INTERSECTIONS BETWEEN PARTICLE AND NUCLEAR PHYSICS

2nd CONFERENCE on the INTERSECTIONS between PARTICLE and NUCLEAR PHYSICS
May 26–31, 1986

**LAKE LOUISE
CANADA**

AIP CONFERENCE PROCEEDINGS 150

RITA G. LERNER
SERIES EDITOR

INTERSECTIONS BETWEEN PARTICLE AND NUCLEAR PHYSICS

LAKE LOUISE, CANADA 1986

EDITOR
DONALD F. GEESAMAN
ARGONNE NATIONAL LABORATORY

AMERICAN INSTITUTE OF PHYSICS   NEW YORK 1986

Copy fees: The code at the bottom of the first page of each article in this volume gives the fee for each copy of the article made beyond the free copying permitted under the 1978 US Copyright Law. (See also the statement following "Copyright" below.) This fee can be paid to the American Institute of Physics through the Copyright Clearance Center, Inc., 21 Congress Street, Salem, MA 01970.

Copyright © 1986 American Institute of Physics

Individual readers of this volume and non-profit libraries, acting for them, are permitted to make fair use of the material in it, such as copying an article for use in teaching or research. Permission is granted to quote from this volume in scientific work with the customary acknowledgment of the source. To reprint a figure, table or other excerpt requires the consent of one of the original authors and notification to AIP. Republication or systematic or multiple reproduction of any material in this volume is permitted only under license from AIP. Address inquiries to Series Editor, AIP Conference Proceedings, AIP, 335 E. 45th St., New York, NY 10017.

L.C. Catalog Card No. 86-72018
ISBN 0-88318-349-8
DOE CONF-860575

Printed in the United States of America

Foreword

At Kicking Horse Pass near Lake Louise, the water from one stream is divided into two as a visible manifestation of the Continental Divide. To the individual droplet of runoff, the division is as arbitrary as the political boundary between Alberta and British Columbia which follows the Divide.

Those who study the physics of leptons and hadrons, in free space, in atomic nuclei, and in stellar objects, are subject to similar arbitrary boundaries. The Second Conference on the Intersections Between Particle and Nuclear Physics emphasized the common basis and scientific goals of these two "provinces". The success of this conference reflects the vitality and significance of this intersection region.

As chairman and co-chairman of the organizing committee, Alan Krisch and Erich Vogt deserve the lion's share of the credit for the conference's success. The coordinators of the parallel sessions independently developed their own excellent programs. And the superb professional staff, Sheila Dodson, Kathy Einfeldt, Lorraine King, Margaret Lear, Frances Mann, Roberta Marinuzzi, Sue Marsh and Karen Poelakker, made everything work so well that a third conference of this series is being planned.

As editor, I wish to add a special thanks to Barbara Weller for her assistance in preparing these proceedings. I also want to thank each speaker and participant for enlivening the spirit of the conference

The mountains don't need to be thanked, but it is important to know that they are there when you need them.

Donald F. Geesaman
July 1986

INTRODUCTION

A. D. Krisch
Randall Lab. of Physics
The University of Michigan
Ann Arbor, MI 48109

Looking at the view out of the window, I am sorry that I cannot welcome you to Lake Louise as a Canadian like Erich. But I do want to welcome you to the Second Conference on the Intersections Between Particle and Nuclear Physics, on behalf of the Organizing Committee.

The main purpose of the conference is to exchange scientific information about particle and nuclear physics between 1 and 100 GeV. It's second purpose is to foster this exciting and diverse area of physics by bringing together particle and nuclear physicists in a beautiful spot which is so isolated that they must interact with each other.

This secret formula of beauty and isolation worked very well at Steamboat Springs. Thus, here in Lake Louise we should have extraordinary success. Recently, there has been a significant increase in the activity and excitement of particle and nuclear physics in the 1 to 100 GeV range. I hope that these conferences have made some contribution to this growth.

To maximize the probability of scientific progress, I believe that we must maintain active and exciting research programs on as many frontiers as we possibly can. We should have as much diversity as possible. The SSC is aimed toward the fascinating high energy frontier. This conference is focused toward the exciting high intensity and high precision frontiers in the GeV range.

From time to time we serve on committees to recommend choices between different experiments or different research areas. Perhaps such choices must be made and certain research areas must be excluded for financial reasons. But we should make such choices as reluctantly as possible. The scientific method states that truth can only be discovered by the reproducibility of experimental observations, not by the deliberations of any learned group of scholars. Rejecting the possibility of making such experimental observations is inherently non-scientific. When we are forced to make such rejections we should be sad at what we are losing.

This conference is a manifestation of the beautiful diversity which I consider essential to the future strength of particle and nuclear physics. We have many interesting plenary and parallel lectures by some distinguished scientists; some are senior and distinguished, some are young and soon to be recognized as distinguished.

CONTENTS

Foreword	v
Introduction	vii
A. D. Krisch	
Organizing Committee and Sponsors	xix
Parallel Session Coordinators	xx

PLENARY SESSIONS

Symmetry Tests in Particle and Nuclear Physics	1
D. Wilkinson	
CP Violation in the Neutral Kaon System	24
B. Winstein	
Superconducting Accelerator Technology	37
H. A. Grunder and B. K. Hartline	
Future Accelerator Technology	53
A. M. Sessler	
High Intensity Proton Synchrotrons	63
M. K. Craddock	
Electromagnetic Structure of Nuclei	83
R. G. Arnold	
Strange Matter and Dihyperon Physics	99
P. D. Barnes	
Anomalous Positron Production	112
J. S. Greenberg	
A Review of the Physics of the Neutrino	115
R. G. H. Robertson	
Hypernuclear Interactions	127
A. Gal	
The Standard Model and Beyond	142
P. Langacker	
The EMC Effect	165
E. L. Berger	
High Precision Parity Tests in Proton-Proton (Nucleon) Scattering	185
M. Simonius	
Quarks in Spectroscopy	200
G. Karl	
Relativistic Heavy Ion Facilities—Worldwide	205
L. S. Schroeder	
Cooling of Stored Beams	226
F. E. Mills	
Quark Model of the NN Interaction	238
M. Oka	
High Energy Spin Physics	257
E. Leader	
Antinucleon Physics	272
C. B. Dover	

Non-Accelerator Experiments 293
 D. H. Perkins
Summary of Accelerator Physics Presentations 303
 P. F. M. Koehler
Antiproton Physics Summary 307
 D. A. Axen
Hadron Scattering: A Biased Summary 310
 N. Isgur
Hadron Spectroscopy Summary 313
 J. F. Donoghue
Summary of the Heavy Ion Physics Sessions at Lake Louise 318
 J. W. Harris
Summary of Hypernuclear Sessions 325
 R. E. Chrien
Kaon Decay Session Summary 334
 M. P. Schmidt
Summary: Neutrino Physics 339
 B. H. J. McKellar
Closing Remarks 346
 R. Hofstadter

ACCELERATOR PHYSICS

On the Possibility of Achieving a Condensed Crystalline State in Cooled
Particle Beams 354
 J. P. Schiffer and A. Rahman
Storage Rings for Heavy-Ion Atomic and Nuclear Physics 357
 G. R. Young
Toward Multi-GeV Electron Cooling 366
 D. J. Larson et al.
Status of Magnet System for RHIC 371
 P. A. Thompson et al.
Does the Transition to Chaos Determine the Dynamic Aperture? 374
 J. M. Jowett
An Exploratory Gas-Target Experiment at PEP Using the TPC/2 γ Facility 378
 F. S. Dietrich, S. O. Melnikoff, and K. A. Van Bibber
Storage Ring Magnet for a Proposed New Precision Measurement of the Muon
Anomalous Magnetic Moment 382
 V. W. Hughes
Use of the Fermilab Antiproton Source for Experiments 391
 J. E. Griffin
Is There a Need for a Lear-Like Facility in North America? 395
 G. A. Smith
LAMPF II 1986: Matching the Machine to the New Physics 400
 H. A. Thiessen
Microwave Instability Criterion for Overlapped Bunches 401
 K-Y. Ng

ANTIPROTON PHYSICS

$N\bar{N}$ Annihilation and the Quark Rearrangements Model 406
    J. A. Niskanen
$\bar{N}N$ Annihilation Experiments 412
    S. M. Playfer
Strangeness Production with Antiprotons 418
    P. D. Barnes and C. J. Maher
Nucleon-Antinucleon Spin Physics 429
    R. D. Ransome
Experimental Evidence for Quantum Gravity? 434
    T. Goldman, R. J. Hughes, and M. M. Nieto
Proposed Measurement of the Gravitational Acceleration of the Antiproton .. 436
    R. E. Brown
Measurement of the Antineutron-Proton Annihilation Cross Section Extremely Close to the $\bar{N}N$ Threshold 445
    B. Bassalleck
Antiproton Spin Physics at LEAR 449
    Y. Onel
Antiproton-Proton Annihilation into Collinear Charged Pions and Kaons 456
    G. M. Marshall et al.
Search for Narrow Lines in Gamma Spectra From $\bar{p}p$ Annihilation at Rest 458
    S. M. Playfer
A Search for Narrow States in Antineutron-Proton Total and Annihilation Cross-Sections Near $\bar{N}N$ Threshold 460
    L. S. Pinsky et al.
Heavy Quarks in Antiproton-Proton Interactions 468
    R. Cester
Physics with Antihydrogen 480
    H. Poth
Antiprotonic Atoms 490
    H. Koch
CP-Odd Asymmetries at Low Energy 495
    J. F. Donoghue
$\bar{P}$-Nucleus Interaction 499
    J. C. Peng
Production of High Energy Density in $\bar{N}$-Nucleus Interactions 505
    W. R. Gibbs
A Report on the First Fermilab Workshop on Antimatter Physics at Low Energy 515
    L. S. Pinsky
Studies of Nuclear Structure in Antinucleon Charge-Exchange Reactions 520
    N. Auerbach
Neutral Strange Particle Production in $\bar{p}$ Ne Annihilations 526
    F. Balestra et al.

ELECTRON AND MUON PHYSICS

A Review of the EMC Effect 530
    G. N. Taylor

Nuclear Medium Effects on the Photon-Proton Vertex Investigated with the $(e,e'p)$ Reaction 535
   L. Lapikás
Large Momentum Components of the Deuteron Wave Function 540
   R. Dymarz and F. C. Khanna
Continuum Shell Model Analysis of $^{16}O(\gamma, p)^{15}N$ at Medium Energies 542
   L. D. Ludeking and S. R. Cotanch
Electromagnetic Structure of ^3H and ^3He 547
   D. H. Beck
Measurements of the Deuteron Magnetic Form Factor at High Q^2 554
   P. E. Bosted
Tensor Polarization of the Deuteron in the Elastic e^-d Scattering 559
   F. C. Khanna and R. Dymarz
Integral Asymmetry Parameter in Muon Decay and Non-Standard Weak Interaction 561
   R. D. von Dincklage
Present Status of QED 567
   J. Sapirstein
"The LAMPF Workshop on Fundamental Muon Physics" 575
   V. W. Hughes
Supersymmetry Corrections to Muon g-2 582
   R. Arnowitt and P. Nath
Thermal Muonium in Vacuum 587
   A. Olin
Leptonic Decays of Heavy Leptonium 595
   J. Malenfant
Muon and Neutrino Production in Proton-^{12}C Scattering 597
   S. L. Mintz
A Large Acceptance Detector for Electromagnetic Nuclear Physics at CEBAF 601
   B. A. Mecking
Polarization in Electron Scattering Experiments at CEBAF 604
   V. Burkert

HADRON SCATTERING
Partial Wave Analysis of $pp \to NN\pi$ and Dibaryon Resonances 609
   A. B. Wicklund
Model Independent Exchange Currents 618
   D. O. Riska
A Precision Test of Charge Independence 627
   K. K. Seth et al.
Eta-Meson Production Experiments at LAMPF 630
   J. C. Peng
Quark Clusters in Pion Absorption 636
   M. A. Moinester
The Splitting of the Roper Resonance 640
   B. M. K. Nefkens et al.
The Sequential Nature of Pion Double Charge Exchange 644
   M. Bleszynski and R. Glauber

Determination of the Real Part of the Isospin-Even Forward Scattering Amplitude of Low Energy Pion-Nucleon Scattering as a Test of Low Energy Quantumchromodynamics 646
    K. Goring et al.
Elastic Meson-Nucleon Partial Wave Scattering Analyses 650
    R. A. Arndt
Models of Multiquark States 657
    H. J. Lipkin
Multi-Quark States in a QCD-Like Potential Model 672
    K. Maltman
Flux Tubes in Multi-Quark Systems 676
    D. Robson
K^+-N, K^{+*}-N Channel Coupling in a Quark Potential Model 686
    R. K. Campbell
Anomalous Dimensions of Multiquark Bound States 688
    C. R. Ji
Dirac vs. Schroedinger Approach in Proton-Nucleus Scattering: A Simple Model 692
    M. Thies
Dudley's Dilemma: Magnetic Moments in Relativistic Theories 696
    J. A. McNeil
(n, p) Reaction Studies 710
    W. P. Alford
Three-Body Forces 719
    B. M. K. Nefkens
Multipole Matrix Elements for ^{208}Pb 727
    D. K. McDaniels et al.
The Chou-Yang Model and p-^{4}He Elastic Scattering from 45–393 GeV ... 729
    M. Kamran and I. E. Qureshi

HADRON SPECTROSCOPY

Interacting Boson Model: Selected Recent Developments 732
    A. B. Balantekin
Fundamental Interaction Studies in Nuclei 738
    W. C. Haxton
Are Nucleons and Deltas Deformed? 749
    M. Bourdeau, R. Davidson, and N. C. Mukhopadhyay
Spin Observables at Intermediate Energies: A Tool in Viewing the Nucleus ... 751
    J. B. McClelland
Spectroscopy without Quarks: A Skyrme-Model Sampler 762
    M. Karliner and M. P. Mattis
Light Quark Pairs in Heavy Quarkonia 770
    N. Byers and V. Zambetakis
Nuclear-Like States of Quark Matter 773
    T. Goldman, K. E. Schmidt, and G. J. Stephenson, Jr.
Recent Mark III Results on J/Ψ Hadronic and Radiative Decays 776
    W. S. Lockman

Hadron Structure from Lattice QCD 788
    R. M. Woloshyn
Lattice Gauge Theory as a Nuclear Many-Body Problem 791
    G. J. Mathews, S. D. Bloom, and N. J. Snyderman
Wigner-Weyl and Nambu-Goldstone Realizations of Chiral Symmetry in
Quantum Chromodynamics 796
    R. Acharya and P. Narayana Swamy

HEAVY ION PHYSICS

Direct Lepton Production in Hadronic and Nuclear Collisions at Low PT and
Low Pair Mass 806
    G. Roche
Subthreshold Production of Strange Hadrons in Relativistic Heavy Ion
Collisions 814
    S. Trentalange et al.
Production of Metastable Strange-Quark Droplets in Relativistic Heavy-Ion
Collisions 822
    G. L. Shaw
Positrons from Heavy Ions: A Puzzle for Physicists 827
    B. Müller
The Nuclear Matter Equation of State from Relativistic Heavy Ions to
Supernovae 835
    J. W. Harris
Collective Flow and the Stiffness of Compressed Nuclear Matter 844
    D. Keane et al.
Phase Transitions in QCD from the Lattice Perspective 848
    R. V. Gavai
Observation of KNO Scaling in the Neutral Energy Spectra from $\alpha\alpha$ and pp
Collisions at ISR Energies 858
    M. J. Tannenbaum
High $p_T \pi^0$ Production in $\alpha\alpha$, dd and pp Collisions at the CERN ISR 872
    M. Tanaka
Finite Temperature Mean Field Calcualtions with Isobars 877
    H. G. Miller et al.
Production Mechanism for a Light PseudoScaler Boson and the Anomalous
Positron Spectrum 879
    M. S. Zahir and D. Y. Kim

HYPERNUCLEAR PHYSICS

Theoretical Aspects of Dibaryons 886
    M. E. Sainio
Future Dibaryon Research at BNL 894
    P. H. Pile
The Reactions $K^-d \to N\Lambda\pi$ and $N\Sigma\pi$ 901
    M. Torres, R. H. Dalitz, and A. Deloff
K-Nucleon Scattering and the Cloudy Bag Model 906
    B. K. Jennings

Relativistic Wave Functions in Kaon Electroproduction from Nuclei 914
    C. Bennhold and L. E. Wright
Hypernuclear Experiments at KEK 917
    J. Chiba
Quasifree Process in Hypernuclear Formation 921
    T. Kishimoto
Correlations in Hypernuclei 924
    J. Speth et al.
The Eta-Mesic Nucleus: A New Nuclear Species? 930
    L. C. Liu and Q. Haider
Weak Decay of Λ Hypernuclei 934
    J. J. Szymanski
Lifetimes for Non-Mesonic Decay in Hypernuclei 940
    L. S. Kisslinger
The Non-Mesonic Decay of Hypernuclei 946
    J. Dubach
Weak Decays of the H Dibaryon 952
    E. Golowich

KAON DECAY PHYSICS

KMC Unitarity and $K^+ \to \pi^+ \nu \bar{\nu}$ 956
    W. J. Marciano
Rare Decays 961
    T. Numao
Search for Rare Muon and Pion Decay Modes with the Crystal Box Detector ... 966
    L. E. Piilonen et al.
Lepton Number and Lepton Flavor Violation is SUSY Models 972
    G. K. Leontaris, K. Tamvakis, and J. D. Vergados
Status of the Standard Electroweak Model in Muon Decay 976
    D. P. Stoker
Review of the Decay of the Tau Lepton 981
    P. R. Burchat
Study of CP Violation in a Tagged Neutral Kaon Beam 987
    J. R. Fry
Search for T and CPT Violation in the Neutral Kaon System 993
    B. L. Roberts et al.
CP Violation: Past and Present 998
    B. R. Holstein
A Progress Report on the Measurement of ϵ'/ϵ at Fermilab 1004
    G. J. Bock

NEUTRINO PHYSICS

Double Beta Decay in ^{82}Se, ^{128}Te and ^{130}Te 1012
    M. K. Moe
^{76}Ge $\beta\beta$-Decay Experiments and Their Analyses—an Update 1017
    F. T. Avignone et al.

Neutrinoless Double β-Decay and Lepton Flavor Violation 1025
    G. K. Leontaris and J. D. Vergados
Upper Limit on the Neutrino Mass from the Los Alamos Free Molecular
Gaseous Tritium Experiment 1031
    D. Knapp et al.
An Upper Limit for the Electron Antineutrino Mass 1036
    M. Fritschi et al.
Coherent Scattering of Neutrinos and Antineutrinos by Quarks in a Crystal .. 1038
    J. Weber
Coherent Scattering of Low Energy Neutrinos from Macroscopic Objects 1040
    R. C. Casella
Neutrino-Oscillation Experiments at Reactors 1042
    Z. D. Greenwood
Recent Results on $\nu_e\, e^-$ Scattering 1050
    R. C. Allen et al.
Neutrino-Nucleus Interactions: Charged Current Interactions at Line E of
LAMPF 1056
    R. Fisk

NON-ACCELERATOR PHYSICS

Ultra-High Energy Point Sources of Cosmic Rays 1070
    J. W. Elbert
Cygnus Experiment at Los Alamos 1078
    B. L. Dingus et al.
New Evidence from SOUDAN 1 for Underground Muons Associated with
Cygnus X-3 1083
    D. S. Ayres
The Search for Solar Neutrinos at Kamioka 1087
    B. Cortez
A D_2O Cerenkov Detector for Solar Neutrinos 1094
    E. D. Earle et al.
A New Force in Nature? 1102
    E. Fischbach et al.
Acoustic Detection of Low-Energy Radiation 1119
    C. J. Martoff, B. Cabrera, and B. Neuhauser
Modern Implications of Neutron β-Decay 1125
    S. J. Freedman
Comparison of Beta-Ray Polarizations in Fermi and Gamow-Teller
Transitions and $SU(2)_L \times SU(2)_R \times U(1)$ Models 1131
    P. Herczeg
Status of the SOUDAN 2 Nucleon Decay Experiment 1134
    E. N. May
Searches for Monopoles: Past and Future 1137
    S. P. Ahlen
A Search for Anomalously Heavy Isotopes of Low Z Nuclei 1143
    D. F. Nitz et al.

SPIN PHYSICS

Polarized Protons at the AGS and High $P_\perp^2$ Spin Effects .... T. Roser	1148
Effects of Spin Asymmetries of Special Effects of 90° .... H. J. Lipkin	1153
High Energy Polarized Beams from Hyperon Decays .... D. G. Underwood	1161
Spin Effects in Exclusive Reactions at High $P_\perp$ .... Y. Makdisi	1166
Spin Observables in Proton-Neutron Scattering at Intermediate Energy .... H. Spinka	1171
Parity and Time-Reversal Violation in Nuclei and Atoms .... E. G. Adelberger	1177
Parity Violation in Proton-Proton Scattering at Intermediate Energies .... V. Yuan et al.	1189
Test of Time Reversal Symmetry with Resonance Neutron Scattering .... J. D. Bowman	1194
Spin Dependence of ρ° Production in $\pi^+ n_\uparrow \to \pi^+\pi^- p$ .... M. Svec, A. de Lesquen, and L. van Rossum	1198
Spin Transfer in Hyperon Production .... J. Kruk	1200
Charge Symmetry Breaking in the n-p System .... L. G. Greeniaus	1202
Spin Observables in Proton Deuteron Elastic Scattering .... E. Bleszynski, M. Bleszynski, and T. Jaroszewicz	1208
Recent Topics on Spin Physics at KEK .... A. Masaike	1214
Tensor Analyzing Power in πd Elastic Scattering .... G. R. Smith	1219
Radiative Capture of Polarized Protons and Deuterons by Hydrogen Isotopes .. W. K. Pitts	1224
A Spin-Splitter for Antiprotons in Low Energy Storage Rings .... Y. Onel, A. Penzo, and R. Rossmanith	1229
Measurement of the Vector Analysing Power in the πd-Breakup .... W. Gyles et al.	1232
Program	1235
List of Participants	1259
Author Index	1267

Organizing Committee

A. D. Krisch (Chairman)
G. M. Bunce
P. A. Carruthers
O. Chamberlain
R. H. Dalitz
G. R. Farrar
G. Fidecaro
D. F. Geesaman
V. W. Hughes
S. E. Koonin
T. D. Lee
T. W. Ludlam

E. W. Vogt (Co-Chairman)
M. H. MacFarlane
J. S. McCarthy
D. F. Measday
F. E. Mills
R. E. Mischke
E. J. Moniz
T. A. O'Halloran
L. S. Schroeder
L. C. Teng
J. D. Walecka

Sponsors

U. S. DEPARTMENT OF ENERGY
U. S. NATIONAL SCIENCE FOUNDATION
ARGONNE NATIONAL LABORATORY
LAWRENCE BERKELEY NATIONAL LABORATORY
BROOKHAVEN NATIONAL LABORATORY
LOS ALAMOS NATIONAL LABORATORY
UNIVERSITY OF MICHIGAN
TRIUMF
E. G. & G. (CANADA)

Session Coordinators

Accelerator Physics	P. Koehler
	R. E. Pollock
Antiproton Physics	D. A. Axen
	G. A. Smith
Electron and Muon Physics	T. Kinoshita
	S. B. Kowalski
Hadron Scattering	G. J. Igo
	N. Isgur
Hadron Spectroscopy	J. F. Donoghue
	A. McDonald
Heavy Ion Physics	O. Hansen
	J. W. Harris
Hypernuclear Physics	R. E. Chrien
	R. H. Dalitz
Kaon Decay Physics	D. A. Bryman
	M. P. Schmidt
Neutrino Physics	W.-Y. Lee
	S. P. Rosen
Non-Accelerator Physics	D. S. Ayres
	D. A. Sinclair
Spin Physics	J. B. Roberts
	W. T. H. van Oers

SYMMETRY TESTS IN PARTICLE AND NUCLEAR PHYSICS

Denys Wilkinson
University of Sussex, Brighton, England

ABSTRACT

Some symmetries are reviewed

INTRODUCTION

The field of symmetries is so vast, even if one eliminates those that proliferate speculatively but of which there is no present sign that Nature is aware, that a full review requires a book, not a short talk. I shall therefore not attempt such a review the more so since: (a) many topics directly or indirectly relating to symmetries are being treated in detail elsewhere in this conference; (b) I have very recently published a partial review of the subject[1] (referred to as I) and should not repeat that here. I will rather pick out a few present trends within more-or-less well established parts of the field where interesting syntheses may be found, where important progress may be anticipated, or where we find great mystery. There will be some overlap with I and with matters presented elsewhere at this conference but in such cases I shall rely for references largely on those other treatments. Nor shall I give references to material that is the object of frequent current review such as the various "supertheories". Conversely, I shall be more liberal with what may be less familiar references.

I shall not mention at all, or only in passing, many important matters, some of them touched on in I: time reversal in strong, electromagnetic and (most) weak processes; charge independence, charge symmetry and isospin; general questions of symmetry and coupling in weak interactions; conserved and partially-conserved weak currents; induced weak couplings; neutrino mixing and oscillation; parity-non-conservation in hadronic interactions and nuclear structure; baryon non-conservation; nucleon/anti-nucleon oscillations; anomalous lepton-hadron couplings; monopoles; the nature of the pion, chiral symmetry and soft pion theorems; etc.

UNIFICATION

The present hope of most, but not all, natural philosophers, is that the four* forces of Nature, the strong, weak, electromagnetic and gravitational, which differ so much in their manifestations under

---

* We may not yet be aware of all of the forces and pseudo-forces, i.e. those simulated by the imposition of the dictates of unknown symmetry requirements on known or unknown forces, of which Nature disposes. For example we may note the long-range neutral vector massless field that may couple to baryon number[2] and the recent re-opening[3] of the question of an intermediate-range non-Newtonian "gravitation".

normal conditions, may be understood as deriving from what at very
high energies (temperatures) is a single force that splits
successively into the everyday forces by processes of spontaneous
symmetry breaking as the temperature is lowered. The processes of
symmetry breaking must each be associated with particle masses or
energy scales. The great challenge is not only to discover, and, one
might hope, to understand the uniqueness of, the master symmetry and
its successive breakings but also to understand the relationship
between the mass scales of the symmetry breaking and of the particles
between which the various forces act and of those that are associated
with the propagation of those forces or are associated with the
breaking (and making) of the symmetries.

We now believe that, starting from the low-temperature end, the weak
and electromagnetic forces are unified through the $W^{\pm}$ and Z^0 bosons of
mass about 100 GeV. This unification appears to be described by $SU(2)_L \otimes U(1)$. The implication, in colloquial terms, that the W and Z bosons
are intrinsically left-handed is thought by some to be aesthetically
unpleasing and serious consideration must be given to the possibility[4]
of the more-symmetrical $SU(2)_L \otimes SU(2)_R \otimes U(1)$ with a right-handed
boson of significantly greater mass than the left; but we would not
then have any immediate understanding of that residual, weaker,
left-right asymmetry.

This electro-weak unification, $SU(2)_L \otimes U(1)$, contains the important
parameter, the Weinberg Angle θ_W, that relates the masses of the W and
Z bosons by $m_W = m_Z \cos\theta_W$. θ_W is an arbitrary parameter at this
stage and to seek an understanding of it we must go to the next stage
of unification.

At low temperature we also now appear to have a possibly full
understanding, in principle, of the strong interaction through QCD,
the non-Abelian local gauge theory of quarks interacting through the
exchange of massless but self-interacting gluons described by $SU(3)_C$.
Our low-temperature world therefore seems, at least approximately, to
work through the "standard model" $SU(3)_C \otimes SU(2)_L \otimes U(1)$.

If unification is to continue we must now absorb QCD and gravitation
at higher temperatures. There is at present no well-established
theory of quantum gravity and so it would be, at the least, convenient
if the QCD absorption were to take place first; that is to say if the
fusion of $SU(3)_C \otimes SU(2)_L \otimes U(1)$ into some single group of "grand
unification", the GUT transition, were to take place at an energy
scale m_x below that at which we know that we cannot escape reckoning
with the quantum aspects of gravity.

This latter energy scale, at which gravity and the quantum theory must
collide, and, presumably, find their reconciliation, is the Planck
scale. The Planck Time is simply the time after the Big Bang (which
is implicity assumed as the starting point for our present discussion)
at which the energy content of the visible universe was equal to the
Heisenberg uncertainty in that energy. Noting the general

relativistic relationship between time and density namely $32\pi G\rho t^2 = 3$ we can write down the Planck Time $t_p = (G\hbar/c^5)^{1/2}$, about 5×10^{-44} sec. Immediately deriving from t_p are the Planck Mass $m_p = (c\hbar/G)^{1/2}$, about 10^{19} GeV, that we need for our present discussion, and the Planck Length $l_p = (G\hbar/c^3)^{1/2}$, about 10^{-33} cm, that we shall need shortly.

It would therefore be convenient if we could persuade ourselves that the GUT unification scale that melds $SU(3)_c \otimes SU(2)_L \otimes U(1)$ into some higher symmetry were such that $m_x \ll m_p$ because then we could work on the GUT symmetry without too much fear of interference from gravity.

We may get some rough idea as to m_x from, for example, our attempts to understand θ_W which is arbitrary within $SU(2)_L \otimes U(1)$ but which should be specified through the GUT unification. The essential point here is that, as with the coupling constants that relate to the forces considered individually, θ_W depends on the energy at which it is measured so that its low-temperature value of $\theta_W \simeq 28°$ is substantially less than its value above m_x. It turns out that the simplest GUT groups suggest $m_x \simeq 10^{14-15}$ GeV for a fit between their theoretical high temperature values of θ_W and that experimentally measured.

Since $10^{14-15} \ll 10^{19}$ this may grant us the desired licence to hunt for the GUT group without regard for gravity or, alternatively, to hunt for the unification above m_p that will include gravity such that it will break down into a suitable GUT for ultimate generation of $SU(3)_c \otimes SU(2)_L \otimes U(1)$.

Searches for an acceptable GUT have not been uniquely successful. The simplest GUT, namely minimal $SU(5)$, appears to be unacceptable because: (i) it predicts a lifetime for the proton against $e^+ + \pi^0$ decay of about 10^{30} years against a present lower limit of 10^{32} years or more; (ii) it leads to galaxy-seeding fluctuations that are too large by a factor of about 10^5; (iii) it probably has unacceptable difficulty in explaining the baryon/photon ratio of the universe. (We may note in passing that another GUT candidate, $SO(10)$, has as a subgroup $SU(2)_L \otimes SU(2)_R \otimes U(1)$ that we have already mentioned and that we shall be considering at several points).

Many other possible GUT symmetries exist as also many possible further symmetries that incorporate gravity the simpler of which, we may note, break down into low-temperature universes of lower complexity than our own so that Nature has evidently decreed that we do not live in the simplest of all possible worlds. (I do not dwell on the anthropic observation that even if such a simple world had been decreed we could not, in fact, have come about in it to make the above remark[5]).

THE COSMOLOGICAL CONSTANT

Before becoming more particular we should pause before one of the great mysteries, namely the cosmological constant, because that may require the recognition of a most powerful overriding symmetry that we have yet to fathom.

The Big Bang naturally locates us in an expanding universe that, on the gross scale, seems to be uniform and isotropic and can therefore be characterized by a "radius" R. General relativity then specifies the equation for the time-behaviour of this universe: $\ddot{R}/R = -4\pi G(\rho + 3p/c^2)/3 + \Lambda c^2/3$ The first terms on the right hand side of this equation are identical with the expectation of Newtonian mechanics for a universe of density ρ plus the mass-density-equivalence of the radiation pressure p. The last term on the right hand side is the novel and inescapable consequence of general relativity: there must be an additional, expansive, term, the cosmological constant Λ. General relativity does not, of itself, specify the numerical value of the constants that appear in its equations and so we must form our <u>a priori</u> ideas as to the likely values of unknown constants from the <u>de facto</u> values taken by others related to them. The cosmological constant has the dimensions, in the above notation, of length^{-2}. The dimension of length naturally specified by the constants that govern the development of the universe is the Planck Length already noted namely $l_p \simeq 10^{-33}$ cm. so that, offhand, we should expect $\Lambda \simeq 10^{66}$ cm^{-2}. In fact[6] $\Lambda \leq 10^{-55}$ cm^{-2} so that we are faced with an <u>a priori</u> discrepancy of a factor of 10^{120} or so. Some very powerful symmetry principle must be presumably at work to quench our <u>a priori</u> expectation. Our argument so far has not, of course, specified any mechanism through which our <u>a priori</u> expectation as to Λ might be generated. We should therefore note that the fact that the energy of the (true) vacuum minimizes at a non-zero expectation value of the Higgs fields associated, as we now understand it, with certain forms of spontaneous symmetry breaking, can be parameterized through an effective cosmological constant and that the Higgs field associated with the electroweak unification alone gives[7] $\Lambda_{WS} \simeq 10^{-4}$ cm^{-2}, at least 10^{50} times greater than Λ_{exp}, while higher, GUT and other, transitions would presumably result in larger effective contributions to Λ.

Is there indeed some powerful symmetry principle at work that decrees $\Lambda \equiv 0$ or may Λ turn out to have a small but finite value? As we shall see immediately the answer to this question might have profound consequences for other aspects of our cosmological view.

Consider the Hubble Constant H_o that measures the present rate of "expansion of the universe". This appears to lie within the range $50 < H_o < 100$ km sec^{-1} mpc^{-1} with some intermediate value, say 75 km sec^{-1} mpc^{-1}, presently favoured[8]. Very evidently, the relationship between H_o, the present mean gravitating density ρ and the age of the universe must involve Λ which could, therefore, be inferred from a confident determination of those other quantities.

The age of the universe is not itself susceptible of any direct assessment, but we may consider measures that constitute lower limits to this quantity. One of them is the age of the globular clusters, chiefly within our own galaxy but including some in the Magellanic Clouds. The most recent[8] estimates favour $[(16-17) \pm 3] \times 10^9$

years. The other lower limit, namely the so-called actinide chronometer, reflects directly upon the topic of this conference. The heavy elements have, following our present understanding, been built up by the r-processes (and, to some degree, the s-processes) in the stars. The present relative abundances of these heavy elements, as determined within our solar system, can be combined with estimates of their relative production rates by the r- and s-processes to give a starting-time for the production of these elements in these parts; this is presumably less than the age of the universe. The earlier "standard value" for this time was about 12×10^9 years ago[9], which is on the low side of the above-reported range of values deriving from studies of globular clusters, but recent developments in our understanding of the decay and reaction mechanisms that lead to the formation are favouring the significantly longer times of $(21^{+2}_{-5}) \times 10^9$ years[10]. These longer times are more consistent with the above recent estimates from globular clusters and, together with them, may seem to favour (a lower limit of) perhaps, 20×10^9 years for the time since the Big Bang. (Lower limit since, in the case of both the measures, some time must have elapsed since the Big Bang and the epoch to which the dating refers).

The clear statement may now be made[8,10] that for $H_o \simeq 75$ km sec^{-1} mpc^{-1} and with the (lower limit) to the present mean density of the universe as inferred from intra- and inter-galactic dynamics this lower limit to the age of the universe cannot be reconciled with $\Lambda = 0$ within the standard model of the Hot Big Bang. With the above value of H_o the best value[10] for Λ is about 2×10^{-56} cm^{-2} for the baryonic density best representing the nucleosynthesis of the lightest elements in the Big Bang; for $H_o = 50$ km sec^{-1} mpc^{-1} $\Lambda = 0$ might just be squeezed in.

We may now examine the implications of a non-vanishing value of Λ. This is most simply done by reference to the geometry of the universe. If we stay with standard general relativity and the idea of a homogeneous and isotropic universe we may discuss the geometry in terms of the Robertson-Walker metric. Within that metric Euclidean geometry, which corresponds to a just-closed, just-open universe, is given at k=0 such as is prescribed by standard inflationary scenarios. For $\Lambda = 0$, k=0 corresponds to a density of gravitating stuff of $\Omega = 1$ in units of the closure density. However, for $\Lambda \neq 0$, k=0 is given by: $\Omega + \Lambda/3H_o^2 = 1$. Note that for $\Lambda \simeq 2 \times 10^{-56}$ cm^{-2}, such as may be implied by the age of the universe derived from globular clusters and from the actinide chronometer, and with $H_o \simeq 75$ km sec^{-1} mpc^{-1}, $\Lambda/3H_o^2 \simeq 1$ so that the value of Ω corresponding to a Euclidean universe would remain very open.

The question of Ω is a fascinating one. The amount of "shining baryons" implied by visible starlight corresponds to $\Omega \simeq 0.01$; intra- and inter-galactic dynamics, particularly the application of the virial theorem to galactic clusters, suggests that gravitating stuff associated with galaxies corresponds to $\Lambda \simeq 0.1$ or a little more although very recent data from IRAS[11] have suggested values as large as 0.5 - 0.8; agreement between the observed cosmic relative abundances of the lightest nuclei and expectation based on the Hot Big Bang also suggests[12] $\Omega \simeq 0.1$ although arguments may be made that it could be significantly less but not significantly more. It is indeed astonishing that Ω should be so close to unity, the more so when one remarks that the expansion of the universe implies that for Ω to be within an order of magnitude of unity today it must have been infinitesimally close to unity shortly after the Big Bang itself. It would then seem to be additionally astonishing if Λ were to have a value that makes $\Lambda/3H_o^2$ also of order unity.

All of this leaves open the question as to where the missing mass, not associated with galaxies, may be if $\Omega = 1$, $\Lambda = 0$. The difference between the $\Omega \simeq 0.01$ from shining baryons and the $\Omega \simeq 0.1$ required for galactic motions and, probably, for nucleosynthesis shortly after the Big Bang, could plausibly be made up by silent baryons in the form of bricks, Jupiters etc or taken into black holes at later stages of galactic evolution. However, if $\Omega = 1$ we must find non-baryonic non-shining missing mass that does not clump with galaxies since the latter event would seem to be ruled out by the application of the virial theorem to galactic clusters. This makes unlikely the intervention of "cold slow matter" such as axions, supersymmetric particles or other WIMPs (weakly interacting massive particles) since they would clump with galaxies even though such matter might have led to a gross distribution of galaxies not unlike that observed. On the other hand "hot light matter" such as massy neutrinos would not clump with galaxies and could therefore top up the mass to $\Omega = 1$ without affecting the $\Omega \simeq 0.1$ galactic contribution inferred from inter-galactic dynamics; however, hot light matter does not lead to the observed form of gross galactic distribution.

Very recent results on the gross distribution of galaxies has revealed a scale-invariance in the Harrison-Zeldovich sense[13] that accords with the expectations of a cosmic string model. Cosmic strings form following a (non-inflationary) transition to the true vacuum, with its associated spontaneous but degenerate symmetry breaking, in the very early stages of the universe. They can seed the formation of galaxies which would then be expected to cluster in chains with great voids between them such as are very recently reported on the scale of 40 mpc or so. The relevant point here is that if the universe is indeed constructed in this way in the gross owing to, for example, such an influence of cosmic strings, then hot light matter such as massy neutrinos might fill the voids to top up to $\Omega = 1$ without having any responsibility for the gross galactic distribution. In other words cosmic strings can let back into contention massy neutrinos as candidates for the missing mass needed to achieve the closure density $\Omega = 1$. We may note that cosmic strings, although not to be confused with the superstrings, that I mention shortly, might be generated by them (through the latter's GUT transition.)

I have, at some length, exposed this problem of the cosmological constant not so much because I think that we can reach any present resolution as, on the one hand, to demonstrate that there are possibly symmetry principles at work that are worth a factor of more than 10^{120} in respect of our primitive expectation and, on the other hand, that nuclear physics considerations as "earthy" as the beta-decay properties and reaction mechanisms of the heaviest nuclei may be of critical importance for a development of our understanding of the cosmos.

(Before leaving this question of the cosmological constant an historical observation might be in order: If something has a certain <u>a priori</u> value but is in fact at least 120 orders of magnitude smaller than this it may be natural to suppose that there exists some over-riding principle that says that it is identically zero. However we may recall the infinities that, because of ultra-violet divergences, plagued quantum electro-dynamics in the 30s and early 40s whenever calculations beyond the first order were carried out. It was then thought that these infinities must really be zero and it was a considerable shock, following the discovery of the Lamb Shift and the subsequent development of renormalizable QED, to find that they were small but finite.)

FLAVOUR ETC

I now move towards what I might term the more conventional mysteries. The first of these, towards which but little present contribution can be made, is that of flavour or generations. Nature seems to be organized, in terms of QCD, into flavour doublets: (u,d), (c,s), (t,b) constituting the three presently-recognized quark generations. Why are there so many flavours? Why are there so few? The GUTs, to which we look forward to link quarks and leptons, would associated these quark generations with the lepton generations, respectively; (e,ν_e), (μ,ν_μ), (τ,ν_τ).

There may well be underlying quantum numbers of a novel kind that label and distinguish the generations or it may be that quark and lepton substructures are at work with their internal re-organizations, that would then give rise to the successive generations, those re-organizations themselves being of the conventional kind, not requiring novel quantum numbers, but rather such as we are familiar with in nuclear and hadron (quark) physics.

Alternatively, the generations could be related by a spontaneously broken global symmetry with its own Goldstone boson of the symmetry breaking, the massless familon (Wilczek) which might show up in, for example, $\mu \rightarrow e f$.

Alternatively, a new gauge symmetry, extending the standard model, could be at work the gauge bosons of which (dubbed "horizontal") being able to bring about neutral current flavour-changing processes such as $K_L^o \rightarrow \mu e$. One might then look forward to recognizing a grand gauge group that would contain the generations (quark plus lepton) in one of its irreducible representations[14].

Suggestions that the three known generations may be all there are come, at present, chiefly from the considerations of Big Bang nucleosynthesis which, within the standard model, understands the observed cosmic ^{4}He abundance in terms of 3 or, at most, 4 species of light neutrino[12]. (Light in this sense means of mass below about 1 MeV so that the neutrinos remain relativistic for at least as long as do electrons in the processes that drive neutrons into protons in the seconds following the Big Bang.) It may, by various manoeuvres, be possible to arrange the baryon density and arguments to do with the relative abundance of the lightest elements, particular deuterium which is more susceptible than most to destruction in later stellar processes, so that more (neutrino) generations might be admitted. But such manoeuvres are very limited in their possibilities if light neutrinos continue to be in question.

Considerations as to number of neutrino species that are not limited by $m_\nu \lesssim m_e$ are available e.g. (i) the width of the Z^0; (ii) the rate of $K^+ \to \pi^+ \nu \bar{\nu}$ *. None has approached a significant limit but current developments in both (i) and (ii) will.

This generation question recalls, of course, the historical development of the (old style) SU(3)/quark hypothesis which relieved us of the need to consider several score hadronic states as "elementary". If the generations proliferate we may find ourselves in an analogous position vis-à-vis the quarks and leptons such that we should welcome the identification of a corresponding "preon" substructure, a compositeness for those particles that within our present classification of both quarks and leptons we are prepared to consider as elementary[15], such as was referred to above. Such models might then enable us to understand fermion masses. All that we can say now is that because of the point-like behaviour of both quarks and leptons down to 10^{-16} cm or so the scale of possible preon compositeness must be beyond 1 TeV.

MASS

The masses of known particles (excluding the photon whose zero mass is understood from gauge principles) cover the range from $\lesssim 30$ eV (the electron neutrino) to 100 GeV (W and Z). As we have already noted, mass scales that are determined by relationships between fundamental constants extend to $m_P \simeq 10^{19}$ GeV. What determines/permits this vast range of mass scales within what we hope eventually to understand as a unified universe? This is another great mystery. The problem is essentially one of understanding how the various mass scales can be decoupled from each other or, better, be derived from each other and this I will now, if only illustratively, examine.

Consider the mass of the W-boson. From general arguments m_W cannot be very different from the mass, m_H, of its associated Higgs particle of

---

* This is a decay process of great richness for new physics that I do not here have occasion to explore in detail.

the electro-weak unification (say 10^{2-3} GeV) if our present understanding of renormalizable symmetry breaking and the generation of mass is correct. This immediately identifies two major problems:

(i) The Hierarchy Problem: what is the origin of the enormously different mass scales $m_H \ll m_P$?

(ii) The Naturalness Problem: how is the hierarchy, once established maintained?

Addressing the second problem first: corrections to m_H arise from quadratically-divergent loop diagrams such that one might expect their cutoff at some higher mass scale to result in:

(i) the correction to m_H from the coupling to m_P through the "foaming of space-time" (Hawking) being of order m_P ($\simeq 10^{16-17} m_H$);

(ii) the correction to m_H from the coupling of the Higgs of the electro-weak unification to that (those) of the putative GUT unification being of order $m_X \simeq 10^{14-15}$ GeV;

(iii) even if correction (ii) above were to be cancelled by <u>fiat</u> at tree level there remaining radiative corrections to m_H of order 10^{13-14} GeV deriving from m_X.

Where might solutions to this mass problem be found?

(i) "Technicolor" solutions regard the Higgs as composite rather than elementary and dissociate it at about 1 TeV. This works but faces many problems to do with the generation of quark and lepton masses, the absence of flavour-changing neutral interactions and the absence of light charged scalar bosons.

(ii) The <u>deus ex machina</u> of supersymmetries cancel all loop corrections to m_H because the corrections from fermions and from their associated bosons have opposite signs. However, these corrections depend upon the differences between the boson and fermion masses (squared) so that to restrict the nett correction to m_H to an admissible figure (say 1 TeV) would demand that the supersymmetric partners of known particles shown up at or below the TeV range. Where are they? This search is a major justification for the next generation of high-energy accelerators.

If supersymmetry exists and can protect low-lying masses against shifts due to interference from higher regions of the mass spectrum the Naturalness Problem may find its resolution. This, however, would not resolve the prior Hierarchy Problem: the mass scale itself. Such resolution might come through supergravities that link gravity with supersymmetry.

Supersymmetry, if it exists, is evidently a broken symmetry because the masses of particle and supersymmetric partner are certainly different. Such symmetry breaking occurs in supergravities with local supersymmetry. The essential point here is that in supergravities the supersymmetry-breaking parameters can be renormalized by the electro-weak interactions (cf gauge couplings). We can <u>arrange that</u>[*] m_H^2 be renormalized from its <u>a priori</u> expectation of $\simeq m_P^2$ to become <u>negative</u> at a certain value simply related to m_P and to the t-quark Yukawa coupling. This we may take to mean that m_H ($\simeq m_W$) should be of about this magnitude which is easily achieved for a t-quark mass of about 40 GeV or a bit more. (We may note that no-scale super-models have all their mass scales determined by such dimensional transmutations so that those mass scales are <u>automatically</u> much smaller than m_P.)

Supergravities/supersymmetries therefore <u>may</u>, although the stress is to be emphasized, constitute a complete solution to the hierachy/naturalness problem.

SUPERSTRINGS

So far, in my review, no truly "top-down" model that, rather than synthesizing the known forces, has generated them from a prior principle, has been suggested. A current very promising candidate is superstrings. Crudely speaking, in this model particles are the normal-mode excitations of (open or closed) strings in a 10-dimensional space-time that, on a scale of l_P or so, compactifies into 4 dimensions. (Shades of Kaluza-Klein). Interactions are the splitting or joining of strings and familiar path integrals derive from summation over topological string configurations.

Superstrings holds a number of attractive promises:

(i) At low energy it reduces to a (no-scale) supersymmetry;
(ii) The number of generations is determined by topology;
(iii) It is a finite theory of gravity;
(iv) Yukawa couplings arise as overlaps of wavefunctions;
(v) It <u>may</u> naturally give a zero cosmological constant.

The possibility that superstrings may be the "theory of everything" makes it interesting to consider seriously those GUT symmetries that may constitute their acceptable "low energy" limit such as E_6 and to ask how such low energy limits differ from the present standard model; for example E_6 implies extra Z^0 bosons the possible consequences of which must be borne in mind at appropriate junctures as one now reads on.

---

[*] I must be forgiven for permitting my administrative preoccupations to obtrude.

AXIONS

A nagging problem in the strong force arising from QCD is the P- and CP-breaking effects in the QCD Lagrangian associated with instantons. The Peccei-Quinn solution is to impose a global U(1) symmetry on the total Lagrangian (QCD plus electroweak) following which the P- and CP-violating terms can be transformed to zero. The extra U(1) symmetry could be imposed in various ways but would always be broken by the spontaneous symmetry-breaking in the electroweak sector; this must imply the existence of a Goldstone-boson-like particle, the axion, which must acquire a small mass (undetermined) through anomaly effects. We must therefore look broadly for such particles.

The simplest implementation of the axion namely a two-doublet extension of the standard model is looking improbable; for example it fails to show up as expected in $K^+ \to \pi^+ a$ [16]. Alternatively we could increase the Peccei-Quinn scale above the weak which results in the "invisible axion"[17] which may nevertheless be put into evidence by its conversion to microwave photons in a magnetic field or by other means[18]. More generally we might break universality and generalize axion-quark and axion-lepton couplings[19].

Such latter axions may perhaps be (extremely tentatively) identified with the monokinetic electron-positron pairs observed in heavy ion collisions at GSI Darmstadt that might be manufactured by photon (laser/wiggler) collisions with relativistic electrons[20].

The high importance that surrounds the axion requires us to examine all places where it might manifest itself. One such is in association with electromagnetic transitions between states of complex nuclei. In particular, axions might be emitted as an alternative to such normal electromagnetic processes when their e^+e^- decay would be superimposed upon the internal pair conversion of the regular electromagnetic decay. This superposition would distort the e^+e^- angular correlation expected for the normal multipole or multipole mixture of the regular decay and so might so manifest itself. In the language of the trade, such axion emission would spoil the analysis of "multipole meter" experiments in complex nuclei; such effects should be seriously pursued[21].

T-INVARIANCE

The recent warnings[22] that a non-null result in traditional T-invariance tests might be due not to T-violation but to a breakdown of rotational Lorentz invariance are adding significance to tests of such rotational invariance; this is, at present, most sharply checked through the precessional frequency of the (J = 3/2) nucleus of ^{201}Hg which sets a limit of about 3×10^{-21} on a breakdown of fundamental Lorentz invariance if that is a $(v/c)^2$ consequence of motion relative to the $3°K$ background radiation[23].

CP IN K_L^0 DECAY

This topic is treated in detail elsewhere at this conference but its central position as the only definite failure of CP-invariance so far detected in Nature merits a word here. I want just to direct attention to two points: (i) the "automatic" violation of CP prescribed by the Kobayashi-Maskawa matrix[24]; (ii) the relevance of K_L^0-decay for CPT.

(Note in passing that the effect of CP-violation in the QCD Lagrangian, to which reference has been made in the section on axions, is limited by, for example, the electric dipole moment of the neutron – see below – to a value so small as to be negligible for CP-violation in K_L^0 decay as compared with experiment.)

As a preliminary consider $K_L^0 \to 2\pi$ which may proceed by the (unavoidable) $\Delta S=2$ mechanism which leads to identical amplitudes for $\pi^+\pi^-$ and $\pi^0\pi^0$ final states: $\eta_{+-} = \eta_{00} = \varepsilon$ and through a $\Delta S = 1$ mechanism such that $\eta_{+-} = \varepsilon + \varepsilon'$ and $\eta_{00} = \varepsilon - 2\varepsilon'$.

"Superweak" hypotheses[25] have $\varepsilon' = 0$; others have $\varepsilon' \neq 0$ as we shall shortly see. The present result, to be improved, is[26] $\varepsilon'/\varepsilon = -0.0025 \pm 0.0047$.

The Kobayashi-Maskawa (KM) observation[24] was that for 3 (or more) generations the weak $d', s', b'...$ quark eigenstates are related to the mass eigenstates $d, s, b...$ through a matrix that couples the W-bosons to the weak current and that must involve an irreducible CP-violating phase that could be zero only by <u>fiat</u> i.e. by some over-riding symmetry principle not required elsewhere so far as we know. To this degree the standard $SU(3)_c \otimes SU(2)_L \otimes U(1)$ model "automatically" violates CP. It also "automatically" entrains a finite value of ε' such that[25] if it accounts correctly for the experimental value of ε, $7 \times 10^{-3} > |\varepsilon'/\varepsilon| > 10^{-3}$ consistent with present data.

This raises the question as to whether or not all K_L^0 phenomena are consistent with such a minimal KM parameterization. If they were so to be, this would in no way prove that other influences were not additionally in play but it would certainly place the onus on such as may wish to promote those other influences to demonstrate their specific relevance.

The consistency or otherwise of the KM account must accommodate:

    (i) super-allowed Fermi beta-decay;
   (ii) hyperon decay;
  (iii) the B-meson's lifetime and its branches;
  (iv) the t-quark mass (effectively a logarithmic dependence);
   (v) η_{+-}/η_{00} ;
  (vi) ε'/ε.

It appears[27] that all these data may be consistent within the KM scheme but that, if they are, CP-violation is probably near maximal (which may be a recommendation). It has recently been pointed out[28] that a constant that multiplies ε in the minimal theoretical expression may, on account of rather tricky chiral loop corrections, be significantly larger than the value deriving from naive chiral perturbation theory; this would ease accommodation with KM.

The importance of studying CP-violating effects in K_S^o is emphasized elsewhere in this conference.

Should KM be found to be violated there are many directions in which we might turn:

- (i) Superweak theories (that would involve a new gauge boson of mass greater than 10^3 TeV);
- (ii) A generalization of the KM matrix through 4 or more generations; (this then carries no definite prediction as to ε');
- (iii) $SU(2)_L \otimes SU(2)_R \otimes U(1)$ which seems to be quite acceptable;
- (iv) Multiple Higgs doublets for the electro-weak spontaneous symmetry breaking;
- (v) Supersymmetric theories (who knows?);
- (vi) GUTs (who knows?).

An immediate remark in relation to (iv) of the above listing is that it is often reported that the multiple Higgs doublet solution is not acceptable because it predicts (in the simplest relevant model of 3 Higgs doublets) $\varepsilon'/\varepsilon \simeq -1/20$ which conflicts with the experimental value reported above. However, it has been observed[29] that proper respect for chiral invariance reduces the theoretical expectation to $\varepsilon'/\varepsilon \simeq -0.006$ (with an uncertainty of a factor of 2 or 3 either way) which is consistent with the present experimental figure. (We must, however, note the possibly unfavourable position in respect of the electric dipole moment of the neutron as we shall do shortly).

CP-VIOLATION AND NUCLEAR EFFECTS

If the KM model for CP-violation in K_L^o is the sole source of that violation then effects outside that system will probably be undetectably small e.g.

- (i) the neutron electric dipole moment;
- (ii) T-odd effects in beta-decay.

However $SU(2)_L \otimes SU(2)_R \otimes U(1)$ has T-odd effects in first order[30], unlike the extended standard model with, for example, multiple Higgs so that the T-odd correlation in beta-decay $D\vec{J}\cdot(\vec{p}_e \times \vec{p}_\nu)/E_e E_\nu$ might have a D-value of 10^{-3} or 10^{-4}. Present values of D are, for the neutron[31], $-(0.5 \pm 1.5) \times 10^{-3}$ and, for ^{19}Ne [32], $(1 \pm 6) \times 10^{-4}$ which by no means exclude the L-R symmetric form.

Expectation for the neutron electric dipole moment (in e.cm) are:

Superweak	$10^{-29} - 10^{-30}$
Kobayashi-Maskawa	$10^{-29} - 10^{-31}$
$SU(2)_L \otimes SU(2)_R \otimes U(1)$	$10^{-26} - 10^{-27}$
Super-symmetric	$O(10^{-26})$
Multiple Higgs	$O(10^{-24})$
GUTs	$O(10^{-25})$

to be compared with experimental values of $-(3.3 \pm 3.5) \times 10^{-25}$ [33], $-(2 \pm 1) \times 10^{-25}$ [34] and $<4 \times 10^{-25}$ [34]. The only one of the listed possibilities that may be being questioned by these figures is, as remarked above, multiple Higgs.

There is considerable developing interest in the use of neutral atoms for electric dipole tests for the nucleus[35]. Mixing of adjacent nuclear states can give considerable enhancement of the intrinsic nucleonic effect (by a factor of 100 or even more[35]). The present best atomic limit of $-(0.3 \pm 1.1) \times 10^{-26}$ e.cm for ^{129}Xe [36] is not much weaker, when referred to the nucleon, than is the above limit for the neutron but may be susceptible of improvement by as much as a factor 10^4 whereas the limit from the neutron will be very difficult to sharpen by more than an order of magnitude.

Although one most naturally thinks of simple systems in the context of conservation tests it may be that macroscopic experiments can be more sensitive. Thus it is suggested[37] that the magnetization of a paramagnet by an electric field, illustratively using 50 gm of EuS near the Curie Point, could reach a limit of 10^{-28} e.cm for the electric dipole moment of the electron; this should be compared with the present limit of about 10^{-24} e.cm that may be alternatively inferred from the ^{129}Xe experiment referred to above.

Some theoretical possibilities for the electron electric dipole moment are, in e.cm:

Scalar-lepton couplings	$< 4 \times 10^{-25}$
Massive Dirac neutrinos	$< 10^{-27}$
Family symmetries	$< 10^{-27}$
Higgs models	$\simeq 10^{-32}$

CPT-TESTS

Much attention has focussed in recent years on C-, P-, CP- and T-tests but comparatiely little on CPT itself which has come to assume an unwarranted sanctity chiefly because we should be at a loss to know how to respond should it be found to fail, stemming as it does from very basic precepts of relativistic field theory. (See, however, e.g. ref.38). Indeed, some checks look very impressive for example that the difference of the g-factors of electron and positron is measured as less than 1 part in 10^{11} of the average value[37] while it is a standard result that the $K^0-\bar{K}^0$ mass difference may be inferred to be no more than 6×10^{-19} (90% CL) of the average value. However, such numbers have diagnostic weight only in relation to some postulated mechanism through which the symmetry might be broken (a remark general to all symmetry tests of course); such a mechanism might leave certain CPT-based equalities intact while subverting others so that tests at all possible points are in order.

The only present suggestion of possible CPT-breaking comes from the CP-violating 2π decays of K_L^0 [40]. With the usual definitions we have the phase angles for $\pi^+\pi^-$ and $\pi^0\pi^0$ decay of ϕ_{+-} =$(44.6 \pm 1.2)°$ and ϕ_{00} = $(54.5 \pm 5.3)°$ respectively so that the two angles appear to be perhaps significantly different from one another. Now there is no fundamental reason why ϕ_{+-} and ϕ_{00} should be equal but if CPT is valid they should split either side of the superweak (ϕ_{SW}) expectation for both ϕ_{+-} and ϕ_{00} namely ϕ_{SW} = arc tan $(2\Delta m \tau_s/\hbar)$ where Δm is the K_L^0 - K_S^0 mass difference and τ_s is the K_S^0 mean lifetime, such that $\phi_{00} - \phi_{SW} = 2(\phi_{SW} - \phi_{+-})$. Now $\phi_{SW} = (43.67 \pm 0.14)°$ so that experimentally $\phi_{SW} - \phi_{+-} = -(0.9 \pm 1.2)°$, $\phi_{00} - \phi_{SW} = (10.8 \pm 5.4)°$ and "CPT is violated at the 2 standard deviation level". Improvement, particularly of ϕ_{00} is obviously most desirable. (As also of ε', the $\Delta S = 1$ contribution referred to above, (some) knowledge of which is important for this discussion of CPT).

(We may note that a theory of higher dimensionality such as superstrings does not display C,P,T as normally considered until after compactification. These "low temperature" symmetries then emerge on the m_p/m_x scale and it is perhaps not inconceivable that the CPT of our own world might carry some small memory of the phase transitions that gave rise to it. Even Lorentz invariance might be subject to this same remark.)

CONSISTENCY OF $SU(2)_L \otimes U(1)$

Consistency of the standard electro-weak model is referred to determinations of the Weinberg Angle by various methods. Current values of $\sin^2\theta_W$, radiative corrections being taken into account, are:

0.221 ± 0.016   (ed)
0.228 ± 0.009 (ν,eN)
0.223 ± 0.030 $(m_W \& m_Z)$
0.223 ± 0.006 $(m_W \& \tau_\mu)$

(for references see I and ref. (41))

Alternatively we may note that the experimental $m_W = 82.0 \pm 1.1$ GeV combines through the standard model with the muon lifetime for the expectation $m_Z = 93.0 \pm 1.2$ GeV against the experimental $m_Z = 92.8 \pm 1.6$ GeV.

Another important set of tests relates to the neutral weak predictions:

$$g_A^e g_A^{\mu,\tau} = 1/4$$
$$g_V^e g_V^{\mu,\tau} = (1-4\sin^2\theta_W)/4 \simeq 3.6 \times 10^{-3}$$

Forward-backward asymmetry measurements in $e^+e^- \to \mu^+\mu^-, \tau^+\tau^-$ give:

$$g_A^e g_A^\mu = 0.25 \pm 0.03 \quad [42]$$
$$g_A^e g_A^\tau = 0.24 \pm 0.05 \quad [42,43]$$
$$g_V^e g_V^\mu = -0.02 \pm 0.10 \quad [42]$$
$$g_V^e g_V^\tau = 0.06 \pm 0.10 \quad [43]$$

while νe scattering gives[44] 0.250 ± 0.013 for the axial and -0.019 ± 0.024 for the vector couplings.

These e^+e^- measurements were at $\sqrt{s} = 29$ GeV. Measurements at $\sqrt{s} = 34.5$ GeV[45] have given, for $\mu^+\mu^-$ production, a forward-backward asymmetry of -0.095 ± 0.010 as against the standard model's expectation of -0.076 ± 0.001 so some small discrepancy may be suggested. Other measurements[46] up to $\sqrt{s} = 44.6$ GeV show good agreement with the standard model, at a somewhat lower level of accuracy, so there is no suggestion of a run-away breakdown.

It is interesting to note that the taxonomy of hadronic jets permits the sorting-out of events in e^+e^- annihilation that originate in $e^+e^- \to b\bar{b}$. These events have[47] an asymmetry (at $\sqrt{s} = 34.5$ GeV) of $-22.7 \pm 6.5\%$ as against the expectation of -25.2%. (The asymmetry goes as the reciprocal of the charge of the product particles so is expected to be 3 times greater for $b\bar{b}$ than for $\mu^+\mu^-$ or $\tau^+\tau^-$).

It must be most strongly emphasized that these consistency tests are no more than that: they merely test the standard model as a continuing-valuable benchmark.

V-A AND WEAK CHARGED UNIVERSALITY

V-A does not come out of the standard electroweak model: it is put in. This makes it the more important to check for V-A against, in particular, signs of a possible right-hand current but more generally also for other possible couplings. The situation was reviewed in I but here we might notice the four standard parameters of muon decay and their comparisons with the expectation of V-A:

Parameter	V-A	Exp.
ρ	3/4	0.7517 ± 0.0026
δ	3/4	0.7499 ± 0.0043
η	0	-0.003 ± 0.030
ξ	1	0.9989 ± 0.0023

The superficial comparison is very close but, in fact, checks V-A only to within several percent, $|g_A/g_V| = 1.028^{+0.051}_{-0.086}$, although the helicity of the muon neutrino is established to be within 0.4% of unity. Muon precession studies[48] give the best "direct" limit yet on the mass of a possible right-hand W-boson: $m_R > 5m_L$ although one may perhaps infer a W_R/W_L mass ratio of more than 20 from the $K_L^o - K_S^o$ mass difference[49].

Somewhat similarly, parity experiments in nuclear beta-decay are consistent with V-A but do not exclude V+A at the 10% level (see I).

If we assume V-A for the muon (and nucleon) we have, after allowance for outer radiative corrections:

$$g_\mu = (1.4357 \pm 0.0001) \times 10^{-49} \text{ erg cm}^3$$
$$g_{\beta V} = (1.41271 \pm 0.00032) \times \text{ "}$$
$$g_{NV} = (1.4519 \pm 0.0011) \times \text{ "}$$

where the change between $g_{\beta V}$ and g_{NV} reflects the Cabibbo Angle as to which the standard model offers no clue. (In the Kobayashi-Maskawa parameterization, using standard nomenclature, nucleon decay is proportional to $\cos\theta_1$, while hyperon decay is proportional not to $\sin\theta_1$, as in Cabibbo theory but to $\sin\theta_1 \cos\theta_3$. We have: $\sin^2\theta_3 \simeq [4 \times 10^{-14} \sec/\tau_B] \times (b \to u)$ where τ_B is the lifetime of the B-meson and $(b \to u)$ is its $b \to u$ branching ratio. Putting in experimental values we have $\cos\theta_3 > 0.9995$ so we may effectively equate θ_1 with the classical Cabibbo Angle). Some unquantified uncertainty still attaches to the Cabibbo Angle owing to the effects of SU(3) symmetry-breaking. Experimentally, therefore $g_{NV} - g_\mu = (1.12 \pm 0.08)\%$. The expectation of the standard model for this quantity, the difference of the inner radiative corrections of nucleon and muon, is 1.05%.

Universality in pion decay is established to about 1% by comparison of the experimental branching ratio[50] for $(\pi \to e\nu)/(\pi \to \mu\nu)$ of $(1.218 \pm 0.014) \times 10^{-4}$ against the expectation, including radiative corrections (which amount to 3.5% in this case), of 1.233×10^{-4}.

Universality in tau decay is established to about 10% by the measurement[51] of the branching ratio $(\tau \to \mu\nu\bar{\nu})/(\tau \to e\nu\bar{\nu})$ of 0.989 ± 0.080; the expectation is 0.973. (The tau lifetime itself[52] of $(3.17 \pm 0.38) \times 10^{-13}$ sec compares with the expectation of $(2.64 \pm 0.15) \times 10^{-13}$ sec the uncertainty in which is due to the many hadronic decay modes.)

(Note that these comparisons involve charged-current couplings. The $e^+e^- \to \mu^+\mu^-, \tau^+\tau^-$ measurements earlier reported[42,43] enable one to make a test of universality in the neutral-current couplings: $g_A^\tau/g_A^\mu = 0.97 \pm 0.25$.)

PARITY NON-CONSERVATION IN ATOMS

The Z^0 can mediate the interaction between atomic electrons and the nucleons of the nucleus and so imposes a weak parity-non-conserving component upon the parity-conserving electromagnetic interaction predominantly responsible for the atom's structure. This can manifest itself in a circular polarization of the light coming from unpolarized atoms and in other ways. This matter was addressed in I since which time certain of the measurements have been sharpened[53] and are summarized below:

Atom	Transition	Experiment	Theory
Cs	$6S_{1/2} \to 7S_{1/2}$	8.5 ± 0.6	9.3 ± 1
Tl	$6P_{1/2} \to 7P_{1/2}$	13 ± 2	10 ± 3
Pb	$6P^2$; $J=0 \to 1$	-9.9 ± 2.5	-10 ± 3
Bi	$6P^3$; $J=3/2 \to 3/2$	-10 ± 1.5	-10 ± 3
	$6P^3$; $J=3/2 \to 5/2$	-9.3 ± 1.4	-12 ± 3

In this listing experiment and theory have been reduced to common units which it is not necessary to specify for our present purposes of comparison.

Relevant remarks are: (i) the overall accuracy of the present set of measurements is effectively about ±5%; (ii) ±1% in these atomic measurements corresponds to about ±0.003 in $\sin^2\theta_W$; (iii) present comparison between theory and experiment is limited by uncertainties in the electronic structure (at the nucleus) of the heavy atoms that are used (heavy because of the Z^2 and Z^3 coherent effects involved - see I). From these remarks it is seen that if we can gain appropriately enhanced confidence in the (electromagnetic) atomic calculations and if we can, in a new generations of experiments, as should certainly be possible, sharpen the experimental data by a modest factor then we should be determining $\sin^2\theta_W$ within the standard model to a precision comparable with that of other approaches as displayed above. The significance of this would be not so much as simply to add one more determination of $\sin^2\theta_W$ to the present list as to effect a test of the standard model through a measurement at zero momentum transfer to be compared with the results at high-momentum-transfer of other approaches; extension of the standard model would affect low and high-momentum-transfer measurements differently.

It similarly goes without saying that these atomic measurements are sensitivity to the existence of a right-handed Z-boson and to other possible neutral weak bosons of any kind.

Before leaving the realm of the atomic electrons we may notice the possibility of "neutral beta decay"[54]. The almost-completely misnamed phenomenon of gamma-ray internal conversion is due chiefly to the electro-magnetic interaction with the atomic electrons of the nucleons in an excited nuclear state which, on making the transition to the final nuclear state may eject an atomic electron as an alternative to emitting a gamma-ray. Such ejection can obviously also take place by virtue of the mediation of the electron-nuclear interaction by the Z^o: "neutral beta-decay". Normally neutral beta-decay would be only a tiny addition to the regular internal conversion but in circumstances, such as indeed arise in practice, where the gamma-decay is very heavily retarded neutral beta-decay and electromagnetic internal conversion might well compete to the degree that the apparent internal conversion coefficient would assume an anomalous value. Cases of anomalous internal conversion coefficients are well recognized but it would be premature to identify them with the influences of neutral beta-decay. Further study might be rewarding.

LEPTON NUMBER

The apparent excellent conservation of lepton family number is a great mystery, presumably connected with the generation mystery and the interconnections, or otherwise, in the lepton sector. That the quark sector interconnects via the Kobayashi-Maskawa matrix may be due to finite mass effects while similar connections in the lepton sector may be kinematically forbidden by mass degeneracy of the neutrinos. But we do not know and, of course, the standard model is by definition silent since it says nothing about generations while neither the Z^o nor the Higgs particle have inter-generation lepton couplings. The following listing, slightly updated from I, gives present limits on lepton-non-conservation.

Process	Branch	
$\mu^+ \to e^+ + \gamma$	$< 5 \times 10^{-11}$	[55]
$\mu^+ \to e^+ + e^+ + e^-$	$< 2.4 \times 10^{-12}$	[56]
$\mu^- + Ti \to e^- + Ti$	$< 4 \times 10^{-12}$	[57]
$K^o_L \to \mu e$	$< 1 \times 10^{-8}$	
$K^+ \to \pi \mu e$	$< 1 \times 10^{-8}$	

Although many of these limits are very impressive we cannot make anything of them except in the context of specific models that subvert, explicitly or implicitily, lepton conservation. Such possibilities abound:

   (i) Additional Higgs bosons;
  (ii) Flavour-changing neutral gauge bosons;
 (iii) Extended technicolor gauge bosons: M > 75 TeV;
  (iv) Composite models: mass scale > 500 TeV;
   (v) Light lepto-quarks;
  (vi) Supersymmetric partners of $SU(2)_L \otimes U(1)$ gauge bosons;

(vii) New electroweak interactions;
(viii) GUTs

and so on.

NEUTRINO MASSES

I am not going to discuss neutrino masses or oscillations (free or induced) except to remark that if a mass of about 30 eV for the electron neutrino were correct then the present upper limits of 0.25 MeV and 50 MeV (both 90% CL) for the masses of the muon and tau neutrinos respectively would exclude models in which the neutrino masses scale as the square of the masses of their charged lepton partners since such scaling would result in masses of 1.28 MeV and 366 MeV for ν_μ and ν_τ respectively. Finite neutrino masses arise naturally in many GUTs e.g. SO(10), as also in $SU(2)_L \otimes SU(2)_R \otimes U(1)$.

DIRAC AND MAJORANA NEUTRINOS

Before the discovery of parity non-conservation we thought that we knew that the neutrino was a Dirac particle, distinct from its own anti-particle, and not a Majorana particle, identical with its own anti-particle. This was because double beta-decay, which with Dirac neutrinos involved the "simultaneous but independent" conversion of two neutrons into two protons with the emission of two electrons and two anti-neutrinos, was at least 10^4 times slower than expected on the basis of Majorana neutrinos where the (anti)-neutrino emitted (virtually) in the first $n \to pe\nu$ process could have been absorbed (as a neutrino) in the second $\nu n \to pe$ process in the same nucleus resulting in neutrinoless double beta-decay so far as external appearances (and phase space) were concerned. But with the discovery of parity non-conservation and the possibility that the (zero mass) anti-neutrino associated with negative electron emission is an absolutely (right) handed particle the argument collapsed because a neutrino of one chirality would be emitted in the first (decay) process but one of the opposite chirality would be demanded for the second (absorption) process to achieve neutrinoless double beta-decay with Majorana neutrinos. So, far from being much faster than two-neutrino double beta-decay, neutrinoless double beta-decay would be absolutely forbidden for Majorana as well as for Dirac neutrinos.

However, if neutrinos have a finite mass they cannot have absolute chirality; we should then expect that neutrinoless double beta-decay, while still being of zero probability for Dirac neutrinos, would proceed for Majorana neutrinos at a rate of about $(m_\nu c^2/E)^2$ of that expected in pre-parity-non-conservation days where E is the transition energy.

So far no cases of neutrinoless double beta-decay has been found. Limits that exist for decay rates in ^{76}Ge, ^{82}Se, ^{128}Te and ^{130}Te give together a face value upper limit of about 1 eV on the Majorana neutrino mass. However, caution must be exercised because calculations, of which there are many, of the rates of two-neutrino double beta-decay, agree only in predicting liftimes shorter by at least an order of magnitude than the well-established values or limits (geochemically determined) for ^{82}Se, ^{130}Te and ^{150}Nd. It would seem best, at this juncture, to say, conservatively, no more than that a Majorana neutrino mass must be less than about 10-20 eV.

With respect to the possibility that the mass of the electron neutrino might be 30 eV it may seen tempting to say that the above upper limit on a Majorana neutrino mass demonstrates that the neutrino cannot be a Majorana mass eigenstate. However, this could not be inferred with certainty owing to the possibility of conspiracy associated with the mixing with neutrinos belonging to other generations.

Of course, these considerations are also affected by the possible existence of right-hand currents but these lead us into a further realm of model dependence and I will not discuss them here.

REFERENCES

1. Denys Wilkinson J. Phys. Soc. Jpn. 55, 347 (1986) Suppl.
2. T.D. Lee and C.N. Yang Phys. Rev. 98, 1501 (1955).
3. E. Fischbach, D. Sudarsky, A. Szafer, C. Talmadge and S.H. Aronson Phys. Rev. Lett. 56, 3 (1986).
4. P. Herczeg, Phys. Rev. D28, 200 (1983).
5. see e.g. J.D. Barrow and F.J. Tipler The Anthropic Cosmological Principle (Clarendon Press, Oxford, 1986).
6. S.A. Bludman and M.A. Ruderman Phys. Rev. Lett. 38, 255 (1977).
   H.J. Blome and W. Preister Naturwiss. 71, 456, 515 and 528 (1984).
7. P. Langacker Phys. Rep. 72, 185 (1981).
8. see R.J. Tayler (Quarterly Journal; Royal Astronomical Society, in course of publication) for a recent review.
9. see e.g. D.D. Clayton in Nucleosynthesis eds. W.D. Arnett and J.W. Truran (Chicago University Press 1986) p.65.
10. H.V. Klapdor and K. Grotz Astrophys. J. Lett. 301, L39 (1986).
11. A. Yahil, D. Walker and M. Rowan-Robinson Astrophys. J. 301, L1 (1986).
    A. Meiksin and M. David Astron. J. 91, 191 (1986).
12. see e.g. D.N. Schramm in Essays in Nuclear Astrophysics eds. C.A. Barnes, D.D. Clayton and D.N. Schramm (Cambridge University Press 1982) p.325.
13. E.R. Harrison Phys. Rev. D1, 2726 (1970).
    Y.B. Zeldovich Mon. Not. R. Astron. Soc. 160, 1 (1970).
14. see e.g. J. Bagger and S. Dimopoulos Nucl. Phys. B244, 247 (1985).
15. L. Lyons Prog. Part. Nucl. Phys 10, 227 (1983).
16. Y. Asano et al Phys. Lett. 107B, 159 (1981).
17. M. Dine, W. Fischler and M. Srednicki Phys. Lett. 104B, 199 (1981).
18. L. Krauss, J. Moody, F. Wilczek and D.R. Morris Phys. Rev. Lett. 55, 1797 (1985).

19. L. Krauss and F. Wilczek unpublished.
20. S.J. Brodsky, E. Mottola, I.J. Muzinich and M. Soldate Phys. Rev. Lett. 56, 1763 (1986).
21. F.P. Calaprice private communication.
22. M.J. Moravcsik, Phys. Rev. Lett. 48, 718 (1982).
    F. Arash, M.J. Moravcsik and G.R. Goldstein, Phys. Rev. Lett. 54, 2649 (1985).
23. R.J. Raab, private communication.
24. M. Kobayashi and K. Maskawa, Prog. Theor. Phys. 49, 652 (1973).
25. L. Wolfenstein Phys. Rev. Lett. 13, 569 (1964) and Ann. Rev. Nucl. Part. Sci. to be published.
26. R.H. Bernstein et al., Phys. Rev. Lett. 54, 1631 (1985).
    J.K. Black et al., Phys. Rev. Lett. 54, 1628 (1985).
27. B.R. Holstein Nucl. Phys. A434, 525c (1985).
    L. Wolfenstein, Comments Nucl. Part. Phys. 14, 135 (1985).
28. J. Bijnens, H. Sonoda and M.B. Wise Phys. Rev. Lett. 53, 2367. (1984)
29. J.F. Donoghue and B.R. Holstein Phys. Rev. D32, 1152 (1985).
30. P. Herczeg in Neutrino Mass and Low Energy Weak Interactions, Telemark, 1984, ed. V. Barger and D. Clive (World Scientific, Singapore, 1985).
31. R.I. Steinberg et al., Phys. Rev. D13, 2469 (1976).
    B. Erozolimski et al., Sov. J. Nucl. Phys. 28, 48 (1978).
32. A.L. Hallin et al., Phys. Rev. Lett. 52, 337 (1984).
    R.M. Baltrusaitis and F.P. Calaprice, Phys. Rev. Lett 38, 464 (1977).
33. J.M. Pendlebury et al., Phys. Lett. 136B, 327 (1984) and private communication.
34. I.S. Altarev et al., Phys. Lett. 102B, 13 (1981).
    V.M. Lobashev et al., Proc. Int. Conf. Neutrino 82 (Balaton, Hungary 1982), vol. 2, p.107.
35. W.C. Haxton and E.M. Henley, Phys. Rev. Lett. 51, 1937 (1983).
    V.V. Flambaum, I.B. Khriplovich and O.P. Sushkov, Phys. Lett. 146B, 367 (1984).
36. T.G. Vold et al., Phys. Rev. Lett. 52, 2229 (1984).
37. W. Bialek, J. Moody and F. Wilczek Phys. Rev. Lett. 56, 1623 (1986).
38. R.M. Wald Phys. Rev. D21, 2742 (1980).
    I.I. Bigi Zeits. f. Phys. C12, 235 (1982).
39. R.S. Van Dyck, P.B. Schwinberg and H.G. Dehmelt in Atomic Physics 9 (World Scientific, Singapore, 1984).
40. V.V. Barmin et al., Nucl. Phys. B247, 293 (1984).
41. D. Bogert et al., Phys. Rev. Lett. 55, 1969 (1985).
    K. Abe et al., Phys. Rev. Lett. 56, 1107 (1986).
42. W.W. Ash et al., Phys. Rev. Lett. 55, 1831 (1985).
43. E. Fernandez et al., Phys. Rev. Lett. 54, 1620 (1985).
44. W. Krenz reported in W. Bartel et al., Zeits. f. Phys. C30, 371 (1986).
45. M. Bohm, W. Hollik and H. Spiesberger, Zeits. f. Phys. C27, 523 (1985) and see ref. 44.
46. B. Adeva et al., Phys. Rev. Lett. 55, 665 (1985).
47. W. Bartel et al., Phys. Lett. 146B, 437 (1984).

48. D.P. Stoker et al., Phys. Rev. Lett. $\underline{54}$, 1887 (1985).
49. G. Beall, M. Bander and A. Soni Phys. Rev. Lett. $\underline{48}$, 848 (1982).
    A. Gangopadhyaya, Phys. Rev. Lett. $\underline{54}$, 2203 (1985).
50. D.A. Bryman et al Phys. Rev. Lett. $\underline{50}$, 7 (1983).
51. R.M. Baltrusaitis et al. Phys. Rev. Lett. $\underline{55}$, 1842 (1985).
52. J.A. Jaros et al., Phys. Rev. Lett. $\underline{51}$, 955 (1983).
    E. Fernandez et al., Phys. Rev. Lett. $\underline{54}$, 1624 (1985).
53. S.L. Gilbert et al., Phys. Rev. Lett. $\underline{55}$, 2680 (1985).
    C.E. Tanner and E.D. Commins Phys. Rev. Lett. $\underline{56}$, 332 (1986) and D.N. Stacey private communication.
54. W.C. McHarris and J.O. Rasmussen unpublished.
55. R.D. Bolton et al., Phys. Rev. Lett. $\underline{56}$, 2461 (1986).
56. W. Bertl et al., Nucl. Phys. $\underline{B260}$, 1 (1985).
57. D.A. Bryman et al., Phys. Rev. Lett. $\underline{55}$, 465 (1985); and unpublished.

CP VIOLATION IN THE NEUTRAL KAON SYSTEM

Bruce Winstein[†]
Department of Physics, Stanford University, Stanford, CA 94305
and
Stanford Linear Accelerator Center, Stanford, CA 94305

ABSTRACT

The phenomenology of CP-nonconservation in the neutral kaon system is presented independent of phase convention. Current progress towards a measurement of "direct CP violation" is described. Finally a test of CPT violation within the neutral kaon system is discussed.

$K^0 - \bar{K}^0$ MIXING

As a result of the violation of strangeness conservation in the weak interaction, the phenomenon of neutral kaon mixing occurs. Then, if CP is a good symmetry, we would expect two orthogonal eigenstates of the weak Hamiltonian which are themselves CP eigenstates.

The mixing phenomenon, even in the absence of CP conservation, can be described by the time-dependent Schroedinger equation

$$H\psi = (id/dt)\psi \qquad (1)$$

where H can be written as:

$$H = M - i\Gamma/2 \qquad (2)$$

While H itself is non-Heritian, both M (the mass matrix) and Γ (the decay matrix) are in general separately Hermitian so that the Hamiltonian can then be written as:

$$H = \begin{pmatrix} M - i\Gamma/2 & M_{12} - i\Gamma_{12}/2 \\ M_{12}^* - i\Gamma_{12}^*/2 & M - i\Gamma/2 \end{pmatrix} \qquad (3)$$

[†] On leave of absence from Enrico Fermi Institute and The Department of Physics, The University of Chicago, Chicago, Il 60637

The two component wave function ψ is the amplitude to be a K^0, $\bar{K}^0$. $M = <K^0|M|K^0> = <\bar{K}^0|M|\bar{K}^0>$ where the last equality follows from CPT invariance. Similarly,
$\Gamma = <K^0|\Gamma|K^0> = <\bar{K}^0|\Gamma|\bar{K}^0>$. $M_{12} = <K^0|M|\bar{K}^0>$; $\Gamma_{12} = <K^0|\Gamma|\bar{K}^0>$.

The weak eigenstates are found by diagonalizing H:

$$K_{L,S} = \frac{K^0 \pm \alpha \bar{K}^0}{\sqrt{1+|\alpha|^2}} \quad (4)$$

where

$$\alpha = \sqrt{\frac{M_{12}^* - i\Gamma_{12}^*/2}{M_{12} - i\Gamma_{12}/2}} \quad (5)$$

The $K_L - K_S$ mass difference is given by

$$\Delta m = 2|M_{12}| \quad (6)$$

while the width difference is

$$\Delta \Gamma = -2|\Gamma_{12}|; \quad (7)$$

the choice of the minus sign results from the empirical observation that the heavier of the two states is the longer lived.

It is clear that the above measurable quantities, Δm and $\Delta \Gamma$, are expressed independent of whatever phase convention is adopted. This is not the case for the mixing parameter α. It is easily seen that the phase of α can be chosen at will by the appropriate adjustment of the relative K^0 and $\bar{K}^0$ phase (which is not in itself measurable) so that in fact only its modulus is physically relevant.

The modulus of α is equal to one only if M_{12} and Γ_{12} are relatively real. In this case, the weak eigenstates are also CP eigenstates so that the measure of CP violation in $K^0 - \bar{K}^0$ mixing is the phase <u>difference</u> between the off-diagonal elements of the mass and decay matrices. Simplifying the above expression for α, we have

$$|\alpha| = 1 + \frac{|\Gamma_{12}|^2/2}{|M_{12}|^2 + \frac{1}{4}|\Gamma_{12}|^2} \text{Im}\left(\frac{M_{12}}{\Gamma_{12}}\right)$$

$$= 1 + \frac{x}{2(x^2 + \frac{1}{4})} \phi_{\Delta S=2}$$

$$= 1 + \frac{\phi_{\Delta S=2}}{2} \qquad (8)$$

where we have used[1] $x \equiv \frac{\Delta m}{\Delta \Gamma} = 0.477$, and

$$\phi_{\Delta S=2} \equiv \arg\left(\frac{M_{12}}{\Gamma_{12}}\right).$$

The parameter $\phi_{\Delta S=2}$ (its departure from 0 or π) is a measure of the degree of CP nonconservation in the mixing. It is most easily determined from a measurement[1] of the semileptonic charge asymmetry in K_L decays:

$$\delta_L = \frac{\text{Rate }(\pi \ell \bar{\nu}) - \text{Rate }(\pi \bar{\ell} \nu)}{\text{Rate }(\pi \ell \bar{\nu}) + \text{Rate }(\pi \bar{\ell} \nu)}$$

$$= 3.30(12) \times 10^{-3}$$

$$= \frac{1 - |\alpha|^2}{1 + |\alpha|^2} \simeq -\frac{\phi_{\Delta S=2}}{2} \qquad (9)$$

Thus, from the experimental result, we have

$$|\alpha| = 0.99671(13); \text{ and} \qquad (10)$$

$$\phi_{\Delta S=2} = -6.58(26) \times 10^{-3}. \qquad (11)$$

The experimental knowledge of the off-diagonal matrix elements can then be expressed as:

$$\frac{M_{12}}{\Gamma_{12}} = -0.4773(23)(1 - i6.58(26) \times 10^{-3}) \qquad (12)$$

DYNAMICS

What is the origin of the phase difference between the off-diagonal elements of the mass and decay matrices? To get some insight into the possibilities, we express both to second order in perturbation theory:

$$M_{12} = <K^0|H_W|\bar{K}^0> + \sum_n \frac{<K^0|H_W|n><n|H_W|\bar{K}^0>}{E_n - M_{K^0}}, \qquad (13)$$

where the first term represents a possible $\Delta S = 2$ term in the weak Hamiltonian and the second term is the sum over all virtual intermediate states, and,

$$\Gamma_{12} = 2\pi \sum_F \rho_F <K^0|H_W|F><F|H_W|\bar{K}^0> \qquad (14)$$

where ρ_F is the density of final states for the state F $(2\pi, 3\pi, \gamma\gamma, ...)$.

From the above expressions, we can delineate some possibilities for the origin of the ~ 6.5 mr phase difference:

A. All real and virtual K meson transition elements are relatively real. In this case, the phase difference arises from the first term in expression (13). This hypothesis is known as the Superweak Model[2]; if true, it has little consequence outside the neutral K system. While not in favor, this hypothesis is nevertheless consistent with all experimental information in that CP violating effects have yet to be seen in the K meson transitions. Discarding this term, we have

B. All virtual matrix elements are relatively real. In this case, the 6.5 mr phase could arise from a difference in phase between, for example, the K→ 2π matrix elements with I=0 and I=2 final states. This possibility can be ruled out as a result of the small amount of $\Delta I = 1/2$ violation.

C. All real K meson decay matrix elements are relatively real. In this case, the phase difference could arise from some heavy state(s) with orthogonal contributions, an example being the state of top quark-anti top quark. This hypothesis would have no consequence within the kaon system; however the heavy states could be produced in higher energy collisions, or they might have significance in the mixing and decays in other systems.

D. The phase difference arises from heavy virtual states (again, for example, $t\bar{t}$) with small contributions to K meson decays. With this possibility, one has the hope of studying CP violation in the K meson decays themselves (eg the 2π and even the 3π and γγ modes). The Kobayashi-Maskawa model[3] is one with the promise of such studies.

K→ 2π TRANSITIONS

We now consider the transitions $K_{L,S} \rightarrow \pi^+\pi^-$ and $\pi^0\pi^0$. There are two decay amplitudes, A_0 and A_2, depending upon the isotopic spin of the final state pions.

A. Suppose A_0 and A_2 are relatively real (no direct CP violation).

In this case, we have, after some algebra[4],

$$\eta_{+-} \equiv \frac{\text{amp}(K_L \to \pi^+\pi^-)}{\text{amp}(K_S \to \pi^+\pi^-)} = \frac{-ix}{2x+i}\phi_{\Delta S=2} \qquad (15)$$

From the known values for x and $\phi_{\Delta S=2}$, we get the following prediction:

$$\eta_{+-} = 2.27(8) \times 10^{-3}\, e^{i43.7°(2)} \qquad (16)$$

which is in excellent agreement with the experimental value[1]

$$\eta_{+-} = 2.279(26) \times 10^{-3}\, e^{i44.6°(1.2)} \qquad (17)$$

B. Consider now the case of a small phase difference between A_0 and A_2.

We let $\phi_{\Delta S=1} \equiv \arg(A_2/A_0)$; then,[5]

$$\eta_{+-} \to \eta_{+-}\left(1 - 0.1\frac{\phi_{\Delta S=1}}{\phi_{\Delta S=2}}\right) \qquad (18)$$

$$\frac{\eta_{+-}}{\eta_{00}} = 1 - 0.3\frac{\phi_{\Delta S=1}}{\phi_{\Delta S=2}} \qquad (19)$$

$$\left|\frac{\eta_{+-}}{\eta_{00}}\right|^2 \simeq 1 - 0.6\frac{\phi_{\Delta S=1}}{\phi_{\Delta S=2}}. \qquad (20)$$

The current experimental information comes from two experiments

$$\left|\frac{\eta_{+-}}{\eta_{00}}\right|^2 = 0.972(35) \quad \text{Chicago Saclay}^6 \qquad (21)$$

$$= 1.010(49) \quad \text{BNL} - \text{Yale}^7 \qquad (22)$$

Averaging these results gives

$$\left|\frac{\eta_{+-}}{\eta_{00}}\right|^2 = 0.985(28) \qquad (23)$$

with the following value for the ratio of the two angles measuring the degree of CP violation in the $\Delta S=1$ and $\Delta S=2$ interactions:

$$\frac{\phi_{\Delta S=1}}{\phi_{\Delta S=2}} = 0.025(47) \qquad (24)$$

Thus there is as yet no evidence for a "direct" CP non conservation, the sensitivity to a non-zero phase angle between A_0 and A_2 being now at the level of a few hundred micro-radians.

RELATION TO THE "STANDARD" PARAMETERIZATION

In the phase convention of Wu and Yang,[8] one chooses A_0, the $K \to 2\pi$, I=0 transition amplitude, to be real. Then the long lived state is written

$$K_L \sim (1+\epsilon)|K^0> - (1-\epsilon)|\bar{K}^0> \qquad (25)$$

where ϵ is a small, complex mixing parameter. In this convention, one has

$$\text{Arg}(\epsilon) \simeq \pi/4 \text{ and} \qquad (26)$$

$$\text{Re}\epsilon = -\frac{\phi_{\Delta S=2}}{4}. \qquad (27)$$

The parameter ϵ' is defined by

$$\epsilon' \equiv \frac{i}{\sqrt{2}} e^{i(\delta_2 - \delta_0)} \frac{\text{Im}(A_2)}{A_0} \qquad (28)$$

and its relation to $\phi_{\Delta S=1}$ is approximately given by (see ref. 5 and 13)

$$\epsilon'/\epsilon \simeq \frac{1}{10} \frac{\phi_{\Delta S=1}}{\phi_{\Delta S=2}} \qquad (29)$$

STANDARD MODEL EXPECTATIONS

In the standard model of the weak interactions, CP nonconservation is accounted for by the Kobayashi-Maskawa (KM) model[4]. The authors pointed out that with 6 quark flavors, an irreducible complexity to the charged current coupling matrix among the quarks would provide for CP non conservation. Not only is the phenomenon of CP violation then unified with that of quark mixing but as well CP violation is predicted among the heavier quark systems, particularly in B^0 decays. While such effects may be observable in the future, the experimental effort has so far been concentrated upon the observation of the very small direct CP violating kaon decays which also arise in the KM model. The 2π decays can sense the heavy quark intermediate states through the $\Delta I = 1/2$ "penguin" amplitude which then interferes with a $\Delta I = 3/2$ amplitude involving only light quarks. These amplitudes are diagrammed in Figure 1.

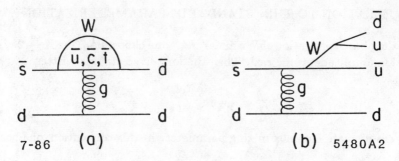

Figure 1. Diagrams contributing to $K^0 \to 2\pi$. (a) is the "penguin" which is pure $\Delta I = 1/2$; in the KM model, it has a phase difference with respect to the $\Delta I = 3/2$ part of (b)

The calculations for $\phi_{\Delta S=1}$ involve considerable uncertainty; they have been refined[9] since the publication of the results above and it now appears that, with 3 generations,

$$|\phi_{\Delta S=1}| \simeq (0.02 \text{ to } 0.08)|\phi_{\Delta S=2}| \simeq 100 \text{ to } 400 \mu r \qquad (30)$$

The sign is uncertain, and, with more than 3 generations, lower values are possible.

EXPERIMENTS

At present there are two experiments nearly ready to take data with sensitivities, to $\phi_{\Delta S=1}$ in the range of 50 μr. These are, first, the Chicago-Fermilab-Princeton-Saclay experiment at FNAL (E731) which is an outgrowth of the Chicago-Saclay experiment cited above, and, second, the CERN-Edinburgh-Maintz-Orsay-Pisa-Siegen experiment at CERN (NA31). These efforts will be briefly treated here; more details about the Fermilab and CERN experiments will be found in the contributions to this conference by G. Bock and M. Calvetti, respectively.

Both experiments need to measure the four rates $K_{L,S} \to \pi^+\pi^-, \pi^0\pi^0$ with high statistics, low backgrounds, and minimal systematic errors, the latter representing the greatest challenge.

The detector for the Fermilab experiment is shown in Figure 2. For that experiment, a new beam-line was constructed making full use the 800 GeV primary protons, the 20 sec beam pulse, large targeting angles, very clean collimation, and low Z absorbers to provide two closely spaced, clean, high intensity K_L beams. One of the beams passes through a thick regenerator so that both K_L and K_S decays to a given channel are studied simultaneously

31

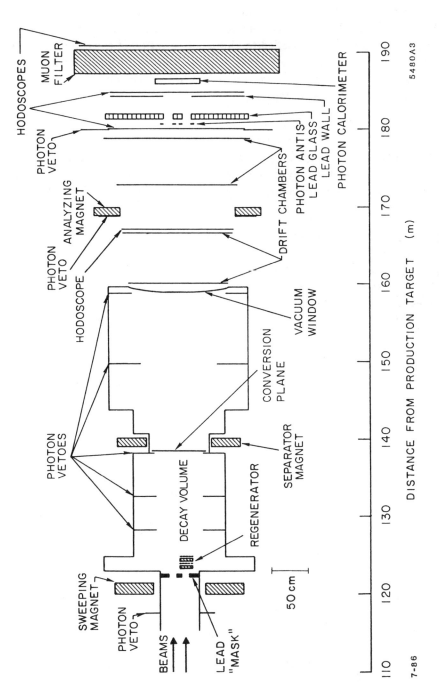

Figure 2. Elevation View of the FNAL (E731) Experiment

thereby significantly reducing one important source of systematic error. A new high rate drift-chamber spectrometer was constructed and many electromagnetic veto planes were installed to aid in the rejection of the copious $K_L \to 3\pi^0$ mode. The regenerator itself was "active" in order to reject inelastically regenerated events with particle production. Photons are detected in a large finely-segmented lead-glass array.

The experiment alternates between two phases – the charged mode and the neutral mode. For the charged mode running, only a simple two-track trigger is needed and statistics are readily obtained. For the neutral mode, a thin lead counter is inserted at the end of the decay region and one converted photon is required in the trigger together with more than 25 GeV deposit in the lead glass array.

This experiment will run in 1987 and will collect in excess of 10^5 $K_L \to 2\pi^0$ events. The experiment had a test run in 1985; the quality of the data can be seen in Figure 3 where, for a sample from this test run, is plotted the $\pi^+\pi^-$ and $\pi^0\pi^0$ invariant mass distributions. The latter distribution is also shown for the earlier Chicago-Saclay experiment[6]; the background has been reduced by about a factor of five.

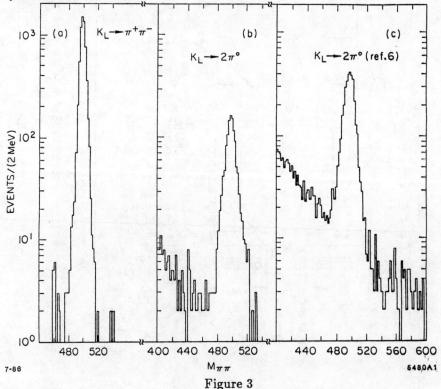

Figure 3

33

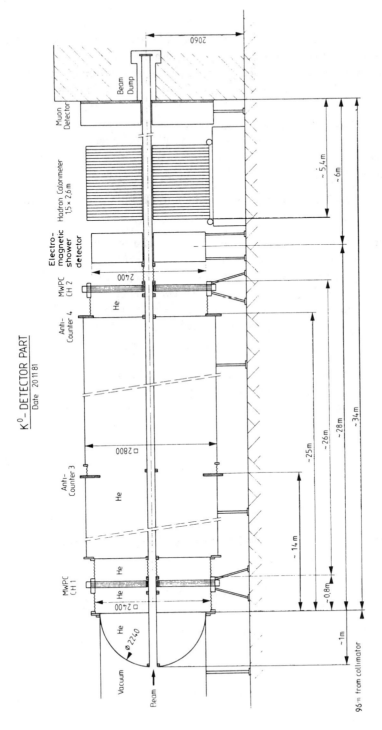

Figure 4. Elevation View of the CERN (NA31) Experiment

The detector for the CERN experiment is shown in Figure 4. Again the detector and beam line are new, and the technique is different from the Fermilab experiment. Both K_S decays modes are simultaneously recorded with a close target which is moved throughout the decay region. A far target is used for recording the K_L modes. The electromagnetic detector is a large liquid argon calorimeter and the charged pions are detected with MWPC's and a hadron calorimeter; no magnet is employed. No conversion is required and the acceptance is very high. Four planes of anti counters are employed to help reject $K_L \to 3\pi^0$ decays.

This experiment will run in 1986 and should collect about 2.5×10^5 $K_L \to 2\pi^0$ events.

As mentioned above, it is likely that systematic uncertainties will ultimately determine the sensitivities of the two experiments. The FNAL experiment is relatively immune to changes in detector response as a function of time or intensity; the CERN experiment is not very dependent upon an accurate Monte-Carlo simulation of the detector response. Both are critically dependent upon knowledge of the absolute energy scale for each decay mode.

TEST OF CPT INVARIANCE[10]

Both groups mentioned in the previous section have proposed to test[11] CPT invariance in the decay of the neutral kaon by measuring the phase difference $(\phi_{+-} - \phi_{00})$ between η_{+-} and η_{00} to a precision of 1^0. We note that in the absence of any direct CP violation, η_{+-} and η_{00} should have identical phases (See (15) above):

$$\eta_{+-} = \eta_{00} = \frac{-ix}{2x+i}\phi_{\Delta S=2} \qquad (31)$$

$$\phi_{+-} = \phi_{00} = \arctan 2x = 43.7^0(2) \qquad (32)$$

As noted earlier the measured value for ϕ_{+-} is in good agreement with the prediction; however the best measurement[12] of ϕ_{00}

$$\phi_{00} = 54.5^0(5.3) \qquad (33)$$

differs by about two standard deviations.

Possible explanations for this deviation are

1. Actually $\phi_{+-} \simeq \phi_{00} \simeq 43.7^0$ and the value for ϕ_{00} represents a fluctuation.

2. CPT is violated in $K^0 \to 2\pi$. This will account for $\phi_{+-} \neq \phi_{00}$. However, CPT must as well be violated in the K^0 - $\bar{K}^0$ mixing since otherwise one would expect that $\phi_{+-} \neq \arctan 2x$.

3. CPT is a good symmetry; however the assumption that direct CP violation is small - which leads to the equality of ϕ_{+-} and ϕ_{00} - is incorrect. The experiments described above which look for a direct CP violation are in fact only sensitive to the component of the amplitude which is in the direction of arctan 2x. The phase of ϵ' is in the direction $\delta_2 - \delta_0 + \pi/2$ which, according to the results of several consistent but somewhat indirect determinations[13], is near 45^0. However, if we ignore these determinations and if the phase of ϵ' were - 45^0, we would then except a splitting between ϕ_{+-} and ϕ_{00}.

A 10^0 difference would then result from $|\epsilon'/\epsilon| \simeq 0.06$. However, one would then expect that ϕ_{+-} would differ from arctan 2x by about 3^0.

Thus the 1st scenario is the most likely; nevertheless the 1^0 measurements of the phase difference will provide the most accurate test of CPT symmetry in a decay amplitude.

CONCLUSION

The conclusion of this talk can be very simply stated: The neutral kaon system continues to show the promise of revealing a deeper understanding of the phenomenon of CP non conservation. As experiments become ever more precise, models are accordingly constrained. The experiments described above on direct CP violation together with studies[14] of the 3π decay mode and future efforts[15] at the LEAR facility at CERN show that this "laboratory" will be exploited for at least another 10 years.

REFERENCES

1. Particle Data Group, RMP56, No. 2, Part II, April 1984. The extraction of $\phi_{\Delta S=2}$ from the semileptonic charge asymmetry assumes CPT symmetry and the validity of the $\Delta S = \Delta Q$ rule.

2. L. Wolfenstein, Phys. Rev. Lett. 13, 562, 1964.

3. M. Kobayashi and T. Maskawa, Prog. Theor. Phys. Japan 49, 652, 1973.

4. Some approximations are made which are justified in the context of current experimental errors but which may be important for very precise future experiments. These will be brought out in a review by the author presently in preparation. In particular, the phase of η_{+-} differs from actan 2x by about $0.2°$, even in the case of no direct CP violation.

5. As is well known, the degree to which the measurable parameters η_{+-} and η_{00} differ depends upon the direct CP violating amplitude (here parameterized by $\phi_{\Delta S=1}$) and upon the difference in the strong interaction phase shifts for two pions in the I=0 and I=2 final states. (See (28)). Our knowledge of these phases is discussed briefly in ref. 13; they are such that the phase of η_{+-} (and of η_{00}) does not change as a function of the magnitude of $\phi_{\Delta S=1}$; this is evidenced in (18); the numerical factor results from the relation $\text{Re}(A_2/A_0) \sim 0.05$.

6. R. H. Bernstein et. al. Phys. Rev. Lett 54; 1631 (1985)

7. J. K. Black et. al. Phys. Rev. Lett 54: 1628(1985)

8. T. T. Wu and C. N. Yang, Phys. Rev. Lett. 13, 380 (1964);

9. See the review article by L. Wolfenstein, CMU-HEP86-3, to be published in Annual Review of Nuclear and Particle Science, and references therein.

10. See the excellent article by V.V. Barmin et. al., Nuclear Phys. B247 (1984) 293.

11. Fermilab E773 (G. Gollin spokesman); CERN NA31 Addendum

12. J. H. Christenson et.al., Phys.Rev.Lett. 43, 1209 (1979).

13. A large set of references can be found in the article by Thomas J. Devlin and Jean O. Dickey, Reviews of Modern Physics, Vol. 51, No. 1, (1979). J. P. Baton et.al. (Phys. Lett. 33B, No. 7, 528 (1970)) determine δ_0 and δ_2 by means of a Chew-low extrapolation using the reactions $\pi^- p \to \pi^- \pi^0 p, \pi^+ \pi^- n$. The resulting phases vary slowly and smoothly over the $\pi\pi$ invariant mass range of 500 MeV to 700 MeV with $\delta_0 = 36°(7)$ and $\delta_2 = -4°(4)$ at $M\pi\pi = 500$ MeV. L. Rosselet et. al. (Phys. Rev. D Vol 15, No. 3, 574 (1977)) use a partial wave analysis to extract δ_0 from threshold to $M\pi\pi \sim 360$ MeV using the (real) final state pions produced in the decay $K^+ \to \pi^+\pi^- e^+ \nu$. Their results are consistent with a slow and smooth variation of δ_0 over the extended kinematric range in that they "tie on" nicely to the results of several pion production experiments. From these experiments, we estimate that, at the K mass, $\delta_2 - \delta_0 = -43°(5)$ where the error is purely systematic.

14. See the contribution from G. Thomson to this conference.

15. See the contribution from J. Fry to this conference.

SUPERCONDUCTING ACCELERATOR TECHNOLOGY

Hermann A. Grunder and Beverly K. Hartline
Continuous Electron Beam Accelerator Facility
Newport News, VA 23606

ABSTRACT

Modern and future accelerators for high energy and nuclear physics rely increasingly on superconducting components to achieve the required magnetic fields and accelerating fields. This paper presents a practical overview of the phenomenon of superconductivity, and describes the design issues and solutions associated with superconducting magnets and superconducting rf accelerating structures. Further development and application of superconducting components promises increased accelerator performance at reduced electric power cost.

INTRODUCTION

When H. Kamerlingh Onnes discovered the phenomenon of superconductivity 75 years ago[1] (Figure 1), he recognized that its exploitation could give rise to highly efficient electrical apparatus operating with no resistive losses, exotic new devices, and the production of intense magnetic fields. Although Onnes was frustrated in his attempts to produce high magnetic fields, because the field itself drove his superconductors normal, the technological capabilities he envisioned have largely been achieved today.

For application to particle accelerators, superconducting technology has entered into an extremely rewarding phase. Its benefits are increased performance at reduced operating cost. Present and planned accelerators that rely on superconducting technology -- Tevatron, HERA, CEBAF, RHIC, SSC, LEP, KEK and others -- provide exciting experimental opportunities for high energy and nuclear physicists, that could not be obtained or afforded using conventional accelerator technology. The primary applications are in the

Fig. 1. H. K. Onnes (seated) with J. van der Waals with the first helium liquifier.

production of magnetic fields to steer the beam and electric fields to accelerate it.

This paper provides a brief background on superconductivity, in particular its properties that underlie its use in accelerators, and discusses design issues and approaches related to superconducting accelerator magnets and rf accelerating structures. The essential supporting technologies, namely cryogenics and superconductor development and processing are not covered, except where they bear directly on component design and performance.

SUPERCONDUCTIVITY

Onnes[1] discovered the phenomenon of superconductivity by measuring the D.C. resistance of mercury as he cooled it to the low temperatures achievable with the helium liquifier he developed (Figure 2). At about 4.2 K, the resistance of mercury effectively vanished! Further study of this remarkable phenomenon revealed that it was a thermodynamic phase transition, and that several metals, in fact, became superconducting at a sufficiently low temperature that depended on the material and the ambient magnetic field. Figure 3 is a "generic" superconductor phase diagram, which introduces two critical parameters: the critical temperature (T_c), and the critical magnetic field (H_c). The critical temperature is the temperature below which a material becomes superconducting in the absence of a magnetic field. The critical magnetic field is the field above which a superconductor at 0 K becomes normal. Note that the phase transition in the absence of a magnetic field, is second order: there is no latent heat associated with it. In the presence of a magnetic field the phase transition is first order, and the transition temperature is less than T_c.

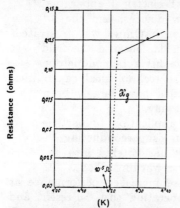

Fig. 2. Onnes' data on the D.C. resistance of mercury showing the transition to superconductivity.

In 1933, W. Meissner and R. Ochsenfeld[2] discovered that superconductors exclude magnetic fields. This so-called Meissner effect is perfect for cylindrical superconductors aligned parallel to the applied magnetic field. In this geometry, the magnetic field is expelled completely from a normal conductor as it is cooled below its transition temperature and becomes superconducting. In 1934, F. London and H. London[3] developed a phenomenological theory to explain some properties of superconductivity, including the Meissner effect. They proposed that a current is induced to flow in a shallow

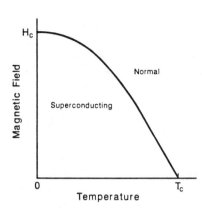

Fig. 3. Generic phase diagram for a superconductor

layer at the surface of the superconductor to produce a magnetic field inside the superconductor which exactly counterbalances the external or applied field there. The applied magnetic field penetrates the superconductor, but its magnitude drops exponentially over the London penetration depth (λ), which is typically on the order of 5×10^{-2} μm. The surface currents induced in a material as it becomes superconducting persist indefinitely without degradation. To avoid such persistent currents, it is important to ensure that devices pass through the superconducting transition in the absence of a magnetic field.

Another length scale, the coherence length (ξ), is also important to a phenomenological understanding of superconductors. The coherence length is a measure of the distance over which the superconducting properties can vary significantly, and it is related to the mean free path of normal electrons. Based on the relationship between λ and ξ, superconductors display distinct behavior (Figure 4).

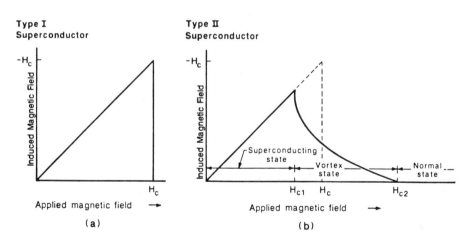

Fig. 4. Magnetic behavior of (a) Type I superconductor and (b) Type II superconductor.

Type I or "soft" superconductors have $\xi \gg \lambda$. For these materials, such as lead and mercury and most other elemental superconductors, the Meissner effect is close to perfect. They tend to have a low critical temperature (T_c) and a low critical magnetic field. The induced magnetic field increases to exactly compensate the applied magnetic field until H_c is exceeded (Figure 4a), and the material becomes normal.

Type II or "hard" superconductors have $\xi \ll \lambda$. In these superconductors, the B field is excluded completely only up to a field, $H_{c1} < H_c$, and it penetrates completely above $H_{c2} \gg H_c$ (Figure 4b). In magnetic fields between H_{c1} and H_{c2}, the magnetic field is only partially excluded, and the material exists in a so-called vortex state. Magnetic flux lines or fluxoids pass through the superconductor, and in the microscopic core of each fluxoid, the material is in a normal state. The supercurrent, J_s, flows around the perimeter of each fluxoid, thereby supporting the flux lines.

The major advantage of Type II superconductors is that superconducting electrical properties prevail up to H_{c2}. Type II superconductors are usually compounds or alloys. Examples include NbTi, Nb_3Sn, and V_3Ga. Typically they have a high critical temperature, and H_{c2} can be high enough to be useful in high-field magnets (Table 1). Ginzburg and Landau[4] developed a theory based on a two-fluid model (normal electrons and super electrons) which is useful for understanding Type II superconductors at temperatures near T_c.

Table 1
Properties of Some Superconductors

Material	Crystal structure	T_c (K)	Energy Gap ($k_B T_c$)	$H_{c2}(0)$ (Tesla)	$H_c(0)$ (Tesla)	Availability*
Pb	fcc	7.2	2.17	Type I	0.08	RF
Nb	bcc	9.2	1.90	.38	.19	RF
Nb_3Al	A15	18.8	2.15	29.5	.65	--
Nb_3Ga	A15	20.3		33.5		Exp
Nb_3Ge	A15	23.6	2.1	37.0		Exp
NbN	B1	16.1	1.74	15.3	.22	Exp
Nb_3Sn	A15	18.3	2.25	22.5	.53	Wire
NbTi	bcc	9.8	1.9	14.8	.24	Wire
V_3Ga	A15	15.0		35.0	.74	Wire
V_3Si	A15	17.1	1.78	15.6	.60	Exp
$Mo_{.57}Re_{.43}$		14	1.8	2.8	.16	Exp
$PbMo_6S_8$		14		55		--

*RF = rf structures; Wire = wire or cable; Exp = experimental use for wire or rf
From H. Piel (Wuppertal)

In 1956 Bardeen, Cooper, and Schrieffer[5] developed a successful microscopic, quantum mechanical theory of superconductivity, which underlies present understanding. According to this theory, the superconducting state is a zero-momentum state. A lattice-mediated interaction binds the electrons into "Cooper pairs", which have zero momentum and carry the supercurrent.

Several superconductors have demonstrated or potential application in accelerator magnets and rf structures. Properties of some promising ones are tabulated in Table 1, and materials with high T_c and H_{c2} are particularly attractive. To be useful, however, these materials must have satisfactory mechanical properties, and be able to be produced in quantity with the required purity in wire and/or sheet.

ACCELERATOR MAGNETS

Accelerator magnets are designed to produce the desired magnetic field and field quality within the beam aperture. There are three basic approaches. In an iron-dominated magnet, the shape of the iron controls the shape of the field, and the field is limited to about 2 Tesla by iron saturation. In a conductor-dominated magnet the placement of the conductor controls the shape of the field, and high currents made possible by superconducting coils make high fields achievable. A hybrid design uses superconducting coils to push the field strength beyond iron saturation; hence these magnets are called "superferric". The shape of the iron and conductor placement together determine the shape of the field. Superconducting accelerator magnets are reviewed in detail by Palmer and Tollestrup[6].

Superconducting accelerator magnets can provide higher magnetic fields at significantly lower power usage than can be achieved with conventional accelerator magnets. Table 2 shows the experience with FNAL Main Ring and Tevatron magnets (Figure 5). With the 4.4-Tesla superconducting Tevatron[7] magnets, the beam energy is doubled

Table 2
Magnet Power Consumption at FNAL

	FNAL Main Ring (1982)	Tevatron (1984)	Ratio Tevatron / Main Ring
Beam energy (GeV)	400	800	2.0
Electric power per flat-top time (KW hr/sec)	175	38	0.2

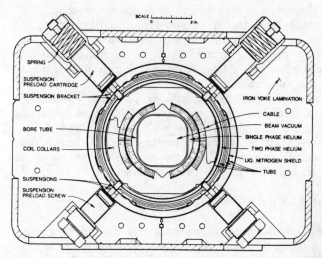

Fig. 5. Cross section of the 4.4-Tesla Tevatron[7] dipole magnet.

and the electric power it takes to produce each second of beam time useful for experiments (flat-top time) is only one-fifth that of the conventional Main Ring magnets.

There are several important aspects to the design of superconducting accelerator magnets:

o Field strength
o Field quality
o Quench protection
o Cost

The magnetic field in a conductor-dominated magnet is determined by the current in the coils and the coil geometry. To achieve a high field requires a high current density, J_c, in the coils, and coil placement close to the aperture. Note that the achievable magnetic field is limited by the magnetic field tolerance of the superconductor. Operation at low temperatures and use of materials with high H_{c2} bring clear advantages. Currently a NbTi alloy (46% Nb, 54% Ti) is the center of development, because it has superior mechanical properties, and can be formed into wire and cable. A major R&D challenge is to develop processes for preparing wire and cable made of Nb_3Sn, which is brittle, and other materials with high H_{c2} and T_c.

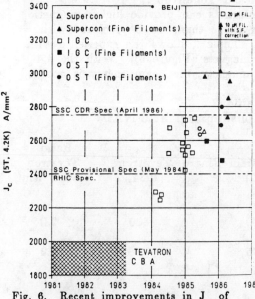

Fig. 6. Recent improvements in J_c of commercial NbTi magnet wire.[10]

With the focus on NbTi, recent R&D in support of magnet development for the Superconducting Super Collider (SSC) has resulted in a 70% improvement in J_c since 1983 in industrially produced wire (Figure 6).

This improvement is due to increased homogeneity in the NbTi alloy, an improved heat treatment and processing schedule, and the use of diffusion barriers to allow highly uniform, fine filaments to result from the drawing process[8]. Multifilament superconducting magnet wire is drawn from large billets containing many rods of NbTi imbedded in a copper matrix. In succeeding steps the billet is drawn to smaller diameters until each NbTi rod becomes a wire filament 5 to 20 μm in diameter. In this process, the NbTi tends to react with the copper matrix, forming a titanium-copper intermetallic phase that disfigures the filaments (Figure 7). The intermetallic phase interferes with the flow of supercurrent, thereby limiting J_c. The use of a niobium jacket around each NbTi rod provides a diffusion barrier that prevents the intermetallic phase from forming (Figure 8). The resulting filaments are fine, smooth, and uniform (Figure 9).

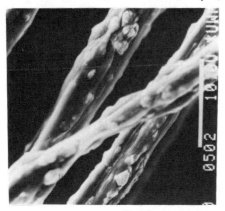

Fig. 7. SEM micrograph of NbTi filaments with large nodules of intermetallic. Source: Lawrence Berkeley Laboratory

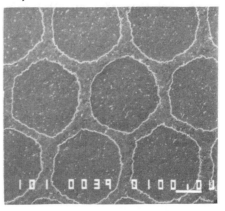

Fig. 8. Niobium-clad NbTi filaments in a copper matrix. Source: Lawrence Berkeley Laboratory

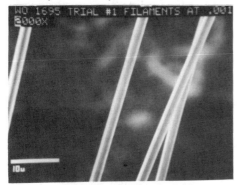

Fig. 9. SEM micrograph of niobium-clad NbTi filaments. Source: Lawrence Berkeley Laboratory

The field quality in a superconducting accelerator magnet is set by the tolerances on conductor placement and mechanical stability, and the compensation or suppression of higher order field multipoles, especially sextupole components due to persistent currents. Superconducting magnets are designed carefully to achieve and maintain the very precise conductor placement, despite the strong forces acting on the coils when the current is on.

In addition, magnets must be designed to be protected against damage during a quench. In a quench, a region of superconductor goes normal and can no longer carry the high J_c. Two alternative approaches are viable in principle: design the cable so that it can carry the current without damage through the copper matrix; or design the magnet so that the quench propagates sufficiently rapidly to avoid damage. The first alternative is not practical for accelerator magnets, because it requires a high ratio of copper to superconductor (~20:1). Therefore it can handle only a limited average current density and achieve a limited field in the beam aperture. The second alternative uses a low ratio of copper to superconductor (1.3:1 to 1.8:1) and a coil design that allows the quench to propagate longitudinally and radially. This approach ensures that a large volume of the magnet goes normal, thereby averting severe local heating and damage. Quench propagation velocities of the order of 20m/sec are required.

Cost issues are dominated by material requirements and assembly considerations. Since over two-thirds of a magnet's cost is in material, it is cost effective to design the magnet with the smallest aperture that is sufficient for the beam. The superconducting cable represents about 30% of the magnet's cost, so economical cable fabrication and high J_c, which reduces the amount of superconductor required, are desirable. Table 3 compares the actual cost of Tevatron dipoles with the estimated production cost of the 6.6-T SSC dipoles. Real progress has been made in this area.

Table 3
Costs of Superconducting Accelerator Dipole Magnets

	Superconductor Cable	All Other Material	Labor	Unit Cost (1986 $)
Tevatron (4.4 T, 6.4m long, $J_c = 1800$ A/mm^2)	26%	40%	34%	65K
	28%	41%	31%	62K*
SSC Conceptual Design (6.6 T, 17m long, $J_c = 2750$ A/mm^2)	27%	48%	25%	106K
	30%	53%	17%	97K*

*Excluding design and inspection.

Two recent dipole designs bear mention.

The magnet designed at Brookhaven National Laboratory (BNL) for their proposed Relativistic Heavy Ion Collider (RHIC) has a single-layer NbTi $\cos\theta$ coil (Figure 10)[9]. With a single layer of superconductor, this magnet is quite cost effective. It is designed to operate at 3.44 Tesla at 4.5K, well below its quench field of 4.65 Tesla.

The SSC Central Design Group[10] has developed a 6.6 Tesla dipole with two layers of superconducting coil (Figure 11). The aperture radius for this magnet is 2 cm, half that of the RHIC dipole, and its operating temperature is 4.35K. A stainless steel collar holds the coils firmly in position. Economical production was a major factor in the development of this design, as the SSC will require nearly 7700 units.

To achieve higher fields with NbTi superconductors, operation at 2K or below is required, and designs have three or more layers of coils. To date, both Japan (KEK) and the U.S. (LBL) have developed and tested short high-field accelerator dipoles that have achieved 9.4 Tesla and 9.3 Tesla respectively. In the future, R&D is likely to focus on increasing J_c and on superconductors, such as Nb_3Sn, with high T_c and H_{c2}, to allow higher fields to be reached. Prototype Nb_3Sn magnets have been made and tested already. Work will continue on cabling and assembly procedures to reduce magnet cost and improve reproducibility and reliability of magnets fabricated in quantity by industry.

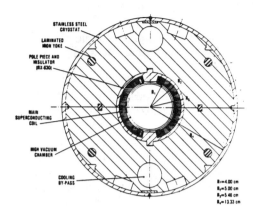

Single-layer NbTi Cos θ coll
Cu: Superconductor 1.8:1
Superconductor J_c spec. 2400 A/mm² at 5T
Length 9.7m
Temperature 4.5K
Quench Field 4.65 T

Fig. 10. Cross section of the 3.44-Tesla dipole for RHIC.

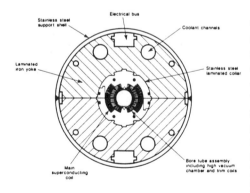

Fig. 11. Cross section of the 6.6-Tesla SSC dipole.

RF ACCELERATING STRUCTURES

Radio frequency accelerating structures produce the electric field that accelerates the particle beam. Design goals for accelerating structures include achieving the desired accelerating field, with the desired frequency and frequency stability. Multipacting, field emission, and other processes that limit or degrade the field are to be avoided or minimized. Superconducting accelerating structures offer the advantages of very low rf losses. Thus, they are energy efficient, and attractive or essential for very high energy e^+e^- storage rings and colliders, TeV-scale linear e^+e^- colliders, and free electron lasers. The low losses also make CW operation feasible, both for electron linacs and heavy-ion linacs. Accelerating structures are not exposed to high magnetic fields; therefore Type I superconductors, as well as Type II superconductors can be used. The most common material employed today is niobium. R&D on rf superconductivity is underway at many laboratories and universities around the world[11], and much progress has been made within the past few years[12].

For low-β accelerators, such as heavy-ion linacs, the resonator normally operates at around 100 MHz. The geometry and size of the resonator determine the resonant frequency. Typically field emission and mechanical stability limit the achievable gradient. Several heavy-ion linacs are currently in operation or under construction (Table 4)[13]. Most serve as boosters to increase the beam energy available at tandems.

Table 4
HEAVY ION SUPERCONDUCTING
BOOSTER LINACS (low β)[13]

System	Superconductor	Resonator type	f_{rf} (MHz)	Active length (m)	Number of Resonators
Operating					
ANL Atlas	Nb	split ring	97	13.3	42
SUNY Stony Brook	Pb	split ring	150	7.5	40
Weizmann Institute	Pb	quarter wave	162	0.7	4
Under construction					
Saclay	Nb	helix	135	12.5	50
Florida State	Nb	split ring	97	4.5	13
Oxford	Pb	split ring	150	2.1	9
U. of Washington	Pb	quarter wave	150	8.6	36
Under development					
Canberra	Pb	quarter wave	150	0.8	4
Kansas State	Nb	split ring	97	3.5	16

Both niobium and lead-on-copper resonators have been used successfully. Niobium resonators have lower rf losses than lead; they therefore are more economical to operate. Lead-on-copper structures are extremely inexpensive to manufacture, however. Resonators made of both materials achieve gradients of about 3 MV/m in operation. Three ATLAS resonators (Argonne) are shown in Figure 12. Note that these three resonators, from the beginning, middle, and end of ATLAS are different. In a low-β linac the resonators must be matched to the β of the particles, which increases along the length of the machine. In addition, the resonators are designed to have a broad velocity acceptance.

Fig. 12. Resonators for the heavy-ion linac ATLAS. Source: Argonne National Laboratory.

For $\beta=1$ structures, resonant cavities are used[11] (Figure 13). The cavity size and shape determine the resonant frequency. Surface defects, multipacting, and field emission must be controlled to achieve high gradients. Higher order modes must be suppressed to achieve beam stability.

Major R&D efforts have been underway for several years at Stanford, Karlsruhe, Cornell, CERN, DESY, Wuppertal, Orsay, and KEK to develop $\beta=1$ superconducting rf structures[11]. Until recently, achieved gradients were limited to about 3 MV/m. Now gradients in excess of 5 MV/m are achieved routinely, at laboratories and by industry (Table 5). The theoretical gradient limit is ~50 MV/m for Nb and ~80 MV/m for Nb_3Sn. Recent progress has been very rewarding, and there are firm plans now to employ these structures in several planned accelerators and major upgrades (Table 6).

Fig. 13. Five-cell, 1500-MHz cavity designed by Cornell and adopted by CEBAF.

Table 5

PERFORMANCE OF $\beta=1$ SUPERCONDUCTING RF CAVITIES
(From H. Piel, Wuppertal)

Laboratory	CERN			KEK	DESY	Cornell	CEBAF**	Darmstadt/Wuppertal	
Accelerator	LEP II			TRISTAN	HERA	CESR	CEBAF	130 MeV	Recyclotron
Material	Nb	Nb	Nb/Cu	Nb	Nb	Nb	Nb	Nb	Nb_3Sn
Frequency in MHz	350	500	500	500	1000	1500	1500	3000	3000
Operating Temperature	4.2K	4.2K	4.2K	4.2K	4.2K	1.8K	2.0K	1.8K	4.2K
Best Single-Cell Results									
E_A (MV/m)	10.8	13.0*	10.8	7.6*	5.5	22.8*	-	23.1*	7.2
Q At E_A (x 10^9)	1.8	0.7	0.4	0.6	0.5	2.5	-	1.2	1.1
Best Multicell Results									
No. of Cells	4	5	-	3	9	5	5	5/20	5
E_A (MV/m)	7.5*	5.0	-	5.8	5.5	15.3*	7.9*	12.3/7.4	4
Q At E_A (x 10^9)	2.2	0.7	-	0.6	0.5	2.2	6.0	3.5/1.2	4.5

* Cavities Fabricated From High Thermal Conductivity Niobium
** Cornell Cavity design

Table 6
PLANNED APPLICATION OF $\beta=1$ SUPERCONDUCTING CAVITIES

Laboratory	Accelerator	Frequency MHz	Active Length (m)
CERN	LEP I, 50 GeV	350	20
CERN	LEP II, 100 GeV	350	650
CEBAF	4 GeV Linac	1500	200
DESY	HERA e, 30 GeV	500	20
KEK	TRISTAN, 33-35 GeV	508	60
KEK	TRISTAN, 40 GeV	508	180-216
Wuppertal/ Darmstadt	130 MeV Linac	3000	12
HEPL/TRW	Free Electron Lasers	1300/ 500	5-6
?	TeV Linear Collider	~1000-3000	~8×10^4

From H.Padamsee

The major design issues for rf cavities are:
o Frequency selection and stability
o Cavity shape
o Superconductor selection, processing, and cleanliness
o Cost

Prototype $\beta=1$ rf cavities have been built to operate at several frequencies between 350 MHz and 8000 MHz. For low frequencies, the cavities are big, awkward to handle, and have a high probability of having a serious defect somewhere on the cavity surface. However, intrinsic RF losses (BCS losses) increase with frequency squared, and deflecting impedances that limit beam current increase as frequency cubed. Thus, the optimum frequency for a specific accelerator will depend on that accelerator's specifications.

Cavity shape is an important factor for both cost and performance. Within the past few years, several improvements in this area have been developed. Spherical or elliptical cell shape has been shown to reduce multipacting. Couplers for fundamental power and for higher order mode (HOM) suppression attach to the beam pipe to minimize field enhancement and multipacting. The number of individual resonant rf cells (half wavelength) in a cavity is limited to control HOMs. Very early rf cavity development at Stanford's High Energy Physics

Laboratory (HEPL) resulted in cavities with 55 cells, and serious HOM problems. Current designs (Table 5) call for five cells or fewer. Cavity design and HOM suppression are aided now by the availability of computer codes, such as URMEL[14].

The most common superconductor currently in use for $\beta=1$ accelerating structures is niobium. Since only a very thin surface layer on the inside of the cavity is active in the formation of the accelerating field, the quality and cleanliness of that surface layer is of the utmost importance to cavity performance. In addition, the surface layer must be kept below the superconducting transition temperature: cooling must be adequate to remove the heat generated by rf dissipation in dust and defects.

Recent developments in niobium processing and cavity fabrication have resulted in real progress in these areas. Niobium suppliers have developed the capability to produce niobium sheet with high thermal conductivity (RRR). High thermal conductivity helps to stabilize the cavity against being driven normal by resistive heating at a defect. Yttrification has been developed as a process to increase the thermal conductivity[15]. The availability of clean rooms and clean manufacturing protocols avoids the introduction of dust or dirt on the active surface. Refined electron beam welding methods help achieve weld smoothness.

Another recent development is thermometric mapping to locate hot spots caused by defects or dirt on the superconducting surface. A cavity can be tested and the factor limiting its gradient can be located and repaired. With the assistance of Cornell University's SRF Group, CEBAF has been working with four vendors to fabricate prototype cavities for CEBAF's 4-GeV CW electron linac. Each delivered cavity is tested to ensure that it meets or exceeds CEBAF's specifications for gradient (5 MV/m) and Q (3×10^9). To date four cavities have been completed by three vendors. All four cavities exceed the specifications (Figure 14). Cavity Number 3 has been tested twice, and undergone "repair" to remove the defect that limited the gradient to 6.8 MV/m on the first test.

Although gradients of 5 to 10 MV/m are achieved by industry today in prototype cavities, the real attraction of rf structures is their potential to achieve gradients in excess of 20 MV/m with very low rf losses (high Q) and high beam-current capacity. Single-cell cavities fabricated at Cornell and Wuppertal have achieved such high gradients (Table 5). According to Sundelin[16], TeV-scale linear colliders become economically feasible at a gradient of about 25 MV/m and a quality factor (Q) of 5×10^{10}.

Experimentation is underway with Nb_3Sn, niobium on copper, and other superconductors as cavity materials. Nb_3Sn offers the potential for lower rf losses and operation at higher temperatures. Niobium on copper would have a very high thermal conductivity, hence excellent stabilization of submicroscopic defects and dust.

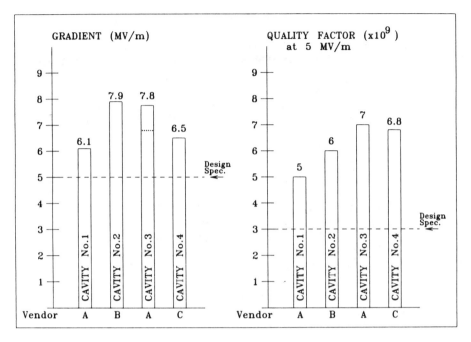

Fig. 14. Preliminary results from CEBAF's cavity prototyping program

CONCLUSIONS

Both in accelerator magnet development and in rf structure development, superconducting technology has reached a sufficient level of performance and reliability to become practical and cost effective. In both areas, the technology is young, and foreseeable improvements promise far greater capabilities in the future, that will serve as the basis for new accelerators and upgrades required by high energy and nuclear physicists to advance the frontiers of knowledge.

REFERENCES

1. H. K. Onnes, Commun. Phys. Lab. Univ. Leiden Suppl. 34b, 55 (1913).
2. W. Meissner and R. Ochsenfeld, Naturwissenschaften 21, 787 (1933).
3. F. London and H. London, Proc. R. Soc. London, Ser A, 149 (1935).
4. V. L. Ginzburg and L. D. Landau, J. Exp. Theor. Phys. (USSR) 20, 1064 (1950).
5. J. Bardeen, L. N. Cooper, and J. R. Schrieffer, Phys. Rev. 108, 1175 (1957).

6. R. Palmer and A. V. Tollestrup, Ann. Rev. Nucl. Part. Sci. <u>34</u>, 247 (1984).
7. H. T. Edwards, Ann. Rev. Nucl. Part. Sci. <u>35</u>, 605 (1985).
8. R. Scanlan, Lawrence Berkeley Laboratory, personal communication (May 1986); see also D. Larbalestier, Proc. 9th Intl. Conf. Magnet Tech, SIN, 453 (1985).
9. Proposal for a Relativistic Heavy Ion Collider, BNL 51932 (1986).
10. Superconducting Super Collider Conceptual Design, SSC-SR-2020 (1986).
11. H. Lengeler, ed., Proc. 2nd Workshop on RF Superconductivity, CERN (1984).
12. H. Piel, in Proc. 1985 Part. Accel. Conf., IEEE Trans. Nucl. Sci. <u>NS32</u> (1985).
13. L. M. Bollinger, Nucl. Instr. and Meth. <u>A244</u>, 246 (1986).
14. T. Weiland, DESY Rept. 82-015 (1982).
15. H. Padamsee, in Proc. 1985 Part. Accel. Conf., IEEE Trans. Nucl. Sci. <u>NS32</u> (1985).
16. R. Sundelin, Cornell University, CLNS 85/709 (1985).

Superconducting accelerator technology is under development at many laboratories. This work was supported by the Department of Energy under contract DE-ACO-84ER4015, and by the Commonwealth of Virginia.

FUTURE ACCELERATOR TECHNOLOGY*

Andrew M. Sessler
Lawrence Berkeley Laboratory
University of California
Berkeley, CA 94720

ABSTRACT

A general discussion is presented of the acceleration of particles. Upon this foundation is built a categorization scheme into which all accelerators can be placed. Special attention is devoted to accelerators which employ a wake-field mechanism and a restricting theorem is examined. It is shown how the theorem may be circumvented. Comments are made on various acceleration schemes.

INTRODUCTION

We know the high-energy accelerators of the present: TEV I, TEV II, SPC, CERN Collider, PEP, PETRA, CESR, etc. and we look forward to the accelerators of the near future: TRISTAN, SLC, LEP, HERA, SSC, UNK, etc. What about the distant future? Can we continue to build ever-larger machines? The circumference of LEP is 27 km; the SSC is projected to have a circumference of 83 km.

An examination of the Livingston graph, Fig. 1, accelerator energy as a function of time, shows that the envelope of accelerator types is a straight line on the semi-log plot. Available energy is ever-increasing, but even more important is the fact that any one technology (a squiggly line on Fig. 1) saturates in its capability; it is only the envelope of lines which continues to rise.

The message is clear: we must develop new technologies, new squiggly lines, if we are to stay on, or anywhere near, the exponential rise in energy which we have experienced in the past. In this paper we survey various technologies which may, someday, contribute squiggly lines to the Livingston graph.

A general survey of new accelerator technologies has been given recently,[1] while much more detailed papers can be found in the proceedings of the two conferences, one held in 1982 and one held in 1985, devoted precisely to the very subject of novel acceleration techniques.[2,3]

CATEGORIZATION

All accelerators use the electromagnetic force. Depending upon the frequency employed one has acceleration by a DC

---

*This work was supported by the Division of High Energy Physics, U.S. Department of Energy under Contract No. DE-AC03-76SF00098.

potential drop, by rf waves, or by lasers. The last are particularly potent having fields at a focus of $10^4 - 10^6$ MV/m (this is with "present - day" lasers). The last should be compared with the accelerating gradient of the Stanford Linear Collider (SLC) which is the largest gradient of any practical accelerator and is 17 MV/m.

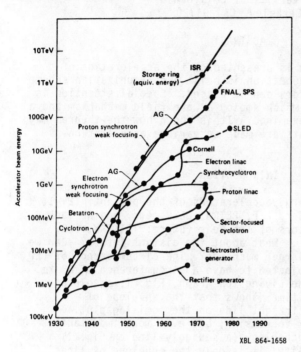

Fig. 1 The Livingston graph of accelerator energy as a function of time. At the same time as energy has been increasing exponentially cost per unit energy has been decreasing exponentially.

However, there are difficulties; namely (1) the field is in the wrong direction, (2) the field is only intense at a focus and a focus isn't very deep so that a particle quickly passes out of the strong field region, and (3) there isn't synchronism between a material particle and a luminous wave.

A particular accelerator scheme must overcome all three of these problems. Infact, as we know, for accelerators do exist, all three problems can be overcome. Before one goes more deeply into the various schemes it is useful to put the above observations on a more formal basis.

A theorem-form of the argument has been formulated by R. Palmer.[4]

Assume:
1. The interaction with light takes place in a vacuum.
2. The interaction takes place far from all dielectrics and conductors.
3. The accelerated particle is sufficiently relativistic that its motion is in a straight line and with constant ($\simeq c$) velocity.

Then:
    There is no acceleration.

We shall not give a proof of this theorem here, but--in a way--it is obvious. Thus all accelerators must violate one, or more, of the hypotheses of the theorem. This allows us to conveniently categorize all accelerators as we have done in Table 1.

Although the categorization of Table I is very useful, people have not devised various schemes with the theorem and categorization in mind. It is, therefore, useful to list all of the pratical high-energy accelerator schemes, of which I am aware, by type. (Of course these could be easily categorized and it is useful to do that.) The result is presented in Table II. One notes in this list that essentially all of the schemes employ the large peak power of lasers or of particle beams. This is quite natural for there is need for large peak power and there are very few other possible choices.

How do the schemes of Table II address the three problems, which we noted at the start of this section, with acceleration by an electromagnetic wave? In diverse ways, and it is a good test of ones understanding of each scheme to see how this is accomplished. For example, take the Surfatron. (See Refs. 2 and 3.) Here the transverse waves of two laser beams are employed, via the beat wave and the non-linear mechanism of a ponderamotive well, to resonantly excite a plasma wave. The plasma wave is longitudinal; i.e. the electric field is now longitudinal; that is, it is "pointing in the right direction." The plasma is expected to provide self-focusing of the light; i.e. to make a channel for the light so that the focus is very deep and not "outrun" by the particle. Finally, synchronization is maintained by having a (small) vertical magnetic field so that the particle gains transverse velocity, and mass, as it is accelerated, but its longitudinal velocity is unaltered and "synchronism is maintained" with the plasma wave.

As a second example consider the laser excitation of an open structure. In this configuration the incident laser light excites a surface wave which has a major component of field along its direction of motion. This is possible within a wavelength of surface, but not--of course--in free space where the solutions of Maxwell's equations must be transverse waves. The wave can be made to run along the surface; i.e. to maintain acceleration over long distance. Synchronisms is maintained for-- as we all know--slow wave structures can be built by (suitably loading a longitudinally smooth metal structure so that the phase velocity of the wave is less than c).

WAKE-FIELD ACCELERATORS

In a wake-field accelerator a large number of particles, N_1, of rather low energy, E_1, are used to accelerate a small

TABLE I

Accelerator Categorization

1. <u>Slow wave down</u> (and let particle go in straight line)
    a) Up frequencies from the 3 GHz at SLAC to (say) 30 GHz and use a <u>slow wave</u> structure. (Two-Beam Accelerator) (Violates 2)
    b) Use a single-sided cavity (i.e., a grating) or droplets as a slow wave structure. (Now one can go to 10 μm of a CO_2 laser or 1 μm of a Nd glass laser) (Violates 2)
    c) Use dielectric slabs (Violates 2)
    d) Put wave in a <u>passive</u> media (Inverse Cherenkov Effect Accelerator) (Violates 1)
    e) Put wave in an <u>active</u> media (Laser Plasma Accelerator) (Violates 1)
2. <u>Bend particles</u> continuously and periodically (and let laser wave go in straight line)
    a) Wiggle particle and arrange that it goes through one period of wiggler just as one period of the electromagnetic wave goes by. (Inverse Free Electron Laser) (Violates 3)
    b) Wiggler particle with an electromagnetic wave rather than a static wiggler field. (Two-Wave Accelerator) (Violates 3)
    c) Use cylotron motion of particle to do the bending. (Cyclotron Resonance Accelerator) (Violates 3)

TABLE II

Accelerator Concepts

1. Plasma Accelerators (Beat-Wave, Surfatron)
    a) Laser excited
    b) Particle beam excited

2. Inverse Cerenkov Accelerator

3. Inverse Free Electron Laser
    a) Regular kind
    b) Gas loaded
    c) Two-wave, Three-wave, ...

4. Droplets, Gratings, Open Structures
    a) Laser excited
    b) Particle-beam excited

5. Plasma Focus

6. Two-Beam Accelerator

7. Wake-Field Accelerators
    a) Electron-Ring Excited
    b) Electron-Beam Excited
    c) Proton Excited
    d) Intense Electron Beam and Laser

8. Switched Linac

9. Collective Radial Implosion

10. Improved Power Sources
    a) Multi-Beam klystrons
    b) Lasertron
    c) Gyrotron
    d) Power multiplying devices

11. Ionization Front Accelerator

number of particles, N_2, to a very high energy. Conservation of energy yields a simple relation between the energy gain of the second group of particles, ΔE_2, and the energy loss of the first group. For a passive structure clearly

$$N_2(\Delta E_2) \leq N_1 E_1,$$

since the first group of particles can't lose more energy than they have.

We thus obtain a restriction on the energy gain of a particle of the second group,

$$\Delta E_2 \leq \frac{N_1}{N_2} E_1 .$$

We can make N_1 very large and thus make the "transformer ratio," defined by

$$\Delta E_2 = R E_1,$$

very large.

The wake-field theorem[5] puts a restriction on the transformer ratio R:

Assume:
1. The bunches of particles can be approximated by delta functions.
2. The two bunches move on the same straight line through the device.
3. The device is arbitrary (It could be made of metal, or contain a plasma, etc.), but passive.

Then:
The transformer ratio is less than or equal to two; i.e.
$$R \leq 2.$$

We shall not prove this theorem here; I urge the reader to attempt to construct his own proof. (Hint: No new physics needs to be put in! One only employs conservation of energy, but recalls that it applies for all values of N_1 and N_2.)

In order to make an interesting wake-field accelerator one has to devise a scheme in which R is much larger than two. Thus the assumptions of the theorem need to be violated and all wake-field accelerators can be categorized by which of the assumptions they violate.

The electron-ring device, the wake-field transformer, has the second group moving along the axis of the electron ring thus violating the second assumption. The expected transformer ratio, here a consequence of the radial implosion of the electromagnetic wave, is between 10 and 20.

Bunch "shaping"; i.e. giving the bunch finite extent, and properly shaping it, is the basis of two recently proposed devices. The first is the resonant excitation of plasma waves by means of bunches of electrons[6] while the second scheme uses the first group to "charge up" a plasma and then creates a radial current implosion by a triggering laser.[7] (This scheme violates not only the first, but also the third, assumption.) In both schemes the expected transformer ratio is very large indeed.

COMMENTS

Little purpose would be served by my describing, here, some-- or even all-- of the acceleration schemes listed in Table II. Most of them have been described superficially in various review articles and comprehensively in the cited references. Rather I will make some comments, perhaps more of an editorial nature, on a number of the proposed approaches. The reader should understand that this section contains "opinions", as contrasted with "hard fact" and that, furthermore, these are the opinions of only one person and not even that of a committee of people!

Work on future accelerators can be divided into three broad sections. The first consists of improved power sources, the second of developments which might impact the next collider (The one after the SLC.), and the third of really far-out developments. Let us take them in turn.

The SLC will begin to provide experience with colliders next year. The run-in time is planned to be a number of years because there are many novel aspects to this device. Hopefully, it will work as predicted and, if so, there will be very large "user pressure" to as quickly as possible build another collider.

What form would such a collider take? It would, naturally, be built very similar to the SLC, but perhaps, to improve efficiency, cut power costs, and increase the accelerating gradient (so as to cut the length) it would operate at a higher frequency than SLC (10 cm wavelength rf). Maybe it would operate at a wavelength of 3 cm or 5 cm.

The development of power sources for such a collider is a major effort of a good number of laboratories. There is work both in the US and Japan, on the lasertron. There is work being done on gyrotrons in the US and (presumably) in the Soviet Union. There is some effort being put on Free Electron Lasers in the US and in Japan. And there is work on pulse multiplying, and on the use of intermediate superconducting linacs, in the US and in CERN.

The major effort is being put into the lasertron and it is hoped that the development effort will "pay off"; ie that the next collider will be built with them.

Turning, now, to schemes that might impact the next collider we find three. The first is the Switched Linac (SL) which is being pursued most vigorously at CERN and to some degree in the US. This scheme depends upon the development of laser switches and has associated with it many questions of jitter, lifetime, alignment, etc. Nevertheless it is a most interesting idea, directly matched to the requirements of a collider, and novel in its approach.

The second is a Wake-Field Transformer (WFT), perhaps with electron rings as the driver. This is being studied at DESY and, to a small degree, in Japan. The primary problem seems to be to provide a proper driving beam, although there are, also, questions of alignment, etc.

The third is the Two-Beam Accelerator (TBA). Work on this is being done in the US. That a free electron laser (FEL) is a prodigious source of power has already been demonstrated, as has the ability of a high-gradient slow wave structure to hold an accelerating gradient at least 10 times that of the SLC. Major problems, such as phase control of the rf and "steady-state" operation of the FEL, remain to be addressed experimentally.

We note that all three of these schemes aim for an accelerating gradient in the range of a few hundred MeV/m. All of them are devices in which one is "close to a conductor"; i.e. within a wavelength of the surface. All three have an effective wavelength in the 1 cm - 2 cm range. The SL and WFT are devices which use shock excitation of the accelerating structure while the TBA uses resonant filling. All three devices are specially designed to the needs of a collider. (Note that in a collider the bunch is only, about, 1 mm long which is quite different than the pulse train in most rf linacs. In SLAC it is 500 m long.) Finally, all three schemes hold out the hope for more efficient operation than the conventional linac. Needless to say, each these three devices would be quite a departure from the usual accelerator structure and, therefore, bring new problems to the accelerator physicist. We can't predict what problems will arise, but they will surely cost in time, money, and reliability. Very soon, in a few years, all three of these approaches, provided they still look attractive, will be ready for scale-up to the next level of experiments.

Turning, now, to really far-out developments the field is -- by definition -- wide open. However there are only two approaches which, so far, have received significant attention; namely, droplets, grating, and open structures or Near Field Accelerators (NFA) and Plasma - Laser Accelerators (PLA).

In both of these approaches one is seeking very high acceleration gradients; about a few GeV/m; namely another order - of - magnitude above the near term devices WFA, SL and TBA.

The NFA is being pursued in the US. It is still in the conceptual stage in that no experiments have yet been done at short wavelengths (like 10 μm). Some experimental work on open structures has been done in the microwave range. No one doubts that the electromagnetic properties of structures can be predicted. But the device depends on the inexpensive and reproducible construction of structures (such as making and properly firing droplets) as well as upon the properties of materials under intense radiation. Presumably the obtaining of a large luminosity in a collider will require very different parameters than one usually contemplates. (For example, much reduced charge per bunch to reduce image charge wake-field effects, but many bunches per unit time.)

The use of plasmas in high energy physics has been pioneered by the UCLA Group, who are still in the vanguard of effort on the Plasma-Laser Accelerator. This work has attracted the attention of a number of group and there is now effort in the US, Canada, Britain, and France. Quite a lot of progress has been made theoretically and a gradient of (about) 1 GeV/m has been demonstrated experimentally. Of course the acceleration length was only a few millimeters, but further work is in progress. Questions of phasing, transverse focusing, pump depletion, etc. must still be studied while the best geometrical configuration is also under study.

In addition, and very importantly, the efficiency of such as accelerator must be studied. (Lasers are notoriously expensive, if one is seeking high average power. They are also rather inefficient.) This accelerator requires the generation of an electromagnetic wave, transfer of the photon energy to plasma motion and, then, transfer of this energy to the high energy particles. But, the PLA is the most promising approach for obtaining real high gradients. (Perhaps, today, the economic minimum of a collider does not require very high gradients, but someday I suspect that we will want very high gradients.)

For the long-term; i.e. beyond the next collider and out into the next century we shall need some novel acceleration concepts. We can't start then; we better start now.

All of the work on new accelerator techniques is "table top" stuff. It is a long way from here to there, i.e., to what we need for high energy physics. One must build 3 meter models (currently the goal is 0.3 m scale models), then 30 meter devices and, then 300 meter machines before one could -- seriously -- consider making a 3 km accelerator.

But even a 30 m scale experiment is non-trivial in its cost. Capitol construction, operation for a few years, theoretical studies, and a research staff would -- very roughly, of course -- cost 10 M$.

A device at the 300 m scale would give high energy particles, (say) 100 GeV, given that one is aiming for a 1 TeV collider at full scale. This energy would be at interest to nuclear physicists, but not to high energy physicists. The machine, since its purpose is for high energy physics, is primarily being built to learn about accelerator physics. And the cost would be non-trivial (say) 10 times 10 M$ or 100 M$.

Do we -- the HEP community - world wide -- have this kind of money for accelerator R&D? I don't know, but I _do_ know that without this kind of effort no really novel (I am excluding evolutionary changes to present power sources and linacs.) idea will even be brought to the point where it will contribute a squiggly line to the Livingston graph. And without new squiggly lines the envelope is going to droop over, and high energy physics--as we have known it -- marching at an exponentially increasing pace to new energy frontiers, will change--slow down--maybe, even, cease being an experimental science.

REFERENCES

1. A. M. Sessler, "The Quest for Ultra-High Energies", Am. J. Phys. 54 (1986) (to be published).
2. P. J. Channell, Editor, Laser Acceleration of Particles, American Institute of Physics Conf. Proc. #91, New York (1982).
3. C. Joshi and T. Katsouleas, Editors, Laser Acceleration of Particles, American Institute of Physics Conf. Proc. #130, New York (1985).
4. R. Palmer, "Laser Driven Accelerators," in Proc. of the Particle Accelerator Conference 1981, IEEE Trans. on Nucl. Sci. NS-28, 3370 (1981).
5. R. Ruth, A. W. Chao, P. L. Morton, and P. B. Wilson, Particle Accelerators 17, 171 (1985).
6. P. Chen, J. Dawson, R. W. Huff, and T. Katsouleas, Phys. Rev. Lett. 54, 693 (1985).
7. R. J. Briggs, Phys. Rev. Lett. 54, 2588 (1985).

HIGH INTENSITY PROTON SYNCHROTRONS

M.K. Craddock
Physics Department, University of British Columbia
and TRIUMF, 4004 Wesbrook Mall
Vancouver, B.C., Canada V6T 2A3

ABSTRACT

Strong initiatives are being pursued in a number of countries for the construction of "kaon factory" synchrotrons capable of producing 100 times more intense proton beams than those available now from machines such as the Brookhaven AGS and CERN PS. Such machines would yield equivalent increases in the fluxes of secondary particles (kaons, pions, muons, antiprotons, hyperons and neutrinos of all varieties) — or cleaner beams for a smaller increase in flux — opening new avenues to various fundamental questions in both particle and nuclear physics. Major areas of investigation would be rare decay modes, CP violation, meson and hadron spectroscopy, antinucleon interactions, neutrino scattering and oscillations, and hypernuclear properties. Experience with the pion factories has already shown how high beam intensities make it possible to explore the "precision frontier" with results complementary to those achievable at the "energy frontier".

This paper will describe proposals for upgrading the AGS and for building kaon factories in Canada, Europe, Japan and the United States, emphasizing the novel aspects of accelerator design required to achieve the desired performance (typically 100 µA at 30 GeV).

INTRODUCTION

"There is no excellent beauty
That hath not some strangeness in the proportion"
Francis Bacon, Essays[1]

With this farsighted remark, whose full meaning has become apparent only after four hundred years, Bacon stands revealed as the true prophet of kaon factories! Coming from a professional lawyer working in Renaissance times, we may expect Bacon's statement to be understandable on more than one level. Of course, it reveals remarkable prescience of two of the more exotic flavours of the standard model, and extols the virtues of strangeness in particular. But from a champion of the experimental method (a cause for which he eventually gave his life[2]) we may also expect a message for the machine builder. And this is clearly that the strangeness factory will itself be unconventional in design.

In case there should be any doubt as to these transparent interpretations, we should recall another of Bacon's claims to fame as the author of "Shakespeare's" plays. The clearest evidence of this lays in various anagrams hidden in the "Shakespearian" text.[3] So it is no surprise to find that Francis Bacon's own name is carefully constructed to immortalize his (or his parents'?) invention:

CAON F-'S BRAIN C.

Amplifying the abbreviations characteristic of the art of cryptograms[4] we find the explicit claim:

KAON FACTORY - HIS BRAIN CHILD.

In more recent times the term "kaon factory", or rather «каонная фабрика», in the sense of a machine to produce high fluxes of kaons, possibly first appeared in print in a paper by Basargin, Komar et al.[5] in 1973 and became familiar only after a Brookhaven meeting[6] in 1976.

This paper will describe high intensity proton synchrotrons designed for kaon factories, i.e. 20-45 GeV machines providing 30-100 µA proton beams. Such machines should provide proton beams 30-100 times more intense than those of the Brookhaven AGS and CERN PS, and correspondingly larger, or cleaner, fluxes of secondary particles:

Kaons	$K^{\pm}$, K^0, $\overline{K}^0$,
Antinucleons	$\overline{p}$, $\overline{n}$
Hyperons	$\Lambda, \Sigma, \Xi, \Omega$
Pions, muons	$\pi^{\pm}$, π^0, $\mu^{\pm}$
Neutrinos	ν_{μ}, $\overline{\nu}_{\mu}$, ν_e, $\overline{\nu}_e$

Two frontiers may be recognized in uncovering new phenomena in subatomic physics — those of high energy and high intensity. The race to higher energies has always left ground behind it only partially explored — a factor which the pion factories have successfully exploited over the last few years with their 200-800 MeV high intensity accelerators. These have made possible high precision comparisons with theory and the investigation of very rare processes, for instance, taking only examples from TRIUMF:

- A muon decay experiment has set a lower limit of 380 GeV on the mass of any right-handed W^+ boson.[7]
- Searches for flavour-changing neutrinoless muon decay have so far proved fruitless, the current branching ratio record being 4×10^{-12} for $\mu A \to eA$, setting a lower bound on the mass of any scalar Higgs of 25 TeV.[8]
- The first observation of charge symmetry breakdown in n-p scattering.[9]

Similar opportunities now arise for the kaon factories, with even greater potential, because of the greater diversity of particles and richer selection of processes available. Indeed there is additional justification in that the grand unified theories are predicting particles with rest energies far above the practical aspirations of any laboratory — and therefore discernible only indirectly through higher-order effects (like those responsible for rare decay modes) in experiments utilizing the largest possible number of particles.

The kaon itself remains one of the most fascinating of the elementary particles. Over the past thirty years its behaviour has led to a number of crucial discoveries in particle physics: strangeness,

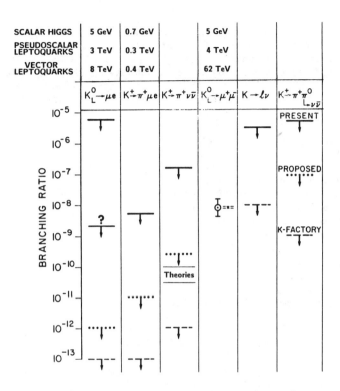

Fig. 1. Branching ratios of several rare kaon decays. Limits achievable with a kaon factory are given by the broken lines. Mass bounds set by present measurements are also shown.

parity violation in the weak interaction, violation of CP invariance and the existence of a fourth "charmed" quark, first suggested by the suppression of decays such as $K_L^0 \rightarrow \mu^+\mu^-$. Today the kaon, as the carrier of a strange quark, continues to promise fundamental insights, not only into particle physics but also into nuclear physics. In the former case the kaon has been referred to as the analogue of the hydrogen atom in atomic physics.[10] Its numerous decay modes provide a rich laboratory for testing higher-order effects. Figure 1 illustrates the improved branching ratios that might be achieved with a kaon factory, together with the lower mass limits for some exotic particles already set by present measurements. In nuclear physics the kaon offers a means of inserting a strange quark into the nucleus, ostensibly tagging one nucleon and freeing it from the Exclusion Principle. A measurement of how far this actually occurs should provide a test of quark deconfinement within the nucleus. Another route opened up would be via Drell-Yan processes.

In spite of this potential the beams of kaons available at present are frustratingly weak ($\sim 10^5$ K$^-$/s) and heavily contaminated with pions (~ 10 π/K). Many desirable experiments are just not feasible, and the same situation holds for the neutrinos, antiprotons, hyperons and other secondary particles produced by GeV accelerators. Consequently there has been a strong interest over the last ten years in building kaon factories to produce beams much more intense than those available at present or, at the sacrifice of some intensity, much less contaminated. The physics case for kaon factories has been made at a number of recent conferences: Brookhaven[6] (1976), TRIUMF[11,12] (1979, 1981), LAMPF[13-15] (1981, 1982), Freiburg[16] (1984), and Mainz[17] (1986). The experimental topics of particular interest at kaon factories can be listed as follows:

Rare kaon and hyperon decays;
CP violation;
Neutrino scattering and oscillations;
Meson and baryon spectroscopy;
Hyperon production, scattering and reactions;
Hadron-nucleon interactions (πN, KN, NN, YN);
Antinucleon interactions;
Charmonium production;
Hypernuclear physics;
K^+-nucleus scattering;
Resonance propagation in nuclei;
Exotic atoms;
Muon physics (muon fluxes will be an order of magnitude higher than at the pion factories).

ACCELERATOR SPECIFICATIONS

In considering the maximum intensity which can be accelerated in a synchrotron two parameters are of particular importance, the number of particles per pulse N and the circulating current I. N is critical because it determines the incoherent (Laslett[18]) space charge tune shift $\Delta \nu$, the decrease in the betatron tune (or number of oscillations per turn) due to the defocusing effects of space charge:

$$\Delta \nu = \frac{r_p \; F \; G \; H \; N}{\pi \; B_f \; \varepsilon_N \; \beta \gamma^2} \; . \tag{1}$$

Here r_p is the classical radius of the proton, B_f is the bunching factor, ε_N is the normalized emittance, β and γ are the relativistic speed and energy parameters, and F, G and H are form factors describing the effect of image forces, the transverse density distribution and the aspect ratio of beam width to height, respectively. Of course, a drop in tune by itself could be compensated by adjusting the quadrupole magnets: the problem here is that the shift can vary across and also along the bunch, so that $\Delta \nu$ also represents a spread in tunes. In order to avoid coming too close to lower-order resonances it is generally agreed that $\Delta \nu$ should be kept below 0.2. The $\beta \gamma^2$ factor makes this condition most critical near injection. It was to take advantage of this energy dependence that the injection energies of the Brookhaven AGS and CERN PS were raised from the original 50 MeV to 200 MeV and 800 MeV, respectively.

The circulating current I is important through its involvement in beam stability. There are many potential instabilities to be considered but one important criterion is that for onset of longitudinal microwave instability, due to Keil, Schnell and Boussard:

$$\left|\frac{Z_\parallel}{n}\right| < \frac{V_p}{I}\left(\frac{\Delta p}{p}\right)^2 B_f |\eta| \beta^2 \gamma . \qquad (2)$$

Here $Z_\parallel$ represents the effective impedance of the beam pipe, n is the mode number, $V_p = m_p c^2/e = 938$ MV, $\Delta p/p$ is the fractional momentum spread in the beam, and the parameter $\eta \equiv \gamma_t^{-2} - \gamma^{-2}$. The energy dependence of this expression is somewhat complicated because of energy-dependent factors η and $\Delta p/p$.

The energy and intensity parameters, including N and I, are listed in Table I for existing and proposed high energy proton synchrotrons. The existing higher energy machines achieve average beam currents of ~1 µA. These currents are limited both by the slow-cycling rate (<1 Hz) and by their low injection energies (<200 MeV

Table I. High-intensity proton synchrotrons.

	Energy (GeV)	Average current (µA)	Rep. rate (Hz)	Protons/ pulse N (× 10^{13})	Circulating current I (A)
Slow Cycling[a]					
KEK PS	12	0.32	0.6	0.4	0.6
CERN PS	26	1.2	0.38	2	1.5
Brookhaven AGS	28.5	0.9	0.38	1.6	0.9
– with Booster		(3)		(5)	(3)
Fast Cycling[b]					
Argonne IPNS	0.5	8	30	0.17	2.3
Rutherford ISIS	0.55(0.8)	40(200)	50	(2.5)	(6.1)
Fermilab Booster	8	7	15	0.3	0.3
AGS Booster[19]	(1)	(20)	(10)	(1.25)	(3)
Proposed Boosters[b]					
TRIUMF	3	100	50	1.2	2.7
European HF	9	100	25	2.5	2.5
LAMPF	6	144	60	1.5	2.2
KEK Booster	1–3	100	15	4	8
Kaon Factories[a]					
TRIUMF[20]	30	100	10	6	2.8
European HF[21]	30	100	12.5	5	2.5
LAMPF[22,23]	45	32	3.33	6	2.2
Japan – Kyoto[24]	25	50	30	1	0.5
– KEK[25]	30	30	1	20	7

[a]Slow extraction
[b]Fast extraction

into their first synchrotron stages) which restrict N to $\sim 2\times 10^{13}$. The circulating current $I \approx 1$ A.

Higher intensities have been achieved in machines using faster cycling rates (10-50 Hz). A record current (for a synchrotron) of 40 µA was recently achieved at the Rutherford ISIS spallation neutron source, and this will be raised to 200 µA when commissioning is completed. The number of protons per pulse N will then be 2.5×10^{13}, only a little more than in the slow-cycling machines, but the circulating current I will rise to 6 A.

The proposed kaon factories aim at energies in the 25-45 GeV range with proton currents of 30-100 µA. Proposals have come from all three existing pion factories at LAMPF, SIN and TRIUMF, these laboratories being unique in already possessing operating machines with adequate energy and current to act as injectors. All the proposals also involve intermediate booster synchrotrons with energies in the 3-9 GeV range. These have a dual purpose. In the first place they raise the injection energy into the main ring and therefore through Eq. (1) the charge per pulse that can be accelerated to $N \sim 6\times 10^{13}$ ppp. This enables the desired current of 100 µA to be achieved with only moderately fast cycling rates ~10 Hz. The second reason for using a booster synchrotron concerns the radio-frequency acceleration requirements. In a fast-cycling machine the much more rapid acceleration requires a much higher rf voltage than has been conventional in slower-cycling machines — about 2 MV for a 10 Hz 30 GeV machine. At the same time a large frequency swing (20-30%) is required when starting from pion factory energies of 500-800 MeV. The use of a booster enables these demands to be handled separately. Almost the entire frequency swing can be provided in the booster at relatively low rf voltage while the main ring provides the 2 MV with only a few per cent frequency swing. Being smaller the booster must cycle faster (15-60 Hz) in order to fill the circumference of the main ring. The charge per pulse would be $N \sim 10^{13}$, comparable to existing machines injected in the same energy range. The circulating current in both booster and main ring would be $I < 3$ A, a level which is not expected to present any problems.

A comparison[23] of the secondary particle fluxes which would be available from the kaon factories is given in Table II, normalized to the yield from the AGS in 1984. We see that in spite of the rather

Table II. Normalized yield from existing and proposed kaon factories.

	K^-/yr (1 GeV/c)	K^-/yr (7 GeV/c)	$\bar{p}$/yr (7 GeV/c)	ν_μ/yr integrated
AGS (1984)[a]	1	1	1	1
TRIUMF/EHF[a]	139	161	176	68
LAMPF II[b]	100	187	290	61

[a] Assumes 50% slow extraction, 50% fast extraction.
[b] Simultaneous slow extraction and fast extraction.

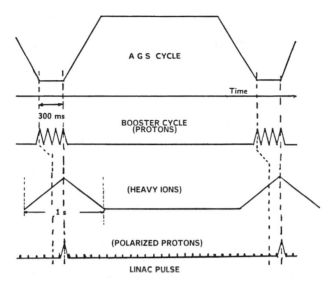

Fig. 2. Injection programs for the AGS.

different energy and current parameters of the LAMPF and TRIUMF & EHF proposals, the flux improvements are comparable at about two orders of magnitude; the LAMPF proposal does provide significantly more higher-momentum antiprotons.

Many of the design features required for these high energy and intensity machines are common to all the proposals; to avoid repetition and because I am most familiar with the TRIUMF design, I will describe its rationale in some detail, reporting only distinctive features in the other cases. All these proposals, however, must yield pride of place to the funded project at Brookhaven to enhance the AGS performance by the addition of a booster synchrotron.[19]

BROOKHAVEN AGS BOOSTER

Funding for the booster began in October 1985 and the project is expected to be complete by the end of 1989. This is in fact a multi-purpose project aimed at the acceleration of heavy ions as well as polarized and unpolarized protons. The booster ring is one-quarter the circumference of the AGS and is located in the angle between the linac tunnel and the AGS ring. The modes of operation for various particles are illustrated in Fig. 2, where the time scale covers the 2.8 s of a single slow extracted pulse. For unpolarized protons four booster pulses would be injected at a 10 Hz repetition rate within a 300 ms flat bottom, enabling the present 1.6×10^{13} ppp to be increased to 5×10^{13} ppp. Initially protons would be accelerated to 1 GeV although the bending capability provided for heavy ions would eventually allow protons to be accelerated to 2.5 GeV. For heavy ions a slower acceleration time is required in the booster, and only one pulse would be injected into the main ring. For polarized protons there is the option of stacking up to 28 pulses in the booster ring before injecting them into the AGS. Further improvements beyond this program include the possibility of adding a 30 GeV stretcher ring and of making modifications to the AGS (rf, etc.) to accommodate $>5\times10^{13}$ ppp and increase the beam intensity to as much as 2×10^{14} p/s (32 µA).

TRIUMF KAON FACTORY

The TRIUMF proposal is based on a 30 GeV main "Driver" synchrotron 1072 m in circumference accelerating 10 µC pulses at a 10 Hz repetition rate to provide an average beam current of 100 µA. For the reasons explained above a Booster synchrotron is used to accelerate protons from the TRIUMF cyclotron at 450 MeV to 3 GeV: this machine is 1/5 the radius of the main ring (Fig. 3) but cycles five times faster at 50 Hz. The Booster energy is chosen to minimize the total cost of the project. This depends mainly on magnet costs, and in particular on the magnet apertures. The minimum cost condition occurs when the emittances set by the space charge tune shift formula [Eq. (1)] are the same for both machines.[26]

Each of the three accelerators is followed by a dc storage ring to provide time-matching and finally a slow extracted beam for coincidence experiments. These are relatively inexpensive, accounting for only 25% of the total cost. Thus the TRIUMF cyclotron would be followed by a chain of five rings, as follows:

A Accumulator: accumulates cw 450 MeV beam from the cyclotron over 20 ms periods
B Booster: 50 Hz synchrotron; accelerates beam to 3 GeV
C Collector: collects 5 booster pulses and manipulates beam longitudinal emittance
D Driver: main 10 Hz synchrotron; accelerates beam to 30 GeV;
E Extender: 30 GeV storage ring for slow extraction

As can be seen from the energy-time plot (Fig. 4) this arrangement allows the cyclotron output to be accepted without a break, and the B

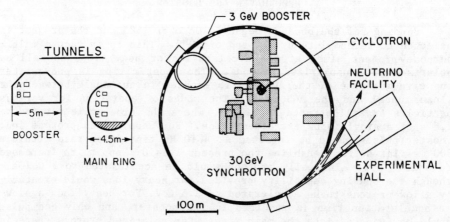

Fig. 3. Proposed layout of the accelerators and cross sections through the tunnels.

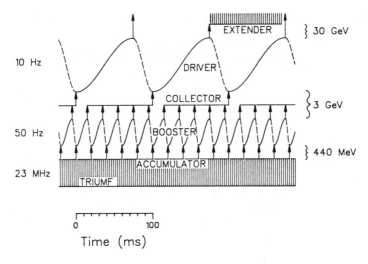

Fig. 4. Energy-time plot showing the progress of the beam through the five rings.

and D rings to run continuous acceleration cycles without wasting time on flat bottoms or flat tops; as a result the full 100 μA from the cyclotron can be accelerated to 30 GeV for either fast or slow extraction. Figure 4 also illustrates the asymmetric magnet cycles used in both synchrotrons. The rise time is three times longer than the fall, reducing the rf voltage required by 1/3, and the number of cavities in proportion. Full-scale power supplies providing such a cycle have been developed by W. Praeg[27] at Argonne, with the encouragement of Los Alamos.

Figure 3 shows the location of the Accumulator directly above the Booster in the small tunnel, and of the Collector and Extender rings above and below the Driver in the main tunnel. Identical lattices and tunes are used for the rings in each tunnel. This is a natural choice providing structural simplicity, similar magnet apertures and straightforward matching for beam transfer. The practicality of multi-ring designs has been thoroughly demonstrated at the high-energy accelerator laboratories, and the use of bucket-to-bucket transfer at each stage rather than the traditional coasting and recapture should keep transfer spills to a minimum.

Separated function magnet lattices are used with a regular FODO quadrupole arrangement, but with missing dipoles arranged to give superperiodicity 6 in the A and B rings and 12 in the C, D and E rings. This automatically provides space for rf, beam transfer and spin rotation equipment. It also enables the transition energy to be driven above top energy in both synchrotrons, avoiding the beam losses usually associated with crossing transition in fast-cycling machines and the difficulties anticipated in making that jump under high beam-loading conditions. It will be recalled that in crossing transition the synchronous phase moves from the rising to the falling side of the rf wave. Right on transition the synchronous phase is at the peak of the wave, but the stable bucket area is reduced to zero.

The transition energy may be expressed roughly in terms of the momentum dispersion $\eta_x = \Delta x/(\Delta p/p)$ by $\gamma_t \simeq \sqrt{R/\langle\eta_x\rangle}$, where R is the machine radius. In conventional alternating gradient proton synchrotrons with regular dipole lattices, $\eta_x \simeq$ constant and $\gamma_t \simeq \nu_x$. In a missing dipole lattice the dispersion function will oscillate and its average value $\langle\eta_x\rangle$ can be driven down towards zero ($\gamma_t \to \infty$) or even to negative values (making γ_t imaginary). This effect can be enhanced, without perturbing the other lattice functions too strongly, by bringing the horizontal tune value ν_x towards, but not too close to, the integer superperiodic resonance. Values of $\nu_x \simeq 5.2$ for the S = 6 Booster and $\nu_x \simeq 11.2$ for the S = 12 Driver prove to be quite convenient. Associated choices of $\nu_y = 7.23$ and 13.22, respectively, keep the working points away from structural resonances and allow room for the anticipated space charge tune spreads, $\Delta\nu_y = 0.18$ and 0.11.

In fast-cycling machines the synchrotron tune ν_s is relatively large and care must be taken to avoid synchro-betatron resonances of the form

$$\ell\nu_x + m\nu_y + k\nu_s = n . \qquad (3)$$

Here k, ℓ and m are integers whose sum gives the order of the resonance, while n is an integer giving the Fourier harmonic of the field component driving it. Only the lowest order of these resonances, satellites of the integer betatron resonances, are of serious concern, but in the Booster ν_s is as large as 0.04, so that these could significantly reduce the working area available for the tune spread. Fortunately it is possible to mimimize their ill effects by suppressing the driving terms for the nearby integer resonances. This can be achieved by placing the rf accelerating cavities symmetrically with the magnet superperiodicity S = 6. The synchro-betatron resonances near $\nu_x = 5$ and $\nu_y = 7$ are then eliminated except for fifth harmonic variations in the cavity voltages and seventh harmonic errors in vertical dispersion.

The superperiodicity of the magnet lattice also drives depolarizing resonances. Compared to the AGS the superperiodicity is stronger but passage through the resonances is more rapid, so that the overall depolarizing effects are comparable in magnitude. At the higher energies it becomes impractical to build pulsed quadrupoles fast enough and strong enough to jump these resonances. Instead it is proposed to use helical "snakes" as proposed by E.D. Courant.[28] These would fit into the 10 m long drift spaces and cause a closed orbit distortion of <4 cm at 3 GeV. Pulsed quadrupoles would be practical for the lower momenta resonances in the Booster.

At injection into the A and B rings the rf accelerating system will operate at 46.1 MHz, twice the radio frequency of the TRIUMF cyclotron. So that every other rf bucket is not missed the radius of these rings is made a half-integer multiple (4.5) of that of the last orbit in the cyclotron. The Booster cavities are based on the double gap cavities used in the Fermilab booster. They will develop a voltage of 26 kV at each gap for a frequency swing of 46-61 MHz. Twelve cavities will develop the required voltage of 580 kV. The cavities for the other rings are based on the single gap cavities designed for the Fermilab main ring. For the main synchrotron 18 cavities each developing 140 kV give a total of 2520 kV with a frequency swing from 61.1-62.9 MHz. To keep the rf power requirements within a factor two of the 3 MW beam power and to provide stability under high beam loading conditions, a powerful rf control system has been designed, based on experience at CERN, including fast feedback around the power amplifiers, and phase, amplitude, tuning, radial and synchronization feedback loops. One-turn-delay feedforward is used to control transient loading effects.

The Accumulator ring is designed to provide the matching between the small emittance cw beam from the TRIUMF cyclotron and the large emittance pulsed beam required by the Booster. The Accumulator will stack a continuous stream of pulses from the cyclotron over a complete 20 ms Booster cycle. Injection and storage over this 20,000 turn long cycle is only possible through the use of the H^- stripping process. This enables Liouville's theorem to be bypassed and many turns to be injected into the same area of phase space. In fact it is not necessary or desirable to inject every turn into the same area; the small emittance beam from the cyclotron (2π mm·mrad)^2 × 0.0014 eV·s) must be painted over the much larger three-dimensional emittance 31 × 93 (π mm·mrad)^2 × 0.048 eV·s needed in the Booster to limit the space charge tune shift [Eq. (1)]. Painting also enables the optimum density profile to be obtained and the number of passages through the stripping foil to be limited. The stripping foil lifetime is estimated to be > 1 day. The painting is achieved by a combination of magnetic field ramping, vertical steering and energy modulation using additional rf cavities in the injection line. A group of 5 rf buckets (out of 45) will be left empty to provide ~100 ns for kicker magnet rise and fall times during subsequent beam transfers.

At present, of course, the H^- ions are stripped in the process of extracting them from the cyclotron. To extract them whole from the cyclotron a new extraction system will be required, and elements of this have been under test for the last year. The first element is an rf deflector operating at half the fundamental frequency: it deflects alternate bunches in opposite directions. Because the cyclotron operates on an odd harmonic (h=5) and the $\nu_R = 3/2$ resonance is nearby, this produces a coherent radial betatron oscillation. Although successive turns are not completely separated a radial modulation in beam density is produced. An electrostatic deflector is then placed so that the septum is at a density minimum. In fact the septum is protected by a narrow stripping foil upstream so that no particles hit it; instead they are safely deflected out of the machine. The particles entering the electrostatic deflector eventually pass into magnetic channels and then into the Accumulator injection line. The magnetic channels have only just become available but tests with the first two elements have demonstrated 85% extraction efficiency.

Successful operation of a high-intensity accelerator depends crucially on minimizing beam losses and the activity they produce. Where some loss is expected, near injection and extraction elements, collimators and absorbers will be provided and equipment will be designed for remote handling. The beam current and any spill will be carefully monitored, and in case of the beam becoming unstable at any time through component failure or power excursions, each of the five rings is equipped with a fast-abort system which would dump the entire beam safely within one turn.

Several processes which give rise to losses in existing machines have been avoided entirely in this design. The use of H^- ions for injection into the Accumulator ring will almost entirely eliminate injection spill. The use of bucket-to-bucket transfer between the rings will avoid the losses inherent in capturing coasting beams. The buckets will not be filled to more than 60% of capacity; this should avoid beam losses while providing a sufficient bunching factor to minimize the space charge tune spread at injection and sufficient spread in synchrotron tune to give effective Landau damping. Magnet lattices are designed to place transition above top energy in all the rings, thus avoiding the instabilities and losses associated with that passage. Moreover, with the beam always below transition, it is no longer advantageous to correct the natural chromaticity, so that sextupole magnets are needed only for error correction, and geometric aberrations in the beam are essentially reduced to zero.

Beam instabilities will be suppressed or carefully controlled. Although all five rings have large circulating currents, rapid-cycling times give the instabilities little time to grow to dangerous levels. Coupled-bunch modes, driven by parasitic resonances in the rf cavities and by the resistive wall effect, will be damped using the standard techniques (Landau damping by octopoles, bunch-to-bunch population spread and active damping by electronic feedback). The longitudinal microwave instability is a separate case because of its rapid growth rate. It will be avoided by making the longitudinal emittance sufficiently large at every point in the cycle and by minimizing the high frequency impedance in the vacuum chamber as seen by the beam. At this stage of the design it is not possible to make accurate estimates of beam blow-up due to instabilities or non-linear resonances, but to be safe, the magnet apertures have been designed to accommodate a 50% growth in the horizontal, and 100% growth in the vertical beam emittance.

The proposal was submitted to TRIUMF's funding agencies, the National Research Council and the Natural Sciences and Engineering Research Council, in September 1985. The total cost, including salaries and E,D&I, but not contingency, is estimated to be M$427 (1985 Canadian dollars). Two review committees have been set up. The first, an international "Technical Panel" of particle, nuclear and accelerator physicists, met in February and has produced a very favourable report. The second "Review Committee" consisting of Canadian industrialists and scientists from various disciplines met recently and is expected to finalize its report later this summer.

LAMPF II

The Los Alamos proposal[22,23] for a 45 GeV, 32 µA facility is described in more detail elsewhere in this volume.[29] It aims at significantly higher energies but lower currents than the other schemes. It is argued that higher energy protons will be more suitable for Drell-Yan studies of quark confinement in the nucleus, besides providing higher momentum secondary beams. In fact 60 GeV could eventually be achieved by running the main ring bending magnets

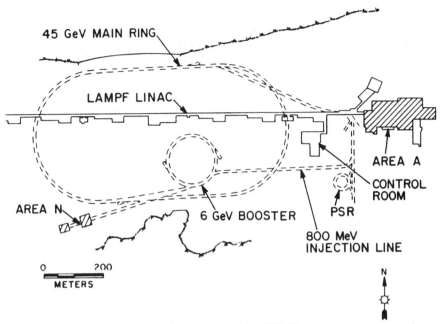

Fig. 5. Site plan of the proposed LAMPF II accelerators and experimental areas.

at 2.0 T. Figure 5 shows the proposed layout, with the existing 800 MeV linac injecting every other H⁻ pulse into a 6 GeV circular booster synchrotron of circumference 330.8 m, operating at 60 Hz and providing an average current of 144 µA. Out of every 18 pulses, 14 (112 µA) are directed to Area N for neutrino studies, while 4 are transferred to the main ring of circumference 1333.2 m for acceleration to 45 GeV. With the magnets flat-topped for 50% of the cycle for slow spill to Area A, and a 3.33 Hz repetition rate, the average proton current is 32 µA. There are two modes of slow spill, either fully debunched with 100% microscopic duty factor, or with rf on giving 1.75 ns pulses (FWHM) every 16.7 ns. Fast extraction without the flat-tops would provide 64 µA currents. Storage rings have not been included in the initial proposal, but space is available for later installation of collector and stretcher rings. This will enable the main ring cycle rate to be increased to 10 Hz and the average fast- or slow-extracted current to be increased to 96 µA.

In the latest design both synchrotrons use separated-function magnets. The booster lattice has a superperiodicity S = 6, created by omitting the dipoles from every other focusing cell and modulating the lengths of the cells. The choice of working point (ν_x = 5.22, ν_y = 4.28) drives transition above the top energy (γ_t = 14.5). The six short empty cells are then used to accommodate 18 rf cavities, while the long empty cells are available for beam transfer elements.

The main ring race-track lattice consists of several functionally different sections — two 144° bending arcs, four 18° missing magnet dispersion suppressors, four short matching sections, and two

90 m long straight sections to accommodate rf cavities, beam transfer elements, and Siberian snakes. The 28 m long high-β section will reduce the beam spill on the extraction septum. The working point ($\nu_x = 7.45$, $\nu_y = 6.45$) was chosen close to a half-integer resonance for slow extraction. The high average dispersion in the bending arcs ($\overline{\eta}_x \simeq 6$ m) keeps the transition energy below the acceleration range ($\gamma_t = 6.4$, 5.06 GeV).

The Los Alamos group is pioneering a number of potentially important technical developments. First among these is the use of an asymmetric magnet cycle[27] (see above) with slow rise and fast fall to reduce the rf voltage requirements. In the latest design this scheme is applied only to the booster synchrotron. For the main ring a linear ramp is proposed with equal 50 ms rise and fall times. Instead of a resonant power supply system this would use SCR bridge rectifiers operating from three phase 60 Hz, supplied by a generator-flywheel set.

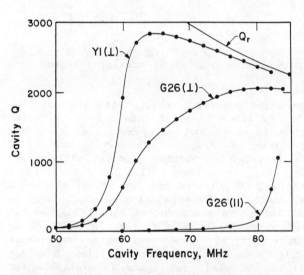

Fig. 6. Variation of the Q-factor of test cavities with frequency for perpendicular- and parallel-biassed ferrite.

To achieve tunable high-power rf cavities economically, the Los Alamos group are proposing to bias the ferrite with magnetic field perpendicular to that of the rf, rather than parallel, as has been conventional. Figure 6 shows how with TDK G26 ferrite this results in cavities with much higher Q values. Tests of a full-scale cavity under high-power conditions are beginning in the laboratory, and will be continued in an existing machine such as the Los Alamos Proton Storage Ring under high beam-loading conditions.

The third development concerns the design of suitable vacuum chambers for the fast-cycling magnets. There must be a conducting surface on the inside of the chamber to prevent the build-up of electrostatic charge and also to provide a low-impedance path for the high-frequency image currents involved in maintaining beam stability. On the other hand eddy currents must be suppressed to minimize heating and magnetic field distortion. The only present example of such a system is at the Rutherford ISIS synchrotron,[30] where a ceramic vacuum chamber is used, fitted with an internal cage of longitudinal wires to provide an rf shield. The Los Alamos group is

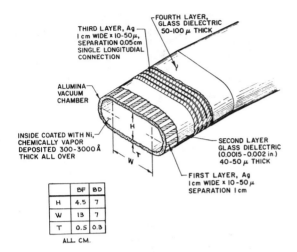

building test sections of a ceramic vacuum chamber with the conducting surfaces provided by an internal 1 μm copper coating and external layers of copper strips (Fig. 7). The system offers a smaller magnet aperture but will require careful design of the end connections.

Fig. 7. Los Alamos design for booster dipole magnet vacuum chamber.

The initial proposal[22] appeared in December 1984. The second edition[23] has appeared hot from the press at this conference. The cost has been reduced somewhat by eliminating one experimental area. This has been done without cutting down the number of secondary channels by using C. Tschalär's MAXIM scheme[31] to produce beams to the left and right of each target. The total cost, including ED&I but not contingency, is estimated to be M$328 (FY 1985 US dollars).

EUROPEAN HADRON FACILITY

The first European scheme for a kaon factory came from SIN. This was for a 20 GeV 80 μA synchrotron fed from the 590 MeV SIN cyclotron and the proposed ASTOR isochronous storage ring.[32] Rapidly growing support led to the formation of an international study group. The conceptual design has been discussed at a number of workshops during the last nine months, and the reference design was agreed on in March. This was described by F. Bradamante[21] at the recent Mainz conference.[18]

The layout of the proposed EHF is shown in Fig. 8. To accelerate 100 μA to 30 GeV the main ring cycles at 12.5 Hz; it is fed by a 9 GeV booster of half the circumference and a specially designed 1.2 GeV linac, both cycling at 25 Hz. Collector (here called "accumulator") and stretcher rings are included in the design to follow the booster and main ring, respectively. The choice of a relatively high 9 GeV for the booster energy raises the cost of the entire project by about 10% over the minimum at ~4 GeV but offers a number of advantages. Not only does it provide greater opportunities for interesting physics at an intermediate construction stage but it brings the booster radius to half that of the main ring, allowing the collector to be placed in the booster tunnel, halving its length and equalizing the number of rings in each tunnel. The desire to accelerate polarized protons plays an important role in this design, and with 9 GeV injection Siberian snakes in the main ring will cause less closed orbit distortion.

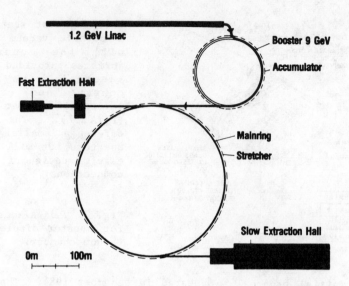

Fig. 8. Schematic layout of the European Hadron Facility.

Although the design is officially "siteless" the circumference of the main ring has been chosen equal to that of the CERN ISR for reference purposes. The advantages of tunnel-stuffing are clearly as obvious in Europe as they have been in the USA. Besides the tunnel, CERN's site offers other advantages — the existence of interim injectors, the availability of the West Hall after the experimental program has moved to LEP, and the infrastructure (even if only as a model for an independent EHF laboratory). The most obvious advantage to CERN would be the provision of a back-up PS. But these are mere speculations and at the moment it seems more likely that the EHF would be located elsewhere, say in Italy or at SIN.

All the rings use separated-function magnets and superperiodic lattices with transition energy out of the acceleration range. The booster uses an interesting doublet lattice with S = 6 and nine cells in each superperiod. Dipole magnets are omitted from three cells, two of which can then be made dispersionless and therefore ideal for the location of rf cavities. The quadrupole doublets give low β-functions and high tunes (ν_x = 13.4, ν_y = 10.2); the transition energy γ_t = 12.7. The main ring has eight superperiods with seven regular FODO cells in each; dipoles are omitted in four of the half-cells, bringing the dispersion close to zero in three of them grouped together. These provide 25 m long straight sections, four of which are used for rf cavities, two for Siberian snakes and two for beam transfer.

The 1200 MeV linear injector consists of a series of linacs of different types (Fig. 9). Following the H$^-$ ion source and dc acceleration to 30 keV, two RFQs operating at 50 MHz and 400 MHz take the beam to 200 keV and 2 MeV, respectively. A 400 MHz drift tube linac then accelerates the ions to 150 MeV and finally a 1200 MHz side-coupled linac to 1200 MeV. Figure 9 also illustrates the time structure, showing how only two out of 24 SCL buckets are filled.

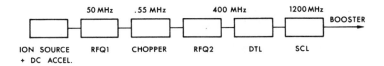

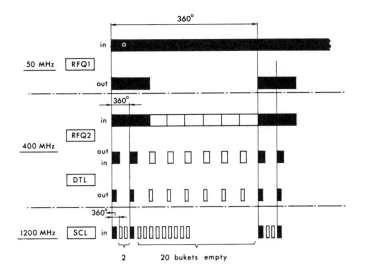

Fig. 9. Schematic diagram of the E.H.F. linear accelerators and of the phase-compression of the dc beam from the ion source into the 2:8 bucket scheme for optimal injection into the Booster.

These are then painted over the central 50% of the 50 MHz booster bucket to provide the desired density distribution.

Work is continuing with the aim of producing a formal proposal within about a year. Some temporary office space has been made available at CERN and the study group plans to make this their headquarters. The total cost is estimated to be MDM867, not including controls, contingency or inflation.

JAPANESE PROPOSALS

There is considerable interest in kaon factories in Japan, with schemes being promoted by Kyoto University and jointly by KEK and the University of Tokyo.

The most fully developed scheme is that from Kyoto, where for some time there has been a proposal for an 800 MeV linac pion factory. The proposed kaon factory complex[24] would add an 800 MeV compressor ring, a 25 GeV fast-cycling synchrotron, a stretcher ring, and antiproton accumulation and storage rings (Fig. 10). The linac would operate at 60 Hz, feeding every other pulse to the synchrotron, operating at 30 Hz with an average current of 50 µA. The magnet lattice has a superperiodicity S = 4 with regular FODO cells and dispersion-free straight sections. The betatron tunes are $\nu_x = 11.2$, $\nu_y = 10.2$; the transition $\gamma_t = 7.9$ GeV.

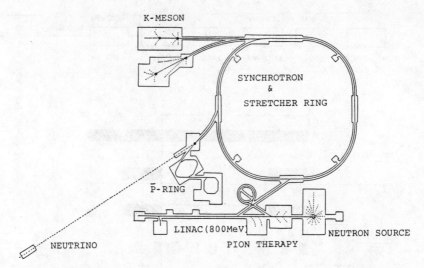

Fig. 10. Layout of the Kyoto accelerator complex based on a 25 GeV, 50 µA synchrotron.

At KEK and the University of Tokyo there has been a longstanding interest[33] in increasing the intensity of the KEK 12 GeV PS by using a higher energy booster (1-3 GeV). There is also the independent GEMINI project,[34] a spallation neutron source and pulsed muon facility, based on an 800 MeV synchrotron operating at 50 Hz to produce 500 µA proton beams. There is now a move to combine these schemes into a single coherent project which might also act as the injector of a kaon factory. One scheme which is being considered would involve a 30 GeV, 30 µA proton synchrotron cycling at 1 Hz fed by a 3 GeV, 100 µA booster cycling at 15 Hz. The booster might be a new fast-cycling synchrotron or it could be a combination of GEMINI with a 3 GeV FFAG after-burner. The booster would form the core of a multi-purpose laboratory, feeding not only the main ring but also a storage ring for spallation neutrons and pulsed muons. The acceleration of heavy ions is also being considered.

A meeting is expected to be held shortly by the proponents of all these schemes to settle the parameters of an accelerator complex — referred to as the "big hadron project" — satisfying their various needs.

CONCLUSIONS

A variety of important questions in both particle and nuclear physics could be attacked using the more intense and/or clean beams of kaons, antiprotons, neutrinos and other particles that high-intensity proton synchrotrons could supply. In response to these rich possibilities a booster is under construction at the Brookhaven AGS and proposals for kaon factories have been completed at LAMPF and TRIUMF and are in preparation in Europe and Japan. Moreover, there is a large body of enthusiastic users numbering well over 1000 worldwide. The first of these proposals to be officially reviewed, the

TRIUMF KAON factory, has been enthusiastically endorsed by an international panel. Prospects seem good that at least one of these projects will be funded within the next few years.

ACKNOWLEDGEMENTS

The author is delighted to acknowledge the information and material that have been generously supplied to him for this talk by Drs. F. Bradamante, M. Inoue, T. Ludlam, A. Masaike, H. Sasaki, T. Suzuki, H.A. Thiessen, and T. Yamazaki.

REFERENCES

1. F. Bacon, Essayes or Counsels, Civill and Morall, XLIII, "On Beauty" (J. Haviland at the King's Head, London, 1625).
2. J. Aubrey, Brief Lives, ed. A. Clark (Oxford, at the Clarendon Press, 1898); ed. O.L. Dick (Penguin Books, London, 1972).
3. E. Bormann, Der Shakespeare-Dichter: Wer War's? (Leipzig, 1902).
4. See e.g., R.A. Knox, Essays in Satire, "The Authorship of 'In Memoriam'" (Sheed & Ward, London, 1928); (gives numerous demonstrations that Tennyson's poem was in fact written by Queen Victoria).
5. Yu.G. Basargin, E.G. Komar, V.M. Lobashev, I.A. Shukeilo, Doklady $\underline{209}$, 819 (1973); Sov. Phys. Dokl. $\underline{18}$, 229 (1973).
6. Proc. Summer Study Meeting on Kaon Physics & Facilities, Brookhaven, June 1976, ed. H. Palevsky, BNL-50579 (1976).
7. J. Carr et al., Phys. Rev. Lett. $\underline{51}$, 627 (1983).
8. T. Numao, "Rare decays", in these proceedings.
9. R. Abegg et al., Phys. Rev. Lett. $\underline{56}$, 2571 (1986).
10. K. Nishijima, Proc. Kyoto Int. Symp. "The Jubilee of the Meson Theory", ed. M. Bando et al., Prog. Theor. Phys. Suppl. $\underline{85}$, 43 (1985).
11. Proc. Kaon Factory Workshop, Vancouver, 1979, ed. M.K. Craddock, TRI-79-1.
12. Proc. 2nd Kaon Factory Workshop, Vancouver, 1981, ed. R. Woloshyn and A. Strathdee, TRI-81-4.
13. Proc. Workshop on Nuclear and Particle Physics at Energies up to 31 GeV, Los Alamos, 1981, ed. J.D. Bowman et al., LA-8775-C.
14. LAMPF II Workshop, 1982, ed. H.A. Thiessen, LA-9416-C.
15. Proc. Second LAMPF II Workshop, ed. H.A. Thiessen et al., LA-9572-C.
16. Proc. Workshop on Future of Intermediate Energy Physics in Europe, Freiburg, April 1984, ed. H. Koch and F. Scheck, KEK Karlsruhe (1984).
17. Proc. Int. Conf. on the European Hadron Facility, Mainz, March 1986 (to be published).
18. J.L. Laslett, Proc. Brookhaven Summer Study on Storage Rings, BNL-7534 (1963), p. 324.
19. AGS Booster Conceptual Design Report, BNL-34989 (1984); Y.Y. Lee, IEEE Trans. Nucl. Sci. $\underline{NS-30}$, 1607 (1985).
20. KAON Factory Proposal, TRIUMF ($\underline{1985}$).
21. F. Bradamante, "Conceptual Design of the EHF", INFN/AE-86/7 (1986).

22. A Proposal to Extend the Intensity Frontier of Nuclear and Particle Physics to 45 GeV (LAMPF II), LA-UR-84-3982 (1984).
23. The Physics and a Plan for a 45 GeV Facility that Extends the High-Intensity Capability in Nuclear and Particle Physics, LA-10720-MS (1986).
24. K. Imai, A. Masaike et al., Proc. 5th Symposium on Accelerator Science and Technology, KEK, September 1984, p. 396.
25. T. Suzuki, private communication.
26. U. Wienands and M.K. Craddock, TRIUMF internal report TRI-DN-86-7 (1986).
27. W.F. Praeg, IEEE Trans. Nucl. Sci., NS-30, 2873 (1983).
28. E.D. Courant, private communication.
29. H.A. Thiessen, "LAMPF-II 1986: Matching the Machine to the Physics Requirements", in these proceedings.
30. J.R.J. Bennett and R.J. Elsey, IEEE Trans. Nucl. Sci. NS-28, 3336 (1981).
31. C. Tschalär, "Multiple Achromatic Extraction System", Nucl. Instrum. Methods (in press).
32. W. Joho, IEEE Trans. Nucl. Sci. NS-30, 2083 (1983).
33. T. Yamazaki, Proc. KEK Workshop on Future Plans for High Energy Physics, March 1985, preprint UTMSL-116.
34. H. Sasaki et al., Proc. Int. Collaboration on Advanced Neutron Sources VII, Chalk River, September 1983, AECL-8488, 50 (1984).

ELECTROMAGNETIC STRUCTURE OF NUCLEI[*]

R. G. ARNOLD

The American University, Washington, D.C. 20016

and

Stanford Linear Accelerator Center, Stanford, CA 94305

ABSTRACT

A brief review is given of selected topics in the electromagnetic structure of nucleons and nuclei for which significant progress was obtained in the two years since the last meeting in this series at Steamboat Springs Colorado.

INTRODUCTION

Electromagnetic probes of nucleon and nuclear targets continues to provide crucial information for understanding the internal structure of hadronic matter. The experimental and theoretical progress in the last two years has extended our understanding in several key areas. On the experimental side, new results have been obtained in part because high current electron beams in the GeV energy range are now available for nuclear scattering experiments. On the theoretical side there is much activity trying to formulate a consistent picture of nucleon and nuclear structure including quark degrees of freedom. This task is far from complete, but we can report progress in the following areas.

NUCLEON FORM FACTORS

We know from much evidence that nucleons are extended composite particles that participate in the strong, electromagnetic, and weak interactions. Measurements of the e.m. form factors

$$G_E = F_1 - \frac{Q^2}{4m^2} F_2 \qquad (1.a)$$

$$G_M = F_1 + F_2 \qquad (1.b)$$

where F_1 and F_2 are the Dirac and Pauli form factors, gives information about the distribution of charge and current in the nucleons and about the nature of the virtual photon-nuclon interaction. At low energy the data are interpreted to give charge and current distributions, but the constituents are not resolved. In the GeV energy range the vector meson dominance model (VMD)[1] pictures the interaction of the virtual photon with the nucleon to be composed of two

---

[*] Work supported in part by the Department of Energy, contract DE-AC03-76F00515 (SLAC) and by the National Science Foundation, Grant PHY85-10549 (American University).

parts: interaction of a bare photon, and interaction of the hadronic (vector meson) components of the virtual photon. At higher energy the hadronic VMD interaction is expected to give way to hard scattering from the nucleon quark constituents described[2,4] by perturbative QCD (PQCD). A central question today is: Where is perturbative QCD applicable? There are new experimental and theoretical results bearing on that question.

Form Factors in Perturbative QCD

The hadron electromagnetic form factors are calculated[2,4] in PQCD as a special case of exclusive reactions (kinematics of all initial and final particles specified). The amplitude for scattering is predicted to factorize into a product of a hard scattering amplitude, containing the pointlike interaction of n valence quarks, times a probability amplitude for finding the n quarks in the initial and final wave functions. The hard scattering is governed by the laws of QCD with quark-gluon coupling given by $\alpha_S(Q^2)$. The quark wave functions for color singlet hadrons can be written as a sum of components starting with the lowest n quark valence component and summing over higher states containing extra quarks and gluons (ocean components). The form factors are predicted[2,3] to have power law falloff at large Q^2

$$F(Q^2) \simeq \left(\frac{1}{Q^2}\right)^{n-1}. \tag{2}$$

The contributions from the non-valence quarks decrease faster with increasing Q^2 due to the penalty for transfering momentum to extra constituents. The quark helicity is conserved in the interactions of the vector photon with massless spin one-half quarks.[5] This leads to the suppression of the Pauli (spin flip) terms compared to the Dirac terms by an extra power of $1/Q^2$

$$F_1 \simeq \frac{C_1}{Q^4}; \quad F_2 \simeq \frac{C_2}{Q^6} \tag{3}$$

where the numbers C_1 and C_2 are determined by the wave functions. Explicit PQCD calculations[4] give

$$G_M^p = \frac{\alpha_s^2(Q^2)}{Q^4} \sum_{n,m} d_{nm} \left(\ln \frac{Q^2}{\Lambda^2}\right)^{-\gamma_m - \gamma_n} \tag{4}$$

If G_M^p falls like $1/Q^4$ at large Q^2, this could be evidence that PQCD is working and scattering from the three valence quarks determines the form factor. There is also a smaller logarithmic dependence on Q^2 contained in the factor $\alpha_s^2(Q^2)$ and in the factors d_{nm} from the quark wave functions.

The early hope was that the wave function dependence would be small allowing a clean test of PQCD. The factor $\alpha_S^2(Q^2)$ could cause G_M^p to decrease faster than $1/Q^4$, with the amount depending on the size of Λ_{QCD}, and give direct evidence for the running of the coupling constant. Recently that hope has been questioned,[6] and the role of the wave functions has been vigorously investigated[7-9]

New Data for ep Scattering

A new experiment[10] on ep elastic scattering at large Q^2 was recently performed at SLAC to obtain high quality data at large Q^2 to measure the slope of G_M^p versus Q^2. The high current SLAC beam up to 20 GeV and a 60 cm long liquid hydrogen target and the 8 GeV/c spectrometer were used to measure from $Q^2 = 2.9$ to 31.3 $(\text{GeV/c})^2$ (Fig. 1). The results show $Q^4 G_M^p$ attaining a constant value between 5 and 12 $(\text{GeV/c})^2$ and then slowly decreasing with increasing Q^2. This shape is consistent with the predictions of PQCD, but this interpretation must be made with some caution.

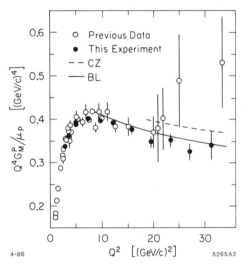

Fig. 1. New results for the proton form factor G_M^p from Ref. 10. The perturbative QCD curves are: BL (Ref. 4), CZ (Ref. 7).

The first PQCD calculations[4] used symmetric wave functions (all quarks have equal probability for carrying some fraction x of the proton momentum). It turns out that the lowest order PQCD diagrams are exactly zero for symmetric wave functions. Chernyak and Zhitnitsky[7] have derived a set of asymmetric wave functions which satisfy the constraints from QCD sum rules and also give fair agreement with the size and shape for G_M^p. If this work survives further tests, it could lead to the rather surprising conclusion that the valence quarks in a nucleon do not more or less equally share the momentum. If this were true it would have important consequences in other areas of physics.

The Neutron

One place to test these questions is in the data for the neutron form factors. For convenience we discuss the neutron in terms of the ratio to the proton. The ratio is also useful for comparison to some calculations because many factors affecting the normalizations cancel. Recall that for small scattering angles where the terms in the cross section proportional to $\tan^2(\theta/2)$ are small,

$$\frac{\sigma_n}{\sigma_p} = \frac{F_{1n}^2 + \frac{Q^2}{4m^2} F_{2n}^2}{F_{1p}^2 + \frac{Q^2}{4m^2} F_{2p}^2}. \tag{5}$$

Several possibilities can be considered. If F_{1n} is zero, or small compared to F_{2n}, which would happen if the nucleon wave function were spatially symmetric, then σ_n would be due only to the higher order term F_{2n}. At large Q^2 the ratio would become

$$\frac{\sigma_n}{\sigma_p} \Rightarrow \left(\frac{C_{2n}}{C_{1p}}\right)^2 \frac{4m}{Q^2} \tag{6}$$

and decrease with increasing Q^2. If F_{1n} is comparable to F_{2n}, then σ_n would eventually be due to F_{1n} at large Q^2, and the ratio σ_n/σ_p would be some number determined by the wave functions

$$\frac{\sigma_n}{\sigma_p} \Rightarrow \left(\frac{C_{1n}}{C_{1p}}\right)^2. \tag{7}$$

If the struck quark has the same flavor as the nucleon (u for the proton, d for the neutron), then $\sigma_n/\sigma_p \to 1/4$; if the struck quark has the same helicity, $\sigma_n/\sigma_p \to 3/7$.

The data[11] for σ_n/σ_p (Fig. 2) shows a nearly constant value from 1 to 6 $(\text{GeV/c})^2$ and then a decrease at higher Q^2. This shape is consistent either with $F_{1n} \simeq 0$, or $F_{1n} \neq 0$ and the same flavor for the struck quarks.

A recent theoretical development that may provide some guidance for interpreting the form factor data comes from Gari and Krümpelmann.[12] They have constructed a phenominological merger of a VMD model with the asymptotic constraints from QCD. The usual VMD expressions[1] for nucleon form factors (without ϕ mesons) are augmented with a momentum

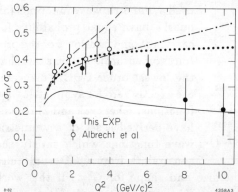

Fig. 2. The ratio of elastic neutron and proton cross sections, data from Ref 11. The dashed and solid curves are from VMD models by Höhler et al and Blatnik and Zovko (Ref. 1). The dotted curve is form factor scaling: $G_M^n/\mu_n = G_M^p/\mu_p = G_E^p$ and $G_E^n = 0$. The dashed-dot curve is the dipole law for G_M^n with $G_E^n = 0$.

dependent factor containing a new scale parameter Λ_2. The meson-nucleon and the Pauli and Dirac e.m. form factors are written as

$$F_1 = \frac{\Lambda_1^2}{\Lambda_1^2 + \hat{Q}^2}\frac{\Lambda_2^2}{\Lambda_2^2 + \hat{Q}^2} \qquad (8.a)$$

$$F_2 = F_1 \frac{\Lambda_2^2}{\Lambda_2^2 + \hat{Q}^2}. \qquad (8.b)$$

The value of $\Lambda_1 = 0.8$ GeV is determined from meson nucleon coupling; the effective momentum transfer $\hat{Q}^2$ contains the logarithmic variation $\log(Q^2/\Lambda_{QCD}^2)$. The parameter Λ_2, to be determined from fits to the form factor data, adjusts the transition from VMD to QCD. For $Q^2 \ll \Lambda_2^2$ the momentum dependence follows the monopole form factors F_1 and F_2 plus the vector meson propagators. For $Q^2 \gg \Lambda_2^2$ the meson propagators die away and $F_1 \sim 1/Q^4$ while $F_2 \sim 1/Q^6$.

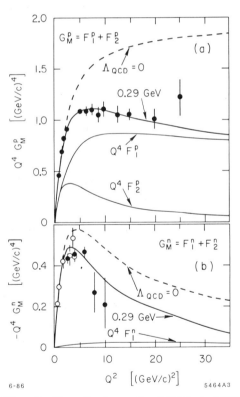

Fig. 3. Results from the VMD + QCD model of Ref. 12. a) the proton G_M^p, b) the neutron G_M^n.

Gari and Krümpelmann fit their model simultaneously to all the available nucleon form factor data (results from Ref. 10 not included). At large Q^2 the G_M^p is mostly due to F_{1p} (Fig. 3a). However, F_{2p} makes a substantial contribution, and we have to be careful when comparing the data to calculations which do not include F_2. The available perturbative QCD calculations[4,7] are ambiguous on this point. They calculate G_M^p, which is identical with F_{1p} asymmptotically. No perturbative calculation of F_{2p} has been done.

The situation for G_M^n is quite different (Fig. 3b). In the Gari and Krümpelmann model F_{1n} is small, near zero, compared to F_{2n}. This result follows from the near cancelation of terms from ρ and ω mesons, and is driven by the requirement to fit the high Q^2 data for σ_n/σ_p in Fig. 2. This phenomenological model could be providing the clues that we need to answer key questions. The value of $\Lambda_2^2 = 5.15$ GeV^2 indicates the transition to the perturbative regime may take place around 5 (GeV/c)^2. The value $\Lambda_{QCD} = 0.29$ GeV is consistent with results determined from scaling violations in deep inelastic scattering.[13] The problem remains how to understand the relative magnitudes of F_1 and F_2 determined by the quark wave functions.

The consequences of $F_{1n} \simeq 0$ are several:

a) From the definition Eq. (1) it follows that G_E^n and G_M^n will be comparable in size around $Q^2 = 4$ (GeV/c)^2 (Fig. 4). It could be that the fokelore that G_E^n is small, maybe zero, is true at low Q^2, but at higher Q^2 the situation is completely reversed (Fig. 4). This prediction can be tested in a standard Rosenbluth measurement of quasielastic ed scattering.[14]

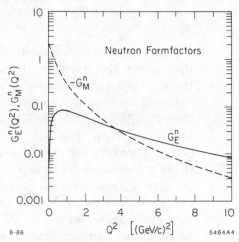

Fig. 4. The neutron form factors G_E^n and G_M^n from Ref. 12.

b) σ_n is determined by the higher order helicity flip term F_2^n and it does not make sense to compare first order PQCD results to the neutron data. The neutron cross section at high Q^2 may provide a useful testing ground for the higher order terms.

c) Knowledge of the neutron form factors is essential for interpretation of other electromagnetic data. One key example is the deuteron forward angle elastic form factor $A(Q^2)$, which in the impulse approximation is mostly proportional to the isoscalar charge form factor

$$A \sim (G_E^p + G_E^n)^2 \phi_{body}^2. \qquad (9)$$

The smaller G_E^n beats against the larger G_E^p and small changes around zero give big effects in $A(Q^2)$. It is possible that a long standing inability of the impulse calculations[15] to give large enough values for $A(Q^2)$ could be traced to using the wrong neutron form factor (Fig. 5).

THE DEUTERON

Measurements of the deuteron form factors at large Q^2 provide important tests of reaction mechanisms and the nature of the nucleon constituents at short distance. The deuteron magnetic form factor $B(Q^2)$ is expected to be especially sensitive to ingredients in the description.[16-19] There is now available very preliminary data from a new experiment[20] at SLAC, with data taking still underway at the time of this conference. The previous data[21] and examples of the range of predictions are shown in Fig. 6. The $B(Q^2)$ is expected to be small compared to $A(Q^2)$ and therefore measurements at large angles are necessary. The new experiment measured ed scattering in coincidence around 180 degrees. The calculations range from impulse models[15-17] which all predict sharp diffraction features, to quark scaling predictions[4,18] with smooth power law fall off. Some models also include scattering from meson exchange currents,[17] delta components in the wave function, or 6-quark components at the core of the wave function.[19] These mechanisms can shift or in some cases totally obliterate the diffraction features from the impulse approximation.

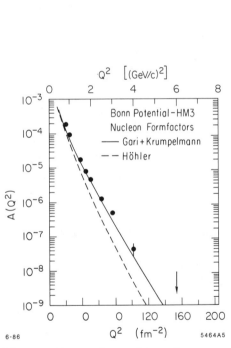

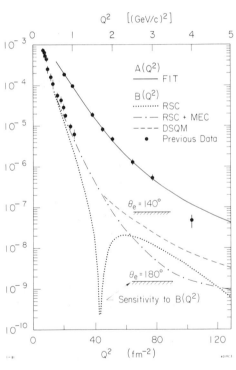

Fig. 5. Comparison of impulse aproximation calculations (Ref. 15) for the deuteron $A(Q^2)$ using nucleon form factors from Refs. 1 and 12.

Fig. 6. Deuteron form factors $A(Q^2)$ and $B(Q^2)$. The curves for $B(Q^2)$ are impulse approximation calculations using reid soft core wave functions, with and without meson exchange currents (Ref. 17) and the dimensional scaling quark model (Ref. 3), arbitrarily normalized at $Q^2 = 1.75$ $(GeV/c)^2$.

The new data (Fig. 7) fall very quickly with Q^2 and seem to indicate a deviation from a smooth decrease around 2 to 2.5 $(GeV/c)^2$. The cross sections at high Q^2 are very low ($\sim 10^{-42}$ cm^2/sr), leading to counting rates of events per week. This new data nearly doubles the range in measured Q^2 and will help narrow the choices for short range deuteron structure.

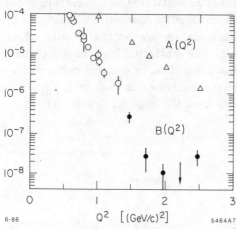

Fig. 7. Preliminary results for $B(Q^2)$ (Ref. 20) and previous data (Ref. 21).

TRITIUM

Two recent experiments, one at Saclay,[22] and one at MIT-Bates,[23] have given a large increase in the experimental information on tritium e.m. structure. These are the long awaited results of major efforts to build difficult liquid[22] and gaseous[23] tritium targets for use in high powered electron beams. The Saclay experiment measured elastic scattering. The charge and magnetic form factors were separated out to Q^2 around 1 $(GeV/c)^2$ and the diffraction features in each have now been revealed (Fig. 8). This data, taken together with similar data already aviailable for ^{3}He, will provide important information needed to sort out the three-body wave functions, the role of meson currents, isobar contributions, and off-shell nucleon form factors. As Figure 8 shows, F_{mag} is particulary sensitive to the meson currents and the choice of nucleon form factors.

The MIT-Bates experiment,[23] which completed data taking only a few weeks ago, was optimized for careful comparison of inelastic scattering from ^{3}He and ^{3}H. Extensive data were obtained for longitudinal and transverse separations in the quasielastic and delta excitation region over the kinematic range accesible with 700 MeV beam (Fig. 9). This data provides a high quality look at the inelastic response function in the lightest nuclei where explicit multi-nucleon effects can appear. The MIT-Bates experiment also measured three points on the charge form factor at Q^2 around 1 $(GeV/c)^2$ that confirm the Saclay results and extends the Q^2 range slightly.

QUASIELASTIC SCATTERING — y-SCALING

Inelastic electron scattering from nuclei in the kinematic region of quasi-free scattering on bound nucleons has attracted a lot of attention recently. The main interest stems from the suggestion by West[27] that such data might be interpreted to yield universal momentum distributions for nucleons in nuclei.

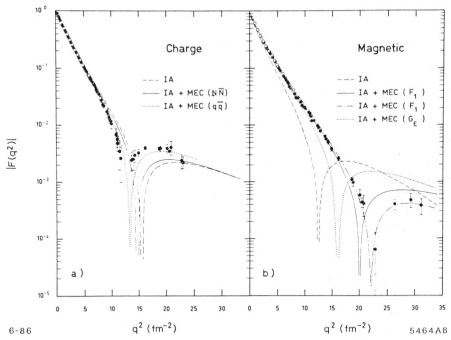

Fig. 8. Results for the tritium charge and magnetic form factors from Saclay (Ref. 22). The theoretical curves are impulse approximation models, some with meson exchange currents, and for various nucleon form factors (Refs. 25 and 26).

The scattering mechanism in this region of the response function is pictured to be quasi-free knockout of single nucleons. The chance for scattering is factorized into a probability $F(y)$ for finding nucleons with mass m moving at momentum y in the initial nucleus, times the cross section for e-nucleon scattering assuming the same form factors as for free nucleons. The $F(y)$ can be deduced from the experimental cross section using

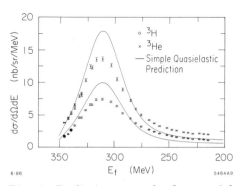

Fig. 9. Preliminary results for quasielastic electron scattering from ^{3}H and ^{3}He from MIT-Bates (Ref. 23).

$$F(y) = \frac{\left(\frac{d\sigma}{d\Omega dE'}\right)_{exp}}{\left[Z\frac{d\sigma}{d\Omega} + N\frac{d\sigma}{d\Omega}\right]} \frac{dE'}{dy} \quad (10)$$

where y is obtained from energy and momentum conservation

$$E = E' - \bar{\epsilon} + \sqrt{(\vec{p}+\vec{q})^2 + m^2} - m + \sqrt{p^2 + (A-1)^2 m^2} - (A-1)m. \quad (11)$$

Various approximations and assumptions are usually applied to the scaling analysis. In most formulations[28,29] for large Q^2 the initial nucleon transverse momentum $k_\perp$ is ignored and then y can be identified with the component of $\vec{k}$ parallel to $\vec{q}$. This approximation also simplifies the kinematical factor dE'/dy which can affect the way the data scale.[30]

Another problem is what to do about the final state interaction of the knocked out nucleon with the (A-1) system and the excitations of the residual nucleus. The term $\bar{\epsilon}$ in Eq. (11), representing an average nucleon separation energy, has sometimes been included [28] to approximately account for these effects for analysis of data for the light nuclei (^{3}He), but this may not be adequate for heavier nuclei.[31] There is also some question about whether it is safe to assume that bound nucleons have the same form factors as free ones, especially at large y (sometimes as large as 600 to 800 MeV/c), where the nucleons are clearly strongly interacting with their neighbors.

Notwithstanding these questions, the y-scaling analysis provides an important method for synthesizing large amounts of data. By virtue of the fact that it works (ie. the data scale), it leads people to accept the basic assumption of single nucleon knockout and continue to regard the derived $F(y)$ as a source of information about the high momentum components.

Preliminary results were presented at this conference from a new experiment[32] that measured quasielastic scattering from a series of nuclei at forward angles (15° to 30°) and beam energies from 2 to 3.6 GeV, corresponding to momentum transfer in the range 0.2 to 2.2 $(\text{GeV}/c)^2$. An example of one series of spectra for ^{12}C is shown in Fig. 10. The original data spanning six orders of magnitude in the cross section are reduced to a universal scaling function in the region of negative y. Comparing similar data for nuclei throughout the periodic table will give insight into the nuclear dependence of the scaling hypothesis.

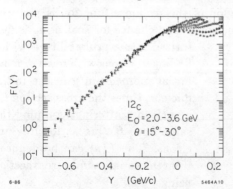

Fig. 10. Preliminary results for quasielastic electron scattering from ^{12}C (Ref. 32).

There presently exists a puzzle in quasielastic scattering which could be giving hints of important new physics. When quasielastic data[33] taken at various angles are analyzed[34] to give separate longitudinal and transverse response functions, there is a large difference between the $F(y)$ for longitudinal and transverse scattering even at the quasielastic peak. The ratio of longitudinal to transverse response for ^{56}Fe is about 0.55, for example. This result contradicts the basic assumption that there should be a universal $F(y)$ for each nucleus.

One suggestion is that these results are obtained because the wrong (free) nucleon form factors were used in the analysis, and that perhaps this effect is another manifestation of modified (swollen) nucleons that is seen also in the EMC effect. To advance the study of this problem, a new experiment[35] is being prepared at SLAC to extend the longitudinal-transverse separation for several nuclei out to Q^2 around 1 $(GeV/c)^2$.

In a related area of physics, two recent experiments[36] measured inelastic scattering in the nucleon resonance region at Q^2 below 1 $(GeV/c)^2$ to study the nuclear dependence of delta excitation.

DEEP INELASTIC SCATTERING — THE EMC EFFECT

The discovery[37] by the European Muon Collaboration that deep inelastic scattering per nucleon from iron was not the same as for deuterium is widely regarded as an important signature for the modification of the quark distributions for bound nucleons. Subsequent data[38] from SLAC agreed with the EMC for $x > 0.3$ and showed that the effect increases with A proportional to the average nuclear density. There remained a discrepancy between EMC and SLAC data for x below 0.3 (Fig. 11).

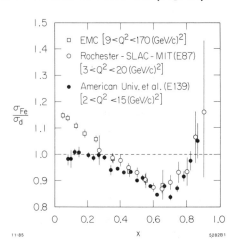

Fig. 11. The ratio of deep inelastic muon and electron cross sections for iron and deuterium from EMC (Ref. 37) and SLAC (Ref. 38).

The suggested explanations for the EMC effect have been many,[39] but most ascribe it to a softening (shift to lower momentum) of the valence quark distribution for bound nucleons, accompanied by an increase in the momentum carried by ocean quarks at low x. This phenomenon can be viewed as caused by an effective increase in the confinement radius for the quarks in bound nucleons, either by nucleon swelling, overlap into clusters, creation of extra pions, or excited nucleons. The region of x below 0.3 is expected to contain complex physics in which shadowing of the virtual photons competes with scattering from pointlike quarks or clusters of quarks (higher twist terms in PQCD) to give effects varying with x, Q^2, A, and ϵ (the virtual photon polarization). The large discrepancy between EMC and SLAC data at low x has helped fuel this discussion, and has stimulated models[40] which depend on enhanced pion content at low x for redistribution of the momentum.

A new measurement[41] by the BCDMS collaboration at CERN of muon scattering on nitrogen and iron has confirmed the effect for $x > 0.3$ (Fig. 12). They only have data at lower x for nitrogen which agree with the trend of the SLAC data.

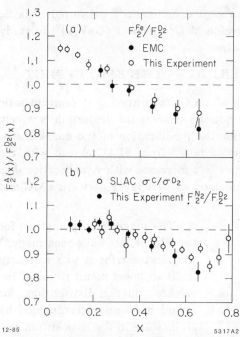

Fig. 12. (a) The ratio of the structure functions F_2 for iron and deuterium as measured by BCDMS (Ref. 41) and EMC (Ref. 37). (b) BCDMS nitrogen data compared to SLAC data (Ref. 38) for carbon. Only statistical errors are shown.

It is conceivable that part of the discrepancy between EMC and SLAC could be due to a difference for $R = \sigma_L/\sigma_T$ for heavy nuclei compared to deuterium.[42] This might be caused, for example, by more higher twist contributions from spin-zero components in nuclei (diquarks or quasi pions) that generates a larger σ_L. The EMC measures near $\epsilon = 1$ and their cross sections are mostly due to $F_2(x, Q^2)$, whereas the SLAC data are measured at ϵ in the range 0.4 to 0.9. Extraction of $F_2(x, Q^2)$ from the cross sections is sensitive to R. A new SLAC experiment[43] has measured R for deuterium, Fe, and Au at $x = 0.2, 0.35, 0.5$; when results are available they should help settle this question.

Meanwhile new important results[44] from the EMC have been presented at this conference which indicate the EMC effect may not be as large at low x as previously thought. Data were taken in such a way as to reduce many of the systematic errors from uncertainty in the acceptance by measuring with different targets (d, He, C, Cu, Sn) in the same geometry. Preliminary results from a partial data sample for Cu (Fig. 13) show a ratio σ_{Cu}/σ_d at low x which never goes above 1.1 and falls to less than one at x below 0.1. The new EMC data are within errors of the SLAC data, though there are small deviations that could be due to variations with Q^2 or ϵ that needs to be sorted out. The large rise at small x in the original EMC data is apparently not real. The new EMC data are consistent with the original values when the systematic errors are included.

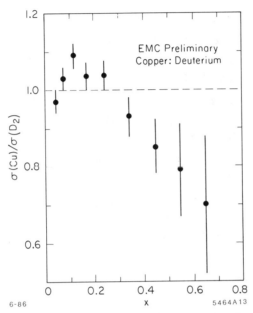

Fig. 13. Preliminary results for the ratio of muon scatering cross sections for Cu and d (Ref. 44).

FUTURE PROSPECTS

A new experiment[45] measuring inclusive muon scattering at CERN using an improved EMC apparatus will begin data taking soon. This experiment will yield high statistics data for a series of nuclei that will be extremely useful for understanding the physics in the low x region. An experiment[46] is being prepared at Fermilab to measure muon scattering using the new high energy beam from the Tevatron. This experiment will complement the CERN efforts by concentrating on measurements of the final state. Possibilities for internal target experiments at the PEP storage ring at SLAC are now being investigated[47]. This could lead to a new generation of electromagnetic probes of nucleon and nuclear targets at GeV energies measuring multiparticle final states, perhaps with polarized beams and targets.

REFERENCES

1. G. Höhler et al., Nucl. Phys. B114, 505(1976); S. Blatnik, N. Zovko, Acta Phys. Austriaca 39, 62(1974), F. Iachello, A. Jackson, A. Lande, Phys. Lett. 43B, 191(1973).
2. S. Brodsky, G. Farrar, Phys. Rev. D11, 1309(1975).
3. S. Brodsky, B. Chertok, Phys. Rev. D14, 3003(1976).
4. S. Brodsky, G. Lepage, Phys. Rev. D22, 2157(1980).
5. G. Farrar, D. Jackson, Phys. Rev. Lett. 35, 1416(1975); I. A. Vainshtein, V. I. Zakharov, Phys. Lett. 72B, 368(1978).
6. N. Isgur, C. Llewellyn Smith, Phys. Rev. Lett. 52, 1080(1984).
7. V. L. Chernyak, I. R. Zhitnitsky, Nucl. Phys. B246, 52(1984), V. L. Chernyak, A. R. Zhitnitsky, Phys. Rep. 112, 173(1984).
8. C. Carlson, Proceedings of the Nato Advanced Study Inst., Banff, Canada, Aug 22 - Sept 4 1985.
9. M. Gari, N. G. Stefanis, Ruhr Universität Bochum Print RUB TPII-85-16; RUB TPII-85-23 (1986).
10. R. Arnold et al., SLAC PUB 3810 (1986); Phys. Rev. Lett. 57, 174(1986).
11. S. Rock et al., Phys. Rev. Lett. 49, 1139(1982); R. J. Budnitz et al., Phys. Rev. 173, 1357(1968); W. Albrecht et al., Phys. Rev. 26B, 642(1968).
12. M. Gari, W. Krümpelmann, Z. Phys. A322, 689(1986); Ruhr Universtät Bochum Print RUB TPII-86-5 (1986).
13. J. J. Aubert et al., Phys. Lett. 114B, 291(1982).
14. R. Arnold et al., "A Proposal to Separate the Charge and Magnetic Form Factors of the Neutron at Momentum Transfer $Q^2 = 2$ to 5 Q^2 ", SLAC-NPAS Proposal NE11 (1986).
15. Examples of impulse approximation results for deuteron form factors including relativistic effects are given in R. Arnold, C. Carlson, F. Gross, Phys. Rev. C21, 1426(1980). The effect of assuming $F_{1n} = 0$ was noted.
16. R. S. Bhalerao, S. A. Gurvitz, Phys. Rev. Lett. 47, 1815(1981).
17. M. Gari, G. Hyuga, Nucl. Phys. A264, 409(1976), Nucl. Phys. A278, 372(1977).
18. S. Brodsky, C. Ji, G. P. Lepage, Phys. Rev. Lett. 51, 83(1983).
19. A. P. Kobushkin, V. P. Shelest, Sov. J. Part. Nucl. Phys. 14, 483(1983).
20. P. Bosted, this conference, preliminary results of SLAC-NPAS experiment NE4, "Measurements of electron deuteron scattering at 180 degrees at large momentum transfer."

21. S. Auffret, et al., Phys. Rev. Lett. 54, 649(1985); R. Cramer, et al., Z. Phys. C29, 513(1985).
22. F. P. Juster et al., Phys. Rev. Lett. 55, 2261(1985).
23. B. Beck, this conference, preliminary results of elastic and inelastic electron scattering from ^{3}H and ^{3}He at MIT-Bates.
24. R. Arnold et al., Phys. Rev. Lett. 40, 1429(1978); J. M. Cavedon et al., Phys. Rev. Lett. 49(1982).
25. W. Strueve, C. Hajduk, P. U. Sauer, Nucl. Phys. A405, 620(1983); C. Hajduk, P. U. Sauer, W. Strueve, Nucl, Phys. A405, 581(1983); E. Hadjimichael, B. Goulard, R. Bornais, Phys. Rev. C27, 831(1983).
26. M. Beyer et al., Phys. Rev. Lett. 122B, 1(1983).
27. G. B. West, Phys. Rep. 18C, 263(1975).
28. I. Sick, D. Day, J. S. McCarthy, Phys. Rev. Lett. 45, 871(1980).
29. P. Bosted et al., Phys. Rev. Lett. 49, 1380(1982).
30. C. Ciofi degli Atti, E Pace, G. Salame INFN (Rome) Print INFN-ISS 85/7 (1985), to be published in Czech. J. Phys.
31. S. A. Gurvitz, J. A. Tjon, S. J. Wallace, U. Md Print 86-103(1986).
32. D. Day, this conference, preliminary results from SLAC-NPAS experiment NE3, "Inclusive electron scattering measurements from nuclei."
33. Z. Meziani et al., Phys. Rev. Lett. 52, 2130(1984); 54, 1237(1985).
34. Z. Meziani, SLAC PUB 3939 (1986).
35. Z. Meziani et al., SLAC-NPAS Proposal NE9 (1985),
36. S. Thornton et al., SLAC-NPAS Experiment NE1 (1985); H. Jackson et al., SLAC-NPAS Experiment NE5 (1985).
37. J. J. Aubert et al., Phys. Letters 123B, 275(1983).
38. A. Bodek et al., Phys. Rev. Lett. 50, 1431; 51, 534(1983); R. Arnold et al., Phys. Rev. Lett. 52, 727(1984).
39. N. N. Nikolaev, "EMC Effect and Quark Degrees of Freedom in Nuclei: Facts and Fancy", Oxford Print TP-58/84, Invited Talk at VII Int. Seminar on Problems of High Energy Physics–Multiquark Interactions and Quantum Chromodynamics, Dubna USSR, 19-23 June 1984; R. R. Norton,"The Experimental Status of the EMC Effect", Rutherford Print RAL-85-054, Topical Seminar on Few and Many Quark Systems, San Marino, Italy 25-29 March 1985; H. J. Pirner, "Deep Inelastic Lepton-Nucleus Scattering", Proc. Int. School of Nucl. Phys. Erice, to be pub. Progress in Particle and Nuclear Physics (1984); R. L. Jaffe, "The EMC effect today", Proc. Workshop on Nuclear Chromodynamics," Santa Barbara, Aug (1985), pub. by World Scientific.

40. E. L. Berger, F. Coester, Phys. Rev. D32, 1071(1985).
41. G. Bari et al., Phys. Lett. 163B, 282(1985).
42. S. Rock, 22nd Int. Conf. on High Energy Phy., Leipzig East Germany, July 19-25(1984); J. Gomez, SLAC PUB 3552(1985).
43. R. Arnold et al., SLAC Experiment E140, "Measurement of the x, Q^2, and A-Dependence of R'" (1985).
44. G. Taylor, this conference, preliminary results from the EMC (1986).
45. D. Allasia et al, the New Muon Collaboration (NMC), CERN Proposal SPSC/P210, "Detailed Measurements of Structure Functions from Nucleons and Nuclei", February 1985.
46. Fermilab experiment E665, T. B. Kirk, (FNAL), V. Eckardt (MPI) spokesmen; D. F. Geesaman, M. C. Green, Argonne print PHY-4622-ME-85 (1985).
47. F. Dietrich, this conference, results of a test experiment using a gas target in the PEP ring at SLAC.

STRANGE MATTER AND DIHYPERON PHYSICS

Peter D. Barnes
Carnegie Mellon University, Pittsburgh, Pa. 15213

ABSTRACT

A short review of the properties of Strange Matter is followed by a dicussion of dihyperon physics. Calculations of the mass, lifetime and decay modes of the H particle are discussed, along with a review of experiments designed to search for the H Dibaryon.

The subject of Strange Matter is not new, but recently has received renewed attention. I will review briefly some of its history, its properties, and will indicate places that have been suggested to look for manifestations of this exotic material. In a related subject I will review the properties of and motivation for looking for the Dihyperon and will discuss experimental methods proposed for detecting it.

I. Strange Matter

Normal matter consists of sets of three u and d quarks combined with gluons to form nucleons, which can in turn be combined to form nuclei. Such nuclear matter is found in a large range of masses consisting of a few nucleons up to cosmic masses, as for example in neutron stars. It has strangeness S = 0 and an energy per baryon of about 932 MeV when the binding energy is taken into account.

Strange Matter (SM), on the other hand, is a plasma of approximately equal numbers of u, d, and s quarks in dynamic equilibrium under the influence of the weak interaction. This idea has been around for about ten years since Collins and Perry[1] in 1975 and later Baym and Chin[2] discussed it in the context of neutron stars. Bjorken and McLerran[3] considered it

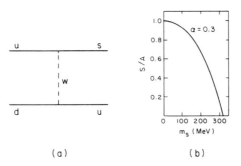

Fig. 1. a) The weak interaction which brings the s, u, and d quarks into equilibrium. b) The strangeness per baryon for various strange quark masses.

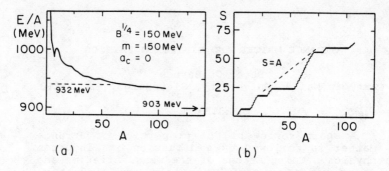

Fig. 2. a) The energy per nucleon, E/A, in strange matter[7] as a function of baryon number. b) Strangeness for A less than 100.

in connection with the Centauro events, while its role in high energy collisions was discussed by Chin and Kerman[4]. A related system with a quite different motivation, with baryon number two and strangeness minus two, was proposed by Jaffe[5] in 1977.

The subject of SM received renewed interest in 1984, when Witten considered seriously that it might be stable[6]. This was followed by the detailed investigation of Farhi and Jaffe[7] of its general properties. The latter authors considered large groups of interacting quarks in the Fermi gas model, for very large systems, and in a bag model, for smaller systems. They find that stable solutions exist for certain mass regions, as for example, in the baryon number range A = 10^2 to 10^{57} (bulk matter). It can have a density approximately equal to nuclear matter and can exist at zero temperature and pressure.

The role of strangeness is apparent in the Fermi gas model where a reduction in the average energy per quark is achieved by going from a two flavor to a three flavor system.

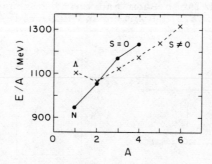

Fig. 3. Energy per nucleon for A < 6 in both strange and nonstrange systems.

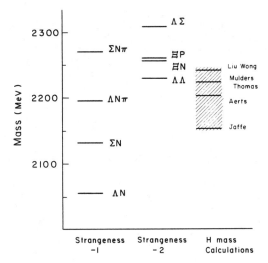

Fig. 4. Comparison of calculations of the H Dibaryon mass with S = - 1 and - 2 systems.

Equilibrium is maintained through the influence of the weak interaction in converting u to s quarks, as indicated in Fig. 1a. The detailed population depends on the mass of the strange quark. Farhi and Jaffe found that the strangeness per baryon number varied from 1.0 to ≈ 0.8 for strange quark masses in the range zero to 150 Mev (see for example Fig. 1b taken from ref. 7). This implies that SM is neutral or slightly positive and that neutral SM can be achieved with an electron cloud distributed through SM volume and extending somewhat outside.

Stability is achieved whenever the energy per baryon falls below the nuclear matter value of 932 MeV/A. Depending on the choice of parameters, Farhi and Jaffe find (Fig. 2a) that this occurs for A > 80 and decreases to 903 MeV/A for bulk matter. The stability against fission comes from the small coulomb energy for SM. In the hadronic bag model, systems with A less than about 80 would appear to be unstable with strangeness as indicated in Fig. 2b.

The very light systems are calculated as multiquark states in the bag model. A comparison of E/A for nonstrange and strange system is given in Fig. 3 for A < 6. For S = 0 systems, E/A increases monotonically with A, and thus gives no bound multiquark states except for the nucleon. For strange systems it is found that E/A is smallest for the special case of S = - 2 , A = 2 and otherwise rises with A. This is just the flavor and spin saturated system discussed by Jaffe[5] in 1977 as the " Dihyperon" or the " H particle", which I will turn to below.

What is the impact of these ideas on experimental physics? Many possibilities have been suggested of which I can only mention a few. In the laboratory, searches for SM impurities in normal matter have been attempted using chemical and mass spectroscopic methods. An interesting accelerator search[32] involves using the coulomb barrier as a filter to distinguish normal matter from SM. For example, bombardment of matter with heavy ions, at energies below the coulomb barrier, might lead to the release of large numbers of photons if "nuggets" of SM were present.

Several astrophysical implications have been mentioned. In the context of dark matter in the universe, a large chunk of strange matter could close the universe. However since it would be unstable against neutron evaporation, for temperatures greater than about 30 MeV, it would be expected to have evaporated away before the universe cooled below this temperature, and thus to no longer exist. Neutron stars are an obvious candidate for SM. Since it can exist at low pressures the star would be expected to consist of SM nearly out to the surface. The possibility that the pulsar, Cygnus X3, might consist of SM and emit Dihyperons has been suggested[8]. This has been proposed as an explanation for the observation of muon showers from the approximate direction of Cygnus X3 with a similar time structure. I will return to this below.

This review is by no means complete but is intended to give the flavor of some of the discussion. We now turn to the related, but independently motivated, possibility of the existence of a dihyperon, the H particle.

II. The Dihyperon

A. Theoretical Expectations.

In 1977 Jaffe[5] reported calculations for systems of six quarks in an MIT Bag model. In particular he emphased the properties of a system of two u, two d, and two s quarks bound in a bag of roughly one fermi, all quarks in relative S waves with a total spin/parity of 0^+ and isospin zero. Under the influence of the spin dependent quark-quark interactions, he found that the mass was 2150 MeV. This places it below the lightest known double strange mass, that of two lambda hyperons. Thus it is stable against strong decay with a binding energy, $B_H = M_{\Lambda\Lambda} - M_H = +80$ MeV. He predicted that it would decay with a life time of about 10^{-10} sec into $\Sigma^- p$, $\Sigma^0 n$, and Λn in the ratios 5 : 3 : 2 .

The origins of this small mass is clearly in the hyperfine splitting as discussed by many authors. I follow here the arguments given by Rosner[9] in the constituent quark model. The hyperfine splitting is:

$$\Delta E = -\omega \Sigma \frac{(\lambda_i * \lambda_j)(\sigma_i * \sigma_j)}{m_i \, m_j}$$

where the scale is determined by the factor ω, which depends on the amplitude of the quark wave function at the origin and the strong coupling constant α_s. In the SU(3) flavor symmetric

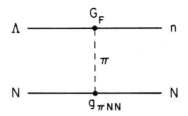

Fig. 5. The ΛN nonmesonic weak interaction as represented in a meson exchange model.

limit this gives, for the Λ hyperon in which the hyperfine interaction is only between the u and d spin zero quark pair, $\Delta E_\Lambda = -8\,\omega/m_u^2$. This can be estimated from the known delta - nucleon splitting which gives $\Delta E_\Lambda = -150$ MeV.

One can build the H particle by superimposing two Λ's and recoupling their spins. The desired wave function is that six quark flavor singlet which gives the largest possible hyperfine splitting, as found by Jaffe. In the SU(3) flavor symmetric case this gives $\Delta E_H = -24\,\omega/m_u^2 = -450$ MeV and a binding energy of $B_H = 2\,\Delta E_\Lambda - \Delta E_H = +150$ MeV. Rosner[9] also considers the SU(3) flavor nonsymmetric case with effective quark masses of $m_s = 538$, and $m_u = m_d = 363$ MeV. This reduces the splitting to $\Delta E_H = -18.8\,\omega/m_u^2 = -353$ MeV and $B_H = 53$ MeV, which corresponds to Jaffe's value of 80 MeV in the bag model. Adopting a heavy strange quark mass, shifts B_H but only changes the H particle wave function in second order.

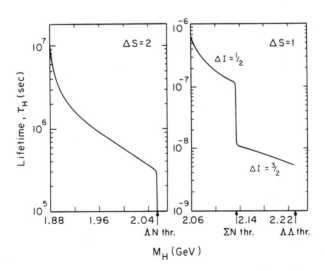

Fig. 6. Lifetime of the H dihyperon as calculated by Donoghue et al[16] for various assumed masses.

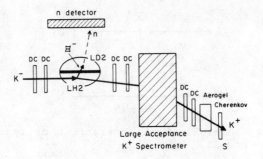

Fig.7. Schematic arrangement of detectors in AGS exp. E-813.

The magnitude of B_H has now been calculated in many models. Calculations in chiral bag models[10] and estimates of center of mass corrections[11] tend to raise the predicted H particle mass above Jaffe's initial estimate of 2150 MeV (see Fig. 4). Soliton solutions to Skyrmion models[12] give larger values of B_H while a recent lattice gauge calculation[13] gives a negative binding energy. The life time of the Dihyperon and its decay modes clearly depend strongly on its mass. In the region above 2230 MeV it can always decay by a strangeness conserving transition into two Λ's. In the region below 1880 MeV it would be lighter than two nucleon masses and hence stable. Of special interest is the weak decay region for which $\Delta S = 1$ transitions are required above the ΛN mass (2056 MeV) and the even slower $\Delta S = 2$ processes required below this mass.

The weak decay of the H particle might be discussed in analogy to the nonmesonic decay of the ΛN system which can be modeled in terms of t channel meson exchange weakly coupled to one of the baryons, Fig. 5, or alternatively by W boson exchange between the s and u quarks, Fig. 1a. These types of calculations give $\Delta I = 1/2$ (enhanced) and 3/2 contributions and agree with the Λ hypernuclear weak decay measurements to about a factor of two (see reference 14 and J. Szymanski's paper at this conference[15]). In this scheme one expects $\Delta S = 1$ and $\Delta S = 2$ H particle decay rates to scale with G_F^2 and G_F^4, respectively. A detailed calculation based on the mechanism of Fig 1a has been reported by Donoghue, Golowich and Holstein[16]. Fig. 6 shows their calculated life time as a function of mass, M_H. They find, in the region above the ΣN threshold, that the $\Delta I = 1/2$ transition, that is dominate in hyperon decay, is suppressed in H particle decay as indicated by structure arguments. Thus for small B_H, decay into the ΣN system through $\Delta I = 3/2$ is available, but gives the rather long life time of $\approx 10^{-8}$ seconds. Below 2135 MeV only $\Delta S = 1$, $\Delta I = 1/2$ is possible and the estimated lifetime increases to $\approx 10^{-7}$ seconds. Below 2056 MeV, the $\Delta S = 2$ transitions to NN require a double W boson exchange and the lifetime jump to $\approx 10^{+6}$ sec (≈ 12 days)[16]. This would seem to rule out the suggestion that Cygnus X3 is a source of H particles with $\gamma \cong 1000$ and lifetime of 12 years, as required by some muon shower data.

III. Experimental Searches for the Dihyperon.

In the literature one finds reports of individual events which both support[17] the existence of the H particle, as well as, evidence for a few double lamda hypernuclear decays[18] which seem to rule out certain mass regions for the H particle. These have been discussed at earlier conferences[19]

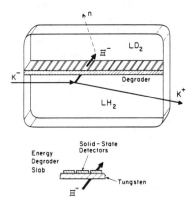

Fig.8. Cryogenic target with tungsten degrader and silicon transmission detectors used in AGS exp. E-813.

and I will not review them here. A major search for the H particle in associated production has been carried out by Carrol et al[20] at the AGS using the reaction:

$$p + p \rightarrow H + K^+ + K^+.$$

They were able to establish an upper limit < 40 nb for H formation. However, recent calculations[21] put the expected cross section at less than 1 nb.

Three reactions have been proposed which attempt to produced the H particle in $\Delta S = 2$ transitions. These may be thought of as intermediate production of a Ξ hyperon which combines with a nucleon to form the Dihyperon. The cross section for Ξ production in the reaction:

$$K^- + p \rightarrow \Xi^- + K^+$$

is about 100 μb.

In 1982 a technique for using the reaction:

$$\Xi^- + D \rightarrow H + n$$

was proposed[22] and is now approved at the AGS as experiment E-813 subject to construction of a 1.8 GeV/c beam line[23]. In this approach a tagged beam of Ξ 's are incident on a liquid deterium target. In the two body final state, a sharp peak in the neutron time of flight (TOF) spectrum is the signature of H particle production and gives a measure of its mass. In order to efficiently use the tagged Ξ beam it is brought to rest in a stopping experiment mode, giving $(\Xi^- D^+)_{atom}$ formation and subsequent decay into the H n channel. The apparatus, which is shown schematically in Fig. 7, consists of three main systems   a) a composite liquid hydrogen/deuterum

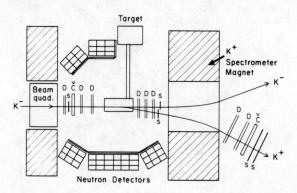

Fig. 9.   Experimental layout of target area, neutron detectors, and tagging K^+ spectrometer in E-813.

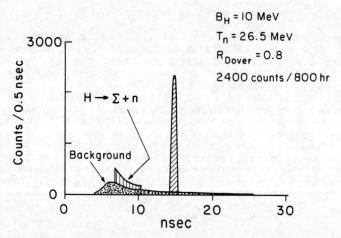

Fig. 10. Time-Of-Flight neutron spectrum showing the narrow peak predicted from H particle formation and the broad peak expected for the subsequent H Particle decay.

target with a tungsten degrader for rapid deceleration of the Ξ^- (see Fig. 8), b) a K^+ tagging spectrometer, and c) a neutron TOF detector. The layout on the floor of the AGS, shown in Fig. 9, is dominated by the very large acceptance ($\approx$ 60 msr) 48D48 spectrometer magnet. A redundant Si strip detector, Fig. 8, down stream of the tungsten degrader gives confirmation of the direction and rate of energy loss of the Ξ's in order to suppress background.

The useful range of this technique is $B_H = -20$ to $+100$ MeV which corresponds to neutron energies of 10 - 130 MeV. In estimating the H particle production rate we assume a K^- flux of $.7 * 10^6$/sec at 1.8 GeV/c, a 30 cm hydrogen target, and a branching ratio, R, for $(\Xi^- D^+)_{atom}$ decay into H n relative to all other channels of R = 0.1 to 1.0, as estimated by Aerts and Dover[25]. This yields 2400 counts in an 800 hour run in the H particle pesk of a TOF neutron spectrum, see Fig. 10. The background from $\Lambda\Lambda$, and H particle decay, as shown, is small.

It is tempting to combine the two targets of this method into one ^{3}He target[22,26] as shown in Fig. 11. In the former case a real Ξ is produced, detected, and degraded in momentum before reabsorption on a proton. In the latter case an on or off shell Ξ is produced and reabsorbed in the same nuclear volume with a momentum transfer of about 400 MeV/c. This technique is being discussed at the AGS as a target change for the E-813 measurement[24] and in the context of a new proposal, E-830 (ref. 27). At KEK it has been proposed as experiment E-127 (ref. 28).

The reaction in this case would be:

$$K^- + {}^3He \rightarrow H + K^+ + n.$$

A coincident momentum measurement of the K^+ and the neutron would permit a missing mass measurement of the H particle. Dover[29] has suggested that if the neutron is truly a spectator in this procress it may be sufficient to measure the inclusive K^+ spectrum in a one arm spectrometer measurement. As shown in Fig. 12 the H appears as a peak at about 1.43 GeV/c K^+ momentum with a width related to the Fermi momentum in ^{3}He. By integrating over the neutron momentum, Dover estimates[29] that the inclusive cross section rises from 0.35 to 0.60 mb/sr as B_H increases from + 30 to 260 MeV. Thus, an advantage of this approach is that the yield rises slowly as the binding energy increases.

A serious disadvantage is due to the large backgrounds that threaten to obscure the H particle signal as shown in Fig. 13. These arise from the quasifree production of Ξ's which then decay, and from the misidentification of pions as kaons either in the entrance or the exit channel. These can give yields 10^{+4} times larger than the estimate H particle peak. A particle identification scheme designed to suppress this background, developed at CMU[24] for E-836, is shown in Fig. 14. It utilizes three Aerogel Cerenkov detectors, a time of flight measurement, a $\Delta p/p \approx 0.5\%$ spectrometer, and good track vertex reconstruction.

These two methods are compared in Table I and are complementary in their region of mass sensitivity. For the spectrometer in E-813/836 they give comparable yields and basically only require a target change in the experimental set up. However, the large background in the ^{3}He case will always be a concern in a search for an exotic particle. The main obstacle to pursuing these two reactions is finding the required funding for the 1.8 GeV/c beam line.

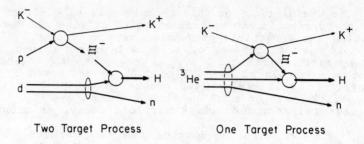

Two Target Process   One Target Process

Fig. 11 Comparison of the composite hydrogen/deuterium target method of E-813 and the ^{3}He target method used in E-836.

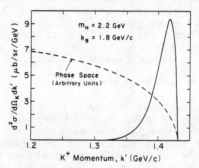

Fig. 12. The inclusive K^+ spectrum calculated by Dover[29] in a neutron spectator model of the K^+ ^{3}He reaction.

A third $\Delta S = 2$ reaction being discussed at the AGS has been proposed by Val Fitch[31]:

$$K^- + D^+ \rightarrow H^0 + K^{0*} \rightarrow H^0 + (K^+ \pi^-).$$

The K^{0*}(892) has $J^\pi = 1^-$, $I = 1/2$, and $S = +1$, so that this reaction has the correct spin and isospin changes to form the H. This can be interpreted as intermediate production of a Ξ (see Fig. 15), in a fashion very similar to the ^{3}He reaction in Fig. 11. In this case the elementary process:

$$K^- + N \rightarrow \Xi^{0-} + K^{0*}$$

has a momentum transfer of q = 900 MeV/c at threshold (1.93 MeV/c). At about 3 GeV/c where the q for the backward going Ξ's is lower, the cross section is $\approx$ 0.6 µb/sr. Dover[30] has estimated the overall cross section for the deuterium target case at 3 GeV/c as $\approx$ 1/10 the ^{3}He target yield at 1.8 GeV/c as discussed above. Nevertheless, the central issue in many of these reactions is not so much the overall yield as it is the role of competing background reactions which, in this case,

I have not reviewed. Overall a 3-4 Gev/c beam line and an imaging detector are required for this measurement.

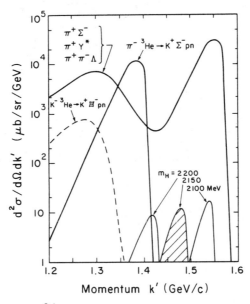

Fig.13. Comparison[24] of possible H particle peaks in the K^+ inclusive spectrum with the background induced by quasifree Ξ formation and misidentification of pions as kaons in the incident or exit channel.

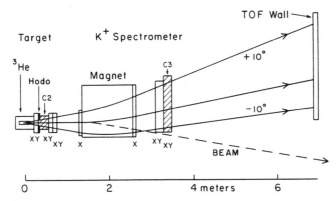

Fig. 14. Detector arrangement for suppression of the background in Fig. 13 as utilized in E-836.

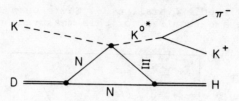

Fig. 15. H particle formation using K^{o*} production and decay in a deuterium target as proposed by Fitch[31].

Table I Comparison of the H/D and ^{3}He Target Methods

	H / D Target	^{3}He Target
Mass Region:	B_H = -20 to +100	+ 10 to + 350 Mev
Events/ 800 hours:	3000 to 300	≈ 3000 ($\Delta\Omega$ = 60 msr)
Background:	Small	Sig/Noise = 1 x 10^4

III. Conclusions

In summary the concept of Strange Matter is certainly intriguing and has astrophysical implications. Various accelerator tests have been proposed and they should be pursued. The specific suggestion of the existence of a Dihyperon is very interesting both for its high degree of symmetry and as a test of confinement models. The question of its mass and decay properties can and should be subjected to test at a modern accelerator facility.

References

1) J. C. Collins and M. J. Perry, Phys. Rev. Lett. 34 1353 (1975).
2) G. Baym and S. Chin, Phy. Lett. 62B 241 (1976).
3) J. D. Bjorken and L. McLerran, Phys. Rev. D20 2353 (1979).
4) S. Chin and A. Kerman, Phys. Rev. Lett. 43 1292 (1979).
5) R. L. Jaffe, Phys. Rev. Lett. 38 195 (1977).
6) E. Witten, Phys. Rev. D30 272 (1984).
7) E. Farhi and R. L. Jaffe, Phy. Rev. D30 (1984) 2379.
8) G. Baym, E. W. Kolb, L. McLerran, T. P. Walker, and R. L. Jaffe, Phys. Lett. 160 181 (1985); SM.V. Barnhill et al, MAD/PH/243/252 S(1985); also see Phy. Lett. 169B 275 (1986).

9) J. L. Rosner, Phy. Rev. D33 2043(1986).
10) K. Saito, Prog. Theor. Phys. 72 674 (1984) also see L.A. Kondratyuk, et al ZHETP 43 10 (1986).
11) K. F. Liu and C. W. Wong, Phys,. Lett. 113B 1 (1982).
12) A. P. Balachandran et al , Phys. Rev. Lett. 52 887 (1984); H. Gomm, R.Lizzi, and G. Sparano, Phys. Rev. D31 226 (1985).; R. L. Jaffe and C. L. Korpa, Nuc. Phy. B258 468 (1985).
13) P. B. Mackenzie and H. B. Thacker, Phy. Rev. Lett. 55 2539 (1985).
14) P. D. Barnes, Nuc. Phy. A450 (1986) 97.
15) J. Szymanski, paper included in these proceedings.
16) J. F. Donoghue, G. Gololich, B.R. Holstein, U. Mass. report UMHEP-259, 1986. and this proceedings; also see M.I. Krivoruchenko and M.G.Shchepkin, Sov. J. Nucl. Phys. 36 769 (1982).
17) B. A. Shanbazian and A. O. Kechechyan, Joint Inst. for Nuc. Res. report, (1984).
18) M. Danysz et al , Nucl. Phys. 49 121 (1963); D.J. Prowse, Phys. Rev. Lett. 17 782 (1966); also see A.R. Bodmer et al, Nucl Phys. 422B 510 (1984) .
19) G.B. Franklin, Proc. of the Int. Symposium on Hypernuclear and Kaon Physics, (1985), Nuc. Phy. A450 (1986)
20) A. S. Carroll et al, Phys. Rev. Lett. 41 777 (1978).
21) A.M. Badalyan and Yu. A.Simonov, Yad. Fiz. 36 1479 (1982), Sov. J. Nucl.Phys. 36 860 (1982).
22) P. D. Barnes, Proc. of the Second LAMPF II Workshop, Los Alamos, 1982, edited by H.,A. Thiessen et al., Los Alamos Nat. Lab. report No.LA-9752-C Vol. I, p 315.
23) AGS Experiment E- 813, Spokesmen: G.B. Franklin and P.D.Barnes., 1985.
24) AGS Experiment 836, Spokesmen: G.B. Franklin and P.D. Barnes, 1986.
25) A.T.M. Aerts, C.B. Dover, Phys,. Rev. D29 443 (1983).
26) C. Dover, proceedings of the International Symposium on Hypernuclear and Kaon Physics-Heidelberg, 1982
27) AGS Proposal E-830, Spokesmen : H. Piekarz, P. H. Pile, 1986.
28) Proc. of the KEK Int. Workshop on Nuc. Phys. in the GEV Region, 1984, KEK report No. KEK-84-20, p 116.; KEK experiment E - 127.
29) A.T. M. Aerts and C.B. Dover, Phy. Rev. Lett. 49 1752 (1982).
30) C. B. Dover, Proc. of the Int. Symposium on Hypernuclear and Kaon Physics, (1985), Nuc. Phy. A450 (1986)
31) V. L. Fitch, Proceedings of the Hawaii Conf. (1985).
32) see references in paper by E. Farhi, Nuc. Phy. A450 (1986) 505;H. Liu and G. L. Shaw, Phy.Rev. D30 1137 (1984).

ANOMALOUS POSITRON PRODUCTION*

Jack S. Greenberg

Yale University, A. W. Wright Nuclear Structure Laboratory

The investigation of positron production from superheavy collision systems was originally motivated by a search for spontaneous positron emission. This process is expected to signature the spontaneous decay of the QED vacuum in the presence of a supercritical electric field (1). A surprising development that emerged from these studies was the discovery (2-6) of prominent narrow peaks in the positron spectra superimposed on the continuous spectrum which is expected from the dynamics of the collision. In particular, in a systematic study (4) of five supercritical collision systems with total nuclear charge Z_u between 180 and 188 that can be formed from the available Th and U projectiles and Th, U, and Cm targets, it was found that a prominent peak appears in all these systems.

Because there is no common collision partner shared by all systems, it was especially striking to find that the peak energies are similar, clustering about a mean value of ~340 keV (4), and that the peak widths also exhibit a near common value of ~80 keV. It appears that the production of the peaks is associated with a narrow interval of projectile energies near the Coulomb barrier where the nuclear surfaces may come in contact, and that the width of the peak can be accounted for entirely by Doppler broadening produced by a source moving with approximately the magnitude of the center-of-mass velocity (2,4). The latter observation particularly indicates that the individual nuclei cannot be the source of the peaks since their motion would produce larger Doppler widths. Moreover, the intrinsic width of the peaks may be much narrower than 80 keV corresponding to a source lifetime long compared to the Coulomb scattering times of ~2×10^{-21} sec.

To date, attempts to associate the peaks with established atomic and nuclear mechanisms for producing low energy positrons have not been successful. The possibility of attributing the peaks to spontaneous positron emission also appears to encounter serious difficulties. Some initial success had been achieved in providing a consistent description of both the peak energy and its intensity in the U+Cm (2) system by incorporating this process into a phenomenological framework (7) which assumes the formation of a long-lived, giant nuclear system. However, the study of the Z_u-dependence (4) has made it clear that the near common peak energies observed for all the systems with Z_u=180 to 188 stand out in marked contrast to the unusual Z_u^{20} expected for spontaneous positron emission from a giant nuclear system composed of normal nuclear density. This scenario is further challenged by preliminary reports of similar peaks found in systems with total nuclear charges below the threshold for generating supercritical fields (8).

Indeed, the mere occurence of low energy, narrow positron peak spectra is anomalous, suggesting an unusual source. Particularly, the near constant energy of the peaks and their narrow widths have led us to suggest (4) that they may be the signature for the two-body, collinear decay into an e+e- pair of a previously undetected source. A follow-up experiment (6) has confronted this possibility by detecting the coincidence emission of electrons and positrons from U+Th, Th+Th, and Th+Cm collisions with a detector geometry that is especially suited to observe back-to-back e+ and e- emission.

These measurements show that an e+ peak in each system is correlated with the simultaneous emission of electrons in a narrow peak of similar energy and width. Particularly significant are the additional observations of a narrow peak in the distribution of the sum of the e- and e+ energies at the position of the sum of the individual e+ and e- peak mean energies, and a peak in the coincidence intensity plotted as a function of the e+ and e- energy difference for a constant sum of their energies centered on the sum-energy peak. The width of the sum-energy peak is consistent with the resolution of the detectors.

The presence of the coincidence-electron peak additionally excludes spontaneous positron emission as a viable source for the positron peaks, as indicated previously by the Z_u-dependence of the peak energy. The coincidence peak intensities and especially the combined appearance of the sharp peak in the sum-energy spectrum and the peak in the difference-energy spectrum cannot be accounted for by known internal conversion processes. A Monte Carlo simulation of a two-body decay of a source moving with approximately the center-of-mass velocity displays a marked resemblance to the features found in the e+e- coincidence spectra, especially the narrow sum-energy peak which reflects the correlated concellation of Doppler shifts expected in a two-body decay. The possibility that there is multiple structure present also becomes evident in this experiment from the added information supplied by the electron channel. These may be related to possible positron-peak energy differences present in the data of Ref. 4, which may also reflect more than one structure. It should be noted in this connection that the sharp sum-energy peak provides a new sensitivity to resolve neighboring structures in the positron spectra.

The central issue of whether we are observing a two-body decay, which is reflected in the narrow width of the sum-energy line and in the presence of a difference-energy peak, as well as the question of the presence of more that one structure, has been pursued in our new measurements. Preliminary analysis of comprehensive data on the U+Th system appears to confirm the previous observations on both these points. How the sharp sum-energy peaks are related and what are their origins are open questions to be addressed. It appears presently difficult to explain the very narrow widths of the sum-energy lines in the context of known mechanisms without invoking a two-body decay process.

* Supported by US DOE Contract No. DE-AC02-76ER03074 and the BFT of the Federal Republic of Germany. Experiments with the EPOS spectrometer were carried out in collaboration with H. Backe, K. Bethge, H. Bokemeyer, T. Cowan, H. Folger, K. Sakaguchi, D. Schwalm, J. Schweppe, and P. Vincent.

1. W. Pieper and W. Greiner, Z. Physik **218**, 327 (1969); S.S. Gershtein and Ya.B. Zel'dovich, ZETF **57**, 654 (1969) (Sov. Phys. JETP **30**, 358 (1970)); Lett. Nuovo Cimento **1** 835 (1969).
2. J. Schweppe et al., Phys. Rev. Lett. **51**, 2261 (1983).
3. M. Clemente et al., Phys. Lett. **137B**, 41 (1984).
4. T. Cowan et al., Phys. Rev. Lett. **54**, 1761 (1985).
5. H. Tsertos et al., Phys. Lett. **162B**, 273 (1985.).
6. T. Cowan et al., Phys. Rev. Lett. **56**, 444 (1986).
7. U. Muller et al., Z. Physik **A313**, 263 (1983); P. O. Hess et al., Phys. Rev. Lett. **53**, 1535 (1984).
8. H. Bokemeyer et al., GSI Report 85-1, 177 (1985); J Schweppe, ICPEAC, Palo Alto, CA, 1985; P. Kienle, GSI Preprint 85-31, 1985.

A REVIEW OF THE PHYSICS OF THE NEUTRINO

R. G. H. Robertson
Physics Division, Los Alamos National Laboratory
Los Alamos, NM 87545 U.S.A.

ABSTRACT

The properties of neutrinos have been partially deduced in a series of elegant and difficult experiments spanning 50 years. Nevertheless, it seems more is unknown than known. Although the basic quantum properties, such as spin, electric charge and helicity, are known, the extent and nature of lepton number conservation and the masses of the neutrinos are still largely open questions. A strong experimental and theoretical effort continues in neutrino oscillations, double beta decay, and neutrino mass measurement to try to illuminate some of these issues. In this review the current situation with specific regard to neutrino properties is considered, but not the wider field of neutrino interactions in general.

INTRODUCTION

It is remarkable how little we know about the neutrino, more than 50 years after its existence was first postulated by Pauli. To be sure, the basic quantum properties such as spin, charge and helicity are known, but (for example) we do not even know if the neutrino is its own antiparticle. The neutrino plays a vital role in nuclear physics, particle physics, and astrophysics, and an intensive experimental effort continues to elucidate its properties.

NEUTRINO PROPERTIES

The Number of Flavors

There are three apparently distinct types (flavors) of neutrino, the electron, mu and tau, and there may well be others. At present the best information on this matter comes from cosmology, by the following simple argument: The primordial mass abundance Y_p of ^{4}He in the universe depends on the rate of expansion during the period of nucleosynthesis because there is competition between the capture of neutrons by protons and the decay of free neutrons. The rate of expansion depends on the energy density of the universe, which in turn depends on the number of species of light neutrino N_ν. The calculations of Yang et al.[1] can be summarized in their equation:

$$Y_p = 0.230 + 0.011 \ln(\eta_{10}) + 0.013(N_\nu - 3) + 0.014(t_{1/2} - 10.6) \quad ,$$

The major uncertainties are in the abundance of ^{4}He, the baryon-to-photon ratio η_{10}, and the half-life $t_{1/2}$ (in minutes) of the free neutron. Figure 1 shows the calculated abundances of ^{4}He, D, ^{3}He,

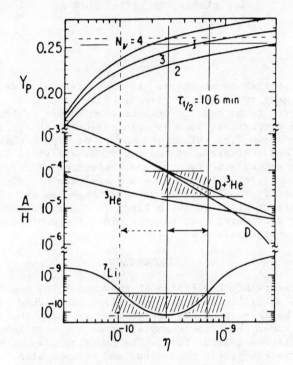

Fig. 1. Abundances as a function of the baryon-to-photon ratio η (Ref.1). The horizontal solid lines are limits derived by Yang et al., and the dashed lines are the upper limits for ^{4}He and (D+^{3}He) suggested by Ellis et al. (Ref.2)

and ^{7}Li as a function of η. Considering the range of concordant values for η, Yang et al. deduced an upper limit of 4 for N_ν. Recently Ellis et al.[2] have concluded that larger abundances of ^{4}He and D are allowed by the data, and although the most probable number of species of light (<100 eV), left-handed neutrino is three, as many as six are possible. (Stable neutrinos between 100 eV and a few MeV are excluded by observational limits on the age of the universe [see Steigman[3]].) However, this proposition implies such a low baryon density that the existence of massive, non-baryonic matter would be essential to explain the dynamics of galaxies.

The total number of neutrino types having a mass less than half that of the Z^0 can be determined in principle from the measured width of the Z^0. In practice the experimental resolution in the UA1 and UA2 detectors is not well enough known, and a better, although model-dependent, method is to compare the decays of the W and the Z. If new charged leptons, but not the associated new neutrinos, are heavier than half the Z^0 mass, then decay channels are accessible to the Z^0 that are closed to the W, and the following ratio may be defined:

$$R = \sigma_Z^e/\sigma_W^e = (\sigma_Z/\sigma_W)(\Gamma_Z^{ee}/\Gamma_Z)(\Gamma_W/\Gamma_W^{e\nu}) \quad ,$$

where $\sigma_{Z(W)}$ is the inclusive cross section for Z(W) production and the Γ's are the partial and total decay widths (Gaillard[4]). A QCD calculation of the cross-section ratios is necessary, and leads to an upper limit on N_ν of 5.6 at 90% C.L., with an additional theoretical uncertainty of 1.7.

Majorana or Dirac?

The question of whether the neutrino is its own antiparticle or not is rendered unreasonably difficult by the left-handed nature of the weak interaction, and is intimately connected to both lepton-number conservation and neutrino mass. Experimentally, it is known that antineutrinos and neutrinos do not induce the same reactions, but is this merely a consequence of the weak interaction that always is involved in the birth of a neutrino, and that guarantees opposite helicities for neutrinos and antineutrinos? Possibly neutrinos are ordinary Dirac particles with attendant antiparticles and two helicity states, but theorists are unhappy with such a situation because it implies neutrino masses similar to those of the charged leptons. Hence the alternative, "Majorana", hypothesis is attractive because of the very different mass terms that can appear for the charged and neutral leptons[5].

Thus, one of the most interesting unanswered questions is the masses of the neutrinos. Before the development of the Standard Model (SM), it was thought that the masses were likely to be zero because the "handedness" of the weak interaction was carried by neutrinos. To insure Lorentz invariance, then, neutrinos would have to move at the speed of light, and therefore be massless. However, the discovery that the neutral current was as left-handed as the charged current demolished this idea. In the SM, the handedness is part of the weak interaction itself. Nevertheless, it is true that neutrinos are massless in the minimal SM.

Lacking a fundamental understanding of the origin of mass for the elementary particles, one generally describes the mass by an arbitrarily normalized self-energy term in the Lagrangian. These terms acquire a vacuum expectation value through coupling to the Higgs. The electron mass term is, for example,

$$\mathcal{L}_e = m\bar{e}e \quad ,$$

or, more properly,

$$\mathcal{L}_e = (m/2)\bar{e}e + h.c. \quad ,$$

where m_μ is here the known mass of the electron. One can rewrite this in terms of separate left- and right-handed field components $e_L = 1/2\ (1-\gamma_5)e$ and $e_R = 1/2\ (1+\gamma_5)e$:

$$\mathcal{L}_e = (m/2)(\bar{e}_L e_R + \bar{e}_R e_L) + h.c.$$

In the Standard Model there are right-handed electron singlets, but one can see immediately that there can be no such mass term for the neutrinos, as they lack a right-handed field. Simply adding right-handed singlets would allow neutrino mass and would do no violence to the successes of the SM, but it would then be hard to understand why the charged and neutral leptons have very different masses. Perhaps the lepton masses are mainly electromagnetic in origin, but this is unlikely, given the enormous differences between the families. A way out of this difficulty is to note that the antineutrino fields are right-handed, and could give us a mass term if neutrinos and antineutrinos are otherwise indistinguishable. Since this option is not open to the charged leptons, it gives a fundamentally different mass term. (It should be noted that, the weak isospin structure of Majorana mass terms require the addition of a Higgs isotriplet.) The most direct evidence for this Majorana structure would be the observation of neutrinoless double β decay. More generally, observation of mass would be important in indicating that, as everyone suspects, there is something beyond the Standard Model.

NEUTRINO MASS - THE EXPERIMENTAL SITUATION

There are three general classes of experiment to consider - ν oscillations, double β decay, and direct mass measurements.

Neutrino Oscillations

If neutrinos have mass, and if lepton family number (L_e, etc.) is not conserved, then mass eigenstates and flavor eigenstates need not be the same. A flavor ("current") eigenstate prepared at some instant, e.g. by beta decay, will evolve into a mixture of flavor components at some later time. The flavor state may be expanded in the mass basis,

$$|\nu_l\rangle = \sum U_{lk} |\nu_k\rangle \qquad (l=e,\mu,\tau;\ k=1,2,3),$$

where U is a unitary matrix. The expression[6] for the probability of observing ν_l when ν_k was initially prepared is

$$P_{ll'} = \sum U_{lk} U_{lk'} U_{l'k} U_{l'k'} \cos(2.54\ \Delta m^2_{k'k} L/E),$$

where $\Delta m^2_{k'k} = |m^2_{k'} - m^2_k|$ is measured in eV^2, L is in m, and E in MeV. In general, then, the transition depends on all mixing matrix elements that share a row or column with the initial and final flavors.

Neutrino oscillation experiments may be divided in an obvious way into appearance and disappearance experiments. The latter are general in that no assumptions must be made about the particular channels involved, but are also not as sensitive to small mixing angles.

The solar neutrino problem may be thought of as a disappearance experiment with a positive result. Because of the great distance (1.5×10^{11} m) from the earth to the sun, the experiment is sensitive to very small mass differences, but, until the recent work by Mikheyev and Smirnov[7], it seemed that nearly maximal mixing between three species would be required to explain the data. Their work showed that the difference in forward scattering amplitudes for ν_e-e and ν_x-e scattering caused by the additional (destructive) charged-current amplitude in the former led to an enhanced effect when neutrinos propagated through dense matter. For certain ranges of mass-squared differences, this effect is substantial.

Physically, the ν_e has a reduced interaction with electrons compared with ν_μ and ν_τ. (That the interference is destructive has recently been verified in the ν_e-e scattering experiment[8] at LAMPF.) Thus ν_e waves see a lower refractive index than the other neutrinos in matter, and tend to get out of phase. If the $\nu_\mu(\nu_\tau)$ is more massive than ν_e, then the refractive index effect acts to enhance the mass effect. (The enhancement was at first thought to require an inverted mass heirarchy for neutrinos, but that is clearly not the case.) Furthermore, the enhancement is maximized resonantly for certain mass differences for a given density and solar radius. Rosen and Gelb[9] have carried out a detailed analysis (as have Hampel and Kirsten[10]), and Fig.2 shows the regions in $\sin^2 2\theta$ ($\sin\theta = U_{ex}$) allowed by the ^{37}Cl result, as well as possible outcomes for a ^{71}Ga experiment. Mixing angles as small as 10^{-3} may be responsible for the solar neutrino deficit. At long last, the crucial Ga detector that is sensitive to neutrinos from the basic p-p reaction is now under construction in Europe, and data-taking should begin early in 1989. It is also clear, in light of the Smirnov-Mikheyev discovery, that further solar neutrino experiments sensitive to other spectral features will be needed (Fig.2). Particularly noteworthy in this regard is a proposed D_2O experiment[11] which could not only measure the shape of the ^{8}B neutrino spectrum, but may also be able to register the neutral-current disintegration of deuterium. Thus the presence of oscillations may be revealed unambiguously in two separate aspects of one experiment.

From time to time, other experiments have indicated the presence of neutrino oscillations. Most recent is the work[12] at the reactor at Le Bugey in France that seems to indicate a statistically significant loss in flux at 18 m as compared to 13 m. However, the measured spectra at the two locations do not appear to be consistent with each other, and the results are (it is claimed) in conflict with measurements made[13] by a different collaboration at Goesgen in Switzerland. Recently, Blümer and Kleinknecht[6] have pointed out that both experiments have been analyzed only for two-flavor oscillations, and that the disagreement is reduced when a more general three-flavor analysis is carried out. This analysis has the effect of weakening the exclusion limits and bringing the two experiments into better agreement. However, the Bugey group also recognize the internal inconsistencies in their data and are repeating some measurements. Very recent results from Rovenskaya in the USSR also indicate a problem with the spectral shape obtained by the Bugey group[13].

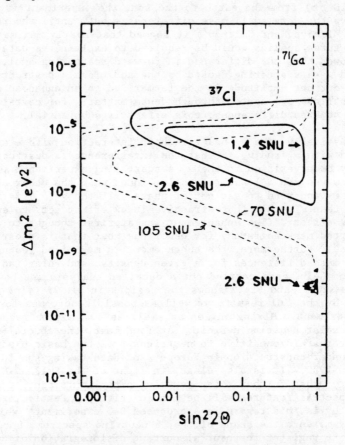

Fig. 2 Correspondence between detected rate (SNU) and oscillation parameters when matter oscillations can occur, for ^{37}Cl and ^{71}Ga detectors. (Hampel and Kirsten, ref.10)

Accelerator experiments have also indicated the existence of neutrino oscillations[14]. Experiment PS191 at CERN (July 1984) searched for electrons produced in a detector exposed to a beam that was 91% ν_μ, 8.6% $\bar{\nu}_\mu$, 0.6% ν_e, and 0.2% $\bar{\nu}_e$. Three times as many electrons were seen as expected, at the 3σ level, and the resulting oscillation parameters, taking into account exclusion limits set on this channel in other experiments, are in the vicinity of $\Delta m^2 = 5$

eV^2 and $\sin^2 2\theta = 0.03$. Competitive experiments at the AGS should be able to shed further light on this exciting possibility.

Figure 3 shows one channel ($e\mu$) of the three-flavor analysis of Blümer and Kleinknecht. It is clear that a region of agreement

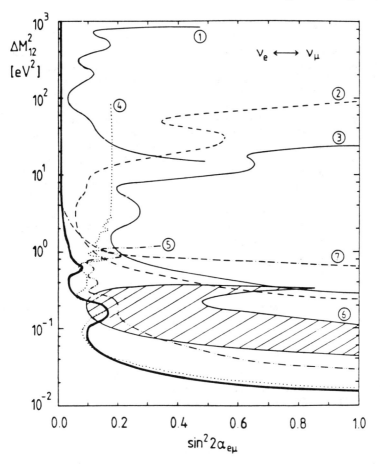

Fig. 3. Limits on mixing parameter $\sin^2 2\alpha_{e\mu}$ vs. neutrino mass difference Δm^2_{12}. The thin lines are 90% C.L. upper limits on $\sin^2 2\alpha_{e\mu}$ from individual experiments (see ref. 6) <u>assuming</u> $\alpha_{e\tau} = \alpha_{\mu\tau} = 0$, the shaded area is the allowed range from the Bugey experiment (Ref. 12). The broad line is the 90% C.L. upper limit on $\sin^2 2\alpha_{e\mu}$ from the three-flavour oscillation analysis, allowing the other two mixing angles to vary over the whole range $0 \le \sin^2 2\alpha_{e\tau} \le 1$ and $0 \le \sin^2 2\alpha_{\mu\tau} \le 1$.

exists between Bugey and Goesgen, although this analysis was carried out before the Goesgen 65-m data became available. The Blümer-

Kleinknecht analysis gives two minima in χ^2 ($\chi^2 = 78$ for 84 D.O.F.) with $\Delta m^2_{21} = 0.2$ eV^2 and $\Delta m^2_{31} = 38$ eV^2 or vice-versa. These minima are about 2σ from a no-oscillation fit ($\chi^2 = 85$ for 87 D.O.F.). It will be most interesting to see what the next round of experiments produces.

Double Beta Decay

The search for neutrinoless double β decay has continued with ever-increasing sensitivity. Most of the experimental effort has been devoted to ^{76}Ge because high-resolution, low-background detectors can be fabricated. The half-life for 0ν decay in this case is now known[15] to be in excess of 4×10^{23} y. The 2ν mode has definitely been observed in two nuclei, ^{130}Te and ^{82}Se, by geochemical methods[16]. For many years there was a discrepancy between the laboratory and geochemical measurements on ^{82}Se, but this appears now to have been conclusively resolved in favor of the latter. The new Irvine TPC data[17] set a 1σ lower limit, 1×10^{20}y, that is close to the geochemical measurement of Kirsten[16], $1.45(15) \times 10^{20}$y, and an order of magnitude longer than the old cloud-chamber result[18]. Observation of the 2ν mode is most important because it has indicated that theoretical calculations of nuclear matrix elements are in error by as much as a factor of 10. It is thus very hard to translate the non-observation of the 0ν mode into an upper limit on the Majorana mass or on the presence of right-handed currents[19]. If that difficulty could be ignored, limits on the Majorana mass would be in the range of 3 eV, but it must also be remembered that the neutrino may be a conventional Dirac particle with mass. The 0ν mode does not occur then.

Tritium Beta Decay

As is well known, the ITEP group in Moscow has reported[20] since 1980 that the electron antineutrino has a mass of about 35 eV (now revised to 30). The method is a careful study of the beta spectrum from tritium decay

$$\frac{dN}{dE} = G_F^2 \frac{m^5 c^4}{2\pi^3 \hbar^7} \cos^2\theta_c |M_{if}|^2 F(Z_f, E) pE[(W_0 - E)^2 - m_e^2 c^4]^{1/2} \quad [1]$$

where M_{if} is the nuclear matrix element and $F(Z_f, E)$ is the Fermi Coulomb distortion factor, a smoothly varying function of energy. The total energy is W_0.

This expression applies to the decay of a bare nucleus and when atomic electrons are present it is necessary to express the total decay probability as a sum of individual branches to all possible final states, which may be done to adequate accuracy by replacing W_0 by

$$W_n = W_0 + V_n,$$

where $V_n = E_i - E_{fn}$, the difference between atomic electron binding energies in the initial and final states.

For the free tritium atom, one can calculate both the branching and the excitation energies to better than 0.1% accuracy. Calculations of comparable accuracy can be done for the T_2 molecule[21], although even in this elementary molecule the situation is vastly more complex because excited states in the ^{3}He daughter are now in the continuum and have a finite width. For a molecule as complicated as valine ($C_5H_{10}NO_2$), used by the ITEP group, a detailed calculation is very difficult, especially for the continuum. The reason why an accurate final state calculation is so important may be seen by considering the case of $m_e = 0$, and evaluating the sums over final states.

$$dN/dE = ApE\ (W_0 - E + V_n)^2 [1 + \sigma^2/(W_0 - E + V_n)]\quad,$$

where F and other constants have been lumped into A, and σ^2 is the variance of the final state distribution. This spectrum looks like a simple Fermi spectrum with an endpoint $W_0 + V_n$ and an "m_e^2" of $-2\sigma^2$. An overestimate of this variance leads to an overestimate of the neutrino mass. Of course, σ^2 must always be ≥ 0, and initially ITEP data gave a finite m_e even when σ^2 was set to zero. However, in the most recent data that model-independent test for the presence of neutrino mass does not give a finite result, which is to be expected if the neutrino mass is smaller than approximately twice the actual variance. Exactly the same arguments can be applied to the instrumental resolution, whose variance must also, therefore, be known accurately.

The development of the ITEP data over the last five years has been reviewed frequently[22]. Many criticisms and suggestions have been made[23], and in each case the ITEP group has responded diligently with new experimental checks and new analyses. We will not repeat the historical material, but report only briefly on the new "ITEP-85" results presented[22] at the Rencontres de Moriond early this year.

The optical response of the spectrometer, a subject of much debate, is now measured directly by observing a convenient conversion line with acceleration and with deceleration. Since the spectrometer resolution is proportional to momentum and all other effects are independent of it, the optical contribution can be unfolded. The ionization loss contribution is also measured directly. A conversion line source is covered with a range of thicknesses of the valine and the various thicknesses are normalized through the total β activity. Backscattering has been examined with conversion line sources on thick and thin backings. Some new tritium data have been taken and the data of ITEP-84 re-analyzed with the result that the mass of ν_e is 30(2) eV when the Kaplan et al.[24] final state spectrum is assumed. Recognizing that this spectrum is somewhat uncertain (the introduction of electron-

electron correlations changed its variance by 50%[25]), ITEP allow a range of variation that gives an endpoint energy still compatible with the known T-^{3}He mass difference. The neutrino mass then is found to lie between 17 and 40 eV (C.L. not assigned).

It is testimony to the considerable ITEP head start that tritium data from other laboratories is only now, 5 years later, becoming available. Zurich[26], Tokyo[27], and Los Alamos[28] now have data, and Beijing[29] is taking data.

The Zurich spectrometer is a toroidal device similar to the ITEP instrument, but electrons are decelerated before entering and re-accelerated on emerging in order to improve the resolution. The source is a carbon film implanted with tritium. (For further details on the Zurich and Los Alamos experiments see the reports at this conference.) The Zurich group find no evidence of mass and an upper limit of 18 eV. It is interesting that, despite the fact that their source is similar in thickness and atomic weight to the ITEP source, the Zurich group find a very different energy-loss spectrum. The ITEP spectrum, on the other hand, agrees very well with results recently obtained by the INS group at Tokyo. Had the Zurich data been analyzed with the ITEP energy-loss spectrum, the results would, remarkably, have agreed quite well with ITEP. Conversely, if the ITEP data were analyzed with a Zurich loss spectrum, no significant neutrino mass would appear. It is thus important to understand why these energy-loss spectra are so different, because the evidence for or against neutrino mass presently rests there.

The Tokyo spectrometer is of the "$\pi\sqrt{2}$" type, and incorporates a non-equipotential source and a multielement detector to improve the count rates. Initial data indicate superb resolution (10 eV FWHM), low backgrounds and high count rates. Pending a detailed analysis of the energy loss and resolution contributions, no result has yet been reported. The source is a complex tritiated organic acid.

The Los Alamos experiment avoids these uncertainties by working with gaseous tritium, for which the final state spectrum and energy-loss effects can be calculated precisely. Instrumental resolution is determined with the aid of a gaseous source that emits a conversion line, ^{83m}Kr, and is presently 35 eV. Figure 4 shows a Kurie plot from one data set (3 days). An upper limit of 30 eV at 95% C.L. has been obtained from 3 data sets.

Masses of ν_μ and ν_τ

Experiments on the other neutrinos have not yielded evidence of finite mass. For ν_μ the best limit comes from measurement[30] of the momentum of muons emitted by π^+ decaying at rest and the upper limit for the ν_μ mass is 250 keV. For ν_μ, strong cosmological arguments require its mass to be less than 100 eV if it is less than 250 keV. Several investigations of the ν_τ mass have been carried out, the one giving the lowest limit (70 MeV) being the ARGUS measurement[31] of the decay of the τ to three charged pions and ν_τ.

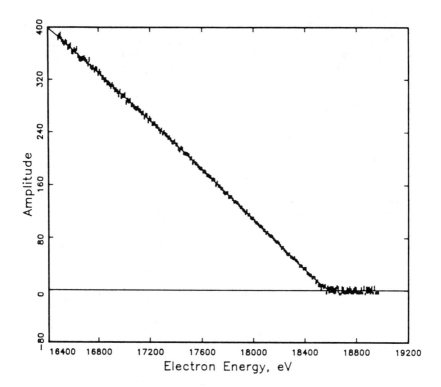

Fig. 4 Kurie plot for gaseous T_2 from the Los Alamos experiment.

REFERENCES

1. J. Yang, M. S. Turner, G. Steigman, D. N. Schramm, and K. A. Olive, Ap. J. 281, 493 (1984).
2. J. Ellis, K. Enqvist, D. V. Nanopoulos, and S. Sarkar, Phys. Lett. 167B, 457 (1986).
3. G. Steigman, Proc. 4th Moriond Workshop on Massive Neutrinos in Astrophysics and in Particle Physics, ed. J. Tran Thanh Van (Editions Frontières, Gif-sur-Yvette, 1984), p.295.
4. J. -M. Gaillard, CERN preprint EP/86-07 (1986).
5. For a lucid review, see W. J. Marciano, Comments Nucl. Part. Phys. 9, 169 (1981).
6. H. Blümer and K. Kleinknecht, Phys. Lett. 161B, 407 (1985).
7. S. P. Mikheyev and A. Yu Smirnov, Yad. Fiz. 42, 1441 (1985); Nuovo Cim. A (to be published).
8. R. C. Allen et al., Phys. Rev. Lett. 55, 2401 (1985).
9. S. P. Rosen and J. M. Gelb (to be published).
10. T. Kirsten, 6th Moriond Workshop on Massive Neutrinos in Astrophysics and in Particle Physics, Tignes, France, January 1986 (to be published).
11. D. Sinclair et al., Proc. 1st Symposium on Underground Physics, Val d'Aosta, Italy (1985), to be published in Nuovo Cim.

11. D. Sinclair et al., Proc. 1st Symposium on Underground Physics, Val d'Aosta, Italy (1985), to be published in Nuovo Cim.
12. J. F. Cavaignac et al., Phys. Lett. 148B, 387 (1984).
13. V. Zacek et al., Phys. Lett. 164B, 193 (1985); see also J. L. Vuilleumier; J. Bouchez, 6th Moriond Workshop (ibid.).
14. F. Vannucci, 6th Moriond Workshop (ibid.).
15. E. Fiorini, (ibid.).
16. T. Kirsten, H. Richter, and E. Jessberger, Phys. Rev. Lett. 50, 474 (1983); T. Kirsten and H. W. Muller, Earth and Planetary Sci. Lett. 6, 271 (1969); K. Marti and S. V. S. Murti, Phys. Lett. 163B, 71 (1985).
17. A. Hahn, 6th Moriond Workshop (ibid.).
18. M. K. Moe and D. Lowenthal, Phys. Rev. C22, 2186 (1980).
19. P. Vogel, 6th Moriond Workshop (ibid.).
20. S. Boris et al. Phys. Lett. 159B, 217 (1985); V. A. Lyubimov, 6th Moriond Workshop (ibid.)
21. See, for example, R. L. Martin and J. S. Cohen, Phys. Lett. 110A, 95 (1985).
22. See, for example, R. G. H. Robertson, Ann. N.Y. Acad. Sci. 461, 565 (1986).
23. See, for example, K. E. Bergkvist, Phys. Lett. 154B, 224 (1985); Phys. Lett. 159B, 408 (1985).
24. I. Kaplan et al., Dokl. Akad. Nauk. USSR 279, 1110 (1984).
25. I. Kaplan et al., Sov. Phys. JETP 57, 483 (1983).
26. M. Fritschi et al., Phys. Lett. B (to be published) 1986.
27. H. Kawakami et al., preprint INS-Rep.-561 (unpublished) 1986.
28. J. F. Wilkerson et al., preprint LA-UR-86-1054, 6th Moriond Workshop (ibid.).
29. Liang, D.-Q., preprint (unpublished) 1986.
30. R.Abela et al., Phys. Lett. 146B, 431 (1984).
31. ARGUS Collaboration, Phys. Lett. 163B, 404 (1985).

HYPERNUCLEAR INTERACTIONS

A. Gal*

TRIUMF, 4004 Wesbrook Mall, Vancouver, B.C., Canada V6T 2A3

ABSTRACT

New developments in extracting strong and weak interaction properties of Λ hypernuclei are reviewed. These include the elucidation of Λ-nuclear spin-dependent interaction parameters from the observation of hypernuclear γ rays in K^- reactions on ^{7}Li, ^{9}Be, ^{10}B and ^{16}O, and the measurement of nonmesonic lifetimes and their branches for $^{12}_\Lambda$C and $^{11}_\Lambda$B ground states. As for Σ hypernuclei, the theoretical difficulties of relating the structures observed in stopped K^- reactions on ^{12}C to Σ-nuclear interaction parameters are reviewed, and production of continuum states is discussed.

INTRODUCTION

The study of hypernuclei is geared to extend the vast body of knowledge of nucleon-nucleon interactions and nuclear phenomena into the strange sector. Λ hypernuclei, in particular, may provide an arena for studying the change in QCD deconfinement scale by "tagging" the s-quark due to a deeply bound hyperon.[1] Whereas the experimental study of the ground-state configuration of heavy Λ hypernuclei is still to be advanced, perhaps through the reactions (π^+,K^+), $(K^-_{stopped},\pi^-)$ or $(e,e'K)$, several possible signatures of quark structure in Λ hypernuclei have been earmarked. These include: (i) underbinding of Λ hypernuclei, beginning with $^5_\Lambda$He, owing to the Pauli exclusion principle for the nonstrange u and d quarks "of the Λ" which forbids them of occupying the lowest 1s(q) quark shell,[2] and (ii) narrowing of $(1s^{-1}(N),1s(\Lambda))$ continuum excitations in light hypernuclei, say in the nuclear p shell, since the most effective strong-interaction deexcitation mode $1p(N) \to 1s(N)$ is partly blocked by the u and d quarks "of the Λ" occupying the lowest 1s(q) shell.[3]

Such signatures are not readily identified through well-defined calculations performed for the existing data because hypernuclei, like ordinary nuclei, are expected to resemble closely baryon gas rather than quark gas. Nonetheless, it is interesting to explore these and other quark signatures in ordinary hypernuclei in the hope of finding clues to and connections with strange quark matter,[4] an extreme form of hypernuclei composed of roughly equal numbers of u, d and s quarks, as recently conjectured to provide a more stable matter than everyday nuclear matter.

In this report, some recent developments in Λ- and Σ-hypernuclear physics during the last two years are reviewed, in accordance with the plan sketched in the abstract.

---

*Permanent address: The Hebrew University, Jerusalem 91904, Israel

SPIN DEPENDENCE OF THE ΛN EFFECTIVE INTERACTION

The spin dependence of the ΛN effective interaction for 1s(Λ) states in the nuclear 1p(N) shell is determined by three parameters (matrix elements) according to the following classification:[5,6]

spin-spin $\quad\Delta \vec{s}_\Lambda \cdot \vec{s}_N$, (1)

spin-orbit $\quad S_\Lambda \vec{s}_\Lambda \cdot \vec{\ell}_N$, (2)

tensor $\quad 6T(3\vec{s}_\Lambda \cdot \hat{r}_N \vec{s}_N \cdot \hat{r}_N - \vec{s}_\Lambda \cdot \vec{s}_N)$. (3)

The theoretical expectations for these parameters, based on the Nijmegen YN potentials of Model D,[7] were discussed by Millener et al.[6] who chose the following values for use in the phenomenological analysis:

$$\Delta = 0.50, \quad S_\Lambda = -0.04, \quad T = 0.04 \quad (\text{MeV}) .$$ (4)

Experiments designed to determine these parameters have been carried out at the AGS of BNL by detecting hypernuclear γ emission, using sodium iodide or germanium detectors, in coincidence with the production pion in the $(K^-,\pi^-\gamma)$ reaction on ^{7}Li, ^{9}Be (Ref. 8) and ^{10}B, ^{16}O (Ref. 9). The γ rays observed in the former are shown in Fig. 1 by solid lines. The $5/2^+$ state of $^7_\Lambda$Li had been predicted[5] to be the most copiously produced in the (K^-,π^-) reaction among the particle-stable excited states of this hypernucleus, and its dominant γ decay to the ground state (g.s.) was indeed observed in the experiment on ^{7}Li thus providing an accurate determination of the excitation energy

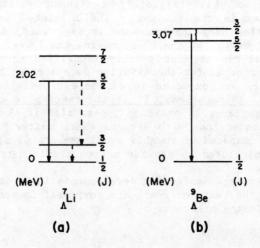

Fig. 1. Observed[8] hypernuclear γ rays (solid lines).

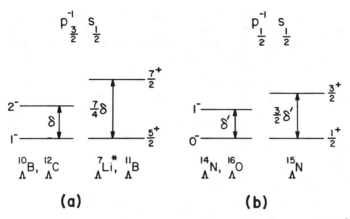

Fig. 2. Doublet splittings in p-shell hypernuclei.[6]

involved. The ultimate goal of a ^{7}Li experiment, however, remains the detection of the $3/2^+ \to 1/2^+$ g.s. intradoublet γ ray, since a determination of the g.s. doublet splitting would provide an excellent measure of the spin-spin interaction (1). Unfortunately, direct excitation of the $3/2^+$ level in the (K^-,π^-) reaction requires spin flip, strongly suppressed at forward angles, and indirect population from higher levels was too weak to be detected in the experiment.[8]

The observation[8] of the 3.1 MeV γ line in $^9_\Lambda$Be (Fig. 1b), with an upper limit of about 100 keV placed on its splitting into two components, stringently constrains the magnitude of the spin-orbit effective interaction which gives the main contribution to the splitting of the $^9_\Lambda$Be excited doublet:

$$|S_\Lambda| \lesssim 0.04 \text{ MeV} . \tag{5}$$

This provides the best limit[6] so far on the strength of the related one-body Λ-nuclear spin-orbit splitting

$$\varepsilon_p \equiv \varepsilon(p_{1/2}) - \varepsilon(p_{3/2}), \qquad |\varepsilon_p| \lesssim 0.25 \text{ MeV} . \tag{6}$$

It is clear that the determination of the spin dependence of the ΛN effective interaction requires detection of several additional intradoublet γ transitions. Typical examples, mostly of g.s. doublets, are shown in Fig. 2 and classified according to their approximate $(1p_j^{-1}, 1s_{1/2})$ $N^{-1}\Lambda$ nature.[6] This classification means that the three parameters of Eqs. (1)-(3) are to be disentangled from the following two combinations

$$1p_{3/2}^{-1}, 1s_{1/2} : \qquad \delta = \frac{2}{3}\Delta + \frac{4}{3}S_\Lambda - \frac{8}{5}T , \tag{7}$$

$$1p_{1/2}^{-1}, 1s_{1/2} : \qquad \delta' = -\frac{1}{3}\Delta + \frac{4}{3}S_\Lambda + 8T . \tag{8}$$

With the expected values given in (4), the main contribution to δ comes from the spin-spin interaction, so a detection of any of the intradoublet γ lines of (a) would provide an approximate measure of Δ, particularly since S_Λ has been constrained by the ^{9}Be measurement (Eq. (5)) and the phenomenologically unknown T was shown[6] to be insensitive to the theoretical model of its determination. Of the four possibilities denoted in (a), ^{10}B provides the best target choice and the most straightforward interpretation[5] of an observation. A preliminary result[9] for the measurement of the $2^- \rightarrow 1^-$ g.s. γ line in $^{10}_\Lambda$B quotes $E_\gamma \sim 157$ keV, compared with the prediction[6] $E_\gamma \sim 170$ keV; in excellent agreement.

The phenomenologically unknown tensor interaction enters significantly in the splitting expression δ', Eq. (8). This suggests measuring one of the γ lines of (b). A measurement of the $1^- \rightarrow 0^-$ g.s. γ line in $^{16}_\Lambda$O, of energy predicted to be $E_\gamma \sim 84$ keV, was undertaken at BNL and the analysis of the data is in progress.[9]

WEAK DECAY OF Λ HYPERNUCLEI

The weak decay of Λ hypernuclei is dominated, except for the very light hypernuclei, by the nonmesonic decay

$$\Lambda N \rightarrow NN . \qquad (9)$$

The free Λ decay, $\Lambda \rightarrow N\pi$ is strongly suppressed in hypernuclei owing to the Pauli principle which excludes the low-energy recoil nucleon. The decay mode (9) cannot be studied in free space. Direct measurement of Λ-hypernuclear lifetimes, therefore, is of a considerable interest, providing for a unique signature of this weak process. Recently, a measurement[10] was undertaken at BNL by detecting relatively high-energy ($E \gtrsim 30$ MeV) protons and neutrons originating from the decay (9) in coincidence with production pions from the reaction

$$^{12}C(K^-,\pi^-)^{12}_\Lambda C . \qquad (10)$$

The proton time of flight (TOF) spectrum from tagged hypernuclear decays was compared with a prompt TOF spectrum measured simultaneously for the inclusive reaction

$$\pi^- + {}^{12}C \rightarrow p + X \qquad (11)$$

induced by the π^- component of the beam. This allows a direct determination of the hypernuclear lifetime. The $^{12}_\Lambda$C excitation spectrum obtained in coincidence with those decay protons used in the lifetime measurement is shown in Fig. 3. The three mass regions shown correspond to the excitation of:

a. $^{12}_\Lambda$C g.s. configuration $(1p^{-1},1s)$, followed by the decay of the 1^- g.s.;
b. $^{12}_\Lambda C^*(0^+,2^+_1,2^+_2)$ belonging to the $(1p^{-1},1p)$ configuration, known[11] to decay by emitting a low-energy proton to the $5/2^+$ g.s. of $^{11}_\Lambda$B, the latter decaying nonmesonically;

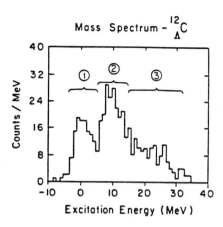

Fig. 3. $^{12}_\Lambda$C spectrum obtained in coincidence with decay protons used in the lifetime measurement.[10]

c. $^{12}_\Lambda$C*(0⁺) belonging to the (1s⁻¹,1s) component of the $^{12}_\Lambda$C continuum, probably decaying by nucleon emission to excited states of $^{11}_\Lambda$B and $^{11}_\Lambda$C (unknown) which then decay nonmesonically through their ground states.

The lifetimes determined, for each of these regions, are given in Table I. Γ_Λ is the free Λ total decay rate. These lifetimes are dominated by the nonmesonic decay (9) to the extent that the π^- decay was hardly noticed (about 4% with an error of a similar magnitude). The measured neutron-to-proton ratio is about 1, which indicates that the one-pion-exchange (OPE), with its characteristically strong $^3S_1 \rightarrow {}^3D_1$ transition contribution to the process (9), cannot be the leading mechanism. Whereas the published calculations[12,13,14] are rooted in nuclear matter, basically ignoring the $^3P_1 \rightarrow {}^3P_1$ transition available to a final nn configuration, a debate between Dubach and Kisslinger concerning the effectiveness of this transition in the decay of $^{12}_\Lambda$C occurred during a parallel session of this conference and resulted in a tie. The above quoted calculations, while differing in their methodology -- OBE[12,14] or OPE plus 6q clusters[13] -- result in

Table I. Hypernuclear lifetime results[10]

hypernucleus	$\tau(10^{-12}s)$	$\Gamma_{tot}/\Gamma_\Lambda$
$^{12}_\Lambda$C	211±31	1.25±0.18
$^{11}_\Lambda$B	192±22	1.37±0.16
$^{11}_\Lambda$B, $^{11}_\Lambda$C, ?	201±30	1.31±0.20

a fair agreement with the lifetimes of Table I. A more stringent test of these theoretical calculations will hopefully emerge from the current measurements[15] in $^5_\Lambda$He and $^4_\Lambda$He.

Σ-HYPERNUCLEAR PRODUCTION WITH STOPPED KAONS

K^- capture from rest has been a standard, and for many years the only method for producing Λ hypernuclei. It is only recently that its potential promise to the production of Σ hypernuclei was realized.[16] It is useful to demonstrate this point by plotting the rate per hyperon as a continuous function of q, the production-pion momentum in the ($K^-_{stopped}, \pi^-$) reaction. Figure 4 shows a plane-wave impulse approximation (PWIA) calculation[17] for stopped K^- in ^{12}C, a situation known to be dominated (~80%) by capture from d orbits. The separate regions of q corresponding to Λ and Σ production are denoted by arrows. As a function of q, the dominant production for small values of q is to $J^\pi = 2^+$ states with relatively strong rates in both Σ and Λ regions, a feature in contrast to small-angle in-flight production where the 0^+ rate dominates. The other transitions, vanishing in the limit $q \to 0$, reach maximum rate for finite values of q and then fall off. Although the rate of the 1^- g.s. transition $1p(N) \to 1s(Y)$ is weaker than the combined $1p(N) \to 1p(Y)$ rate to the 0^+ and 2^+ excited configuration, by about a factor 3 in the Σ region, it still represents a measurable rate. The departure from the PWIA calculations is

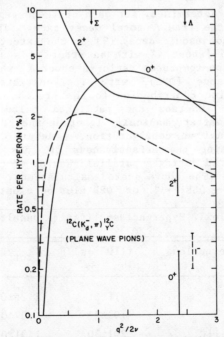

Fig. 4. Hypernuclear production rates in K^- capture from atomic d orbits in ^{12}C, calculated[17] in PWIA.

moderate (less than 20% off) for Σ production when the various approximations are removed, but substantial for Λ production because of the strong pion-nuclear interaction in the (3,3) vicinity. The DWIA calculations for Λ-hypernuclear production are marked by bars in the figure. The net outcome is that Σ-hypernuclear production rates are higher by about an order of magnitude than Λ-hypernuclear production rates.

Yamazaki et al.[18] recently reported a striking Σ-hypernuclear structure in the π^+ spectrum for the reaction

$$^{12}C(K^-,\pi^+)^{12}_\Sigma Be \qquad (12)$$

with stopped kaons at KEK (Fig. 5). The mass resolution of this experiment was 1.3 MeV, allowing for a clear observation of three narrow peaks ($\Gamma \lesssim 4$ MeV) in the momentum range q=150-170 MeV/c. The formation rate of these peaks is fairly large, of order 1% per stopped kaon. None of the peaks observed is believed to correspond to the 1s(Σ) g.s. configuration which should have been excited with a calculated rate of about 0.5% per stopped kaon, a sufficient intensity to be observed if its width were small. One possibility is that the 1s(Σ) level is too broad to be uniquely observed, in agreement with the phenomenological analysis of Refs. 19, 20. It is remarkable that the observed narrow $^{12}_\Sigma Be$ states all stand out in the Σ continuum, particularly for the π^0 tagging shown in the figure (this tagging ensures that the states decay by conversion $\Sigma^- p \to \Lambda n$ and subsequently by the weak decay $\Lambda \to n\pi^0$, not by the quasi-free mode $\Sigma^- \to n\pi^-$). Similar findings have been preliminarily reported by Hungerford[21] for the reaction (12) in flight, at p(K⁻) = 720 MeV/c.

Yamazaki et al.[18] argued that the energy difference ΔE = 4.6 MeV between the two lowest $^{12}_\Sigma Be$ peaks $[(1p^{-1}_{3/2}, 1p_{3/2})$ at q=164 and

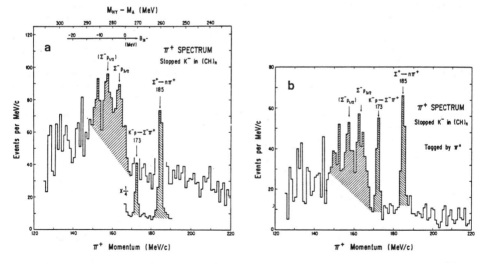

Fig. 5. π^+ spectrum from K⁻ stopped in (CH)$_n$ as taken in KEK.[18]

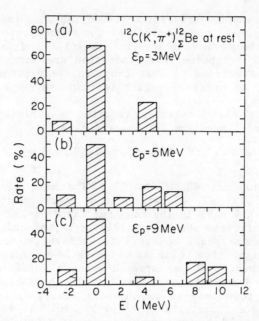

Fig. 6. Relative intensities calculated[22] for $^{12}C(K^-,\pi^+)$ at rest, for three assumed values of the Σ-nuclear spin-orbit splitting ε_p.

$(1p_{3/2}^{-1},1p_{1/2})$ at 158 MeV/c] be attributed almost entirely to the Σ-nuclear spin-orbit splitting ε_p (c.f. Eq. (6) for the corresponding situation in Λ hypernuclei). The shell-model calculations of Dover et al.[22] raise some serious reservations about such a conclusion. Figure 6 gives the relative intensities calculated, for ΣN residual interaction $V(\Sigma N)$ determined from Model D,[7] as a function of ε_p. It is clear from (b) that although the choice ε_p=5 MeV does reproduce the two lowest peaks in $^{12}_\Sigma Be$, it does not leave room for a substantial $1p(P) \to 1p(\Sigma^-)$ intensity in the 9 MeV region corresponding to the third observed peak (q=152 MeV/c). In fact, it has not proved possible to find a reasonable $V(\Sigma N)$ which would displace sufficient intensity to this region without over-depleting the intensity calculated for the 5 MeV region. On the other hand, increasing ε_p to about 9 MeV (Fig. 6c), as favored by straightforward interpretations[23,24] of the somewhat ambiguous $^{16}O(K^-,\pi^+)$ data[23] at $p(K^-)$=450 MeV/c, leads to a splitting between the two main peaks which varies linearly with ε_p, but the remaining intensity around 5 MeV is depleted, again, leaving only two peaks explained. Of course, if ε_p were as large as 9 MeV, the second observed peak could tentatively be assigned[22,24] to the $d_{5/2}(\Sigma)$ configuration; but it is far from clear why a $^{11}B \otimes d_{5/2}(\Sigma^-)$ state should remain unfragmented and narrow, and this reservation would become even more valid if, for $\varepsilon_p \sim 5 MeV$, one attempted to ascribe the origin of the third peak to an excited $d(\Sigma)$ state.

It is worth noting, Fig. 6a, that for $\varepsilon_p < \Delta \sim 5$ MeV the energy splitting of the two calculated main peaks no longer varies linearly with ε_p. Different choices of $V(\Sigma N)$ produce radically different spectra, the one in (a) corresponding to a particular choice that fits the symmetry energy[25] of the $^{12}_\Sigma C$ system.[26] Hence, the tentative identification made by Yamazaki et al.,[18] besides leaving out the third peak, overlooks the observation that the energy splitting between the main two peaks of the $1p(\Sigma)$ configuration is close to ε_p only if $\varepsilon_p \gtrsim \Delta$, where Δ is related to the matrix elements of $V(\Sigma N)$ and to nuclear-core excitation energies which cannot be ignored in picking up a proton from the nuclear target.

COMMENTS ON QUASI-FREE PRODUCTION AND NARROW Σ-NUCLEAR STATES

The problem of understanding why narrow Σ-hypernuclear continuum excitations ($\Gamma \lesssim 5$ MeV) have been observed, implying the suppression of escape width and a substantial degree of quenching for the strong $\Sigma N \to \Lambda N$ conversion process, is not yet solved. I will not review here the various unsuccessful attempts to explain this remarkable phenomenon, but would like to make a few comments on mechanisms mentioned recently which control production of continuum states.

No final state interactions

Consider the process

$$K^- + N \to \pi + Y \qquad (13)$$
$$(\vec{p}_i) \qquad (\vec{p}_f)(\vec{p})$$

occurring on a nucleon with momentum-space form factor $S(P) = \overline{|\tilde{\psi}(\vec{P})|^2}$, where the bar marks spin average. The momenta of the initial and final mesons, and the momentum of the outgoing free hyperon, are as denoted above. The laboratory differential cross section for the nuclear (K^-,π) reaction, within the quasi-free (QF) assumption, is written as follows:

$$d\sigma_L = \frac{1}{v_i} \overline{|T_L|^2} \, 2\pi \, \delta(E_i - E_f + m_N - m_Y - B_N - T_Y) d^3 p_f d^3 p/(2\pi)^6 , \qquad (14)$$

where B_N is the struck nucleon separation energy and $T_Y \sim p^2/2m_Y$ is the kinetic energy of the produced hyperon. Here we neglected the relatively small nuclear-recoil energy; this recoil can absorb any sizable momentum which is why only the energy-conserving δ-function appears in (14), but the kinematics remains essentially two-body (the QF assumption). The many-body T matrix is approximated correspondingly by $T_L = t_L(0°)\rho$, where the transition density is computed in PWIA:

$$\rho = \langle e^{i\vec{p}\cdot\vec{r}} | e^{i\vec{q}\cdot\vec{r}} | \psi(\vec{r}) \rangle = \tilde{\psi}(\vec{q}-\vec{p}), \qquad (\vec{q} = \vec{p}_i - \vec{p}_f) . \qquad (15)$$

The two-body t matrix is related to the two-body differential cross section of (13) by an expression similar to (14), but with an additional momentum-conserving δ-function:

$$d\sigma_L^{K^-N \to \pi Y}(0°)/d\Omega \equiv |f_L(0°)|^2 = \frac{p_f E_f}{(2\pi)^2 v_i} |t_L(0°)|^2 \ . \qquad (16)$$

Substituting Eqs. (15) and (16) in Eq. (14) and integrating over T_Y yields

$$\frac{d\sigma_L}{d\Omega dE_f} = \frac{d\sigma_L^{K^-N \to \pi Y}(0°)}{d\Omega} R(p) \ , \qquad (17)$$

where $R(p)$ is the appropriate nuclear response function:

$$R(p) = \frac{m_Y p}{2\pi^2} \int \hat{S}(\vec{p}) d\Omega_p/4\pi \ , \qquad \hat{S}(\vec{p}) \equiv S(|\vec{q}-\vec{p}|) \ . \qquad (18)$$

By taking q to be fixed along the spectrum, which is closely satisfied in practice, one obtains the sum rule

$$\int R(p) dT_Y = \int \hat{S}(\vec{p}) d^3p/(2\pi)^3 = 1 \ . \qquad (19)$$

It follows from Eqs. (17) and (18) that the QF pion spectrum reflects the momentum distribution of the struck nucleon. The narrowest spectrum is expected for $q \to 0$,

$$R(p) \xrightarrow[q \to 0]{} \frac{m_Y p}{2\pi^2} S(p) \ . \qquad (20)$$

A useful representation for $q \neq 0$ follows by replacing $d\Omega_p/4\pi$ in Eq. (18) by $QdQ/2qp$, where $\vec{Q} \equiv \vec{q} - \vec{p}$,

$$R(p) = \frac{m_Y}{2\pi^2} \frac{1}{2q} \int_{|p-q|}^{p+q} QS(Q) dQ \ , \qquad (21)$$

from which one derives the small p behavior (for $q \neq 0$)

$$R(p) \xrightarrow[p \to 0]{} \frac{m_Y}{2\pi^2} S(q) p \ . \qquad (22)$$

As a practical illustration we take a 1p harmonic-oscillator (HO) form factor:

$$S(Q) = \frac{16\pi^{3/2}}{3} b^5 Q^2 e^{-b^2 Q^2} \qquad (23)$$

which gives

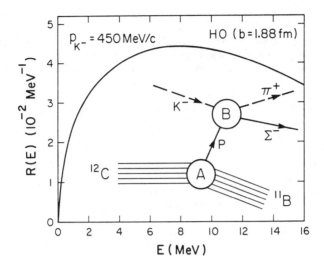

Fig. 7. Hypernuclear response function calculated from Eq. (24) (no Σ interaction) for $^{12}C(K^-,\pi^+)$ at p_{K^-}=450 MeV/c. The harmonic-oscillator (HO) parameter b was chosen to fit the extracted[27] (e,e'p) form factor which corresponds to vertex A of the "pole" diagram.

$$R(p) = \frac{2m_Y b}{3\pi^{1/2}q} \left\{ [1+b^2(p-q)^2]e^{-b^2(p-q)^2} - [1+b^2(p+q)^2]e^{-b^2(p+q)^2} \right\} . \quad (24)$$

This expression shows explicitly that, as q increases, the maximum of R(p) moves closer and closer to p=q, yielding the characteristic QF law $T_Y^{max} \simeq q^2/2m_Y$.

Expression (24) is plotted in Fig. 7 for q=90 MeV/c (corresponding to the in-flight data[26] at p(K$^-$)=450 MeV/c) and b=1.88 fm, corresponding to a maximum of S(Q) at $Q_{max}=b^{-1}$=105 MeV/c as given by the form factor derived from the $^{12}C(e,e'p)$ data.[27] The (K^-,π^+) spectrum is seen to be rather broad and its maximum at 8.5 MeV lies quite apart from the Σ^- excitation observed in ^{12}C at about 3 MeV (5.5-6 MeV if the Σ^- Coulomb energy is subtracted). This maximum cannot then be loosely identified[28] with the peak or enhancement observed in the data.[26] In fact, the consideration of a smaller value of b, b ~ 1.65 fm as appropriate to the ^{12}C size, plus the effect of meson absorption, are expected to shift the maximum of R(p) to even higher energies, about 10-12 MeV. For larger values of q, say q ~ 160 MeV/c as appropriate to the stopped-kaon reactions, the QF shape will become even broader. The conclusion of this paragraph is that the "pole" diagram[28] shown in Fig. 7, which yields essentially the same result as the development above (the two factors on the r.h.s. of Eq. (17) correspond to vertices B and A, respectively, of the diagram), produces QF shapes too broad to explain the relatively narrow excitations seen so far in (K^-,π^+) reactions on light nuclei.

Final state interactions

It is well known that replacing the free Σ wave $\exp(i\vec{p}\cdot\vec{r})$ in Eq. (15) by a distorted wave generated in a nuclear optical potential $V_\Sigma = -(U+iW)$ produces peaks corresponding to bound states or resonances of U when W is relatively weak. However, one has to consider non-perturbative effects of the $\Sigma N \to \Lambda N$ conversion, here represented by the imaginary part W of the optical potential when the latter is strong. Morimatsu and Yazaki[29] showed that, for realistic values of the coupling between the Σ- and Λ-hypernuclear channels, the pion spectrum is dominated by the conversion process which peaks near the Σ threshold with width roughly of order W_0. We review here a derivation[30] of their result.

The double differential cross section for the nuclear (K^-,π) reaction is written, in close analogy to Eq. (14), as

$$\frac{d^2\sigma_L}{d\Omega dE} = |f_L(0°)|^2 \sum_f \delta(E-E_f)|\langle\Psi_f^{(-)}|\Sigma F|i\rangle|^2 \qquad (25)$$

where F is the baryonic transition operator $\left(\exp(i\vec{q}\cdot\vec{r}) \text{ in PWIA}\right)$ onto the Σ channels (the notation Σ is used for the corresponding projector) and Ψ solves the many-channel Hamiltonian:

$$\Psi = \begin{pmatrix}\psi_\Sigma \\ \psi_\Lambda\end{pmatrix}, \quad H = \begin{pmatrix}H_{\Sigma\Sigma} & H_{\Sigma\Lambda} \\ H_{\Lambda\Sigma} & H_{\Lambda\Lambda}\end{pmatrix}, \quad H\Psi = E\Psi. \qquad (26)$$

Neglecting the dependence of F on the final-state energy E_f, the summation over the complete set of final states $\Psi_f^{(-)}$ can be done in a closed form, resulting in

$$\frac{d^2\sigma_L}{d\Omega dE} = |f_L(0°)|^2 \frac{1}{\pi} \operatorname{Im} \langle i|F^+ G_{\Sigma\Sigma}^{(-)}(E) F|i\rangle, \qquad (27)$$

where

$$G_{\Sigma\Sigma}^{(-)}(E) = \sum_f |\Sigma\Psi_f^{(-)}\rangle \frac{1}{E-i\varepsilon-E_f} \langle\Psi_f^{(-)}\Sigma|$$

$$= \Sigma \frac{1}{E^{(-)}-H} \Sigma = \frac{1}{E^{(-)}-H_\Sigma^+}, \qquad (28)$$

$$H_\Sigma = H_{\Sigma\Sigma} + H_{\Sigma\Lambda} \frac{1}{E-H_{\Lambda\Lambda}} H_{\Lambda\Sigma} \equiv T_\Sigma + V_\Sigma(E), \qquad (29)$$

in terms of a complex optical potential V_Σ. Morimatsu and Yazaki,[29] assuming that the (K^-,π) transition is localized on the surface R due to the strong absorption of the mesons, or the peripherality of the kaon atomic orbit in capture from rest, evaluated the pion spectrum as proportional (by (27)) to $\text{Im} G^{(-)}_{\Sigma\Sigma}(R,R;E)$. They noted the following identity

$$\text{Im} G^{(-)}_{\Sigma\Sigma} = (1 + G^{(-)}_{\Sigma\Sigma} V^*_\Sigma)(\text{Im} G^{(-)}_0)(1 + V_\Sigma G^{(-)\dagger}_{\Sigma\Sigma}) + G^{(-)}_{\Sigma\Sigma}(\text{Im} V^*_\Sigma) G^{(-)\dagger}_{\Sigma\Sigma} \qquad (30)$$

which follows from the relationship

$$G^{(-)}_{\Sigma\Sigma} = G^{(-)}_0 + G^{(-)}_0 V^*_\Sigma G^{(-)}_{\Sigma\Sigma} . \qquad (31)$$

Expression (30) defines a splitting of the spectrum into an escape component ($\sim \text{Im} G_0$) and a conversion component ($\sim \text{Im} V^*_\Sigma$). To appreciate the significance of such a splitting we show how to devise the appropriate contributions to the summation over $\Psi^{(-)}_f$ in the first line of Eq. (28).

a. escape component: the Σ distorted waves χ_Σ are generated by applying the optical potential to plane waves ϕ_Σ, while the Λ wave function does not have an outgoing or ingoing part

$$\psi_\Lambda = \frac{1}{E - H_{\Lambda\Lambda}} H_{\Lambda\Sigma} \psi_\Sigma , \qquad \psi_\Sigma \to \tilde{\chi}^{(-)}_\Sigma = (1 + G^{(-)}_{\Sigma\Sigma} V^*_\Sigma) \phi_\Sigma . \qquad (32)$$

The sum

$$\text{Im} \, \Sigma_f |\tilde{\chi}^{(-)}_\Sigma\rangle_f \frac{1}{E - i\epsilon - E_f} {}_f\langle \tilde{\chi}^{(-)}_\Sigma| \qquad (33)$$

then yields exactly the "escape component" term in (30) because

$$\Sigma_f |\phi_\Sigma\rangle_f \frac{1}{E - i\epsilon - E_f} |\phi_\Sigma\rangle_f = G^{(-)}_0(E) . \qquad (34)$$

b. conversion component: here the Λ wave function acquires an inhomogeneous part χ_Λ

$$\psi_\Lambda = \chi_\Lambda + \frac{1}{E - H_{\Lambda\Lambda}} H_{\Lambda\Sigma} \psi_\Sigma , \qquad (E - H_{\Lambda\Lambda}) \chi_\Lambda = 0 \qquad (35)$$

where χ_Λ is a Λ distorted wave. The Σ wave function does not have an outgoing or ingoing part

$$\psi^{(-)}_\Sigma = G^{(-)}_{\Sigma\Sigma} H_{\Sigma\Lambda} \chi_\Lambda . \qquad (36)$$

The sum

$$\text{Im} \, \Sigma_f |\psi^{(-)}_\Sigma\rangle_f \frac{1}{E - i\epsilon - E_f} {}_f\langle \psi^{(-)}_\Sigma| \qquad (37)$$

is evaluated by substituting (36) and noting that

$$\Sigma_f |\chi_\Lambda\rangle_f \frac{1}{E-i\varepsilon-E_f} {}_f\langle\chi_\Lambda| = \frac{1}{E^{(-)}-H_{\Lambda\Lambda}} , \qquad (38)$$

so that this sum (37) becomes

$$G^{(-)}_{\Sigma\Sigma} \left(\text{Im}(H_{\Sigma\Lambda} \frac{1}{E^{(-)}-H_{\Lambda\Lambda}} H_{\Lambda\Sigma}) \right) G^{(-)\dagger}_{\Sigma\Sigma} \qquad (39)$$

reducing by (29) to

$$G^{(-)}_{\Sigma\Sigma} (\text{Im} V^*_\Sigma) G^{(-)\dagger}_{\Sigma\Sigma} . \qquad (40)$$

This agrees with the "conversion component" term in (30).

In Fig. 8, taken from Ref. 29, we show the spectrum resulting from Eq. (30) for

$$V_\Sigma = -(U_o + iW_o)\theta(R-r) , \qquad (41)$$

with R=3 fm and U_o=16 MeV, W_o=6 MeV. This value of W_o is not far from the one expected,[20] after renormalizing the square-well shape to a common density, from an unquenched conversion. It is clear from the figure that the conversion component (dash line for E > 0) dominates the calculated spectrum, giving rise to a structure of width $\gtrsim W_o$ centered just above threshold. That the calculated spectrum no longer reflects the momentum distribution of the struck nucleon may be understood by recalling the restriction to the surface value r=R which corresponds to an infinitely uniform momentum-space form factor.

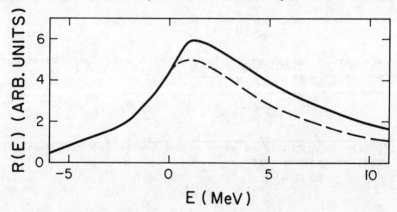

Fig. 8. Hypernuclear response function (solid line) calculated[29] for $^{12}C(K^-,\pi^+)$, assuming the production is localized at R=3 fm. The Σ-nuclear interaction (41), with U_o=16 MeV and W_o=6 MeV, was used. The ordinate is in arbitrary units. The dash line gives the conversion component (40).

REFERENCES

1. T. Goldman, in Intersections Between Particle and Nuclear Physics (Steamboat Springs, 1984), ed. R.E. Mischke, AIP Conference Proceedings, No. 123 (New York, 1984), p. 799;
   T. Yamazaki, Nucl. Phys. $\underline{A446}$, 467c (1985).
2. E.V. Hungerford and L.C. Biedenharn, Phys. Lett. $\underline{142B}$, 232 (1984);
   T. Goldman, in Hadronic Probes and Nuclear Interactions (Tempe, 1985), ed. J.R. Comfort, W.R. Gibbs and B.G. Ritchie, AIP Conference Proceedings, No. 133 (New York, 1985), p. 203.
3. A. Gal, in Hadronic Probes and Nuclear Interactions (Tempe, 1985), ed. J.R. Comfort, W.R. Gibbs and B.G. Ritchie, AIP Conference Proceedings, No. 133 (New York, 1985), p. 30.
4. E. Witten, Phys. Rev. $\underline{D30}$, 272 (1984);
   E. Farhi and R.L. Jaffe, Phys. Rev. $\underline{D30}$, 2379 (1984).
5. R.H. Dalitz and A. Gal, Ann. Phys. $\underline{116}$, 167 (1978).
6. D.J. Millener, A. Gal, C.B. Dover and R.H. Dalitz, Phys. Rev. $\underline{C31}$, 499 (1985).
7. M.M. Nagels, T.A. Rijken and J.J. deSwart, Phys. Rev. $\underline{D12}$, 744 (1975); $\underline{15}$, 2547 (1977).
8. M. May et al., Phys. Rev. Lett. $\underline{51}$, 2085 (1983).
9. M. May, Nucl. Phys. $\underline{A450}$, 179c (1986).
10. R. Grace et al., Phys. Rev. Lett. $\underline{55}$, 1055 (1985).
11. R.H. Dalitz, D.H. Davis and D.N. Tovee, Nucl. Phys. $\underline{A450}$, 311c (1986).
12. B.H.J. McKellar and B.F. Gibson, Phys. Rev. $\underline{C30}$, 322 (1984).
13. C.Y. Cheung, D.P. Heddle and L.S. Kisslinger, Phys. Rev. $\underline{C27}$, 335 (1983);
    D.P. Heddle and L.S. Kisslinger, Phys. Rev. $\underline{C33}$, 608 (1986).
14. J.F. Dubach, Nucl. Phys. $\underline{A450}$, 71c (1986).
15. J. Szymanski, this volume.
16. T. Yamazaki, T. Ishikawa, K. Yazaki and A. Matsuyama, Phys. Lett. $\underline{144B}$, 177 (1984).
17. A. Gal and L. Klieb, Phys. Rev. C, in press (1986);
    A. Gal, Nucl. Phys. $\underline{A450}$, 23c (1986).
18. T. Yamazaki et al., Phys. Rev. Lett. $\underline{54}$, 102 (1985);
    Nucl. Phys. $\underline{A450}$, 1c (1986).
19. A. Gal and C.B. Dover, Phys. Rev. Lett. $\underline{44}$, 379 and 962(E) (1980).
20. C.J. Batty, A. Gal and G. Toker, Nucl. Phys. $\underline{A402}$, 349 (1983).
21. E.V. Hungerford, Nucl. Phys. $\underline{A450}$, 157c (1986).
22. C.B. Dover, A. Gal, L. Klieb and D.J. Millener, Phys. Rev. Lett. $\underline{56}$, 119 (1986).
23. R. Bertini et al., Phys. Lett. $\underline{158B}$, 19 (1985).
24. R. Hausmann and W. Weise, Z. Phys., in press (1986).
25. C.B. Dover, A. Gal and D.J. Millener, Phys. Lett. $\underline{138B}$, 337 (1984).
26. R. Bertini et al., Phys. Lett. $\underline{136B}$, 29 (1984).
27. M. Bernheim et al., Nucl. Phys. $\underline{A375}$, 381 (1982);
    S. Frullani and J. Mougey, Adv. Nucl. Phys. $\underline{14}$, 1 (1984).
28. T. Kishimoto, Nucl. Phys. $\underline{A450}$, 447c (1986).
29. O. Morimatsu and K. Yazaki, Nucl. Phys. $\underline{A435}$, 727 (1985).
30. A. Gal, Nucl. Phys. $\underline{A450}$, 343c (1986).

THE STANDARD MODEL AND BEYOND

Paul Langacker
Department of Physics, University of Pennsylvania
Philadelphia, Pennsylvania 19104-6396

INTRODUCTION

In the last fifteen or twenty years there has been incredible progress in our understanding of the elementary particles and their interactions. We now have a theory - the standard model of the strong, weak and electromagnetic interactions - which is almost certainly a correct description of Nature to some degree of approximation. With the possible exception of the anomalous positron events from Darmstadt, the standard model either predicts or is compatible with all known facts. It describes the physics that has been observed under normal terrestrial and astrophysical conditions, and at existing accelerators up to energies of about 100 GeV. Despite the great success of the standard model, however, it is certainly not the ultimate theory of Nature. This is because it is too complicated: it has many free parameters, and there are many aspects of Nature which are input and not explained in any fundamental way. Therefore, there is almost certainly something more fundamental. In this talk I will describe one particularly promising line of development of ideas for physics beyond the standard model. This line incorporates grand unified theories (GUTs), in which the strong, weak, and electromagnetic interactions are all components of a single underlying unified interaction. A further development involves supersymmetry and supergravity. Supersymmetry is a particular type of symmetry in which fermions and bosons are related to each other. Supersymmetry was introduced mainly to help explain why the weak interaction scale can be so much smaller than the Planck scale (the scale at which quantum gravity becomes important). Most supersymmetric models do not explain the origin of this scale, but at least they stabilize it with respect to higher order corrections. In addition to this technical but very important problem, supersymmetry brings gravity into the theory in a non-trivial way. Unfortunately, it does not explain how to generate a sensible quantum theory of gravity. Finally, in the last couple of years there has been a great deal of excitement about superstring theories. These very speculative theories suggest that physics beyond the Planck scale may be very different than what we see at low energies. In particular, the basic entities are extended one-dimensional objects called strings rather than point-like particles. Superstring theories apparently lead to finite quantum theories of gravity, and, moreover, they give the hope of being the ultimate theory of everything. However, we are very far from knowing whether Nature indeed chooses these very elegant theories.

THE STANDARD MODEL

The standard electroweak theory is a gauge field theory. Gauge field theories are mathematically consistent, renormalizable field theories based on an underlying symmetry principle called gauge invariance. Gauge invariance predicts the existence of spin-1 gauge bosons which mediate the underlying interactions. The gauge symmetry not only predicts the existence of these gauge bosons, but it prescribes the form of their interactions with each other and with the fermions and other particles of the theory, up to a constant known as the gauge coupling constant. In the standard model the basic symmetry is described by the $SU_3 \times SU_2 \times U_1$ group. This is the group of transformations of the fundamental fields of the theory under which the equations of motion are left invariant. A gauge invariance means that the symmetry transformation can be performed independently at different space-time points. The strong interactions are associated with the SU_3 group (with gauge coupling constant g_s) of transformations of the different colors of quarks (and antiquarks) into each other; red and green quarks can be interchanged, for example, without changing the form of the equations of motion. The electroweak interactions are based on the $SU_2 \times U_1$ group, with two coupling constants g and g'. The charged current weak interactions are associated with transformations of up quarks into down quarks, and neutrinos into electrons. The neutral current and electromagnetic interactions are associated with phase transformations on the fields of all of the quarks, anti-quarks, leptons and anti-leptons.

THE STRONG INTERACTIONS (QCD)[1]

It is now generally believed that the strong interactions are based on a gauge theory known as quantum chromodynamics (QCD). This is a gauge theory based on the group SU(3) which describes transformations of the three colors (red, green, and blue) of quarks into one another. The SU(3) group has 8 generators; therefore, there are 8 massless gauge bosons or gluons associated with it. The underlying force involves the emission of a gluon by a quark. The quark then changes into a quark of a different color but the same flavor and electric charge; the gluon can then be absorbed by another quark. The amplitude for emission or absorption is proportional to the strong coupling constant g_s. In addition, there are self-interactions amongst the gluons because, unlike tha analog of electrodynamics, the gluons themselves carry color charge. There are three and four point vertices, the form of which are also prescribed by the gauge symmetry and have strengths related to the strong coupling constant. The

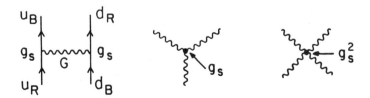

Fig. 1. Fundamental QCD interactions.

strong coupling constant is large (i.e. of order one) and for that reason the strong interactions are strong. It also means that the force is non-perturbative. It is very hard to do calculations, and therefore QCD has not been subjected to the same degree of rigorous quantitative testing as the other interactions. It is thought, though not rigorously proved, that the large value of the coupling constant and the existence of the gluon self-interactions lead to quark confinement (i.e. the non-observation of isolated quarks). It is also believed that all colored objects are confined and unable to propagate freely. This is the reason that the strong force is short range even though the gluon is massless.

QCD has been extremely successful qualitatively. It has all the right general features, correct limits, and so forth. For example, a) QCD incorporates the one-boson exchange potential model of the strong force in the appropriate limit, with the exchange of a pion or other meson interpreted as a higher order effect.

b) Current algebra and $SU_3 \times SU_3$ chiral symmetry are naturally incorporated into the theory. The success of the soft pion theorems is associated with the smallness of the up and down quark masses.

c) QCD naturally explains parity conservation to order α.

d) The approximate scaling observed in deep inelastic electron-nucleon scattering is incorporated in the theory. No other realistic theory has this property.

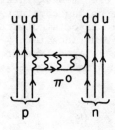

Fig. 2. The one pion exchange diagram.

Because of the strength of the QCD coupling constant it is very hard to do the necessary calculations to subject the theory to rigorous, quantitative tests; however, to the extent that such tests have been carried out, it has been generally successful. In particular, the logarithmic scaling violation observed in deep inelastic scattering of electrons, muons and neutrinos from nucleons is described in a satisfactory way. Secondly, hard scattering and jet production observed in e^+e^-, $\bar{p}p$, and pp scattering is consistent with the QCD predictions, although again it is hard to do the calculations in detail.

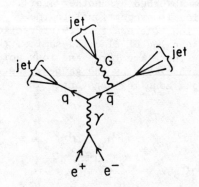

Fig. 3. Jet production in e^+e^- annihilation.

There is a great deal of current activity, both theoretical and experimental. Much of the ultimate motivation is to provide the final definite proof that QCD is the correct theory. For example, there is much theoretical effort trying to develop non-perturbative calculational techniques; e.g. lattice calculations of the hadronic spectrum, phase transitions, attempts to prove quark confinement, etc. Another area of interest both theoretically and experimentally is the search for glueballs, which are predicted bound states of gluons only. There is much effort on the possibility of new forms of matter, such as quark matter or quark-gluon plasma, which could occur in conditions of sufficient density and energy. Finally, there is a lot of continuing activity in jet physics, deep inelastic scattering, heavy quark systems and so forth.

QCD is probably correct. It has all the right limits, and those tests that we can make are successful. However, the most compelling argument is that there just doesn't seem to be any competing candidate within the general framework of renormalizable field theories.

ELECTROWEAK INTERACTIONS[2]

The standard model incorporates the electroweak theory of Glashow, Weinberg and Salam. This is based on the gauge group $SU_2 \times U_1$, which has gauge couplings g and g' for the two subgroups. The SU_2 transformations relate up and down quarks, neutrinos and electrons, and so forth, whereas the neutral generator of SU_2 and the U_1 generator involve phase transformations on all of the quarks and leptons. The charged current weak interactions are associated with the exchange of a charged boson called the W. The Fermi constant is related by

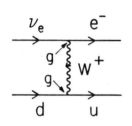

Fig. 4. The charged current interaction.

$G_F/\sqrt{2} \simeq g^2/8 M_W^2$. In addition, there is a neutral current interaction associated with the exchange of a massive electrically neutral boson called the Z. The couplings of the Z involve g and also $\sin^2\theta_W = g'^2/(g^2+g'^2) = 1 - M_W^2/M_Z^2$. Finally, the electromagnetic interactions (quantum electrodynamics) is incorporated in the theory as well. The amplitude for photon emission or absorption is proportional to the electric charge of the positron, given by $e = g \sin\theta_W$.

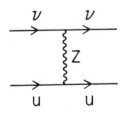 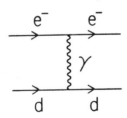

Fig. 5. The neutral current and electromagnetic interactions.

The $SU_2 \times U_1$ model is a renormalizable field theory. It incorporates the Fermi theory of the charged current weak interactions (as modified later by others to include parity violation and strangeness changing effects). In addition, it predicted the existence of the neutral current interaction. In the 1970's the neutral current interaction was observed and in the last decade it has been intensively studied in many types of reactions and over a very wide range of momenta (and therefore distances). For example, the parameter $\sin^2\theta_W$ has been measured in such diverse reactions as parity violation in atoms, neutrino-electron elastic scattering, polarization asymmetries in electron and muon scattering from hadrons, deep

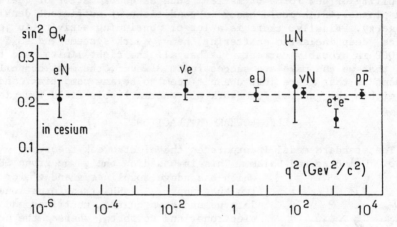

Fig. 6. $\sin^2\theta_W$ determinations from a variety of processes. From M. A. Bouchiat et al.[3]

inelastic neutrino scattering, forward-backward asymmetries in electron-positron annihilation, and indirectly in the measured masses of the W and Z bosons. The consistency of these results indicates that the $SU_2 \times U_1$ model is approximately correct down to a distance scale of 10^{-16}cm, which corresponds to about 100 GeV. The value of the weak angle determined from these measurements is $\sin^2\theta_W = 0.223 \pm .004$ (the quoted error is experimental only; there is an additional theoretical error of approximately 0.005. This average represents all of the available data as of the summer of 1985). In addition to the prediction of the neutral current interaction, the $SU_2 \times U_1$ model predicted the existence of the W and Z bosons and of their masses. The W and Z were observed by the UA1 and UA2 groups at the CERN $\bar{p}p$ collider a few years ago. By now their masses and other properties have been measured in detail. The measured masses are compared with the theoretical prediction in Table 1. The agreement is remarkable.

	UA1	UA2	Theory
M_W(GeV)	$83.1^{+1.3}_{-0.8} \pm 3$	$81.2 \pm 1.1 \pm 1.3$	$A/\sin\theta_W = 82.0 \pm 1.7$ (77.2 without rad. corr.)
M_Z(GeV)	$93.0 \pm 1.6 \pm 3$	$92.5 \pm 1.3 \pm 1.5$	$2A/\sin 2\theta_W = 93.0 \pm 1.4$ (88.2 without rad. corr.)

Table 1. Measured and predicted values for M_W and M_Z. $A \equiv (\pi\alpha/\sqrt{2}G_F)^{\frac{1}{2}}$ $(1-\Delta r)^{-\frac{1}{2}}$, where $\Delta r = 0.0696 \pm 0.002$ is due to radiative corrections (other corrections affect the value of $\sin^2\theta_W$ extracted from neutral current experiments).

CURRENT WORK

There is still a great deal of effort on the standard electroweak model. Some of the major activies are in the following directions:

a) A great deal of experimental work is going on to continue studying the neutral and charged current interactions to a high degree of precision, and to measure the masses and properties of the W and Z to higher accuracy. One of the motivations for this work is that one would like to test the theory at the level of the radiative corrections. At present the measured value of the W mass is about two standard deviations away from the prediction that one would have without including the corrections and it is in perfect agreement with the prediction that includes the radiation corrections. One is therefore on the verge of being able to test these effects, and with the advent of LEP and SLC it should be possible to make these tests precise and conclusive; unfortunately, it will probably never be possible to test the radiative corrections to the same degree of precision that has been done with quantum electrodynamics. Another motivation for these precision tests is to search for and put limits on possible new physics beyond the standard model, such as on the possible existence of additional heavy W and Z gauge bosons. At present the limits on additional neutral bosons are rather weak. For bosons with the couplings predicted by some of the more popular extensions of the standard model, one could tolerate masses as low as 100-200 GeV. At present the most stringent limits come from the indirect neutral current constraints, but as soon as the Fermilab Tevatron starts running the direct production limits will quickly become more important. The most popular theory predicting additional charged W's are the $SU_2 \times SU_2 \times U_1$ models. In such models the additional charged bosons couple to right-handed rather than to left-handed currents. Because of this new type of coupling it turns out that the extra W's can give very important contributions to the K_L-K_S mass difference. It is therefore likely that any such additional W's should be more massive than 2-3 TeV. However a great deal of theory goes into this limit and it should therefore be regarded as suggestive, but not rigorous.

b) Another area of activity is the study of flavor mixing between quarks. This refers to the determination of the mixing angles which govern the rate of weak decays of heavy quarks into lighter quarks. These mixings are free parameters in the standard model; however, we would like to know them in detail because they are useful for computing new weak processes and also because they may give us some insight into the origins of the fermion mass.

c) CP violation. We are still not certain whether CP violation is due to higher order weak interactions or whether it requires the introduction of new physics or new interactions.

d) Rare Decays, such as $K_L \rightarrow \mu e$. The search for rare decays, such as those which violate quark or leptonic flavor, is very important. The predicted rates for such decays in the standard model (via higher order effects) are very low or zero. However, most extensions of the standard model predict one or another type of rare decay at some level. Therefore it is important to continue such searches as far as is feasible as a probe of new physics.

e) The Top Quark. The observed decays of the bottom quark indirectly indicate that it must have a partner, the top quark. So far, the t has not been discovered. It must be more massive than about 23 GeV because of direct searches at PETRA. Theoretical arguments strongly suggest that the t should be less than about 200 GeV.

f) The Higgs Scalar. The masses of the W and Z bosons require that the $SU_2 \times U_1$ gauge symmetry be spontaneously broken down to the U_1 of electromagnetism. The simplest mechanism for accomplishing this breaking is the Higgs mechanism. One introduces a scalar field called the Higgs field. If the lowest energy state (the vacuum) has a non-zero value for this Higgs field, then the symmetry will be spontaneously broken. As the W and Z propagate, they constantly interact with this field and acquire effective masses. The quantum of this field is known as the Higgs scalar. The Higgs has not been observed - it is the last important ingredient of the $SU_2 \times U_1$ which needs to be verified. Unfortunately, the theory does not predict the Higgs mass. Plausible theoretical arguments suggest that it should be somewhere between 10 GeV and 1 TeV, but these are not rigorous. Unfortunately, the Higgs scalar is relatively weakly coupled, it is hard to produce, and it does not have a very dramatic signature. As we will see, the Higgs sector is not only the least understood, but is also the most unattractive feature of the standard model. It is possible that the Higgs mechanism is right, or it is possible that there is some other mechanism for spontaneous symmetry breaking. Investigations in this direction are one of the major motivations for the desire to build larger accelerators.

g) Neutrino Mass. One of the most interesting areas of investigation is the search for a possible non-zero mass for the neutrino. The neutrino is predicted to be massless in the minimal standard model, but almost all extensions predict a non-zero mass at some value. For some years the ITEP group in Moscow has reported evidence from tritium β decay for a non-zero mass for the electron neutrino somewhere in the 17 to 40 eV range. This result has not yet been confirmed by other experiments and in fact it is on the verge of being contradicted by the Zurich experiment. If true, however, the ITEP result would be extremely important, not only for particle physics but for cosmology as well. Relic neutrinos left over from the early Universe are expected to still fill space ($n_\nu \sim 50/cm^3$ for each neutrino type) and if the neutrino has even a tiny mass, it could dominate the energy density of the Universe and perhaps provide an explanation for the dark matter. There are various current neutrino mass experiments. In addition to Zurich, there are experiments at Los Alamos, Tokyo, Livermore and elsewhere which should confirm or refute the ITEP result in the near future. Results are eagerly awaited.

Another interesting development concerns the possibility of much smaller neutrino masses. Recently Mikheyev and Smirnov[4] have developed an old idea of Wolfenstein that may well explain the missing solar neutrinos. The idea is that electron neutrinos produced at the core of the sun interact with electrons on their way out. For appropriate sets of parameters, an adiabatic conversion of the electron neutrinos into muon or tau neutrinos is possible, explaining the results of the

Davis experiment. This is a very elegant and plausible sounding solution to the solar neutrino problem. If true, it would suggest neutrino masses of order 10^{-2} eV or so for the ν_μ or ν_τ and even smaller for the ν_e. Unfortunately, it would be very difficult to verify neutrino masses of this range in the laboratory. Therefore further solar neutrino experiments, including gallium and heavy water experiments, are urgently needed.

THEORETICAL PROBLEMS

The standard $SU_3 \times SU_2 \times U_1$ model, combined with classical general relativity, is almost certainly correct at some level. However it contains far too much arbitrariness to be the final story of Nature. One way of seeing this is that the minimal version with massless neutrinos has 21 free parameters, and nobody believes that the ultimate theory of Nature could have so much arbitrariness. I have classified the problems or unexplained features of the standard model under five headings.

a) The Gauge Problem

The standard model gauge group is a complicated direct product of three factors with three independent coupling constants. Charge quantization, which refers to the fact that the magnitudes of the proton and of the electron electric charges are the same, is not explained and must be put into the standard model by hand.

b) Fermion Problem

The standard model involves a very complicated reducible representation for the fermions. Ordinary matter can be constructed out of the fermions of the first family (ν_e, e^-, u, d). However, in the laboratory we observe at least two additional fermion families – (ν_μ, μ^-, c, s), (ν_τ, τ^-, t, b) – which appear to be identical with the first except that the quarks and leptons are heavier. We have no fundamental understanding of why there should be these heavy families or how many there are. In addition, the standard model does not explain or predict the pattern of fermion masses, which are observed to vary over five orders of magnitude, and it does not explain the various mixings.

c) Higgs/Hierarchy Problem

The next problem is a fine-tuning problem. The spontaneous breakdown of the $SU_2 \times U_1$ symmetry in the standard model is accomplished by the introduction of a Higgs field. Consistency requires that the mass of the Higgs be not too much different from the weak interaction scale; that is, it should be equal to the W mass within one or two orders of magnitude. The problem is that there are higher order corrections involving loop effects (or even tree-level corrections in grand unified theories) which change (renormalize) the value of the square of the Higgs mass by $\delta m_H^2 / M_W^2 \gtrsim 10^{24}$–$10^{34}$. Therefore the bare value of m_H^2 in the original Lagrangian must be adjusted or fine-tuned to 24 to 34 decimal places to achieve a satisfactory theory. Such a fine-tuning is possible, but extremely unattractive.

d) Strong CP Problem

Another fine-tuning problem is the strong CP problem. It is possible to add an additional term $(\theta/32\pi^2)g_s^2 F\tilde{F}$ to the QCD Lagrangian. This term breaks P, T, and CP invariance. Limits on the electric dipole moment of the neutron require θ to be less than 10^{-9}. However, weak interaction corrections change or renormalize the lowest order value of θ by about 10^{-3} - that is, 10^6 times more than the total value is allowed to be. Again, one must fine-tune the bare value against the correction to a high degree of precision.

e) Graviton Problem

Finally, I come to the graviton problem. This really has several aspects. First of all, gravity is not unified with the other interactions in the standard model in a fundamental way. Secondly, even though general relativity can be incorporated into the model by hand, we have no idea how to achieve a mathematically consistent theory of quantum gravity: attempts to quantize gravity within the standard model framework lead to horrible divergences and a non-renormalizable theory.

Finally there is yet another fine-tuning problem associated with the cosmological constant. The vacuum energy density associated with the spontaneous symmetry breaking of the $SU_2 \times U_1$ model generates an effective renormalization of the cosmological constant $\delta\Lambda = 8\pi G_N <V>$, which is about 50 orders of magnitude larger than the observed value. One must fine-tune the bare cosmological constant against the correction to this incredible degree of precision.

EXTENSIONS OF THE STANDARD MODEL

It is therefore almost certainly true that there must be new physics beyond the standard model. Some of the possible types of new physics that have been discussed extensively in recent years are shown in Table 2. I do not have time to discuss all of these, so I will concentrate on one line of development which I find particularly promising, namely grand unification, supersymmetry/supergravity, and superstrings.

GRAND UNIFIED THEORIES (GUTs)[5]

In the standard model the apparent symmetry of Nature at energies less than 100 GeV is $SU_3 \times U_1^{elm}$ (as well as classical gravity). At the electroweak unification scale ($\simeq$ 100 GeV) one sees a larger symmetry group, namely $SU_3 \times SU_2 \times U_1$ combined with classical gravity. In the standard model this symmetry continues up to the Planck scake (10^{19} GeV), at which point we know that the theory breaks down because quantum gravitational effects become important. In grand unified theories one embeds the $SU_3 \times SU_2 \times U_1$ standard model into a larger group, which is not a direct product of factors, such as SU_5, SO_{10} or E_6. There is therefore a single underlying gauge coupling constant and a single underlying interaction. In a very real sense all of the

Extension	Typical Scale (GeV)	Motivation	Comments
extra W's, Z's extra Fermions extra Higgs	10^2-10^{19}	–	Generally remnants of something else
Family Symmetry Composite Fermions	10^2-10^{19}	Fermion problem	No compelling models
Composite Higgs Composite W, Z	10^2-10^3	Higgs problem	No compelling models
Grand Unification	10^{14}-10^{19}	Gauge problem	strong↔electroweak
Supersymmetry/ Supergravity	10^2-10^4	Higgs, graviton problems	Fermion↔boson
Kaluza-Klein	10^{19}	Graviton	(spin-0)↔(spin-1)↔ ↔(spin-2) extra dimensions
Superstrings	10^{19}	all?	Gauge + Gravity Fermion↔boson extra dimensions

Table 2. Extensions of the standard model.

interactions are unified: they are merely different aspects of this larger underlying symmetry group. The larger group has additional generators and corresponding gauge bosons, which acquire a large mass of order of the grand unification scale. If the theory is observed or probed at an energy that is much higher than that scale, one sees the full symmetry; all interactions look alike and there is only one observed gauge coupling. However, at energies much lower than the unification scale the strong and electroweak interactions look very different because of symmetry breaking effects, and the effective (momentum-dependent) coupling constants split apart. Therefore we observe completely different features and strengths for the strong and electroweak interactions.

In the simplest grand unified theories one can predict the unification scale in terms of the observed coupling constants at low energy, using the theoretical expression for the momentum dependence of the effective couplings. From the ratio of electromagnetic to strong interaction couplings one predicts the unification scale $M_X \sim 1.3 \times 10^{14 \pm .5}$ GeV, where the uncertainty is mainly due to the fact that it is hard to determine the strong coupling constant precisely. One can also use the fact that there are three couplings which come together at M_X to predict the weak angle, which is defined as a ratio of coupling constants. In the simplest models one predicts $\sin^2\theta_W = .218 \pm .007$, which is remarkable agreement with the experimental number $.223 \pm .004 \pm .005$. This agreement strongly suggests that grand unified theories may indeed have something to do with the real world. However, it is possible that the agreement is fortuitous.

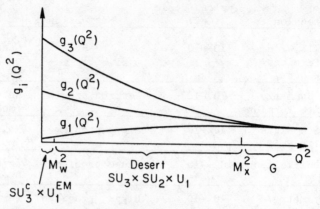

Fig. 7. The momentum dependent coupling constants $g_3 \equiv g_s$, $g_2 \equiv g$, and $g_1 \equiv \sqrt{5/3}\ g'$.

In addition to unifying the interactions, the additional symmetries within a grand unified theory relate quarks, antiquarks, leptons and antileptons. For example, the simplest grand unified theory - the SU_5 model of Georgi and Glashow - assigns the fermions of the first family to a reducible $5^* + 10$ dimensional representation:

$$W^\pm \updownarrow \begin{pmatrix} \nu_e & & \\ & & \bar{d} \\ e^- & & \end{pmatrix}_L \quad \begin{pmatrix} & u & \\ e^+ & & \bar{u} \\ & d & \end{pmatrix}_L$$

$$\overset{5^*}{\leftarrow X,Y \rightarrow} \qquad \overset{10}{\overset{\leftrightarrow \quad \leftrightarrow}{X,Y}}$$

The neutrino and electron can be rotated into the anti-down quark by the new symmetry generators, as can the positron, the up and down and the anti-up quark. Because of this relation between the different types of fermions, the grand unified theories naturally explain charge quantization; that is, quark and lepton electric charges are related, and the fact that the electron and the proton have the same magnitude of electric charge is predicted by the theory. In addition, there are new gauge bosons associated with these extra symmetries. The SU_5 model contains new bosons which carry both color and electric charge, known as the X and Y bosons, with masses $M_X \sim M_Y \sim 1.3 \times 10^{14}$ GeV. These can mediate proton decay. For example, a diagram for $p \to e^+ \pi^0$ is shown in Figure 8. The proton lifetime can be estimated in the simplest SU_5 model: one has for the partial lifetime into $e^+\pi^0$ the expression $\tau_{p \to e^+\pi^0} \simeq 6.6 \times 10^{28 \pm 0.7} (M_X/1.3 \times 10^{14} \text{ GeV})^4$ yr, where the

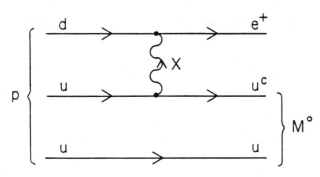

Fig. 8. A typical proton decay diagram.

uncertainty in the coefficient is due to the difficulty of computing the hadronic matrix elements of the quark operators derived from the diagram. For the expected range of unification scale, this implies $\tau_{p \to e^+ \pi^0} < 8.4 \times 10^{30}$ yr. However, the uncertainties in the calculation both of the coefficient and unification scale are large and difficult to quantify. It is therefore useful to take a more conservative limit $\tau_{p \to e^+ \pi^0} < 1.4 \times 10^{32}$ yr for the partial lifetime in SU(5) and similar models.

A number of experiments have been searching for proton decay for some time. There is no definitive evidence for proton decay and at present the lower limit is $\tau_{p \to e^+ \pi^0} > 3.3 \times 10^{32}$ yr. Therefore SU(5) and similar unified theories are excluded. However, grand unification itself is alive and well. Many variations and extensions of the simplest theories give longer lifetimes. For example, some models invoke extra gauge boson, fermion, or scalar thresholds; the new particles can lead to a larger M_X and a longer lifetime. Supersymmetric grand unified theories, for example, automatically lead to a larger unification scale. Some models invoke large mixing effects, which can suppress the decay amplitude, and some theories predict the existence of higher dimensional effective operators which have the ultimate effect of raising the unification scale. It should be cautioned, however, that many of the models which give a longer proton lifetime also lose much of their predictive powers for the weak angle.

To summarize the status of grand unified theories:
a) Grand unified theories solve the gauge problem. The strong, weak and electromagnetic interactions are unified. Charge quantization is explained naturally and $\sin^2\theta_W$ is correctly predicted, at least in the simplest theories.
b) Another attractive feature of grand unified theories is that they can naturally explain the cosmological baryon asymmetry. It is an observational fact that our part of the Universe consists of matter and not antimatter: there is approximately one baryon for every 10^{10} microwave photons, but essentially no anti-baryons. In grand unified theories, this asymmetry does not need to be invoked as an initial condition on the Big Bang; rather, it can be generated dynamically in the first instant (the first 10^{-35} sec. or so) after the Big Bang, by baryon number violating interactions related to those which lead to proton decay.
c) On the other hand, grand unified theories give no help with the fermion, strong CP, Higgs (the Higgs problem is even aggravated), or

graviton problems.

d) The non-observation of proton decay rules out the simplest grand unified theories, but many more complicated schemes are still possible.

e) There are other implications of grand unified theories as well. For example, all grand unified theories predict the existence of superheavy magnetic monopole states with masses $M_M \sim M_X/\alpha_G \sim 10^{16}$ GeV. These may have been produced during phase transitions during the early Universe and could still be left over as relics today. However, we have no firm prediction as to what the flux of these monopoles should be. Early estimates suggested that far too many of these monopoles should be left over and that they should contribute much more to the energy density of the present Universe than is allowed by observation. This monopole problem motivated the development of the inflationary cosmology. If the Universe did undergo a period of inflation subsequent to the grand unified period, there might be very low numbers of relic monopoles. Anything in between is also possible.

f) Another implication of the simplest grand unified theories is that they often predict the b quark mass in terms of the observed τ^- mass. However, the models which correctly predict m_b almost always have trouble with similar predictions for the d and s masses. However, small perturbations on such theories could solve the latter problem without significantly affecting the m_b prediction.

g) Finally, some grand unified theories (and even some analogous theories which are not truly grand unified) allow the oscillation of neutrons into anti-neutrons. The interactions which mediate such neutron to anti-neutron oscillations are generally driven by extra Higgs bosons in the theory.

In some ways then, the simplest grand unified theories are probably too naive. They have some successes, but there are many problems that they too leave unanswered. It is likely that some aspects of grand unification are correct, but they are almost certainly not the final story.

SUPERSYMMETRY (SUSY)[6]

Supersymmetry, which can be implemented either with or without grand unification, is motivated by the Higgs problem and by the graviton problem. It involves a new type of symmetry in which fermions are related to bosons. In order to solve the Higgs problem, this new symmetry must be broken at a scale not too much larger than the weak scale - that is between 10^2 and 10^4 GeV.

The primary motivation for supersymmetry is the Higgs problem. To generate the weak interaction scale by the Higgs mechanism, one needs to introduce a scalar field, the quantum of which (the Higgs boson) has a mass not too much larger than the weak scale. However, loop diagrams associated with Higgs fields, gauge fields, or fermions yield quadratically divergent renormalizations of the Higgs mass. One finds $\delta m_H^2 = 0 \ (\lambda, \ g^2, \ h^2) \Lambda^2$, where the cutoff Λ is interpreted as the Planck scale $G_N^{-\frac{1}{2}} \simeq 10^{19}$ GeV or as the grand unification scale $M_X \simeq 10^{14}$ GeV.

In the simplest type of theory with exact supersymmetry, fermions and bosons occur in degenerate pairs. For example, the spin-1/2 quark

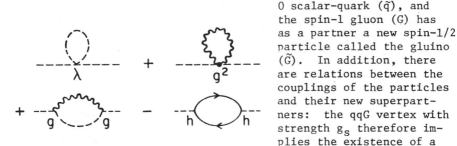

Fig. 9. Diagrams contributing to the renormalization of the Higgs mass.

(q) has a partner, a spin-0 scalar-quark ($\tilde{q}$), and the spin-1 gluon (G) has as a partner a new spin-1/2 particle called the gluino ($\tilde{G}$). In addition, there are relations between the couplings of the particles and their new superpartners: the qqG vertex with strength g_s therefore implies the existence of a $\tilde{q}\tilde{q}G$ vertex with the same strength. If the supersymmetry is exact, then there is a cancellation between the boson loops and the fermion loops, which occur with opposite sign, and the renormalization of the Higgs mass becomes zero.

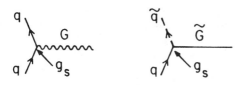

Fig. 10. Ordinary and new vertices.

The real world is not exactly supersymmetric. We do not have bosons and fermions in degenerate pairs. However, if there is an approximate supersymmetry which is broken by mass terms (and/or by cubic scalar interactions) then the renormalization effects remain finite though nonzero, yielding $\delta m_H^2 = O(\lambda, g^2, h^2) |m_B^2 - m_F^2|$. Therefore, as long as the splitting between bosons and fermions is less than about 10 TeV, the Higgs problem is at least partially solved. There will be no unacceptable large renormalization of the Higgs mass and no large fine tunings will be necessary. I should emphasize that the origin of the weak scale and why it is so much smaller than the Planck and GUT scales is not explained in the simplest supersymmetric models. It is merely stabilized with respect to higher order corrections.

IMPLICATIONS OF SUPERSYMMETRY

The major prediction of supersymmetric theories is the existence of a large number of new particles. For various reasons, no known fermion-boson pairs can be identified as superpartners of each other. Therefore, it is necessary to assume a doubling of the spectrum; for every known fermion one must introduce a new boson and vice versa, as shown in Table 3. The new particles, denoted with a tilde, are generally given funny names. For example, associated with the W and Z there are new fermionic partners known as the wino and zino, respectively. Similarly, the photon and gluon have partners known as the photino and the gluino. The quarks and leptons have new spin-0 partners, known as scalar-quarks (or squarks) and scalar-leptons (or sleptons), and the Higgs (there must be at least two Higgs doublets in

J = 1	J = 1/2	J = 0
$W^{\pm}$	$\tilde{w}^{\pm}$ (wino)	
Z	$\tilde{z}$ (zino)	
γ	$\tilde{\gamma}$ (photino)	
G (gluon)	$\tilde{G}$ (gluino)	
	q_L, q_R (quarks)	$\tilde{q}_L, \tilde{q}_R$ (squarks)
	ℓ_L, ℓ_R (leptons)	$\tilde{\ell}_L, \tilde{\ell}_R$ (sleptons)
	$\tilde{H}_1, \tilde{H}_2$ (Higgsinos)	H_1, H_2 (Higgs)

Table 3. The low energy spectrum in the minimal supersymmetric model.

a supersymmetric model) have fermionic partners known as Higgsinos. The masses of these particles are very model dependent. All that we really know is that the masses should be larger than existing limits and less than 1-10 TeV. The photino, the zino, and the neutral Higgsinos are fermions which interact very weakly - very much like neutrinos. The signature of supersymmetric theories is therefore unobserved momentum. This is associated with the escape from the detector of the lightest supersymmetric particle (e.g. the photino), which is typically produced in the decays of heavier superpartners such as the gluino or scalar quarks. The existing limits from the nonobservation of such missing momentum events at the CERN collider are $m_{\tilde{q}}$, $m_{\tilde{G}} \gtrsim 40$ GeV. The limits on the charged scalar leptons, the winos, and the charged Higgsinos are around 20 GeV because they have not been observed at the e^+e^- colliders PEP or PETRA. The limits on the photino, zino, and neutral Higgsinos are very weak; they could easily be as light as a GeV or so. In addition to the limits from direct searches at accelerators, there are a number of indirect limits on the new particles due to the fact that higher order diagrams involving these particles could lead to CP violation, to large parity violating effects, or to rare decays, such as $\mu \to e\gamma$.

SUSY - GUTs

One can also consider supersymmetric grand unified theories. In such theories one has supersymmetric partners not only for the ordinary light particles, but also for the superheavy gauge and Higgs bosons introduced in GUTs. An important feature of supersymmetric GUTs are that all of the new light superparticles, that is the scalar quarks, scalar leptons, Higgsinos, gauginos, etc., modify the theoretical predictions for the momentum dependence of the effective coupling constants and therefore for $\sin^2\theta_W$ and M_X. The $\sin^2\theta_W$ is about .02 larger than in standard GUTs, but this is still probably within an experimentally acceptable range. The unification scale is about 30 times larger than in the standard grand unified theory. The proton

	Ordinary Gut	SUSY-GUT	Experiment
M_X (GeV)	1.3×10^{14}	5×10^{15}	-
$\sin^2\theta_W$	~0.22	~0.24	$0.223 \pm 0.004 \pm 0.005$
$\tau_{p \to e^+ \pi^0}$ (yr)	$6.6 \times 10^{28 \pm 2}$	10^{35}	$\geq 3 \times 10^{32}$

Table 4. Predictions for M_X, $\sin^2\theta_W$. and $\tau_{p \to e^+\pi^0}$ (via X, Y exchange) in ordinary and supersymmetric GUTs.

lifetime due to ordinary heavy gauge boson exchange is therefore increased by a factor of approximately 10^6 to an observable rate of about 10^{35} yr.

However, there are new proton decay mechanisms (so-called dimension-5 operators) which can lead to much more rapid proton decays. For example, a typical diagram involves the exchange of the fermionic partner of a superheavy color Higgs scalar. This diagram leads to a proton lifetime that is proportional to the square of the unification scale and not to the fourth power as in traditional GUTs. There are many suppression factors, but nevertheless in many supersymmetric GUTs the proton lifetime is predicted to be in the range 10^{26} to 10^{31} yrs. The preferred decay mode in most such models is $p \to \bar{\nu}K$, rather than into $e^+\pi^0$. Observational limits exclude proton lifetimes into this mode in this range, so many supersymmetric GUTs are in trouble. However, there are many unknown masses and parameters in the prediction, so one should keep pushing the proton decay experiments into these and other modes as far as possible.

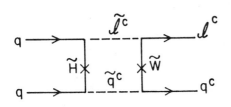

Fig. 11. A diagram leading to a dimension-5 operator.

SUPERGRAVITY

So far I have been describing global supersymmetry in which the symmetry transformations between fermions and bosons are the same at all points in space and time. In complete analogy to gauge theories, in which symmetry transformations are allowed to take place independently at different space-time points, one can also consider local (gauge) supersymmetry in which the fermion-boson transformations can be performed independently at different space-time points. Such theories, known as supergravity theories, necessarily require the introduction of a spin-3/2 particle, called the gravitino, which is analogous to the spin-1 boson that is needed for a local gauge symmetry. In addition, the gravitino has a spin-2 partner, the graviton, which generates gravity. Supergravity therefore incorporates gravity in a non-trivial way. However, it does not solve the problem of quantum gravity; supergravity theory is still non-renormalizable. The spontaneous breaking of supergravity generates a mass for the gravitino in much the same way that gauge bosons acquire a mass when an ordinary gauge symmetry is spontaneously broken.

SUSY BREAKING

We have seen that supersymmetry must be broken, with the typical breaking scale somewhere in the range from 10^2 to 10^4 GeV. The simplest possibility would be to simply introduce some symmetry breaking mechanism in this mass range. However, for a variety of reasons all such schemes fail. The so-called hidden sector models lead to an acceptable theory, but they are rather complicated. In these theories one introduces a set of particles which are almost totally decoupled from the ordinary particles and their superpartners. Supersymmetry breaking presumably occurs in this hidden sector and is transmitted only indirectly via weak couplings to the ordinary particles and their superpartners. This weak coupling may be via gravity or it may be via very high order gauge interactions. If the scale of symmetry breaking is F in the hidden sector, then the result is that the gravitino and all scalars in the ordinary sector acquire a mass of order $m_{3/2} \sim F^2/m_p \ll F$. Since the Higgs scalar has a positive mass^2 of this value, $SU_2 \times U_1$ is not spontaneously broken at tree level. However, large Yukawa couplings associated with the top quark mass may lead to $SU_2 \times U_1$ breaking via radiative corrections in such models. The upshot is that the electroweak and effective supersymmetry breaking scales are naturally of the same order of magnitude, namely the gravitino mass. Since this magnitude must be around 10^2 to 10^4 GeV, one has that the hidden sector SUSY breaking scale is of order 10^{11} GeV.

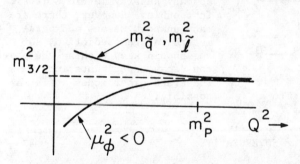

Fig. 12. Running scalar masses, leading to radiative $SU_2 \times U_1$ breaking in hidden sector models.

SUSY SUMMARY

a) Supersymmetry is a very elegant idea.
b) Supersymmetry helps to solve the Higgs Problem. It stabilizes the electroweak scale against unacceptably large radiative corrections as long as the effective breaking scale is less than about 10 TeV.
c) Supersymmetry does not help to solve the fermion problem.
d) Gravity is incorporated into supergravity theories, but quantum

gravity is still nonrenormalizable.
e) There is not a shred of experimental evidence for supersymmetry. However, since the breaking scale cannot be arbitrarily large (and still solve the Higgs problem) it should be possible to produce and detect supersymmetric particles at new high energy supercolliders.
f) Phenomenologically viable supersymmetric models are extremely complicated.

SUPERSTRINGS[7]

Superstrings are a very exciting development which may yield finite theories of all interactions with no free parameters. Superstring theories introduce new structure above the Planck scale. The basic idea is that one changes the fundamental entities in a superstring theory. Instead of working with (zero-dimensional) point particles as the basic quantities, one considers one-dimensional objects known as strings, which may be open or closed. The quantized vibrations of these strings lead to an infinite set of states, the spectrum of which is controlled by the string tension.

Strings were introduced in the early 1970's as a model of the strong interactions and of the hadronic states. However, the strong interaction strings never quite worked out, and since then they have been replaced by QCD as the most likely theory of the strong interactions. More recently, strings have re-emerged with the new interpretation that they are a theory of everything, not just the strong interactions. The string tension is now given by the square of the Planck scale $T \sim m_P^2$. When one probes or observes a string at energies much less than the Planck scale, one sees only the "massless" modes and these represent ordinary particles. The physical size of the string is given by the inverse of the Planck scale $\simeq 10^{-33}$cm, and at larger scales a string looks like a point particle.

Fig. 13. Open and closed strings.

There are various types of string theories. The most promising is the heterotic string, which is a theory with closed strings only in which the gauge symmetry charge densities are distributed along the length of the string. The basic interaction in the closed string theory involves the joining of two strings to form one or the breaking of one closed string into two. This should be compared to the three point vertex in ordinary field theory. The various string configurations correspond to ordinary matter particles such as fermions or their scalar particles, to gauge bosons and their partners, and to the graviton and the gravitino. String theories therefore incorporate both gauge interactions and gravity - they are unified. The enormously exciting development is that string theories apparently lead to completely finite theories of quantum gravity and of the other interactions; not just renormalizable but completely finite. No known ordinary field theory leads to a mathematically consistent, renormalizable theory of quantum gravity, let alone to a finite theory.

Fig. 14. Interactions in string and field theories.

An interpretation is that the string represents an infinite number of physical states and in field theory language one has an infinite set of diagrams and exchanges involving this hierarchy of states. The couplings and masses of these states are constrained in such a way that the divergences cancel, leading to a completely finite theory.

There are two important constraints on the string theory. The first is that the string theories are only consistent in ten dimensions of space and time, that is nine space and one time dimension. In different dimensionalities, one has unacceptable negative norm (ghost) states. The fact that we apparently live in four and not ten dimensions can be accounted for by the assumption that six of the dimensions compactify into a small manifold. That is, six of the dimensions curl up to a small closed manifold of radius $r_{comp} = 1/m_{comp} \simeq 10^{-32\pm1}$ cm. This idea of compactification was borrowed from Kaluza-Klein theories, which are an alternative class of models that also gave some promise of unifying gravity (but which have run into serious problems). Another constraint is that string theories tend to be plagued with anomalies, which are pathologies or inconsistencies which can occur in a theory, leading to breaking of gauge and other symmetries that are needed for the consistency of the theory. In four-dimensional field theories, anomalies are generated by certain triangle diagrams, and one generally insists that the fermion content of a theory be such that the anomalies cancel. In a ten-dimensional string theory there are hexagon-type diagrams which lead to unacceptable anomalies. However, the anomalies cancel if the underlying gauge group is either $G = E_8 \times E_8$ or SO_{32}. The SO_{32} theories are not physically realistic, so the underlying gauge symmetry of a superstring theory is constrained to be the group $E_8 \times E_8$.

The scenario that one expects is the following: At high energies (large compared to the string tension or Planck scale) the theory of Nature would be an $E_8 \times E_8$ ten-dimensional string theory. However, six of the dimensions compactify into a small manifold at the compactification scale $m_{comp} \sim 10^{18\pm1}$ GeV. At energies much lower than m_{comp} one sees an effective four-dimensional particle theory. It is possible and expected that some of the gauge flux of the $E_8 \times E_8$ theory becomes trapped in the manifold so that one sees an effective gauge symmetry smaller than the original $E_8 \times E_8$ group. In order to have a chiral low energy supersymmetry below the compactification scale, it is necessary that the extra six dimensions compactify into what is known as a Calabi-Yau manifold. In such theories the low energy gauge group is either $E_6 \times E_8$ or a subgroup of $E_6 \times E_8$. The E_8 factor corresponds to a hidden sector, while the E_6 factor (or the subgroup which survives the compactification) is associated with the ordinary particles. Because the E_6 is broken to a subgroup such as $SU_3 \times SU_2 \times U_1 \times G'$, the compactification scale plays the role of the GUT unification scale; compactification is a new method of spontaneous symmetry breaking that does not rely on an ordinary Higgs mechanism. Consistency of the entire approach requires $M_X \sim m_{comp} \sim 10^{18 \pm 1}$ GeV.

One of the most exciting features of the superstring is that essentially all low energy physics, that is all physics below the compactification scale, is in principle determined by the topology of the Calabi-Yau manifold. The nature of the low energy gauge group, the number of fermion families, and even the Yukawa couplings (and therefore the fermion masses and mixings) are in principle not free parameters but are determined by the topology of the compactification manifold. Unfortunately, there are many (10^4) possible Calabi-Yau manifolds. At present we are not capable of calculating which manifold is chosen or even why it is a Calabi-Yau manifold, why six dimensions are compactified, or what the compactification scale is. Therefore although everything is determined in principle, we are far from able to do the calculations.

CONSEQUENCES OF STRINGS

a) Proton decay is probably unobservable in superstring theories. This is because the basic unification scale is typically of order 10^{18} GeV so that gauge boson mediated decays are unobservably slow. The necessary Yukawa interactions can in principle be calculated in terms of topological considerations, but we are unable to do that at present. Models typically have proton decays via Higgs that are either much too fast, in which case those models are excluded experimentally, or much too slow to observe. However, it is always possible that in some models proton decay may occur at an observable rate. One should keep looking.

b) Baryon Asymmetry. It is simply too early to make meaningful statements about such cosmological implications as the baryon asymmetry.

c) Most Calabi-Yau manifolds imply the existence of extra Z bosons (i.e., the symmetry group below the GUT scale generally contains an extra factor G'. This is typically U_1 or $U_1 \times U_1$, although there are other possibilities. These may be very light; e.g., in the range from 10^2 to 10^3 GeV. Present experimental limits are very weak, typically allowing extra Z's as light as 100 to 200 GeV. One should certainly pursue the search for such bosons vigorously.

d) Matter fields. Most manifolds imply the existence of either three or four families of fermions and their scalar partners. (The simplest manifolds imply far too many families and are therefore excluded.) The fermions and their partners occur in 27-dimensional representations of the E_6 group. Therefore, it is predicted that for each family there should be the ordinary fermions $\begin{pmatrix} u \\ d \end{pmatrix}_L \bar{u}_L \bar{d}_L \begin{pmatrix} \nu_e \\ e^- \end{pmatrix}_L e^+_L$

(and their CPT conjugates) as well as a class of exotic new fermions $D_L, \bar{D}_L \begin{pmatrix} E^+ \\ E^0 \end{pmatrix}_L \begin{pmatrix} \bar{E}^0 \\ E^- \end{pmatrix}_L N_{1L}, N_{2L}$. These exotic new fermions and their scalar partners may be in the hundred GeV range, or possibly much more massive.

e) $\sin^2\theta_W$. The predictions for the weak angle in superstring theories are similar to those in other supersymmetric GUTs, except that one must take the exotic particles into consideration. The correct value $\sin^2\theta_W$ is therefore a constraint on model building and on assumptions as to whether there are intermediate thresholds in the desert, at which the extra Z's or exotic fermions could possibly acquire masses.

f) Neutrino mass. The neutrino mass turns out to be too large in most superstring theories. It is typically comparable to the charged lepton mass. There are possible ways of evading this problem, but nothing particularly elegant or attractive has emerged so far.

g) SUSY breaking. The origin of supersymmetry breaking in the superstring model is somewhat obscure. It is thought that a gluino condensate in the hidden E_8 sector may lead to the original supersymmetry breaking. However, the details of how this is transmitted to the observed sector are somewhat obscure and it seems that the superstring models do not directly imitate the superstring breaking schemes that have been previously worked out for non-string theories.

STRING SUMMARY

a) Superstrings are very elegant.

b) They yield a completely finite theory of gravity and of all other interactions with no free parameters; therefore, the entire structure of physics below the compactification scale is, in principle, determined.

c) However, the details of the low energy physics depend on the topology of the manifold. We are not, at present, able to calculate what the manifold is, the number of compactified dimensions, the compactification scale, etc. Therefore we are unable at present to predict the structure of the low energy world.

d) If the superstring idea is correct, it is likely that superstring models will incorporate some, but by no means all, aspects of the standard supergravity-GUT picture.

SUMMARY

a) The standard model of the strong, weak, and electromagnetic interactions is almost certainly approximately correct up to around 100 GeV.
b) However, the standard model with classical gravity involves at least 21 free parameters and is plagued by many unexplained features,

such as the gauge, fermion, Higgs/hierarchy, strong CP, and graviton problems. Therefore new underlying physics is necessary.
C) Grand Unified Theories solve some, but not all of the problems of the standard model. They elegantly explain the gauge problem and charge quantization. However, proton decay limits exclude the simplest GUTs. Some aspects of grand unification are probably true, but it is unlikely that a simple grand unified theory is the final story either.
d) Supersymmetry and supergravity models stabilize the weak scale and therefore solve much of the Higgs problem, but they do not in general explain it in the first place. Supergravity theories incorporate gravity, but they are still nonrenormalizable. Supersymmetric models do not help with the fermion problem.
e) Superstring models are conceivably the ultimate theory of everything. They yield finite theories of quantum gravity, and in principle they lead to a completely determined theory, with no arbitrariness or free parameters. However, we are very far from being able to do the calculations needed to test superstring theories or to predict the structure of the low energy world. If superstrings are correct, they probably incorporate some of the aspects of the supergravity-GUT picture, but probably not all. I don't want to give the impression that the superstring theories are obviously correct. It may well be that we are on the wrong track. They are very exciting and promising, but whether Nature chooses to use these elegant theories remains to be seen.
f) There are many other possible extensions of the standard model, such as Kaluza-Klein theories or models in which fermions, Higgs bosons, or gauge bosons are composites of still more elementary particles. It is not clear what direction physics beyond the standard model will ultimately take. In the end it will be an experimental question. In the future we expect to probe physics beyond the standard model with a new generation of accelerators and with new proton decay experiments, neutrino mass experiments, monopole searches, etc. The future promises to be very exciting indeed.

ACKNOWLEDGMENT

It is a pleasure to thank Alan Krisch and the organizers of this meeting for their invitation and travel support.

REFERENCES

1. For reviews of QCD, see W. Marciano and H. Pagels, Phys. Rep. 36, 137 (1978) and A. H. Mueller, Proc. of the 1985 Int. Symp. on Lepton and Photon Interactions at High Energies, ed. M. Konuma and K. Takahashi (Nisha, Kyoto, 1986), p. 162.
2. For reviews of the standard electroweak model, see P. Langacker, 1985 Int. Symp. on Lepton and Photon Interactions, p. 186; W. Marciano, First Aspen Winter Physics Conference, ed. M. Block, (N.Y. Acad. Sci., N.Y., 1986) p. 367.
3. M. A. Bouchiat, J. Guena, and L. Pottier, CNRS preprint.
4. S. P. Mikheyev and A. Yu Smirnov, INR preprint, Moscow (1985).

5. For reviews of GUTs, see P. Langacker, ref. 2; Comm. on Nucl. and Part. Phys. 15, 41 (1985); and Phys. Rep. 72, 185 (1981).
6. For reviews of supersymmetry, see H. E. Haber and G. L. Kane, Phys. Rep. 117, 75 (1985); H. P. Nilles, Phys. Rep. 110C, 1 (1984); S. Dawson, E. Eichten, and C. Quigg, Phys. Rev. D31, 1581 (1985); J. Ellis, 1985 Int. Symp. on Lepton and Photon Interactions, p. 850.
7. For reviews of superstrings, see M. B. Green, 1985 Int. Symp. on Lepton and Photon Interactions, p. 372; G. Segre, NATO Advanced Study Inst. on Particle Physics, Cargese, France, 1985; V.S. Kaplunovsky and C. R. Nappi, Princeton preprint.

THE EMC EFFECT

Edmond L. Berger
Argonne National Laboratory, Argonne, IL 60439

ABSTRACT

A review is presented of data and theoretical interpretations of the nuclear dependence of quark and antiquark distributions as observed in the deep inelastic scattering of neutrinos and charged leptons from nuclei. After a summary of the experimental situation and a survey of the broad spectrum of proposed explanations, I concentrate on the Q^2-rescaling approach and on interpretations in terms of conventional nuclear physics. The review concludes with a list of desirable future experiments.

I. INTRODUCTION

The European Muon Collaboration (EMC) Effect is the observation[1-4] that deep-inelastic structure functions of nuclei, $F_2^A(x,Q^2)$, differ from those of free nucleons, $F_2^N(x,Q^2)$. In this paper I review[5] the data and interpretations of the effect, concluding with a summary of desirable future investigations.

I begin with a definition of the structure functions per nucleon of a nucleus of baryon number A, $F_{2,3}^A(x,Q^2)$. In Sec. III, the quark model description of $F_{2,3}^A(x,Q^2)$ is reviewed. Data are summarized in Sec. IV along with general statements regarding their interpretation. In Sec. V, I review the spectrum of models[5] proposed to explain the data, concentrating specifically on the Q^2 rescaling approach[6] and on explanations in terms of nuclear binding.[7-9] Massive lepton-pair production -- the Drell-Yan process -- is discussed in Sec. VI for two reasons.[10] First, it is important to establish the universality of the nuclear dependence of the quark and antiquark densities. Second, the reaction $pA \to (\mu\bar{\mu})X$ is an especially valuable source of information on the antiquark density of nuclei, $\bar{q}^A(x,Q^2)$.

One of my conclusions, expressed in Sec. VII, is that current data on the ratios $F_2^A(x,Q^2)/F_2^D(x,Q^2)$ and $\bar{q}^A(x,Q^2)/\bar{q}^D(x,Q^2)$ are fully compatible with an interpretation in terms of conventional nuclear binding. There is no hard evidence in these data for new quantum chromodynamics (QCD) effects such as a change in the confinement size of nucleons. However, more precise data on $F_2^A(x,Q^2)$ and on $\bar{q}^A(x,Q^2)$, especially at small x, $x \lesssim 0.3$, are of critical importance for future quantitative tests of all interpretations.

II. KINEMATICS AND DEFINITIONS

For $\ell A \to \ell' X$, q is the four vector momentum transfer from the initial lepton ℓ to the final lepton ℓ', and ν is the energy transfer measured in the reference frame in which target A with mass M_A is at rest. Symbol X represents an inclusive sum over all final states. It is conventional to define $Q^2 = -q^2$, and two scaling variables x and y; $x = Q^2A/2M_A\nu = (Q^2/2M_N\nu) \cdot (AM_N/M_A)$ and $y = \nu/E$; M_N is the nucleon mass, and E is the laboratory energy of the initial lepton. The deep inelastic domain is that in which ν and Q^2 are both "large" (i.e. greater than a few GeV^2). For scattering from a free nucleon, x is bounded by 1, but for scattering from a nucleus, the range of x extends above 1.

In $\ell A \to \ell'X$, with ℓ = e, μ, or ν, the structure of the lepton vertex is well known, and one particle (photon or vector boson) exchange is a good approximation. The deep inelastic data determine the total current-nucleus cross sections: $\sigma(\gamma^*A)$, $\sigma(W^\pm A)$,.... This physics is usually expressed in terms of dimensionless structure functions, designated $F_i(x,Q^2)$ in the spin averaged case.[11] I use structure functions <u>per nucleon,</u> meaning that the cross section has been divided by A. For $E \gg M_N$,

$$\mu A \to \mu'X \qquad (1)$$

$$\frac{d^2\sigma^{\mu A}}{dxdy} = \frac{4\pi\alpha^2}{Q^4} \frac{2M_A E}{A} \left[F_2^{\mu A}(x,Q^2)(1-y) + 2xF_1^{\mu A}(x,Q^2)\left(\frac{y^2}{2}\right) \right] .$$

$$\binom{\nu}{\bar{\nu}} A \to \binom{\mu^-}{\mu^+} X \qquad (2)$$

$$\frac{d^2\sigma^{\nu A}}{dxdy} = \frac{G^2 M_A E}{\pi A} \left[F_2^{\nu A}(x,Q^2)(1-y) + 2xF_1^{\nu A}(x,Q^2)\left(\frac{y^2}{2}\right) \pm xF_3^{\nu A}(x,Q^2)y(1-y/2) \right] .$$

One may also replace the set (F_1, F_2) by the set (F_2, R), where $R \equiv \sigma_L/\sigma_T \equiv (F_2 - 2xF_1)/2xF_1$; σ_T and σ_L are virtual γ^* nucleon cross sections for absorption of transversely and longitudinally polarized virtual γ^*'s. Because data[12] show that $R \approx 0$, within, say, $\pm 10\%$, one practical advantage of this replacement is a simplification of Eqs. (1) and (2). If $R \equiv 0$ is inserted in the equations, one may use data on $\mu A \to \mu'X$ to determine $F_2^{\mu A}$ directly, and use data on $\nu A \to \mu^- X$ and $\bar{\nu}A \to \mu^+ X$ to isolate $F_2^{\nu A}$ and $F_3^{\nu A}$. The current data sample[13] extends in Q^2 up to about 200 GeV^2.

III. QUARK MODEL DESCRIPTION

The fact that $F_2(x,Q^2)$ is nearly independent of Q^2 (approximate scaling) led to the hypothesis that deep inelastic scattering occurs at large Q^2 from pointlike constituents of hadrons, called partons.[14] A number of crucial properties of partons and parton densities may be determined directly from deep inelastic data.[11] Those which refer to parton properties such as spin and fractional charge confirm the identification of the charged partons as the quarks and antiquarks of the spectroscopic quark model.

We may define probabilities $q_f^A(z,Q^2)$, $\bar{q}_f^A(z,Q^2)$ and $G^A(z,Q^2)$ which represent the quark, antiquark, and gluon number densities in a nucleus, A. These are densities per nucleon, just as is $F_{2,3}^A(x,Q^2)$, meaning that a factor of A has been removed. (These densities "per nucleon" should not be assumed to be the parton densities of "nucleons within nuclei".) Subscript f on q_f and $\bar{q}_f$ labels the flavor of the quark or antiquark: u,d,s,c,b,t.

The light front fraction of the momentum of the target A carried by a parton of type i is $z \equiv A(E + p_L)_i/(E + p_L)_A$, where E and p_L are the energy and the longitudinal component of momentum. The longitudinal direction is the direction of the momentum of the exchanged boson.

In the parton model, z, $q_f^A(z,Q^2)$, $\bar{q}_f^A(z,Q^2)$, and $G^A(x,Q^2)$ are measurable quantities. Indeed, $z \equiv x$, with x determined from the lepton kinematics, as defined above. Furthermore,

$$F_2^{\mu A}(x,Q^2) = \sum_f e_f^2 x [q_f^A(x,Q^2) + \bar{q}_f^A(x,Q^2)] . \qquad (3)$$

Thus, the observable $F_2^{\mu A}$ measures the fraction of momentum of the target A, per nucleon, carried by quarks and antiquarks, weighted by the squares of the fractional quark charges e_f.

Invoking isospin symmetry, one may refer quark and antiquark densities to those of the proton: e.g. for the up and down flavors, $u(x) = u_p(x) = d_n(x)$ and $d(x) = d_p(x) = u_n(x)$; subscripts p and n denote proton and neutron. For the strange quark ocean, $s_p(x) = s_n(x) = s(x) = \bar{s}(x)$. In terms of these,

$$F_2^{\mu p}(x) = x[\tfrac{4}{9}(u(x) + \bar{u}(x)) + \tfrac{1}{9}(d(x) + \bar{d}(x)) + \tfrac{2}{9} s(x)] ;$$

$$\qquad (4)$$

$$F_2^{\mu n}(x) = x[\tfrac{4}{9}(d(x) + \bar{d}(x)) + \tfrac{1}{9}(u(x) + \bar{u}(x)) + \tfrac{2}{9} s(x)] .$$

I have ignored small contributions from the charm, bottom, and top quark oceans.

For neutrino and antineutrino charged current processes ($\nu A \to \mu^- X$, $\bar{\nu} A \to \mu^+ X$) on an "isoscalar" target A containing an equal number of protons and neutrons, the structure functions per nucleon are

$$F_2^{\nu A} = x[u + d + \bar{u} + \bar{d} + 2s] ;$$
$$xF_3^{\nu A} = x[u + d - \bar{u} - \bar{d} + 2s] ; \qquad (5)$$
$$xF_3^{\bar{\nu} A} = x[u + d - \bar{u} - \bar{d} - 2s] .$$

Note that structure function F_2 provides information on the sum

$q_f(x) + \bar{q}_f(x)$, whereas F_3 yields the difference. Together, the two may be used to isolate the ocean $\bar{q}_f(x)$ from $q_f(x)$.

For each quark flavor, a decomposition may be made of the form $q(x) = q_V(x) + q_O(x)$, where subscripts V and O denote "valence" and "ocean", with $u_V(x) = u(x) - \bar{u}(x)$ and $d_V(x) = d(x) - \bar{d}(x)$. Data[15] indicate that at $Q^2 \simeq 20$ GeV^2, the valence components are dominant for $x \gtrsim 0.1$, with the ocean vanishing for $x \gtrsim 0.3$. Below $x \simeq 0.1$, the ocean quark densities are substantial. Information on the strange quark ocean is derived from heavy flavor production.

IV. NUCLEAR DEPENDENCE

Turning to the principal subject of this review, we wish to compare $q^A(x,Q^2)$ with $q^N(x,Q^2)$; $\bar{q}^A(x,Q^2)$ with $\bar{q}^N(x,Q^2)$; and $G^A(x,Q^2)$ with $G^N(x,Q^2)$. Densities with superscript N refer to those of free nucleons.

The standard prejudice before 1983 held that at large Q^2, nucleons contribute incoherently to $F_2(x,Q^2)$, at least for $0.05 \lesssim x \lesssim 0.7$. For quark densities, this statement may be expressed in the form

$$Aq^A(x,Q^2) = Zq^p(x,Q^2) + (A-Z) q^n(x,Q^2) \quad . \qquad (6)$$

At small enough x or Q^2, shadowing[16] is expected to invalidate Eq. (6). Because the photon does not resolve individual nucleons within the nucleus, the cross section cannot grow as rapidly as linearly with A. For large enough x, Fermi smearing[17] of the nucleon's momentum distribution also invalidates Eq. (6). In this case, the fact that the range of x extends beyond 1 for scattering from a nucleus means that $q^A(x,Q^2)$ remains finite for $x > 1$ whereas $q^N(x,Q^2)$ vanishes for $x > 1$.

Except for shadowing at very small x and Fermi smearing at large x, it was generally believed[18] that Eq. (6) would hold to a good approximation. Moreover, prior to 1983, experiments were usually not of sufficient statistical or systematic precision to discern more than gross deviations from Eq. (6). The first "test" was published by the European Muon Collaboration.[1] Since it is known that the x dependences of up and down quark densities differ, it would be inappropriate to compare heavy target data with hydrogen data. The EMC group compared data from iron with data from deuterium. The fact that there is a slight neutron excess in Fe leads to only a small correction. The data were published in the form of a ratio:

$$R_{EMC}(x,Q^2) \equiv \frac{F_2^{Fe}(x,Q^2)}{F_2^D(x,Q^2)} \quad . \qquad (7)$$

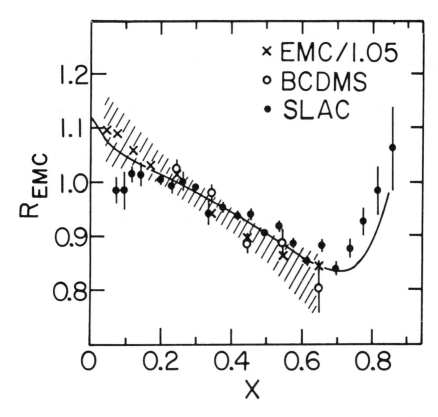

Fig. 1. A compilation of data on the ratio of structure functions $R_{EMC}(x,Q^2) \equiv F_2^{Fe}(x,Q^2)/F_2^D(Q^2)$ for deep inelastic electron and muon scattering. Shown are published results from the EMC Collaboration (Ref. 1), divided by 1.05, as well as data from the BCDMS Collaboration (Ref. 3), and from SLAC experiments (Ref. 2). The shaded band indicates the EMC group's estimate of experimental systematic uncertainties. The solid curve is calculated from the pion exchange model of Ref. 7.

A compilation is presented in Fig. 1. The data from the CERN-EMC[1] and CERN-BCDMS[3] collaborations are taken at "large Q^2" in the sense that $Q^2 \gtrsim 10$ GeV^2 for all x. On the other hand, in the SLAC-E139 data[2] sample, Q^2 descends to rather small values at small x (i.e. $Q^2 \simeq 1$ GeV^2). No marked Q^2 dependence of $R_{EMC}(x,Q^2)$ is observed within either the CERN or SLAC data samples. Nevertheless, Q^2 dependence of $R_{EMC}(x,Q^2)$ might be responsible for the discrepancy between the CERN and SLAC data in Fig. 1. Another potential resolution[5] lies in a possible dependence of σ_L/σ_T on A at low Q^2. These possibilities will be examined in future experiments.

The "naive" expectation expressed in Eq. (6) implies that

$R_{EMC}(x,Q^2) \simeq 1$, for $0.05 \lesssim x \lesssim 0.7$. The data in Fig. 1 show a ±15% deviation from this expectation. The "EMC effect" is precisely this deviation, especially the excursion above $R_{EMC} = 1$ at small x.

Although it is necessary to try to understand the x dependence in Fig. 1 for all x, hopefully in a unified fashion, it is still useful to begin with an examination of the data in three distinct regions of x:

-- $x \lesssim 0.2$ "small x". In this region the antiquark density is important, and shadowing presumably influences results in a Q^2 dependent fashion. Neutrino data[4], shown in Fig. 2, indicate that the antiquark density is increased at small x by at most a modest amount. While of limited precision, the neutrino data contradict proposed interpretations[19] in terms of a 40 to 60% increase in the ocean.

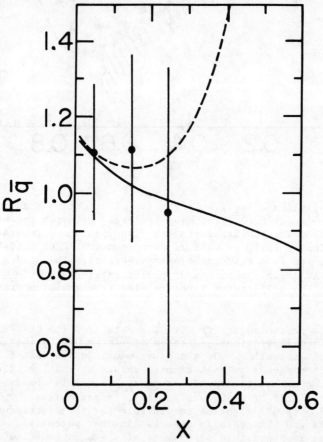

Fig. 2. Data of Abramowicz et al. (Ref. 4) on the ratio $R_{\bar{q}}(x) = \bar{q}^{Fe}(x)/\bar{q}^P(x)$ from neutrino scattering. Here $\bar{q} = (\bar{u} + \bar{d} + 2\bar{s})$. The dashed curve illustrates predictions of the pion exchange model (Ref. 7), whereas the solid curve shows expectations of the rescaling model (Ref. 6).

The greatest source of confusion about the small x region is whether one should focus on the EMC results which show a clear rise of $R_{EMC}(x)$ above unity, or on the SLAC data which shows no rise. At this meeting new EMC data were presented[20] on the ratio $F_2^{Cu}(x)/F_2^D(x)$. These data confirm that $R_{EMC}^{Cu}(x)$ exceeds unity for $0.05 < x < 0.3$. Bearing in mind that $A_{Cu} = 63$ but $A_{Fe} = 56$, one would expect a slightly larger "EMC effect" in Cu than in Fe at the same Q^2. However, the magnitude of the deviation of $R_{EMC}^{Cu}(x)$ from 1 at small x appears to be 3 to 5% smaller than that of $R_{EMC}^{Fe}(x)$. These new data lend further justification to the 5% renormalization of the original EMC data made in Fig. 1.

My working hypothesis is that there is indeed an excursion of $R_{EMC}(x)$ above unity for $0.05 < x < 0.2$ and $Q^2 \gtrsim 10$ GeV^2. The magnitude of the excursion is in the range $R_{EMC}(x) \simeq 1.05$ to 1.10. I surmise that the absence of an excess in the SLAC data is to be attributed to the fact that the SLAC data at small x are taken at very small Q^2.

-- $0.2 \lesssim x \lesssim 0.7$. In this intermediate region of x all experiments agree that $R_{EMC}(x) < 1$. The ratio $R_{EMC}(x,Q^2)$ shows no pronounced Q^2 dependence. Since the valence quarks are dominant in this region, the data show directly that the momentum distribution of the valence quark is softened. In order to maintain momentum balance, softening of the valence quark component means that, at a given Q^2, the ocean quark and/or glue components must carry more momentum per nucleon in a nucleus than in a free nucleon.

-- $x \gtrsim 0.7$. In this region of x, 70% of the momentum of the nucleon is carried by a single parton. The structure functions $F_2^A(x,Q^2)$ and $F_2^N(x,Q^2)$ are both very small, and $R_{EMC}(x)$ is the ratio of two small quantities. The behavior of R_{EMC} in this region is controlled by Fermi smearing[17] of the momentum distribution of nucleons in the nucleus and by possible collective phenomena.[21]

To recapitulate, the <u>original</u> ECM effect was defined in terms of

i. a clear, ~15%, enhancement of $R_{EMC}^{Fe}(x)$ above unity for $x \lesssim 0.2$ and $Q^2 \gtrsim 10$ GeV^2;

ii. a depletion, $R_{EMC}(x) < 1$, for $0.2 \lesssim x \lesssim 0.7$; and

iii. a possibly large, ~40 to 60%, enhancement of the antiquark ocean, $\bar{q}^A(x)$.

The data <u>now</u> present a different picture. The effect now consists of

i. a 5 to 10% enhancement of $R_{EMC}^{Fe}(x)$ above unity for $x \lesssim 0.2$ for $Q^2 \gtrsim 10$ GeV^2, but no enhancement for $Q^2 \simeq 1$ GeV^2;

ii. a depletion, $R_{EMC}(x) < 1$, for $0.2 \lesssim x \lesssim 0.7$, independent of Q^2

(for $Q^2 \gtrsim 5$ GeV^2); and

iii. a modest enhancement of $\bar{q}^A(x,Q^2)$ of on the order to 10%:
$\langle R_{\bar{q}} \rangle = 1.10 \pm 0.10 \pm 0.07$.

In attempts to model the EMC effect, various authors have emphasized phenomena in different regions of x. Some have focussed on the enhancement in $R_{EMC}(x)$ at small x, others on the depletion at intermediate x, and others even on the rise in $R_{EMC}(x)$ for $x \gtrsim 0.7$. Regrettably, it is not always made clear which "effect" is being emphasized. As indicated above, my preference is to seek a unified interpretation of the full x dependence. My prejudice is that the most essential aspect of the "effect" is the enhancement of $R_{EMC}(x)$ above unity for $x \lesssim 0.2$. If this enhancement is discredited in future experiments then I suspect most would agree that there is no "EMC effect".

V. INTERPRETATIONS

The data summarized in Fig. 1 and discussed in Section IV show that Eq. (6) is violated. It is clear that $q^A(x,Q^2) \neq q^N(x,Q^2)$. In examining explanations of this difference, we should be careful about how questions are phrased: Are we discussing

-- quark and antiquark densities of a nucleon in a nucleus, or
-- quark and antiquark densities of a nucleus, normalized per nucleon?

In other words, does the way in which the question is phrased lead us to consider that

-- the nucleon's "intrinsic" properties differ in the nuclear medium, or
-- that clustering, binding, or other ingredients in the nucleus affect what one measures?

A bias-free approach begins with the realization that the measurable $q^A(x,Q^2)$, $\bar{q}^A(x,Q^2)$, and $G^A(x,Q^2)$ are parton densities of a nucleus, not necessarily parton densities of nucleons within a nucleus.

Various approaches[5] have been developed to explain the observed nuclear dependence of $F_2^A(x,Q^2)$. In all cases, deep inelastic scattering occurs from quark and antiquark constituents. The approaches differ in the manner in which the constituents are grouped into color singlet degrees of freedom within a nucleus. Interpretations of the EMC effect run the gamut from full deconfinement[22] of partons within a nucleus, to baryon clustering[21] (e.g. dibaryons), to changes in nucleon properties such as mass or size[6,19,23], and to conventional nuclear models[7-10] in which binding effects are responsible for differences between $F_2^A(x)$ and $F_2^N(x)$. All approaches yield fits of roughly similar quality to $F_2^A(x)$. Two curves are shown in Fig. 3.

However, models differ in their predictions for the antiquark density $\bar{q}^A(x)$, as shown in Fig. 2. Experiments designed to measure $\bar{q}^A(x)$ precisely could provide substantial new insight into the short distance structure of nuclei.

In the paragraphs below, I will describe the Q^2 rescaling model[6], sometimes called the QCD approach, and the conventional nuclear physics approach.[7-10]

A. Q^2 Rescaling Model

The structure function $F_2(x,Q^2)$ is known experimentally[11] to vary slowly with Q^2. For $x \lesssim 0.1$, $F_2(x,Q^2)$ __grows__ with increasing Q^2, whereas, for $x \gtrsim 0.1$, $F_2(x,Q^2)$ __decreases__ with increasing Q^2. These deviations from scaling are understood in quantum chromodynamics (QCD) in terms of gluonic radiative corrections.[24]

The scaling deviations imply that the ratio $F_2^N(x,Q_2^2)/F_2^N(x,Q_1^2)$, $Q_2^2 > Q_1^2$, is a decreasing function of x. At typical values of Q_1^2 and Q_2^2, the ratio crosses unity at $x \simeq 0.1$. Focussing on x dependence, one observes a similarity:

$$\frac{F_2^N(x,Q_2^2)}{F_2^N(x,Q_1^2)} \quad \text{resembles} \quad \frac{F_2^A(x,Q_1^2)}{F_2^N(x,Q_1^2)} . \tag{8}$$

On the left-hand-side, the target N is fixed, but Q^2 is changed. On the right-hand-side, Q^2 is fixed, but the target is changed.

The resemblance in x dependences of the two ratios suggests that the effective value of Q^2, the "scale" used to measure Q^2, may not be the same for a nucleus and for a free nucleon. This observation is the motivation for the Q^2 rescaling model.[6]

To proceed towards a possible dynamical understanding of the resemblance, one may begin by examining the source of Q^2 dependence in QCD. Because QCD is a theory of interacting constituents, prescribed changes occur with Q^2 in the densities $q(x,Q^2)$, $\bar{q}(x,Q^2)$ and $G(x,Q^2)$. For the valence quark density in hadron h, $q_v^h(x,Q^2)$, the theory specifies that[24]

$$\frac{dq_v^h(x,t)}{dt} = \frac{\alpha_s(Q^2)}{\pi} \int_x^1 \frac{dy}{y} q_v^h(y,t) P_{qq}(\frac{x}{y}) , \tag{9a}$$

where
$$t = \ln(Q^2/\mu_h^2). \tag{9b}$$

Equation (9) expresses the fact that the valence quark distribution at x is determined by its value in the range $x \leq y \leq 1$, and by the probability $P_{qq}(x/y)$ for the transition $q(y) \to q(x)$ via gluon radiation. In Eq. (9) $\alpha_s(Q^2)$ is the strong coupling strength describing the qqg vertex.

Because the integral over the transverse momentum of the radiated gluon in $q \to qg$ is formally divergent, $\propto \int^{Q^2} dp_T^2/p_T^2$, a scale μ_h^2 is introduced as a lower momentum "confinement" cutoff. This scale μ_h^2 appears in Eq. (9b); its value is unspecified in QCD perturbation theory.

Within the context of Eq. (9), one may propose at least two reasons for the EMC effect, $q^A(x,Q^2) \neq q^N(x,Q^2)$: either $\mu_A^2 \neq \mu_N^2$, or $q_v^A(x,\mu^2) \neq q_v^N(x,\mu^2)$. In other words, the lower momentum confinement cutoff μ_h^2 may be different for a quark in a nucleus, or the initial (non-perturbative) distributions from which one begins the QCD evolution may be different.

In a series of papers[6], Close, Jaffe, Roberts, and Ross have argued that the difference between $q^A(x,Q^2)$ and $q^N(x,Q^2)$ implies that $\mu_A^2 < \mu_N^2$. They state that the softening of the quark distributions in nuclei is due to the radiation of gluons between $Q^2 = \mu_A^2$ and $Q^2 = \mu_N^2$. Implicit in this view are the assumptions that partons are confined within nucleons which, in turn, make up the nucleus and that $q^N(x,Q^2 = \mu_N^2) = q^A(x,Q^2 = \mu_A^2)$. While not inconsistent with what is known in QCD, these important assumptions are not necessarily justified either. Acknowledging that the "details of quark confinement in free or bound nucleons are not yet solved, [Close et al.] believe that when a free nucleon is placed in nuclear matter, the modification to confinement is characterized by a change of a [single] length scale", the confinement size. Close et al. then use $\mu_A^2 < \mu_N^2$ to conclude that the physical size λ of the nucleon is greater in a nucleus, $\lambda_A > \lambda_N$. Within this model, the magnitude of the EMC effect implies $\lambda_A/\lambda_N \simeq 1.15$, a 15% "swelling" of nucleons in iron. The specific rescaling relationship proposed is

$$q^A(x,Q^2) = q^N(x,\zeta Q^2) \quad (10a)$$

with
$$\alpha_s(Q^2) \ln \zeta_{Fe}(Q^2) = 0.15 \quad . \quad (10b)$$

In the Q^2-rescaling model, momentum lost by valence quarks is taken up by an increase in the ocean q and $\bar{q}$ pairs and by the glue. A calculation based on the rescaling approach is compared with data in Fig. 3. Agreement is better with the SLAC data than with the CERN results. This is curious since the Q^2-rescaling approach should be more relevant at large values of Q^2. The rescaling curve crosses $R_{EMC} = 1$ at $x \simeq 0.1$, whereas the crossing point in the CERN data is closer to $x \simeq 0.3$. The model does not incorporate Fermi smearing and, correspondingly, fails to describe the data for $x \gtrsim 0.7$.

Swelling of nucleons in nuclei has been proposed[23] in nuclear physics for reasons other than the EMC effect. However, I am far from persuaded that the large physical size (~ 1 fermi) of a nucleon is at all relevant for determining a short distance quantity such as μ.

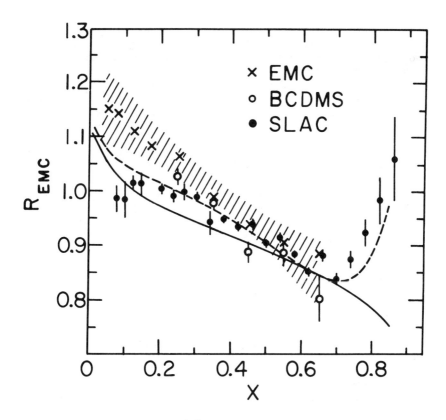

Fig. 3. Calculations of $R_{EMC}(x)$ compared with data. The solid curve shows the expectation of Q^2 rescaling (Ref. 6), and the dotted curve results from the pion exchange model (Ref. 7).

As I mentioned above, another way to maintain Eq. (9) along with $q^A(x,Q^2) \neq q^N(x,Q^2)$ is to assert that boundary conditions $q^A(x,Q_0^2)$ and $q^N(x,Q_0^2)$ are different. In the conventional nuclear physics approach to which I now turn, the initial boundary conditions are indeed different.

B. "Conventional" Nuclear Physics Approach

There are several models which may be grouped under this heading. I will first list the assumptions which are common to them and then turn to specific differences.

The nucleus is viewed as a system whose relevant degrees of freedom are A bound nucleons plus an indefinite number of mesons. The mesons are associated with nuclear binding.

The essential assumptions[25] are:
i. A nucleus is a bound system of interacting mesons and nucleons

with partons confined within these hadrons.

ii. The current density operator j^μ is a sum of one-body operators. Since a "one-body operator" cannot depend on the spectators or on the state on which it operates, this formal assumption implies the physical assumption that the structure functions (and hence the quark distributions) of the nucleons and mesons are not affected by the nuclear medium. They are the same as those measured on free nucleons and mesons.

iii. The contribution of interference terms $j_i^\mu(0) j_k^\nu(0)$, $i \neq k$, to the tensor $F_A^{\mu\nu}$ can be neglected. Intuitively, this means that hadrons in a nucleus contribute incoherently in deep inelastic lepton scattering.

iv. In the sum over final states, final state interactions between the residue of the struck hadron and the spectator hadrons in the nucleus do not affect the inclusive cross section.

v. $Q^2 > M_N^2$, where M_N is the mass of the proton or neutron.

In this approach, which is an extension of successful nuclear theory, the role of quantum chromodynamics (QCD) is restricted to the description of the structure of single hadrons. The approach cannot be "derived" from QCD since even bound-state wave functions of single hadrons cannot, at present, be obtained in this manner. As is true also for the Q^2-rescaling approach, the fundamental validity of the model cannot be judged until a successful quantitative calculation has been completed of parton densities and confinement. If asymptotic freedom extends over regions of the order of 1 fm or more, then some of the assumptions above are almost certainly invalid. For the present, the best one can do is to explore quantitatively the consequences of the assumptions and confront the data.

In the conventional nuclear physics approach, one defines[7] fractions of the (light-front) momentum of the nucleus carried by nucleons and mesons. These are fractions per nucleon:

$$y = \frac{A(E + p_L)_\pi}{(E + p_L)_A} \quad , \quad \text{for } \pi\text{'s} ; \tag{11a}$$

and

$$z = \frac{A(E + p_L)_N}{(E + p_L)_A} \quad , \quad \text{for nucleons} . \tag{11b}$$

Number densities per nucleon of mesons and nucleons are also defined, $f_\pi(y)$ and $f_N(z)$. I will describe below how they are computed. Baryon number conservation is expressed as $\int f_N(z) \, dz = 1$. The mean number of pions per nucleon is given by the integral $\int f_\pi(y) dy = \langle n_\pi \rangle$. Momentum conservation takes the form

$$\int z \, f_N(z) \, dz + \int y \, f_\pi(y) \, dy = 1 . \tag{12}$$

Given the assumptions above, one may derive[25] directly a "convolution" formula for the structure function $F_2^A(x, Q^2)$:

$$F_2^A(x,Q^2) = \int_x f_\pi(y) F_2^\pi\left(\frac{x}{y}, Q^2\right) dy + \int_x f_N(z) F_2^N\left(\frac{x}{z}, Q^2\right) dz \quad . \quad (13)$$

The structure function of a nucleus, per nucleon, $F_2^A(x,Q^2)$ is expressed in terms of the structure functions $F_2^N(x,Q^2)$ and $F_2^\pi(x,Q^2)$ measured on unbound nucleons and pions. Expressions analogous to Eq. (13) may be derived for the other structure functions, $F_1^A(x,Q^2)$ and $F_3^A(x,Q^2)$, and by implication, therefore, for the quark and antiquark densities per nucleon, $q^A(x,Q^2)$ and $\bar{q}^A(x,Q^2)$, as well as for the gluon density $G^A(x,Q^2)$. For example,

$$q^A(x,Q^2) = \int_x f_\pi(y) q^\pi\left(\frac{x}{y},Q^2\right) dy + \int_x f_N(z) q^N\left(\frac{x}{z}, Q^2\right) dz \quad . \quad (14)$$

Since the bound state wave function of a nucleus is known, at least in principle, one should be able to compute the single particle densities $f_N(z)$ and $f_\pi(y)$ explicitly and reliably, and relate their properties to quantities measured in experiments other than deep inelastic lepton scattering. Different conventional nuclear models part company at this point, emphasizing different input information.

1. Binding

The concept of binding as measured by one nucleon separation energies in (e, e'p) is emphasized by Akulinichev et al.[9] and by Birbrair et al.[9] They consider that deep inelastic scattering occurs from a bound nucleon moving in an average potential well and do not treat mesons explicitly. In terms of single particle wave functions $\phi_\lambda(\vec{p})$ and energy levels ε_λ, they derive

$$f_N(z) = \frac{M_N}{(2\pi)^2} \sum_\lambda \int_{|M_N(z-1)-\varepsilon_\lambda|} |\phi_\lambda(\vec{p})|^2 |\vec{p}| d|\vec{p}| \quad (15)$$

Results obtained in this fashion are illustrated in Fig. 4. Average single particle separation energies, $|\langle \varepsilon_\lambda \rangle|$ range from 30 to 40 MeV.

As is evident in Fig. 4, "binding" yields a depletion, $R_{EMC}(x) < 1$, but no rise in $R_{EMC}(x)$ above unity at small x. The approach provides a curve which resembles the behavior of the SLAC data, but it cannot describe the original EMC result. A more serious limitation of the approach is that momentum is not conserved. Indeed,

$$\langle z \rangle \simeq \left(1 - \frac{|\langle \varepsilon_\lambda \rangle|}{M_N}\right) \simeq 0.96 \quad . \quad (16)$$

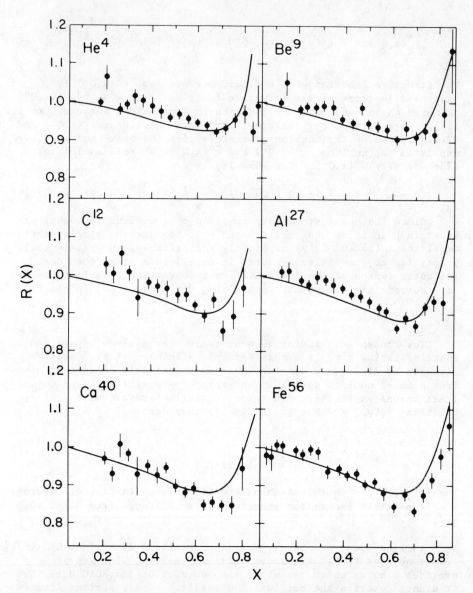

Fig. 4. Curves from a model based on binding are compared with data from a SLAC experiment, Ref. 2. This figure is taken from Akulinichev et al., Ref. 9.

Equation (16) may be exploited to obtain a qualitative understanding of why "binding" leads to a depression of $R_{EMC}(x)$ below unity for $0 < x \lesssim 0.7$. Since the function $f_N(z)$ obtained from Eq. (15) is concentrated fairly sharply near $z = <z>$, one may use the convolution Eq. (13) to show that

$$F_2^A(x) \simeq F_2^N(x/<z>) \qquad (17)$$

for $x \lesssim 0.7$. (Recall that $f_\pi(y)$ is ignored in this model). Because $F_2^N(x)$ is a monotonically decreasing function of x, Eq. (17) implies that $R_{EMC}(x) < 1$ for $0 < x \lesssim 0.7$. To reproduce the observed magnitude of the depression of $R_{EMC}(x)$ below unity, values of $|<\varepsilon_\lambda>|$ in the range 30 to 50 MeV are required. In particular, it would not suffice to use the average binding energy $|<\varepsilon>| \simeq 8$ MeV.

2. Exchange Pions

Soon after the EMC effect was published, it was proposed that deep inelastic scattering from excess pions in the nucleus is responsible for the rise of $R_{EMC}(x)$ above unity at small x. Llewellyn Smith[8] and Ericson and Thomas[8] pursued calculations in terms of a medium modification of the pion propagator. A different approach to incorporating pion exchange effects was developed at Argonne.[7] I will summarize some aspects of the Argonne approach.

We begin with the conventional view in which the nucleon-nucleon potential features an attractive one-pion exchange tail at large distances, an intermediate range attraction dominated by two-pion exchange, and a strong short-range phenomenological repulsion. The expectation value of the pion exchange potential is then calculated in the nuclear ground state.[7] From this, the density of exchange pions in the nucleus is computed by a variational technique.[26] For the momentum density of exchange pions, we derive

$$\rho_\pi(p) = \rho_N^{(2)}(p) \frac{F(p)}{(p^2 + m_\pi^2)^2} \,. \qquad (18)$$

In Eq. (18), $\rho_N^{(2)}(p)$ is the two-nucleon density, and $F(p)$ is the πNN, $\pi N\Delta$ vertex function. A standard value of $\Lambda = 7$ fm^{-1} is used in the vertex function as well as in the determination of the many-body wave function used in computing the expectation value.

Among the advantages of the Argonne approach, one may mention:
i. The single- and two-nucleon densities $\rho_N^{(1)}(p)$ and $\rho_N^{(2)}(p)$, as well as the single pion density $\rho_\pi(p)$ are all computed from the same well-defined many-body bound state formalism. These densities are used to derive $f_\pi(y)$ and $f_N(z)$.
ii. The momentum balance constraint on $f_N(z)$ and $f_\pi(y)$, Eq. (12), arises naturally. It does not have to be imposed "by-hand".
iii. The structure functions $F^N(x,Q^2)$ and $F^\pi(x,Q^2)$ which appear on the right hand side of Eq. (13) are empirical, "on-shell" quantities, not "off-shell" continuations thereof.

The pion exchange model provides a unified description of $R_{EMC}(x)$ for all x. The "books are balanced" in the sense that momentum lost to nucleons through binding, $<z> < 1$, is carried by exchange pions. A comparison with data is shown in Fig. 1. In this calculation, $< n_\pi^{Fe} > = 0.095$, meaning that in an Fe nucleus, there are on average 5 to 6 pions from which deep inelastic scattering occurs. The mean momentum per nucleon carried by those pions is $< y^{Fe} > = 0.052$. Correspondingly, $< z > = 1 - < y > \simeq 0.95$. The value of $< n_\pi >$ controls the size of the enhancement of $R_{EMC}(x)$ above unity at small x, whereas $< y >$ controls the shape and size of the depression below unity for intermediate x. In the model, there is a modest change of $< n_\pi >$ with A. For Aℓ, Fe, and Au, we compute $< n_\pi > =$ 0.089, 0.095, and 0.114.

Comparison with the data in Fig. 1 indicates that the pion exchange model accounts successfully for the data at large Q^2 for all x. It does not reproduce the SLAC data at low Q^2. The model does not yet incorporate shadowing. The success of the conventional pion exchange model in explaining $R_{EMC}(x)$ for all x implies that there is no need to invoke "new physics", such as a change in confinement scale, to deal with the present data.

C. Ocean Quark Density

It is instructive to compare the prediction of the pion exchange model for the ocean quark density $\bar{q}^A(x)$ with that of the Q^2 rescaling model. Both are shown in Fig. 2 along with data from deep inelastic neutrino scattering. In the pion exchange model, the antiquark distribution $\bar{q}^A(x)$ is broadened in a nucleus because exchange pions supply valence antiquarks. However, in the rescaling model, QCD evolution leads to a narrowing of $\bar{q}^A(x)$ relative to $\bar{q}^N(x)$. As indicated in Fig. 2, the two models provide quite different expectations for the antiquark ratio measured in deep inelastic neutrino scattering. Unfortunately, the differences are significant only for $x \gtrsim 0.35$, where there are essentially no results on $\bar{q}(x)$ from neutrino experiments.

Even for $x < 0.3$, very precise neutrino data would be valuable for establishing whether indeed $R_{\bar{q}}$ is greater than unity, as shown in both models in Fig. 2. Another source of information on $R_{\bar{q}}$ is massive-lepton pair production, the Drell-Yan process.

VI. MASSIVE LEPTON-PAIR PRODUCTION

Massive lepton pair production, the Drell-Yan process, offers another means for studying the A dependence of $q^A(x,Q^2)$ and $\bar{q}^A(x,Q^2)$. Experiments are important for at least two reasons[10]: (i) to establish whether the A dependence of $q^A(x,Q^2)$ observed in deep inelastic lepton scattering is a universal, process-independent property of parton densities, and (ii) to determine the A dependence of the ocean, $\bar{q}^A(x,Q^2)$.

In the Drell-Yan model[27] for $hN \to \mu\bar{\mu}X$, the $\mu\bar{\mu}$ pair arises from the decay of a massive photon γ^*. This virtual photon is produced in the annihilation of a quark (antiquark) from hadron h with an antiquark (quark) from the target nucleon: $q\bar{q} \to \gamma^*$. The invariant mass M of the γ^* is related to the fractional longitudinal momenta x_i of the annihilating partons by $M^2 = s x_1 x_2$, and the scaled longitudinal momentum of the pair is $x_F = x_1 - x_2$; $\tau = M^2/s$. The variable s denotes the square of the center of mass energy of the hN process.

As described in Ref. 10, the reaction $pA \to \gamma^* X$, with $x_F > 0$, should be a valuable source of detailed information on the A dependence of the antiquark distribution $\bar{q}^A(x,Q^2)$, notably that of the up flavor. It is noteworthy that the range in x spanned by current lepton pair data[28] is greater than in the neutrino case, e.g., varying from $x = 0.15$ to $x = 0.4$ for $4 < M < 11$ GeV at 400 GeV/c and $x_F = 0$. This range could be extended by studies at larger values of τ. Lepton pair data appear, therefore, to have more discriminating power, and a precise series of experiments on $pA \to \gamma^* X$ seems advisable.[10]

VII. SUMMARY AND FUTURE INVESTIGATIONS

The original EMC effect stimulated broad interest in quark and antiquark distributions of <u>nuclei</u>. The data provide new tests of models of nucleon binding and, perhaps, information on properties of constituent confinement. The pion exchange model yields a unified quantitative description of all features of the present data sample at large Q^2. It reproduces the magnitude and shape of $R_{EMC}(x)$, for all x, and the weak enhancement of the antiquark distribution $\bar{q}^A(x)$ demonstrated by neutrino experiments. Since this conventional nuclear physics approach suffices to explain the data, there is no need to invoke new QCD effects such as a proposed change of the color confinement size of nucleons within nuclei.

The above conclusions are based on an analysis of the current data sample. New data, especially in the small x region, could have a significant impact. The absolute normalization of $R_{EMC}(x)$ at small x is crucial. If an $\sim 15\%$ enhancement of $R_{EMC}(x)$ above unity, as suggested in the original EMC paper, were to be confirmed in future experiments, then none of the current models would survive. This conclusion is obvious from an examination of Fig. 2. Neither the Q^2-rescaling nor the pion exchange model yields an enhancement above unity which would be sufficiently large. Shadowing is another important feature of the small x region. The Q^2-rescaling and the pion exchange models do not incorporate this physics as yet. Better data should stimulate the required effort. Finally, data on $\bar{q}^A(x)$ are most desirable.

As discussed elsewhere in these proceedings, an experiment is underway[29] at SLAC to measure the A dependence of the ratio σ_L/σ_T. Two important new experiments are almost ready to take data on $\mu A \to \mu' X$ at large Q^2: CERN NA37 and Fermilab E-665. These experiments should establish the magnitude of $R_{EMC}(x)$ for $x \lesssim 0.2$ as well as determine the x and Q^2 characteristics of shadowing. Regrettably, I know of no new experiment planned for $\nu A \to \mu A$ which would be sufficiently precise to substantially reduce the size of the error flags on $\bar{q}^A(x)$ in Fig. 2. On the other hand, an important proposal[30] to study the Drell-Yan process pA → $\mu\bar{\mu}$X is being considered at Fermilab. The proponents intend to accumulate enough data in the relevant kinematic region to determine $\bar{q}^A(x)$ precisely at small x.

I have said little about the gluon distribution $G^A(x,Q^2)$. There is no simple process in which $G^A(x,Q^2)$ is probed in the precise way μA → μ'X or hA → $\mu\mu$X probe the quark and antiquark densities. However, inclusive prompt photon production[31] at large p_T, pA → γX, and inclusive chi production, pA → χX, are my favorite candidates. I am certain that good data could be analyzed to yield interesting information on $G^A(x,Q^2)$.

ACKNOWLEDGMENT

I have benefitted from many discussions with Fritz Coester. This work was supported by the U.S. Department of Energy, Division of High Energy Physics, Contract W-31-109-ENG-38.

REFERENCES

1. J. J. Aubert et al., Phys. Lett. 123B, 275 (1983).
2. A. Bodek et al., Phys. Rev. Lett. 50, 1431 (1983) and 51, 534 (1983); R. G. Arnold et al., Phys. Rev. Lett. 52, 727 (1984).
3. G. Bari et al., Phys. Lett. 163B, 282 (1985).
4. H. Abramowicz et al., Z. Phys. C25, 29 (1984); see also M. A. Parker et al., Nucl. Phys. B232, 1 (1984); A. M. Cooper et al., Phys. Lett. 141B, 133 (1984).
5. For previous reviews of both experiment and theory, as well as an extensive list of references, see E. L. Berger, Argonne report ANL-HEP-PR-85-70, Proc. Seminar on Few and Many Body Quark Systems, San Miniato, 1985; O. Nachtmann, Proc. 11th Intl. Conf. on Neutrino Physics and Astrophysics, Nordkirchen near Dortmund, 1984, p. 405; C. H. Llewellyn Smith, Proc. PANIC (Heidelberg) 1984, Eds. P. Povh and G. zu Putlitz, Nucl. Phys. A434, 35c (1985); A. Krzywicki, Proc. 11th Europhysics Conf. "Nuclear Physics with Electromagnetic Probes" Paris 1985, Eds. A. Gerard and C. Samour, Nucl. Phys. A446, 135c (1985).
6. F. E. Close, R. G. Roberts, and G. C. Ross, Rutherford report RAL-85-101; F. E. Close et al., Phys. Lett. 129B, 346 (1983); R. L. Jaffe et al., Phys. Lett. 134B, 449 (1984); F. E. Close et al., Phys. Rev. D31, 1004 (1985).
7. E. L. Berger and F. Coester, Phys. Rev. D32, 1071 (1985); E. L. Berger, F. Coester, and R. Wiringa, Phys. Rev. D29, 398 (1984); E. L. Berger and F. Coester in Quarks and Gluons in Particles and Nuclei, ed. by S. Brodsky and E. Moniz (World Scientific, Singapore, 1986), p. 255.
8. J. D. Sullivan, Phys. Rev. D5, 1732 (1972); C. H. Llewellyn Smith, Phys. Lett. 128B, 107 (1983); M. Ericson and A. W. Thomas, ibid., 128B, 112 (1983); A. W. Thomas, Phys. Rev. 126B,

97 (1983); J. Szwed, Phys. Lett. 128B, 245 (1983).
9. S. V. Akulinichev et al., Phys. Lett. 158B, 485 (1985); S. V. Akulinichev et al., Phys. Rev. Lett. 55, 2239 (1985); B. L. Birbrair et al., Phys. Lett. 166B, 119 (1986).
10. E. L. Berger, Nucl. Phys. B267, 231 (1986).
11. For reviews, consult F. Eisele, Proc. XXI Int. Conf. on High Energy Physics, Paris (1982), eds. P. Petiau and M. Porneuf, p. C3-337; F. Dydak, Proc. 1983 Intl. Symposium on Lepton and Photon Interactions at High Energies, Cornell, eds. D. G. Cassel and D. L. Kreinick, p. 634; J. Drees and H. E. Montgomery, Ann. Rev. Nucl. Part. Sci 1983, 33 p. 383; F. Sciulli, Nevis report R-1345 (1985), Proc. 1985 Int. Symposium on Lepton and Photon Interactions at High Energies, Kyoto, to be published; E. L. Berger, Argonne report ANL-HEP-PR-86-17, Proc. Workshop on Fundamental Muon Physics, Los Alamos, 1986.
12. CHARM, F. Bergsma et al., Phys. Lett. 141B, 129 (1984); EMC, J. J. Aubert et al., Nucl. Phys. B259, 189 (1985); SLAC, M. D. Mestayer et al., Phys. Rev. D27, 285 (1983).
13. See e.g. the EMC data, Ref. 12.
14. J. D. Bjorken and E. A. Paschos, Phys. Rev. 185B, 1975 (1969); R. P. Feynman, Phys. Rev. Lett. 23, 1415 (1969).
15. See, e.g., D. Allasia, Phys. Lett. 135B, 231 (1984), and H. Abramowicz et al., Z. Phys. C25, 29 (1984).
16. G. Gramer and J. Sullivan in Electromagnetic Interactions of Hadrons, ed. by A. Donnachie and G. Shaw (Plenum, New York, 1978), Vol. 2, p. 195; A. Mueller, in Quarks, Leptons, and Supersymmetry, Vol. 1 of Proceedings of the XVIIth Recontre de Moriond, Les Arcs, France 1982, edited by J. Tran Thanh Van (Editions Frontieres, Gif-sur-Yvette, 1982), and references therein; L. V. Gribov, E. M. Levin, and M. G. Ryskin, Phys. Rep. 100, 1 (1983).
17. A. Bodek and J. L. Ritchie, Phys. Rev. D23, 1070 (1981); D24, 1400 (1981).
18. Exceptions include N. N. Nikolaev and V. I. Zakharov, Phys. Lett. 55B, 397 (1975); A. Krzywicki, Phys. Rev. D14, 152 (1976).
19. R. L. Jaffe, Phys. Rev. Lett. 50, 228 (1983).
20. G. Taylor, these proceedings.
21. I. A. Schmidt and R. Blankenbecler, Phys. Rev. D15, 3321 (1977) and D16, 1318 (1977); A. M. Baldin, Nucl. Phys. A434, 695C (1985); H. Faissner and B. R. Kim, Phys. Lett. 130B, 321 (1983); H. Faissner et al., Phys. Rev. D30, 900 (1984); G. Berlad, A. Dar, and G. Eilam, Phys. Rev. D22, 1547 (1980); L. L. Frankfurt and M. I. Strickman, Phys. Rep. 76, 215 (1981); Nucl. Phys. B181, 22 (1981); H. J. Pirner and J. P. Vary, Phys. Rev. Lett. 46, 1376 (1981); C. E. Carlson and T. J. Havens, Phys. Rev. Lett. 51, 261 (1983); A. I. Titov, Sov. J. Nucl. Phys. 38, 964 (1983); M. Chemtob and R. Peshanski, Journal of Phys. G10, 599 (1984); J. Dias de Deus et al., Phys. Rev. D30, 697 (1984) and Z. Phys. C26, 109 (1984); B. C. Clark et al., Phys. Rev. D31, 617 (1985).
22. O. Nachtmann and H. J. Pirner, Z. Phys. C21, 277 (1984); G. Chanfray et al., Phys. Lett. 147B, 249 (1984); H. J. Pirner,

Comm. Nucl. Part. Phys. 21, 199 (1984); W. Furmanski and A. Krzywicki, Z. Phys. C22, 391 (1984); A. Krzywicki, Phys. Rev. D14, 152 (1976); C. Angelini and R. Pazzi, Phys. Lett. 154B, 328 (1985).

23. J. V. Noble, Phys. Rev. Lett. 46, 412 (1981); M. Jändel and G. Peters, Phys. Rev. D30, 1117 (1984); L. S. Celenza et al., Phys. Rev. Lett. 53, 892 (1984); T. Goldman and G. J. Stephenson Jr., Phys. Rev. Lett. 146B, 143 (1984); M. Staszel et al., Phys. Rev. D29, 2638 (1984); M. Rho, Phys. Rev. Lett. 54, 767 (1985); I. Sick, Nucl. Phys. A434, 677c (1985).
24. Consult, e.g. A. Buras, Rev. Mod. Phys. 52, 199 (1980).
25. See E. L. Berger and F. Coester in Quarks and Gluons in Particles and Nuclei, Ref. 7.
26. F. Coester, in Recent Progress in Many Body Theories, Proc. of the Third International Conference, Odenthal-Altenberg, Germany (Lecture Notes in Physics, Vol. 198), edited by H. Kümmel and M. L. Ristig (Springer, Berlin, 1984), p. 16.
27. S. D. Drell and T. M. Yan, Phys. Rev. Lett. 25, 316 (1970); E. L. Berger in Proc. of the Workshop on Drell-Yan Processes, Fermilab, 1982, pp. 1-62.
28. A. S. Ito et al., Phys. Rev. D23, 604 (1981).
29. R. Arnold, these proceedings.
30. Fermilab proposal P-772, J. Moss et al.
31. E. L. Berger, E. Braaten, and R. D. Field, Nucl. Phys. B239, 52 (1984).

HIGH PRECISION PARITY TESTS IN PROTON-PROTON (NUCLEON) SCATTERING

Markus Simonius
Institut für Mittelenergiephysik, Eidg. Techn. Hochschule,
CH-8093 Zürich, Switzerland

ABSTRACT

Tests of parity violation in proton-proton scattering at low and intermediate energy and of proton H_2O scattering at high energy are reviewed. New measurements at 45 MeV are presented. Theoretical analysis and implications of such experiments and prospects for future measurements are discussed.

INTRODUCTION

Parity violation in the nucleon-nucleon system is at present the only experimentally accessible signature of flavour non changing purely hadronic weak interactions. Its analysis tests our ability to calculate matrix elements of weak interactions for strongly interacting systems.

The **theory** is difficult and open problems persist.
The **experiments** are extremely challenging.

Since measurements are due to interference between weak (parity violating) and strong amplitudes, both have to be known or understood, including their phases, for a useful analysis. For the weak amplitudes this implies that one has to understand the strong wavefunctions (or equivalents) at the quark as well as the nucleon-nucleon (or 6 quark?) level.

The quantity measured in scattering experiments is

$$A_z = A_L = \frac{\sigma^+ - \sigma^-}{\sigma^+ + \sigma^-}$$

where $\sigma^\pm$ denotes the crossection for positive and negative helicity. There are two different ways to obtain this quantity, shown schematically in Fig. 1.

Fig. 1a represents a **transmission experiment** in which the helicity dependence of the total crossection is infered from the helicity dependent of the absorption in the target. This is the preferred method at high energy but cannot be used at low energy, since there only small absorption can be reached since energy loss of the beam in the target does not permit one to use a thick enough target.

Fig. 1b represents a **scattering experiment** in which the scat-

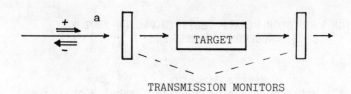

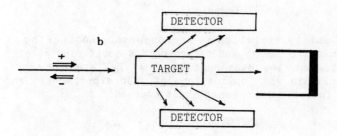

Fig. 1. Schematic arrangement for transmission (a) and scattering (b) experiment.

tered particles are detected. Unlike usual scattering experiments, however, one does not actually count the number of scattered particles ($> 10^{10}$ sec!) but measures a total integrated current produced by them in the detector. As a consequence one in general measures some complicated angular average of A_z. Only for low energy p-p scattering (< 50 MeV) this is essentially equivalent to a total crossection measurement since in this case A_L is angle independent up to a few percent due to the dominance of the lowest partial wave (in the weak and the strong amplitude).

The most significant experimental results for nucleon-nucleon scattering obtained so far at different Lab energies are:

p-p scattering:
15   MeV A_L = $(-1.7\pm0.8)\times10^{-7}$ ref. 1
45   MeV A_L = $(-1.5\pm0.22)\times10^{-7}$ ref. 2
800 MeV A_L = $(2.4\pm1.1\pm0.1)\times10^{-7}$ ref. 3

p-H$_2$O scattering:
5.3 GeV A_L = $(26\pm7) \times 10^7$ ref. 4

Note the difference by an order of magnitude between the low and

the high energy experimental data. In the following I shall
first present the new 45 MeV experiment in some detail. I will
then discuss the analysis and significance of p-p scattering
experiments below ~ 400 MeV (pion-threshold) and then turn to
the high energy experiment. I will not discuss the 800 MeV
experiment which was presented at this conference in a separate
session: it lies between the two energy domains focussed upon.

THE NEW MEASUREMENTS OF A_L FOR 45 MeV p-p SCATTERING
W. Haeberli, S. Kistryn, J. Lang, J. Liechti, Th. Maier, R.
Müller, F. Nessi-Tedaldi, M. Simonius, J. Smyrski, J. Sromicki.

This experiment was resumed after first measurements in p-p
scattering at 45 MeV[5] and a similar investigation in p-α scattering[6] in order to reduce the experimental uncertainty and provide more significant constraints for theoretical analyses.

The principle of the experiment is as shown in Fig. 1b. The
incoming polarized protons are scattered in a 100 bar H_2 target
and detected in a cylindrical ionization chamber of 20 cm outer
radius and 2 cm active thickness coaxial with beam and target.
A_L is obtained from the ratio

$$\frac{(N_s/N_p)^+ - (N_s/N_p)^-}{(N_s/N_p)^+ + (N_s/N_p)^-}$$

where N_s is the integrated ionization chamber current and N_p the
integrated beam current in the Faraday cup. The basic integration time is 20 msec in order to suppress line frequency (50 Hz)
modulations. In between integrations there is a roughly 10 msec
deadtime, used for beam monitoring. At its beginning the beam
polarization is switched between $p_z = \pm 84\%$ by RF transitions at
the SIN groundstate atomic beam polarized ion source according
to one of the following patterns

```
|+-+-+-+--+-+-+|
|-+-+-+-++-+-+-|
```

alternatively selected by a pseudorandom sign generator. With
a typical current of 1-4 μA the statistical accuracy achieved is

$(0.5 - 1) \times 10^{-4}$ in 20 msec
$(2.5 - 5) \times 10^{7}$ in 20 min.

20 min is the length of our standard data runs. Obviously the
main problem is systematics.

The general outlay of the experiment is shown in Fig. 2.

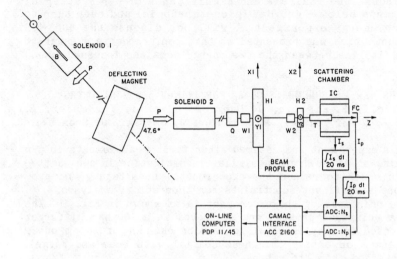

Fig. 2. Schematic view of beam line and experimental set-up. The polarization vector P is shown for positive p_z. After the elastic scattering in the high pressure (100 bar) H_2 gas target (T) the protons are detected in the cylindrical ionazation chamber (IC). The direct beam is measured in the Faraday cup (FC) for calibration. H1 and H2 are the intensity and polarization profile monitors. Q, W1 and W2 are small magnets used for artificial beam modulations.

Note the two precession solenoids. The first one is used in conjunction with the 47.6° deflection to obtain longitudinal polarization at the target from the vertical one coming from source and cyclotron. Reversing its field reverses the phase between the longitudinal polarization at the target and its switching at the ion source which is important for the suppression of some systematic effects. This solenoid was therefore reversed after every second 20 min data run. Solenoid 2 is used to produce horizontal transverse polarization at the target needed to measure the sensitivity of the detection system to this polarization component.

An essential part of our arrangement are the two beam profile monitors[7] H_1 and H_2 placed about 95 and 10 cm respectively, in front of the entrance window to the target. Their crucial feature is that besides the usual intensity profiles they measure also the polarization profiles of the beam. This is done by mooving carbon strip targets as analyzers through the beam and measuring the two transverse polarization components as a function of their position with a left-right and a down-up scintillation counter pair. One of the 4 profiles is scanned in each 10 ms deadtime between current integrations. The eight polarization profiles so obtained in some sample 20 min run with nominally longitudinal polarization is shown in Fig. 3. The

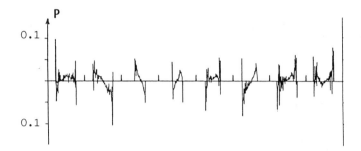

Fig. 3. Polarization profiles from the two beam profile monitors H1 and H2. Fluctuations, at the edges of the profiles are due to lack of statistics.

large variations at the edge of each profile are insignificant statistical fluctuations since the intensity goes to zero. The important feature is the (skew) inhomogeneous distribution in the center of the profiles which can lead to corrections of the order of a few times 10^{-7} in A_L. Its potential danger is easily understood: Transverse polarization of the beam — say vertical for concreteness — leads to a left-right asymmetry in the number of scattered protons due to the $\lesssim$ 3% regular analyzing power of p-p scattring at our energy. If the detection system is left-right symmetric, and that is how it was built, this does not matter. However, for beam portions not in the center of the detection system this symmetry is broken and in fact broken with opposite sign for portions left or right of the symmetry axis. Thus a false signal can be produced if the beam has transverse polarization components which have opposite sign on both sides of the beam axis even if the average transverse polarization is zero (or small as is seen to be the case in Fig. 3). While it is relatively easy to adjust the beam so that it has small average transverse polarization, it is difficult to eliminate these inhomogeneities which are produced almost inevitably for instance in deflection dipoles[8]. However, from the profiles, measured with these monitors, a reliable correction can be determined[8] if the sensitivity of the detection system is measured as a function of beam position relative to the target axis with transverse beam polarization. Such and similar sensitivity measurements are repeated from time to time between parity measurements.

Table I contains the list of corrections to the final result and limits of systematic effects. Though corrections to individual runs for transverse polarization components are typically 1-4 times 10^{-7} their contribution to the final result averages out almost completely. The only other effect for which a nonzero correction has been found actually is intensity modu

Table I: Systematic effects and corrections in the final result for A_L in units of 10^{-7}.

Runwise corrections (stat. errors)

Transverse polarization comp. (average + first moments)[a]	0.05 ± 0.03
Intensity modulations[a,b]	0.00 ± 0.00
Position modulation[a]	0.01 ± 0.04

Limits on systematic effects:

Transverse polarization comp. (sensitivities, 2nd and 3rd moments)	± 0.04
Emittance modulation[a]	± 0.03
Energy modulation[b]	± 0.03
β-decay asymmetry[c]	± 0.05
Double scattering[c]	± 0.02
Empty target background	± 0.05
Electronic crosstalk etc.[b]	0.00
Uncertainty of beam polarization ($\pm 2\%$)	± 0.03

a) Sensitivities measured with artificially enhanced modulations. Coherent modulations monitored during measurement.

b) suppressed by solenoid reversal.

c) depends on the incoming helicity itself, cannot be made small. Separate investigation needed.

lation of the beam in step with polarization reversal. This also can be corrected run-wise from the measured intensity modulation and experimentally determined sensitivity to it. In addition it completely cancels, even without correction, due to solenoid 1 reversal as discussed above. For all other effects no nonzero contribution has been seen above the statistical accuracy of the tests. I cannot go through a detailed discussion of all other effects listed in Table I (see ref. 5). Two of them, however, should be mentioned. The beta decay asymmetry and the double scattering effect. Unlike all other effects these could produce an asymmetry which is proportional to the helicity of the beam itself which obviously cannot be held small during the measurements as the other modulations which could lead to a fake asymmetry.

Beta decay asymmetry: The impinging protons produce β-active nuclei in various parts of the apparatus whose decay give a contribution to the measured currents. If these nuclei inherit some of the beam polarization and retain it until they decay the usual parity violating beta decay asymmetry can lead to a false effect.

Double scattering effect: In p-p scattering in the target part of the longitudinal polarization is transfered to transverse polarization of the final protons. If these rescatter, for instance in the target wall, this leads to a regular asymmetry in the rescattered flux which, however, can contribute to the measured asymmetry only if the beam-target detector arrangement has no intact symmetry plane. With perfect rotational symmetry of the apparatus this, however, is not the case even if the beam is not centered ideally. Retaining rotational symmetry is therefore important to suppress such a contribution.

Both these effects were studied in separate investigations. Our new measurements contain 350 20 min data runs acculmulated in 6 separate series (beam periods). Their normalized distribution after applying the runwise corrections mentioned in table 1 is given in Fig. 4 together with the theoretical distribution.

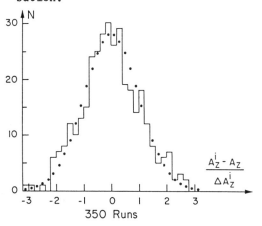

Fig. 4. Normalized distribution of the 350 20 min runs compared to gaussian.

The new result obtained is

$$A_L = (-1.46 \pm 0.22) \times 10^{-7}.$$

The error is the rms sum of
purely statistical error $\quad 0.19 \times 10^{-7}$
statistical error of corrections 0.05×10^{-7}
total of systematic uncertainties 0.09×10^{-7}.
Including the old measurements[5] yields

$$A_L = (-1.50 \pm 0.22) \times 10^{-7}.$$

Some statistical investigations of the data are still in progress.

This is the actually measured result without correction for angular dependence. It differs by < 5% from the asymmetry in the total crossection[5]. Using the so-called DDH best value coupling constants[9] (which have a large uncertainty) and old Reid soft-core potential calculations of the matrix elements[10] yields the prediction $A_L = (-1.3 \times 10^{-7})$.

ANALYSIS OF LOW ENERGY p-p EXPERIMENTS

The aim of this section is to present some general, essentially kinematical, features.

Only even total angular momentum (J) singlet (s=0)-triplet (s=1) transition can contribute to the parity violating p-p scattering amplitude because of the Pauli principle. The two lowest ones, $f_0 = f(^1S_0 - ^3P_0)$ and $f_2 = (^3P_2 - ^1D_2)$ dominate even at 400 MeV due to the short range of the parity violating p-p interaction which cannot contain π-exchange if CP is conserved. At 45 MeV even f_2 is negligible. Fig. 5 shows the energy dependence of $f(^1S_0 - ^3P_0)$ and $f(^3P_2 - ^1D_2)$, calculated with Reid soft-core potential, toghether with their contribution to the asymmetry $\bar{A}_L$ in the total nuclear crossection, neglecting the Coulomb interaction[10]. The scale chosen is such that the curve for the $^1S_0 - ^3P_0$ transition (but not the other) represents the DDH best value prediction. The remarkable feature of Fig. 5 is the zero crossing of $A_L(^1S_0 - ^3P_0)$ around 230 MeV. This is a pure strong interaction phase effect as seen from the following formula:

$$A_L(^1S_0 \; ^3P_0) = 2f(^1S_0 - ^3P_0) \sin(\delta(^1S_0) + \delta(^3P_0))/\sigma k^2 \quad (1)$$

$$= 2.9 \; f(^1S_0 - ^3P_0) \text{ at 45 MeV}$$

where $\delta(\;)$ are p-p scattering phases and $f(^1S_0 - ^3P_0)$ is defined in terms of the S-matrix by

$$f(^1S_0 - ^3P_0) = e^{-i(\delta(^1S_0) + \delta(^3P_0))} \langle ^3P_0 | \frac{1-S}{2i} | ^1S_0 \rangle \quad (2)$$

and is real by unitarity and time reversal invariance below pion

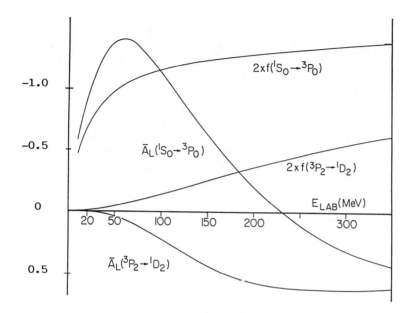

Fig. 5. Energy dependence of the amplitudes f and their contribution to the asymmetry $\bar{A}_L$ of the total nuclear cross section for the two lowest parity violating transitions in p-p scattering. The total asymmetry $\bar{A}_L$ is the sum of such contributions. The scale is chosen such that the $^1S_0-^3P_0$ contribution (but not the other) represents the DDH prediction.

threshold. The zero crossing of $\bar{A}_L$ is due to

$$\delta(^1S_0)+\delta(^3P_0) = 0.$$

As mentioned earlier, below ~ 50 MeV the angular distribution of A_z is practically constant and completely dominated by the $^1S_0-^3P_0$ transition. The energy dependence of the f's does not contain much new information since it is essentially determined by the strong p-p scattering states. On the other hand, as will be discussed below, $f(^3P_2-^1D_2)$ contains different and independent physics and its determination, best done around 230 MeV, would therefore be complementary to the 45 MeV experiment.

Unlike at 45 MeV, the angular dependence has to be considered if one goes to higher energy. It can be expressed as follows:

$$A_L(\theta) = \sum_{J^t even} f_{J^\pm} K_{J^\pm}(\theta) \qquad (3)$$

where $K_{J\pm\theta}$ is determined by the known strong interaction p-p scattering amplitudes. The amplitudes f are connected to the parity violating partial wave S-matrix elements according to the following generalization of eq.(2):

$$\langle J, L\pm 1, s=1 | S | J, L=J, S=0 \rangle = (\cos\varepsilon_J \, f_{J\pm} - i \sin\varepsilon_J \, f_{J\mp}) e^{i(\delta_{J^0} + \delta_{J^t})}$$

where δ_J, δ_J and ε_J are usual p-p phaseshift and mixing angles. In this way the f's are real below pion threshold due to unitarity and symmetry (time reversal invariance) of S where it should be noted that for the states the phase conventions of Jabob + Wick[11] have to be chosen which differ from the usual ones by a factor i^L. $K_{0+}(\theta)$ for the $^1S_0 - ^3P_0$ and $K_{2-}(\theta)$ for the $^3P_2 - ^1D_2$ transition are shown for 250 MeV in Fig. 6. As one sees,

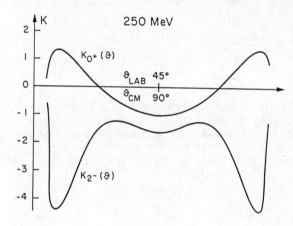

Fig. 6. Angular dependence of the two dominating contributions f_{0+} and f_{2-} to A_L for 250 MeV Lab energy.

the average of K_0 is close to zero corresponding to the zero crossing in Fig. 5.

I now turn to the conventional analysis in terms of vector meson exchange in order to reveal the origin of the different physical contents of $f_{0+} = f(^1S_0 - ^3P_0)$ and $f_{2-} = f(^3P_2 - ^1D_2)$. Only ρ_0 and ω exchange can contribute since π-exchange is CP-forbidden and 2π exchange is confined to the ρ^0-channel. The effective couplings are defined as follows:

weak: $-h_{\rho,\omega}^{pp} \, \overline{\psi}_p \, \gamma_\mu \, \gamma_5 \, \psi_p \, \varphi_{\rho,\omega}^\mu$

strong: $g_{\rho,\omega} \, \overline{\psi}_p (\gamma_\mu + \frac{i \chi_{\rho,\omega}}{2M} \sigma_{\mu\nu} \, q^\nu) \psi_p \, \varphi_{\rho,\omega}^\mu$

where $g_\rho = \frac{1}{3} g_\omega \sim 2.6$ represents the strong coupling constants and $\chi_\rho = \chi_V = 3.7$ and $\chi_\omega = \chi_S = -0.12$ are the isovector and isoscalar nucleon magnetic moments. The difference between χ_V and χ_S will be seen to be crucial. The weak ρpp and ωpp coupling constants introduced here are connected to the usual DDH[9] definition and best values by

$$h_\rho^{pp} = h_\rho^0 + h_\rho^1 + \frac{h_\rho^2}{\sqrt{6}} \qquad \text{DDH: } -15.4 \times 10^{-7}$$

$$h_\omega^{pp} = h_\omega^0 + h_\omega^1 \qquad \text{DDH: } -3.0 \times 10^{-7}.$$

The weak coupling constants h are the central parameters of this approach. They should be determined experimentally in order to compare them to theoretical predictions which, however, are still plagued with large uncertainties[9].

The meson exchange parity violating pp interaction obtained from above couplings is

$$V_{pp}^{PV} = -\left[h_\rho^{pp} g_\rho + h_\omega^{pp} g_\omega\right](\vec{\sigma}_1 - \vec{\sigma}_2)\left\{\vec{p}, F_m(r)\right\} +$$

$$-\left[h_\rho^{pp} g_\rho(1+\chi_V) + h_\omega^{pp} g_\omega(1+\chi_S)\right](i\vec{\sigma}_1 \times \vec{\sigma}_2)\left[\vec{p}, F_m(r)\right]_-$$

$$F_m(r) = \frac{e^{-mr}}{4\pi M} \qquad m = m_\rho = m_\omega.$$

Using the fact that the spin matrix elements obey

$$\langle \text{triplet}|\vec{\sigma}_1-\vec{\sigma}_2|\text{singlet}\rangle = \langle \text{triplet}|i\vec{\sigma}_1 \times \vec{\sigma}_2|\text{singlet}\rangle$$

it is convenient to split f_J into two parts of the form

$$f_J^s \propto \langle J\ L=J\pm1\ s=1|(\vec{\sigma}_1-\vec{\sigma}_2)\ F_m(r)\vec{p}|J\ L=J\ s=0\rangle$$

$$f_J^t \propto \langle J\ L=J\pm1\ s=1|(\vec{\sigma}_1-\vec{\sigma}_2)\vec{p}\ F_m(r)|J\ L=J\ s=0\rangle$$

which represent the difference (proportional to χ) and sum (proportional to $2+\chi$) of the anticommutator and commutator part, respectively, of the matrix element of V_{pp}^{PV}. The superscripts s and t stand to remind that the operator p acts on the singlet (L=J) or triplet L=J±1) side of the matrix element. Now p is responsible for the transition between different L values and since $F_m(r)$ has short range ($m^{-1} \sim 0.25$ fm) its matrix element in the lowest possible L state dominates at energies considered here. For a semiquantitative analysis we therefore find

$$f({}^1S_0 - {}^3P_0) \sim \left[h_\rho^{pp} g_\rho(2+\chi_V) + h_\omega^{pp} g_\omega(2+\chi_S)\right] f_{o^+}^t \sim \left[15\ h_\rho^{pp} + 15\ h_\omega^{pp}\right] f_{o^+}^t$$

$$f({}^3P_2 - {}^1D_2) \sim -\left[h_\rho^{pp} g_\rho \chi_V + h_\omega^{pp} g_\omega \chi_S\right] f_2^s \sim -\left[9.7\ h_\rho^{pp} - 0.94\ h_\omega^{pp}\right] f_2^s.$$

Fig. 7 shows new calculations of $f_{o^+}^t$ and f_2^s done using the Paris potential to describe the p-p scattering states[12].

Referring to the 45 MeV measurement one sees that this rough analysis yields $h_\rho^{pp} + h_\omega^{pp} \sim 18 \times 10^{-7}$ for $f_{o^+}^t$ calculated with the Paris potential while the DDH value is 18.4×10^{-7}. With the Reid potential $h_\rho^{pp} + h_\omega^{pp} \sim 21 \times 10^{-7}$.

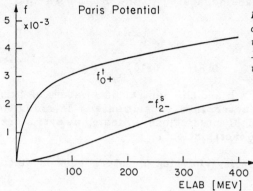

Fig. 7. Energy dependence of the dominating amplitudes f_{0+}^t and f_{2-}^s calculated with Paris potential p-p wavefunctions.

A measurement around 230 MeV could determine h_ρ^{pp} alone[13].

It must be emphasized that this analysis is based on the strong interaction properties of ρ and ω, i.e. the values of their anomalous magnetic coupling to nucleons. The value used is due to vector meson dominance of the electromagnetic formfactors of the nucleon (at low q^2).

p-NUCLEUS SCATTERING AT HIGH ENERGIES

Recall that at 6 GeV/c beam momentum an effect in the p-H_2O total crossection has been observed which is an order of magnitude larger than the ones discussed so far. Theoretical analysis at this high energy is much less consolidated than at low energies. Even the strong N-N scattering system, needed for the analysis of parity violating effects, is not understood as well.

The first theoretical calculations have been done on parity violating meson exchange, as at low energies and on simple $W^\pm$ and Z exchange. In spite of the different approaches, predictions are all $<(1$ to $2)\times 10^{-7}$ or even much smaller and thus miss the 6 GeV result by at least an order of magnitude[14].

An explanation of the large high energy effect was proposed by Nardulli and Preparata[15]. They found that a parity violating wavefunction admixture in the outer nucleons together with an effective Lagrangian describing Regge-behaviour of the strong scattering amplitudes is able to reproduce the large high energy data. The effect is predicted to be proportional to P/E and thus constant at high energy. However, it has been pointed out[16] that introducing the same parity violating wavefunction admixture into the low energy N-N system leads there also to similarly large effects, in clear contradiction with the data, unless the strong meson-nucleon coupling constants are reduced drastically[17] (~ factor 3) from the values generally accepted (for low energy physics).

The idea of a wavefunction effect in the nucleon has been taken up again and integrated into an explicit quark model calculation by Goldman and Preston[18]. They consider diquark-quark scattering where parity violation is generated by weak interaction acting within the diquark which is part of the (polarized) beam nucleon (Fig. 8a). They also find that it is possible to reproduce the large experimental value of A_L at high energy. In addition they predict a rise by an additional factor 60 in going to 500 GeV (FNAL). This large prediction is based on interference between the weak and strong amplitudes shown in Fig. 8a and 8c, respectively, devided by the total crossection calculated from Fig. 8b. But this means that the weak-strong interference contains an amplitude which is omitted in the calculation of the crossection used to normalize it. In my opinion this is not consistent and the question arises whether treatment of strong interaction to the same order in the numerator and the denominator of A_L, would still lead to the large prediction. Numerical calculations show[19] that inclusion of the amplitude of Fig. 8c in the total crossection indeed reduces the prediction considerably if the same parameters are used as in the calculation as in ref. 18. On the other hand Fig. 8b alone gives a good description[18] of the high energy total crossection of ~ 40 mb, roughly constant, and this agreement is completely destroyed if Fig. 8c is added[19]. No consistent picture therefore

Fig. 8. Graphs for diquark-quark model of N-N scattering. The point denotes weak interaction and wavy lines gluons.

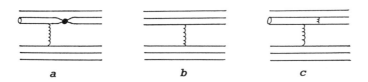

emerges neither way and probably this should also not be expected since what deermines the total crossection is the foward scattering amplitude which is dominated by difraction and low momentum transfer for which pertubative QCD is hardly expected to work.

Thus it is still not clear whether the large experimental result can be understood.

Additional experiments above 6 GeV would certainly be desirable, and could, in principle be done at the new polarized proton facility at the Brookhaven AGS. What are the prospects of such a measurement? The equivalent of the β-decay processes discussed in connection with the 45 MeV measurement is hyperon production (on the target itself) and its parity violation decay. In the 6 GeV experiment hyperon decays have been eliminated by a spectrometer with a pair of deflection dipoles. The problem with such an arrangement is that it lowers the symmetry of the arrangement and thus makes it more sensitive to double scattering effects discussed also in connection with the 45 MeV experiment. The hyperon production crossection is expected to grow if the energy is increased so that the problem may get worse at BNL energies (~20 GeV). If this can be overcome in a satisfactory way, I think an experiment at a level of one to a few times 10^{-7} is feasible.

CONCLUSION

Parity violation in p-p scattering below pion threshold is amenable to a reliable analysis in terms of two parity violating meson-nucleon coupling constants h_ρ^{pp} and h_ω^{pp} which in turn can be used to test methods to treat weak matrix elements in quark model calculations. The new high precision data at 45 MeV gives significant restriction for $h_\rho^{pp}+h_\omega^{pp}$. A measurement at ~ 230 MeV could determine h_ρ^{pp} alone but very high precision will be needed.

For high energy p-N scattering no similarly well established method of analysis is available yet and I do not think that the 6 GeV/c result is understood. Additional experiments at the same or higher energy would be important.

I wish good luck to whoever accepts the challenge to do one or the other of these measurements.

REFERENCES

1. D.E. Nagle, J.D. Bowman, C.M. Hofman, J.L. McKibben, R.E. Mischke, J.M. Potter, H. Frauenfelder, L. Sorenson, in High Energy Physics with Polarized Beams and Polarized Targets Argonne-1978, AIP Conf. Proc. <u>51</u> (1979), ed. by G.H. Thomas (AIP New York, 1979) p. 224
2. S. Kistryn, J. lang, J. Liechti, Th. Maier, R. Müller, F. Nessi-Tedaldi, M. Simonius, J. Smyrski, W. Haeberli, J. Sromicki, to be published
3. V. Yuan, this conference
4. N. Lockyer, T.A. Romanowski, J.D. Bowman, C.M. Hofman, R.E. Mischke, D.E. Nagle, J.M. Potter, R.L. Talaga, E.C. Swallow, D.M. Alde, D.R. Moffett, and J. Zyskind, Phys. Rev. Lett. <u>45</u>, 1821 (1980) 1 and Phys. Rev. <u>D30</u>, 860 (1984)
5. R. Balzer, R. Henneck, Ch. Jacquemart, J. Lang, F. Nessi-Tedaldi, Th. Roser, M. Simonius, W. Haeberli, S. Jaccard, W. Reichart, and Ch. Weddigen, Phys. Rev. <u>C30</u>, 1409 (1984)
6. J. Lang, Th. Maier, R. Müller, F. Nessi-Tedaldi, Th. Roser, M. Simonius, J. Sromicki, and W. Haeberli, Phys. Rev. Lett. <u>54</u>, 170 (1985)

7. W. Haeberli, R. Henneck, Ch. Jacquemart, J. Lang, R. Müller, M. Simonius, Ch. Weddigen, and W. Reichart, Nucl. Instr. and Methods 163, 403 (1979)
8. M. Simonius, R. Henneck, Ch. Jacquemart, J. Lang, W. Haeberli, and Ch. Weddigen, Nucl. Instr. and Methods 177, 471 (1980)
9. B. Desplanques, J.F. Donoghue, and B.R. Holstein, Ann. Phys. (NY) 124, 449 (1980)
10. M. Simonius, in Interaction Studies in Nuclei, eds. H. Jochim anbd B. Ziegler (North Holland, Amsterdam, 1975) p.3
11. M. Jacob and G.C. Wick, Ann. Phys. (NY) 7, 404 (1959)
12. F. Nessi-Tedaldi and M. Simonius, to be published
13. R. Abegg et al., Triumf proposal
14. E.M. Henley and R.F. Krejs, Phys. Rev. D11, 603 (1975)
    A. Barroso and D. Tadic, Nucl. Phys. A364, 194 (1981)
    T. Ika, Prog. Theor. Phys. (Japan) 66, 977 (1981)
    P. Chiappetta, J. Sofer and Tai Tsun Wu, J. Phys. G8, L93 (1982)
    S.K. Singh and I. ahmand, Phys. Lett. 143B, 10 (1984)
15. G. Nardulli and G. Preparata, Phys. Lett. 117, 445 (1982)
16. M. Simonius, in High Energy Spin Physics-1982, ed. G.M. Bunce, AIP Conf. Proc. 28 (AIP, New York, 1983) p. 139
17. G. Nardulli and G. Preparata, Phys. Lett. 137B, 111 (1984)
18. T. Goldman and D. Preston, Nucl. Phys. B217, 61 (1983) and Phys. Lett. 168B, 415 (1986)
19. M. Simonius and L. Unger, in preparation

QUARKS IN SPECTROSCOPY[$]

Gabriel KARL

Guelph Waterloo Program
for Graduate Work in Physics
Guelph, Ontario, N1G 2W1

Abstract: I discuss current issues in hadron spectroscopy and their connection to Quantum Chromodynamics and to more general issues of hadron physics. The discussion is in elementary terms and is directed to non-experts.

When Alan Krisch asked me to give this lecture he stressed that it should be extremely elementary, at the level of a Colloquium talk. So if you find all of this too elementary you should recall that I am following instructions. There is not much advanced or original stuff in this talk, all of it is well known to the experts. I have attempted to describe in simple terms the present situation of the field of hadron spectroscopy with the hope that it may encourage some of you to do new and better experiments which is what the subject is in desperate need of.

What is spectroscopy? Spectroscopy is the branch of physics in which we study systems through their response to different frquencies. Thus we find the spectrum of excitations of the system. From the spectrum we learn:
(a) the ´moving parts´ or ´degrees of freedom´ which are called ´constituents´ of the system in hadron spectroscopy, and
(b) the ´restoring´ forces which are called ´forces between constituents´ in hadron spectroscopy.

The classic example of spectroscopy is ATOMIC spectroscopy which studies atomic excitations usually in the electron-volt (eV) range. The degrees of freedom are electrons and the forces are electromagnetic. The situation is ideal because the constituents are point-like and the forces are very simple. In MOLECULAR spectroscopy (rotational and vibrational) the energy scale is even smaller, fractions of an eV, and the moving parts are atoms (or their nuclei). The forces between atoms are residuals after most of the Coulomb forces cancel out.

[$]: Lecture at the 2nd Conference on the Intersections between Particle and Nuclear Physics, Lake Louise, Canada, May 26-31, 1986

NUCLEAR spectroscopy is in the MeV range, the degrees of freedom are nucleons and the forces between them are also residuals after most inter-quark forces are cancelled out. HADRON spectroscopy is in the GeV range, the constituents are quarks and the forces between them are strong, transmitted through gluon exchange or hadronic strings. Because quarks are point-like and the forces between them stem from gluon exchange, there is hope that hadron spectroscopy may reach the simplicity and precission attained by atomic spectroscopy.

In the period between 1950-1970 hadron spectroscopy was an important branch of high energy physics. By 1975, when the quark structure of hadrons became widely accepted, hadron spectroscopy was handed over to intermediate energy physics, and in this decade there are signs that it is considered a branch of nuclear physics. By extrapolation it follows that in another ten years hadron spectroscopy will join atomic spectroscopy as a branch of Chemistry !

Traditional hadron spectroscopy deals with the spectrum of of mesons (quark-antiquark systems) and baryons (three quark systems). These simple SU_3-color singlet systems have to be constructed from the ultimate theory of hadrons, Quantum Chromodynamics, (Q.C.D.). The entities of Q.C.D. are color triplet quarks and color octet gluons. The path from the world of quarks and gluons to the world of mesons and baryons is not yet well charted although a great deal of effort has already been spent. I guess that the best path will turn out to be : QCD---> LATTICE---> STRING---> POTENTIAL---> HADRON SPECTRUM. The connection of potential models to hadron spectra is already well developed, as is the relation of lattice models to QCD. It is likely that strings will provide the physical explanation of of potential models and that the strings will be derived from lattice models. It seems unlikely that computations of the spectrum will be made directly in lattice models especially in cases like the upsilon spectrum where there are many bound states. There are also other paths from QCD to the world of mesons and baryons, like the BAG MODEL, but these are less likely to be useful in spectroscopy on account of unphysical degrees of freedom. A path which has been studied recently, connecting QCD to baryons is the SKYRMION MODEL which is valid in the limit of N_c very large. Many baryon properties are remakably close in this limit to their physical values.

MESONS: Heavy quarkonia are simpler than their light counterparts, both because nonrelativistic approximations are valid, and also because for heavy systems the quark antiquark distance is small and at short distances the potential is dominated by Coulomb forces just as in atoms. At larger distances it has been found that the potential between quark and antiquark is approximately linear. A linear potential is easy to understand in terms of strings since strings have energy proportional to their lemgth. A very nice physical analogy was proposed ten years ago by Parisi. A magnetic string is produced when a magnetic monopole and antimonopole are immersed in a superconductor. Because of the Meissner effect the magnetic field does not penetrate in the superconductor and the field lines are collimated in a narrow string between the monopole and the antimonopole. The field strength inside the string is large enough to effect the transition of the superconductor to normal phase, and this critical value determines the diameter of the string. The energy of the system is proportional to the distance between the monopole and antimonopole-- a linear potential. It is generally believed that the vacuum of QCD inside a hadron differs from the vacuum outside the hadron, and this notion was first understood in the context of the bag model. The difference between bags and strings is purely quantitative: for if the diameter of the string is large compared to the quark-antiquark separation we have a bag, and in the opposite limit we have a string. The diameter of the string is typically the penetration depth of the "magnetic" field, which we can approximate by the distance over which single gluon exchange dominates. From fits to the charmonium spectrum with the Cornell potential we can take this distance to be of the order of 0.2 fermi whereas the typical distance between quark and antiquark in light quarkonia is about 0.6 fermi. Thus a string picture seems more appropriate than a bag picture. Note also that this way of looking at hadrons looks very similar to the understanding of phase transitions between normal and superconducting states of metals. In this light the progress of QCD is not so slow when compared with the 30 years required to reach the BCS milestone from QED. Moreover the understanding of superconductivity did not orriginate in extensive numerical work but from the physical insight provided by ingenious experiments on isotopic effects. The spectroscopy of mesons has provided already information about quark spins, from hyperfine and spin-orbit splittings. The spin independent forces between quarks and anti-quarks also appear to be universal, the same for any quark antiquark system consistent with the color origin of this potential. It seems very likely that in not too long a time lattice models will give us this universal quark-antiquark interaction.

The physics of quarks and strings at short distances is revealed most directly in HADRON DECAYS. These processes correspond to string breaking with the creation of quark-antiquark pairs. A very important question is whether at distances where the decays take place we should consider individual gluons or we are already in the non-perturbative regime of strings. The issue is not yet solved. Of particular interest in this connection are the so-called OZI-forbidden decays, such as the decay of charmonium into mesons composed of light quarks only. Examples are the decy of J/psi to pi-rho or to omega-f. In these decays the heavy quark and antiquark disappear and light quarks are created instead. In a perturbative description the annihilation of the heavy quarks is into three gluons which then proceed to create light quark-antiquark pairs. It is not obvious whether at the distance scale of annihilation, which is the Compton wavelength of the charmed quark, about 0.15 fermi we should consider gluons or strings . One relevant question (for which I am indebted to R.Petronzio) is whether the separation of graphs into singly OZI-suppressed and doubly OZI-suppressed is gauge invariant. There are indications that the decays into light quark pairs proceed sequentially, with strings between quark antiquark pairs breaking. This mechanism will suppress omega-f´ relative to omega-f since the f´ contains only strange quarks and antiquarks and cannot be produced in conjunction with an omega by a sequential mechanism. The dominance of sequential production indicates that even at the distances where the decay takes place strings and their breaking are relevant rather than gluons.

The spectrum of BARYONS can be derived in potential models in a very satisfactory way and agreement with experiment is very good. Since baryons are made of three quarks they have two normal modes and their spectrum is somewhat richer:"baryon" = "two-mesons". The most interesting recent development involving baryons is the derivation of their properties in the Skyrme model which is valid in the limit of a large number of colors. At this Coference there is a discussion of the derivation of pion-nucleon phase-shifts from a Skyrme model.

The main question of hadron spectroscopy is whether one can excite additional degrees of freedom in hadrons in addition to quarks. In terms of strings this is formulated in two ways:
(A) Are there closed strings , without quarks ? These states are the so-called gluonia, which in perturbative language are bound states of a few gluons.
(B) Can one excite transverse vibrations of the open strings ? These states are called hybrids and in a perturbative language correspond to bound states of quarks and (valence) gluons.

Gluonia are excitations of the vacuum whereas hybrids are excitations of ordinary mesons and baryons. Because our understanding of the vacuum is rather unsatisfactory it is probable that the predictions of hybrids are more reliable than the prediction of gluonia.

The experimental situation regarding these states is still confused and what we need is higher statistics data in the 1-3 GeV region as emphasized in the talk of Chanowitz at this Conference.

There are other exotic hadrons like meson-molecules and dibaryons like the H discussed at the Coference by Barnes. Although these states are very important in many ways their impact on hadron spectroscopy is less dramatic than that of gluonia and hybrids.

In Summary hadron spectroscopy is still not finished. There is room both for new qualitative and quantitative advance. Even as a branch of Nuclear Physics there is hope for further progress in this field!

Acknowledgements I am greatly indebted for the opportunity to give this talk to Allan Krisch and the organizers of this Conference. I also thank Dean J.R.Macdonald and the Sheriff of the County of Wellington for making this talk possible. Conversations with M.Chanowitz,J.Donoghue,N.Isgur,H.Lipkin, M.Mattis and other participants at the Conference helped me in preparing both the talk and this manuscript.

REFERENCES:
1. N.Isgur and G.Karl, Physics Today, November 1983,p.36
2. M.Chanowitz,These Proceedings.
3. J.Donoghue,These Proceedings.
4. P.Barnes,These Proceedings.
5. M.Mattis,These Proceedings.
6. S.Koonin,These Proceedings
7. W.Lockman,These Proceedings.

RELATIVISTIC HEAVY ION FACILITIES--WORLDWIDE*

L.S. Schroeder

Nuclear Science Division, Lawrence Berkeley Laboratory
University of California, Berkeley, California 94720

ABSTRACT

A review of relativistic heavy ion facilities which exist, are in a construction phase, or are on the drawing boards as proposals, is presented. These facilities span the energy range from fixed target machines in the 1-2 GeV/nucleon regime, up to heavy ion colliders of 100 GeV/nucleon on 100 GeV/nucleon. In addition to specifying the general features of such machines, I will also outline the central physics themes to be carried out at these facilities, along with a sampling of the detectors which will be used to extract the physics.

INTRODUCTION

The aim of this talk is to provide you with an updated review (May 1986) of relativistic heavy ion facilities throughout the world. As my working definition, "relativistic" will refer to those machines capable of providing beams of heavy ions at kinetic energies ≥ 1 GeV/nucleon. Furthermore, I will include those facilities which are either in the real (operating or under construction) or virtual (proposal) state. Seven facilities survive this classification scheme and are shown on the world map in Fig. 1. They are:
1) operating--Bevalac, Dubna, Saturne II, AGS[1] and SPS[2]
   (these latter two are set to operate in late 1986)
2) construction phase--SIS 18 at GSI Darmstadt[3]
3) proposal stage--Bevalac Upgrade,[4] Holifield Upgrade,[5]
   Saturne II + MIMAS,[6] Nuclotron[7] and RHIC.[8]
It is interesting to note that four of the seven either are or were in a previous life involved only in particle physics research. A fifth, namely RHIC, has the potential of rising out of the ashes of the CBA high energy physics project.

In this talk I will also briefly cover the physics issues that are being or will be addressed at these facilities. Being an experimentalist, I think of a facility as being composed not only of the accelerator but also of the detectors that are used to isolate the interesting physics. So I want to provide you with some idea of the devices that are already operating or will be called upon at future machines. Finally I will end with a time

---

*This work was supported by the Director, Office of Energy Research, Division of Nuclear Physics of the Office of High Energy and Nuclear Physics of the U.S. Department of Energy under Contract DE-AC03-76SF00098.

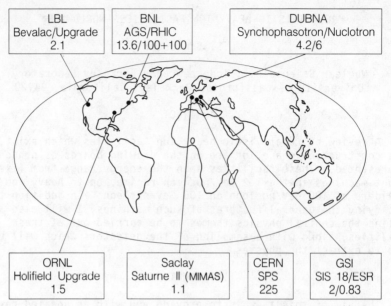

Fig. 1. World map showing location of the relativistic heavy ion facilities. The maximum kinetic energy for each is indicated.

table indicating the fondest dreams of proposers as to when we might expect the new accelerator complexes to be ready for action, and some personal observations.

PHYSICS GOALS

The upper portion of Fig. 2 shows the now familiar phase diagram of nuclear matter (T vs. ρ/ρ_0), which can be probed in heavy ion collisions. However, there are other degrees of freedom which can also be studied in such collisions. As an example, one can cite the isospin (I) degree of freedom, where one studies the limits of proton and neutron number in nuclei. This area is indicated by the chart of the nuclides in the lower portion of Fig. 2. In terms of physics to be studied with the relativistic heavy ion facilities under discussion here, a convenient division into two parts is possible:
 1) intermediate energy machines (E < 10 GeV/nucleon)--used to probe nuclear matter under extreme conditions (T,ρ,I,J,S)
 2) high energy (quark-matter) machines (E > 10 GeV/nucleon)-- used to achieve high energy density and thereby produce and study the quark-gluon plasma (QGP).[9,10]

A. Physics with Intermediate Energy Machines

The general program to be carried out at these machines (Bevalac/Upgrade, Holifield Upgrade, Saturne II + MIMAS, SIS 18 and Synchrophasotron/Nuclotron) includes:

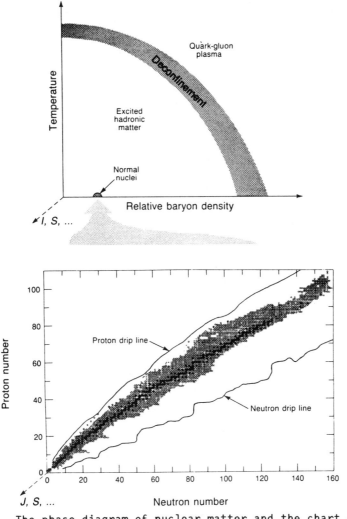

Fig. 2. The phase diagram of nuclear matter and the chart of the nuclides.

1) extension of existing studies--à la the present Bevalac program
2) new opportunities provided by much higher beams currents (typical increases of 10^2-10^3)
3) exploiting new techniques such as cooler-storage rings for studying a wide range of phenomena.

It should be noted that by using modern strong focusing synchrotrons, experiments will be greatly enhanced through improved beam quality, duty factor and allow for more flexible machine operation. More specifically, the essential elements of a physics program will include:

1) nuclear matter equation of state via measurements of
   - π, K, e^+e^- probes (excitation functions)
   - collective flow
   - composite yields (sensitive to entropy in system)
2) liquid-gas phase
3) expanding domain of nuclei
   - exotic nuclei
   - radioactive beams
4) nuclear dynamics
   - cooperative effects (subthreshold π,K)
   - 1-2 body forces
   - transfer reactions
5) nuclear structure
   - decay modes
   - giant resonances
6) nuclear astrophysics
7) other areas
   - atomic physics
   - biomed applications
   - technology applications

It is clear that a rich and varied program of nuclear physics is accessible with such facilities.

B. Physics with High Energy Machines

The general thrust of these facilities (AGS, SPS, RHIC) will be the exploration of high energy and baryon density in central nucleus-nucleus collisions. This program will include:

1) quark-gluon plasma (QGP)--present calculations[10] indicate that energy densities of ~1-2 GeV/fm^3 are sufficient to produce the QGP
2) pushing the equation of state of nuclear/hadronic matter to higher T, ρ
3) studying the properties of highly excited hadronic matter (as a necessary by-product of isolating the QGP).

As pointed out in T. Ludlam's talk,[10] at high energies in nucleus-nucleus collisions there are two areas of interest. Somewhere in the range of 1-10 GeV/nucleon in the c.m. frame one expects to achieve maximum baryon density or "nuclear stopping." At energies above 30 GeV/nucleon in each beam, "transparency" sets in and two separate regions can be identified--the fragmentation regions carrying the net baryon number of the colliding systems and a central region mostly occupied by mesons. These two scenarios are sketched in Fig. 3.

Signatures for the formation of quark matter are actively debated. Standard arguments include:[10-16]

1) studying $<p_T>$ as a function of energy density ($\propto dN/dy$)-- a first order transition could lead to a flat $<p_T>$ curve (analogous to the ice $\rightarrow$ H$_2$O transition)
2) large fluctuations in quantities like dE_T/dy signaling explosions or deflagrations from the QGP
3) enhanced yields of strange particles (particularly antihyperons) due to $gg \rightarrow s\bar{s}$ processes in the QGP

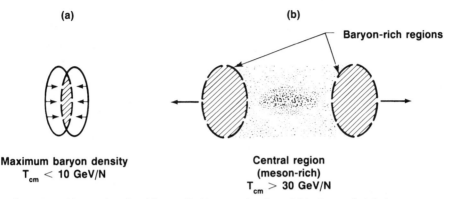

Fig. 3. Characterization of the central collision of high energy nuclei showing a) "nuclear stopping" and b) "transparency" regimes.

4) directly studying the plasma phase with weakly interacting probes (γ, e^+e^-, $\mu^+\mu^-$)--these can carry information from the hot, compressed stage of the collision.

To date, no single signature for the QGP has emerged as the prime candidate. In all cases, one must be concerned with the yield of particles from the hadronization phase also, as each experiment will measure contributions from both (hadronic + plasma) phases. In general, this means that each experiment will have to be capable of measuring several observables within a given event and studying their correlations. Thus, some form of global analysis will be required. In addition, experiments will need to study collisions as a function of energy (adjusting energy density) and mass (for light systems only expect production from excited hadronic gas) of the colliding partners. These considerations have profound effects on detector systems.

THE FACILITIES (MACHINES, DETECTORS)

In this section I want to outline the basic parameters of each facility and point out their unique features. In addition, some discussion of detector requirements for nuclear physics experiments will be included. As earlier, I divide this section into two parts--one dealing with intermediate energies, the other higher energies.

A. Intermediate Energy Facilities

Five facilities come under this category: Bevalac/Upgrade, Holifield Upgrade, Saturne II + MIMAS, SIS 18 and the Synchrophasotron/Nuclotron. Figs. 4-8 show a plan view of each. Table I

Table I MACHINE PARAMETERS

Site	Facility	KE (GeV/N) Ne	KE (GeV/N) U(chg.st.)	Estimated intensity* Ne/sec	Estimated intensity* Ne/cycle	Estimated intensity* U/sec	Estimated intensity* U/cycle
LBL	Bevalac	2.1	$0.96(68^+)$	$\sim 10^9$	---	$\sim 3 \times 10^5$	---
	Upgrade	1.92	$0.86(68^+)$	$\sim 10^{11}$	---	$\sim 4 \times 10^8$	---
ORNL	Holifield Upgrade	1.5	$0.45(54^+)$	---	$\sim 8 \times 10^{10}$	---	$\sim 2 \times 10^{10}$
Saclay	Saturne II + MIMAS	1.18	$0.58(?)$	---	$\sim 5 \times 10^{10}$	---	$\sim 2 \times 10^7$
GSI	SIS 18	2.0	$1.0(78^+)$	$\sim 2 \times 10^{10}$	---	$\sim 4 \times 10^{10}$	---
	ECR	0.83	$0.57(92^+)$	(see Ref. 3)			
Dubna	Synchro.	4.1	---	---	$\sim 10^5$	---	---
	Nuclotron	6.0	$3.5(82^+)$	---	$\sim 10^{10}$	---	$\sim 10^9$

*Intensities are quoted either as particles/sec or particles/cycle. For definition of individual machine <u>cycles</u> see Refs. 5, 6 and 7.

lists a typical beam energy for two nuclei (Ne and U) and the corresponding intensities that are expected. <u>One needs to bear in mind that Table I is not meant as a critical evaluator of the overall performance features of these facilities--but rather as a rough comparator of two specific parameters.</u> The reader is strongly urged to consult Refs. 3-7 for more detailed information.

I now briefly describe each of these facilities--starting in the west and moving east.

1) <u>Bevalac/Upgrade (Fig. 4)</u>: The Bevalac at LBL is at present the only relativistic heavy ion machine capable of accelerating the complete periodic table (p to U) for nuclear physics research. It came on-line in 1974 with beams of A ≤ 56. In 1982, after installation of the new vacuum liner (providing pressures ~few x 10^{-10} torr), the heavier beams up to uranium became available. Significant studies of compressed nuclear matter have resulted from the uranium upgrade. However, the Bevatron is a weak focusing synchrotron and must be replaced if higher currents of the heavier ions (A > 100), as needed by the physics, are to be achieved. In addition, the Bevatron cannot function as an adequate injector for other devices, such as a storage ring. With this in mind, LBL is looking at the possibility of replacing the Bevatron, with a modern strong-focusing synchrotron injected by the existing SuperHILAC. This

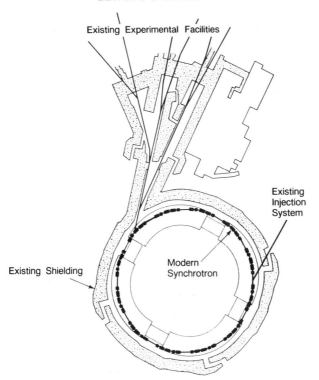

Fig. 4. View of Bevalac Upgrade showing replacement synchrotron and existing experimental hall.

machine would have $(B\rho)_{max} = 18$ Tm. Existing shielding and experimental halls would be used to keep costs down. A future addition of a storage ring with the potential of cooling is being studied. Such a device would be useful for a variety of experiments with radioactive beams, both in and out of the ring, stretcher device for large 4π detector experiments, stripping and re-injection into the main ring to provide higher energies, atomic physics, etc.

2) <u>Holifield Upgrade (Fig. 5)</u>: The ORNL group has proposed a major upgrade to their existing Tandem facility. This involves a veritable 3-ring circus, including: a 4 Tm accumulator/booster, a 15 Tm main ring and a 10 Tm storage/cooler ring to provide duty factor, resolution ($\Delta p/p \sim 10^{-4}$) and brightness. The full range of projectiles would be available for studies at both high and low energies at high intensities. The facility would use the 25 MV Tandem as its injector, with construction of the synchrotron facility and experimental hall taking place on a "green site" next to the Tandem building. A broad physics program is envisioned.

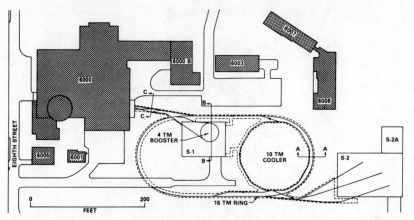

Fig. 5. Site plan for the Holifield synchrotron facility. Existing buildings denoted by cross hatching.

3) <u>Saturne II + MIMAS (Fig. 6)</u>: The Saturne II facility will be upgraded (by 1987) with the addition of a strong focusing booster synchrotron ring (MIMAS) and a new heavy ion source (Dione). MIMAS will replace the present LINAC. This combination will allow the acceleration of heavier ions and increase intensities of polarized protons and deuterons for nuclear physics research. The heavy ions will be extracted into the existing experimental halls for research in a wide variety of existing detectors.

4) <u>SIS 18 + ESR (Fig. 7)</u>: GSI has embarked on a major new project (this one is funded!) to produce a modern synchrotron facility injected by their existing UNILAC heavy ion linac. In addition to the main 18 Tm synchrotron ring, they are planning to include an experimental 10 Tm storage ring (ESR) which will incorporate both electron and stochastic cooling—a true state-of-the-art device. They envision a major physics program with the ESR. The thrust of the program with SIS 18 will be to provide beams up to ^{238}U beams at 1-2 GeV/nucleon at high intensity. These beams will be delivered to a new experimental hall (generally using slow extraction) or to the ESR (fast extraction). In the ESR facilities for storing and cooling completely ionized stable and radioactive (produced via projectile fragmentation) beams will be available. The ESR will also have internal target facilities for studies with the cooled circulating beams. They also plan a mode of operation of the ESR in which two beams of completely stripped uranium with different energies can be made to merge in order to study effects of large coulomb fields as discussed at this conference.[17] This facility is expected to turn on for physics research in 1989 and will represent a major new tool for the heavy ion community to exploit.

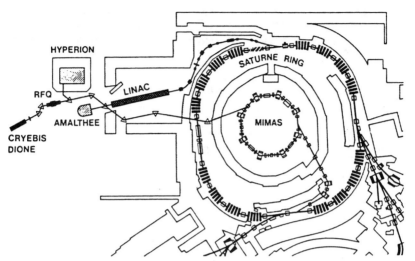

Fig. 6. MIMAS inside the main Saturne II ring together with three ion sources (Hyperion, Cryebis, Dione) and the RFQ pre-injector.

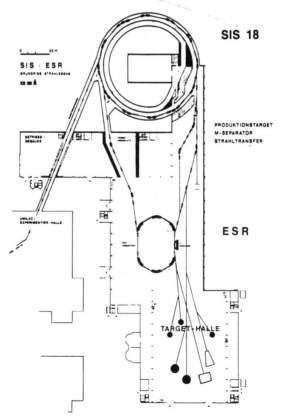

Fig. 7. Plan view of SIS 18 and the experimental storage ring (ESR).

5) Synchrophasotron/Nuclotron (Fig. 8): The Synchrophasotron at Dubna is the largest of the weak focusing synchrotrons. Since the early 1970's they have been engaged in a program of research at 3-4 GeV/nucleon with relatively weak beams of light ions (A < 30). They have been studying the possibility of placing a new strong focusing synchrotron (Nuclotron) in the tunnel below the Synchrophasotron. The Nuclotron would employ a superconducting design based around a superferric magnet. When coupled with a new ion source they expect to achieve energies up to about 6 GeV/nucleon for q/m = 1/2 ions and much higher intensities than presently available. Beams up to uranium would be available for research in the existing experimental halls.

The new facilities being discussed will place heavy demands on detectors, particularly those involving full event measurements. As we already know from Bevalac experience a single streamer chamber photograph will show ~100-200 charged particles being emitted in central collisions of large nuclei at 1-2 GeV/nucleon. For electronic experiments covering a large solid angle this implies the need for a high degree of segmentation to avoid multiple hits and allow full event analysis. An example of such a detector is the GSI/LBL Plastic Ball/Wall shown in Fig. 9. This device contains over 650 elements in the Ball and over 150 elements in the Wall and provides almost 4π coverage. Figure 10 shows the result of a multiplicity measurement with it. Tracking devices such as TPC's and streamer chambers with CCD read-outs are actively being pursued to study interactions at these energies. Of course, not all experiments will involve ~4π measurements. Magnetic spectrometers with limited apertures will be required to study a variety of phenomena (e.g., excitation functions for subthreshold production, production and delivery of radioactive beams, etc.). A large arsenal of detectors will be needed to exploit the full potential of these facilities.

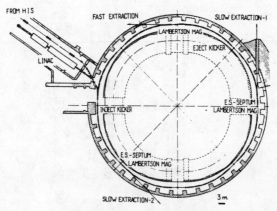

Fig. 8. General layout of Nuclotron at Dubna.

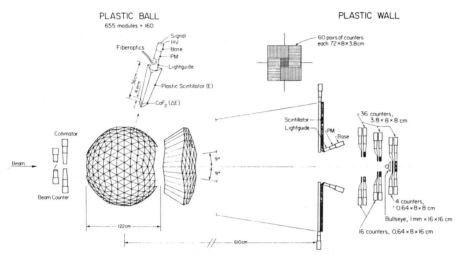

Fig. 9. The GSI/LBL Plastic Ball detector.

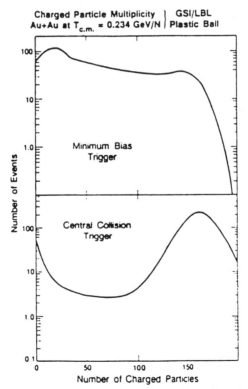

Fig. 10. Charged particle multiplicity distribution for Au + Au collisions studied by GSI/LBL Plastic Ball collaboration at the Bevalac.

B. High Energy Facilities

We next consider the three high energy facilities (AGS/RHIC, SPS). Since we finished the last section on the European side of the Atlantic, we start there and move west.

1) SPS at CERN (Fig. 11): In 1980 a proposal was submitted by a GSI/LBL collaboration to accelerate light ions (^{16}O) in the CERN PS for delivery to two experiments--involving the Plastic Ball and a Streamer Chamber. Since that time the program has expanded and now involves a large experimental effort (involving ~300 nuclear and particle physicists) to conduct several major experiments at the CERN SPS.[2] The initial round of experiments will be carried out in Nov.-Dec. 1986, with a follow-up run in Sept.-Oct. 1987 (with possibly improved ECR source giving ions up to Ca).

To provide light ions to the SPS, a new ECR source built by Geller has been coupled to an RFQ (built by GSI/LBL) and used to inject LINAC 1 at the PS.[18] These components are now in place. ^{16}O ions will be accelerated in the PS, transferred and accelerated in the SPS, where they will be extracted for high energy heavy ion physics experiments at 60-225 GeV/nucleon in both the North and West areas. As indicated earlier, the central theme will be the exploration of high energy and baryon density and the possible production of the QGP. In terms of the diagrams in Fig. 3 one can characterize the SPS (and AGS) program as probing the "nuclear stopping" regime. Table II lists the present makeup of the CERN program.

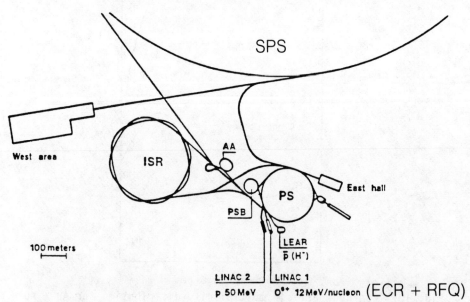

Fig. 11. Layout of the CERN accelerator complex (PS → SPS → heavy ion experiments in West and North (not shown) areas).

Table II SUMMARY OF CERN (SPS) HEAVY ION EXPERIMENTS

Experiment	Main detectors	Partial list of observables
NA34 (HELIOS)	~4π calorimetry, tracking, dimuon spectrometer	1,2 particle inclusive, E_T, direct γ, e^+e^-, $\mu^+\mu^-$
NA35 (Str. Ch.)	2m streamer chamber, calorimeters, π^0 detector (PPD)	Λ, $\bar{\Lambda}$, K, $<n_{\pi^-}>$, $dN/d\eta$
NA36 (TPC)	TPC, forward calorimiters, multiplicity	strange-antistrange baryons ($\Lambda,\ldots,\Omega$)
NA38 (Quark Search)	Hg (to be analyzed in levitometer)	fractional charge
NA39 (Dimuons)	NA10 dimuon spectrometer, neutral calorimetry	$\mu^+\mu^-$, $\dfrac{dE_T^0}{dy}$
WA80 (Plastic Ball)	Plastic Ball, forward calorimeters, π^0/γ detector	$<M>$, $dN/d\eta$, neutral and charged energy flow, π^0, direct γ
EMU 1,2,3	Emulsions, plastics	$dN/d\eta$, survey various distributions

At these energies the detectors not only have to respond to much higher fluxes of particles (expect individual events to contain ~100-1000 particles) but also the strong kinematic focusing which throws everything into a narrow forward cone at a fixed target machine. This imposes strict constraints on the segmentation of detectors. Figure 12 shows a schematic of the NA35 streamer chamber experiment. Shown in plan view is the streamer chamber, along with the downstream bank of calorimeters. These calorimeters can be used to trigger the streamer chamber cameras (UA5 cameras with image itensifiers) on events of interest, e.g., those involving large hadronic or electromagnetic energy deposition. Almost all of the SPS experiments are employing major pieces of high energy physics apparatus used in earlier CERN experiments.

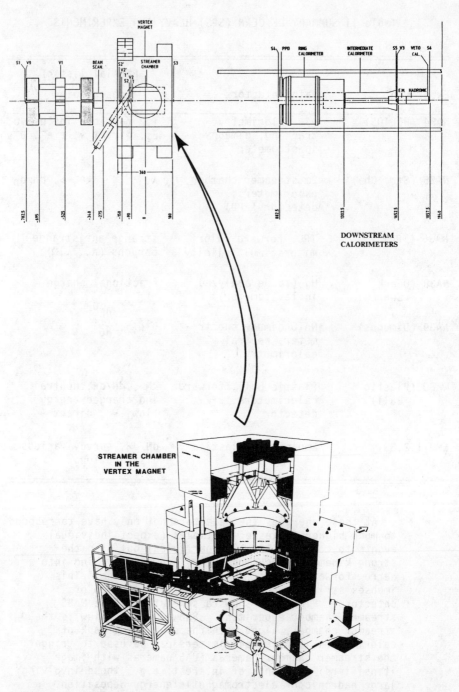

Fig. 12. Plan view of the NA35 streamer chamber experiment at the CERN SPS. An isometric view of the streamer chamber is included (note scale).

2) **AGS/RHIC (Fig. 13)**: A major effort has been underway at Brookhaven to provide a new high energy heavy ion capability. This includes:
   - completion of the transfer line connecting the BNL Tandem to the AGS to start a light ion (A $\leq$ 32) program beginning in late 1986 (energies up to 14 GeV/nucleon)
   - building a booster synchrotron (funds actually provided by high energy physics to increase proton current in AGS) which will allow acceleration of heavy ions up to Au in AGS
   - BNL has submitted a proposal to fill the CBA tunnel with a lattice of superconducting magnets to operate as a Relativistic Heavy Ion Collider (RHIC) at energies up to 100 GeV/nucleon on 100 GeV/nucleon for Au-Au collisions (modest R&D for RHIC was included in the President's FY87 budget submission to Congress).

A major milestone was passed in the week just prior to this Conference when ^{16}O ions were injected and circulated in the AGS. In FY87 about 10 weeks of heavy ion running is expected. Of course, the AGS continues with a strong program of proton experiments. Table III indicates the first round of heavy ion experiments at the AGS.

Now we come to the facility which has the potential of becoming the "crown jewel" of the U.S. heavy ion program--RHIC. The present design[8] calls for a flexible machine, one capable of providing colliding beams of both symmetric (AA) and unsymmetric (pA) partners. The BNL design employs superconducting magnets (B = 3.5 T for dipoles) to reach Au-Au collisions at 100 GeV/nucleon + 100 GeV/nucleon. At this energy, one should have a fully developed central plateau region of about ±1 units of rapidity in which the meson-rich environment of Fig. 3 can be explored. The rate (counts/sec), R, for any process is given by R = $\sigma \mathscr{L}$, where σ is the cross section of interest (generally interested in central collisions) and $\mathscr{L}$ is the luminosity. The design $\mathscr{L}$ for RHIC is shown in Fig. 14 for four different masses. At top energy one is able to sample ~10-50 central collisions/sec--quite a comfortable rate. The machine has been designed to operate for ~10 hours for energies >30 GeV/nucleon in each beam. Below this energy, storage time falls rapidly. The major culprit responsible for beam loss is expected to be intrabeam scattering.[19] Note also that RHIC can be operated in a fixed target mode--circulating beam on a gas target/fiber--to achieve lower c.m. energies.

For Au-Au collisions at RHIC energies one anticipates literally thousands of particles in each central collision. Workshops[20,21] on heavy ion collider detectors have been held to study what the appropriate techniques are in such an environment. Many ideas have emerged from these workshops and it is not possible to cover them here. An example of a collider detector is

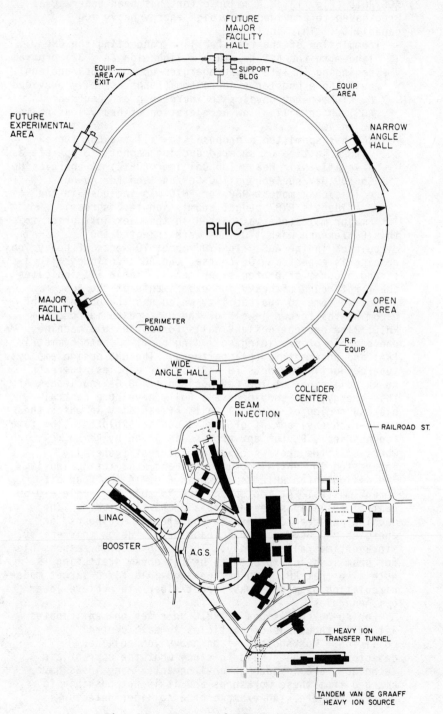

Fig. 13. Layout of AGS/RHIC at BNL.

Table III 1st ROUND HEAVY ION EXPERIMENTS AT AGS

Experiment	Main detectors	Partial list of observables
E-801	Hg (to be analyzed in levitometer)	fractional charge
E-802	large angle magnetic spectrometer, aerogel, cerenkov arrays	semi-inclusive: $\pi^\pm$, $K^\pm$, $p^\pm$, d, α, $\phi \to K^+K^-$, multiplicity
E-810	TPC inside MPS, CCD cube	strange-antistrange, baryons ($\Lambda,\ldots,\Omega$), mult.
E-814	forward spectrometer, calorimetry	energy-flow, proj. fragmentation
E-815	emulsion, $\pi°$-Pb calorimeter	survey
E-825	foils and off-line counting	radiochemical studies
E-793,804, 806,808, 819,826	emulsion/plastic	survey various distributions

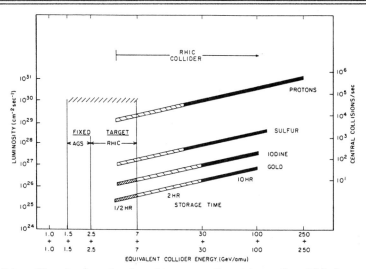

Fig. 14. The design luminosity as a function of collision energy over the energy interval spanned by AGS and RHIC. On left-hand scale, central collisions correspond to impact parameters <1 fm.

shown in Figs. 15-16. The general notion is to have
~4π calorimeter coverage with a limited number of ports
(e.g., at midrapidity) to sample particles. Thus, the main
calorimeter could be used to signal interesting events
(e.g., large transverse energy flow) and one would
correlate this with observations of particles in the
several limited aperture ports surrounding the
calorimeter. Additionally, groups are looking at the
possibilities of tracking particles--both small and large
numbers. Granularity of detectors and questions on
particle ID are also being pursued. It is clear that one
needs to draw on the experience of the particle physics
community to help answer some of those questions.

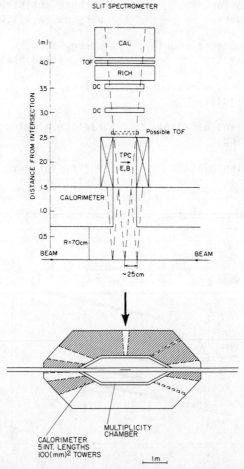

Fig. 15. Concept of 4π calorimeter with small aperture ports
instrumented with special-purpose spectrometers such as the
midrapidity tracking spectrometer above it (this would provide
Δθ = Δφ = 10° at y = 0 coverage).

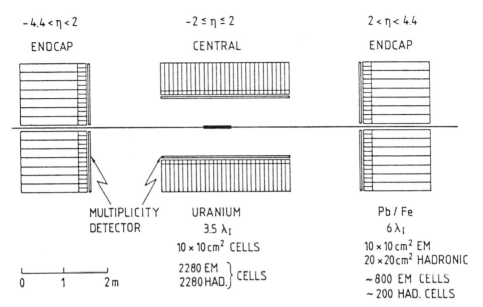

Fig. 16. Exploded view of the 4π calorimeter showing essential elements. Central part would have full azimuthal coverage.

SUMMARY

As we have seen there are five relativistic heavy ion facilities which will be in operation by the end of 1986, with a sixth in the construction phase. Additional new or upgraded facilities are anxiously waiting in the wings for a nod from their friendly funding agency to come on stage. If each were to realize their fondest dreams (and this will surely not happen) then the earliest that we can expect these new facilities is shown below:
1) 1987--Saturne II + MIMAS
2) 1989--AGS Booster (Au in AGS), SIS 18 + ESR
3) 1991--Bevalac Upgrade
4) 1992--RHIC
5) 1995--Holifield Upgrade

One final note. The era of relativistic heavy ion physics is still in the formative stage, having just begun in the early 1970's. But in that time, we have made tremendous strides in coming to grips with the complexity of these reactions, and are now extracting the interesting physics from them. One measure of the advancement of the field is shown in Fig. 17--a "Livingston curve" for relativistic heavy ion machines. Note that the general trend is similar to that found for high energy proton synchrotrons[22]-- except with a time displacement of 17-20 years. New and exciting facilities are becoming available to probe both old areas of nuclear physics with high intensity and improved machine performance, and to strike out for the new territory at high energy and baryon density.

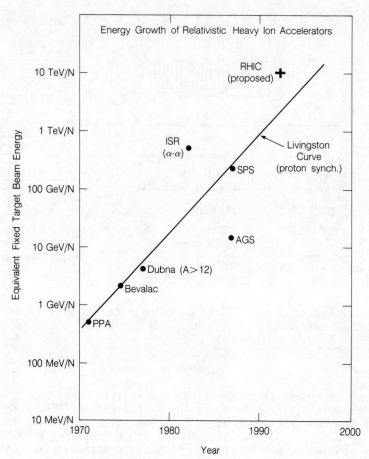

Fig. 17. Energy growth of relativistic heavy ion accelerators.

ACKNOWLEDGMENTS

I wish to thank J. Ball, J. Bartke, P. Kienle, T. Ludlam and G. Young for providing me with material for this report.

REFERENCES

1. R. Palmer, Proceedings of the Meeting on the Intersections Between Particle and Nuclear Physics (Steamboat Springs, 1984), AIP Conference Proceedings No. 123, edited by R. Mischke, p. 197.
2. R. Stock, Nucl. Phys. A447, 371 (1985).
3. P. Kienle, Nucl. Phys. A447, 419 (1985).
4. The Bevalac Upgrade, LBL PUB-5166 (March 1986).
5. Future Directions in Intermediate Energy Heavy Ion Physics (A Proposed Expansion of the Holifield Facility), ORNL (January 1986).

6. P. Radvanyi, Nucl. Phys. $\underline{A447}$, 435 (1985).
7. A.M. Baldin et al., IEEE Transaction on Nucl. Sci., Vol. NS-30, No. 4 (1983), p. 3247.
8. RHIC--Proposal for a Relativistic Heavy Ion Collider, BNL 51932 UC-28 (March 1986).
9. A. Goldhaber, Proceedings of the Meeting on the Intersections Between Particle and Nuclear Physics (Steamboat Springs, 1984), edited by R. Mischke, AIP Conference Proceedings No. 123, p. 21.
10. See talk of T. Ludlam elsewhere in these Proceedings.
11. M. Gyulassy, Nucl. Phys. $\underline{A418}$, 59 (1984).
12. L. McLerran, Proceedings of the 4th International Conference on Ultra-Relativistic Nucleus-Nucleus Collisions, Helsinki (1984), edited by K. Kajantie (Springer-Verlag), p. 1.
13. See section on "Signatures of the Quark-Gluon Plasma" to be published in the Proceedings of the 5th International Conference on Ultra-Relativistic Nucleus-Nucleus Collisions, Asilomar (1986), edited by M. Gyulassy and L. Schroeder (North-Holland), to be published as a special edition of Nucl. Phys. $\underline{A}$.
14. L. Van Hove, Nucl. Phys. $\underline{A447}$, 443 (1985).
15. L. Van Hove, Proceedings of the 5th International Conference on Ultra-Relativistic Nucleus-Nucleus Collisions, Asilomar (1986), edited by M. Gyulassy and L. Schroeder (North-Holland), to be published as a special edition of Nucl. Phys. $\underline{A}$.
16. P. Koch, B. Mueller and J. Rafelski (GSI-86-7), to be published in Physics Reports.
17. See talk of J. Greenberg elsewhere in these Proceedings.
18. H. Huseroth, Proceedings of the Bielefeld Workshop--Quark Matter Formation and Heavy Ion Collisions (1982), edited by M. Jacob and H. Satz (World Scientific), p. 253.
19. G.R. Young, Proceedings of the Meeting on the Intersections Between Particle and Nuclear Physics (Steamboat Springs, 1984), edited by R. Mischke, AIP Conference Proceedings No. 123, p.169.
20. Proceedings of the Workshop on Detectors for Relativistic Nuclear Collisions, LBL (March 1984), edited by L. Schroeder, LBL-18227.
21. Proceedings of the RHIC Workshop on Experiments for a Relativistic Heavy Ion collider, BNL (April 1985), edited by P. Haustein and C. Woody, BNL 51921.
22. See for example, W.K.H. Panofsky, IEEE Transactions on Nucl. Sci., Vol. NS-30, No. 4 (1983), p. 3689.

COOLING OF STORED BEAMS

F. E. Mills
Fermi National Accelerator Laboratory, Batavia, Illinois*

ABSTRACT

Beam cooling methods developed for the accumulation of antiprotons are being employed to assist in the performance of experiments in Nuclear and Particle Physics with ion beams stored in storage rings. The physics of beam cooling, and the ranges of utility of stochastic and electron cooling are discussed in this paper.

INTRODUCTION: MIXING AND HEATING

Since we are dealing with particles circulating in storage rings, we need to describe the properties of the orbits of the particles. We will install devices in the ring, which the particles will pass at their revolution frequencies $f = V/C$, where V is the particle velocity, and C is the length of the orbit. If we consider particles of differing momenta ΔP, then both the particle velocities and the particle orbit lengths will differ, the latter due to the dispersion in the magnetic system which confines the particles. There will be a spread of revolution frequencies given by

$$\Delta f/f = \Delta V/V - \Delta C/C \equiv (1/\gamma^2 - 1/\gamma_t^2)\Delta P/P \equiv \eta \Delta P/P \qquad 1.1$$

If f_0 is the central revolution frequency of the beam particles, and we detect the beam current at frequencies in the neighborhood of nf_0, where n is an integer, then we will see a spread of frequencies

$$\Delta f = nf_0 \eta \Delta P/P \qquad 1.2$$

where $\Delta P/P$ is the fractional momentum spread of the beam. This signal is called the nth longitudinal Schottky line of the beam. At sufficiently high frequency, $\Delta f \rightarrow f_0$, the bands overlap, and the signal becomes uniform in frequency.

* Operated by Universities Research Association Inc., under contract with the U.S. Department of Energy.

In order that the particles will stay in the ring, we configure the magnetic system such that particles not on their closed orbit will oscillate about it with a "betatron oscillation". Both the particle's position x and its angle x' (with respect to its closed orbit) will oscillate with a frequency νf_0. If we now detect the beam position, or more precicely its dipole moment at some location in the ring, we see, for each particle, a periodic δ-function modulated at frequency νf_0. Then the transverse Schottky bands appear at frequencies $(n \pm \nu) f_0$. In most cases of interest, the magnetic structure will be configured such that ν is independent of P ("chromaticity" $\xi = 0$), and that the dependence of ν on betatron oscillation amplitude is small. Then the width of the nth transverse Schottky line is just the width of the nth longitudinal line. We now need to introduce new variables (I,σ) to describe the betatron oscillations. Let

$I = (\gamma x^2 + 2\alpha x x' + \beta x'^2)/2$, $\tan \sigma = \alpha + \beta x'/x$

$x = (2I\beta)^{1/2}\cos \sigma$, $x' = (2I/\beta)^{1/2}[\sin \sigma - \alpha \cos \sigma]$ 1.3

where (α,β,γ) are the lattice parameters determined by the confinement sustem, and are periodic functions of position s along the orbit[1]. The equations of motion are

$I' = 0$, $\sigma' = 1/\beta$ 1.4

We infer that $\nu = (\oint ds/\beta)/2\pi$.

We now inquire about beam heating, that is, we subject the beam to a deflection Θ(t) at some location, where Θ is a random function of time and has zero mean. Then we find on each passage

$\delta I = (2\beta I)^{1/2}\sin \sigma\, \Theta + \beta \Theta^2/2$, $\delta \sigma = (\beta/2I)^{1/2}\cos\sigma\, \Theta$ 1.5

We next sum up the contributions from successive passages ℓ at times $t_\ell = t_0 + \ell/f_0$, and $\sigma_\ell = 2\pi f_0 \nu t_\ell + \sigma_0$ during a long time interval T and find the long term rates of change $dI/dt \equiv \Sigma \delta I_\ell / T$ and $dI^2/dt \equiv \Sigma (\delta I_\ell)^2/T$ for a particle averaged over initial phases and times. We are not interested in the changes in σ. We find[2]

$dI/dt = \beta f_0 \int P(f) df$

$dI^2/dt = \beta I f_0^2 \Sigma P[f_0(n \pm \nu)]$ 1.6

Here P(f) is the spectral power density of Θ (in the sense that Θ^2 represents power)[3]. The sum is over all transverse Schottky lines. The first moment, wide band heating, is typical of gas and target scattering, dipole magnet ripple, etc. The second moment will be important in stochastic cooling, where the beam itself provides noise at the frequencies of the Schottky lines. At this point we could introduce the Fokker-Planck equation to determine the evolution of the distribution of betatron amplitudes, but we will wait until we have introduced the cooling terms. The second moment term is typical of results we will obtain, where the effect on the beam particles is obtained by adding up the contributions of all Schottky bands in the bandwidth of the system in question.

STOCHASTIC COOLING

Consider the arrangement shown in Fig. 1. A single particle is circulating in a storage ring. A beam pickup detector sends its signal to

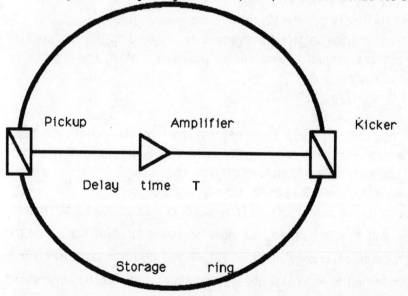

Figure 1. Stochastic cooling system schematic

an amplifier, which drives a dipole kicker. The time delay of the signal T is adjusted so that it equals the transit time of a particle with central momentum from pickup to kicker. The betatron phase shift between pickup and kicker is chosen to be near π/2. As the particle passes through the system on successive revolutions, it will, on the average, receive a kick which reduces its betatron amplitude in proportion to its amplitude. The amplitude will fall exponentially. Amplifier noise will heat the particle at the same time, and the particle will approach a state of equilibrum between the heating and the cooling[4].

The pulse sent to the kicker will have a nonzero time width because of the limited bandwidth of the system. If the particle has a momentum differing from the central momentum, it will arrive at the kicker at a different time than the signal, and will receive less kick, so its cooling rate will be reduced. The higher the bandwidth, the shorter the pulse, and the more important the effect. This is called "bad mixing" or "mixing between pickup and kicker". The effect of the high frequency Schottky bands can be reduced or even turned into heating by this effect. If we define Y(f) to be the ratio, at frequency f, of the kick θ to the dipole moment at the kicker, then we can calculate the long term rate of change of the amplitude I of a particle due to the effect of the system (ignoring noise). It is,

$$dI/dt = f_0 I \sqrt{(\beta_1 \beta_2)} \sin\phi \Sigma [Y\cos(2\pi n f_0 \Delta t)]/2\pi \qquad 2.1$$

The amplitude of Y, which contains the pickup sensitivity, the amplifier gain, and the deflection strength/volt of the kicker, has been taken to be constant in the bandwidth. ϕ is the betatron phase shift, the sum is over all longitudinal Schottky lines, and β_1, β_2 are the lattice functions at the pickup and kicker. Δt is the particle transit time difference $T\eta\Delta P/P$, but can include unwanted amplifier phase shifts as well.

Now let us add more particles to the ring, say N of them. Each particle will feel its own signal, and receive cooling as in 2.1. In addition it will feel the signal of all the others, and be heated according to 1.6. To estimate the heating, we need the power spectrum of the Schottky signal at the kicker. To obtain this, we find the power spectrum of the dipole moment of the beam P_D at the pickup and multiply by Y^2. We find

$$P_D(f) = \beta_1 J N \Sigma \Psi[(f \pm \nu f_0)/n]/n(2\pi)^2 \qquad 2.2$$

The sum is over all transverse Schottky lines, J is the mean value of I in the beam, and Ψ is the distribution function of revolution frequencies f_0 in the beam. This is just the beam signal described above. As the frequency increases, the power density decreases until the bands overlap, and then remains constant. When all bands overlap, the situation is called "good mixing" in the sense that a fresh sample of beam is presented to the pickup on each revolution. We might further comment that in most situations of interest to this audience, namely intense beams, we can neglect amplifier noise. To get an estimate of the combined effect of the cooling and heating we employ the Fokker-Planck equation for the evolution of the distribution function F(I) of amplitudes. The Fokker-Planck equation is essentially a continuity equation in which the particle flux Φ includes both the flux due to the cooling and the diffusive flux due to the heating[5].

$$\partial F/\partial t + \partial \Phi/\partial I = 0; \quad \Phi = (dI/dt)F - (dI^2/dt)(\partial F/\partial I)/2 \qquad 2.3$$

We multiply this equation by I and integrate over I to find the rate of change of J, the mean value of I. Historically the result has been written in the following useful way[6];

$$J'/J = -(W/N)[g - g^2 M/2] \qquad 2.4$$

Here W is the bandwidth of the cooling system, g can be interpreted as the fractional correction applied by the system on each revolution when it measures an apparent offset of the beam (-gW/N is just the right hand side of eq. 2.2), and M is the "mixing factor". The mixing factor is equal to one for good mixing, i.e. all bands overlap. If no bands overlap,

$$M = [\Sigma f_0/\Delta f_n]/n_\ell \qquad 2.5$$

Δf_n is the width of the nth Schottky band and n_ℓ is the number of Schottky lines in the bandwidth W included in the sum. Clearly M describes the effect of frequency spreads on the ratio of the heating term to the cooling term.

From equation 2.4 we see that optimum cooling takes place when g = 1/M, in which case the shortest cooling time is

$$T_{opt} = 2MN/W \qquad 2.6$$

Another effect appears when systems operate near optimum gain. The kicker signal induces coherent motion in the beam. This signal tends to reduce the spontaneous beam signal at the pickup, and to slow down the cooling. As a practical matter, this does not pose a much greater limitation than the heating term, and is intimately related to it, since both effects depend critically on the mixing. On the other hand, if the gain is set high, and the system has improper phase adjustment, instability can take place in the improperly adjusted bands.

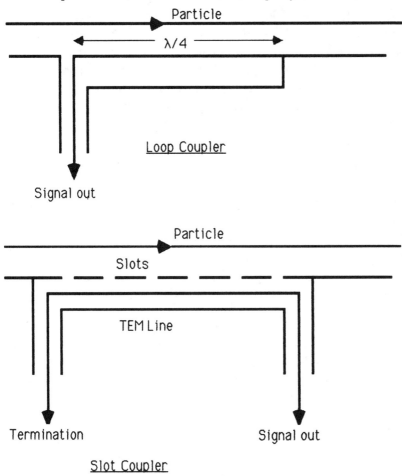

Figure 2. Pickup and Kicker Geometries

Concerning hardware, loop couplers of length $\lambda/4$ in midband seem to offer the maximum sensitivity for relativistic particles, although TEM lines coupled through slots to the beam find application at lower velocity, or in situations in which there is large signal. These pickups are shown pictorially in Fig. 2. Kickers tend to be the same structure, except for power handling capability, because of a reciprocity relationship. Special slowwave structures (helices, lumped lines) have been used in experiments. The signal from many loops can be combined to improve the signal to noise ratio. In the TeV-1 systems, the signals of 128 loops are combined. Travelling wave tubes with octave bandwidth have been employed for frequencies up to 2-4 gHz, and solid state amplifiers may soon be available. Output power requirements can be reduced by splitting the high level signal and driving many kickers. Again, TeV-1 systems drive 128 loop couplers[8].

Stochastic cooling can be employed to cool momentum spread as well. The principal difference is that as the momentum spread is reduced, the Schottky heating power density is increased, and the cooling slows down. Special filters have been employed to shield the beam, in frequency, from its dense core and to allow cooling of hotter particles[9].

We can use Eq. 2.6 to estimate the utility of stochastic cooling as the primary cooling means for a ring such as the Indiana Cooler. For high precision experiments it would be desireable to have $\Delta P/P = 10^{-4}$, say $\eta = .5$, $f_0 = 10^6$Hz, $N = 10^{10}$, and $W = 10^9$Hz. Then $M = 20$, and $T_{opt} = 400$ sec. The momentum cooling system to obtain $\Delta p/p = 10^{-4}$ would be very difficult, with many pickups and kickers, which would have to be delay-adjusted to cool different particle velocities. Nevertheless, stochastic cooling will probably find use in such machines. As we shall see below, cooling of large amplitudes is very slow for electron cooling systems. Scattering from internal targets will, in some circumstances, create a beam halo which will cause backgrounds which are undesireable. Betatron cooling systems are being contemplated for LEAR in which the pickup electrodes have spatially varying sensitivity so that the system responds to the halo and not to the dense core[10]. Similar systems are under study at Fermilab for the Tevatron collider, where normal cooling systems would have cooling times of days for the intense bunched beams. In colliders, the beam-beam tune shift provides some frequency separation between the core and the halo. This aids the situation, since the halo signal will not heat the core.

ELECTRON COOLING

Consider a system of spatially coincident ion and electron beams which have the same mean velocity vectors. Of course this cannot be true for the whole circumference of the storage ring, so we will be speaking of what happens in some fraction η of the ring, called the cooling region. The means for bringing the two beams together will be discussed below. Now consider the two beams from the point of view of an observer moving at the beams mean velocity. He will see an ion gas and an electron gas. If the ion gas is hotter than the electron gas, the ions will lose energy to the electrons by Coulomb collisions and tend to come into temperature equilibrium with the electron gas[11,12]. Kilovolt ions come to rest in condensed matter in times of the order of 10^{-13} sec, so that in an electron beam whose electron density is of order 10^{-11} that of condensed matter, we might expect that equilibrium would come in some milliseconds, or seconds considering the fraction η of the time the ion is actually in the electron beam.

Since the process is Coulomb scattering, the rates will be determined by the accessible range of impact parameters and the relative velocities of the two species. Electron guns for this purpose have operated in a longitudinal magnetic field of about 1 kilogauss, and either provide a beam of constant diameter, or increase the field as the electrons move from the cathode, in which case the beam diameter decreases and the temperature increases. For our purposes, let us consider a beam of 5 Amp, 5 cm diameter, T_e = 1 eV, and kinetic energy of 110 keV. Some important parameters are then;

density	n	$1 \cdot 10^8 \text{cm}^{-3}$
plasma frequency	ω_p	$6 \cdot 10^8 \text{Hz}$
cyclotron frequency	ω_c	$2 \cdot 10^{10} \text{Hz}$
electron $\perp$ velocity	$\beta_{e\perp} \approx \sqrt{(2T_e/mc^2)}$	$2 \cdot 10^{-3}$
Debye length	$\lambda = \sqrt{(T_e/4\pi n e^2)}$	$7 \cdot 10^{-2} \text{cm}$
electron gyroradius	ρ_e	$3 \cdot 10^{-3} \text{cm}$
minimum approach	b_{min}	$1.4 \cdot 10^{-7} \text{cm}$

When the electrons are accelerated, their transverse momenta, and hence their transverse temperature, tend to remain constant or to be regulated by the magnetic field as noted above. In the longitudinal direction, however, the electron temperature is compressed by the acceleration process, ($\Delta E = mv\Delta v$), and the electron parallel velocity becomes $\beta_{e\parallel} = T_{cath}/cP_e = 3\cdot 10^{-6}$. Then the electron velocity distribution in the moving system is far from Maxwellian, rather resembling a disc. For impact parameters above λ the electric field of the ion is screened by the intervening electrons. Collisions taking place for impact parameters between b_{min} and ρ_e are normal free-free Coulomb collisions, exactly the same as those considered in stopping power in normal matter. Collisions for impact parameters between ρ_e and λ are adiabatic with respect to the electron gyromotion, and no momentum can be transferred perpendicular to the field lines. This means that the only electron velocity important in the collision is the parallel one, and it is very small. If the ion velocity is small also, the cooling can be very fast. For the nonadiabatic collisions, the electron $\perp$ velocity normally determines the cooling rates. Then there are two Coulomb logarithms Λ, the nonadiabatic one of about 10, and the adiabatic one of about 3.

Space charge also contributes to the electron velocity spreads. The space charge potential of the e-beam $V = 30\cdot I/\beta \approx 200$ eV causes a spatially correlated $\parallel$ electron velocity spread. The same radial electric field, combined with the magnetic field, causes the e-beam to rotate, typically .05-.1 rad/m.

The ion velocities are determined by the emittance and momentum spread of the ion beam. Since the $\perp$ momentum is invariant, the ion $\perp$ velocity is given by the ion angle, $\beta_{i\perp} = (\beta\gamma)x'$. To have the same $\beta_\perp$ as the e-beam the ion emittance would be [$(\beta\gamma) = .7$, lattice $\beta = 20$m] $\epsilon/\pi = 100$ mm mr, which is not restrictive. In the $\parallel$ direction, $\beta_\parallel = \beta\Delta P/P$, so 10^{-3} is typical. Two other transfomations between laboratory and moving system are required by Lorentz invariance, affecting clock rates and densities: $n_{moving} = n_{lab}/\gamma$, and $\Delta t_{lab} = \Delta t_{moving}/\gamma$. Both of these are in the sense to reduce cooling rates as the beams become relativistic.

In order to calculate cooling rates, we must add up the momentum tranfers of collisions of all impact parameters and all electron velocities as an ion moves through the electron gas. There is an analogy, for the nonadiabatic collisions, between the ion friction force due to the electron velocity distribution, and the electric field due to a similar charge distribution. Because of the multiparameter nature of the process, it is customary to treat the ∥ and ⊥ cooling separately. For the ∥ cooling, because of the disc nature of the electron distribution, the friction force is slowly varying for $\beta_e < \beta_{i\parallel} < \beta_{e\perp}$ and then falls as $1/\beta_i^2$ for larger velocities (as does the transverse friction force). The ∥ friction force is a good measure of the ability to cool a beam, and to restore energy losses due to targetry. For the above parameters, with the fractional cooling length $\eta = .04$ we can estimate the rate of reduction of momentum spread $\delta \equiv \Delta P/P$ to be

$$\delta' = -8\pi n c m_e c^2 r_e^2 \Lambda \eta / [(m_i/m_e)\beta \gamma^2 T_e] \approx 10^{-3} \text{Hz} \qquad 3.1$$

For transverse cooling, when $\beta_{\perp i} < \beta_{\perp e}$ the friction force is proportional to velocity so there is a characteristic cooling time

$$T = (T_{e\perp}/m_e c^2)^{3/2} \gamma^3 / [2\pi n r_e r_i c \Lambda \eta] \approx 12 \text{ sec} \qquad 3.2$$

For smaller ion emittances the adiabatic or magnetic cooling can reduce the cooling time appreciably. There is a substantial premium to be gained by starting with the smallest possible emittance and momentum spread. The magnetic friction forces are;

$$F_\perp = -[8\pi n e^4 \Lambda \eta / m_e][v_\perp^2 - 2v_\parallel^2]v_\perp/v^5$$
$$F_\parallel = -[6\pi n e^4 \Lambda \eta / m_e]v_\perp^2 v_\parallel/v^5 \qquad 3.3$$

Here $(v, v_\perp, v_\parallel)$ refer to the ion velocities in the moving system, as do the friction forces F, and the density n. Λ is here the adiabatic Coulomb logarithm, equal to about 3. Note that the parallel force is zero when the ion moves parallel to the field, and that the ion velocity vector is first rotated to a more ⊥ direction before its magnitude is reduced.

There is no experience in the electron cooling of very intense beams, say for more than 10^9 particles. One can observe, however, that the cooling will reduce the frequency spreads in the beams and render them susceptible to instabilities. In the past means have usually been available to control these instabilities when they occur.

It is beyond the scope of this report to describe completely the technological basis of cooling systems. Suffice it to say that guns for the currents and temperatures required have been constructed, and that newer designs incorporate significant improvements. The use of toroidal fields to bring the ion and electron beams together works well, and efficient electron collection systems have been built with 97% energy recovery efficiency and 10^{-4} electron recovery inefficiency. Adequate methods have been developped to sweep unwanted trapped ions from the electron beams, and suitable vacuum practices now exist to be consistent with long beam storage times[13,14,15].

CONCLUSION

Modern cooling techniques are directly applicable to Nuclear and Particle Physics Experiments utilizing cooled stored beams and internal targets. Altogether about 8-10 cooling rings are being planned in Europe, North America, and Asia for these purposes and for Atomic Physics experiments. This experimental medium promises high resolution, low backgrounds, and adequate luminosity for many types of experiments. To achieve all these goals, probably both electron cooling and stochastic cooling will be required.

REFERENCES

1. Courant, E., Snyder, H., Annals of Physics 3, 1 (1958)
2. Cole, F.T., Mills, F.E., Ann. Rev. Nucl. Part. Sci. 31, 295 (1981)
3. Wiener, N., Time Series, MIT Press (Cambridge 1949), p. 37
4. van der Meer, S., Stochastic Damping of Betatron Oscillations in the ISR. CERN/ISR-PO/72-31
5. Chandrasekhar, S., Stochastic Problems in Physics and Astrometry. In " Selected Papers on Noise and Stochastic Processes ", Dover (New York 1954), p. 3-93
6. Sacherer, F., Stochastic Cooling Theory, CERN-ISR-TH/78-11
7. Lambertson, G.R., et. al., Stochastic Cooling of 200 MeV Protons, Proc. 11th Int. Conf. High Energy Accelerators, Birkhauser (Basel 1980), p.794
8. Design Report Tevatron 1 Project, Fermilab (Batavia 1984), p. 5-12
9. Carron, G., Thorndahl, L., Stochastic Cooling Of Momentum Spread by Filter Techniques, CERN/ISR-RF/78-12
10. Mohl, D., Private Communication, 1986

11. Budker, G.I. et. al, Proc. Int. Symp. Electron and Positron Storage Rings, Atomnaya Energiya 22, 346 (1967)
12. Derbenev, Y.S., Skrinsky, A.N., Preprint INP-225, (Novosibirsk 1968)
13. Budker, G.I., et. al., Part. Accel. 7, 197 (1976)
14. Krienen, F., Proc. 11th Int. Conf. High-Energy Accelerators, Birkhauser (Basel 1980), p. 781
15. Ellison, T., et. al., Electron Cooling and Accumulation of 200 MeV Protons at Fermilab, IEEE Trans Nuc. Sci., NS-30, 2636 (1983)

QUARK MODEL OF THE NN INTERACTION

Makoto Oka

Department of Physics, University of Pennsylvania
Philadelphia, PA 19104

ABSTRACT

The quark cluster model approach to the baryon-baryon interaction is reviewed. The model incorporates full antisymmetrization among valence quarks and takes into account the exchange interaction completely. It explains the strong short-range repulsion between two nucleons. State dependence of the short-range baryon-baryon interaction is understood as being based on the spin-flavor symmetry structure. Various extention of the model for the realistic nuclear force are discussed. Application to the spin-orbit interaction is briefly reviewed. Nucleon size modification in the quark cluster model is presented and related to a simple argument on the relation between the size change and the nucleon interaction.

INTRODUCTION

The nuclear force, or the nucleon-nucleon NN interaction, is one of the most fundamental subjects of nuclear physics. The conventional picture of the nuclear force involves meson exchanges and has been established with a long history since Yukawa proposed the pion. The meson exchange potential, especially the one-pion exchange, is quite successful in the study of the nucleus including the simplest two-nucleon system (bound and scattering).

According to the standard model, strong interactions among hadrons can be described by quantum chromodynamics (QCD). Mesons and baryons are composite particles of quarks and gluons. The nucleus provides us with a unique opportunity for exploring the low energy behavior (soft processes) of QCD, which has not yet been well understood compared with the solid evidence for hard processes such as inclusive lepton scattering. In fact, the size of the nucleon (0.8–0.9 fm) is of the same order as the range of the two-pion exchange, which is most responsible for the nuclear binding, and is larger than the range of the short-distance NN repulsion as well as heavy meson exchanges. Therefore it is not surprising that the conventional picture of the nucleus is modified by effects of the nucleon substructure.

Nucleon-nucleon interaction has been studied in this context for a while.[1-8] The main purpose of this report is to review recent attempts to understanding the baryon-baryon interactions based on a quark model. A simple valence quark model with nonrelativistic kinematics is employed because it is quite successful in the study of single baryon spectroscopy.[9-11] We present the quark cluster model, which is suitable for the study of quark exchange process.[2,3] The model takes into account full antisymmetrization among the valence quarks, which induces exchange interactions between two baryons. Various baryon–baryon (BB) interactions are studied as well as the most interesting NN interaction. It is pointed out that the symmetry structure in the spin and flavor space plays an important role.

In the following sections, we first present the quark cluster model, discussing several possibilities for choosing quark confining potential. Next we review some results on the short range BB force, stressing the importance of the symmetry structure. Possible approaches to the realistic NN interaction are discussed in the next section. Recent effort on the spin-orbit interaction is briefly reviewed. Finally, modification of the nucleon size in the nuclear medium is studied based on the quark cluster model approach as well as a more general model independent argument.

QUARK CLUSTER MODEL

We outline the quark cluster model in the two-baryon system in general.[3]

(a) Hamiltonian

It is well known that nonrelativistic quark hamiltonian with long-range confining potential and short-range spin-dependent interaction describes low-energy hadron states very well.[9–11] The key ingredients there are balance of the constituent quark kinetic energy and the confining potential, which determines the overall size of hadrons and the scale of the excitation energy, and the short-range color-magnetic potential, $(\lambda_i \cdot \lambda_j)(\sigma_i \cdot \sigma_j)$ term, generated by a gluon exchange. The latter is crucial to describe the N-Δ and Λ-Σ mass difference, the neutron charge radius, etc., because it breaks the $SU(6)$ symmetry.

In making the effective hamiltonian in the six-quark system, we need a quark confining potential. The quark confinement is totally in the nonperturbative region of QCD, which is not well understood so far. A conventional choice of the confining potential with overall color symmetry is a two-body potential,[12]

$$V_{\text{conf}} = -a \sum_{i<j} (\lambda_i \cdot \lambda_j) r_{ij}, \qquad (1)$$

which prevents isolation of colored subsystems including a single quark, while it saturates in a color-singlet baryon and releases it from the confinement. The linear r dependence is motivated by a color-electric string confinement expected in the nonperturbative QCD.[13]

However, the potential (1) produces a long-range attraction known as the color van der Waals force between two color-singlet hadrons.[14] This attraction is unphysical, at least at large distances, say $R > 2$ fm, because the confining color flux would break (creating a $q\bar{q}$ pair) before getting that long. Remember the string tension $\sigma \simeq 1$ GeV/fm is quite large compared with typical meson ($q\bar{q}$) energy. This is a pathology in the nonrelativistic approach without $q\bar{q}$ creation or annihilation mechanism. In order to avoid this flaw, several attempts have been made to construct a many-body confinement model based on strong coupling expansion of lattice QCD, and various string or flux-tube type models have been proposed.[15,16]

One of them proposed by Lenz et al.[15] is a simple potential model (without dynamical strings), which is applicable to the quark cluster model for hadron-hadron interactions. The idea is taking the adiabatic limit of the color string (or the color electric field) configuration among valence quarks. For well-separated two baryons (color-singlet, 3 quarks each), the string saturates within the individual baryon and the dynamics is identical to that for a free 3-quark system. There exists no interaction between two baryons (fig. 1). No additional long range force is induced in any higher order effects. The string configuration for two overlapping baryons is determined so as to minimize the string energy. We assume that the rearrangement of the gluon field is fast enough to justify the adiabatic approach. Then the scattering of two baryons occur via a recombination of the quarks as is shown in fig. 1. However, the recombined three-quark system can be color nonsinglet, and then it cannot be isolated. Extra strings are necessary in order to confine such a color nonsinglet 3-quark cluster. We have an ambiguity in choosing the strength of this extra string, which is totally independent of the spectroscopy of a single hadron. We introduce at least one (free) parameter to specify this strength, which is, in principle, determined in studying multihadron systems.

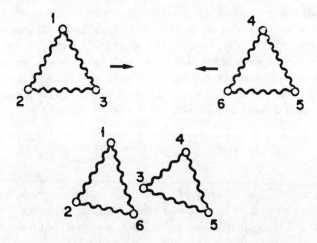

Figure 1 Baryon-baryon scattering into a quark exchange channel in the string quark model.

A choice of such a potential, applied in the study of the NN interaction,[17] is

$$V_{\text{conf}} = \frac{v_0}{12} \left[\sum_{i<j} r_{ij}^2 - 9 R_{\max}^2 \left(1 - \epsilon(1 - P_{\max})\right) \right]. \qquad (2)$$

Here the maximum in the second term is chosen among ten independent groupings of three quark clusters so that it gives the maximum distance $R_{\max}$ between two clusters. The color projection operator $P_{\max}$ is 1 (0) when the chosen clusters are color singlet (nonsinglet). The parameter ϵ ($0 < \epsilon \leq 1$) determines the confining strength for color-nonsinglet clusters and the choice of ϵ does not affect the dynamics of single baryons. The nonsinglet confinement is stronger for larger ϵ, i.e., the maximum strength is at $\epsilon = 1$, when every pair of quarks is connected by a string. A quadratic confinement is taken here for simplicity. The overall strength v_0 is determined by the spectrum of low lying baryons, by reducing the potential to

$$V_{\text{conf}}(\text{three} - \text{quark}) = \frac{v_0}{6} \sum_{i<j} r_{ij}^2. \qquad (3)$$

Different confining models would result in different baryon-baryon interactions even when they are identical in the single baryon system. We will later discuss the difference among the models.

The short-range interquark potential is taken from the static limit of the one-gluon exchange (oge) process,[10] i.e.,

$$V_{\text{oge}} = \sum_{i<j} (\lambda_i \cdot \lambda_j) \frac{\alpha_s}{4} \left\{ \frac{1}{r_{ij}} - \frac{\pi}{m_i m_j} \left(1 + \frac{2}{3}(\sigma_i \cdot \sigma_j)\right) \delta(\vec{r}_{ij}) - \frac{1}{m_i m_j r_{ij}^3} S_{ij} \right\}. \qquad (4)$$

This potential has been extensively used in hadron spectroscopy.[9-11] The most important terms in eq. (4) are spin-dependent ones, i.e., the color-magnetic gluon exchange interaction, $(\lambda_i \cdot \lambda_j)(\sigma_i \cdot \sigma_j)$, and the tensor interaction, S_{ij} (only for the mixing of orbital angular momentum). It should be stressed again that the color-magnetic gluon exchange plays a crucial role in describing the S-wave baryon spectrum. We have not included the spin-orbit force here, which will be discussed later.

The total hamiltonian is given by

$$H = K + V$$
$$K = \sum_i \frac{p_i^2}{2m_i} - K(\text{center of mass}) \qquad (5)$$
$$V = V_{\text{conf}} + V_{\text{oge}}.$$

Much work has been done with considerable success for the single hadron energy and other properties.[9-11] Also numerous modifications of the hamiltonian including the use of the relativistic kinematics was proposed for better understanding of the baryon properties.[18]

(b) Six quark Schrödinger equation (Resonating Group Method)

Our aim is to study the quark exchange force between two nucleons, or in general two baryons, under the hamiltonian given above. Fig. 2 shows one of the exchange diagrams. This is analogous to the Heitler-London force in H_2 molecule, where the exchange property of two electrons plays a crucial role. The quark exchange interaction is nonlocal and in general has a complicated color, spin and flavor structure.

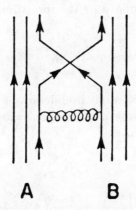

Figure 2   An example of the quark exchange between two baryons, A and B.

In order to solve this problem, a particular form of the six-quark wave function (i.e., cluster wave function) is introduced,

$$\Psi_\beta(1,2,\ldots,6) = \mathcal{A}\left[\{\phi_B(1,2,3)\phi_{B'}(4,5,6)\}_\beta \chi_\beta(R_{123-456})\right], \qquad (6)$$

where β stands for (B, B') combined into a definite spin-flavor state, ϕ_B is a wave function for a single baryon B, χ_β for a relative motion between B and B' and $R_{123-456} \equiv R_{123} - R_{456}$ is the relative coordinate. $\mathcal{A}$ denotes the antisymmetrizing operator acting on the quark labels. This wave function incorporates both the full antisymmetrization among six valence quarks and the correct asymptotic form when the two baryons are well separated. Although the identity of the baryons B and B' becomes ambiguous in the overlapping region, the form (6) is valid for any separation as a form of six-quark wave function. In fact, we have proved that the form (6) with all possible combinations (B, B') of *color singlet* baryons give a complete set of states for a totally color-singlet six-quark systems.[3] Especially, linear combinations of (6) with excited *color-singlet* baryon states describe hidden color states with two color nonsinglet baryon configurations as well as six-quark shell model states.[19]

The Schrödinger equation $(H - E)\Psi = 0$ can be expressed as a coupled integral equation (called resonating group method (RGM) equation[20])

$$\sum_{\beta'} \int dR' \left[H_{\beta\beta'}(R,R') - E N_{\beta\beta'}(R,R') \right] \chi_{\beta'}(R') = 0 \qquad (7)$$

where the hamiltonian (normalization) integral kernel, $H_{\beta\beta'}$ ($N_{\beta\beta'}$) is given by an integral involving the internal wave functions ϕ_B. Eq. (7) takes care of all the exchange interactions completely. It can be applied to the scattering problem as well as the bound state problem thanks to the correct asymptotic behavior of the wave function (6). Clearly, the antisymmetrization makes the BB interaction nonlocal.

In the actual calculation, we have to limit the β' sum mostly to the ground state baryon configurations, sometimes including coupling of several different spin-flavor(-color) states. The individual baryon wave function ϕ_B is taken as a harmonic oscillator shell model wave function for the ground state baryons. The color van der Waals force for the two-body confinement (1) does not appear in the results shown in the following sections because we limit β and β' to the ground state S-wave baryons. Solving this equation for appropriate boundary conditions, we may obtain the binding energy and the wave function for bound states, and the phase shifts or the S-matrix for scattering problems.

SHORT RANGE BARYON-BARYON INTERACTION

Fig. 3 shows calculated S-wave NN scattering phase shifts. (See ref. 3 for parameter values and details for the results in this report.) They are always negative and behave like scattering from a repulsive core. The repulsion for $^1S_0$ is slightly stronger than for $^3S_1$. Equivalent radius of the repulsive core is about $0.4 - 0.5$ fm, obtained from the slope of the phase shift $d\delta/dk$ in the energy region $E_{\rm cm} = 100 - 200$ MeV. This repulsion is quite consistent with the short-range behavior of the nuclear force observed in the NN scattering. It has been shown that the choice of the confining potential $V_{\rm conf}$ does not affect the short-range NN repulsion, i.e., the many-body string-type potential (2) with the short-range force $V_{\rm oge}$ shows qualitatively the same NN phase shift as is shown in fig. 4.[17] It turns out that the effective strength and the range of the NN repulsion are determined only by two parameters: the size of the nucleon b and the strength of the color-magnetic gluon exchange $\alpha_s/m^2 b^3$. These parameters are less ambiguous than the others because they are directly related to observables, i.e., the nucleon radius and the N–Δ mass difference.

The strong repulsion observed above is not necessarily universal to other two-baryon systems. The quark model approach allows us a unified description for various two-baryon systems. We have applied the quark cluster model to $N\Delta$, $\Delta\Delta$, ΛN, ΣN, $\Lambda\Lambda$ etc. The results for the S-wave scattering for those systems are summarized in table I. We can categorize the effective B-B' potential into three classes: (I) strong repulsive core with a large radius $r_c \approx 0.7 - 0.8$

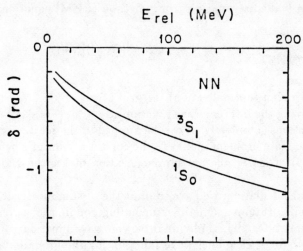

Figure 3 S-wave NN scattering phase shifts for the two-body confining potential (1).

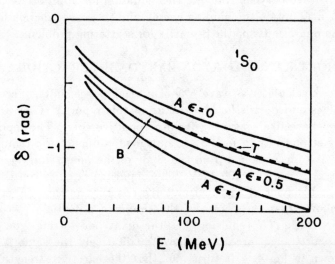

Figure 4 $^1S_0$ NN scattering phase shifts for various confining potentials.[17] T denotes the two-body confinement (1) and the curves labelled A are for the potential (2) with various ϵ. The curve B corresponds to another choice of the many-body confining potential.

fm, (II) medium repulsion with a range $0.3-0.45$ fm, and (III) weakly repulsive or attractive potential. A symmetry argument of two-baryon systems clearly explains the difference among them. The strong repulsion (class I) is generated purely due to the Pauli principle or antisymmetrization in the channels with [51] symmetry of SU(6) (spin-flavor). There the totally symmetric ([6]) orbital wave function is forbidden by the Pauli principle and therefore the relative wave function $\chi(R)$ has a node almost independent of the energy. The effective BB' potential is strongly repulsive at the node of $\chi(R)$. The repulsion is not dynamical but structural. The similar situation is known to occur in interactions between two nuclei, such as ^4He-^4He.

Table I Baryon-baryon interaction

BB'	(S, I)	core radius	class
$\Delta\Delta$	(3, 2)	0.81 fm	I
$\Delta\Delta$	(2, 3)	0.83 fm	I
$N\Delta$	(1, 1)	0.80 fm	I
$N\Delta$	(2, 2)	0.82 fm	I
$\Delta\Delta$	(3, 0)	attractive	III
$\Delta\Delta$	(0, 3)	weakly repulsive	III
$NN - \Delta\Delta$	(1, 0)	0.44 fm	II
$NN - \Delta\Delta$	(0, 1)	0.50 fm	II
$N\Delta - \Delta\Delta$	(2, 1)	repulsive	II
$N\Delta - \Delta\Delta$	(1, 2)	repulsive	II
$N\Lambda$	(0, 1/2)	0.44 fm	II
$N\Sigma$	(0, 1/2)	0.72 fm	I
$N\Lambda$	(1, 1/2)	0.37 fm	II
$N\Sigma$	(1, 1/2)	0.30 fm	II
$N\Sigma$	(0, 3/2)	0.40 fm	II
$N\Sigma$	(1, 3/2)	0.77 fm	I
$\Lambda\Lambda - N\Xi - \Sigma\Sigma$	(0, 0)	see text	III

By contrast, no strong repulsion is seen in the channels with [33] SU(6) symmetry (class III). Among them, $\Delta\Delta$ ($S = 3, I = 0$) has a shallow bound state,[2] which could be experimentally observed as a dibaryon resonance. $\Lambda\Lambda$-$N\Xi$-$\Sigma\Sigma$ ($S = 0, I = 0$) system has been noted because the MIT bag model predicts a deeply bound flavor singlet six-quark state due to a strong colormagnetic attraction in this channel.[21] The quark cluster model does not give a bound state but shows a sharp $\Lambda\Lambda$ resonance state just below the $N\Xi$ threshold ($E_{\Lambda\Lambda} \simeq 27$ MeV).[22] The resonance wave function coincides with the flavor singlet combination qualitatively. These bound or resonance states seem very different from the bound nucleus, say the deuteron. The nucleus is a very extended bound state, where the quark wave function of each nucleon does not overlap much with that of the others. For instance, the deuteron 6-quark wave function is far from the 6-quark shell model wave function. On the contrary, in the above bound $\Delta\Delta$ or resonance $\Lambda\Lambda$ states, the quark wave function is quite similar to the shell model one because of the lack of the short-range repulsion.

The NN states are within the class II. There the two symmetries, [33] and

[51], are mixed because of the SU(6) breaking interaction. The NN repulsion observed above is not totally due to the Pauli principle, but has different origin. In fact, a simple symmetry argument shows that the color-magnetic interaction (CMI) is most responsible for the repulsion. CMI makes the energy of the totally symmetric ([6]) orbital state higher relative to that of the mixed symmetry state ([42]) and therefore induces effective short-range repulsion. We can relate the central NN repulsion to the N–Δ mass difference by evaluating the CMI matrix element in the symmetry basis states, i.e.,

$$V(R=0;\,^1S_0) \simeq \frac{3}{2}(M_\Delta - M_N) \simeq 450 \text{MeV}, \qquad (8)$$

which is consistent with the adiabatic NN potential in the quark cluster model. A careful analysis of the NN wave function in the symmetry basis has confirmed the above argument.[7] We should stress again that the class II repulsion with dynamical origin is very different from the class I with structural origin.

The $N\Lambda$ and $N\Sigma$ interactions are all repulsive as is seen in table I, which seems consistent with N-Hyperon potential models based on the one-boson exchange. One of such models proposed by Nijmegen group is quite successful.[23] The repulsions for $N\Sigma(0,1/2)$ and $N\Sigma(1,3/2)$ are structural (class I) and much stronger than the others. Therefore the state dependence of the short-range repulsion is significantly different from the conventional models based on flavor SU(3) symmetry. The coupling between the $N\Lambda$ and $N\Sigma$ channels due to the quark exchange is very weak and negligible. In other words, the short-range transition potential is much weaker than the diagonal one.

NN INTERACTION

Apparent lack of medium and long range attraction in the NN potential shows the limitation of the quark cluster model approach. In fact, the range of the quark exchange interaction is determined by the size of the nucleon because the process requires that the quark wave functions of the two nucleons overlap with each other. We need another mechanism beyond that range. Here we discuss several attempts to make the quark model approach more realistic.

(a) Medium and long range interaction

The excited spectra of hadrons suggest that the color degree of freedom is strongly confined in a small region, typically within ≤ 1 fm. Because the gluon carries a color charge, it is also confined in a color-singlet system, and therefore gluon exchange among valence quarks seems not to describe the long-range hadronic interactions. It is, instead, color-singlet hadrons, mostly pion, that are exchanged between nucleons. The pion exchange interaction has been well established in few-nucleon systems. In principle, meson exchange can also be described in terms of quarks and gluons. There is an attempt by Fujiwara and Hecht[24] that couples $(q\bar{q})$ creation and annihilation to three valence quarks with one-gluon exchange process. The model reproduces qualitative features of the one-boson-exchange NN interaction at large distances, although it fails to explain enough medium-range attraction. This would be an interesting direction to explore.

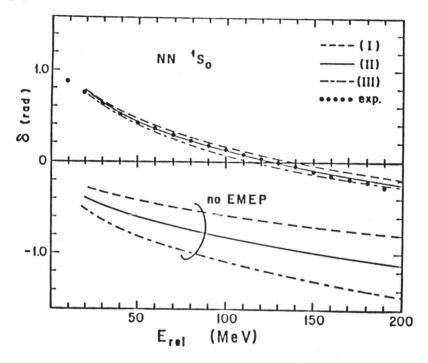

Figure 5 $^1S_0$ NN scattering phase shifts with a long range potential (the effective meson exchange potential or EMEP).[25] The curves I–III correspond to different choices of the quark model parameters. The phase shifts calculated without the long range potential are also shown for comparison.

Less ambitious alternatives are hybrid models. Abondoning the hope of expressing meson exchange in terms of quarks and gluons, the meson fields could be introduced as fundamental degrees of freedom. This approach is familiar in the baryon models with chiral symmetry, such as the chiral bag models and the topological soliton models. We made the most primitive attempt to construct a realistic model by hybridizing a phenomenological attractive NN potential with the quark exchange repulsion.[25] Figs. 5–7 show the results compared with experimental phase shifts. The parameters of the attractive potential (the range and the strength) have been fixed by the low energy scattering parameters, i.e., the scattering length and the effective range. We see that such a simple model can work really well, which indicates the quantitative success of the short-range quark exchange interaction.

More elaborate analyses by similar hybrid models have been done by several

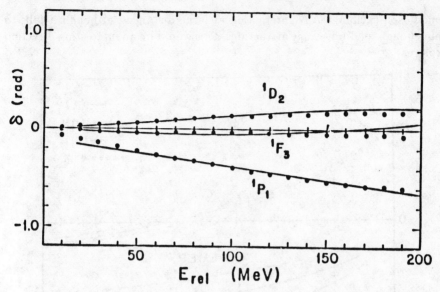

Figure 6 Spin-singlet NN phase shifts with a long range potential.

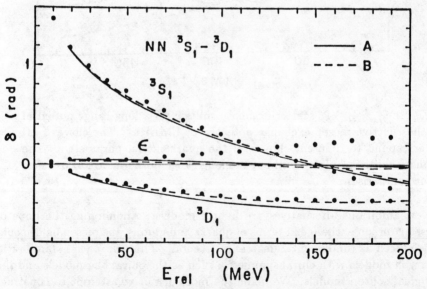

Figure 7 $^3S_1 - {}^3D_1$ NN phase shifts with a long range potential. The curves A and B correspond to different choices of parameters in the long range potential.

groups.[26-30] Among them, Wakamatsu et al.[26] have proposed superposing to the quark cluster model one- and two-pion exchanges with a short range cutoff (similar to the Paris potential). They successfully reproduced the NN scattering phase shifts as well as the deuteron properties. Shimizu[27] and Faessler et al.[28] have introduced a surface type meson-quark coupling form factor and have superposed the quark exchange interaction. They have studied the modification of the conventional one-boson exchange potential due to the quark existence.

Most of those hybrid models can reproduce the NN scattering phase shifts ($E_{cm} \leq 200$ MeV) by choosing appropriate parameters. More interesting are off-shell properties, such as electro-weak form factors, response functions and disintegration cross sections, where the short range behavior of the wave function becomes important. Takeuchi et al.[31] studied the quark exchange effects in detail on the electroweak form factors and the response functions by constructing quark cluster model wave functions for the deuteron and ^4He. They found that the form factors would hardly tell the quark effect when the parameters are adjusted so as to reproduce the observed r.m.s. radius of the nucleus, while the response functions for $\alpha \to p + t$ would have $5 - 16\%$ effects.

(b) Color polarization

Another possibility for the medium range attraction was pointed out by Maltman and Isgur.[8] They evaluated the effect of color-dipole excited nucleon (and Δ) (p-wave, color-octet) mixing using the two-body confining potential (1) and obtained a substantial attraction. The origin of this attraction is in fact the same as the long-range color van der Waals force associated with the potential (1), which is recognized unphysical at large distances because of the string breaking with $q\bar{q}$ creation. Without a reliable picture of the string breaking, we have to make an arbitrary assumption on where the attraction should be cut off. It is also highly dependent on the confining model, i.e., the many-body string confining potential, for instance, given by eq. (2), does not have the color van der Waals attraction. Maltman and Isgur took a cutoff at $R \simeq 2$ fm in the calculation of an adiabatic potential and connected the result with the one-pion exchange long-range part. No nonadiabatic (resonating group method) calculation has been done including the color dipole polarization.

In the string like confining potential, the arbitrariness of the strength of the color nonsinglet confinement suggests another possible color polarization. If the color nonsinglet confinement is sufficiently weak, the energy of cluster states with two color polarized baryons (so called hidden color state[6]) become low enough to couple to the color singlet channels. We might have low lying hidden color dominant bound or resonance states.[32] Those states would manifest themselves as dibaryon resonances, which again have a six-quark shell model like wave function. These hidden color dominant resonances are entirely dependent on the strength of the color nonsinglet confinement. We have so far no evidence for such a state.

(c) Spin-orbit interaction

The quark cluster model approach to the spin-orbit (SO) interaction seems quite interesting. Unlike the central force, the SO force due to the meson exchange is short-range ($\leq (2m_\pi)^{-1}$) and therefore we may expect a significant contribution from the quark exchange process. However, the situation is not yet clear. The main ambiguity comes from the qq SO interaction.[33] It originates both from the one-gluon exchange (oge) process and from the quark confining potential (conf). Furthermore, they both have a (so called) symmetric SO potential (sym-oge or sym-conf), proportional to $(\vec{r}_1 - \vec{r}_2) \times (\vec{p}_1 - \vec{p}_2) \cdot (\vec{\sigma}_1 + \vec{\sigma}_2)$, and an antisymmetric one (anti-oge or anti-conf), $(\vec{r}_1 - \vec{r}_2) \times (\vec{p}_1 + \vec{p}_2) \cdot (\vec{\sigma}_1 - \vec{\sigma}_2)$. The latter vanishes in the $q\bar{q}$ system but survives in $3q$ system because $\vec{p}_1 + \vec{p}_2 = \vec{p}_3 \neq 0$. It is well known that the spin-orbit force is suppressed relative to the spin-spin (color-magnetic) force in the (nonstrange) baryon spectrum. If we neglect the antisymmetric terms, then the suppression can be explained by the cancellation between the sym-conf and sym-oge contributions.[11] However, the anti-conf and anti-oge have the same sign and therefore do not cancel, though their magnitude is considerably less than the symmetric terms. Thus the understanding of the spin-orbit splitting in the single baryon system might not be sufficient to make the NN SO problem conclusive. It is still an open problem.

Several groups[34] have studied the NN and N–nucleus SO interaction in the quark cluster model. Mostly the ^3P_J NN phase shifts have been calculated including all or a part of the qq SO potentials stated above. It is found that the sym-oge term gives a significant SO force between two nucleons, though it is greatly reduced by the sym-conf term due to the two-body confining potential (1). Koike et al.[35] generalized the approach to the string-like confining potential, similar to eq. (2). They only took the symmetric SO terms and adjusted the parameters so as to cancel the SO force exactly for the P-wave nonstrange baryon system. They found that in the NN system the sym-conf contribution due to the string-like confinement (unlike the two-body confinement) does not cancel the sym-oge and the net result is of the correct order for the observed NN SO force.

We would like to stress that the study of the BB SO force is important in order to distinguish different quark models, because it has less ambiguity due to the long-range meson exchange effects than the central force in comparing to experiment. So far, it is not yet conclusive because of the confusion in the qq SO interaction. The consistency between the three-quark and the six-quark systems is important.

An application to the hyperon–nucleon SO interaction seems interesting. Pirner (and Povh)[36] proposed an explanation of the difference of the magnitude between the Λ–nucleus and the Σ–nucleus SO potentials in an additive quark model. It was pointed out that the Λ SO potential is significantly weaker because the strange quark is relatively inactive due to the fact that the qq SO potential is proportional to $1/m_i m_j$. This qualitative conclusion has been confirmed in a quark cluster model analysis by Morimatsu et al.[37] They have compared the oge SO contribution with the Nijmegen potentials[23] and found a rough agreement except in an $N\Sigma$ channel.

NUCLEON SIZE AND INTERACTION

Modification of nucleon properties in nuclear medium is a current interesting subject. We have several experimental indications (though not "evidences") that the electromagnetic and weak properties are modified.[38,39] In general, an interacting nucleon could have any polarization. Among them size change of the nucleon is the simplest.

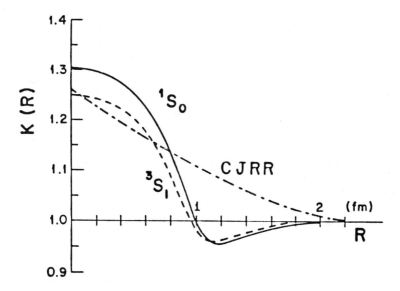

Figure 8 $K(R) \equiv \lambda(R)/\lambda_{\text{free}}$ in a two-nucleon system.[41] The curve CJRR shows the ratio used in the rescaling model analysis of the EMC effect, ref. 38.

The quark cluster model has been applied to the study of the nucleon size in the two nucleon systems.[40,41] The adiabatic approximation is taken and the six-quark energy H_6 is minimized changing the nucleon size parameter b for a fixed distance R between the two nucleons, i.e.,

$$\left. \frac{\partial H_6}{\partial b} \right|_{\text{fixed} R} = 0. \tag{9}$$

The ratio of the quark confinement scale λ is given as a function of R by

$$K(R) \equiv \frac{\lambda(R)}{\lambda_{\text{free}}} = \left(\frac{\langle k^2 \rangle_{\text{fixed} R}}{\langle k^2 \rangle_{\text{free}}} \right)^{-1/2}. \tag{10}$$

Fig. 8 shows $\lambda(R)/\lambda_{\text{free}}$ calculated for the $^1S_0$ and $^3S_1$ NN systems, which is compared with the curve used in the rescaling model analysis of the EMC effect.[38] We see an enhancement at short distances, but a slight reduction for large R, which deviates from the curve for the rescaling model analysis. It was shown that these effects come mainly from change of b, the size of the individual nucleon, rather than from the quark exchange between the nucleons.

The size of the interacting baryon has also been studied in the skyrmion-skyrmion system,[42] where again the size decrease for medium and large R region is observed. However, neither model reproduces the correct sign of the long range NN interaction. In fact, we showed recently that the nucleon size change is directly related to the sign of the interaction by a quite general argument independent of model.[43] It has been proved that the nucleon size increases (decreases) when the N-N interaction is attractive (repulsive) at large distances (or for low density). One of the two proofs we provided is based on first order perturbation theory for a quantum system in an external potential. We showed that the direction of the size change by the first-order mixing of the monopole excitation mode N^* is determined only by the sign of the external potential, or more precisely, the sign of the second derivative, $\nabla^2 V$, which is equivalent to the former for potentials falling faster than $1/r$. (Remember that $\nabla^2 V = \mu^2 V$ with the range μ of the potential.) We showed that the ratio of the sizes for the perturbed and unperturbed states is given by

$$\frac{r^2}{r_0^2} = 1 - \alpha(\mu r_0)^2 \frac{V(R)}{\Delta E_M}, \qquad (11)$$

where ΔE_M is the excitation energy of the monopole state and α a positive numerical constant, which depends on the dynamics. We found that for reasonable parameter choice the nucleon is expected to grow as much as *a few* − 10 % at around $R \simeq 1-1.5$ fm. Note that this result is not inconsistent with small probability of N^* in a nucleus because it is the interference term that contributes to the size increase. For a nucleon in a nucleus, the potential V to use in eq. (11) is not the mean nuclear potential $\langle V \rangle$, which is constant in the nucleus and therefore has $\nabla^2 \langle V \rangle = 0$, but the fluctuations around that mean due to the presence of other nucleons. If V is the sum of pairwise forces acting on a particular nucleon, we write $V = \langle V \rangle + (V - \langle V \rangle)$. Then $\nabla^2(V - \langle V \rangle) = \nabla^2 V$, which is the contribution from the two-body forces. Since that contribution is monopole, it does not average to zero when averaged over all particles.

Our other proof is appropriate for any model of the nucleon which has a single dimensional parameter, like QCD. In such a model, because every quantity should scale according to its dimension, the size change is also directly related to the change of the effective nucleon mass. The decrease of the effective nucleon mass due to an attractive interaction makes the nucleon bigger. We can relate the nucleon size increase to the single nucleon energy in the nucleus.

The above connection of the nucleon size, or more generally the nucleon's internal structure, and the nuclear interaction, or the global motion of the nucleon, is important in understanding the effect of nucleon modifications in the nuclear medium. For instance, the EMC effect is known to be explained by (at least) two completely different mechanisms, one of which is the increase of the quark confinement size in nuclei and the other is the effect of the off shell kinematics of the nucleon.[44,45] We now have a picture with both of these. The nucleon is in a bound state, which modifies the kinematics, and its size is enhanced accordingly. However, these two effects are not totally independent, because the modification of the nucleon structure couples to the nucleon motion. More specifically, the convolution formula with a modified (rescaled) nucleon structure function,

$$F_{q/A}(x) = \int F_{q/N}^{(rescaled)}(x/y) f_{N/A}(y) dy, \qquad (12)$$

is valid only if one takes an average confinement size neglecting the coupling of the nucleon motion and the size increase. One may take into account the mixing of the monopole excitation N^* explicitly in calculating the structure function. Then an interference term between N and N^*, shown in fig. 9, contributes as a twist-2 diagram. This interference contribution cannot be represented by a convolution form.

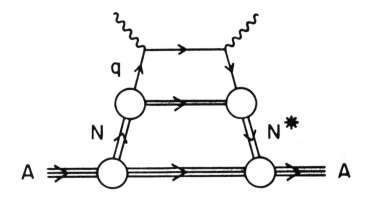

Figure 9   Interference of N and N^* in the structure function.

CONCLUSION

We have reviewed the quark cluster model approach to the baryon-baryon interaction. Some conclusions are:

(1) Short-range BB interactions due to quark exchange process are qualitatively determined by the spin-flavor symmetry structure of the state.

(2) The color-magnetic gluon exchange plays a dominant role in the NN short-range repulsion. Ambiguity in the confining potential does not change the qualitative conclusion.

(3) Various ellaborations of the basic quark cluster model have been discussed. Hybrid models with quark exchange and meson exchange show quantitative success in the NN scattering phase shifts. In order to demonstrate

the quark effect in the nucleus, however, more extensive study of the offshell properties will be necessary.

(4) Generalization of the model including $q\bar{q}$ creation and annihilation would be an interesting future direction. The possible contribution from the color dipole polarized state in the medium range NN force is not conclusive yet.

(5) There are many applications and sophistications of the quark cluster model, which have not been discussed in this report. Various other models of the baryon, like topological or nontopological soliton models, chiral bag models, ...etc., have also been applied to multibaryon systems. Relations among the models are not clear yet.

(6) Nucleon size change in the NN system is related to the NN interaction at large distances. We expect a size increase in the nuclear medium.

This work is supported in part by the National Science Foundation.

REFERENCES

1. D.A. Liberman, Phys. Rev. D16, 1542 (1977); C. DeTar, Phys. Rev. D17, 302, 323 (1978).
2. M. Oka and K. Yazaki, Phys. Lett. 90B, 41 (1980); Prog. Theor. Phys. 66, 556, 572 (1981).
3. M. Oka and K. Yazaki, in *Quarks and Nuclei*, ed. by W. Weise (World Scientific, 1985).
4. J.E.T. Ribeiro, Z. Phys. C5, 27 (1980).
5. C.S. Warke and R. Shanker, Phys. Rev. C21, 2643 (1980); M. Cvetic, B. Golli, N. Mankoc-Borstnik and M. Rosina, Phys. Lett. 93B, 489 (1980); 99B, 486 (1981); Nucl. Phys. A395, 349 (1983); H. Toki, Z. Phys. A294, 173 (1980).
6. M. Harvey, Nucl. Phys. A352, 301, 326 (1981); M. Harvey, J. LeTourneux and B. Lorazo, Nucl. Phys. A424, 428 (1984).
7. A.F. Faessler, F. Fernandez, C. Lübeck and K. Shimizu, Phys. Lett. 112B, 201 (1982); Nucl. Phys. A402, 555 (1983).
8. K. Maltman and N. Isgur, Phys. Rev. Letters 50, 1827 (1983); Phys. Rev. D29, 952 (1984).
9. A.J.G Hey and R. Kelly, Phys. Rep. 96C, 71 (1983).
10. A. DeRujula, H. Georgi and S.L. Glashow, Phys. Rev. D12, 147 (1975).
11. D. Gromes and I. Stamatescu, Nucl. Phys. B112, 213 (1976); W. Celmaster, Phys. Rev. D15, 1391 (1977); N. Isgur and G. Karl, Phys. Lett. 72B, 109 (1977); 74B, 353 (1978).
12. H.J. Lipkin, Phys. Lett. 45B, 267 (1973); N. Isgur, in *The New Aspects of Subnuclear Physics*, ed. by A. Zichichi (Plenum, 1980).

13. J. Kogut and L. Susskind, Phys. Rev. D11, 395 (1975).
14. P.M. Fishbane and M.T. Grisaru, Phys. Lett. 74B, 98 (1978); T. Appelquist and W. Fischler, Phys. Lett. 77B, 405 (1978); S. Matsuyama and H. Miyazawa, Prog. Theor. Phys. 61, 942 (1979).
15. F. Lenz, J.T. Londergan, E.J. Moniz, R. Rosenfelder, M. Stingl and K. Yazaki, preprint SIN 85-09; J.T. Londergan, in *Hadron Substructure in Nuclear Physics*, ed. by W.Y.P. Hwang and M.H. Macfarlane (AIP, 1984); K. Yazaki, Nucl. Phys. A416, 87c (1984);
16. I. Barbour and D.K. Ponting, Z. Phys. C4, 119 (1980); J. Carson, J.B. Kogut and V.R. Pandharipande, Phys. Rev. D27, 233 (1983); D28, 2807 (1983); N. Isgur and J. Paton, Phys. Lett. 124B, 247 (1983); Phys. Rev. D31, 2910 (1985); L. Heller and J.A. Tjon, preprint LA-UB-84-3219; O.W. Greenberg and J. Hietarinta, Phys. Rev. D22, 993 (1980); D. Robson, Florida State Univ. preprints; in *Hadron Substructure in Nuclear Physics*, ed. by W-Y.P. Hwang and M.H. Macfarlane (AIP, 1984).
17. M. Oka and C.J. Horowitz, Phys. Rev. D31, 2773 (1985).
18. See ref. 9, and references therein. Recent papers on a relativized potential model are: S. Godfrey and N. Isgur, Phys. Rev. D32, 189 (1985); S. Capstick and N. Isgur, preprint UTPT-85-34.
19. I.T. Obukhovsky, V.G. Neudatchin, Y.F. Smirnov and Y.M. Tchuvil'sky, Phys. Lett. 88B, 231 (1979); P.J. Mulders, A.T. Aerts and J.J. deSwart, Phys. Rev. D21, 2653 (1980); S. Ohta, M. Oka, A. Arima and K. Yazaki, Phys. Lett. 119B, 35 (1982).
20. J.A. Wheeler, Phys. Rev. 32, 1083 (1937); 32, 1107 (1937); K. Wildermuth and Th. Kanellopoulos, Nucl. Phys. 7, 150 (1958); I. Shimodaya, R. Tamagaki and H. Tanaka, Prog. Theor. Phys. 27, 793 (1962).
21. R.L. Jaffe, Phys. Rev. Lett. 38, 195 (1977).
22. M. Oka, K. Shimizu and K. Yazaki, Phys. Lett. 130B, 365 (1983).
23. M.M. Nagels, T.A. Rijken and J.J. deSwart, Phys. Rev. D12, 744 (1975); D15, 2547 (1977); D20, 1633 (1979).
24. Y. Fujiwara and K.T. Hecht, Nucl. Phys. A444, 541 (1985), 451, 625 (1986); Phys. Lett. B 171, 17 (1986).
25. M. Oka and K. Yazaki, Nucl. Phys. A402, 477 (1983).
26. M. Wakamatsu, R. Yamamoto and Y. Yamauchi, Phys. Lett. 146B, 148 (1984); Y. Yamauchi, R. Yamamoto and M. Wakamatsu, Phys. Lett. 146B, 153 (1984).
27. K. Shimizu, Phys. Lett. 148B, 418 (1984).
28. K. Bräuer, A. Faessler, F. Fernandez and K. Shimizu, Z. Phys. A320, 609 (1985).
29. A. Faessler and F. Fernandez, Phys. Lett. 124B, 145 (1983); Y. Suzuki and K.T. Hecht, Phys. Rev. C27, 299 (1983); C28, 1458 (1983).

30. D. Robson, Florida State Univ. preprints; in *Hadron Substructure in Nuclear Physics*, ed. by W-Y.P. Hwang and M.H. Macfarlane (AIP, 1984).
31. S. Takeuchi and K. Yazaki, Nucl. Phys. A438, 605 (1985); S. Takeuchi, K. Shimizu and K. Yazaki, Nucl. Phys. A449, 617 (1986).
32. M. Oka, Phys. Rev. D31, 2274 (1985).
33. F.E. Close and H. Osborn, Phys. Rev. D2, 2127 (1970); F.E. Close, L.A. Copley, Nucl. Phys. B19, 477 (1970); L.J. Reinders, J. Phys. G4, 1241 (1978); in *Baryon 80, IVth International Conference on Baryon Resonances*, ed. by N. Isgur, Toronto, 1980.
34. Y. Suzuki and K.T. Hecht, Nucl. Phys. A420, 525 (1984); O. Morimatsu, S. Ohta, K. Shimizu and K. Yazaki, Nucl. Phys. A420, 573 (1984); O. Morimatsu, K. Yazaki and M. Oka, Nucl. Phys. A424, 412 (1984); J. Burger, H.M. Hofman, Phys. Lett. 148B, 25 (1984); Y. He, F. Wang and C.W. Wong, Nucl. Phys. A438, 620 (1985); A451, 653 (1986); Y. Fujiwara and K.T. Hecht, U. Michigan preprint.
35. Y. Koike, O. Morimatsu and K. Yazaki, Nucl. Phys. A449, 635 (1986); Y. Koike, to be published.
36. H.J. Pirner, Phys. Lett. 85B, 190 (1979); H.J. Pirner and B. Povh, Phys. Lett. 114B, 308 (1982).
37. O. Morimatsu, S. Ohta, K. Shimizu and K. Yazaki, in ref. 34.
38. F.E. Close, R.G. Roberts and G.G. Ross, Phys. Lett. 129B, 346 (1983); R.L. Jaffe, F.E. Close, R.G. Roberts and G.G. Ross, Phys. Lett. 134B, 449 (1984); F.E. Close, R.L. Jaffe, R.G. Roberts and G.G. Ross, Phys. Rev. D31, 1004 (1985).
39. G. Karl, G.A. Miller and J. Rafelski, Phys. Lett. 143B, 326 (1984); T. Yamazaki, Phys. Lett. 160B, 227 (1985); J. Noble, Phys. Rev. Lett. 46, 412 (1981); P.J. Mulders, Phys. Rev. Lett. 54, 2560 (1985).
40. K. Bräuer, A. Faessler and K. Wildermuth, Nucl. Phys. A437, 717 (1985).
41. M. Oka, Phys. Lett. 165B,1 (1985).
42. M. Oka, K.F. Liu and H. Yu, UPR 0289-T, appear in Phys. Rev. D.
43. M. Oka and R.D. Amado, preprint UPR-295T.
44. C.H. Llewellyn Smith, Phys. Lett. 128B, 107 (1983); M. Ericson and A.W. Thomas, Phys. Lett. 128B, 112 (1983); E.L. Berger, F. Coester and R.B. Wiringa, Phys. Rev. D29, 398 (1984); S.V. Akulinichev, S.A. Kulagin and G.M. Vagradov, Phys. Lett. 158B, 485 (1985); S.V. Akulinichev, S. Shlomo, S.A. Kulagin and G.M. Vagradov, Phys. Rev. Lett. 55, 2239 (1985); B.L. Birbrair, A.B. Gridnev, M.B. Zhalov, E.M. Levin and V.E. Starodubski, Phys. Lett. 166B, 119 (1986); R.P. Bickerstaff and G. Miller, Phys. Lett. 168B, 409 (1986).
45. D.V. Dune and A.W. Thomas, Phys. Rev. D33, 2061 (1986); F.E. Close, R.G. Roberts and G.G. Ross, preprint RAL-85-101.

HIGH ENERGY SPIN PHYSICS

Elliot Leader
Birkbeck College, University of London
Malet Street, London WC 1E 7HX

ABSTRACT

The confrontation between Theory and Experiment in high energy spin physics is already tantalizing. Future experiments will provide a severe test for QCD.

1. INTRODUCTION

I shall argue that Spin Physics at high energies provides and will continue to provide an enormous challenge, to both theory and experiment in Elementary Particle Physics.

We believe that we have uncovered the theory of the strong interactions QCD. It is aesthetically beautiful and deserves to be the theory, but it has not been critically tested. Many experimental results point towards it, but no single one is absolutely convincing. In the past, spin measurements have often been known to make or break a theory (V-A weak interactions ; g-2 ; Regge Poles...) and it is possible that they will provide a decisive test for QCD. In truth, we could claim that there is already disagreement between theory and experiment, but, as will become clear, the theory may be too naive and the experiments may require higher energies and larger momentum transfers. The challenge is then to improve the calculations and to extend the experiments. If, thereafter, the conflict persists, we shall have to reassess our trust in QCD.

In our current state of wisdom we are forced to divide the physics into two regimes, the perturbative and the non-perturbative.

2. THE NON-PERTURBATIVE REGIME

Ultimately QCD must provide us with the wave-function of the quarks inside a hadron H, of helicity h,

$$\psi_h(\underline{r}_1, \underline{\sigma}_1, \underline{\tau}_1; \underline{r}_2, \underline{\sigma}_2, \underline{\tau}_2; \underline{r}_3, \underline{\sigma}_3, \underline{\tau}_3)$$

where the position, spin and iso-spin of the quarks is indicated. Meanwhile we have to be content with the "distribution functions" $q_i(x)$ [the number density of quarks of type i with momentum $\underline{p} = x\underline{P}$, inside the hadron of momentum $\underline{P}$], which we deduce from lepton-hadron deep inelastic scattering (D.I.S.) experiments. (See Fig. 1)

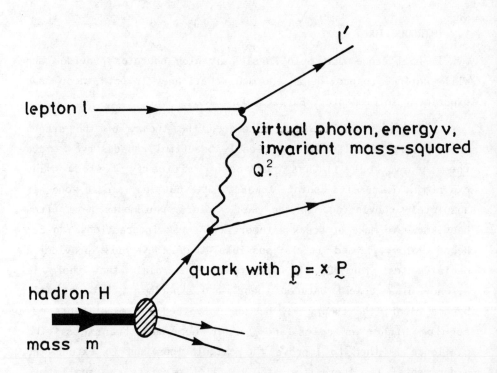

Fig.1 : Physical interpretation of deep inelastic scattering.

According to the naive quark model what we measure is $q_i(x)$ at $x = Q^2/2m\nu$. According to QCD the distribution functions must depend on Q^2 also, in a well-defined way controlled by the Altarelli-Parisi evolution equations[1].

Equally important, though less well known, are the spin-dependent quark distribution functions $q_i^+(x)$, $q_i^-(x)$, which give the number density of quarks with momentum fraction x and with helicity either parallel (+) or antiparallel (-) to the helicity of the parent hadron[2]. The usual $q_i(x)$ is given by

$$q_i(x) = q_i^+(x) + q_i^-(x) \qquad ---(1)$$

and

$$P_i(x) = \frac{q_i^+(x) - q_i^-(x)}{q_i^+(x) + q_i^-(x)} \qquad ---(2)$$

is the polarization of the ith quark inside a 100 % polarized hadron.

The $q_i^\pm(x)$ are best measured in lepton-hadron D.I.S. using a polarized lepton beam and polarized hadron target[3], but they can also be obtained from the decay of W's produced in pp or p$\bar{\text{p}}$ collisions with one beam polarized[4]. Again QCD predicts a specific Q^2 dependence.

A remarkable result of the SLAC-Yale polarized D.I.S. experiment[5] is that one high momentum "up" quark carries nearly all of the spin of a polarized proton.

Most of what has been said above about quarks applies equally well to the gluons inside a hadron except that the spin-dependent gluon distribution functions $G^\pm(x, Q^2)$ are much more difficult to measure and are not yet well determined.

As will be seen in Section 4, the $q_i^{\pm}(x, Q^2)$ are vital ingredients in the calculation of the spin parameters in a hadronic reaction. But they also have an intrinsic interest, in particular as regards their Q^2 dependence. For example consider a proton of helicity + 1/2 moving along OZ. Naively the sum of the z-components of the spins of the constituents should add up to + 1/2. But the constituents may have orbital angular momentum as well, (though this is ignored in taking $\underline{p} = x \underline{P}$) so there will be a sum rule[6]

$$\langle S_z \rangle_{quarks} + \langle S_z \rangle_{gluons} + \langle L_z \rangle = \frac{1}{2} \quad \text{--(3)}$$

where

$$\langle S_z \rangle_{quarks} = \frac{1}{2} \sum_i \int_0^1 dx \left\{ q_i^+(x, Q^2) - q_i^-(x, Q^2) \right\} \quad \text{--(4)}$$

and

$$\langle S_z \rangle_{gluons} = \int_0^1 dx \left\{ G^+(x, Q^2) - G^-(x, Q^2) \right\} \quad \text{--(5)}$$

Now it can be shown (on the basis of the helicity theorem in Section 3) that $\langle S_z \rangle_{quarks}$ does NOT vary with Q^2. But $\langle S_z \rangle_{gluons}$ grows like $\ln Q^2$ and we are forced to conclude that $\langle L_z \rangle$ must become very important as Q^2 grows. It is important to test these statements experimentally.

3. THE PERTURBATIVE REGIME

Because of the simple γ_μ coupling and because the quarks are effectively massless in QCD one can prove the following powerful theorem : If in an <u>arbitrary</u> Feynman diagram a quark (helicity λ) enters in the initial state and eventually emerges (helicity λ') in the final state, then $\lambda' = \lambda$, i.e. the quark preserves its helicity no matter how many interactions it undergoes. [Analogously, for a quark (helicity λ) and an antiquark (helicity λ') belonging to the <u>same</u> fermion line in any diagram, the amplitude is non-zero only if $\lambda' = -\lambda$]. (See Fig. 2)

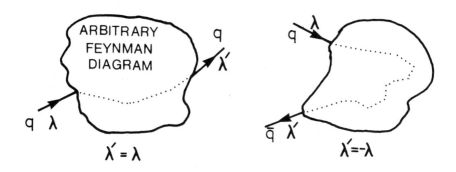

Fig.2 : Helicity rules for quarks in QCD.

Thus in perturbative QCD one cannot flip the helicity of a quark, though one can reverse the helicity of a gluon.

It is believed that for present-day values of Q^2 the quarks are much more sensitive than the gluons are to the spin state of their parent hadron. So to induce a hadron to flip its helicity (and thereby to produce interesting spin effects) it would be most efficacious to flip the helicity of one or more of its quark constituents, which, precisely, perturbative QCD is unable to do.

This does not mean that no hadron can flip its helicity. By re-routing quarks from one hadron to another (quark interchange) one can cause helicity-flip but then it has to happen to at least two hadrons in the reaction. An example of such a transition, $|1/2, -1/2\rangle \rightarrow |-1/2, 1/2\rangle$ in NN $\rightarrow$ NN, is illustrated in Fig. 3.

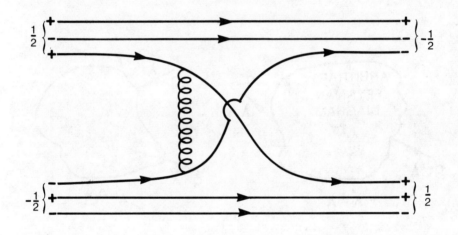

Fig. 3 : Quark interchange yielding double helicity-flip.

In the world of partons only, the powerful helicity rules lead to very clear predictions for spin dependent parameters. Some illustrative examples are :

i) Polarization in $q_i q_j \rightarrow q_i q_j$. The polarization P is zero because

$$P \propto \text{Im}\, \phi_5 \times \{-----\}$$

where ϕ_S is the single flip amplitude

$$\left|\tfrac{1}{2},\tfrac{1}{2}\right\rangle \longrightarrow \left|\tfrac{1}{2},-\tfrac{1}{2}\right\rangle$$

and is therefore zero. (In lowest order P would in any case be zero because the Born terms are real.)

ii) The partonic reaction at the heart of the Drell-Yan process : $q\bar{q} \to e^+e^-$. It is highly dependent on the helicities of the q and $\bar{q}$. Indeed

$$d\sigma_{++} = d\sigma_{--} = 0$$

The most challenging question is how to translate this sort of dramatic statement into the world of hadrons.

4. THE HADRONIC WORLD

The treatment of hadronic reactions requires the merging of the non-perturbative element describing how the partonic constituents are distributed inside the hadron (which information is largely taken from experiment), and the perturbative aspect concerning the dynamical interaction of the partons (which is calculated using perturbative QCD). One has then a more or less convincing prescription for calculating both inclusive and exclusive hadronic reactions which is expected to be valid for high energies and large transverse momenta.

a) <u>Inclusive reactions</u>. We consider reactions of the generic type

$$A + B \to C + \text{anything}$$

in which A and/or B may be polarized and we may or may not measure the polarization of C. The canonical approach is perhaps best illustrated by a concrete example $pp \to \pi^{\pm} X$ which is visualised as in Fig. 4.

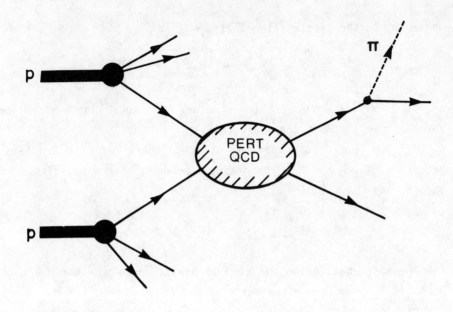

Fig.4 : Inclusive production in the parton-QCD model.

In this simple picture the polarized hadron beam simply provides a beam of polarized quarks.

An interesting prototype asymmetry, when both protons are longitudinally polarized [either parallel ($\rightrightarrows$) or antiparallel ($\rightleftarrows$)] is

$$A_{LL} \equiv \frac{d\sigma^{\rightrightarrows} - d\sigma^{\rightleftarrows}}{d\sigma^{\rightrightarrows} + d\sigma^{\rightleftarrows}} \qquad -- (6)$$

Intuitively the result will be

$$A_{LL} \approx \langle P_1 \rangle \langle P_2 \rangle A_{LL}^{Partonic} \quad --(7)$$

where the $\langle P_j \rangle$ are the mean polarization of the quarks in the protons. The partonic A_{LL} is very large because of the helicity rules discussed in Section 3 but the hadronic A_{LL} is diluted by the moderate mean quark polarizations. Some results of ref.(6) are in Fig. 5 below :

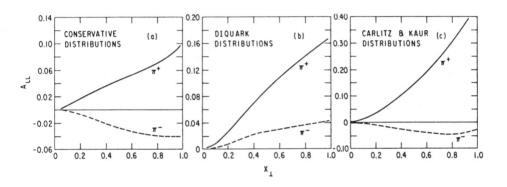

Fig.5 : Asymmetry A_{LL} for the reactions $pp \rightarrow \pi^{\pm} X$ as a function of $x_{\perp} = 2p_T/\sqrt{s}$ for (a) conservative SU(6) distributions, (b) diquark distributions, and (c) Carlitz-Kaur distributions.

It is seen that significant asymmetries are expected. Experimental results are awaited with interest !

Similar results can be obtained for many other two-spin asymmetries in a host of inclusive reactions[7]. The experiments are not easy but the results will be of immense value.

Single-spin asymmetries, either with a polarized target, or with an unpolarized initial state producing a particle whose polarization is monitored, are much easier. Indeed the reactions

$$pp \rightarrow (\Lambda, \Sigma, \Xi) + \text{anything}$$

with unpolarized protons, have a very healthy cross-section and enjoy the benefit that the hyperons are self-analyzing, in that their polarization can be deduced from their decay pattern.

The situation at the moment is tantalizing. Much data exists and it spans a huge range in energy (p_{LAB} = 24 GeV/c up to CERN ISR energies) and extends out to $p_T \approx$ 2-3 GeV/c. A sample of this data (plotted as function of LAB momentum for fixed LAB production angle) is shown in Fig. 6.

The polarization P is significantly non-zero and seems to grow with p_T. The theoretical prediction is unequivocally

$$P = 0 \qquad \qquad -- (8)$$

What are we to make of this ? Is p_T still too small to justify the perturbative calculation ? We shall discuss this further in Section 5.

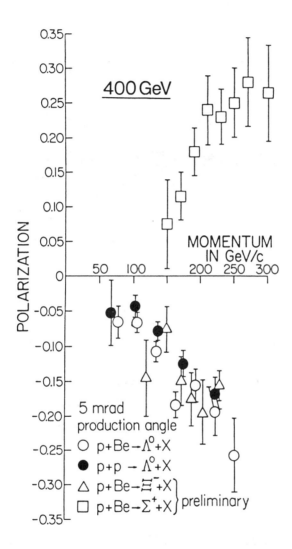

Fig.6 : Polarization in inclusive hyperon production at 400 GeV at fixed LAB. angle.

b) <u>Exclusive processes</u>. Here we need the actual wave-function of the quarks that make up the hadron, i.e. we require the amplitude

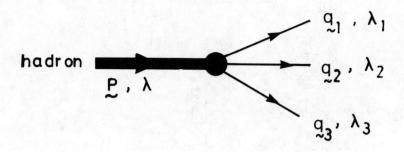

for the hadron, momentum $\underset{\sim}{P}$, helicity λ to break up into quarks of momentum $\underset{\sim}{q}_j$ and helicity λ_j. We consider only reactions with large p_T.

It is believed that the most efficient way to produce large p_T is for each hadron to produce a beam of essentially parallel quarks which then get a high-p_T kick via a perturbative QCD interaction. In that case the only non-perturbative input is the amplitude where $\underline{P}$ and each $\underline{q}_j$ lie along one direction, say OZ. Although we can't compute this amplitude we can deduce an important piece of information. Since all momenta are along OZ any orbital angular momentum must be perpendicular to OZ. Thus the only angular momentum along OZ is spin angular momentum. Conservation of J_z then implies

$$\lambda = \lambda_1 + \lambda_2 + \lambda_3 \qquad -- (9)$$

for each hadron in the reaction.

Since each quark that interacts perturbatively conserves its helicity we end up with a remarkable result due to Brodsky and Lepage[8]. In any exclusive reaction

$$A + B \longrightarrow C + D + E + \ldots$$

one has

$$\lambda_A + \lambda_B = \lambda_C + \lambda_D + \lambda_E + \cdots \qquad -- (10)$$

that is, total initial helicity equals total final helicity.

Consequences abound ! Here are some examples :

(i) The polarization in NN $\longrightarrow$ NN is zero because it is proportional to the single-flip amplitude $|1/2\ 1/2\rangle \rightarrow |1/2\ -1/2\rangle$ for which initial helicity (=1) is not equal to final helicity (=0). Experimentally the polarizations measured at BNL are not small[9]. (See Fig. 7) On the other hand p_T is not very large in these experiments. But there is certainly no sign of the polarization decreasing with p_T. Again a tantalizing situation crying out for experiments at larger p_T.

(ii) For $\pi^- p \longrightarrow \rho^0 p$, with unpolarized target, the helicity density matrix elements ϱ_{ij} for the produced ρ meson are zero if $i \neq j$. A recent BNL experiment[10] at p_{LAB} = 10 GeV/c with ρ's emerging at 90° in the C.M, corresponding to $p_T \approx 2$ GeV/c, finds a significant $\sin^2\phi$ dependence in the ρ decay

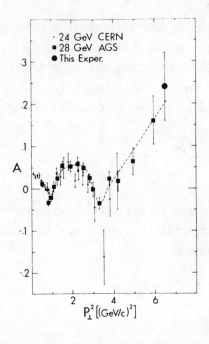

Fig.7 : Polarization asymmetry in proton proton scattering.

distribution, implying a large value for $S_{1,-1}$ (0.29 ± 0.07) instead of zero. More data for this and several related reactions should soon appear. The problem as always is that p_T may be too small.

5. THE CONFRONTATION BETWEEN THEORY AND EXPERIMENT

With a fairly convincing, but certainly not rigorous, blending of perturbative and non-perturbative elements of QCD one arrives at dramatic predictions for the spin dependent parameters in exclusive and inclusive hadronic reaction at high energies and large p_T. Without exception every one of these predictions that has been tested is contradicted by present day experiments !

It would be premature to conclude that QCD is disproved. The experiments run out to $p_T \approx$ 2-3 GeV/c, which may well be too small for the validity of the perturbative calculations. But the worrying aspects are i) that other perturbative QCD predictions, such as the behaviour of electromagnetic form factors and fixed angle elastic scattering do seem to work for p_T of this order, and ii) there seems to be no hint in the spin data for a change of behaviour in the required direction as p_T increases[10]. It may well also be that the calculations are too naive.

Whatever the true answer it is clear that the confrontation between theory and experiment is tantalizing. There is a great challenge to improve the theoretical calculations and to push the experiments to larger p_T.

ACKNOWLEDGEMENT : The author is grateful for the hospitality of the Institut de Physique Nucléaire, Université Claude Bernard, Lyon, where part of this study was carried out.

REFERENCES

1) G.Altarelli and G. Parisi, Nucl. Phys. B126 (1977) 298.

2) Their Q^2 - dependence has been studied numerically in P.Chiappetta and J. Soffer, Phys. Rev. D31 (1985) 1019.

3) For an elementary discussion see E. Leader and E. Predazzi, "An Introduction to Gauge Theories and the 'New Physics' ", Cambridge University Press (1985), Section 12.6 .

4) E. Leader, Phys. Rev. Lett. 56 (1986) 1542.

5) G. Baum et al., Phys. Rev. Lett. 51 (1983) 1135.

6) J. Babcock, E. Monsay and D. Sivers, Phys. Rev. D19 (1979) 1483.

7) N.S. Craigie, SACLAY preprint DPhPE 83-01 (1983).

8) G.P. Lepage and S.J. Brodsky, Phys. Rev. D22 (1980) 2157.

9) S. Heppleman et al., Phys. Rev. Lett. 55 (1985) 1824.

10) For a more sanguine interpretation, see G.R. Farrar, Rutgers Preprint U-85-46.

ANTINUCLEON PHYSICS

Carl B. Dover
Brookhaven National Laboratory*, Upton, NY 11973

ABSTRACT

We review and interpret some of the recent data from LEAR, Brookhaven, and KEK on low and medium energy interactions of antinucleons ($\bar{N}$) with nucleons (N). Our emphasis is on elastic and charge exchange scattering, total cross sections, and studies of $\bar{N}N$ annihilation, with particular focus on the emerging evidence for broad resonances and/or bound states of the $\bar{N}N$ system and the selection rules which reveal the quark-gluon dynamics of the annihilation process.

INTRODUCTION

Since the advent of the LEAR (Low Energy Antiproton Facility) at CERN, there has been a veritable deluge of new data pertaining to the interactions of antinucleons with nucleons and nuclei. Rather than attempting to catalog these results (as well as recent data from Brookhaven and KEK) in a systematic fashion, I have selected only a few topics for which I can offer some theoretical interpretation.

The first topic to be treated is antiproton-proton ($\bar{p}p$) elastic scattering. Recent LEAR data[1] extend to rather low lab momenta. Even at the lowest momentum $p_L \approx 180$ MeV/c, the anisotropic angular distributions demand a large p-wave contribution. The ratio ρ of real to imaginary parts of the 0° elastic amplitude displays more than one zero in the low momentum region, suggestive of a resonance phenomenon.

There is new data on the charge exchange reactions[2-4] $\bar{p}p \to \bar{n}n$ and $\bar{p}p \to \bar{\Lambda}\Lambda$. For $\bar{n}n$, the small angle "dip-bump" structure seen at higher momenta[2] is found to disappear[3] for $p_L < 300$ MeV/c. This phenomenon is interpreted as the interference of the one pion exchange (OPE) amplitude with a momentum dependent background term sensitive to the range of the annihilation potential.

Recent searches[5-11] for long-lived "baryonium" states are reviewed. Earlier evidence for such objects has not been confirmed by a series of high statistics experiments, which include $\bar{p}p$ and $\bar{n}p$ cross sections[5-9], and γ, π^0, $\pi^\pm$, $K^\pm$ inclusive spectra[10,11].

There is much new data on $\bar{p}p$ and $\bar{p}n$ annihilation into specific mesonic channels[12-15]. The data on $\bar{p}p \to \pi^+\pi^-\pi^0$ from initial states of orbital angular momentum $\ell = 0,1$ provide evidence for a <u>dynamic selection rule</u>: the production of $\pi\rho$ events is strongly favored from $^{2I+1,2S+1}\ell_J = {}^{13}S_1$ and $^{11}P_1$ $\bar{p}p$ states, and the πf channel is most strongly populated from the $^{33}P_1$ state. This surprising result implies a strong constraint on the <u>quark-gluon dynamics</u> of the annihilation process. We offer an explanation in terms of the $^3P_0$ model in which quark-antiquark ($Q\bar{Q}$) pairs annihilate or are created with vacuum quantum numbers $(J^{\pi C}(I^G) = 0^{++}(0^+))$. Other models are also discussed.

---

*The submitted manuscript has been authored in part under contract DE-AC02-76CH00016 with the U.S. Department of Energy.

Realistic $\bar{N}N$ potential models, including the effects of annihilation, predict a spectrum of bound states and resonances[16-19]. Except in special cases, these are expected to be rather broad, and hence difficult to observe in $\bar{N}N$ total cross sections and inclusive spectra. However, such states, as well as broad $Q^2\bar{Q}^2$ or $Q\bar{Q}g$ mesons[20,21], for instance, may be visible in certain specific annihilation channels. We offer a possible interpretation of a broad structure recently observed[14] in the $\bar{p}n \to 3\pi^-2\pi^+$ channel as the $^{13}P_2 - {^{13}F_2}$ quasinuclear (QN) bound state of the $\bar{N}N$ system. Appropriate channels for the production of other members of the predicted $0^{++}, 1^{--}, 2^{++}$.... natural parity isospin zero QN band are discussed.

Future prospects for the field of antinucleon physics are evaluated, with emphasis on some novel aspects of meson spectroscopy.

ELASTIC SCATTERING

The angular distributions for $\bar{p}p \to \bar{p}p$ elastic scattering at low momenta were recently measured by the Heidelberg group[1] at LEAR. Selected data are shown in Fig. 1. Even at the lowest momentum of 181 MeV/c, the angular distribution is anisotropic, indicative of a strong p-wave component. This is in contrast to the situation for pp scattering at comparable momenta, which is s-wave dominated. The "precocious p-wave" in $\bar{p}p$ scattering arises naturally from the very strong attractive central and tensor forces due to meson exchange (coherence of scalar and vector contributions in the central potential, for instance, unlike the pp case).

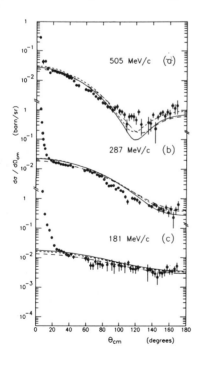

Fig. 1. Differential $\bar{p}p \to \bar{p}p$ elastic cross sections at lab momenta p_L = 181, 287, and 505 MeV/c. The solid, dashed, and dash-dotted curves are optical model predictions from refs. (25,23,24), respectively. The data are from Brückner et al[1].

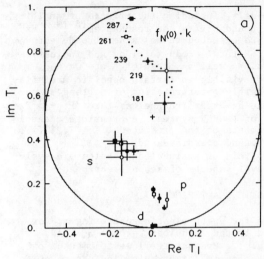

Fig. 2. The complex partial wave amplitudes T_ℓ for $\bar{p}p$ elastic scattering in the momentum range 181–287 MeV/c, from Ref. 1. The forward amplitude $kf_N(0)$ is also shown. Only s and p wave contributions are important at these low momenta.

The data were subjected to a partial wave analysis, neglecting the spins of the $\bar{p}$ and p. The usual dimensionless T-matrix elements $T_\ell = (\eta_\ell e^{2i\delta_\ell}-1)/2i$ for orbital angular momentum ℓ are plotted in Fig. 2 in the momentum range 181-287 MeV/c. In this region, $\ell \geq 2$ waves were found to make a negligible contribution. The $\ell=0$ amplitude T_0 does not display a clear momentum dependence, while Im T_1 grows with p_L and Re T_1 tends to zero from positive values. Note that Re $T_0 < 0$, corresponding to an effective repulsion. This is the signal of $\ell=0$ bound states in the $\bar{N}N$ system.

For $\ell=1$, one expects Re $T_1 > 0$, based on optical model calculations[22-25]. Here, only the $^{13}P_0$ and $^{13}P_2$ states are expected to be strongly bound, whereas other configurations are weakly bound or unbound. To obtain T_1, one must perform a spin-isospin average over $\ell=1$ channels, which leads to considerably more model dependence than for $\ell=0$.

The ratio ρ of the real to imaginary parts of the $\bar{p}p$ forward amplitude[26-29] is shown in Fig. 3, as a function of p_L. The surprise is that ρ rises above zero at low momenta. This result has been confirmed by the PS172 collaboration[7] at LEAR. Existing optical model calculations[22-25] are consistent with the data for $p_L > 400$ MeV/c, but fall smoothly to values $\rho \approx -1$ for $p_L \approx 0$. Data from $\bar{p}p$ atoms[30] indeed give $\rho \approx -1$.

The rise of ρ to positive values at low p_L has been interpreted[31] as an effect of the coupling of elastic scattering to $\bar{p}p \to \bar{n}n$ charge exchange, the threshold for which lies at 98 MeV/c. Although the coupling to the $\bar{n}n$ channel may be important in a quantitative analysis, it seems unlikely as the sole explanation for the structure in ρ, which occurs in a region of p_L where the $\bar{p}p \to \bar{n}n$ cross section does not vary rapidly[3,32]. Another possible explanation would be offered by a p-wave $\bar{N}N$ resonance in the region $p_L \approx 250$ MeV/c. A reasonable model for the forward $\bar{p}p$ amplitude $f(0^0)$ consists of the sum of background and p-wave resonance

$$f(0°) = a + Ak^2 + \sqrt{2J+1} \; x\Gamma_T \, k^2/2k_r^3 \cdot \frac{(E-E_r+i\Gamma_T/2)}{(E-E_r)^2+(\Gamma_T/2)^2} \quad (1)$$

Here a and A are the complex s and p-wave scattering length and volume, $E_r = 2(k_r^2 + m_N^2)^{1/2}$ is the c.m. resonance energy, $x = \Gamma_{EL}/\Gamma_T$ and Γ_{EL}, Γ_T are the elastic and total widths of the $\ell=1$ resonance of spin J.

In the absence of a resonance, the smooth trend from $\rho \approx -1$ at zero energy to $\rho \approx 0$ near 500 MeV/c, typical of many optical model calculations, is easily understood in terms of the cancellation of repulsive $\ell=0$ and attractive $\ell=1$ contributions to Re $f(0°)$. For typical values $a \approx (-1+i)$fm and $A = (1+i)$fm^3, we would obtain a zero of Re $f(0°)$ around $p_L \approx 2k \approx 400$ MeV/c. Such a model, however, does not explain the rise of ρ at low momenta seen in Fig. 3.

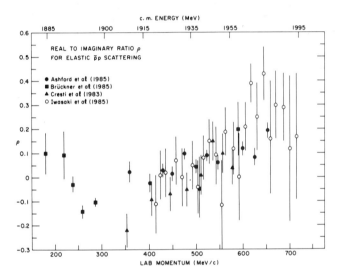

Fig. 3. The ratio ρ = Re $f_N(0)$/Im $f_N(0)$ of real to imaginary forward amplitudes for $\bar{p}p$ elastic scattering. The data points, as indicated, are from references 26-29.

Phenomenological optical potentials can lead to p-wave resonances near threshold. For instance, the Paris potential predicts[18,23] a 3^1P_1 resonance at $E_r = 1880$ MeV with $\Gamma_T \approx 65$ MeV. However, the elasticity x of broad optical model resonances is generally small (x < 0.5), and so they do not produce a sharp structure as in Fig. 3, which seems to require a fairly narrow resonance with $x \approx 1$.

From the behavior of ρ alone, one cannot deduce the quantum numbers of a possible resonance. Data on the momentum dependence of exclusive annihilation channels would be very useful. For instance, a 3^1P_1 $\bar{N}N$ resonance should show up as a peak in the $\bar{p}p \to \pi^0\omega$ channel, $1^3P_{0,2}$ resonances in $\pi^+\pi^-$, etc. The very low momentum region

$p_L < 180$ MeV/c is well worth study, since ρ must display at least one more zero in this region.

THE $\bar{p}p \to \bar{n}n$ CHARGE EXCHANGE REACTION

The most recent data on $\bar{p}p \to \bar{n}n$ differential and total cross sections are due to Nakamura et al[2] (KEK) and Brückner et al[3] (LEAR). The KEK measurements at higher momenta (particularly at 490 and 690 MeV/c) show a striking dip-bump structure at forward angles, while the LEAR data at low momenta (183 and 287 MeV/c) are forward peaked but do not display this structure. The LEAR data at 287 and 590 MeV/c are plotted in Fig. 4, together with theoretical predictions[22-24,33] of various optical models. The LEAR and KEK data at 590 MeV/c are seen to agree rather well, except possible for $\theta < 20^0$. The Nijmegen[24] and boundary condition[33] models are seen to describe the data well, while the Paris model[23] predicts an exaggerated dip-bump structure which is not observed. The Dover-Richard[22] predictions also disagree with the data below 300 MeV/c, although to a lesser extent.

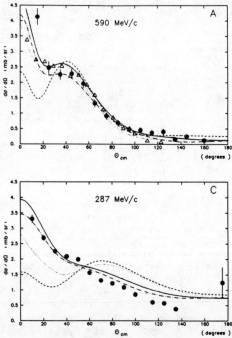

Fig. 4. Differential $\bar{p}p \to \bar{n}n$ charge exchange cross sections at $p_L=287$ and 590 MeV/c. The full circles are the data from Brückner et al[3], while the open triangles represent the data of Nakamura et al[2]. The solid, dashed, dotted, and dash-dotted lines are optical model predictions of the Nijmegen[24], Paris[23], Dover-Richard[22] and boundary condition[25] models, respectively.

It is possible to highlight the essential physics of the $\bar{p}p \to \bar{n}n$ reaction at low momentum transfer, i.e., in the dip-bump region, in terms of the dominant long-range one pion exchange (OPE) interfering with a coherent background amplitude. Recall that the $\bar{p}p$ elastic scattering and charge exchange cross sections are given in terms of the usual helicity amplitudes[33,34] ϕ_i by

$$\frac{d\sigma}{dt} = \frac{\pi}{2m^2 p_L^2} \{|\phi_1|^2 + |\phi_2|^2 + |\phi_3|^2 + |\phi_4|^2 + 4|\phi_5|^2\}, \quad (2)$$

where $\phi_1 = \langle ++|M|++\rangle$, $\phi_2 = \langle ++|M|--\rangle$, $\phi_3 = \langle +-|M|+-\rangle$, $\phi_4 = \langle +-|M|-+\rangle$, $\phi_5 = \langle ++|M|+-\rangle$. Now we note that OPE gives a contribution to ϕ_2 and ϕ_4 only, viz.

$$\phi_2^\pi = \phi_4^\pi = \frac{g_{\pi NN}^2}{4\pi} \frac{t}{t - \mu_\pi^2} \quad (3)$$

where $g_{\pi NN}^2/4\pi \approx 14$, t is the invariant four momentum transfer ($-2k^2(1-\cos\theta)$ for elastic scattering), and μ_π is the pion mass. In contrast to the π, the exchange of a ρ contributes to all ϕ_i. Since $\phi_{2,4}^\pi$ vanishes at t=0, it is clear that the observed forward peak in $\bar{p}p \to \bar{n}n$ and the subsequent dip-bump structure (if present) requires the addition of a background contribution B to ϕ_2, which interferes destructively with ϕ_2^π. Such a constant background cannot be present in ϕ_4, which must rigorously vanish at t=0 (flip of two units of helicity). Thus a simple model, valid for small t, is

$$\left(\frac{d\sigma}{dt}\right)_{\bar{p}p \to \bar{n}n} \approx C_0[(\tilde{t}-B)^2 + \tilde{t}^2] + C_1 \quad (4)$$

where $C_1 > 0$ is an assumed constant (incoherent) background from ϕ_1 and ϕ_3, and $\tilde{t} = |t|/(\mu_\pi^2 + |t|)$. The above representation has a peak at $\tilde{t} = 0$ (t=0), a minimum at $\tilde{t} = B/2$, and, for $\tilde{t} = B$, returns to its value at t=0, i.e., the desired dip-bump structure. According to Nakamura et al[2], the positions of the dip at lab momenta p_L = 390, 490 and 690 MeV/c are $|t|/\mu_\pi^2 \approx 0.9$, 0.8 and 0.5, respectively. This implies corresponding values $B \approx 0.94$, 0.87 and 0.68, i.e., the movement of the dip with p_L implies an energy dependence of B, which we may roughly parametrize in the 400-700 MeV/c region as

$$\frac{1}{B} \approx 0.36 + 1.61 \, p_L (\text{GeV/c}). \quad (5)$$

If we now extend this result to the lower momenta of 287 and 183 MeV/c where LEAR measurements exist, we get $B \approx 1.22$ and 1.53, respectively. However, for B>1, we no longer expect to see the dip-bump structure clearly, since the dip moves to larger angles, and the width of the dip region becomes too large. This is consistent with the absence of structure in the data. In fact, one notes that the kinematic constraint that the minimum must occur for $\theta < 180°$, is given by $B < 2(2k/\mu_\pi)^2/(1+(2k/\mu_\pi)^2)$. For $p_L \approx 2k \approx 183$ MeV/c, this gives $B < 1.26$, so our deduced value $B \approx 1.53$ implies that $(d\sigma/dt)_{\bar{p}p \to \bar{n}n}$ has no minimum. Note that the $\bar{p}p \to \Lambda\bar{\Lambda}$ reaction[4] does not display a dip-bump structure, since the longest range transition potential arises from K exchange, already a short range phenomenon compared to OPE.

Mizutani et al[33] have studied the helicity amplitudes ϕ_i in detail. They find that for elastic scattering, ϕ_1 and ϕ_3 are the most important amplitudes, while for charge exchange, ϕ_2 and ϕ_4 prevail. The simple model of Eq. (4) captures the essence of the physics of the dip-bump region, but it does not offer a quantitative description. The amplitudes ϕ_1 and ϕ_3, treated as constants in Eq. (4), actually decrease fairly rapidly near t=0 (note ϕ_5 vanishes here). Different optical models vary significantly in their predictions for ϕ_1 and ϕ_3. The incorrect prediction of the Paris model[23] of a significant dip-bump structure in $\bar{p}p \to \bar{n}n$ even at low momenta may be due to having amplitudes ϕ_1 and ϕ_3 which are too small to fill in the dip, or to having the wrong momentum dependence for B.

Where does B come from? The lack of significant t variation imples that it is a manifestation of the short range dynamics. As shown by Leader[34] (at least for high energies), the ρ exchange contributions to ϕ_2 and ϕ_4 are imaginary, so they do not interfere <u>coherently</u> with $\phi_{2,4}^\pi$. Thus B must be related to the annihilation potential, in particular to the spin-isospin dependence, energy dependence and range of W. The models which successfully reproduce the low momentum $\bar{p}p \to \bar{n}n$ data (i.e. the Nijmegen[24] and boundary condition[25,33] models) have a longer range and less energy depencence than the Paris model[23].

TOTAL CROSS SECTIONS

There has been considerable focus on total cross section measurements[5-9] in the quest for narrow "baryonium" states. Earlier evidence[35] for such structures has not been confirmed in recent high statistics experiments[5-11]. For instance, recent $\bar{p}p$ annihilation (σ_A) and total cross section (σ_T) measurements[5-7] show no structure in the region of the putative S(1930) meson. The $\bar{n}p$ cross sections[8] σ_A and σ_T were measured at Brookhaven in the very low energy region E = 1880-1940 MeV/c^2, with tight upper limits on the production of narrow states, particularly near threshold. The $\bar{n}p$ data constrain the properties of a possible low momentum isospin I=1 resonance which might be introduced to explain the behavior of ρ (see previous section).

The measured low momentum $\bar{n}p$ cross sections (p_L < 300 MeV/c) lie considerably below the expected $\bar{p}p$ cross sections, extrapolated from A + B/p_L fits to the $\bar{p}p$ data[5,6] at momenta above 300 MeV/c. If confirmed by direct measurements of the low momentum $\bar{p}p$ cross sections, this would be an interesting effect. In the context of potential models, the observation that $\sigma_{\bar{p}p} > \sigma_{\bar{n}p}$ could signal the effect of tensor forces which are particularly strong and attractive[36] for I=0, J = $\ell \pm 1$ states (the coherence of contributions from π, ρ and ω exchange). Such coherent tensor forces operate for $\bar{p}p$, but not for $\bar{n}p$, since the later system has I=1, for which the total tensor potential is expected to be much weaker. Note that differences in the $\bar{p}p$ and $\bar{n}p$ total cross sections could also arise from an isospin dependence of the absorptive potential, which is expected in various microscopic quark models.

$\bar{N}N$ ANNIHILATION: GENERAL CONSIDERATIONS

The $\bar{N}N$ system has a total baryon number of zero, and hence couples directly to multi-meson final states. This offers a variety of possibilities for studies of <u>meson spectroscopy</u>. For instance, there is copious production of the usual s-wave quark-antiquark ($Q\bar{Q}$) mesons s = $\{\pi, \eta, \rho, \omega\}$ or p-wave mesons p = $\{f, B, A_2 \ldots\}$ in reactions of the type $\bar{N}N \to$ ss, sp, sss .... . The relative branching ratios for such modes provide tests of the dynamics of the reaction mechanism, in particular the choice of an effective operator for $Q\bar{Q}$ annihilation, as well as the topology ($Q\bar{Q}$ annihilation vs quark rearrangement) of the annihilation process. We return to these questions in the next section.

In addition to the usual s and p mesons, one could imagine the production of new types of mesons X via $\bar{N}N$ annihilation, either as scattering resonances ($\bar{N}N \to X \to \bar{N}N$) or, for masses $M_X < 2m_N$, as annihilation products in reactions like $\bar{N}N \to \pi X$ or γX, followed by decays $X \to M_1 M_2$. The meson X could be an $\bar{N}N$ bound state[15-19] or a $Q^2\bar{Q}^2$ "baryonium" state[20]. In the absence of some special selection rules[37], such structures are expected to be very broad, decaying readily into ordinary $Q\bar{Q}$ mesons. A spectrum of mesons containing "dynamic" gluons, i.e., the hybrid $Q\bar{Q}g$ states[21] or the glueballs[38] (quarkless mesons) has been predicted using several different approaches. These objects are expected to be very unstable, but they may have unusual decay modes which provide a distinct signature. For example, in a strong coupling approach to QCD, Isgur and Paton[39,40] have predicted hybrid mesons of masses 1800-1900 MeV/c^2 with exotic quantum numbers (ex. $J^{\pi C} = 1^{-+}$). These decay via $X \to sp$ rather than $X \to ss$, and thus would not have been seen in conventional analyses.

Since the $\bar{N}N$ system can provide a rich source of gluons through the (non-perturbative) annihilation of any one of its constituent $Q\bar{Q}$ pairs, one would expect measurable branching ratios for the production of mesons with gluonic content. Promising reactions include $\bar{p}p \to X \to \phi\phi$ (ref. 41) or

$$\bar{p}p \to \pi X \atop {\downarrow \atop {\pi B \atop {\downarrow \atop {\pi\omega \atop {\downarrow \atop \pi\gamma}}}}} \qquad (6)$$

The latter reaction involves the detection of more than one neutral particle, and can be studied with the proposed "Crystal Barrel" detector[42] at LEAR.

As a final example, we mention the study of charmonium spectroscopy via direct s-channel formation. With $\bar{N}N$, one can access ψ/χ states with quantum numbers other than 1^{--}; the widths of these states can also be directly measured. The first such experiment[43] was recently completed at the CERN ISR, and a continuation is being readied at Fermilab[44].

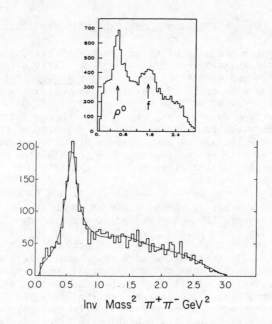

Fig. 5. Invariant mass distribution for $\pi^+\pi^-$ in the reaction $pp \to \pi^+\pi^-\pi^0$, in gaseous hydrogen[12] (top) and liquid hydrogen[45] (bottom). The figure is taken from the Crystal Barrel proposal[42].

$\overline{N}N$ ANNIHILATION INTO THREE PIONS

The reaction $pp \to \pi^+\pi^-\pi^0$ was originally studied in bubble chambers[45]. The data, corresponding primarily to $\ell=0$ absorption, show a very strong $\pi\rho$ signal and little sign of πf production (bottom half of Fig. 5). From the approximate equality of the number of events N for the $\pi^+\rho^-$, $\pi^-\rho^+$ and $\pi^0\rho^0$ charge states, and also from an analysis of ρ decay angular distributions, one concludes

$$N(\overline{p}p(^{13}S_1) \to \pi^\pm\rho^\mp) \gg N(\overline{p}p(^{31}S_0) \to \pi^\pm\rho^\mp) \tag{7}$$

Since the transition $^{31}S_0 \to \pi^0\rho^0$ is forbidden by C-parity, any appreciable production of $\pi^\pm\rho^\mp$ from $\overline{p}p(^{31}S_0)$ would imply $N(\overline{p}p \to \pi^0\rho^0)$ is less than $N(\overline{p}p \to \pi^+\rho^-)$.

Recently, the ASTERIX collaboration[12] has measured the $\overline{p}p \to \pi^+\pi^-\pi^0$ reaction using a gaseous hydrogen target. In this case, the absorption from $\ell=1$ and $\ell=0$ states is comparable. The data (top part of Fig. 5) clearly indicate a strong ℓ dependence of the $\overline{p}p \to \pi^+\pi^-\pi^0$ reaction: the $\pi^0 f$ signal from $\ell=1$ is now very prominent. Again, the rates for $\pi^0\rho^0$, $\pi^+\rho^-$ and $\pi^-\rho^+$ seem to be about equal, implying

$$N(\overline{p}p(^{11}P_1) \to \pi^\pm\rho^\mp) \gg N(\overline{p}p(^{33}P_{1,2}) \to \pi^\pm\rho^\mp) , \tag{8}$$

whereas for the $\pi^0 f$ channel, one finds

$$N(\bar{p}p(^{33}P_1) \to \pi^0 f) \gg N(\bar{p}p(^{33}P_2) \to \pi^0 f) . \tag{9}$$

The above relations provide examples of dynamical selection rules in $\bar{N}N$ annihilation. In my opinion, such selectivity will be the rule rather than the exception. Indeed, there is evidence that $K^*\bar{K}$, $\bar{K}^*K$ production[46] from $\ell=0$ is dominated by the $^3S_1$ rather than the $^1S_0$ channel. Recent $\bar{p}p$ data[47] also point to a suppression of the $(K^+K^-/\pi^+\pi^-)$ ratio for $\ell=1$, i.e.,

$$(K^+K^-/\pi^+\pi^-)_{\ell=0} \approx 0.3$$
$$(K^+K^-/\pi^+\pi^-)_{\ell=1} \approx 0.05 - 0.07 . \tag{10}$$

Similarly, the $\bar{p}p(\ell=1) \to K_S K_S$ process is suppressed[47] relative to $\bar{p}p(\ell=0) \to K_L K_S$.

Note that the above inequalities involve the number of events N, not the decay widths Γ. Since N contains a statistical factor $2J+1$ favoring the formation of $^3S_1$ over $^1S_0$ atoms, it is clear that the $\bar{p}p$ atom is not the optimum situation for studying the relative strength of transitions from the $^1S_0$ state. Rather, one studies[14,15] the reaction $\bar{p}d \to p_S +$ mesons in the kinematic regions (momentum ≤ 150 MeV/c) where the recoil proton p_S is a spectator, in order to isolate the process $\bar{p}n \to$ mesons. For $\bar{p}n \to 2\pi^-\pi^+$, for example, this has the advantages that the initial state has unique quantum numbers ($^{31}S_0$ for for $\ell=0$), and is free of the complication of isospin mixing $\bar{p}p \leftrightarrow \bar{n}n$), a significant effect[48] for $\bar{p}p$ atoms. The $\bar{p}n \to 2\pi^-\pi^+$ reaction at rest has been investigated by Kalogeropoulos and collaborators[15]. They find clear peaks in the $(\pi^+\pi^-)$ mass spectrum corresponding to the transitions $\bar{p}n(^{31}S_0) \to \pi^-\rho^0$, π^-f. The ρ^0 and f peaks are also evident[49] in the same reaction "in flight." This shows that the relations (7, 8, 9) do not imply an absolute selection rule forbidding $\pi\rho$ production from the $^{31}S_0$ state, as predicted in ref. (50). The analysis of refs. (15, 49, 51) yielded several fits to the data, all of which have the property that

$$N(\bar{p}n(^{31}S_0) \to \pi^-\rho^0) < N(\bar{p}n(^{31}S_0) \to \pi^-f) . \tag{11}$$

QUARK-GLUON DYNAMICS OF $\bar{N}N$ ANNIHILATION

In a conventional picture one considers the $\bar{N}N$ annihilation process to be driven by a baryon exchange mechanism. This concept (i.e., a u-channel pole approximation) is useful in many high energy reactions (for instance, in the description of large angle processes involving numerous partial waves). However, for $\bar{N}N$ interactions near threshold, one studies interactions in non-peripheral partial waves ($\ell=0,1$), where the overlap between the N and $\bar{N}$ "bags" is significant. Under these circumstances, the concept of a very short range force (a $\approx 1/2m_N \approx 0.1$ fm), generated by baryon exchange between point particles, is not very useful. The finite spatial extent of the N and $\bar{N}$ introduces another effective range parameter into the problem, related to the bag radius R. For $r \gg R$, the $\bar{N}N$ absorptive potential presumably heals to the nucleon exchange form $e^{-r/a}$ predicted long ago for point particles by A. Martin[52]. However, the relevant region for phenomenological analyses is $r \approx R$.

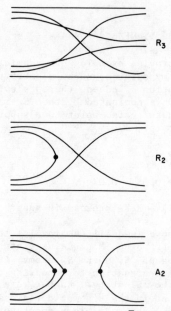

Fig. 6. Quark diagrams corresponding to $\bar{N}N$ annihilation into three meson (R_3) or two meson (R_2, A_2) channels. The dot represents an effective operator O for $Q\bar{Q}$ creation/annihilation.

In the quark model, the processes leading to $\bar{N}N$ annihilation are shown in Fig. 6. The "rearrangement" graph R_3 at the top leads to direct production of three meson final states. The so-called rearrangement model, consisting solely of this process, has been used to analyze mesonic branching ratios[53,54,55]. The basic assumption of such analyses is that direct three body (3B) modes dominate, with perhaps 10-20% of annihilation proceeding via quasi-two-body (QTB) reactions. This would support the idea of a hierarchy of processes, in which one pays a penalty for each $Q\bar{Q}$ annihilation vertex, i.e., the strengh λ associated with the vertex is presumed to be a small parameter. In such a picture, the leading correction to the pure rearrangement amplitude R_3 would be the process R_2 with one $Q\bar{Q}$ vertex. This is the basic assumption of the Helsinki and Osaka groups[54,55].

Experimentally, it is difficult to separate QTB and 3B contributions, particularly if the QTB mode involves two broad mesons. For modes of multiplicity n=3 and 4, for instance $\pi^+\pi^-\pi^0$ and $2\pi^+2\pi^-$, the analysis[46] of the early CERN data is in fact consistent with QTB dominance: $\rho\pi$ exhausts most of the $\pi^+\pi^-\pi^0$ strength, while πA_2 and ρf make significant contributions to $2\pi^+2\pi^-$. The purported 3B branches $\pi^0\pi^+\pi^-$ and $\rho^0\pi^+\pi^-$ could also be interpreted in terms of the QTB modes $\pi^0\varepsilon$ and $\rho^0\varepsilon$, while a large $\rho\rho\pi$ signal in the 5π channel may be consistent with QTB $\rho A_{1,2}$ production. So far, so consistent analysis of $\bar{N}N$ annihilation has been done using a QTB hypothesis. This was presumably due to the sizable number of QTB modes (many involving two

broad mesons) which contribute for n>4, and the lack of theoretical
guidance as to their relative branching ratios. Given a choice of
the effective $Q\bar{Q}$ operator which enters in graphs R_2 and A_2 of Fig. 6,
one can now calculate these ratios (see next section). The strong
possibility exists that QTB processes in fact dominate $\bar{N}N$ annihilation, and that graph A_2 is the major process. This would imply that
there is no small parameter λ which characterizes the $Q\bar{Q}$ vertex,
i.e., one is in the strong coupling limit of QCD. Note that in
ordinary meson and baryon resonance decay, there is in fact little
indication of direct transitions to three body final states. The
observed decays are consistent with a QTB hypothesis, in which a
multiparticle final state is produced by a chain of two-body decays
(ex. $B \to \omega\pi$, $\omega \to \rho\pi$, $\rho \to \pi\pi$ gives $B \to 4\pi$). Since the $\bar{N}N$ system
shares the quantum numbers of ordinary $Q\bar{Q}$ mesons (differing of course
in its quark content), one might expect that its decays would display
a similar QTB character.

THE EFFECTIVE OPERATOR FOR $Q\bar{Q}$ ANNIHILATION

To perform dynamical calculations of QTB processes, one must
choose a form for the $Q\bar{Q}$ vertices in processes R_2 and A_2. Attention
has been focused on the perturbative one or two gluon[50,56] and non-
perturbative $^3P_0$ models[57]. In its simplest form, the effective
operator O for $Q\bar{Q}$ annihilation in the $^3P_0$ model is

$$O(^3P_0) = \lambda_p \chi_f \chi_c^{(1)} \chi_m (1m1-m|00) \mathcal{Y}_{1,-m}(\underset{\sim}{k}) \qquad (12)$$

where $\chi_f = (u\bar{u} + d\bar{d} + s\bar{s})/\sqrt{3}$ is an SU(3) flavor singlet combination,
$\chi_c^{(1)}$ is an SU(3) color singlet, χ_m is a spin 1 wave function with z
projection m, and the spherical harmonic $\mathcal{Y}_{1,-m}(\underset{\sim}{k})$ expresses the p-
wave character of the vertex. The Clebsch-Gordon coefficient
$(1m1-m|00)$ couples S=1, ℓ=1 to $J^\pi = 0^+$. Overall, the $^3P_0$ vertex
corresponds to the quantum numbers of the vacuum $(0^{++}(0^+))$. The
strength constant λ_P has been determined from an analysis of meson
and baryon resonance decays. In this context, a single value of λ_p
is able to account for decay widths over a considerable mass scale
($\rho \to \pi\pi$ to charmonium), and also for other delicate features such as
relative phases of decay amplitudes and polarization of vector mesons
produced in decay processes (see ref. 59 for details). The $^3P_0$ model
has recently been "derived" in a strong coupling approach to QCD by
Isgur and Paton[39]. In this formulation, quarks and flux tubes are
the relevant degrees of freedom. Gluonic degrees of freedom are condensed into collective string-like flux tubes. This leads to the
idea of "constituent gluons" and hybrid mesons containing a $Q\bar{Q}$ pair
and an excited flux tube, while reducing to the usual constituent
quark model for heavy quarks. Creation of $Q\bar{Q}$ pairs occurs by breaking a flux tube with equal amplitude anywhere along its length, and
leads one to a version of the $^3P_0$ model in the limit where the size
of the flux tube becomes large. A detailed comparison of the full
strong coupling results and the simplified $^3P_0$ limit has been given
by Isgur and Kokoski[39].

In the weak coupling limit of QCD, $Q\bar{Q}$ pair creation/annihilation
is coupled to a single gluon. The simplest form of O is then

$$O(\text{one gluon}) = \lambda_s \chi_f \chi_c^{(8)} \vec{S} \cdot (\vec{\sigma}_i \times \vec{q}) \tag{13}$$

where χ_f is a flavor singlet as in (12), $\chi_c^{(8)}$ is a color octet, $S=1$ is the gluon spin, $\vec{\sigma}_i$ is the Pauli spin of the quark i struck by the gluon, and $\vec{q}$ is the momentum transfer imparted to this quark. In the non-relativistic reduction of the full relativistic form of O(one gluon), one also finds[56] recoil corrections to (13) which resemble the $^3P_0$ form.

There is as yet no comprehensive study of meson and baryon resonance decay processes using O(one gluon). Rubinstein and Snellman[60] have suggested that the weak coupling approach may be appropriate for heavy quark pair production (including $s\bar{s}$), but regard the production of light quark $u\bar{u}$ and $d\bar{d}$ pairs to be essentially non-perturbative. The one gluon approximation has been shown[48] to fail to describe vector meson polarization in the decays $B \to \omega\pi$ and $A_1 \to \rho\pi$, whereas the $^3P_0$ model[59] works well. It seems unlikely that the one gluon approximation has much relevance to the problem of $\bar{N}N$ annihilation into non-strange mesons in the low energy regime. The distance scales involved can be estimated from the momentum q released in the final state in a reaction $\bar{N}N \to M_1M_2$ at rest. We find

$$q^{-1}(\text{fm}) = \begin{cases} 0.2 & (\pi\pi) \\ 0.26 & (\pi\rho) \\ 0.37 & (\rho\rho) \\ 0.4 & (\pi f) \\ 0.43 & (\pi A_2) \end{cases} \tag{14}$$

For prominent sp modes, such as πf or πA_2, $q^{-1} \approx R/2$, suggesting that one is probing the confinement regime (strong coupling, with many soft gluons) rather than the perturbative regime ($q^{-1} \ll R$).

The quark graphs of Fig. 6 are subject to selection rules which are testable experimentally. For instance, the rearrangement process R_3 conserves intrinsic spin S and allows only relative s-waves between mesons M_i in the final state. Thus we have for R_3:

Allowed: $\ell = 0 \to sss$
$\ell = 1 \to ssp$

Forbidden: $\ell = 0 \to ssp$
$\ell = 1 \to sss$

(15)

If a three-body component can be identified in the ASTERIX data[12], Eq. (15) can easily be tested. This analysis has not yet been completed.

For the processes R_2 and A_2, the selection rules depend on the choice of the effective $Q\bar{Q}$ operator O. For the one gluon case, these have been discussed by Henley et al[56]. For the $^3P_0$ model, we have for R_2:

Allowed: $\ell = 0 \to sp(\ell_f=0)$
$\ell = 1 \to ss(\ell_f=0)$

Forbidden: all transitions with $\ell_f \geq 1$,

(16)

where ℓ_f is the relative orbital angular momentum of the two mesons in the final state. Since the partial wave content of R_2 is strongly restricted, it is clear that this process by itself cannot account for the data. For instance, R_2 in the $^3P_0$ model gives no amplitude for $^{13}S_1 \to \pi\rho(\ell_f=1)$ or $^{11}S_0 \to \pi A_2(\ell_f=2)$, both of which are observed to be prominent transitions. For A_2, we have

<u>Allowed</u>: $\ell = 0 \to sp(\ell_f=0,2)$
$\ell = 0 \to ss(\ell_f=1)$
$\ell = 1 \to sp(\ell_f=1)$
$\ell = 1 \to ss(\ell_f=0)$ (17)

In a "no recoil" approximation[61], in which the effects of finite meson size are neglected in certain angular factors, we find for A_2

<u>Forbidden</u>: $\ell = 1 \to ss(\ell_f=2)$,
all transitions with $\ell_f \geq 3$ (18)

The various choices for O and the topology of annihilation give rise to dramatically different selection rules, as per Eqs. 16-18. An analysis of the experimental data which can separate the numerous transitions $\ell=0,1 \to ss, sp(\ell_f=0,1,2)$ is thus of paramount importance.

$\bar{N}N \to \pi\rho, \pi f$ TRANSITIONS IN THE $^3P_0$ MODEL

To obtain a transition amplitude for $\bar{N}N \to M_1 M_2$, one embellishes the quark lines in A_2 or R_2 with all possible spin-flavor combinations and then assigns weights based on the SU(6) wave functions of the external lines. We focus here on the $\pi\rho$, πf channels, for which only A_2 contributes. More details are found in ref. 61. The width Γ_i for a channel $i = \{M_1, M_2\}$ populated from an initial $\bar{N}N$ state has the form

$$\Gamma_i = W_i |F_i|^2 \qquad (19)$$

where W_i is the spin-flavor weight and F_i is a kinematical form factor:

$$F_i(\ell,\ell_f,k,q) = N_i(\ell,\ell_f) \, (kR)^\ell \, (qR)^{\ell_f} \, e^{-R^2(k^2/9 + q^2/4)} \qquad (20)$$

In Eq. (20) the usual penetrabilities $(kR)^\ell$ and $(qR)^{\ell_f}$ for the initial and final states are modified by Gaussian factors arising from wave function overlap (R = bag radius). For fixed ℓ, F_i is maximum when ℓ_f and q are well matched., i.e.

$$\ell_f = (R^2 q^2)/2 \qquad (21)$$

This kinematical matching of q and ℓ_f is a crucial ingredient in understanding the relative annihilation branching ratios. We find that $\bar{p}p(\ell=0,1) \to sp(\ell_f=2)$ transitions are well matched, whereas the ss, $sp(\ell_f=0)$ cases are suppressed. As an example, we find

$$\left|F_{\pi\rho}(\ell=0,\ \ell_f=1)/F_{\pi f}(\ell=0,\ \ell_f=2)\right|^2 \approx 0.03$$

$$\left|F_{\pi\rho}(\ell=1,\ \ell_f=0)/F_{\pi f}(\ell=1,\ \ell_f=1)\right|^2 \approx 0.08 \qquad (22)$$

for $R = 0.8$ fm. Thus, if the $\pi\rho$ and πf channels compete for flux from the same initial $\overline{N}N$ state, πf is expected to dominate. The weights W_i do not upset this conclusion. Up to an overall normalization, we find[61] in the $^3P_0$ model

$$W_i = \begin{cases} 1 & (^{13}S_1 \to \pi^\pm \rho^\mp) \\ 1 & (^{31}S_0 \to \pi^\pm \rho^\mp) \\ 1/2 & (^{31}S_0 \to \pi^0 f) \end{cases} \qquad (23)$$

for $\ell = 0$, and

$$W_i = \begin{cases} 1 & (^{11}P_1 \to \pi^\pm \rho^\mp) \\ (5/3)^2 & (^{33}P_1 \to \pi^\pm \rho^\mp) \\ 500/27 & (^{33}P_1 \to \pi^0 f) \end{cases} \qquad (24)$$

for $\ell = 1$.

Eqs. 22-24 enable us to explain the "$\pi\rho$ puzzle", namely the absence of $\pi^\pm\rho^\mp$ production from the $^{31}S_0$ and $^{33}P_1$ channels. We obtain

$$\frac{N(^{31}S_0 \to \pi\rho)}{N(^{13}S_1 \to \pi\rho)} \approx \frac{1}{3} \frac{\Gamma(^{31}S_0 \to \pi\rho)}{\Gamma(^{31}S_0 \to \pi\rho + \pi f)} \approx \frac{1}{50}$$

$$\frac{N(^{33}P_1 \to \pi\rho)}{N(^{11}P_1 \to \pi\rho)} \approx \frac{1}{10} \qquad (25)$$

$$\frac{N(\ell=1 \to \pi\rho)}{N(\ell=1 \to \pi f)} \approx 1 ,$$

consistent with the data[12]. Note that the transition $^{33}P_2 \to \pi^\pm \rho^\mp (\ell_f = 2)$ is suppressed due to Eq. (18). Observe that W_i is in fact the same for $^{13}S_1 \to \pi^\pm\rho^\mp$ and $^{31}S_0 \to \pi^\pm\rho^\mp$. Thus the apparent suppression in $^{31}S_0 \to \pi^\pm\rho^\mp$ is not due to a selection rule which forbids this transition in Born approximation[50], but rather to a strong competition from the $\pi^0 f$ channel. The same comment applies for $\ell=1$. The effect of the $\pi^0\varepsilon$ channel has been estimated[61] and found to be small.

The $^3P_0$ model suggests that dynamical selection rules, and a strong ℓ dependence of branching ratios are rather general phenomena. The search for these among the multiplicity $n \geq 4$ annihilations often requires the detection of modes with two or more neutral pions (ex. $\rho^+\rho^-$ vs $\rho^0\rho^0$). This is a task for the "crystal barrel" detector[42], which should be a premier device for annihilation studies in the ACOL era of LEAR.

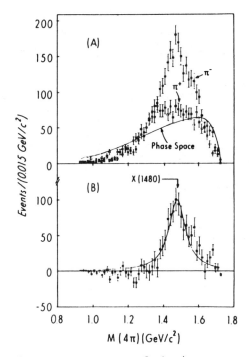

Fig. 7. The invariant mass spectrum of the 4π system recoiling against a π^+ or π^- in the reaction $\bar{p}n \to 3\pi^- 2\pi^+$ is shown in (A). In (B) the $\pi^- - \pi^+$ difference spectrum is displayed. The figure is taken from Bridges et al[14].

QUASINUCLEAR BOUND STATES OF THE $\bar{N}N$ SYSTEM

Recent data[14] on the reaction $\bar{p}n \to 3\pi^- 2\pi^+$ have demonstrated the existence of a broad structure $X^0(1480)$, produced via

$$\bar{p}n \to \pi^- X^0(1480) \quad\quad (26)$$
$$\hookrightarrow \rho^0 \rho^0$$

The π^- and π^+ spectra are shown in Fig. 7(A). The $X^0(1480)$ shows up as a peak in the difference spectrum $N(\pi^-)-N(\pi^+)$ displayed in Fig. 7(B). The study of $N(\pi^-)-N(\pi^+)$ helps to separate the primary π^- produced in $\bar{p}n \to \pi^- X^0$ from a π^- background arising from decays. The $3\pi^- 2\pi^+$ channel has the advantage that only a few QTB modes are possible, namely $\bar{p}n \to \rho^0 A_{1,2}^-$, which do not lead to an understanding of Fig. 7(B).

In ref. 14, a width $\Gamma = 116\pm9$ MeV was found for $X^0(1480)$, a dominant decay mode $X^0 \to \rho^0 \rho^0 (\ell_f = 0)$, and a product branching ratio $BR(\bar{p}n \to \pi^- X^0) \cdot BR(X^0 \to \rho^0 \rho^0) \approx 3.5\%$. A quantum number assignment $J^{PC}(I^G) = 2^{++}(0^+)$ was given[14].

A structure at a comparable mass and width, dubbed the f_2', was seen by Gray et al[15] in the reaction

$$\bar{p}n \to \pi^- f_2' \quad\quad (27)$$
$$\hookrightarrow \pi^+ \pi^-$$

The data are shown in Fig. 8. It is tempting to identify $X^0(1480)$ and f_2' as the same object, seen in two different decay modes.

The $X^0(1480)$ appears near the $\rho\rho$ threshold, so it could possibly be interpretable in terms of a final state s-wave rescattering of the $\rho\rho$ system. A broad bump was also seen[62] in the process $\gamma\gamma \to \rho^0\rho^0$ near threshold, whereas in the $\gamma\gamma \to \rho^+\rho^-$ reaction[63] this structure is suppressed. This behavior was interpreted in refs. 64 and 65 as an interference between $2^{++}(I=0,2)$ $Q^2\bar{Q}^2$ resonances. The partial wave analysis of the $\gamma\gamma \to \rho^0\rho^0$ data[62] suggested mostly 0^{++} strength near threshold and 2^{++} strength at higher energies, but the soundness of the analysis has been questioned[65]. The relationship between the structures seen in $\bar{p}n \to \pi^-\rho^0\rho^0$ and $\gamma\gamma \to \rho^0\rho^0$ is not clear. If an I=2 $Q^2\bar{Q}^2$ state is indeed involved, it should be seen in $\bar{p}n \to \pi^+X^{--}$, $X^{--} \to \rho^-\rho^-$. This corresponds to the $\pi^+2\pi^-2\pi^0$ channel, again a task for the "crystal barrel"[42].

A plausible interpretation[66] of $X^0(1480)$ is that it corresponds to a quasinuclear (QN) bound state of the $N\bar{N}$ system, namely the $^{13}P_2$-$^{13}F_2$ configuration. The mass and width are reasonable: for instance the Paris optical model[23] predicts a mass of 1500 MeV and $\Gamma \approx 65$ MeV

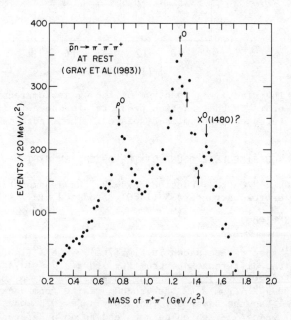

Fig. 8. The invariant $\pi^+\pi^-$ mass spectrum for the reaction $\bar{p}n \to 2\pi^-\pi^+$ at rest, from Gray et al[15]. Note the peak near $M(\pi^+\pi^-) \approx 1480$ MeV/c^2, where a comparable structure[14] is seen in $M(4\pi)$ in the $\bar{p}n \to 3\pi^-2\pi^+$ reaction (see Fig. 7).

for the $2^{++}(0^+)$ bound state. The mechanism for strong binding is provided by the coherent and attractive tensor potential[36] ($\pi+\rho+\omega$) which operates for I=0 natural parity $\overline{N}N$ states. Production branching ratios of a few percent are expected[48]: the $\bar{p}p$ atom finds it rather easy to radiate a pion and drop into a QN bound state having the same quark content. Assuming that $X^0(1480)$ and f_2' are indeed the same state, the preference for the $\rho\rho$ decay mode, i.e.,

$$\Gamma(X^0 \to \rho\rho) > \Gamma(X^0 \to \pi\pi) \tag{28}$$

is understandable in terms of the approximate selection rule of Eq. (18), which suppresses the decay $2^{++}(0^+)(\ell=1) \to \pi\pi(\ell_f=2)$.

If the interpretation of $X^0(1480)$ as the $2^{++}(0^+)$ QN state is sensible, one may also expect to find the $1^{--}(0^-)$ and $0^{++}(0^+)$ members of the isoscalar, natural parity $\overline{N}N$ band at lower mass. For the J=0 state, one expects to see the $\pi^+\pi^-$ decay mode more strongly than for J=2, so the reactions

$$\bar{p}n \to \pi^- X^0(0^{++}) \quad , \quad \rho^- X^0(0^{++}) \tag{29}$$
$$\hookrightarrow \pi^+\pi^-(\ell_f=0) \quad \hookrightarrow \pi^+\pi^-(\ell_f=0)$$

are promising. There is some evidence for a peak around 1100 MeV/c^2 in both of these reactions[15,67], but this has not yet been confirmed. The J=1 state should appear in the $2\pi^-\pi^+\pi^0$ channel:

$$\bar{p}n \to \pi^- X^0(1^{--}) \tag{30}$$
$$\hookrightarrow \pi^0\rho^0, \pi^\pm\rho^\mp$$

This process is difficult to disentangle from $\bar{p}n \to \pi^- A_2^0$, $A_2^0 \to \pi\rho$ without a full partial wave analysis, since $X^0(1^{--})$ and A_2 are expected to lie in the same mass region.

FUTURE PROSPECTS

In this talk, I have only addressed a few of the areas under active investigation in antinucleon physics. This field is in a stage of rapid evolution. The first round of experiments at LEAR has almost been completed, and one looks forward to the ACOL Era[68], when further improvements in $\bar{p}$ beam intensity will permit the explorations of fundamental questions[68] such as CP, T, CPT violation and the $\bar{p}$ gravitational mass, as well as the logical continuation of $\bar{p}$ strong interaction phenomenology (spin physics, for example).

The different aspects of meson spectroscopy, as probed through $\overline{N}N$ annihilation, have received emphasis here. We have argued that $\overline{N}N$ annihilation is very promising as a spectroscopic tool in the search for new mesons. However, such non-$Q\overline{Q}$ objects are likely to be highly unstable, with widths generally in excess of 100 MeV. Data of high statistical quality are required to permit a full amplitude analysis. Bump hunting in total cross sections has proved disappointing, and is likely to remain so. Reactions of the type $\overline{N}N \to M_1 M_2$ act as quantum number filters, but one needs to measure neutral as well as charged modes in order to further constrain C and I. Inclusive spectra ($\pi^\pm, \pi^0, K^\pm, \gamma$, etc.) have yielded negative results in the search for new narrow mesons, and they are not well suited to the study of new broad states because of a generally large background of conventional quasi-two-body modes. It is necessary to go beyond inclusive measurements, and to study reactions such as $\bar{p}p \to \pi X$, where the decay products of X are studied in coincidence with the "direct" π (or other narrow meson). Measurements of particular $\bar{p}d$ or $\bar{n}p$ annihilation modes is necessary as a supplement to the $\bar{p}p$ results. The search for exotic mesons X with masses in the 2 GeV/c^2 region via the $\bar{N}N \to \pi X$ reaction, for instance, requires high intensity $\bar{p}$ beams with lab momenta well in excess of the threshold around 1 GeV/c. The need for such high momentum, high intensity $\bar{p}$ beams for a variety of physics goals was discussed in a recent workshop[69] at Fermilab.

I would like to thank C. Amsler, P. Barnes, T. Kalogeropoulos, H. Koch, R. Landua, S. Pleyfer, H. Poth, and G. Smith for extensive and illuminating discussions of the experimental $\bar{N}$ data and their interpretation.

REFERENCES

1. W. Brückner et al, Phys. Lett. 166B, 113 (1986).
2. K. Nakamura, Phys. Rev. Lett. 53, 885 (1984).
3. W. Brückner et al, Phys. Lett. 169B, 302 (1986).
4. P. Barnes, these proceedings (PS185 collaboration).
5. T. Brando et al, Phys. Lett. 158B, 505 (1985).
6. A.S. Clough et al, Phys. Lett. 146B, 299 (1984).
7. PS172 collaboration at LEAR (Y. Onel, private communication).
8. T. Armstrong et al, Penn State preprint PSU HEP/86-03, to appear in Phys. Lett.
9. W. Fickinger et al, submitted to Phys. Rev. D.
10. A. Angelopoulos et al, Phys. Lett. 159B, 210 (1985), and Penn State preprint PSU HEP/86-02, to appear in Phys. Lett.
11. T. Brando et al, Phys. Lett. 139B, 133 (1984);
    S. Ahmad et al, Phys. Lett. 152B, 135 (1985);
    A. Angelopoulos et al, Proc. Int. Conf. on Meson Spectroscopy, College Park, Maryland, April 1985.
12. S. Ahmad et al, Proc. VII European Symposium on Antiproton Interactions; Durham, England, July 1984 (Inst. of Physics Conf. Series No. 73, Adam Hilger Ltd, Bristol, 1985, Ed. M.R.Pennington), p. 287.
13. S. Ahmad et al, Proc. Third LEAR Workshop; Tignes, France, January 1985 (Editions Frontières, Gif-sur-Yvette, 1985, Eds. U.Gastaldi et al), p. 347;
    L. Adiels et al, ibid, p. 359.
14. D. Bridges et al, Phys. Rev. Lett. 56, 211 and 215 (1986).
15. L. Gray et al, Phys. Rev. D27, 307 (1983).
16. I.S. Shapiro, Phys. Rep. C35, 129 (1978);
    W. Buck, C.B. Dover, and J.M. Richard, Ann. Phys. (NY) 121, 47 (1979);
    C.B. Dover and J.M. Richard, Ann. Phys. (NY) 121, 70 (1979).
17. J.A. Niskanen and A.M. Green, Nucl. Phys. A431, 593 (1984).

18. M. Lacombe, B. Loiseau, B. Moussallam, and R. Vinh Mau, Phys. Rev. C29, 1800 (1984);
    R. Vinh Mau, in Proc. Int. Symp. on Medium Energy Nucleon and Antinucleon Scattering; Bad Honnef, Germany, June 1985 (Lecture Notes in Physics, Vol. 243, Springer Verlag, Berlin, 1985, Ed. H.V. von Geramb), p. 3.
19. J.C.H. Van Doremalen, Yu.A. Simonov, and M. Van der Velde, Nucl. Phys. A340, 317 (1980).
20. R.L. Jaffe, Phys. Rev. D15, 281 (1977) and D17, 1445 (1978).
21. F.E. Close, Nucl. Phys. A416, 55 (1984);
    M. Chanowitz and S. Sharpe, Phys. Lett. 132B, 413 (1983).
22. C.B. Dover and J.M. Richard, Phys. Rev. C21, 1466 (1980).
23. J. Coté et al, Phys. Rev. Lett. 48, 1319 (1982).
24. P.H. Timmers, W.A. van der Sanden, and J.J. deSwart, Phys. Rev. D29, 1928 (1984).
25. O.D. Dalkarov and F. Myhrer, Nuovo Cim. 40A, 152 (1977).
26. V. Ashford et al, Phys. Rev. Lett. 54, 518 (1985).
27. W. Brückner et al, Phys. Lett. 158B, 180 (1985).
28. M. Cresti, L. Peruzzo, and G. Sartori, Phys. Lett. 132B, 209 (1983).
29. H. Iwasaki et al, Nucl. Phys. A433, 580 (1985).
30. S. Ahmad et al, Phys. Lett. 157B, 333 (1985);
    T.P. Gorringe et al, Phys. Lett. 162B, 71 (1985).
31. B.O. Kerbikov and Yu.A. Simonov, preprint ITEP-38 (Moscow, 1986);
    O.D. Dalkarov and K.V. Protasov, Lebedev Institute preprint No. 34 (Moscow, 1986).
32. R.P. Hamilton et al, Phys. Rev. Lett. 44, 1179 (1980).
33. T. Mizutani, F. Myhrer, and R. Tegen, Phys. Rev. D32, 1663 (1985).
34. E. Leader, Phys. Lett. 60B, 290 (1976).
35. B. Richter et al, Phys. Lett. 126B, 284 (1983);
    W. Brückner et al, Phys. Lett. 67B, 222 (1977).
36. C.B. Dover and J.M. Richard, Phys. Rev. D17, 1770 (1978).
37. G.C. Rossi and G. Veneziano, Phys. Rep. 63, 149 (1980).
38. T. Barnes in Proc. Int. School of Exotic Atoms; Erice, Italy, 1984 (Plenum Publ. Co., NY, 1985, Ed. P. Dalpiaz et al), p. 191;
    P.M. Fishbane and S. Meshkov, Comments on Nucl. and Part. Phys. 13, 325 (1984).
39. N. Isgur, R. Kokoski, and J. Paton, Phys. Rev. Lett. 54, 869 (1985);
    N. Isgur and J. Paton, Phys. Rev. D31, 2910 (1985).
40. R. Kokoski and N. Isgur, Toronto preprint UTPT-85-05 (1985).
41. G. Bassompierre et al, JETSET proposal, CERN/PSCC 86-23, PSCC/P97, March 1986.
42. E. Aker et al, CRYSTAL BARREL proposal (PS197), CERN/PSCC/85-56, PSCC/P90, October 1985.
43. C. Baglin et al, submitted to Phys. Lett. B;
    R. Cester, these proceedings.
44. Fermilab proposal P760 (approved 12/85), Fermilab, Ferrara, Genoa, Irvine, Northwestern, Penn State, Torino collaboration.
45. M. Foster et al, Nucl. Phys. B6, 107 (1968).
46. R. Armenteros and B. French, in "High Energy Physics", Ed. E.H.S. Burhop, Academic Press, NY (1969), p. 284.
47. S. Ahmad, in ref. 13, p. 353.

48. C.B. Dover, J.M. Richard, and M. Zabek, Ann. Phys, (NY) 130, 70 (1980).
49. S.N. Tovey et al, Phys. Rev. D17, 2206 (1978).
50. J.A. Niskanen and F. Myhrer, Phys. Lett. 157B, 247 (1985).
51. P. Anninos et al, Phys. Rev. Lett. 20, 402 (1968).
52. A. Martin, Phys. Rev. 124, 614 (1961).
53. H. Rubinstein and H. Stern, Phys. Lett. 21, 447 (1966).
54. M. Maruyama and T. Ueda, Nucl. Phys. A364, 297 (1981) and Phys. Lett. 149B, 436 (1984).
55. A.M. Green and J.A. Niskanen, Nucl. Phys. A412, 448 (1984) and Nucl. Phys. A430, 605 (1984);
   A.M. Green, V. Kuikka, and J.A. Niskanen, Nucl. Phys. A446, 543 (1985);
   a recent review is A.M. Green and J.A. Niskanen, Helsinki preprint HU-TFT-85-60, to appear in Progress in Particle and Nuclear Physics, Pergamon Press (1986).
56. E.M. Henley, T. Oka, and J. Vergados, Phys. Lett. 166B, 274 (1986);
   M. Kohno and W. Weise, Phys. Lett. 152B, 303 (1985) and Nucl. Phys. A454, 429 (1986).
57. C.B. Dover and P.M. Fishbane, Nucl. Phys. B244, 349 (1984).
58. S. Furui, Orsay preprint (1985).
59. A. LeYaouanc, L. Oliver, O. Pene, and J.C. Raynal, Phys. Rev. D8, 2223 (1973), D9, 1415 (1974) and D11, 1272 (1972).
60. H.R. Rubinstein and H. Snellman, Phys. Lett. 165B, 187 (1985).
61. C.B. Dover, P.M. Fishbane, and S. Furui, preprint (1986).
62. M. Althoff et al, Z. Phys. C16, 13 (1982);
   H.J. Behrend et al, Z. Phys. C21, 205 (1984).
63. H. Kolanoski, Proc. 5th Int. Workshop on Photon-Photon Interactions; Aachen, Germany, 1983 (Springer Publ, Co., Berlin, 1983, Ed. Ch. Berger).
64. B.A. Li and K.F. Liu, Phys. Rev. Lett. 51, 1510 (1983).
65. N.N. Achasov, S.A. Devyanin, and G.N. Shestakov, Phys. Lett. 108B, 134 (1982) and Z. Phys. C27, 99 (1985).
66. C.B. Dover, Brookhaven preprint (1986).
67. T. Daftari, Syracuse preprint (1986).
68. see the Tignes proceedings (ref. 13).
69. Proc. 1st Workshop on Antimatter Physics at Low Energy; Fermilab, April 1986 (to appear as Fermilab report).

NON-ACCELERATOR EXPERIMENTS

D.H. Perkins
Department of Nuclear Physics, University of Oxford,
Keble Road, Oxford, England

1) PROTON DECAY SEARCHES

Let us first remind ourselves of the reasons that have prompted proton decay searches. First is that given by Lee & Yang in 1955, following the extension of the principle of local gauge invariance to non-Abelian fields by Yang and Mills in 1954. If the baryon number B were absolutely conserved, then in the context of local gauge invariance, one expects that a new long-range field coupled to B should exist, just as conservation of electric charge implies the existence of a field (the electromagnetic field) coupled to charge. Eötvos-type experiments with different materials (i.e. slightly different baryon number per unit mass) do not find any evidence for such a coupling, and the result can be expressed as a limit to $K_B/K < 10^{-9}$, where K is the Newtonian constant and K_B is the analogous coupling of the baryon number field. While not proving that such a field does not exist, it is clear that a new field with such a weak coupling would make enormously more difficult the problems that we already have, in seeking to unite the different fundamental interactions.

A second reason for believing that protons are unstable at some level was given in 1966 by Sakharov, following the discovery of CP violation in 1964. In the context of the big bang model, Sakharov emphasized three conditions needed to account for the baryon-antibaryon asymmetry of the universe:- baryon number violation, CP violation (unequal production of left-moving quarks and right-moving antiquarks) and an "arrow of time" (non-equilibrium expansion). Sakharov emphasized the inevitability of proton decay (conversion of baryon to lepton plus mesons) and took for his energy scale the Planck mass, estimating a proton lifetime of 10^{50} years or more.

In grand unified theories, characterized by a mass $M_X \sim 10^{15}$ GeV for the GUT bosons (where the electroweak and strong couplings, α_g, merge together), the proton lifetime via X-exchange is

$$\tau = \frac{A\, M_X^4}{\alpha_g^2 M_p^5}$$

where A is a dimensionless number of order unity and contains the matrix elements for the conversion of a quark pair to a lepton and quark $(Q + Q \to X \to l^+ + \bar{Q})$. In minimal SU(5), $A \simeq 1$ and $\tau(p \to e^+\pi^\circ) = 10^{29\pm1}$ years. This prediction is in definite disagreement with the recent result (Bionta et al 1985) from the IMB water Cerenkov experiment, yielding

$$\tau(p \to e^+\pi^\circ) > 2.10^{32} \text{ years.}$$

Other versions of GUT make less exact predictions for the lifetime and cannot be excluded. Those GUTS incorporating supersymmetry involve supersymmetric Higgs exchange and therefore predict decay to the heaviest possible quarks and leptons i.e. the decay modes $p \to \mu^+ K^\circ$, $\bar{\nu}_\tau K^+$.

Table 1 summarizes the results on total event rates in the different detectors: the tracking-type detectors consisting of steel plates and proportional counters (Kolar Gold

Field, KGF), streamer tubes (NUSEX) or flash chambers triggered by Geiger planes (FREJUS); and the water Cerenkov detectors (Kamioka, IMB and HPW). The event rates observed are in good agreement with those calculated by Gaisser et al (1983) assuming they are due entirely to interactions of atmospheric neutrinos. (The uncertainty in expected rates is of order 30%).

Practically all the experiments report proton decay candidates, but no-one actually claims to have observed a clear signal. A small fraction of neutrino reactions (up to 5%) can simulate some modes of proton decay rather closely. In fact, the main task of the experimenters is to understand the background in sufficient detail to make meaningful subtractions and hopefully find a clear signal (if there is one).

Table 1:   Event Rates in Proton Decay Detectors

Group	KGF	NUSEX	FREJUS	KAMIOKA	IMB
Method	Fe/PWC	Fe/ST	Fe/FC	Water-Cerenkov	
$\neq$ Events	17	31	65	141	401
Kiloton years*	0.22	–	0.46	1.11	3.77
Rate/Kty	77±19	151±28	143±18	127±11	142±7
Prediction (Gaisser) of Neutrino event rate	85	132	132	99	132

* Including factor for efficiency

Two different approaches have been made to the estimation of neutrino background for event configurations simulating proton decay. Fig. 1 shows IMB results on the asymmetry, A, versus Cerenkov energy E_c. The anisotropy is defined as the vector sum of unit vectors drawn from the effective event vertex to each of the hit photomultipliers, divided by the number of hits. For a single straight track, A will be ~ 0.7 (the cosine of the Cerenkov angle, 42°), while for two back-to-back tracks, $A \simeq 0$. Fig. 1 shows, in (a), the observed values for a sample of 169 events; in (b), the values expected from a computer simulation of the decay $p \to \mu^+\pi^0$, taking account of Fermi motion and scattering; and in (c), the distribution for simulated neutrino events, corresponding to the rates expected for the period of observation (204 days). The neutrino event simulation is based on the topologies of neutrino interactions observed in accelerator experiments using bubble chambers (BNL Ne/H data, ANL D_2 data, Gargamelle freon data). Clearly, the observed underground event distribution is compatible with the expected neutrino background. In particular, no events are observed which could correspond to the decay $p \to e^+\pi o$, giving the limit previously quoted. From this experiment, the 90%CL limits for various decay modes are:-

Table 2:   IMB Limits.

	($\tau_{min} \times$ branching ratio)
$p \to e^+\gamma, \mu^+\gamma, e^+\pi^\circ, \mu^+\pi^\circ$	2-3.10^{32} years.
$n \to e^+\pi^-, \mu^+\pi^-$	5-8.10^{31} years.
$p \to \mu^+\rho^\circ, \mu^+\omega, \nu\rho^+, \nu K^+, \mu K^\circ$	0.8-3.10^{31} years.

The Kamiokande experiment, with a smaller water volume but better photoelectron yield than IMB, currently achieves similar limits.

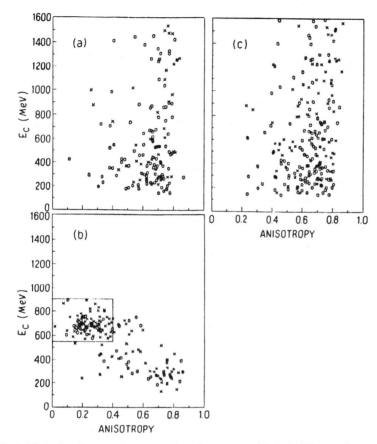

Fig.1 Plot of anistropy A versus Cerenkov energy E_c in IMB experiment.
  (a) 169 contained events in 204 day run in underground detector
  (b) simulation of a sample of proton decay evants in the mode $p \to \mu^+ \pi^0$.
  (c) simulation of 204 days of atmospheric neutrino interactions.

The second approach to the background estimation is that adopted in the NUSEX experiment—to expose a module of the underground detector to an accelerator neutrino beam. From comparison of these genuine neutrino interactions in the detector with those from the underground exposure we obtain a very direct background estimate. Fig. 2 shows a proton decay candidate from the NUSEX experiment, with the two interpretations: proton decay $p \to \mu^+ K^0$, or a neutrino reaction $\nu N \to \mu \pi N$, with backscattering of the pion. While, in my opinion, the simulation of neutrino interactions in water from the configurations of accelerator neutrino interactions in deuterium and heavy liquids is dubious at best, the Monte-Carlo simulation of neutrino reactions in iron is virtually impossible—hence the desirability of a sample of accelerator neutrino interactions. From analysis of 400 such interactions the experimenters measured the probability that the neutrino background could contain an event like that in Fig. 2. This **measured** background was 0.15 events (Battistoni et al 1986).

As an example of statistical results from tracking detectors, Fig. 3 shows a plot

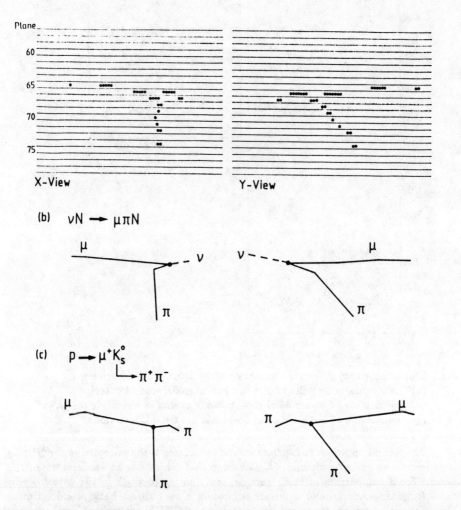

Fig.2 (a) The orthogonal views of a NUSEX event that is a candidate for proton decay.
(b) Interpretation of event in terms of a neutrino interaction $\nu N \to \mu \pi N$, with large-angle scattering of the pion.
(c) Interpretation according to the decay $p \to \mu^+ K_s^\circ$, $K_s^\circ \to \pi^+ \pi^-$.

of the net vector momentum of secondaries, divided by visible energy, $|\Sigma \vec{p}_s|/E_{vis}$, for the Frejus experiment (Berger et al 1986). One might suppose that, apart from Fermi motion effects, this ratio should be one, if there is a charged current reaction due to a massless neutrino. Any scattering of secondary particles in the nucleus, or misidentification of protons as pions, is going to reduce the ratio, so a mean value of 0.75 is not surprising. Equally, re-interaction of secondaries in a proton decay event will increase the value of $|\Sigma \vec{p}_s|$ above the zero expected (apart from Fermi motion effects). So, proton decays in iron will look more like neutrino interactions, and conversely. Fig. 3 shows events in the Frejus detector, with the results expected from a MC simulation of neutrino interactions. Apart from one candidate event (which is however complicated and difficult to analyse unambiguously) all is compatible with neutrino background.

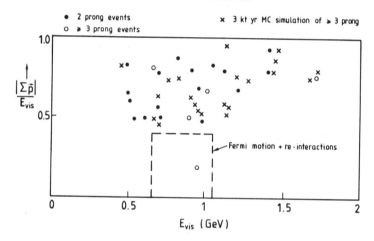

Fig. 3

What are the future possibilities for the proton decay experiments? First, one has to remark that proton decay, if it is ever observed, will be a most important happening in its own right, as well as a crucial test of grand unified theories. Second, it is clear that, for particular decay modes—those in the first line of Table 2—the water Cerenkov method has no equal and appears not to be background limited. Therefore, in time the lifetime limits for these modes can be pushed to 10^{33} years and beyond. If multibody decays to non-relativistic particles are important, then some improvements are certainly possible from the tracking detectors, in particular from improved spatial resolution and ionization information. At the moment, modest improvements are certainly worthwhile, and the use of several different detectors with different technology ensures that any claims of a signal can be quickly and independently corroborated or refuted.

2) MONOPOLE SEARCHES

Two methods have been used to search for GUT monopoles, expected to be of mass $\sim 10^{16}$ GeV and of cosmic velocities $\beta > 10^{-4}$. First, induction experiments use SQUID magnetometers to detect the change in magnetic flux $\Delta\phi$ when a monopole traverses a (superconducting) loop, where

$$\Delta\phi = 4\pi g$$

and the monopole charge g is an integer times the Dirac charge

$$g = n(\hbar c/2e)$$

SQUIDS are certainly sensitive enough to detect such a flux change, and the method has the great advantage that it is independent of monopole mass or velocity—but has the disadvantage that the sensitive area is restricted. Present limits on monopole flux, with detectors of area up to 1m^2, are $F < 10^{-11}$ monopoles cm^{-2}sr^{-1}s^{-1}.

The second type of experiment relies on monopole interactions with matter. For velocities $\beta > 10^{-2}$, the monopole acts like an electric charge $g\beta$ and the usual Bethe-Bloch ionization loss formula applies. For $10^{-4} < \beta < 10^{-3}$, the energy-loss mechanism is expected to be by atomic excitation via interaction with the magnetic moments of atomic electrons (the Drell effect). De-excitation is by photon emission (e.g. fluorescence), or ionization if the energy is transferred to molecules of low ionisation potential (Penning effect). For example in a helium-methane mixture,

$$\text{He} \xrightarrow{M} \text{He}^* + \text{CH}_4 \rightarrow \text{He} + \text{CH}_4^+ + e^-.$$

A large number of experiments have been carried out with scintillation or proportional counters, with areas upto 1000m^{2}sr. It is usual to have at least 2 counter layers and require the time delay appropriate to a slow monopole traversal. Present flux limits by these techniques are $F < 10^{-13} - 10^{-15}$ monopoles cm^{-2}sr^{-1}s^{-1}, from the UCSD, Texas A and M, and Baksan experiments.

There is a third class of experiments which requires for their interpretation assumptions about monopole interactions with matter. As an example, it has been proposed that even a slow monopole could attach to a heavy nucleus (Al) in traversing the earths' crust and therefore leave "fossil" tracks in mica, and this approach leads to very low $(F < 10^{-17})$ flux limits (Price, 1984). At the other extreme, it has been proposed that monopoles may calatyze nucleon decay, leading to flux limits from proton decay experiments of $F < 10^{-14}$ (depending somewhat on monopole velocity and cross-section). It goes without saying that these two competing processes are mutually exclusive.

In summary, the most stringent present limits on GUT monopole fluxes are in the region of $F < 10^{-13} - 10^{-15}$cm^{-2}st^{-1}s^{-1}, that is a factor $10 - 10^3$ larger than the "Parker bound" (Turner et al 1982) of $F < 10^{-16}$ for the maximum galactic flux which could be tolerated by the need to maintain the observed galactic magnetic field. This shortcoming will be remedied by forthcoming experiments with large area ionization detectors.

3) POINT COSMIC SOURCES

Great interest has arisen over the last years in stellar point sources of gamma-rays. These are interesting because of the enormous energy concentrated in a relatively small number of sources, and because it has been claimed that underground muons are associated with one such object, Cygnus X3. Any correlation of muons with a distant source like Cygnus, at the rates observed could almost certainly indicate completely new physical phenomena; hence all the excitement.

Cygnus X3 is an X-ray binary discovered in 1966, distant $\geq 30,000\,ly$, with right ascension $\alpha = 307.6°$ and declination $\delta = 40.8°$ (i.e. it is in the Northern Sky and only dips below the northern horizon by $\sim 10°$ at $50°$ latitude). It is invisible in the optical region (absorption) but detected at radio, infra-red, X-ray and γ-ray wavelengths. The X-ray and IR data show a clear period of 4.8 hours between sharp minima. This period is assumed to be related to orbital motion and eclipse of a compact, active star (N star, pulsar ...) by a close and diffuse companion. Accurate measurements (Van der Klis et al 1981) of the X-ray period give a value $P_0 = 0.1996830$ days, which is lengthening by about $1.2.10^{-9}$ parts per revolution.

A prominent feature of Cygnus X3 is its extreme variability—factors of 10 in X-ray emission on a timescale of months/years and 20-30% fluctuations on a timescale of seconds. This variability is observed in an extreme form for occasional violent radio bursts.

We are mostly concerned here with γ-ray and underground muon signals. The low energy γ's (~ 1 GeV) have been detected in shower counters carried in balloons and satellites. High energy γ's ($1-10^3$ TeV) will produce extensive showers in the atmosphere, and have been observed by (a) twin or multiple ground based mirrors detecting the Cerenkov light emitted as the shower particles traverse the atmosphere (b) ground arrays to detect charged particles in the shower. Underground muons have been detected in large proton decay experiments. From about 30 experiments on γ-rays and muons, slightly more than half claimed a signal originating from Cygnus X3; the others had not seen anything, or were not claiming a significant effect. Part of this result is surely because some of the experiments had poor angular resolution or statistics, but there is equally no question that the photon flux from Cygnus fluctuates very widely. It should also be borne in mind that some types of detector—the air-shower Cerenkov light detectors—can only operate occasionally, in moonless, cloudless nights, that is with an efficiency of a few per cent only.

Fig. 4(a) shows, as examples, the result of plotting the Cerenkov light signal against phase of Cygnus X3 in the experiment of Danaher et al (1981), while fig. 4(b) shows the results of Samorski and Stamm (1983) on air-showers at sea-level (Kiel) also plotted against phase. In the first experiment, the (on-source)/(off-source) signal ratio integrated over time (phase) is 1.20 ± 0.05 (3.75σ) and the shower energies are ~ 1 TeV; while in the second the ratio is 2.15 ± 0.4 (5σ) and energies are $10^3 - 10^4$ TeV. The enhancements in the phase plots are at 0.75 for the air Cerenkov data and at about 0.25 for the ground air-shower data (if corrected to the Van der Klis ephemeris). An interesting reported feature of the Kiel data is that the showers are muon-rich. Since it is established that Cygnus X3 emits in the radio, infra-red, X-ray and few GeV region of the electromagnetic spectrum, it is not too difficult to believe in a signal in the 1–100 TeV region also. If it assumed that the air-showers are photon-induced (although there is little direct evidence to support this), the rate indicates that the spectrum is quite

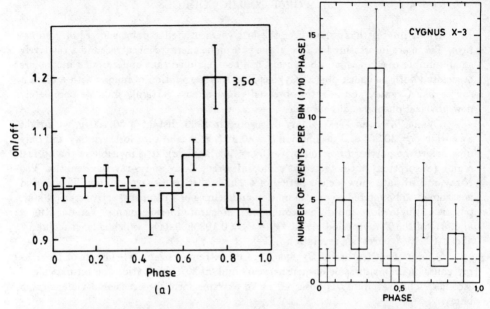

Fig.4(a) Results of 95 30-minute scans (10 mins off, 10 mins on, 10 mins off source) of Cygnus X3 by Danaher *et al* (1981) using twin mirrors to detect Cerenkov light from air-showers of order 1 TeV energy.

(b) Results of Samorski and Stamm (1983) on air showers of $> 10^5$ particles recorded by Kiel array (28 1m^2 scintillators). The plot is for events within 1.5° of Cygnus X3, over 3800 hours. Dashed line is off-source background. Primary energies $\sim 2000 - 20,000$ TeV.

hard, $f(E)dE \sim E^{-2}dE$, instead of $E^{-3}dE$ for the general cosmic-ray spectrum.

Coming to the underground muon data, **no** experiment sees a significant access of muons when pointing at Cygnus X3 and integrated over time (see Table 3). All the arguments about possible signals rest entirely on **enhancements in phase plots**.

Table 3: $R =$ On-source/Off-source ratios for Cygnus X3 muons

	Depth mwe	E_μ (min) TeV	R
Soudan I	1800	0.7	$1.03 \pm .03$
Nusex	6200	5.0	$1.16 \pm .10$
Frejus	5000	3.2	$0.93 \pm .09$

Fig. 5 shows (a) Soudan I data (Marshak *et al* 1985) for muons within 3° of Cygnus (b) Nusex data (Battistoni *et al* (1986) for muons inside a 10° × 10° bin centred on Cygnus. These angular bins were chosen to maximize the signal. Both data sets were collected over a 1½-3 year period. The random probability of the enhancements in the phase bins $\phi = 0.6 - 0.8$ are given as $10^{-3} - 10^{-4}$ in both cases. On continuing the observations from Feb. '85-Feb. '86, with a total of 50 muons in the angle bin (compared with 150 in Fig. 5(b)), NUSEX observe no enhancement (in fact 2 events of

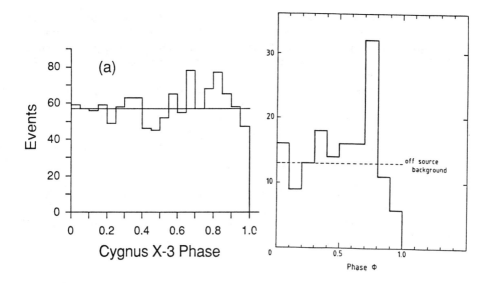

Fig.5(a) Soudan I results (Marshak *et al.* 1985) on underground muons (minimum energy ~ 1 TeV at sea-level) recorded in proton decay detector. The phase plot is for muons pointing within 3° half-angle of Cygnus X3: off source background shown by line.
(b) Nusex data (Battistoni *et al.* 1986) on underground muons (E_{min} ~ 5 TeV) in Mont Blanc tunnel, within a 10° × 10° bin in declineation and right ascension centred on Cygnus X3.
Both plots cover a period of observation of 1–2 years (1983–85).

$\phi = 0.7 - 0.8$ compared with 12.5 ± 2.3 expected).

The Frejus experiment finds no significant phase enhancements with the same cuts as NUSEX, and the HPW experiment a completely flat phase distribution. Both experiments use integration times comparable with (but not identical with) the two other experiments. To make life even more complicated, Soudan I have presented a phase plot for muons arriving within a 3° cone of Cygnus during the period 3–12 Oct. 1985, which is that of the largest ever radio flare from Cygnus. On the basis of off-source measurements, they expect 14 events in this period and observe 20; **but** 6 of them are concentrated in the phase bin $\phi = 0.7 - 0.75$.

What can be said about this data? Let us repeat that (integrated over phase) no underground experiment has seen an excess of muons associated with Cygnus, but 2 experiments have seen enhancements in phase plots in the region ($\phi \sim 0.7$) where they are also seen in air-showers. It is important to emphasize the different energy thresholds (Table 3) and the episodic nature of Cygnus X3 in all regions of the electromagentic spectrum (radio, X-ray etc.). It is clear that in order to clarify the situation, much more detailed observations over long periods, with careful comparison between experiments, is necessary. Since we believe we are dealing with point sources, the angular resolution is crucial in enhancing the signal/noise ratio. It appears that both ground-based air-shower detectors and those for underground muons should be able to achieve angular resolutions of 5 mrad (determined by timing of the shower fronts in the case of air-showers, and by Coulomb scattering in the rock overburden, for the muons). This will allow improvements of signal/noise ratios by 1–2 orders of magnitude. Hence, over the next few years, there is a good chance of cleaning up this somewhat messy field.

REFERENCES

Ch. Berger, *et al* 1986 (Proc. 7th Workshop on Grand Unification, Toyama, Japan, April 16–18, 1986)

G. Battistoni, *et al* 1986 (Proc. 7th Workshop on Grand Unification, Toyama, Japan, April 16–18, 1986)

R.M. Bionta, *et al Phys. Rev. Lett.* **54**, 22 (1985)

C. Danaher, *et al Nature* **289**, 568 (1981)

T. Gaisser, *et al Phys. Rev. Lett.* **51** 223 (1983)

T.D. Lee and C.N. Yang, *Phys. Rev.* **98** 1501 (1955)

M.L. Marshak, *et al Phys. Rev. Lett.* **54**, 2079 (1985)

P.B. Price, CERN EP/84-28 (1984)

A. Sakharov, *JETP Letters* **5**, 24 (1967)

M. Samorski and W. Stamm, *Astr. J.* **268** L17 (1983)

M.S. Turner, *et al Phys. Rev.* **D26** 1296 (1982)

M. Van der Klis and J.M. Bonnet-Binaud, *Astron. & Astroph.* **95** L5 (1981).

SUMMARY OF ACCELERATOR PHYSICS PRESENTATIONS

P.F.M. Koehler
Fermi National Accelerator Laboratory, Batavia, IL 60510*

INTRODUCTION

The organizers of this conference have offered the participants a rich menu of presentations on accelerator physics topics; in addition to the twelve talks which were scheduled during the three accelerator physics parallel sessions, six of the lectures given during the two evening plenary sessions were devoted to accelerator physics. There appears to be no shortage of plans for new or upgraded accelerator facilities to serve the needs of the nuclear and particle physics communities. Significant advances in accelerator technology have been achieved and new ways of utilizing facilities for experiments are being developed. Here the cross-fertilization between the subfields of particle and nuclear physics is particularly fruitful. In the following paragraphs I will sketch only at a rather superficial level the contents of the various accelerator physics presentations, grouped by broad topic areas. For the details of the talks I refer the reader to the individual papers themselves.

NEW FACILITIES

The present status of the planning for the Superconducting Super Collider was summarized by J.D. Jackson. A great deal of work has been accomplished by the SSC Central Design Group. The magnet selection has been made and confirmed and a design report and cost estimate based on that choice have been presented to and reviewed by DOE. The project is now making its way through the decision-making chain in the DOE and above. The outcome is uncertain.

The plans for high-intensity proton accelerators, particularly TRIUMF II and LAMPF II, were presented by M.K. Craddock and H.A Thiessen, respectively. The LAMPF II design has recently been revised in an effort to match the machine better to the physics requirements; the design now opts for higher energy at the expense of lower beam current and the experimental area has been simplified. Since it is unlikely that both of these "kaon factories" will be built, it is important that the one which may be ultimately built has the correct parameters to meet the programmatic needs.

The capabilities of the existing heavy ion facilities and the many plans for upgrades and new projects were surveyed by L.S. Schroeder and G.R. Young. These machines range from rings with rigidity less than 2 Tesla•m to RHIC with 835 Tesla•m. It is an area

───────────────
*Operated by the Universities Research Association, Inc. under contract with the U.S. Department of Energy.

of rapid growth not only in the U.S. but also in Western Europe and Japan, motivated by the broad range of physics studies which such machines make possible. At this time there exists a great deal of overlap and duplication among the proposals which will surely be eliminated before a subset of these facilities is funded.

TECHNOLOGY

Better accelerator facilities are made possibly by advances in technology. This theme was illustrated in H.A. Grunder's presentation of the recent dramatic advances in the technology of superconductors (e.g. cable and rf cavities) and how they have influenced the design of the SSC and CEBAF. A.M. Sessler gave us a glimpse of some of the latest developments in advanced accelerator technology which might become the basis for the next generation of accelerators. Given the time constants in this business we need to invest in such R&D work now.

On a more specialized topic, B.E. Norum described methods for producing longitudinally polarized electrons of energy up to 10 GeV for experiments in intermediate nuclear physics. For both linear and circular accelerators the present limitation in producing external beams with longitudinal polarization is imposed mainly by the source. For storage rings, the electron beam is self-polarizing, with a time constant which decreases as E^{-5}. Longitudinal polarization can be achieved by the insertion of a Siberian snake or, at lower energy, by a resonant snake. Thus, while in principle polarization of electrons up to 12 GeV is possible, the implementation is non-trivial.

In his theoretical talk, J. Jowett reviewed the notion of dynamic aperture in a storage ring and related this to the general ideas of dynamical instability, notably the transition to chaos. He compared practical approaches to the problem of selecting stability criteria and pointed out that KAM theory and resonance selection rules suggested a novel quantitative guide to the old problem of choosing machine tunes.

COOLING

In his tutorial talk F.E. Mills reviewed the various methods which have made possible the cooling of stored beams and discussed their relative merits.

The latest developments toward achieving electron cooling in the multi-GeV energy range were presented by D. Larson. This work is motivated by the desire to improve present antiproton sources by reducing the transverse and the longitudinal emittances by a factor close to 20 each. The equipment for the recirculating electron system has been assembled and adjusted. Recirculation tests will follow next.

J.P. Schiffer discussed what might happen to charged particles in a storage ring when they are cooled to the point that in the moving frame their random, thermal energies are comparable to the Coulomb repulsion between adjacent particles. Making several simplifying assumptions his calculations indicated the possibility that a phase transition takes place from "liquid" correlations to a frozen "solid" bcc lattice.

EXPERIMENTS

Plans for a new precision measurement of the muon g-2 value were discussed by V.W. Hughes. Almost 10 years ago a series of three CERN experiments measured $a_\mu^{ex} = (1\ 165\ 924 \pm 8.5) \times 10^{-9}$, where $g_\mu = 2(1+a_\mu)$. This compares to a theoretical value of $a_\mu^{th} = (1\ 165\ 920.1 \pm 2.0) \times 10^{-9}$.

In these experiments most of the error (7.0 out of 7.3 ppm) was due to statistical uncertainty in the determination of the precession frequency. The new experiment aims to determine a_μ to within 0.35 ppm. The proposal calls for the construction of a 3.1 GeV/c muon storage ring at the AGS, using conventional magnets. The required intensity of 5×10^{13} protons/pulse can only be achieved after completion of the AGS Booster.

F.S. Dietrich reported on an exploratory experiment at PEP which used a gas target in conjunction with the TPC/2γ facility to investigate the possibility of carrying out an internal-target physics program, e.g. to study exclusive final states. Data were collected with targets of D_2, Ar, and Xe gas and are now being analyzed. In terms of technical feasibility the results of this first try are encouraging. But the trigger needs to be refined to avoid serious deadtime problems.

The possibility of using the Fermilab antiproton source for experiments was discussed by J. Griffin. He did so in the context of the requirements imposed by the heavy quark spectroscopy experiment which has received approval. Some modifications of the Accumulator will be needed to achieve, for example, deceleration of the cooled antiproton beam through transition and to provide continuous stochastic cooling. But such modifications are judged technically feasible and can be implemented after investing some time in machine studies.

Finally, G.A. Smith tried to answer the question whether there is a need for a LEAR-type facility in North America. After reviewing the varied and exciting physics opportunities presented by the LEAR program (pre- and post-ACOL) and the proposed use of the Fermilab Accumulator for experiments he concluded that any proposal for a new facility dedicated to the study of the low-energy antiproton sector should first await the results of these ongoing efforts. In order to have a chance of receiving funding it should be cleverly designed to be as inexpensive as possible. In the meantime, a working group (AMPLE) has been set up to explore the possibilities.

CONCLUSION

The presentations on accelerator physics have provided the conference participants with a good overview of what lies ahead in terms of technology, facilities, and novel means of utilization. The two subfields of nuclear and particle physics are obviously interacting strongly with each other. No doubt the number of projects will be reduced when choices are imposed by the restrictions of funding and manpower available for implementation. The latter limitation should be of particular concern since the scale of the facilities and the experiments is increasing rapidly.

I would like to acknowledge gratefully the valuable advice and assistance I received from Robert E. Pollock, the co-coordinator for the Accelerator Physics presentations at this conference.

ANTIPROTON PHYSICS SUMMARY

D.A. Axen
University of British Columbia, Vancouver, B.C. Canada V6T 2A6

Antiproton physics is the intersection of particle and nuclear physics with atomic physics. Activity in antiproton physics has increased dramatically following successful operation of the Low Energy Antiproton Ring (LEAR) at CERN in December 1983. Twenty papers covering atomic interactions, $\bar{p}p$ annihilation at rest, $\bar{N}N$ interactions above threshold, $\bar{N}$ nucleus interactions and finally antimatter gravitational effects were presented in four parallel sessions. I would like to review briefly the developments reported and where possible mention the future plans of the groups presenting results or describing new proposals. I wish to apologize in advance for any errors or misquotations.

Atomic physics was reviewed by H. Koch. K and L series X-rays have been observed from $\bar{p}p$ atoms formed by stopping antiprotons in gaseous hydrogen. At atmospheric pressure the L X-ray (≈ 1.7 KeV) yield is approximately .13 and K X-ray (≈ 9 KeV) yield 0.0026 of all annihilations. In heavier nuclei the energies are higher and the backgrounds reduced. Antiproton X-ray data from lead comparable in quality to muonic X-ray data were presented, from which a new value for the antiproton magnetic momentum has been determined.

H. Poth described an interesting proposal to produce a neutral e^+p^- beam in the LEAR ring. Calculations indicate that sufficient intensities could be achieved for measuring physical constants such as the Rydberg constant, the Lamb shift and the hyperfine splitting of this exotic system. Such a low energy atomic system could possibly lead to a source of polarised antiprotons.

The theory of $\bar{p}p$ annihilation to two mesons was reviewed by J. Niskanen and again by Carl Dover in a plenary talk, so I will not discuss it further here. The experimental results from LEAR were reviewed by S. Playfer. The significant feature of $\bar{p}p$ annihilation is the non observation of narrow states. Annihilation dynamics however has been found to depend strongly on the angular momentum of the atomic $\bar{p}p$ state from which the annihilation proceeds. An example was described by Glen Marshall. In liquid, annihilation from the S state predominates due to Stark mixing, and the branching ratio R for K^+K^- with respect to pi^+ and pi^- was of the order of 30%. In gas only 30% of the annihilation is from the S state and the branching ratio R is reduced to approximately 6%.

The next generation of detector for studying $\bar{p}p$ annihilation at rest will be the crystal barrel detector at LEAR which is being optimised for the detection of neutral annihilation products. The DM1 magnet from ORSAY will be upgraded with a superconducting coil and a liquid or gaseous target. Detector apparatus will include an X-ray drift chamber, a charged particle drift chamber and a segmented cesium iodide calorimeter.

The NN interaction up to ≈ 1 GeV had been described phenomonologically using a meson exchange model. A description of the N̄N system must include the annihilation channel. An amplitude analysis of the N̄N system will require a data base of at least the same size as the existing NN data base. For instance, the p̄p system is a linear combination of I = 0 and I = 1 states whereas the n̄p system is pure isospin 1. Parameterisation of the N̄N interaction is likely to require n̄p and p̄p spin observables. B. Bassalleck and L. Pinsky presented new n̄p total and annihilation cross-section data from Brookhaven. Near threshold $\beta\sigma_{np}$ = 35 mb. Above threshold neither of these cross sections shows any structure.

Two proposals for polarising antiprotons in the LEAR ring were presented. Ron Ransome described a technique in which the antiproton beam is passed through a polarised neutral atomic hydrogen beam. The p̄p total cross-section can be written as

$$\sigma_{tot} = \sigma_0 + \sigma_1 P_B \cdot P_T$$

and initially 1/2 the circulating beam particles are spin up and 1/2 spin down. If $\sigma_1 \neq 0$ in time one spin component will be selectively scattered out of the beam. Estimates are $\sigma_1/\sigma_0 \approx 0.05$ and the corresponding polarisation time constant 10 hours. The second proposal described by Y. Onel is to use the strong gradient in the focussing quadrupoles to separate the two spin components. Superconducting solenoids are required between quadrupole pairs to rotate the spin direction through 180°. The calculated separation is 3 mm/hr.

Peter Barnes presented preliminary data on the excitation function and Λ polarisation in the reaction p̄p → Λ̄Λ. The lambas are polarised. If the Λ and Λ̄ polarisations are different this would be the first evidence of CP violation in a system other than kaon decay. A test at the 10^{-4} level requires reconstruction of 10^8 events. This program is underway. Tests of CP invariance in the pp → Λ̄Λ interaction were discussed by J. Donnoghue.

At higher energy all the J/ψ states can be formed in p̄p collisions. The 1^{--} photon quantum number for the intermediate state characteristic of e^+e^- collision is not present. Dr. R. Cester presented new ISR results for masses and widths of the ψ(3096), χ_1(3511) and χ_2(3556) states. Further work is proposed for Fermilab.

One session was devoted fully to N̄ nucleus interactions. J. Peng presented p̄ nuclear elastic scattering data obtained with the SPES II spectrometer at LEAR. In general p̄ nucleus elastic scattering is much more diffractive than p nucleus scattering. P. McGaughey showed that N̄ nucleus annihilation data could be reproduced very well using Intra-Nuclear Cascade calculations. W. Gibbs showed very convincingly that the production of high energy density in nuclei is better done with antiprotons than any other projectile and suggested looking for evidence of phase transitions to a quark gluon plasma with existing p̄ beams. N. Auerbach also showed very convincingly that isovector states could be strongly excited in charge exchange reactions with anti-nucleons. Finally Terry Goldman and Ron Brown discussed an interesting proposal to study gravitational effects with antiprotons. The suggestion is that the gravitational interaction could have scalar and vector terms which have a strong cancellation for matter but could be very large for antimatter. The approved experiment is to measure the difference of the gravitational effect for antiprotons and negative hydrogen ions. The requirement is to cool the $\bar{p}$'s to 1 meV in a series of traps. Tests with H^- ions are in progress at Los Alamos.

HADRON SCATTERING: A BIASED SUMMARY

Nathan Isgur
Department of Physics
University of Toronto
Toronto, Canada M5S 1A7

The four parallel sessions George Igo and I organized were roughly divided into four topics: the baryon-baryon interaction, the meson baryon interaction, multiquark theory, and relativistic nuclear physics. Names appearing in parentheses in what follows are those of contributors to these or related sessions of this conference.

I would claim that there is no more fertile ground on the intersection of particle and nuclear physics than hadron scattering. I base this claim on the equation

$$\text{QCD : hadrons : nuclei} = \text{QED : atoms : molecules} \qquad (1)$$

One practical application of (1), which, though still in its infancy, illustrates its potential impact, is to the study of the nature of the short-range NN interaction. The traditional picture of two nucleons at a separation of $m_\omega^{-1} \approx 0.25$ fm is

• •

while in the quark model the picture is

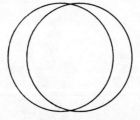

where the circles represent the nucleon rms radii. On the basis of this simple "argument" alone it seems to me that either some justification for the traditional pointlike picture must be found or it should be discarded.

At the same time that the traditional picture has been on the defensive, major progress has been made in the quark description: the six quark potential model gives a natural explanation of the repulsive core and of the very existence of nucleonic clusters (Oka, Maltman, Lipkin, Robson). Lipkin aptly described the physics behind these results as the "Pauli-Fermi-Heisenberg effect" since the repulsive core arises from an interplay of the Pauli exclusion principle, the short-range Fermi spin-spin colour magnetic interaction, and the Heisenberg uncertainty principle. Because the physics behind this source of short-range repulsion is structural, it is quite model insensitive and therefore well established. I conclude that ω exchange probably has little or nothing to do with the core.

This (many would say premature) conclusion is testable since the quark model of the NN force leads to what I consider some of the most interesting predictions on the interface of particle and nuclear physics:

    1) it predicts the absence of dibaryons in most channels (Wicklund, Maltman, Oka),

    2) it predicts the dominant character of low energy baryon-baryon interactions (Moinester, Maltman, Oka),

    3) it predicts weakly bound dibaryons in certain as yet unexplored channels (Maltman, Oka),

    4) it predicts the low energy KN interaction and the existence of weakly bound resonances in certain channels (Campbell, Maltman),

    5) it predicts N-body interhadron forces (Nefkens, Maltman).

The confrontation of these predictions with experiment constitutes the first major challenge of the quark picture.

A second major challenge is more theoretical. Considerable effort is now required to clarify the role of <u>long-range</u> gluon dynamics in multi-hadron systems. The analog of the meson-meson gluonic flux transition

leads to a medium-range attractive van der Waal's-like force between nucleons. This is a pure gluonic effect with no $(q\bar{q})$ mesonic counterpart (Koonin). There is a significant possibility that this effect, not σ exchange, is the real source of nuclear binding (Maltman, Robson, Lipkin).

A third major challenge, which involves a mixture of theory and experiment, has been implicit in the whole of the preceeding discussion: the role of meson exchange must be clarified. Since $m_\pi^{-1} > r_N$ it is certain that the pion tail is real and important. However, this inequality is anomalous: it is the result of the pion being the nearly massless Goldstone boson of broken chiral symmetry. For all other mesons $m^{-1} \ll r_N$ so that their importance as exchange particles is, to say the least, theoretically murky. The success of the meson exchange <u>parameterization</u> of the NN potential (or of Regge theory) is not an argument for its reality since the quantum numbers of the mesons are in one-to-one correspondence with those of quark exchange: mesons have quark-antiquark quantum numbers while quark exchanges have quark-quark hole quantum numbers. A better understanding of the model independence of "meson exchange currents" might help to sort out this ambiguity (Riska).

Despite this drift toward a quark-based model of the NN interaction, there are some strong countercurrents. (Note the biased labelling). The chiral quark model (Banerjee) is a quark-meson theory which has some of the characteristics of each of the traditional and quark-based approaches. The chiral soliton model (Mattis) is a pure meson theory which in its idealized form is probably equivalent to QCD. It is difficult, however, to see how to map it onto either picture. Finally, perhaps the strongest countercurrent is that produced by the practicioners of quantum hadrodynamics and relativistic nuclear physics. There has been continuing success in these extensions of the traditional picture. For example, a solution has now been proposed to the old problem with magnetic moments in the presence of strong scalar potentials (McNeil). There has also been significant new work done on understanding quasielastic scattering in this framework (Horowitz). At the same time there has been progress in understanding the basic mechanisms responsible for some of the phenomenal successes of the relativistic approach. The association of this mechanism with $N\bar{N}$ pair terms has led in turn to some criticism of the approach, and to the claim that non-relativistic models can give the same good results in a physically better motivated way (Thies).

There were also many topics discussed in our parallel sessions that don't fit neatly into any of the above "boxes": charge independence tests (Seth), the (π,η) reaction (Peng), π-nucleus scattering above the Δ region (Baer), a new determination of the πN σ-term (Kluge), double charge exchange (Bleszynski), the (n,p) reaction (Alford), evidence for two resonances in the Roper region in πN scattering (Nefkens), and a study of ^{208}Pb multipoles (McDaniels). I apologize to these speakers for taking a slant in this summary that didn't allow me time to discuss their contributions.

My main conclusion echoes my opening claim: keep your eyes on this section of the interface. There is a lot of friction here, and therefore heat, so that things are cooking.

HADRON SPECTROSCOPY SUMMARY

John F. Donoghue
University of Massachusetts, Amherst, Mass. 01003

The sessions on Hadron Spectroscopy provided a wide range of talks on the nuclear and particle spectra and the uses of such studies in regions of common interest of both fields. I do not feel that it is appropriate to summarize each and every talk (they are available in these Proceedings), but rather I will highlight some developments which are of such general interest that all participants should be aware of them. These are the Skyrme model (discussed at the Conference by M. Mattis[1]) and the future directions of hadron spectroscopy (primarily related to the talks by W. Lockman[2] and M. Chanowitz[3]). Included as an example in the latter section is a discussion of the $\theta(1700)$, which is most likely the first identified non $Q\bar{Q}$ meson.

Of the tools available to a theorist studying the low energy region of hadronic physics the only two (aside from the obvious isospin and SU(3) symmetries) which could be called rigorous are chiral symmetry and lattice techniques. Chiral symmetry provides the structure of the low energy interactions of pions (and, in its SU(3) extension, K's + η's). This structure can be summarized by effective Lagrangian such as the one for the strong interactions

$$L = \frac{F_\pi^2}{4} \operatorname{Tr}(\partial_\mu u \partial^\mu u^+) - \frac{m_\pi^2 F_\pi^2}{4} \operatorname{Tr}(u + u^+ - 2) \\ + \frac{1}{32e^2} \operatorname{Tr}([(\partial_\mu u) u^+, (\partial_\nu u^+) u]^2) + \ldots \quad . \tag{1}$$

where $u = \exp(\frac{i\vec{\tau}\cdot\vec{\phi}}{F_\pi})$. The expansion is in the number of derivatives, and the omitted terms have four or more derivatives (or m_π^2 times two or more derivatives). At low energy (typically $q^2 \ll 1$ Gev^2) terms with high numbers of derivatives are small and the leading two terms generate the predictions of chiral symmetry. Lattice calculations are rigorous in the limit of zero lattice spacing as $g \to 0$. However we seem to be far from this ideal at present. Both these techniques have well defined (even if difficult) methods for improvements in accuracy without any fundamental limitations in principle.

Other tools are less rigorous but are often more useful. These include potential models, bag models, string models, ITEP sum rules, and the Skyrme model.[1,4] The Skyrme model has been developed somewhat more recently than the rest and deserves some discussion. It is almost orthogonal to the other theories because, instead of building pions and baryons out of quarks, it builds baryons out of pions. It starts from the effective chiral Lagrangian written in Eq. 1 (dropping the higher order terms not explicitly specified), and applies it to baryons. This is no longer a rigorous theory because it is applied at high energy and hence is out of the region of validity for the chiral expansion. However, as a model, it has a soliton of the form

$$u_0 = \exp\{i\, F(r)\, \bar{\tau}\cdot\hat{r}\}$$

i.e. with the isospin and spatial directions correlated. Upon quantization, states of definite spin and isospin are projected out and these become identified with the nucleons, deltas, etc. Surprisingly the model does reasonably well. The mass of the proton can be roughly given in terms of quantities known from pion interactions.[5] Some of the static properties of nucleons can be calculated.[6] Mattis presented work[7] which studies pion-Skyrmion scattering and extracts phase shifts and resonances. The results roughly resemble the real world to a degree which is remarkable given the approximate nature of the model. There are applications of the Skyrme model to the intersections of particle and nuclear physics, for example to the internucleon potential. Another application, which I have not seen previously mentioned, is a direct pion production potential in $NN \to NN\pi$. This arises when one adds pions to the Skyrme soliton, using[8]

$$u = \exp(i\vec{\tau}\cdot\vec{\pi}/2F_\pi)\, u_0 \exp(i\vec{\tau}\cdot\vec{\pi}/2F_\pi)$$

When this is expanded to first order in the pion field, the same techniques that yield the potential in $NN \to NN$ will produce a direct pion producing potential in $NN \to NN\pi$. The analysis of pion production will then involve not only pion radiation off of nucleons and isobars, but also this direct potential. I would think that this would be of significant interest in the nuclear community.

There was at this Conference a good deal of hype for the potential models and some good matured "bag-bashing" and Skyrme-slapping". I feel compelled to reply in the same spirit. The potential models are useful these days as a method for calculating the background (i.e. known $Q\bar{Q}$ and QQQ states) to the more interesting physics which we seek to uncover. In this use they may be even more useful than the Particle Data Tables as the latter has significant gaps. However, the models are filled with parameters (of order two dozen) and are useful because they have basically managed to parameterize the regularities of the quark model. It is to be expected that potential models may fail to be applicable for new physics outsdie of the range of validity of the parameterization. Moreover the prime focus of contemporary hadron spectroscopy (such as the connection to QCD and the nature of glue related states) is beyond the reach of potential models. If we are to address the truly interesting problems we must outgrow the use of potentials and nonrelativistic quantum mechanics. It is not just a question of the velocity of the particles, but rather has to do with the richness of field theory and the real world and the complexity of the strong interactions compared to the naive assumption of a potential. As an example, consider the

QCD trace anomaly, which requires the nucleon mass to be given, in the chiral limit, by

$$m_N = \langle N | \frac{\beta}{2g} F^A_{\mu\nu} F^{A\mu\nu} | N \rangle$$

This seems unusual in that one would naively expect that gluonic operators would at the least not determine the entire nucleon mass. However this relation can be understood[9] in the bag model to be exactly equivalent to the bag model virial theorem $m_N = 4BV$ if one uses the relation of B to $\langle 0 | F^A_{\mu\nu} F^{A\mu\nu} | 0 \rangle$. Another example is in the $\bar{s}s$ content of the nucleon[10], which appears nonzero given the value obtained via the Gell-Mann-Okubo relation

$$\langle N | m_u \bar{u}u + m_d \bar{d}d - 2m_s \bar{s}s | N \rangle = 26 \text{ MeV}$$

and the πN sigma term[11]

$$\langle N | m_u \bar{u}u + m_d \bar{d}d | N \rangle = 57 \pm 7 \text{ MeV}$$

which implies

$$\frac{\langle \bar{u}u + \bar{d}d \rangle}{\text{TOT}} = .79 \pm .03 \qquad \frac{\langle \bar{s}s \rangle}{\text{TOT}} = .21 \pm .03$$

While this appears surprising it is predicted pretty accurately in the Skyrme model[10], where the $\bar{s}s$ content comes when one extends the chiral symmetry to SU(3) (i.e. roughly, it is kaons in the soliton). The effect also arises[10] naturally in the bag model but is less accurately given (here it is related to $\langle 0 | \bar{s}s | 0 \rangle \neq 0$). These are examples of nontrivial effects which are best addressed outside of potential models.

This said, it is best to stress the advantage of knowing a multiplicity of models. All of the models are wrong. They are not approximations to QCD but are perhaps caricatures of it. There is no equivalent of perturbation theory where one can calculate the next order in a well defined approximation. The closest thing which approaches a "quark model perturbation theory" is to calculate a quantity in a variety of models. The "next order" would be equivalent to using another model. To the extent that magnetic moments, for example, can be understood in essentially all models indicates that our understanding of this aspect is quite good.

One example where models do not agree is on the issue of glueball masses. As far as I can see the scorecard is

    bag model - glueball masses are light
    strings   - glueball masses are heavy
    sum rules - some masses are light - others are heavy
    potentials -pretend glueballs don't exist
    Skyrmions - "What is a gluon?"

The result appears to be a failure of theory - we cannot trust the models. In this case we must rely on experiment.

Unfortunately, experimental answers are not yet satisfactory either - even for the nonexotic aspects of the spectra. Chanowitz[13] presented evidence that, despite 30 years of experimental spectroscopy, only three complete $Q\bar{Q}$ nonets are unambiguously known. This of course has important consequences in looking for new exotic states, as one has difficulty comparing these to usual $Q\bar{Q}$ states when the latter are not mapped out.

An example of the kind of comparative data required in this subject was given by W. Lockman[1] in his discussion of the Mark III results. To discuss just one example, we can consider the θ(1700), a novel $J^{PC} = 2^{++}$ state seen in $J/\psi \to \gamma\theta$ decaying into $K\bar{K}$ and $\eta\eta$ strongly and $\pi\pi$ slightly. The Mark III has measured the related final states

$$J/\psi \to \begin{Bmatrix} \gamma \\ \omega \\ \phi \end{Bmatrix} + \begin{Bmatrix} K\bar{K} \\ \pi^+\pi^- \end{Bmatrix}$$

The 2^{++} f(1270), a $u\bar{u} + d\bar{d}$ state appears in $\gamma\pi\pi$, $\omega\pi\pi$ while the $s\bar{s}$ state, f'(1515), is strong in $\gamma K\bar{K}$, $\phi K\bar{K}$ and θ is strong in $\gamma K\bar{K}$, $\omega K\bar{K}$ and may be seen in $\gamma\pi\pi$.

The θ appears most likely to be our first non $Q\bar{Q}$ state.[3,12] It falls between the ground state nonet which includes f and f', and the radial excited 2^{++} nonet which experiment[13] and theory[14] both have starting at 1.8 GeV for the $(u\bar{u}+d\bar{d})_R$ state, with the $s\bar{s}$ state expected at 2.0 GeV. The low mass makes it clear that it is not the radially excited $(s\bar{s})$ state while the dominant $K\bar{K}$ coupling (and to some extent its mass) tells us that it is not the radial $(u\bar{u} + d\bar{d})$ state. However, we can be less certain exactly what it is. The Mark III data will play a crucial role in this determination.

When asked about the future, experimenters tend to say that more theory is required to guide them. Theorists tend to say that more experimental work is needed to guide them. For once, the theorists are right. A more complete knowledge of the spectrum is required, in order to allow more comparative studies. In his talk Chanowitz stressed the importance of high statistics. Theorists will not be in a position to guide experiment reliably for many years but, if led by at least partially complete information, they may be able to help sort out the spectrum. The complicated issues will not be resolved quickly, but determining the spectrum of QCD is a valuable enterprise which deserves attention.

REFERENCES

1. M. Mattis, this conference.
2. W. Lockman, this conference.
3. M. Chanowitz, this conference.
4. T. H. R. Skyrme, Proc. R. Soc. London A260, 127 (1961).
5. J. F. Donoghue, E. Golowich and B. R. Holstein, Phys. Rev. Lett. 53, 747 (1984).
6. G. Adkins, C. Nappi, and E. Witten, Nucl. Phys. B228, 552 (1983).
7. M. Karlinger and M. Mattis, Phys. Rev. D31, 2833 (1985); M. Mattis and M. Peskin, Phys. Rev. D32, 58 (1985).
8. H. Schnitzer, Phys. Lett. 139B, 217 (1984).
9. J. F. Donoghue, in "Theoretical Aspects of Quantum Chromodynamics," ed. by J. Dash (CRNS-LUMINY-1981), p. 136.

10. J. F. Donoghue and C. Nappi, Phys. Lett. (to be published).
11. T. P. Cheng and R. Dashen, Phys. Rev. Lett. 26, 594 (1971).
12. J. F. Donoghue, in Proc. of the Int. Conf. on High Energy Physics, Bari, Italy, ed. by L. Nitti and G. Preparata (LATERZA BARI-1985), p. 326.
13. N. Cason et al., Phys. Rev. Lett. 48, 1316 (1982).
14. S. Godfrey and N. Isgur, Phys. Rev. D32, 189 (1985).

SUMMARY OF THE HEAVY ION PHYSICS SESSIONS AT LAKE LOUISE

John W. Harris
Nuclear Science Division, Lawrence Berkeley Laboratory
University of California, Berkeley, CA 94720 USA

In the last several years the intersections between particle and nuclear physics have grown rapidly to the point where many particle and nuclear physicists now speak the same language and even work together on the same experiments. In heavy ion physics this interdisciplinary communication ranges from describing ground state nuclei and highly excited nuclear matter in terms of quark degrees of freedom to searching for quark deconfinement in high energy nucleus-nucleus interactions utilizing nuclear beams from high energy physics accelerators. Presentations in the Heavy Ion Physics Sessions encompass this range and can for the most part be divided into two subcategories. The first is the study of different phases of <u>nuclear matter</u>. This typically refers to understanding the nuclear equation of state from low baryon densities and temperatures (where a liquid-gas phase transition should occur) through nuclear structure, neutron star formation and supernova explosion to the high densities and temperatures where the onset of deconfinement of quarks in nuclear matter is predicted. Beyond this is the second region, that of the study of <u>quark matter</u>.

In the quest to understand the nuclear equation of state considerable advances have been made on several fronts. A semi-empirical equation of state has been extracted from pion production data in the Streamer Chamber[1] in relativistic nucleus-nucleus collisions at the Bevalac. This is shown in Fig.1 along with theoretical equations of state labelled FP,[2] BCK,[3] and VUU [4] which are used in nuclear matter calculations, supernova simulations and microscopic calculations for relativistic nucleus-nucleus collisions, respectively. The FP and BCK equations of state are considerably "softer" than that extracted from the data and VUU. This apparent discrepancy is assuaged by the fact that relativistic nucleus-nucleus collisions, supernova explosions and nuclear structure effects are governed by the nuclear equation of state in different regions of temperature and density. Nuclear structure studies and nuclear matter calculations near the ground state of nuclei reflect the equation of state at zero temperature near saturation density ρ_0 while neutron star formation and supernova explosions are governed by the nuclear equation of state at low temperatures (T ~ 15 MeV) and densities ρ ~ 2 ρ_0. In contrast relativistic nucleus-nucleus collisions are sensitive to the nuclear equation of state at high temperatures (T ~ 60 - 100 MeV) and densities ρ ~ 3-4 ρ_0. The behavior of the nuclear equation of state under these different conditions has yet to be fully understood. However, it should be stiffer at higher temperatures partly due to a decrease in the effective nucleon mass and an increase in prominence of the repulsive ω-interactions at high temperatures.

Another important result from the Bevalac is the recent observation and systematic study of collective sidewards flow of matter. Shown in Fig.2 are

319

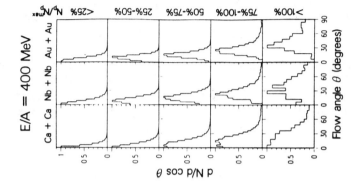

Fig.2. Flow angular distributions for three systems as a function of the fractional multiplicity.

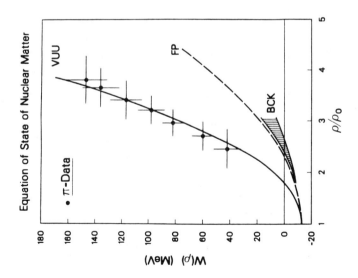

Fig.1. Internal energy per baryon as a function of nuclear density extracted from pion multiplicity data (points). Also displayed are three theoretical equations of state (curves). See text for details.

results from the Plastic Ball[5] for three different systems at fixed incident energy. Plotted as a function of fractional multiplicity N_p/N_p^{max} (which is inversely related to the impact parameter) are the distribution of flow angles, defined as the angle of the major axis of the event ellipsoid with respect to the beam axis, as determined from a sphericity analysis. The flow angle is observed to increase with the mass of the system and the multiplicity. These data provide strong evidence for the collective flow of matter. A rather stiff nuclear equation of state is necessary to describe these and other flow data[6] along with the pion production data.

An interesting result[7] from the Bevalac and possibly another approach[8] to the equation of state is the recent study of subthreshold kaon production. Whereas the threshold for associated production of K^- in nucleon-nucleon reactions occurs at $E_{lab}(NN \rightarrow NNK^{+0}K^-) = 2.5$ GeV, K^-'s have been observed in Si + Si collisions at incident energies as low as $E_{lab}=1.3$ GeV/n. In an attempt to understand the production mechanism, predictions have recently been made for the K^- yields assuming various theoretical models. An estimate assuming thermal and chemical equilibrium, which is unreasonable below threshold, overpredicts the yields by a factor of 20 as would be expected. Quite surprisingly, a single collision Fermi momentum model[9] underpredicts the yield by a factor of 20. Two more sophisticated microscopic models, the hadrochemical model[10] and semianalytic transport theory[11] predict the energy dependence of the data but not the correct magnitude of the K^- yields or all the systematics observed in the data. These models and experiments are still somewhat in their infancy. However, subthreshold kaon production, which should be more sensitive than pion production to the available energy and in turn to the compressional degrees of freedom, may provide new information on the nuclear equation of state as the experimental techniques become more refined and the models better developed.

On the way to studying the properties of nuclear collisions in the quark matter regime, several interesting experiments have already taken place. In T. Ludlam's talk[12] he discussed the mechanism by which gammas and dileptons should probe the early stages of relativistic nucleus-nucleus collisions, and specifically may be signatures for the formation of a quark-gluon plasma. Single leptons, namely electrons, have been measured over a large incident energy range from KEK in Japan to the ISR at CERN and the e/π ratio has been found to be relatively independent of projectile-target-energy combinations.[13] Experiments on the lepton and dilepton production in pp, pA and AA reactions are underway[13] at the Bevalac and should provide valuable and necessary information for understanding the lepton production in future searches for the quark-gluon plasma.

Results from the ISR light ion running periods have also shed first light on nuclear effects in very high energy nuclear reactions. Displayed in Fig.3 are results for the total neutral energy spectra from pp and αα by the R110-BCMOR collaboration.[14] The pp data are well fit by a single gamma function over nine orders of magnitude in cross section as represented by the solid curve. The fit

321

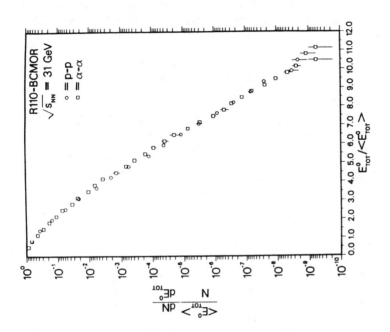

Fig.4. Total neutral energy spectra of Fig.2 plotted as a function of the scaled energy variable.

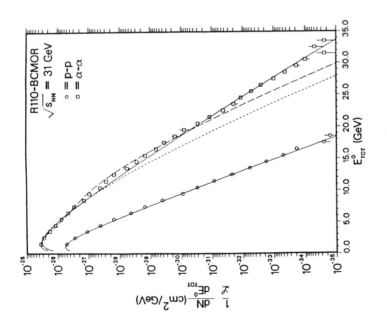

Fig.3. Total neutral energy spectra for pp and αα interactions with fits described in the text.

corresponds to fitting the function $f(E) = a(aE)^{p-1}e^{-aE}/\Gamma(p)$ to the data with parameters a and p. The dotted curve on the $\alpha\alpha$ data corresponds to the best fit using the n-collision probabilities derived by the AFS collaboration[15] from particle multiplicity data. This only fits the data at the lowest neutral energies. The dashed curve corresponds to the wounded nucleon model in the extreme case where all nucleons are assumed to participate in the collision. It still underpredicts the high neutral energy part of the spectrum. Only the solid curve where a gamma function was fit to the $\alpha\alpha$ data with all parameters free can fit the data over the entire range of neutral energies. It was found that the parameters p for the pp and $\alpha\alpha$ cases were identical thereby obeying KNO scaling.[16] This can be seen by plotting the pp and $\alpha\alpha$ data as a function of the scaling variable $E_0^{tot}/<E_0^{tot}>$ where $<E_0^{tot}>$ is the average value of the total neutral energy E_0^{tot}. This is shown in Fig.4 where the pp and $\alpha\alpha$ data overlay each other identically. A fundamental description for this simple phenomenon has not been presented but the results strongly suggest that nucleons in the helium nuclei are strongly correlated. A final comment on these data. The tail of the $\alpha\alpha$ data above 20 GeV (at a level of 10^{-5} of the interaction cross section) exhibits tremendous energy density, 15 GeV per unit rapidity. This corresponds approximately to an energy density of 1.3 GeV/fm^3 which is already quite near the predicted critical energy density for formation of the quark-gluon plasma.

Insight into the quark matter regime has come recently from advances in QCD lattice gauge calculations. A consistent set of results appears to be emerging from calculations which employ different approximation schemes for solving finite temperature QCD on the lattice. Displayed in Fig.5 are results from three different approaches[17] each indicating two coupled, weak, first order phase transitions. Plotted as a function of $\beta = 6/g^2$, where g is the coupling constant, are the order parameter for chiral symmetry $<\bar{\psi}\psi>$ and the quantity $<L>$, the expectation value of the thermal Wilson line (or Polyakov loop). $<L>$ is a measure of the change in free energy of the system. If the free energy F of an isolated quark diverges, i.e. $e^{-F/T} = <L> \rightarrow 0$, then a deconfinement transition appears as a sudden increase in $<L>$ to some nonzero value. This is observed in all three calculations. The order parameter for chiral symmetry $<\bar{\psi}\psi>$, which is a measure of the quark-antiquark mass, is also observed to abruptly change at this same value of β approaching zero where chiral symmetry restoration occurs. It appears that the deconfinement transition and chiral symmetry restoration may be related as they appear coincidentally. To better understand this and the effects of the various approximation schemes, more physical observables will have to be investigated through the use of faster calculations, requiring large amounts of supercomputer time (hundreds of Cray CPU hours).

Of considerble recent interest and uncategorizable in the present heavy ion physics summary of intersections between particle and nuclear physics are the positron lines observed at the GSI in heavy ion collisions. J. Greenberg[18] presented the experimental results in the plenary sessions. Theoretically there is still no consistent explanation which adequately describes the systematics of the experimental measurements.[19] The singles measurements rule out internal nuclear conversion ($A^* \rightarrow A\ e^+ e^-$) primarily from the line width ($t_{source} >$

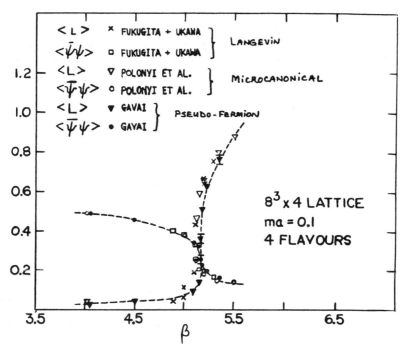

Fig.5. Expectation values of thermal Wilson line <L> and chiral symmetry order parameter <YY> plotted as a function of B=6/g² from QCD lattice gauge calculations.

3 x 10⁻²⁰ s) and gamma and electron spectra. In the coincidence measurements the narrow energy sum peak and similar energies of the e^+ and e^- peaks leave open the possibility of two-body decay of a slowly moving particle (momentum < 100 KeV/c) in the c.m. system. The narrow energy difference peak for e^+ and e^- implies decay far away from the nuclei (distance > 13000 fm). From the energy of the peaks, E_{e^+} and E_{e^-}, the mass of such a particle would be $M = 2m_e + E_{e^+} + E_{e^-} = 1.78$ MeV. A problem arises from the coincidence lines observed in Th + Th. At incident energy $E_{lab} = 5.70$ MeV/n the e^+ and e^- peaks are observed at approximately 310 KeV and at $E_{lab} = 5.75$ MeV/n they appear at around 370 KeV. Is this one line that moves abruptly as the incident energy is increased slightly or are there two different lines? There still seems to be a large sentiment among experimentalists and theoreticians alike that a "simple" explanation may still loom in the future explanation of these lines, but for now there are none. The more exotic explanations[19,20] involve the decay of a pseudoscalar particle produced from the nuclear current where the presence of multiple lines could be explained by a particle with internal structure.

In conclusion, the intersection between particle and nuclear physics research is entering a very interesting period where in the next year there will be high energy nucleus-nucleus experiments[21] on both the European (CERN-SPS) and American (BNL-AGS) continents which for the first time may have the capability to transcend the nuclear matter to quark matter barrier. Furthermore, a proposal to build a relativistic heavy ion collider (RHIC) at even higher energies in the US has received strong support from NSAC. These experiments will all involve a large number of high energy and nuclear physicists working together to understand the common physics of nuclear and quark matter.

REFERENCES
1. R. Stock et al., Phys. Rev. Lett. 49 (1982) 1236; J.W. Harris et al., Phys. Lett. 153B (1985) 377 and J.W. Harris, "The Nuclear Matter Equation of State from Relativistic Heavy Ions to Supernovae", presented at this conference.
2. B. Friedman and V.R. Pandharipande, Nucl. Phys. A361 (1981) 502.
3. E. Baron, J. Cooperstein and S. Kahana, Nucl. Phys. A440 (1985) 744.
4. H. Kruse, B.V. Jacak and H. Stöcker, Phys. Rev. Lett. 54 (1985) 289; J.J. Molitoris, D. Hahn and H. Stöcker, MSU Preprint MSUCL-530, June 1985.
5. H.G. Ritter et al., Nucl. Phys. A447 (1985) 3c.
6. H.A. Gustafsson et al., Phys. Rev. Lett. 52 (1984) 1590; R.E. Renfordt et al., Phys Rev. Lett. 53 (1984) 763; P. Danielewicz and G. Odyniec, Phys. Lett. 157B (1985) 146; D. Keane, "Collective Flow and the Stiffness of Compressed Nuclear Matter", presented at this conference; also see comparisons in J.J. Molitoris and H. Stöcker, Phys. Rev. C32 (1985) 346.
7. S. Trentalange, "Subthreshold Production of Strange Hadrons in Relativistic Heavy Ion Collisions" presented at this conference.
8. J. Aichelin and C.M. Ko, Phys. Rev. Lett. 55 (1985) 2661.
9. A. Shor et al., Phys. Rev. Lett 48 (1982) 1597.
10. H.W. Barz et al., Z. Phys. A311 (1983) 311.
11. W. Zwermann and B. Schürmann, Phys. Lett. 145B (1984) 315.
12. T. Ludlam, " Relativistic Heavy Ion Collisions" presented at this conference.
13. G. Roche, " Universality of Lepton Production in pp, pA and AA Collisions", presented at this conference.
14. A.L.S. Angelis, Phys. Lett. 168B (1986) 158 and M. Tannenbaum, " Observation of KNO Scaling in the Neutral Energy Spectra from aa and pp Collisions at ISR Energies", presented at this conference.
15. T. Akesson et al., Phys. Lett. 119B (1982) 464.
16. Z. Koba, H.B. Nielsen and P. Olesen, Nucl. Phys. B40 (1972) 317.
17. R. Gavai, "Phase Transitions from Lattice QCD Perspectives", presented at this conference.
18. J.S. Greenberg, " Anomalous Positron Production", presented at this conference.
19. B. Muller, " Positron Lines - New Particle or Decaying Vaccum?", presented at this conference.
20. M.S. Zahir, " Production Mechanism for a Light Pseudoscalar Boson (m = 1.6 MeV) and Anomolous Positron Spectra", presented at this conference.
21. L.S. Schroeder, " Relativistic Heavy Ion Facilities", Presented at this conference.

SUMMARY OF HYPERNUCLEAR SESSIONS

R. E. Chrien
Brookhaven National Laboratory
Upton, New York 11973-5000

Fifteen papers were presented in the four hypernuclear sessions. Since the Steamboat Springs meeting of two years ago, it is easy to discern a change in emphasis from nuclear spectroscopy to the question of possible quark clustering in nuclei. It is also obvious, and should be stressed, that hypernuclear studies, especially experimental ones, are presently severely hampered by a lack of suitable facilities and sufficient running time at existing accelerators. It is hard to be optimistic about significant amelioration of this situation in the near future.

The study of nuclei begins properly with the study of the simplest nuclear system; namely deuterium and its strangeness analogues, the YN systems.

The first figure shows a missing mass spectrum of the Λp system from the Λ threshold to above the Σ-N threshold near 2130 MeV. The most striking feature here is the appearance of a large

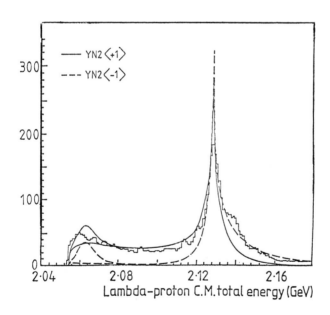

Fig. 1. Invariant mass spectrum for (K^-, π^-) on deuterium in the Λp channel.

spike or a threshold cusp in the Λp spectrum very near the onset
of Σ production. This has led to intense studies over the past
few years by Dalitz and his collaborators. They interpret the
spike as a final state interaction among the Λ, Σ, and N channels,
as originally suggested by Rodberg and Karplus. At this meeting
Torres presented a detailed analysis of the reaction using the
Faddeev method and incorporating all the known two-body NN and YN
potentials. The graph shows fits to the stopped kaon data of
Tan. A satisfactory fit requires the introduction of a pole in
the ΣN channel quite near to the Σ threshold, a state which could
be interpreted as an SU_3 partner of the deuteron.

Of particular interest is the appearance of a shoulder on
these data, near an energy of 2140 MeV, some 10 MeV above
threshold. This shoulder has been suggested by Piekarz et al. as
a candidate for a six quark state, an S = -1 dibaryon resonance.

The status of the searches for Jaffe's H-particle have
already received some attention at this conference. However, it
should be stressed that quark bag models also predict the
appearance of narrow S = -1 states, at mass values near the Σ
production thresholds. Sainio reviewed the status of the bag
model predictions and pointed out that the presence of the s quark
allows flavor representations of lower dimensionality in the

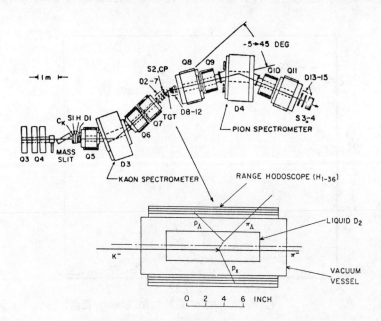

Fig. 2. Apparatus for the dibaryon search in (K^-, π^-) on
deuterium. The range hodoscope allows emphasis in
the Λ-P exit channel.

six-quark system, maximizing the color-magnetic attraction. The predictions thus lie below the thresholds for pion emission from these systems. Hence a strange dibaryon could be expected to be narrow and actually observable as a peak in the mass spectrum.

The data on the Brookhaven experiment of Piekarz allow the examination of the Λp channel in detail by a trick which suppresses the Σ exit channels. A deuterium target is surrounded by a range hodoscope which requires the detection of three charged particles in the final state--two from Λ decay and one from the proton partner (Fig. 2). The BNL experiment indicates the presence of this 2140 MeV peak and its p-wave character, which is demanded by the bag model predictions. There is no calculation based on the known two-body potentials which can produce this peak for the boson exchanges; hence it is apparently our best dibaryon candidate.

Phil Pile discussed planned dibaryon searches in future AGS experiments. The singlet partner of the triplet dibaryon just discussed would be predicted to be 40 MeV below the triplet--that is near 2100 if the 2140 candidate is indeed a dibaryon reconance. It could not be excited in (K^-,π^-) on deuterium because of the smallness of the spin flip amplitude. However, it could be seen in the $(K^-\pi^+)$ reaction on ^{3}He, in a region distant from the complication of the Σ threshold and the attendant quasi-free Σ production. If this peak is seen in the experiment--scheduled for the fall of this year at the BNL hypernuclear

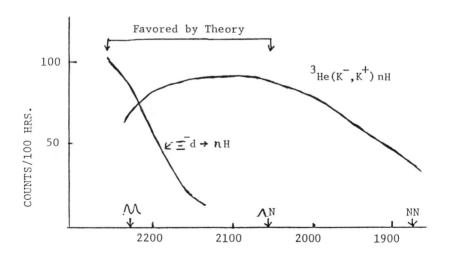

Fig. 3. Count rate predictions for two proposed H-particle searches.

spectrometer--I believe a very strong case can be made for the reality of this 6 quark cluster.

Pile also discussed suggested H-particle experiments. There are two varieties--the single target ^{3}He method and the two target method involving cascade production in hydrogen and a subsequent cascade-deuteron capture. An atomic state is formed which decays to the H--should it lie near the $\Lambda\Lambda$ threshold. Since this was discussed in the plenary session I will make only some general comments. The sensitivity of these experiments over the mass range is shown in Fig. 3. Both kinds of experiments require a separated kaon beamline operating near 1.8 GeV/c, and both experiments have been analyzed by Aerts and Dover who predict cross sections well within the kaon intensities provided by the AGS and the proposed line. Thus the results of these experiments, whether they are positive or negative, would provide a definitive test of the model prediction. It indeed would be a pity if these experiments--which are among the most significant test of the role of quarks in nuclei--could not be mounted at the AGS because of lack of funds for this line. The line could be called the 1-cent beamline--it could be built at a U.S. per capita cost of one penny.

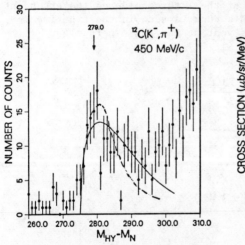 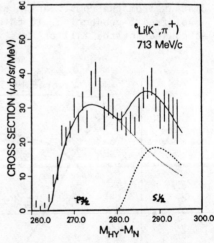

Fig. 4. At left are shown CERN ^{12}C(K^-, π^+) data compared to quasi free calculations with a constant Σ production amplitude (solid line) and including the Y* (1520) resonance (dashed line). At the right are shown the quasi free calculation for ^{6}Li(K^-, π^+) where the $P_{3/2}$ and $S_{1/2}$ hole strength distributions are accounted for.

Turning to that great mystery of hypernuclear spectroscopy--
the observation of narrow Σ states--I need not repeat the arguments given earlier by Gal, who indicated that it is difficult to
find a theoretical justification for narrow Σ states 80 MeV above
the N threshold. There are several new and important bits of
evidence concerning those states and one rather provocative
suggestion.

It will be recalled that the observed Σ structures lie above Σ
binding, where the quasi free sigma-production cross section rises
rapidly. No bound Σ states have ever been observed. Kishimoto
presented calculations of the quasi free Σ production by the pole
graph method in which several important effects are included--
namely the experimental momentum distribution from (e, e'p), the
effect of Y* resonances, and the distribution of hole strength in
the recoiling nucleus. The illustration, Fig. 4a, shows data from
CERN at 400 MeV/c ($K^-\pi^+$) on ^{12}C and the quasi free
calculation. Shown is the effect of including the Y* resonance at
1520. Figure 4b shows a quasi free fit to ^{6}Li(K^-, π^+), BNL
data at 713 MeV/c. The effect of the $P_{3/2}$ and $S_{1/2}$ hole
strengths is indicated. It is clear that a proper calculation of

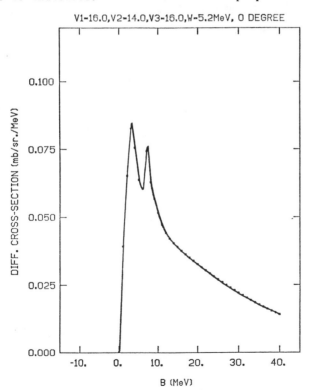

Fig. 5. A calculation of final state interaction peaks in
^{12}C(K^-, π^+) at 720 MeV/c by Hungerford and Tang.

the quasi free production must be taken into account before narrow
Σ-state claims can be validated. In other cases, such as the
(K^-, π^+) ^{12}C data from KEK, the evidence for Σ-states is
stronger. Hungerford, however, has raised the question of whether
final state interactions, like those producing the prominent cusp
in the deuterium system, could be producing similar effects in
^{12}C. He also points out that the π^0 tagging would also enhance
the final state interactions, inasmuch as they emphasize the Λ exit
channel. Figure 5 shows he has been successful in reproducing at
least two of the peaks in the (K^-, π^+) reaction on ^{12}C. I hope
that further data from experiments in progress at KEK as reported
by Chiba, and more refined calculations, will clarify these issues.

The lifetime and decay brancing ratio measurements,
appropriate to the weak decay of Λ hypernuclei, performed at the
AGS were described in some detail by Gal in his plenary session
talk. John Szymanski presented a status report on the extension of
these measurements to the A=5 system from a study of $(K^-\pi^-)$ on
^6Li. Here we are studying the transition from the almost purely
nucleonic weak decay mode in the mass 12 system to a case in which
mesonic decay is presumably more evident. Figure 6 shows the
prompt and delayed timing curves. Although a lifetime number is
not yet available, the delayed time distribution is obviously

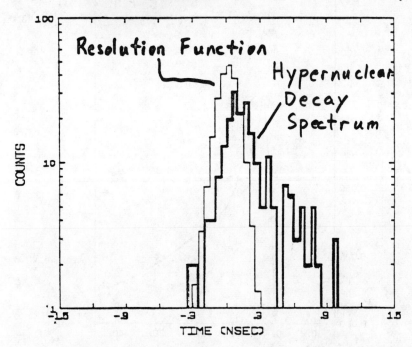

Fig. 6. The decay timing curve for ^5He ground state weak decay
comparted to the resolution function (σ = 95 ps).

distinct from the prompt timing curve. A remarkable resolution
figure of 95 picoseconds (σ) has been obtained for the prompt
distribution, indicating the excellent timing resolution achieved
over a two-week running period. Calculation of the branching ratios
is in progress.

Theoretical calculations for the nucleon-stimulated weak decay
modes of Λ-hypernuclei were presented by Kisslinger and Dubach.
Kisslinger uses a hybrid quark model which uses meson exchange for
the long range part of the nuclear force, while treating the short
range part with quark gluon exchange. Dubach uses a meson exchange
treatment incorporating the $(SU_6)_W$ symmetry. Both emphasize the
necessity of separating nuclear structure effects in order to
understand the fundamental transition operator. A calculation of
decay in the nuclear matter limit is inappropriate to a discussion
of real nuclei. Both indicated the importance of experiments for A
= 4 and Λ = 5 systems where the Λ-N are in relative S-states for
this separation.

A weak decay calculation for the H particle using the
constituent quark model was presented by Golowich. He considered
$\Delta S = 2$ and $\Delta S = 1$ transitions, appropriate to a more or less
strongly bound H. The important result for cosmology is that the

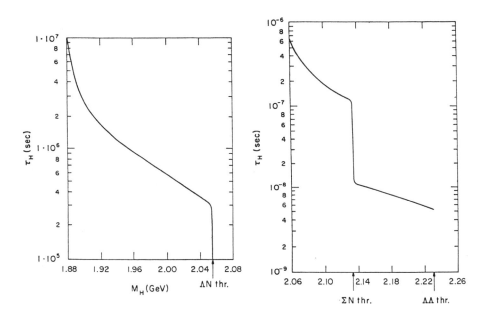

Fig. 7. Calculations of the H-particle lifetime for $\Delta S = 2$
transitions (left) and $\Delta S = 1$ transitions (right).

estimated lifetime for the $\Delta S = 2$ transition is about 10^6 seconds. This rules out the H as a possible source of the Cygnus X-3 signal. For $\Delta S = 1$ transitions, a lifetime of 10^{-8} and 10^{-7} seconds is suggested, and these lifetimes have relevance to possible accelerator based searches for the H. The results are shown in Fig. 7.

A cloudy quark-bag model calculation of the kaon-nucleon interaction was presented by <u>Byron Jennings</u>, who pointed out that in his model the $\Lambda(1405)$ is predominantly an anti-kaon nucleon bound state, rather than a three-quark object. <u>Joseph Speth</u> presented results from his extension of the Bonn-potential to the Λ-nucleon system. Speth presents curves which illustrate the effects of the nuclear medium in altering the real two-body forces to effective two-body forces. In particular the large difference in the tensor forces between the NN and ΛN are reduced by polarization effects in nuclear matter (Fig. 8). These differences suggest that interpreting spin-orbit splittings in N and Σ hypernuclei, for example, in terms of a simple additive quark model, is not possible. <u>C. Bennhold</u> presented calculations for kaon photoproduction based on relativistic wave functions.

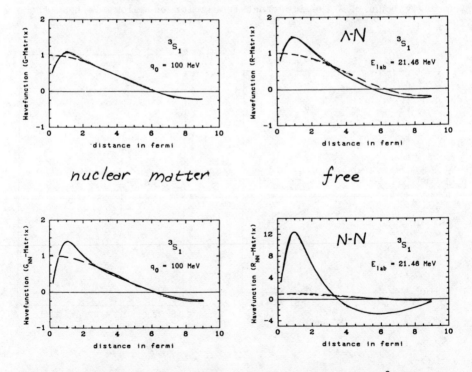

Fig. 8. Correlated (—) and uncorrelated (--) ΛN and NN $^3S_1$ wavefunctions in free space (left) and in nuclear matter (right).

Finally L. C. Liu presented a proposal to search for an eta-mesic bound nuclear system—a system which, if not strange, is certainly unusual. The ejection of a proton by a pion in ^{16}O and the simultaneous creation of an η at rest in the nucleus is kinematically possible at a pion incident momentum near 800 MeV/c. The idea is based on a suggestion made by J. C. Peng several years ago. The calculations of Liu indicate η binding starts near mass 12 and is possible for heavier systems (Fig. 9). An experimental search for η-bound nuclei will start at the AGS in January.

A wealth of nuclear physics problems involving strange quarks in nuclei exists. We need only the accelerator, facilities, and people to exploit this field.

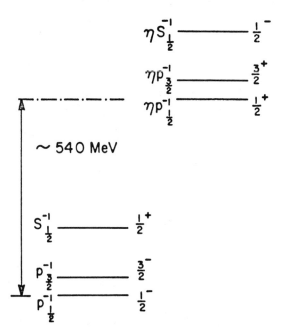

Fig. 9. Diagram for η-hole states formed in ^{16}O(π^+, p).

KAON DECAY SESSION SUMMARY

M.P. Schmidt
Physics Department, Yale University
Box 6666, New Haven, Connecticut 06510

INTRODUCTION

In keeping with the interdisciplinary nature of the conference, the parallel sessions on kaon decay physics were richly augmented by the presentation of new results on muon and tau lepton decays.

RARE K DECAY

An assessment of unitarity constraints on the Kobayashi-Maskawa (KM) matrix elements and the implication of these constraints for Standard Model predictions for the decay $K^+ \to \pi^+ \nu \bar{\nu}$ was given by W. Marciano. From measurements determining the weak charged couplings of the up quark to the down, strange, and bottom quarks, one finds that there is still room for amplitudes of transitions involving a fourth generation at about the 5% level. In addition, under the assumption of only three generations of quarks, the unitarity constraints imply limits of about 6% for the amplitudes of transitions of the top quark to down or strange quarks. This limited top quark mixing balances against its relatively large mass (~ 40 GeV/c^2 ?) to make the charm and top quark constributions to the $K^+ \to \pi^+ \nu \bar{\nu}$ decay rate comparable. The situation is considerably altered if m_t is larger than suggested by the CERN data, with BR($K^+ \to \pi^+ \nu \bar{\nu}$) increasing from $\sim 7 \times 10^{-11}$ to $\sim 2 \times 10^{-10}$ as m_t increases from ~ 40 GeV/c^2 to ~ 70 GeV/c^2. Even more dramatic is the effect of a possible fourth generation, where an order of magnitude increase in the $K^+ \to \pi^+ \nu \bar{\nu}$ decay rate could be due either to the contributions of a very massive top' quark, or to a substantially increased coupling of the top to light quarks resulting from a relaxation of the unitary constraints of the (3 × 3) KM mixing matrix.

This is good news to the proponents of the search for $K^+ \to \pi^+ \nu \bar{\nu}$, E787 at Brookhaven, who have the goal of pushing the present BR limit from 1.4×10^{-7} to 10^{-10}. In this experiment the full $K^+ \to \pi^+ X$ ($X = \nu \bar{\nu}$), $\pi^+ \to \mu^+ \nu_\mu$, $\mu^+ \to e^+ \nu_e \bar{\nu}_\mu$ decay chain is detected, with the K^+ tracked to rest in an active scintillating fiber target. In the progress report by T. Numao we learned that construction is well underway and the detector will take first beam this coming winter.

STANDARD LEPTON DECAYS

The decay systematics of muons, and now the tau leptons, provide a testing ground for the standard Model free of the vaguaries associated with the low momentum strong interactions. A brief but comprehensive review of standard muon decay ($\mu \to e\nu\nu$) was presented by D. Stoker. A recent series of measurements, including in particular those on transverse and longitudinal e^+ polarization at SIN, allow the form of the interaction to be determined up to the ambiguities associated with the unmeasured neutrino helicities. A general analysis leads to the statement that > 79% (90% CL) of the muon total decay rate is due to the V-A interaction, with the remaining uncertainty dominated by the limit on a scalar-pseudoscalar interaction as set by the CERN (CHARM) measurement of the inverse muon decay rate. A specific limit on the existence of right-handed W bosons (M > 432 GeV/c^2 for arbitrary left-right mixing) has been obtained from the elegant endpoint measurements at TRIUMF of the muon polarization and decay rate for helicity unfavored positrons.

The situation with tau leptons was likewise reviewed by P. Burchat, naturally with somewhat reduced statistical significance. In short, the measured tau lifetime (τ_τ = 0.285 ± 0.018 psec), the relative decay mode branching ratios, and the measured leptonic momentum spectra ($\to \rho_\tau$ = 0.71 ± 0.09 vs ρ_μ = 0.7518 ± 0.0026), all agree well with the Standard Model predictions. Still to be solved is the mystery associated with the observation that 11% of the single charged particle topological branching fraction is as yet unaccounted for by specific decay channels. Continued progress can be expected at CESR and DORIS where the cross section to background ratio is most advantageous.

LEPTON NUMBER NONCONSERVATION

Returning to muon decays, this time of the rare, that is forbidden by the Standard Model, variety, we heard from L. Piilonen on results from the LAMPF Crystal Box. The experimenters achieved excellent performance from the detector with its large NaI array ($\Delta E/E$ = 6.5% FWHM @ 130 MeV, Δt = 1.2 nsec FWHM, and Δx, Δy = 3 cm FWHM). Maximum likelihood analyses of the data allow the following (90% CL) limits to be obtained: BR($\mu \to e\gamma$) < 4.9 × 10^{-11}, BR($\mu \to e\gamma\gamma$) < 1.5 × 10^{-10} (preliminary), and BR($\mu \to eee$) < 3.1 × 10^{-11}. Only the latter limit is surpassed by the SIN result of BR($\mu \to eee$) < 2.4 × 10^{-12}. Further progress on the $\mu \to e\gamma$ limit (down to ~ 10^{-13}) will result from the project MEGA now beginning at LAMPF (E969). New results were also presented by T. Numao on limits on muon-electron conversion in titanium obtained at TRIUMF with a TPC. The (90% CL) limit obtained $\Gamma(\mu Z \to eZ)/\Gamma(\mu Z \to \nu_\mu(Z-1))$ < 4 × 10^{-12} is especially important for models of lepton nonconservation in which the coherent interaction with the nucleus is important.

CP VIOLATION

Well primed by B. Winstein's review talk in an earlier plenary session, we had an excellent session on the recent progress in studies of CP violation in kaon decays. B. Holstein brought the theoretical perspective together succinctly and suggested that with the next round of experiments we may well be able to determine which, if any, of the 'favored' models of CP violation nature has chosen. Our current uncertainty about the mass of the top quark, the relative couplings of the b quark to the up and charm quarks, and difficulties with calculations for weak decays of quarks residing in light hadrons prevents us as yet from ruling out the standard KM model, or the not always so popular multi-Higgs models of CP violation.

The observation of $|\varepsilon'/\varepsilon| > 10^{-3}$ is expected in either case with the Higgs models distinguished by an observable (?) electric dipole moment for the neutron, $d_n > 10^{-27}$ ecm. Superweak models would preclude such observations, and the remaining possibility (ε'/ε observable, d_n not) would then point to other possibilities such as a variant of the left-right symmetric models.

Progress reports on the CERN (NA-31, M. Calvetti) and FNAL (E731, G. Bock) measurements of ε'/ε were presented, including some detailed discussion of the analyses of data collected during preliminary running of the two experiments last summer. A precise determination of ε'/ε involves a measurement of the ratio of the CP violating decays $K^0_L \to \pi^0\pi^0$ and $K_L \to \pi^+\pi^-$, normalized by the CP allowed $K_S \to 2\pi$ decays: $Re(\varepsilon'/\varepsilon) = 1/6(1 - |\eta_{00}|^2/|\eta_{+-}|^2$ with $|\eta_{00}|^2 = \Gamma(K_L \to \pi^0\pi^0)/\Gamma(K_S \to \pi^0\pi^0)$ and $|\eta_{+-}|^2 = \Gamma(k_L \to \pi^+\pi^-)/\Gamma(K_S \to \pi^+\pi^-)$. Each of the experiments have already collected on the order of 10,000 of the difficult $K^0_L \to 2\pi^0$ decays. This, in principle, allows for an improvement on the limit $|\varepsilon'/\varepsilon| < 0.01$ established by the recent Chicago-Saclay (FNAL E671) and Yale-BNL (BNL E749) experiments.

Analyses underway are focused on establishing control over possible systematic effects which differ considerably for the two experimental techniques. The Fermilab experiment compares the 2π decay rates (first $\pi^0\pi^0$, then $\pi^+\pi^-$) in a twin K^0_L beam arrangement with K^0_S produced in one of the beams by regeneration. The twin beam technique allows for the cancellation of many systematic effects associated with differences in instrumental backgrounds from a single K_L or K_S beam. However, care must be taken in correcting for contributions due to other than <u>coherently</u> regenerated K_S, and Monte Carlo techniques must be employed to compensate for differences in the K_S and K_L decay distributions along the two beams. The CERN experiment compares the $\pi^0\pi^0$ to $\pi^+\pi^-$ decay rates in a single beam arrangement, first with $K^0{}_L$, and then with $K^0{}_S$ produced <u>directly</u> in target station which can move along the beam line. The larger acceptance of the detector (no spectrometer magnet and no photon conversion requirements) will presumably allow for a larger raw statistical sample on which the systematic corrections can be based. Both groups are upgrading their data acquisition electronics (triggers, Fastbus, etc.) to cut down on the dead time suffered. The next running period begins in June for the CERN experiment and in January for the Fermilab experiment and both groups expect to collect ~ 100K $K^0{}_L \to 2\pi^0$ events and determine ε'/ε to ~ 0.001.

Recent progress in the test for CP violation in K_S decay was reported on by G. Thomson. E621 at Fermilab aims to measure the K_L, K_S interference in the $\pi^+\pi^-\pi^0$ decay mode and thus determine the CP violating parameter $\left|\eta_{+-0}\right|^2 = \Gamma(K_S \to \pi^+\pi^-\pi^0)/\Gamma(K^0{}_L \to \pi^+\pi^-\pi^0)$. A twin beam technique is employed in which the interference is searched for downstream of a near target in one beam and acceptance biases are removed without recourse to Monte Carlo calculations by using decays from $K^0{}_L$ originating from a far target in the other beam. With a typical momentum of 150 GeV/c, the K_S travel 7.8 m in one lifetime and the experimental resolution in determining the decay position along the beam line of $\Delta Z <$ 1 m corresponds to $\Delta t = \pm \tau_S/8$. Analysis of data collected in 1984 yield the result $\left|\eta_{+-0}\right| = 0.023 \pm 0.029$, and ongoing analysis of 1985 data should allow a determination of η_{+-0} to 0.003. Of course $\left|\eta_{+-0}\right| \sim \left|\eta_{+-}\right| \sim \left|\varepsilon\right| \sim 0.002$ for the superweak or standard (KM) model of CP violation, although the difference of $\left|\eta_{+-0}/\eta_{+-}\right|$ from unity might be 50 to 100% in other less constrained models.

A new era of CP violation studies will commence with the completion of ACOL at CERN (1987 ?). Two groups intend to exploit the copius rate of $\bar{p}p$ annihilation at rest then available at LEAR and use channels such as $\bar{p}p \to K^+\pi^-\bar{K}^0$ or $K^-\pi^+K^0$ in order to study the evolution and decays of tagged K^0 or $\bar{K}^0$ mesons. J. Fry (for Pavlopoulos et al.) and B. Roberts (for Tanner et al.) reviewed the two proposals. The emphasis for the first group is the measurement of the time dependent asymmetries of the form

$$A_f(t) = \left.\frac{R(K^0 \to f) - R(\bar{K}^0 \to f)}{R(K^0 \to f) + R(\bar{K}^0 \to f)}\right|_t$$

for $f = \pi^+\pi^-$, $\pi^0\pi^0$, $\pi^+\pi^-\pi^0$ and $\pi^0\pi^0\pi^0$ in order to determine ε'/ε to $\sim 2 \times 10^{-3}$, the phase difference of η_{+-} and η_{00} ($\phi_{+-} - \phi_{00}$) to $\sim 2^0$, and $|\eta_{+-0}|$ and $|\eta_{000}|$ to $\sim 7 \times 10^{-4}$ among other things. The thrust of the second group is to look at large proper times (after $K^0 - \bar{K}^0$ mixing is complete) for the asymmetry

$$A = \frac{R(\bar{K}^0_{tag} \to \pi^- e^+ \nu_e) - R(K^0_{tag} \to \pi^+ e^- \nu_e)}{R(\bar{K}^0_{tag} \to \pi^- e^+ \nu_e) + R(K^0_{tag} \to \pi^+ e^- \nu_e)} \bigg|_\infty$$

and establish new limits on $\Delta S \neq \Delta Q$ interactions, CPT violation and a first measure of the expected T violation manifest in the difference in the mixing rate of $K^0 \to \bar{K}^0$ vs. $\bar{K}^0 \to K^0$. These are extremely challenging experiments requiring $\sim 1\%$ determination of small integrated asymmetries ($\sim 10^{-3}$, requiring 10^8 K decays to a given final state) and commensurate control of systematic errors, which however are of a totally different nature than those of current experiments on ε'/ε.

CONCLUSIONS

The activity in the field promises to deliver important new results in CP violation by the next conference of this series. Results will also be forthcoming from rare K decay experiments, i.e. $K^+ \to \pi^+ \nu \bar{\nu}$, as well as those not specifically reviewed at this conference, $K^0_L \to \mu e$, ee and $K^+ \to \pi^+ \mu^+ e^-$. Indeed two groups YALE-BNL-WASHINGTON-SIN (E777, $K^+ \to \pi^+ \mu^+ e^-$) and YALE-BNL (E780, $K^0_L \to \mu e$ and ee) are expected to take useful physics data in the coming month.

This work is supported in part by the U.S. Department of Energy under Contract No. DE-AC02-76ER03075.

SUMMARY: NEUTRINO PHYSICS

Bruce H.J. McKellar
School of Physics, University of Melbourne,
Parkville, Victoria, Australia 3052

INTRODUCTION

Much of the present interest in neutrino physics is directed towards the question of the mass of the neutrino(s). While there is no reason in the standard $SU(3) \times SU(2) \times U(1)$ model for a non zero neutrino mass, almost any extension of the standard model allows a non zero mass, and many extensions require a non zero mass. When the neutrino mass is non zero the weak interaction need no longer be diagonal in the mass eigenstate basis, so that the search for neutrino mixtures or neutrino oscillations is an indirect search for evidence of finite neutrino masses. Nine of the thirteen papers presented in the neutrino physics parallel sessions were related to neutrino masses, and this question will occupy much of the summary.

Of the other papers in the neutrino physics sessions, two were related to tests of the standard model, and two to a proposed macroscopic effect of neutrino interactions.

While I will not report on them in this summary I draw your attention to the papers on solar neutrinos in the Non-accelerator physics session, the discussion of that subject by Haxton in the hadron spectroscopy session, the paper on acoustic detection of neutrinos in the non accelerator physics session, and the discussions of lepton flavour violation in the Kaon Decay sessions.

One notable feature of the neutrino physics sessions was the controversy generated, which I will attempt to describe.

DIRECT OBSERVATION OF THE ELECTRON-NEUTRINO MASS

Results of the Zurich and Los Alamos experiments to measure the mass of the electron neutrino from the tritium Kurie plot were reported by Kundig[1] and Knapp[2] respectively. It should be emphasised that the mass parameter which enters the fit to the data is m_ν^2 rather than m_ν, and this should be kept in mind in assessing confidence limits placed on m_ν in quoted results.

The two experiments reported here differ in their choice of the tritium source. The Zurich group used tritium implanted in a carbon foil, which has the advantage that an intense source can be produced in this way, but the disadvantage that the analysis depends on solid state effects - on the ^{3}H-C bonds which influence the electronic final states after the decay, and on the energy loss spectrum of the β particles. The Los Alamos group uses atomic and molecular tritium as a source, avoiding the difficulties with the energy loss, and allowing a calculation of the distribution of atomic and molecular final states, but at a cost in intensity.

Both groups presented impressive data which are consistent with zero neutrino mass. From the Zurich data the χ^2/df for $m_\nu=0$ is 1370.2/1370. The Los Alamos group used a modified statistic, Ξ^2, to determine the quality of the fits to the data. While it is computed in a slightly different way Ξ^2 is distributed much like χ^2 and for $m_\nu=0$ the $\Xi^2/d.f.$ is 714/709. Both groups thus find no reason to reject a zero neutrino mass, and report 95% confidence limits (which include an allowance for systematic errors)

and
$$m_\nu < 18 eV \quad (Zurich)$$
$$m_\nu < 29.7 eV \quad (Los\ Alamos),$$

which are to be compared to the "realistic mass estimates" reported by the ITEP group[3]

$$20 eV < m_\nu < 34 eV\ .$$

The ITEP results appear to be just inconsistent with the results reported at this meeting. It would be useful to have the χ^2/df for an $m_\nu=0$ fit to the ITEP data, but this is not at present available.

The first controversy was raised in the discussion of the Zurich results. It was claimed by Robertson[4] that the electron energy loss curve used by the Zurich group differed significantly from that used by the ITEP group, and that this difference is the principal explanation of the different neutrino masses reported by these two groups. The Zurich energy loss spectrum was obtained by profiling the ^3H distribution in the carbon foil using nuclear techniques, and folding this distribution with electron energy loss spectra measured for electrons incident on carbon at energies of 23-70 keV and extrapolated down to the required energies using the plasmon excitation theory of electron energy loss. From the discussion it was not possible to resolve this conflict.

LEPTON FLAVOUR MIXING

If at least one of the neutrinos has a non zero mass, then the analogy between the quark and lepton sectors is much closer, and just as strangeness and charm are not conserved in the weak interactions we would no longer expect individual muon and tauon neutrino numbers to be conserved. For example the charged current will now transform the electron into an electron neutrino ν_e which is a linear combination of the mass eigenstates ν_1, ν_2, ν_3, ...

$$\nu_e = L_{e1}\nu_1 + L_{e2}\nu_2 + L_{\mu 3}\nu_3 + \ldots \qquad (1)$$

This flavour mixing can be detected in a variety of ways, through distortion of β decay spectra, through neutrino oscillations, though processes such as $\mu \to e\gamma$, and so on. Most of these processes were first discussed by the Nagoya group[5], and have been reanalysed many times since then.

With the advent of explicit extension of the standard model, we now have a number of models which make quantitative predictions of neutrino masses and of the rates for these lepton flavour violating processes. Many of these predictions were reviewed and compared with the available data by Vergados[6]. At present only a few of the model predictions are within reach of experiment, and further experiments should be encouraged in the hope of testing those models. Moreover there is always the possibility that unexpected results will be observed - nature is not always limited by the imagination of the theorists!

One such unexpected result was the kink in the ^{3}H β spectrum found by Simpson[7], which corresponds to a neutrino of mass 17.1 keV weakly coupled to the electron neutrino. Simpson conveyed a paper to the neutrino physics session[8] describing a reanalysis of experiments which did not see the knik in other spectra and presenting new data of his own on ^{3}H, all of which showed evidence for the elusive neutrino. He emphasised that, in looking for structure in a part of the spectrum making a fit to the whole spectrum may obscure the local structure, and that it was not reasonable to show that a two component spectrum does not fit the data by first fitting a one component spectrum and superposing the additional component.

A lively discussion ensured on the validity of the local method of searching for kinks in the spectrum. One needs to be convinced that further kinks would not be uncovered when such an analysis is applied section by section to the whole spectrum.

NEUTRINO OSCILLATIONS

We saw in equation (1) that the neutrino coupled through the weak current to a charged lepton is a superposition of mass eigenstates, and thus that the different components will evolve in time with different frequencies. After evolving in this way the state is in general no longer uniquely coupled to one lepton, and thus for example $P(\nu_e \to \nu_\mu)$, oscillates as a function of time, or as a function of distance from the source.

Experiments are usually analysed assuming the mixing of two species for simplicity, but the real situation could involve mixing of more than two species, so discrepencies which may appear in results reported from experiments at different energies or distances may be artifacts of the two species analyses. In the simple two family analysis the oscillation pattern is described by such probabilities as

$$P(\nu_\mu \to \nu_e; \ell) = \sin^2 2\theta \sin^2\left(1.27 \frac{\Delta m^2 \ell}{E}\right)$$

where θ is the mixing angle, Δm^2 is the difference in the squares of the masses of the species in eV^2, ℓ is the distance from the source in m, and E is the neutrino energy in MeV.

In the neutrino physics sessions O'Halloram[9] reviewed accelerator experiments on neutrino oscillations, most of which have searched for oscillations from the ν_μ and $\bar{\nu}_\mu$ states, and Greenwood[10] reviewed experiments from reactors which are all disappearance experiments, searching for oscillations in $P(\bar{\nu}_e \to \bar{\nu}_e)$. Typically results are presented by plotting the region in the ($\sin^2 2\theta$, Δm^2) plane which represents the range of parameters which are allowed by the experiment, as illustrated in refs. 9 and 10. Typically the accelerator experiments reach lower values of $\sin^2 2\theta$ than the reactor experiments, whereas the reactor experiments reach lower values of Δm^2.

The two experiments, the experiment PS191 at CERN[11], and the Bugey reactor experiment[12] have reported evidence for oscillations. The PS191 experiment investigated $\nu_\mu \to \nu_e$ oscillations and found results consistent with $\sin^2 2\theta \sim 2\text{-}3\%$, $\Delta m^2 \sim 5 eV^2$. O'Halloran reported that the region of parameter space allowed by the PS191 contained in the <u>forbidden region</u> of BNL and BEBC experiments.

The Bugey reactor experiment is consistent with

$$\sin^2 2\theta \sim 0.2 \qquad \Delta m^2 \sim 0.2 eV^2 ,$$

and these points are just outside the region of parameter space excluded by other reactor experiments.

In summary, no non-controversial evidence for oscillations is available at present and future results are awaited with interest.

DOUBLE BETA DECAY

Neutrinoless double beta decay can occur when the neutrino is a Majorana particle, so it is transformed into itself by particle-antiparticle (CPT) transformations, and it has a non zero Majorana mass[13] M which couples ν_L to ν_R. In general with a number of families, neutrinoless double beta decay is possible for non-zero values of a number of distinct combinations of masses, coupling parameters, and L-R interactions, as described in detail by Haxton[14] and Vergados[6] in their contributions, but in the absence of any evidence for neutrinoless double beta decay it is adequate to describe the results in terms of the simplest parameter M.

The experimental situation has been complicated in the past by a discrepency between the rates of two neutrino double beta decay as observed by geochemical and laboratory methods. In his paper Moe[15] showed new laboratory rsults for the $^{82}Se \rightarrow ^{82}Kr$ $\beta_1 + \beta_2$ spectra which were in agreement with the geochemical results of the Missouri and Heidelberg groups. However the discrepency between the calculated and the observed rates for 2ν $\beta\beta$ decay persists (which represents a factor of 3 in the matrix element for ^{82}Se) and was discussed in some detail by Haxton[14]. This discrepency has survived several attempts to do the nuclear physics calculation in different nuclear models. It shows that the 2ν $\beta\beta$ rate is sensitive to details of the nuclear wavefunction which are not well determined in our existing models, and must cast doubt on our ability to calculate the expected 0ν $\beta\beta$ decay rate accurately.

Avignone[16] reviewed the substantial effort that has gone into the $^{76}Ge \rightarrow ^{76}Se$ experiments. The effort is devoted to reducing the background counts in the region of 2 MeV in the summed β spectrum where the 0ν decay spectrum would show a peak. Running the experiments deep underground, with incredibly pure materials used in the construction of the spectrometer, and with shielding of old lead has reduced the background to the extent that the present world limit is

$$T^{0\nu}_{1/2}(^{76}Ge) \geq 4.4 \times 10^{23} \text{ yr},$$

which corresponds to an upper bound for the Majorana neutrino mass which ranges from 0.8eV to 2.3eV depending on which nuclear physics calculation is used in the analysis. The lifetime for 2ν $\beta\beta$ decay of ^{76}Ge has not yet been measured in these experiments, but it is expected that a value for the lifetime will be obtained after another year or two of running. However, without further background reduction, 20 years of observations will be required to reduce the limit on M by a factor of 2.

TESTS OF THE STANDARD MODEL

Two Los Alamos experiments were reported which test neutrino interactions gainst the expectations of the standard model. Fiske[17] reported on the progress of the ν_μ-nucleus interaction experiment, which will take data this summer, and Lee[18] reported results from the ν_e-e scattering experiment. The latter experiment is significant in that it probes the interference between W and Z exchange processes in lepton-lepton scattering. The results presented were consistent with the standard model and correspond to

$$\sin^2\theta_W = 0.24 \, {}^{+0.09}_{-0.10} \, {}^{+0.05}_{-0.06}$$

which is consistent with the present world average for the weak angle.

DETECTION OF NEUTRINOS BY MACROSCOPIC EFFECTS ON CRYSTALS

Weber presented a paper[19] escribing his proposal for a neutrino detector based on the forced exerted by a neutrino beam on a crystal. He claimed that

(i) a cross section essentially proportional to N_0^2 (N_0 is the number of nuclei in the crystal lattice) will result from neutrino scattering from a perfectly rigid crystal

(ii) this N_0^2 cross section persists as long as the neutrino energy is less than the binding energy of a nucleus in the crystal.

(iii) that the effect had been observed in crystals exposed to a β source, and a reactor neutrino flux.

Casella[20] presented a paper which claimed that the condition for the N_0^2 cross section was coherent scattering, i.e. that the neutrino wavelength be large compared to the dimensions of the crystal, and that the momentum transfer to the crystal would then be too small to be observed.

A vigorous discussion followed these presentations. It seems to me that the Weber effect, if it occurs, could be expected to influence the transmission of neutrinos by microcrystalline materials, e.g. the material shielding reactors may be expected to obserb energy from the neutrino flux, and the spectrum of upward going cosmic ray neutrinos and muons could be distorted.

CONCLUSION

In spite of heroic efforts, there is still no uncontroversial evidence, direct or indirect, that the mass of any of the neutrinos is non zero. The efforts will continue, and we look forward with interest to further results at the next meeting on the intersections of nuclear and particle physics.

ACKNOWLEDGEMENT

It is a pleasure to thank TRIUMF for their hospitality and support while this paper was written.

REFERENCES

1. W. Kundig, these proceedings.
2. D. Knapp, these proceedings.
3. S. Boris et al., Phys. Lett. <u>159B</u>, 217 (1985).
4. See also R.G.H. Robertson, these proceedings.
5. Z. Maki, M. Nakagawa and S. Sakata, Prog. Theor. Phys. (Kyoto) <u>28</u>, 870 (1962);
   M. Nakagawa et al., ibid <u>30</u>, 258 (1963).
6. J. Vergados "Double Beta Decay and Lepton Flavour Violation", these proceedings.
7. J.J. Simpson, Phys. Rev. Lett. <u>54</u>, 1891 (1985).
8. J.J. Simpson, these proceedings.
9. T. O'Halloran, these proceedings.
10. Z. Greenwood, these proceedings.
11. F. Vanucci, Proc. 20th Rencontre de Moriand (1985).
12. J.F. Cavaignac et al. Phys. Lett. <u>148B</u>, 387 (1984).
13. When the Majorana neutrino mass vanishes there is no physical distinction between the 2 component Majorana neutrino and the 2 component Weyl neutrino.
14. W. Haxton, "Double Beta Decay and Nuclear Theory", these proceedings.
15. M. Moe, these proceedings.
16. F. Avignone, these proceedings.
17. R. Fiske, these proceedings.
18. W.P. Lee, these proceedings.
19. J. Weber, these proceedings.
20. R. Casella, these proceedings.

CLOSING REMARKS

Robert Hofstadter
Department of Physics, Stanford University
Stanford, California 94305

When Alan Krisch twisted my arm in inviting me to give this talk, I told him I had a couple of problems. One was that I had been out of nuclear physics for over twenty years. This didn't phase him for he said, "You'll be able to say something." Then I told him I would have to miss the Thursday of the meeting because I had a long-standing commitment in Seattle. He said, "You'll be able to say something." But what he didn't tell me was that Eric Vogt would be assiduously handling the transparency projector for the benefit of those sitting in the back of the room. As you know, the transparencies were mostly in motion, and what I wanted to see on them was gone whenever I looked up from my note-pad. Not only were the transparencies constantly moving, but some were masked in all the critical places, even diagonally masked in one case, and besides, many speakers often shadowed some of the data that I wanted. So there will be no transparencies in my talk!

In any case, I have really been happy to participate in this exciting conference at Lake Louise. It is not often that one can top Sir Denys Wilkinson, who told us he was here 40 years ago, but I had been at Lake Louise 46 years ago! The beauty and splendor of the surroundings were just as I remembered them. It has also been a great pleasure to renew acquaintances with old friends and to meet new ones.

In the last week we've been able to hear new physics and older physics too. You have already heard lots of great summaries, better than any I can give. I'm going to try to talk about some of those topics from impressions I gained, and also through a choice of subjects I liked. I won't try to be encyclopedic and I expect to make some mistakes and maybe treat topics superficially. I'm sorry but I'll just have to plead "finiteness." I learned I was finite a long time ago. So, here are my impressions and likings. Broadly speaking, I think we experienced a consolidation and extension of previously known material at this conference. We did not hear of revolutionary new developments, or breakthroughs, which almost by definition are very rare anyway. Let me first talk about theory from the point of view of an experimenter. Theory took some pummeling at a few of the sessions--from theorists--but also had vigorous supporters. Our basic tool is now QCD, the quark-gluon-color picture that has slowly developed over the past couple of decades. In previous years, nuclear physics was presented with a more or less phenomenological view of nuclear forces and a scheme of how nuclei were built employing those forces. But now we have a solid, well tested, and even elegant basis in QCD which promises to explain the nucleon-nucleon force and much more. In principle we also have the basis for understanding how nucleons may be modified, as in the EMC effect for example, when coexisting in various nuclear ground states, in excited states, and even in the presence of strange matter, as in hypernuclei. Previously, we didn't know much about the repulsive core, or about the intermediate range of the nuclear force, and certainly not as much as we knew about the one pion theory of the outer range of the nuclear force. Perhaps we still do not know it all but the basis is there. For example, we can put together a six quark configuration and make QCD perturbative calculations for the nucleon-nucleon potential. We can also construct dibaryon complexes like the H-particle for which there is not a wide separation between the sets of quarks, as in the deuteron, or as an p-p, p-$\bar{\text{p}}$, p-n, etc., scattering situations. The methods used are probably still clumsy and one needs to be an excellent bookkeeper to be sure that some terms in setting up wave functions are not omitted. We hear phrases like the Heitler-London model, quark molecules, and Van der Waals forces, phase transitions and lattice simulations of finite chemical potential. An innocent bystander wandering into this meeting might wonder whether this is a conference on chemistry. Perhaps we don't know it, but maybe it is! We have one pion effects, two pion effects, meson exchange, quark exchange, quark-gluon exchanges, gluon-gluon exchanges, etc., etc. We really have to learn to be good bookkeepers to keep all these effects in place.

However, it is obvious that what is new, disregarding the obvious complexities of six quarks in even a relatively simple dibaryon assembly, is that we are dealing with self-interacting boson entities we call gluons. This is what a non-abelian gauge theory introduces. My hunch is, and it is a rather obvious thought, that that is where the new physics lies. Glue balls of various kinds will, when understood, enlighten us thoroughly in a new craft of calculation. Perturbative techniques in QCD are now hard enough, but how do we handle the non-pertubative situations? I hope that new clues provided by gluon interactions may suddenly make the calculations elegantly simple. While we now have a theory that at best gives 10 or 20 percent agreement with experiment, what we all want is agreement to 1 percent or better. Then we may have a real theory!

Polarization experiments, spin physics at medium or low P_T and polarization theory, based on QCD, are in a rather puzzling state. We don't even know whether theory and experiment should be expected to agree at present P_T values. At higher P_T we shall have a chance to make a better check of QCD, and at this conference experimenters have been thoroughly alerted to this deep question.

We also heard presentations on lattice gauge theory of QCD, and the numerical approximation methods used in it to make non-perturbative calculations. LGT successfully gives confinement and flux tubes, asymptotic features, reproduction of hadron masses, though within wide limits, glueball mass predictions, pion and proton form factors, etc. But there is a real question about whether the grid is too coarse and there are still many unresolved questions, not to speak of the difficulties of using huge computers for the approximations. Hamiltonian methods are also being examined but only time will tell how successful any of these techniques may be. One would like to know where the boundary lies between perturbative and non-perturbative methods.

The SU(3)S(U2)U(1) standard model came out unscathed from this conference, at least when we think of 10-20% agreement with experiment. Calculated nucleon form factors, as we shall see, do give the "asymptotic" dependence on Q^2 for the proton, neutron, and pion in terms of the valence quarks. But what about the sea quarks? We should know the answer before long, when HERA at DESY starts operating. I think this is where the action will begin and, if I were younger, I would try to participate in the colliding beam e-p experiments to be done at HERA. My guess is that these experiments will be path-breaking and will lead to a new understanding of QCD and nucleon structure. I would even guess that HERA will produce simplifications and make really new discoveries before things cloud up again as precision and detail increase and when the SSC comes along. In the meantime, the Tevatron collider will also open up new ways of dealing with QCD.

The standard model, although unifying the strong and electroweak interactions, cannot be expected to be an ultimate theory because of the many arbitrary parameters (some say 21) it must assume. Furthermore, where is the Higgs particle? QED is now good to 10^{-16} cm or so. How long will this hold up? LEP, SLC, and HERA will once again tell us the answer to this question. Grand unified theory now makes predictions at enormously high energy, not reachable in today's accelerators, so that we have no present ways of testing it except by studying proton decay, looking for monopoles, or finding neutrino masses, etc. On the other hand, at the one TeV scale we may be able to test supersymmetry by finding the new particles it predicts. And maybe we will see preons. In spite of difficulties with testing

GUT, supersymmetry, technicolor, and superstring theory, these imaginative constructs offer delights to theorists that they cannot resist. But I'm betting that when SLC, LEP, the Tevatron Collider, Tristan, HERA etc., get into routine operation the deluge of new experimental facts will bring theory and theorists closer to reality and we shall hope for a lot of fun and success in spite of mirages far away. I'm not a believer of deserts in physics.

Let me turn to the experimental side of the Conference. Again, there is no way I can be complete. On CP violation we are witnessing what I would call heroic efforts at CERN, FNAL and elsewhere to measure the ratio ϵ'/ϵ in kaon decay. This quantity is related to the difference between the ratio of amplitudes of two pion decays in K_L and K_S for charged pions and the corresponding ratio for neutral two pion decays. Up to now, experiment fails to assign a definite value of ϵ'/ϵ that is different from zero. The ratio is probably less than 0.005, but the efforts I referred to aim at reducing this limit to 0.001 or better. When ϵ'/ϵ is measured, the result will tell which among the following choices is relevant to the understanding of CP violation. The choices are the superweak hypothesis, the Kobayashi-Maskawa scheme, Higg's particles, or a left-right symmetrical model.

In accelerator physics, various particle accelerators of many different types are now being built all over the world, some with superconducting magnets. Except for SLC these machines use conventional techniques. We have heard that a new method of obtaining very high gradients in linear machines may come from a two beam approach, using a free electron laser in one beam as the power-producing element for the other beam's acceleration. Microwave power exceeding present klystron behavior seems to have been achieved, but efficiencies of operation are not known. But I think something more radical is still needed, and I have suggested thinking about excited atoms or ions in crystals as a replacement for accelerator cavities.

Several proposals now exist for producing kaon factories such as at Los Alamos, KEK, Triumf and the European Hadron Facility. All have ambitious plans for producing proton accelerators in the 30-60 GeV range with currents perhaps two orders of magnitude above those presently achieved in meson factories. New heavy ion facilities are also being proposed.

The electromagnetic structures of nucleons and nuclei have now been extended by new electron scattering experiments at SLAC and at BATES and SACLAY. The new results show that the vector dominance model of nucleon form factors, so successful in its day, may be suitable for lower Q^2 values but at the highest Q^2 studied, about 30 $(GeV/c)^2$, where elastic cross sections as small as 8×10^{-40} cm^2/sr have been measured, QCD behavior takes over. At such large Q^2 the proton form factors scale as $(1/Q^2)^{n-1}$ where n is the number of valence quarks. For the proton this gives F_1 varying as $1/Q^4$ and F_2 as $1/Q^6$. Perturbative QCD can account nicely for these results, as remarked earlier. However, at the largest values of Q^2 where $G_M^p Q^4$ should remain constant, the observed gradual decline of this product may show that α_s, the strong coupling constant, may be exhibiting an expected decrease. For $G_M^n Q^4$ this product decreases strongly with increasing Q^2, and F_{1n} is probably not zero above $Q^2 > 4$ $(GeV/c)^2$.

Nuclei such as the deuteron, tritium and He^3 have been studied, with elastic cross sections as low as 10^{-42} cm^2/sr. Electric and magnetic form factors have been separated by a Rosenbluth analysis and yield beautifully clear diffraction minima. Quasielastic spectra are also obtained that correspond to the emission of one or more nucleons or other particles. Scaling of these spectra has been accomplished successfully. New <u>coincidence</u> studies with the TPC detector at the PEP ring with an argon gas jet may uncover <u>exclusive</u> channels in nuclear disintegration processes by high energy electrons. Such new techniques may give radically new information on nuclear breakup into various quark combinations such as pions, kaons, or nucleons etc.

I found the planned search for strange matter particularly intriguing. Jaffe's H particle, a six quark combination of two sets of uds quarks would be one of the objects of such a search. This strange combination has even been connected with the possible detection of hadrons incident on the earth that could have been emitted by Cygnus X-3. Large air showers with very large area scintillation and Cerenkov detector assemblies are the coming thing for detecting cosmic ray hadrons. While I find the evidence for Cygnus X-3 radiation in underground muon detectors very ambiguous, it is probably true that high-energy gamma radiation in the $10^{14}-10^{16}$ eV range is identifiable and measurable.

Elaborate experiments at the AGS and at KEK will search for the H-particle as evidence of strange matter, whether stable, or decaying. For these searches negative kaons on deuterium or ^3He have been proposed. On the other hand, multiquark droplets have also been suggested as being metastable or even stable, and an experiment has been proposed to look for their existence in fixed target high-energy heavy ion facilities.

The possible show stopper at this conference is the anomalous positron production in heavy ion collisions. The stubborn appearance of one or more narrow positron peaks with kinetic energy about 300 keV, on top of the positron production by known processes, has led to the assumption that possibly a new metastable entity may have been discovered. This entity decays by a two-body process into a positron and an electron of equal energies and emitted back to back in coincidence. The energy sum peak of the e^+e^- pair appears to be quite narrow, on the order of 25-50 keV wide, while the difference in energies also centers near zero. This object is produced at very low velocities, about 1/20 the speed of light. It is not clear whether we are observing the discovery of a new particle or whether there is still some normal explanation which is perfectly adequate. Perhaps the object is a composite of several e^+e^- pairs, but no one is sure at present what it is, or indeed if it is real at all. Only time will tell and new experiments using other methods of looking for this object are being discussed.

The value of the neutrino mass is obviously one of the most important parameters of nuclear physics, particle physics, and cosmological theory. New measurements were presented at this conference differing from earlier published Russian experiments. The shape of the beta decay curve at the 18.6 KeV end point of tritrium determines the mass value of the electron antineutrino Results of experiments at SIN and Los Alamos give upper limits of 18 eV and about 25 eV respectively. In the SIN measurement the best fits occurred at zero mass. These results do not agree with the Russian measurement of 35 eV ± (approximately 15 or 20 eV) which, of course, represents a non-zero mass. Some controversy still exists about details of the new measurements, but in my opinion these experiments point again to the traditional zero mass, agreeing with the value used successfully in many previous years of experience with the weak interaction. I think it is also fair to say that neutrino oscillations have not been observed.

In hadron spectroscopy we heard beautiful new results in J/Ψ decay observed with the Mark III spectrometer at SPEAR at SLAC. The iota (1460) and the θ (1700) are still considered to be glueball candidates but one cannot be sure because there are so many peaks in the same neighborhood among which these states are seen. What is called for is much greater statistics and higher resolution. Other new data such as those on the ξ (2230) are too detailed to be discussed here but all kinds of quark combinations, hybrids $\bar{q}qg$ called meiktons and possible quark molecules now seem to be on the verge of discovery. LEAR and SUPER LEAR and perhaps a high luminosity J/Ψ factory may be what is needed in the future.

On the subject of spin physics, we have already mentioned the possible failure of QCD to explain the results. I do not think this is really likely since QCD has been so successful everywhere else. Nevertheless, the strikingly large polarization effects observed in pp scattering are still somewhat mysterious and more experimentation at higher momenta and at $90°$ in the CM system are highly desirable.

These remarks are incomplete and my summation certainly does not represent a piece of scholarship. But I hope I have given you a feeling for the conference proceedings. I will be happy to correct any errors or omissions.

I want to thank the Conference organizing committee for inviting me to give this talk. I also want to thank the Committee for producing a splendid meeting in a splendid environment. I look forward to another successful meeting two years hence, whichever beautiful venue is finally selected. Thank you for your attention.

ACCELERATOR PHYSICS

Coordinators: P. Koehler
              R. E. Pollock

ON THE POSSIBILITY OF ACHIEVING A CONDENSED CRYSTALLINE STATE IN COOLED PARTICLE BEAMS

John P. Schiffer
Argonne National Laboratory, Argonne, IL 60439
and
A. Rahman
Supercomputer Institute and School of Physics & Astronomy,
University of Minnesota, Minneapolis, MN 55455

ABSTRACT

Calculations have shown that when a plasma of one kind of particle (one-component plasma or OCP) is below a certain temperature, it will undergo a phase transition and the particles will form a crystalline (bcc) array.[1] The relevant parameter is $\Gamma = (q^2/a)/kT$ (where a is the average spacing and T the temperature) and the transition takes place at $\Gamma = 170$. The properties of proposed storage rings for heavy ions with $q \gg e$, are such that an ordered array may actually be realized.[2]

We have performed molecular dynamics calculations for an idealized system in which an average (constant in time) focusing force was assumed, perpendicular to the x-axis: $F = Kr$, ($r^2 = y^2 + z^2$). The calculations were done as if in the moving frame; the particles had no velocity other than that arising from their thermal motion. No effects of the circular orbits were included. Under these circumstances, with $\Gamma = 180$, the particles form cylindrical shells around the "beam axis" as is shown in figure 1, with hexagonal two-

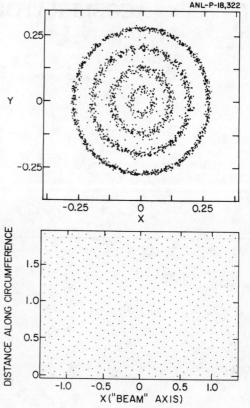

Figure 1:
Upper part: Projection of 2000 particles in a molecular dynamics calculation onto the plane perpendicular to the beam (x-axis) for $\Gamma = 180$. Lower part: distribution of particles in the outer shell with the shell unfolded into a plane. All but the innermost shell show a similar pattern.

dimensional order within the surface of each shell, except the innermost one. The pair correlation function in figure 2 is plotted against the coordination number n(r): the number of neighbors up to r. The pronounced peaks contain 14 and 44 nearest neighbors, indicating approximate three-dimensional bcc order. The two-dimensional correlation function shows peaks with 6 and 12 particles that are characteristic of triangular order. All distances are in units of $\xi \equiv (q^2/K)^{1/3}$ where q is the charge and K the average focusing constant mentioned above.

When the number of particles per unit length is reduced, only what was previously the innermost shell remains, as is shown in figure 3, and when it is further reduced, to the point where the two-particle Coulomb repulsion can be overcome by the focusing force, the particles in this shell line up in a straight line along the beam (x-axis).

For higher temperatures ($\Gamma = 40$) the particles still form an outer shell, but the inner shells become blurred. Further calculations are planned to simulate somewhat more realistically the condition in a storage ring: the effective shear that is needed to represent the circular motion and the periodicity that is more likely to approximate the strong-focusing conditions that are used at most storage rings.

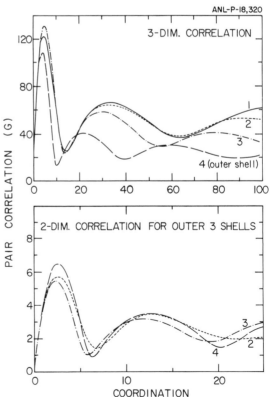

Figure 2:
Pair correlation functions g(r) starting with particles in particular shells, plotted as a function of coordination n(r) (total number of particles up to that radius). The upper curves show the three- dimensional pair correlation where the first peak has a coordination of 14 characteristic of bcc structure, except for the outer shell that clearly has no particles outside it. The lower curves are the two-dimensional pair correlation within each of the three outer shells, with the coordination of 6 for the first peak, characteristic of the triangular order in two dimensions.

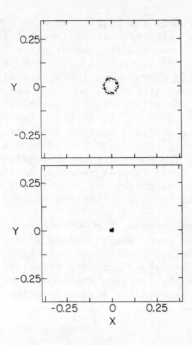

Figure 3: Same as the upper part of figure 1 but with only 100 particles (above) and 50 particles (below).

REFERENCES

1. E. L. Pollock and J. P. Hansen, Phys. Rev. A8, 3110 (1973).
2. J. P. Schiffer and P. Kienle, Z. Phys. A, 321, 181 (1985); J. P. Schiffer and O. Poulson, Europhys. Lett. 1, 55 (1986).

Work supported by the U. S. Department of Energy, Nuclear Physics Division, under Contract W-31-109-ENG-38.

> The submitted manuscript has been authored by a contractor of the U. S. Government under contract No. W-31-109-ENG-38. Accordingly, the U. S. Government retains a nonexclusive, royalty-free license to publish or reproduce the published form of this contribution, or allow others to do so, for U. S. Government purposes.

STORAGE RINGS FOR HEAVY-ION ATOMIC AND NUCLEAR PHYSICS

G. R. Young
Oak Ridge National Laboratory,* Oak Ridge, Tennessee 37831

ABSTRACT

A brief review is given of the physics goals and a few of the accelerator physics issues relevant to the new generation of small storage rings being built for atomic and nuclear physics use.

The last few years have seen a rapid growth in the number of projects to build storage rings designed for use in atomic and nuclear physics. All these projects involve some type of beam cooling. A table listing several of the projects and their status as of May 1986 is given below.

Table I

Type	Location	Bending Limit (T·m)	Status (May 1986)
"Granddaddy"	Indiana	3.6	Constr/Operating
"Small"	MPI, Heidelberg (TSR)	1.6	Constr
	Aarhus (ASTRID)	1.6	Constr
	Stockholm (CRYRING)	1.5	Constr
	ORNL (HISTRAP)	2.0	FY 1988 - DOE
	INS, Tokyo (TARN I)	2.0	Operating
"Medium"	GSI (ESR)	10	Constr
	Uppsala (CELSIUS)	7	Constr
	INS, Tokyo (TARN II)	7	Constr
	ORNL	10	Study
	LBL	9	Study
"Large"	BNL (RHIC)	835	FY 1988 - DOE

In 1985 there was one operating small- or medium-sized ring. In 1992 there will be nine to ten of these devices. In the following we attempt to give some reasons for the rapid growth in interest, physics to be pursued with these devices, and the accelerator physics issues and constraints which arise.

---

*Operated by Martin Marietta Energy Systems, Inc., under contract DE-AC05-84OR21400 with the U.S. Department of Energy.

The principal areas of physics of interest for the atomic physics rings (the "small" rings) can be broken into four broad areas. One area is basic studies of interest in atomic physics. This includes studies of "Lamb-shifts" in hydrogenlike ions of heavy atoms, which provide tests of perturbative and strong field QED, and of single and dielectronic recombination rates in multiply-charged ions, which provide sensitive tests of atomic structure calculations. A second area is the study of atomic exotica, such as very high-n Rydberg states, so-called "giant atoms," and multiphoton ionization. A third area is the study of electronic transitions in highly ionized heavy ions, which is of interest to the plasma physics program. This yields useful information helpful in identifying impurities in magnetically confined plasmas and for studying ways in which these impurities affect the plasma by absorbing energy from it. A fourth area is the (somewhat speculative) search for new devices. An interesting example is the proposal to optically pump a stored, highly charged ion beam. This could result in a "tuneable" (possibly lasing) light source in the ultraviolet or shorter wavelength region, depending on the bending limit of the storage ring.

Atomic physicists are led to the use of storage rings, not by center-of-mass energy considerations as are high-energy physicists, but because of the desire for improved beam currents, useful beam lifetime, and improved beam phase-space density. Heavy atoms are most easily put into high charge states by first accelerating them to energies of a few MeV/nucleon and then passing them through a thin ($10~\mu g/cm^2 \approx 1000$ Å) carbon foil to strip off outer shell electrons. The higher the energy, the more inner shell electrons are removed; a good estimate of the charge state obtained can be had from the Bohr velocity-matching criterion. Unfortunately, such a group of ions forms a beam which then passes only once through a target and/or apparatus. Storing the ions in a ring alleviates this problem and then allows the possibility of beam phase-space cooling. Cooling, especially electron cooling, can alleviate several of the problems of phase-space dilution resulting from accelerating and stripping very heavy nuclei. Once the atoms are stored in a ring, several advantages occur for the atomic physics program. Obviously, the magnetic elements can be adjusted to accommodate a wide variety of ions and/or different charge states for different experiments, and the magnets can be ramped to decelerate the ions to low energies (<1 MeV/nucleon) of interest in atomic collision physics. By storing the beam in a ring, the effective currents for, e.g., merged photon-ion or electron-ion or crossed electron-ion experiments can be increased by four to five orders of magnitude over those obtainable in fixed-target experiments. In addition, storing the ions in a ring means time can be allocated to allow the metastable states created by stripping to decay before beginning a measurement. Finally, if one is interested in using laser beams to probe a stored beam, and if the lifetime of the excited state of interest exceeds the beam revolution time in the ring, an exposure occurring over only part of a ring circumference appears to the ion to be effectively continuous. This permits pumping a stored beam into an excited state, for example.

A graph of peak kinetic energy per nucleon vs. atomic mass number for the Oak Ridge HISTRAP project, a 2.0-Tm ring, is shown in Fig. 1.

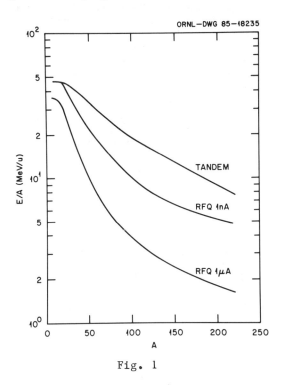

Fig. 1

Kinetic energy per nucleon is shown for three possible injectors: the existing ORNL 25-MV tandem, which could produce beams from $^{12}C^{6+}$ to $^{197}Au^{40+}$; an electron cyclotron resonance ion source (ECRIS) coupled to an RFQ running high charge states at low current ("RFQ 1 nA") from $^{12}C^{6+}$ to $^{197}Au^{33+}$; and an ECRIS-RFQ running lower charge states at high current, representative ions being $^{12}C^{5+}$ to $^{197}Au^{20+}$ in this case. The ordinate is computed from

$$E/A \text{ (MeV/nucleon)} = [(299.79 \cdot B_\rho \cdot Q/A)^2 + 931.5^2]^{1/2} - 931.5 ,$$

where B_ρ is in tesla meters, Q is the charge state, and A is the mass number in amu.

For the nuclear physics rings ("middle" in the table), the physics issues include a broad range of topics of interest in heavy-ion nuclear physics. (The "large" project at BNL has, of course, as its goal the discovery of a quark-gluon plasma, which is discussed by Tom Ludlam in these proceedings.) One area needing cooled beams is the study of inelastic scattering, giant resonance and delta resonance excitation. These studies can be pursued cleanly using high resolution heavy-ion beams with kinetic energies in the range of 300-1000 MeV/nucleon (energies that are high enough that single-step excitation works); these beams need energy resolutions in the range of $\Delta E/E = 10^{-4}$ to 10^{-5}, clearly requiring phase-space cooling. A second area needing considerable beam energy for heavy nuclei is the study of the equation of state for nuclear matter via the production of kaons at energies well below the free nucleon-nucleon threshold for strangeness production. Differing equations of state for nuclear matter give predictions for kaon yields that differ sharply, making the measurement a good discriminator among theories. A third area of study, which requires a sophisticated production target plus charge and mass separators, is the production, storage, and cooling of exotic nuclei far from the valley of β stability. This also requires an accelerator ring for primary beam production and a second, storage ring to act as the catcher, cooler and (possibly) decelerator. Being able to produce energetic heavy nuclei also leads to the possibility of studying "kinematically reversed" reactions, wherein a high-mass projectile bombards a low-mass target, resulting in a quite large dynamic solid angle for particle correlation experiments. Finally, a large enough accelerator ring allows producing the heaviest elements with enough energy that all or nearly all electrons can be removed by a post-stripping, meaning that the QED tests and optical pumping studies suggested above could be carried out for any element. Table II lists possible experiments, typical "heaviest beams" that might be employed, kinetic energies of interest, and finally the resulting B_ρ for the associated accelerator ring. Figure 2 summarizes the peak energy per nucleon performance of several proposed and existing machines; the top four curves are for synchrotrons, while the lower three curves are for cyclotrons and the "GSI" line is for the heavy-ion linac (UNILAC) there.

Table II

Topic	Beam	MeV/Nucleon	B_ρ (T·m)
"Subthreshold" kaon, hyperon production	$^{100}Mo^{32+}$	500-1000	11.3-17.6
Δ resonance excitation via transfer	$^{12}C^{6+}$, $^{40}Ca^{20+}$	1000-1500	11.3-15.0
Bremsstrahlung photons	$^{238}U^{43+}$	50-500	5.7-20.0
Elastic, inelastic scattering, GMR excitation	$^{40}Ca^{20+}$	300-1500	5.4-15.0
Exotic nuclei	$^{238}U^{43+}$ $^{40}Ca^{20+}$	500-800 500-800	20.0-26.9 7.3-9.7
"Atomic physics"	$^{238}U^{43+}$ $^{238}U^{75+}$	200-900 200-900	11.9-29.0 6.8-16.6

Table II continued

Topic	Beam	MeV/Nucleon	B_ρ (T·m)
Inverse kinematics experiments	$^{238}U^{43+}$	5-200	1.8-11.9
Pion interferometry	$^{20}Ca^{20+}$	700-1800	8.9-17.1
	$^{150}Nd^{50+}$	700-1800	13.4-25.7
Inverse (d,p) reactions for astrophysics	$^{238}U^{43+}$	20-200	3.6-11.9
$\gamma\gamma$ Hanbury-Brown/Twiss	$^{238}U^{43+}$	20-200	3.6-11.9
Subthreshold π^0 production	$^{238}U^{43+}$	20-200	3.6-11.9

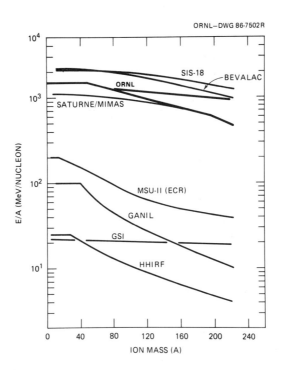

Fig. 2

Considerations on beam brightness also motivate the construction of the medium-sized facilities. The only available facility in this range at present is the Bevalac, which is a relatively slow-cycling weak-focussing machine. Figure 3 shows desired vs. existing beam brightness as a function of mass number, with the proposed GSI, LBL, and ORNL machines' projected performance lying near the solid line shown.

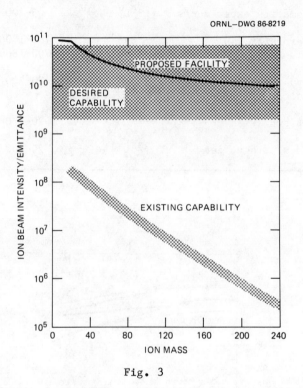

Fig. 3

A number of issues confront the lattice designer for these small storage rings. Since they are designed for hadrons at moderate energies ($\gamma < 3$), no synchrotron radiation damping occurs as for the small synchrotron light sources. Thus, the maximum beam emittances (before cooling) tend to be large, of the order of 40–100 π mm mrad for various projects. Injection into the small rings has to be by multiturn injection, charge-exchange injection being ruled out because of repeated charge state fractionation and multiple scattering in the stripping foils.

Many of the designs provide dispersion-free straights in which to locate electron-beam type coolers; whereas, others locate these in straight sections with dispersion from 1–3 m. There is an ongoing discussion as to which solution provides the best cooling rates. However, some of the atomic physicists wish to use the electron beams also as electron sources for electronic recombination studies. Since changing the charge state of an ion in a straight with nonzero dispersion would result in inducing a large amplitude betatron motion, via $X_{eq} \Delta q/q$, a lattice to be used for such studies must have a dispersion-free straight for the cooling section. Usually 3-5 meters of free space is needed for the cooling apparatus. Beta functions of usually 2-10 m are chosen, there being the usual tradeoff between cooling rate (favoring large β) and ultimate lower limit for the emittance (favoring small β). The nuclear physics projects are also complicated by the need to allow for stochastic cooling to be able to handle the "hot" secondary beams produced by fragmentation of the heavy-ion beams from the primary accelerator.

The desire to accommodate multiple charge states also means that the ring needs a stable acceptance extending over $\Delta(p/q)/(p/q)$ of ±2-6%. Orbit stability calculations and tracking must also include effects of magnet ends and the $1/\rho^2$ terms for these small rings. These same calculations are complicated, when sextupoles are introduced, due to the rather "lumpy" nature of the β functions and the desire to keep $\hat{X}_{eq}$ small, leading to relatively large normalized strengths. All these rings operate below γ_{tr}, meaning sextupoles are not needed for the negative mass instabilities. However, the chromaticities ξ_x/ξ_y tend to be much larger than the tunes Q_x/Q_y, exacerbating the sextupole strength problem. Most groups are looking at lattices which use as a basic building block a doublet or triplet achromatic bend. The triplet achromat proposed for the ORNL HISTRAP project is shown in Fig. 4a. The lattice functions (focussing structure OBFDFBO) are as shown in Fig. 4b.

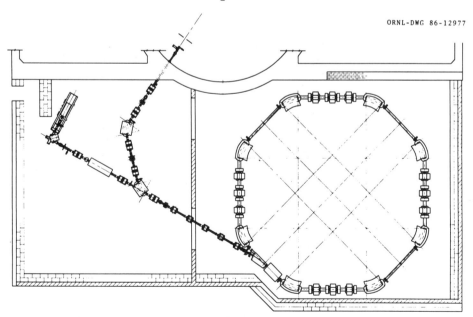

Fig. 4a

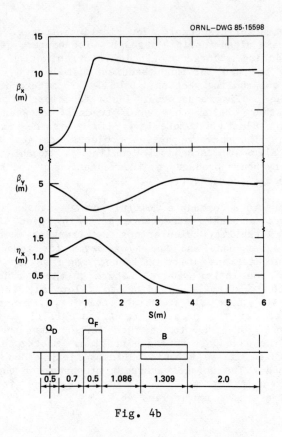

Fig. 4b

Many of the atomic physics proponents talk about quite low numbers of ions, from 10^6–10^{10}, in the ring for many experiments, whereas the nuclear physics projects emphasize the highest attainable currents, for example, needing 10^{11} ions/second hitting a production target to produce a useful number of radioactive secondary nuclei, 10^5–10^7/s, for subsequent storage, cooling, and measurement. The atomic physics projects usually also involve deceleration of the ions to energies as low as tens of keV/nucleon. At this point, the Keil-Schnell limit strongly determines the number of ions which can be handled, given its scaling as $(\Delta p/p)^2 \beta^2 \gamma^3 (A/Q^2)$. How effective cooling is at low energies will also be strongly affected by intrabeam scattering, which is severe for very highly charged ions due to the Q^4/A^2 dependence of the Rutherford cross section. Otherwise, the number of particles which can be injected is usually governed by the incoherent space-charge limit, $\beta^2\gamma^3 A/Q^2$, and/or the injector + ion source performance. The proposed injectors are a bewildering array of ECRIS + RFQ's, tandem electrostatic generators, heavy-ion linacs, and isotope separators injecting at energies from 100 keV to 20 MeV per nucleon, depending specifically upon the project considered.

Although the magnet and rf systems are straightforward for these projects (except for the 10:1 frequency swing often needed for the rf system), the vacuum requirements tend to be severe. To avoid charge-changing collisions with residual gas molecules requires pressures in the range of 10^{-10} to 10^{-12} torr (nitrogen equivalent) to attain storage times of even a few seconds in the worst cases (e.g., 1-MeV/nucleon $^{238}U^{43+}$ has a single electron capture cross section of $\sim 10^{-16}$ cm^2, or 100 megabarns, on atomic hydrogen). The vacuum systems proposed include high temperature bakes (up to 350°C), generous use of titanium sublimation and sputter ion pumping in the magnet chambers, and NEG pumping near serious gas loads such as the anode and collector for the electron beam. Pump spacings of 3-4 meters appear to be necessary, as well as extensive chamber wall preparation, in order to preserve the possibility of an ultimate pressure of 10^{-12} torr. Due to the high bakeout temperatures, thermally insulating jackets, such as the ceramic wool type used at LEAR and the $\bar{p}$ sources, are necessary to prevent damage to magnet yokes and coils. Thin copper cooling sheets will probably have to be inserted between the thermal jackets and the yokes to keep the yoke temperatures low enough during bakeout. To allow fast (~2 T/s) pulsing of the dipoles, these will have to be slotted in the magnet gaps to minimize eddy current distortions of the magnet field. The whole process is made somewhat easier for the dipoles, as they often must have C-shaped yokes to allow experimenters access to the ring vacuum.

The next five years promises to be a busy period in this area of accelerator construction. The potential experiments for such rings are still being invented by the atomic and nuclear physicists awaiting the completion of the rings. Let us hope many fruits await the labors of the hundreds of people involved.

TOWARD MULTI-GEV ELECTRON COOLING

D.J. Larson, D.B. Cline, D.R. Anderson
The Institute for Accelerator Physics, Physics Dept.,
University of Wisconsin, 1150 University Ave., Madison, WI, 53706

J.R. Adney, M.L. Sundquist
National Electrostatics Corporation, 7540 Graber Road, Middleton, WI 53562

F.E. Mills
Fermilab, P.O. Box 500, Batavia, Il, 60510

ABSTRACT

We discuss progress being made in the development of an ampere intensity MeV recirculating electron beam system. The system is presently intended for the upgrading of antiproton sources, but is also ideally suited for ion beam cooling in the GeV energy range. We present results of a theoretical study applying intermediate energy electron cooling to the Fermilab antiproton source, a brief overview of the design of the electron cooler, and discuss progress on the assembly and test of the system.

INTERMEDIATE ENERGY ELECTRON COOLING APPLIED TO THE FERMILAB ANTIPROTON SOURCE

One possible use of the electron beam system described herein is in the cooling of antiprotons. The Fermilab antiproton source (after stochastic cooling) has transverse emittances of 2π mm-mr and a longitudinal emittance of $\Delta p/p = 2 \times 10^{-4}$.[1] A detailed study[2] of the cooling time indicates that the Fermilab antiproton beam should cool to equilibrium in about twenty minutes. The study found that the effects of intrabeam scattering (IBS) and finite electron beam temperature were unimportant in determining the cooling time, while betatron oscillations of the antiproton beam enhanced the cooling by about 50%. The calculations assumed an electron current of four amps, and an electron velocity distribution much narrower than the antiproton velocity distribution.

At equilibrium the rate of electron cooling is equal to the rate at which the beam is heated. The heating may arise from several factors, such as intrabeam scattering, finite electron temperature, residual gas scattering, internal targetry, and wall impedance. For the case of the Fermilab accumulator a study[2] has determined that intrabeam scattering dominates the heating. (The wall effects are important as well – these are treated separately as beam instabilities.) The equations for the cooling rates are compared to the IBS heating rates as determined by Bjorken and Mtingwa.[3] The study indicates that the equilibrium transverse emittance should be 0.12π mm-mr and the equilibrium longitudinal emittance should be $\Delta p/p = 1 \times 10^{-5}$. Work must be done on the accumulator to reduce instability problems for these emittances to be realized.

DESIGN OF THE ELECTRON COOLER

Electron cooling of antiproton sources in the several GeV range will require electron beam energies in the range of three to ten MeV in order to match the antiproton velocity. High quality electron beams of this energy range can be produced by Pelletron accelerators manufactured by National Electrostatics Corporation (NEC) of Middleton, Wisconsin. To obtain currents in the order of amperes highly efficient recovery of the electron beam is required since the Pelletron charging capability is limited to about 500 microamps. The collection of these nonmagnetized electron beams requires a highly efficient collector just above terminal potential as well as electron optics designed to keep the beam within apertures throughout the system.

In the Fermilab electron cooling experiment[4] a collector was operated that reached a collection efficiency in excess of 99.99%. The collector designed for the intermediate energy electron cooling effort (figure 1) has been designed using the principles of the successful FNAL model. The collector uses a Pierce geometry for the initial electrodes allowing the beam to be slowed to an energy of about one KeV without space charge blowup. The beam next enters a solenoidal field region where the space charge of the electron beam is neutralized by the formation of an ion cloud from residual gas in the beam pipe. Finally, the beam is accelerated into the collector cone where the accelerating electric field serves to suppress secondary electron emission. The ions within the solenoid are trapped radially by the solenoidal field and longitudinally by the electrostatic field.

One of the major design considerations is the power dissipation in the collector. The collector supply is a 3 KV, 5 A supply. The 400 Hz generator supplying power to this part of the terminal produces 10 KVA, thus the collector will not be able to supply 3 KV and 5 A simultaneously. The 10 KW of heat that can be produced in the collector will require the presence of water cooling and a $H_2O - SF_6$ heat exchanger.

The planned testing of the electron beam system will use the design shown in figure 2. The beam will be accelerated, turned around by two dipoles, decelerated, and collected. The beamline consists largely of standard NEC components. In order from the electron gun at the upper right: nine high gradient accelerating tube sections; pumping tee with 120 l/sec ion pump; high power Faraday cup, magnetic quadrupole singlet lens; pumping tee with 220 l/sec ion pump; beam profile monitor; 90 degree double focusing dipole;

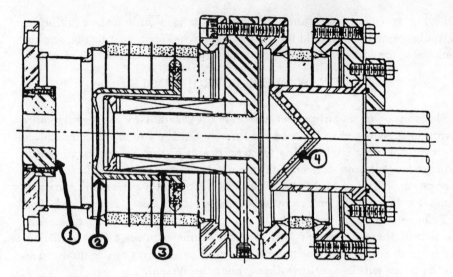

Figure 1 – Schematic of the electron beam collector: (1) entrance aperture, (2) focus aperture, (3) solenoid windings, (4) Collector cone.

and single slit assembly. All components then repeat in reverse order to the collector at the upper left to produce a symmetrical beamline.

The electron gun design is similar to that used in early FEL work done by Luis Elias.[5] However, in order to be able to slowly increase the current in the system, a cathode consisting of three concentric rings and a center spot has been fabricated. The optics calculation of the gun and initial section of the Pelletron was treated using the program EGUN developed by Herrmannsfeldt. The entrance to the Pelletron accelerating column presents an extremely strong focusing lens to the beam. Once the electric field of the Pelletron no longer has any significant radial component a separate treatment of the problem[2] shows that the beam should have no trouble passing the periodic one inch apertures of the Pelletron.

MeV ELECTRON COOLER – PROGRESS MADE TO DATE

After the gun was assembled it was put under vacuum and baked. Optical pyrometry was used to determine the cathode temperature as a function of the four filament heating currents. Each emitting portion of the cathode was heated to the same temperature, and the currents required for each filament were recorded. In this way we will be able to keep the entire cathode at the same temperature during operation. The calibration was done using a combination of one, two, three and four rings as the emitting surface. A logarithmic amplifier used to measure recirculated electron current over a wide dynamic range was installed in the collector power supply and calibrated.

Upon completion of calibration the gun was bolted directly to the collector in order to operate in air at potentials between 0 and 40 KeV. This bench test provides a test of the power supplies and control systems, allows for a measurement of the beam current losses occurring in the gun and collector, and tests the collector cooling system.

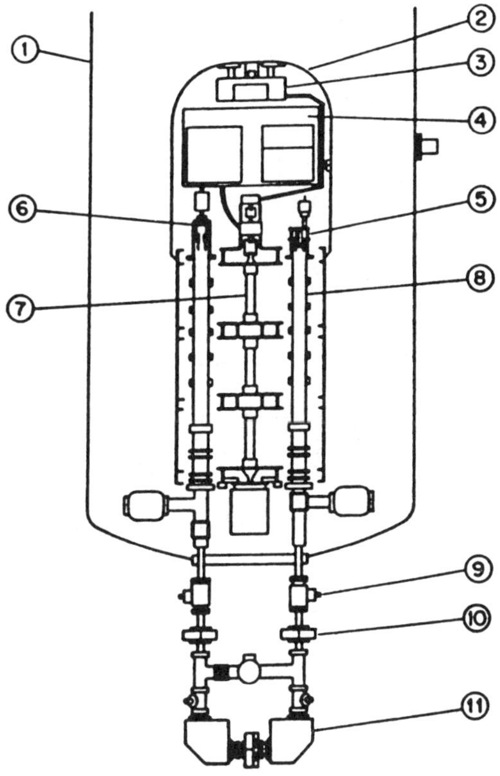

The initial runs of the bench test allowed the upgrading of several components that were easily spark damaged. EGUN simulations of the bench test configuration predicted that space charge blowup would exceed the collector electrode diameter for currents greater than 0.5 A. This behavior was experimentally confirmed – as currents approached 0.5 A. power supplies were overloaded and the beam transmission failed.

Figure 2 – Schematic of the recirculation test set up: (1) Tank, (2) Terminal, (3) Terminal Potential Electronics, (4) Cathode Potential Electronics, (5) Electron Gun, (6) Collector, (7) Rotating Shaft, (8) Accelerating Tube, (9) Faraday Cup, (10) Quadrupole Singlet, (11) Bending Dipole.

In order to test collector losses in a configuration like the Pelletron account must be made of the strong focusing of the tube entrance lens. This can be simulated by installing a solenoid lens between the gun and collector (Figure 3). Operation of the system with the solenoid will permit the study of space charge limited currents as well as investigations of beam halo.

CONCLUSION

With the bench test now complete we plan in the next few months to place all electronics in the 3MeV terminal and begin recirculation tests. Successful

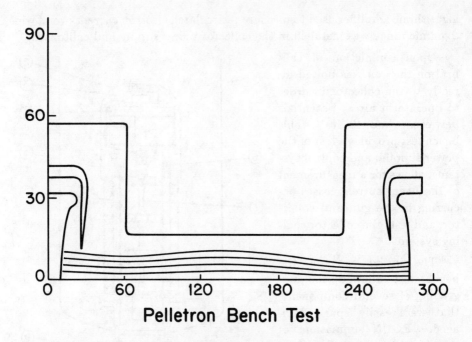

Figure 3 – EGUN simulation of bench test optics.

operation of the system will lead to an opportunity for GeV electron cooling of (anti)hadron beams.

REFERENCES

1. "Design Report - Tevatron I Project" Fermi National Accelerator Laboratory, September 1983.

2. D.J. Larson, "Intermediate Energy Electron Cooling for Antiproton Sources", PhD thesis, University of Wisconsin, 1986.

3. J.D. Bjorken and S.K. Mtingwa, "Intrabeam Scattering", Particle Accelerators Vol. 13 pp. 115-143, 1983.

4. T. Ellison, et al., "Electron Cooling and Accumulation of 200 MeV Protons at Fermilab", IEEE Trans. Nuc. Sci., Vol. NS-30, No. 4, August 1983, 2636-2638

5. Infrared free electron laser uses electrostatic accelerator, Physics Today, Nov. 1984, p. 21-23.

STATUS OF MAGNET SYSTEM FOR RHIC*

P.A.Thompson, J.Cottingham, P.Dahl, R.Fernow,
M.Garber, A.Ghosh, C.Goodzeit, A.Greene, H.Hahn, J.Herrera,
S.Kahn, E.Kelly, G.Morgan, S.Plate, A.Prodell, W.Sampson,
W.Schneider, R.Shutt, P.Wanderer, E.Willen
Brookhaven National Laboratory, Upton NY, 11973

Introduction

A Relativistic Heavy Ion Colliding beam accelerator[1] (RHIC) has been proposed at Brookhaven National Laboratory. The machine would generate colliding beams of energies up to 100 GeV/amu of ions as heavy as ^{197}Au. The facilities neccessary to accelerate these ions up to 11 GeV/amu are either already operational or under construction at BNL. This paper will discuss the magnet system for the actual collider ring itself, which will further accelerate the particles to beam energies of between 7 and 100 GeV/amu, store them, and provide interaction regions. This magnet system will consist of two rings of superconducting magnets placed in an existing 3.8 km tunnel.

Because of the much larger charge, the emittance of heavy ion beams is greatly increased by intra-beam scattering[2]. This requires a larger aperture than an equivalent proton machine. Much of the interesting physics requires the collision of ions of differing masses, with protons striking gold being the extreme case. To accomplish this, the two rings must run at magnetic fields differing by a factor of 2.5. This is a comparatively small machine and must maximize the use of available technology to minimize R&D.

RHIC MAGNET INVENTORY

Magnet Type	Length (m)	Bore (mm)	Field (T)	Number	Status
Regular Arc					
Dipoles	9.5	80	3.5	288	4 full size prototypes under construction 5 successful tests
Quadrupoles	1.2	80	67/m	276	Engineering Design
Sextupoles	0.75	80	1300/mxm	276	Engineering Design
Correctors	0.5	80	..	276	Magnetic Design
Insertions					
Dipoles	3.5/5.5	80	3.5	24	Same as Regular Arc
	4.4	100	4.4	24	Magnetic Design
	3.3	200	3.3	12	Conceptual Design
Quads	1.1/1.7	80	67/m	144	Same as Regular Arc
	1.1/2.2	130	57/m	72	Conceptual Design
Sextupoles	0.75	80	1300/mxm	12	Conceptual Design
Correctors	0.5	80	..	144	Conceptual Design
	0.5	130	..	96	Conceptual Design

---

* Work supported by the U.S. Department of Energy

Magnet Design

These constraints produce a design which is significantly different from the superconducting proton machines built or under construction: large bore -- 80 mm, modest dipole field -- 3.5 Tesla, short, strong focussing cell structure -- 1 dipole/half cell, independence of the two rings -- both magnetic and cryogenic. The lattice consists of 6 arcs of 12 cells each plus six interaction regions.

The design of the quadrupole, which uses the same concepts as the dipole is shown in Fig. 1. The conductor is a 30 strand Rutherford cable (Cu-63%, NbTi-37%) which is partially keystoned. The molded coil fits into the precision molded insulator which is forced against alignment steps in the iron lamination. The iron laminations are compressed and welded to establish the 10 kpsi prestress. The helium is contained with a stainless steel jacket welded around the yoke.

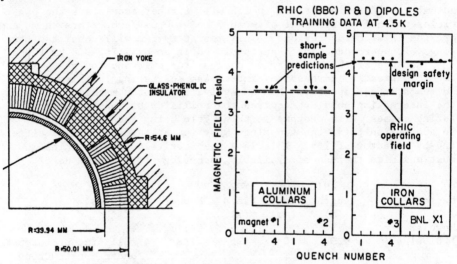

Fig. 1 Quadrupole Cross Section   Fig. 2 Quench History Experimental Magnets.

Prototypes

Three classes of prototypes are complete or under construction. The first class consists of five experimental magnets to test the basic properties of the one layer iron clamped design. The second consists of 4.5 meter long field quality magnets (only one built to date). The last consists of full length (9.5 meter) pre-production prototypes. (Four of these are under construction). The experimental magnets were intended to investigate construction techniques and quench performance. These were built with coils wound and molded by FNAL. BNL constructed one magnet using these coils and a modified CBA yoke. No difficulties were experienced in the construction and the test results are shown in Fig. 2. In addition to the complete absence of training apparent in this figure, quench velocity measurements indicated that for a full scale magnet the maximum temperature would be 700 K, comfortably below the measured 1050 K damage threshold. The test results for three additional magnets assembled by Brown Boveri et Cie are also shown. These magnets demonstrated that the technology could be transfered successfully to private industry. The variations of collar materials verified that the iron collared magnets performed exactly as predicted. The performance shown in this figure is unprecedented for a new magnet design.

The field quality magnets are being constructed to verify the calculations and to check the details of the final design geometry. The first of these was constructed of the same cable as intended for the final design. This cable was fully keystoned. This resulted in excessive compression of the inner edge of the cable which is believed to be the cause of the quench performance shown in Fig. 3. Because of this training, the coil was redesigned as shown in Fig. 1 to use partially keystoned cable. The field measurements are shown in Fig. 4; the agreement with calculations is within the uncertainties.

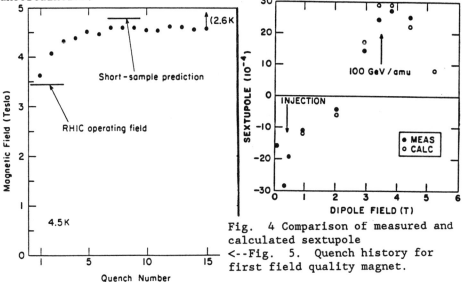

Fig. 4 Comparison of measured and calculated sextupole

<--Fig. 5. Quench history for first field quality magnet.

Presently full length magnets with the revised cable are under construction (one at BNL and three at BBC). These are intended to be full scale production models and will be connected together for a string test, as well as individual quench and field quality testing.

References

1. "RHIC and Quark Matter: Proposal for a Relativistic Heavy Ion Collider at Brookhaven National Laboratory", BNL 51932, UC-28,TIC-4500(1985)
   P.A. Thompson, et al.,"Superconducting Magnet System for RHIC",IEEE Trans. Nucl. Sci. NS-32, No. 5,3698
2. G. Parzen, "Strong Intra-beam Scattering in Proton and Heavy Ion Beams",IEEE NS-32,No. 5,2326,(1985)

DOES THE TRANSITION TO CHAOS DETERMINE THE DYNAMIC APERTURE?*

JOHN M. JOWETT[†]
Stanford Linear Accelerator Center
Stanford University, Stanford, California 94305

ABSTRACT

We review the important notion of the dynamic aperture of a storage ring with emphasis on its relation to general ideas of dynamical instability, notably the transition to chaos. Practical approaches to the problem are compared. We suggest a somewhat novel quantitative guide to the old problem of choosing machine tunes based on a heuristic blend of KAM theory and resonance selection rules.

INTRODUCTION

Nowadays, machine designers are much exercised by the *dynamic apertures* of their storage rings. Roughly speaking, the dynamic aperture, or non-linear acceptance, is the region around the central closed orbit in which single particle motion is stable. The term itself only recently came into use but neatly encapsulates the essence of what it is meant to describe.

Formerly, the aperture *tout court* of a synchrotron or storage ring was what is now variously described as the physical or mechanical acceptance or aperture; that is, the aperture determined by the vacuum chamber or other material obstruction presented to the beam.

To explain the distinction, we recall that, in terms of action-angle variables of linearised motion (R.D. Ruth in Ref. 1), $\mathbf{J} = (J_x, J_y, J_s)$, $\boldsymbol{\phi} = (\phi_x, \phi_y, \phi_s)$, the radial displacement of a particle at some azimuth $\theta = s/R$ may be written

$$x(\theta) = \sqrt{2J_x \beta_x(\theta)} \cos(\phi_x + \psi_x(\theta)) + \eta_x(\theta) \underbrace{\sqrt{2J_s} \cos(\phi_s + \psi_s(\theta))}_{\delta}. \qquad (1)$$

The vertical displacement y is similar and the Hamiltonian has the form

$$H(\boldsymbol{\phi}, \mathbf{J}) = \boldsymbol{\nu} \cdot \mathbf{J} + \{\text{nonlinear terms in } \boldsymbol{\phi} \text{ and } \mathbf{J}\}. \qquad (2)$$

The actions, $\mathbf{J}$, are exact invariants only insofar as the oscillations are linear, although perturbation methods may be used to find approximate invariants in certain other cases. The average of $J_{x,y}$ over all the particles in the beam is the corresponding emittance, $\varepsilon_{x,y}$. Part of the displacement is attributed to the instantaneous momentum deviation from a central value, $\delta = (p - p_0)/p_0$,

---

* Work supported by the Department of Energy, contract DE-AC03-76SF00515.
† Permanent address: CERN, CH-1211 Geneva 23.

through the dispersion function, η_x. The amplitude of the betatron part of the oscillation is given by the β-function. If the beams are bunched, δ undergoes synchrotron oscillations and is also a dynamical variable, as indicated in (1). Often, however, synchrotron oscillations are so slow that they can be regarded as a parametric modulation of betatron motion. The β-, η- and ψ-functions are determined by the focusing structure, independently of any particle.

Generally, the phases ϕ may take all values independently with equal probability so that the mean square beam size is

$$\langle x(\theta)^2 \rangle = \beta_x(\theta)\langle J_x \rangle + \eta_x(\theta)^2 \langle J_s \rangle = \beta_x(\theta)\epsilon_x + \eta_x(\theta)^2 \sigma_\delta^2. \qquad (3)$$

Similarly for $\langle y^2 \rangle$. To avoid loss of particles in the tails of the distribution, these dimensions have to be significantly less than the physical aperture. How much less is different for e^+e^- storage rings, where the emittance is determined by radiation effects, and hadron colliders, where it depends essentially on the injection system. The beam size must also stand in a similar relationship to the dynamic aperture and, particularly in the new generation of large colliders and synchrotron light sources, this can turn out to be the more stringent requirement. The dynamic aperture is a 6-dimensional subset of $(\phi, \mathbf{J})$ space.

The number of betatron or synchrotron oscillations per turn (as $\mathbf{J} \to 0$) is called the *tune* of the machine and is also determined by the lattice, although small adjustments are easily made. The tune vector $\boldsymbol{\nu} = (\nu_x, \nu_y, \nu_s)$ plays a fundamental rôle in determining single-particle stability. Instability is associated with resonance conditions $\mathbf{k} \cdot \boldsymbol{\nu} = p$ where $\mathbf{k} = (k_1, k_2, k_3) \in \mathbf{Z}^3$ is an integer vector and $p \in \mathbf{Z}$ is an integer related to the harmonic of the revolution frequency at which nonlinear terms drive the resonance. A working point diagram like Figure 1(a) is a useful aid in the avoidance of resonances.

Some resonances do not lead directly to instability but instead cause beating of amplitudes. Nonetheless it is important to avoid these too since the beating may lead to particle loss at the physical aperture. Thus, the two concepts of aperture are not completely distinct.

If the machine were perfect, the betatron oscillations could be made linear (with pure quadrupole focusing). The dynamic aperture would be infinite in both betatron degrees of freedom but the momentum acceptance would be almost zero. Since $\sigma_\delta \neq 0$, the natural chromaticity (*i.e.* betatron tune dependence on momentum) has to be compensated by the introduction of sextupoles. Careful arrangements of sextupole families are used in large machines to cancel the linear and quadratic parts of the chromaticity. A few other harmful effects, *e.g.* systematic excitation of low order resonances, can be removed at the same time. However the interactions of one sextupole with others (or itself on later turns) generate driving terms for resonances of arbitrarily high order and there remain the field errors (especially in superconducting magnets).

The oscillations are then necessarily non-linear and the betatron tunes, where they can be defined at all, are functions $\boldsymbol{\nu}(\mathbf{J})$ with many singularities and folds related to resonances and chaotic layers. It is impossible to avoid this, since the resonance planes form a dense web in $\boldsymbol{\nu}$-space and, generically, there is no way of eliminating the tune-dependence on amplitude. It is now a commonplace that the interactions of many resonances will inevitably lead to chaotic motion in some regions of the phase space.

THE PRACTICAL APPROACH: TRACKING

In practice, most estimates of dynamic aperture are made by computer tracking of particles through models of the lattice. Simulated random errors in magnetic fields or magnet positions may or may not be included.

Limited computer time forces a number of compromises on us: fast algorithms (*e.g.* thin lens approximation) have to be preferred to more accurate but slower methods; only a 2- or 3-dimensional section of initial condition phase space can be sampled; the number of turns of the machine must be limited, typically to a few hundred; the sampling of possible errors has to be extremely limited; although tracking results are very strongly affected by the tune values, only a few of these can be tested, *etc.*

In the face of such overwhelming difficulties, the practical stability criterion in fairly general use might be baldly paraphrased as: *If, for a given lattice design, with a few typical sets of random errors, and a particular set of tune values, a particle with initial betatron emittances ε_x, ε_y, an initial momentum deviation δ and initial phase zero in all three degrees of freedom survives for (say) 400 turns, then particles with the same amplitude and arbitrary phases will be stable for the same lattice with most other set of errors and most reasonable tunes.* A "reasonable" tune satisfies a number of well-known criteria concerning the avoidance of the resonances one can expect to be driven in a given lattice. For hadron colliders particularly, one may add the further caveat that the tracked orbit must remain essentially indistinguishable from a linear motion.

The dynamic aperture is then taken to be the set of all initial conditions generated in this way from the largest connected set of stable conditions (usually this includes the origin) in the sampling section.

Clearly this stability criterion is far removed from anything which can be found in the mathematical theory of dynamical systems. From a strictly logical point of view, it also sounds like wild optimism since we know that even millions of turns of apparently regular, stable motion may be just the initial segment of a chaotic orbit which will eventually find its way to large amplitude—such orbits can sometimes be found more quickly by finely scanning the sampling section.

Yet this approach seems to work quite well!

It is difficult to believe that the survival of the *last* invariant torus is an adequate criterion for stability of a particle beam. With or without the help of external perturbations, particles may easily jump or bypass such a flimsy barrier. One cannot avoid including some narrow chaotic regions in the dynamic aperture. On the other hand, regular orbits, particularly those associated with resonant beating, may have to be rejected.

It is not easy to distinguish regular from chaotic motion in systems with three degrees of freedom. Two-dimensional phase projections usually show only a cloud of points, but much more information can be obtained from watching a "movie" of the motion. Otherwise, computational means of making the distinction include Lyapounov exponents, reversal tests (A. Wrulich in Ref. 1) and

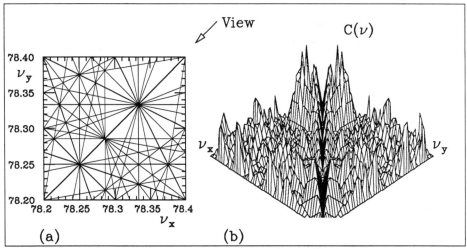

Fig. 1 Tune diagram and $C(\nu)$ for $K = \{\mathbf{k} \in \mathbf{Z}^3 : \|\mathbf{k}\| \leq 7, k_3 = 0\}$

fractal dimensions of orbit power spectra (J.M. Jowett in Ref. 2).

A GUIDE TO CHOOSING ν

KAM theory suggests the definition of a function

$$C(\nu) = \min_{\substack{\mathbf{k} \in K \\ p \in P}} \frac{|\mathbf{k} \cdot \nu - p|}{\|\mathbf{k}\|^4}, \qquad \|\mathbf{k}\| \stackrel{\text{def}}{=} |k_1| + |k_2| + |k_3| \qquad (4)$$

which, in a well-defined sense, measures the distance of ν from the resonance most likely to influence it *a priori*; large values of $C(\nu)$ indicate a better chance of stability. The sets of integer vectors $K \subset \mathbf{Z}^3$ and harmonics $P \subset \mathbf{Z}$ are chosen with the help of selection rules reflecting the best available judgment of which resonances are important, *e.g.*, if the lattice has superperiodicity N, then p must be a multiple of N for the systematic resonances. Placing an upper bound on the order $\|\mathbf{k}\|$ limits $\mathbf{k}$ to an octahedron. Figure 1(b) shows $C(\nu)$ in a small portion of the ν_x-ν_y plane. In one dimension, maximising $C(\nu)$ leads, in principle, to a ν value related to the "golden mean" but its 3-dimensional analogue is not yet fully understood. If detailed information on resonance widths were available, it could be incorporated into (4). The width of the peaks of $C(\nu)$ is also important. In other words, to accommodate the distribution of amplitudes, and therefore tunes, we should look for regions, rather than points, in ν space where $C(\nu)$ is large.

REFERENCES

1. J.M. Jowett, M. Month and S. Turner (eds.), *Nonlinear Dynamics Aspects of Particle Accelerators*, Springer-Verlag, Berlin, 1986.
2. W. Busse and R. Zelazny (eds.), *Computing in Accelerator Design and Operation*, Springer-Verlag, Berlin, 1984.

AN EXPLORATORY GAS-TARGET EXPERIMENT AT PEP USING THE TPC/2Y FACILITY

F. S. Dietrich, S. O. Melnikoff, and K. A. Van Bibber
Lawrence Livermore National Laboratory, Livermore, CA 94550

ABSTRACT

We have performed a pilot nuclear-physics experiment at the PEP ring by bleeding deuterium, argon, and xenon gases into the interaction region containing the TPC/2Y facility. The purpose of the experiment was to obtain information for the design of a dedicated nuclear-physics interaction region at PEP, as well as to obtain a first view of the physics that could be studied at such a facility.

During the past few years major interest has arisen in the observation of exclusive final states in inelastic electron scattering. Coincidence measurements at low duty-factor accelerators below 1 GeV (MIT/Bates, Saclay, NIKHEF) have yielded important information on single-nucleon momentum distributions in nuclei. The CEBAF accelerator will feature a program of coincidence measurements with electrons up to at least 4 GeV. Exclusive final-state measurements at even higher energies (up to 14.5 GeV) are possible with internal targets at the PEP electron-positron storage ring at the Stanford Linear Accelerator Center. This energy is sufficient to reach the range in which Bjorken scaling is well established in inclusive measurements.

The present exploratory experiment was undertaken as part of an effort to design an internal-target facility in one of the PEP interaction regions. It was performed in the interaction region that houses two major detector systems, the Time Projection Chamber (TPC) and 2Y facilities. The latter is an assembly of detectors very close to the beam axis on either side of the TPC that includes NaI and Pb-scintillator shower counters suitable for tagging electrons scattered at angles of roughly $1°$ to $10°$. The experiment was motivated by the observation that even at the low pressures in the PEP ring (in the range 10^{-9} Torr in the interaction regions), approximately 20% of the TPC triggers are due to collisions between the beams and the background gas.

Approximately 50000 events were recorded for each of the three gases in a running period totaling 24 hours. This sample contains a few hundred events in the deep-inelastic kinematic range $Q^2 \sim 1-5$ $(GeV/c)^2$ and $x \sim 0.1-0.5$. By analyzing these events we hope to obtain a first glimpse of the topology of the products of deep-inelastic scattering; i.e., whether there is evidence for jet-like structure in this kinematic regime, and how the exclusive processes that sum up to the inclusive cross section may be characterized. Comparison of events from the heavy gases with those from deuterium may yield information on the hadronization process in the nuclear medium; we note that a softening of the energy spectrum of the hadrons produced in C and Cu targets relative to that in hydrogen has been inferred from earlier deep-inelastic experiments at SLAC[1]. Other reactions of interest for the present experiment as well as for future planning include the knockout reactions (e,e'p), (e,e'π), (e,e'K), and the ejection of correlated pairs (e.g., (e,e'2p)) that may yield evidence on multibody currents in nuclei.

From the technical point of view, the present experiment was performed to ensure that usage of gas targets is compatible with the PEP ultrahigh vacuum system, and that the circulating beams are not affected by the target in an unexpected manner. As an aid in understanding constraints on future experiments in other interaction regions, singles count rates in the various components of the TPC/2Y facility with and without gas-target injection were recorded.

Target gases were introduced into the PEP vacuum chamber at a point 9 m north of the e^+-e^- intersection via a controlled leak. The gas composition in the PEP ring and the degree of cleanliness of the gas-handling system were monitored with residual gas analyzers. Vacuum gauges on the ring showed that the pressure rise was localized to the neighborhood of the TPC/2Y interaction region. At the pressures utilized during the runs (mid 10^{-8} Torr range) no effects were noticed on either the beam lifetime or on the performance of detectors at other interaction regions. The luminosity, estimated at 10^{28} $cm^{-2}sec^{-1}$, was sufficient to yield event rates in the neighborhood of 5 sec^{-1} and significant dead time with the loose trigger employed in this experiment. By contrast, the beam-lifetime limitation on the luminosity is expected to be near 10^{33} $cm^{-2}sec^{-1}$.

Fig. 1 shows the longitudinal distribution of the vertices of the events observed in the TPC with the argon target. The spike near z=0, superposed on the broad distribution of beam-gas events,

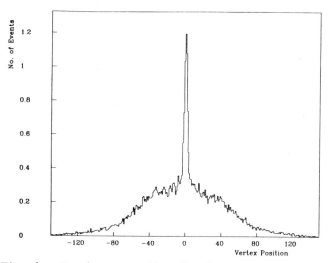

Fig. 1. Track vertex distribution, argon target.

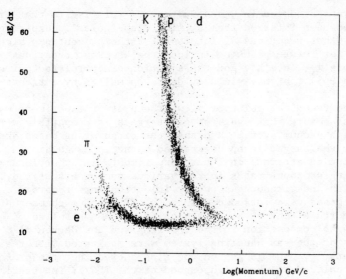

Fig. 2. Scatter plot of dE/dx vs. ln(p) for TPC tracks in events with argon target.

is due to e^+-e^- collisions. The TPC extends from -100 to +100 cm; the shape of the vertex distribution is a complicated function of the geometry and trigger efficiency. Fig. 2 is a scatter plot of energy loss vs. momentum for the events in the argon run. This figure illustrates the excellent particle-identification capability of the TPC. Bands corresponding to electrons, pions, protons, and deuterons are clearly visible; there is also evidence for a number of events producing kaons.

Figs. 3 and 4 show two typical events recorded with the deuterium target. Fig. 3 is a deep-inelastic event, in which an electron entering the interaction region from the north side is tagged in the south shower counter. The energy and angle of the scattered electron identify the kinematics of this event as Q^2=3.7, x=0.44. Energy-loss measurements in the TPC identify the particle exiting at the upper-right corner of the central detectors as a proton, and the remaining four charged particles as pions. Events of this type will be examined for jet-like structure by studying the momentum distribution transverse to the momentum-transfer axis. Quite a different type of event is shown in Fig. 4, in which a scattered positron is tagged in the north shower counter, and a proton and pion produced at back angles are detected in the TPC. The invariant mass of the proton and pion is close to the delta mass. The event represents the probable backward production of a delta with very high momentum (1.8 GeV/c).

This work was performed in part under the auspices of the U. S. Department of Energy by the Lawrence Livermore National Laboratory under contract number W-7405-ENG-48. We also gratefully acknowledge the participation and advice of members of the TPC and 2Y collaborations.

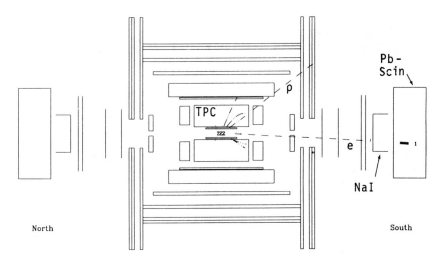

Fig. 3. Deep inelastic event. Electron from north, tagged in south shower counter. Unlabeled tracks are pions.

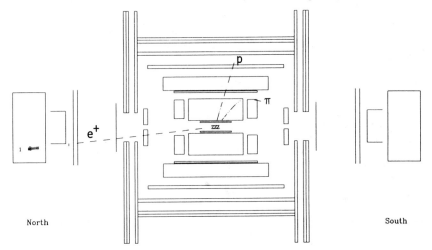

Fig. 4. Backward delta production. Positron from south, tagged in north shower counter. Proton and pion observed in TPC.

REFERENCE

1. H. E. Montgomery, in Intersections Between Particle and Nuclear Physics, AIP Conf. Proc. No. 123, p. 444 (1984).

STORAGE RING MAGNET FOR A PROPOSED NEW PRECISION
MEASUREMENT OF THE MUON ANOMALOUS MAGNETIC MOMENT

Vernon W. Hughes
J.W. Gibbs Laboratory, Physics Dept.
Yale University, New Haven, CT 06520

(with G. Danby, J. Jackson, E. Kelly, A. Prodell, R. Shutt, W. Stokes, Brookhaven National Laboratory; S.K. Dhawan, A. Disco, F.J.M. Farley, Y. Kuang, H. Orth, G. Vogel, Yale University; W. Williams, University of Michigan; F. Krienen, Boston University; M. Lubell, City College of New York; P. Marston, J. Tarrh, Massachusetts Institute of Technology.)

ABSTRACT

A brief discussion is given of a proposed new precision measurement at the BNL AGS of muon g-2 at the level of 0.35 ppm which would represent an improvement by a factor of 20 over the famous CERN experiment. As at CERN a muon storage ring will be employed. This paper gives the parameters for the experiment and emphasizes features of the storage ring magnet and of the NMR system required to achieve the requisite precision of 0.1 ppm in the mean magnetic field.

I. INTRODUCTION

The physics motivation for a more precise measurement of the muon anomalous magnetic moment a_μ or equivalently of the muon g-2 value has been reviewed recently.[1] With increased precision in a_μ the weak interaction contribution to a_μ of 1.7 ppm could be determined. This contribution analogous to the Lamb shift in QED arises from single loop Feynman diagrams involving the W±, Z and Higgs particles in the unified electroweak theory and results from the as yet untested renormalization prescription of the theory. More generally muon g-2 serves as a sensitive calibration standard for particle physics against which the existence of new phenomena and new particles can be tested.

The present experimental value for a_μ has a quoted accuracy of 7.3 ppm and was determined in a well-known experiment at CERN[2] which involved the storage of polarized muons with momentum $p_\mu \approx 3$ GeV/c in a storage ring having a homogeneous magnetic field with $B \approx 1.5$ T and a quadrupole electrostatic field for focussing. The g-2 precession frequency $\omega_a = \frac{eB}{mc} a$ was measured from the modulation in time of the number of high energy decay electrons, and hence with a knowledge of B a_μ was determined. The principal error of 7.0 ppm in the CERN experiment was the statistical counting error and the main systematic error of 1.5 ppm was due to inaccuracy in knowledge of B.

II. NEW AGS PROPOSAL

Our group has recently proposed[3] a new precision measurement of a_μ to 0.35 ppm - a factor of 20 improvement over the CERN value - at the Brookhaven AGS. A number of different approaches were considered including the following:
(1) Use of an intense beam of surface muons (e.g. at LAMPF) ($p_\mu \simeq$ 30 MeV/c) in a magnetic bottle in an experiment similar in principle to the Michigan electron g-2 experiment.[4]
(2) Use of muons with high momentum ($p_\mu \sim$ 30 GeV/c) in a strong focussing alternating gradient machine (e.g. ISR) having momentum compaction of 1 and with magnetic field measurement with a deuteron beam.
(3) Use of a storage ring with homogeneous B and quadrupole E for muons with the magic $\gamma(\gamma = 29.3)$ for which ω_a is determined only by B and not by the focussing E field.
   (i) B = 1.5 T, π injection (CERN case)
   (ii) B = 1.5 T, μ injection
   (iii) B = 5 T (π or μ injection)

The decision of the group was that approach (3i) - the CERN case - was the most certain, inexpensive and rapid one, and our proposal is based on this approach.

The modern possibility of improving on the CERN accuracy rests on two principal factors. A higher primary proton beam intensity will be available at the AGS with its new Booster Ring by a factor of about 100 as compared to the proton beam from the PS used in the CERN experiment and hence the statistical counting error can be reduced greatly. Secondly, a new and improved storage ring magnet with B = 1.47 T will be based on a superferric approach, i.e. an iron magnet with a superconducting coil, and on several shimming techniques, together with much improved NMR measurement. It is designed to achieve a field homogeneity over the storage volume of about 1 ppm and a field stability and measurement accuracy of about 0.1 ppm. These characteristics would improve on the CERN storage ring magnet[5] by a factor of 10. The general scheme of the proposed AGS experiment is shown in Fig. 1. Its parameters and projected errors are shown in Tables I and II.

III. STORAGE RING MAGNET AND NMR

Design of the storage ring magnet can be separated into two parts: (1) Basic iron and superconducting coil structure, (2) Shimming. A preliminary design of the basic structure has been completed and is indicated in Figs. 2 and 3 where an overall view of the storage ring magnet and the pole face configuration are shown. The magnet is C-shaped as dictated by the requirement of the experiment that decay electrons be observed inside the ring, and it will have azimuthal continuity. The overall weight of the steel is about 2×10^6 lbs. The cross section of the return yoke, which is to be made of 1010 or 1020 steel, is large enough to provide good field

TABLE I. PRINCIPAL PARAMETERS OF AGS EXPERIMENT

MAGNET
- Orbit radius — 7.0 m
- Central magnetic field — 1.47 T

STORAGE SYSTEM
- Storage region, circle with diameter — 9 cm
- Vertical focusing by electric quadrupole field (pulsed 1 ms) — 40 kV
- Field index (average), n — 0.12

PARTICLE INJECTION
- Pulsed magnetic inflector
- π-μ decay

ELECTRON DETECTORS
- (Number, each segmented into 4 units) — 20

KINEMATICS
- Muon momentum — 3.094 GeV/c
- Gamma — 29.3
- Lifetime — 64.4 μsec
- Revolution frequency — 6.81 MHz
- Time — 147 nsec
- (g-2) frequency — 0.2327 MHz
- Period — 4.3 μsec

AGS WITH BOOSTER — 5×10^{13} protons/pulse

OVERALL PRECISION — 0.35 ppm

TABLE II. ERRORS IN AGS EXPERIMENT

COUNTING RATES AND STATISTICAL ERRORS

Storage aperture diameter	90 mm
Protons per RF bunch with booster	4.2×10^{12}
Bunches ejected per ring fill	2
Protons per fill	8.4×10^{12}
Pion $\Delta p/p$	$\pm 0.6\%$
Pions at inflector exit per fill (1.28×10^7 per 10^{12} protons)	1.07×10^8
Muons stored per fill (at 134 ppm capture efficiency)	14×10^3
Electrons counted above 1.6 GeV per fill (20% of the decays)	2.8×10^3
Fills per AGS cycle (1.4 s)	3
Fraction of AGS protons used	50%
Fills per hour	7714
Electron counts per hour	22×10^6
Running time for 0.3 ppm (1 std. dev.)	1288 hours

SYSTEMATIC ERRORS

Source	Comments	Error (ppm)
Magnetic field	Includes absolute calibration of NMR probes and averaging over space, time, and muon distribution.	0.07
Electric field correction	0.7 ppm correction	0.03
Pitch correction	0.4 ppm correction	0.02
Particle losses		0.05
Timing errors		0.01
	TOTAL	0.09

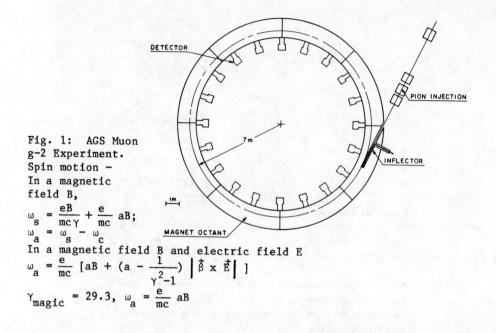

Fig. 1: AGS Muon g-2 Experiment.
Spin motion —
In a magnetic field B,
$\omega_s = \dfrac{eB}{mc\gamma} + \dfrac{e}{mc} aB$;
$\omega_a = \omega_s - \omega_c$
In a magnetic field B and electric field E
$\omega_a = \dfrac{e}{mc} [aB + (a - \dfrac{1}{\gamma^2 - 1}) |\vec{\beta} \times \vec{E}|]$
$\gamma_{magic} = 29.3$, $\omega_a = \dfrac{e}{mc} aB$

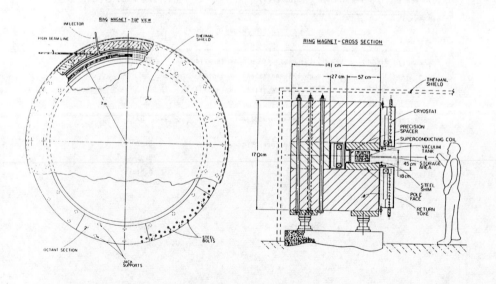

Fig. 2. Storage Ring Magnet Plan Drawings

uniformity and to reduce the effect of variations in the B/H curve due to non-uniformity in the steel. Machining tolerances on the yoke need only be about 1/8" except for some mating surfaces where flatness to about 10 mils is needed. The critical pole pieces will be made of low carbon 1006 forged steel and the pole pieces will be machined to 1 mil tolerances with pole face flatness to 0.1 mil. Nonmagnetic spacers precision ground to 0.1 mil tolerance will be used to maintain the gap spacing. The air gaps of 1 cm between the pole pieces and the return yoke serve to decouple the field in the storage region from that in the return yoke and will be used to insert iron shims as needed. With the pole face configuration shown calculations with a Poisson code assuming steel with uniform characteristics and the exact dimensions indicated predict that the field uniformity over the storage volume (a circular cross section 9 cm in diameter centered in the gap) would be better than 10 ppm. (This design exercise neglected the quadrupole component caused by the left-right asymmetry of the flux return, but this should not present any problem in the actual design.) Without extensive shimming we expect to achieve field uniformity to several parts in 10^5 over the storage volume from a storage ring constructed to our preliminary design. On the basis of our drawings and specifications quotes are being obtained from manufacturers.

For shimming a variety of techniques will be used including iron shimming on the return yoke, iron shimming in air gaps provided between the pole faces and the return yoke, circumferential current loops near the pole faces, and small current loops or iron shims on the pole faces. Calculational techniques have been developed using a two-dimensional Poisson code together with harmonic expansion of the solution so that the differential effects of shims can be calculated to the requisite accuracy in magnetic field of about 1 ppm.

Four ring-shaped superconducting coils, each in its own helium vessel, provide the excitation for the magnet. Each coil consists of 30 turns of insulated copper-stabilized NbTi conductor. Existing CBA cable soldered in a copper substrate with an overall copper-to-superconductor ratio of 20 to 1 will be used. The operating current will be 4000 A. A helium refrigerator with a capacity of about 100 W at 4.5 K is needed. Field stabilization will be accomplished with feedback to the magnet power supply controlled by an NMR probe.

Measurement of the magnetic field to an accuracy of 0.1 ppm can be made by NMR[6]. We plan to measure the magnetic field using an NMR trolley which will be movable within our vacuum system at a pressure of about 10^{-7} Torr. (Figs. 4 and 5). The trolley will contain 25 NMR probes, and electrical power and signals will be transmitted via the sliding contacts on the trolley rails. Within 8 hours a field map can be obtained around the circumference of the ring at 500 different locations (12,500 points). In addition to the NMR trolley, there will be 176 fixed NMR probes equally distributed

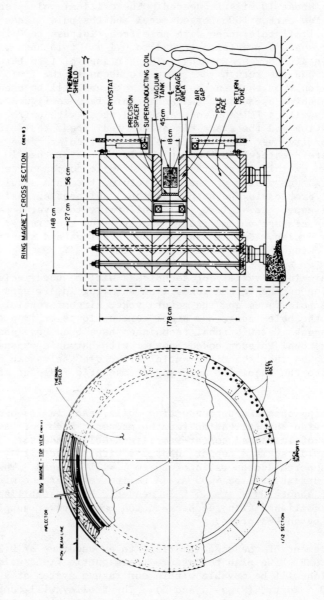

Fig. 2: Storage ring magnet plan drawings.

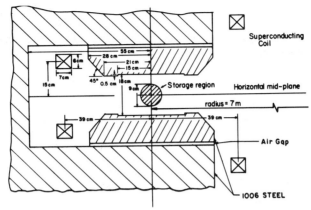

Fig. 3: Preliminary profile of magnet pole face-computer simulated.

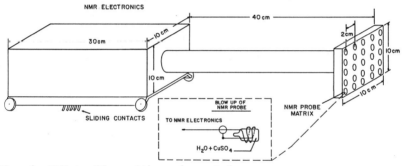

Fig. 4: NMR trolley with probe matrix.

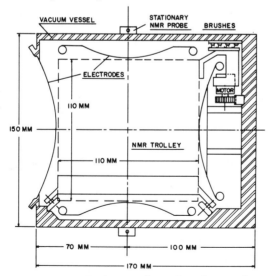

Fig. 5: Configuration of the quadrupole electrodes showing the arrangement for movement of the NMR trolley.

above and below the vacuum chamber for continuous monitoring of the magnetic field. Absolute calibration of the trolley NMR probes will be made with a standard probe which can be inserted into the vacuum chamber through an airlock.

Several R&D studies are now being planned and pursued as follows:

(1) Shimming techniques will be investigated experimentally at BNL using an existing dipole magnet. Ground pole pieces will be made and inserted in the magnet gap with the provision of an air gap between each pole piece and a face of the dipole magnet. Shimming with iron in the air gaps and with current loops and iron shims on our pole faces will be studied at the several ppm level using differential NMR measurements.

(2) Design and construction of a prototype trolley is in progress to test mechanical movement and electrical transmission.

(3) An NMR system involving rf frequency modulation and a proton probe is being designed for 0.1 ppm magnetic field measurements. It will probably be tested with a Yale superconducting solenoid now located at BNL.

Research supported in part by the U.S. Department of Energy.

REFERENCES:
1. V.W. Hughes and T. Kinoshita, Comments Nucl. Part. Phys. $\underline{14}$, 341 (1985).
2. J. Bailey et al., Nucl. Phys. $\underline{B150}$, 1 (1979).
3. "A New Precision Measurement of the Muon g-2 Value at the Level of 0.35 ppm.", AGS Proposal 821, November, 1985, V.W. Hughes, spokesman. J.P. Miller, B.L. Roberts – Boston University; H.N. Brown, E.D. Courant, G.T. Danby, C.R. Gardner, J.W. Jackson, M. May, M. Month, P.A. Thompson – Brookhaven National Laboratory; M.S. Lubell – City College of New York; A.M. Sachs – Columbia University; T. Kinoshita – Cornell University; W.P. Lysenko – Low Alamos National Laboratory; L.R. Sulak, W. Williams – University of Michigan; J.T. Reidy – University of Mississippi; F. Combley – Sheffield University; F. Krienen – Stanford University; D. Joyce, R.T. Siegel – College of William & Mary; S.K. Dhawan, A.A. Disco, F.J.M. Farley, V.W. Hughes, Y. Kuang, J.K. Markey, H. Orth – Yale University.
4. J. Wesley and A. Rich, Phys. Rev. $\underline{A4}$, 1341 (1971); Rev. Mod. Phys. $\underline{44}$, 250 (1972).
5. H. Drumm et al., Nucl. Instr. Meth., $\underline{158}$, 347 (1979).
6. K. Borer, Nucl. Instr. Meth., $\underline{143}$, 203 (1977); F.G. Mariam et al., Phys. Rev. Lett. $\underline{49}$, 993 (1982); B.N. Taylor and W.D. Phillips, ed., Precision Measurement and Fundamental Constants II, Nucl. Bur. Stand. Spec. Pub. 617 (U.S. Govt. Print, Wash. D.C., (1984).

USE OF THE FERMILAB ANTIPROTON SOURCE FOR EXPERIMENTS

James E. Griffin and Fermilab Antiproton Source Dept. Members
Fermilab, Box 500, Batavia, IL 60510, USA

ABSTRACT

A brief description of the Fermilab antiproton source is presented. The primary purpose of this facility is to provide a bright source of antiprotons for $\bar{p}p$ physics in the TeV energy range (Fermilab Tevatron). During periods of scheduled Tevatron fixed target physics it will be possible to operate the antiproton source at reduced efficiency in a parasitic mode. During such periods it appears feasible to use the antiproton Accumulator Ring for low energy (below 8.9 GeV) antiproton experiments. One such program is outlined. An additional small ring with good stored beam lifetime might possibly be operated in conjunction with the high energy $\bar{p}p$ program.

INTRODUCTION

On Oct. 13, 1985, the first series of twenty-three proton-antiproton collisions with C.M. energy 1.6 TeV were observed in the colliding beam detector facility (CDF) at Fermilab.[1] During the initial experiment a series of antiproton bunches, each containing about 10^9 particles, was delivered to the Tevatron via the Fermilab Main Ring accelerator from a dense, stochastically cooled beam stored in the Antiproton Source Accumulator Ring. Successful operation of the entire complex of production, capture, cooling, storing, and delivery of antiprotons to the Tevatron marked completion of the first phase of commissioning of the Tevatron I colliding Beam Facility.[2,3]

The Antiproton Source, Figure 1, may be divided into two main facilities. The first of these is the Main Ring Accelerator (with its Linac and Booster injector) operated in a specially dedicated manner. The second major component is the compound of beam transport lines, production target, particle focusing lens, and two additional Booster-sized rings for capture, cooling, and storage of antiprotons. The design goals of the project require that the Main

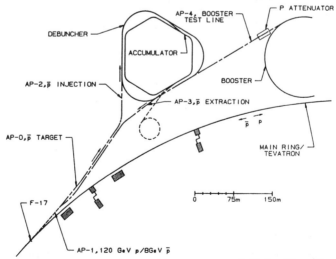

Fig. 1. Antiproton Source rings with segment of Main Ring/Tevatron.

Ring, operating at a 2 second repetition rate, deliver bursts of 2×10^{12} protons per cycle to an antiproton production target via a specially designed extraction channel. The same channel is used for reinjection of cooled antiprotons into the Main Ring for acceleration to 150 GeV and injection into the Tevatron. Antiprotons from the production target are delivered to the outer "Debuncher" ring where the momentum spread and transverse emittance are reduced by about a factor of ten by an rf manipulation[5] and transverse stochastic cooling[6,7] during a 2 second interval. This allows the beam to be transferred to the injection orbit of the smaller "Accumulator" ring just before a new burst of particles arrives. In the Accumulator ring the freshly injected beam is captured by a 53 MHz (harmonic number 84) rf system, moved away from the injection orbit by deceleration, and deposited at a momentum near the range of the "stack tail" momentum cooling system. During the 2 second interval this cooling system moves the particles away from the deposit point toward a core of cooled particles while at the same time increasing their phase space density. Additional core cooling systems increase the transverse and longitudinal phase space density of the stored antiprotons. The design accumulation rate is 10^{11} antiprotons per hour with a goal of 4.5×10^{11} particles in a roughly Gaussian distribution with a peak density of 10^5 per eV after 4.5 hours.

The antiproton accumulation rate on Oct. 11, 1985 was 10^9 per hour and a peak density of 10^4/eV was achieved. These deficits resulted from a variety of sources. Not all of the stochastic cooling systems were operating. Misalignment of some components in the Debuncher ring reduced the transverse acceptance. The lithium lens could be pulsed to only half of its required 5×10^5 A due to a peripheral component failure. Finally, the Main Ring was delivering protons at less than the design intensity and it was operating at a 4 second repetition rate. The lower repetition rate was a carryover from the operating mode which existed during the previous months during which commissioning of the source coexisted with the Tevatron fixed target program.

The manner in which the antiproton source can coexist with the Tevatron fixed target program is shown in Figure 2. A single Tevatron cycle occupies 65 seconds. Each Tevatron ramp requires a single 150 GeV Main Ring injection ramp of about 6 seconds. A large fraction of the remaining time is available for 120 GeV antiproton production ramps. Figure 2 shows eleven 120 GeV cycles within the Tevatron cycle although Main Ring performance is not uniform on all such cycles. Fine details of Main Ring tuning must, of course, be optimized for the 150 GeV cycle, and this precludes certain optimizations of the parasitic cycles. Nevertheless, it is clear that the antiproton source can be operated at perhaps one third efficiency during periods of Tevatron fixed target physics.

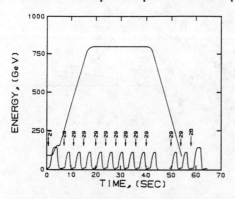

Fig. 2. Tevatron and Main Ring ramps for fixed target physics and parasitic antiproton production.

PROSPECTS FOR LOW ENERGY ANTIPROTON PHYSICS

Fermilab plans to schedule fixed target physics for five to six months per year. During these periods the antiproton source will be operated for development studies. Since these studies are expected to require 3 or 4 shifts per week, the availability of the source for antiproton physics during the remaining time is very promising. Detailed discussions of the problems associated with such a program appear in the proceedings of two recent workshops on low energy antiproton physics.[8,9]

There is presently an approved experiment[10] to study the formation of charmonium states using a hydrogen gas jet target with a decelerated and cooled antiproton beam in the Accumulator ring. The physics goals of the experiment are reviewed by R. Cester in this proceeding.[11] Accelerator aspects of this proposal are discussed here inasmuch as they apply generally to similar proposals.

The masses of the charmonium resonances range from the η_c (2980) through about 3700 for the higher angular momentum excited states. This corresponds to incident antiproton momenta between 5.8 and 6.3 GeV/c. Deceleration and subsequent cooling of antiprotons in the Accumulator presents several problems. The large momentum aperture at 8.9 GeV, where the guide field is 16.8 kG and substantial saturation occurs, is achieved by pole face shaping which is not effective at lower fields. The useful aperture is therefore reduced and the beam must be held near the center. Since the stochastic core cooling pickups are located off axis they are no longer useful. Also the particle velocity is sufficiently different at lower momentum so that existing cable delays are no longer correct. A design has been completed for an array of pickups and cables capable of the required cooling, and there is space in the lattice for insertion of the required equipment.

The transition energy of the ring, 5.4 GeV, presents another problem because it falls within the range of energies where data, and hence cooling, are required. The cooling rate is proportional to $\eta = \gamma_t^{-2} - \gamma^{-2}$. At energies very near γ_t the value of η is too small for effective cooling. A modified lattice with $\gamma_t = 3.8$ has been calculated. This can be achieved by moderate changes in quadrupole and sextupole currents at the expense of relaxing the requirement for zero dispersion at the longitudinal cooling kickers. The resulting betatron heating has been shown to be insignificant.

A circulating beam current of 1.5×10^{11} antiprotons colliding with a gas jet of density 10^{14} atoms/cm will provide a luminosity of 10^{31} cm^{-2} sec^{-1} and an acceptable event rate. At this beam current the modified cooling system will provide a transverse emittance sufficiently small so that the beam remains within the 0.6 mm gas jet but sufficiently large so that beam heating resulting from intrabeam scattering can be counteracted by the cooling system. Under these conditions the rms momentum width can be cooled to less than 1 MeV. This translates to a mass resolution for charmonium states of less than 300 kV. Higher resolution, limited eventually by microwave instability, might be achieved by electron cooling.[12]

Once the energy has been lowered near a resonant state of interest further incremental changes may be made by slowly changing the magnet fields and allowing the stochastic cooling system to recenter the beam on its pickups. Very precise measurements of the energy increments are possible simply by measuring the rotation frequency of the centroid of the cooled distribution. In this manner all resonance energies can be related to that of the ψ/J.

VERY LOW ENERGY PHYSICS--ADDITIONAL RING

The proposed experiment may be capable of examining the $\xi(2200)$ since this is near the lower limit of the rf frequency available. How far below this energy (1.64 GeV total or 701 Mev kinetic) the energy can be reduced remains to be determined. The h-84 rf system may have to be augmented with a lower harmonic system with broader bandwidth and some effort will be required to learn how to cross transition energy with minimal beam degradation. Nevertheless one might expect to reach kinetic energy of 500 MeV with reasonable beam quality. Since the beam will probably be confined to the center of the aperture and the extraction channel is off axis, some development will be required if the beam is to be extracted. Nevertheless, it is reasonable to consider a smaller storage ring similar to the LEAR ring at CERN.[13] Such a ring might be filled with 10^{11}-10^{12} antiprotons on a daily basis even during periods of colliding beam physics since each fill would require only a few hours of accumulation. An appropriate location for such a ring is noted on Figure 1 by a dashed circle between the Accumulator extraction line and the Main Ring.

CONCLUSIONS

With the modifications to the antiproton accumulator outlined here, the prospects for a program of low energy antiproton physics at Fermilab are promising. Because the primary function of the laboratory is to serve the high energy user community, the program would have to be essentially non-intrusive from the point of view of machine time and particularly manpower and funding. If personnel and funding are made available, such a program is viable and should produce exciting physics.

REFERENCES

1. CERN Courier, 25, 419 (Dec. 1985)
2. J. Peoples, The Fermilab Antiproton Source, IEEE Tran. Nucl. Sci. NS-30 1970 (1983).
3. G. Dugan, Tevatron I, Energy Saver and P Source, IEEE Proc. Nucl. Sci. NS-32, 1582 (1985).
4. B.F. Bayanov et al., A Lithium Lens for Axially Symmetric Focusing of High Energy Particle Beams, Nucl. Instr. and Methods, 190, 9 (1981).
5. J.E. Griffin, J. MacLachlan, A. Ruggiero and K. Takayama, Time and Momentum Exchange for the Production and Collection of Intense Antiproton Beams at Fermilab, IEEE Proc. Nucl. Sci NS-30, 2360 (1983).
6. D. Mohl et al., Physics and Technique of Stochastic Cooling, Physics Reports, 58, 73 (1980).
7. B. Autin, J. Marriner, A. Ruggiero and K. Takayama, Fast Betatron Cooling in the Debuncher Ring for the Fermilab TeV-I Project, IEEE Trans. Nucl. Sci. NS-30, 2593 (1983).
8. Workshop on the Design of a Low Energy Antimatter Facility in the USA, Univ. of Wisconsin, Madison (Oct. 1985).
9. Fermilab Low Energy Antiproton Facility Workshop, Fermilab (Apr. 1986).
10. V. Bharadwaj et al., A Proposal to Investigate the Formation of Charmonium States Using the pbar Accumulator Ring, Fermilab Expt. E760 (1985).
11. R. Cester, Heavy Quark Physics in $\bar{N}N$ Interactions, This proceeding.
12. D. Larson, Towards Multi-GeV Electron Cooling, This proceeding.
13. G. Giannini et al., Low Energy Antiprotons at the CERN PS, Proc. 12th Internat. Conf. on High Energy Accelerators, 20 (Fermilab 1983).

IS THERE A NEED FOR A LEAR-LIKE FACILITY IN NORTH AMERICA?

Gerald A. Smith*
Department of Physics
The Pennsylvania State University
University Park, PA 16802 USA

ABSTRACT

Considerations on the usefulness of a new LEAR-like facility in North America for low energy antiproton physics are presented.

I. INTRODUCTION

Recently there have been discussions about the feasibility of a LEAR-like facility in North America. These discussions have included the BNL AGS [1], and possible proton synchrotrons at Los Alamos [2] and TRIUMF [3]. Last month the Low Energy Antiproton Facility Workshop on Antimatter Physics was held at Fermilab. Preceding that was the Antimatter Facility Workshop held at Madison, Wisconsin. These activities demonstrate clear evidence of strong community interest in the physics which can be done with low energy antiprotons.

In the brief time alloted to me, I will attempt to give my personal assesment of the physics available to a new facility and a recommendation for possible future developments. I will address three topics in the following order: (1) I will review the LEAR program for the post-ACOL era ($\geq$ 1987); (2) I will discuss other experiments that I think are important and should be pursued; and (3) I will attempt to develop a recommendation for a North American facility that could support a vigorous program of physics.

II. FUTURE LEAR PROGRAM

The program for experiments at LEAR after the ACOL upgrade ($\geq$1987) is presently being developed at CERN [4]. To date, seven experiments (Table 1) have been approved. Five other proposals are in various states of review (Table 2). In addition, several current LEAR experiments will continue into the post-ACOL era (Table 3). I indicate in Table 4 the broad areas of physics interest and machine requirements represented by these experiments. The number of experiments (15) rivals the first-round LEAR program in number (16), which is already known for its high density.

There appear to be several clear differences between the pre-ACOL and post-ACOL programs. First, the emphasis on very low momenta (<200 MeV/c) is much greater in the post-ACOL program. Second, the internal experiment is becoming increasingly popular. And third, there seems to be a greater emphasis on experiments involving symmetry tests (categories 3-5) than before. Overall, I see an exciting future with tremendous variety, but not free of scheduling conflicts which may result from demands for intensity and extracted (internal) energy

---
* Work supported in part by the U.S. National Science Foundation.

Table 1 - Presently Approved Post-ACOL Experiments at LEAR

Title (Groups)	Physics Goals
PS189 - Precision Comparison of Antiproton and Proton Masses in a Penning Trap: (Washington-Mainz-Fermilab)	Precision comparison of antiproton and proton masses ($1/10^9$). (a)
PS195 - Tests of CP Violation with $\bar{K}^°$ and $K^°$ at LEAR; (Athens-Demokritos-Basel-CERN-Fribourg-Liverpool-Saclay-SIN-Stockholm-Thessaloniki-Zurich)	Measure CP-violating parameters: $\epsilon'/\epsilon \sim 2\times10^{-3}$ (different systematics from regeneration experiments); η_{+-0}, η_{000} improved by 3 orders of magtude; $\|\phi_{+-}-\phi_{00}\| < 2°$ (presently 14±7°). (b)
PS196 - High Precision Mass Measurements with a Radiofrequency Mass Spectrometer - Application to the Measurement of the $\bar{p}p$ Difference: (CERN-Orsay)	Precision measurement of antiproton-proton mass difference ($1/10^9$). (a)
PS197 - The Crystal Barrel: Meson Spectroscopy at LEAR with a 4π Neutral and Charged Detector: (Karlsruhe-Zurich-Berkeley-Queen Mary-Surrey-Mainz-Strasbourg-Vienna-Munich)	Searches for glueballs and hybrids; radiative and rare decays of mesons; $\bar{p}p$ dynamics from initial L=0,1 states; $\bar{p}p$ bound states-baryonium; $\bar{p}p$ resonances into exclusive channels. (a)
PS198 - Measurement of Spin Dependent Observables in the $\bar{p}N$ Elastic Scattering from 300 MeV/c to 700 MeV/c: (Saclay-Karlsruhe-Lyon-SIN)	Measure differential cross-sections and polarization for $\bar{p}$-nucleus scattering using SPESII high resolution spectrometer. (a)
PS199 - Study of the Spin Structure of the $\bar{p}p \to \bar{n}n$ Channel at LEAR: (Cagliari-Geneva-Karlsruhe-Trieste-Turin)	Measure differential cross-sections and polarization for charge-exchange with polarized target; scan for resonances in the direct channel. (a)
PS200 - A Measurement of the Gravitational Acceleration of the Antiproton: (Pisa-Los Alamos-Rice-Texas A&M-Genova-Kent State-Case Western-CERN-NASA/Ames)	Measure gravitational acceleration of antiprotons and protons using RFQ for deceleration. (c)

(a) Ref. 4.   (b) J. Fry, this conference.   (c) R. Brown, this conference.

Table 2 - Post-ACOL Proposals Presently Under Consideration at LEAR

Title (Groups)	Physics Goals
P86 - Feasibility Study for Antihydrogen Production at LEAR: (CERN-Heidelberg-Karlsruhe)	Production of antihydrogen in LEAR ring using corotating beams of electron-cooled antiprotons and positrons; tests of CPT in antihydrogen. (a)
P88 - Tests of Time Reversal and CPT Invariance at LEAR: (Zurich-William & Mary-Oxford-New Mexico-Ljubljana-Delft-Coimbra-Birmingham)	Tests of CPT to $\sim 6\times10^{-4}$ in $K°,\bar{K}°$ mass difference and K^+,K^- lifetime difference; test of T to $\sim 4\times10^{-4}$ from CPT and present level of CP violation; CP and CPT tests to $\sim 2\times10^{-4}$ in $K_{\ell3}$. (b)
P92 - Spin-Dependence in $\bar{p}p$ Interaction at Low Momenta: (Heidelberg-Houston-Rice-Karlsruhe-Wisconsin-Marburg-Mainz-Munich-Rutgers)	Creation and analysis of polarized antiproton beam in LEAR ring using a polarized atomic beam target. (c)
P95 - Antinucleon Annihilations at LEAR with OBELIX, a Large-Acceptance and High-Resolution Detector Based on the Open Axial Field Spectrometer: (Brescia-Cagliari-CERN-Frascati-Geneva-Dubna-Legnaro-Orsay-Padua-Pavia-Trieste-Torino-Udine-British Columbia)	Glueball, hybrid, 4 quark meson spectroscopy; $\bar{N}N$ dynamics, $\bar{p}p$ atoms and strong interaction effects, quark-gluon aspects of nuclear matter, search for highly excited states of nuclear matter. (d)
P97 - JETSET: Physics at LEAR with an Internal Gas Jet Target and an Advanced General Purpose Detector: (Annecy-CERN-Freiburg-Geneva-Genoa-Julich-Oslo-Uppsala-Texas-Trieste-Minnesota)	Study of $\bar{p}p \to \phi\phi, K^°_s K^°_s$ using unpolarized and polarized gas jets in the LEAR ring; observation of $\xi(2230)$ with high mass resolution. (e)

(a) H. Poth, this conference.   (c) R. Ransome, this conference.   (e) Y. Onel, this conference.
(b) L. Roberts, this conference.   (d) Ref. 4.

operations.

III. OTHER PHYSICS OPPORTUNITIES

LEAR has a maximum $\sqrt{s}$ of 2430 MeV as a fixed target machine. In a collider mode one could study $\bar{p}p$ collisions up to $2\times2(GeV)^2$, or $\sqrt{s}$ = 4420 MeV. However, the maximum luminosity is estimated to be

Table 3 - Pre-ACOL LEAR Experiments Which May Continue into the Post-ACOL Era

Title (Groups)	Physics Goals
PS170 - Precision Measurement of the Proton Electromagnetic Form Factors in the Time-like Region and Vector Meson Spectroscopy: (Ferrara-Padova-Saclay-Torino)	Study of $\bar{p}p \to e^+e^-$, $V^\circ \to (e^+e^-)$ + neutrals; EM form factors, vector mesons. (a)
PS177 - A Search for Heavy Hypernuclei at LEAR: (Orsay-Saclay-Uppsala-Warsaw)	Search and measurement of lifetime of heavy hypernuclei. (a)
PS185 - Study of Threshold Production of $\bar{p}p \to Y\bar{Y}$ at LEAR: (Carnegie-Mellon-Erlangen-Freiburg-Illinois-Julich-Rice-Saclay-Uppsala-Vienna)	Measure differential angular distributions and polarization for $\bar{p}p \to Y\bar{Y}$ near threshold; $\bar{p}p \to \xi(2230)$, meson states, CP violation. (b)

(a) Ref. 4.   (b) P. Barnes, this conference.

Table 4 - Broad Physics Categories and Beam Requirements for Approved and Proposed Post-ACOL LEAR Experiments.

	External Momenta (MeV/c)			Internal Momenta (MeV/c)
	≤ 100	100-200	> 200	
1. Meson Spectroscopy		PS197(b) P95(b)	PS185(a)	P97 (gas jet, 600-1900)
2. Polarization Effects			PS198(a) PS185(a) PS199(a)	P92 (polarized gas jet, <2000)
3. Tests of Discrete Symmetries (CP,CPT,T,ΔS=ΔQ)	PS189(c) PS196(c)	PS195(a) P88(a)	PS185(a)	
4. Gravity and Inertial Mass	PS200(c)			
5. Formation of Antimatter				P86 (with positron beam, 300)
6. EM Form Factor			PS170(a)	
7. Hypernuclei	PS177(b)			
Totals	4	4	4	3 = 15

(a) ≥ 10^6 $\bar{p}$/s   (b) ~ 10^4-10^5 $\bar{p}$/s   (c) < 10^3 $\bar{p}$/s

only 1.5×10^{29} cm^{-2}sec^{-1} [5]. Therefore, I take the viewpoint that any new physics requiring $\sqrt{s}$ in excess of 2430 MeV and high luminosity must be done at a machine other than LEAR. What are the possibilities? I discuss two candidates in the following sections.

(a) Charmonium

The recent success of the R704 experiment at the ISR points out the importance and practicality of forming charmonium states in $\bar{p}p$ collisions. At Bari Susan Cooper provided a rather succinct review of the R704 results [6]:

"The last ISR experiment R704 pioneered a new technique of observing $c\bar{c}$ states. They produce them directly in $p\bar{p}$ annihilation, scanning the beam energy over the expected mass region. Their scans of the χ_c $^3P_2$ and $^3P_1$, shown in Figure 1, are quite impressive. The χ_c states are identified by their decay to $\gamma\psi$, $\psi \to e^+e^-$. The fitted masses are 3511±0.6 and 3556±0.6 MeV. The ISR beam gives a mass resolution of ~ 0.3 MeV, allowing them to measure the total widths of these states. These results were

obtained with only a few weeks running time. More data
would certainly improve the width measurements."

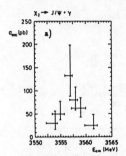
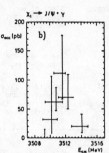

Fig. 1: R704 P states. (a) $p\bar{p} \to \gamma\psi$ cross sections in $^3P_2$ region. The plot contains 56 candidates, with an estimated background < 2. (b) $^3P_1$ region, with 32 candidates, < 2 background.

In the collider mode, LEAR has just sufficient c.m. energy for a full charmonium study. However, the anticipated luminosity falls short by two orders of magnitude of that which will be available in the Fermilab E760 experiment utilizing a hydrogen gas jet in the accumulator [7]. A more detailed account of E760 can be found in ref. [8]. I also refer the reader to a recent review talk by R. Jaffe [9]. In summary, $\bar{p}p$ charmonium physics seems to fit nicely into Fermilab capabilities and plans.

Unfortunately, extension of this work to the upsilon region appears out of range. The $BR_{had}(b\bar{b})$ is predicted in QCD to scale as $(m_c/m_b)^8 \cdot BR_{had}(c\bar{c})$ for $J^{PC} = 1^{--}$, 1^{++}, 2^{++} states and $(m_c/m_b)^{10} \cdot BR_{had}(c\bar{c})$ for $J^{PC} = 0^{-+}$, 0^{++}, 1^{+-} states [10]. This drives down rates for bb by a huge factor (10^4-10^5).

(b) <u>CP Violation in $\bar{p}p \to \Lambda°\bar{\Lambda}°$</u>

It is important to search for evidence for CP violation in non-$K°$ systems. At this conference J. Donoghue discussed possible CP violations in the reaction $\bar{p}p \to \Lambda°\bar{\Lambda}°$ [11]. Under CP, the polarization of the lambda and antilambda should be equal. He estimates that a violation may appear at the level of $\sim 10^{-4}$. The threshold for this reaction is ~ 1.5 GeV/c, so a window between 1.5 and 2.0 GeV/c exists at LEAR to do this experiment [12]. An alternative approach is to utilize the Fermilab accumulator as a source of antiprotons incident upon a gas jet target. The cross section peaks around 3 GeV/c at ~ 130 μb. An effective cross section of ~ 13 μb combined with a luminosity of 10^{31}cm^{-2}sec^{-1} for 10^7sec (one year run) would yield $\sim 10^9$ events, corresponding to a statistical accuracy in the asymmetry parameter of $\sim 3 \times 10^{-5}$. This looks very attractive, although systematic errors would have to be given careful consideration.

IV. CONCLUSIONS

I have attempted to review plans for low energy antiproton physics at LEAR and Fermilab, as well as suggest additional experiments for the future. The variety of these programs is truly amazing, ranging over twelve orders of magnitude in energy (meV to GeV) and covering fundamental symmetry tests, spectroscopy, QCD, polarization phenomena, etc.

Therefore, it seems that we should be anticipating the need for another facility for low energy antiproton physics. The most obvious option is a dedicated storage ring at Fermilab. Designs for such a ring should include high luminosity ($> 10^{31} cm^{-2} s^{-1}$), a large dynamic range in energy (100 MeV/c - 10 GeV/c) and excellent energy resolution. The option of multi-GeV electron cooling ($\Delta p/p = 10^{-5}$ with transverse emittance of 0.12 mm-mrad) discussed at this conference [13] should be considered seriously. Every effort should be made to incorporate existing facilities into this scheme, since new funding will be hard to come by as the demands for new major facilities in North America intensify. A working group (called AMPLE ≡ Antiproton Matter Physics at Low Energies) has been formed at Fermilab and this is the logical starting point for such studies. Naturally, options for antiproton facilities at the proposed machines at LAMPF and TRIUMF should be left open.

REFERENCES AND FOOTNOTES

[1] Report of the AGSII Task Force (Chairman, G.A. Smith), Feb. 15, 1984.
[2] A Proposal to Extend the Intensity Frontier of Nuclear and Particle Physics to 45 GeV (LAMPFII), Los Alamos National Laboratory, Los Alamos, NM, LA-UR-84-3982.
[3] M.K. Craddock, these proceedings, and KAON Factory Review, Report of Technical Panel, May 9, 1986.
[4] Physics With Antiprotons at LEAR in the ACOL ERA, Third LEAR Workshop, Tignes, France, 1985, ed. U. Gastaldi, R. Klapisch, J.M. Richard and J, Tran Thanh Van, Editions Frontieres; Physics at LEAR with Low-Energy Cooled Antiprotons, ed. U. Gastaldi and R. Klapisch, Plenum Press (1984).
[5] D. Möhl, "Technical Implications of Possible Future Options for LEAR," ref. 4, p. 65.
[6] S. Cooper, International Europhysics Conference on High Energy Physics, Bari, Italy, July 18-24, 1985; SLAC-PUB-3819, Oct. 1985.
[7] P-760: A Proposal to Investigate the Formation of Charmonium States Using the PBAR Accumulator Ring, Fermilab-Farrara-Genoa-Irvine-Northwestern-Penn State-Torino Collaboration, March 1985 (approved December 1985).
[8] R. Cester, these proceedings.
[9] R. Jaffe, proceedings of the Fermilab Low Energy Antiproton Facility Workshop, 10-12 April 1986.
[10] W. Buchmüller, ref. 4, p. 321.
[11] J. Donoghue, these proceedings.
[12] P. Barnes, these proceedings.
[13] D. Larsen, these proceedings.

LAMPF II 1986: MATCHING THE MACHINE TO THE NEW PHYSICS

Henry A. Thiessen, MP-14, MS H847
Los Alamos National Laboratory, Los Alamos, NM 87545

We discuss the measurement of quark structure functions of nuclei using the Drell-Yan process. These experiments require proton energy of at least 45 GeV, and 60 GeV if it is desired to use kaons as the projectile. In addition, these and other coincidence experiments require that the 60 MHz microstructure of the beam be eliminated and replaced with a fully debunched slow-extracted beam. We present a plan for a 45-GeV, 32-µA proton facility, upgradeable to 60 GeV and 100 µA. The improved 1986 version produces twice as many kaons per proton as does the previously proposed machine.

This talk was a summary of the 1986 update to the LAMPF II proposal. A report of this work, LA-10720-MS, entitled "The Physics and a Plan for a 45-GeV Facility that Extends the High-Intensity Capability in Nuclear and Particle Physics" may be requested by writing to

                Mrs. Carol Harkleroad
                Mail Stop H847
                Los Alamos National Laboratory
                Los Alamos, New Mexico 87545

MICROWAVE INSTABILITY CRITERION FOR OVERLAPPED BUNCHES

King-Yuen Ng
Fermi National Accelerator Laboratory[1], Batavia, IL 60510

ABSTRACT

Debunching can be a method to measure Z/n of a storage ring by timing the start of microwave instability. However, if this instability begins to show up when two or more bunches overlap each other, the situation becomes more complex, because one is confused of which local current and energy spread should be used. An analysis shows that exactly the same microwave instability criterion should be used as if there is only one bunch.

INTRODUCTION

During debunching, the energy spread of a bunch becomes smaller and smaller. Eventually, Landau damping fails and microwave instability starts. By measuring the time when instability starts, the Z/n of the storage ring can be inferred. However, this instability may start when two or more bunches overlap each other. One may wonder whether one should take the total energy spread of the bunches or the RMS energy spread of one bunch in the Keil-Schnell criterion. Also, one is not sure whether the total local current of the overlapped bunches or the local current of a single bunch should be used in the criterion. This problem is solved in this paper[2].

THE DISPERSION RELATION

Consider two overlapped Gaussian bunches as shown in Fig. 1. At any azimuthal point in the overlap, the dispersion relation is

$$1 = -(\Delta\Omega_o/n)^2 \int F'(\omega)/(\Delta\Omega/n-\omega) d\omega, \qquad (1)$$

where $\Delta\Omega_o/n = [ie\eta\omega_o^2 I_t (Z/n)/(2\pi\beta^2 E)]^{1/2}$ is the growth without Landau damping, η the frequency dispersion coefficient, ω_o the revolution frequency, β the velocity of a bunch particle of energy E in unit of c, and $\Delta\Omega/n$ is the coherent frequency per revolution harmonic of the perturbing wave in excess of ω_o. Note that we have used the total local current I_t which is equal to the sum of the local currents I_1 and I_2 of the two individual bunches. The normalized frequency distribution function is

$$F(\omega) = (\sqrt{2\pi}\sigma)^{-1}\{a_1 \exp[-(\omega-\omega_1)^2/2\sigma^2] + a_2 \exp[-(\omega-\omega_2)^2/2\sigma^2]\}, \qquad (2)$$

where σ is the RMS revolution frequency spread of each bunch which is considered to be Gaussian, ω_1, ω_2 are respectively the mean deviations of revolution frequencies of the bunches from that of a synchronized particle (we take $\omega_1 < 0$ and $\omega_2 > 0$). The fraction of each bunch in the overlap is represented by $a_i = I_i/I_t$, $i = 1,2$.

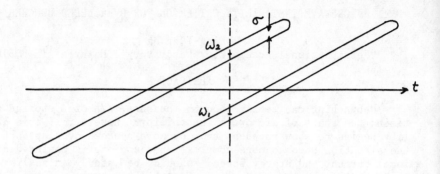

Figure 1

Let us consider the case when Z/n is imaginary; i.e., $(\Delta\Omega_o/n)^2$ is real. Then the thresholds are given by

$$1 = -(\Delta\Omega_o/n)^2 \int F'(\omega)/[\text{Re}(\Delta\Omega/n)-\omega]d\omega, \tag{3}$$

where $\text{Re}(\Delta\Omega/n)$ is any of the 3 zeros of $F'(\omega)$ which are ω_1, ω_2 and another one in between. Equation (3) can be solved exactly:

$$(\Delta\Omega_o/n)^{-2} = -\sigma^{-2}[1-i\sqrt{\pi}a_1 u_1 w^*(u_1) - i\sqrt{\pi}a_2 u_2 w^*(u_2)], \tag{4}$$

where $u_i = (\Delta\Omega/n - \omega_i)/\sqrt{2}\sigma$ and $w(u)$ is the complex error function. Then, at one zero, for example, $u_1=0$, Eq. (4) becomes

$$(\Delta\Omega_o/n)^{-2} = -\sigma^{-2}[1-2a_2 K\exp(-K^2)\int_0^K \exp(t^2)dt], \tag{5}$$

where $K=\Delta\omega/\sqrt{2}\sigma$ and $\Delta\omega=|\omega_1-\omega_2|$. During debunching, we always have $2K\gg 1$; Eq. (5) can therefore be simplified to

$$(\Delta\Omega_o/n)^{-2} = -\sigma^{-2}\{1-a_2[1+(\sigma/\Delta\omega)^2]\} = -a_1/\sigma^2 + a_2/\Delta\omega^2, \tag{6}$$

Neglecting the last term and putting in the relation between σ and the RMS energy spread σ_E of a bunch, we get

$$ie\eta\omega_o^2 I_t(Z/n)/(2\pi\beta^2 E) = -(\eta\omega_o^2\sigma_E/E)^2/a_1. \tag{7}$$

Recalling that $I_1 = a_1 I_t$, this is just the same stability criterion of a single Gaussian bunch with RMS energy spread σ_E and local current I_1. Similarly, with $u_2 = 0$, we obtain the same stability criterion with σ_E and I_2 for the second bunch.

This result can also be visualized as follows. Consider two coasting beam with frequencies $\omega_o+\omega_{1,2}$ and each has a RMS spread of σ. Imagine a small perturbing current wave of the form $\exp(in\theta-i\Omega t)$ where θ is the azimuthal angle around the accelerator ring. If the coherent frequency $\Omega \sim n(\omega_o+\omega_1)$, it will set the particles in the first beam to oscillate with harmonic n and eventually lead to a growth if σ is not large enough to destroy the coherency. If $\sigma \ll |\omega_1-\omega_2|$, the particles in the second beam will not be affected. On the other hand, if $\Omega \sim n(\omega_0+\omega_2)$, it can only drive a growth of harmonic n in the second beam while the first one will not affected. Thus the stability criterion applies to each beam individually. In debunching, the bunches are long and resemble coasting beams so we expect the same reasoning applies to overlapped bunches as well.

THE STABILITY CURVE

The stability curve in the $(\Delta\Omega_0/n)^2$-plane is shown in the Fig. 2 with $(\omega_1/\sqrt{2}\sigma)^2=10$, $|\omega_1|=\omega_2$ and $a_1=a_2$. It wraps around the origin twice in two Riemannn sheets as the real part of the coherent frequency shift $\Delta\Omega/n$ increases, the cut being the positive imaginary axis. The real coherent frequency shift $Re(\Delta\Omega/\sqrt{2}n\sigma)$ is marked along the curve. For the shake of clarity, only one half of the curve is plotted. The other half is just an mirror image about the cut. The two identical intercepts it makes with the negative imaginary axis in two the different sheets correspond to $Re(\Delta\Omega/n) = \omega_{1,2}$ for the two bunches if Z/n is capacitive. The intercept it makes with the positive imaginary axis corresponds to the threshold criterion of Eq. (4) using the third zero of $F'(\omega)$ and corresponds to substituting $\sim\Delta\omega$ in the stability criterion and is therefore $(\Delta\omega/2\sqrt{2}\sigma)^2$ farther away from the origin than the two other intercepts. Any Z/n corresponding to a point inside the center region of the curve is completely stable. Thus, different from the situation of a single bunch, a big enough inductive Z/n above transition can also lead to instability.

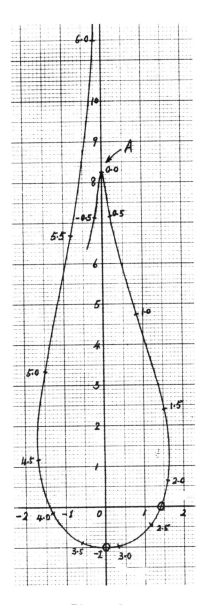

Figure 2

1. Operated by the Universities Research Association, Inc., under a contract with the U.S. Department of Energy.
2. For the debunching experiment and other analysis, see K. Y. Ng Fermilab Report TM-1389.

ANTIPROTON PHYSICS

Coordinators: D. A. Axen
               G. A. Smith

$N\bar{N}$ ANNIHILATION AND THE QUARK REARRANGEMENT MODEL

J.A. Niskanen

Research Institute for Theoretical Physics
University of Helsinki, Siltavuorenpenger 29 C,
SF-00170 Helsinki, Finland

ABSTRACT:

A short review is given on the status of the quark rearrangement model of $N\bar{N}$ annihilation.

INTRODUCTION

There are good reasons to study low energy $N\bar{N}$ annihilation dynamics today. To start with, annihilation is one of the most spectacular phenomena in modern physics, but as yet, it is not well known in the case of strongly interacting particles like baryons and, at a deeper level, quarks. There are data as well as valid theoretical arguments for high energies, but apparently there is little doubt that the process also at low energies is intrinsically a quark phenomenon - anyway it is the most likely candidate for the explicit appearance of quarks at low energies. The basic mechanism is expected to be some kind of rearrangement of the quarks and antiquarks of the nucleon and antinucleon into a number of mesons along with a number of $q\bar{q}$ pair annihilations and creations. Undoubtedly at low energies theoretical difficulties may well arise from the largeness of the basic strong coupling constant α_s, rendering invalid perturbation approaches which are valid at high energies. This leads inevitably to a need for some phenomenological "effective" couplings, which may, in principle, be taken at least roughly from hadron spectroscopy. Even with these reservations there could be expected important constraints and advances in strong interaction dynamics from $N\bar{N}$ (or more generally baryon antibaryon) annihilations at low energies and also from reactions $N\bar{N} \to B\bar{B}'$. In addition to these theoretical reasons, the LEAR facility at CERN has a potential to raise the experimental knowledge to a new level giving further impetus to the interest in low energy antiproton physics. For reviews see refs.[1,2].

In the following a short summary of the so called rearrangement model, supplemented by one $q\bar{q}$-pair annihilation, is presented. A reason for this concentration is given above and also by the fact that, in spite of some success of the most conventional baryon exchange model[3] for $p\bar{p} \to \pi^+\pi^-$, this is very dependent on the form factors of the particles. In fact, the size of the nucleon is much larger than the range $\simeq 0.1$ fm of the effective optical potential from the latter model, whereas the range of the quark processes is directly related to the sizes of the particles (rather of the mesons than of baryons). Although the baryon exchange might be claimed to be necessarily the longest range effect[4] (asymptotically the finite range of confined quarks is indeed shorter than the Yukawa type of baryons exchange), in practice it is the $r \lesssim 1$ fm region presumably dominated by quarks that is important, not the weak tail.

THE REARRANGEMENT MODEL

Historically the rearrangement model started from an attempt to understand some three meson branching ratios in terms of simply the SU(6) recoupling coefficients $(qqq)(\bar{q}\bar{q}\bar{q}) \to (q\bar{q})(q\bar{q})(q\bar{q})$ and the phase space[5]. Later several phenomenologically adjustable parameters were introduced by Maruyama and Ueda to simulate different effective meson-$q\bar{q}$ couplings and initial state distortions[6]. In fact, these parameters can be qualitatively understood by differences in the meson wave functions (e.g. not all of the π and η is of light quark $q\bar{q}$ configuration), form factor effects and meson exchange potential effects[1,7,8].

The philosophy of the Helsinki approach is to attempt first to understand the bulk of the annihilation cross section with a minimum of free phenomenological parameters. In this pursuit first the rearrangement into three mesons (fig. 1a) was studied for initial S and P waves[9]. The simplifying assumptions of mutually non-interacting mesons and harmonic oscillator internal wave functions for the hadrons are presumably qualitatively reasonable. The resulting optical potential from this mechanism is both energy dependent and separable and of the form

$$V = -\frac{\hbar^2}{M} \lambda\, I(E,B\bar{B}')\, f(r)f(r')\, e^{-\frac{3}{4}\beta(r^2+r'^2)}. \tag{1}$$

Here M is the nucleon mass, β is the meson wave function oscillator parameter $\beta = 2/\sqrt{3}/b^2$, b = RMS radius of the nucleon, and λ is a parameter for the strength of the rearrangement. It can be given a rough theoretical estimate (14 MeV fm)[10], although in practice it has been given some freedom around this value due to the uncertainties in quark dynamics. The coefficient $I(E,B\bar{B}')$ is a combination of the SU(6) recoupling coefficients from a baryon state $B\bar{B}'$ and intermediate meson state phase space integrals. The inclusion of the meson widths

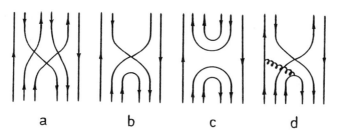

Fig. 1. Mechanisms in $N\bar{N}$ annihilation: a) rearrangement, b) $q\bar{q}$ annihilation into vacuum ($^3P_0$ model), c) one meson intermediate state effect, d) $q\bar{q}$ pair annihilation into a gluon.

was found to be essential especially below the nominal three meson thresholds - a case very typical of a P-wave annihilating into two s-wave mesons and one p-wave meson. Also I(E) has a very attractive real part. Additional attraction in the annihilation potential was found to improve significantly the fit with experiment in the phenomenological approach of Dover and Richard[11]. This result, which was contrary to contemporary expectations, gets now an explanation from quark rearrangement. Also it is important to note that a potential of type (1) can only yield mesons in relative s states - setting strong constraints. The functions f depend on the initial L value and the type of final state mesons (the low case letters denote the internal ℓ value of the mesons):

$$f(r) = 1 \qquad \text{for } S \to sss$$
$$f(r) = \sqrt{\beta}\, r \qquad \text{for } P \to ssp$$
$$f(r) = \beta/\sqrt{5}\, r^2 \qquad \text{for } D \to ssd \qquad (2)$$
$$f(r) = \beta(r^2 - \frac{8/3}{\alpha+\beta}) \qquad \text{for } S \to spp$$
$$f(r) = \sqrt{2/5}\, \beta\, r^2 \qquad \text{for } D \to spp$$

Of these, the effect of the high mass states with a d-wave meson or two p-wave mesons is rather negligible, although the principal importance of the latter is high due to the close analogy of the $\varepsilon(0^{++})$ meson with the vacuum, i.e. in the simplest model for annihilation into two mesons one ε-meson may be replaced by the vacuum.

In addition to the annihilation a meson exchange potential, similar to that of ref.[11] is included. Also it is found important to incorporate $N\bar{\Delta} \pm \Delta\bar{N}$ and $\Delta\bar{\Delta}$ isobaric state admixtures as a doorway for annihilation. The result - omitting the negligible high mass states - is shown in fig.2 for the S wave at the laboratory energy 100 MeV. This is far below the unitarity limit 26 mb for S waves. Even adding the P and D waves does not yield more than 60 % of the experimental $\sigma_{ann} \simeq 100$ mb.

A further mechanism necessary is annihilation into two mesons by destroying one $q\bar{q}$ pair. In ref.[10] this is performed using the so called $^3P_0$ model where a $q\bar{q}$ in a 1^3P_0 state annihilates into the vacuum (fig. 1b). A very similar result is achieved for the optical potential as in eq. (1) except that in the exponential $3\beta/4 \to 3/8(\alpha+3\beta)$, $\alpha = 1/b^2$ and

$$f(r) = \frac{3}{2}\sqrt{3\beta}\, (r^2 - \frac{4/3}{\alpha+\beta}) \qquad \text{for } S \to sp$$
$$f(r) = \sqrt{3}\, r \qquad \text{for } P \to ss \qquad (3)$$
$$f(r) = 3\sqrt{3/10\,\beta}\, r^2 \qquad \text{for } D \to sp$$

It is assumed that the overall strength can now be obtained from meson and baryon resonance decays[12]. Except for P → ss again the real parts are strongly attractive. Even with this addition the annihilation

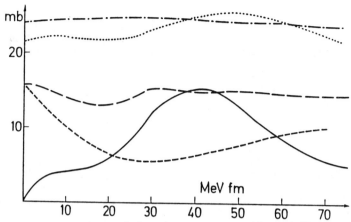

Fig. 2. The S-wave annihilation cross section at E_{lab} = 100 MeV. Solid: pure rearrangement into sss (fig. 1a); long dashed: the sp two meson contribution (fig. 1b) added (the short dashes show the portion of the sp final state); dotted: shows the effect of phenomenological local repulsion of ref.[8] with the isospin dependence omitted; dash-dot: a strong effect of two s-wave mesons in a relative p wave added to dotted, ie. sss+sp+(ss)$_{relp}$. The unitarity limit is 26 mb.

cross section remains too low, although much of the annihilation does go into two meson channels - see fig. 2.

POSSIBLE IMPROVEMENTS

To understand the above failure to obtain arbitrarily large cross sections with the increasing strength parameter, one has to realize that for a separable annihilation potential with a single term and baryon channel

$$\sigma_{ann} \propto \frac{\lambda |F_0|^2}{|1-\lambda F_1|^2}, \tag{4}$$

where the F's are complex integrals depending on the potential and $B\bar{B}'$ wave functions $u(r)$[1]. In the multiterm, multichannel case the result is similar though more complicated[10]. Clearly (4) has a peaking structure with $\sigma_{ann} \to 0$ as $\lambda \to \infty$. Further insight is obtained by noting that the S states and some P waves tend to have nodes inside the annihilation region[1] making the annihilation integrals small. This is due to the extremely strong attraction in the meson and annihilation potentials. (In the above model - Re $V_{ann} \gg$ -ImV_{ann}. Dover and Richard[11] require Re$V_{ann} \simeq$ ImV_{ann}.) Therefore, it seems that some repulsion would improve matters at least in S-waves. This was shown to be true by Maruyama and Ueda[8], who added a phenomenological repulsive term - see also ref.[13]. The overall effect is, however, not enough, since, after all, the weight of the S states is quite small and only $\simeq$ 10 mb at most are gained in this way. Therefore, Maruyama and Ueda still add a phenomenological local annihilation term in D waves[8].

The Helsinki group has attempted to justify these modifications by possible microscopic mechanisms. Ref.[10] introduced one meson intermediate states (fig. 1c). This is again separable, contains itself nodes in the S state and is of extremely short range, and has consequently a small effect. In ref.[13] various forms of repulsion suggested by the Skyrmion model of baryons are studied as a possible mechanism to increase annihilation. Ref.[14] suggests a relativistic extension of the $^3P_0$ model yielding an interaction resembling in form the one gluon mechanism[15] (fig. 1d). However, in S-states the relativistic "small component" effect yielding two s-wave mesons <u>in a relative p-state</u> - a repulsive contribution! - has the same radial form as (3) involving nodes and helps little[16]. It is remarkable, nevertheless, that taking a radial form for $B\bar{B}' \to ss$ suggested by a long range gluon exchange (over the confined quark region) without nodes and arbitrarily with the same strength as $B\bar{B}' \to sp$ (except for the $I(E,B\bar{B}')$ which is calculated) yields nearly the unitary limit for the s states virtually independent of λ^{16} - see fig. 2. It is possible that the nodes in the $^3P_0$ model are due to an incorrect structure of the basic phenomenological vertex[17] which is a one body operator incapable of any momentum or quantum number transfer. One should also note that it may not be legitimate to neglect higher order mechanisms, where e.g. two $q\bar{q}$ are annihilated and one pair created, to form two mesons[18]. It seems that after the first steps of infancy the models of annihilation are finally closing in on the basic quark interaction vertices themselves also in the bulk of annihilation. This, of course, is the ultimate aim of the whole exercise.

This talk is an outcome of collaboration with A.M. Green.

REFERENCES

1) A.M. Green and J.A. Niskanen, Int. Rev. of Nuclear Physics, Vol.I, Quarks and Nuclei, ed. W. Weise (World Scientific, Singapore 1985) p. 569
2) A.M. Green and J.A. Niskanen, Helsinki preprint HU-TFT-85-60, to be published in Progress in Nuclear and Particle Physics, ed. A. Faessler (Pergamon Press, 1986)
3) B. Moussallam, Nucl. Phys. <u>A407</u>, 413 (1983), <u>A429</u>, 429 (1984)
4) Coté et.al, Phys. Rev. Lett. <u>48</u>, 1319 (1982)
5) H. Rubinstein and H. Stern, Phys. Lett. <u>21</u>, 447 (19866)
6) M. Maruyama and T. Ueda, Nucl. Phys. <u>A364</u>, 297 (1981), Phys. Lett. <u>124B</u>, 121 (1983), Prog. Theor. Phys. <u>73</u>, 1211 (1985)
7) J.A. Niskanen, V. Kuikka and A.M. Green, Nucl. Phys. <u>A443</u>, 69 (1985)
8) M. Maruyama and T. Ueda, Phys. Lett. <u>149B</u>, 436 (1984), Prog. Theor. Phys. <u>74</u>, 526 (1985)
   M. Maruyama, Osaka University thesis (1984)
9) A.M. Green and J.A. Niskanen, Nucl. Phys. <u>A412</u>, 448 (1984), <u>A430</u>, 605 (1984)

10) A.M. Green, V. Kuikka and J.A. Niskanen, Nucl. Phys. A446, 543 (1986)
11) C. Dover and J.M. Richard, Phys. Rev. C21, 1466 (1980)
12) A. Le Youanc et al, Phys. Rev. D8, 2223 (1973)
    J.P. Ader, B. Bonnier and S. Sood, Nuovo Cim. 68A, 1 (1982)
13) A.M. Green and J.A. Niskanen, Physica Scripta, to be published
14) A.M. Green, J.A. Niskanen and S. Wycech, Phys. Lett. B (1986) to be published
15) M. Kohno and W. Weise, Regensburg preprint TPR-85-25
    E.M. Henley, M. Oka and J. Vergados, Phys. Lett. 166B, 274 (1986)
16) A.M. Green and J.A. Niskanen, in preparation
17) R. Koniuk and N. Isgur, Phys. Rev. D21, 1868 (1980)
    N.A. Törnqvist and P. Zenczykowski, Helsinki preprint HU-TFT-85-10
18) C.B. Dover and P. Fishbane, Nucl. Phys. B244, 349 (1984)
    C.B. Dover, Proc. Int. Symp. on Medium Energy Nucleon and Antinucleon Scattering, Bad Honnef 1985, Springer Lecture Notes in Physics 243, ed. H.V. von Geramb, p. 80

$\overline{N}N$ ANNIHILATION EXPERIMENTS

S.M. Playfer
The Pennsylvania State University
University Park, PA 16802

ABSTRACT

General features of $\overline{N}N$ annihilation at rest are discussed, with particular emphasis on new results on two meson annihilation channels from experiments at LEAR. In the search for baryonium and other exotic states nothing has been seen in recent experiments. It is argued that good neutral detection is important for future annihilation experiments.

ANNIHILATION INTO TWO MESONS

According to a variety of theoretical models of $\overline{N}N$ annihilation [1], the final states should contain two or three mesons, with subsequent decays of broad meson resonances giving the higher multiplicities of charged pions (kaons), and gammas (from $\pi^°(\eta)$ decays), which are observed experimentally [2]. To test theoretical models one would like to distinguish two and three meson final states, but this is very difficult because of the large widths of some mesons (e.g. A_1, $\epsilon(1300)$). Table 1 summarizes the known yields of two meson final states from $\bar{p}p$ annihilation at rest in liquid hydrogen. Only 14 channels are listed out of ~ 100 possible channels, representing 15% of all annihilations. As yet there is little information on multiple $\pi^°$ events which comprise 57% of all annihilations. It can be concluded that a substantial fraction of all annihilations give two meson final states.

Table 1 - Yields of Two Meson Final States in Liquid Hydrogen at Rest

Channel	CERN[2]	Columbia[2]
$\pi^+\pi^-$	0.37±0.03	0.32±0.03
$\pi^\pm\rho^\mp$	}5.8±0.3	2.7±0.4
$\pi^°\rho^°$		1.4±0.2
$\pi^\pm B^\mp$	0.7±0.1	0.7±0.2
$\pi^\pm A_2^\mp$	2.9±0.4	-----
$\pi^° f$	0.24±0.07	-----
$\rho^°\rho^°$	0.12±0.12	0.4±0.3
$\rho^°\omega$	2.1±0.2	0.7±0.3
$\rho^° f$	0.9±0.2	-----
ωf	1.7±0.2	-----
K^+K^-	0.096±0.008	0.11±0.01
$K^°K^°$	0.081±0.005	0.072±0.010
$K^\pm K^{*\mp}$	0.104±0.011	0.086±0.018
$K^°K^{*°}$	0.158±0.015	0.128±0.024

Some interesting results on two meson final states are coming from the ASTERIX detector at LEAR which studies $\bar{p}p$ annihilation at rest in a hydrogen gas target. In a gas target stark mixing effects are reduced, so the atomic cascade is more likely to reach a low-lying P-state where P-wave annihilation can occur. The ASTERIX detector observed an L X-ray yield of $(13\pm2)\%$ and a K_α yield of $\sim 10^{-3}$, showing that almost all of the 2p level annihilates. The observed L X-rays are used to tag P-wave annihilations which differ from S-wave annihilations through selection rules based on angular momentum, parity and G-parity (Table 2). An experiment measures the yield of a channel as a fraction of all annihilations. To understand this yield it is necessary to establish the initial atomic state populations, and then for each initial state to consider the competition between the final states selected by that initial state. From Table 2 one sees that these patterns of competition differ significantly between P-states

Table 2 - Selection Rules for Two Meson Final States

Initial $\bar{N}N$ State	$\pi^+\pi^-$	$\pi^0\pi^0$	$\pi^0 n$	$\pi^{\pm}\rho^{\mp}$	$\pi^0\rho^0$	$\pi^0\omega$	$\pi^{\pm}B^{\mp}$	$\pi^0 f$	$\pi^{\pm}A_2^{\mp}$
$^{11}S_0$	no	no	no	no	no	no	no	no	$\ell_f=2$
$^{13}S_1$	no	no	no	$\ell_f=1$	no	no	$\ell_f=0,2$	no	no
$^{31}S_0$	no	no	no	$\ell_f=1$	$\ell_f=1$	no	no	$\ell_f=2$	no
$^{33}S_1$	$\ell_f=1$	no	no	no	no	$\ell_f=1$	no	no	$\ell_f=2$
$^{11}P_1$	no	no	no	$\ell_f=0,2$	$\ell_f=0,2$	no	$\ell_f=1$	no	no
$^{13}P_0$	$\ell_f=0$	$\ell_f=0$	no	no	no	no	no	no	no
$^{13}P_1$	no	no	no	no	no	no	no	no	$\ell_f=1$
$^{13}P_2$	$\ell_f=2$	$\ell_f=2$	no	no	no	no	no	no	$\ell_f=1$
$^{31}P_1$	no	no	no	no	no	$\ell_f=0,2$	no	no	$\ell_f=1$
$^{33}P_0$	no	no	$\ell_f=0$	no	no	no	$\ell_f=1$	no	no
$^{33}P_1$	no	no	no	$\ell_f=0,2$	no	no	$\ell_f=1$	$\ell_f=1$	no
$^{33}P_2$	no	no	$\ell_f=2$	$\ell_f=2$	no	no	$\ell_f=1$	$\ell_f=2$	no

(ℓ_f = Orbital Angular Momentum between Final State Mesons).

and S-states.

The ASTERIX group uses yields for $\bar{p}p \to K_S K_S$ (from $^3P_{0,2}$ only), and $\bar{p}p \to K_S K_L$ (from $^3S_1$ only), to determine a $(54\pm10)\%$ P-wave annihilation probability with no X-ray tag [3][+]. They also obtain the surprising result:

$$\frac{(\bar{p}p \to K_S K_S)}{\text{all P-wave annihilations}} = (0.052\pm0.015) \frac{(\bar{p}p \to K_S K_L)}{\text{all S-wave annihilations}}$$

In another contribution to this conference [4], you will hear that:

$$\left(\frac{\bar{p}p \to K^+K^-}{\bar{p}p \to \pi^+\pi^-}\right)_P = (0.15\pm0.05) \left(\frac{\bar{p}p \to K^+K^-}{\bar{p}p \to \pi^+\pi^-}\right)_S$$

These two cases of order of magnitude suppressions of kaon channels from the P-wave look like the same effect. It would be interesting for ASTERIX to also measure the yield of $\bar{p}p \to K^*K$ from the P-wave.

As a final example of the ASTERIX results I shall discuss $\bar{p}p \to \pi\rho$. From Table 2 one sees that $\pi\rho$ can be produced from initial states $^{13}S_1$, $^{31}S_0$, $^{11}P_1$, $^{33}P_1$ and $^{33}P_2$. The different ρ polarizations lead to different decay angle distributions for $\rho \to \pi\pi$. A comparison of predicted angular distributions with the data, gives 83% of all P-wave $\pi\rho$ from $^{11}P_1$ [5]. From bubble chamber data it is known that $^{13}S_1 > 97\%$ of all S-wave $\pi\rho$ [6]. There is an apparent selection rule for I=0, but this cannot be absolute since $\bar{p}n \to \pi^-\rho^0$ is also seen in bubble chamber experiments.

[+] The observation of one $K_S K_S$ event in liquid suggests $\sim 10\%$ P-wave annihilation in liquid.

Table 3 - Searches for Narrow States in $\bar{p}p$ Annihilation

Experiment	Ref.	Measurement	States Seen	Signal or 90% C.L.
Bruckner et al	[7]	σ_A	1936 MeV	(26±6)mb-MeV
Brando et al	[8]	σ_A	--	<8mb-MeV
Clough et al	[9]	σ_T	--	<2mb-MeV
Pavlopoulos et al	[10]	$\bar{p}p \to \gamma X$	1395	$(8.5\pm2.0)\times10^{-3}/\bar{p}$
			1646	$(6.0\pm1.9)\times10^{-3}/\bar{p}$
			1684	$(7.2\pm1.7)\times10^{-3}/\bar{p}$
Richter et al	[11]	$\bar{p}p \to \gamma X$	1210	$(1.0\pm0.3)\times10^{-3}/\bar{p}$
			1638	$(3.0\pm0.8)\times10^{-3}/\bar{p}$
			1694	$(1.6\pm0.4)\times10^{-3}/\bar{p}$
			1771	$(1.8\pm0.4)\times10^{-3}/\bar{p}$
Angelopoulos et al	[12]	$\bar{p}p \to \gamma X$	--	$<5\times10^{-4}/\bar{p}$
Angelopoulos et al	[13]	$\bar{p}p \to \pi^{\pm}X$	--	$<5\times10^{-4}/\bar{p}$

SEARCHES FOR EXOTIC STATES

A major goal of recent $\bar{N}N$ annihilation experiments has been the search for exotic states. $\bar{N}N$ potential models predict a large spectrum of bound states ("baryonium"), clustered about the $\bar{N}N$ threshold. Models of QCD confinement suggest that in addition to the known $q\bar{q}$ mesons, there should also be four quark states ($qq\bar{q}\bar{q}$), hybrid states ($q\bar{q}g$), and glueballs (gg),

A decade ago signals were reported for a narrow state above threshold in σ_A, σ_T (S-meson), and for narrow states below threshold in $\bar{p}p \to \gamma X$. Attempts to reproduce these signals in recent experiments have failed (Table 3). It now appears that there are no new narrow states ($\Gamma \lesssim 10$ MeV) with significant coupling to $\bar{p}p$ below 2 GeV. The existence of new broad states is still possible, but the experiments are difficult, and require measurements of exclusive final states. No convincing evidence for new broad states exists at this time.

THE IMPORTANCE OF NEUTRAL DETECTION

The lack of information on annihilations with several neutrals is an important obstacle to understanding the mechanisms of $\bar{N}N$ annihilation. From Table 2 it can be seen that neutral two meson channels are more selective than charged channels (e.g. $\pi^\circ\pi^\circ, \eta\pi^\circ$ are forbidden from the S-wave, whereas $\pi^+\pi^-$ is allowed). This should make their yields easier to understand. All neutral events are a good place to try and find possible three meson final states, since many possible two meson contributions are forbidden (e.g. $\bar{p}p \to \pi^\circ\pi^\circ\pi^\circ$ contains no $\pi^\circ\rho^\circ$ contribution).

A new 4π detector known as the Crystal Barrel is to be installed at LEAR in 1988. The design of the detector is shown in Figure 1. It includes gamma detection over 98% of 4π by CsI crystals, and charged particle detection over 99% of 4π using a precision cylindrical drift chamber in a 1.5T magnetic field. It is possible to select events in which the measurement of energy (momentum) and direction of all the annihilation products gives the maximum number of constraints for the kinematic fitting. The detector is specifically optimized

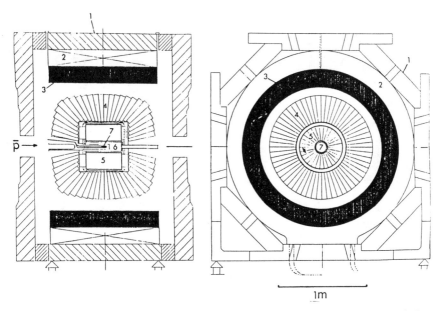

Fig. 1: The Crystal Barrel Detector: (1) Yoke, (2) Old Coil, (3) New Coil, (4) CsI Barrel, (5) Jet Drift Chamber, (6) XDC/PWC, (7) LH$_2$ Target.

for the detection of the decays $\eta \to \gamma\gamma$, $\omega \to \pi^0\gamma$ and $K_S \to \pi^+\pi^-$, where the constraints are important to isolate narrow peaks on a combinatorial background (Fig. 2(a)).

A HYBRID MESON SEARCH

Table 4: The dominant decays of the low-lying exotic meson hybrids.

Hybrid state	(Decay mode)$_{L \text{ of decay}}$	Partial width
x_2^{+-} (1900)	$(\pi A_2)_P$	450 (MeV)
	$(\pi A_1)_P$	100
	$(\pi H)_P$	150
y_2^{+-} (1900)	$(\pi B)_P$	500
x_1^{-+} (1900)	$(\pi B)_{S,D}$	100, 30
	$(\pi D)_{S,D}$	30, 20
y_1^{-+} (1900)	$(\pi A_1)_{S,D}$	100, 70
	$[\pi\pi(1300)]_P$	100
	$(\overline{K}Q_2 + \text{c.c.})_S$	~100
x_0^{+-} (1900)	$(\pi A_1)_P$	800
	$(\pi H)_P$	100
	$[\pi\pi(1300)]_S$	900
y_0^{+-} (1900)	$(\pi B)_P$	250

In the search for exotic states, Isgur et al [14] have found an ingenious way of hiding hybrid mesons. In their flux-tube model the hybrid mesons decay to P-state mesons, with large widths (Table 4). Hybrid mesons are particularly interesting to search for because they can have exotic (non $q\bar{q}$) J^{PC} such as 0^{+-}, 1^{-+}, 2^{+-}, which make their identification unambiguous. In Table 4 the relatively narrow x_1^{-+} (1900) state decaying to πB, πD looks the most accessible to experiment. Consider $\bar{p}p \to \pi X$ at a $\bar{p}$ momentum of 2 GeV/c. The final state $\pi\pi B$ gives $\pi\pi\pi\omega$, while $\pi\pi D$ gives

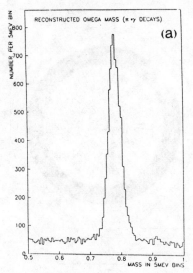 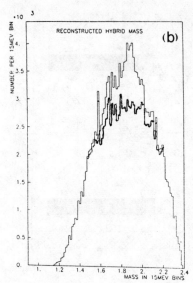

Fig. 2: Simulation of Hybrid Meson Search in $\bar{p}p \to \pi X$ at 2 GeV/c.
(a) $\omega \to \pi^\circ \gamma$ signal and background, (b) $X \to \pi B$ signal and background.

$\pi\pi\pi\pi\eta$, $\pi\pi\pi KK$. These high multiplicity channels are difficult to identify, and almost all contain several neutrals.

The Crystal Barrel will search for exotic J^{PC} resonances in these channels using its ability to isolate η, ω and K_S decays. As an example take the channel $\bar{p}p \to \pi^+\pi^-\pi^\circ\pi^\circ\gamma$, where the decay $\omega \to \pi^\circ\gamma$ is identified (Figure 2(a)). Figure 2(b) shows a simulated signal for a 10^{-4} yield of $\bar{p}p \to \pi X$, including estimated background channels. The yield of 10^{-4} is at the limit of sensitivity for this experiment. For a more typical hadronic yield of 10^{-2} (like $\pi B, \pi A_2$), the signal for the hybrid meson would be clear, and a J^{PC} assignment could be made.

CONCLUSIONS

After two years of experimental work at LEAR, our knowledge of $\overline{N}N$ annihilation remains limited. Data on P-wave annihilations at rest show surprising differences from S-wave annihilations, but earlier evidence for narrow baryonium states has not been confirmed by the new high statistics experiments. In order to make a comprehensive study of $\overline{N}N$ annihilation, final states with several neutrals should be studied. The new Crystal Barrel detector is designed to do this, and also to search for hybrid states with exotic J^{PC}.

ACKNOWLEDGEMENTS

I am indebted to C. Dover for guidance on theoretical aspects of $\overline{N}N$ annihilation, and to C. Amsler for discussions about the ASTERIX experiment and the Crystal Barrel project.

REFERENCES

[1] J.A. Niskanen, C.B. Dover, this conference.
[2] R. Armenteros and B. French, in "High Energy Physics," ed. E.H.S. Burhop, Academic Press (1969).
[3] S. Ahmad et al, Proc. of the Third LEAR Workshop (Tignes), Editions Frontieres (1985), p. 353.
[4] G. Marshall, this conference.
[5] S. Ahmad et al, Proc. of the VII European Symposium on Antiproton Interactions (Durham), Adam Hilger (1984), p. 287.
[6] M. Foster et al, Nucl. Phys. B6, 107 (1968).
[7] W. Bruckner et al, Phys. Lett. 67B, 222 (1977).
[8] T. Brando et al, Phys. Lett. 158B, 505 (1985).
[9] A.S. Clough et al, Phys. Lett. 146B, 299 (1984).
[10] P. Pavlopoulos et al, Phys. Lett. 72B, 415 (1978).
[11] B. Richter et al, Phys. Lett. 126B, 184 (1983).
[12] A. Angelopoulos et al, to be published; S. Playfer, this conference.
[13] A. Angelopoulos et al, Proc. Conference on Hadron Spectroscopy (Maryland), AIP Conf. Proc. No. 132 (1985), p. 144.
[14] N. Isgur, R. Kokoski and J. Paton, Phys. Rev. Lett. 54, 869 (1985).

STRANGENESS PRODUCTION WITH ANTIPROTONS

Peter D. Barnes and Christopher J. Maher
Carnegie-Mellon University, Pittsburgh, Pa. 15213

PS - 185 Collaboration*

ABSTRACT

New data from LEAR on the production of hyperon-antihyperon pairs in strong interactions is reported. The possibility of checking CP invariance in the $\bar{\Lambda}\Lambda$ system is reviewed together with plans to search for the $\xi(2230)$ in the $p + \bar{p} \to K_S + \bar{K}_S$ reaction.

The antiproton-proton interaction is being investigated extensively at the LEAR facility and includes measurements of leptonic, mesonic, and baryonic annihilation channels. I will restrict my remarks to antiproton-proton annihilation channels into two body strange/anti-strange final states. Recent studies of $\bar{\Lambda}\Lambda$ production will be compared with new theoretical models, the possibility of looking for CP violation in the $\bar{\Lambda}\Lambda$ system will be discussed, and finally a planned search for the $\xi(2230)$ resonance, originally reported by the Mark III detector group, will be reviewed.

LAMBDA ANTI-LAMBDA PRODUCTION

A detailed investigation of hyperon-antihyperon production in the reaction $\bar{p} + p \to \bar{Y} + Y$ is in progress at LEAR[1-4]. Threshold measurements of $\bar{\Lambda} + \Lambda$ and $\bar{\Sigma} + \Lambda$ ($\bar{\Lambda} + \Sigma$) are in progress, while a study of $\bar{\Sigma} + \Sigma$ is planned. I will report preliminary results from an investigation of total $\bar{\Lambda} + \Lambda$ cross sections, $\sigma_{\bar{\Lambda}\Lambda}$, differential cross sections, $d\sigma/d\Omega$, and polarization measurements, $P_{\bar{\Lambda}}$ and P_{Λ} near 1.5 GeV/c. These were made by the PS185 collaboration[1] using an imaging detector for event reconstruction and taking advantage of the self analyzing feature of the weak Λ decay to extract information on the spin variables.

*PS 185 Collaboration: P.D. Barnes, R. Besold, P. Birien, B. Bonner, W.H. Breunlich, G. Diebold, W. Dutty, R.A. Eisenstein, G. Ericsson, W. Eyrich, R.v. Frankenberg, G. Franklin, J. Franz, N. Hamann, D. Hertzog, A. Hofmann, T. Johansson, K. Kilian, C.J. Maher, R. Mueller, W. Oelert, S. Ohlsson, H. Ortner, P. Pawlek, B. Quinn, E. Roessle, H. Schledermann, H. Schmitt, J. Seydoux, J. Szymanski, P. Woldt.

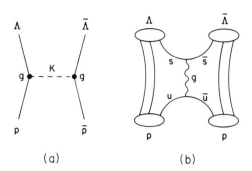

Fig. 1. Models of Hyperon production in $\bar{p}p$ interactions: a) t-channel kaon exchange, b) one-gluon exchange.

These measurements have stimulated several theoretical investigations[5-8] of this reaction in terms of t-channel meson exchange, Fig. 1a, and one-gluon exchange, Fig. 1b. Exchange of the K(495 MeV, 0^-) meson gives the transition potential[8]:

$$V_K(r) = (g^2_{KN\Lambda} m_K^2 / 48 \pi M_N M_\Lambda) [\sigma_1 \cdot \sigma_2 Y_0 + S_{12} Y_2]$$

which has a strong spin dependence through the tensor interaction, S_{12}. While Kohno and Weise[8] have utilized only one-kaon exchange, other authors have also included exchange of more massive mesons, namely, Niskanen[7] : K^*(892 MeV, 1^-), Tabakin and Eisenstein[5] : K^*(892) and K^{**}(1430 MeV, 2^+), and Dillig and Frankenberg[6] : K^*(892) and κ(1350 MeV, 0^+).

Various authors[8-10] have interpreted this reaction in terms of the quark flow diagram of Fig. 1b which gives a one-gluon exchange potential[8]:

$$V(q\bar{q} \to s\bar{s}) = -(4\pi \alpha_s/24m_G^2) S^2 (\lambda_{q\bar{q}} - \lambda_{s\bar{s}})^2 \delta^3_{q\bar{q}} \delta^3_{s\bar{s}} \delta^3_{q\bar{s}}$$

where λ are Gellmann's color matrices and S is the total qq spin. Kohno and Weise[8] take as the coupling strength, $\alpha_s/m_G^2 \sim 0.13$ fm^2 and use gaussian quark wave functions to derive an overall transition potential of:

$$V^{SI}(r) = 16 \pi \alpha_s \delta_{S1} \delta_{I1} (3 m_G^2)^{-1}$$
$$* [3/(4\pi <r^2>)]^{3/2} \exp(-3r^2/4<r^2>),$$

where $<r^2>$ is the mean square radius associated with the quark distributions.

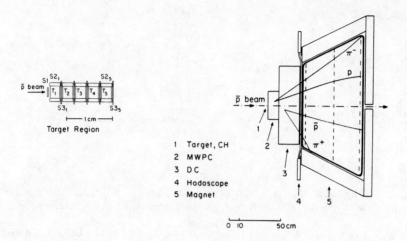

Fig. 2. Experimental layout for experiment PS185.

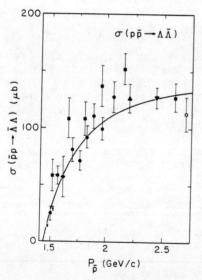

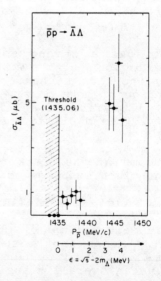

Fig.3. Data[21] and calculation[8] of the momentum dependence of $\sigma_{\bar{\Lambda}\Lambda}$.

Fig.4. Preliminary Lear measurements (PS185) of $\sigma_{\bar{\Lambda}\Lambda}$ near threshold.

These various transition potentials are then incorporated in the calculations, with different schemes for handling the initial and final state interactions being employed by the different authors. Also, different choices are made for the distortion potentials, such as the $\bar{p}p$ potential of Dover and Richard[11] for the incident channel and the Nijmegen potential[12] for the exit channel.

The PS185 technique for study of the reaction

$$\bar{p} + p \rightarrow \bar{\Lambda} + \Lambda \rightarrow (\bar{p}\,\pi^+)(p\,\pi^-)$$

has been described previously[1-4] and is shown in Fig. 2. Briefly, the antiproton beam is incident on a small target followed by MWPC and drift chamber planes, a scintillator hodoscope, and a magnet for baryon number identification. The target consists of one to five cells of CH_2 or graphite (2.5 cm dia. x 2.5 cm thick) each surrounded by a scintillator veto box as part of the neutral trigger. Measurements have been made in the region from below threshold (1435.06 MeV/c) up to 1550 MeV/c. Preliminary cross section results, $\sigma_{\bar{\Lambda}\Lambda}$, are compared to previous measurements and to the one kaon exchange calculations[8] in Fig. 3. with detailed results for the threshold region shown in Fig. 4. The differential cross sections obtained at 1476.5 and 1507.5 MeV/c are shown in Fig. 5 and are compared to the

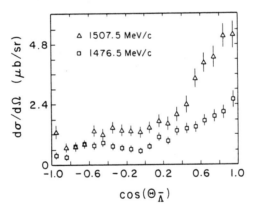

Fig. 5. Preliminary differential cross sections at 1508.5 and 1476.5 GeV/c.

kaon and gluon exchange calculations in Figs. 6a and 6b. The agreement of both the Tabakin[5] and the Kohno[8] meson exchange calculations with the data is quite good as is the effective one-gluon exchange representation of Kohno[8].
    The polarization, transverse to the production plane, P_y, measured at 1508 MeV/c is shown in Fig. 7. and demonstrates good agreement between $P_{\bar{\Lambda}}$ and P_{Λ} as expected from C invariance in strong interactions. These results are also compared with the kaon exchange calculations in Fig. 7. Only the one-kaon exchange calculations of Kohno and Weise reproduce the magnitude and the apparent sign reversal of the measured polarization.

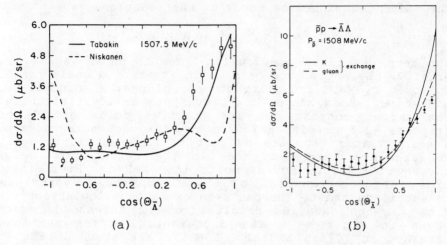

Fig. 6. Comparison of $d\sigma/d\Omega$ with a) kaon exchange calculations of Tabakin[5] (solid) and Niskanen[7] (dashed) and b) with one-gluon and one-kaon exchange calculations of Kohno and Weise[8].

Additional runs very near threshold and a high statistics run at 1550 MeV/c have now been completed and are in the process of analysis. These will provide even more severe tests of both the t-channel meson exchange calculations and of the effective one-gluon exchange treatments of this strangeness production reaction.

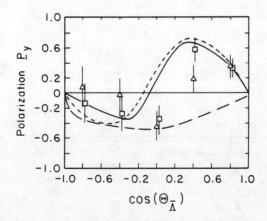

Fig. 7. Comparison of PS185 polarization data with calculations: Tabakin[5] (long dashes), Kohno and Weise[8] one-kaon (solid) and one-gluon (short dashed). The triangles (squares) are $\bar{\Lambda}$ (Λ) data.

CP NONCONSERVATION IN THE $\bar{\Lambda} + \Lambda$ SYSTEM

The predominant mesonic and leptonic decay modes of the Λ hyperon have been well studied. The current value[13] of the total decay rate is $\Gamma_\Lambda = 3.7994 \pm 0.0003 \times 10^9$ / sec and of the weak decay asymmetry parameter is, $\alpha_\Lambda = 0.647 \pm 0.013$. CPT invariance implies $\Gamma_{\bar{\Lambda}} = \Gamma_\Lambda$, while CP invariance implies $\alpha_{\bar{\Lambda}} = -\alpha_\Lambda$.

The possibility of testing CP invariance by comparing Λ and $\bar{\Lambda}$ decay has been discussed in the past[14,15] and will be explored in new calculations by John Donoghue at this conference[16]. I will give a brief review of estimates of CP noninvariance in the $\bar{\Lambda} + \Lambda$ system and discuss some of the implications for the design of an experiment.

The weak decay amplitudes can be catalogued in terms of orbital angular momentum and isospin change as:

	$\Delta I = 1/2$	$\Delta I = 3/2$
S Wave:	$A_S(1) \, e^{i\delta_1}$	$A_S(3) \, e^{i\delta_3}$
P Wave:	$A_P(1) \, e^{i\delta_{11}}$	$A_P(3) \, e^{i\delta_{31}}$

where δ is the πN phase shift induced by the final state interaction. Two possible CP violating phases, θ_1 and ϕ, can be defined by the ratios:

$$A_S(1) / A_P(1) = | A_S(1) / A_P(1) | \, e^{+i\theta_1}$$

and

$$A_S(3) / A_S(1) = | A_S(3) / A_S(1) | \, e^{-i\phi}$$

These phases can be related to the measured decay rates and weak decay asymmeteries using measured values of the strong πN phase shifts, δ, as:

$$\delta\Gamma = [\Gamma - \bar{\Gamma}]/[\Gamma + \bar{\Gamma}] = (7 \times 10^{-3}) \sin\phi$$

and

$$\delta\alpha = [\alpha + \bar{\alpha}]/[\alpha - \bar{\alpha}] = (1 \times 10^{-1}) \tan\theta_1.$$

The present limit on CP invariance is $(\alpha p)_{\bar{\Lambda}} / (\alpha p)_\Lambda = -1.04 \pm 0.29$, obtained from the R608 experiment at the ISR[17]. Insofar as ϕ and θ_1 are of comparable magnitude, then $\delta\alpha$ is the more sensitive experimental parameter.

To obtain estimates of these CP violating phases one can turn to the $K^0 \bar{K}^0$ system. The only known examples of CP noninvariance come from three branches of K_L decay: $K_L \to \pi^+\pi^-$, $\pi^0\pi^0$, and $\pi l \nu$. They are described in terms of the two parameters, ε and ε':

and
$$\varepsilon = (2.27 \pm 0.002) \times 10^{-3}$$
$$\varepsilon' = -(3 \pm 6) \times 10^{-3}.$$

In the superweak model of Wolfenstein[18] where a $\Delta S = 2$ force is introduced, the CP violation is in the mass matrix which describes the $K^0 \bar{K}^0$ mixing in which case, $\varepsilon' = 0$. In milliweak models, as for example the Kobayashi-Maskawa model[19], ε' describes the $\Delta S = 1$, CP violating amplitude and is expected to have a magnitude given by:

$$\varepsilon/\varepsilon' = -(1.0 \text{ to } 7.0) \times 10^{-3}.$$

In this context several authors have made predictions for the $\bar{\Lambda} \Lambda$ CP violating phases. Wolfenstein and Chang[14] find

$$\delta\Gamma = \varepsilon'/5 = -(1 \pm 2) \times 10^{-6}$$
and
$$\delta\alpha = 3\varepsilon' = -(2 \pm 4) \times 10^{-5}.$$

Donoghue[15] finds that $\Delta S = 1$ CP violation interactions contribute dispersion corrections to the mass matrix and obtains:

$$\delta\Gamma = -(0.5 \text{ to } 8.0) \times 10^{-6}$$
and
$$\delta\alpha = -(0.1 \text{ to } 0.8) \times 10^{-4}.$$

An experimental search for these CP violating effects requires a detailed study of the reaction:

$$\bar{p} + p \rightarrow \bar{\Lambda} + \Lambda \rightarrow (\bar{p}\pi^+)(p\pi^-)$$

illustrated in the center of mass frame in Fig. 8. At the

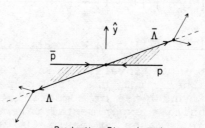

Production Plane (c.m.)

Fig. 8. Production plane for $\bar{\Lambda}\Lambda$ production relative to which the asymmetery test is applied.

$\bar{\Lambda}\Lambda$ production vertex C invariance in strong interactions requires that for the polarization transverse to the production plane: $P_{\bar{\Lambda}} = P_{\Lambda}$. In the subsequent weak decay of these hyperons, the proton(antiproton) recoil direction is correlated with the $\Lambda(\bar{\Lambda})$ spin direction:

$$I(\theta_p) = I_0 [1 + \alpha P \cos \theta_p]$$

where P is typically about 0.5. Thus a nonzero value of $\delta\alpha$ will be manifest as an asymmetry in proton (antiproton) emission about the production plane as measured by :

$$\frac{[N_p(\text{up}) - N_p(\text{down})] + [N_{\bar{p}}(\text{up}) - N_{\bar{p}}(\text{down})]}{N_p + N_{\bar{p}}}$$

In a discussion of the feasibility of this measurement some primary considerations are:

a) Statistics-- Assuming about 5×10^6 antiprotons/sec, a production cross section of 120 μb, an LH_2 target of 10 cm thickness, then of the order of 10^8 useful $\bar{\Lambda}\Lambda$ four prong events will be produced in 35 LEAR days.

b) Trigger-- It is neccessary to adopt a tight target geometery which will permit a clean charge-neutral-charge trigger.

c) Systematic errors-- The relative advantages of colliding beam, jet target, and fixed target geometeries must be evaluated for removal of detection asymmetries of the $\bar{\Lambda}$ and Λ decays.

d) Computer reconstruction time-- The rate for full tracking analysis on $\bar{\Lambda}\Lambda$ four prong events in PS185 is currently about 300 events/VAX-780 hour. Full analysis of 10^8 events at this rate would require 3×10^6 VAX hours or about two Cray years. More efficient programing could significantly shorten this but will still leave one with a major data reduction problem.

In conclusion, a search for CP nonconservation in the $\bar{\Lambda}\Lambda$ system is probably a feasible experiment but would require about 10^{13} to 10^{14} antiprotons and a major computer effort.

SEARCH FOR THE $\xi(2230)$ IN $K_S K_S$ DECAY AT LEAR

The MARK III collaboration at SPEAR has reported evidence[20] for a narrow resonance (ξ) observed in the radiative decay of the J/ψ into both the K^+K^- and $K_S K_S$

channels, see Fig. 9. Analysis of the latter decay channel gives a mass of 2232 ± 10 MeV, $J^{PC} = 0^{++}$, 2^{++}, ..., and a width of 18 +23/-15 ± 10 MeV. It was not observed in $\pi^+\pi^-$, and $\mu^+\mu^-$ channels. It has been discussed in terms of a new L= 3 $s\bar{s}$ state, a glue ball, a light Higgs scalar, an exotic $s\bar{s}s\bar{s}$ state, and a Meikton (q q g) as sumarized in Table I. It is not known whether it couples to the $\bar{p}p$ system. In order to provide confirmation of its existence and additional information about its character and width, three experimental groups at LEAR are making plans to look for it in the $K\bar{K}$ decay channels (PS185, PS170, and PS172). Its mass corresponds to ~ 1435 MeV/c momentum for

Table I Explanation for the $\xi(2230)$

Interpretation	Evidence For:	Evidence Against:
Higgs Scalar	0^{++}, very narrow $\xi \to K\bar{K}$ expected.	B.R. too high, $\mu^+\mu^-$ not seen.
Glueball	2^{++}, $\xi \to K\bar{K}$	Too narrow? $\xi \to \pi\pi$ not seen
L = 3 meson	2^{++}, mass ok	width ≈ 60 MeV
Exotic $s\bar{s}s\bar{s}$	0^{++}, narrow	$\xi \to \eta\eta$, $\eta\eta'$, $\varphi\varphi$ should be substantial

which the nonresonant $K\bar{K}$ background is 55 ± 5 µb in the K^+K^- channel and 3 µb in the $K_S K_S$ channel.

The PS185 collaboration is making preparations to use its imaging detector to look for the $K_S K_S \to (\pi^+\pi^-)(\pi^+\pi^-)$ decay. In order to enlarge the acceptance for large angle decay products, three layers of streamer tube tracking planes are being mounted around a thicker production target. It is difficult to estimate in advance the yield for this measurement. However using the η_c decay for guidance, we find that if
$$BR[\bar{p} p \to \xi] = 0.01$$
and
$$BR[\xi \to K_S K_S] = 0.05$$
a peak cross section of 1.2 µb can be expected for J = 0 and 6.0 µb for J = 2. The PS185 detector sensitivity is such that a 0.4 µb cross section would give a five standard deviation effect for a double branching ratio of about 1×10^{-4}. This measurement is scheduled to take place in August, 1986.

In conclusion, the strange annihilation channels of the $\bar{p}p$ system continue to offer a rich environment to investigate both strong and weak interactions in antiproton physics.

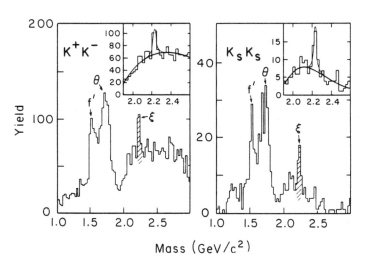

Fig. 9. Narrow resonance (ξ) reported by the Mark III collaboration[20] in the decay of the J/ψ into $K\bar{K}$ channels.

References:

1) PS 185 Collaboration: P.D.Barnes, R.Besold, P.Birien, B.Bonner, W.H.Breunlich, G.Diebold, W.Dutty, R.A.Eisenstein, G.Ericsson, W.Eyrich, R.v.Frankenberg, G.Franklin, J.Franz, N.Hamann, D.Hertzog, A.Hofmann, T.Johansson, K.Kilian, C.Maher, R.Mueller, W.Oelert, S.Ohlsson, H.Ortner, P.Pawlek, B.Quinn, E.Roessle, H.Schledermann, H. Schmitt, J.Seydoux, J.Szymanski, P.Woldt, CERN/PSCC/81-69 (1981)
2) B. Bonner, Proceedings of the Workshop on Antiproton Interactions, Durham, U.K., 1983.
3) P. D. Barnes, Proceedings of the Conf.on Antinucleon and Nucleon-Nucleus Interactions, Plenum Press, 1985.
4) N. Hamann, Proceedings of the International Symposium on Medium Energy Nucleon and Antinucleon Scattering, Bad Honnef, Germany, 1985.
5) F. Tabakin and R. A. Eisenstein, Phys. Rev.C31(1985)1857.
6) M. Dillig and R.v.Frankenberg, Proceedings of the Conf.on Antinucleon and Nucleon-Nucleus Interactions, Plenum Press, 1985.
7) J. Niskanen, Helsinki preprint HU-TFT-85-28,(1985); A.M.

Green and J. Niskanen, review paper, Helsinki preprint (1986).
8) M.Kohno and W. Weise, Regensburg preprint, TPR-86-9, (1986).
9) B. Anderson
10) H.R. Rubinstein and H. Snellman, Phys. Lett. 165B 187(1985).
11) C.B Dover and J.M. Richard, Phy. Rev C21 (1980).
12) M.M. Nagels, T.A. Rijken and J.J. de Swart, Phys. Rev. D15 2547(1977).
13) Particle Data Book, Reviews of Modern Physics, B56 (1984).
14) L. Wolfenstein and D. Chang, CMU preprint CMU-HEP83-5 and Proc. of the Symposium on Intense Medium Energy Sources of Strangeness, Univ. of Calif. at Santa Cruz, 1983; L.L. Chau, BNL Report 31859-R, Phys. Reports
15) J. F. Donoghue,Proceedings of the Workshop on Antimatter Interactions at Low Energy, Fermi Lab. 1986.
16) J. F. Donoghue, E. Golowich and B. R. Holstein, Invited talk delivered at this conference.
17) P.Chauvat et al,CERN preprint CERN-EP/85-185, submitted to Phy Lett B.
18) L. Wolfenstein, Phys. Lett. 13 562 (1964).
19) M. Kobayashi and K. Maskawa, Prog. Theor. Phy. 89 282 (1972); for a review see ref. 14.
20) R. M. Baltrusaitis et al,Phy Rev. Lett. 56 107(1986).
21) B. Jayet et al., Nuovo Cim. 45A 371(1978),

NUCLEON-ANTINUCLEON SPIN PHYSICS

R.D. Ransome
Rutgers University, Piscataway, NJ 08854

ABSTRACT

The current status spin dependent measurements in the nucleon-antinucleon system is briefly reviewed. A proposed method for producing a polarized beam is discussed, along with the experiments that could be done with such a beam.

INTRODUCTION

During the past 30 years a great deal effort has been invested in the study of the spin dependence of the of the nucleon-nucleon interaction. This has been important both for the understanding of the fundamental interaction between nucleons and the understanding of nuclear structure.

Today a nearly complete phenomenological description of nucleon-nucleon scattering exists up through to 800 MeV.[1] This was made possible by the existence of polarized proton targets and polarized proton and neutron beams over a large energy range.

In contrast, the nucleon-antinucleon interaction is much less well understood, both theoretically and experimentally.

Theoretically, one would like to be able to use models of the NN interaction to predict the $\overline{N}N$ interaction. With meson-exchange models one can relate the $\overline{N}N$ potential to the NN potential via a G-parity transformation plus the addition of an imaginary potential to describe the annihilation part of the cross section. If this technique does not give a good description of the $\overline{N}N$ scattering, this could be evidence for quark-quark interactions at low energies. One would also like to know the spin dependence of annihilation, both the total and for specific channels, as this more directly gives information on the spin dependence for quark-quark interactions at low energies. A complete partial wave analysis could also indicate resonances which are too broad or weak to be seen easily in the total cross section measurements.

Before the startup of the Low Energy Antiproton Ring (LEAR) at CERN, there were few measurements below 400 MeV/c. At higher momenta there were measurements of only differential and total cross sections, and a few analyzing power measurements. Recent experiments at LEAR have extended the measurements to lower momenta. LEAR experiment PS173 has made measurements of the differential elastic and charge exchange cross sections down to 180 MeV/c.[2] The SING collaboration (PS172) has made measurements of the total cross section down to 220 MeV/c,[3] the analyzing power and Donon[4] to about 350 MeV/c.[5] Although these are important measurements, a complete description of the interaction requires spin dependent measurements.

The scattering of two spin one-half particles can be described by a matrix of the form[4]

$$M = 0.5[(a + b)+(a - b)(\sigma_{1n}\sigma_{2n}) + (c + d)\sigma_{1m}\sigma_{2m} \qquad (1)$$
$$+ (c - d)\sigma_{11}\sigma_{21} + e(\sigma_1 + \sigma_2)\sigma n + f(\sigma_1 - \sigma_2)n]$$

assuming invariance under spatial rotations, time reversal, and parity transformations. For nucleon-nucleon scattering, the last term, f, is zero under isospin invariance. It also zero for nucleon-antinucleon scattering under charge conjugation invariance. In principle, 9 experiments are required for each energy and angle for a complete determination of the scattering amplitude. This is simplified by use of the phase shift analysis technique, which is especially at low energies where only a few partial waves contribute.

A phase shift analysis for $\overline{N}N$ scattering is somewhat more complicated than for NN. All partial waves are allowed for each isospin state and the reaction has inelastic channels from zero energy, which makes it essential to measure as many spin dependent parameters as possible.

POLARIZATION TECHNIQUES AND EXPERIMENTS

Up to now, the major obstacles have been the lack of polarized antiproton beams and low rates. To a large degree LEAR has overcome the rate problem. Several suggestions have been made for producing polarized antiproton beams and it now seems probable that the spin dependence of the $\overline{N}N$ system will be explored within the next few years.

The following methods have been suggested for polarizing antiprotons: double scattering, making anti-Hydrogen and using optical pumping[6] or the Lamb-shift,[7] using the Stern-Gerlach effect,[8] and the spin-filter method (SFM).[9,10] Antiproton beam intensities are still too low to make double scattering a feasible method. The other methods all potentially allow polarization of the antiprotons without unacceptably high losses of the antiprotons. The SFM will be described here, along with the feasible first-generation experiments.

The spin filter method was first suggested by Csonka for the ISR, and by Kilian and Mohl for LEAR.[10] A collaboration Heidelberg, Houston, Mainz, Munich, Rice, Rutgers, and Wisconsin has proposed to test the method at LEAR.[11]

The SFM uses the spin dependence of the $\overline{N}N$ interaction to polarize the beam. A polarized hydrogen target is inserted in a storage ring, such as LEAR. The intially unpolarized beam can be thought of as two beams polarized in opposite directions. Assuming a 100% polarized target, the total cross section for each beam is

$$\sigma = \sigma_0 + \sigma_1 \text{ and } \sigma = \sigma_0 - \sigma_1 \qquad (2)$$

where σ_0 is the total cross section for unpolarized beam and target.

If we define a time constant $\tau = (\sigma_0 n f)^{-1}$ where n is the target thickness and f the circulation frequency, the intensity and polarization as a function of time will be given by

$$I = I_0 e^{-t/\tau} \cosh(\sigma_1 t/\sigma_0 \tau) \qquad (3)$$

$$P = \tanh(\sigma_1 t/\sigma_0 \tau)$$

For $\bar{p}p$ at 300 MeV/c, σ_0 is about 250 mb and f is $.5 \cdot 10^6$, Assuming a target thickness of 10^{14} H/cm² gives a value for tau of 10 hours. The ratio of σ_1/σ_0 is completely unknown but various models give values in the neighborhood of 0.05.[11] The maximum for np scattering is about 0.1 and it seems unlikely a much higher value will be found for pp. After 10 hours, one could expect an antiproton beam with a polarization of about 5%, and intensity of 37% the original intensity, about 10^{10} stored antiprotons.

The major questions for the filter method are: can a target with 10^{14} H/cm² be developed, can a low momentum beam be stored for 10 hours or more without depolarizing, and is the ratio σ_1/σ_0 at least 0.05 for some momentum in the LEAR range.

The proposed target is a storage bottle, suggested by Haberli.[13] This consists of a cylinder with the axis along the beam and open at each end. Polarized hydrogen is pumped into the cylinder via an opening on one side. The gas fills the cylinder, giving an effective target thickness of about 10 cm, rather than the few millimeters of a jet target. With current technology, densities of a few 10^{13} H/cm² can be acheived. The collaboration[10] is currently working to improve this to 10^{14} H/cm².

The question of beam storage for long periods at low momenta will be explored at Heidelberg. A test storage beam (TSB) is currently being built. This will allow tests of the filter method to be made with protons. The prototype target will also be tested. If tests of the method are successful at the TSB, the experiment will be done at LEAR, with antiprotons. The final question of σ_1/σ_0 will be known only when the experiment is performed at LEAR.

Given a polarized beam, the following parameters might be measured: Donon, Dosos, Dosol, Knoon, Ksoos, Ksool, Aoonn, Aooss, Aoosl. The D parameters require measurement of the scattered proton polarization, the K's measurement of the scattered antiproton polarization, and the A's only the asymmetry of the scatter. Only the A parameters require a polarized antiproton beam. Neglected here are those parameters requiring longitudinally polarized beam, which is more difficult to produce with this method. Assuming a polarized beam of 5% is acheived, what experiments could be done? Assume one has 10^{10} polarized $\bar{p}$ (or $5 \cdot 10^{10}$ unpolarized $\bar{p}$), a target thickness of 10^{14} H/cm^2, a detector solid angle of 10 msr, and cross section for elastic scattering cross section of 1 mb/sr, about the minimum cross section for momenta less than 800 MeV/c. For those parameters requiring measurement of the scattered particles polarization, assume a mean analyzing power of 0.2 for protons and 0.05 for antiprotons, with 0.1% usefully scattered, and that parameters will be measured with a statistical uncertainty of 0.05.

Measurement of the A parameters requires about 80,000 scatters to achieve the desired precision. The rate is $(10^{10}\ \bar{p})\ (10^6/\text{sec})\ (10^{14}$ H/cm$^2)\ (10^{-27}$ cm$^2)\ (.01\ \text{sr}) = 10/\text{sec}$, for a required time of 2.2 hours. A similar calculation for the D parameters gives 10,000 scatters required with a rate of .05/sec, for a total time of 55 hours. For the K parameters 160,000 scatters are required with a rate of .05/sec for a total time of 880 hours.

From a rate viewpoint, both the A and D parameters are easily measured. The K's are difficult with the lowest cross sections. However, it should be noted the lab scattering cross sections for the antiprotons reach nearly 80 mb/sr for forward scattering, which brings the times required here to less than a day.

The detector for this experiment is still being designed. I would just like to mention here a few of the constraints which limit the energy and angular range of the measurements.

Particle identification is required for two reasons. Determination of the center of mass scattering angle requires separation of protons from antiprotons. At all energies forseen annihilation is a dominant part of the cross section requiring the distinction between pions and kaons from protons/antiprotons. It would be difficult to use a magnetic field due to the proximity of the beam. The most practical design seems to be a dE/dx and total energy measurement, which would distinguish protons/antiprotons from mesons. A segmented total energy detector is needed to detect the mesons coming from the antiproton annihilation, which will distinguish antiprotons from protons. The detector needs to be thick enough to stop the protons/antiprotons, which will probably limit the first experiments to momenta less than about 800 MeV/c.

The lowest energies depend on the type of measurement. For the D and K parameters, the particle needs enough energy to pass through the detectors and analyzer needed for the polarization measurement, while the A parameters require only one measurement of the position before the particle is stopped. We estimate this will limit the lowest momentum for the A's to about 200 MeV/c while the D's and K's will be limited to about 300 MeV/c. These lower limits can only be reached if the detectors are in the vacuum.

It will also be possible to measure the A parameters for the charge exchange reaction ($\bar{p}p$ to $\bar{n}n$). Because the final state contains only neutral particles it is somewhat more difficult to measure polarization in the final state. Those experiments will probably be done only when antiproton beams of higher intensities are available.

CONCLUSIONS

For the elastic case it seems possible to measure at up to 9 spin dependent parameters, in addition to the differential cross section and analyzing power. Three spin dependent parameters could be measured for the CEX reaction. This allows a complete determination of the scattering amplitude for at least some angles in the momentum range of 300-800 MeV/c.

REFERENCES

1. Bugg, D. V., Nucl. Phys. A416, 227 (1984).
2. Brückner, W., et al., Phys. Lett. 158B, 180 (1985);
   Brückner, W., et al., Phys. Lett. 166B, 113 (1986);
   Brückner, W., et al., Phys. Lett. 169B, 302 (1986).
3. Clough, A. S., et al., Phys. Lett. 146B, 289 (1984).
4. Notation of J. Bystricky et al., J.Physique, 39, 1 (1978).
5. Onel, Y., private communication.
6. Imai, K., et al., J. Phys. Soc. Jpn., 55, 1136 (1986).
7. Neumann, R., et al., Z. Physik, A313, 253 (1983);
   Poth, H., et al., CERN PSCC/P86.
8. Niinikoski, T. O. and R. Rossmanith, J. Phys. Soc. Jpn., 55, 1134 (1986);
   Onel, Y., et al., CERN PSCC/86-22.
9. Csonka, P., NIM, 63, 247 (1968).
10. Kilian, K. and D. Mohl, Physics at LEAR with Low Energy Antiprotons (Plenum, NY, 1984) p.701.
11. H. Döbbeling, et al., CERN PSCC/P92.
12. Dover, C. B., preprint BNL 37502 (1986).
13. Haberli, W. and T. Wise, J. Phys. Soc. Jpn., 55, 483 (1986);
    Haberli, W. and T. Wise, J. Phys. Soc. Jpn., 55, 1092 (1986).

EXPERIMENTAL EVIDENCE FOR QUANTUM GRAVITY?

T. Goldman, R. J. Hughes and M. M. Nieto*
Theoretical Division, Los Alamos National Laboratory
Los Alamos, New Mexico 87545

ABSTRACT

A broad range of quantum field theories of gravity predict the existence of vector and scalar components of gravity in addition to the usual tensor component. These may produce effects observable at the classical level. We refer to experimental results suggestive of the appropriate effects and show that a clear, low-energy test of these theories may be made by measuring the gravitational acceleration of antiprotons in the Earth's field.

Modern quantum field theories of gravity grew out of gauge theories[1]. These ideas have peaked in 26-dimensional superstring theories[1] which grew out of local supersymmetry.

In virtually all of these theories, spin-one and spin-zero partners of the graviton appear, sometimes in great profusion. However, it is difficult to observe their effects since, like any gravitational strength interaction, the scattering amplitudes they induce are about 34 orders-of-magnitude smaller than weak amplitudes. The impetus to test for these effects arises from the improvements these theories bring to the renormalization properties of theories of gravity.

The effects of these partners, the graviphoton and graviscalar, can be represented[2] as a modification of the gravitational interaction energy

$$I(r) = - \frac{G_\infty m_1 m_2}{\gamma_1 \gamma_2 r} \left[\langle 2(u_1 \cdot u_2)^2 - 1 \rangle \mp q_1^v q_2^v (u_1 \cdot u_2) e^{-r/v} + q_1^s q_2^s e^{-r/s} \right] \quad (1)$$

where u_i is the four-velocity of the ith particle, the q's are vector and scalar charges in units of $G_\infty^{1/2}$, and v,s are the inverse masses of the vector and scalar bosons.

---

* and Niels Bohr Institute, Copenhagen, Denmark

It has been suggested[3] that v,s may be as large as ~km in scale which would allow classical observable effects. In fact, there may be evidence for a scale dependence

$$G(r) \neq G_{\infty} \tag{2}$$

from geophysical and astrophysical measurements[4] and a material dependence of G also[5]. Both of these small effects could arise from (1) due to inexact cancellation of the two new terms.

Note, however, that the minus sign for the vector term in (1) applies to like particle (matter-matter) interactions. Such a cancellation would be completely undone if one studies the matter-antimatter interaction, which requires the plus sign in (1). Ron Brown will talk next about LEAR experiment PS-200 which proposes to do just such a measurement.

The question may also be phrased completely generally as a study of the weak equivalence principle for antimatter. The conflict between trajectories needed for general relativity and the tenets of quantum mechanics prevent the use of CPT to "prove" this principle. We conclude that the antimatter gravity experiment is of great value to the study of quantum gravity.

REFERENCES

1. S. Ferrara, et al., eds., Unification of the Fundamental Interactions, (Plenum, New York, 1981).
2. T. Goldman, et al., Phys. Lett. 171B, 217 (1986).
3. I. Bars and M. Visser, "Feeble Intermediate Range Forces from Higher Dimensions", USC preprint 86/05, Feb. 1986.
4. S. C. Holding, et al., Phys. Rev. D, in press; P. Hut, Phys. Lett. 99B, 174 (1981).
5. E. Fischbach, et al., Phys. Rev. Lett. 56, 3 (1986).

PROPOSED MEASUREMENT OF THE GRAVITATIONAL ACCELERATION
OF THE ANTIPROTON

Ronald E. Brown
Los Alamos National Laboratory, Los Alamos, NM 87545

ABSTRACT

A collaboration has been formed to measure the acceleration of the antiproton in the gravitational field of the earth. The technique consists in obtaining antiprotons of the lowest energy possible from the LEAR facility at CERN, decelerating them further in the external beam line, trapping and cooling them to ultralow energy, and measuring their gravitational acceleration by time-of-flight methods. The experiment has been granted CERN approval (PS-200). Present plans and initial development efforts are described.

INTRODUCTION

The suggestion by Goldman and Nieto[1] that it would be of great interest to measure the effect of the earth's gravitational field on antimatter has stimulated the formation of a collaboration, listed in Table I, to measure the gravitational acceleration of antiprotons. A proposal[2] for this measurement has been approved (experiment PS-200) to be performed at the CERN Low Energy Antiproton Ring (LEAR). Some details of this proposal have been discussed before.[3-7] Previously, the effect of the earth's gravity on photons,[8] electrons,[9] neutrons,[10] and atoms[11] has been detected; however, no experimental test of the gravitational interaction of antimatter with either matter or antimatter has ever been made.

Modern theories of quantum gravity suggest that the equivalence principle embodied in general relativity is violated to some extent, and in particular suggest that antimatter is affected differently from matter by the earth's gravity. Such theories, which try to unify gravity with the other known forces, predict deviations from the inverse square law,[12-15] and in fact geophysical studies[16] show evidence for a repulsive Yukawa component in the earth's gravitational potential with a range between a few m and 10 km. A reanalysis[17] of the original Eötvös data[18] claims to verify this. Theories based on local supersymmetry generate such components[14,15,19,20] via spin-1 (vector) and spin-0 (scalar) partners of the graviton. Furthermore, these components can violate the equivalence principle. For example, the vector component repels a proton from the earth, but attracts an antiproton.

Table I. The PS-200 collaboration.

CASE WESTERN RESERVE UNIV. R. M. Thaler	NASA/AMES RESEARCH CENTER F. C. Witteborn
CERN M. Weiss	RICE UNIVERSITY B. E. Bonner
KENT STATE UNIV. P. C. Tandy	TEXAS A&M UNIV. D. A. Church D. J. Ernst
LOS ALAMOS NATIONAL LAB. J. H. Billen R. E. Brown L. J. Campbell K. R. Crandall T. Goldman D. B. Holtkamp M. H. Holzscheiter S. D. Howe R. J. Hughes M. V. Hynes N. Jarmie N. S. P. King M. M. Nieto A. Picklesimer W. Saylor E. R. Siciliano J. E. Stovall T. P. Wangler	A. L. Ford R. A. Kenefick J. Reading UNIV. DI GENOVA V. Lagomarsino G. Manuzio UNIV. DI PISA N. Beverini L. Bracci G. Torelli

Recently, Goldman, Hughes, and Nieto[21] have put forth a model in which they invoke the scalar and vector terms to show how the anomalous part of the earth's potential could be quite large for antimatter, even though it produces[17] effects of only a few parts in 10^8 for matter. This could arise from a strong cancellation between these terms for matter, whereas for antimatter the change in sign of the vector part allows them to add. Independent of such ideas, however, the proposed measurement is important because it is a new, fundamental test of the equivalence principle in a matter-antimatter context.

The experiment will use extensions of several techniques pioneered by Witteborn and Fairbank[9] in their electron gravity measurement. Our experiment has two main advantages over theirs: (1) The larger antiproton mass allows the toleration of stray electric fields 1800 times higher, and (2) The use of H^- ions allows an accurate comparison measurement. We will use antiprotons from the Low Energy Antiproton Ring (LEAR) at the CERN laboratory, and after deceleration, trapping, and cooling, the confined ultralow-energy antiprotons will be gradually launched upward into a 1-m long, cylindrical, conducting drift tube containing an axial, guiding magnetic field. The flight times would then be measured and compared with those obtained for H$^-$ ions.

THE SLOWING AND TRAPPING OF ANTIPROTONS

LEAR will supply bursts of about 10^7 antiprotons at an energy possibly as low as 2 MeV. Their average energy will be lowered further to about 10^{-3} eV ($\approx$10 K temperature) by several deceleration and cooling stages. The first stage, deceleration to 20 keV, will employ a buncher and radiofrequency quadrupole.[22] Next, electrostatic deceleration to a few keV will take place just prior to the particles' entering an ion trap. In this first trap, they will be cooled to about 50 eV and then transfered to a second trap, where they will be cooled to 10 K prior to launching. Our system will also contain an ion source for H$^-$ injection into the traps.

The ion traps are Penning traps,[5,23] in which confinement is achieved by static electric and magnetic fields. The first trap is elongated in the beam direction ($\approx$15 cm long) and uses a strong ($\approx$6 T) axial magnetic field. The particle burst will be captured by rapidly varying the trap-electrode voltages. After capture, the particle energy will be lowered (cooled) to around 50 eV. Further cooling in this relatively large trap becomes inefficient, and so the bunch will be transferred to a smaller trap, the launching trap, to cool the particles to the 10-K region. This trap can release several particles at a time into a vertical drift tube where their time-of-flight will be measured. We are considering three possible types of cooling in the traps: resistive, stochastic, and electron. Cooling techniques, vacuum requirements, and details of trapping, intertrap transfer, and launching have been discussed by M. H. Holzscheiter.[5]

THE GRAVITY EXPERIMENT

The proposal is to compare the acceleration of the antiproton to that of the H$^-$ ion in the earth's gravitational field. The H$^-$ ion moves in electromagnetic fields very similarly to the antiproton, and therefore its use will allow a precise comparison to be made, even though we may not be able to eliminate stray fields completely. Several particles will be launched at a time from a 10-K thermal distribution upward into a drift tube of length L≈1 m, and the distribution of arrival times at the top of the drift tube will be measured, thereby determining the particle cutoff time $(2L/g)^{1/2}$ (about 0.45 sec). A comparison will be made of the cutoff times for antiprotons and H^- ions. A computer study[24] shows that about 10^7 particles must be launched from a 10-K Maxwell-Boltzmann distribution to attain enough data near the cutoff time to achieve a statistical accuracy for g of 0.5%. A considerable number of run sequences of this type will be performed in the experiment to study systematic effects in the H^--antiproton comparisons.

The concept of the experiment is simple — its carrying out is difficult. For example, extreme care must be taken to control vertical electric and magnetic forces on the drifting particles; a vertical electric field of 10^{-7} V/m will give a force on the antiproton equal to that given by the earth's gravity. In the electron experiment[9] it was found that stray electric fields can be lowered to about 10^{-11} V/m. The technique is to employ a cylindrical, conducting drift tube at liquid-helium temperature with a uniform axial magnetic field to keep the particle motion close to the axis of symmetry.

Sources of systematic errors arising from electric and magnetic effects are listed in Table II. Many of these have been discussed by Witteborn and Fairbank.[9]

Table II. Sources of systematic errors.

1. End effects	7. Magnetic field uniformity
2. The patch effect	8. Alignment and tube uniformity
3. Thompson emf	9. Off-axis orbits
4. Electron sag	10. Collisions with residual gas
5. Lattice compression	11. Neighbor interactions
6. Thermal fluctuations	

(1) End effects — The regions at the ends of the drift tube may be at varying potentials, causing some distortion of the time-of-flight spectrum. Contact potentials, the particle detector, and the launching trap can contribute to this. Shielding grids will alleviate the problem, but more importantly, the effect will cancel to high order in the comparison between the antiproton and the H^- ion. This cancellation also occurs for other items in Table II, but will not be reiterated below.

(2) The patch effect — The surface of even the best of conductors is not at a uniform potential because it consists of many crystal faces (patches) that can have differing work functions.[25] The rms field produced by a random patch distribution on the surface of a typical

conductor is expected to be considerably larger than the gravitational equivalent.[26] However, we expect to coat[27] the inner surface with an amorphous material to reduce this field significantly. Furthermore, it has been discovered[26] that the patch field is masked strongly at liquid-helium temperatures, although the reason for this suppression is not well understood.

(3) Thompson emf — A temperature gradient along the drift tube causes a potential variation of a few µV/K, resulting in a requirement that the temperature difference be less than 10^{-2} K. Immersing the entire tube in liquid helium cannot accomplish this, because the change in boiling point with pressure causes a temperature gradient of 0.3 K/m. However, it is possible[9] to reduce the gradient to <10^{-5} K/m by having the tube in contact with the helium bath at only one location.

(4) Electron sag — This was the limiting effect in the electron experiment[9]. Schiff and Barnhill[28] pointed out that the force of gravity on the Fermi gas of electrons in the tube material would produce an electric field of $mg/e = 5.6 \times 10^{-11}$ V/m in the drift region so as to cancel the expected gravitational force. In fact, it was found[9] that the net acceleration of the electrons in the drift region was consistent with zero. In our experiment this effect will be quite small because of the larger antiproton mass.

(5) Lattice compression — A vertical tube undergoes a differential compression of the positive-ion lattice. Dessler et al.[29] estimated that this should yield an electric field in the drift region much larger than that from electron sag. Yet, such a strong field was not observed in the electron work.[9] Later, the field in the drift region was studied[26] as a function of tube temperature. Fields of 10^{-6} to 10^{-7} V/m were observed as the temperature was decreased from 300 K toward 4.2 K, with the field plummeting into the 10^{-11} range below 4.5 K. Attempts to explain this observation have been unsatisfactory. We will study this in tests with H$^-$ ions.

(6) Thermal fluctuations — The rms field in the drift tube caused by thermal fluctuations was calculated by Maris[30] and found to be small enough to be ignored here.

(7) Magnetic field uniformity — A field of 6 T in a 1-m tube must be uniform to one part in 10^4 to control vertical magnetic forces on the particles. This assumes antiprotons or H$^-$ ions at 10 K, resulting in the forces being due to the orbital magnetic moment. Such uniformities are available in commercial magnets. In addition, although a field of 6 T is necessary for the trapping at 20 keV, we might use a smaller field in the drift tube, which will relax the uniformity requirement.

(8) Alignment and tube uniformity — A charged particle moving precisely along the axis of a perfectly fabricated conducting cylinder will experience no force from its

image charge distribution. However, if there is some misalignment of the magnetic axis with the tube axis, or if the tube wall is nonuniformly constructed, the particle will find itself in an electric field tending to pull it radially toward the wall. Of course, the guiding magnetic field will resist this radial motion; however, two effects should be considered. The particle, being in crossed electric and magnetic fields, will develop, via magnetron motion, an orbital magnetic moment that can interact with any magnetic field gradients present, and in addition, a vertical component of the image field could act on the particle. The calculations of Ref. 9 indicate that these effects can be held well within bounds with standard fabrication and alignment techniques.

(9) Off-axis orbits — If the antiproton is not launched directly along the tube axis, the crossed-field effect again results in the development of an orbital magnetic moment. An estimate as in item (8) shows that, even with relatively large magnetic gradients, the particle trajectory could be off-axis by a significant fraction of the tube's diameter without causing a problem.

(10) Collisions with residual gas — The background gas density in the drift tube must be low enough for antiproton collisions to be unlikely to cause energy transfers of 10^{-7} eV or more. The interaction of greatest importance in a cryogenic system such as we envisage is the antiproton's attraction to the induced dipole moment of the helium atom. Following the calculation in Ref. 9, we conclude that the pressure in the tube must be less than 10^{-9} Torr. This pressure is far greater than is readily achievable in cryogenic systems, and furthermore is much larger than that needed to keep the antiproton annihilation rate at an acceptable level.[5]

(11) Neighbor interactions — The gravitational force on an antiproton at the earth's surface is equal to the electrostatic force between two antiprotons 12 cm apart. However, in a conducting tube the antiprotons are partially shielded from each other. One finds that the effect is small compared to gravity if the antiprotons are separated by at least two to three times the tube's radius. The effect of Coulomb forces on the velocity distribution of the launched particles is currently being studied by computer simulation. These questions can be studied experimentally in tests with H⁻ ions.

PRESENT STATUS

At the Ion Beam Facility of the Los Alamos National Laboratory we have constructed a test beam line to begin developing the apparatus for the antiproton gravity experiment. An H⁻ beam is obtained from an existing CW ion source[31] and is sent through a horizontal section consisting of valves, cold traps, four-way slits, cryopumps, magnetic steerers, and electrostatic lenses. The beam is then turned into the vertical direction by a 90° magnet mounted in a large vertical support stand. Above the 90° magnet, the beam is steered and focused into a Penning trap situated in the 6-T field of a superconductiong solenoid magnet. A chopping system will soon be installed to more nearly reproduce the LEAR beam characteristics and to study beam diagnostic methods.

Our ongoing tests with this apparatus will allow us to study vacuum isolation and bakeout procedures, ultimate vacuum capability with a room-temperature ion trap, trap pulsing to capture protons or H^- ions, the features needed in subsequent trap designs, and the type of H^- source to be taken to LEAR. In a later phase we will be able to test a variety of drift-tube designs to study the suppression of the patch effect and lattice compression, and other criteria for use in the gravity experiment. The present tube is simply a commercial stainless steel tube with no special care taken in its fabrication, and the present solenoid magnet surrounds only the trap region and does not extend the length of the tube.

In our initial tests we have run H^- beams of 5 to 20 keV energy and have obtained beam currents of 3 to 12 μA through the trap's 3-mm diameter apertures, the current being measured at the top of a 1-m tube with the superconducting solenoid magnet energized to 6 T. We have obtained pressures in the trap region as low as 3×10^{-10} Torr, which should allow a sufficient captured-ion lifetime to demonstrate trapping. The planned demonstration technique is to run the beam into the trap with enough voltage on its upper end cap to repel the beam, and then to pulse the lower cap to sufficient voltage to trap the ions. Several hundred msec later the voltage on the upper cap will be dropped to release the ions upward to be detected in a microchannel plate at the top of a short drift tube.

CONCLUSION

Development of an experiment is underway by the collaboration of Table I to measure the gravitational acceleration of the antiproton in the earth's field. No measurement of the gravitational interaction between matter and antimatter has ever been carried out. Current attempts to unify gravity with the other forces of nature indicate the possibility that the matter-antimatter interaction could show significant gravitational anomalies. A test system for this experiment has been assembled at the Los Alamos National Laboratory and development of the apparatus and techniques is in progress.

ACKNOWLEDGMENTS

Besides expressing appreciation to my colleagues in Table I for discussions and help, I would like to thank D. C. Lizon, R. Martinez, R. R. Showalter, and C. B. Webb for their expert help in constructing and operating the test system at the Los Alamos Ion Beam Facility.

REFERENCES

1. T. Goldman and M. M. Nieto, Phys. Lett. $\underline{112B}$, 437 (1982).
2. N. Beverini, et al., CERN proposal P-94 (Los Alamos National Laboratory report LA-UR-86-260).
3. M. V. Hynes, in Physics with Antiprotons at LEAR in the ACOL Era, Proc. Third LEAR Workshop, Tignes, Savoie, France, January, 1985, edited by U. Gastaldi, R. Klapisch, J. M. Richard, and J. Tran Thanh Van (éditions Frontières, Gif sur Yvette, 1985), p. 657.
4. R. E. Brown, in Workshop on the Design of a Low Energy Antimatter Facility in the USA, Madison, Wisconsin, October, 1985 (American Institute of Physics).
5. M. H. Holzscheiter, in the Workshop of Ref. 4.
6. T. Goldman, in Antimatter Physics at Low Energy, Fermilab Low Energy Antiproton Facility Workshop, April, 1986 (to be published by the Fermi National Accelerator Laboratory).
7. M. V. Hynes, in the Workshop of Ref. 6.
8. F. W. Dyson, A. S. Eddington, and C. Davidson, Philos. Trans. R. Soc. London $\underline{A220}$, 291 (1919); R. V. Pound and G. A. Rebka, Phys. Rev. Lett. $\underline{4}$, 337 (1960); R. V. Pound and J. L. Snider, Phys. Rev. $\underline{140}$, B788 (1965).
9. F. C. Witteborn and W. M. Fairbank, Phys. Rev. Lett. $\underline{19}$, 1049 (1967); Rev. Sci. Instrum. $\underline{48}$, 1 (1977), this ref. discusses the design criteria and sources of error for the electron gravity experiment; F. C. Witteborn, Ph. D. thesis, Stanford University (1965).
10. A. W. McReynolds, Phys. Rev. $\underline{83}$, 172, 233 (1951); J. W. T. Dabbs, J. A. Harvey, D. Paya, and H. Horstmann, Phys. Rev. $\underline{139}$, B756 (1965); L. Koester, Phys. Rev. D $\underline{14}$, 907 (1976).
11. I. Easterman, O. C. Simpson, and O. Stern, Phys. Rev. $\underline{71}$, 238 (1947).
12. Y. Fujii, Nature Phys. Sci. $\underline{234}$, 5 (1971); Gen. Relativ. Gravit. $\underline{6}$, 29 (1975); J. O'Hanlon, Phys. Rev. Lett. $\underline{29}$, 137 (1972).
13. A. Zee, Phys. Rev. Lett. $\underline{42}$, 417 (1979); K. I. Macrae and R. J. Riegert, Nucl. Phys. $\underline{B244}$, 513 (1984).
14. J. Scherk, Phys. Lett. $\underline{88B}$, 265 (1979).
15. J. Scherk, in Unification of the Fundamental Particle Interactions, edited by S. Ferrara, J. Ellis, and P. van Nieuwenhuizen (Plenum, New York, 1981), p. 381.

16. S. C. Holding and G. J. Tuck, Nature 307, 714 (1984); F. D. Stacey, G. J. Tuck, S. C. Holding, A. R. Maher, and D. Morris, Phys. Rev. D 23, 1683 (1981); S. C. Holding, F. D. Stacey, and G. J. Tuck, submitted to Phys. Rev. D.
17. E. Fishbach, D. Sudarsky, A. Szafer, C. Talmadge, and S. H. Aronson, Phys. Rev. Lett. 56, 3 (1986).
18. R. v. Eötvös, D. Pekár, and E. Fekete, Ann. Phys. (Leipzig) 68, 11 (1922).
19. C. K. Zachos, Phys. Lett. 76B, 329 (1978).
20. P. Fayet, Phys. Lett. 95B, 285, (1980).
21. T. Goldman, R. J. Hughes, and M. M. Nieto, Phys. Lett. B 171, 217 (1986); see also Ref. 6.
22. J. H. Billen, K. R. Crandall, T. P. Wangler, and M. Weiss, in *Physics with Antiprotons at LEAR in the ACOL Era*, Proc. Third LEAR Workshop, Tignes, Savoie, France, January, 1985, edited by U. Gastaldi, R. Klapisch, J. M. Richard, and J. Tran Thanh Van (éditions Frontières, Gif sur Yvette, 1985), p. 107.
23. J. Byrne and P. S. Farago, Proc. Phys. Soc. (London) 86, 801 (1965); L. S. Brown and G. Gabrielse, Rev. Mod. Phys. 58, (1986).
24. M. V. Hynes, private communication.
25. C. Herring and M. H. Nichols, Rev. Mod. Phys. 21, 185 (1949).
26. J. M. Lockhart, F. C. Witteborn, and W. M. Fairbank, Phys. Rev. Lett. 38, 1220 (1977).
27. R. Mah, private communication; S. Takayama, J. Mat. Sci. 11, 164 (1976).
28. L. I. Schiff and M. V. Barnhill, Phys. Rev. 151, 1067 (1966).
29. A. J. Dessler, F. C. Michel, H. E. Rorschach, and G. T. Trammell, Phys. Rev. 168, 737 (1968).
30. H. J. Maris, Phys. Rev. Lett. 33, 1594 (1974).
31. N. Jarmie, R. E. Brown, and R. A. Hardekopf, Phys. Rev. C 29, (1984); erratum 33, 385 (1986).

MEASUREMENT OF THE ANTINEUTRON-PROTON
ANNIHILATION CROSS SECTION EXTREMELY
CLOSE TO THE $\bar{N}N$ THRESHOLD

B. Bassalleck, University of New Mexico[*]
Department of Physics and Astronomy
Albuquerque, NM 87131

[*](on behalf of the AGS-E795 collaboration)[†]

ABSTRACT

In an experiment at the Brookhaven AGS the $\bar{n}p$ annihilation has been studied very close to threshold, i.e. down to $T_{\bar{n}} \lesssim 1$ MeV. The antineutrons were produced by antiprotons via charge exchange $\bar{p}p \to \bar{n}n$. The $\bar{n}$ annihilation was detected in the same liquid hydrogen target in which they were produced. Very slow antineutrons were tagged by recording the associated, forward going neutron in a separate neutron time-of-flight detector. The $\bar{n}p$ annihilation cross section is extracted from the measured time dependence of the $\bar{n}$ annihilation in the target. Preliminary results indicate values around $\beta_{\bar{n}}\sigma_A \approx 35$ mb ($\bar{n}$ velocity * annihilation cross section).

The quantity of interest in this experiment, $\beta_{\bar{n}}\sigma_A$, is proportional to the imaginary part of the $I = 1$ $\bar{N}N$ s-wave scattering length, the $\bar{n}p$-system being a pure isospin $I = 1$ state. An attractive feature of using antineutrons is their lack of dE/dx losses, which allowed this experiment to investigate a yet unexplored energy region very close to threshold ($T_{\bar{n}} \lesssim 1$ MeV).

[†]D. I. Lowenstein - BNL; B. Mays, L. Pinsky, L. S. Vinson, Y. Xue - University of Houston. H. Poth - University of Karlsruhe; B. Bassalleck - University of New Mexico; T. Armstrong, A. Hicks, R. A. Lewis, W. Lockstet, G. A. Smith - Pennsylvania State University; J. Clement, J. Kruk, G. Mutchler, B. Moss, W. von Witsch - Rice University; M. Furic - University of Zagreb.

Figure 1 shows the experimental set-up used at the AGS LESBII beam line. Also schematically indicated is a typical event of interest.

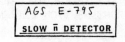

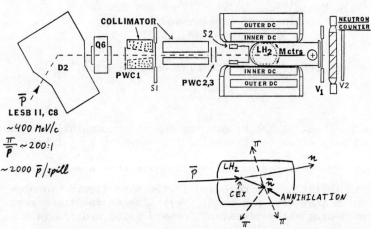

Fig. 1: Experimental set-up and schematic of a typical event of interest.

A total of 4×10^9 antiprotons were incident on a 60 liter liquid hydrogen target (40 cm diameter, 50 cm length). The $\bar{p}$ momentum was adjusted so that they stopped near the center of the target. All timing was done relative to the scintillator S2. The target was surrounded by a barrel arrangement of 12 scintillators ("M-counters") viewed by phototubes on both ends. Outside of these scintillators were layers of drift chambers used to reconstruct the annihilation vertex. In order to accept only $\bar{n}$ annihilations, and not the large number of $\bar{p}$ annihilations, the trigger was only sensitive to annihilation products that were at least several nsec late relative to a "prompt" $\bar{p}$ annihilation. Downstream of the target was an array of plastic scintillator neutron detectors (total area 1 m^2, thickness 20 cm) that were used to detect the neutron associated with the antineutron. The relatively unique signature of a charge exchange followed by $\bar{n}$ annihilation was the detection of late annihilation products coupled with the neutron detected. Position and timing information from the neutron detectors was used in the CEX kinematics reconstruction.

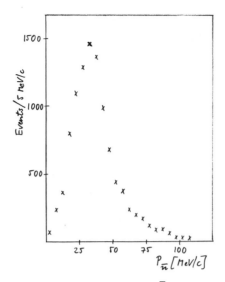

Fig. 2: Reconstructed n̄ momentum distribution

Fig. 2 shows the reconstructed n̄ momentum distribution and demonstrates the very low energy antineutrons involved in this experiment. The n̄p annihilation cross section σ_A is related to $\tau_{\bar{n}}$, the effective antineutron lifetime against annihilation in liquid hydrogen. This lifetime is of course directly measured by the M-counters in our experiment. The actual extraction of σ_A is based upon a comparison of the data with a detailed Monte Carlo code with σ_A as one of the input parameters.

A preliminary comparison, using $\beta_{\bar{n}} \sigma_A = 35$ mb, is shown in Fig. 3. The extensive Monte Carlo effort includes, amongst other things, a parametrization of the known p̄p → n̄n cross section, geometry, vertex reconstruction and the neutron detection efficiency. Another important ingredient is the effect of the unknown (at these low energies) n̄p elastic cross section, which moderates in energy and direction the antineutrons as they propagate through the target. The sensitivity of our results to different extreme assumptions on the n̄p elastic scattering cross section is carefully being studied, and turns out to be relatively small.

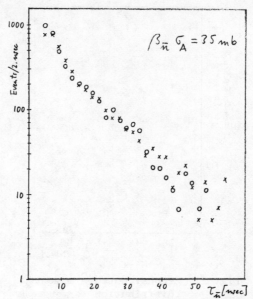

Fig. 3: n̄ lifetime data (x's) and Monte Carlo with $\beta_{\bar{n}}\sigma_A$ = 35 mb (circles). Both are integrated over the n̄ momentum spectrum of Fig. 2.

In conclusion, we have shown that the technique of producing a "tagged n̄-beam" by detecting the forward neutron works. Our final results will contain the n̄p annihilation cross section both integrated over our n̄ momentum distribution and for selected n̄ momentum ranges very close to threshold, $T_{\bar{n}} \lesssim 1$ MeV.

Work supported in part by the U. S. Department of Energy, the U.S. National Science Foundation and the Federal Ministry for Research and Technology, FRG.

ANTIPROTON SPIN PHYSICS AT LEAR

Y. Onel

DPNC, University of Geneva, 24, quai Ernest-Ansermet, CH-1211 Geneva

ABSTRACT

Present and future experiments at LEAR for investigating spin effects in the $\bar{p}p$ system will be discussed.
PS 172 is using a polarized target to investigate the two body channels $\bar{p}p \to \bar{p}p$, $\pi^+\pi^-$, K^+K^-. PS 185 is studying $\bar{p}p \to \bar{\Lambda}\Lambda$ production from which it extracts spin information by the weak decay distributions. PS 172 has also measured the asymmetry in $\bar{p}$'s doubly scattered at small angles from hydrogen and carbon, but only a small effect was observed.
For the post-ACOL period, experiments using external frozen spin and internal polarized jet targets have been proposed to measure polarization observables and asymmetries in the $\bar{p}p \to \bar{p}p$ and $\bar{n}n$ channels. A recent proposal includes channels like $\phi\phi$, $K_S K_S$, $Y\bar{Y}$, $\pi^0\pi^0$, $\eta\eta$ and $\eta\pi^0$ using a polarized jet target.

POLARIZATION AT SMALL ANGLES IN ELASTIC $\bar{p}p$ AND $\bar{p}C$ SCATTERING

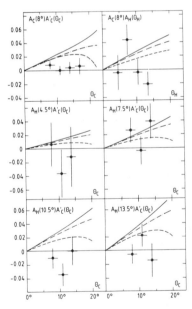

Fig.1 Results for the three asymmetry measurements. Also shown are predictions from Frahn and Venter (full lines), Richard (dashed lines) and "Paris" potential model (dash-dotted lines).

As a tool for investigating proton-antiproton elastic scattering at LEAR it is important to know the analyzing power of carbon for antiprotons. A large analyzing power would open up the possibility of producing a polarized $\bar{p}$ beam by scattering, and also of analyzing the $\bar{p}$ polarization in a carbon polarimeter. At energies of about 150 MeV, pp and pC scattering are characterized by large values of the polarization (about 0.2 and 0.4, respectively, at $10°$ Lab. angle), but the corresponding information for antiprotons is extremely scanty with proton targets, and is non-existent for nuclei. Calculations[1] indicate small values for the $\bar{p}C$ analyzing power, mainly because of the strong absorption due to the annihilation channel.
In this spirit, PS 172 has carried out a double scattering measurement on elastic $\bar{p}p$ and $\bar{p}C$ scattering at small angles using a 610.8 MeV/c $\bar{p}$ beam and a 5.22 g/cm^2 carbon scatterer.

Three asymmetries were measured: $A_C(8°).A_H(\theta_H)$, $A_C(8°).A_C'(\theta_C)$ and $A_H(\theta_H).A_C'(\theta_C)$. The first two were recorded simultaneously, using $\bar{p}$'s scattered at 8° from the first carbon target, and alternating the sign of the beam production angle at each one-hour spill. The momentum resolution of the beam line ensured that the first scattering on carbon was elastic, with analyzing power A_C. A total of 4.5×10^6 events were written on tape, 0.6×10^6 of which were with no hydrogen in the target (ET data) so as to measure the background from the target flask.

The third asymmetry was measured sending the direct $\bar{p}$ beam into the liquid-hydrogen target, looking for $\bar{p}$'s scattered both from hydrogen and from the carbon stab in the polarimeter. The second scatter is mostly, but not entirely elastic, and we denote the analyzing power by A_C'. The azimuthal asymmetries measured in the various conditions are reported in fig. 1. The lines show predictions based on three different models used to generate the $\bar{p}p$ and $\bar{p}n$ amplitudes, namely: the optical model developed by Daum et al.[2] from an earlier model of Frahn and Venter[3] to produce their $\bar{p}p$ data; the model of Dover and Richard[4]; and the "Paris" potential model[5]. The $\bar{p}C$ amplitudes have been evaluated using the model of Osland and Glauber[6] as in ref. 1. In this model the polarization A_C depends mainly on the isospin averaged forward slope of the elementary spin-orbit scattering amplitude.

Measured values are clearly smaller than predicted. Since the overall set of data is compatible with the hypothesis $A_C = A_C' = 0$ ($x^2 = 19.8$ for 20 degrees of freedom), the disagreement could be mainly in A_C rather than A_H. Indeed a different nuclear model calculation indicates smaller values for A_C[7]. On the other hand, if the Glauber theory is reliable, it is also possible that both A_H and A_C are small, or that there is a cancellation between $\bar{p}p$ and $\bar{p}n$ amplitudes.

TWO BODY ELASTIC AND ANNIHILATION CHANNELS AT LEAR

It has long been known that there are broad structures between 1 and 2 GeV/c in the total cross section, both in hydrogen and deuterium. Effects attributable to these structures have been seen more clearly in the two-body annihilation channels $\pi^+\pi^-$ and $\pi^0\pi^0$.

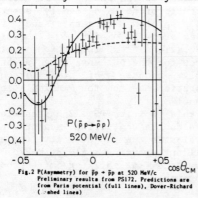

Fig.2 P(Asymmetry) for $\bar{p}p \to \bar{p}p$ at 520 MeV/c Preliminary results from PS172. Predictions are from Paris potential (full lines), Dover-Richard (dashed lines).

The energy dependence of the angular distribution in these annihilation channels are very striking and indicate that there is more going on in these channels. These are the so-called T and U mesons. There is a claim from the latest phenomenological analyses that the angular distributions require resonances in nearly all partial waves that contribute to the two final states. All of these new structures appear to be broad (> 100 MeV) and the absence of complete data to threshold prevents having a reliable phase shift analysis in these channels. PS 172 at LEAR has, as the main objective, the measurement of $d\sigma/d\Omega$ and the polarization P of $\bar{p}p \to \pi^+\pi^-$ (1), $\bar{p}p \to K^+K^-$ (2), $\bar{p}p \to \bar{p}p$ (3) in the momentum range 420-1550 MeV/c at 20 momenta using a conventional polarized target. In reaction 1 and 2 the complete angular range on 180° are covered. Reaction 3 is studied over the angular range where p and $\bar{p}$ have sufficient energy to escape from the target. At those energies and angles where the proton from reaction 3 has sufficient energy, a measurement of its polarization (Wolfenstein parameter D) is also made parasitically. The data taking is now completed (May 86). The amount of data collected is 1-2 million events per momentum. Only 10% of the total number of events are true two body events and the ratio of pp : $\pi^+\pi^-$: K^+K^- is roughly :70:23:7. The experiment has taken data at the following momenta: 420, 450, 500, 530, 550, 600, 700, 800, 900, 1000, 1100, 1200, 1300, 1360, 1400, 1425, 1450, 1475, 1510, 1550 MeV/c. The preliminary results of the polarization parameter $P(\bar{p}p \to \bar{p}p)$ at 550 MeV/c is shown in fig. 2 (using only 30% of the available data at this momentum). The dip and maximum are clearly visible. This experiment will also be able to give additional information about the following new interesting results. The first one concerns the measurement of K. Nakamura et al. at KEK[8] which has measured $d\sigma/d\Omega$ for $\bar{p}p \to K^+K^-$, $\pi^+\pi^-$ and finds evidence for a resonance in the K^+K^- and $\pi^+\pi^-$ channels. This state has a mass of 1940 ± 20 MeV, a width of < 40 MeV. (PS 172 will have the following advantages over this experiment, firstly they have collected 10K K^+K^- events at each momentum around this structure and secondly the polarization measurements will help enormously to get definitive quantum numbers if this state exists).

The second result of interest is that of the MARK III Spectrometer at SPEAR which has reported convincing evidence for a narrow state, called the ξ, decaying strongly into K^+K^- and $K_S K_S$. The reported mass and width are M = 2.230 ± 0.006 ± 0.014 GeV/c², Γ = 0.026 ± 0.020 ± 0.017 GeV/c². In view of the narrow width, there has been speculation that it might be a Higgs boson, a glueball or some unusual state. It lies conveniently within the mass accessible at LEAR. PS 172 has also taken data around the ξ region, measuring $d\sigma/d\Omega$ and P in $\bar{p}p \to \pi^+\pi^-$, K^+K^- and $\bar{p}p$.

THE $\bar{p}p \to Y\bar{Y}$ CHANNEL

The experiment PS 185 studies the reaction $\bar{p}p \to Y\bar{Y}$ near threshold and the results are expected to be especially informative about the nature of the $\bar{p}p$ annihilation process, the production of strange quarks, and, in certain cases, about the spin dynamics of the strange quark itself. The experiment has taken data for $\Lambda\bar{\Lambda}$ production at 1509 and 1480 MeV/c. Since symmetry consideration requires the light (ud) diquark in the Λ to have zero spin, all the spin information in the Λ is

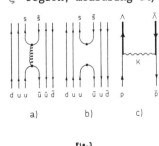

Fig.3

Mechanism for $\bar{p}p \to \Lambda\bar{\Lambda}$; a) Annihilation of a uu quark pair into a gluon and recreation of an ss quark pair, b) Quark pair annihilation into vacuum and recreation, c) Kaon exchange

carried by the strange quark. The acceptance of the apparatus at threshold is such that in 41% of the cases the four charged decay $\Lambda\bar\Lambda \to \bar{p}p\pi^+\pi^-$ is detected, so that not only the polarization vectors are measured from the recorded tracks, but also the $\Lambda\bar\Lambda$ spin-correlations. Hence one gets detailed information on the spin of the $\bar{s}s$ quark pair. Further understanding of this reaction will probably require more complete information on the spin correlation parameters and will be essential for comparing several theoretical models which were proposed as a mechanism for $\bar{p}p \to \Lambda\bar\Lambda$ (fig. 3).

FUTURE PROSPECTS AT LEAR

It is unlikely that one can understand the $\bar{N}$-N interaction mechanisms and meson spectroscopy, without polarization input. Ideally one wants both polarized beam and target to map out in detail the spin structure of the $\bar{N}$-N system and/or allow spin-parity determination of new states.

For the post-ACOL period, there are two new experiments which use external polarized target. The first one plans to measure P, D, A, A', R, R' in the $\bar{p}p$ elastic channel using a recoil polarimeter (proton) between 300 and 700 MeV/c[10]. The second one plans to measure P and D parameters in $\bar{p}p \to \bar{n}n$ at two momenta[11]

POLARIZED ANTI-PROTON BEAM

A method originally proposed for the ISR, and often discussed for LEAR[12], consists in selectively absorbing one of the spin components of a stored beam, via interaction with a polarized gas internal target, if at some energy the interaction rates for parallel and antiparallel beam and target polarizations are appreciably different, one spin state would be enhanced in the stored beam. This is called the 'spin-filter' method. In order to achieve reasonably short polarization build-up times (few hours), it will also need the development of a high density polarized gas target. This method is presently considered in a LEAR proposal[13] and could be implemented if preliminary feasibility tests at the new cooler ring at Heidelberg would be successful.

A much more appealing scheme, which does not involve, as in the previous ones, a large reduction of flux in the course of polarization, has been proposed recently[15] and is based on the spatial separation of circulating particles with opposite spin directions, by means of the Stern-Gerlach effect in the field gradients of quadrupoles in the machine. Although the effect of the field gradient on the magnetic moment is extremely small compared to the effects due to the charge of the circulating particle, the choice of specific coherence conditions between the orbit and spin motions, allows a constructive build-up of spatial separation, leading to a mechanism of self-polarization.

When used in conjunction with a polarized (internal jet or external solid state) target the polarized $\bar{p}$ beam would allow an extensive program of double-spin measurements in the $\bar{p}$-p elastic scattering (with a recoil proton polarimeter) allowing a direct experimental reconstruction of the amplitudes. Analogously the spin structure of the charge-exchange process $\bar{p}$-p $\to$ $\bar{n}$-n could be fully mapped out. In this case the meson exchange potentials for N-$\bar{N}$

could be understood and compared to the N-N system; a deeper insight could be gained in the long range part of the nucleon interaction, and the short range description in terms of QCD inspired models would be further extended by investigating the relative roles of $\bar{q}$-q exchange versus annihilation mechanisms. Indications from hadron spectroscopy are that spin effects are important and mainly governed by a spin-spin short range interaction, characterized by vector gluon exchange, and that the spin orbit term is negligible.

SPIN STUDIES WITH POLARIZED JET TARGET (PJT)

Obviously an ideal situation would be to use a polarized $\bar{p}$ beam circulating in LEAR in conjunction with a PJT, as in this case a complete spin dependent description of the N-$\bar{N}$ interaction would be possible, thus allowing the determination of amplitudes. In this case the meson exchange potentials for N-$\bar{N}$ could be understood and compared to the N-N system.

A lot can be achieved already with the PJT alone in the study of annihilation processes. Here the major problem is to disentangle the relative importance of quark diagrams and conventional meson-baryon intermediate states[16].

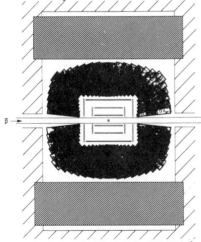

Fig.4 Micro detector for LEAR gas jet (vertical section)

There is experimental evidence that a large negative polarization is found for many reactions where Λ are produced and this extends to other hyperon inclusive production over an energy interval of few orders of magnitude. This effect must be related directly with the spin dynamics in the creation process of the s-s pair and the recombination process leading to the final state hadrons.

The measurement of the polarization asymmetry in the study of the $\bar{p}p$ annihilation dynamics and the search for new exotic states is likely to complete the cross-section measurements in a decisive way, allowing in many cases the determination of the quantum numbers of these new states and the amplitude structure of the relevant processes. From this point of view annihilation channels into two spinless mesons $\pi\pi$, KK, $\eta\eta$, $\eta\pi$ have definite advantages as compared with $\bar{p}p$ elastic scattering.

As a first step to achieve this goal of complete spin studies in LEAR, there is a new proposal by the JETSET group[17] which uses an internal atomic polarized jet target. They have proposed an advanced detector with large acceptance and complete information on charged and neutral tracks in its final implementation (fig. 4) however, the immediate interest is to study of $\phi\phi$ and $K_S K_S$ production, on both unpolarized and polarized hydrogen targets in the momentum range 600-1900 MeV/c. The measurement of P will have two interesting aspects:

i) As stressed by S. Cooper[18], the exclusive reaction and the two spin directions allow partial wave analysis of the $\phi\phi$ mass plot, and $\bar{p}p \to \phi\phi$ is simpler than $\pi^-p \to \phi\phi n$ (BNL experiment). The polarization parameter P will provide additional constraints on the partial wave analysis.

ii) The polarization will allow to study the correlation of the spin directions in the initial and final states, in a process involving purely gluonic intermediate states. The specific nature of the spin-spin short range interaction characteristic of vector gluon coupling to $\bar{q}$-q pairs, involves a correlation in the helicity of the $\bar{q}$-q both in the initial and final states.

The channel $\bar{p}p \to K_S K_S$ represents 25% of $K^o K^o$ events. The series of quantum numbers accessible to both $\bar{p}p$ and $K_S K_S$ are $J^{PC} = 0^{++}, 2^{++}, 4^{++}...$ in a pure isospin system I = 1. The initial $\bar{p}p$ spin state is also constrained to have spin equal to one. It is foreseen to measure $d\sigma/d\Omega$ and P at the same momenta as the K^+K^- annihilation measurements done by PS 172. That will allow a simultaneous amplitude analysis from the two isospin coupled channels.

Concerning spin amplitudes, scattering in these channels can be expressed in terms of two amplitudes F^{++}, F^{+-} which refer to helicity non-flip and helicity flip parts, as $\bar{p}p \to \pi^+\pi^-$ annihilation. The differential cross section and polarization are then defined as:

$$d\sigma/d\Omega = |F^{++}|^2 + |F^{+-}|^2$$

$$P = 2Im[\{(F^{++})(F^{+-})\}/\{|F^{++}|^2 + |F^{+-}|^2\}].$$

These measurements combined with the PS 172 data could provide significant insight into the annihilation mechanism of $\bar{p}p \to KK$. In association with an adequate electromagnetic calorimeter the apparatus would be capable of measuring $\bar{p}p \to \pi^o\pi^o$, $\eta\eta$, $\eta\pi^o$, which gives access to $T^G = 0^+$, 0^+, 1^- s-channel states with a simple 2-amplitude structure.

REFERENCES

1. G. Alberi et al., Proc. 5th European Symposium on Nucleon-Antinucleon Interactions, Bressanone, (CLEUP, Padua, 1980), p. 51, (1980).
2. C. Daum et al., Nucl. Phys. B6, 617 (1968).
3. W.E. Frahn and R.H. Venter, Ann. Phys. 27, 135 (1964).
4. C.B. Dover and J.M. Richard, Phys. Rev. C21, 1466 (1980).
5. J. Cote et al., Phys. Rev. Lett. 48, 1319 (1982).
6. P. Osland and R.J. Glauber, Nucl. Phys. A 326, 255 (1979).
7. H.V. von Geramb, Third LEAR Wokshop, Tignes, p. 531 (1985)
8. T. Tanimori et al., Phys. Rev. Lett. 55, 1835 (1985).
9. R.M. Baltrusaitis et al., SLAC-PUB-3786 (1985).
10. R. Bertini et al., CERN/PSCC/85-71.
11. M.P. Macciotta et al., CERN/PSCC/85-85.
12. K. Killian, D. Mohl, CERN/PS/LEAR 82-11.
13. H. Dobbeling et al., CERN/PSCC/85-80.
14. K. Imai, Preprint KUNS 801, July 1985.

15. Y. Onel et al., contributed paper to Workshop on Low Energy Antiproton Facility, Fermilab, April (1986).
16. J.A. Niskanen, Third LEAR Workshop, Tignes, p. 193 (1985).
17. G. Bassompierre et al., CERN/PSCC 86-23.
18. S. Cooper, SLAC-PUB-3819 (1985).

ANTIPROTON-PROTON ANNIHILATION INTO COLLINEAR CHARGED PIONS AND KAONS

ASTERIX Collaboration

S. Ahmad, C. Amsler, R. Armenteros, E. Auld, D. Axen, D. Bailey,
S. Barlag, G. Beer, J.-C. Bizot, M. Botlo, M. Caria, M. Comyn,
W. Dahme, B. Delcourt, M. Doser, K.-D. Duch, K. Erdman, F. Feld-Dahme,
U. Gastaldi, M. Heel, B. Howard, J. Jeanjean, H. Kalinowsky,
F. Kayser, E. Klempt, C. Laa, R. Landua, G.M. Marshall, H. Nguyen,
N, Prévot, J. Riedlberger, L. Robertson, C. Sabev, U. Schaefer,
R. Schneider, O. Schreiber, U. Straumann, P. Truöl, H. Vonach,
B. White, W.R. Wodrich and M. Ziegler
(Presented by Glen Marshall)

CERN-Mainz-Munich-Orsay-TRIUMF-Vancouver-Victoria-Vienna and Zurich

ABSTRACT

An analysis is presented of two body final states of collinear charged pions or kaons from antiproton-proton annihilation at rest in the ASTERIX spectrometer at LEAR. The relative branching ratio of kaons to pions, which is sensitive to the dynamics of quark-antiquark annihilation and rearrangement, is shown to differ for P and S wave initial states.

The sensitivity of antiproton-proton annihilation observables to details of quark dynamics has been recently emphasized by many authors.[1] Annihilation into two mesons has been treated in terms of planar (annihilation and creation) and rearrangement (following annihilation) quark line graphs. Different assumptions have been invoked such as the $^3P_0$ model (the vertex has vacuum quantum numbers) and the gluon exchange model (the vertex has gluonic properties). One observable often calculated is the relative branching ratio R for production of K^+K^- with respect to $\pi^+\pi^-$, since the former channel is not accessible in the rearrangement picture due to the presence of strange quarks.

The ASTERIX spectrometer at LEAR[2] provides the possibility to enhance at the trigger level and to select in the analysis the annihilation from P wave ($\ell=1$) orbital angular momentum states of antiprotonic hydrogen. Whereas in liquid H_2 annihilation from S states ($\ell=0$) dominates, due to density-dependent mixing at higher atomic levels, in gaseous H_2 at NTP a large fraction of antiprotons cascade to the 2P level[3] from which annihilation is strongly favoured over a radiative transition to the 1S state. The nd-2P L x rays are detected with high efficiency in an x ray drift chamber (XDC) and indicate P wave annihilation. The distinctive ionization from x ray absorption is normally collected on only one of ninety XDC wires with no signal on neighbouring wires, providing an x ray trigger ('one gap').

For the present analysis two data samples were used, one with no x ray enhancement and the other identical except for the 'one gap' demand. Charged collinear tracks were selected by fitting the coordinates for both tracks (negative and positive) for candidate events to a single arc. The momentum spectrum for events well repre-

sented by the single arc was then fit to a function describing two
gaussian peaks on a constant background (Fig. 1). This procedure

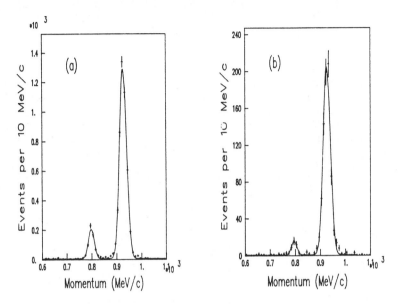

Fig. 1 Fits to collinear momentum spectra for annihilation in gas
(a) and for P wave annihilation (b).

improves resolution by more than a factor of three, so that kaon and
pion peaks are well separated. The ratio of events in the kaon to
pion peaks R, after correction for kaon decay in flight by Monte Carlo
simulation, is obtained for each data set. In addition a value was
obtained for 'one gap' events which were subjected to a further demand
of x rays with pulse shape, pulse length, position and energy
appropriate to L x rays (L_α energy 1.74 KeV). With no x ray
enhancement, R=0.155(12), while the 'one gap' data sample yields
R=0.097(9)). For events in which an L x ray was selected,
R=0.064(12); however, known backgrounds (from argon fluorescence and
inner bremsstrahlung) exist in the L region so that the latter value
is an upper limit. The quoted errors are statistical only.

The extraction of an absolute branching ratio for annihilation to
$\pi^+\pi^-$ is foreseen shortly which, coupled with improved S wave
measurements from current experiments in liquid H_2, should provide
further important tests of the annihilation mechanism.

REFERENCES

1. For reviews of the theory, see contributions to this conference by
   C. Dover and J. A. Niskanen.
2. The ASTERIX collaboration, CERN proposal CERN/PSCC/80-101 (1980)
3. T.B. Day, G.A. Snow, and J. Sucher, Phys. Rev. <u>118</u>, 864 (1960)

SEARCH FOR NARROW LINES IN GAMMA SPECTRA FROM p̄p ANNIHILATION AT REST

S.M. Playfer, Pennsylvania State University
(on behalf of the PS183 Collaboration)[+]

ABSTRACT

The gamma ray spectrum from p̄p annihilation at rest has been measured with good energy resolution and high statistics using a magnetic pair spectrometer. We find no narrow lines with widths comparable to the resolution. We obtain upper limits (90% C.L.) on the yield of narrow baryonium states of $2-5 \times 10^{-4}/\bar{p}$ for states below 1700 MeV, and $5-10 \times 10^{-4}/\bar{p}$ for states between 1700 and 1800 MeV.

Measurements of the inclusive gamma spectrum from p̄p annihilation at rest showed narrow lines with a typical significance of 3σ [1], which were interpreted as evidence for narrow baryonium states below threshold. We describe here new measurements of the inclusive gamma spectrum in which we use a pair spectrometer to improve on the energy resolution of previous experiments.

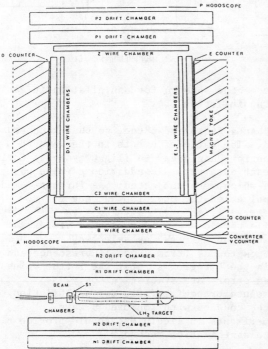

Fig. 1: The PS183 Spectrometer.

Figure 1 shows the PS183 magnetic spectrometer. An antiproton beam of 300 or 460 MeV/c from the LEAR facility enters from the lower left and stops in a liquid hydrogen target 7 cm in diameter and 70 cm in length at a typical rate of 10^5/s. The annihilation vertex is measured by drift chambers on either side of the target.

Inside the magnet aperture a 3.5 kG field acts perpendicular to the plane of Fig. 1. Gammas are converted in a 10% L_R lead converter at the entrance to the magnet, and the electron-positron pair are momentum analyzed by proportional wire chambers.

[+] A. Angelopoulos, A. Apostolakis, P. Papaelias, H. Rozaki, L. Sakelliou, M. Spyropoulou-Stassinaki - University of Athens; T.A. Armstrong, J. Biard, R.A. Lewis, S.M. Playfer, G.A. Smith, M.J. Soulliere - The Pennsylvania State University; B. Bassalleck, P. Denes, N. Graf, R. Hill, N. Komninos, D.M. Wolfe - University of New Mexico; G. Bueche, H. Koch, W. Rohrbach, D. Walther, K. Willuhn - University of Karlsruhe; M. Fero, M. Gee, M. Mandelkern, R. Ray, D. Schultz, J. Schultz, T. Usher - UC-Irvine

Depending upon the momentum of the particles, they are either reflected back out of the magnet (R tracks), trapped within the magnet (T tracks) or passed through the magnet (P tracks). Gamma events can have many different topologies (RR,RT,TT,PT,RP and PP), which are triggered by combinations of the A hodoscope (R tracks), D and E counters (T tracks), and the P hodoscope (P tracks). In this paper only RT,TT and PT gamma spectra are analyzed since other topologies have rather small acceptances. From a total of 3×10^{10} antiprotons, 2.1×10^7 gamma triggers have been recorded on magnetic tape. Of these, 4.8×10^6 events have sufficient wire chamber hits to be fully reconstructed by the analysis procedures (Fig. 2).

The energy resolution of the spectrometer has been determined from calibration processes to be $\sigma/E=1\%$ at 100 MeV, increasing gradually to 2% by 700 MeV [2]. A search for narrow structures consistent with the experimental resolution has been made in the spectra of Fig. 2. A gaussian peak is fitted on top of a sixth order polynomial background. Over the entire range of the fitted data spectra (75-700 MeV) we find peaks which are consistent with statistical fluctuations. Figure 3 shows 90% confidence limits for the yield of narrow structures. Also shown in Fig. 3 are the yields of the four lines observed by Richter et al [1]. Our data definitely exclude the interpretation of their lines at 176, 222 and 549 MeV as transitions to narrow states. At 103 MeV we would expect a 4.1σ peak, and we actually see a 2.5σ peak at 98 MeV, which is just inside the ±5% calibration uncertainty of Ref. 1. While we cannot completely exclude this line, we do not support it with a statistically significant peak.

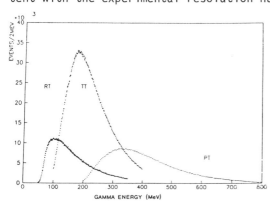

Fig. 2: Gamma Ray Spectra.

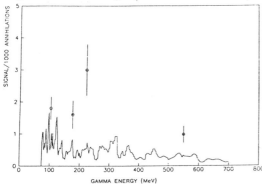

Fig. 3: 90% C.L. Limits for Signals.

Work supported in part by the Greek Ministry of Research and Technology; U.S. Department of Energy; Federal Ministry for Research and Technology, FRG; and the U.S. National Science Foundation.

REFERENCES

[1] B. Richter et al, Phys. Lett. 126B, 184 (1983).
[2] A. Angelopoulos et al, submitted to Physics Letters (1986).

A Search for Narrow States in Antineutron-Proton Total and
Annihilation Cross-Sections Near $\overline{N}N$ Threshold[1]

Brookhaven[2]-Houston[3]-Pennsylvania State[4]-Rice[3]

T. Armstrong[a], C. Chu[b], J. Clement[c], C. Elinon[a], M. Furic[b]
K. Hartman[a], A. Hicks[a], E. Hungerford[b], T. Kishimoto[b], J. Kruk[c],
R. Lewis[a], D. Lowenstein[d], W. Lochstet[a], B. Mayes[b], B. Moss[c],
G. Mutchler[c], L. Pinsky[b], G. A. Smith[a], L. Tang[b], W. von Witsch[c],
and Y. Xue[b]

[a] Department of Physics, The Pennsylvania State University, University Park, PA 16802, USA

[b] Department of Physics, University of Houston, Houston, TX 77004, USA

[c] T. W. Bonner Nuclear Physics Laboratory, Rice University, Houston, TX 77001, USA

[d] AGS Department, Brookhaven National Laboratory, Upton, NY 11973, USA

ABSTRACT

The $\overline{n}p$ total and annihilation cross-sections have been measured from near $\overline{N}N$ threshold (1880 MeV) to 1940 MeV with RMS resolution ranging from 0.08 MeV (1880 MeV) to 6.7 MeV (1940 MeV). No significant narrow meson structures were seen, with 90% C.L. upper limits of 40-180 mb-MeV on σ_T for states with width less than our resolution. Combined with increasing unitarity bounds on σ as one approaches threshold, these limits confine widths of possible predicted states below 1900 MeV to less than ~ 1 MeV.

During the past ten years much experimental effort has been devoted to the search for narrow ($\Gamma \lesssim$ 10 MeV) NN states above threshold. For instance, evidence for a structure in the total and annihilation cross-sections has been reported in several $\overline{p}p$ formation experiments in the S(1936) region [1-6]. However, other experiments

---

[1] A paper on this work has been submitted to Physics Letters B
[2] Work supported by the U.S. Department of Energy.
[3] Work supported in part by the U.S. Department of Energy.
[4] Work supported in part by the U.S. National Science Foundation.

have found negative results [7-12]. Such states have been predicted to exist by both potential models [13-14] and quark models [15-18] based on $q^2\bar{q}^2$ baryonium states.

Due to problems associated with low intensities from conventional $\bar{p}$ sources and energy loss and coulomb scattering in low energy $\bar{p}$ beams, cross-sections have not been published below ~ 1900 MeV mass. Independently, we have pursued this problem by developing a source of $\bar{n}$'s for measuring $\bar{n}p$ cross-sections for incident momenta 100-500 MeV/c (1880-1940 MeV mass). This method allows especially good resolution due to time-of-flight techniques employed to measure the $\bar{n}$ momentum. For instance, at 1880 (1900) MeV the mass resolution is 0.08 (1.4) MeV RMS. Hence, in the heretofore unexplored mass region of 1880-1900 MeV, we have excellent sensitivity to narrow states. Furthermore, the $\bar{n}p$ system is pure I=1. This is especially important, as the potential models [14] predict a clustering of I=1 states close to threshold.

The experiment was performed in the low energy separated beam (LESBII) at the Brookhaven National Laboratory AGS. The apparatus is shown in Figure 1. Antiprotons were identified by time-of-flight (TOF) between a scintillation hodoscope (BH, not shown) 3.89 m upstream of scintillation counter S1, along with pulse height (PH) discrimination in counters S1 and S2. The beam intensity was typically 2500 $\bar{p}$/s and $5\times10^5\pi^-$/s, with less than 1% pion contamination at the trigger level. The direction of the beam immediately in front of the detector was monitored by two (x,y) proportional chambers (PWC1,2). A collimator around the beam eliminated most pion and muon background. Antiprotons were triggered by the logic
$\bar{p} \equiv BH \cdot S1 \cdot TOF(S1-BH) > 20 \text{ ns} \cdot PH(S1) > 3\times$ minimum ionizing.

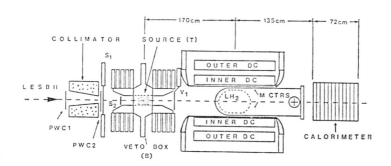

Figure 1

Schematic layout of apparatus

The beam was transported at 505 and 520 MeV/c and focussed onto a stack of twenty thin (6 mm) scintillation counters (T) of dimensions 5 x 13 cm^2, referred to as the "source". Its overall thickness was sufficient to degrade the $\bar{p}$'s from the transport momentum to rest. About 2% of the $\bar{p}$'s produced $\bar{n}$'s via the charge exchange process in the source, the threshold for which is 98 MeV/c. Surrounding the source on four sides parallel to the beam was a veto box (B) for rejecting events in which the $\bar{p}$ annihilated. Each side was constructed of three layers of lead and scintillator each for identifying charged pions and gamma rays. Its efficiency for rejecting annihilations was measured to be ~ 98%. Just downstream of the source was an additional veto counter (V_1), which was used to reject triggers associated with beam pion accidentals. Antineutrons were triggered by the logic $\bar{n} \equiv \bar{p} \cdot (\Sigma B < 2)$, where one hit in the veto box was allowed to avoid rejecting charge exchange events in which the neutron interacted in the box.

Further downstream (170 cm) from the source was a cylindrical liquid hydrogen target of 50 cm length and 40 cm diameter. It was surrounded by a twelve element barrel-stave scintillation hodoscope (M) for triggering the $\bar{n}$ annihilations. Each counter, 100 cm long and 15 cm wide, was read out on both ends to obtain good timing resolution. Immediately surrounding these counters were four modules of planar drift chambers which subtended a solid angle of ~ 70% of 4π steradians from target center. Each module measured 151 x 163 x 30 cm^3, and consisted of an inner and outer drift chamber with 4 planes each.

The drift chambers permitted the reconstruction of $\bar{n}p$ annihilation vertices in the target. The vertex resolution was determined by replacing the target with a thin scintillation counter and irradiating it with beam antiprotons. The resultant distribution of reconstructed vertices due to annihilation in the scintillator had an RMS width of 1 cm. Events with a unique vertex in hydrogen, reconstructed from two or more tracks and located at least 2.5 cm inside the target walls, were used in the final analysis. The overall efficiency for vertex reconstruction was calculated to be 29±2% and is nearly independent of the antineutron momentum. The main source of error arises from uncertainties on the charged particle multiplicity distribution for $\bar{n}p$ annihilation. A target annihilation was defined at TA $\equiv \bar{n} \cdot (\Sigma M > 2)$.

Approximately 135 cm downstream of the target was the calorimeter, designed to measure the yield of antineutrons transmitted through the target. It consisted of twelve modules, each module containing a 4 element (x or y) scintillation hodoscope (C), two (x,y) planes of aluminum proportional drift tubes, and a 2 cm aluminum thick absorber in all but the first three modules. The total active volume of the calorimeter was 60 x 60 x 72 cm^3 and represented ~ 2.5 interaction lengths to an antineutron of average energy.

The vertex of an annihilation in the calorimeter was obtained from drift tube hits. Based on a Monte Carlo simulation, the vertex resolution was estimated to be 5 cm RMS, limited by the granularity of the drift tubes and multiple coulomb scattering. Cuts on the quality of the reconstructed

vertex and total energy deposited in the calorimeter provided a clean separation of $\bar{n}$ events from background, comprised largely of beam pion accidentals. The overall efficiency for reconstructing an antineutron event has been determined by Monte Carlo calculations and is 50±8%. The main source of error arises from insufficient data on $\bar{n}$ or $\bar{p}$ cross-sections in aluminum and carbon in our momentum region. However, such errors should be smooth functions of momentum and do not affect our search for narrow structures. A calorimeter annihilation trigger was defined as
$CA \equiv \bar{n} \cdot (\Sigma C > 0) \cdot TOF(V_1-C) > 10$ ns.

As noted previously, TOF is the key ingredient in achieving good mass resolution. Care has been given to understanding timing and position uncertainties in all scintillation counters. One source of error considered was timing jitter among the T, M and C counters. These errors were found to be 0.45 ns RMS and 0.55 ns RMS for CA and TA events respectively. Another source of error was position accuracy at the point of creation and interaction of the antineutron. Due to the fine segmentation of the source, the charge exchange point was located to 3 mm RMS. These uncertainties, along with previously quoted errors on vertex reconstruction, have been combined to give the overall mass resolutions quoted in Table 1. Final antineutron momentum distributions are given in Figure 2.

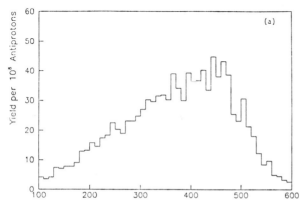
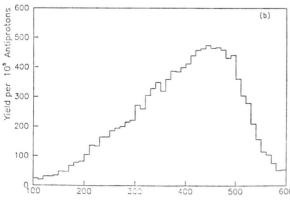

Reconstructed antineutron momentum distributions for (a) TA events and (b) CA events

Figure 2

TABLE 1 - Mass Resolutions in this Experiment.

Mass (MeV)	TA Event (MeV RMS)	CA Event (MeV RMS)
1880	0.08	0.08
1890	0.5	0.4
1900	1.4	0.9
1910	2.5	1.4
1920	3.8	2.0
1930	5.2	2.7
1940	6.7	3.2

We have made cross-section measurements utilizing two different methods. The first involved measuring the rate of CA events with the target full and empty. The ratio of these rates, normalized to an antiproton flux monitor, is directly related to $\sigma_{tot}(\bar{n}p)$. Systematic errors associated with the calorimeter cancel in this method. About 20% of $\sigma_{el}(\bar{n}p)$ escaped detection, corresponding to $\bar{n}$'s which scatter in the hydrogen and fail to hit the calorimeter. No correction has been made for this loss. The resultant cross-sections are shown in Figure 3(a). A search was made for narrow structures by fitting these data to a simple background expression plus a gaussian line shape with widths given in Table 1. No significant structures were found. The fit to the background shape (solid curve) without structures is excellent ($\chi^2/DF = 29/38$). The 90% confidence level (C.L.) upper limits for signal area ($\sigma_{tot}\Gamma$) are shown in Figure 3(b).

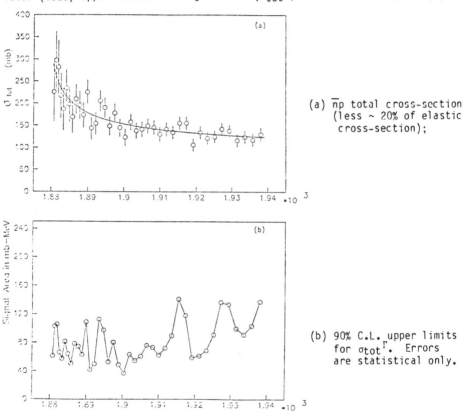

(a) $\bar{n}p$ total cross-section (less ~ 20% of elastic cross-section);

(b) 90% C.L. upper limits for $\sigma_{tot}\Gamma$. Errors are statistical only.

Figure 3

The second method involved measuring the ratio of TA rates with target full to CA rates with target empty, again normalized to an antiproton flux monitor. This measured $\sigma_{ann}(\bar{n}p)$ directly. In this method detector efficiencies do not cancel, and cross-section values are subject to the previously discussed errors associated with determining TA and CA rates. The resultant cross-sections are shown in Figure 4(a). As before, we fit to simple background plus a gaussian line shape. Again, no significant structures are found, with a good fit $(\chi^2/DF) = 42/38$ to the background term alone. Upper limits for $\sigma_{ann}\Gamma$ are given in Figure 4(b). The structure at 1905 MeV, the largest of any observed in both measurements, has a significance of 2.8σ and is not correlated with any structure at this mass in Figure 3(a).

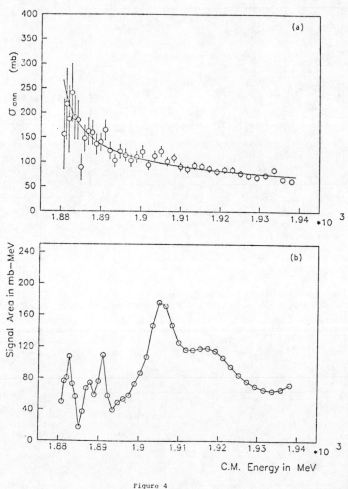

Figure 4

(a) $\bar{n}p$ annihilation cross-sections; (b) 90% C.L. upper limits for $\sigma_{ann}\Gamma$. Errors are statistical only.

In summary, the two measurements have permitted us to establish limits on narrow structures below 1900 MeV, a previously unexplored region. Combined with unitarity bounds given by $\sigma_{tot} = \pi\lambda^2 (2J+1)\Gamma_{el}/\Gamma$, stringent upper bounds can be placed on Γ as one approaches threshold. For example, let us assume a state produced from $\overline{N}N$ with quantum numbers

$$^{2I+1, 2S+1}L_J = {}^{33}P_1 \text{ or } {}^{31}P_1$$

with $\Gamma_{el}/\Gamma = 0.5$. Such states appear frequently near threshold in potential model predictions [14]. Under these assumptions, our data establish upper limits on Γ of 0.1(0.6) MeV at 1880(1900) MeV. These limits are well below predicted values [14].

REFERENCES

[1] A. S. Carroll et al, Phys. Rev. Lett. 32, 247 (1974).
[2] V. Chaloupka et al, Phys. Rev. Lett. 61B, 487 (1976).
[3] W. Bruckner et al, Phys. Lett. 67B, 222 (1977).
[4] S. Sakamoto et al, Nucl. Phys. B158, 410 (1979).
[5] R. P. Hamilton et al, Phys. Lett. 44, 1182 (1980).
[6] C. Amsler et al, Physics at LEAR With Low-Energy Cooled Antiprotons eds. U. Gastaldi and R. Klapisch (Plenum Pub., New York, 1984), p. 375.
[7] E. Jastrzemski et al, Phys. Rev. D23, 2784 (1981).
[8] D. I. Lowenstein et al, Phys. Rev. D23, 2788 (1981).
[9] T. Sumiyoshi et al, Phys. Rev. Lett. 49, 628 (1982).
[10] K. Nakamura et al, Phys. Rev. D29, 349 (1984).
[11] A. S. Clough et al, Phys. Lett. 146B, 299 (1984).
[12] T. Brando et al, Phys. Lett. 158B, 505 (1985).
[13] I. S. Shapiro, Phys. Rep. 35C, 129 (1978).
[14] W. W. Buck, C. B. Dover and J. M. Richard, Ann. Phys. 121, 47 (1979); C. B. Dover and J. M. Richard, Ann. Phys. 121, 70 (1979); C. B. Dover J. M. Richard and M. C. Zabek, Ann. Phys. 130, 70 (1980).
[15] R. L. Jaffe, Phys. Rev. D17, 1445 (1978).
[16] C. Hong-Mo and H. Hogaasen, Nucl. Phys. B136, 401 (1978).
[17] I. M. Barbour and D. K. Ponting, Z. Phys. C5, 221 (1980).
[18] L.A.P. Balazs and B. Nicolescu, Z. Phys. C6, 269 (1980).

HEAVY QUARKS IN ANTIPROTON-PROTON INTERACTIONS

R. Cester

Istituto Nazionale di Fisica Nucleare - Sezione di Torino

ABSTRACT

A new experimental method to study $(q_H, \bar{q}_H)$ states formed in $\bar{p}p$ annihilations is discussed and compared to similar experiments at e^+e^- colliders. Results on Charmonium states from a pioneering experiment at the ISR are summarized and plans for a future experiment at Fermilab are presented. The possibility of extending the method to the study of $(b, \bar{b})$ states is evaluated.

In this paper I discuss the resonant formation of bound states of one heavy quark ($q_H = c, b$) and the corresponding antiquark in the reaction:

$$\text{Antiproton} + \text{Proton} \rightarrow (q_H, \bar{q}_H) \rightarrow \text{final state}$$

This type of process is interesting both from the theoretical and the experimental point of view.

In the framework of Quantum Cromo Dynamics it is described at the lowest order by the graphs:

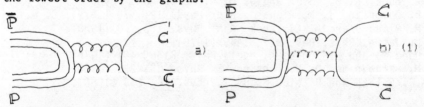

$$(1)$$

The coupling of heavy quarkonium to the Antiproton-Proton state probes therefore the structure of gluonic intermediate states in the energy region ($3. < \sqrt{s} < 10.$ GeV/c^2) where overlap between the confinement and the perturbative regimes takes place and calculations become model dependent. In this energy range it is essential to have precise measurements to constrain the theoretical models.

In a formation experiment the parameters which characterize the energy levels (mass and total width) can be determined from the study of the excitation curve obtained by changing the center of mass energy of the initial state. If this energy is accurately controlled

it is possible to obtain very precise measurements of the resonance
parameters independent of the characteristics of the final state
detector which only function is to tag the formation of a resonance
through the detection of its final states.

This paper will focus on the formation of Charmonium states; in
the last part I will attempt an extrapolation of the difficulties and
challenges of applying this method to the higher energy regime where
(b,b̄) states can be formed.

Table I shows the level structure of Charmonium states as
calculated with a non relativistic potential model.[1]

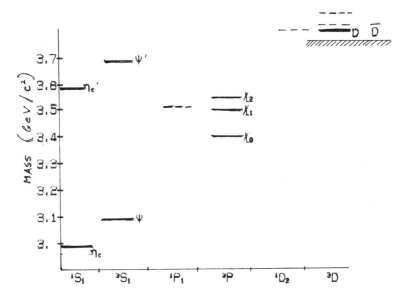

Table I Charmonium spectrum

It must be noted that:

a) Only states with the quantum number of the photon ($J^{PC} = 1^{--}$) can be formed directly from an e^+e^- initial state through the first order QED graph:

Since most of the early information on Charmonium states comes from experiments at e^+e^- accumulation rings, the only energy levels measured with great accuracy are the 1^{--} (ψ) states. Typically the mass of the $n = 1$ $^3S_1$ state has been determined to ± 100 keV and its width to ± 9 keV.

b) Other states with quantum numbers different from those of the photon have been detected at e^+e^- machines through radiative decays of the ψ states. In this case the precision in the measurement of the level parameters is limited by the ability of the detector to reconstruct accurately the decay products.

c) Three narrow states, the $^1P_1$, $^1D_2$ and the $^3D_2$, have never been detected at e^+e^- machines. The ψ' radiative decay to $^1P_1$ violates C, while the D states are predicted at masses higher than those of the narrow ψ states.

In 1974, immediately after the discovery of the ψ, it occurred to many that starting from a $\bar{p}p$ initial state all charmonium levels, independent of their quantum numbers, could be directly formed through intermediate gluon states. For example an $^1S_2$ $J^{PC} = 0^{-+}$ state could be reached through the 2 gluons graph 1 a, while the $^1P_1$ 1^{+-} state could be reached through 3 gluon exchange (1b).

To clarify why only 10 years after, this type of experiment was attempted and successfuly carried out at the ISR at CERN, I will compare the values of relevant parameters (cross sections, background level and source characteristics) for the e^+e^- and $\bar{p}p$ experiments.

Cross sections: elementary scattering theory predicts for the resonant cross-section the well-known Breit-Wigner behavior:

$$\sigma_R(\sqrt{s}) = \frac{A\Gamma_R^2}{\Gamma_R^2 + 4[\sqrt{s}-M_R]^2}$$

with M_R and Γ_R the mass and width of the resonance and A a constant fully determined by the resonance characteristics (mass, spin, total and partial widths etc.).

At the Breit-Wigner maximum, summing over all decay channels:

$$\sigma_R(M_R) = \frac{4\pi(2J+1)(\hbar c)^2}{(M_R^2 - 4M_P^2)} \; BR(R \to \text{initial state})$$

In Table II a direct comparison is given for the formation cross section of ψ, ψ' respectively from e^+e^- and $\bar{p}p$ initial states. Also compared are the formation cross section of η_c and χ_2 from $\bar{p}p$ to the production cross section of these states in e^+e^- through the radiative decay of the ψ and ψ' respectively. In the last column the measured branching ratios $BR(R \to \bar{p}p)$ are also given.

Table II Cross section at the resonance peak (summed over final states)

R	$e^+e^- \to R$ (μb)	$e^+e^- \to \psi(\psi') \to R$ (μb)	$\bar{p}p \to R$ (μb)	$BR(R \to \bar{p}p)$
ψ	113.		5.3	.0022
ψ'	9.7		.3	.00019
η_c	---	1.5	1.	.0011
χ_2	---	.72	.24	.00009

Background levels: in the energy range $2.5 < \sqrt{s} < 4.5$ GeV/c² which brackets the Charmonium mass region, the total hadron production cross section in the continuum varies from 36.5 to 25. nb for e^+e^- and from 70 to 50 mb for $\bar{p}p$ interactions. In the latter case, therefore, the resonant signal must be extracted from a background which is $\sim 2 \cdot 10^6$ times higher. This forces the choice of electromagnetic and O.Z.I forbidden final states which are suppressed in strong interactions.

The peak cross section for detected states is given in Table III where the branching ratios for the decay channels (last column) have been folded in.

Table III Cross section at the resonance peak (for detected final states)

R	Decay channel	$e^+e^- \to R \to \ldots$ (μb)	$\bar{p}p \to R \to \ldots$ (nb)	$BR(R \to \ldots)$
ψ	e^+e^-	8.4	392.	.074
ψ'	e^+e^-	.09	2.7	.009
η_c	$\phi\phi \to 2K^+2K^-$		2.	.002
	$\gamma\gamma$		.8	.0008
χ_2	$\psi\gamma \to e^+e^-\gamma$		2.7	.0011

It should be apparent at this point that $\bar{p}p$ and e^+e^- experiments complement each other; starting from a $\bar{p}p$ initial state precise measurements of the line parameters (mass and total width) for all the energy levels can be obtained along with detailed information on specific final state channels. e^+e^- collider experiments, on the other side, have given precise measurements of the 1^{--} line parameters and allow for a systematic study of all decay channels of these states, a field which has turned out to be surpisingly rich (see W. Lockman talk at this conference).

Given the peak cross section, the event rate can be determined by the convolution of the Breit-Wigner and a function which describes the energy profile of the source instantaneous luminosity. The event rate can be conveniently parametrized by:

$$N(\sec^{-1}) = L_0 (cm^{-2}\ \sec^{-1}) \times \sigma_{effective}(cm^2)$$

where $\sigma_{effective} = \sigma_{PEAK} \times K(\Gamma_R/\Gamma_{BEAM})$ and Γ_{BEAM} measures the width of the luminosity distribution in center of mass energy. For Γ_{BEAM}:

$$\sigma_{effective} \simeq \sigma_{PEAK} \times \Gamma_R/\Gamma_{BEAM}$$

In conclusion, the relevant parameter for the source, if we want to study narrow Charmonium states, is the luminosity density $DL_0/d\sqrt{s}$. Typical values for the source parameters at SPEAR are: $L_0 \sim 5\ 10^{29}$ and $\sigma_E \sim .7$ MeV. The calculated event rate for $e^+e^- \to \psi \to e^+e^-$ is then:

$$N(\sec^{-1}) \sim 8.4\ 10^{-30} \times 5\ 10^{29} \times .063/2.3 = .11\ \text{events/s}$$

with a rate suppression of ~35 due to the beam and resonance width mismatch.

Clearly given the intrinsic difficulties of the $\bar{p}p$ experiment (low cross sections and high background) a highly performing annihilation source was required to carry out a competitive experiment. It was only when the antiproton source facilities became available first at CERN and now also at FNAL that the construction of a $\bar{p}p$ annihilation source with adequate performance became possible.

The method adopted both in the pioneering experiment (R704) carried out at the CERN ISR and in experiment E760 now in preparation at FNAL uses the antiproton beam circulating inside the vacuum pipe of an accumulator ring and for the (fixed) target a molecular Hydrogen Jet continuously intersecting the beam inside the vacuum pipe. Figure 1 shows the transfer line from the Antiproton Accumulator (AA) through the PS to ring 2 at the ISR where 3.5 GeV/c antiprotons were stored for experiment R704.

The RF system and cooling equipment of ring 2 at the ISR provided great flexibility of beam control operations, while the well localized, high density target granted the small size, high luminosity source required for this experiment.

The parameters that characterize the source performance are given in Table IV. With the values quoted the event rate for the reaction $\bar{p}p \to \psi \to e^+e^-$ was:

$$N(sec^{-1}) = .07 \text{ events}$$

to be compared to $N(sec^{-1}) = .11$ events estimated for the SPEAR experiment.

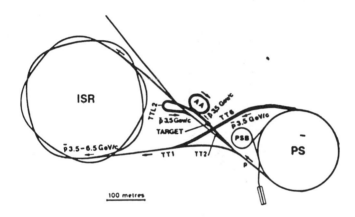

Fig. 1 - TRANSFER LINE FOR THE ANTIPROTON BEAM.

Table IV Antiproton annihilations source for experiment R704

Circulating antiprotons:	$N_p \leq 1.1 \cdot 10^{11}$
Jet target density :	$\rho_H \simeq 10^{14}$ atoms/cm²
Instantaneous Luminosity:	$L_o(cm^{-2}sec^{-1}) \leq 3 \cdot 10^{30}$
Momentum byte (R.M.S.) :	$\Delta P/P \simeq 4 \cdot 10^{-4}$
Luminosity density :	$\dfrac{L_o}{\Gamma_{BEAM}} = \dfrac{3 \cdot 10^{30}}{1.1}$ cm⁻²sec⁻¹MeV⁻¹

The design of the final state detector (Fig. 2) for R704 emphasized $e^{\pm}/\pi^{\pm}$ and γ/π° separation and had a two arm topology to preferentially select high mass two body final states. Two important additional elements completed the detector system: a) a guard counters array to veto on both the charged and the γ ray components outside the two arms and b) a silicon counter telescope measuring large angle recoil protons from $\bar{p}p$ elastic scattering to provide an on-line measurement of the source luminosity. Details on the source and detector systems can be found in Ref. 3.

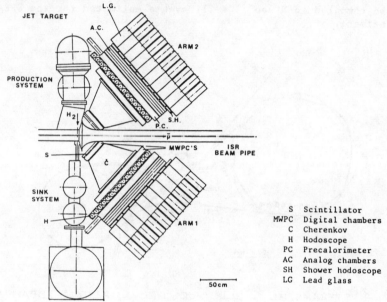

Fig. 2 - TARGET - DETECTOR SYSTEM FOR R704

Fig. 3 shows the χ_1, χ_2 excitation curves obtained at the ISR experiment in three weeks of data taking. The reaction studied was:

$$\bar{p}p \rightarrow \chi_{1,2} \rightarrow \psi + \gamma \rightarrow e^+e^- + \gamma$$

The upper limit to the residual background level as drawn in Fig. 3 was actually measured only in the $\sqrt{s}$ range between 3.520 and 3.531 GeV/c^2.

In Fig. 4a the $\chi_{1,2}$ excitation curves are compared to the entire sample of events accumulated for the same final state reaction by the Crystal Ball experiment at SPEAR. Here the advantage of the formation experiment is obvious: in spite of the small statistics the ISR experiment gives a more precise measurement of the total width of the χ states. Table V summarizes the results of experiment

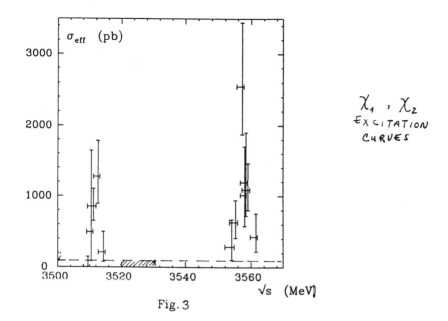

χ_1 , χ_2
EXCITATION
CURVES

Fig. 3

Table V Summary of results from experiment R704

1) New measurement of the ψ mass:
   $M_\psi = 3096.95 \pm .1 \pm .3$ MeV

2) New set of measurements of the parameters of χ_1 and χ_2

	Mass (MeV)	Γ_{TOT} (MeV)	$\Gamma_{\bar{p}p}$ (eV)
χ_1	$3511.3 \pm .4 \pm .4$	< 1.3	$57 ^{+13}_{-11} \pm 11$
χ_2	$3556.9 \pm .4 \pm .5$	$2.6 ^{1.4}_{-1.0}$	$233 ^{+51}_{-45} \pm 48$

3) Search for:
   $$\bar{p}p \rightarrow {}^1P_1 \rightarrow \psi + \ldots$$
   5 events at $\sqrt{s} \sim 3525$ MeV (a 2.3 σ effect)

4) Search for:
   $$\bar{p}p \rightarrow \eta_c \rightarrow \gamma\gamma$$
   preliminary result: $\Gamma \simeq 5$ KeV (2σ effect) or
                       upper limit < 7 KeV (94 % C.L.)

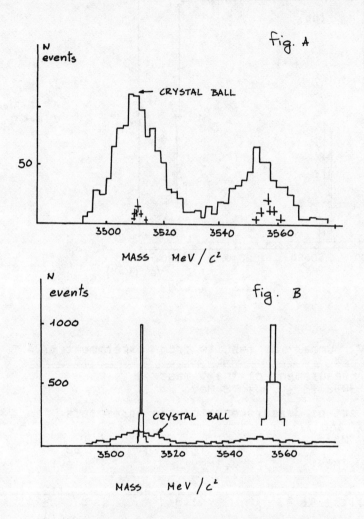

Fig. 4 - COMPARISON OF χ_1, χ_2 EXCITATION CURVES WITH CRYSTAL BALL DATA SAMPLE
A) R 704 DATA   B) PROJECTION FOR E 760

R704.[3-6] The scheduled close down of the ISR in the summer of 1984 interrupted the data taking when the experimental program of R704 was still far from being completed. The experiment at CERN has, however, demonstrated the validity of the method and a new experiment to continue and extend the study of the Charmonium spectrum with $\bar{p}p$ is scheduled to run in the Accumulator of the Antiproton source at Fermilab with an H_2 Jet target similar in design and performance to the one used for the ISR experiment.

Experiment E760, which will start data taking in spring 1987, will run parassitically to the fixed target program. Up to 14 antiproton production cycles can be obtained from the main ring in the 58 sec interval of each supercycle when the beam for fixed target operation is accelerated in the superconducting ring and delivered to the experimental areas. Without interferring with the high energy program it will be possible in a few hours to store the required Antiproton current in the Accumulator. The Antiproton beam will then be cooled in the accumulator ring and decelerated to the required momentum (for details on machine operation see J. Griffin talk at this conference). Figure 5 shows a section of the target/detector system in the horizontal plane through the beam. The detector, which has a cylindrical symmetry around the beam axis, is a scaled up

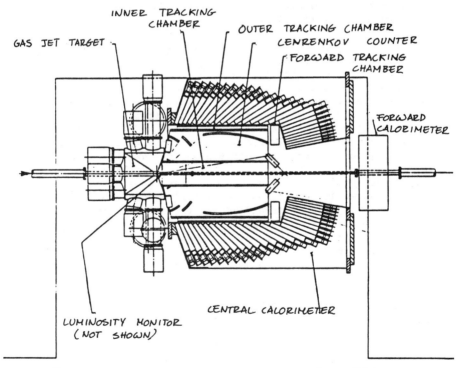

FIG. 5 - TARGET-DETECTOR SYSTEM FOR E 760.

Table VI Performance of source/detector system for R704 and E760

	R704	E760
Instant. Luminosity (cm^{-2}sec^{-1})	3 10^{30}	10^{31}
R.M.S. beam momentum spread: $\Delta P/P$	4 10^{-4}	10^{-4}
Event rate for $\bar{p}p \to \psi \to e^+e^-$	.07 sec^{-1}	.9 sec^{-1}
Detector acceptance (for $\bar{p}p \to \psi \to e^+e^-$)	11%	54%

R704 collaboration: Annecy (LAPP)-CERN-Genova-Lyon
Oslo-Roma-Strasburg-Torino
E760 collaboration: Fermilab-Ferrara-Genova-Irvine
North Western-Penn.State-Torino

version of R704 designed to give a much higher performance both in particle identification and energy measurement. A comparison of the performance of source and detector for the two experiments is given in Table VI. To exemplify the improvement expected from the Fermilab experiment we show in Fig. 4b the excitation curves for $\chi_{1,2}$ as can be obtained in three weeks of data taking. The data are again compared to the Crystal Ball sample.

With the program to study Charmonium states formed in $\bar{p}p$ annihilations well under way, the question naturally arises as to the applicability of the method to the higher energy regime where formation of $(b,\bar{b})$ states could be observed. The predicted level diagram is given in Table VII. Also in this case, experiments at e^+e^- colliders have provided valuable information but many states predicted to be narrow are still undetected. To analyze the feasibility of these experiments we compute the cross section at the peak of the resonance for the formation of the n = 1, $^2S_1$ state respectively from e^+e^- and $\bar{p}p$. In the e^+e^- case the decay ratio of the resonance to e^+e^-, which enters the cross-section determination, has been measured to be 2.5% only a factor ~3 down from the $\psi \to e^+e^-$ decay ratio. When the $\sqrt{s}$ dependence is taken into account one computes a cross section of 4 μb for $e^+e^- \to \gamma$. Unfortunately the branching ratio of $\gamma \to \bar{p}p$ has not been measured and we must rely on theoretical estimates to evaluate the cross section for the process $\bar{p}p \to \gamma$ at the resonance peak is 50 pb, a factor 10^{-5} down from the ψ case.

With a source luminosity of 10^{32} cm^{-2} sec^{-1} and a narrow beam ($\Gamma_{BEAM} < \Gamma_{RES}$ or $\Delta P/P < 10^{-5}$) one expects a maximum formation rate of:

$$10^{32} \times 5 \ 10^{-35} \times 8.64 \ 10^4 \approx 450 \ \text{day}^{-1}$$

Table VII (b,b) mass spectrum

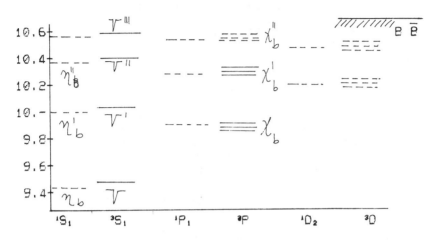

over a strong interaction background rate 10^9 times higher.

It is not inconceivable that, with the advent of high intensity accelerators and continuous progress in cooling techniques (see for example F. Mills talk at this conference) a source with the required characteristics could be built in the future (the construction of $\bar{p}p$ colliders in this energy range has been already discussed at CERN and Fermilab). A major problem will remain, however that of extracting the signal from the large background of strong interactions.

REFERENCES

1) See for example: T. Appelquist, R.M. Barnett and K. Lane, Ann. Rev. Nucl. Part. Sci., Vol. 28, 387 (1978).

2) Einsweiler Ph.D. thesis, Stanford University, SLAC (1984).

3) J/ψ resonant formation and mass measurement in antiproton proton annihilations. C. Baglin et al. (to be published in Nuclear Physics).

4) Limits on the $p\bar{p} \to \eta_c \to \gamma\gamma$ formation/decay process. C. Baglin et al. (to be published).

5) Formation of the χ_1 and χ_2 resonances in Antiproton-Proton Annihilations and new measurements of their Masses and Total Widths. C. Baglin et al. Submitted for publication to Phys. Lett.

6) Search to $^1P_1$ Charmonium state in $p\bar{p}$ annihilations at the CERN ISR. C. Baglin et al. Submitted for publication to Phys. Lett.

PHYSICS WITH ANTIHYDROGEN

H. Poth
Kernforschungszentrum Karlsruhe, Institut für Kernphysik,
Postfach 3640, D-75 Karlsruhe, Fed. Rep. Germany.

ABSTRACT

A cooled and stored antiproton beam of high intensity promises to be the best method for forming an intense anti- hydrogen beam. How this can be accomplished using the electron cooling technique will be discussed and eventual experiments with antihydrogen described.

ANTIHYDROGEN PRODUCTION

It was pointed out some time ago that antihydrogen can be produced through spontaneous radiative capture of a positron by an antiproton

$$e^+ + \bar{p} \rightarrow \bar{H} + h\nu$$

in an arrangement similar to electron cooling of proton beams[1]. Spontaneous electron capture by protons was observed in electron cooling experiments performed with stored beams of protons[2]. The measured capture rate agreed rather well with theoretical calculations[1,3,4].

A specific experimental arrangement to form anti-hydrogen through spontaneous capture and various methods of producing a positron beam of suitable properties were discussed in Ref. 5. They proposed to use a dedicated linac to produce a positron beam which would be injected into a storage ring, cooled stochastically at high energy, decelerated to low energy, cooled with electrons and then merged with an antiproton beam. An antihydrogen rate of a few per second was estimated to be achievable.

Recently it was shown theoretically that induced capture

$$n \cdot h\nu + e^+ + \bar{p} \rightarrow \bar{H} + h\nu + n \cdot h\nu$$

may be used to enhance the antihydrogen formation by more than two orders of magnitude[3], making it possibile to produce sufficient antihydrogen for future precision experiments.

In contrast to Ref.5 a scheme to produce antihydrogen based on laser-induced capture was elaborated in Ref. 6. It makes use of well established technologies. Positron beams of a quality equal to that of cooling electron beams (except for the intensity) can be obtained by moderating positrons from a β$^+$ emitter or from an

accelerator produced beam[7]. The technique to achieve the
required positron density for an initial antihydrogen
production experiment is summarized in Ref. 8.

Recently a proposal[9] has been submitted to CERN for
an exploratory experiment to produce for the first time
antihydrogen through spontaneous capture as a first step
towards an abundant production of antimatter atoms. In
this proposal the measurement of the enhancement through
induced capture, initially to be done with electrons and
protons, is also planned. The outcome of this pilot
experiment could provide the necessary information for
the design of an antihydrogen factory.

The experimental set-up for proposal P86 is shown[9]
schematically in Fig.1. In straight section 3 (SL3) of
the low energy antiproton ring (LEAR) an electron cooler
will be installed which will cool the circulating proton
beam. To study the stimulated electron capture by protons
a pulsed laser beam of 496 nm wavelength will be directed
on to both charged particle beams. Moreover the laser
pulse is stored for about 20 passages in an optical
cavity in order to in- crease the useful laser power. In
the cooling region capture into the n=2 levels will be
stimulated. The rate of hydrogen atoms emerging from SL3
will be measured during the laser pulse duration and
compared to the spontaneous capture rate. From this the
enhancement will be determined.

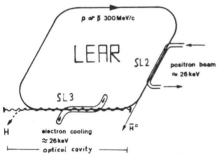

Fig. 1. Schematic set-up for antihydrogen formation

Table I: experimental parameters for proposal P86[9]

production of	hydrogen	antihydrogen
hadron beam	protons	antiprotons
momentum (MeV/c)	300	300
stored beam intensity	5 10¹⁰	5 10¹⁰
positron source		²²Na
positron source activity (Ci)		10
lepton energy (keV)	26	26
lepton current(A)	2.3	5 10⁻¹¹
lepton beam diameter (cm)	5	0.1
beam velocity (c)	0.3	0.3
length of interaction region (m)	1.5	2.5
spontaneous neutral atom rate (s⁻¹)	2 10⁴	0.01
laser pulse length (ns)	20	
laser repetition rate (Hz)	25	
laser pulse intensity (MW/cm²)	20	
laser beam diameter (mm)	2	
stored usefull laser pulses	10	
induced capture rate (s⁻¹)	10	

In straight section 2 (SL2) the positron beam will be
installed to produce antihydrogen through spontaneous
capture. Antihydrogen atoms escape from SL2. Outside the
bending magnet the positron is stripped off in the vacuum
chamber window. The antiprotons will be identified by
signals of two scintillation counters and the annihilation in a calorimeter. In table I the experimental
parameters are summarized.

Before coming to the eventual experiments with antihydrogen, we will first discuss the antihydrogen production in more detail.

The rate for the neutral atom production through spontaneous capture is given by[*]:

$$R_o = N_p n_e \, a_r \, \eta \, \gamma^{-2} \qquad (1)$$

where N_p, n_e, a_r and η is the number of stored (anti-) protons, the (anti-)electron density, the recombination coefficient (about $2 \cdot 10^{-12}$ cm^3 s^{-1} for most cases) and the fraction of the ring circumference occupied by the interaction region. The relativistic factor γ comes from the fact that the rate was initially deduced in the particle rest frame and then transformed to the laboratory system. Hence one γ comes from the Lorentz-transformation of time and the other from the transformation of the lepton density. This formula assumes a coasting hadron and a continuous lepton beam. It shows that for the case of electrons and protons ($n_e = 10^8$ cm^{-3}, $\eta = 0.05$, $N_p = 10^{11}$, $\gamma = 1$) about 10^6 s^{-1} atoms can be produced in a single pass arrangement (electron cooling) which means that the entire proton beam can be transformed into neutral atoms within approximately one day. In the case of antiprotons and positrons the situation is less favourable because of the low positron densities achievable. Here one has to look into further possibilities to increase the rate.

The maximal stored antiproton intensity is ideally given by the space charge limit and hence scales as $\beta^2 \gamma^3$ but will hardly exceed 10^{13} in the near future (accumulation rate limited).

Continuous positron beams can be produced from radioactive sources, by moderating the positrons and accelerating them to the desired velocity. The achievable positron rate R_e is:

$$R_e = 3.3 \cdot 10^{11} \, C_i \, e \, q \, f_\beta \qquad (2)$$

where C_i, e, q, and f_β is the source activity (Ci), the moderation efficiency (about 10^{-3}), the remodera- tion efficiency (about 0.3) and the β^+ branching ratio. One could expect to achieve source activities of roughly 100 Ci for reasonably long lived β^+ emitters.

On the other hand slow positrons can be produced in batches from a pulsed high energy (200MeV) electron beams

$$R_e = 10^{-5} \, I_e / e \qquad (3)$$

where e is the elementary electric charge and I_e the current of the incident electron beam. This tells us, that a 100 Ci source produces about as many positrons continuously as a 200 MeV electron linac of 0.1 mA average current. The pulsed production of positrons permits their accumulation in a storage ring and the increase of the useable positron beam intensity until one reaches the space charge limit or the limit given by the beam lifetime. The space charge limit scales as for antiprotons like $\beta^2 \gamma^3$ and the same is true for the lifetime.

With a pulsed positron beam one can take full advantage of stimulated capture which enhances the formation rate by a gain factor G, defined[3] as the ratio of the stimulated capture into a particular state to all spontaneous captures. The luminosity is then:

$$L_{ind} = L_{spon} \; G \; x \qquad (4)$$

where x is the effective duty cycle (pulse length * repetition rate * number of useful photon passages) for the stimulating pulse. The gain factor depends on the radiation power P as follows:

$$G = 2 \; P \; n^5 \; (\hbar c/13.6eV)^3/(a^2 m_e c \; \Delta_\parallel \Delta_\perp) \qquad (5)$$

where n is the principal quantum number, a is the radius of the photon beam, $\Delta_\parallel$ and $\Delta_\perp$ is the longitudinal and the transverse velocity spread of the leptons. The photon wavelength λ has to be

$$\lambda = 911 \text{ Å } n^2/(1+\beta \cos \theta)\gamma \qquad (6)$$

where θ is the angle between the photon and the charged particle beam ($\theta=180°$ for head-on collisions).

For $L_{ind} \gg L_{spon}$ we need to have $G \; x \gg 1$. The gain factor can in principal be made very large by increasing the radiation power. However, in practice the maximum admissible power is given by the re-ionization limit[3]. High radiation power can be obtained from pulsed lasers operating in the visible and infrared at the expense of a moderate repetition rate. Hence one has to look into possibilities of storing and recirculating the photon pulses. By synchronizing all three beams one can then achieve a duty cycle close to unity. A few examples are now discussed.

We consider first antihydrogen production through spontaneous capture in a single pass experiment. Let us take an antiproton storage ring of 80m circumference where $5 \; 10^{10}$ cooled antiprotons circulate at $\beta=0.3$. Positrons are obtained from a β^+ emitter of, say of 20 Ci (e.g. ^{48}V, $f_\beta=0.5$). They are moderated with an efficiency of 10^{-3}. The thermalized positrons are accelerated to a few keV and focussed to $1mm^2$ on a remoderator to increase the brightness. The efficiency of this process is assumed to be 30% (q=0.3). Then the positron beam is accelerated to 26 keV and merged with the equally large antiproton beam over an interaction region of 3m. The positron density (Eq. 2) is

$$n_{e+} = R_o/\bar{F}\beta c = 1.2 \text{ cm}^{-3} \qquad (7)$$

Hence we get a spontaneous antihydrogen rate (Eq. 1) of

$$R_o = 4.4 \cdot 10^{-3} \text{ s}^{-1} \qquad (8)$$

This rate is rather low for atomic spectroscopy but high enough to study the details of the production process. The achievement of such a rate would be a milestone towards an intense antihydrogen beam.

Next we consider an antiproton ring of 30m circumference where $5 \cdot 10^{10}$ antiprotons are circulating with $\beta=0.5$. Bunching the antiproton beam at the k=2 harmonics of the revolution frequency (5 MHz) could produce bunches of 3m length. Hence every 100ns an antiproton pulse would pass through the interaction region of 3m length were it is merged with a positron beam. The latter is produced from an electron linac (0.1A instantaneous current) yielding pulses of cold positrons 20 ns long and containing $1.2 \cdot 10^5$ particles (after moderation). The positrons are accelerated electrostatically to $\beta=0.5$ and injected into a recirculator of 15m length.

At the same time, when the particle beams pass through the interaction region the 20ns long photon pulse is sent through that section. The photon pulse is stored in an optical cavity of 30m length where it bounces back on forth between mirrors. With a correctly chosen cavity length it passes through the interaction region in the right sense and synchronism with the particle beams. Now if the photon pulse is stored for 10000 passages and the positron pulse is recirculated an equal number of times one has covered 1ms. If the linac and the photon source operates at a repetition rate of 1kHz one achieves a duty cycle of unity. For captures in the n=2 levels and a laser beam of $P/F = 20$ MWcm^{-2} a gain factor of about 100 was estimated[3]. Charged particle beam sizes of 1mm^2 can be achieved for the above described conditions when beam cooling techniques are applied. Hence in this arrangement ($N_p=5 \cdot 10^{10}$, $n_e=1.2 \cdot 10^5/(1\text{mm}^2 \ast 3\text{m})=4 \cdot 10^4 \text{cm}^{-3}$, $\eta=0.1$, $G \times = 100$) one can produce about $3.2 \cdot 10^4$ s^{-1} antihydrogen atoms. Even still higher rates are conceivable[10].

Finally we want to point out that one day heavier antinuclei will certainly be collected (e.g. antideuterons, antialphas, etc.). At that stage, the above described method will most likely become a key-technique to accumulate antimatter. For instance an antialpha beam could be down-charged with a positron beam through radiative capture and then decelerated to very low energies where it could be laser-cooled to subthermal energies, neutralized and stored in a magnetic field (acting on the magnetic moment) as antihelium gas. This might pave the way for the production of microscopic amounts of antimatter in very distant future.

Unfortunately this technique cannot be applied for antihydrogen, since once produced it is neutral an cannot be decelerated. Starting with trapped antiprotons might be a possibility. However, we are presently unable to estimate how many antihydrogen atoms per second could be produced in this way, since it is not known yet how effective antiprotons can be decelerated to very low momenta and cooled to subthermal energies. It is difficult, without decreasing the phase space density, to decelerate many charged particles and to store them at low energy because of space charge effects. The decrease of phase space density leads, however, to a reduction of the production rate.

PHYSICS WITH ANTIHYDROGEN

It is well accepted that the laws of physics are valid for particles and their antiparticles, except for the known case (kaons) where CP symmetry is violated. In spite of this we expect that in the Universe as a whole the CPT symmetry holds. Yet the visible abundance of antibaryons is to low as to support this theorem. We are left with a puzzle. We have either to explain an obvious asymmetric Universe through a violation of fundamental symmetries or to understand how antimatter in the Universe can hide itself from our observation. If there is a violation of basic symmetries which can cause a asymmetric world we should look for it in the most fundamental antimatter system, the antihydrogen. Our conclusions about an asymmetric Universe are based on the observation of cosmic antiprotons and the shape of the high energy gamma spectrum. We know, however, that antprotons can be produced from energetic particles and we also know from antiproton annihilations observed in laboratory experiments that most of the gamma rays are produced from meson decays. On the other hand mesons can be produced in any high energy reaction and they are not a specific indicator of the presence of antibaryons. Only the detection of antiatoms can tell us whether there is a significant amount of antimatter in the Universe. Yet, we know of no means to distinguish radiation emitted from ordinary atoms from that emitted from antiatoms, because we believe in the validity of our fundamental principles. With the production of antihydrogen on laboratory scale we have, however, the possibility to verify some of these assumptions to a certain extent.

The abundant production of antihydrogen opens a plethora of new physics: spectroscopy of antihydrogen; study of matter-antimatter interaction; technical applications (polarized antiproton beams); gravitational experiments; solid state physics; chemical applications; storage of antihydrogen and accumulation of macroscopic quantities; production of heavier antimatter systems.

<u>Spectroscopy of antihydrogen</u>. The aim of these experiments is to measure very precisely observables in antihydrogen and to compare them to the corresponding quantities in hydrogen. The hydrogen atom is one of the best measured and understood systems. Any measurement in antihydrogen which can approach a similar precision as achieved in hydrogen provides a sensitive test of the validity of fundamental symmetries. Moreover we cannot exclude that an antimatter atom may change its properties slightly in the overwhelming presence of matter not necessarily violating fundamental symmetry principles.

The basic properties of antihydrogen which one wants to measure are the binding energy and lifetime of the atomic levels. Some of these properties are: Lamb shift of the ground state; ground state hyperfine splitting; the $2s_{1/2} - 2p_{1/2}$ Lamb shift; hyperfine splitting of the 2s and 2p states; Rydberg constant; lifetime of the 2s, 2p, 3d, etc. states.

The lifetime of the excited states can be measured by populating a state at a spacially well defined position and then determine its the decay length. In some cases (e.g. 3s and 4s states) one might be able to achieve a precision of say 1mm/100m, i.e. 10 ppm by chosing the energy of the antihydrogen beam accordingly.

The Rydberg constant can be determined from the ionization threshold or from a precise measurement of a transition energy. The precision achievable here is essentially limited by the velocity spread of the beams and hence the Doppler profile. However, eventually a Doppler-free measurement could be done.

The hyperfine splitting of the n=2 levels could be measured in a level crossing experiment, where the 2s level is quenched with either one of the 2p levels. Another possibility is building-up a population in the 2s level and then inducing microwave transitions to the 2p state which rapidly decays. The measurement of the K_α-line or the remaining 2s population as a function of the microwave frequency would yield the 2s-2p Lamb shift. The precision to which the frequency could eventually be measured is at best 10^{-4} of the 2p-width (100 MHz) hence 10 ppm. This requires a flight path of at least 30m ß in order to keep the transition broadening much below the lifetime related width.

The ground state hyperfine splitting is one of the best measured quantities in physics. It was determined from the hydrogen maser. The precision with which the frequency of the famous 21 cm line could be measured in antihydrogen depends mainly on the size of the transition broadening. As in the case of the Lamb shift measurement the achievable precision should be in the order of 10 ppm unless one finds a method to store a large amount of antihydrogen at very low energy.

Matter-Antimatter Interaction. The interaction of a hydrogen with an antihydrogen atom leads eventually to the annihilation of the particle-antiparticle pairs. The question whether there is a stable bound state between H and $\bar{H}$ prior to annihilation let to controversal discussion in the literature[11]. It could be deduced from the H-$\bar{H}$ total cross section measurement at very low energies. The required relativ H-$\bar{H}$ energies are below 1 eV. Cross sections at such a low energy are difficult to measure as it requires cold gases of hydrogen and antihydrogen. However one could imagine to merging beams of hydrogen (e.g. stripped H^-) and antihydrogen with an adjustable energy difference.The attenuation of the antihydrogen in the hydrogen beam and vice versa would be a measure and of the cross section. With a very intense hydrogen beam (10mA stripped H^-) one could achieve annihilation rates of about $1\ s^{-1}$.

It might also be of interest to measure the stripping cross section of antihydrogen in thin foils or in gas curtains. In contrast to hydrogen we expect the stripping cross section varying differently with the target thickness since once the positron is stripped off it is very unlikely that it can be recaptured. With usual matter projectiles one observes, however, an equilibrium between ionisation and capture. Hence the antihydrogen may allow to test particular aspects here.

Gravitational experiments. The gravitational interaction has gained recently considerable interest. We refer here also to the contibution to this conference where the measurement of the gravitational mass of the antiproton is described and the theoretical interest is outlined[12]. Antihydrogen provides in principal also this possibility with the advantage of having a neutral system. However, for this purpose the antihydrogen atoms have to be of subthermal energies. As mentioned earlier this can probably achieved for a few atoms starting from antiprotons stored at very low energy in a trap.

Antiproton-Polarization. The production of intense antihydrogen beams has the practical aspect that it can be used to polarize antiprotons. This can be achieved in several ways[13,14].

The population of the magnetic levels of the ground state of antihydrogen can be changed through optical pumping with circularly polarized laser light[14]. Stripping off the positrons leaves the antiproton in a well defined state of polarization.This produces an external polarized antiproton beam.

Another way to polarize the antiprotons is by making use of the principle of classical Lamb shift spin filter source[13]. Antihydrogen is formed in the 2s state. The magnetic substates are quenched in a magnetic field to the ground state. Only one 2s magnet sublevel remains stable. Stripping off the positron produces again polarized antiprotons.

The described techniques do not waste any antiprotons, since all the unpolarized antiprotons could be recirculated.

<u>Storage of antihydrogen</u>. Because of its magnetic moment the neutral antihydrogen atoms can be stored in a magnetic field similar to that used for the storage of slow neutrons. In order to achieve that very low energy antihydrogen is needed. This can eventually be obtained if antihydrogen is formed from antiprotons stored in a trap.

<u>Production of heavier antimatter systems</u>. We can imagine that other systems of antimatter can be formed once antinuclei can be produced with a sufficient rate. Then through the same process as described above negative anti-ions can be formed. This gives us finally the possibility to decelerated such ions down to very low momenta. In contrast to naked ions we can then apply laser-cooling to reach subthermal energies. Having achieved that, the ions can be neutralized and stored in a magnetic bottle. Apart from that we can imagine of forming the anti-ion to H^- or the anti-molecule to H_2^+. In order to accomplish the latter we need, however, very high densities of antiprotons and positrons.

1) G.I. Budker and A.N. Skrinsky, Sov. Phys. Usp. 21, 277 (1978).

2) H. Poth, Review of electron cooling experiments, Proc. Workshop on Electron cooling and related applications (ECOOL 84), Karlsruhe 1984, ed. H.Poth, KfK 3846, 1985, p. 45

3) R. Neumann, H. Poth, A. Wolf and A. Winnacker, Z. Phys. A313, 253 (1983).

4) M. Bell and J.S. Bell, Part. Accelerators 12, 49 (1982).

5) H. Herr, D. Möhl and A. Winnacker, Proc. 2nd Workshop on Physics with Cooled Low Energy Antiprotons at LEAR, Erice, 1982(eds. U. Gastaldi and R.Klapisch) (Plenum Press, NY, 1984), p. 659.

6) A. Wolf, H. Haseroth, C.E. Hill, J.-L. Vallet, C. Habfast, H. Poth, B. Seligmann, P. Blatt, R. Neumann, A. Winnacker, G. zu Putlitz,

Electron cooling of low-energy antiprotons and the production of fast antihydrogen atoms, Proc. Workshop on the design of a low-energy antimatter facility in the USA, Madison, Wisconsin, 1985, ed. D. Cline

7) A. P. Mills, in Positron scattering in gases (eds. J.W. Humbertson and M.R.C. McDowell)(Plenum Press, NY, 1981), p. 121.
R. Howell, R.A. Alvarez and M. Stanek, Appl. Phys. Lett. 40, 751 (1982).
G. Gräff, R. Ley, A. Osipowicz, G. Werth and J.Ahrens, Appl. Phys. A33, 59 (1984).
A. Vehanen, K.G. Lynn, P.J. Schultz and M. Eldrup, Appl. Phys. A32, 2572 (1983).

8) R.S. Conti and A. Rich, The status of high intensity low energy positron sources for antihydrogen production, ibid. Ref. 6.

9) J. Berger et al., Feasibility Study for Antihydrogen Production at LEAR, Proposal to CERN CERN-Heidelberg-Karlsruhe Collaboration, CERN/PSCC/85-45, PSCC/P86, CERN/PSCC/86-21, Add.1 and CERN/PSCC/86-37 Add.2 (extended collaboration includes Ann Arbor and Brookhaven National Laboratory)(spokesman H. Poth).

10) H. Poth, On the production of antihydrogen, in preparation.

11) W. Kolos, D.L. Morgan, D.M. Scrader, L. Wolniewicz, Phys. Rev. A11, 1792(1975).
B.R. Junker, J.N. Bardsley Phys. Rev. Lett. 28, 1227 (1972).

12) R.E. Brown, this conference
F. Goldman, this conference

13) H. Poth and A. Wolf, Antiproton polarization through antihydrogen, KfK 4098 (1986).

14) K. Imai Proc. 6th Int. Symposium on Polarization Phenomena in Nuclear Physics, Osaka, August 1985, eds. M. Kudo et al., Supplement to J. Roy. Phys. Soc. of Japan, 55,302 (1986).

ANTIPROTONIC ATOMS
H. Koch
Kernforschungszentrum Karlsruhe, Institut für Kernphysik and
University of Karlsruhe, Institut für Experimentelle Kernphysik,
W.-Germany

ABSTRACT

The results of five LEAR experiments looking for $\bar{p}$-atomic X-rays are reviewed. In the $\bar{p}p$-system, for the first time, transitions to the ground state were observed, and its strong interaction shift and width could be derived. Experiments on heavier atoms produced very accurate data, from which a new best value for the magnetic moment of the antiproton was obtained. For the first time a spin-dependence of the $\bar{p}$-nucleus interaction was observed.

Antiprotonic atoms are formed, when a slow antiproton kicks an electron off an atom and is subsequently bound by the Coulomb field of the nucleus. This process can happen on Hydrogen or on all heavier nuclei. The energy levels of an antiprotonic atom are characterized by the main quantum number n and the total angular momentum j. After its capture the antiproton finds itself mainly in states with high n,j-values and then cascades down to states with higher binding energies emitting Auger electrons (low energy cascade transitions) and X-rays. In addition to the electromagnetic force the antiproton feels the strong $\bar{p}$-nucleon interaction, when its Bohr orbit approaches the nucleus. This strong force gives rise to shifts (ε) (compared to the pure QED-case) and natural widths (Γ) which can be measured in the last observable transition of the X-ray spectrum. Only in the case of Hydrogen (Deuterium) the $\bar{p}$ reaches eventually the 1s ground state (2p-1s = last observable transition), in heavier atoms the $\bar{p}$-nucleus annihilation is so strong, that the X-ray cascade ends at higher levels. In $\bar{p}$-^{16}O, e.g., the 4f-3d transition is the last observable transition.

The availability of the high-quality low energy $\bar{p}$-beams at Lear has stimulated a new round of measurements on antiprotonic X-rays the first results of which are reviewed in the following.

1. $\bar{p}$-Hydrogen (Deuterium)

Measurements on these systems are difficult because of (i) the low energies of the X-rays (3d-2p: 1.7 keV; 2p-1s ≈ 9 keV) and (ii) the low X-ray yields due to the Day-Snow-Sucher mechanism. It means that a $\bar{p}p(d)$-system, while it is in energy levels with n ≈ 10-3, penetrates the strong electric field of a neighbouring Hydrogen-atom, and thus a strong Stark-mixing happens.

This effect mixes atomic levels of the same n-, but different l-quantum numbers, giving rise to an s-or p-state admixture in all states leading to larger than normal annihilation rates from these levels. It is the strongest in liquid Hydrogen (Deuterium) and weakens in gas with decreasing pressure and is responsible for the fact, that the 2p-1s X-ray transition is only observable in gaseous

Hydrogen.

Three experiments at Lear are looking for $\bar{p}p(d)$-X-rays using different targets and detectors:

-PS 171 (Asterix collaboration)
Target: 1 atm gas (NTP); Detector: X-ray drift chamber with moderate energy resolution
Tag: Annihilation products and X-ray coincidences

- PS 174 (Birmingham, Rutherford, William and Mary, Nikhef-Collaboration)
Target: Gas (partly low temperature), corresponding to pressures of 10, 2, 0.25 atm
Detector: Si(Li) with NaI-Compton suppression

- PS 175 (Kernforschungzentrum and University, Karlsruhe)
Target: Gas with pressures as low as 60 mbar (Cyclotron-trap: Antiprotons spiralling in a 4.5 T superconducting magnetic field)
Detector: Si(li) with guard-ring supression

- X-ray yields (pressure dependence)

All three experiments have identified various members of the X-ray L-series and have determined the absolute yields which are pressure dependent. At NTP the total yield of the L-X-rays is 13% [1] while it increases to 40% at 60 mbar gas pressure [2]. The pressure dependence of the X-ray yields can be well reproduced by appropriate cascade calculations [3], the parameters of which could be determined for the first time.

- Strong interaction effects (ε_{1s}, Γ_{1s}, Γ_{2p})

The components of the K-series X-rays have been seen in two experiments until now [1,4]. The yield of the K_α-line at NTP, e.g., is down to 2.6%o [1] because of the strong 2p-annihilation width Γ_{2p} which is about 40 meV [1,2]. Recently, Asterix was able to produce pure K_α- and L_α-lines by requesting coincidences with the respective higher transitions [5]. The measured shifts (ε_{1s}) and widths (Γ_{1s}) range from -500 to -700 eV and 800-1000 eV, respectively. They are compatible with earlier theoretical predictions, but the measurements are not yet accurate enough to allow a selection between the different approaches. However, the large width Γ_{1s} seems to preclude the existence of narrow $\bar{N}N$-states very near to the threshold.

2. Heavier $\bar{p}$-atoms

The X-rays of $\bar{p}$ ^{4}He-[1,2,4] and $\bar{p}$ ^{3}He[2]-atoms have been measured using gaseous targets, while all heavier atom-data were taken on liquid or solid targets. These measurements were performed at Lear from two collaborations:

- PS 176 (Kernforschungzentrum and University, Karlsruhe; University of Basel; Institute for Nuclear Physics, Stockholm;

CNR Strasbourg; University of Thessaloniki)

- PS 186 (Technical University, Munich, University of Mississippi)

In short running periods high statistics spectra with excellent peak/background ratios were collected. Up to eight detectors were run simultaneously measuring electromagnetic radiation in the range from 2 keV to 1 GeV and neutrons. The evaluation of the data is still in progress and only some aspects of the measurements are discussed here:

- Magnetic moment of the antiproton

From the fine-structure splitting of the 11-10 X-ray transition of $\bar{p}$-^{208}Pb a new best value for $\mu_{\bar{p}} = 2.8007(91) \cdot \mu_N$ [6]) was derived which is two times better than the present world average. Within the errors it is compatible with the CPT-prediction relating μ_p and $\mu_{\bar{p}}$.

- Strong interaction effects

Table I gives a survey on the strong interaction effects measured at Lear:

The widths and shifts of the lower level were derived from an accurate determination of the energies and the natural line widths of the last observable transitions, the width of the upper level using the intensity balance of the X-rays feeding and depopulating the level [10]). In the case of the Mo-isotopes, even a fourth quantity (width of the level below the last observable transition) was measured by taking advantage of a resonance effect between nuclear and atomic levels [9,11]).

The data can be compared with theoretical predictions [12-18]). Parameterfree analysis using the $\bar{p}p$- and $\bar{p}n$-amplitudes at threshold, the nucleon distributions in the nuclear tail and various medium corrections are already available for a few nuclei and reproduce the data within 30%, while phenomenological analysis with one or more adustable parameter do a similar job, but for the whole range of nuclei measured.

Of particular importance for the determination of as yet still uncertain input parameters, like $f_{\bar{p}n}$ (threshold), ρ_n (nuclear tail), spin dependence of the $\bar{p}$-nucleus interaction, are the precise measurements on isotopes (e.g. $^{16/17/18}$O) and the strong interaction effects individually determined for both components of a fine-structure splitted level (^{174}Yb).

- Isotope effects in $^{16/17/18}$O

Isotope effects have been measured on $^{3/4}$He, $^{16/17/18}$O and $^{92/94/95/98}$Mo. They are particularly clear in the case of the Oxygen-isotopes, where the addition of two (one) neutrons gives rise to 8(4)σ-effects. A proper theoretical treatment of these effects is expected to give insight into the isospin-structure of the low energy $\bar{p}$-nucleon amplitude.

Table I Strong interaction effects derived from measurements of
PS 176/186 [7,8,9]. Also included are the data from PS 171/
174/175 [1,2,4] on Hydrogen, Deuterium and Helium. Values
with * are still preliminary.

	ε_{1s} [eV]	Γ_{1s} [eV]	Γ_{2p} [meV]
^{1}H	-500±300 (PS171)	1000* (PS171)	≈ 40* (PS175)
	-730±150 (PS174)	850±390 (PS174)	

	ε_{2p} [eV]	Γ_{2p} [eV]	Γ_{3d} [meV]
^{3}He	-17 ± 4 (PS175)*	25 ± 8 (PS175)*	2.13 ± 0.14 (PS175)*
^{4}He	-16 ± 3 (PS175)*	44 ± 8 (PS175)*	2.38 ± 0.10 (PS175)*
	-7.4 ± 5.3 (PS174)	35 ± 15 (PS174)	2.4 ± 0.4 (PS174)
^{6}Li	-212 ± 65*	685 ± 160*	≈ 150*
^{7}Li	-271 ± 40*	745 ± 120*	

	ε_{3d} [eV]	Γ_{3d} [eV]	Γ_{4f} [meV]	
^{14}N	-18 ± 6	189 ± 15	275 ± 25	
^{16}O	-112 ± 20	495 ± 45	602 ± 24	) Clear
^{17}O	-140 ± 46	540 ± 150	731 ± 37	) isotope
^{18}O	-195 ± 20	640 ± 40	795 ± 25	) effects
^{19}F	-450 ± 66	1397 ± 54	3245 ± 130	
^{23}Na	-2000 ± 200*	2500 ± 800	31020 ± 7020*	

	ε_{4f} [eV]	Γ_{4f} [eV]	Γ_{5g} [meV]
^{40}Ca	-1067 ± 143*	3576 ± 403*	34900 ± 3800*

	ε_{6h} [eV]	Γ_{6h} [eV]	Γ_{7i} [eV]	Γ_{5g} [eV] indir. det. via nucl.res.[9]
^{92}Mo	-460±80 (PS186)	1400±300 (PS186)	19.5±1.2 (PS186)	
^{94}Mo	-640±220 (PS186)	2300±900 (PS186)	68.9±10.2 (PS186)	42500±9200
^{95}Mo	-740±120 (PS186)	1900±400 (PS186)	28.5±2.6 (PS186)	49900±19800
^{98}Mo	-550±160 (PS186)	2300±700 (PS186)	47.9±5.3 (PS186)	45100±10000

	ε_{7i} [eV]	Γ_{7i} [eV]	Γ_{8j} [eV]
^{138}Ba	-350 ± 150*	1800 ± 450*	

- Spin-orbit $\bar{p}$-nucleus interaction

In the case of the 9k-8j X-ray transition in ^{174}Yb an individual determination of the shifts and widths of each fine-structure component of the 8j-level could be performed for the first time. The shifts and widths are definitely dependent (3σ-effect) on $\vec{L}\cdot\vec{S}$ ($\epsilon_{\uparrow\uparrow} - \epsilon_{\uparrow\downarrow}$ = (68.0 $\pm$ 25.0)eV; $\Gamma_{\uparrow\uparrow} - \Gamma_{\uparrow\downarrow}$ = (195 $\pm$ 58)eV), so that a clear spin-dependence of the $\bar{p}$-nuclear interaction shows up, which is surprising having in mind the weak spin-dependence of the elastic $\bar{N}N$-scattering data. A theoretical treatment has to show, if both results are compatible or if the here observed strong spin-effect is due to the annihilation-part of the interaction.

3. Outlook to future measurements

The spectroscopy of $\bar{p}p(d)$-systems would profit from higher $\bar{p}$-fluxes, allowing the use of high resolution detectors, like crystal-spectrometers. With high $\bar{p}p$-densities even the use of optical- or microwave methods could be considered. These detectors would allow the detection of strong interaction effects including FS- and HFS-interaction in levels above the 1s-groundstate, the determination of the polarizability of the antiproton and eventual medium range QCD-effects.

Similar statements hold for heavier atoms, where particularly coincidence measurements between atomic X-rays and particles (γ's) emitted after the $\bar{p}$-nucleus annihilation are attractive.

Needless to say that high intensity, low energy $\bar{K}$, $\bar{\Sigma}$, $\bar{\Xi}$-beams from a Kaon-factory would make similar experiments feasible with particles containing strange quarks and thus give further information on the flavour-dependence of quark-interactions.

REFERENCES

1. S. Ahmad et al., P.L. B157 (1985) 333
2. P. Blüm, D. Gotta, L.M. Simons, Private communication
3. E. Borie, M. Leon, P.R. A21 (1980) 1460
4. T.P. Gorringe et al., P.L. 162B (1985) 71
5. E. Klempt, Private communication
6. A. Kreissl, Thesis, University of Karlsruhe, 1986
7. Th. Köhler et al., submitted to P.L.B., 1986
8. D. Rohmann et al., submitted to Z. Phys., 1986
   H. Barth, Private communication
9. W. Kanert et al., P.R.L. 56 (1986) 2368
10. H. Koch et al., P.L. 29B (1969) 140
11. M. Leon, P.L. 50B (1974) 425
12. J.F. Haak et al., P.L. 66B (1977) 16
13. H. Nishimura, T. Fujita, P.L. 60B (1976) 413
14. W. Kaufmann, H. Pilkuhn, P.L. 62B (1976) 165
15. A.M. Green et al., Nucl. Phys. A399 (1983) 307
16. T. Suzuki, H. Narumi, Nucl. Phys. A426 (1984) 413
17. A. Deloff, J. Law, P.R. C10 (1974) 2657
18. S. Dumbrajs, Proceedings of the Third LEAR Workshop, Tignes, 1985

CP-ODD ASYMMETRIES AT LOW ENERGY

John F. Donoghue
University of Massachusetts, Amherst, MA 01003

ABSTRACT

CP violation can be seen in the reactions $pp \to \Lambda\bar{\Lambda}, \Sigma\bar{\Sigma}, \Xi\bar{\Xi}$. This talk discusses the CP odd observables in hyperon decay and the way that they produce asymmetries in $\bar{p}p$ reactions.

The goal of studies of CP violation is to find a new form of this phenomena. In the superweak model the only form of CP non-conservation is a $\Delta S=2$ interaction mixing K^0 and $\bar{K}^0$. In other models, such as the KM theory, the $\Delta S=2$ interaction is still dominant. The strength of this effect cannot distinguish among models, because each model will adjust the strength of their CP odd parameter in order to fit the experimental number, ε. However many models have CP violation in the $\Delta S=1$ interaction also. It is here that models stand a chance of being distinguished. We need to observe some non-ε effect if the field is to progress. This talk describes one such proposal.[1]

Let us look at an example of $\Delta S=1$ CP violation in order to discuss its strength. In the case of $K_S \to 3\pi$ we can define

$$\eta_{+-0} \equiv \frac{\langle \pi^+\pi^-\pi^0 | H_w^{CP\text{-odd}} | K_S \rangle}{\langle \pi^+\pi^-\pi^0 | H_w^{CP\text{-even}} | K_L \rangle} \equiv \varepsilon + \varepsilon_{3\pi} \qquad (1)$$

The factor ε arises from mass matrix mixing while $\varepsilon_{3\pi}$ comes from direct $\Delta S=1$ effects in the $K_S \to 3\pi$ transition. The present generation of experiments will measure η_{+-0} to $O(\varepsilon)$. Is there any chance of seeing a non-ε signal? The answer is that $\varepsilon_{3\pi}$ may in general be large, but that many specific models will have it less than 0.2ε. Let us label the amount of $\Delta S=1$ CP violation by the parameter χ. If a theory violates parity, there will be a direct CP nonconservation in $K_L \to 2\pi$, leading to an effect $\varepsilon' \sim \chi/20$ (the factor of 1/20 comes from the $\Delta I = 1/2$ rule), so that $\chi \approx 20\varepsilon' < .2\varepsilon$. However if the CP odd part of the theory is parity conserving, no effect is found in $K \to 2\pi$. In this case the bound comes from the dispersive component to the mass matrix, made up of two subsequent $\Delta S=1$ transitions as in $K^0 \to \pi^0, \eta, \eta' \to \bar{K}^0$. An estimate of this indicates that (being somewhat generous) $\chi \leq 4\varepsilon$. Surprisingly, there is no more stringent bound than this at present. This means that the proposed measurements are in fact significant, although in most models a smaller signal is expected (see the review by Holstein[2] in these Proceedings).

Proton antiproton reactions are especially useful for studying CP violation because the initial state has an invariance property under CP. Consider $\bar{p}p$ in their center of mass. We have then

$$PC|p(\vec{p})\ \bar{p}(-\vec{p})\rangle = P|\bar{p}(\vec{p})\ p(-\vec{p})\rangle$$
$$= -|\bar{p}(-\vec{p})\ p(+\vec{p})\rangle \qquad (2)$$
$$= |p(p)\ \bar{p}(-\vec{p})\rangle$$

This means that if CP is valid the final state must be CP symmetric. There has been considerable discussion of using kaons in $p\bar{p}$ (e.g. see the talk by Fry[3] and Winstein[4] at this conference). However it is also possible to exploit this property using hyperons.

An example of a CP odd asymmetry in the reaction

$$p(\vec{p}) + \bar{p}(-\vec{p}) \to \Lambda(\vec{\ell}) + \bar{\Lambda}(-\vec{\ell}) \to p(k) + \bar{p}(\bar{k}) + \pi^+(\bar{q}) + \pi^-(q)$$

is of the form

$$A_\Lambda \equiv \vec{p}\cdot(\vec{q}\times\vec{k} - \vec{\bar{q}}\times\vec{\bar{k}})$$
$$= \vec{p}\times\vec{\ell}\cdot(\vec{k}+\vec{\bar{k}}) \qquad (3)$$

In the latter form $\vec{p}\times\vec{\ell}$ defines the production plane and the asymmetry counts the p and $\bar{p}$ above and below the plane. Another way to discuss this is to use

$$N_\Lambda = \frac{N_p(\text{up}) + N_{\bar{p}}(\text{up}) - N_{\bar{p}}(\text{down}) - N_p(\text{down})}{N_{\text{events}}} \qquad (4)$$

A nonzero value of N_Λ or $\langle A_\Lambda\rangle$ indicates CP violation.

The theory of CP violation in hyperon decay has been worked out elsewhere.[5] The important quantities are the partial widths (e.g. $\Gamma(\Lambda \to p\pi^-)$) and the parameters α and β which govern the decay distribution and spin in the final state

$$W(\theta) = \frac{1}{4\pi}[1 + \vec{\alpha}\cdot\vec{k}]$$
$$\langle\vec{\sigma}\rangle_f = \frac{1}{1+\alpha\vec{s}_i\cdot\hat{p}_f}\{\beta\vec{s}_i\times\hat{p}_f + \ldots\} \qquad (5)$$

The CP violation in α and β occurs through an interference of the s and p wave final states, while in the partial decay rate it is an interference of isospin 1/2 and 3/2 final states. Moreover one needs to compare hyperon and antihyperon decay in order to separate the CP violating phases from phases that occur due to the strong final state interactions. In this game there are two small numbers. One is the final state phase where typically one finds $\sin(\delta_i - \delta_j) \approx 1/10$. The other has to do with the $\Delta I = 1/2$ rule where $\Delta I = 3/2$ amplitudes are small by a factor of $A_3/A_1 \sim 1/20$. In the decay quantities this leads to a hierarchy of signals. A model independent analysis finds (again using χ as the measure of $\Delta s=1$ CP violation)

$$\beta + \bar{\beta} \sim \chi$$

$$\alpha + \bar{\alpha} \sim \chi \sin(\delta_s - \delta_p) \sim \frac{1}{10} \chi \tag{6}$$

$$\frac{\Gamma - \bar{\Gamma}}{\Gamma + \bar{\Gamma}} \sim \chi \sin(\delta_{1/2} - \delta_{3/2}) \frac{A_{3/2}}{A_{1/2}} \sim \frac{1}{100} \chi$$

Explicit calculations in the KM model give[5], for example,

$$(\alpha + \bar{\alpha})_\Lambda = -0.8 \times 10^{-4} \quad , \quad (\beta + \bar{\beta})_\Xi = 2 \times 10^{-4} \tag{7}$$

These explicit calculations are rough and should not be trusted in detail. However they are reasonably conservative in that they use the bag model which tends to give small transition amplitudes. I could, if I wanted to, increase them by a factor of six without being dishonest by 1) normalizing the CP-odd portion to a <u>calculation</u> of the CP-even part rather than to experiment (arguing that matrix element uncertainties cancel better in the ratio of two calculated quantities) and 2) using a larger value of the KM element V_{ub} which <u>may</u> be allowed in recent analyses. However experimenters should be prepared for small signals and need to plan their experiments for the more pessimistic values.

The hyperon decay parameters generate the asymmetries discussed earlier in the following way. When the Λ and $\bar{\Lambda}$ are produced, T invariance of the strong interactions implies that their polarizations are equal and are normal to the production plane. Due to the α parameter the angular distribution, Eq. 5, then populates the upper and lower hemispheres differently. Explicit calculations give

$$N_\Lambda = \frac{1}{2} P(\alpha + \bar{\alpha}) \tag{8}$$

In past experiments polarizations of O(50%) are achieved at least for some range of momentum transfer.

There are also asymmetries which can measure the parameter $\beta + \bar{\beta}$. However these are considerably more complicated because they require the measurement of the final baryon spin. As far as I can see, this requires the study of cascade decay, $\Xi \to \Lambda \pi$, where the final state Λ will analyze its own spin. The asymmetry governed by β is of the form $\vec{s}_\Xi \cdot \hat{p}_\Lambda \times \vec{s}_\Lambda$. The cascade spin will be in the direction $\vec{y} \equiv \vec{p} \times \vec{p}_\Xi$, while the lambda spin will be measured by the outgoing proton momentum. We then can define the asymmetry by the sign of the product

$$p \equiv \hat{y} \cdot \hat{p}_\Lambda \times \vec{p}_{p_f} \tag{9}$$

measured in the cascade frame. Overall (using $\beta + \bar{\beta} > \alpha + \bar{\alpha}$)

$$A_\Xi = \frac{N_p(p>0) + N_{\bar{p}}(p>0) - N_p(p<0) - N_{\bar{p}}(p<0)}{N_{events}}$$

$$= \frac{\pi}{8} \alpha_\Lambda (\beta + \bar{\beta})_\Xi P_\Xi \tag{10}$$

This measurement will be more difficult than that of $\alpha + \bar{\alpha}$, but may be worth considering if the signal is larger.

These asymmetries will not be easy to measure as they require high statistics, good control over systematics and a large amount of computing time. However they may be within reach of the next generation of experiments at LEAR or at a $p\bar{p}$ facility at Fermilab. If such experiments were successful, they would be extremely important in understanding the origin of CP violation. The overall number and SU(3) pattern of the CP odd signals should provide a distinct test allowing one to decide among the various theories. These experiments deserve careful thought to see if they are feasible.

REFERENCES

1. J. F. Donoghue, B. R. Holstein and G. Valencia, UMHEP-257.
2. B. R. Holstein, this conference.
3. J. Fry, this conference.
4. B. Winstein, this conference.
5. J. F. Donoghue, X. G. He and S. Pakvasa, submitted to Phys. Rev.;
   J. F. Donoghue and S. Pakvasa, Phys. Rev. Lett. $\underline{55}$, 162 (1985);
   L. L. Chau and H. Y. Cheng, Phys. Lett. $\underline{B131}$, 302 (1983);
   T. Brown, S. F. Tuan and S. Pakvasa, Phys. Rev. Lett. $\underline{51}$, 1823 (1983).

p̄-NUCLEUS INTERACTION

J. C. Peng
Los Alamos National Laboratory
Los Alamos, NM 87545

ABSTRACT

Status and future prospects of p̄-nucleus scattering experiments are presented.

I. INTRODUCTION

During the past two years, high quality p̄-nucleus scattering data obtained at the LEAR facility have provided extremely valuable information on p̄-nucleus interaction. In this talk, I will summarize the data[1-5] from the LEAR experiment PS 184 and discuss their impact on current knowledge on p̄-nucleus interaction. Some possible future directions for p̄-nucleus scattering experiments will also be mentioned. Other aspects of p̄-nucleus interaction, such as p̄-atoms, p̄ annihilation in nuclei, and (p̄,n̄) reaction on nuclei, are discussed by Koch, McGaughey, Gibbs, and Auerbach in this conference.

II. p̄-NUCLEUS ELASTIC SCATTERING

Using a high resolution, large acceptance spectrometer, the antiproton elastic scattering has been measured on a number of targets at 300 MeV/c and 600 MeV/c. Table I lists the target nucleus and the beam momentum for these measurements. Optical model calculations have been performed[1,2,5] to extract the phenomenological optical model potentials. The results of this analysis are the followings:

(a) Strongly absorptive potentials are required to fit the data. This reflects the importance of p̄-annihilation at low energies.

(b) The strength of the real and imaginary parts of the potential is well determined only at the nuclear surface. The real potential at this distance is found to be attractive and weaker than the imaginary potential; $|W(R)| \geq 2 |V(R)|$. R is about 1.5 $R_{r.m.s.}$ and decreases slightly at higher incident energy.

(c) Despite the ambiguities in the optical potentials, the reaction cross sections calculated by various potentials are in good agreements (± 5%). The extracted p̄-nucleus reaction cross sections are roughly a factor of two larger than the p-nucleus reaction cross sections.

TABLE I ELASTIC SCATTERING DATA FROM PS 184

$\bar{p}$ momentum	target nucleus						
300 MeV/c	-	^{12}C	---	---	^{40}Ca	^{48}Ca	^{208}Pb
600 MeV/c	d	^{12}C	^{16}O	^{18}O	^{40}Ca	---	^{208}Pb

Although the $\bar{p}$ elastic scattering data are only sensitive to the interaction at nuclear surface, they already provide valuable constraints on $\bar{p}$-nucleus interaction. The prediction[6] of a deep real potential in a relativistic mean-field approach is not supported by our finding of a rather weakly attractive real potential. The ambiguity[7] that either a shallow or a deep imaginary potential can fit the $\bar{p}$-atom data[8] is also removed by our finding of the dominance of absorption in $\bar{p}$-scattering. The prediction[9] of orbiting phenomena, which requires $V(r)>W(r)$, is at a variance with our results. The elastic scattering data provide enough constraint on the $\bar{n}$-nucleus potential so that a more reliable $n\bar{n}$ oscillation $t_{nn} \sim 3 \times 10^7$ sec is deduced[10] from the proton decay experiment.

There are several theoretical works to describe the p-nucleus interaction in microscopic models. The ingredient for such models are the nuclear density and the free or medium modified $N\bar{N}$ amplitudes. In the framework of Glauber theory, it was shown[11] that the data can be well reproduced and the value of real to imaginary ratio ϵ for the $\bar{p}$ N amplitude can be extracted. KMT calculations[12] performed with Paris and Dover-Richard $N\bar{N}$ free t- matrix are also very successful. The effects of medium correction is apparently not very important since the strong absorption prevents $\bar{p}$ from penetrating the nuclear interior. Calculations[13] based on relativistic impulse approximation give good agreement with $\bar{p}$ +^{12}C data. The differences between relativistic and non-relativistic calculations are found to be small, again reflecting the fact that antiprotons do not reach the nuclear interior where the relativistic effects are important.

III. ($\bar{P}$,P) REACTION

The analysis of $\bar{p}$-nucleus scattering data and the $\bar{p}$-atom x-ray data suggest that the real potential is sufficiently attractive to accommodate p̄-nucleus bound or resonant states. In order to search for such p̄-nucleus states, we have measured the (p̄,p) reaction on d, ^{6}Li, ^{12}C, ^{63}Cu and ^{209}Bi. By detecting the (p̄,p) reaction at very forward angles, we search for events where antiproton transfers its momentum to the knocked-out proton and gets trapped in possible p̄-nucleus states. Evidence of p̄-nucleus states could then show up as peaks in the energy spectrum of the detected protons.

The (p̄,p) spectra for the ^{6}Li and the scintillator targets are shown in Fig. 1. The narrow peak observed in Fig. 1a comes from the 180° p̄p elastic scattering. In addition to this peak, the proton spectrum exhibits an exponential-like shape falling with increasing proton energy. The dominant process responsible for the (p̄,p) spectra appears to be proton emission following antiproton annihilation in the target. The energetic pions produced in the annihilation could undergo final-state interactions, causing energetic protons to be emitted. The dashed line in Fig. 1 shows the proton spectrum calculated[14] with a intranuclear cascade model.

A broad peak is observed in the ^{6}Li (p̄,p) spectrum. The location and width of this peak suggest that it corresponds to the quasi-free (p̄,p) knock-out reaction. This quasi-free peak is also clearly observed in the d (p̄,p) spectrum. However, on heavier targets such as ^{12}C, ^{63}Cu, ^{209}Bi, the quasi-free peak is not observed[3]. Apparently, the annihilation-knockout cross sections, which has a mass dependence of $A^{0.63}$, dominate the quasi free cross sections for heavier targets.

No evidence for narrow peaks corresponding to bound or resonant p̄-nucleus states could be found in the (p̄,p) spectra. Experimental limits for the production of p̄-^{5}He and p̄-^{11}B states, assuming these states are bound by an energy equal to the binding energy of the ejected proton, are 6 μb/(sr.MeV) and 20 μb/(sr.MeV), respectively. This limit is already about one order of magnitude lower than one theoretical prediction[15].

IV. P̄-NUCLEUS INELASTIC SCATTERING

The p̄-nucleus inelastic scattering is interesting for the following reasons:

(a) The spin-isospin selectivity in the inelastic scattering can improve our knowledge on the poorly known spin-isospin dependence of the NN interaction.

(b) Unlike the proton-nucleus inelastic scattering, the reaction mechanism for p̄-nucleus inelastic scattering is expected to be relatively simple. In particular, there is no exchange process and the contribution from multiple-scattering process is relatively unimportant due to the strong absorption.

(c) Further constraints on p̄-nucleus optical potential can be obtained by requiring a simultaneous description of the elastic and inelastic data.

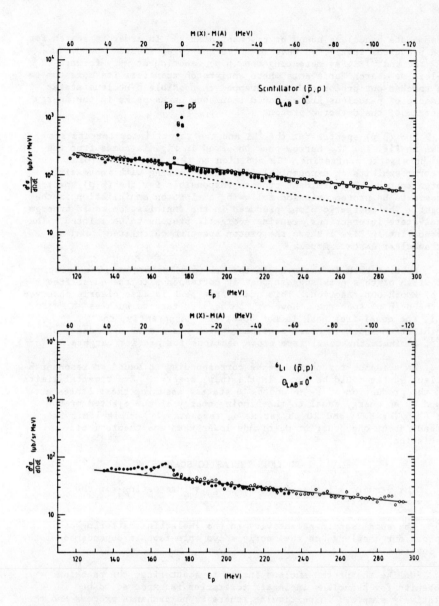

Fig.1. Double differential cross sections for the $(\bar{p},p)$ reaction scintillator and 6Li targets at 600 MeV/c.

Transitions to low lying collective states have been measured[16] on ^{12}C and ^{18}O. Very diffractive angular distribution shape is observed for these transitions. At 300 MeV/c, the ^{12}C 2+ state is excited with a cross section three times smaller for $(\bar{p},\bar{p}')$ than for (p,p'); while the situation is reversed at 600 MeV/c. The energy dependence and the shape of these cross sections are well reproduced by DWIA calculations. This shows that the spin-isospin averaged $N\bar{N}$ amplitude is pretty well understood.

A significant amount of beam time was invested to obtain high statistics on the transitions to the two 1+ unnatural parity states of ^{12}C at 12.7 MeV (T=0) and 15.1 MeV (T=1). It was advocated by Dover et al[18] that the relative strength of exciting the T=0 state versus T=1 state provides a good test of different $N\bar{N}$ models. Unfortunately, the 1.2 MeV energy resolution achieved in our experiment did not allow us to resolve the 15.1 MeV state from the nearby 15.3 MeV natural parity state. The 12.7 MeV state is well resolved and the measured cross sections are in better agreement with the Paris potential prediction.

V. FUTURE PROSPECTS

With the completion of PS 184, it can be said that the first-generation experiments on $\bar{p}$-nucleus scattering have already been performed. Anticipating the arrival of ACOL and other future $\bar{p}$ facilities, it might be worthwhile to mention some possible future experiments. The PS 184 elastic data were limited to scattering angles smaller than 60°. It would be very interesting to extend the measurement to large angles; $180° > \theta > 60°$. The large angle data often bring surprises and provide very stringent constraints on the optical potential. If polarized $\bar{p}$ beam becomes available, then one can separate the central potential from the spin-orbit potential and check the theoretical predictions of very weak spin-orbit potential.

The $(\bar{p},p)$ results show that the annihilation-knockout background greatly reduces the sensitivity for hunting $\bar{p}$-nucleus states. The $(\bar{p},\Lambda)$ reaction, which can be used to search for $\bar{\Lambda}$-nucleus states, merits consideration since the annihilation background is unimportant[19]. Another method to search for $\bar{p}$-nucleus resonant states is to measure excitation functions of $\bar{p}$-nucleus elastic or inelastic cross sections. It is worth noting that single-nucleon resonant states were clearly identified from the excitation functions of proton nucleus scattering.

A significant improvement in $\bar{p}$-nucleus inelastic scattering requires better energy resolutions which can be obtained with a thin target. The $(\bar{p},\bar{n})$ charge exchange reaction is also very interesting for revealing the I=1 $N\bar{N}$ interaction.

The author thanks all collaborators in the PS 184 experiments.

REFERENCES

1. D. Garreta et al., Phys. Lett. <u>146 B</u>, 266 (1984).
2. D. Garreta et al., Phys. Lett. <u>149 B</u>, 64 (1984)

3. D. Garreta et al., Phys. Lett. 150 B, 95 (1985).
4. G. Bruge et al., Phys. Lett. 1699 B, 14 (1986).
5. S. Janouin et al., Nucl. Phys. A451, 541 (1986).
6. A. Bouyssy and S. Marcos, Phys. Lett. 114 B, 397 (1982)
7. C. Y. Wong et al., Phys. Rev. C29, 574 (1984).
8. P. Barnes et al., Phys. Rev. Lett. 29, 1132 (1972);
   P. Roberson et al., Phys. Rev. C16, 1945 (1977);
   H. Poth et al., Nucl. Phys. A294, 435 (1978).
9. E. H. Auerbach et al., Phys. Rev. Lett. 46, 702 (1981).
10. C. B. Dover et al., Phys. Rev. D27, 1090 (1983);
    Phys. Rev. C31, 1423 (1985).
11. G. Dalkarov and V. Karmanov, Phys. Lett. 147 B, 1 (1984).
12. A. Chaumeaux et al., unpublished.
13. B. C. Clark et al., Phys. Rev. Lett. 53, 1423 (1984).
14. M. R. Clover et al., Phys. Rev. C26, 2138 (1982).
15. H. Heiselberg et al., Phys. Lett. 132B, 279 (1983).
16. M. C. Lemaire et al., to be published
17. C. B. Dover and D. J. Millener, preprint BNL-36438.
18. C. B. Dover et al., Phys. Lett. 143, 45 (1984).
19. J. C. Peng, Proc. Third LAMPF II Workshop, Los Alamos, 1983, Report LA-9933-C, p. 531.

PRODUCTION OF HIGH ENERGY DENSITY IN $\bar{N}$-NUCLEUS INTERACTIONS

W. R. Gibbs
Theoretical Division, Los Alamos National Laboratory
Los Alamos, New Mexico 87545

ABSTRACT

The results of an investigation of $\bar{p}$- (and to a lesser extent) $\bar{d}$-nucleus interactions are reported. The technique involves following the classical production and propagation of mesons ($\pi, K^+, K°, K^-, \bar{K}°, K^*, \eta, \omega, \phi$) and baryons (N, Λ, Σ) in nuclei after antiparticle annihilation. It is found that small regions of the nucleus can be raised to sufficiently high energy densities that some predictions of a quark-gluon phase transition can be tested with the use of energetic antiprotons (5-10 GeV/c). The strangeness signal is examined and compared with the amount of strangeness produced in a recent experiment with 4 GeV/c incident antiprotons. A general expression is given for the total amount of strangeness produced which is invariant under intranuclear strangeness exchange reactions.

I. Introduction

It is clear that the $\bar{p}$-nucleus interaction is one of the really new and exciting tools available for the exploration of the nucleus, and hadronic physics in general. In no other system is the production of a particle beam available directly within the nuclear medium. The first question that arises, from a fundamental and pedantic point of view is "When is a hadron (a member of this 'particle beam') a hadron?" This question touches on the time structure of the strong interactions in the most fundamental manner. If one is willing to assume that the time required for hadron formation is small (or that the interactions of the constituents lead to approximately equivalent results) then the problem can be addressed from an hadronic viewpoint. This does not immediately take us to the meson and excited baryon picture -- far from it. Since we made the initial assumption on the basis of lifetimes we will continue to do so. While the contribution to the spectral representation of hadronic interactions of the Δ or ρ may be substantial, the approximation of these short lived systems by a "particle" which moves around in our coordinate space seems inappropriate. On the other hand, long lived entities, such as the eta or omega <u>should</u> be treated as real objects. Note that this has little contact with the modern boson exchange picture.

While this discussion has in no way resolved the fundamental question of the time structure of hadronic interactions, it does suggest a hierarchy for the development of a theory of non-coherent nuclear interactions. This is the type of theoretical structure needed for reactions in which radical changes occur in the nuclear system. In such reactions, where many states are averaged over in the final system, the quantum mechanical phase information is lost and classical mechanics is approximately restored. In this case we may use (relativistic) Newtonian mechanics employing classical probabilities based on bilinear quantum-mechanical calculations or, better yet, measured cross sections.

I shall present the results of such a calculation of two interesting quantities. The first is the energy density in the nuclear medium. This is of fundamental interest because, if it can be made of the same order as that in a nucleon, it is possible that a substantial volume of the nucleus can be transformed into a state resembling that of the nucleonic interior. In this advent we shall have our first macroscopic (on a nuclear scale) view of the true hadronic "soup". This is the closest we are likely to come, in the laboratory, to conditions that existed in the first instants of the universe. Such conditions are very challanging to achieve and great efforts are being made to arrive there in the heavy-ion arena. It would seem that these conditions are quite possible with antiprotons (or antideuterons) and, while the volumes attainable are modest, the control of "external" systems makes the study potentially very interesting.

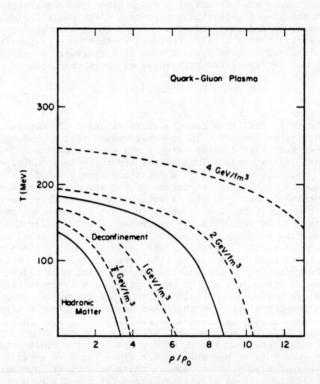

Figure 1. The general view of the possible phases of nuclear matter.

To understand this last comment consider figure 1. It represents a simplified schematic phase diagram of nuclear matter as conjectured by a number of people[1]. While the formation of the universe presumably followed a path originating from a point at very high temperature and low density, along the vertical axis and progressed along this axis to arrive at the cold nuclei which form stable matter, we must start with target nuclei and trace out a heating, as well as cooling, curve. As we shall see shortly, the antiproton annihilation forms a small pocket of high energy density involving several baryons. This volume has a natural center-of-mass velocity determined by the incident antiproton momentum and is usually formed not very far into the nucleus (at about 1 fm depth as estimated from the mean-free-path for annihilation). If a phase transition takes place in this small volume so that quark and gluon, rather than hadronic, degrees of freedom are the relevant ones, then one needs to study the cooling of this system. While the new state is within the nucleus it is cooled by conduction (collisions with the "colder" nucleons surrounding it) and convection (mixing the "cold" nucleons into the "hot" system itself). When it exits the back side of the nucleus it finds itself in free space and will cool radiatively only. By varying the size of the nucleus one can control the time that the hot matter remains in the nucleus, and hence the rate of cooling of the high energy density region. In this way we may study the "signals" of the phase transition as a function of the "temperature" at which the object breaks into free space. In this introduction I have freely used the colorful jargon of thermodynamics; in the actual reactions one should define the finite-particle-number analogue of these concepts and use them.

The second result that I will present concerns the signals that one can use to determine the possible presence of a quark-gluon plasma. The only one I will discuss is that measured by the quantity of strangeness produced, relative to, for example, light q-$\bar{q}$ pairs. If, as has been discussed by Rafelski[2] and co-workers, the gluon-gluon interactions Bremsstrahlung $S\bar{S}$ pairs with the same probability as lighter pairs, then the amount of strangeness produced will increase dramatically over that available from the usual, OZI forbidden mechanisms. As he points out there will already be considerable enhancement in a hot hadronic soup, so some quantitative understanding must be in hand before the significance of an enhancement in strangeness production can be evaluated. I note a point here which will be emphasized later: Strangeness exchange may rearrange significantly the individual strangeness channels so it is not sufficient to look at a single channel (such as only Λ or only K^+, etc.).

It is with these two goals in mind that the calculations are undertaken. Before going on to describe the calculation and present the results, it is useful to obtain an intuitive view of the process.

Energetic antiprotons are much more efficient for energy deposition than very low energy antiprotons for three reasons: 1) The mesonic debris from the annihilation is pushed into the nucleus by the overall motion of the system, 2) The annihilation takes place well within the nuclear system due to the decrease in annihilation cross section and 3) The total energy available is greater, assuming that a reasonable fraction of the energy can be shared among several nucleons. With respect to this last point I note that pions are very efficient as a means of distributing energy among many nucleons since they make ~3-4 collisions with nucleons during the absorption process. A more complete discussion of this energy sharing process can be found in some previous lectures[3].

II. Calculational Technique

The nuclear target is modeled by a system of A nucleons propagating in a Saxon-Woods potential with classical motion. Isotropic (in the CM) NN collisions are governed by an approach distance corresponding to a classical circular cross section of 40 mb. The antiproton is assumed to annihilate 1 Fermi inside the surface of the nucleus on the beam axis. For the antideuteron calculations it is assumed that the two antinucleons annihilate simultaneously ± 1/2 Fermi away from the central axis. The products of annihilation are taken to be pions and $K\bar{K}$ pairs according to the experimentally measured fractions. $\Lambda\bar\Lambda$ production is neglected in the present version of the code.

The mesons thus produced are then propagated within the nucleus with no mean field (in contrast to the case for the nucleons) but with the

Table 1. Channels included in the current version of the code.

NN → NN	$\pi N \to \pi N$	$\pi N \to \phi N$	$\bar{K}N \to \bar{K}N$
	$\pi N \to \pi N$	$\pi N \to K\Lambda$	$\bar{K}N \to \pi\Lambda$
	$\pi N \to \pi\Delta$	$\pi N \to K\Sigma$	$\bar{K}N \to \pi\Sigma$
	↳ πN	$\pi N \to K^*\Lambda$	KN → KN
	$\pi N \to \pi\pi\pi N$	$\pi N \to K^*\Sigma$	
	$\pi N \to \omega N$		
	$\pi N \to \eta N$		

reactions in table I occuring. The pion induced reactions are very important since these deposit most of the energy and produce strangeness as well. Their calculation is implemented by first deciding if there is to be a pion-nucleon collision based on the pion-nucleon distance compared to the pion-nucleon total cross section in the laboratory. The momentum used for the calculation of the cross section is the effective momentum that the π-N system would have **if** the nucleon were at rest, i.e., the Fermi motion of the nucleon is taken into account as if it were on shell. Once it is determined that a pion-nucleon collision will take place a number of branches are possible. These are chosen according to the cumulative probabilities given in Figure 2. The logarithmic graph is used so that the small, OZI forbidden, particle productions can be seen. The π-2π and π-3π reactions are important for an estimate of energy deposition. These probabilities were calculated from data taken from the CERN-HERA[4] reports.

The kaon reactions are treated in a similar, but less exhaustive manner.

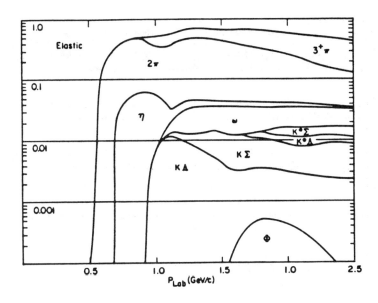

Figure 2. Branching ratios from πN collisions.

III. Results for Energy Densities

To estimate the degree to which the conditions necessary for a phase transition are achieved several approaches are possible. One way is to tabulate the kinetic energies of all of the nucleons and then plot the number of nucleons as a function of their energy. This distribution is seen to consist of two components; one corresponds to the nucleus in its "cold" state and the other to the heated portion. From an analysis of the hot distribution one can obtain both a "temperature" and the number of nucleons involved in the distribution. Since these results have been published elsewhere[3,5] I won't repeat them here, but temperatures in excess of 200 MeV are achieved. This conclusion is in agreement with the results of hydrodynamic calculations as well[5].

Another way to estimate the usefulness of antimatter annihilation to achieve a phase transition is to look at the energy density directly. The brute force method is to choose a set of small volumes and directly calculate the energy contained in the volumes. Since we are not interested in the energy associated with overall translational motion only the density in the volume's rest frame is counted, i.e., the invariant mass density. One might choose to look at the energy contained in all particles (nucleons, pions, kaons, lambdas, etc.) in the volume. This has the disadvantage that the full fireball energy is contained, and even in free

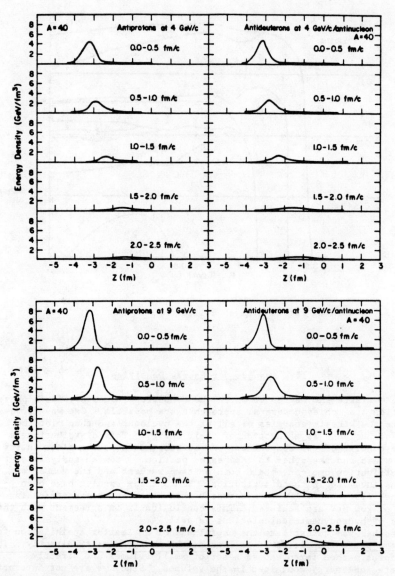

Figure 3. Maximal antiproton and antideuteron induced nuclear energy densities, such densities may be expected in 1-3% of central annihilations.

space one will see a high energy density. From previous work we know that an appreciable amount of energy is transferred to nucleons in a short period of time in a few percent of the cases. Thus this estimate can be used if one is careful to say that it only is applicable in about 3% of annihilations. These results are shown in figure 3 for both antiproton and antideuteron annihilation. It is clear that the maximum energy density is sufficient to cause a phase transition. Clear questions exist regarding whether the space-time extent is sufficient for the required parton equilibrium to be established. Clearly the reactions initiated with antideuterons are superior.

One might also choose to look only at the energy transfered to the nucleons. The problem with this is that this energy transfer to nucleons seems to occur at different places for different events. Since we are interested in the energy deposited in each individual case the average will spread the energy density around and dilute it. In this case we must examine individual events, which forces us to deal with much poorer statistics. The sampling volumes must not be taken too small or the nucleon granularity can cause artificial and arbitrarily large energy densities. Figure 4 shows two cases of single events. The energy densities in this case are more modest but we should remember that there really are mesons in these volumes which are not being counted. It is clear that the decay of the energy density with time is now much slower.

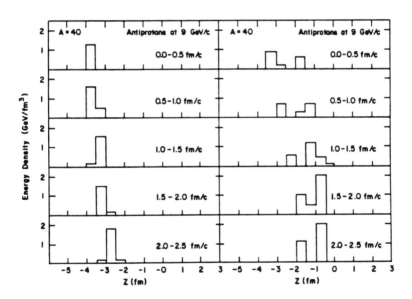

Figure 4. Single events showing energy deposited in nucleons only. This may be expected to be a minimum of the energy density deposited in nuclei by antiparticles.

The relevant parameters may be expected to lie between the two estimates.

Table II. Λ Production

P_{LAB}(GeV/c)	Target	Λ/Ann (%)	Ref
0.0	D	0.36	Bizzarri, et al. Lett. Nuovo Cimento 9, 431 (1969)
≤ .3	C,Ti,Ta,Pb	1.9	Condo, et al. Phys. Rev. C29, 1531 (1984)
0.6	Ne	~2.0	Balestra, et al. Nucl. Phys. A452, 573 (1986)
4.0	^{181}Ta	11.8	Miyano, et al. Phys. Rev. Lett. 53, 1725 (1984)

IV. Strangeness Production

As mentioned earlier an enhancement in strangeness production has been suggested as a signal of the formation of a quark-gluon plasma. Table II gives a brief summary of Λ production as measured in p̄-nucleus collisions. In order to set the investigation of this signal in a realistic framework it is particularly useful to compare with the recent KEK data[6] taken in a regime which approaches the relevant regions. The incident p̄ momentum was

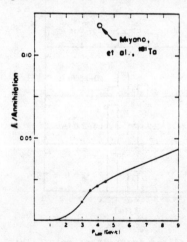 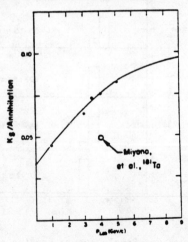

Figure 5. Production of Λ hyperons with p̄-beams.

Figure 6. Production of K-shorts with p̄-beams.

4 GeV/c and the target ^{181}Ta. Observed was the number of K_s, Λ and $\bar{\Lambda}$ which follow antiproton interaction with the nucleus. Few $\bar{\Lambda}$'s were observed, consistent with $\bar{\Lambda}$ annihilation in the nuclear medium. The number of Λ's seen was greatly enhanced, however, as indicated in figure 5. Is this large enhancement to be interpreted as evidence for a phase transition? After all, a very recent estimate[7] based on the non-topological soliton bag model, has predicted the quark-gluon phase transition to occur at a temperature of ~112 MeV. It seems extremely likely that such temperatures were reached in this experiment. Figure 6 points out that a slightly smaller number of K-shorts than expected were observed.

Running the classical nuclear modeling code previously described leads to predictions agreeing quite well with the experimental results. The code created about 3% of strangeness (.03 S$\bar{\text{S}}$ pairs per annihilation) but this is probably an overestimate since only central collisions were considered. This is not a major effect in any case. The most important reaction was $\bar{K}N \to \pi\Lambda$ creating Λ's in about 7% of the annihilations and depleting the K_s population. Table III gives the strangeness balance corresponding to the results of the calculation.

Table III. Strangeness Balance on ^{181}Ta at 4 GeV/c

	Λ(%)	K_s(%)
$\bar{p}p \to \Lambda\bar{\Lambda}, K\bar{K}$	2.0	6.5
$\bar{K}N \to \pi\Lambda, \pi\Sigma$	7.1	-2.5
$\pi N \to K\Lambda, K\Sigma$	3.1	1.0
	12.2	5.0
Experiment[6]	11.8	5.0

While it is instructive to follow the strangeness exchange in detail to learn about the reaction process, it is not necessary if one only wishes to know the total strangeness produced. Note that the observation of a K_s does not distinguish the presence of an S quark or an $\bar{\text{S}}$ quark, so one cannot just count S or $\bar{\text{S}}$ quarks. However, if one assigns a probability for strange quark exchange between bags, i.e., for $\bar{K}N \to \pi\Lambda$ and $\bar{\Lambda}N \to K +$ anything, as well as a probability for S$\bar{\text{S}}$ formation, then one can eliminate these first two probabilities algebraically from the species transformation equations to find an expression which is invariant under all strangeness exchange reactions. This measure is:

$$S = \frac{1}{2}(4N_{K_s} + N_Y + N_{\bar{Y}})$$

where S is the number of S$\bar{\text{S}}$ pairs produced and N_{K_s}, N_Y, $N_{\bar{Y}}$ are the numbers of K-shorts, hyperons and antihyperons respectively produced. As an example let us apply this equation to the experiment at 4 GeV/c. In free space: $S = \frac{1}{2}(4 \times 0.065 + 0.02 + 0.02) = \underline{0.15}$. From Ref. 6: $S = \frac{1}{2}(4 \times 0.05 + 0.118 + 0.002) = \underline{0.16}$

This leaves 1% to be produced in (OZI forbidden) hadronic interactions. Remember that the classical code (over)estimated 3%, so that, if anything, the experiment observed *less* than expected. Certainly these small differences are in the noise at this point. What is anticipated from a phase change is an enhancement in S by about an order of magnitude. It seems that this experiment could be taken for evidence *against* such a transition, at least at some level.

V. Conclusions

We have seen that energy densities of the order of that required for a phase transition to a quark-gluon plasma are present in energetic antimatter-matter interactions. Whether the space-time extent is sufficient for the parton thermalization to become complete, and for the signals to be generated, is not clear, but the conditions would seem to be of comparable quality to those generated in the heavy-ion collisions. Greater control of the cooling curve could well provide an advantage in interpreting any anomalous behavior in signals at the critical energy density.

Analysis of the strangeness produced in one recent antiproton experiment shows it to be completely consistent with no enhanced strangeness production, both from a detailed model and from a relation derived from invariance under strangeness exchange in strong interactions.

This work was supported by the U. S. Department of Energy.

REFERENCES

1. G. Baym, Nucl. Phys. $\underline{A447}$, 463c (1985); Nucl. Phys. $\underline{A418}$, 433c (1984); Helmut Satz, Nucl. Phys. $\underline{A400}$, 541c (1983).

2. J. Rafelski, Phys. Lett. $\underline{B91}$, 281 (1980); Nucl. Phys. $\underline{A418}$, 215c (1984).

3. W. R. Gibbs and D. Strottman, Proceedings of the "International Conference on Antinucleon- and Nucleon-Nucleon Interactions", Telluride, CO March 18-21, 1985, Plenum Press, New York; W. R. Gibbs, Proceedings of the Fermilab Workshop on Low Energy Antiproton Physics (In Press).

4. CERN-HERA Reports 83-01, 83-02, 84-01.

5. D. Strottman and W. R. Gibbs, Phys. Lett. $\underline{149B}$, 288 (1984).

6. Miyano *et al.*, Phys. Rev. Lett. $\underline{53}$, 1725 (1984).

7. H. Reinhardt, B. V. Dang and H. Schulz, Phys. Lett. $\underline{159B}$, 161 (1985).

A REPORT ON THE FIRST FERMILAB WORKSHOP ON ANTIMATTER PHYSICS AT LOW ENERGY (AMPLE)

Lawrence S. Pinsky
University of Houston, Houston, Texas 77004

ABSTRACT

The existance of the Antiproton Source at Fermilab makes practical the possibility that a new low energy (below 10 GeV) antinucleon physics facility be located at Fermilab. To that end, a workshop was organized to clearly identify the interesting physics that could potentially be investigated at such a facility. Three major physics areas were emphasized as being in the forefront of experimental and theoretical research. The first probes the origins of the Standard Model and includes test of CP, T, and CPT symmetries as well as $\Delta S = \Delta Q$. The other two areas probe the dynamics of confinement in QCD and include Heavy Quark Spectroscopy (e.g. Charmonium) and Detailed Measurements of Exclusive Final States in p̄p Annihilation (e.g. glueballs, CP exotics BB̄ resonances, etc.). In addition, the physics of many other significant fields were discussed. These include, physics with trapped p̄'s, p̄-nuclear physics, QCD plasmas, p̄-atomic physics, annihilation mechanism studies, hyperon physics, and others. A working group has been formed to pursue the goal of formally proposing a facility at Fermilab.

INTRODUCTION

The advent of the Antiproton Source to support the Tevatron Collider at Fermilab provides us with a familiar situation. In 1976,[1] with the Antiproton Accumulator coming into existance at CERN, to support the SppS the suggestion to use the antiprotons at low energy lead to the first LEAR workshop in 1979 and to the LEAR facility. Indeed, from the inception of the Fermilab Antiproton Source, the suggestion has been repeated made that the antiprotons could be used in a dedicated facility at low energies.[2] To some extent this is already happening in the approved experiment (E760) to use an internal gas jet target in the accumulator to do a Charmonium formation experiment as a follow on to the ISR experiment R704.[3] As an idea whose time we hope has come, a workshop was held at Fermilab April 10-12 this year for the purpose of identifying the physics that could be done with cooled antiprotons below 10 GeV/c. In particular, a facility such as considered here would be unique between 2-10 GeV/c and would overlap LEAR below 2 GeV/c. Given the constraints on LEAR operation, it is likely that the timely construction of a facility at Fermilab would find much of the interesting physics below 2 GeV/c still remaining to be done. A proceedings is in preparation and the purpose of this contribution is to present a brief overview of the physics that was presented at the workshop.

ORIGINS OF THE STANDARD MODEL

The very success of the Standard Model has perhaps clouded the questions that exist concerning its origins. For example, the causes of Weak Symmetry Breakdown, the origins of CP violation, the source of Quark and Lepton masses and angles, and the fundamental gauge groups are unknown. At a $\bar{p}$ facility there are a number of possible experiments that can be done to probe such questions. One example is the production of tagged neutral kaons. Consider the reactions:

$$\bar{p}p \rightarrow \begin{array}{c} K^-\pi^+K^o \\ K^+\pi^-\bar{K}^o \end{array}$$

The identification of the charged mesons uniquely tags the K^o's and $\bar{K}^o$'s. One can thus consider experiments such as $K^o \rightarrow \bar{K}^o/\bar{K}^o \rightarrow K^o$ which would be the first observation of expected T-invariance violation. Similarly $\Delta S \neq \Delta Q$ is expected at the CP violation level if CPT is good. Measurements of K^+ and K^- lifetimes would test CPT to more than an order of magnitude better than presently known in the first generation of $\bar{p}$ experiments alone. Similarly $K^o/\bar{K}^o$ masses would provide an alternative test of CPT to a similar level.

It was pointed out by Wolfenstein and Donoghue that the only experimental observations of CP violation have been in ratios of 3 decays of the K_L^o meson. The tagged K^o beams could of course extend these observation at least as well as conventional measurements and with a different collection of systematics. Hope for the future might include even more precise measurements at a $\bar{p}$ facility than is achievable elsewhere.

An interesting possibility for observing CP-violation in another system was discussed by Donoghue and is also presented elsewhere in these proceedings. The suggestion consists of measuring the correlations of decay directions of individual hyperons and antihyperons produced in pairs in $\bar{p}p$ annihilations. Such an observation could uncover mili-weak violation in a $\Delta S = 1$ interaction. It would also be the first CP-violation observed outside of the K^o system.

The importance of pursuing these questions is clear from interest shown by the diverse number of experiments being done at different laboratories all over the world. The understanding of the origins and existance of these symmetries is at the root of our present level of understanding of the structure of the universe. A dedicated $\bar{p}$ facility at Fermilab would allow a strong new method of approach to these investigations.

HEAVY QUARK QCD

The study of the Charmonium and Bottomonium systems hold the promise of providing a spectroscopy that can be used to constrain the dynamics in QCD theories. In particular, Charmonium spectroscopy can be done far more precisely at a $\bar{p}$ machine than at e^+e^- machines. The technique consists of controlling the $\bar{p}$ momentum spread to yield a narrow invarient mass spread in the direct formation of the desired state in a $p\bar{p}$ annihilation. For example a width resolution of ~70 KeV is concievable in the Charmonium region. The production of a desired state is tagged by detecting the e^+e^- pair from the decay of the J/ψ cascade product. A number of unseen but expected states can be measured and a precise splitting and spacing of all states can be obtained. Further, the η_c' can be confirmed and measured with precision.

These measurements could provide a detailed data set to contrain QCD calculations. A more exciting possibility would be to accomplish a similar set of measurements in the Bottomonium system. Since the b quark is heavy enough to allow relavent non-relativistic calculations, that system could provide a simple proving ground for calculable theories. However, the coupling of $b\bar{b}$ to $p\bar{p}$ is expected to go as the eighth power of the mass. As such the bottomonium formation cross sections should be down by 3-4 orders of magnitude from charmonium. The potential reward is sufficient that these estimates should be checked, but if correct, there is not much serious hope of doing bottomonium spectroscopy at a first generation Fermilab AMPLE facility.

MESON SPECTROSCOPY
(VOODOO QCD)

In what may be termed conventional meson spectroscopy (1-2 GeV mass range) there are two general areas of interest. First there are existing states that are not understood. Their identification including detailed measurement of their quantum numbers needs verification. Consider for example the $\iota(1460)$ or the $\theta(1720)$. Possible claims for identification of these and other states as glueballs requires clear careful determination of their J^{PC} quantum numbers (0^{++}, 2^{++}, 0^{-+}, 2^{-+}, etc. are possible glueball candidates) as well as studies of their exclusive final states partial widths and branching ratios.

The second area of interest is the search for the so called exotic states. Since $C = (-1)^{L+S}$ and $P = (-1)^{L+1}$, $J^{PC} = 0^{--}$, 0^{+-}, 1^{-+}, 2^{+-}, 3^{-+}, etc. are forbidden in $q\bar{q}$. The discovery of such a state would herald a new degree of freedom and possibly signal the existance of mixed quark-gluon states (Meiktons).

The prediction of 2 quark-2 antiquark states has lead to a sordid past of fading baryonium. The theoretical arguments remain compelling so one can conclude that the states must be broad and any further experimental attempts in this field must be in the form of detailed amplitude analyses rather than bump hunting.

TRAPPED ANTIPROTONS

The prospects for decelerating antiprotons to low enough energies to allow their capture in a Penning Trap, has lead to proposals to attempt to measure the sign of the gravitational mass of the antiproton. Such a measurement has enormous fundamental importance and will be attempted in a first generation set of experiments at LEAR.

The concept of having trapped cold antiprotons available has led to a number of suggestions for interesting physics experiments of the type not normally associated with accelerators. These include experiments in atomic physics, chemical physics, solid-state physics, and condensed matter physics. As an example, it has been predicted that cold $\bar{p}$'s in superfluid helium would act as condensation centers forming charged "bubbles" and "snowballs". Although these ideas do not argue for the facility by themselves, they show the breadth of possible interest.

FURTHER INTERESTS

There is a large body of work begun at LEAR to study antinucleon-nucleon and antinucleon-nucleus interactions. These include such subjects as studies of annihilation, elastic scattering, charge exchange, and $\bar{p}$ atoms. To these we add the possible studies of quark-gluon plasmas resulting from 6-8 GeV antiprotons annihiliating in heavy nuclei. So far polarized antinucleons have eluded us. Recently several proposals have been put forward to attempt to polarize circulating antiprotons. If these attempts succeed, the entire program becomes possible with polarized antiprotons and maybe polarized antineutrons. One might even be able to eventually accelerate these polarized antiprotons for use in other Fermilab experiments.

PROSPECTS

Given the strong physics case for such a machine the question is how best should this be pursued. To that end, a working group has been assembled and will proceed to write an extensive summary of these physics arguments for submission to the Fermilab Advisory Committee. Continued support from that committee would lead to a formal design effort and ultimately to a proposal. Pursuit of this goal in a timely fashion is the intent of the working group.

REFERENCES

In general the comments in this paper paraphase the contributions the Workshop by the various speakers and they will be published in written form as a proceedings to that workshop.

1. K. Killian, private communications

2. T. Fields, T. Kalogeroupolis, private communications

3. See Fermilab E760 proposal

STUDIES OF NUCLEAR STRUCTURE IN
ANTINUCLEON CHARGE-EXCHANGE REACTIONS

N. Auerbach

Los Alamos National Laboratory, Los Alamos, NM 87545
and
School of Physics and Astronomy
Tel-Aviv University, Tel-Aviv, Israel

ABSTRACT

The antinucleon-nucleus charge exchange reaction is discussed and its use as a probe of isovector excitations in nuclei is described. Attention is drawn to the fact that the ($\bar{p},\bar{n}$) reaction will predominantly excite "pionic" (i.e., longitudinal spin) modes in nuclei. Comparison between ($\bar{p},\bar{n}$) and (n,p) reactions is made. Plans for ($\bar{p},\bar{n}$) experiments in the near future are mentioned.

INTRODUCTION

In recent years charge-exchange reactions have proven to be a very useful means for studying giant isovector resonances in nuclei. The (p,n) reactions at intermediate energies have probed the spin excitations of nuclei and have lead to the observation of the Gamow-Teller (GT) resonance and the study of its properties for a wide range of nuclei.[1] In the pion charge-exchange reactions (π^+,π°) and (π^-,π°), on the other hand, the spin excitations are suppressed and one is able to select the spin-independent electric isovector excitations in nuclei. In this manner the charge-exchange analogs of the giant dipole were observed and the components of the giant isovector monopole were seen for the first time.[2,3]

The ($\bar{p},\bar{n}$) reaction combines some of the features of the nucleon and pion charge-exchange processes. In this sense it is a quite unique reaction. On the one hand, it is able to excite strongly spin states in nuclei but as opposed to nucleons and similarly to pions it is, because of absorption channels, a reaction that takes place at the periphery of the nucleus.

We will now discuss several properties of the ($\bar{p},\bar{n}$) reaction and point out its various applications.

EXCITATION OF UNNATURAL PARITY STATES IN NUCLEI

The most striking feature of the ($\bar{p},\bar{n}$) reaction is that it will preferentially excite unnatural parity ("pionic") states such as the $J^\pi = 0^-, 1^+, 2^-, 3^+$, etc. The reasons are the following: the natural parity spin excitations ($J^\pi = 0^+, 1^-, 2^+, 3^-, 4^+$, etc.) involve only transverse spin modes (i.e., modes in which the spin vibrations are perpendicular to the momentum transfer) whereas the unnatural parity transitions are composed of both transverse and longitudinal modes (i.e., spin vibrations parallel to momentum transfer). (The 0^- is a purely longitudinal mode.) The longitudinal part of the $N\bar{N}$ interaction is due to one pion-exchange. The transverse part is

caused by the exchange of heavier mesons such as the ρ and therefore
the longitudinal part is of longer range than the transverse part of
the NN̄ force. Because of the strong absorption of the antinucleon
the transverse part will be suppressed and the longitudinal part
will dominate. The isovector spin-independent part of the force is
also of short-range and will also be considerably weaker than the
isovector longitudinal, spin-dependent coupling. Figure 1 illustrates in a schematic manner the physical picture. As the antinucleon approaches the nucleus and touches the nuclear density it is quickly annihilated. Among the virtual mesons which are exchanged between the approaching p̄ and nucleons in the nucleus only the lightest one (i.e., the pion) will reach the nuclear interior and interact with

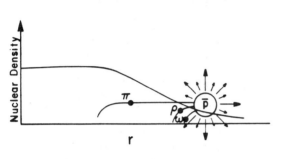

Figure 1

several nucleons. The exchange of a ρ-meson, for example, will
occur only with a very small number of nucleons at the nuclear
surface.

These ideas were outlined some time ago in Ref. 4. Also in
Ref. 4, cross-sections for the various (p̄,n̄) transitions in ^{90}Zr
were calculated using two common types of NN̄ forces.[5,6] The DWIA
calculations have shown that for $E_{p̄}$ = 175 MeV the "pionic"
(unnatural parity) states are excited about 10 times more strongly
than the corresponding (the same L) natural parity states.

For example,[4] for an (n,p) reaction with E_n = 200 MeV the cross
section at θ = 0° for exciting the 1⁻ state is 6.8 mb/sr while for
the 0⁻ and 2⁻ states the cross sections are 3.0 mb/sr and 4.4 mb/sr
respectively. On the other hand, for the (p̄,n̄) at $E_{p̄}$ = 175 MeV the
cross section for the 1⁻ is 0.05 mb/sr while for the 0⁻ and 2⁻
states the corresponding cross sections are 0.34 mb/sr and
0.57 mb/sr, thus the L = 1 unnatural parity states are much more
strongly excited than the giant dipole. For more details see
Ref. 4.

In a recent model calculation[7] in which only the 1f → 1g
transition is considered the differential cross sections for the 1⁺
and 0⁻ isovector excitation in ^{90}Zr(p̄,p̄') at 175 MeV were estimated.
For the 1⁻ state only the L = S = 1 part was included so that a pure
transverse spin transition resulted. The 0⁺ → 0⁻ transition is a
purely longitudinal one. The calculated DWIA cross section at its
maximum for the 0⁺ → 0⁻ transition was 16 times larger than for the
0⁺ → 1⁻ transition at its maximum. The moduli of the longitudinal
and transverse transition potentials for the 0⁻ and 1⁻ as calculated
in Ref. 7 are shown in Fig. 2. One sees clearly that the
longitudinal transition potential dominates the transverse one for
distances larger than the strong absorption radius R_{SA}. It was

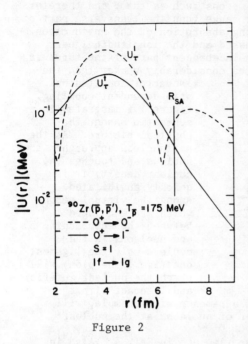

Figure 2

pointed out[7-] that in light nuclei (A ~ 20) the node in the longitudinal transition potential is outside R_{SA} and thus cancellations occur for the $0^+ \rightarrow 0^-$ transition. As a result the dominance of the longitudinal transitions is reduced in light nuclei. One should therefore attempt to measure the pion-like transitions in heavy nuclei.

The dominance of the longitudinal transitions in the antinucleon charge-exchange reaction follows from the very general properties of the $N\bar{N}$ interaction and does not depend on the details of the force. This characteristic feature of the $(\bar{p},\bar{n})$ reaction, we feel may provide exciting possibilities of research. There has been a longstanding question whether the pionic field is enhanced in nuclei[8-10] and whether precursor phenomena[8] to pion condensation in nuclei. These effects should show up in the longitudinal response function of nuclear excitations.[8-10] The $(\bar{p},\bar{n})$ reaction presents some possibilities to study such effects in nuclei. One should keep in mind however that these effects occur at higher momentum transfer in which case the dominance of the longitudinal transitions in $(\bar{p},\bar{n})$ will be reduced.

OBSERVATION OF GAMOW-TELLER STRENGTH

The $(\bar{p},\bar{n})$ reaction induces $\Delta T_z = +1$ transitions in the target nucleus. Therefore it will be possible to measure β^+ strength (S_+) particularly of Gamow-Teller (GT) type.

By measuring the S_+ strength and comparing it to the GT, S_- strength measured[1] in (p,n), one may be able to better understand the problem of missing GT strength.[1] Presently two different explanations have been put forward to account for the missing GT strength. The first suggests that the strength is removed to very high energies due to the coupling of the GT to Δ-particle-nucleon hole states.[11] In the second approach[12,13] the missing GT strength is due to the coupling of the 1p-1h GT state to 2p-2h configurations leading to a fragmentation of GT strength.

By using (n,p) or $\bar{p},\bar{n}$ reactions it may be possible to shed light on this problem and provide more direct experimental evidence concerning one of these possibilities. The $(\bar{p},\bar{n})$ reaction might be even better suited than (n,p) to study GT strength because of the suppression of natural parity excitations.

The inclusion of 2p-2h configurations in the configuration space in which the GT is calculated leads not only to a fragmentation of the strength but also the the appearance of new additional strength.[14] For a β_- transition the strength in a nucleus like ^{90}Zr should be

$$S_- = 3(N-Z) + \Delta S$$

where ΔS is the additional GT strength mentioned above. In the β_+ channel the strength ΔS should appear[14] so that the sum rule

$$S_- - S_+ = 3(N-Z)$$

is obeyed. If indeed the second hypothesis about the missing GT is valid, then one should find some of this strength in the $(\bar{p},\bar{n})$ reaction.

In some light and medium heavy nuclei GT, β^+ transitions of the $0\hbar\omega$ type are allowed. For example, in ^{60}Ni the transition $f_{7/2} \rightarrow f_{5/2}$ is allowed for both β^- and β^+. A comparison of the (p,n) on the one hand and the (n,p) and $(\bar{p},\bar{n})$ on the other might be of interest. Note that in the case of ^{60}Ni RPA correlations[15] do affect the $f_{7/2} \rightarrow f_{5/2}$ part of the GT transitions. Our DWIA-RPA estimate of the ^{60}Ni$(\bar{p},\bar{n})^{60}$Co, $f_{7/2} \rightarrow f_{5/2}$, GT transition gives a forward peaked cross section of about 4.5 mb/sr for $\theta = 0°$.

EXCITATION OF NEW RESONANCES

One will also be able to excite new types of giant resonances in $(\bar{p},\bar{n})$. In particular there is the possibility that the spin isovector monopole[16] will be detected. The transition density for a monopole excitation has a volume part and a surface part of opposite sign which usually cancels to a large extent the contribution of the volume part in the cross section. However, if the interaction is at the surface as is the case with 160 MeV pions or with antinucleons, then mostly the surface part of the transition density contributes and the cross sections could be measured. One would expect, therefore, to observe the isovector monopole state in $(\bar{p},\bar{n})$ and because of the relatively strong spin-isospin component in the $N\bar{N}$ force the spin part of an isovector monopole transition would show up. The cross section for exciting the spin isovector monopole was computed in a DWIA calculation with an RPA transition density. The calculation includes in addition to the $L = 0$, $S = 1$, $J = 1^+$ also the $L = 2$, $S = 1$, $J = 1^+$ component. The forward peaked differential cross section is about 0.8 mb/sr for $\theta = 0°$.

STUDYING THE $N\bar{N}$ INTERACTION

The $(\bar{p},\bar{n})$ reaction will help to determine the spin, isospin components of the $N\bar{N}$ force. In inelastic $(\bar{p},\bar{p}')$ scattering on ^{12}C an attempt was made to measure[17] the cross sections for the transition to the $J = 1^+$, $T = 0$ at 12.7 MeV and the $J = 1^+$, $T = 1$ state at 15.1 MeV. These measurements should then provide essential information that will help determine[18] the v_σ, and $v_{\sigma\tau}$ parts of the

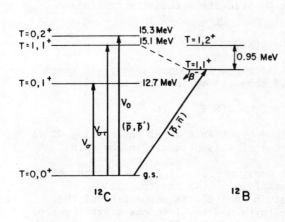

Figure 3

$N\bar{N}$ force. Measurement of the $J = 1^+$, $T = 1$ resonance is strongly impaired however by the closeness of an $J = 2^+$, $T = 0$ (15.3 MeV) level, which because of its isoscalar nature is strongly excited in the reaction. The $(\bar{p},\bar{n})$ reaction of course is a means of avoiding this difficulty. In this reaction one can excite the $\Delta T_z = +1$ component of the $J = 1^+$ isotriplet in the A = 12 nuclei and obtain the same kind of information as in the $(\bar{p},\bar{p}')$ measurement of the $J = 1^+$, $T = 1$ state (see Fig. 3).

THE EXPERIMENTAL PROGRAM. OUTLINE

In the first stages of the experimental program[19] at LEAR, the following reactions will be studied.

(A) The transition $^{13}C(\bar{p},\bar{n})^{13}B$ to the $J = 3/2^-$ g.s. of ^{13}B will be measured. The GT, β^--decay transition between the $J = 3/2^-$ g.s. of ^{13}B and $J = 1/2^-$ g.s. of ^{13}C is well known and its ft value measured. The $^{13}C(p,n)^{13}N$ reaction has also been studied.[20] By measuring this transition in the $(\bar{p},\bar{n})$ reaction and combining it with the known GT strength one will be able to assess the distortion effects entering the reaction mechanism. An (n,p) measurement of this transition was performed recently.[21]

(B) The reaction $^{12}C(\bar{p},\bar{n})^{12}B$ (g.s.) will be studied. The ^{12}B ground state ($J = 1^+$, $T = 1$) is the analog of the 15.1 MeV level in ^{12}C. The importance of studying this transition was mentioned in the previous section. In ^{12}B the first excited level at 0.95 MeV is a $J = 2^+$, $T = 1$ state and with the present limited resolution it could in principle obstruct the analysis of the transition to the g.s. We hope, however, that as discussed above the excitation of the natural parity $J = 2^+$ state will be weaker as compared with transition to the g.s.

(C) Also in the first stage of the $(\bar{p},\bar{n})$ experiments the $^6Li(\bar{p},\bar{n})^6He$ (g.s.) transition will be measured. The ^6Li and ^6He ground states have $J = 1^+$, $T = 0$ and $J = 0^+$, $T = 1$, respectively and thus the transition is a pure GT one. The β^- transition from ^6He to the g.s. of ^6Li is known. All the above reactions will be studied for forward angles.

With these "calibration" experiments in hand one hopes to measure $(\bar{p},\bar{n})$ transition in medium and heavy mass nuclei such as ^{90}Zr and ^{208}Pb in the near future.

The comparison between the ($\bar{p},\bar{n}$) and (n,p) results (anticipated from experiments at TRIUMF and LAMPF) should reveal some new interesting features in the structure and nuclear spin excitations. This comparison will tell us more about transverse versus longitudinal components of various nuclear transitions.

ACKNOWLEDGMENTS

I wish to thank M. A. Franey, C. D. Goodman, A. Klein, W. G. Love, and A. I. Yavin for many helpful discussions and suggestions. This work was supported in part by DOE contract W7405-ENG-36.

REFERENCES

1. C. D. Goodman, Comments Nucl. Part. Phys. 10, 117 (1981).
2. J. D. Bowman et al., Phys. Rev. Lett. 50, 1195 (1983).
   H. W. Baer et al., Phys. Rev. Lett. 49, 1376 (1982);
3. N. Auerbach and A. Klein, Phys. Rev. C29, 2075 (1983).
4. A. Klein, W. G. Love, M. A. Franey, and N. Auerbach, Proceedings of the International Conference on Antinucleon and Nucleon-Nucleus Interactions, eds. G. Walker, C. Olmer, and C. D. Goodman, Plenum Press, New York, 1985, p. 351.
5. J. Cote et al., Phys. Rev. Lett. 48, 1319 (1982).
6. C. B. Dover and J. M. Richard, Phys. Rev. C21, 1466 (1980).
7. A. Klein, W. G. Love, and M. A. Franey, Phys. Rev. Lett. 56, 700 (1986).
8. See for example E. Oset, H. Toki, and W. Weise, Phys. Rep. 83, 281 (1982).
9. W. M. Alberico, M. Ericson, and A. Molinari, Nucl. Phys. A279, 429 (1982).
10. T. A. Carey et al., Phys. Rev. Lett. 53, 144 (1984).
11. M. Ericson, A. Figureau, and D. Thevenet, Phys. Lett. 45B, 19 (1973); A. Bohr and B. R. Mottelson, Phys. Lett. 100B, 10 (1981).
12. K. Shimizu, M. Ichimura, and A. Arima, Nucl. Phys. A226, 282 (1974); G. F. Bertsch and I. Hamamoto, Phys. Rev. C26, 1323 (1982).
13. A. Klein, W. G. Love, and N. Auerbach, Phys. Rev. C31, 710 (1985); F. Osterfeld, D. Cha, and J. Speth, Phys. Rev. C31, 372 (1985).
14. N. Auerbach, A. Klein, and W. G. Love, Proceedings of the International conference on Antinucleon and Nucleon-Nucleus Interactions, eds. G. Walker, C. Olmer, and C. D. Goodman, Plenum Press, New York, 1985, p. 323.
15. N. Auerbach, L. Zamick, and A. Klein, Phys. Lett. 118B, 256 (1982).
16. N. Auerbach and A. Klein, Phys. Rev. C30, 1031 (1984).
17. D. Garreta et al., in Ref. 4.
18. C. B. Dover et al., Phys. Lett. 143B, 45 (1984).
19. Cagliari-Legnaro-Torino-Padova-Indiana-Tel-Aviv Collaboration.
20. C. D. Goodman et al., Phys. Rev. Lett. 54, 877 (1985).
21. P. Alford, these proceedings.

NEUTRAL STRANGE PARTICLE PRODUCTION IN $\bar{p}$Ne ANNIHILATIONS

F.Balestra[7], Yu.A.Batusov[3], G.Bendiscioli[6], S.Bossolasco[7], F.O.Breivik[5], M.P.Bussa[7], L.Busso[7], I.V.Falomkin[3], L.Ferrero[7], C.Guaraldo[4], A.Haatuft[1], A.Halsteinslid[1], T.Jacobsen[5], E.Lodi Rizzini[2], A.Maggiora[4], K.Myklebost[1], J.Olsen[1], D.Panzieri[7], G.Piragino[7], G.B.Pontecorvo[3], A.Rotondi[6], P.Salvini[6], M.G.Sapozhnikov[3], S.O.Sorensen[5], F.Tosello[7], A.Zenoni[6].
Bergen[1], Brescia[2], Dubna[3], Frascati[4], Oslo[5], Pavia[6], Torino[7] -collaboration.

Studying the $\bar{p}$Ne interaction at LEAR with a self-shunted streamer chamber in a magnetic field[1], it has been found that a percentage of $\bar{p}$, increasing with the kinetic energy, annihilates deep inside the nucleus[2]. The detection of deep annihilations is of interest because theoretical works predict that they might develop through mechanisms different from those included in the "standard physics" of intranuclear cascade models, such as the annihilation on more than one nucleon[3] and the excitation of new degrees of freedom of the nuclear matter[4]. It is a common opinion that typical signatures of these exotic phenomena are an enhancement in hyperon and strange-meson production and the emission of a high number of energetic nuclear fragments. We studied at 607 MeV/c the reaction $\bar{p}$Ne → V°+anything (where V°=Λ or $K_s^°$). Preliminary results[2] gave lower limit (without geometrical efficiency corrections) of V° production. In this paper we present the result of complete analysis of 7622 annihilations produced by $9.4 \cdot 10^6$ $\bar{p}$ in Ne. These events yielded (147±27)Λ and (64±13) $K_s^°$ mesons, which give a final fraction of (1.9±0.4)%Λ and (0.8±0.2)% $K_s^°$. The neutral strange particles were detected by their charged particles decays: $\Lambda \to p\pi^-$ and $K_s^° \to \pi^+\pi^-$. The $K_s^°/\Lambda$ ratio is close to that measured[5] in ^{131}Ta at 4 GeV/c and also for the Ne the rapidity values of Λ and $K_s^°$, <y> =(0.08±0.03) and <y> =(0.44±0.08) respectively, show that ($\bar{p}$,4÷11N) and ($\bar{p}$, 1÷2N) systems in the c.m.s. are involved in their production. Very preliminary measurements of Λ and $K_s^°$ production in Ne, for $\bar{p}$ at rest, show that the $K_s^°/\Lambda$ ratio is about 10 times higher than at 607 MeV/c. It seems possible to conclude that when the $\bar{p}$ annihilate in the nuclear soft surface the $K_s^°$ production is higher than of the Λ's. When the annihilation of $\bar{p}$ occurs deep inside the nucleus, nucleon-systems are involved and more Λ are produced. These results show the great interest of the proposal made at CERN to continue the study of $\bar{p}$-nucleus interaction, with a large acceptance and high resolution detector as OBELIX[6], in particular in the following directions:
a) wide survey of $\bar{p}$ annihilation channels in nuclear targets as ^{2}H,

^{3,4}He, ^{14}N,Ne,Ar,Kr,Xe up to the highest $\bar{p}$ momenta (1-2 GeV/c);

b) study of quark substructures and manifestations of quark-gluon plasma in nuclei[7]. Such "exotic" effects could be detected studying:
1) pionless annihilations $\bar{p}A \to p(A-2)N$ as for example $\bar{p}\,^3He \to np$;
2) single pion annihilation $\bar{p}A \to \pi(A-1)$ as for example $\bar{p}\,^2H \to \pi^- p$, or quasideuteron $\bar{p}$ absorption;

c) neutral strange particle production $\bar{p}A \to V^° \text{ anything }$ ($V^°=\Lambda,\bar{\Lambda},K_s^°$) and $\bar{p}A \to \Lambda\Lambda + KK + \text{anything}$ (for $A \geqslant 3$) in the search for bound (H dibaryon) or unbound $\Lambda\Lambda$ states[8].

REFERENCES
1) F. Balestra et al. Nucl. Instr. Meth. A234, 30 (1985).
2) F. Balestra et al. Nucl. Phys. A452, 573 (1986) and Europhys. Lett. (in press).
3) J. Cugnon et al. Phys. Lett. B146, 16 (1984); J. Derreth et al. Phys. Rev. C31, 1360 (1985).
4) J. Rafelski, Phys. Lett. B146, 16 (1984); J. Rafelski et al. Phys. Rev. Lett. 48, 1066 (1982).
5) K. Myiano et al. Phys. Rev. Lett. 53, 1725 (1984).
6) R. Armenteros et al. CERN/PSCC/86-4 (1986).
7) D. S. Koltun et al. Phys. Lett. B172, 267 (1986).
8) G. T. Condo et al. Phys. Lett. B114, 27 (1984); M. P. Locher et al. Adv. Nucl. Phys. (in press).

ELECTRON AND MUON PHYSICS

Coordinators: T. Kinoshita
              S. B. Kowalski

A REVIEW OF THE EMC EFFECT

G.N. Taylor
University of Oxford

and
C.E.R.N.

INTRODUCTION

The result shown in figure 1 created an industry. First presented four years ago [1] this result, obtained by the European Muon Collaboration (EMC), shows the ratio of structure functions, F_2, for iron and deuterium as a function of the scaling variable x_{Bj}. The observed ratio falling approximately linearly from $x_{Bj}=0.1$ to $x_{Bj}=0.65$ was in stark contrast with that expected from Fermi motion effects. It was subsequently confirmed [2] in an analysis of data taken 10 years earlier at SLAC – events taken from a deuterium target were compared with those originating in the deuterium vessel walls. The result implies that nucleon parton distributions are modified by neighbouring nucleons in nuclei. A flood of theoretical papers has ensued as well as a solid experimental programme to pursue the "EMC Effect".

After briefly describing the significance of these results, I will present some new results from the EMC, and then give an indication of what to expect in the future from the EMC and from it's successor, the NMC. The deeper theoretical meaning of the EMC, along with detailed references, is covered elsewhere in this conference [3]. Only brief mention will be made of the several neutrino scattering results. For a summary of these data see [4,5].

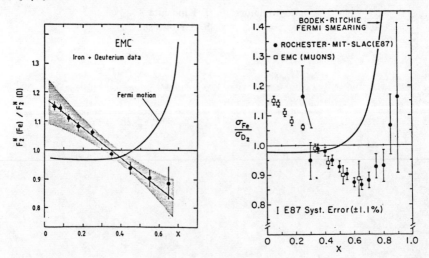

Figure 1
"The EMC Effect" as measured by (a) EMC [1] and (b) Bodek et al., [2]

MEASUREMENT AND INTERPRETATION

To discuss the physics we must first define our terms. The differential cross section for muon or electron scattering is given by

$$\frac{d^2\sigma}{dQ^2 dx} = \frac{4\pi\alpha^2}{Q^4} [(1-y - \frac{Mxy}{2E}) \frac{F_2(x,Q^2)}{x} + \frac{y^2}{2} 2F_1(x,Q^2)]$$

where $-Q^2$ is the 4 momentum transfer from the incoming lepton of energy E to the outgoing scattered lepton, energy E', y is the fraction of beam energy transfered (E-E')/E, M is the proton mass and x_{Bj} is the Bjorken scaling variable, $Q^2/2MEy$. F_1 and F_2 are related via $R=\sigma_L/\sigma_T$, the ratio of longitudinal to transverse cross sections. In the parton model, provided transverse momenta are negligible, this relation reduces to $F_2=2xF_1$. If R is not zero, then the differential cross section and F_2 are not directly equivalent. In the parton model, F_2 describes the momentum distribution of the quarks within the nucleon. The question of A-dependence of R will not be addressed here (see ref. 6).

The EMC effect indicates a change of the parton momentum for nucleons bound in nuclei, beyond that expected from Fermi motion. A pedagogical description of the present understanding of these nuclear effects is most easily given for different regions of x_{Bj}. (See figure 2).

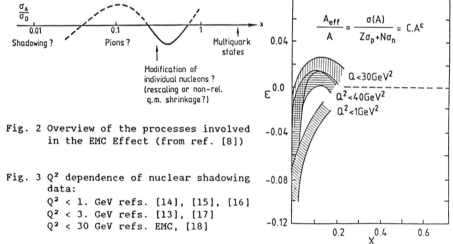

Fig. 2 Overview of the processes involved in the EMC Effect (from ref. [8])

Fig. 3 Q^2 dependence of nuclear shadowing data:
$Q^2 < 1.$ GeV refs. [14], [15], [16]
$Q^2 < 3.$ GeV refs. [13], [17]
$Q^2 < 30$ GeV refs. EMC, [18]

Since x_{Bj} relates to the structure function per nucleon, the proton mass is used in its definition. For nuclei, x_{Bj} can in principle extend to A. An enhancement in F_2 at very high x_{Bj} thus corresponds to the kinematics of multi-nucleon effects, the simplest being the sharing of the Fermi momentum between nucleons. More exotic phenomena such as multi-quark clusters within nuclei have also been widely discussed [3].

In the central x_{Bj} region, a softening of the parton momentum distribution is found in nuclei. This effect has been "explained" in many guises. However, physically, it relates to changes in properties of the individual nucleons [8]. Theoretically this has been described by, for example, swelling of the nucleon within nuclei (rescaling models and color conductivity) or the nuclear binding energy changing the effective nucleon mass within nuclei [3].

In the language of "light cone physics", the small x region corresponds to large longitudinal correlation lengths, well beyond the nucleon charge radius. Effects such as increased virtual pion cloud, parton fusion, extra gluons and sea quarks may play a role as the physical manifestation of these small x_{Bj} correlations. Data from neutrino scattering [9] indicate that no major increase in the sea quark distribution is found in nuclei. Studies of ψ production in hydrogen, deuterium and iron have shown evidence, interpreted as an enhancement in the gluon distribution in nuclei, in the region below $x_{Bj}=0.08$.

The very small x_{Bj} region is where nuclear shadowing is expected. Here the full geometrical cross section of the nucleons is unattainable. The experimental status of shadowing is somewhat confused, however, following an admittedly biased selection of results based on data set consistently [10] the picture of shadowing shown in figure 3 emerges. Below Q^2 of about 1 GeV^2, shadowing is seen at all measured x_{Bj} values. Above this value, antishadowing develops in the region of $x_{Bj}=0.1$.

The low x_{Bj} region where data have not given a clear, consistent picture and where theoretical ideas are fewer, is where the EMC apparatus is well suited. Data were taken by this collaboration in 1984-85 from small parasitic targets mounted during the running of the polarised target phase of NA2. Three targets were used during each data taking period, each distributed over the same volume, being cycled into the beam every few hours. Systematic differences in acceptance were thus eliminated. Effects due to changes in the apparatus also cancelled to negligible levels. This setup was thus optimized for measuring ratios of cross sections of several targets with very small relative systematic errors. Figure 4 shows the ratios of events for carbon/deuterium and copper/deuterium normalised to the same measured incident reconstructuble muon flux. Figure 5 is the same data with radiative corrections applied. These corrections affect the first point by 2% for C/D2 and 4% for Cu/D2, becoming negligible compared with statistical errors by $x_{Bj}=0.1$. No correction has been made for the neutron excess which amounts to less than 3% at $x_{Bj}=0.65$ becoming negligible at smaller x_{Bj}. The ratio of events reconstruction efficiency and acceptance are determined to be unity with a precision, presently limited by the amount of Monte Carlo simulation available, of 4% at small x_{Bj} rising to 10% at large x_{Bj}. These errors, being the present systematic uncertainty will be reduced with further Monte Carlo studies and data

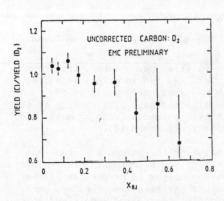
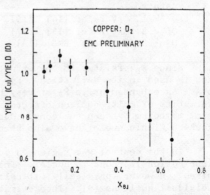

Figure 4
Preliminary, uncorrected event ratios, normalised to incident reconstructible muon flux from the EMC

analaysis. The average Q^2 for the data points shown range from 7 GeV^2 at low x_{Bj} to 50 GeV^2 at the highest x_{Bj} value. A clear tendency for the ratio to rise near $x_{Bj}=0.1$ then to fall at lower values of x_{Bj} is seen in the copper data. The latter effects was not seen in the previous results [1] where the systematic errors at low x_{Bj} were large. This structure is not so pronounced for the carbon to deuterium ratio. In figure 6 the new results are compared with previously reported charged lepton scattering results, on iron, copper, carbon and nitrogen targets.

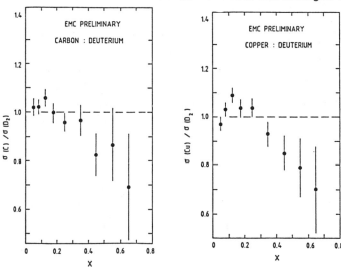

Figure 5
As in figure 4 with radiative corrections applied

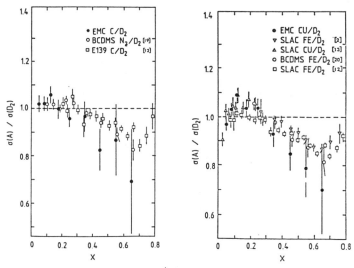

Figure 6
Comparison of new EMC results with previously reported results
(refs. indicated)

THE FUTURE

Remaining to be analysed are data taken from helium and tin targets, as well as dedicated run with segmented and interspersed copper/deuterium targets simultaneously in the beam, providing higher statistics for studying this pair of nuclei.

For the longer term future, a new collaboration has inherited the EMC apparatus [11], designed optimised target arrangements to reduce experimental target to target biases to a minimum, and upgraded the existing equipment for a new round of experiments to compare different targets over the next three years.

Data taking will soon commence with a detailed study of H_2 and D_2, followed by a measurement of F_2 ratios as a function of x_{Bj} for nuclei spanning the periodic table. Finally, using a massive calorimeter target made up of slabs of two different nuclei, widely separated in A (e.g. Al/Pb or Be/Nb), large statistics will be obtained to look for any Q^2 dependence of the F_2 ratios in one experiment. A complimentary programme measuring ψ production from different targets will be carried out simultaneously.

SUMMARY

Without doubt, nuclear physics has staked a claim in the previously exclusive domain of high physics – deep inelastic scattering. The resulting picture, which is becoming clearer, provides a study of quarks in nuclei and nuclear effects on nucleon quark distributions. From shadowing at very small x_{Bj} through to an enhancement of the parton momentum distributions near $x_{Bj}=0.1$, attributable to such effects as pion field enhancement, onto effects of nucleon expansion and nuclear binding energy within nuclei at medium x_{Bj}. Finally, multi-nucleon correlation effects are seen near $x_{Bj}=0.1$. To distinguish between the specific models vying for a place in this general picture, very precise data covering a wide range of nuclei are required. For their part, the EMC and the NMC offer the hope of providing such data.

References

1. EMC, J.J. Aubert et al., Phys. Lett. 123B (1983) 275.
2. A. Bodek, et al., Phys. Rev. Lett. 50 (1983) 1431.
3. E. Berger, this conference.
4. P.R. Norton, RAL-85-054.
5. J. Hanlon, Phys. Rev. D32 (1985) 2441.
6. J. Gomez, SLAC-PUB-3552.
7. R.G. Arnold, this conference.
8. C. Llewellyn-Smith, Proc. 10th Int. Conf. on Particles and Nuclei, Heidelberg, 30th July – 3rd August, 1984.
9. R. Voss, CERN-EP/85-204.
10. P. Grafström, private communication.
11. NMC, CERN/SPSC/85-18, SPSC/p.210.
12. R.G. Arnold, et al., Phys. Rev. Lett. 52 (1984) 727.
13. S. Stein, et al., Phys. Rev. D12 (1975) 1984.
14. G. Huber, et al., Z. Phys. C2 (1979) 279.
15. J. Franzel, et al., Z. Phys. C10 (1981) 105.
16. J. Bailey, et al., Nucl. Phys. B151 (1979) 367.
17. W.R. Ditzler, et al., Phys. Lett. 57B (1975) 201.
18. M.S. Goodman, et al., Phys. Rev. Lett. 47 (1981) 293
19. G. Bari, et al., Phys. Lett. 163B (1985) 282.

NUCLEAR MEDIUM EFFECTS ON THE PHOTON-PROTON VERTEX
INVESTIGATED WITH THE (E,E'P) REACTION

L. Lapikás

NIKHEF-K, P.O.Box 4395, 1009 AJ Amsterdam, The Netherlands

ABSTRACT

A longitudinal-transverse separation has been carried out for the coincident quasi-free proton knockout reaction on ^{12}C in the momentum-transfer range 270<q<460 MeV/c. The deduced ratio of longitudinal and transverse response functions for 1p proton knockout indicates a significant effect of the nuclear medium on the coupling of the virtual photon with the proton. The result is compared with predictions based on the σ-ω model and on models involving an enlarged nucleon radius.

INTRODUCTION

In recent years high-resolution (e,e'p) experiments[1] have become a novel tool for the study of the spectroscopy of valence protons in nuclei. The present experimental technique enables accurate (< 5%) absolute measurements, while the achieved 100 keV energy resolution allows to resolve a multitude of states below, say, 10 MeV excitation energy. In this energy region (around the Fermi level) one probes the single-particle aspects of those nuclear states whose wave functions receive their main contributions from nucleons in the valence shells. The occupation of these shells plays a crucial role in the shell-model picture of the nucleus. Recent estimates, e.g. those based on nuclear matter calculations[2], indicate a depletion of about 30 %.

In fig. 1 we show a survey of the occupation probability of valence shells for a number of nuclei as deduced from (e,e'p) experiments. The displayed data are from recent NIKHEF-K work and from older Saclay measurements[3]. Apart from ^{2}H, all data show a systematic depletion ranging from 30 to 60%. This is to be contrasted with the results from proton pick-up reactions, like (d,τ), which tend to yield completely filled valence shells. However, recent refinements[4] in the analysis method of these hadronic reactions have demonstrated the large model dependence of the deduced occupation numbers.

Before drawing conclusions from a possible discrepancy between results from the two reactions, it is necessary to consider the ingredients in the analysis of the (e,e'p) reaction in more detail. Two aspects are important: i) the assumption that the reaction can be described in Impulse Approximation (IA) and ii) the description of the Final State Interaction (FSI) between the ejected proton and the residual nucleus. Recently we studied[5] the FSI effects by using in the (e,e'p) analysis optical potentials that are fitted to proton scattering in the relevant energy domain. We found a model dependence of the spectroscopic factors due to the uncertainty in the potential of about 10%. This error has been included in fig. 1.

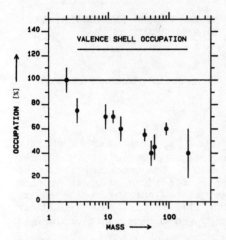

Figure 1. Occupation of valence shells against mass number, as deduced from (e,e'p) experiments

In this paper we describe some experiments that test the IA assumption. If the basic electron-proton cross section σ_{ep}, which enters in the (e,e'p) analysis, were to be modified due to the fact that we do not scatter from free protons (Impulse Approximation), but from protons bound in the nucleus, then the data shown in fig. 1 should be also modified. Indications for such a modification have been discussed on the basis of experimental evidence (EMC effect[7], Longitudinal-Transverse response functions in quasi elastic scattering[8], while some models (bagmodel of the nucleon[9], relativistic $\sigma-\omega$ model[10] predict estimates for the size of these effects.

THE (E,E'P) REACTION IN IMPULSE APPROXIMATION

The usual description of the quasi-free scattering process assumes the validity of the impulse approximation, i.e. the free nucleon current is used[11]. In this approximation the (e,e'p) coincidence cross section can be expressed[11] in terms of four structure functions W_i. In parallel kinematics, when the momentum of the outgoing proton p' is parallel to the momentum transfer q, only two structure functions W_L and W_T remain[12]:

$$d^6\sigma/d\vec{e}'d\vec{p}' = |e'|^2\sigma_{Mott}Q^2(q^2\varepsilon)^{-1}\{\varepsilon W_L(\omega,q)+W_T(\omega,q)\} \qquad (1)$$

where σ_{Mott} is the Mott cross section, $Q^2 = q^2-\omega^2$, and $\varepsilon = [1+(2q^2/Q^2)\tan^2(\theta_e./2)]^{-1}$ is the photon polarization parameter. For scattering from a bound nucleon W_L and W_T depend on the separation energy (E_m) and momentum (p_m) of the nucleon inside the nucleus. In the plane-wave impulse approximation this dependence is the same for W_L and W_T and eq. (1) can be factorized:

$$d^6\sigma/d\vec{e}'d\vec{p}' = |e'|^2\sigma_{ep}S(E_m,p_m), \qquad (2)$$

where σ_{ep} describes the off-shell electron-proton scattering cross section given by $\sigma_{ep} = \sigma_{Mott}Q^2(q^2\varepsilon)^{-1}\{\varepsilon|F_L(Q^2)|^2+|F_T(Q^2)|^2\}$ and thus the longitudinal-transverse (LT) character is determined by the nucleon form factors. For elastic scattering from a free proton the F_i's are equal to the electric and magnetic proton form factors: $F_L^{free}(Q^2) = G_E(Q^2)$ and $F_T^{free}(Q^2) = Q^2/4m^2 G_M(Q^2)$. The spectral function $S(E_m,p_m)$ represents the probability to find a proton with separation energy E_m and momentum $p_m=p'-q$ inside the nucleus.

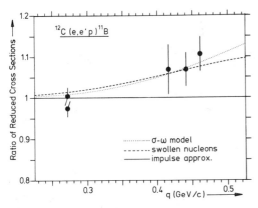

Figure 2. Ratio $\sigma_{red}(E_0,\theta_e)_{back}/\sigma_{red}(E_0,\theta_e)_{forw}$ as a function of q for the $^{12}C(e,e'p)^{11}B$ reaction leading to the ^{11}B ground state.

If the photon-proton coupling is modified inside the nucleus, σ_{ep} will be different from the free electron-proton cross section. Hence the influence of the nuclear environment can be investigated by studying the q and ε dependence of σ_{ep}. Since the absolute value of $S(E_m,p_m)$ is unknown, only the ratio of σ_{ep}'s can be obtained in cases where E_m, p_m and p' are the same. This is done by performing measurements at two values of the initial electron energy E_0 and corresponding values of θ_e, such that q and ω remain constant. This results in a different value of ε, hence it follows from eq. (1) that the two measurements comprise a Rosenbluth separation, which will yield the ratio W_T/W_L.

EXPERIMENT AND RESULTS

The experiment was carried out with the NIKHEF-K coincidence facility[13] at initial energies of 313 and 443 MeV. The outgoing proton momentum was kept constant at 370 MeV/c. The kinematics were centered around four values of q with special emphasis on the points near 460 MeV/c, where the most accurate LT-separation could be made. The q-range chosen corresponds to a p_m-range between -90 and +100 MeV/c. In the present experiment[6] we studied knockout of 1p protons from ^{12}C, leading to the ground state of ^{11}B. Reduced cross sections $\sigma_{red}(E_0,\theta_e)$ were obtained[14] by dividing the cross sections by $|e'|^2\sigma_{ep}$ as calculated by de Forest[11].

In fig. 2 the ratio $\sigma_{red}(E_0,\theta_e)_{backw}/\sigma_{red}(E_0,\theta_e)_{forw}$ is displayed with only statistical errors shown. The systematic error in the ratio is 1.5%. Since the distorted spectral function is the same at both energies, the ratio shown in fig. 2 should be equal to one, if σ_{ep} is not affected by the nuclear medium. At q=270 MeV/c the data and the predictions of the models are close to unity. In this q-region the longitudinal contribution dominates. At large values of q, where the transverse contributions are largest, the data indicate a deviation of 8.4±2.6 % from the impulse approximation.

In order to make a comparison with the inclusive (e,e') data[8] on ^{12}C, a LT-separation has been performed. Using eqs. (1) and (2) the ratio W_T/W_L was derived from the exclusive data of fig. 2. We plotted $R_G = \sqrt{(4m^2/Q^2)W_T/W_L}$ as a function of Q^2 in fig. 3. R_G is the

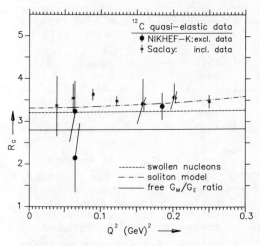

Figure 3. $R_G = \sqrt{(4m^2/Q^2)}W_T/W_L$ as a function of the four momentum squared Q^2 for both inclusive and exclusive quasi-scattering from ^{12}C.

ratio of the form factors G_M/G_E for a free proton. For a bound proton we interpret R_G as the value of G_M/G_E inside the nucleus. The quantity R_G has also been deduced from the inclusive data. Here we have applied corrections[6] for neutron knockout and contributions of momenta perpendicular to q. As can be seen from fig 3. the two results are in good agreement lending support to the assumption of single-nucleon knockout dominance in both reactions. The main observation from fig. 3 is the enhancement of the ratio G_M/G_E inside the nucleus. The 'bound' ratio is 22% larger than the 'free' ratio, which at small values of Q^2 corresponds to μ_p=2.79.

CONCLUSIONS

In figs. 2 and 3 the results of several calculations are shown. The dotted curve represents the prediction of the σ-ω model[10]. In both figures the dashed curves were obtained from an enlargement of the size of the proton[15]. Since the magnetic moment is proportional to the size, e.g. the radius in a quark-bag model[9], the enlargement factor (λ=1.15) simply multiplies μ_p. Also drawn is a curve corresponding to a calculation by Celenza et al.[16] based on many-body soliton dynamics in a relativistic frame work. All models considered give a good description of the experimentally observed breakdown of the impulse approximation. Therefore it would be of interest to study the relation between these models.

A further preliminary conclusion can be drawn from the present data if we indeed trust the relative accuracy of the calculated FSI effect on the momentum distribution (see fig. 4) to the level of 10%. Then we can make a relative comparison of the data at p_m=-90 MeV/c and p_m=+100 MeV/c and obtain the ratio $R_E=G_E(270$ MeV/c$)/G_E(460$ MeV/c) for bound protons. It appears that $R_E = (1.01\pm0.05) \times R_E^{free}$ from which a change in the rms charge radius for a bound proton of 1±6 % is found. This leads to conclusion that both inclusive and exclusive data (see fig.3) can be explained by a 22% increase of the magnetic moment for a bound proton without a change in the electric and magnetic radii. A similar conclusion has recently been obtained by Mulders[17] in an analysis of (e,e') quasi-free elastic scattering on a number of nuclei.

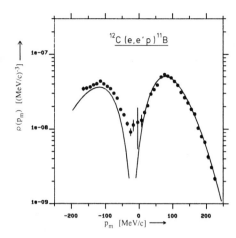

Figure 4. Momentum distribution for the ground-state transition in the reaction $^{12}C(e,e'p)^{11}B$. The data at positive (negative) p_m were obtained at forward (backward) electron scattering angle. The amplitude of the DWIA curve was fitted to the positive p_m data.

Finally we return to the observed systematic depletion of valence orbitals as deduced from the (e,e'p) reaction (see fig. 1). These data have all been obtained at forward electron kinematics, i.e. where the longitudinal contribution to σ_{ep} dominates. As argued above no change in G_E is needed, and hence these data remain unchanged. Thus we are left with an observed reduction of valence shell occupation numbers by about 50%. Such a large effect is not predicted by most current models of nuclear structure.

ACKNOWLEDGEMENT

The work described above has been carried out by the coincidene group at NIKHEF-K in cooperation with the Free University, Amsterdam and the State University, Utrecht. Important support has been received from the theory groups at NIKHEF-K and at the Universitá, Pavia, and from J. Mougey.
This work is part of the research program of the Foundation for Fundamental Research on Matter (FOM), which is financially supported by the Netherlands' Organization for Advancement of Pure Research (ZWO).

REFERENCES

1. L. Lapikás, Proc. 4th Miniconference on Nuclear Structure in the 1p shell, Amsterdam, 1985, p. 13
2. V.R. Pandharipande et al., Phys. Rev. Lett. **53**, 1133(1984)
3. S. Frullani and J. Mougey, Adv. Nucl. Phuys. **14**, 1(1984)
4. G.J. Kramer et al, to be published
5. J.W.A. den Herder et al, to be published
6. G. van der Steenhoven et al., to be published in Phys.Rev.Lett.
7. J.J. Aubert et al., Phys. Lett **123B**, 275(1983)
8. R. Barreau et al., Nucl. Phys. **A402**, 515(1983)
   C. Marchand et al., Phys. Lett. **153B**, 29(1985)
9. T. de Grand et al., Phys., Rev.**D12**, 2060(1975)
10. T. de Forest Jr., Phys. Rev. Lett.**53**, 895(1984)
11. T. de Forest Jr., Nucl. Phys. **A392**, 232(1983)
12. S. Boffi, C. Giusti and F.D. Pacati, Nucl. Phys. **A386**,599(1982)
13. C. de Vries et al., Nucl. Instr. **223**, 1(1984)
14. L. Lapikás and P.K.A. de Witt Huberts, J. Phys. **C45**, C4-57(1984)
15. P.J. Mulders, Phys. Rev. Lett. **54**, 2560(1985)
16. L.S. Celenza,A.Rosenthal and C.M. Shakin,Phys.Rev.**C31**, 232(1985)
17. P.J. Mulders, to be published

LARGE MOMENTUM COMPONENTS OF THE DEUTERON WAVE FUNCTION[*]

R. Dymarz and F.C. Khanna
Theoretical Physics Institute, Physics Department,
University of Alberta, Edmonton, Alberta T6G 2J1, Canada

and TRIUMF, Vancouver, B.C. V6T 2A3, Canada

One-body momentum distribution in nuclei, n(k), is not well known near and above the Fermi surface. Both the theoretical and experimental studies give rather ambiguous information on large momentum components of the nuclear wave function[1]. Among the experimental methods to measure these high momentum components the quasi-free scattering of high-energy electrons is most promising. The inclusive electron scattering on the low-energy transfer side of the quasi-free peak is supposed to be a simple one-step knock-out process with all many-body mechanisms (like final state interaction, meson exchange current, virtual isobar production) strongly suppressed. It was originally suggested by West[2] that under certain conditions (large momentum transfer (q) and small energy transfer (ω)), the inclusive e^--nucleus cross section scales in the variable y defined as a component of the momentum of the struck nucleon ($\tilde{k}$) along the momentum transfer ($y = \tilde{k} \cdot \tilde{q}/|\tilde{q}|$).

$$\sigma^{E(T)}(\omega,q) \, d\omega = (Z\sigma_{ep} + (A-Z)\sigma_{en}) \, F(y)^{E(T)} \, dy, \qquad (1)$$

where F(y) is a scaling function, $\sigma_{ep(n)}$ are the elementary e^--proton (neutron) scattering cross sections, and E(T) refers to experimental (theoretical) quantities. The function F(y) can be related to n(k) enabling then a study of n(k) in inclusive electron scattering.

Scaling hypothesis has been verified in the wide range of y for experimental cross sections in the (e,e') reaction on ^{3}He (Ref.3) and deuterium[4]. In both cases the calculated $F^T(y)$ underestimated $F^E(y)$ at large negative y and it was interpreted as an indication of the lack of high momentum components in the wave function used. In the present contribution we will show that in d(ee') scattering the $\Delta\Delta$ components of the deuteron wave function (DWF) is responsible for the large |y| behaviour of $F^T(y)$ assuring good agreement between theoretical and experimental scaling functions. The $\Delta\Delta$ components of the DWF were obtained within the coupled channel model of the nucleon-nucleon interaction with Δ-isobars included explicitly. The model is briefly described in Ref. 5.

In Fig. 1 we show F(y) evaluated according to Eq. (1) with the experimental cross section taken from Ref. 6 and the theoretical

[*]Work supported in part by NSERC (Canada).

cross section as given by McGee[7]. We adopted approximations to McGee's formulas similar to those in Ref. 4 and then in our model $F^T(y)$ is proportional to the integral over one-body momentum distribution which, in our case, besides the nucleon components contains $\Delta\Delta$ components. The elementary cross sections $\sigma_{ep(n)}$ were calculated with dipole nucleon form factors and with the neutron electric form factor neglected. We evaluated $F^T(y)$ with our model of deuteron wave function (B1) and two other models: Reid soft core (RSC) and Paris potential wave functions which contain only nucleon components. We can see in Fig. 1 that all three models are indistinguishable at small $|y|$ and differ at large $|y|$. The differences are only moderate at $|y| < 0.8$ GeV/c but are substantial at larger $|y|$. The experiment clearly favors model B1 at large $|y|$. The large values of $F^T(y)$ are determined by large momentum components of the DWF which in the model B1 are dominated by the $\Delta\Delta$ part of the DWF.

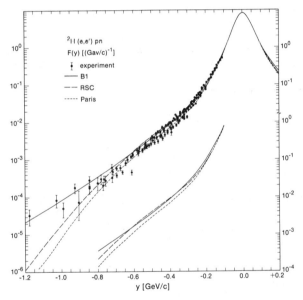

Fig. 1. Scaling function F(y) calculated from the experimental and theoretical cross section according to Eq. (1). Three curves represent calculations with different deuteron wave functions.

REFERENCES

1. M. Jaminon et al., Nucl. Phys. A452, 445 (1986).
2. G.B. West, Phys. Rev. 18C, 263 (1975).
3. I. Sick et al., Phys. Rev. Lett. 45, 871 (1980).
4. P. Bosted et al., Phys. Rev. Lett. 49, 1380 (1982).
5. R. Dymarz and F.C. Khanna, Phys. Rev. Lett. 56, 1448 (1986).
6. W.P. Schütz et al., Phys. Rev. Lett. 38, 259 (1977).
7. I.J. McGee, Phys. Rev. 158, 1500 (1967).

CONTINUUM SHELL MODEL ANALYSIS OF $^{16}O(\gamma,p)^{15}N$ AT MEDIUM ENERGIES[†]

Larry D. Ludeking[*] and Stephen R. Cotanch
North Carolina State University, Raleigh N.C. 27695-8202

ABSTRACT

A coupled-channels calculation, based on a continuum shell model formulation, is reported for $^{16}O(\gamma,p)^{15}N$ at energies between 50 and 400 MeV. The sixteen particle Hamiltonian is diagonalized in a truncated, but realistic model space spanned by products of continuum single-particle states and bound states of the A = 15 system. All non-localities arising from antisymmetrization, Pauli blocking, and exchange are rigorously retained. Using the complete effective two-body interaction which describes the ^{16}O giant dipole resonance, a quantitative description is obtained for the photonuclear cross section for energies spanning the quasi-elastic, Δ isobar region. For photon energies above 300 MeV it is necessary to include all electric and magnetic transitions up to E11 and M9, respectively, to achieve convergences for the multipole expansion. Preliminary calculations involving explicit Δ isobar degrees of freedom are also reported and briefly discussed.

INTRODUCTION

Fostered by several exciting discoveries, such as the EMC effect[1], the intersecting discipline between particle and nuclear physics is rapidly developing. This field, the area of high energy nuclear physics, offers the possibility for a more fundamental description of the nucleus as well as for new and deeper insight into baryon interactions and structure. Indeed, the growing awareness that the nucleus is an excellent laboratory for particle physics and that particle physics concepts, such as strangeness, are important in nuclear studies has finally removed the ill-defined demarcation which previously existed between these two fields.

The current work is representative of this intersecting research area and involves the microscopic description of the photonuclear reaction $^{16}O(\gamma,p)^{15}N$ at medium energies. This paper, which emphasizes conventional nuclear techniques extended to higher energies, is part of a broader investigation which intends to clearly delineate the role of the Δ isobar in nuclear processes. Our philosophy is to exploit the known electromagnetic interaction and relatively simpler excitation structure of the closed-shell nucleus ^{16}O to perform a comprehensive calculation that will provide a solid framework for consistently assessing a variety of effects, including Δ propagation. Our results, using only conventional nuclear degrees of freedom, are encouraging and provide a

[†]Supported in part by the U. S. Department of Energy.

[*]Present address: Mission Research Corporation, Alexandria, Va.

reasonable quantitative, but not complete, description of the data. Our subsequent communication will report if the remaining descrepancies can be meaningfully resolved by including explicit Δ excitations.

DETAILS OF THE CALCULATION

As discussed in our previous low energy analysis[2] of the giant dipole resonance in ^{12}C, we use the continuum shell model[3,4] to describe the p + ^{15}N system and the conventional shell model for ^{16}O. The A = 16 particle Hamiltonian is diagonalized in a truncated model space spanned by products of single-particle continuum states and A = 15 shell model eigenfunctions. Our model includes the ground state, $1/2^-$, and first excited state, $3/2^-$, for both ^{15}O and ^{15}N. The diagonalization generates a set of coupled integro-differential equations for the unknown continuum partial-waves which angular momentum couple with the A = 15 states to form a total J^π. For each J^π there are 12 coupled equations (except for J = 1 for which there are 10) and we have found it necessary to include all J^π contributions up to 11^- for natural parity transitions (E1 to E11) and 9^+ for unnatural parity transitions (M1 to M9) to achieve accurate convergence of the multipole expansion for the electromagnetic field. The total composite wavefunction is rigorously antisymmetrized which leads to complicated exchange (non-local) and Pauli blocking (inhomogeneous) terms in the coupled equations. Further, wavefunction orthogonality is fully preserved. This is important for photonuclear reactions as nonorthogonal wavefunctions can generate sizeable differences[5] between the density and current formulation for electric transition operator. We have carefully checked that both current and density form give identical numerical results. Also, the long wavelength approximation, which is inappropriate for medium energy photons, is not used and physical electric charges and magnetic moments are adopted. Finally, because of the closed shell nature of ^{16}O, we use 1 particle - 1 hole shell model wavefunctions for the bound A = 15 states (^{16}O final ground state is treated as a closed shell).

The effective two-body residual interaction is a modification of a phenomenological particle-hole potential used to describe the giant dipole resonance in ^{16}O[4] and consists of a central

$$v = 58[.0125 + .25\vec{\tau}_1\cdot\vec{\tau}_2 + .0425\vec{\sigma}_1\cdot\vec{\sigma}_2 + .1\vec{\sigma}_1\cdot\vec{\sigma}_2\vec{\tau}_1\cdot\vec{\tau}_2]Y_0(\mu r), \quad (1)$$

and tensor component

$$v^T = 5.1\vec{\tau}_1\cdot\vec{\tau}_2 S_{12}[Y_2(\mu^T r) - (\frac{\mu^c}{\mu^T})^3 Y_2(\mu^c r)]. \quad (2)$$

Here the overall strength is in MeV; S_{12}, $\vec{\sigma}$, and $\vec{\tau}$ are the standard tensor, spin, and isospin operators and Y_0, Y_2 are Yukawa form factors

$$Y_2(x) = (1 + \frac{3}{x} + \frac{3}{x^2})Y_0(x), \qquad (3)$$

$$Y_0(x) = \frac{e^{-x}}{x}, \qquad (4)$$

with range parameters $\mu = .73$ fm^{-1}, $\mu^T = .714$ fm^{-1}, and cut-off parameter $\mu^C = 6$ fm^{-1}. Using this residual interaction we compute both diagonal and off-diagonal coupling potentials but not the Hartree term. Instead, we use a Wood-Saxon and Thomas spin-orbit potential[5] having respective strengths[4] of 63-66 MeV and 5 MeV, diffuseness of .6 fm, and radius of 2.836 fm. Our Hamiltonian remains hermitian as there is no absorption in this calculation.

RESULTS AND DISCUSSION

Using the model outlined above with all parameters determined by the giant dipole resonance, we calculated the photonuclear cross section for three lab angles (45°, 90°, and 135°) and lab photon energies up to 400 MeV. Our results are summarized in Fig. 1 where the theoretical lab cross section is compared to two independent, but complimentary, sets of data (+ is ref. 6, * ref. 7). It is important to stress that our calculation is more in the nature of a prediction as no parameters have been readjusted. While a complete detailed description of the data is still lacking, especially for the 135° measurements which entails larger errors (20-50%), the overall results support the general theoretical framework and should provide a consistent basis for investigating additional effects such as Δ isobar propogation.

To this end we have generalized our code to include specific Δ degrees of freedom. We treat the Δ^+ and Δ^0 on equal footing with the nucleons (Δ-hole model). The Δ introduces several technical difficulties: the additional coupling associated with spin 3/2 triples the number of channels to 36 (26 for J=1); the large width $\Gamma_\Delta \simeq 115$ MeV (we incorporate it by allowing for a complex energy) requires high accuracy and multiple integration regions to accurately accomodate the short range Δ propagation (especially important numerically below the Δ threshold = 309 MeV). Our preliminary calculation using between 22 and 26 total channels and including all multipole contributions through E8/M7 indicates a sizeable (up to an order of magnitude) increase in the cross section throughout the quasi-elastic region (i.e. near the Δ threshold). See refs. 8-12 for further discussion concerning Δ isobar effects in photo-

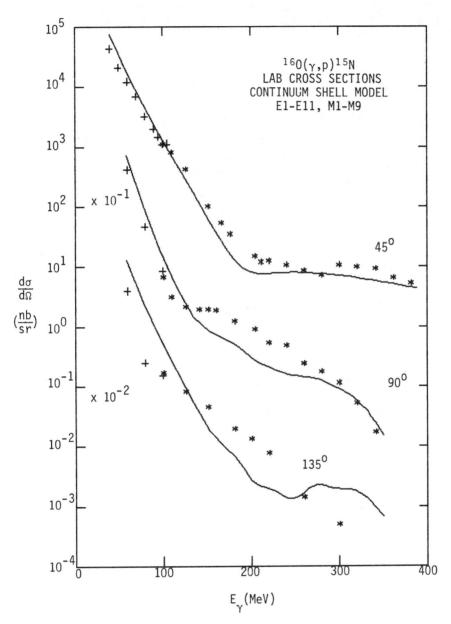

Fig. 1. Theory and experiment for $^{16}O(\gamma,p)^{15}N$. For graphical clarity the $90°$ and $135°$ results have been multiplied by .1 and .01 respectively.

nuclear processes. We have also found that it is very important to include the full Δ width, otherwise one obtains unrealistically large effects. We will report the results of our complete Δ calculation in a future communication.

CONCLUSION

In summary, we have performed a comprehensive continuum shell model calculation using a realistic effective interaction and have obtained a reasonable, although not complete description of the reaction $^{16}O(\gamma,p)^{15}N$ at medium energies. We have also performed preliminary calculations in which about half of the channels describing the propagation of Δ^+ and Δ^0 isobars have been included. While isobar excitation and propagation appears important it presently remains unclear whether this effect will significantly improve the description of the data. The complete calculation is in progress and will be reported in the near future.

REFERENCES

1. J. J. Aubert et al., Phys. Lett. 123B, 275 (1985).
2. L. D. Ludeking and S. R. Cotanch, Phys. Rev. C29, 1546 (1984).
3. B. Buck and A. D. Hill, Nucl. Phys. A95, 271 (1967).
4. J. Raynal et al., Nucl. Phys. A101, 369 (1967).
5. J. M. Lafferty and S. R. Cotanch, Nucl. Phys. A373, 363 (1982).
6. D. J. S. Findlay and R. O. Owens, Nucl. Phys. A279, 385 (1977).
7. M. J. Leitch et al., Phys. Rev. C31, 1633 (1985).
8. J. T. Londergan and G. D. Nixon, Phys. Rev. C19, 998 (1979).
9. M. Gari and H. Hebach, Phys. Rep. 72, 1 (1981).
10. E. Grecksch et al., Z. Phys. A302, 247 (1981).
11. J. Finjord, Nucl. Phys. A274, 495 (1976).
12. C. Y. Cheung and B. D. Keister, Phys. Rev. C33, 776 (1986).

Electromagnetic Structure of ^{3}H and ^{3}He

D. H. Beck
Department of Physics and
Laboratory for Nuclear Science
Massachusetts Institute of Technology
Cambridge, MA 02139

1. Introduction

The current interest in studying the three-nucleon systems with electromagnetic probes is motivated by at least two observations. The three-body system is the heaviest nucleus for which the (non-relativistic) wave equation can in practice be solved with arbitrary precision. A great deal of effort has been directed at this problem by theorists; considerable success has been achieved. Secondly, ^{3}H and ^{3}He form an isospin doublet. One can therefore predict the properties of both systems with essentially the same theory – a change of T_z is all that is required. In addition, the nearly exact isospin symmetry for the nucleon wave functions in the two systems is also applicable to the quark wave functions.

What are the physics issues explored in this system? I see three major areas of interest – structure, non-nucleonic degrees of freedom and the effect of relativity. We are learning mostly about structure – wave functions in general and particular aspects of structure such as correlation functions. A very important contribution to the observed electromagnetic currents comes from non-nucleonic degrees of freedom – reflecting the composite nature of nucleons. Effects of meson exchange currents (MEC) and delta excitation are in fact manifest in both the system wave function and in its response to the electromagnetic probe. Finally, with momentum transfers reaching the scale of the nucleon mass, both the kinematic and dynamic effects of relativity might be expected to be important.

I shall try to review developments on these fronts by discussing three classes of experiments performed recently. Elastic scattering experiments perhaps yield the most information because the theoretical models are most complete for this process. Both inclusive and exclusive inelastic measurements have also been made. They, however, suffer in interpretation from the lack of continuum solutions to the three-body wave equation. I would like take this opportunity to salute our colleagues at NIKHEF-K and especially at Saclay for important contributions to experimental work in this area and on which I will report.

2. Elastic form factors

Let me begin by discussing the elastic scattering response. There is some new evidence emerging to suggest that the ground state wave function is influenced by non-nucleonic degrees of freedom. Inclusion of a so-called three-body force in the hamiltonian by Friar et al.[1] has shown an increase in the binding energy (of the only bound state) of about 1 MeV. This repairs the traditional shortfall of Faddeev solutions to the three-nucleon wave equation – calculations with only two-nucleon potentials give a binding energy of about 7.5 MeV as compared with the experimentally observed

8.47 MeV. Three separate models of the three-nucleon force were employed in the calculation of Friar and all contained important contributions from diagrams containing an intermediate virtual Δ resonance (Figure 1). A different calculation, taking the approach of including Δ's explicitly in the wave function[2], yields only 300 keV of additional binding because the Δ is allowed to interact (unlike the other calculation where the Δ is considered to be static). There is, however, general agreement among theorists that inclusion of the Δ in the ground state results in significant contributions.

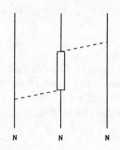

Figure 1. Dominant three-body force diagram.

Recently, two experiments have made elastic scattering measurements on tritium which will help to constrain theory. At Saclay[3] a liquid target with 10 kCi of tritium, operating at 20 K, was used to determine the charge form factor in the range $0.01 \leq q^2 \leq 0.89\,(\text{GeV}/c)^2$ and the magnetic form factor in the range $0.12 \leq q^2 \leq 1.22\,(\text{GeV}/c)^2$. The quoted systematic error (dominant at low q) in the cross sections is 4.5%. At MIT-Bates[4] a gas target with 110 kCi of tritium, operating at 45 K and 225 psia, was used to determine in particular the low momentum transfer form factors with an uncertainty in the cross sections of about 2%. An additional feature of the MIT target system allows ^{3}He cross sections to be measured with a target possessing essentially identical geometry. Along with a few high momentum transfer elastic cross sections, a set of inclusive inelastic measurements of both systems were made, taking advantage of the complete range of incident energies available at Bates, i.e. from 65 to 805 MeV.

Because of the lack of binding, calculated elastic charge form factors were, at low q ($q^2 \leq 0.3\,(\text{GeV}/c)^2$) smaller than those determined experimentally by 5 – 10%. Figure 2 shows the low q form factors of ^{3}H and ^{3}He. The new Saclay ^{3}H charge form factor data appear to disagree with the new results of Friar, though the 4.5% systematic uncertainty in the data should be born in mind. Bates data are still being analyzed.

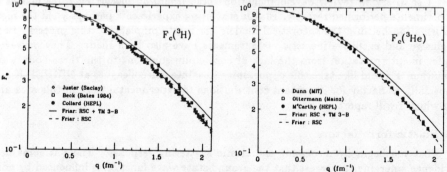

Figure 2. Low q elastic charge form factors. Theory is due to Friar[5]. Data are from Juster[3], Beck[4], Collard[7], Dunn[8], Ottermann[9] and McCarthy[10].

At high momentum transfer ($q^2 \geq 0.5\,(\text{GeV}/c)^2$) the need for non-nucleonic degrees of freedom is made obvious particularly in the magnetic form factors. Whereas the first minimum occurs at about 0.65 $(\text{GeV}/c)^2$, a nucleons-only calculation locates it at 0.35 $(\text{GeV}/c)^2$. Indeed it has been known for many years that the magnetic moments of these nuclei contained a 15% contribution from MEC. In pursuing the analysis of the high momentum transfer form factors it is useful to consider the isospin decomposed form factors defined by

$$F_i^{\,^V_S} = \frac{F_i^{^3He} \pm F_i^{^3H}}{2}.$$

The isovector magnetic form factor, F_m^V, contains meson exchange currents which enter to the same relativistic order, $(p/m)^1$, as the one body currents and thus is consistent in this sense. (The other three 3N form factors have MEC which enter only to higher order in p/m than the one body currents and hence are inconsistent in this same sense.) The agreement between the calculations including MEC and experiment for this quantity are reasonable with discrepancies at the second maximum at the 20% level or smaller (Figure 3). One interesting input to the model which results in some variation is the nucleon form factor used in the MEC. If the Dirac form factor F_1^V is used the agreement discussed above is obtained. If, however, the Sachs form factor G_E^V is used the minimum in F_m^V moves by about 0.2 GeV/c. (As an example, two calculations of the ^{3}H magnetic form factor are shown in Ref. 3.) The particular difficulty with this difference is that the nucleon form factors disagree only in order $(p/m)^2$ – since the Sachs form factor is, in the language of this three nucleon problem, the Dirac form factor with a relativistic correction, the general question about the size of relativistic effects is raised.

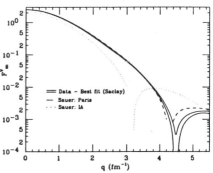

Figure 3. Three-nucleon elastic isovector magnetic form factor. Theory is due to Sauer[2]. Data are from Martino[11].

The three-nucleon wave functions are, at this point, calculated non-relativistically. Microscopic relativistic calculations of wave functions have been made only for the deuteron. The example of the deuteron elastic magnetic form factor is, I think, important to this discussion of the size of relativistic effects (even though, being an isoscalar transition, it does not relate directly to F_m^V discussed above). Tjon and coworkers[12] have found the effect of coupling to the negative energy states significant – about 20% at $q^2 = 0.6$ $(\text{GeV}/c)^2$. The contribution is coincidentally nearly equal in magnitude but opposite in sign to the contribution from the pion pair terms in the non-relativistic calculations and therefore the relativistic calculations do not agree with the data.

What can one say about the charge form factors? Some interesting exploratory work has been done recently by Kisslinger et al.[13] (and also by Vary[14]) using a hybrid quark-hadron model in which, for nucleon separations of less than 1 fm, a simple 6 or 9 quark bag model wave function is used. Calculations using the same values of this separation parameter and bag radius simultaneously fit the deuteron elastic form factors, the deuteron threshold electrodisintegration, and the ^{3}He charge form factor all at q's up to about 1 GeV/c. Unfortunately the calculations for the three-nucleon system are limited to this maximum momentum transfer because, since the probability of two 'nucleons' being in a 6 quark bag state is substantial (~15%), at higher q's some significant unphysical oscillations show up – the result of the sharp boundary condition. At Bates, three high q points have been measured which determine (together with F_m from Saclay) the ^{3}H charge form factor to 1.2 $(GeV/c)^2$ as shown in Figure 4. The model appears to drop off more rapidly than these new, preliminary data. The uncertainty due to the inconsistent treatment of MEC described above is reduced in these models because the short range contributions are eliminated in favor of the simple quark dynamics.

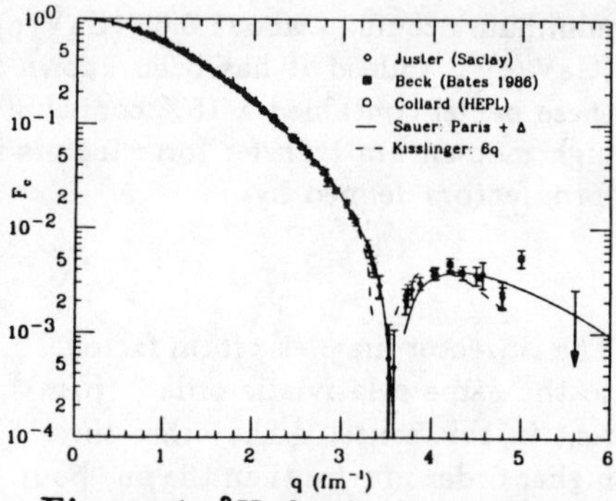

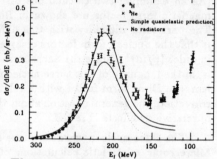

Figure 4. ^{3}H elastic charge form factor. Theories are due to Sauer[2] and Kisslinger[13].

3. Inclusive inelastic scattering

A recent Saclay measurement[15] has determined the inclusive longitudinal and transverse response functions for ^{3}He in the region $0.1 \leq q^2 \leq 0.3$ $(GeV/c)^2$. At Bates essentially the same measurements were performed for both ^{3}H and ^{3}He in such a way as to be able to determine the response functions to the 5% level (and where statistical uncertainties dominate those of systematic origin). In addition, higher momentum transfer (up to 1.2 $(GeV/c)^2$) cross sections and the response of the Δ excitation were also measured at Bates.

One sees in the Laget calculations (described in Ref. 16, shown most clearly with the data in Ref. 15) for the Saclay results overprediction of the transverse response by about 25% at the peak for all momentum transfers. This effect is sometimes blamed on wave functions with the wrong binding energy. If that were so one would expect: 1) the overall strength to be correct and 2) that the effect should also be present in the longitudinal response. Neither of the above statements appear to be correct. In an attempt to understand this phenomenon the on-line Bates data for both 3H and ^{3}He were compared with a simple model calculation[4] using the momentum distribution function of Ciofi degli Atti[17] (Figure 5 – the momentum transfer at the peak is 0.27 $(GeV/c)^2$). Since these data have not been radiatively corrected, the model of Ciofi was also used to calculate responses at lower momentum transfer in order to predict the measured cross section (i.e. including the effects of the known radiators in the target system). The agreement between theory and experiment in this comparison is very good. Similar calculations with the momentum distributions of Sauer[2] and of Wiringa[18], the latter defining the size of three-body effects by fitting the ^{3}H binding energy, give essentially identical results. The model of Laget does, however, contain more physics – further interpretation awaits direct comparison of his calculation with the carefully analyzed data.

Figure 5. Quasi-elastic data for ^{3}H and ^{3}He at $E_i = 370$ MeV, $\theta = 134.5°$. Theory is referenced in text (calculation with no radiators is also shown for ^{3}He).

Note should be made of two additional problems in the transverse response. First, there is the chronic problem of a lack of strength in the so-called 'dip' region, even when making use of a ground state wave function containing reasonable two-body correlations and including final state interactions. Second, the prediction is smaller than the data by a factor of as much as three in the low ω region. This has often been ascribed to a lack of high momentum components in the wave function. However, Laget seems to be able to repair the difficulty in the longitudinal response. One can look more directly for missing high momentum components using the (e,e'p) reaction as will be discussed in the next section.

The longitudinal response of ^{3}He is reasonably well described by the predictions of Laget for higher q^2 whereas they exceed the data by again about 25% at the peak for the lowest q^2. By integrating the longitudinal response for constant $|\vec{q}|$ one can investigate this apparent excess of strength with respect to the non-relativistic coulomb sum rule:

$$\Sigma(|\vec{q}|) = \frac{1}{G_{E_p}^2(q^2)} \int_0^\infty R_L(|\vec{q}|,\omega) d\omega$$
$$= Z + Z(Z-1)C(|\vec{q}|).$$

For low q the two proton correlation function term $C(|\vec{q}|)$ is expected to reduce the strength of $\Sigma(|\vec{q}|)$ below Z. As far as I know there is no published calculation of $C(|\vec{q}|)$ as such, though this quantity by itself is interesting – it provides important input on the question of extensions of the mean field for heavier systems. However, as correlations are, to some extent, included implicitly in the Laget calculation one should perhaps conclude that there is another mechanism reducing the strength of R_L at low q. One can again look at the very preliminary results for ^{3}H for further guidance, since for $Z = 1$, the correlation term vanishes and one should expect to see the same overprediction for ^{3}H as for ^{3}He. Figure 6 shows that there is 5 - 10% more strength in the ^{3}H cross section (relative to the calculation) than in the ^{3}He cross section. Final assessment awaits detailed analysis of the new data.

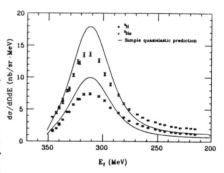

Figure 6. Quasielastic data for ^{3}H and ^{3}He at $E_i = 371$ MeV, $\theta = 54°$. Theory is as for Figure 5.

4. Exclusive inelastic scattering

Three experiments on the ^{3}He(e,e'p) and ^{3}He(e,e'd) reactions have recently been performed. The first measurement[19] was made at Saclay with the electron detected in quasi-elastic kinematics and momentum transfers of 300 and 430 MeV/c. If interpreted in the simple PWIA, i.e. assuming that the residual nucleus (Z-1,N) acts only as spectator throughout, the proton spectral function $S(E_m,p)$ may be factored out of the expression for the cross section. Integrating over E_m (the sum of the proton separation energy and the internal energy remaining in the residual nuclear system) then gives the momentum distribution of the initial proton. The results indicate that whereas predictions of the low ω inclusive transverse response function fall short of the data by as much as a factor of three, out to 300 MeV/c the momentum distribution predicted by the same model agrees with the exclusive data to within about 30%. In fact the calculated strength is larger than that measured. The agreement holds for both the two- and three-body exclusive channels.

An interesting place to, roughly speaking, check these conclusions is to look with electron kinematics defining a reaction away from the quasi-elastic peak – one should see the same momentum distribution. Preliminary results from Marchand et al.[20] at Saclay show that again for both the two- and three-body breakup channels separately the predictions come reasonably close to the data, even for higher proton momenta. Figure 7 shows the results for the two-body breakup. Laget's model of the distribution in the two-body channel agrees with the data at the 20% level up to $p = 500$ MeV/c. Thus these first two measurements suggest that calculations

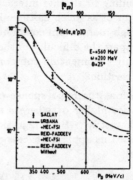

Figure 7. Preliminary data from Marchand et al.[20]. Reid-Faddeev theory is from Ref. 16 and Urbana theory is from Ref. 18.

do contain the correct proportion of high momentum components. Laget suggests that the observed lack of strength at low ω in the inclusive response may be due not to larger high momentum components, but to polarization or deformation induced by the medium. The differences in the longitudinal and transverse responses (see Ref. 15) – coupling to charge and spin, respectively – may guide further investigation of this idea.

There is some suggestion of an enhancement in a high q measurement of the inclusive response at $x = 2$ (corresponding to quasi-elastic scattering from a deuteron). Interestingly Laget's model also predicts an enhancement of the cross section at this point (measurement at 7.26 GeV and 8°, see Ref. 16). At NIKHEF-K an (e,e'd) experiment has been performed[21] which confirms this prediction – the cross section exceeds the 'proton only' calculation by a factor of 10 for initial deuteron momenta of 50 MeV/c. When direct coupling of the photon's four-momentum is made to the deuteron and some account is taken of final state interactions there is good agreement between theory and experiment for $50 \leq p_d \leq 200$ MeV/c. This constitutes quite a direct confirmation of the np correlation structure of current three-nucleon wave functions.

In a recent measurement at NIKHEF-K[22] the value of the extracted proton momentum distribution function in the two body channel and at $p = 100$ MeV/c was found to be independent of the relative kinetic energy of the outgoing proton and deuteron (Figure 8). Essentially no variation is observed at the 5 - 10% level for kinetic energies ranging from 25 to 100 MeV. The value of $\rho(p)$ is, however, only about 70% of that calculated in the simple PWIA. Interpretation of this curious result is important for understanding the validity of the PWIA analysis framework for these exclusive experiments.

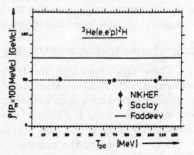

Figure 8. Preliminary data from Keizer et al.[22]. Calculation is due to Meier-Hajduk[23].

5. Summary

The three nucleon system interests us because one can make exact calculations of the ground state using the best physics input and because the two physical nuclei are effectively projections of the same system in isospin space. Elastic scattering experiments show the need for both meson exchange currents and delta isobars (even static properties require both). At higher momentum transfers relativity apparently plays an important role both through kinematics and dynamics. The inclusive transverse inelastic response shows discrepancies among the standard calculations in addition to a general lack of calculated strength above and below the quasielastic peak. Exclusive measurements show that this shortfall is not likely due to a lack of high momentum components in the model wave functions. The integrated inclusive longitudinal response can yield information about the proton-proton correlation function in ^{3}He; this quantity may also be calculated from exact ground state wave functions. Recent exclusive measurements from NIKHEF-K show: 1) a large direct coupling of the photon four-momentum to the deuteron in (e,e'd) and 2) a remarkable insensitivity to the final state p - d kinetic energy in the same reaction.

References

1. J. Friar et al., Phys. Lett. 161B, 241 (1985).
2. P. Sauer, priv. comm. and C. Hajduk, P. Sauer and W. Streuve, Nucl. Phys. A405, 581 (1983).
3. F. Juster et al., Phys. Rev. Lett. 55, 2261 (1985).
4. D. Beck et al., MIT-Bates Annual Report, 1985, p. 69.
5. J. Friar, priv. comm.
6. D. Beck et al., Phys. Rev. C 30, 1403 (1984).
7. H. Collard et al., Phys. Rev. 138, B57 (1965).
8. P. Dunn et al., Phys. Rev. C 27, 71 (1983).
9. C. Ottermann et al., Nucl. Phys. A436, 688 (1985).
10. J. McCarthy, R. Whitney and I. Sick, Phys. Rev. C 15, 1396 (1977).
11. J. Martino, priv. comm.
12. For example, J. Tjon and M. Zuilhof, Phys. Lett. 84B, 31 (1979).
13. L. Kisslinger, Phys. Lett. 112B, 307 (1982) and priv. comm.
14. J. Vary, priv. comm. and H. Pirner and J. Vary, Phys. Rev. Lett. 46, 1376 (1981).
15. C. Marchand et al., Phys. Lett. 153B, 29 (1985).
16. J. Laget, Phys. Lett. 151B, 325 (1985).
17. C. Ciofi degli Atti, E. Pace and G. Salme, Phys. Lett. 141B, 14 (1984).
18. R. Wiringa, priv. comm. and R. Schiavilla, V. Pandharipande and R. Wiringa, Nucl. Phys. A449, 219 (1986).
19. E. Jans et al., Phys. Rev. Lett. 49, 974 (1982).
20. C. Marchand, priv. comm.
21. P. Keizer et al., Phys. Lett. 157B, 255 (1985).
22. E. Jans, priv. comm.
23. H. Meier-Hajduk et al., Nucl. Phys. A395, 332 (1983).

Measurements of the Deuteron Magnetic Form Factor at High Q^2

P. E. Bosted*
The American University, Washington DC 20016

ABSTRACT

We have doubled the range of existing data for the magnetic form factor of the deuteron by making measurements at $Q^2 = 1.5, 1.75, 2.25$, and 2.5 $(GeV/c)^2$. The results show a shallow minimum around $Q^2 = 2$ $(GeV/c)^2$, in qualitative agreement with impulse approximation calculations.

INTRODUCTION

The electromagnetic form factors of the deuteron have long been of interest for the information they contain on the short range nucleon- nucleon interaction and the transition from nucleon to quark degrees of freedom. Predictions for the magnetic form factor, in particular, show great sensitivity to the high momentum components in the deuteron wave fuction, isobar contributions, nucleon form factors, relativistic effects, and the contributions of iso-scalar meson exchange currents. In contrast to perturbative QCD, which predicts a smooth falloff of the cross section with four-momentum transfer Q^2, the models based on the impulse approximation all predict a minimum around $Q^2 = 1.5$ to 2.5 $(GeV/c)^2$. In order to distinguish between all these approaches and provide valuable constraints on nucleon-nucleon models, we have measured the magnetic form factor of the deuteron to the highest possible Q^2.

KINEMATICS

The cross section for elastic scattering can be written as

$$d\sigma/d\Omega = \sigma_{MOTT}[A(Q^2) + B(Q^2)\tan^2(\theta/2)],$$

where σ_{MOTT} is the non-structure cross section and θ is the electron scattering angle. The structure function $A(Q^2)$ is a combination of the charge, quadrupole, and magnetic form factors, and has previously[1] been measure out to $Q^2 = 4$ $(GeV/c)^2$. $B(Q^2)$ is proportional only to the magnetic form factor, and has recently been measured[2,3] out to $Q^2 = 1.3$ $(GeV/c)^2$. Extrapolation of the data and the predictions of most calculations indicated that $B(Q^2)$ would become more than two orders of magnitude smaller than $A(Q^2)$ at higher Q^2. For this reason we decided that a Rosenbluth separation would not work, and that a 180° spectrometer system would be needed.

EXPERIMENTAL SETUP

Figure 1 shows a layout of the spectrometer system that was built in End Station A at the Stanford Linear Accelerator Center. Electrons from the new Nuclear Physics Injector were transported through chicane magnets B1, B2, and splitting magnet B3 to the target. The beam had peak currents of 20 to 50 mA in 1.6μsec long pulses at 150 Hz, with energies from 0.7 to 1.3 GeV. The electrons scattered at 180° were gathered up by the three quadrupoles Q1, Q2, and Q3 and bent by magnets B3 and B4 into the electron detectors. The solid angle was 22 msr for a thin target (less for extended targets), and the momentum acceptance ±5%. Because the anticipated cross sections were very small, we constructed liquid deuterium targets with thicknesses of 10, 20, and 40 cm. The energy loss smearing in these targets was much too large to be able to resolve elastic from inelastic scattering in the electron arm only, so that it was necessary to build a recoil spectrometer at 0° to detect the coincident deuterons. This was accomplished using quadrupoles Q4, Q5, and Q6, and bending magnets B5, B6, B7, and B8. The electron beam was bent in the opposite direction from the deuterons into a well shielded moveable dump by splitting magnet B5.

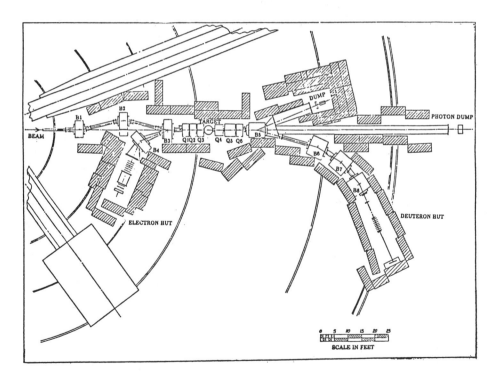

Figure 1. Plan view of the double arm spectrometer system.

Electrons were identified in the 180° spectrometer using a threshold gas Cerenkov counter and a forty segment lead glass shower counter. Taken together, they rejected nearly all of the large flux of pions. The particle trajectories were measured with six planes of wire chambers, and two banks of scintillators provided fast timing. Deuterons were identified in the recoil spectrometer using time-of-flight between two banks of scintillators placed 7 meters apart. As can be seen in Figure 2, they are well-separated from the much more abundant protons. A clear *ed* coincidence signal can also be seen, with only a small contamination from random *ed* coincidences. Trajectories in the recoil arm were reconstructed using eight planes of wire chambers.

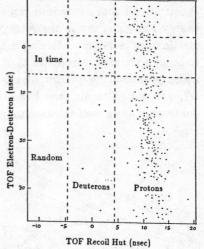

Figure 2. Electron-recoil coincidence data at $Q^2 = 2.5 \ (GeV/c)^2$. The horizontal axis is the time-of-flight between the front and rear scintillators in the deuteron arm. The vertical axis is the time-of-flight between electrons in the 180° spectrometer and particles in the recoil spectrometer. A clear *ed* coincidence signal can be seen.

The entire system was simulated with a Monte Carlo program which used the measured field maps of the magnets. The program was used to predict both forward and reverse matrix elements, and effective solid angles including radiative effects. Good agreement was found with vertical matrix elements as measured using a grid placed between the Q3 and the target, and horizontal matrix elements measured by sweeping peaks from *ep* elastic scattering across the wire chambers. Good agreement was also found for the absolute *ep* elastic cross sections measured at each beam energy with a variety of target thicknesses.

Data were taken in two running periods: one in May-July 1985 and one in April-May 1986. The preliminary results for *ed* elastic scattering for the 1985 running only are shown in Figure 3. An online analysis of the 1986 running at $Q^2 = 1.2, 1.625, 2.25, 2.5$ and $2.75 \ (GeV/c)^2$ is in reasonable agreement with the trend of the 1985 data. All of the 1985 data were taken using the 20 cm target, except for the point at $Q^2 = 2.5 \ (GeV/c)^2$, which used the 40 cm target. During the analysis we discovered an unanticipated source of background, i.e. *ed* coincidences arising from $\gamma d \rightarrow \pi^0 d$ in which one of the photons from the π^0 decay pair-produces in the target to give a high energy electron. This process is kinematically cleanly separated from elastic ed scattering for a thin target, but has some overlap (due to energy loss smearing) for thick targets, especially the 40 cm one. Further analysis using the complete double arm kinematics will improve the separation, which is the principal reason that our quoted results are still preliminary. The point at $Q^2 = 2.5$ $(GeV/c)^2$ was repeated in the 1986 running using the 20 cm target to reduce this background. The other correction applied to the data is the subtraction of the contribution from $A(Q^2)$ due to the finite acceptance of the 180° spectrometer. This correction was largest at $Q^2 = 2$ $(GeV/c)^2$, amounting to some 40%.

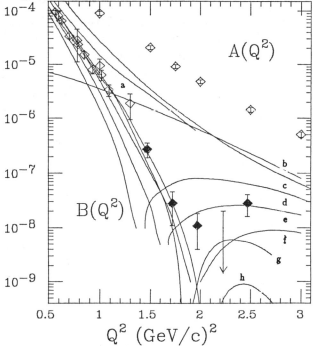

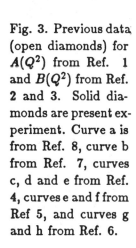

Fig. 3. Previous data (open diamonds) for $A(Q^2)$ from Ref. 1 and $B(Q^2)$ from Ref. 2 and 3. Solid diamonds are present experiment. Curve a is from Ref. 8, curve b from Ref. 7, curves c, d and e from Ref. 4, curves e and f from Ref 5, and curves g and h from Ref. 6.

As can be seen in Figure 3, our new data for $B(Q^2)$ continue to fall steeply with Q^2 up to $Q^2 = 1.75$ $(GeV/c)^2$, then flatten out and rise slightly at the highest Q^2. The ratio of B/A becomes very small, less than 0.003 for both $Q^2 = 1.75$ and 2.0 $(GeV/c)^2$. The simplest approach to making predictions for $B(Q^2)$ is to use the non-relativistic impulse approximation. The calculations of two papers[4,5] agree and are shown in curve e. The Paris potential was used in both cases. The results are very sensitive to the amount of high momentum components in the wave fuctions. This can be seen in curve g for a calculation[6] in which $\Delta\Delta$ isobar components have been included. This calculation also includes iso-scalar meson exchange currents. A major unknown here is the $\pi\rho\gamma$ coupling constant. Increasing this value from .406 to .560 reduces[6] curve g down to curve h. At these large Q^2, relativistic effects should play a major role. The size of this can be gauged from a comparison of curves e (non-rel) and f (relativistic)[4]

Finally, the inclusion of six quark clusters can shift the minimum to lower Q^2, as was shown[5] in the calculation of curves e and d. Although large variations are seen between these calculations in the traditional framework, they have the common features of a diffraction minimum followed by a secondary maximum, in qualitative agreement with our preliminary data. This is not the case for the covariant quark oscillator model of Ref. 7 (curve b), the parton model of Ref. 4 (curve c), or the hybrid model of Ref. 8 (curve a), all of which use a quark description of the deuteron as their basis. The disagreement with experiment is not too surprising, since these models should work best at very high Q^2.

During both the 1985 and 1986 running, inelastic data were taken from threshold through the quasi-elastic region. When analyzed, these data will be used to extract the structure function W_1. Comparison with previously measured W_2 will give R, the ratio for scattering longitudinally and transversely polarized virtual photons. If R is much larger than zero at high Q^2, it might be an indication that quark clusters play an important role in inelastic scattering near threshold[9].

REFERENCES

*Reporting on an experiment performed by R. Arnold, D. Benton, L. Clougher, G. DeChambrier, A. T. Katramatou, A. Lung, G. Petratos, A. Rahbar, S. E. Rock, and Z. M. Szalata, *The American University*, R. A. Gearhart, *SLAC*, B. Debebe, M. Frodyma, R. S. Hicks, A. Hotta, and G. A. Peterson, *University of Massachusetts*, J. Alster and J. Lichtenstadt, *Tel Aviv University*, and J. Lambert, *Georgetown University*. Work supported by U.S. Department of Energy Contract DE-AC03-76F00515 and National Science Foundation Grant PHY85-10549.

1. R. Arnold *et al*, Phys. Rev. Lett. 35, 776 (1975).
2. S. Auffret *et al*, Phys. Rev. Lett. 54,649 (1985).
3. R. Cramer *et al*, Z. Phys. C29, 513 (1985).
4. R. S. Bhalerao and S. A. Gurvitz, WIS-81/32-Ph, 1981.
5. M. Chemtob and S. Furui, SACLAY-SPhT-85/173, 1985.
6. E. Lomon, P. Blunden, P. Sitarski, private communication.
7. N. Honzawa *et al*, Prog. of Theo. Phys. 73, 1502 (1985).
8. A. P. Kobushkin and V. P. Shelest, Sov. J. Part. Nucl. 14, 483 (1983).
9. I. Schmidt and R. Blankenbecker, Phys. Rev. D15, 3321 (1977).

TENSOR POLARISATION OF THE DEUTERON IN THE ELASTIC e⁻d SCATTERING*

F.C. Khanna and R. Dymarz
Theoretical Physics Institute, Department of Physics
University of Alberta, Edmonton T6G 2J1

and TRIUMF, 4004 Wesbrook Mall, Vancouver V6T 2A3

Suggestions[1] have been made that measurement of spin observables in the elastic electron-deuteron scattering may serve as an unambiguous test of the Perturbative Quantum Chromodynamics (PQCD) in the nuclear physics domain at momentum transfer of few GeV/c. Recently[2] a counter argument has been advanced that admixture of ΔΔ-components in the ground state NN-components of the deuteron may provide significant changes in the short distance behaviour of the nucleonic wavefunction such that the tensor polarisation, t_{20}, at large q may be approaching the asymptotic limit of $-\sqrt{2}$ obtained with PQCD. It may be conjectured that additional effects at large q arise from the small components of resonances of N and Δ present in the ground state wavefunction of the deuteron.

A non-relativistic model of coupled nucleons and Δ-isobars has been developed to treat the two nucleon problem in a consistent way. A one-Boson-exchange potential approach is adopted. The coupling constants are obtained from experiment with the inter-relationship between nucleon and isobar couplings taken from the quark model. The vertices have monopole form factors with a parameter Λ that is free. It may be stated that the eigenvalue problem has been converted to an initial value problem with the σNN coupling constant being varied to obtain the deuteron binding energy. In this note we investigate the dependence of t_{20} on the variation of the parameter Λ. This is effectively providing us with a measure of the role of the short range potential in determining the high momentum components in the deuteron wavefunction. We have chosen $\Lambda_\rho = \Lambda_\pi$ and $\Lambda_\sigma = \Lambda_\omega$. These parameters are varied in the range of 0.9 GeV and 1.8 GeV. It should be stressed that since we are solving an initial value problem the coupling constant σNN is found to be different for each choice of the Λ-parameters. The results of such a variation may be summarized as follows:

(i) A large total ΔΔ-probability is obtained for the case when the tensor force is weaker. In such a case the S-state of the ΔΔ-component has a large probability. This leads to the largest change in the wavefunction for the nucleonic components. It is to be emphasized that the large ΔΔ-probability (≳ 1%) has the salutory effect of changing the nucleon wavefunctions at short distances, thus increasing the high momentum components. The ΔΔ-component wavefunction has little or no effect on t_{20} at small values of q but has very important contributions at q ~ 1 GeV/c and larger.

*Work supported in part by the Natural Sciences and Engineering Research Council of Canada.

(ii) The dependence of the overall shape of the curve for t_{20} vs. q on Λ is quite weak. Still the asymptotic value for t_{20} is different in the various cases. This is important for the conclusion that the variation of the wavefunction at short distances is more dependent on the coupling of several channels than on the short distance behaviour of the potential.

(iii) Meson exchange current effects are important at the intermediate range of q but become small for large values of q. This is largely due to the fact that the charge form-factor is small at high q and it is only the ratio of quadrupole and magnetic form-factors that survives. The role of meson exchange currents is minimized and the overall results are insensitive to any variation of parameters.

(iv) The static properties of the deuteron, quadrupole moment and magnetic moment, are well-reproduced. (See Table I.) The form-factors of the deuteron are affected strongly at large momentum transfers (q > 600 MeV/c). Both the meson exchange currents and the ΔΔ-components play an important role. Unfortunately the charge form-factor of the neutron is not known well enough to make any definite statements.

To sum up, a variation of the parameters in the coupled channel calculation indicates clearly the dominant role of the ΔΔ-components in the ground state wavefunction of deuteron energies from the coupling of the six channels and not from a particular set of parameters that determine the short-range behaviour of the potential through the monopole form-factor at the vertices.

TABLE I

$g_{\pi NN}$ = 13.5, $g_{\rho NN}$ = 2.6, K_ρ = 6.6

$g_{\omega NN}$	$g_{\sigma NN}$	$\Lambda_\pi = \Lambda_\rho$ [GeV]	$\Lambda_\sigma = \Lambda_\omega$ [GeV]	NN(%) D-State	ΔΔ(%) Total	Q_D [fm²]	μ n.m.
9.8	3.89	0.9	1.2	5.48	1.26	0.260	0.876
	5.78	0.9	1.8	5.42	0.34	0.271	0.858
	1.54	1.2	1.2	6.04	0.78	0.270	0.863
13.3	4.89	1.2	1.2	5.86	0.37	0.276	0.856
7.6 (K_ρ = 3.7)	2.77	0.9	1.0	5.85	0.33	0.269	0.857
					Experiment:	0.286	0.858

REFERENCES

1. C.E. Carlson and F. Gross, Phys. Rev. Lett. 53, 127 (1984).
2. R. Dymarz and F.C. Khanna, Phys. Rev. Lett. 56, 1448 (1986).

INTEGRAL ASYMMETRY PARAMETER IN MUON DECAY AND NON-STANDARD WEAK INTERACTION

Ralph D. von Dincklage
Institut für Mittelenergiephysik, Eidgenössische Technische
Hochschule Zürich, CH-5234 Villigen, Switzerland

ABSTRACT

The integral asymmetry parameter ξ of muon decay is being studied in view of possible non-standard contributions to the leptonic weak interaction. Transversely polarized muons were produced from pions decaying in flight. A "longitudinal drift chamber" was developed to measure the muon emission angles with high precision thereby determining the polarization to one part in thousand. Two approaches to observing the muon decay asymmetry have been pursued. The first method relies on muon decay in flight and is estimated to yield an integral asymmetry parameter with uncertainties of a few tenths of a per cent. The second method investigates muons stopped in beryllium metal. Here, the integral asymmetry parameter extracted has the virtue of a higher event rate but is limited by a small unknown depolarization.

INTRODUCTION

Ordinary muon decay $\mu \to e\nu\nu$ continues to be a sensitive testing ground for the theory of the weak interactions.

Assuming the most general, lepton number conserving, derivative free, local, four fermion coupling, the decay $\mu \to e\nu\nu$ has recently been shown [1] as being completely determined. In fact, numerical results have been derived from existing experiments for all ten (complex) coupling constants. Although they are compatible with the V-A form, it turns out that sizeable admixtures of non-standard forms are permitted by current data, namely

(i) "V+A interaction", mediated by an additional vector boson, W_R, which couples right-handed muons to right-handed electrons and (if W_L and W_R mix) muons of given handedness to electrons of the opposite handedness [2,3].

(ii) Scalar-pseudoscalar interaction mediated by a heavy scalar lepton which causes the muon to decay into an electron and two (hitherto unobserved) photini (i.e. the supersymmetric partners of the photons) [4].

(iii) Supersymmetric interaction of muons and electrons with the light scalar partners of the neutrini, mediated by a wino, $\tilde{W}$ [5].

The last example is in-so-far particularly interesting, as it constitutes a decay mode different from the above four fermion ansatz. These "exotic" interactions may be studied via deviations of the muon decay parameters from their V-A predictions. The integral asymmetry parameter, ξ, is very sensitive to such admixtures.

Incidentally, the present value [6] of ξ deviates [7] significantly from its V-A predicted magnitude. As an illustration we note that a 0.1 % measurement centered around the "V-A" value $\xi=1$ has the potential of pushing the lower limit for the masses of the hypothetical particles W_R as well as $\tilde{W}$ to about seven times the one of the "familiar" W_L (i.e. $M(W_R,\tilde{W}) \gtrsim 500$ GeV). Consequently it is timely to attempt a re-measurement of ξ.

At the meson beams of the Swiss Institute of Nuclear Research (SIN), a series of experiments was initiated [8] with the objective to ascertain ξ with about 0.1 % accuracy.

These experiments subdivide themselves into two parts, namely the preparation of a polarized muon beam on one hand and the actual observation of the $\mu \to e\nu\nu$ decay asymmetry on the other hand.

The overall set-up is displayed in Fig. 1.

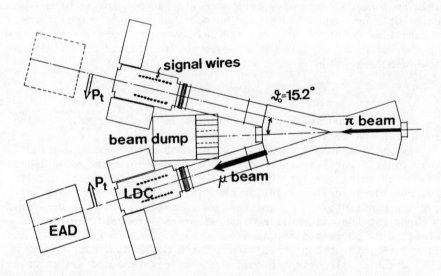

Fig. 1. The overall set-up consists of the X-shaped vacuum tank which is coupled to the longitudinal drift chamber, LDC. This LDC measures the emission angles $\theta \lesssim \theta_0$, of muons from pion decay in flight. From these the transverse polarization, P_t, can be deduced, which is indicated for the case where positive pions are used as primary particles. The muons of interest undergo the $\mu \to e\nu\nu$ decay within the electron asymmetry detector, EAD.

POLARIZED MUONS

A muon beam of high transverse polarization, P_t, was extracted from a 150 MeV/c parallel pion beam. Muons created by pion decay in flight exhibit a sharp cut-off angle, in our case at $\theta_0 = 15.2°$, called the Jacobian edge. It is interesting to note that although for emission angles θ smaller than θ_0 muons of two different energies will contribute, this does not at all impede the experiment since they possess the same transverse polarization P_t, namely

$$P_t = \sin\theta / \sin\theta_0.$$

In addition to this the pion decay probability is strongly enhanced near θ_0 thus making this scheme viable. Numerically we expect an average polarization $<P_t>$ of larger than 97 %, when accepting only muons emitted within 0.020 radians of the Jacobian edge.

Clearly, when striving for a sub-per-cent precision for ξ also P_t must be known to this degree. To this end the individual polarization of muons were measured by comparing their trajectory with the Jacobian edge. For this purpose an especially adapted "longitudinal" drift chamber, LDC, was developed. In this device the muon flight path is parallel to the wire plane (cf. Fig. 1). The final accuracy of $<P_t>$ averaged over the muon sample used thus is better than 0.1 %. The LDC is described in more detail in ref. 9. Figure 2 shows the raw distribution of muons traversing the LDC as a function of their angle θ as analysed with this device. It should be emphasized that for the final analysis software cuts will be applied further reducing the (already) small background.

The measurements used so far positive pions and were performed on each side of the pion beam, that is on each arm of the X-shaped tank of Fig. 1. The observed Jacobian edges emphasize with their mirror symmetry the fact that their spin points always towards the pion axis and so does the transverse polarization, which is also being indicated in Fig. 1.

MUON DECAY IN FLIGHT

Decay electrons were observed in the electron asymmetry detector EAD, which consists of two walls of thin plastic scintillators. These are arranged symmetrically to the left and the right of the θ_0-axis, the Jacobian direction. The decay asymmetry is then assessible via a comparison of the left-right counting rates.

In order to achieve an 0.1 % accuracy in ξ, the geometry of the EAD must be well known. The effect of several small apparative asymmetries was investigated. Their effect can be neglected if measurements with both spin orientations were performed.

For this purpose the EAD is mounted on a movable platform allowing it to be swung from one arm (at +15° relative to the π axis) to the other (at -15° relative to the π axis) thereby reversing

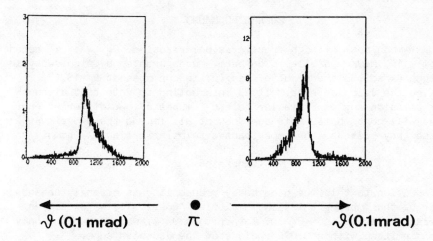

Fig. 2. The muon intensity distribution is a function of emission angle and clearly exhibits the two Jacobian edges (at $\pm\theta_0$) with mirror symmetry relative to the direction of the primary pion axis.

the transverse polarization.

As a further check on potentially remaining apparative asymmetries we also investigated the $\pi \to \mu \to e$ decay of pions at rest, which produces unpolarized muons and hence a symmetric electron distribution.

Furthermore, the apparative asymmetry was carefully measured with a radioactive ^{106}Ru beta decay source, and found to be smaller than 0.5 per cent.

During two weeks of running time about two millions of $\mu^+ \to e^+ + \nu_e + \bar{\nu}_\mu$ decays were detected by the EAD.

The analysis of the data, which includes a large amount of further information from various veto counters and MWPC's has begun. An accuracy of this first experiment in the regime of 0.4...0.7 % is expected.

MUON DECAY AT REST

As an alternative to the above scheme we stopped the polarized muons in 1 cm thick Beryllium metal, for which it is reasonable to expect only a small depolarization [10].

The Be (as well as the counters T_2 and A) were placed inside a 32 G solenoid, thus giving rise to muon spin rotation μSR. The arrangement is shown schematically in Fig. 3.

The two plastic scintillators, T_2 and A sandwich the Be stopper and signal a valid muon-stop, if the telescope T_1T_2 but not the veto counter A detect a traversing muon. The moderator M decelerates the muons in order to optimize their range. The decay electrons finally are identified by the threefold telescope $T_2Z_1Z_2$.

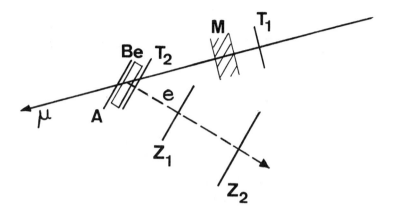

Fig. 3. Set-up for the detection of the decay electron asymmetry via the μSR method. Muons of interest are slowed down in the Aℓ moderator M and stopped in the Be plate. Their electronic signature is $T_1 \cdot T_2 \cdot \bar{A}$. Decay electrons are observed in the telescope $T_2 Z_1 Z_2$. All detectors T_1, T_2, Z_1, Z_2 and A, are plastic scintillators. The solenoid which surrounds T_2 Be A and produces a vertical magnetic field ±B is not shown for clarity.

For moderator thicknesses ranging from 10 to 40 mm Aℓ, muon spin rotation spectra were recorded. In fact, for each case two spectra were accumulated corresponding to both orientations ±B of the vertical magnetic field B.

A typical (M=27 mm Aℓ) pair of spectra is shown in Fig. 4. The difference in shape between a and b near time=0 is clearly seen and reflects the fact that the sense of rotation is reversed by changing from +B to -B, thereby changing the μSR phase.

The analysis of the data is still in progress.

DISCUSSION

The μSR approach to detecting the decay electron asymmetry benefits from the simplicity of the detector system. No apparative asymmetries play a role. On the other hand, the accuracy of any ξ parameter to be extracted is limited by the material's depolarization. Although this quantity is believed to be small, a 0.1 % precision will be difficult to obtain.

Utilizing muon decay in flight is free of these shortcomings but the decay rates are reduced drastically. The intrinsic difficulty of apparative asymmetries was overcome by periodically changing the direction of the muon spin relative to the apparatus. Thus systematic uncertainties on ξ are estimated to be negligible down to the 100 parts per million regime.

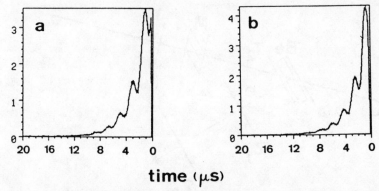

time (μs)

Fig. 4. Two μSR-type spectra show the electron rate observed with the set-up of Fig. 3 as a function of time. The spectra a and b differ in their phase, corresponding to the reversal of the direction of the vertical magnetic field.

In conclusion, one may expect new data on the integral asymmetry parameter ξ of μ decay in the near future with overall uncertainties well below one per cent. While the two proposed detection schemes will contribute at this level of precision, only by utilizing muon decay in flight the goal of one part in thousand accuracy will be reached.

REFERENCES

1. W. Fetscher, H.-J. Gerber and K.F. Johnson, Phys. Lett. 173B, 102 (1986).
2. M.A.B. Beg, R.V. Budny, R.N. Mohapatra, A. Sirlin, Phys. Rev. Lett. 38, 1252 (1979).
3. F. Scheck, Leptons, Hadrons and Nuclei, North-Holland, Amsterdam (1983).
4. S. Barber and E. Shrock, Phys. Lett. 139B, 427 (1984).
5. W. Buchmüller and F. Scheck, Phys. Lett. 145B, 421 (1984).
6. V.V. Akhmanov, I.I. Gurevich, Yu.P. Dobretsov, L.A. Makariyna, A.P. Mishakova, B.A. Nikolskii, B.V. Sokolov, L.V. Surkova, V.D. Shestakov, Sov. J. Nucl. Phys. 6, 230 (1968).
7. "Review of particle properties", Particle Data Group, Rev. Mod. Phys. 56, 2, II (1984).
8. I. Beltrami, H. Burkard, R.-D. von Dincklage, W. Fetscher, H.-J. Gerber, K.F. Johnson, E. Pedroni, M. Salzmann, F. Scheck, E. Ungricht and V. Zacek, "Measurement of the ξ parameter in μ decay", SIN-proposal (1983), R-83-29.
9. C. Witzig, Diplomarbeit, IMP der ETH Zürich (1986), unpublished.
10. J. Carr, G. Gidal, B. Gobbi, A. Jodidio, C.J. Oram, K.A. Shinsky, H.M. Steiner, D.P. Stocker, M. Strovink, R.D. Tripp, Phys. Rev. Lett. 51, 627 (1983).

PRESENT STATUS OF QED

J Sapirstein
Department of Physics
University of Notre Dame
Notre Dame, IN 46556

The range of processes involving Quantum Electrodynamics is extremely large, in principle covering the whole of atomic and molecular physics, as well as scattering processes involving leptons. In this brief review this range will be restricted to certain QED tests with a bearing on the subject of this conference, the intersection of particle and nuclear physics. There are three aspects of QED that are of particular interest in this regard:

1) Regardless of any direct applicablility of QED to the study of the strong, weak, and gravitational interactions, it always serves as an example of what a successful theory should be. The basic framework of the theory, the relativistic quantum mechanics of fields, is shared with the unified theories of weak and electromagnetic interactions and the presently accepted underlying theory of the strong interactions, QCD. Thus it is of intrinsic interest to find out just how successful such a theory can be when applied to systems in which the dominant interactions are electromagnetic. The agreement of theory and experiment at the part per million(ppm) level to be detailed below simultaneously tests a variety of field theory effects such as vacuum polarization, self energy corrections, and the intricate renormalization scheme that isolates finite answers from divergent graphs.

2) Once the validity of the theory has been established at a certain level of confidence, one can turn the process around, and use precision QED tests as a sort of microscope. It is very probable that there is interesting structure in

the mass range of a few Tev connected with a more basic theory of spontaneous symmetry breaking than the simple Higgs model. One important approach to study this mass regime is the construction of the SSC to directly produce any such high mass particles. Very massive particles however also show up already at low energies if they have any couplings to leptons by producing very slight shifts in physical quantities. In the particularly simple systems used for QED tests, such tiny effects can in principle be seen, once the understood QED effects have been accounted for. The hoped for situation is that eventually some real discrepancy will be seen in one of these tests that cannot be accounted for with any known interaction: then one will have discovered new physics just as much as if a previously unknown particle is directly produced at the SSC.

3) A final point of contact with the subject of this conference, which is primarily concerned with the strong interactions, is the merging of QED tests with the subject of the electromagnetic interactions of hadrons. Particularly in tests in hydrogen, where the electromagnetic structure of the proton enters directly, but also in purely leptonic processes, where vacuum polarization graphs involving hadrons cannot be avoided, at some level of precision the fact that an accurate understanding of this feature of the strong interactions is needed cannot be avoided. I will try to highlight some features of this problem in the following, and it will be seen that some tests actually become an unusual probe of strong interactions instead of a pure test of QED.

The highest development of QED is in the magnetic anomaly of the electron, both theoretically because the highest number of loops have been calculated, and experimentally, due to the great precision that a fundamental property of this elementary particle has been measured. The beautiful Penning trap experiment of Van Dyck, Schwinberg, and Dehmelt[1] gives the result

$$a_e = 1\ 159\ 652\ 193(4) \cdot 10^{-12} \tag{1}$$

An important recent development related to this quantity has been the discovery that the standard ohm used by the NBS was drifting: after correcting for this fact the Josephson junction determination of the fine structure constant is[2] $\alpha_J^{-1} = 137.035981(12)$. With this value the one loop contribution to the electron anomaly, $\frac{\alpha}{2\pi}$, is 1515.707ppm higher than experiment, with an error due to uncertainty in α of .090ppm. Addition of the two loop correction, $-.328478966(\frac{\alpha}{\pi})^2$ brings theory 12.601ppm below experiment. The three loop contribution of $1.1765(13)(\frac{\alpha}{\pi})^3$ improves agreement to .114ppm above, and finally the four loop calculation of Kinoshita and Lindquist[3], $-.8(1.4)(\frac{\alpha}{\pi})^4$ takes theory to .094ppm above experiment. Given the uncertainty associated with α, one can conclude that theory and experiment agree to within one standard deviation: the previous value of α was giving a two standard deviation disagreement. Several comments are in order about this very precise QED test: firstly, the main bottleneck for future progress is the relative imprecision in the solid state determination of α: this will hopefully stimulate further work on this difficult metrological problem. Secondly, Brown et.al.[4] have made calculations of the effect of the finite size of the Penning trap on the electron energy levels, and have found that considerable structure can be present that can complicate the extraction of a_e: such structure has not yet been found. The agreement between theory and experiment has ruled out countless models involving exotic couplings of the electron. The sheer size of the calculations is also worth noting: completion of the four loop calculation is a sort of existence proof that given present computer capablities that extremely large scale calculations can be carried out. Finally, the problem of solid state uncertainties in α can be finessed by assuming that QED is working correctly and inferring α^{-1}_{g-2} = 137.035 993(7) . In order to check the internal consistency of QED another system must be found with comparable accuracy. The next best test of QED is in ground state muonium hyperfine splitting(hfs).

In this system theoretical progress has been stimulated by the very accurate measurement[5]

$$\Delta v(exp) = 4\ 463\ 302.88(16)\ kHz \qquad (2)$$

With the use of the new fine structure constant the Fermi splitting is E_F= 4 459 033.4(1.5) kHz, 956.574ppm below experiment. Addition of the electron anomaly adds in a term $a_e E_F$, which takes theory 201.969ppm above experiment. The long established terms of order $\alpha^2 E_F$ and $\alpha\ m_e/m_\mu\ E_F$ reduce this disagreement to 6.121ppm above, and the more recently completed $\alpha^3\ E_F$ and $\alpha^2 m_e/m_\mu\ E_F$ calculations reduce it further to only .072ppm above experiment. It is worth noting that the perturbation expansion in the atomic case has a different kind of difficulty than that encountered in anomaly calculations: a significant problem is that often exchange of Coulomb photons must be considered to all orders, so infinite sets of diagrams must be considered. The main bottleneck to progress in this system is the .336ppm error in E_F arising from the uncertainty in the muon magnetic moment. Given a possible .2ppm contribution from binding corrections to the two loop self energy that are as yet uncalculated, one can infer α^{-1}_{hfs} = 137 .035 991 (20) (15), with the first error experimental and the second theoretical: while this is consistent with the g-2 determination, more experimental and theoretical work is clearly needed.

A third 'QED theory' value of α^{-1} has been available for some time from analysis of the triplet P splitting in Helium: the 2^3P_0-2^3P_1 splitting, v_{01}, when compared with theory[6] gives α_{He}^{-1} = 137. 036 080 (130), consistent with the previous determinations, but an order of magnitude less precise. A more accurate experimental measurement would be a major stimulus to understanding the QED of two electron systems. To get to the level of $m\alpha^7$ will probably require a major streamlining of the previous massive calculations or possibly an entirely different approach. In addition, the methods for calculating recoil corrections to the two body problem, based on the Bethe-Salpeter equation or extensions of it may have to be extended to the three body problem to determine radiative corrections to recoil effects.

Strong interactions affect the above QED tests only very weakly. To see how they can enter even in a purely leptonic system, consider the g-2 of the muon. The experimental measurement[7], a_μ = 11 659 110(110) · 10^{-10} is discrepant with $\alpha/2\pi$ by -3680.598ppm. The two loop effect, .76578171$(\alpha/\pi)^2$ reduces the discrepancy to -316.760ppm, the three loop term, 24.45(6)$(\alpha/\pi)^3$ to -53.961ppm, but the recently calculated 135(63)$(\alpha/\pi)^4$ still leaves theory 50.616ppm below experiment. This disagreement marks the discovery of a 'new' interaction, though of course the strong interactions, which are entering in vacuum polarization loops, were encountered long before. Put in another way, understanding of the muon anomaly at the 50ppm level is equivalent to a probe of structure at the 1Gev level. The hadronic vacuum polarization is given by

$$a_\mu(\text{had}) = (\frac{\alpha m_\mu}{3\pi})^2 \int_{4m_\pi^2}^{\infty} ds/s^2 \; K(s) \; R(s) \quad (3)$$

where K is a kinematical function and R is proportional to the total cross section for e^+e^- annihilation into hadrons. A recent analysis[8] gives a_μ(had) = 702(19)·10^{-10}, or 60.210(1630)ppm, bringing theory and experiment into one standard deviation agreement. The experimental error of 9.435ppm clearly should be reduced, as it is completely obscuring the interesting weak interaction contribution, which enters at the 1.715ppm level. In this sense, getting to 1ppm precision in this system is equivalent to a probe of 100GeV structure. It is important to note, however, that the uncertainty in hadron vacuum polarization is at the same level as the weak interaction effect: therefore for progress in QED to continue for the muon anomaly, progress must be made simultaneously in our understanding of the electromagnetic properties of strongly interacting particles, preferably both experimentally and theoretically.

A similar situation obtains for the Lamb shift. The calculation of terms of order α^3 Ry is 6886.642ppm below the measurement of Lundeen and Pipkin[9]

$$\mathcal{S} = 1057.845(9) \text{ MHz} \qquad (4)$$

While the α^4 Ry calculation takes theory to 147.470ppm below the measurement, the α^5 Ry and $\alpha^3 m_e/m_p$ Ry terms leaves theory 109.660 ppm below. While there are several recoil and two non-recoil QED calculations not yet performed, they should be smaller than 20ppm, so again new physics is being probed, in this case basically the finite size of the proton. A measurement at Mainz[10] of the proton rms radius gives a 137.181ppm effect, in contrast to a 120.151ppm effect from the older Stanford[11] measurement. This leads to the interesting possibility that once the calculations mentioned above are performed, that QED can determine which experiment is correct. However, unless the electron scattering experiments are sorted out, it is clear that advancing the calculations to the few ppm level is not going to provide a sensitive QED test, but rather will be measuring the proton electromagnetic charge radius.

In contrast to the sensitive muonium hfs QED test, ground state hydrogen hfs is a poor QED test because of proton structure uncertainty. Terms of order $\alpha\, m_e/m_\mu\, \ln(m_\mu/m_e)\, E_F$ in muonium that are precisely calculable for the pointlike muon are sensitively dependent on proton structure in hydrogen. Use of experimental form factors bring theory and experiment to within 1.6ppm, with experimental uncertainty at the .9ppm level. There remains a contribution from the same effect where the proton is excited into different states, referred to as the proton polarizablity term. It's value can be inferred as 1.6(9)ppm if one ignores other QED effects and requires theory and experiment to agree: it would be of interest if this property of the proton could also be predicted from some model of hadron structure. Again, as with the g-2 of the muon and the Lamb shift, QED tests are beginning to merge with the study of the electromagnetic structure of hadrons.

While all the QED tests so far discussed are working well, there is one system that may be indicating an unexpected breakdown. The decay rate of orthopositronium has been measured[12] to be $\Gamma = 7.051(5)$ μsec^{-1}. The lowest order calculation of Γ_0 is 22692 ppm above this rate. The large first order correction, $\alpha/\pi\ \Gamma_0$ (-10.282(3)) takes theory 1701.9ppm below experiment. The problem is that, after accounting for a known $\alpha^2 \ln \alpha\ \Gamma_0$ term, that a very large coefficient of 34(13) in the uncalculated $\alpha^2\ \Gamma_0$ term is needed to bring theory and experiment into agreement. Such a large coefficient would be surprising: if the experimental error is reduced enough to indicate a many standard deviation discrepancy, this large scale calculation will have to be undertaken. If a smaller coefficient is found, the resulting discrepancy may mark the first truly unexpected new physics revealed in a QED test, which is one of the fundamental goals of the field.

REFERENCES

1. R.S. Van Dyck, Jr., P.B. Schwinberg, and H.G. Dehmelt, Atomic Physics 9, ed. by R.S. Van Dyck, Jr. and E.N. Fortson, (World Scientific Pub. Co. , Singapore, 1984) p. 53
2. B.N. Taylor, Journal of Research of the National Bureau of Standards 90, #2, 91 (1985)
3. T. Kinoshita and W.B. Lindquist, Phys. Rev. Lett. 47, 1679 (1981)
4. L. Brown, G. Gabrielse, K. Helmerson, and J. Tan, University of Washington preprint 40048-05, 1985
5. F.G. Mariam et.al., Phys. Rev. Lett. 49, 993 (1982)
6. A. Kponou, V.W. Hughes, C.E. Johnson, S.A. Lewis, and F.M.J. Pichanick, Phys. Rev. A24, 264, (1981)
7. J. Bailey et. al., Nucl. Phys. B150, 1 (1979)
8. T. Kinoshita, B. Nizic and Y. Okamoto, Phys. Rev. Lett. 52, 717 (1984)
9. S.R. Lundeen and F.M. Pipkin, Phys. Rev. Lett. 46, 232 (1981)
10. G.G. Simon, Ch. Schmitt, F. Borkowski, and V.H. Walther, Nucl. Phys. A333, 381 (1980)
11 D.J. Dirckey and L.N. Hand, Phys. Rev. Lett. 9, 521 (1962); L.N. Hand, D.J. Miller, and R. Wilson, Rev. Mod. Phys. 35, 335, (1963)
12. D.W. Gidley, A. Rich, E. Sweetman, and D. West, Phys. Rev. Lett. 49, 525 (1982)

THE LAMPF WORKSHOP ON FUNDAMENTAL MUON PHYSICS

Vernon W. Hughes
J. W. Gibbs Laboratory, Physics Department
Yale University, New Haven, CT 06520

ABSTRACT

A Workshop on Fundamental Muon Physics: Atoms, Nuclei and Particles was held at LAMPF from January 20-22, 1986 with 85 physicists from Europe, Japan, Canada and the United States in attendance.[1] Selected topics were examined where there is lively current interest and prospect for future advances, and included muon decay, muon capture, QED and electroweak interactions, laser spectroscopy of muonic atoms, high-energy muon-nucleon and muon-nucleus scattering, muon beams - new developments, and muon catalysis (see Table I). In this short review only the briefest comments on a few selected matters can be made.

I. MUON DECAY AND MUON CAPTURE

There is a large effort in progress to improve our understanding of muon decay and muon capture.

For muon decay, which is one of the simplest and most fundamental of the weak interaction processes, major experiments have been completed recently and others are in progress both on the dominant normal decay mode, $\mu^+ \rightarrow e^+ \nu_e \bar{\nu}_\mu$, and on rare decay modes.

For the normal muon decay mode and its inverse an adequate number of different measurements have been made by now[2] to determine the 19 parameters of the most general four-Fermion interaction which is local, derivative-free and lepton-number conserving. The results of a complete statistical analysis of these data indicate that the results are consistent with the prediction of the GSW electroweak theory which gives the values $\rho = \delta = 3/4$, $\eta = 0$ and $\varepsilon = 1$. A recently completed precise measurement by Möller scattering of the polarization of e^+ from μ^+ decay done at SIN was reported[2] (Fig. 1); further improvements in precision seem possible. Data-taking for a high statistics measurement using a TPC detector of the ρ parameter in normal μ^+ decay has been completed at LAMPF (Fig. 2) and 4×10^7 events are available for analysis.[3]

The desperate search continues for rare muon decays which would violate muon number conservation, driven by the general faith that the lepton generations should communicate but unenlightened by any believable theoretical estimate.[4] A major new experiment is being developed at SIN[5] to search for the $\mu^- Z \rightarrow e^- Z$ decay (Fig. 3) to a BR of 10^{-13} relative to μ^- capture and another at LAMPF[6] to search for $\mu \rightarrow e\gamma$ (Fig. 4), also to a BR of 10^{-13}.

Searches for the rare process muonium → antimuonium are being actively pursued at LAMPF, SIN, and TRIUMF with current sensitivity goals of $G_{M\bar{M}} \sim G_F$ or less.[7]

The basic muon capture reaction is that in hydrogen, $\mu^- + p \rightarrow n + \nu_\mu$, from which the weak form factors at low energy are determined, in particular the axial form factor f_A which is used to test PCAC.[8] Some of the better recent data come from experiments at Saclay in which the time distribution of capture neutrons in liquid hydrogen is measured. The accuracy in determining f_A is at the 10% level. An experiment at the 1% level based on a high μ^- stopping rate in low pressure gaseous H_2 where molecular complications could be avoided would be very valuable, but would require further development of low momentum μ^- beams with high density in phase space.

II. QED AND ELECTROWEAK INTERACTIONS

Muonium and simple muonic atoms provide important systems for testing QED and the behavior of the muon as a heavy electron, and potentially also for studying the electroweak interaction as well. The ultrahigh precision which has been attained in hydrogen spectroscopy, particularly recently by the methods of laser spectroscopy, provide a challenging goal for muon-atom spectroscopy.[9]

For muonium (M)[7] the hyperfine structure and Zeeman energy levels in the ground n = 1 state are shown in Fig. 5a and the energy levels of the n = 1 and n = 2 states in Fig. 5b. For the n = 1 state the hfs interval $\Delta\nu$ and the ratio of muon to proton magnetic moments, μ_μ/μ_p, have been measured with precisions of 36 ppb and 0.36 ppm, respectively. The agreement of theoretical and experimental values of $\Delta\nu$ to 0.4 ppm constitutes one of the most sensitive tests of QED and of the behavior of the muon as a heavy electron. The most precise value for μ_μ and for the muon mass m_μ are derived from this Zeeman effect measurement in the M ground state. With the much improved intensity and purity of the presently available LAMPF μ^+ beam and with the possibility of resonance line-narrowing with a bunched μ^+ beam, it should be possible to improve the precision in $\Delta\nu$ and in μ_μ/μ_p by a factor of 5 to 10.

The Lamb shift transition $2^2S_{1/2} \to 2^2P_{1/2}$ has been observed and the interval has been measured to about 1%. Although the signal intensity is small, an experiment is in progress at LAMPF to improve the precision. If measured with sufficient accuracy, the Lamb shift in muonium will provide a cleaner test of QED than in hydrogen because proton structure contributes to the Lamb shift in H.

One of the more exciting features of the Workshop was the announcement from both KEK[10] and TRIUMF[11] of the production of thermal muonium into vacuum. A surface μ^+ beam of about 25 MeV/c was used in both experiments and copious emission of thermal M downstream of the target foil was observed. In the KEK experiment (Fig. 6) the target was a tungsten foil heated to 2000K and the fraction M/μ^+ was 0.04 with M at a temperature T of 2000K, and at TRIUMF (Fig. 7) the target was SiO_2 at room temperature and the fraction M/μ^+ was 0.20 with M at room temperature. Muonium was identified downstream of the target and its velocity measured by observing the origin in position and time of the decay e^+ from M. A copious source of thermal M should make possible several fundamental measurements on M: (1) the 1S → 2S transition by laser spectroscopy; (2) a more sensitive search for the spontaneous conversion of M to $\bar{M}$.

Using crystal diffraction spectrometers, measurements of transitions in medium Z muonic atoms have attained precisions in the ppm range. These results provide excellent confirmation of muon electrodynamics and sensitive limits to anomalous muon-nucleus interactions.[12]

TABLE I

LA-10714-C
Conference

UC-34
Issued: May 1986

Proceedings of the Workshop on Fundamental
Muon Physics: Atoms, Nuclei, and Particles

Held at Los Alamos National Laboratory
Los Alamos, New Mexico
January 20-22, 1986

Compiled by
Cyrus M. Hoffman
Vernon W. Hughes*
Melvin Leon

*Department of Physics, Yale University, New Haven, CT 06511.

Los Alamos National Laboratory
Los Alamos, New Mexico 87545

TABLE OF CONTENTS

PREFACE...vii
ABSTRACT..1

INVITED PAPERS

SESSION I. MUON DECAY

Herbert Anderson, Los Alamos National Laboratory
   Introduction..2

Howard Georgi - Harvard University
   Theory of Rare Muon Decay.............................4

Hans Juergen Gerber - ETH/SIN
   Positron Polarization in Muon Decay
   and General Analysis..................................8

Andreas Badertscher - SIN
   Search for μe Conversion in Ti.......................14

Douglas Bryman - TRIUMF
   Experiments on Muon Rare Decay Modes at TRIUMF.......24

Martin Cooper - Los Alamos National Laboratory
   LAMPF Experiments Searching for the Rare
   Muon Decay $\mu^+ \to e^+\gamma$.........................27

SESSION II. MUON CAPTURE

Masato Morita - Osaka University
   Muon Capture in Light Nuclei and Induced Pseudoscalar
   Interaction..34

Emilio Zavattini - CERN
   A Comment on Muon Capture in Hydrogen................39

Jules Deutsch - University of Louvain
   Muon Capture in Nuclei: The Case of ^{3}He...........47

SESSION III. QED AND ELECTROWEAK INTERACTIONS

Jonathan Sapirstein - Notre Dame University
   QED and Exotic Atoms.................................56

Herbert Orth - Yale University
   Muonium..62

Hans Jorg Leisi - ETH/SIN
   Muonic X-Rays with Crystal Diffraction Spectrometry..75

John Missimer - SIN
   Electroweak Effects in Muonic Atoms..................81

Vernon W. Hughes - Yale University
   The Muon Anomalous g-Value...........................87

SESSION IV. LASER SPECTROSCOPY OF MUONIC ATOMS

Theodor Hänsch - Stanford University
   High Resolution Laser Spectroscopy of Atomic Hydrogen:
   Advances and Prospects...............................99

Franz Kottman - SIN
   Lambshift Measurements in μp and μHe Atoms and the
   μHe-2S-Lifetime.....................................103

SESSION V. HIGH-ENERGY MUON-NUCLEON AND MUON-NUCLEUS SCATTERING

Edmond Berger - Argonne National Laboratory
   Deep Inelastic Lepton Scattering from Nucleon
   and Nuclei..109

K. Peter Schüler - Yale University
   Deep Inelastic Muon Scattering - Experiment.........132

SESSION VI MUON BEAMS-NEW DEVELOPMENTS

Claude Petitjean - SIN
   Introduction..138

David Taqqu - SIN
   Phase Space Compression of Muon Beams...............141

Leopold Simons - SIN
   On the Use of A Cyclotron Trap......................147

SESSION VII. MUON CATALYSIS

Melvin Leon - Los Alamos National Laboratory
   Brief Summary of the Theory of Muon-Catalyzed Fusion.....151

Steven Jones - Brigham Young University
   Muon-Induced Fusion - Experiments at LAMPF..........157

Wolfgang Breunlich - Austrian Academy of Sciences
   Muon Catalyzed Fusion...............................165

CONTRIBUTED PAPERS

SESSION I. MUON DECAY

V. Wayne Kinnison - Los Alamos National Laboratory
   A High-Statistics Normal Muon Decay Experiment......177

SESSION II. MUON CAPTURE

Georges Azuelos - TRIUMF
   Radiative Muon Capture on Hydrogen..................180

Wolfgang Breunlich - Austrian Academy of Sciences
   Measurement of Nuclear Muon Capture in Deuterium....182

Antonio Bertin - University of Bologna
   Muon Capture in Deuterium: Present Status
   and Future Work.....................................184

Antonio Bertin - University of Bologna
   Muonic Hydrogen and Deuterium:
   Future Work on Selected Topics......................191

Edward McIntyre - Boston University
   Radiative Muon Capture in ^{3}He....................198

Jules G. Deutsch - University of Louvain
   A Comment on T-Violating Triple-Correlations
   in Muon-Capture.....................................201

Kanetada Nagamine - University of Tokyo
   Negative Muon Repolarization with Polarized
   Nuclear Targets.....................................207

SESSION III. QED AND ELECTROWEAK INTERACTIONS

Robert T. Siegel - College of William and Mary
   Metastability of the 2S State of ^{4}He$^-$.........212

SESSION IV. LASER SPECTROSCOPY OF MUONIC ATOMS

Morgan May - Brookhaven National Laboratory
   3d - 3p Transitions in (μ$^-$Be$^+$)$^-$............213

Kanetada Nagamine - University of Tokyo
   Thermal Muonium in Vacuum:
   Its Birth and Future................................219

Glen Marshall - TRIUMF
   Production of Muonium in Vacuum
   from Silica Powders.................................225

SESSION VI. MUON BEAMS-NEW DEVELOPMENTS

Michael A. Paciotti - Los Alamos National Laboratory
   Degraded Muon Beams at LAMPF........................228

Michael A. Paciotti - Los Alamos National Laboratory
   A Muon Facility at Biomed...........................231

CHARACTERISTICS OF VARIOUS STOPPED MUON BEAMS............................234
PROGRAM..235
LIST OF ATTENDEES..238

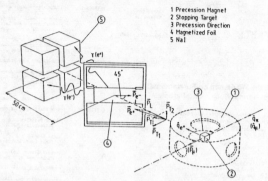

Fig. 1. Apparatus to measure the polarization components P_L, P_{T1} and P_{T2} of the positrons from the decay of polarized and unpolarized (positive) muons.

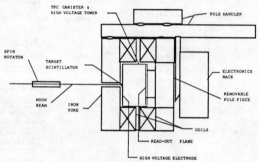

Fig. 2. A schematic view of the TPC detector apparatus for the High Statistics Normal Muon Decay experiment at LAMPF.

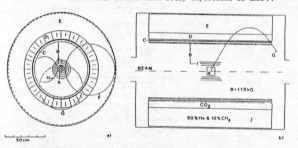

Fig. 3. Front view a) and side view b) of the new μ-e conversion setup. A: Target (e.g. Ti); B: MWPC with cathode strip readout; C: Scintillator hodoscope; D: Drift chamber serving as a 180° spectrometer for a 100 MeV electron; E: He drift chamber; F: 105 MeV track θ = 90° (θ: angle to beam axis); G: 105 MeV track, θ = 60°; H: 53 MeV track, θ = 90°.

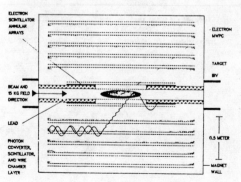

Fig. 4. A sectioned plan view of the MEGA apparatus with an idealized $\mu^+ \to e^+ \gamma$ event shown.

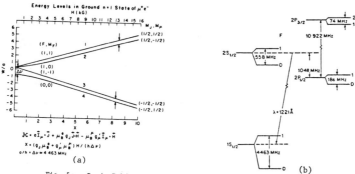

Fig. 5a. Breit-Rabi energy level diagram for the ground state of muonium.
5b. Energy level diagram of the n=1 and n=2 states of muonium.

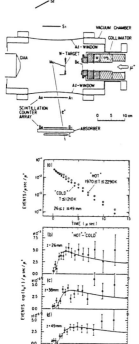

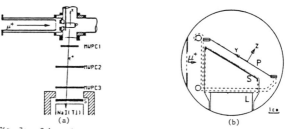

Fig. 6. (A) Apparatus used for the first successful observation of muonium in vacuum (left); A_i, B_i, C and S_i are scintillation counters and CMA is the Auger spectrometer. (B) Time spectra of μ^+ decay events (right); (a) raw spectra binned and normalized, obtained with the W target hot (between 1970 and 2290 K) and cold (below 1210 K); (b-d) "hot" - "cold" difference spectra for **the Z positions of the** decaying Mu from the W-target.

Fig. 7a. Schematic of experimental apparatus. **The powder target is** indicated by P, and S is a stack of three scintillators.
7b. Target region viewed by the wire chambers. The powder layer (P), the thin scintillator (S), and its lucite light guide (L) are indicated, as are the coordinate axes.

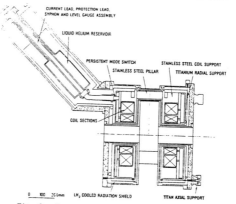

Fig. 8. Superconducting split coil magnet.

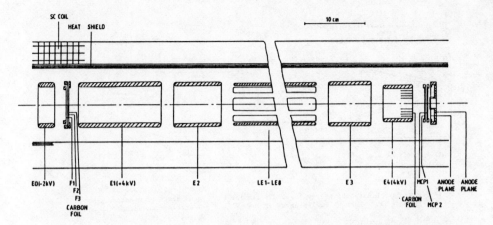

Fig. 9. A method of phase space compression of a low momentum muon beam involving observation of a muon's position and velocity with detectors based on secondary electron emission from C foils and subsequent deceleration and steering with electric fields.

Laser spectroscopy of muonic atoms still remains an exciting and potentially fertile but unplowed field. The only successful experiment to date was the pioneering measurement of the fine structure and Lamb shift in the muonic helium ion, $(^4He\mu^-)^+$ in the n = 2 state by Zavattini et al.[13] Furthermore, there now exists some controversy and puzzle about the lifetime of the metastable $2^2S_{1/2}$ state at the high pressure at which the measurement was made.[14] A long list can easily be made of fundamental measurements of hyperfine structure, fine structure, and Lamb shift intervals in simple muonic atoms including μ^-p, μ^-d, $^4He\mu^-$, and $^3He\mu^-$ which might be measured by laser spectroscopy. The field is certain to blossom when adequate sources of high density μ^- beams are developed.

Although electroweak interference effects are relatively large in muonic atoms as compared to electronic atoms, no observation of an electroweak effect in a muonic atom has yet been made.[15] Again we can expect some successful measurements when more intense, high-density muon beams become available.

III. NEW DEVELOPMENTS IN MUON BEAMS

As indicated above several times, the development of higher intensity muon beams (both μ^+ and μ^-), which will provide a large number of muon atoms at low gas pressure, should extend most significantly possibilities for fundamental experiments on muon atoms. Two of the most interesting developments are in progress at SIN.

One of these involves deceleration in a cyclotron trap[16] (Fig. 8). Muons (or pions) with momenta up to 123 MeV/c can be decelerated in the weak-focusing cyclotron field to obtain stopping densities some factor of 10^2 higher than with ordinary muon beams. Tests of this approach have been made at LEAR with encouraging results.

The second approach is that of phase space compression[17] and is indicated in Fig. 9. The positions and velocities of individual muons are measured and then altered by pulsed electric fields to compress the phase space of the beam. Large compression factors are projected, but due to the pulsing requirements the number of muons that can be handled is limited.

Supported in part by the U.S. Department of Energy.

REFERENCES

1. V. W. Hughes, C. M. Hoffman, and M. Leon, ed., "Proceedings of the Workshop on Fundamental Muon Physics: Atoms, Nuclei and Particles," LA-1074-C (May, 1986).
2. H.-J. Gerber, ibid., p. 8.
3. W. Kinnison, ibid., p. 177.
4. H. Georgi, ibid., p. 4.
5. A. Badertscher, ibid., p. 14.
6. M. Cooper, ibid., p. 27.
7. H. Orth, V. W. Hughes, ibid., p. 62.
8. E. Zavattini, ibid., p. 39.
9. T. Hänsch, ibid., p. 99.
10. K. Nagamine, ibid., p. 219.
11. G. Marshall, ibid., p. 225.
12. H. J. Leisi, ibid., p. 75.
13. A. Bertin et al., Phys. Lett. 55B, 411 (1975).
14. R. T. Siegel, ibid., p. 212.
15. J. Missimer, ibid., p. 81.
16. L. Simons, ibid., p. 147.
17. D. Taqqu, ibid., p. 141.

SUPERSYMMETRY CORRECTIONS TO MUON g-2

R. Arnowitt* and Pran Nath
Northeastern University, Boston MA 02115

ABSTRACT

Supersymmetry corrections to the muon g-2 in N=1 supergravity models are comparable to and generally larger than Standard Model loop corrections due to the fact that Supergravity unified models predict the existence of relatively low-lying superparticles. A discussion is given of the present status of these corrections using current limits on superparticle masses obtained from accelerator experiments.

1. INTRODUCTION

The muon anomalous magnetic moment $g_\mu \equiv 2(1+a_\mu)$ is one of the most accurately determined quantities in physics. The current experimental value from the CERN muon storage ring experiment is[1]

$$a_\mu^{exp} = 1,165,924(8.5) \times 10^{-9} \quad . \tag{1}$$

Theoretical calculations by Kinoshita et al.,[2] yield the value of

$$a_\mu^{th} = 1,165,920.2(2.0) \times 10^{-9} \tag{2}$$

and include QED corrections through α^4 order, the large hadronic contribution, as well as the Weinberg-Salam electroweak contributions:

$$a_\mu^{had} = (70.2 \pm 1.9) \times 10^{-9} \quad ; \quad a_\mu^{WS} = (1.95 \pm 0.01) \times 10^{-9} \tag{3}$$

The possibility exists, however, that new measurements of $e^+e^- \to$ hadrons (e.g., at Novosibirsk) could reduce the error in a_μ^{had} by a factor of $\simeq$ 5-6 and that the error in a_μ^{exp} can be reduced by a factor of $\simeq$ 20, allowing a measurement of the electroweak loop correction.[1]

However, supersymmetry contributions are comparable and often larger than the Standard Model corrections. Thus any experiment designed to test the Weinberg-Salam theory will also test supersymmetry. Further, as will be discussed below, recent accelerator experiments has significantly limited the supersymmetric mass spectrum, thus allowing a more specific theoretical prediction for the supersymmetry contribution.

2. N=1 SUPERGRAVITY MODELS

We limit the discussion to N=1 Supergravity unified models.[3]

---

* Speaker at Conference.

Model dependence arises due to the existence of different mechanisms for SU(2)×U(1) breaking. Since supersymmetric models must possess at least two Higgs doublets, H_α and H'_α, whose VEVs contribute to SU(2)×U(1) breaking, one defines a Higgs mixing angle α_H by

$$\tan \alpha_H = \langle H'_0 \rangle / \langle H_0 \rangle \qquad (4)$$

Models then fall into two classes depending on the size of α_H:
 (1) $\alpha_H \cong 45°$: SU(2)×U(1) breaking occurs at the tree level (T.B. models) or via dimensional transmutation (D.T. models).
 (2) α_H Small ($\lesssim 15°$): SU(2)×U(1) breaking occurs via renormalization group improvement in going from M_{GUT} to M_W (R.G. models).

Associated with each normal particle in a SUSY theory is a super partner. For light photinos, a general result holds:[4]
The $\tilde{W}_{(-)}$ always lies below the W mass and at least one $\tilde{Z}$ lies below the Z mass. (In almost all models the photino is light i.e., $m_{\tilde\gamma} \cong 1$-15 GeV.) Other sparticles have masses of $O(M_W)$, but their masses are not explicitly determined. In general, an experimental determination of $\tilde{m}_-$ (the mass of $\tilde{W}_{(-)}$), $m_{\tilde\nu}$, $m_{\tilde\gamma}$ (and/or the gluino mass $m_{\tilde g}$) and the Polonyi constant A (which is restricted theoretically by $|A| \lesssim 3$) will uniquely determine all masses and coupling constants. The low-lying nature of the superparticle spectrum is what makes Supergravity models phenomenologically interesting. It is also what makes the supersymmetric contributions to a_μ relatively large.

3. SUPERSYMMETRIC CORRECTIONS TO a_μ

The supersymmetric corrections to a_μ arise from the graphs of Fig. 1.[5,6] The results for the R.G. model and the T.B. model[5]

Fig. 1. Diagrams leading to supersymmetric corrections to a_μ.

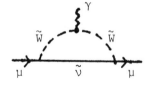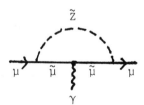

are shown in Figs. 2 and 3 parameterized by the $\tilde{W}_-$ and sneutrino masses. (The supersymmetric curves include the the Weinberg-Salam contribution.) The contributions for the R.G. model are quite large, as can be seen from the solid horizontal line which represents the lower bound on Δa_μ from the CERN g-2 measurement.[1] As can be seen, the suggested new experiments[1] will indeed test the supersymmetric models strongly. The supersymmetric contributions to a_μ for the T.B. model (Fig. 3) is smaller than the R.G. Case. However, it is also considerably different from what one expects from the Standard Model.

4. LIMITS ON SUPERSYMMETRIC MASSES FROM ACCELERATORS

Up to last year, the only limits on superparticle masses were from pair production experiments at Petra ($\tilde{m}_-$, $m_{\tilde{q}}$, $m_{\tilde{\ell}} \geq 20$ GeV). More recently, new limits from other accelerators have been obtained.

(1) <u>CERN Collider</u> (UA1)
Analysis of the monojet events at CERN have led to strong constraints on superparticle masses. This is because supersymmetry represents a "new physics" possibility for the origin of this phenomenon. In supersymmetry, the missing transverse momentum p_T is due to an escaping $\tilde{\gamma}$. The UA1 monojet data is summarized in Table I. Since the 1985 run doubles the integrated luminosity, one may expect ≈60-70 monojet events when the total data sample is analysed.

i) <u>Squark/Gluino Production</u>. Squark and/or gluino production at the collider (via strong interactions) predict monojets from decays of these particles, e.g., $\tilde{q} \to q + \tilde{\gamma}$. Detailed theoretical analysis shows, however, that the event rate is too high which puts a lower bound on the squark

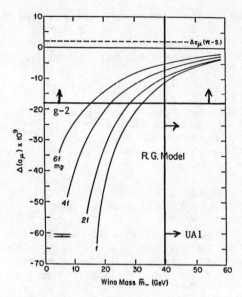

Fig. 2. Supersymmetric correction to a_μ for R.G. model ($m_{\tilde{\gamma}}$=7GeV, A=1.27).

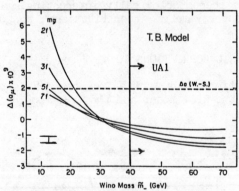

Fig. 3. Supersymmetric correction to a_μ for T.B. model ($m_{\tilde{\gamma}}$=7GeV, A=1.27).

Table I. Monojets events at CERN Collider (UA1)

Year	$\sqrt{s}$ (GeV)	Luminosity nb^{-1}	No. of events
1983	546	113	7
1984	630	275	23
1985	630	350-400	being analysed

and gluino mass $m_{\tilde{q}}$; $m_{\tilde{g}} \geq$ 65-70 GeV to inhibit their production.[7]

ii) <u>Supersymmetric Decays of W and Z</u>. An alternate possible source of monojets is the supersymmetric decays of vector bosons, $W \to \tilde{W} + \tilde{\gamma}$ and $Z \to \tilde{W} + \tilde{W}$ with subsequent Wino decay $\tilde{W} \to \bar{u} + d + \tilde{\gamma}$, etc. One may analyse the UA1 data two ways to get bounds on the Wino mass. Using the high p_T monojet events ($p_T > 34$GeV) gives the result[8]

$$\tilde{m}_- \geq 35\text{-}45 \text{ GeV} \qquad (5)$$

Recently, Mohammadi[9] has analysed the limits monojet data impose on the mass of a fourth sequential lepton arising from W decay. He made use of the monojet events with τ likelihood L<0 (which emphasizes the low p_T events). Adapting this analysis to Wino decay of the W one obtains approximately, $\tilde{m}_- \geq$ 40GeV consistent with the above.

In Figs. 2 and 3, the vertical lines represent the approximate UA1 bounds of $\tilde{m}_- \geq$ 40. Thus in the R.G. model of Fig. 2, the allowed region is restricted to the upper right hand corner. The error flag on the left hand side of the diagram represents the expected size of errors in the improved experiment proposed in Ref. 1. Thus the distinction between the Standard Model and its supersymmetric generalization should be observable. In the T.B. model of Fig. 3, the allowed region is to the right of the vertical line. The expected error of Ref. 1 is also shown there.

iii) <u>Supersymmetric Origin of Monojets</u>. After subtracting monojets expected from the Standard Model,[10] the high p_T monojet data ($p_T \gtrsim 34$GeV) is consistent with $\approx$2 monojets/100 nb^{-1} coming from "new physics" origins. Such event rates arise from Winos in the mass range of Eq. (5).[8] Whether monojets possess a new physics signature is not at present known (though the matter may be clarified when the 1985 data is analysed which should double the number of events). High p_T monojets would require relatively light photinos and so if monojets from Winos do exist, one would also require $m_{\tilde{\gamma}} \lesssim$ 5 GeV.

(2) <u>PEP (ASP,MAC)</u>
The ASP and MAC collaborations have looked for processes $e^+e^- \to \gamma \tilde{\gamma} \tilde{\gamma}$ or $\gamma \tilde{\nu} \tilde{\nu}$ at the PEP accelerator obtaining joint limits on $m_{\tilde{e}}$ and $m_{\tilde{\gamma}}$ (or $\tilde{m}_-$ and $m_{\tilde{\nu}}$) Fig. 4 shows the excluded region in the $m_{\tilde{e}}$-$m_{\tilde{\gamma}}$ plane. These restrictions produce constraints on the muon magnetic moment only if additional assumptions are made. Thus if $m_{\tilde{\gamma}} \lesssim$5GeV, one finds from Fig. 4 $m_{\tilde{e}} \geq$48GeV. Further, for the T.B. model $m_{\tilde{\nu}} \cong m_{\tilde{e}}$. Hence if we assume, as in Sec.

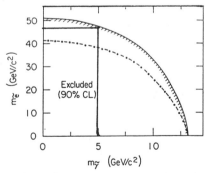

Fig. 4. Limits on $m_{\tilde{e}}$ and $m_{\tilde{\gamma}}$ from ASP data

(1)(iii) above that some monojets are of Wino origin, then one has for the T.B. model

$$\tilde{m}_- \simeq 40\text{-}45 \text{ GeV}, \ m_{\tilde{\gamma}} \lesssim 5 \text{ GeV}, \ m_{\tilde{\nu}} \gtrsim 48 \text{ GeV} \qquad (10)$$

As can be seen in Fig. 3, this greatly restricts the theoretical predictions of a_μ. Of course, further UA1 data is needed to verify the assumptions made above.

5. CONCLUSIONS

A measurement of the muon g-2 which reduces the error by a factor of 20 will not only be able to see the Standard Model contribution but will also test the validity of a wide class of Supergravity models. This is because the supersymmetry corrections are comparable to and often larger than the Standard Model term and generally of opposite sign. (Thus one can even test some Supergravity models without any improvement on the error of the hadronic contribution to a_μ).

Accelerator constraints on superparticle masses have begun to significantly restrict the allowed values of supersymmetric contributions to muon g-2. As discussed above, g-2 depends upon the Wino, sneutrino, photino (and/or gluino) masses, and the Polonyi constant A. One expects further accelerator constraints on these parameters in the relatively near future which would make the supersymmetric a_μ predictions more precise, i.e., (i) the UA1 1985 data should give a better determination of the Wino mass, (ii) Tevatron should be sensitive to higher mass squarks and gluinos and (iii) if $2\tilde{m}_- < M_{Z_2}$ the Wino should show up dramatically at SLC and LEP from the $Z \to \tilde{W}\tilde{W}$ decays.

This work was supported by the Natinal Science Foundation under Grant No. PHY-83-05734.

REFERENCES

1. See e.g., talk by V. Hughes at this Conference.
2. T. Kinoshita, B. Nizic and Y. Okamoto, Phys. Rev. Lett. **52**, 717 (1984).
3. For reviews of Supergravity models see P. Nath, R. Arnowitt and A.H. Chamseddine, "Applied N=1 Supergravity" (World Scientific Pub. Co., Singapore 1984) and HUTP-83/A077-NUB #2588 (1983).
4. S. Weinberg, Phys. Rev. Lett. **50**, 387 (1983); R. Arnowitt, A.H. Chamseddine and P. Nath Phys. Rev. Lett. **50**, 232 (1983).
5. T.-C. Yuan, R. Arnowitt, A.H. Chamseddine and P. Nath, Zeit. f. Phys. C **26**, 407 (1984).
6. D.A. Kosower, L.M. Krauss and N. Sakai, Phys. Lett. **133B**, 305 (1983).
7. R.M. Barnett, H. Haber, and G. Kane, LBL-20102 (1985).
8. A.H. Chamseddine, P. Nath and R. Arnowitt, NUB #2681-HUTP-85 A091 (1985) to be pub. Phys. Lett. B.
9. M. Mohammadi, talk at Workshop on Physics Simulations at High Energy, University of Wisconsin, May 1986.
10. J. Rohlf (UA1 Collaboration) CERN-EP/85-160 (1985) to be pub. Proc. of DPF Meeting, Univ. of Oregon, August 1985.

THERMAL MUONIUM IN VACUUM

Arthur Olin
TRIUMF, 4004 Wesbrook Mall, Vancouver, B.C. V6T 2A3, Canada

ABSTRACT

Two recent measurements have established large rates of emission of thermal muonium into vacuum. These experiments and their implications for future fundamental studies of the muonium system are discussed.

In the past six months groups at TRIUMF and KEK have developed techniques to produce muonium in vacuum moving with thermal velocities. Muonium (Mu) is usually produced by stopping μ^+ in insulators - often noble gases are used - but the interaction with the moderator is a problem for fundamental studies of the purely leptonic muonium system in isolation. Some examples are measurement of muonium conversion to antimuonium ($\mu-e^+$),[1] hyperfine splitting,[2] Lamb shift,[3] or the 1S-2S atomic transition energy. In the measurement of the Lamb shift epithermal beams of muonium were produced using beam foil techniques.[4] In this technique muons are passed through a thin metal foil, a sweeping magnet is used to define the neutrals, and the muonium identified from the decay positrons. The intensity of the epithermal beams achieved at TRIUMF were ~3×10^4 Mu/s, or about 1% of the 29 MeV/c positive muon beam. These beams have typical mean velocities of 0.01 c, so the Mu will travel 6 m in a muon lifetime.

In contrast, thermal Mu is being produced by stopping μ^+ in metal foils or SiO_2 powders, and then the Mu diffuses out into the vacuum. The first attempts to observe this effect by Kendall et al.[5] in Pt and W foils gave positive results, but there were difficulties in duplicating them. I would like now to review the recent work at TRIUMF[6] and KEK,[7] and then discuss its relevance to some of these measurements.

The behaviour of muons in fused silica powders (grain size 3.5 nm radius) has been studied previously at TRIUMF, where it has been established that (61±3)% of the stopping muons form muonium which escapes the silica and moves in the regions between the powder grains.[8] From this observation it was predicted that muonium should be emitted from a thin powder layer with yield approximately 5%. Our experimental arrangement is shown in Fig 1. A beam of 2×10^4 µ/s at 20 MeV/c was incident at 60 to the normal of a 125 µm plastic scintillator, and transmitted muons then struck a silica powder layer target supported by a 2.5 µm mylar film parallel to the scintillator. The targets studied were: two grades of silica powder (3.5 and 7.0 nm particle radius, density 30 mg/cm^3),[9] a powder compressed under 7 tons pressure to a thickness of 3 mg/cm^2, and an aluminum foil of 3 mg/cm^2 for background estimation. Different thicknesses of uncompressed powder (1 and 3 mg/cm^2) were employed, but fragility of the targets prevented accurate thickness measurements and the layers were not always uniform. The target region was evacuated to a pressure of less than 10^{-6} Torr.

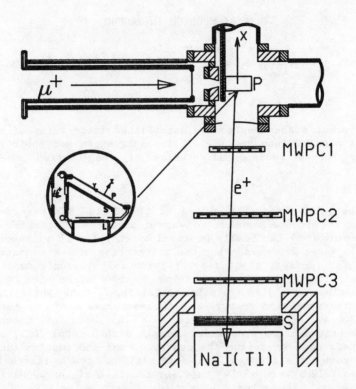

Fig. 1. Schematic of experimental apparatus. The powder target is indicated by P, and S is a stack of three scintillators. Target region (insert) shows powder layer (P), the thin scintillator (S), and its lucite light guide (L). Dashed lines indicate structures out of the plane of the beam.

The scintillator-target region was viewed through a 25 μm stainless steel window by three two-dimensional multiwire proportional chambers, three scintillators, and a large NaI(Tℓ) crystal. The data were analyzed by fitting the MWPC coordinates of positrons above 20 MeV to a straight track. Figure 2 shows a density plot of the trajectories of the decay positrons extrapolated to the vertical plane containing the beam axis. Decays originating from the target and beam scintillator are clearly seen.

A Monte Carlo simulation has been used to obtain a quantitative understanding of the data. Simulated muons are stopped in the powder layer, and 61% are converted to muonium and undergo a random walk using an effective diffusion constant D. The muonium is allowed to escape upon reaching the downstream powder surface with a thermal velocity for 300 K and a cos(theta) angular distribution. The position resolution of the MWPC track extrapolation is modelled by a Gaussian distribution with a width of 6 mm FWHM determined by the relative intensity of counts in region V1 for the Aℓ foil target and for the simulation with D=0. The diffusion constant D is then adjusted to match the ratio of target decays to decays from region V2 of Fig. 2.

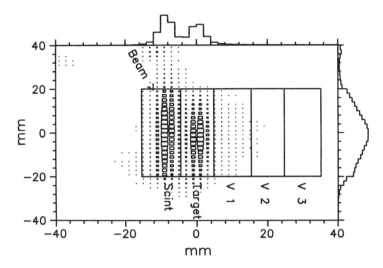

Fig. 2. Density plot of projection of decay positron trajectory to x=0 plane, showing definition of regions for time histograms. The coordinate axes are defined in Fig. 1.

Histograms of the observed time of decay for real events in different spatial regions are shown in Fig. 3 for a 1 mg/cm^2 layer of 3.5 nm powder and for the Aℓ foil target, and are compared to the simulation for D=80 cm^2/s and beam momentum 20.0 MeV/c. Note the nonexponential time dependence introduced by motion of muonium into and out of the different regions. The simulation allows estimates of partial yields for diffusion from the layer followed by decay in each of the four regions. Based on the measured yield for region V2 of 1.6±0.12% per stop in the silica, yields into the vacuum of 6.7%, 5.3% and 0.7% are estimated from the simulation for decays from the target, V1, and V3 regions, with the remainder decaying elsewhere. However, the inferred total yield depends on the input parameters and the model. From simulations with other assumptions, in particular for isotropic emission from the layer, a systematic uncertainty of 30% and a total yield of (19±6)% is indicated. The same analysis for a 1 mg/cm^2 layer of 7.0 nm particles gives a similar result while a 3 mg/cm^2 layer of 3.5 nm particles yields (15±5)%. Figure 4 shows a histogram of the distance of the muon from the layer at decay divided by the time of decay (an average velocity during diffusion and drift) for muons decaying in regions V1 and V2. The agreement with the simulation is again very good, confirming the hypothesis of thermal emission. No evidence was noted for a change in yield with an applied electric field of 2.5 V/cm, which is sufficient to inhibit escape of charged muons from the proximity of the layer. This confirms the neutral nature of the signal, while a muonium-like mass is inferred from the velocity spectrum. No evidence was noted for muonium emission from the compressed target. We conclude that the effect observed from silica is due to muonium moving at velocities consistent with room temperature thermal energies.

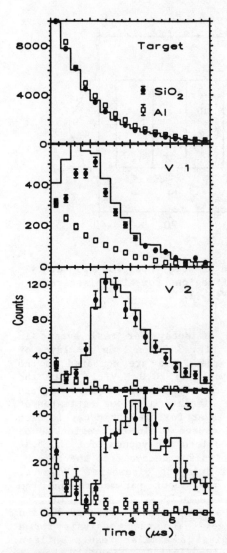

Fig. 3. Muon decay times for events from the spatial regions defined in Fig. 2. Filled circles represent the data from a 1 mg/cm^2 powder, open squares from a 3 mg/cm^2 Al foil, and the histograms result from the simulation calculation with diffusion constant of 80 cm^2/s.

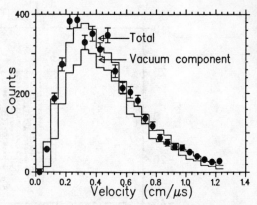

Fig. 4. Velocity distribution of events from spatial regions V1, V2, and V3 for the powder and the simulation as in Fig. 3. Also shown is the component of the simulation from muonium in vacuum.

A thermal diffusion constant D can be calculated as a function of particle number density and size by treating the layer as a gas of powder particles.[10] For 3.5 nm spheres one obtains $D = 7.9$ cm^2/s. However, the simulation requires $D = 80$ cm^2/s, corresponding to a substantially increased yield. Our results indicate even higher values of D for a thicker powder layer. We believe the primary reason is the existence of large chain-like agglomerates in the powder, regions of higher density separated by large empty volumes, which tend to greatly increase the effective diffusion constant.

Muonium emission into the vacuum from hot W foils has been recently observed at KEK.[7] The choice of foils was motivated by the observed emission of positrons and positronium from W foils. The Mu position was determined from the decay positron track, which was measured with a scintillator hodoscope with a resolution of 25 mm. A vane was used to reduce scattering from the target onto the vacuum window which, together with the larger vacuum vessel, led to smaller backgrounds than at TRIUMF, but also meant that the W target could not be viewed directly with the same counters. The μ^+ beam intensity was 1800/s at 23 MeV/c. The foil used was 50 μm thick and resistively heated to 2300-2800 K. The vacuum at 2300 K was 3×10^{-9} Torr, which proved to be important. A CMA Auger analyser was used to monitor the surface cleanliness, which showed only small fraction of a monolayer of C and O.

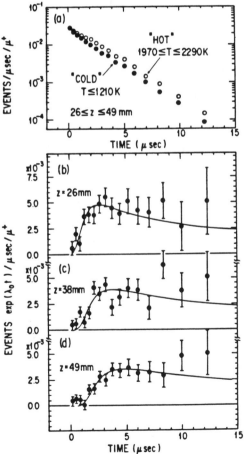

Results are shown in Fig. 5 where there is an obvious hot-cold difference and similar time structures to the TRIUMF work are evident in the difference spectra. The mean velocity of the Mu is 2 cm/μs, corresponding to T=2100 K. The magnetic field from the heater current was sufficient to confine charged particles of these velocities, confirming the neutral nature of the signal. The observed yield is 4±2% per μ^+. Figure 6 shows the temperature dependence of the delayed count rate of Fig. 5. The fit corresponds to

Fig. 5. Time spectra of μ^+ events from KEK run.[7] (a) Raw spectrum, binned and normalized, obtained with the W target "hot"-"cold"; (b-d) "hot"-"cold" difference spectra for three z positions of the counter telescopes.

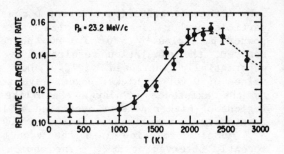

Fig. 6. Temperature dependence of delayed events summed over the time interval from 2 to 6 μs. The fitting curve represents an Arrhenius-type activation with a correction due to μ^+ trapping at thermal vacancies at high temperature.

an Arrhenius-type activation of E=0.66 eV with a drop at high temperatures due perhaps to trapping at thermally activated vacancies. The diffusion constant at high temperatures is in reasonble agreement with the value for H in W.

In which experiments is this muonium flux useful? The measurement of the hyperfine splitting is so far the only measurement on Mu to push QED. There is strong motivation to improve this limit as weak interaction effects will enter as interference terms and therefore go as the inverse square of the relevant mass scale. The experiment was performed in Kr gas to a precision of 0.160 kHz (0.036 ppm), and a correction for the gas density of 35.85(8) kHz had to be applied. At some level the density correction will limit the experimental accuracy and a vacuum measurement will be required. The premium on small size to restrict the volume over which homogenous magnetic field is required suggests a thermalized beam.

The measurement of the 1S-2S transition energy could be done using Doppler-free two-photon absorption, followed by ionization by a third photon. Appropriate lasers have been developed,[11] and the intrinsic linewidth is determined by the 2S state lifetime. In this case thermal muonium from SiO_2 is indicated together with a pulsed muon beam matched to the laser duty cycle. For the Lamb shift measurements the requirement of 2S production indicates the use of epithermal beams.

Muonium-antimuonium conversion, $\mu^+e^- \to \mu^-e^+$, is now a topical experiment, with active proposals at TRIUMF, LAMPF, KEK, and SIN. The process, first discussed by Feinberg and Weinberg, is strongly suppressed by magnetic fields or atomic collisions, and thus must be performed in vacuum. The experiment consists of producing Mu, allowing it to drift in a vacuum field-free environment for 1-2 muon lifetimes, and then detecting a signature of the μ^-. The most intriguing mechanism for this reaction involves the exchange of a doubly charged Higgs which conserves flavour but not lepton number, and whose neutral partner would give Majorana mass to the electrons. The present limit on this branching ratio is ρ=0.01 based on the Marshall et al.[10] measurement revised to use the new muonium yields. I have very little information about the SIN and KEK experiments, but the TRIUMF and LAMPF proposals provide an interesting contrast in experimental technique.

The experimental arrangement envisioned in the LAMPF proposal[12] is shown in Fig. 7. An epithermal beam of 1400 Mu/s is produced at the foil converter. After drifting for 800 ns through the clearing magnets and field-free region 25 Mu/s reach a 1200 cm^2 uranium target centred in the Crystal Box detector. The μ^- are captured into atomic

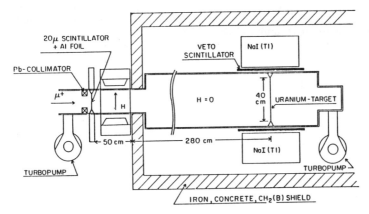

Fig. 7. Experimental arrangement for LAMPF Mu conversion experiment.

orbits in the uranium, and the K and L X-rays in coincidence are recorded in the NaI(Tℓ) crystals. A sensitivity to $\rho \sim 3 \times 10^{-5}$ is anticipated from a run this summer.

The TRIUMF experiment will use thermal muonium from silica powder. With 2000 Mu/s leaving the powder layer, we will have 200 Mu/s arriving 2 cm from the layer after having drifted for ~3 μs. In later runs we hope to substantially increase these rates through use of a multi-layered target. The μ^- will be detected through the muon capture reaction on W, which produces ^{184}Ta. This radionuclide has an 8 h half-life and a distinctive decay signature which will be detected out of beam. We hope to probe $\rho \sim 10^{-5}$ in the initial run.

REFERENCES

1. B. Pontecorvo, Sov. Phys JETP 6(33), 429 (1958); G. Feinberg and S. Weinberg, Phys. Rev. Lett. 6, 381 (1961) and Phys. Rev. 123, 1439 (1961).
2. F.G. Mariam et al., Phys. Rev. Lett. 49, 993 (1982).
3. C.J. Oram et al., Phys. Rev. Lett. 52, 910 (1984);
   A. Badertscher et al., Phys. Rev. Lett. 52, 914 (1984).
4. C.J. Oram et al., J. Phys. B 14, L789 (1981); P.R. Bolton et al., Phys. Rev. Lett. 47, 1441 (1981).
5. K.R. Kendall, Ph.D. thesis, University of Arizona (unpublished) (1972).
6. G.A. Beer, G.M. Marshall, G.R. Mason, A. Olin, Z. Gelbart, K.R. Kendall, T. Bowen, P.G. Halverson, A.E. Pifer, C.A. Fry, J.B. Warren, and A. R. Kunselman, submitted to Phys. Rev. Lett.
7. A.P. Mills, Jr. et al., Phys. Rev. Lett. 56, 1463 (1986).
8. R.F. Kiefl et al., Proc. First International Topical Conference on Muon Spin Rotation, Hyper. Int. 6, 185 (1979); G.M. Marshall, J.B. Warren, D.M. Garner, G.S. Clark, J.H. Brewer, and D.G. Fleming, Phys. Lett. 65A, 351 (1978).
9. The powders were kindly furnished by Cabot Corporation, P.O. Box 188, Tuscola, IL, USA, and are described in their technical report. The grades used were EH-5 and M-5.

10. G.M. Marshall, J.B. Warren, C.J. Oram, and R.F. Kiefl, Phys. Rev. D **25**, 1174 (1982); G.M. Marshall, Ph.D. thesis, University of British Columbia, unpublished (1981).
11. T.W. Hansch et al., in Workshop on Fundamental Muon Physics (1986).
12. V.W. Hughes et al., LAMPF research proposal 985 (1985).

LEPTONIC DECAYS OF HEAVY LEPTONIUM[†*]

Jerome Malenfant
Department of Physics, University of California, Los Angeles, CA 90024

We have calculated order α radiative corrections to the rate for bound μ^+-μ^- or τ^+-τ^- states to decay to e^+-e^- via one photon annihilation.[1] Nonrelativistically this rate is given by the Weisskopf-Van Royen formula[2]

$$\Gamma_{NR} = 16 \pi \alpha^2 |\psi(0)|^2 / 3 M^2 . \qquad (1)$$

Relativistically $\psi(0)$ diverges for potentials that behave like $1/r$ as $r \to 0$.[3] Since the relativistic corrections to the wave function are order α, and we had no doubt the rate as calculated in the relativistic theory is finite, we calculated all the order α radiative corrections to see precisely how the divergence of the wave function at the origin is canceled. We find that it is canceled by the ultraviolet divergence of the wave function renormalization constant Z_2 which belongs to the heavy lepton lines.

We evaluated relativistic corrections to the wave function using the Barbieri-Remiddi perturbation theory.[4,5] It turns out that our final result can be expressed in terms of the classic results of Karplus and Klein[6] and of Källen and Sabry.[7] Including bremsstrahlung we find that the inclusive rate for dimuonium including only order α terms is

$$\Gamma = \Gamma_{NR} \left(1 + (\alpha/\pi) [(4/3) \ln(2m_\mu/m_e) - 221/36] \right) . \qquad (2)$$

Numerically this gives a rate of 5.561×10^{11} sec^{-1} for decays of the ground state.

We have studied these corrections for excited $^3S_1$ states as well as the ground state and found that the order α radiative corrections cancel out of ratios; i.e., the rate for leptonic decay of the n^3S_1 state is simply $1/n^3$ that of the ground state (nonrelativistically $|\psi(0)|^2$ are in the ratio $1/n^3$).

Our calculations apply also to tauonium. Using (1) for τ^+-τ^-, we obtain

$$\Gamma_{NR} = m_\tau \alpha^5 / 6 n^3 . \qquad (3)$$

This gives a rate of 9.4×10^{12} sec^{-1}. To this must be added the hadronic decay rate. It is

$$\Gamma^{(hadronic)} = \Gamma_{NR} \times R \qquad (4)$$

where R is the usual ratio of cross section for $e^+ e^-$ annihilation to hadrons divided by the cross section for $e^+ e^- \to \mu^+ \mu^-$. Using the measured value for R,[8] we find that the total annihilation decay rate is 4.2×10^{13} sec^{-1}. This is about 14 times higher than the decay rate of a free τ.

Vacuum polarization effects due to electron loops give large higher order corrections to the wave functions for the bound states. This is because the electron Compton wavelength is approximately equal to the Bohr radius of dimuonium and considerably larger than the Bohr radius of tauonium. These diagrams contribute differently in the ground and first excited states. Their contributions are characterized by the parameter $\kappa = \alpha m/2m_e$. This number is only slightly less than one for the dimuonium case and large compared to one for tauonium. Consequently, especially for tauonium, higher order in α corrections are needed to obtain reliable estimates of leptonic decay rates of heavy leptonium states.

[†] Presented by Nina Byers.
[*] Work supported in part by Department of Energy, contract DE-AT03-81ER40024.

REFERENCES

1. J. Malenfant, UCLA thesis (1986).
2. R. Van Royen and V. Weisskopf, Nuovo Cimento A 50, 617 (1967).
3. B. Durand and L. Durand, Phys. Rev.D 28, 396 (1983); Phys. Rev. D 30, 1904 (1984).
4. R. Barbieri and E. Remiddi, Nuclear Physics B 141, 413 (1978).
5. W. Buchmüller and E. Remiddi, Nuclear Physics B 162, 250 (1980).
6. R. Karplus and A. Klein, Phys. Rev. 87, 848 (1952)
7. G. Källen and A. Sabry, Kgl. Danske Videnskab. Selskab, Mat. Fys. Medd. 29, No. 17 (1955).
8. M. Perl, Physica Scripta 25, No. 1:2, 172 (1982).

MUON AND NEUTRINO PRODUCTION IN PROTON-^{12}C SCATTERING

S.L. Mintz
Florida International University, Miami, Fla. 33199

ABSTRACT

The cross section for the reaction $p + {}^{12}C \rightarrow {}^{13}C + \mu^+ + \nu$ is calculated for center of mass proton energies from 1060 MeV. to 1080 MeV. by the use of experimental data for the $p + {}^{12}C \rightarrow {}^{13}C + \pi^+$ and $p + {}^{12}C \rightarrow \gamma + {}^{13}N$ reactions. Preliminary calculations for the inclusive reaction $p + {}^{12}C \rightarrow \mu^+ + \nu + X$ are discussed. A possible role of these processes in studying the contributions of anomalous threshold states in nuclear PCAC is also discussed.

INTRODUCTION

The reaction $p + {}^{12}C \rightarrow {}^{13}C + \mu^+ + \nu$ is interesting for a number of reasons. In this process the momentum transfer squared is entirely time-like and it is therefore possible to study a kinematic region not generally available. Processes at time-like q^2 also would allow the possibility of directly observing anomalous threshold contributions to the to the nuclear matrix elements of the axial current divergence.

The axial current matrix element for a $1/2^+ \leftrightarrow 1/2^+$ transition may be written as:

$$<f| A_\mu (0) |i> = \bar{u}_f[\gamma_\mu F_A(q^2) + (q_\mu/m_\pi) F_P(q^2)]\gamma_5 u_i \quad (1)$$

which leads to a matrix element for the divegence of the axial cureent given by:

$$<f| \partial^\eta A_\eta (0) |i> = \bar{u}_f[(m_i + m_f)F_A + (q^2/m_\pi)F_P]\gamma_5 u_i. \quad (2)$$

The quantities, F_A and F_P, the axial current form factor and the psuedoscalar form factor respectively are related in the nucleon case by the formula:

$$F_P(q^2) = -m_\pi(m_n + m_p)F_A(q^2)/(q^2 - m_\pi^2) \quad (3)$$

This result, Eq.(3), represents pion pole dominance. In the nuclear case, we expect pion pole dominance to be less good. One may write Eq.(3) in the form:

$$F_P(q^2) = -m_\pi(m_f + m_i)F_A(q^2)(1 + \epsilon(q^2))/(q^2 - m_\pi^2) \quad (4)$$

where $\epsilon(q^2)$ represents a correction to pion pole dominance which may be expressed[1] as:

$$\epsilon(q^2) = (m_\pi^2/q^2)[1 - a_\pi f_{\pi if}(q^2)/F_A(q^2)] \quad (5a)$$

$$f_{\pi if}(q^2) = f_{\pi if}(m^2) + (1 + q^2/m_\pi^2)I(q^2) \quad (5b)$$

$$I(q^2) = \int \text{Im } f_{\pi if}(m^2)[m^2(m^2 - m_\pi^2)(1+q^2/m^2)]^{-1}dm^2 \quad (5c)$$

where $f_{\pi if}(q^2)$ is the pion-nuclear coupling constant. From Eq.(5c) the anomalous threshold contributions can be noted. Estimates[1] for $\epsilon(q^2)$ yield values of -0.04 for the $^3H \longleftrightarrow ^3He$ transition and -0.15 for the $^{12}B \longleftrightarrow ^{12}C$ transition and a limiting value of -0.29 for large nuclei at q^2 appropriate to muon capture. On the other hand, there is some evidence[2] that is consistent with zero. If F_P were examined via a time-like process, states contributing to the cut in Eq.(5) would become realizable and would be evident in the cross section.

Finally there has been some speculation[3] that F_A may not analytically continue in the $q^2 = m_\pi^2$ region. A time-like process would be useful in testing this idea.

MATRIX ELEMENTS

The reaction $^{12}C + p \longrightarrow ^{13}C_{gs} + \mu^+ + \nu$ is particularly interesting. The matrix element for the axial current $<^{13}C_{gs}| A_\mu^+(0)|^{12}C,p>$ requires eight form factors for a complete description. However, because $^{13}C_{gs}$, ^{12}C, and p are $1/2^-, 0^+$, and $1/2^-$ states respectively, the major contribution from this matrix element comes from the time component, A_0, of the axial current. This component can be described by two form factors so that:

$$<^{13}C_{gs}|A_\mu^+(0)|^{12}C,p> = \bar{u}_f[\gamma_\mu F_1 + q_\mu F_2]u_p \quad (6)$$

The form factors F_1 and F_2 are analogous to F_A and F_P respectively. They can be expressed as functions of two variables, q^2 and n^2 where $n^\delta = \epsilon^{\delta a \eta \kappa} p_a p_{i\eta} p_{f\kappa}$. There exists sufficient data[4] to determine F_1 and F_2 from $p + ^{12}C \longrightarrow ^{13}C_{gs} + \pi^+$ and by making use of a relation similar to Eq.(4) but neglecting ϵ, which is permissible in an approximate calculation such as this one.

The matrix element of the vector current is also described by eight form factors. If CVC is imposed and the largest components are kept one obtains:

$$<^{13}C_{gs}|V_\lambda^+(0)|p,^{12}C> = \bar{u}_f[d_\lambda F_{v3} + Q_\lambda F_{v4} + \sigma_{\lambda\kappa}Q^\kappa F_{v5}]_5 u_p \quad (7)$$

where $F_{v4} = -F_{v3}d \cdot q/Q \cdot q$ with $d = p_f - p$, and $Q = p_f + p$. There does not exist sufficient data to determine the two independent form factors in Eq.(7). However the contribution of this matrix element to the cross section is small. Furthermore, when suitably normalized, the form factors tend to have a similar magnitude. We therefore as an approximation assume that they are equal and determine them at $q^2 = 0$ from data for the reaction $^{12}C + p \rightarrow {}^{13}C + \pi^+$. With this approximation, all form factors are determined and the cross section may be calculated. We show the results in figure 1. An earlier calculation for this process had been undertaken by conventional means by Weiss and Walker[5]. They speculated that the values which they obtained were an order of magnitude smaller then that which would be obtained by a more direct use of PCAC. In the calculation presented here we obtain values for the cross section approximately three times as large as that obtained by Weiss and Walker and in agreement with their expectations. We again stress that we have used experimental data directly in this empirical calculation and that the estimates have been on the conservative side.

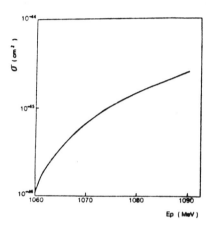

Fig. 1. Plot of the cross section as a function of CM proton total energy.

THE INCLUSIVE PROCESS

It was noted by Weiss and Walker that transitions to a number of positive parity final states gave substantially larger cross sections than the transition to the $^{13}C_{gs}$ final state. We therefore consider an inclusive process $p + {}^{12}C \rightarrow X + \mu^+ + \nu$. We may therefore write:

$$d = \sum_k \int \frac{d^3p_k m_k \, d^3p_\mu m_\mu \, d^3p_\nu m_\nu |M_{ik}|^2 \delta(p_{out}-p_{in})}{E_k(2\pi)^3 \, E_\mu(2\pi)^3 \, E_\nu(2\pi)^3} \quad (8a)$$

$$|M_{ik}| = \langle k|J_\nu(0)|{}^{12}C,p\rangle\langle k|J_\eta(0)|{}^{12}C,p\rangle \mathcal{L}^{\nu\eta} G^2 \cos^2(\theta)/2$$
(8b)

Here $\mathcal{L}^{\nu\eta}$ is the usual lepton trace. We can make use of the results of Mintz and King[6] and assume closure to obtain:

$$|M_{ik}| = M_{\alpha\kappa} \mathcal{L}^{\alpha\kappa} G^2 \cos^2(\theta)/2 \quad (9a)$$

$$M^{\eta\kappa} = \alpha g^{\eta\kappa} + \beta p^{\eta} p^{\kappa}/m^2 + \gamma p_i^{\eta} p_i^{\kappa}/M^2 + \delta <q^{\eta}><q^{\kappa}>/M^2$$
$$+a\, p^{\eta} p_i^{\kappa}/M^2 \ldots + bi\epsilon^{\eta\kappa\delta\iota} p_{\delta} p_{\iota} + \ldots \qquad (9b).$$

However arguments of the type used in reference 6 indicate that the large contributions to the cross section should come from the first four terms of Eq.(9b). We can then calculate the cross section and obtain:

$$\sigma \cong \frac{2\, M_x}{(2\pi)^3 [p_o + P_o]} \int E_{\nu}^2 E_{\mu} p_{\mu} dE_{\mu} \{2\alpha - \beta - \gamma\} \qquad (10)$$

Some data exists for the process $p + {}^{12}C \rightarrow X + \pi^+$ and we are presently attempting to determine a value for the approximate cross section for the inclusive process.

REFERENCES

1. C.W. Kim and H. Primakoff, Mesons in Nuclei,Ed.,M. Rho and D.H. Wilkinson(North Holland,Amsterdam,1979),p.69.
2. H. Primakoff,Nucl.Phys A317,279(1979).
3. B. Bosco, C.W. Kim, and S.L. Mintz,Phys. Rev.C25, 1986 (1982)
4. F. Soga et al.,Phys.Rev C24,570(1981)
5. D.L. Weiss and G.E. Walker,Phys. Rev.C25,991(1982)
6. S.L. Mintz and D.F. King, Phys. Rev.C30,1585(1984)

A LARGE ACCEPTANCE DETECTOR FOR ELECTROMAGNETIC NUCLEAR PHYSICS AT CEBAF

Bernhard A. Mecking
CEBAF
12070 Jefferson Avenue
Newport News, Virginia 23606

ABSTRACT

At CEBAF a large acceptance magnetic detector is planned for the investigation of electromagnetic nuclear reactions. The detector will be used to investigate multiple particle final states with high efficiency and to achieve sufficient count rate in cases of limited luminosity.

In addition to experiments that require high resolution magnetic spectrometers (with relatively small solid angles) there are two important classes of experiments at CEBAF that can only be performed with (or will at laeast benefit from) a large acceptance detector (LAD):

1. The detection of multiple particle final states, like in the photo- (or electro-) excitation of the nucleon resonances in both elementary and nuclear production ($\gamma N \rightarrow N^* \rightarrow \pi\pi N$, $\gamma D \rightarrow \Delta\Delta \rightarrow NN\pi\pi$, $\gamma A \rightarrow \pi NX$, $\gamma A \rightarrow NNX$).

2. Measurements at limited luminosity. The limitation can be due to the target (e.g. polarized target with low radiation resistance, active targets for the detection of low range recoils) or due to the beam (e.g. all experiments using a tagged bremsstrahlung or other secondary beams independent of the number of particles in the final state).

The physics program results in the following requirements for a general purpose large acceptance detector:

1. Homogeneous coverage of a large angular and momentum range for charged particles (magnetic analysis), photons (total absorption counters) and neutrons.
2. Good energy and angular resolution (for all particles).
3. Good particle identification properties.
4. No transverse magnetic field at the beam axis (to avoid sweeping (e^+, e^-) pairs into the detector).
5. Large field free space around the target to allow for the installation of complicated targets (cryogenic, polarized, track sensitive, etc.) or auxilliary equipment (nucleon polarimeter, γ-converter).

6. High count rate capability. The operation in a tagged photon beam ($N_\gamma = 10^7$/sec) should be easy; it will be more challenging to operate the detector in the high background environment caused by a reasonable intensity electron beam (e.g., 10nA on a 0.1 g/cm^2 target → luminosity $\simeq 10^{33}$/cm^2·sec).

The field requirements are best satisfied by a toroidal magnet (see figure 1) consisting of 8 coils arranged around the beam line to produce essentially a magnetic field in ϕ-direction. Charged particles are tracked by drift chambers; scintillation counters are used for the trigger and for time-of-flight; photons are detected by shower counters. A description of the main features of the detector will be given below.

1. <u>Magnetic field</u>. The field is generated by a toroidal magnet (4m diameter, 4 m length) consisting of 8 super-conducting coils. Each coil carries a current of 750 kA. A large volume of constant magnetic field ($B_\phi \simeq 0.5$ Tesla) can be generated by an appropriate radial distribution of the conductors; the $\int B_\phi \cdot dl$ for particles emitted at 90° is $\simeq$0.7 Tesla·meter. The forces that pull the coils towards the axis (about 50 tons/coil) are handled by using a warm support structure around the outside of the coils. About 20% of the ϕ-range is obstructed by the vacuum chamber.

2. <u>Particle tracking</u>. In each of the eight segments charged particles are tracked by planar wire chambers. The segments are instrumented independently to achieve high count rate capability. The estimated momentum resolution is $\simeq$ 1% in the momentum range of interest; the initial ϕ and θ of the track can be determined to $\leq$ 5 mrad.

3. <u>Scintillation counters</u>. The outer planar drift chambers are covered by scintillation counters. In each segment there are 8 barrel counters, each about 400 cm long, 20 cm wide, and 5 cm thick. The counters are viewed by phototubes at both ends for improved timing and position resolution. Each endcap segment is covered by 4 counters viewed by a single photomultiplier. The scintillation counters serve the double purpose of providing the trigger and the time-of-flight information. Also, a fraction of the high energy neutrons ($\simeq$5%) will interact in the outer scintillation counters and will thus be detected.

4. <u>Shower counter</u>. The detector can be surrounded by shower counters for the detection of showering particles, especially for high energy photons from Compton scattering, $\pi°$ and η decays, etc.).

5. Particle identification. The combination of momentum and time-of-flight is used for the identification of charged particles. Pions and kaons can be separated up to 1.5 GeV/c, kaons and protons up to 2.5 GeV/c. π/μ and μ/e separation can be achieved by using the pulse height in the shower counter in addition.

The performance of the LAD can be predicted using Monte-Carlo methods to generate multiple particle final states and to test particle tracking and reconstruction. As an example, the probability to reconstruct the complete $\pi^+\pi^-p$ final state from the F35(1975) → $\pi^+\pi^-p$ decay has been found to be 30% (most of the events are lost because one of the 3 particles hits a coil).

The LAD will occupy the lightly shielded end station B. In addition to low intensity electron and photon beams, it can also be used for experiments with secondary kaon, pion and muon beams. The LAD is complimentary to the high resolution magnetic spectrometers: by using the standard spectrometers for high luminosity and the LAD for low luminosity experiments a wide range physics phenomena can be covered at CEBAF.

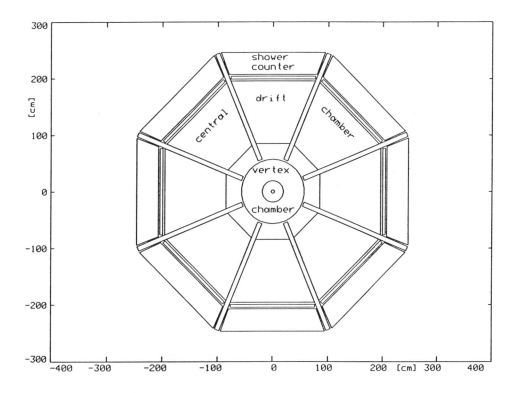

Figure 1: View of the toroidal detector in the direction of the beam

POLARIZATION IN ELECTRON SCATTERING EXPERIMENTS AT CEBAF.

V. Burkert
CEBAF, Newport News, Va. 23606, USA

ABSTRACT

Some aspects of a tentative program for using high current polarized electron beams as well as solid state targets with high nucleon and deuteron polarization in nuclear physics experiments at CEBAF are briefly discussed.

INTRODUCTION

In the long history of electron scattering from nucleons and nuclei little use has been made of intense polarized electron beams and polarized nucleon or nuclear targets. This may partly be due to the lack of appropriate polarized electron sources, partly due to the cumbersome procedure that has been involved in using polarized solid state targets in high current electron beams. In recent years, however there has been enormous progress both, in the development of high current polarized electron injectors, as well as in the development of target materials with high polarization and high radiation resistivity, which will make polarization experiments a powerful tool in the study of the electromagnetic structure of nucleons and light nuclei.

PROGRAM AT CEBAF

Current ideas on using polarized beams and targets at CEBAF include programs to measure photo- and electroexcitation of baryon resonances, the electric formfactor of the neutron, a separation of the electric monopole and quadrupole formfactors of the deuteron, and experiments to study electroweak interference effects in electron scattering from nucleons and nuclei[1]. Due to the limited time for this presentation I want to restrict the discussion to the first two topics.

Virtual photons are an ideal probe of the electromagnetic structure and of the spin structure of the nucleon. At high enough energies and four-momentum transfers, transverse photons are sensitive to spin 1/2 objects like quarks, whereas longitudinal photons are sensitive to spin 0 objects like pions or diquarks. The ultimate aim of electron scattering experiments in the nucleon resonance mass region is to study the $\gamma_v NN^*$ vertex. Since excited states of the nucleon are generally broad and overlap, the identification of single resonances requires a study of the resonance decay and therefore calls for coincidence measurements. Obviously, a high duty factor electron machine of several GeV maximum energy will be ideally suited for this purpose.

Nucleons are composed of valence quarks and sea quark and antiquark pairs, bound by gluon exchange. In the mass region of nucleon resonances the $\gamma_v NN^*$ vertex contains information on quark-gluon interaction at large and intermediate distances (confinement regime) and quark distribution in excited nucleonic systems. The precise knowledge of resonance transitions in free nucleons provides the necessary data base for a study of possible modification in resonance parameters due the nuclear environment in nuclei. Previous experiments have studied single pseudoscalar meson production using unpolarized electron beams and unpolarized hydrogen targets. The cross section is given by

$$d\sigma/d\Omega_\pi = \sigma_u + \epsilon\sigma_l + \epsilon\sigma_t\cos2\phi + \sqrt{2(\epsilon+1)\epsilon}\ \sigma_I\cos\phi$$

where the σ_i are related to the photon wave polarization and are function of Q^2, W, θ_π^*. The process $\gamma_v N \rightarrow \pi N'$ is generally described by 6 parity conserving complex helicity amplitudes and requires at least 11 independent measurements to be fully determined. Measuring the four terms specifying the unpolarized cross section, does therefore not provide sufficient information to conduct a model independent extraction of the contributing amplitudes. Yet, some attempts have been undertaken to extract the transverse helicity 1/2 and helicity 3/2 photocoupling amplitudes for some of the dominant resonances, using additional information from elastic πN scattering. The results of these attempts may be summarized as follows:[2]
The photocoupling amplitudes of the most prominent resonances, the $P_{33}(1232)$, $D_{13}(1520)$, $S_{11}(1535)$, and $F_{15}(1688)$ have been extracted from proton data for Q^2 up to 3 (GeV/c)^2. The excitation of the $P_{33}(1232)$ remains dominantly magnetic up the highest Q^2. The $D_{13}(1520)$ and the $F_{15}(1688)$ exhibit a rapid change from helicity 3/2 to helicity 1/2 dominance in the initial $\gamma_v N$ system, in qualitative accordance with quark model calculations. The $S_{11}(1535)$ shows a strikingly weak Q^2 dependence, which is not well understood within the framework of quarkmodels. The coupling of longitudinal photons to these resonances studied is weak with the possible exception of the $P_{11}(1440)$.

The data are clearly insufficient to extract the photocoupling amplitudes of the weak resonances. Very little information exists on electroexcitation of neutron resonances and there is a complete lack of information on higher mass resonances (W>1700 MeV/c^2). In general the results of existing analyses suffer from systematic uncertainties, due to limited experimental information.

PROSPECTS OF NUCLEON RESONANCE STUDIES AT CEBAF

Some of the features that make CEBAF an ideal laboratory for this kind of experiments are listed in table I. In conjunction with the appropriate experimental equipment high statistics coincidence measurements will become feasible, to cover complete angular distribution in the hadron decay system. In the initial $\gamma_v N$ system a large range in the four momentum transfer Q^2, the invariant mass W of the hadronic system, and of the

Table I. Parameters of the CEBAF and proposed experimental equipment

Maximum Energy	E > 4 GeV
Unpolarized e$^-$ current	I_e > 100μA
Duty Factor	η = 100%
Polarized e$^-$ current	I_p > 100μA
Electron Polarization	P_e > 40%
Polarized Protons	P_p = 50 to 90%
Polarized Deuterons	P_d = 30 to 60%

photon polarization ϵ will be covered. Using polarized beams and polarized targets will enable to perform 13 sensible asymmetry measurements at a given Q^2, W and ϵ, in addition to the 4 unpolarized measurements. Such an experimental program leads to a highly redundant data set and would provide the material for a rather complete and largely model independent analysis.

Polarization measurements are particularly sensitive to interferences of amplitudes from different partial waves. Measurements of polarization asymmetries may therefore be a powerful tool in extracting information on weakly excited resonances if they interfere with stronger resonance amplitudes. Fig.1 shows the sensitivity of the target asymmetry in single π^0 production to the excitation of the $P_{11}(1440)$. Despite the fact that this resonance is clearly visible in photoproduction experiments, it has barely been seen in electroproduction experiment. By choosing a suitable orientation of the target polarization and by carefully selecting the kinematics of the decay particles, interference effects may become large and may exhibit large effects even from weak resonances.

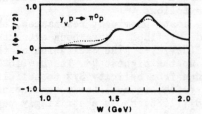 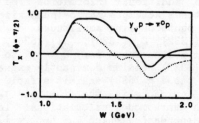

Fig.1 Sensitivity of the target asymmetries T_x (target polarization in electron scattering plane, perpendicular to γ_v) and T_y (target polarization perpendicular to the electron scattering plane) to the excitation strength of the $P_{11}(1440)$ at $Q^2=1.(GeV/c)^2$. The prediction is based on an analysis[3] of DESY data. Solid lines: with P_{11}; dashed lines: without P_{11}.

ELECTRIC FORMFACTOR OF THE NEUTRON

Precise knowledge of the neutron electric formfactor G_E^n is important for testing microscopic models of the nucleon. It would also remove considerable uncertainties in the interpretation of electron nucleus scattering data where G_E^n enters as a fundamental quantity. Present information comes from various sources and is limited to $Q^2 < 1.(GeV/c)^2$: The slope dG_E^n/dQ^2 at $Q^2=0$. was found to be positive in scattering of thermal neutrons off electrons in atoms. Quasielastic electron deuteron scattering has been used to measure the magnetic formfactor but failed to allow for an extraction of the electric formfactor. The most precise values of G_E^n are believed to be extracted from elastic electron deuteron scattering. This method requires, however, the adoption of a specific wavefunction, the choice of which strongly influences the resulting G_E^n. To circumvent this, it has been proposed to measure the polarization transfer in quasielastic scattering of polarized electrons from neutrons in unpolarized deuterium[4]. Measuring the recoil neutron polarization requires a second scattering experiment (neutron polarimeter) with an a priori not well known analyzing power and efficiency. These quantities have to be determined independently.

We consider as an alternative, but otherwise equivalent method quasielastic scattering of polarized electrons from polarized neutrons in vector polarized deuterium[1]. For an orientation of the neutron spin in the electron scattering plane, perpendicular to the direction of γ_v, the polarized cross section writes as

$$\frac{d\sigma}{d\Omega} = \frac{d\sigma}{d\Omega}_0 (1+P_e P_n A^n) \qquad A^n = \frac{2 G_E\, G_M\, \sqrt{\tau/(\tau+1)}\, \mathrm{tg}(\theta_e/2)}{(G_E^2 + \tau G_M^2)/(1+\tau) + 2\tau^2 G_M^2 \mathrm{tg}^2(\theta_e/2)}$$

(P_e = electron polarization, P_n = effective neutron polarization)

Since G_M^n is known, A^n and hence G_E^n can be determined by measuring the cross section asymmetry for opposite spin orientation of the incident electrons. $A^n(Q^2)$ is shown in fig.2 for two electron scattering angles and two parametrizations of G_E^n, both of which are consistent with present data at $Q^2 > 0.3 (GeV/c)^2$. With realistic figures on presently achievable electron and neutron polarization as well as on the electron current which present polarized deuterium targets are able to handle, one obtains the running time for a measurement of A^n with a given accuracy as shown in fig.3. Measurements of G_E^n for Q^2 up to $2(GeV/c)^2$ seem feasible at CEBAF energies. A unique feature of this method is that by measuring the corresponding quantities for free protons and for protons bound in deuterium one may correct the measured neutron formfactor for nuclear effects.

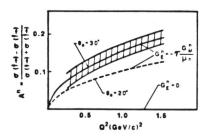
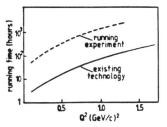

Fig.3 Left: Expected neutron asymmetry for two parametrizations of the electric formfactor and for two electron scattering angles. The hatched band indicates $\delta A^n = {}^+_-0.02$. Right: Expected running time at a given Q^2 at CEBAF energies. The dashed line is based on $P_n=0.4$, $I_e=0.3$ nA, and a 1.6 cm long ND_3 target. The solid line assumes a 6 nA electron current.

(1) RPAC Report CEBAF 1986, V. Burkert CEBAF-R-86-002, CEBAF-R-86-003
(2) F. Foster and G. Hughes ; Rep. Progr. Phys.46, 1445 (1983)
    V. Burkert ; Lecture Notes in Physics 234,pg. 228 (1985)
(3) G. Kroesen; Thesis Bonn Univ. (unpublished)
(4) R. Arnold, C. Carlson, F. Gross ; Phys. Rev. C23, 363 (1981)

HADRON SCATTERING

Coordinators: G. J. Igo
               N. Isgur

PARTIAL WAVE ANALYSIS OF pp → NNπ AND DIBARYON RESONANCES

A. B. Wicklund
Argonne National Laboratory, Argonne, IL 60439

ABSTRACT

This talk summarizes results of a partial-wave analysis (PWA) of $p_\uparrow p \to \Delta^{++} n \to p\pi^+ n$ in the dibaryon resonance region, using comprehensive data on single spin correlations from the effective mass spectrometer at the Argonne Zero Gradient Synchrotron.[1] These results are compared with recent PWA for the bound nucleon final state, $pp \to \pi^+ d$, by Bugg.[2] These reactions are closely related, and neither displays Breit-Wigner resonance behavior in the larger waves.

INTRODUCTION

We discuss in turn the motivation for this study, the relationships between free and bound nucleon final states, the P_{lab} dependent features of the data, the amplitude results, and the conclusions on dibaryon resonances.

MOTIVATION

Elastic nucleon-nucleon phase-shift-analyses[3-6] have clearly indicated Breit-Wigner behavior in some of the proton-proton waves, notably 1D2 and 3F3. As the inelastic data base has improved, many reviewers have advocated analogous PWA on the pp → NNπ reactions.[7] For example, we quote P. Kroll:[7] "According to the usual phenomenological criteria there are resonances in NN scattering (1D2, 3F3 dibaryons). A good quantitative understanding of NN → NNπ is needed in order to settle the interpretation of the dibaryons as conventional or unconventional resonances."

We can categorize the elastic partial waves according to the Argand plots of Figs. 1a-c. The largest waves are "deuteron-like" (similar to the 3S1 wave in np → np) and rotate clockwise with increasing energy (Fig. 1a). The "dibaryon resonance" waves rotate counter-clockwise (in a Breit-Wigner sense), but have very small elasticities (Fig. 1b). Other waves (Fig. 1c) are small and nondescript. None of the proton-proton waves exhibits the behavior of Fig. 1d, that is, Breit-Wigner with strong elastic coupling.

The elastic PSA indicates that the dibaryon waves are strongly coupled to the ΔN and NNπ channels, with partial cross sections of 5-8 mb at the 1D2 and 3F3 peaks (~ 1.2 and ~ 1.5 GeV/c respectively). If these waves were coupled-channel Breit-Wigner resonances, then we would expect the ΔN → ΔN elastic amplitudes to behavior like Fig. 1d, with $\Gamma_{\Delta N}/\Gamma_{tot} \geq 0.7$. On the other hand,

the PSA can measure only the proton-proton phase shifts and
inelasticities, and these parameters would be consistent with a
variety of behaviors for $\Delta N \to \Delta N$, for example "deuteron-like"
behavior as in Fig. 1a. Physically, the "deuteron-like" behavior
might correspond to virtual bound states of ΔN ("conventional
resonances") which decay into pp; Breit-Wigner behavior
("unconventional") might correspond to color-confined QQQ_8-QQQ_8
six-quark states that decay into both pp and ΔN. The only way to
distinguish these possibilities would to isolate the $\Delta N \to \Delta N$ phase
shift, $\delta_{\Delta N}$. For a two-channel case, $\delta_{\Delta N}$ is related to
the pp $\to \Delta N$ transition phase by

$$\phi(pp \to \Delta N) = \delta_{pp} + \delta_{\Delta N}, \qquad (1)$$

and since $\delta_{pp} \sim 0$ for the dibaryon waves, we would expect
dramatically different behaviors for ϕ and $\delta_{\Delta N}$, corresponding to
the "deuteron-like" or Breit-Wigner possibilities. Thus, our
objective in partial wave decomposition of pp $\to$ NNπ is to detect
large phase variations ($\Delta\phi = \pm \pi$) in the 1D2, 3F3 or 3P2 waves,
which are known to have large cross sections (e.g., 10-40% of the
total inelastic cross sections). These goals are modest, compared
with the very detailed knowledge that has been obtained from the
elastic PSA.

COMPARISON OF NNπ AND π^+d FINAL STATES

Deuteron production presumably occurs by final state
interactions, via pp $\to \pi^+$pn $\to \pi^+$d. Thus we would expect
resonances in the primary production process, pp $\to$ NNπ, to show
up in both the free and bound nucleon final states. Specifically,
we would expect the behavior of individual waves (i.e., phase
versus energy) to be the same in these two cases. On the other
hand we might expect the relative intensities of the individual
waves to differ, as is experimentally the case. In comparing
the free and bound nucleon reactions, some complications deserve
note.

(1) There are more partial waves in pp $\to \Delta N$ ($S_{\Delta N}= 1,2$) then
in pp $\to \pi^+$d ($S_{\pi d}= 1$).
(2) There are more observables in pp $\to \Delta N \to p\pi^+$n then
in pp $\to \pi^+$d, due to the $\Delta^{++} \to p\pi^+$ decay distribution. The
observables are listed in Table I.
(3) Non-Δ backgrounds from pp $\to$ n(pπ^+)$_{s-wave}$ are expected
from pion-exchange. These can be separated using the pπ^+ angular
distributions for the NNπ final states; they cannot be isolated
in pp $\to \pi^+$d, where they are expected to contaminate the 3P1
waves.
(4) For the free nucleon case, we select on $M_{p\pi^+}$ to define a
Δ^{++} band. For pp $\to \pi^+$d, $M_{p\pi^+}$ is a function of P_{lab} ($M_{p\pi^+} \sim M_{\Delta^{++}}$
at 1.3 GeV/c). Consequently, pp $\to \pi^+$d reflects the underlying

pp → ΔN process only over a limited energy range (p_{lab} ≲ 1.5 GeV/c). Also, amplitude phases for pp → π^+d are usually quoted including the Δ^{++} Breit-Wigner phase from the ΔN intermediate state. For dibaryon resonance studies we need to remove the common Δ^{++} decay phase in order to isolate the pp → ΔN production amplitude behavior.

Table I. Observables used for amplitude analysis in pp → ΔN and pp → π^+d and their sensitivity to partial wave contributions; here S(T) denote single (triplet) initial pp states. See Ref. 1 for definitions of ρ_{jk} and $P_i \rho_{jk}$.

Spin	pp → ΔN	pp → π^+d	Sensitivity
0-spin	$\rho_{11}, \rho_{33}, \rho_{31}, \rho_{3-1}$	σ	Re[SS'*+TT'*]
1-spin	$P_y\rho_{11}, P_y\rho_{33}, P_y\rho_{31}, P_y\rho_{3-1}$	A_{yo}, iT_{11}	Im[ST*], Im[TT*]
	$P_x\rho_{31}, P_x\rho_{3-1}, P_z\rho_{31}, P_z\rho_{3-1}$		
2-spin	limited data available	$A_{xx}, A_{yy}, A_{zz}, A_{xz}$	Re[SS'*+TT'*], Re[ST*], Re[TT*]

SOME FEATURES OF THE DATA

Is there any interesting energy dependence in the inelastic production observables? Figure 2 depicts the trend of selected spin observables in pp → π^+d and pp → Δ^{++}n, evaluated near 90° in the cm. The kinematics are such that pp → π^+d should proceed mainly from helicity-1/2 Δ^{++} production, and there is qualitative agreement between A_{yo}[pp → π^+d] and A_{yo}[pp → $\Delta^{++}(\lambda=\frac{1}{2})$n], as in Fig. 2b.[1] The helicity-3/2 Δ^{++} production, which does not feed the π^+d final state, exhibits fairly dramatic energy dependence and changes sign between the 1D2 and 3F3 peaks (Fig. 2c). If only 1D2 and 3F3 waves were present, we would have

$$-A_{xz} \propto + \text{Re}[1D2\ 3F3^*] \qquad (2a)$$

$$A_{yo}(\lambda_\Delta = \frac{1}{2} \text{ or } \frac{3}{2}) \propto + \text{Im}[1D2\ 3F3^*] \qquad (2b)$$

Comparison of Figs. 2a and 2b suggests gradual variation in the 1D2 and 3F3 relative phase; comparison of 2b and 2c suggests that 1D2 and 3F3 alone cannot account for the behavior of A_{yo}.

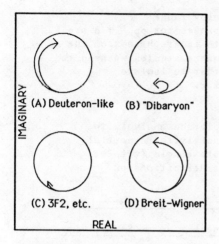

Figure 1. Argand plot behaviors. Examples are 1S0, 3S1, 3P0, 3P1 (A); 1D2, 3P2, 3F3 (B); 3F2 (C).

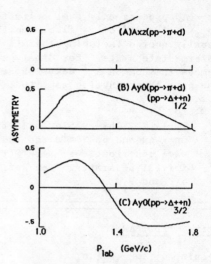

Figure 2. A_{xz} and A_{yo} vs. P_{lab} ($\theta \sim 90°$); B(C) refer to Δ-helicity 1/2 (3/2).

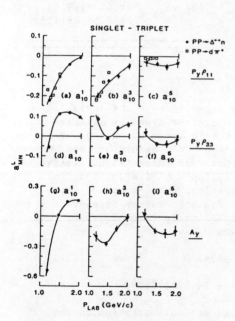

Figure 3. Normalized L=1,3,5 moments from Eq. (3); π^+d data are from Ref. 13.

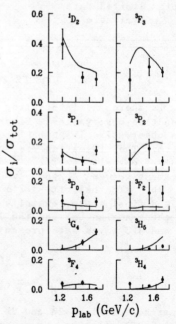

Figure 4. Relative intensity for leading $pp \rightarrow \Delta^{++}n$ waves: curves are PSA fits, data points are OPE fits.

As indicated in Table I, there are twelve 0,1-spin observables in $pp \to \Delta^{++}n$. These can be further expanded in θ_Δ, the cm Δ^{++} production angle,[1]

$$P_i \rho_{jk} \frac{d\sigma}{d\cos\theta_\Delta} = \sum_L a^L_{MN} d^L_{MN}(\theta_\Delta) \qquad (3)$$

where the coefficients a^L_{MN} are related quadratically to the production waves. Figure 3 illustrates the p_{lab} dependence for some of the odd-L coefficients for helicity - 1/2 and 3/2 Δ-production asymmetries (odd-L correspond to singlet-triplet interferences); also shown are corresponding coefficients for $pp \to \pi^+d$. Strong p_{lab} dependence is seen in all of the moments. Altogether, at each p_{lab} we obtain ~ 80 a^L_{MN}'s from the 12 $P_i \rho_{jk}$ for the PWA.

RESULTS

We used an absorbed pion-exchange (OPE) model, fitted to the unpolarized cross sections and dme's from 1.18 to 11.75 GeV/c,[1,8] to estimate the complexity needed for an accurate PWA expansion up to 2 GeV/c. The 59 $pp \to (p\pi^+)n$ transition amplitudes used in our fits are sorted into categories in Table II. The "very high partial waves" and the non-Δ^{++} production waves were fixed by the OPE model. Ten "low" (LPW) and eighteen "high" (HPW) $pp \to \Delta^{++}n$ waves were varied in the fits; the HPW were constrained to be real (except for the common Δ^{++} decay phase, which is not included in our fitted results). The small contributions from $pp \to \Delta^+p \to n\pi^+p$, required by isospin, were not explicitly included in our PWA and are distributed among the various $pp \to (p\pi^+)n$ isobar contributions; it is easily shown that Δ^+-Δ^{++} interference has negligible effect on the spin correlations after averaging over the Δ^{++} mass band.[1] To test for continuity, we carried out the PWA over a fine grid in p_{lab}, by interpolating the data smoothly between the measured momenta (1.18, 1.47, 1.71, and 1.98 GeV/c).

Table II. Listing of $pp \to (p\pi^+)n$ transition amplitudes: LPW (low partial waves), HPW (high partial waves), VHPW (very high partial waves) and S/P isobars (non-Δ^{++} isobars with $J^P = 1/2^\pm$).

LPW	HPW	VHPW	S/P Isobars
1D2->5S2	1D2->3D2	1D2->5G2	1S0->3P0 (S)
3P0->3P0	1D2->5D2	1G4->3G4	3P0->1S0 (S)
3P1->3P1	3P1->5F1	1G4->5G4	3P1->3S1 (S)
3P1->5P1	3P2->3F2	3F3->5H3	1D2->3P2 (S)
3P2->3P2	3P2->5F2	3F4->3H4	3P1->3D1 (S)

3P2->5P2	3F2->3F2	3F4->5H4	3P2->1D2 (S)
3F2->3P2	3F2->5F2	3H4->3H4	3P2->3D2 (S)
3F2->5P2	3F3->3F3	3H4->5H4	3F2->1D2 (S)
3F3->5P3	3F3->5F3	3H5->3H5	3F2->3D2 (S)
1S0->5D0	3F4->3F4	3H5->5H5	3F3->3D3 (S)
	3F4->5F4	3J6->3H6	1S0->1S0 (P)
	1G4->5D4	3J6->5H6	3P0->3P0 (P)
	3H4->3F4	3J7->5H7	3P1->1P1 (P)
	3H4->5F4		3P1->3P1 (P)
	3H5->5F5		3P2->3P2 (P)
	1I6->5G6		3F2->3P2 (P)
	3H6->3H6		1D2->1D2 (P)
	3H6->5H6		1D2->3D2 (P)

We considered many PWA scenarios, of which we show two: (a) "PSA" fits - the partial cross sections for the 1D2, 3F2,3, and 3P0,1,2 were constrained to agree, percentage-wise, with the inelastic partial cross sections from the PSA of Ref. 3. The HPW were fitted separately and were constrained to be smooth and monotonic and roughly consistent with the PSA intensities at 1.7 GeV/c. (b) "OPE" fits-the data on $\Delta\sigma_L$ and $\Delta\sigma_t$ were used to help constrain the partial wave intensities, and the HPW were freely varied simultaneously with the LPW. The values of $\Delta\sigma_L$ and $\Delta\sigma_t$ were obtained from the PSA of Ref. 3 together with spin correlation data on $pp \to \pi^+ d$;[9] the estimates and fit values for $\Delta\sigma$ are listed in Table III. We remark that our fits also predict the $pp \to \Delta N$ contributions to $\Delta\sigma_{L,t}$ which arise from the $^3P_2$-$^3F_2$ coupling;[10] these do not seem to agree in sign or magnitude with the PSA, but they depend on the coupling angle, ε, which may not be well determined above inelastic threshold.

Table III. Contributions to $\Delta\sigma_L$ and $\Delta\sigma_t$; "$\Delta\sigma$(INEL)" are total inelastic contributions from the PSA, and "$\Delta\sigma(\pi d)$" are estimates based on A_{xx}, A_{yy}, and A_{zz} data of Ref. 9. $\Delta\sigma/\sigma$ are ratios expected for the $NN\pi$ final states; $\Delta\sigma/\sigma$(Fit) are the "OPE" fit results for the $pn\pi^+$ data in the Δ^{++} band.

P_{lab} (GeV/c)	$\Delta\sigma_L^{INEL}$	$\Delta\sigma_t^{INEL}$ (mb)	$\Delta\sigma_L^{\pi d}$	$\Delta\sigma_t^{\pi d}$	$\frac{\Delta\sigma_L}{\sigma}$	$\frac{\Delta\sigma_L^{Fit}}{\sigma}$	$\frac{\Delta\sigma_t}{\sigma}$	$\frac{\Delta\sigma_t^{Fit}}{\sigma}$
1.2	3.3	7.9	2.8	3.7	+ 0.06	0.06	0.51	0.55
1.5	-6.7	5.0	~0.2	~1.0	- 0.34	-0.34	0.20	0.19
1.7	-3.9	4.8	~0	~0.5	- 0.20	-0.12	0.22	0.22

The relative intensities for the "PSA" and "OPE" solutions are compared in Fig. 4; the "OPE" fits have large statistical uncertainties and generally give higher intensities for 3F2 and the HPW. The phases of the LPW[11] are compared in Fig. 5, together with smooth curves that show the results of the continuity test described above; for the "OPE" fits, the smooth curves depict a variant solution, in which the HPW were simply fixed at the OPE values. While the (three) solutions are not in precise agreement, the qualitative behavior of the dominant 1D2, 3F3, and 3P0,1,2 waves generally suggests a gradual p_{lab} dependence in all cases.

In Bugg's analysis[2] of $pp \to \pi^+ d$, the HPW were fixed by theory, and only seven LPW ($a_{0...6}$ = 1S0,3P1,1D2,3P1',3P2,3F2, and 3F3 respectively) were fitted; the 1D2 phase was also constrained by theory, and was approximately consistent with the expected Δ^{++} decay phase. To remove the unwanted Δ^{++} Breit-Wigner phase, we use the 1D2 as a reference wave, and compare the relative phases, $\phi_i - \phi_{1D2}$, for $pp \to \pi^+ d$ and $pp \to \Delta^{++} n$ in Fig. 6.[12] There are inherent ambiguities in this comparison, since there are two lowest orbital waves in $pp \to \Delta^{++} n$ for 3P1, 3P2, and 3F2 compared with only one in $pp \to \pi^+ d$. In the case of 3P2 and 3F2, we can ignore the 3P,F2 $\to$ 5P2 $pp \to \Delta^{++} n$ transitions because they contribute very little to helicity-1/2 Δ production. Overall the level of agreement between the two sets of PWA phases is encouraging. Note that neither solution indicates strong p_{lab} dependence for the relative production phases; the backwards motion of the 3P1 - 1D2 relative phases in $pp \to \pi^+ d$ probably reflects the $p\pi^+$ S-wave background in 3P1 noted above.

Finally, Fig. 7 compares Argand plots for the dibaryon waves in $pp \to \Delta N$ and $pp \to pp$.

CONCLUSION

We find that neither the absolute nor the relative phase motions in $pp \to \Delta N$ or $pp \to \pi^+ d$ suggest Breit-Wigner behavior in the production waves. Instead, our solutions indicate "deuteron-like" behavior for the largest ΔN waves, that is, clockwise phase rotation with increasing p_{lab}. This behavior is similar to that of the large waves (1S0, 3P0,1, 3S1) in elastic nucleon-nucleon scattering, and suggests "conventional" interpretations of the dibaryon resonances, such as bound states or strong hadronic interactions in the $\Delta N \to \Delta N$ channels.

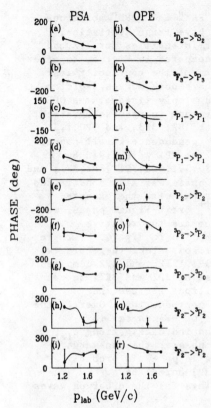

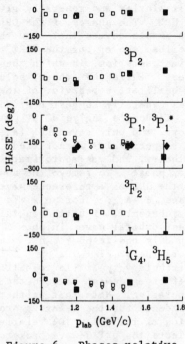

Figure 5. LPW phases for pp → Δ^{++}n for PSA and OPE fits, with interpolation as described in text.

Figure 6. Phases relative to 1D2 for pp → π^+d (open points, Ref. 2) and pp → ΔN (solid points, PSA fits).

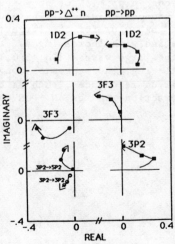

Figure 7. Argand plots for 1D2, 3F3, and 3P2 waves in pp → Δ^{++}n (OPE solution) and pp → pp. From arrows, points are 1.7, 1.5, 1.2 GeV/c.

REFERENCES

1. A. B. Wicklund et al., Argonne preprint, ANL-HEP-PR-86-37 (1986).
2. D. V. Bugg, Nucl. Phys. A437, 534 (1985); J. Phys. G: Nucl. Phys. 10, 717 (1984).
3. R. A. Arndt et al., Phys. Rev. D28, 97 (1983).
4. J. Bystricky, C. Lechanoine-Leluc, and F. Lehar, Saclay preprint, DPhPE 82-12 (Rev. Feb. 1984).
5. N. Hoshizaki, Prog. Th. Phys. 61, 129 (1979).
6. R. Dubois et al. Nucl. Phys. A377, 554 (1982).
7. M. P. Locher, Nucl. Phys. A416, 243C (1984); J. P. Auger, C. Lazard, R. J. Lombard, and R. R. Silbar, Nucl. Phys. A442, 621 (1985); P. Kroll, Journal de Physique C2, 493 (1985).
8. A. B. Wicklund et al., Argonne preprint ANL-HEP-PR-85-81; to be published in Phys. Rev. D.
9. E. Aprile-Giboni et al., Nucl. Phys. A415, 365 (1984); G. Glass et al., Phys. Rev. C31, 288 (1985); W. B. Tippens et al., Few Body Problems in Physics, Vol. II p. 17, B. Zeitnitz (ed.), Elsevier Science Publishers B.V. (1984).
10. M. Arik and P. G. Williams, Nucl. Phys. B136, 425 (1978).
11. The 1S0 wave (not shown) is very small and not well determined, and was set to zero for the PSA fits.
12. To correct for phase conventions we have applied an overall $180°$ phase to the odd-parity pp $\to$ ΔN waves.
13. A. Saha et al., Phys. Rev. Lett. 51, 759 (1983).

MODEL INDEPENDENT EXCHANGE CURRENTS

D.O.Riska
Department of Physics, University of Helsinki
00170 Helsinki, Finland

Abstract
It is shown that the exchange current operators of main importance for the electron scattering observables of the few-nucleon systems can be determined almost uniquely in a gauge invariant way. The most important isovector exchange current operator can be obtained directly from the spin- and isospin-dependent components of the nucleon-nucleon interaction once these have been decomposed into terms that are associated with exchange of objects with pseudoscalar and vector quantum numbers. The main isoscalar exchange current operator is uniquely determined by the anomalous baryon current operator in the description of the baryons as topological solitons of meson fields a la Skyrme.

1. Introduction

Many of the electron scattering observables for light nuclei at low energy but large values of momentum transfer reveal large contributions from two-body exchange current operators [1,2]. Although reasonably successful models for these exchange current operators have been constructed on the basis of simple boson exchange mechanisms, such involve considerable numbers of more or less phenomenological parameters as masses and coupling constants and ad hoc high-momentum cut-off factors. Hence it has remained as a desireable goal to find ways of constructing the exchange current operators that are consistent with the model for the nucleon-nucleon interaction and which respect fundamental constraints as gauge- and chiral invariance. Some recent work in this direction will be reviewed here.

The isovector exchange current operator of main importance in the few-nucleon systems is that which is associated with the isospin independent tensor and spin-spin components of the nucleon- nucleon interaction. In section 2 a consistent and gauge invariant method [3,4] to construct this exchange current operator directly from the interaction is described. In this method the interaction components are separated into terms that are due to the exchange of systems with pseudoscalar (0^-) and vector (1^{-+}) quantum numbers. The resulting exchange current operator is completely determined by the interaction and in this sense it is model independent.

The isoscalar exchange current operators that are most important for the electron scattering observables in light nuclei are purely transverse operators, which are unrelated to the nuclear force. For these one has hitherto had to rely on explicit meson exchange mechanisms - the most important being the $\rho\pi\gamma$ -mechanism [5,6] - that are totally model dependent. It has however now been shown [7] that all these exchange current operators are effectively included in the isoscalar exchange current operator that is given by the anomalous baryon current in the description of baryons as topological solitons of meson fields [8]. This exchange current operator is then determined by the chiral anomaly and the topology and it is therefore essentially model independent. The construction of this exchange current operator, and its manifestations in the magnetic form factors of the nucleons and the deuteron, are described in section 3.

In addition to this isoscalar exchange current operator there are isoscalar exchange current operators that are associated with the explicitly velocity dependent components of the nucleon-nucleon interaction. In section 4 we describe the construction of these operators directly from the corresponding interaction components by minimal substitution.

2. Model independent isovector exchange currents

As the isospin dependent components of the nucleon-nucleon interaction contain the isospin factor $\tau_1.\tau_2$, which does not commute with the single-nucleon charge operator, the continuity equation requires the presence of exchange current operators, the divergence of which cancel the isospin commutators. Although the interaction term in the continuity equation only specifies the longitudinal component of the exchange current operator the form of all its components will be closely related to the form of the corresponding interaction. Hence it is not possible to construct a model for the exchange current operator separately from the interaction in a consistent way. The commonly employed single meson exchange current operators are e.g. not consistent with the realistic semiphenomenological models for the nucleon-nucleon interaction and do not satisfy the continuity equation.

One can nevertheless use the known forms for the single pion and ρ-meson exchange interactions and exchange current operators, which are mutually consistent, if unrealistic, as suggestive for the form of the consistent exchange current operators for realistic interaction models. For this purpose one has to separate the isospin dependent spin-spin and tensor components of a given interaction into terms that arise from the exchange of objects with the quantum numbers of pseudoscalar and vector mesons [3,4].

If the spin-spin and tensor parts of a given phenomenological nucleon-nucleon interaction model are

$$V = [v_{SS}(r)\sigma_1.\sigma_2 + V_T(r)S_{12}]\tau_1.\tau_2, \qquad (2.1)$$

the components of e.g. the tensor potential v_T that are due to pseudoscalar and vector exchange mechanisms are (in momentum space)

$$v^{PS}(k) = [2v_T(k) - v_{SS}(k)]/3, \qquad (2.2a)$$

$$v^V(k) = [v_T(k) + v_{SS}(k)]/3. \qquad (2.2b)$$

Here $v_{SS}(k)$ and $v_T(k)$ are the following Hankel transforms of $v_T(r)$ and $v_{SS}(r)$:

$$v_{SS}(k) = \frac{4\pi}{k^2} \int_0^\infty dr\, r^2 [j_0(kr) - 1] v_{SS}(r), \qquad (2.3a)$$

$$v_T(k) = \frac{4\pi}{k^2} \int_0^\infty dr\, r^2 j_2(kr) v_T(r). \qquad (2.3b)$$

The tensor potentials $v^{PS}(k)$ and $v^V(k)$ may be viewed as "generalized pion" and "generalized vector meson exchange potentials".

If these generalized pseudoscalar and vector exchange tensor interactions are substituted in place of the single pion and ρ-meson pole terms in the expressions for the single pion and ρ-meson exchange current operators one obtains isovector exchange current operators that when combined satisfy the continuity equation with the general interaction (2.1). The explicit forms for the pseudoscalar and vector exchange current operators are [3]

$$\vec{j}_{PS} = -i\,(\vec{\tau}^1 \times \vec{\tau}^2)_3 \,\{\, v_T^{PS}(\vec{k}_2)\,\vec{\sigma}^1(\vec{\sigma}^2 \cdot \vec{k}_2) - v_T^{PS}(\vec{k}_1)\,\vec{\sigma}^2(\vec{\sigma}^1 \cdot \vec{k}_1) $$
$$ - \frac{\vec{k}_1 - \vec{k}_2}{k_1^2 - k_2^2}\,\vec{\sigma}^1 \cdot \vec{k}_1\,\vec{\sigma}^2 \cdot \vec{k}_2\,[v_T^{PS}(\vec{k}_2) - v_T^{PS}(\vec{k}_1)]\,\}, \qquad (2.4a)$$

$$\vec{j}_V = -i\,(\vec{\tau}^1 \times \vec{\tau}^2)_3 \,\{\, v_T^V(\vec{k}_2)\,\vec{\sigma}^1 \times (\vec{\sigma}^2 \times \vec{k}_2) - v_T^V(\vec{k}_1)\,\vec{\sigma}^2 \times (\vec{\sigma}^1 \times \vec{k}_1) $$
$$ + \frac{v_T^V(\vec{k}_2) - v_T^V(\vec{k}_1)}{k_1^2 - k_2^2}\,[(\vec{k}_1 - \vec{k}_2)\,\vec{\sigma}^1 \times \vec{k}_1 \cdot \vec{\sigma}^2 \times \vec{k}_2 + (\vec{\sigma}^1 \times \vec{k}_1)\,\vec{\sigma}^2\cdot(\vec{k}_1 \times \vec{k}_2) + (\vec{\sigma}^2 \times \vec{k}_2)\,\vec{\sigma}^1\cdot(\vec{k}_1 \times \vec{k}_2)]\,\}. \qquad (2.4b)$$

These current operators should finally be multiplied by the isovector form factor $F_1^V(q)$ of the nucleon [3].

The exchange current operators (2.4) are model independent in that they are completely determined by the nucleon-nucleon interaction. A particularly nice feature of these operators is that their short range behaviour is determined by the interaction and thus indirectly by the nucleon-nucleon phase shifts without any need for introduction of ad hoc high-momentum cut-off factors.

In Fig.1 we show the exchange current dominated cross section for backward electrodisintegration of the deuteron near threshold as obtained with the exchange current operators (2.4) and the wavefunctions given by the parametrized Paris potential [9]. For the particular case of that potential which is given as a superposition of Yukawa functions the exchange current operators (2.4) were also obtained in Ref.[4]. The agreement with the empirical values shown in Fig.1 is quite satisfactory even at large values of momentum transfer.

In addition to these model independent isovector exchange current operators there are naturally, in spite of what might be concluded on the basis of Fig.1, an infinity of transverse model dependent exchange current operators that cannot be obtained from the nuclear force. Of these however only those that are associated with the excitation of virtual intermediate Δ_{33}-resonances are significant. While these play no noticeable role at large momentum transfer in electrodisintegration of the deuteron near threshold [10], they are important at low momentum transfer. In the inverse reaction of radiative np-capture the Δ-excitation current operators contribute roughly on third of the total exchange current correction [11]. These operators also play a very interesting role for the quenching of the effective isovector spin magnetic moment and M1 strength in heavy nuclei [12,13].

3. Model independent Transverse isoscalar exchange currents

The isoscalar exchange current operators can be divided into terms that have longitudinal components, which satisfy the continuity equation with the velocity dependent components of the nucleon-nucleon interaction [6,14] and into terms that are purely transverse and unrelated to the interaction. In the few-nucleon systems the latter ones are the most important ones [5,14]. These have hitherto been constructed form completely model dependent meson exchange current mechanisms. We shall here describe the recent suggestion [7] that these can in fact be constructed in a model independent way from the anomalous baryon current that appears in Skyrme's description of the baryons as topological solitons of meson fields [8]. This description may be viewed as an effective bosonization of QCD in the large color limit at low energies. The baryon current current is topological and in effect determined by the hidden Wess-Zumino term that is required by the chiral anomaly.

The anomalous baryon current has the form

$$B^\mu = \frac{\epsilon^{\mu\nu\alpha\beta}}{24\pi^2} Tr\{U^\dagger \partial_\nu U \, U^\dagger \partial_\alpha U \, U^\dagger \partial_\beta U\}, \qquad (3.1)$$

where U is the SU(2) field of the Skyrme model or generalizations thereof. The spatial integral of B gives the topological charge, which is interpreted as the baryon number[8,15].

For the fields of the single nucleons one uses Skyrme's ansatz

$$U(r) = exp(i\hat{r}.r\theta(r)), \qquad (3.2)$$

where the chiral angle θ satisfies the Euler-Lagrange equations of motion for the Lagrangean.

For the two-nucleon system one employs the product ansatz

$$U(r_1, r_2, r) = U(r - r_1)U(r - r_2). \qquad (3.3)$$

Here r_1 and r_2 are the centers of the two solitons and r is the point of interaction with the electromagnetic field. With this ansatz the baryon current splits into a sum of two single nucleon current operators and an exchange current operator as

$$B^\mu(r_1, r_2, r) = B_1^\mu(r - r_1) + B_1^\mu(r - r_2) + B_{ex}^\mu(r_1, r_2, r) \qquad (3.4)$$

Here B_1^μ is the baryon current for a single nucleon and B_{ex}^μ the two-nucleon exchange current operator.

In order to construct the e.m. current operator with the proper isoscalar properties from the baryon current (3.4) the soliton fields U in (3.3) have to be rotated so as to split the spin-isospin degeneracy [15]. Rotation of the single baryon current operators in (3.4) leads to the following expressions for the isoscalar current operator for a single nucleon

$$j_0^S(\vec{r}) = -\frac{1}{4\pi^2} \frac{\sin^2\theta(r)}{r^2} \theta'(r), \qquad (3.5a)$$

$$\vec{j}^S(\vec{r}) = \frac{1}{8\pi^2 \lambda} \frac{\sin^2\theta(r)}{r^2} \theta'(r) \, \vec{\sigma} \times \vec{r} \qquad (3.5b)$$

Here $\theta(r)$ is the chiral angle and λ the moment of inertia of a single soliton [15]. These expressions imply the following interesting relation between the isoscalar electric and magnetic form factors of for a single nucleon:

$$G_M^S(q) = -\frac{2m}{\lambda}\frac{d}{dq^2} G_E^S(q), \tag{3.6}$$

which is in rather good agreement with the empirical data on the isoscalar form factors.

The explicit expression for the isoscalar exchange current operator for the deuteron is given in Ref.[7].

Although the structure of the exchange current operator is completely determined by the baryon current operator (3.1) the amplitudes of its components depend on the chiral angle, which is of course dependent on the explicit version of the nonlinear Lagrangean. One can however also determine the chiral angle from the isoscalar electric form factor of the nucleon by taking the Fourier transform of eq.(3.5a). In the Skyrme approach the isoscalar exchange current operator is then completely determined by the isoscalar form factor of a single nucleon.

In Fig.2 we show the magnetic form factor of the deuteron as it is obtained using the Skyrme approach for both the single nucleon and exchange current operators. For the chiral angle I have used the result of Ref.[16], which gives reasonably good predictions for the isoscalar form factors of the nucleon. The deuteron wavefunctions used were those obtained with the parametrized Paris potential [9]. The result in Fig.2 should be compared with the similar results that are obtained with the help of various transverse meson exchange current contributions - mainly that of the $\rho\pi\gamma$-mechanism [5,6]. Although the results obtained with conventional meson exchange models can be brought into good agreement with the empirical values by suitable parameter adjustments and multiplication by additional e.m. formfactors as suggested be the vector-meson dominance model they are nevertheless completely model dependent and therefore have little predictive power. In the Skyrme approach in its original form on the other hand the topology of the solitons completely determine the isoscalar form factors of both nucleons and nuclei. It does however seem to be difficult to explain both the isoscalar for factors of the nucleons and the deuteron well with the same chiral angle in the original Skyrme model. This suggests that one should build in the vector meson degrees of freedom already at the level of the Lagrangean as in Ref.[23]. This actually does not bring in additional model dependence as in that case the solitons are stabilized by the vector mesons instead of by Skyrme's quartic term in the Lagrangean.

4. Velocity dependence and exchange currents

The nucleon-nucleon interaction contains explicitly velocity dependent terms, which by the continuity equation, require the presence of corresponding exchange current operators. Although one may construct isoscalar exchange current operators directly from these terms by minimal substitution that satisfy the continuity equation with the interaction, those current operators are in general inconsistent with microscopic models [6,17]. The reason is that the phenomenological interactions are available only in the center-of-mass

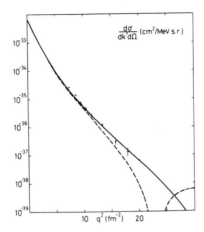

Fig.1 The backward cross section for ed-enp at 1.5 MeV above threshold. The dashed line is the impulse approximation and the solid line the prediction when the exchange current correction is included. The data points are from Ref.[21].From Ref.[3]

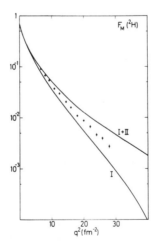

Fig.2 The magnetic form factor of the deuteron. The curve I represents the impulse approximation result as obtained with the Skyrme model and the curve II shows the result as obtained with inclusion of the exchange current correction. The data points are from ref.[22].

frame, whereas the construction of the proper exchange current by minimal substitution requires the form of the interaction valid in a general frame. A method for the generalization of a given phenomenological spin-orbit interaction to a general frame has however recently been developed in Ref.[17], and by that method a nonrelativistically adequate model independent exchange current operator may be constructed. This method will be reviewed here.

The usual form of the spin-orbit interaction in the c.m. frame is

$$V_{LS} = \frac{1}{2} \frac{1}{r} \frac{d}{dr} v_{SO}(r) \ (\vec{\sigma}^1 + \vec{\sigma}^2) \cdot \vec{\ell} \ . \tag{4.1}$$

In momentum space this interaction becomes

$$V_{LS} = \frac{i}{2} v_{SO}(k) \ (\vec{\sigma}^1 + \vec{\sigma}^2) \cdot (\vec{p}' \times \vec{p}) \ , \tag{4.2}$$

where p and p' are the initial and final nucleon momenta and $v_{SO}(k)$ is the Fourier transform of $v_{SO}(r)$. In a general frame (4.2) would be replaced by the expression

$$V_{LS} = \frac{i}{2} \Big\{ v_a(k) [\vec{\sigma}^1 \cdot \vec{p}_1' \times \vec{p}_1 + \vec{\sigma}^2 \cdot \vec{p}_2' \times \vec{p}_2] \\ + v_b(k) [\vec{\sigma}^1 \cdot (\vec{p}_1' - \vec{p}_1) \times \frac{\vec{p}_2' + \vec{p}_2}{2} + \vec{\sigma}^2 \cdot (\vec{p}_2' - \vec{p}_2) \times \frac{\vec{p}_1' + \vec{p}_1}{2}] \Big\}, \tag{4.3}$$

which reduces to (4.2) in the c.m. frame with $v_{SO} = v_a + v_b$. To determine the two separate components v_a and v_b of the spin-orbit interaction one may proceed in the following way [17]. One first determines the components of the phenomenological interaction in terms of 5 relativistic spin amplitudes (combinations of γ-matrices) in place of the nonrelativistic spin amplitudes in terms of which the interaction is given. A subsequent nonrelativistic reduction carried out in a general frame yields the spin-orbit components v_a and v_b as linear combinations of all the components of the phenomenological interaction[17].

By minimal substitution in the spin-orbit interaction (4.3) one obtains an exchange current operator that satisfies the continuity equation with the isospin independent spin-orbit interaction when the nonrelativistic single nucleon charge operator is used in the equation. Furthermore this exchange current agrees with the seagull-type exchange current operators that arise when nonrelativistic meson-nucleon coupling Lagrangeans are used [17].

A similar exchange current operator naturally also arises from the isospin dependent spin-orbit interaction which contains a factor $\tau_1 \cdot \tau_2$. This current operator can be obtained by a simple replacement in the previous one as shown in Ref.[17].

In addition to the linearly velocity dependent spin-orbit interaction realistic nucleon-nucleon interaction models contain terms with quadratic velocity dependence. The central and spin-spin interactions contain small velocity dependent terms of the form [9,18]:

$$V(r) = \frac{p^2}{m} v(r) + v(r) \frac{p^2}{m}. \tag{4.4}$$

The quadratic spin-orbit interaction is of the same order in v/c as the velocity dependent central interaction (4.4). The complete for of this interaction is [6]

$$\begin{aligned} V_{LS2} &= -\frac{1}{2} \frac{1}{r} \frac{d}{dr} \frac{1}{r} \frac{d}{dr} v(r) \, (\vec{\sigma}^1 \cdot \vec{\ell} \, \vec{\sigma}^2 \cdot \vec{\ell} + \vec{\sigma}^2 \cdot \vec{\ell} \, \vec{\sigma}^1 \cdot \vec{\ell}) \\ &+ \frac{i}{2} \frac{1}{r} \frac{d}{dr} \frac{1}{r} \frac{d}{dr} v(r) \, (\vec{\sigma}^1 \times \vec{r} \cdot \vec{\sigma}^2 \times \vec{p} + \vec{\sigma}^2 \times \vec{r} \cdot \vec{\sigma}^1 \times \vec{p}) - \frac{1}{r} \frac{d}{dr} v(r) \, \vec{\sigma}^1 \times \vec{p} \cdot \vec{\sigma}^2 \times \vec{p}. \end{aligned} \tag{4.5}$$

In most phenomenological interaction models only the first term in this expression is included, although consistency with boson exchange models requires the inclusion of the complete expression.

By minimal substitution in the velocity dependent interaction (4.4) and quadratic spin-orbit interaction (4.5) one can generate associated isoscalar exchange current operators [6]. These will however only be consistent with dynamical models if the interaction is consistent with a relativistic model for the scattering amplitude that has proper spin structure. The boson exchange models for the interaction usually satisfy this criterion, but most phenomenological interaction models do not.

The effect of the isoscalar exchange current operators that are associated with the velocity dependent components of the parametrized Paris potential [9] on the magnetic form factor of the deuteron were investigated in Ref.[9]. The results show that these exchange current corrections reduce the predicted form factor from the impulse approximation prediction. As the velocity dependent components of the potential represent relativistic corrections it is in fact natural that inclusion of the corresponding exchange current corrections reduce the predicted form factor as this brings the prediction closer to that obtained by the relativistc potential model in Ref. [19]. The remaining large discepancy between the predicted and the measured form factor values can only be explained by inclusion of the transverse exchange current corrections discussed in the previous section.

5. Discussion

The exchange current contributions to nuclear electron scattering observables are most notable in the few-nucleon systems. In large nuclei the exchange current effects are often hidden by corrections of comparable magnitude from small components of the wave functions although inclusion of exchange current corrections usually improves the agreement with measured data [20]. Hence from the point of view of determining the form of the exchange current operators by comparison to data, the electromagnetic form factors of the few nucleon systems provide the most fruitful field for study.

The empirical cross-section for backward electrodisintegration of the deuteron near threshold (Fig.1) shows that the isovector exchange current operator that has been constructed directly from the nucleon-nucleon interaction is quite adequate. This operator does of course contain far more physics than the usual simple boson exchange current operators and it is essentially unique once the interaction has been specified.

The situation regarding the isoscalar exchange current operators has also now been improved in the sense that there is now a good theoretical framework for their construction.

Those isoscalar exchange currents that are associated with the velocity dependent parts of the nucleon- nucleon interaction can be constructed from the corresponding interaction components by the methods discussed in section 4. More importantly the Skyrme approach to the description of the baryons as topological solitons of meson fields has made it possible to construct the previously poorly known transverse isoscalar exchange currents directly from the anomalous baryon current without even specifying the form of the nonlinear meson Lagrangean. The isoscalar exchange current operator depends on the Lagrangean only indirectly through the chiral angle. The Skyrme approach also allows one to construct the hitherto rather uncertain isoscalar exchange charge operators by a theoretically satisfactory way.

References
1. Progress in Nuclear and Particle Physics,Vol.11,D.S. Wilkinson ed.,Pergamon Press (1984)
2. Mesons in Nuclei, M.Rho and D.S.Wilkinson eds., North Holland, Amsterdam, Vol.2 (1979)
3. D.O.Riska, Physica Scripta 31, 471 (1985)
4. A.Buchmann, W.Leidemann and H.Arenhovel, Nucl.Phys.A443,726 (1985)
5. M.Gari and H.Hyuga, Phys.Rev.Lett. 36, 345 (1976)
6. D.O.Riska and M.Poppius, Physica Scripta 32, 581 (1985)
7. E.M.Nyman and D.O.Riska, "Static electromagnetic properties of the deuteron in the Skyrme model", Preprint HU-TFT-85-47 (1985)
8. T.H.R.Skyrme, Proc.Roy.Soc. A460, 127 (1961)
9. M.Lacombe et al., Phys. Rev.C21,861 (1980)
10. J.Hockert et al., Nucl.Phys.A217,14 (1973)
11. D.O.Riska and G.E.Brown, Phys.Lett.38B,193 (1972)
12. W.Knupfer, M.Dillig and A.Richter, Phys.Lett.95B,349 (1980)
13. D.O.Riska, Prog.Nucl.and Part.Phys.11,199 (1984)
14. D.O.Riska, Physica Scripta 31,107 (1985)
15. G.S.Adkins, C.R.Nappi and E.Witten, Nucl.Phys.B228,251 (H1984)
16. G.S.Adkins and C.R.Nappi, Nucl.Phys.B233, 251 (1984)
17. D.O.Riska, Physica Scripta 31, 107 (1985)
18. K.Holinde and R.Mandelius, Nucl.Phys. A256, 479 (1976)
19. R.Arnold, C.E.Carlson and F.Gross, Phys.Rev.C21 1426 (1980)
20. B.Desplanques and J.F.Mathiot, Phys.Lett. 116B, 82 (1982)
21. M.Bernstein et al.,Phys.Rev.Lett. 46, 402 (1981)
22. S.Auffret et al., Phys.Rev.Lett. 54, 649 (1985)
23. U.G.Meissner and I.Zahed, Phys.Rev.Lett. 56, 1035 (1986)

A PRECISION TEST OF CHARGE INDEPENDENCE*

Kamal K. Seth, M. Artuso, D. Barlow
B. Parker and R. Soundranayagam
Northwestern University, Evanston, IL 60201, USA

There is renewed interest in isospin invariance, or equivalently, charge independence (CI) and charge symmetry (CS) in hadronic interactions. It is well known that SU_2 isospin symmetry is substantially broken at the level of the current algebra masses of the up and down quarks[1] $m_u/m_d = 4.3 MeV/7.5 MeV = 0.57$. ($SU_3$ is even more badly broken : $m_s/m_d = 20$). SU_2 is happily restored to a remarkably good degree at the level of the constituent quark masses: $m_d^{con} - m_u^{con}/m_d^{con} \approx 1\%$. It is to this happy circumstance, which has been sometimes called accidental, that we owe isospin invariance in hadronic interactions. The basic purpose of all experimental tests is to measure deviations from CS and CI predictions in appropriately constructed observables and to see how well they can be explained in terms of the hadronic mass differences originating from the quark mass differences, and electromagnetic effects, which, of course, violate CI and CS.

Most experiments to date test the weaker invariance, charge symmetry. Precision tests of charge independence are few and far between. The best known of these is the difference between the $^1S_0$, T=1 scattering lengths determined in low energy experiments :

$$a_{np} - \langle a_{nn}, a_{pp} \rangle = -5.8 \pm 1.0 \ fm$$

While this clearly establishes CI breaking, it tests a very special sector of the nucleon-nucleon interaction. For example it sheds no light on class IV charge independence breaking forces[2] of the type :

$$V_4 = D[\tau_3(i) - \tau_3(j)][\vec{\sigma}(i) - \vec{\sigma}(j)] \cdot \vec{L}_{ij} + E[\vec{\tau}(i) \times \vec{\tau}(j)]_3 [\vec{\sigma}(i) \times \vec{\sigma}(j)] \cdot \vec{L}_{ij}$$

which may be particularly important at medium energies. Notice that the above interaction is spin dependent and its effects may be best seen in polarization observables.

Since tests of CI necessarily involve comparison of two different measurements, it is necessary to design an experiment in which the parameters of the two measurements are as identical as possible, and uncontrolled systematic errors cancel rather than add. With these criteria in mind, we have made measurements of differential cross sections ratio R, and analyzing power differences ΔA_{y_0} for the two reactions :

$$\vec{p} + d \to {}^3H + \pi^+$$
$$\vec{p} + d \to {}^3He + \pi^0$$

The two observables are defined as :

$$R(\theta) = \sigma(\theta, {}^3H)/\sigma(\theta, {}^3He)$$
$$\Delta A_{y_0}(\theta) = A_{y_0}(\theta, {}^3H) - A_{y_0}(\theta, {}^3He)$$

Charge independence predicts $R(\theta)=2$, and $\Delta A_{y_0}(\theta)=0$, for all angles θ.

The experiment was done at the High Resolution Spectrometer facility at LAMPF. A 730 MeV beam of transversely polarized protons and a liquid deuterium target were used, and measurements were made at $\theta_{(lab)} = 12°$ by detecting the recoil particles, ^3H and ^3He in the same detection system. The outgoing particle identification was done by means of both pulse height measurement in a trigger scintillator located beyond the focal plane and time of flight measurement between the first scintillator and another located about 1.9 meters beyond it.

In order to make the conditions for the detection of $^3H^+$ and $^3He^{++}$ similar, voltages of drift chambers and scintillation counters were adjusted to give approximately the same pulse height in each detector. Total count-rates were kept as nearly the same for the two measurements as possible. Target pressure, beam position and beam polarization were continuously monitored, and measurements for the two reactions were alternated many times to average over long-term variations in beam, target and detector characteristics. Measurements of p-^{208}Pb elastic scattering, p-d elastic scattering, and p-d continuum scattering were done to determine beam energy, exact scattering angle, momentum acceptance variations, and detector efficiency variations across the focal plane. A Monte-Carlo program was written to simulate all aspects of the experiment (incident beam energy spread, dE/dx, straggling, and multiple Coulomb scattering in the target, transport through the spectrometer, and energy loss, straggling and multiple Coulomb scattering in the detectors). The program was thoroughly tested for the measured quantities and then used to make the small background and differential efficiency corrections to the data.

The final results are :

$$R = 2.193 \pm 0.007 \pm (0.025)$$
$$\Delta A_{y_0} = -0.007 \pm 0.004 \pm (0.002)$$
$$\Delta A_{y_0}/\langle A_{y_0}\rangle = [-1.7 \pm 1.0 \pm (0.5)]\%$$

(The numbers in parantheses are estimates of systematic errors). The present result for R is almost an order of magnitude more precise than the earlier results[3]. The measurement of ΔA_{y_0} is the first of its kind.

A definitive interpretation of these results requires a reliable theory of pion production in p-d collisions. No such theory exists at present. In absence of such a theory one can only attempt to make models for calculating relative differences between the two reactions. For $T_{(p)}$=600 MeV, Kohler[4] has made estimates of the various 'external' corrections which must be applied to the pure charge independence prediction of $R = 2$. Kohler finds that the major corrections arise from differences in the ^{3}H, ^{3}He wave functions (+8.5% at $\theta_{cm}(\pi) = 128°$) and phase space differences (−0.7%). Coulomb distortion corrections are negligible. Corrections for momentum transfer differences and θ_{cm} differences (both due to differences in π^+, π° masses) depend on the experimental knowledge of the variation of the differential cross sections with incident energy and outgoing particle angle. Using the latest results for wave function differences[6] we obtain a correction factor of +6.6%. For our experiment the phase space difference is −0.5%. We find that the corrections for q and θ_{cm} differences are negligible and thus obtain a correction factor of +6.1%, or $R = 2.12$. Laget and LeColley[5] have recently attempted to calculate the so called 'internal Coulomb' corrections to our data. This is done by taking account of mass differences between the different charge states of nucleons, pions and deltas in all the diagrams for the reactions. They find that this results in an overall additional correction factor of −5%. If this is added to Kohler's external corrections, we have a net correction of +1% left, i.e., $R = 2.02$. Obviously, our experimental result, $R = 2.19(3)$ is significantly different. Laget also obtains $\Delta A_{y_0}/\langle A_{y_0}\rangle \approx -0.3\%$, which, although smaller, is not inconsistent with our result.

It is quite clear that a complete theoretical calculation, which consistently includes all the 'internal' and 'external' corrections − including distortions in the exit channel, is sorely needed.

* Research supported in part by the U.S. Department of Energy.

References

1) S. Weinberg, Trans. N.Y. Acad. Sci. Ser.II 38 (1977) 185.
2) E. M. Henley, in Isospin in Nuclear Physics, ed. D. H. Wilkinson (North Holland Publishers, Amsterdam) 1969, Chap. II.
3) D. Harting et al., Phys. Rev. 119 (1960) 1716; B. H. Silverman et al., Nucl. Phys. A444 (1985) 621.
4) H. S. Kohler, Phys. Rev. 118 (1960) 1345.
5) J. M. Laget and J. F. LeColley, Proc. PANIC-X, Heidelberg, Vol.I (1984) E21; also private communication.
6) T. Sasakawa, T. Sawada and Y. E. Kim, Phys. Rev. Lett. 45 (1980) 1386.

ETA-MESON PRODUCTION EXPERIMENTS AT LAMPF

J.C. PENG

LOS ALAMOS NATIONAL LABORATORY
LOS ALAMOS, NM 87545

ABSTRACT

Results of recent experiments on the (π,η) reactions in nuclei are reported.

I. INTRODUCTION

The existing meson factories have greatly advanced our knowledge on pion-nucleus interactions. In contrast, the roles of heavy mesons (η, μ, ρ, σ, etc.) in nuclear physics have received very little attention. Very little information exists on the interaction between these heavy mesons with nucleons, and almost nothing is known about their interaction with nuclei. About two years ago, we proposed to study the (π,η) reaction on nuclear targets at LAMPF. A detailed discussion on the physics objectives of the (π,η) reaction can be found in Ref. 1. In this talk, I will present the results obtained very recently on the (π,η) reaction.

II. THE EXPERIMENTS

We have made the following measurements:

(a) ^{3}He(π^-,t)η reaction

Tritons are detected with a magnetic spectrometer at forward angles ($5° < \theta_{lab} < 15°$) with 620 MeV/c and 680 MeV/c π^- beam.

(b) ^{3}He(π^-,η)t reaction

The LAMPF π°-spectrometer[2] and the University of Virginia ^{3}He target[3] have been used to measure this reaction through the detection of $\eta \to 2\gamma$ decays. Cross sections at forward angles ($0° < \theta_{lab} < 30°$) have been measured at six beam momenta between 590 MeV/c and 700 MeV/c.

(c) (π^+,η) reaction on ^{7}Li, ^{9}Be and ^{13}C

Energy resolution in these measurements is optimized to search for discrete states.

(d) Inclusive (π^+,η) reaction on ^{3}He, ^{4}He, ^{7}Li, ^{9}Be, ^{12}C, ^{27}Al, ^{40}Ca, ^{120}Sn and ^{208}Pb at 620, 650 and 680 MeV/c.

The data are currently being analyzed. Some of the preliminary results are presented here.

III. EXPERIMENTAL RESULTS AND DISCUSSION

Fig. 1 shows the triton spectrum obtained in the ^{3}He(π^-,t) measurement using the Large Aperture Spectrometer[4] (LAS). A peak is observed in the triton spectrum at a location expected for the ^{3}He$(\pi^-,t)\eta$ binary reaction. Note that this peak stands out rather clearly above some backgrounds such as the ^{3}He$(\pi^-,t)\pi^+\pi^-\pi^0$ reaction. Preliminary analysis gives a cross section of 1μb/sr for the ^{3}He$(\pi^-,t)\eta$ reaction. Since tritons are detected at forward angles, the measurement corresponds to backward angle in the ^{3}He$(\pi^-,\eta)t$ reaction. The momentum transfer at this angle is ~ 800 MeV/c. This is the first time that a discrete final state is observed in the (π,η) reaction on nuclear targets.

To measure the forward angle cross sections of the ^{3}He$(\pi^-,\eta)t$ reaction, the LAMPF π^0 spectrometer was used. Fig. 2 shows that η mesons are clearly identified from the invariant mass plot. The η energy spectra are shown in Fig. 3 for the ^{3}He(π^-,η) reaction at three pion beam momenta. At 680 MeV, which is near the free p(π^-,η)n threshold, the ^{3}He$(\pi^-,\eta)t$ reaction appears as a shoulder of the quasi-free ^{3}He(π^-,η) reaction. As the beam momentum is lowered to 620 MeV/c, the separation between the ground state transition and the quasi-free continuum improves greatly. At 590 MeV/c pion momentum, which is just above the absolute threshold of the ^{3}He$(\pi^-,\eta)t$ reaction and below the threshold of any other ^{3}He(π^-,η) channels, the ^{3}He$(\pi^-,\eta)t$ reaction is umambiguously detected. It is intriguing that the subthreshold (π,η) reaction has such an appreciable cross sections. Several theoretical efforts[5-7] to calculate the (π,η) cross section on nuclei are underway. In particular, a DWIA calculation[6] by Liu describes the shape of the ^{3}He$(\pi^-,\eta)t$ angular distribution rather well, but underestimates the magnitude by a factor of ~5.

Fig. 4 shows the energy-integrated (π^+,η) inclusive cross sections on a number of nuclei measured at three pion momenta. In an attempt to understand the observed mass dependence of the (π^+,η) inclusive cross sections, we use a version of the Glauber theory developed by Kolbig and Margolis[8]. In this model, the inclusive cross section is proportional to the effective nucleon number N_{eff}. The ingredients for calculating N_{eff} are the nucleon density distribution, the hadron-nucleon total cross sections for the incident and the outgoing particles. The only unknown input in this calculation is the total cross section for ηN. The various curves on Fig 4 correspond to calculations using various values of $\sigma(\eta N)$. A good description of the mass dependence at all beam momenta is obtained with $\sigma(\eta N)$ = 15 mb. The extracted ηN total cross section is rather small compared to the πN total cross section of 34 mb at this energy. This is in qualitative agreement with the theoretical expectation[1] that ηN interaction is weaker than the πN interaction.

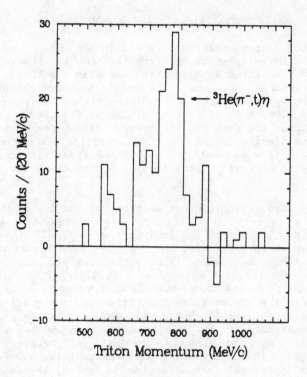

Fig.1. Energy spectrum of the ^{3}He(π^-,t) reaction at 680 MeV/c.

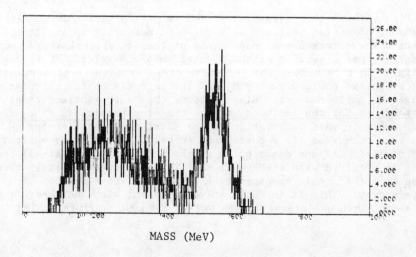

Fig.2. Invariant mass plot for the two photons detected in the $\pi^- + {}^3$He reaction at 680 MeV/c using the LAMPF π^0- spectrometer.

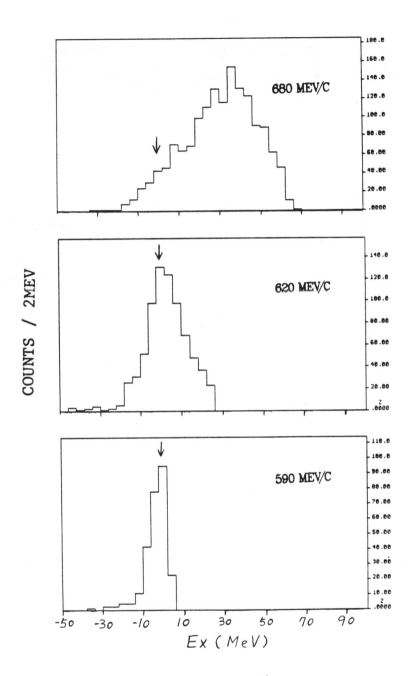

Fig. 3. η energy spectra measured in the ^{3}He(π^-,η) reaction at three π^- momenta. The arrows indicate the location of the ^{3}He(π^-,η)t transition.

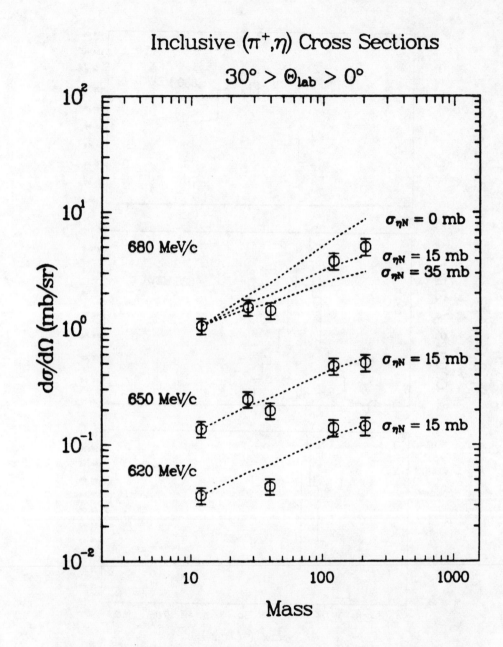

Fig.4. Inclusive (π^+,η) cross sections on several target nuclei. The dashed curves are Glauber calculations using various ηN total cross sections.

ACKNOWLEDGEMENT

I wish to thank sincerely the excellent efforts by all the collaborators in this experiment. They are M. J. Leitch, J. Simmons, J. Kapustinsky, C. Smith, T. K. Li, F. Irom, D. Bowman, J. Moss, Z. F. Wang, C. Lee and R. Whitney.

REFERENCES

1. J. C. Peng, AIP Conference Proceedings No. 133, 255 (1985).
2. H. W. Baer et al., Nucl. Instr. and Meth. 180, 445 (1981).
3. J. S. McCarthy, I. Sick and R. R. Whitney, Phys. Rev. C15, 1396 (1977).
4. E. P. Colton et al., Nucl. Instr. and Meth. 178, 95 (1980).
5. Li Yang-guo and Chiang Huan-Ching, preprint, BIHEP-TH-85-17.
6. L. C. Liu, private communication, 1986
7. J. A. Johnstone, private communication, 1986
8. K. S. Kolbig and B. Margolis, Nucl. Phys. B6, 85 (1968)

QUARK CLUSTERS IN PION ABSORPTION

Murray A. Moinester
Raymond and Beverly Sackler Faculty of Exact Sciences
School of Physics and Astronomy,
Tel Aviv University, 69978 Ramat Aviv, Israel.

ABSTRACT

Experimental differential cross sections for two-body π^- absorption on the diproton ($^1S_0$ proton pair in ^{3}He) are described. A partial wave analysis of the 63 MeV data demonstrates that 88% of the cross-section proceeds through the p-wave non-Δ T=0 channel. It is shown how a partial wave analysis including polarization data can determine the physical amplitudes for this process. We describe how these amplitudes can test six quark cluster descriptions of the short distance pp wave function, and compare the cross-section data to six quark cluster calculations.

An important question involving intersections between particle and nuclear physics is whether or not nucleons near enough together form six quark (6q) clusters. A possible testing ground[1,2,3] with hadron beams for 6q cluster models is the two-body pion absorption reaction on the diproton in ^{3}He (1S0 proton pair with T=1,S=0). This is studied with the ^{3}He(π^-,pn)n reaction, measuring the angular distribution of the differential cross-section for a proton detected in coincidence with a neutron. The data are taken at two-body kinematic conditions, where the spectator neutron recoil momentum equals its value in ^{3}He. The two-body mechanism π^-+pp $\to$ pn is a fundamental process of pion absorption, and can give information on states not accessible through absorption on the well studied T=0, S=1 deuteron. Considering the large momentum transfer to the two active nucleons, one expects that pp correlations at short distances (or 6q degrees of freedom) should play an important role.

Recently, pion-diproton absorption differential cross-sections have become available from experiments at TRIUMF[4] at 37,63,83 MeV, at SIN[5] at 120 MeV, and at LAMPF[6] at 165,250,350,500 MeV. Based on published and preliminary results, these studies yield cross-sections for the reaction which are roughly constant ($\approx$ 0.7 mb) over the entire energy region. In addition, the differential cross-section at 63 MeV (and most other energies) is asymmetric about 90° in the π-pp center-of-mass. The low-energy pion-diproton absorption data have already been described in detail[4] in the literature. The 63 MeV data given in Fig.1 are representative of the data more generally, and are shown together with a recent 6q cluster calculation (discussed later) by G. A. Miller.[2,3] The coefficients A_0, A_1, A_2 shown on the figure describe a Legendre Polynomial fit to the data.

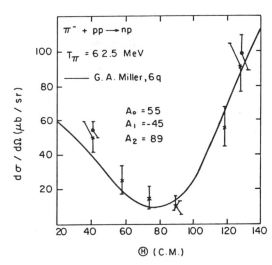

Fig. 1. Differential cross-sections for π^- absorption on the diproton in ^{3}He at 62.5 MeV from Ref. 4. The coefficients A_i are the Legendre polynomial coefficients expressed in μb/sr. The angular distribution is plotted in the π-pp center-of-mass system. The curve is the six quark cluster calculation of G. A. Miller from Refs. 2, 3.

Partial wave analyses[4,7] of the 63 MeV angular distribution (discussed later) demonstrate that at low energies, the process is dominated by the absorption of p-wave pions. In that case, the intermediate state has $J^\pi = 1^+$, which can only decay to a T=0 np state, ruling out intermediate states involving the Δ. This suppression of the usually dominant long range Δ mechanism increases the sensitivity to the weak and more exotic short range 6q cluster contributions. By contrast, $\pi^+D \to pp$ absorption cross-sections[8] at low energies are some 10-20 times larger, follow the Δ resonance, and are characterized by symmetric angular distributions. The π^+D absorption in the Δ region is dominated by the 1D2 final state formed following p-wave absorption that produces an intermediate state ($J^\pi = 2^+$) with a Δ - N S-state. The strong Δ channel makes πD absorption less sensitive to the weaker 6q cluster effects, although this sensitivity has also been studied.[9]

We now consider a partial wave analysis of the 63 MeV angular distribution. At low energies, assuming that only s- and p-wave pions contribute, only three complex amplitudes are needed to describe the differential cross section, and also polarization of the outgoing proton. These amplitudes (A,B,C) describe respectively, the transition from the initial 1S0 diproton state to the final p-n states characterized by 3P0, 3S1, 3D1. They can be described by five variables; three magnitudes and two relative phases. The cross-section data determine three Legendre coefficients, which are functions of the five variables. Without additional data it is impossible to solve these three equations to deduce the five unknown amplitude variables. However, it is nonetheless possible to get information[4,7] regarding their magnitudes. Consider all possible pairs of values for the unknown phases. For each choice we solve the three equations for the magnitudes of the amplitudes. The probability for p-wave absorption depends on the relative magnitudes of B and C compared to A; for p-wave absorption, the probability that the p-n final state will be 3S1 or 3D1 depends on the relative magnitudes of B and C. Considering all the allowed solutions, we find[4,7] a cross section probability of 88 ± 5% for p-wave absorption. The probability for the 3S1 final state is less well determined to be between 10 and 65%. In the recent analysis of Piasetzky et al.,[7] the phases of the amplitudes are fixed using n-p elastic scattering phase shifts for 3P0,3S1,3D1 to describe the final state. This more strongly constrains the allowed solutions.

Further experimental information is required in order to unambiguously determine the amplitudes. We consider spin-observables and measurements in which say the polarization of the final state proton is measured at each angle. The polarization is an axial vector, and must be perpendicular to the reaction plane. The angle-dependence of the polarization is described[7] by two parameters. These two parameters together with the three Legendre coefficients can be used to fix the amplitudes A,B,C. This is a crucial point, since different amplitudes can be consistent with the differential cross section, and yet give[7] radically different polarizations. The true test of dynamical models requires comparison then of the experimental and theoretical amplitudes (effectively $d\sigma/d\Omega$ AND polarization).

Finally, we consider the 6q cluster calculation of G. A. Miller shown in Fig. 1. This figure corrects the one given in Ref. 2 to take into account the most recent value[3] of 6.4% for the 6q probability in ^{3}He, and uses also the latest[4] cross-section values. A π^- is absorbed by any of the quarks in the 6q cluster representing the short distance pp wave function. The π-quark interaction is used to calculate the three amplitudes involving the transition from the initial 1S0 quark state to the three final 6q states (3P0, 3S1, 3D1). The calculation so far has only been carried out at low energies with only s-and p-wave pions considered. In this calculation,[2] the 3S1 probability following p-wave absorption is large, since the final state 6q probability is high for an S-state, and lower for a D-state due to centrifugal barrier effects.

The agreement at 63 MeV of the calculation with the cross-section data is good; but as noted previously does not adequately test the theoretical amplitudes. It is also not clear how well the 6q calculation can describe the energy dependence of all the pion-diproton amplitudes. Standard calculations[10,11] for this process at low energies have not been very successful to date. It will only be possible to accept the 6q interpretation if the 6q amplitudes agree with the experimental amplitudes at the same time that the amplitudes from improved standard calculations do not. This reminds one of some initial enthusiasm in which low energy pion single and double charge exchange data at small angles were described[12] by 6q cluster calculations. However, a number of standard long-range calculations[13] since claimed to describe the data without resorting to a 6q ansatz. In the case of low energy DCX, it was anyhow not intuitively obvious that such a low momentum transfer reaction should necessarily probe NN correlations at short distances. Perhaps then the pion-diproton absorption reactions (and possibly photodisintegration of the diproton) will turn out to be a more natural testing ground for 6q cluster models.

REFERENCES

1. E. Piasetzky, Proc. LAMPF Workshop on Pion Double Charge Exchange, LA-10550-C, 1985. H. W. Baer, M. J. Leitch, eds.
2. G. A. Miller, Proc. Workshop on Nuclear Chromodynamics, Inst. Theoretical Physics, Santa Barbara, Cal., E. Moniz, S. Brodsky,eds.,World Press, 1986; and private communication.
3. V. Koch and G. A. Miller, Phys. Rev. C$\underline{31}$, 602 (1985); Phys. Rev. C$\underline{32}$, 1106(E) (1985).
4. M. A. Moinester et al., Phys. Rev. Lett. $\underline{52}$, 1203 (1984); K. A. Aniol et al., Phys. Rev. C$\underline{33}$, 1714 (1986).
5. H. J. Weyer et al., Proc. European Symposium on Few Body Physics, Bechyne, unpublished (1985).
6. D. Ashery et al., Proc. PANIC Conf. , F.Gutther, B. Povh, G. Zu Putlitz, eds., Heidelberg, p. E20 (1984); D. Ashery, S. Mukhopadhyay, C. Smith, private communication.
7. E. Piasetzky, D. Ashery, M. A. Moinester, G. A. Miller, A. Gal, to be published.
8. D. Ashery and J. P. Schiffer, Ann. Rev. of Particle and Nucl. Science, 1987, to be published.
9. G. A. Miller and L. S. Kisslinger, Phys. Rev. C$\underline{27}$, 1669 (1983).
10. R. E. Silbar and E. Piasetzky, Phys. Rev. C$\underline{29}$, 1116 (1984); C$\underline{30}$, 1365(E) (1984).
11. O. V. Maxwell and C. Y. Cheung,TRIUMF preprint TRI-PP-85-69, 1985, to be published.
12. G. A. Miller, Phys. Rev. Lett. $\underline{53}$, 2008 (1984).
13. T. Karapiperis and M. Kobayashi, Phys. Rev. Lett. $\underline{54}$, 1230 (1985).

THE SPLITTING OF THE ROPER RESONANCE

B.M.K. Nefkens, S. Adrian, A. Eichon, J. Engelage,
G. Kim, Y. Ohashi, and H. Ziock
University of California, Los Angeles, Ca. 90024

J. Arends and D. Sober
Catholic University of American, Washington, D.C. 20064

W. Briscoe, C. Seftor, and M. Taragin
George Washington University, Washington, D.C. 20052

S. Graessle and M. Sadler
Abilene Christian University, Abilene, Texas 79699

The Particle Data Tables list 44 pion-nucleon resonances. The isospin 1/2 states with a mass less than 2 GeV are listed in Table I. Except for a few historic cases where the resonances have been discovered as peaks in the πN total cross section, each resonance was found as the appropriate extremum of the partial wave in an Argand diagram.

Partial waves cannot be measured directly. They are the outcome of a Partial Wave Analysis, PWA, which encompasses the simultaneous analysis of all elastic scattering data. The current PWA data bank lists over 10,000 entries. The PWA is complicated by the fact that most resonances are broad and they overlap each other, and many have a sizeable inelasticity.

There are three modern PWA's: 1) Karlsruhe-Helsinki,[2] K-H; 2) Carnegie-Mellon University - Lawrence Berkeley Laboratory,[3] C-L; 3) Virginia Polytechnique Institute,[4] VPI. An important difference between them concerns the properties of the $P_{11}(1440)$ or Roper resonance: C-L concludes that the P_{11} is broad, $\Gamma \sim 340$ MeV; K-H believes it to be narrow, $\Gamma \sim 135$ MeV; VPI finds two nearby poles in the complex plane at (1359-100i) and (1410-80i) MeV. The P_{11} is particularly interesting for testing the validity of the many quark bag and skyrmion model calculations; the predictions of such models are given in Table II, Refs. 5-15.

A sensitive way to test the three PWA's is via a measurement of the left-right asymmetry in $\pi^-+p\uparrow \rightarrow \pi^0+n$, CEX, in the region of the Roper resonance as illustrated in Figs. 1d-f. A few selected but accurate measurements will be sufficient. We have carefully evaluated, for this purpose, the "background" CEX data from our recent experiment[16] on the measurement of the asymmetry in the related reaction, $\pi^-+p\uparrow \rightarrow \gamma+n$. Details of this counter experiment are given elsewhere so we shall proceed directly to the results.

The experiment covered three energy regions:
a) The measurements at $p_\pi = 301$ and 316 MeV/c cover the area just

Table I Isospin 1/2 πN Resonances with Mass Less Than 2 GeV

Resonance	Symbol	Star status
N(1440)	P_{11}	4
N(1520)	D_{13}	4
N(1535)	S_{11}	4
N(1540)	P_{13}	1
N(1650)	S_{11}	4
N(1675)	D_{15}	4
N(1680)	F_{15}	4
N(1700)	D_{13}	3
N(1710)	P_{11}	3
N(1720)	P_{13}	4
N(1990)	F_{17}	2

around the peak of the Δ(1232) resonance. As one can see in Figs. 1a and 1b, both the VPI and K-H PWA do a good job in predicting the overall shape of the asymmetry curve; our data show a slight preference for K-H.
b) The measurement at p_π = 471 MeV/c is in a region where there is no πN resonance. Our data is in agreement with the overall shape of VPI and K-H with a slight preference for VPI, see Fig. 1c.
c) The region of the greatest interest is at p_π = 547, 586, and 625 MeV/c. Our data overwhelmingly favors VPI over K-H and C-L, see Figs. 1d, e, f. We recall that VPI finds two poles in the P_{11} in contrast to the harmonic oscillator quark model prediction by Isgur and Karl[5] which prefers one P_{11} around 1440 MeV and another one at 1700 MeV. The recent skyrmion calculations favor two nearby P_{11} poles, namely the SLAC work by Mattis et al.,[16] on the exact

Table II Mass of Low Lying P_{11} Resonances as Predicted by Recent Models

Mass (MeV) P_{11} ("Roper")	P_{11} ("Arndt")	Model	Authors
1405	1705	Harmonic Oscil. Pot.	Isgur-Karl[5]
1543	1646	MIT-bag, radial excit.	Bowler-Hey[6]
1410	1603	MIT-bag breathing mode	De Grand-Rebbi[7]
1416	1617	Perturbed MIT bag	Close-Horgan[8]
1418	1533	Cloudy bag	Umland et al.[9]
1580		Colored flux tubes	Carlson et al.,[10]
~1270		Skyrmion-breath.	Breit-Nappi[11]
~1340		Skyrmion-adiab.	Hayashi et al.[12]
	1427	Skyrmion	Mattis et al.[13]
	~1575	q^3b - hybrid	Barnes-Close[14]
1436	1565	q^3b - hybrid	Golowich et al.[15]

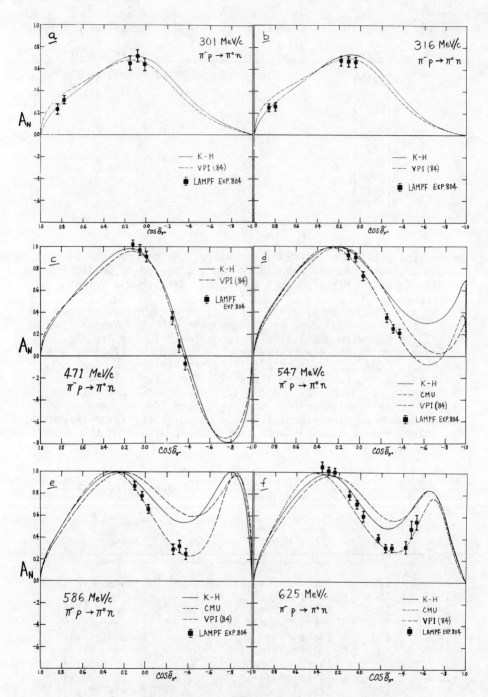

Fig. 1. Left-right asymmetry, A_N, measured with a transversely polarized target in $\pi^-p \rightarrow \pi^0 n$.

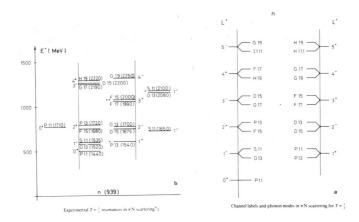

Fig. 2. Mass hierarchy predicted by skyrmion calculations of the Siegen group.[17]

masses, as well as the results of the Siegen group[17] on the hierarchy of masses of the πN resonances as illustrated in Fig. 2.

This work was supported in part by the U.S. D.O.E.

REFERENCES

1. Particle Data Group, Rev. Mod. Phys. 56, S1 (1984).
2. G. Hohler et al., Handbook of Pion-Nucleon Scattering (Karlsruhe, Germany, 1978); R. Koch and E. Pietarinen, Nucl. Phys. A336, 331 (1980).
3. R.E. Cutkosky et al., Phys. Rev. D20, 2839 (1979).
4. R.A. Arndt et al., Phys. Rev. D32, 1085 (1985).
5. I. Isgur and G. Karl, Phys. Rev. D19, 2653 (1979).
6. K.C. Bowler and A.J.G. Hey, Phys. Lett. 69B, 469 (1977).
7. T. DeGrand and C. Rebbi, Phys. Rev. D17, 2358 (1978).
8. F.E. Close and R.R. Horgan, Nucl. Phys. B164, 413 (1980).
9. E. Umland et al., Phys. Rev. D27, 2678 (1983).
10. J. Carlson et al., Physl Rev. D28, 2807 (1983).
11. J.D. Breit and C.R. Nappi, Phys. Rev. Lett. 53, 889 (1984).
12. A. Hayashi et al., Phys. Lett. 147B, 5 (1984).
13. M.P. Mattis and M. Karliner, Phys. Rev. D31, 2833 (1985).
14. T. Barnes and F.E. Close, Phys. Lett. 123B. 89 (1983).
15. E. Golowich et al., Phys. Rev. D28, 160 (1983).
16. G.J. Kim et al., Phys. Rev. Lett. 56, 1779 (1986).
17. G. Eckhardt et al., Nucl. Phys. A448, 732 (1986).

THE SEQUENTIAL NATURE OF PION DOUBLE CHARGE EXCHANGE
M. Bleszynski[+] and R. Glauber[*]
[+]Physics Department, University of California, Los Angeles, CA90024
[*]Lyman Laboratory of Physics, Harvard University, Cambridge, MA02138

Recently several detailed measurements of angular distributions in the low energy positive pion induced single and double charge exchange reactions on ^{14}C have been performed at TRIUMF[1] and LAMPF.[2,3] Of particular interest are the reactions leading to the isobaric analog state in ^{14}N and the ground state of ^{14}O which the represent simplest charge exchange processes since they involve only the target isospin rotation and essentially elastic nuclear transitions. The double charge exchange reactions are particularly attractive at low energies (≤ 50 MeV), where the πN interaction is relatively weak. Consequently a theoretical treatment based on the lowest (second) order term in a multiple scattering expansion should describe such processes quite adequately. Due to this feature the double charge exchange processes may provide us with a simple and relatively unobscured way of selecting information on the collision of a pion with the system of 2 target nucleons.

An interesting aspect of the data taken at 50 MeV is the occurence of a very deep minimum in the differential cross section in the reaction $\pi^+ \, ^{14}C(0^+) \rightarrow \pi^0 \, ^{14}N(0^+)$ at forward scattering angles and the presence of a strongly forward peaked angular distribution in the process $\pi^+ \, ^{14}C(0^+) \rightarrow \pi^- \, ^{14}O(0^+)$.

The observed strong suppression of the differential cross section for the single charge exchange reaction is related in a well known way to the behavior of the differential cross section for the charge exchange reaction $\pi^+ n \rightarrow \pi^0 p$. Because of the destructive interference between the s- and p-wave partial wave amplitudes the latter cross scetion has a deep minimum near the forward direction at 50 MeV.

The enhancement of the angular distribution for the reaction $\pi^+ \, ^{14}C(0^+) \rightarrow \pi^- \, ^{14}O(0^+)$ at small angles[1] has created a kind of puzzle. There is considerable controversy about the underlying physical mechanism making a charge exchange process being almost forbidden for small momentum transfers, when it takes place twice over, to be particularly favored for the same small momentum transfers.

In particular this apparently anomalous behavior of the double charge exchange cross section has prompted Miller[4] to attempt the examination of the double charge exchange reaction between the isobaric analog states of ^{14}C and ^{14}O at 50 MeV as a possible means of detecting the presence of six-quark clusters in the nucleus. The presence of such structures, it was argued, would lead to the strong forward peak in the double charge exchange cross section and an enhanced probability for the occurrence of the process.

We show that the above mentioned apparent puzzling behaviour of angular distributions of charge exchange reactions can be easily explained by the careful multiple scattering analysis. We have constructed a simple model based on the first and the second order impulse approximations which allows us to discuss both single and double charge exchange processes by means of most elementary approximations. The model succesfully describes pion charge exchange reactions between the isobaric analog states of ^{14}C and ^{14}O and ^{14}N at 50 MeV (see Fig. 1) and enables us to explain even the more puzzling aspects of the single and double charge exchange reactions in terms of the well known parameterizations of the πN charge exchange amplitude and the well established nuclear shell model wave functions. Our approach, due to its analytical simplicity, enables us to understand which features of the nuclear structure and πN interaction are responsible for the shape and the magnitude of the angular distributions of low energy pions undergoing the double charge exchange.

We find that the forward peak in the angular distribution is determined by nuclear structure properties of the two-particle wave function of the valence nucleons participating in the double charge exchange reaction. The shape of the peak can be well represented with the square of the Fourier transform of their center-of-mass coordinate $\vec{R} = (\vec{r}_1 + \vec{r}_2)/2$, $\rho_+(R)$:

$$\frac{d\sigma}{d\Omega}(Q) \simeq \frac{d\sigma}{d\Omega}(0) |\int d^3R \rho_+(R) e^{i\vec{Q}\cdot\vec{R}}|^2,$$

where $\vec{Q}$ is the momentum transfer.

The value of the cross section at Q=0 turns out to be proportional to the square of the average of the pion Green function $\tilde{G}(r)$ with the distribution of the coordinate of the relative distance of the valence nucleons $\vec{r} = \vec{r}_1 - \vec{r}_2$, $\rho_-(r)$:

$$\frac{d\sigma}{d\Omega}(0) \simeq |\int d^3r \rho_-(r) \tilde{G}(r)|^2.$$

The above expression is very sensitive to the spatial dependence of the πN interaction at short range distances. The distributions $\rho_+(R)$ and $\rho_-(r)$ depend strongly on the amount of admixture of the S- and P-states in the 2-particle density of the nucleons on which the charge transfers take place.

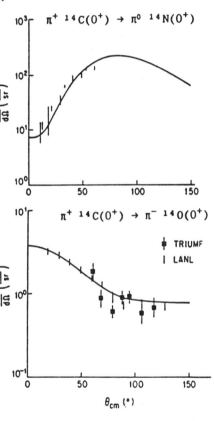

Fig 1. Comparison of the predictions of the model (solid curves) with tha data.

References
1. I. Navon, et al., Phys. Rev. Lett. 52, 105(1984).
2. M. J Leitch et al., Phys. Rev. Lett. 54, 1482 (1985).
3. M. J Leitch et al., Phys. Rev. C31, 278 (1985).
4. G. A. Miller, Phys. Rev. Lett. 53, 2008 (1984).

DETERMINATION OF THE REAL PART OF THE ISOSPIN-EVEN FORWARD
SCATTERING AMPLITUDE OF LOW ENERGY PION-NUCLEON SCATTERING
AS A TEST OF LOW ENERGY QUANTUMCHROMODYNAMICS

K. Göring, U. Klein, W. Kluge, H. Matthäy, M. Metzler, U. Wiedner
Kernforschungszentrum Karlsruhe, Institut für Kernphysik und
Universität Karlsruhe, Institut für Experimentelle Kernphysik
Postfach 3640, D 7500 Karlsruhe 1

E. Pedroni
Schweizerisches Institut für Nuklearforschung, CH 5234 Villigen

W. Fetscher, H.J. Gerber
Eidgenössische Technische Hochschule Zürich, Institut für
Mittelenergiephysik, CH 5234 Villigen

ABSTRACT

The real part of the isospin-even forward scattering amplitude
Re $D^+(t=0)$ of the pion-nucleon scattering has been determined at an
incident pion kinetic energy of 55 MeV by measuring the elastic
scattering of positive and negative pions on protons within the
Coulomb-nuclear interference region (t being the four momentum
transfer). The value of Re $D^+(t=0)$ confirms the prediction of the
Karlsruhe-Helsinki phase shift analysis[1] for that energy. These
phases can be used to determine the σ-term of pion-nucleon
scattering by means of forward dispersion relations. Koch[2] obtained
in such an analysis $\sigma = (64 \pm 8)$ MeV at the Cheng-Dashen point
$t = 2 m_\pi^2$, $\nu = 0$, which differs significantly from the value
determined by chiral perturbation theory of QCD[3] $\sigma = (35 \pm 5)$ MeV
at $t = \nu = 0$.

INTRODUCTION

The strong interaction is believed to be described by quantum
chromodynamics (QCD), the Lagrangian of which is given by:

$$L_{QCD} = \sum_{q=u,d,s,\ldots} i\bar{q}\, \gamma^\mu D_\mu q \; - \frac{1}{4} F^a_{\mu\nu} F^{\mu\nu a} - \sum_{q=u,d,s,\ldots} m_q \bar{q} q + \ldots \quad (1)$$

Two limiting cases can be investigated: at high four momentum
transfer t the perturbative expansion in the strong coupling
constant α_s can be tested and at low energies predictions can be
made within the framework of chiral perturbation theory (CPT),
which treats the current quark masses m_q as small perturbations.
The chiral limit itself is defined by setting the masses of the
light quarks equal zero: $m_u = m_d = m_s = 0$. All the well known low
energy (soft pion) theorems (e.g. Goldberger-Treiman relation,
Adler-Weisberger relation, Adler consistency condition), in the
past derived from current algebra and PCAC, are preserved in the
CPT of QCD.

One of the most important testgrounds of current algebra and PCAC and consequently of CPT is low energy pion-nucleon scattering. A basic number is the σ-term of πN scattering:

$$\sigma = \frac{m_u + m_d}{4m_N} < N | \bar{u}u + \bar{d}d | N > \qquad (2)$$

(m_N being the nucleon mass) which is a measure of the size of chiral symmetry breaking of QCD due to the quark mass term in L_{QCD}. The σ-term has been calculated by Gasser and Leutwyler[3] from the hadronic mass spectrum to be $\sigma = (35 \pm 5)$ MeV. The σ-term is also connected with the isospin even πN scattering amplitude D^+ at an unphysical point of the complex ν-t-scattering plane ($\nu = \frac{s-u}{4m_N}$) by a low-energy theorem of current algebra

$$\sigma = \frac{f_\pi^2}{2} [D^+ (t=2m_\pi^2, \nu = 0) - \frac{g^2}{m_N}], \qquad (3)$$

g being the πN-coupling constant.

It is therefore possible to determine the σ-term from experimental pion-nucleon scattering data by means of forward dispersion relations [2,4], extrapolating from the physical region of the ν-t-plane to the unphysical Cheng-Dashen point $t = 2m_\pi^2$, $\nu = 0$. Koch obtained in a recent dispersion analysis $\sigma = (65 \pm 8)$ MeV. This "experimental" value is almost a factor of 2 bigger than the CPT-value of Gasser and Leutwyler, even if one takes into account, that the CPT-value has not been calculated for the Cheng-Dashen point, but for $t = \nu = 0$, which will reduce Koch's value by about 4-5 MeV [3].

The inconsistency of both values is not understood and is regarded as a serious problem which hints either at a deficiency of CPT or at an unsufficient πN data basis.

RESULTS

Unfortunately only a few measurements of πN scattering or pion charge exchange exist below 100 MeV, which are even partly contradictory. To improve this unsatisfactory situation we have measured at the πM3 channel of SIN angular distributions of the elastic scattering of 55 MeV pions on protons in the region of the Coulomb-nuclear interference (between 7.5° and 27.5°). The goal of the measurement is the determination of the real part of the D^+ amplitude Re $D^+(t)$ at $t = 0$, which is an important constraint for πN phase analyses at low energies. The experimental set-up and the data analysis have been described elsewhere[5]. In order to get Re D^+ ($t=0$) it is sufficient to plot the expression

$$A = \frac{\frac{d\sigma^+}{d\Omega} - \frac{d\sigma^-}{d\Omega}}{\frac{d\sigma^+}{d\Omega} + \frac{d\sigma^-}{d\Omega}} \cdot \frac{1}{t}$$

as a function of t and to extrapolate it to $t = 0$, because:

$$\lim_{t \to 0} \text{Re } D^+(t) = 4\pi\alpha\omega A(t) \tag{4}$$

(ω being the total laboratory pion energy, and α the Sommerfeld constant). Building the ratio A out of the differential cross-sections $\frac{d\sigma^{\pm}}{d\Omega}$ of $\pi^{\pm}$ p scattering reduces the systematical errors of the value of Re D^+ (t=0). Fig. 1 shows the angular distributions for the elastic scattering of pions ($\pi^{\pm}$) on protons together with the predictions of the Karlsruhe-Helsinki phase shift analysis (solid lines). The agreement is very good as well for the differential cross-sections as for the asymmetry A, which is displayed in Fig. 2.

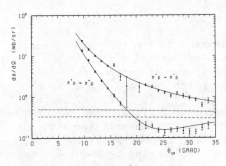

Fig. 1. Elastic $\pi^{\pm}$p scattering at 55 MeV. The solid lines represent predictions by means of the Karlsruhe-Helsinki phase shift analysis . The dashed-dotted and dashed curves denote the pure hadronic cross-sections for π^+p and π^-p, respectively.

Fig. 2. The asymmetry A as given in the text as function of t. Re D^+ (t=0) is obtained by equation (4).

For Re D^+(t=0, 55 MeV) a value of (14.6 ± 0.5) GeV^{-1} is obtained. We conclude from our results, which have been measured in a kinematical region never explored experimentally before, and which agree with the Karlsruhe-Helsinki-phase shift analysis, that Koch's value for the σ-term is heavily supported by our investigation. The disagreement between the CPT-value of Gasser and Leutwyler derived from the hadronic spectrum and the value from the dispersion analysis of πN scattering of Koch obviously persists.

REFERENCES

1. R. Koch, E. Pietarinen, Nucl. Phys. A336, 331 (1980)
2. R. Koch, Z. Physik C15, 161 (1982)

3. J. Gasser, M. Leutwyler, Phys. Reports C87, 77 (1982), and private communications
4. G. Höhler, Landolt-Börnstein Vol. I/9b: Pion-Nucleon Scattering, ed. H. Schopper (1983), Springer-Verlag Berlin-Heidelberg-New York
5. Karlsruhe-SIN-ETH Zürich collaboration, SIN Newsletter No. 17, p. NL 25 (1985);SIN Newsletter No. 18, p. NL 36 (1986)

Elastic Meson-Nucleon Partial Wave Scattering Analyses

Richard A. Arndt
Department of Physics
Virginia Polytechnic Institute and State University
Blacksburg, Virginia 24061

ABSTRACT

Comprehensive analyses of π-n elastic scattering data below 1100 MeV(Tlab), and K+p scattering below 3 GeV/c(Plab) are discussed. Also discussed is a package of computer programs and data bases (scattering data, and solution files) through which users can "explore" these interactions in great detail; this package is known by the acronym SAID (for Scattering Analysis Interactive Dialin) and is accessible on VAX backup tapes, or by dialin to the VPI computers. The π-n, and k+p interactions will be described as seen through the SAID programs. A procedure will be described for generating an interpolating array from any of the solutions encoded in SAID; this array can then be used through a fortran callable subroutine (supplied as part of SAID) to give excellent amplitude reconstructions over a broad kinematic range.

INTRODUCTION

The method used by the VPI group for analysing elastic meson-nucleon scattering at intermediate energies consists of combined energy-dependent and "single-energy" representations which are evolved in combination with each other[1]. This produces a set of solutions at about 50 MeV intervals with meaningful error matrices, and a "global", smoothed solution which has been parameterized in analytically "proper" forms which can be continued into the complex energy plane for assessment of the poles, and branch cuts associated with observed partial wave structures. Our method differs from that of the Karlsruhe, and Carnegie-Mellon-Berkeley[2] analyses in our encoding of a more restricted energy range, and in our focus on direct channel analyticity and unitarity requirements; except for forward dispersion information, we use only real scattering data.

A package of programs and data files known as SAID is used to encode these analyses as well as recent analyses of n-n data below 1 GeV. The programs are run interactively on computers at VPI&SU and on a large number of VAX11 systems throughout the world. The programs use one of a large number of solutions, including possible user input, to calculate a multitude of quantities predicted by the solution; amplitudes, observables, or partial waves. These, in turn, are used to plan experiments, examine the data base, and ascertain disparities and uncertainties in the solutions. While the system supports a large number of terminal types for graphics output, a feature which we consider of great value, any terminal can be used to obtain numerical output. Most of the plots presented in this report were generated through SAID. Copies of SAID are available upon request on VAX backup tapes which are exceedingly easy to transport.

EXPLORING THE DATA BASE

Elastic scattering data bases used in our analyses can be viewed graphically through the "GO11" option of SAID. A kinematic domain (Energy, Angle) is specified and the data can be diplayed by charge-channel (pi+, pi-, or cxs), type (cross section, pol,..), and date (1980-1986). In figure 1 we plot the most recent data (1985-86) for pi-n scattering; these are LAMPF measurements which have yet to be published.

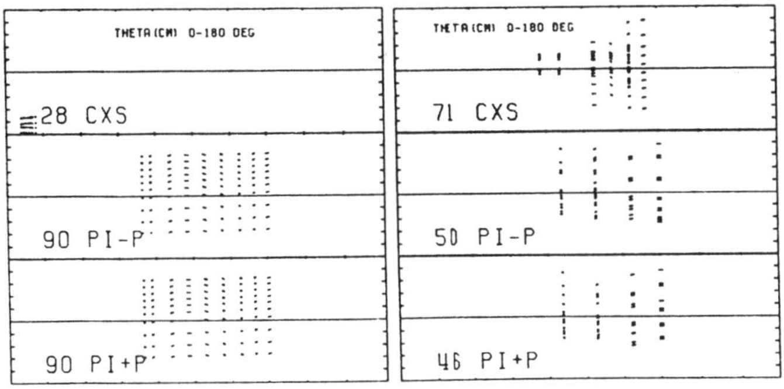

Fig. 1
$\pi-n$ LAMPF data for Tlab=0-800 MeV, Theta(cm) =0-180 deg.

Numerical exploration is possible with the "GO5" option where a solution is used to predict and compare to experimental numbers; data is selected by Energy range, charge channel, data type, date, short reference, or in the case of $\pi-n$ data by "flag" rating. The "flag" rating of π-n experiments was implemented in the summer of 1985 in collaboration with the Nefkens UCLA group; experiments below 800 MeV were rated for quality (and age), on a 0 to 3 scale. Our present analyses use only 2, and 3 star data, but all data remain accessible through SAID. Data which has been scratched from analysis is clearly labeled in any output from SAID.

PREDICTING AND PLOTTING OBSERVABLES

Data can be calculated, and plotted, through SAID in a variety of representations. Independent variables are Tlab, Plab, Wcm, Theta(cm), or Momentum transfer (-t). Experimental data can be selected for superimposing on predicted quantities as illustrated in figure 2. If a "single-energy" solution is used, the error matrix will be used to provide errors on the calculated observable. Normalization of the experimental data is an allowed option at this step.

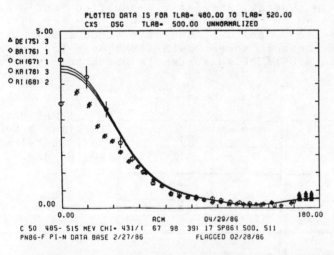

Fig. 2 Some π-η observables plotted through SAID.

ASSESSING NEW DATE, PLANNING FUTURE MEASUREMENTS

The "GO8" option of SAID uses a solution and error matrix to represent data at some energy. To this representation one can add or subtract real or hypothetical data and calculate a one-step adjustment of the phase parameters of the solution. This enables the user to determine sensitivity of the phase parameters to particular data types, and to evaluate compatibility between various experiments. Output from this step is x^2, Normalization, Phase parameters, and predicted observables with errors before, and after adding/subtracting the new experiment. It has been used as an effective planning tool by groups at LAMPF and TRIUMF.

PARTIAL WAVE AMPLITUDES

Partial wave amplitudes can be plotted, using the "GO7" option of SAID in a variety of representations; Argand, T(re, im), $\delta(1-\eta^2)$, σ(tot, react). They can be plotted vs. Tlab, Plab, or Wcm. The I = (1/2) S, P, and D waves from our most recent solution (SP86) are shown in figure 3. It is

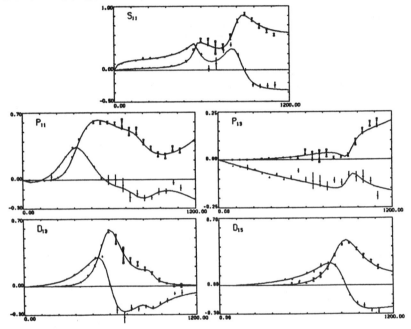

Fig. 3 S, P, D wave amplitudes from solution SP86.

important to note the degree to which the "single-energy" values agree with the "global" solution, SP86. Very recently we have attempted to encode the secondary structure in P_{11} between 700 MeV and 1 GeV which is suggested by the single energy solutions. In figure 4 we illustrate P_{11} as determined

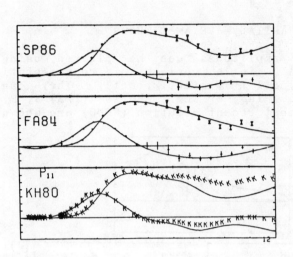

Fig. 4
P_{11} from SP86, FA84, and KH80 from 0-1200 MeV (Tlab).

by our published solution, FA84, and by the Karlsruhe solution, KH80. These results are preliminary and will be carefully examined in the future.

CALCULATIONAL USE OF GLOBAL SOLUTIONS

A very accurate, and efficient encoding of the on-shell amplitudes from our analyses is available with the "GO15" option of SAID which creates a (BCD) file for use with a FORTRAN calleable subroutine, PNAMP, to produce on-shell amplitudes at desired kinematic (energy, angle) points. This allows our results to be used in calculational ways such as in π-nuclear reactions. The package of subroutines, and the interpolating arrays are easily transported and are contained on the VAX backup tapes which are used to transport the SAID programs.

INTERPRETATION, RESONANCES

Our objective is primarily to obtain the most accurate two-body amplitudes available from elastic scattering data, and secondarily to provide some interpretation of the structures which emerge from our analyses. The forms used to express our global solutions allow analytic continuation into the complex energy plane where we can identify the poles, and cuts underlying the observed structures. In figure 5, we illustrate the complex plane mapping of

Fig. 5. P_{11}, and D_{13} from Wcm=1300-1800 Mev, Im Wcm=-200 to 0 MeV. The $\pi-\Delta$ branch cut is run in opposite directions in the P_{11} bottom plot. The bottom D_{13} plot is the residue function $T*(W_p-W)$.

P_{11}, and D_{13} from Wcm=1300 MeV to 1800 MeV. The plotted quantity is Log(T**2) and it clearly reveals the poles, and zeroes associated with on-shell structures. If the T-matrix is multiplied by a factor (Wp-W), Wp=pole energy, we obtain a residue function illustrated as the bottom plot under D_{13}. Table 1 contains the positions, and residues so obtained from our most recent analyses; it does not yet reflect the complexity in P_{11} which we are working to understand.

Table 1. Resonance poles for π-n amplitudes.
$T*(W_p-W)=G*\exp(i*\phi)$
Where: T=Tmatrix, W=cm energy (MeV), Wp=pole position

State	Wp(MeV)	G(MeV)	Phi(deg)
S_{11}(P1)	1464-i*75	40	136
S_{11}(P2)	1656-i*54	34	126
P_{11}(P1)	1355-i*100	62	72
P_{11}(P2*)	1416-i*78	118	176
P_{13}(P1)	1690-i*33	3.7	42
D_{13}(P1)	1508-i*62	40	171
D_{13}(P2)	1676-i*24	2	-137
D_{15}(P1)	1658-i*68	32	160
F_{15}(P1)	1668-i*55	33	162
S_{31}(P1)	1592-i*54	13	63
P_{33}(P1)	1211-i*51	56	150
P_{33}(P2)	1588-i*149	37	73
D_{33}(P1)	1674-i*168	32	156
F_{35}(P1)	1872-i*114	23	167
F_{37}(P1)	1864-i*108	50	160

OTHER SAID FEATURES

Amplitudes from the VPI isobar analysis[3] of π-π-n reactions is obtainable through SAID, as are a number of other features. Correllations of the error matrices can be explored, "movies" of angular distributions evolving with energy, and utiltiy programs to do two body kinematic transformations are also included. A number of special functions such as low energy scattering length parameters have also been encoded for particular users.

ACKNOWLEDGEMENTS

This work was sponsored by the United States Department of Energy Contract DE-AS05-76-ER04928.

REFERENCES

1. R. Arndt and L. D. Roper, Phys. Rev. <u>D32</u>, 1085 (85). "Pion-nucleon partial wave analysis to 1100 MeV".
2. G. Hohler et al, Physics Data Handbook of Pion Nucleon Scattering, 1979. R.E. Cutkosky et al, Phys Rev Letters 37,645(76).
3. D. Mark Manley, Richard A. Arndt, Yogesh Goradia, and Vigdor Teplitz, Phys. Rev. <u>D30</u>, 904(84)

MODELS OF MULTIQUARK STATES*

Harry J. Lipkin
Argonne National Laboratory, Argonne, IL 60439
and
Department of Physics, Weizmann Institute of Science, Rehovot, Israel

ABSTRACT

The success of simple constituent quark models in single-hadron physics and their failure in multiquark physics is discussed, emphasizing the relation between meson and baryon spectra, hidden color and the color matrix, breakup decay modes, coupled channels, and hadron-hadron interactions via flipping and tunneling of flux tubes. Model-independent predictions for possible multiquark bound states are considered and the most promising candidates suggested. A quark approach to baryon-baryon interactions is discussed.

1. INTRODUCTION AND DEDICATION--"WHERE IS THE PHYSICS?"

Gabriel Karl[1] has shown how results in remarkable agreement with with experimental hadron spectroscopy have been obtained from simple constituent quark models using effective two-body interactions with properties motivated by QCD although not rigorously derived.[1,2,3] But this approach has broken down completely in the multiquark sector, where not a single prediction has been confirmed by experiment.

Why does Gabriel Karl's beautiful picture fall apart at N > 3? One answer was given in another context by Y. Yamaguchi's response in 1960 to the question "Has there been any thought about the problem of...?" "No! Many calculations, no thought!"

One of the earliest and still valid theoretical investigations at the intersections between particle and nuclear physics used a "nuclear physics approach to hadrons". The matrix elements for effective interactions are obtained from the 2-body problem. These are then used for predictions in the N-body problem.[4,5] This approach has been very successful for predicting the baryon spectrum from the meson spectrum. In 1966 Sakharov and Zeldovich[4] obtained

---

*Work supported by the U.S. Department of Energy, Division of High Energy Physics, Contract W-31-109-ENG-38.

**1985-86 Argonne Fellow on Leave from Weizmann Institute of Science, Rehovot, Israel.

the mass relation

$$M_\Lambda - M_N = \frac{3}{4}\lfloor M_{K^*} - M_\rho \rfloor + \frac{1}{4}\lfloor M_K - M_\pi \rfloor .$$

177 MeV   180 MeV

This paper is dedicated in honor of the 65th birthday of a pioneering contributor to the intersections between particle and nuclear physics--unable to attend this meeting--Andrei D. Sakharov.

Hadron physics is very different from electroweak physics, where there has always been a standard model, and experiments either test reliable predictions or look for new physics beyond the standard model. Even though we now believe the correct theory for hadron physics to be QCD, nobody knows how to use QCD to calculate the hadron spectrum. A collective effort by theorists and experimentalists is needed with experimental data guiding the theorists in constructing QCD-motivated models, and with the predictions of these models as guides to future experiments. A successful implementation of this program will at least teach us how to use QCD for hadron physics. It may also lead to the discovery of new types of hadrons suggested by QCD, like glue balls, hybrids or multiquark exotics. It may give us insight into the early universe or astrophysical puzzles like Cygnus-X3. It may even lead to evidence for new physics beyond the standard model.

It is interesting to look at Standard Model Physics and Hadron Physics with a "burger model". These days in America everything has been burger-ized. There are beefburgers, fishburgers, pizzaburgers, shrimpburgers, etc. etc. Even the U.S. Supreme Court has been burgerized, with Chief Justice Earl Warren replaced by Warrenburger. Fast-Food outfits have put so much other junk into their burgers that one very popular TV commercial showed a lady asking "Where's the beef?" This tradition has been followed by the Fast-Physics calculators, who have put so much other junk into their physicsburgers that one can ask "Where's the physics?".

There are two kinds of physicsburgers. The electroweak burger has a thick slice of solid predictions on a base of a well defined standard model, covered with a reliable calculation, and garnished with data, Monte Carlo, computer programs and χ^2 fits. The physics is clear. The hadron burger has a base of ad hoc assumptions, covered with free parameters and nothing else and garnished with "reliable" data, Monte Carlo, computer programs and χ^2 fits. There is usually a nearby waste basket filled with rejected "unreliable" data. One can well ask "Where is the physics?"

There are two approaches to using QCD for hadron physics: the southern fundamentalist approach and the northern iconoclastic approach.

The fundamentalists believe that "In the beginning God created the Bag", and follow the implications of the Bag with religious fervor. The lunatic fringe believe that the "n" in Big Bang cosmology is a typographical error and that all multiquark physics is describable with a Big Bag. They lose all contact with the real world as they follow their religion and send experimentalists on wild goose chases for nonexistent objects like narrow baryonium states.

The iconoclasts are atheists (or asakists - from the Greek ΣΑΚΚΟΣ) who refuse to believe anything and are always looking for alternative models in case their favorite model is wrong. Even when they have invented the great standard model for which they eventually get the Nobel Prize, they do not browbeat experimentalists into looking for the phenomena predicted by their model, like charm and weak strangeness-conserving neutral currents. Instead they produce a plethora of alternative models with five quarks, six quarks, eight quarks, new unobserved heavy leptons, etc. to explain all possible disagreements of their right standard model with wrong experiments.

North and south in this context refer of course to locations of the two great centers of particle physics on Massachusetts Avenue, Harvard and M.I.T. (Nit-picking purists may point out that they are really Northwest and Southeast). The correct approach for experimentalists is to recognize that all these diverse types of theorists contribute to our understanding of physics. It is good that we have them, rather than one party line. But just as any good experimenter is very careful to look for all kinds of biases and acceptance criteria before drawing conclusions from a particular set of experimental data, he should also be aware of all the biases and acceptance criteria that go into any theoretical paper before drawing conclusions from their predictions. The key question is "Where is the physics?"

Two examples of these two approaches are the H-dibaryon predicted by Jaffe[6] (M.I.T.) and the prediction of the Λ magnetic moment by DeRujula et al (Harvard).[3]

Jaffe's six-quark-bag calculation predicted the existence of the H and estimated its mass. Where is the physics? Solid general QCD-symmetry arguments show that the H should be the most stable dibaryon. The mass prediction clearly does not include all the right physics. Any bound state near the Λ-Λ threshold must have a Λ-Λ piece in its wave function that decreases exponentially and continues well outside the boundary of any bag. This has been pointed out by Jaffe, but overlooked by others who use his result. This exponential tail reduces the kinetic energy and lowers the mass. Experimental seaches for the H should have high priority, but no mass prediction should be taken seriously unless it manifestly contains all the right physics. For example, any calculation which says that the lowest dibaryon state with the H quantum numbers has a mass greater than the mass of two Λ's must be missing some physics.

DeRujula et al predicted the Λ magnetic moment by using the Δ-nucleon and Σ^*-Σ splittings to estimate flavor-SU(3) symmetry breaking and predicted μ_Λ = -0.61 n.m. This was later confirmed with surprising precision by experiment which found exactly the same value, μ_Λ = -0.61 n.m.,

Where is the physics? It is in the natural assumptions that (1) The Λ moment is entirely due to the strange quark. (2) The SU(3) prediction, μ_Λ = -(1/3) μ_p, must be multiplied by the ratio of the strange quark moment to the down quark moment. (3) Hyperfine splittings are due to "color-magnetic" quark-quark interactions which are proportional to the product of quark "color-magnetic moments". (4) The electromagnetic magnetic moments of the quarks are proportional to the color magnetic moments; thus the ratio of the strange quark moment to the down quark moment is given by the ratio of the Σ^*-Σ and Δ-nucleon mass splittings.

Assuming that assumed quark magnetic moments are Dirac moments with a scale determined by some effective quark mass,[7] and using the Λ-nucleon mass difference as the mass difference between the strange and nonstrange quarks,[4,8] gives a completely different prediction for μ_Λ with exactly the same value, μ_Λ = -0.61 n.m.

The physics input here is that the same "effective quark mass" which may include all kinds of complicated quark-gluon interactions appears both in the quark magnetic moment and in the hadron masses. Why this should be so is an open question, left to be solved by QCD theorists. But the simple constituent quark model, with its manifestly simple physics, appears here as a bridge between the experimental data and the fundamental QCD description.[1]

2. THE NUCLEAR APPROACH TO MULTIQUARK HADRONS--"WHERE'S THE PHYSICS?"

2.1 Beyond Gabriel Karl's Standard Model

To go beyond N=3 we must recognize the new physics arising in multiquark configurations[12], where confinement of the color field no longer requires confinement of the system and different properties arise in different domains. Any model for N>3 must pass the following two tests.

1. Can a given model predict N=3 spectroscopy from N=2? If not, throw it away. It won't predict N>3. This kills the bag and skyrmion.

2. Can a model good for N=2,3 handle the new physics beyond N=3? If not, throw it away. It will predict nonsense. This kills the naive potential models.
   Needed! <u>One</u> prediction that works!

Multihadron states at large distances have color flux confined within hadrons and no color field nor long range interactions between them. At intermediate distances interactions between separated hadrons can occur when flux tubes flip from one configuration to another. At short distances there are many ways of connecting the constituents with flux tubes and no obvious optimum configuration. So far there is no tractable model with all the proper characteristics in these three domains and which matches smoothly between them. The problem resembles nuclear reactions which go via a "compound nucleus" intermediate state, but with the additional complications of the color degree of freedom, confinement and flux tubes.

Some of these difficulties would be avoided in a strongly bound multiquark state where only the small distance behaviour is relevant. The experimental discovery of such a strongly bound state would provide valuable information about how QCD works in multiquark systems. So far none have been found, and theoretical guidance directing searches for promising candidates is of great interest.

2.2 From N=2 to N=3

In the quark-antiquark and three-quark systems the coupling of the colors of the constituents to a color singlet is unique, and the color degree of freedom factorizes out to leave a single Schroedinger equation in the space, spin and flavor variables. Mesons and baryons are made of the same quarks with an effective two-body interaction V satisfying the relation[9,10,11,12]:

$$\langle q(x_1)q(x_2);3^*|V|q(x_1)q(x_2);3^*\rangle = (1/2)\langle q(x_1)\bar{q}(x_2);1|V|q(x_1)\bar{q}(x_2);1\rangle \tag{1.1}$$

where $|q(x_1)q(x_2);3^*\rangle$ and $|q(x_1)\bar{q}(x_2);1\rangle$ denote respectively quark-quark and quark-antiquark states with the color antitriplet (3*) and color singlet (1) couplings at the points x_1 and x_2.

Potential models which replace the interactions via the color field by static confining potentials have recently been justified by QCD arguments and lattice gauge calculations for heavy quarkonium states. The relation (2.1) for additive two-body interactions in baryons is easily derived by noting that the color field and the interaction obtained from lattice gauge calculations in a color-singlet baryon become the same as in the quark-antiquark configuration in mesons whenever two of the three quarks in the baryon are at the same point.[9,12,13] However, no QCD argument yet shows that baryons can be described by an effective potential with only two-body forces, nor justifies a potential model for hadrons composed of light quarks. The commonly used "color-exchange" force with the color dependence of one-gluon exchange in all channels satisfies the relation (2.1), but a very different spatial dependence from that of one gluon exchange is obtained from lattice gauge calculations or phenomenological fits to hadron spectra. Thus any justification from QCD for the use of additive two-body effective interactions with realistic radial dependence in hadron spectroscopy must include appreciable multigluon contributions and explain why they do not give appreciable three-body forces. The use of the relation (2.1) has recently led to new successful prediction of baryon masses with meson masses as input.[13] It therefore seems reasonable to extrapolate the approach for larger N. But there are new problems.

2.3 The New Physics Beyond N = 3.

The essential features of multiquark physics are exhibited in the simple example of meson-meson scattering with the possibility of quark exchange; e.g.

$$K^-(s\bar{u}) + K^0(d\bar{s}) \rightarrow \pi^-(d\bar{u}) + \phi(s\bar{s}). \qquad (2.2)$$

It is also interesting to note that if there were only two colors, so that a diquark can also be a color singlet, analogous to the baryon in the three-color case, an additional reaction could occur:

$$K^-(s\bar{u}) + K^0(d\bar{s}) \rightarrow B(ds) + \bar{B}(\bar{u}\bar{s}). \qquad (2.3)$$

where B and $\bar{B}$ denote diquark baryons and antibaryons in two-color QCD.

The analog of these reactions in abelian QED is fermion exchange in positronium-muonium scattering:

$$(e^+e^-) + (\mu^+\mu^-) \rightarrow (e^+\mu^-) + (\mu^+e^-) \qquad (2.4)$$

In Abelian QED the dynamics of the reaction (2.4) can be described by a static two-body coulomb interaction between each pair, having the form

$$V_{ij} = q_i q_j V(r_{ij}) \qquad (2.5)$$

where q_i and q_j are the charges of the particles and $V(r_{ij})$ is just the coulomb interaction. The value of the interaction is completely determined by the positions of the particles, and the coulomb field of each particle extends throughout all space. Thus even when the two bound states are separated by a large distance, there are long range Van-der-Waals forces between them.

In nonabelian QCD the color flux is confined to flux tubes within hadrons and is not determined completely by the positions of the particles. There is no color field nor Van-der-Waals force between separated hadrons.

Consider the case where the four particles in the reactions (2.2) and (2.3) are at the corners of a tetrahedron. The forces between the particles will depend upon the location of the flux tubes connecting them. In three-color QCD there are two equivalent ways of drawing the flux tubes, as each quark has two possible antiquarks with which it can be connected. In two-color QCD there are three equivalent ways of drawing the flux tubes, since the baryon-antibaryon configuration is also allowed. Thus in contrast to QED, the interactions are not completely determined by the positions of the particles with all necessary information about the field included in the static potential. Additional information about the field configuration is necessary, because the fields are not determined by the positions of the particles alone; there are several ways of drawing the flux tubes for a given configuration.

Attempts to express this additional degree of freedom using the color variable have had some apparent success in the four-body system.[10] The two ways of drawing the flux tubes have been placed in one-to-one correspondence with the two ways of coupling the colors of the quarks and antiquarks to make a color singlet. A "color-exchange" potential of the form (2.5) with color matrices replacing charges is a 2 X 2 matrix in color space and its eigenvectors can be interpreted as corresponding to the two ways of drawing flux tubes. However, this correspondence is a numerical accident occurring in the three-color-four-body system. With only two colors, there are three ways of drawing the flux tubes, including the "baryon-antibaryon" coupling (2.3), but the potential matrix is still only 2 x 2 in color space. A similar situation occurs in the six-quark baryon-baryon system with three colors, where the color matrix is 5 x 5, but there are 10 ways to draw flux tubes.

The new effects in multiquark configurations have been summarized in ref. 12. We list them briefly here:

A. "Hidden-Color" and the Color Matrix. Multiquark systems contain pairs which are neither in color singlet nor color anti-triplet states. The interactions in such hidden color states are not defined by eq. (1.1) and completely unknown, with no theoretical basis nor experimental information available. The color dependence no longer factorizes, since the colors of the constituents can be coupled in different ways to form an overall color singlet, and the Schroedinger equation is a nontrivial matrix equation in color space.

B. Breakup Decay Modes and Coupled Channels. A multiquark system is not confined and can break up into separated color singlet clusters. Multiquark dynamics involve coupled channels with the size of the S matrix determined by the number of ways in which asymptotic states can be defined as two color singlet clusters. This is particularly important for the H dibaryon which can break up into a $\Lambda\Lambda$ state.

C. __Motion of Flux Tubes -- Flipping and Tunneling__. The Coulomb fields of muonium and positronium spread out through all space, cancel exactly only when the proton and electron are exactly at the same point and give rise to long range power-law Van der Waals forces. The QCD color field is confined to flux tubes within hadrons and gives no long range interactions between hadrons. Forces between hadrons arise in QCD from the motion or flipping of flux tubes from one configuration to another through a domain of energetically less favorable configurations.[14,15] This motion is a tunneling process which gives an interaction decreasing exponentially with the distance between color singlet hadrons.

Any model with additive two-body interactions like (2.5) between hadron constituents gives power-law forces between separated hadrons, rather than exponentially decreasing forces, if it has confining forces within hadrons. Flux-tube physics can be introduced to give confining forces within clusters and no forces between clusters.[14,15] This requires defining how to draw flux tubes in the physically interesting domain where the distance between hadrons is comparable to the hadron size. Placing the flux tubes in the spatial configuration which has minimum energy for a given spatial configuration of the quarks implies instantaneous flipping of flux tubes each time the quarks move through a configuration where two flux tube configurations become degenerate.[14]

At very short range, one can imagine motion of the multiquark system in some kind of mean color field or bag which does not change violently as the quarks move. How to connect the well defined asymptotic states at large distances with a bag picture at short distances is not known. Uncertainties in what happens at intermediate range where flipping and tunneling occur have not been resolved, neither by theoretical derivations nor by experimental tests of phenomenological models. The physics resembles that of nuclear reactions, where a "compound nucleus" basis of states is used when all the particles are within a small volume, a basis of asymptotic states describes the breakup channels and there is no simple relation between the two bases. The P-matrix formalism of Jaffe and Low[16] follows the nuclear example.

Despite numerous attempts to include flux-tube dynamics, no treatment has yet made contact with either experimental data or rigorous theory.

3. THE SEARCH FOR MULTIQUARK BOUND STATES USING KNOWN HADRON PHYSICS

The experimental discovery of a convincing candidate for a multiquark bound state would be a significant breakthrough for our understanding of multiquark spectroscopy and of how QCD operates in mutiquark systems. So far no such states have been found, although the δ and S^* mesons are possible candidates[15,16] with insufficient experimental evidence either for or against. Theoretical predictions suggesting experimental searches for multiquark bound states are therefore of particular interest.

3.1 Two approaches to model-independent predictions.

Attempts to construct "nearly model-independent" bound multiquark wave functions whose properties can be reasonably well predicted from conventional hadron spectroscopy have been discussed in Refs. 12 and 13. Reliable estimates of multiquark binding are possible only when unknown hidden color contributions can be neglected. We note the following two approaches:

A. <u>Hyperfine binding</u>. The observation that hyperfine (color magnetic) energies are much larger than binding energies of multihadron states suggests that the energy difference between bound states and breakup channels in multiquark systems is dominated by the hyperfine interaction. This hyperfine dominance is expressed quantitatively by noting that the N-Δ mass difference $M(\Delta)-M(N)$ is much larger than the deuteron binding energy $M(n) + M(p) - M(d)$,

$$M(\Delta)-M(N) \gg M(n) + M(p) - M(d) \qquad (3.1)$$

B. <u>Heavy Quark Binding</u>. Hidden color interactions may also be negligible in systems like two heavy antiquarks and two light quarks. If the relative motion and structure of the two-heavy-antiquark wave function can be separated from the motion of the two light quarks relative to the heavy antidiquark, the four-body problem is then broken up into a two-body problem and a three-body problem with only color triplet and antitriplet two-body couplings.

3.2. Systems Dominated by the Hyperfine Interaction

3.2.1 <u>Meson "Alpha-Particle" Configurations</u>. The hyperfine energy has been calculated for systems of two quarks and two antiquarks in a spatially symmetric s-wave state with a radial wave function such that each pair has the relative radial dependence of a corresponding meson wave function. Such "alpha-particle" meson wave functions have been shown not to give bound multiquark states.[12,18]

3.2.4 <u>Dibaryons and the H Dibaryon</u>. The same approach applied to the six quark system shows that one of the most promising candidates for a bound multiquark state is the H dibaryon predicted by Jaffe[16] with strangeness -2 and spin zero. The recent suggestion connecting the H with Cygnus X3 events[19] requires it to be stable against the first-order-weak Λ-nucleon decay.

$$M(H) < M(\Lambda) + M(N). \qquad (3.2)$$

Thus

$$M(N) + M(\Xi) - M(H) > M(\Xi) - M(\Lambda) > M(\pi), \qquad (3.3)$$

and the mass of the H is below the threshold for pion production by stopped Ξ's in the reaction

$$\Xi + N \to H + \pi. \tag{3.4}$$

The pion produced in the reaction (3.4) provides a distinctive signature against a very low background.[12] For stopped Ξ's or a low momentum Ξ beam below the conventional pion production threshold no other open channel can give prompt pion production.

Theoretical calculations[6] predict higher dibaryon states in an octet of SU(3) flavor including an (I=1, S=-2) Ξ-nucleon resonance which can decay into H-π. This resonance could enhance the cross section for the reaction (3.4), and also the reaction

$$K^- + d \to H^{*-} + K^+ \to H + K^+ + \pi^- \tag{3.5}$$

This reaction has been proposed[20] for an H search with the K^+ and π^- forming a K^{*0}. It is also of interest to look for the H^{*-}.

The inequality (3.2) increases the energy release and therefore reduces the cross section in the reaction suggested for producing the H in deuterium[20]

$$\Xi + d \to H + n. \tag{3.6}$$

In this reaction the momentum of the final neutron is generally believed to come from its Fermi momentum in the wave function of the initial deuteron state. The cross section is therefore suppressed for high values of the neutron momentum by a deuteron form factor at the nucleon (not quark) level. It might be of interest to cover the low H-mass region in this experiment by looking for the reactions

$$\Xi^- + d \to H + n + \pi^0 \tag{3.7}$$

$$\Xi^- + d \to H + p + \pi^-. \tag{3.8}$$

No convincing theoretical argument places the mass of the H either above or below the N-Λ threshold. Searches for this particle have so far been inconclusive and the question is still open. Experiments including both a pion detector sensitive to the reactions (3.4), (3.7) or (3.8) as well as a neutron detector sensitive to the reaction (3.6) can detect the H over a wide mass range.

3.2.2 <u>Baryon "Alpha-Particle" Configurations</u>. Consider the system of twelve nonstrange quarks in a spatially symmetric s-wave state with a radial wave function such that each pair has the relative radial dependence of a quark pair in the nucleon wave function. This configuration has the spin, isospin and baryon number quantum numbers of an alpha particle, but does not describe the physical alpha particle. This "closed-shell" configuration has been shown[12] to have the color-spin couplings of four Δ's with a high hyperfine energy, of the order of four times the N-Δ splitting. This high energy could be the source of the repulsive core in the nucleon-nucleon force as discussed below.

3.2.3 Meson "Deuteron-like" Configurations. Meson-meson states can be loosely bound by short-range interactions barely strong enough to bind a single state with tails in the wave functions going beyond the range of the potential as in simple models of the deuteron. Qualitative predictions for such states indicate that the δ and S* mesons might indeed be K$\bar{\text{K}}$ bound states.[17,18]

3.3 Heavy Diquark Meson Configurations.

In systems containing heavy quarks the hyperfine interaction is much lower and no longer dominates completely over the color electric interaction. The higher mass particles are allowed by the uncertainty principle to come much closer together than light quark pairs, and therefore to be much deeper in the coulomb-like potential at short distances. In such systems the color-electric force also becomes important in binding.[12,13,21]

The first states where color-electric binding may be important are configurations with two light quarks and two heavy antiquarks; e.g. (ud$\bar{c}\bar{c}$). Because the two heavy particles both have the same baryon number, either both quarks or both antiquarks, the allowed breakup channels have the two in separated mesons. Thus the additional binding produced by the possibility of having them very close together increases stability against breakup.

The masses of such states can be estimated from the masses of known hadrons, and indicate that an axial vector (ud$\bar{c}\bar{c}$) meson is on the bordeline of stability, while the (ud$\bar{b}\bar{b}$) state is found to be stable against strong decays. The experimental production and detection of such states appears to be very difficult.

4. THE QUARK APPROACH TO BARYON-BARYON INTERACTIONS

The interactions between nucleons are generally described as a short-range repulsive core, an intermediate range attraction, and a long range tail attributed to one-pion exchange. Attempts to obtain these properties from a more fundamental quark picture have focused primarily on the short range repulsion, explained as an effect of the Pauli principle. The intermediate range attraction might be attributed to gluon exchanges. The pion exchange is generally put in by hand at this stage, since there does not seem to be any hope of describing this part of the interaction before QCD provides a good description from first principles of the pion and its emission and absorption. Most treatments assume certain specific models and obtain results by detailed calculations. We examine here some normally overlooked simple and general results which are model-independent and follow from symmetry principles.

Consider a system of six nonstrange quarks at very short distances, with a wave function totally symmetric in space and having a finite value when all six quarks are at the same space point. The allowed color singlet states for six nonstrange quarks with this space symmetry have the following spin-isospin quantum numbers:

($I=3;S=0$) and ($I=0;S=3$). These quantum numbers occur in the Δ-Δ system.
($I=2;S=1$) and ($I=1;S=2$). These quantum numbers occur in the Δ-Δ and Δ-N systems.
($I=1;S=0$) and ($I=0;S=1$). These quantum numbers occur in the Δ-Δ and N-N systems.

States with the following quantum numbers do not occur:

($I=3;S=2$) and ($I=2;S=3$) found in the Δ-Δ system.
($I=2;S=2$) and ($I=1;S=1$) found in the Δ-N system.

The following states which are not found in any dibaryon system also do not occur:

($I=3;S=3$), ($I=3;S=1$), ($I=1;S=3$) and ($I=0;S=0$).

From these results the following general conclusions can be drawn:

1. No states allowed for the six-quark system are forbidden for the dibaryon system. Thus <u>there are no exotic six-quark states at short distances with quantum numbers forbidden for a dibaryon system</u> and no bound states of quarks at short distances which are forbidden by known selection rules to couple to the dibaryon system.

2. The four "exotic" states ($I=3;S=3$), ($I=3;S=1$), ($I=1;S=3$) and ($I=0;S=0$), which are not found at short distances and do not occur for asymptotic dibaryon systems have quantum numbers which would occur in the Δ-Δ and/or nucleon-nucleon systems if they were not equivalent fermions, but are forbidden by the Pauli principle at the baryon level. These states therefore play no role in dibaryon physics.

3. The states ($I=3;S=2$) and ($I=2;S=3$) which are allowed for the Δ-Δ system and the states ($I=2;S=2$) and ($I=1;S=1$) which are allowed for the Δ-nucleon system are not allowed states for six quarks at short distances. These states would not be Pauli forbidden if the Δ and nucleon were elementary fermions. However, they are forbidden at short distances for composite baryons made of the same quarks, as the Pauli principle forbids placing two quarks with the same quantum numbers in a spatially symmetric state.

For example, a Δ-Δ state with charge +3, spin 2 and spin projection +2 on the z-axis is a perfectly good state for two elementary Δ's. One can be in a spin state with m=3/2 and the other with m=1/2. However, if each Δ is made of three u-quarks, the m=2 Δ-Δ state has four u-quarks with spin up and two with spin down. Since there are only three colors, the Pauli principle forbids the four u-quarks with parallel spins from being in a symmetric spatial state. Similar arguments hold for the other states allowed by Pauli for two elementary baryons and forbidden for six quarks. These dibaryon states must always see a repulsive core in any phenomenological description at the baryon level, independent of the dynamical interactions between the baryons.

4. The states (I=3;S=0) and (I=0;S=3) allowed for the six quark system and having only one allowed dibaryon state; namely Δ-Δ, have exactly the same spin-isospin couplings both at short range and as asymptotic states. There is no effective repulsion due to the Pauli principle at short distances because these six-quark states have no pair of quarks with the same quantum numbers. For example the Δ-Δ system with charge 3, with one Δ having m=3/2 and the other m=-3/2, is simply a color singlet state of three u-quarks with "spin up" and a color singlet state of three u-quarks with "spin down". This state of the Δ-Δ system feels no repulsive core due to Pauli effects. Thus a relatively weak attractive interaction at short range might produce binding. These channels offer the best candidates for possible bound dibaryons.

5. The remaining states (I=2;S=1) and (I=1;S=2), allowed for the Δ-Δ and Δ-N systems, and (I=1;S=0) and (I=0;S=1), allowed for the Δ-Δ and N-N systems, each have two allowed asymptotic dibaryon states but only a single allowed six-quark state at short distances. Here there is a complicated interplay of the Pauli principle and dynamics. Consider the nucleon-nucleon state with I=0;S=1, m=1; i.e. the quantum numbers of a deuteron with spin up. Each nucleon has m=1/2. The proton has a piece in its wave function with both u-quarks in a state with m=1/2. The neutron has a piece in its wave function in which the u-quark is in a state with m=1/2. The six-quark description of this asymptotic dibaryon state thus has a piece in its wave function with three u-quarks all having m=+1/2. When the two nucleons are very close together, the Pauli principle allows these three u-quarks to be at the same point only if they are in the antisymmetric color singlet state. But the u quark in the neutron is coupled to a color singlet with the two d-quarks in the neutron and has no color correlation with the two u-quarks in the proton. Thus this piece of the wave function has the three u-quarks in a mixture of color singlet and color octet states, with the probability of color singlet being only 1/9. Thus we see that when the two nucleons come close together, only certain pieces of the wave function are allowed; other pieces are forbidden or "Pauli blocked". The same is true for the Δ-Δ system with these quantum numbers.

The two channels are therefore necessarily coupled at short distances. If the dynamics allow the two baryon clusters to come close enough together to remove the "Pauli-blocked" pieces of the asymptotic wave function, the remaining wave function must have both nucleon-nucleon and Δ-Δ components and there will be transitions between these two asymptotic states as a result of the "wave-function-rearrangement" imposed by the Pauli principle at the quark level.

A repulsive core in the nucleon-nucleon interaction can be produced by combining the Pauli blocking effect with the hyperfine interaction dominant at short distances. Two nucleons can be pushed together so that their constituent quark wave functions overlap only by modifying the wave function to exclude the Pauli-blocked pieces. This automatically introduces a Δ-Δ component which has a higher interaction energy by at least twice the nucleon-Δ mass splitting. The contribution of this Δ-Δ energy to the total energy depends upon the volume in which this component of the wave function exists. This volume cannot be made arbitrarily small, even though the Pauli-blocking occurs only when the two particles are at the same point, because the Heisenberg uncertainty principle prevents the Δ-Δ piece of the wave function from being confined to a small volume without paying a large price in kinetic energy. This combination of statistics, hyperfine interactions and kinetic energy has been called the "Pauli-Fermi-Heisenberg" repulsive core and may be quantitatively responsible for the observed repulsion in nucleon-nucleon interactions.

REFERENCES

1. G. Karl, these proceedings; N. Isgur and G. Karl, Phys. Lett. $\underline{74B}$, 353 (1978); Phys. Rev. $\underline{D18}$, 4187 (1978); $\underline{D19}$, 2653 (1979), Phys. Rev. $\underline{D20}$, 1191 (1979); Phys. Rev. $\underline{D21}$, 3175 (1980).
2. H. J. Lipkin, in Common problems in low- and medium-energy nuclear physics, Proc. NATO Advanced Study Institute/1978 Banff Summer Institute on Nuclear Theory (Banff Canada, 1978), eds. B. Castel, B. Goulard and F. C. Khanna NATO ASI Ser. B. Physics, Vol. 45, (Plenum, New York) p. 175; H. J. Lipkin, in the Quark Structure of Matter, (Proceedings of the Yukon Advanced Study Institute Whitehorse, Yukon, Canada, 1984) edited by N. Isgur, G. Karl and P. J. O'Donnell (World Scientific Publishing Co., 1985) p. 108.
3. A. De Rujula, H. Georgi and S. L. Glashow, Phys. Rev. $\underline{D12}$, 147 (1975).
4. Ya. B. Zeldovich and A. D. Sakharov, Yad. Fiz. $\underline{4}$, 395 (1966); Sov. J. Nucl. Phys. $\underline{4}$, 283 (1967).
5. P. Federman, H. R. Rubinstein and I. Talmi, Phys. Lett. $\underline{22}$, 203 (1966).
6. R. L. Jaffe, Phys. Rev. Lett. $\underline{38}$, 195 (1977).
7. H. J. Lipkin, Phys. Rev. Lett. $\underline{41}$, 1629 (1978).
8. H. J. Lipkin, Phys. Lett. $\underline{74B}$, 399 (1978).

9. H. J. Lipkin, in Short-Distance Phenomena in Nuclear Physics, edited by David H. Boal and Richard M. Woloshyn (Plenum Publishing Corporation, 1983), p. 51.
10. H. J. Lipkin, Phys. Lett. 45B, 267 (1973).
11. O. W. Greenberg and H. J. Lipkin, Nucl. Phys. A179, 349 (1981).
12. H. J. Lipkin, Proceedings of the Workshop on Nuclear Chromodynamics, Santa Barbara, August 12-23, 1985, ed. by S. Brodsky and E. Moniz, p. 328.
13. H. J. Lipkin, Phys. Lett. 171B, 293 (1986).
14. F. Lenz, Proceedings of the Workshop on Nuclear Chromodynamics, Santa Barbara, August 12-23, 1985, ed. by S. Brodsky and E. Moniz, p. 289
15. D. Sivers, Proceedings of the Workshop on Nuclear Chromodynamics, Santa Barbara, August 12-23, 1985, ed. by S. Brodsky and E. Moniz, p. 322.
16. R. L. Jaffe and F. E. Low, Phys. Rev. D19, 2105 (1979).
17. R. L. Jaffe, Phys. Rev. D15, 267 and 281 (1977).
18. Nathan Isgur and Harry J. Lipkin, Phys. Lett. 99B, 151 (1981); Kim Maltman and Nathan Isgur, Phys. Rev. Lett. 50, 1827 (1983); Phys. Rev. D29, 952 (1984).
19. G. Baym et al., Phys. Lett. 160B, 181 (1985).
20. P. D. Barnes, these proceedings.
21. Leon Heller, Proceedings of the Workshop on Nuclear Chromodynamics, Santa Barbara, August 12-23, 1985, ed. by S. Brodsky and E. Moniz, p. 306.
22. H. J. Lipkin in Intersections Between Particle and Nuclear Physics, (Steamboat Springs, 1984), Edited by R. E. Mischke, AIP Conference Proceedings No. 123, p.346 (1984).

MULTI-QUARK STATES IN A QCD-LIKE POTENTIAL MODEL

Kim Maltman
T5-B283 Los Alamos National Laboratory, Los Alamos, N.M., 87545

ABSTRACT

The requirements of quark antisymmetry and the color-spin structure of the quark hyperfine interaction generate effective short range NN repulsion, in agreement with observation. Multi-quark channels where these same aspects of the underlying quark physics favor formation of exotic states are identified. Narrow resonances are not expected. The non-existence of these states would most likely require unexpected confinement physics in the multi-quark sector and a re-evaluation of the apparent succeses of the quark picture of the origin of the short range structure of the NN force.

INTRODUCTION

If, as widely believed, QCD is the correct theory of the strong interactions and confines color non-singlet objects, baryons and mesons, whose valence structures, q^3 and $q\bar{q}$, contain no color singlet sub-units, must necessarily occur as either stable particles or (albeit potentially broad) resonances in the strong interaction spectrum. More complicated color-singlet multi-quark configurations ($q^2\bar{q}^2$, $q^4\bar{q}$, q^6,...), on the other hand, necessarily admit color singlet sub-units, the question of whether such degrees of freedom manifest themselves as true resonances or only as apparently broad background contributions to hadron-hadron scattering, therefore, being one of dynamics. In principle one would like to simply apply QCD directly to the multi-quark sector. At present, however, it is only feasible to employ phenomenological models which, as much as possible, incorporate the dominant feature of QCD.

Much of the interest in multi-quark exotics was spurred by early calculations in the MIT bag model[1-4]. If one assumes that bag model wavefunctions provide an approximate representation of the short distance region, one may study the influence of even artificially confined bag model "states" (primitives) on hadron-hadron scattering[5]. Primitives lying below fall-apart thresholds should, of course, appear as resonances. As in any model in which confinement is implemented phenomenologically, one assumes that the structure and parameters of the hadronic confinement mechanism can be taken over unchanged into the multi-quark sector, despite the appearance of new color structure in the confinement region. In addition, the non-linear (pressure balance) boundary condition is generally only tractable in the static, spherical cavity approximation. As is well-known[6], the color-spin structure of the quark hyperfine interaction is such as to favor lower spatial symmetries, and this suggests that important hyperfine binding effects may be missed in the bag model.

In what follows we will employ a QCD-inspired non-relativistic quark potential model in the strangeness 0 and -1 positive parity di-baryon and Z^* sectors, all of which are free of phenomenologically unconstrained annihilation effects. In this model there are no technical impediments to the inclusion of lower spatial symmetries. In addition, since the lowest lying two hadron state of a given channel is not artificially confined, one may approach the short distance region from the two-hadron side. Channels in which the residual inter-hadron forces induced by the requirements of quark antisymmetry are attractive will then be favorable to deep binding through mixing to available short range hidden color excitations. Note that calculations involving similar models show that short range NN forces can be understood as a residual quark effect[7-11]. The physics underlying these calculations motivates the present approach. In the NN channels the exchange hyperfine interaction generates a dynamical, non-local repulsion which shields the system from access to short distance hidden-color excitations. Since this repulsion depends on the hyperfine color-spin structure and the channel color-spin-isospin couplings, one anticipates the existence of channels in which the induced interaction is attractive.

METHOD AND DISCUSSION

The approach procedes in two phases, the first identifying channels in which the quark-induced interactions of ordinary hadrons are either attractive or weakly repulsive. The qualitative features here are dominated by the exchange hyperfine interaction and quark antisymmetry, and presumably reliable, barring radical changes in the effective confinement interactions. Phase II implements mixing of available excited states in the channels identified in Phase I. Since calculated hidden-color energies are, of necessity, phenomenologically unconstrained, the magnitude of the resulting induced binding is subject to large uncertainties, despite verifying the plausibility of typically hadronic-sized transition matrix elements.

Our method is essentially that of the resonating group. For a state of two hadrons, A and B, in channel α, we write

$$|AB\rangle_\alpha = \frac{1}{N_A} A | [\phi^A_{\{k_i\}} \phi^B_{\{l_i\}}]_\alpha X(R_{AB}) \rangle. \tag{1}$$

N_A is a normalization factor, $\{k_i\}$ and $\{l_i\}$ sets of quark and/or anti-quark labels, $\phi^{A,B}$ the corresponding internal wavefunctions, and A the quark antisymmetrizer. The inter-cluster "wavefunction", $X(R_{AB})$, is a variational degree of freedom, to be determined by minimizing the expectation of the quark Hamiltonian, H. Using permutational symmetry, the matrix elements required for Phases I and II can be cast in the form

$$\langle [\phi^A \phi^B]_\alpha X^{AB} | H A | [\phi^C \phi^D]_\alpha X^{CD} \rangle / N_{AB} N_{CD}. \tag{2}$$

For further details the reader is referred to Refs. 12-14.

Let us take for H the Isgur-Karl Hamiltonian[15]

$$H_{IK} = \sum_i (m_i + p_i^2/2m_i) + \sum_{i<j} \vec{F}_i \cdot \vec{F}_j (h^c_{ij} + h^{hf}_{ij}) \tag{3}$$

$$h^c_{ij} = -(C + \tfrac{1}{2} k r_{ij}^2 + U(r_{ij}))$$

$$h^{hf}_{ij} = -\frac{\alpha_s}{m_i m_j} \left(\frac{8\pi}{3} \vec{S}_i \cdot \vec{S}_j \delta^3(r_{ij}) + S_{ij} r_{ij}^{-3} \right). \tag{4}$$

The $\vec{F}_i$ are the usual quark (anti-quark) color charges $\vec{\lambda}_i/2$ ($-\vec{\lambda}_i^*/2$), and U represents departures from the harmonic limit, including the spin-independent pieces of one-gluon exchange (OGE). h_{ij}^{hf} contains the contact (hyperfine) and tensor interactions of the non-relativistic reduction of OGE. H_{IK} provides a good fit to both the baryon spectrum[15] and baryon decay amplitudes, including a resolution of the problem of "missing resonances"[16]. See Refs. 11,12 for a discussion of H_{IK} in the context of multi-quark systems.

The singular nature of the contact term in (4) necessitates a perturbative treatment of hyperfine effects. A simultaneous fit to K and K^* masses is not possible within this framework. While one might suppose this to be a result of the association of the K with chiral symmetry breaking, note that s-wave K^+N scattering can be well-accounted for in a constituent quark picture[17]. Recent work employing a regularized modification of (4)[18] demonstrates that the meson spectrum can, indeed, be fit, though with a smaller value of the effective strong coupling constant and smaller meson cluster sizes than associated with H_{IK}. This modified Hamiltonian appears also to provide a good account of the baryon spectrum[19], again with smaller cluster sizes than those associated with H_{IK}. Since multi-quark binding effects are driven by antisymmetrization and the exchange hyperfine interaction, and both are reduced by the above changes, we must investigate whether or not our qualitative results are altered by such changes. To this end we consider a simplified version, H_{mod}, of the Hamiltonian of Ref. 18, in which h^c of (4) is replaced by a linear-plus-Coulomb-plus-constant form, the Coulomb piece involving a strong coupling constant, α_s, which implements saturation at large, and one-loop asymptotic freedom at small, distances, this change in α_s also serving to regularize the hyperfine interaction. See Refs. 13,18 for details. H_{mod} has also been applied to the strangeness -1 dibaryon channels, but not, as of yet, to the non-strange sector. The results of the calculations are presented in Table 1. We include only those channels predicted to support multi-quark resonances. Note that the $IJ^P=30^+$ state may be subject to downward corrections of O(100 MeV) due to pionic effects[20] which have not been included here. Given the existence of open meson decay channels, none of the predicted states are expected to be narrow. Partial widths to two-body final states will also be small, in the case of the high spin states, as a consequence of the weakness of the quark tensor force. The $IJ^P=03^+$ state, for example, has a $^3D_3$ NN width of ~5 MeV[12].

Table 1. Predictions for multi-quark resonances[a]

IJ^P	hadron content	Phase I_{IK}	Phase II_{IK}	Phase I_{mod}	Phase II_{mod}
30^+	$\Delta\Delta$	wr	BE=30	—	—
03^+	$\Delta\Delta$	BE=3	BE=260	—	—
$\frac{1}{2}3^+$	$\Delta\Sigma^*$	BE=2	BE=60	att	BE=20
$1\frac{5}{2}^-$	ΔK^*	BE=26	—	att	BE=56

a) wr=weakly repulsive, att=attractive but insufficient to bind, BE= binding energy (MeV)

While we have stressed the multi-quark resonance aspect of the calculations, one should bear in mind that many of the channels whose induced interactions are strongly repulsive are, nonetheless, of interest viz a vis hadron-hadron scattering. For example, as also noted by Oka and Yazaki[7], the short distance quark-induced $N\Lambda$-$N\Sigma$ transition potential is negligible, in consequence of which the full conversion process is predicted to be peripheral and dominated by pion exchange. This may have some bearing on hypernuclear structure.

In conclusion, let us stress the qualitative nature of the predictions of Table 1; should these states not exist, a revamping of the quark picture of NN forces would be required. In addition, we re-iterate the uncertainties associated with hidden color mixing in the Phase II numbers therein. More extensive lattice calculations in the multi-quark sector, of the type contained in Ref. 21, would provide useful constraints for confinement phenomenology and be most welcome in this regard.

REFERENCES

1. R.L. Jaffe, Phys. Rev. Lett. 38, 195 (1977); Phys. Rev. D 15, 267 (1977).
2. P.J. Mulders, A.T. Aerts and J.J. deSwart, Phys. Rev. D 21, 2653 (1980) and references therein.
3. D. Strottman, Phys. Rev. D 20, 748 (1979).
4. C. Roiesnel, Phys. Rev. D 20, 1646 (1979).
5. R.L. Jaffe and F.E. Low, Phys. Rev. D 19, 2105 (1979).
6. Yu. F. Smirnov et al., Sov. J. Nucl. Phys. 27, 456 (1978); I.T. Obukhovsky et al., Sov. J. Nucl. Phys. 31, 269 (1980).
7. M. Oka and K. Yazaki, in "Quarks in Nuclei", ed. W. Weise (World Scientific, Philadelphia, 1984), p. 489.
8. A. Faessler et al., Nucl. Phys. A 402, 555 (1983).
9. Y. Suzuki and K.T. Hecht, Phys. Rev. C 27, 299 (1983); C 28, 1453 (1983); Nucl. Phys. A 420, 525 (1984).
10. M. Harvey, J. LeTourneux and B. Lorazo, Nucl. Phys. A 424, 428 (1984).
11. K. Maltman and N. Isgur, Phys. Rev. D 29, 952 (1984).
12. K. Maltman, Nucl. Phys. A 438, 669 (1985).
13. K. Maltman and S. Godfrey, Nucl. Phys. A 452, 669 (1986).
14. K. Maltman, Los Alamos Preprint LA-UR-86-242, 1986.
15. N. Isgur, XVIth Int. School of Subnuclear Physics, Erice, 1978, ed. A. Zichichi, (Plenum, N.Y., 1980).
16. R. Koniuk and N. Isgur, Phys. Rev. D 21, 1868 (1980); D 23, 818 (1981)(E).
17. I. Bender et al., Nucl. Phys. A 414, 359 (1984).
18. S. Godfrey and N. Isgur, Phys. Rev. D 32, 189 (1985).
19. S. Capstick and N. Isgur, U. of Toronto UTPT-85-34, 1985.
20. P.J. Mulders and A.W. Thomas, J. Phys. G 9, 1159 (1983).
21. S. Ohta, M. Fukugita and A. Ukawa, KEK-TH 118, 1986.

FLUX TUBES IN MULTI-QUARK SYSTEMS

D. Robson[†]
Department of Physics, Florida State University
Tallahassee, Florida 32306

ABSTRACT

The many-body properties of flux tubes are outlined and found to lead to a modified Pauli Principle for quarks in different hadrons. Incorporation of flux tube ideas into nucleon-nucleon potential models including one pion exchange yields a repulsive core without invoking one gluon exchange and provides a reasonable description of the low energy properties of the nucleon-nucleon system with only four parameters. The size of nucleons in A=2 and 3 nuclei are found to increase by a few percent at small nucleonic separations.

1. INTRODUCTION

One of the major driving forces in non-perturbative quantum chromodynamics (QCD) is the concept that gluonic fields form into localized flux tube configurations whenever the quarks and/or antiquarks are <u>all</u> separated beyond some critical range r_{cr}. Phenomenological fits to hadron spectra[1] coupled with the predicted strong coupling limits[2] for QCD suggest that r_{cr} is small compared to the observed size of hadrons (~1F). Consequently we expect that flux tubes play a dominant role in describing the behavior of hadrons at low and intermediate momentum transfer ($Q^2 \lesssim 1$ GeV^2/c^2).

At high enough momentum transfer we know that quarks behave according to the predictions of perturbative QCD. This regime of "asymptotic freedom" occurs when the dominant amplitudes for high momentum transfer processes arise from quark-gluon dynamics within a small hypersphere of radius r_{cr}, i.e., where <u>all</u> quarks and antiquarks are close enough so that a minimal set of one gluon exchanges becomes the dominant process. We emphasize the region of asymptotic freedom as a small <u>multi</u>-quark volume because it is easy to erroneously think that small hadron-hadron separation (r_{HH}) is a necessary and sufficient condition for applying perturbative QCD. Making the separation between two hadrons less than r_{cr} is not sufficient to guarantee correspondingly small interquark separations. In fact if $r_{cr} \ll 1$F as we are proposing here then just requiring $r_{HH} < r_{cr}$ will still allow almost all of the multi-quark hypervolume to be occupied with at least one interquark separation being larger than r_{cr}.

The above is our rationale for stressing the need to consider flux tubes not only for peripheral hadronic interactions but also for understanding the nature of hadronic interactions at small hadron-hadron separations. In the present approach the commonly accepted idea that the repulsive core in the nucleon-nucleon interaction arises[3] from one-gluon exchange interactions (magnetic) is found to

---

[†]Supported in part by the U.S. Department of Energy.

be inadequate because of the important properties of multi-quark flux tubes. In multi-quark systems the flux tubes are compact objects (except at very high Q^2) and the integration over each field variable has a finite norm. This property of flux tubes leads to the most interesting result that quarks in "different" hadrons (i.e., belonging to different flux tube configurations) are "distinguishable". Such a result was first proposed by Greenberg and Hietarinta[4] and recently[5] has been more directly related to QCD in lattice gauge theory. This "modified Pauli Principle" has recently been incorporated into a calculable quark model of the nucleon-nucleon interaction. Here we show briefly some of the qualitative aspects of flux tubes and the basic model results for the nucleon-nucleon system at low energies.

2. FLUX TUBE PROPERTIES

Our discussion here has to be qualitative. We imagine we can take instantaneous snapshots of multi-quark configurations in momentarily frozen positions wherein the gluonic fields have just a long enough exposure time to form their flux tube connections. In this Born-Oppenheimer type of approach the gluon degrees of freedom adjust rapidly to the relatively slowly changing quark positions so that the energy residing in the gluon fields plays the role of a multi-quark potential energy. In obtaining the properties of flux tubes in QCD it is essential to impose <u>local</u> gauge invariance on the state vectors of the system. This is most conveniently done in a Hamiltonian lattice gauge approach[2] wherein quarks are placed at lattice sites and gluon fields are placed on the "links" connecting neighboring sites.

In QCD local gauge invariant states satisfy the requirement that they are unchanged by SU(3) color gauge transformations at <u>any</u> lattice site. A quark at x and an anti-quark at y ≠ x will have a flux tube state made out of a superposition of gauge invariant states of the form[2],

$$U_c(x,y) = \bar{\psi}(x)U(x,n)U(x+n,m)...U(y-\ell,\ell)\psi(y)|0>, \quad (1)$$

in which the linkages $U(i,j)$ are all connected to form a piecewise continuous (ordered) path (c) from x to y. Now if $\bar{\psi}$ and ψ are three color component (fundamental) representations of SU(3) and the linkages $U(i,j)$ behave like SU(3) "rigid" rotors then under local gauge transformations,

$$\bar{\psi}(x) \xrightarrow{G} \bar{\psi}(x)G^\dagger(x),$$

$$U(i,j) \xrightarrow{G} G(i)U(i,j)G^\dagger(j)$$

$$\psi(y) \xrightarrow{G} G(y)\psi(y), \quad (2)$$

with G being an arbitrary 3x3 SU(3) matrix operating in color space such that $G^\dagger(z)G(z) = I$ for all z. As pointed out by Greenberg and Lipkin[6] conventional quark models (often referred to as additive potential models) which only use quark degrees of freedom explicitly and ignore local gauge invariance have unphysical and erroneous predictions, particularly for multi-hadron systems.

Introducing linkage operators $L_0^\dagger(x,y)$ which create the lowest energy flux tube eigenstates for fixed values of x and y, then Eq. (1) summed over c can be used to represent a gauge invariant meson state[4]:

$$|M(p)\rangle = (2\pi)^{-3/2} \int d^3x\, d^3y\, \psi(x-y) e^{ip \cdot \frac{1}{2}(x+y)}$$
$$\times \bar\psi(x) L_0^\dagger(x,y) \psi(y) |0\rangle \qquad (3)$$

in which $L_0^\dagger(x,y)$ involves[5] an integral over all link variables and satisfies the only non-vanishing commutation rules:

$$[L_0(x,y)_{\alpha\beta}, L_0^\dagger(x',y')_{\alpha'\beta'}] = \frac{1}{9} \delta_{\alpha\alpha'} \delta_{\beta\beta'} \delta_k(x-x') \delta_k(y-y') \qquad (4)$$

with the Kronecker delta function,

$$\delta_k(z) = 1\,,\ z=0$$
$$= 0\,,\ z \neq 0$$

and α,β are color labels. The unusual Kronecker delta functions in position space appear[5] to be a consequence of the fact that strong coupling eigenstates of the gluonic sector are orthogonal for $x' \neq x$ and/or $y' \neq y$ and have a finite norm for $x=x'$, $y=y'$.

The overlap between any two meson states now involves the usual non-vanishing terms

$$\langle 0 | L_0(x_2,y_2) L_0(x_1,y_1) L_0^\dagger(x_1,y_1) L_0^\dagger(x_2,y_2) |0\rangle = 1,$$

and

$$\langle 0 | L_0(x_1,y_1) L_0(x_2,y_2) L_0^\dagger(x_1,y_1) L_0^\dagger(x_2,y_2) |0\rangle = 1, \qquad (5)$$

but terms with different linkages

$$\langle 0 | L_0(x_2,y_1) L_0(x_1,y_2) L_0^\dagger(x_1,y_1) L_0^\dagger(x_2,y_2) |0\rangle, \qquad (6)$$

given no contribution due to the linkage commutation properties in Eq. (4). In conventional quark models the meson states $[\bar\psi(x_1)\psi(y_1)] \otimes [\bar\psi(x_2)\psi(y_2)]$ and $[\bar\psi(x_2)\psi(y_1)] \otimes [\bar\psi(x_1)\psi(y_2)]$ do not have zero overlap. These states clearly violate local gauge invariance and allow erroneous predictions[6] of long range Van der Waals interactions.

For identical quarks in different hadrons such as nucleons the above orthogonality corresponds to an orthogonality between quark configuratons with different quark labellings which arise via quark or linkage exchange between hadrons. Such orthogonalities correspond to using a modified Pauli Principle for quark dynamics once the gluon degrees of freedom have been eliminated. Such a "Pauli exclusion" between quarks in different hadrons leads to repulsive interactions between the hadrons at small separations between color singlet centers in much the same way as it does in polyatomic physics. It also means that the hadrons must become larger as the hadronic separations become smaller due to quarks being forced into orthogonal orbits about the different hadronic centers. These Pauli effects arise even when one gluon exchanges are totally neglected. In fact in the specific case of nucleon-nucleon interactions the Pauli effect arising from the properties of QCD flux tubes causes almost total suppression[7] of any repulsion from one gluon exchange terms in the Hamiltonian.

Finally in this section we note that the three quark baryons do not have flux tubes connecting pairs of quarks. Instead, as pointed out by Greenberg and Hietarinta, there is only one end of a flux tube to each quark and the other ends must be joined together via a color singlet junction denoted by $\varepsilon_{\alpha\beta\gamma}$. In attempting to derive the formalism of Greenberg and Hietarinta we find that the minimum energy flux tube configurations are a linear superposition of strong coupling eigenstates with different junction positions so that the junction position is not a simple dynamical variable as assumed by Greenberg and Hietarinta. However minimum energy linkage operators $L_0^\dagger(x_1, x_2, x_3)$ connecting three quarks are found which yield similar non-vanishing commutators to those found for mesonic systems. Excited "linkages" $L_n^\dagger$ with n>0, are also possible since orthogonal linear combinations of strong coupled eigenstates can be constructed. We assume here that these are at high excitation and essentially decouple from the low energy region of interest here.

3. OCTET COLORED HADRONS

Much has been written about the importance of hidden color excitations in multi-hadron systems. Such ideas arise because a color singlet made out of more than three quarks (or one quark-anti-quark) can have several "colored coupling schemes". For two nucleons with three quarks each it is possible to couple three quarks to either a symmetric or an antisymmetric octet color state and then couple the two octets back to a color singlet. However with flux tubes obeying local gauge invariance these octet colored baryons must have additional flux connections compared to the flux configurations for two singlets. The leading strong coupled state for 6 quarks involves two junctions, one for each baryon in a color singlet state. If we ignore four tube junctions the leading strong coupled state for 6 quarks with hidden color substructure involves at least four junctions as schematically shown in Fig. 1, but these are assumed to be at much higher energies than the color singlet pair with only two junctions. Since such hidden color configurations are expected to be much higher in energy for all hadronic separation distances when flux tube ideas are invoked we do not believe that earlier predictions about the importance of hidden color excitations should be taken as seriously as they have been in the low energy regime.

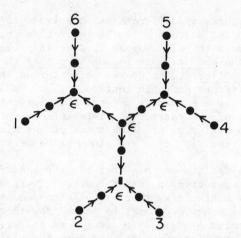

Fig. 1. Example of a hidden color six quark configuration.

4. TRACTABLE MODEL

The model Hamiltonian used here is based on flux tube philosophy discussed above. After integrating out gauge fields the Hamiltonian for n-nucleons or 3n-quarks is taken to be

$$H_{NN} = P_{NN}^2/M_N + \langle H_q \rangle \quad,$$

with
$$H_q = \sum_{i=1}^{3n}(P_i^2/2m_i + V_i) + \sum_{i=1}^{3n}\sum_{j>i}(V_{ij}^\pi + V_{ij}^{LE}) \quad (7)$$

in which $\sum_i V_i$ is a simplified confinement potential which models the gluonic field energy associated with all the quark positions x_i and V_{ij}^π, V_{ij}^{LE} are residual quark-quark interactions for pion exchange and linkage exchange respectively. For V_i we assume a two-center harmonic oscillator as discussed in an earlier presentation[7]. The two-center model has the advantage that we can construct orthogonal quark orbits for all separation distances between the two centers. It is a simplification of the more realistic many-body confinement models wherein for six quarks.

$$V = \min_{\{S_6\}}[\sum_{i=1}^{3}b_i(\vec{r}_i-\vec{\omega}_1)^2 + \sum_{j=4}^{6}b_j(\vec{r}_j-\vec{\omega}_2)^2 \quad (8)$$

and minimization is carried out for all labellings in S_6. Models with this "flip-flop" between different flux tube connections depending upon all the quark positions have been used[8] but they are difficult to use. In particular the orthogonality of the quark orbits in such potentials at finite momenta have not been considered.

The one particle approximation for V given in Eq. (7) is made plausible by constructing 6 quark solutions wherein three quarks are localized around a center at $\omega_1 = -\frac{R}{2}\hat{z}$ and the other three around $\omega_2 = +\frac{R}{2}\hat{z}$. This is achieved by using orthogonal linear combinations of the two lowest eigenfunctions $\phi_\pm$ of the two-center oscillator at each separation R:

i.e., $\phi_a(\vec{r}_i, -\frac{R}{2}) = \frac{1}{\sqrt{2}} [\phi_+(\vec{r}_i, R) - \phi_-(\vec{r}_i, R)]$

$$\phi_b(\vec{r}_j, +\frac{R}{2}) = \frac{1}{\sqrt{2}} [\phi_+(\vec{r}_j, R) + \phi_-(\vec{r}_j, R)] \qquad (9)$$

in which $\langle\phi_a|\phi_b\rangle$ is zero at every R because $\langle\phi_+|\phi_-\rangle$ is zero at every R and $\langle\phi_+|\phi_+\rangle = \langle\phi_-|\phi_-\rangle$ is unity. The particular weight factors of $\pm 1/\sqrt{2}$ are derivable because the quark energy associated with each nucleon is expected to be identical and at large R it gives the correct asymptotic results. To a very good approximation we find

$$\phi_\pm \simeq \hat{\phi}_\pm(\vec{r}, R) \simeq N_\pm \{\chi(|\vec{r} - \frac{\vec{R}}{2}|) \pm \chi(|\vec{r} + \frac{\vec{R}}{2}|)\} \qquad (10)$$

where $N_\pm$ are normalization factors (R-dependent) and $\chi(\xi)$ are Gaussian wave functions centered at $\xi=0$ with a width given by the oscillator parameter a.

A Born Oppenheimer estimate for the nucleon-nucleon potential at each separation R is obtained by integrating out the quark variables $\vec{r}_i$, i.e.,

$$U_{NN}^{ST}(R) \simeq \prod_i \int d^3 r_i \, \phi_{ST}^*(\vec{r}_1 \ldots \vec{r}_6, \vec{R}) \, H \, \phi_{ST}(\vec{r}_1 \ldots \vec{r}_6, \vec{R}) - U_{NN}(\infty) \qquad (11)$$

in which

$$\phi_{ST}(\vec{r}_1 \ldots \vec{r}_6, \vec{R}) = \mathcal{A} \prod_{i=1}^{3} \phi_a(\vec{r}_i, -\frac{R}{2}) \prod_{j=4}^{6} \phi_b(\vec{r}_j, +\frac{R}{2}) \chi_{ST} \xi_a \xi_b \qquad (12)$$

is an antisymmetrized six quark state if χ_{ST} is constructed by vector coupling the spin-isospin states $(S_a T_a)(S_b T_b)$ of each nucleon to total spin-isospin values of (ST). The states ξ_a, ξ_b are both singlet (antisymmetric) states in their respective color spaces labelled (123) and (456) respectively.

A more accurate approach to the potential involves angular momentum projection using the Hill-Wheeler approach and the results given below have used this for partial waves up to L=2.

The "residual" interactions V_{ij}^π, V_{ij}^{LE} in Eq. (7) are included at the quark-quark interaction level. The one-pion exchange (OPE) between quarks is assumed to be the same form as the regularized OPE interaction between nucleons in the static limit but with a coupling constant chosen to give the same large separation behavior for OPE in the Paris potential   The use of an elementary pion field rather than a more microscopic quark-anti-quark pair produced by flux tube breaking is justified a posteriori because the major contribution of pion exchange in our approach is at distances which are large compared to the regulator distance $1/\Lambda_\pi \sim 1$ GeV^{-1} wherein Λ_π is the parameter used in the vertex regularization function[9]

$$F_\pi^2(Q^2) = (\Lambda_\pi^2 - m_\pi^2)^2 / (\Lambda_\pi^2 + Q^2)^2 \qquad (13)$$

The linkage exchange term is taken to act at the interface of the two oscillator wells, i.e.,

$$V_{ij}^{LE} = \text{const} \times \delta(z_i)\delta(z_j)\vec{F}_i \cdot \vec{F}_j \tag{14}$$

with $\vec{F}_i, \vec{F}_j$ being the octet of color generators for each quark. Simple flux tube considerations suggest that short range magnetic (plaquette terms in the lattic gauge approach) interactions will occur near interfaces between competing flux tube configurations (which become degenerate in energy at interfaces). The simple assumption of $\delta(z_i)\delta(z_j)$ will be adequate for our single interface model provided the range of plaquette interaction terms are not larger than the length scale of quark wave functions in hadrons. The linkage exchange interaction given by Eq. (14) is easily calculated if we ignore angular momentum projection and calculate with $\hat{\phi}$ in place of ϕ. We find

$$U_{NN}^{LE,ST}(R) = [4K_{NN}Z^2(R)/(1+Z(R))^2]f_{ST} \tag{15}$$

with spin-isospin factors given by

$$f_{00} = 7/9, \quad f_{11} = 31/81, \quad f_{01} = f_{10} = -1/27.$$

and

$$Z(R) = \exp[-R^2/4a^2]. \tag{16}$$

To allow for spin-dependent interactions in the nucleon-nucleon interaction we have assumed they arise from the linkage exchange mechanism. Assuming a Lorentz vector in the <u>nucleon</u> spinor space for U^{LE} we can find spin-dependent terms via the usual Fermi-Breit prescription without adding any new parameters.

i.e. $U_{NN}^{LE,ST}(\text{spin-dependent}) = \dfrac{3}{2M_N^2} \dfrac{1}{R} \dfrac{dU_{NN}^{LE,ST}(R)}{dR} + \text{tensor} + \text{spin-spin etc.}$

(17)

5. COMPARISON WITH DATA

The various components of the effective nucleon-nucleon potential $U_{NN}^{ST}(R)$ involve four parameters: i) oscillator size and ii) quark mass (which are already roughly known[10] from a consideration of excited states of the nucleon and delta to be in the ranges 250 MeV $\leq m_q \leq$ 450 MeV, 0.38F $\leq a \leq$ 0.47F.) plus iii) a regulator cut-off Λ_π for pion exchange and iv) the strength K_{NN} of the link-exchange term. We use physical masses for m_π and M_N and take $g_{\pi qq}^2$ to fit the long range part of the Paris potential.

The data considered here are the static properties of the bound deuteron and the n-p phase shifts over the range 0-300 MeV bombarding energy. Consideration of charge dependent effects have been considered recently[11] and inclusion of the simple Coulomb interaction between quarks yields an even better description of the T=1 proton-proton data than the T=1 neutron-proton data. The values of the parameters used to produce an eyeball fit to all the n-p data are $a=.4F$, $m_q=0.38$ GeV, $\Lambda_\pi=1.15$ GeV and $K_{NN}=2.80$ GeV. The importance of the linkage exchange contribution lies primarily in the odd partial waves where all attempts with $K_{NN}=0$ yield a very poor description. Of course other mechanisms than linkage exchange could be involved. However our attempts to include heavy meson exchanges such as ρ,ω with reasonable coupling constants yield very small effects for the same reason that OPE and one gluon exchange effects are suppressed at small hadronic separations. The orthogonality of quark orbits ϕ_a, ϕ_b is very effective in reducing the effects of boson exchange at distances below 1F so that only the light mass of the pion allows it to be the significant contributor at larger separations.

Our results for the static properties of the bound deuteron are shown in Table 1.

Table I Static Properties of the Deuteron

	E_D	$P_D^\%$	D/S	$Q_D(f^2)$	μ_D(n.m.)	$a_{np}(f)$
Expt	2.2249	3→6	0.0271	0.272	0.8574	5.38
Present Model	2.2249	4.01	0.0240	0.254	0.857	5.43
Paris Potential	2.2249	5.79	0.0260	0.279	0.853	5.427

Also shown for comparison are the results of the Paris group[12] and the experimental results. The binding energy E_D is used to fix the value of Λ_π. The major difference of the present approach compared to the Paris model is the prediction of a smaller D-state component (P_D). This four parameter model is clearly capable of giving a reasonable description of the deuteron using values of a, m_q which are in line with those values expected[10] from fits to the baryon spectra.

The calculated phase shifts for all partial waves not involving coupling to $L \geq 3$ are shown in Fig. 2. Also shown are the 1977 and 1983 phase shifts extracted from experiment by Arndt et al[13]. The calculated singlet scattering length is -9.7F which is to be compared with the value of -23.7F extracted from data. Other significant deviations occur for $^3D_2$, $^1P_1$ and E_1 for $E_{LAB}>100$ MeV, although E_1 is very sensitive to parameter changes and even "experimentally" keeps changing by significant amounts. Overall the model is quite successful in describing the low energy nucleon-nucleon data. It should be kept in mind that m_q, a and Λ_π are already constrained by

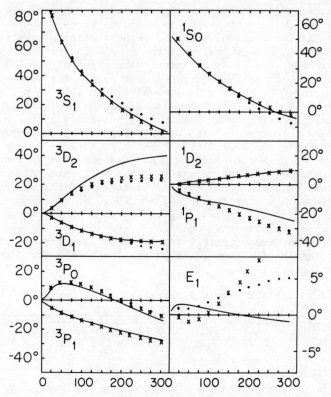

Fig. 2. Comparison between the low energy n-p phase shifts (degrees) calculated (full curves) and those extracted from experiment[13] in 1977 (crosses and 1983 (circles) as a function of bombarding energy (MeV).

the baryon spectrum and E_D and the curves in Fig. 2 represent fits obtained by freely varying K_{NN} with only small changes being allowed in the other parameters. Since the long range tail here is the same as the Paris potential we do expect the higher partial waves will be well described for E_{LAB}<300 MeV. However we also expect the Born-Oppenheimer estimate to become less applicable as E_{LAB} is increased and it is perhaps surprising that the present model works as well as it does up to 300 MeV.

Finally we note that the size of a "nucleon" in the present approach will be a function of the nucleonic separation distance. Using the simple approximate solutions $\hat{\phi}$ we can estimate the ratio ρ of the size of a nucleon at R compared to its size when $R\to\infty$ using $<\hat{\phi}|r^2|\hat{\phi}>^{\frac{1}{2}}$ as the measure. Table 2 shows our results for ρ for the deuteron and for mass three in a triple oscillator with an equilateral triangle configuration.

Table 2 Size ratio of "nucleons" in A=2,3 nuclei.

R(F)	0	.1	.2	.3	.4	.5	.6	.7
ρ(A=2)	$2/(3)^{1/2}$	1.11	1.07	1.05	1.03	1.02	1.01	1.00
ρ(A=3)	$(13)^{1/2}/3$	1.15	1.11	1.08	1.05	1.03	1.02	1.01

Clearly ρ here is a function of R which translates into ρ being a function of Q^2. We expect that ρ for $Q^2 \sim 1$ GeV^2/c^2 will be a few percent which is in reasonable agreement with the recent analysis[14] of ρ for A=3 quasi-elastic data at this Q^2. Such size changes as a function of Q^2 may be observable in more careful measurements of quasi-elastic data. The present type of model is only qualitative with respect to high Q^2 data (such as the EMC effect) and a relativistic treatment will be needed to provide a realistic model for all Q^2.

ACKNOWLEDGEMENTS

It is a pleasure to thank Chris Long and Mario Encinosa for checking many of the calculations and for several discussions. This work was supported in part by the U.S. Department of Energy.

REFERENCES

1. See for example W. Buchmuller and S.-H.H. Tye, AIP Conf. Proc. 74, 370 (1981).
2. J. Kogut and L. Susskind, Phys. Rev. C11, 395 (1975).
3. K. Maltman and N. Isgur, Phys. Rev. C29, 952 (1984) and refs. therein. D. Robson, Progr. in Particle and Nuclear Phys. 8, 259 (1982), D.H. Wilkinson, ed.
4. O.W. Greenberg and J. Hietarinta, Phys. Rev. D22, 993 (1980).
5. D. Robson, submitted to Phys. Rev. D.
6. O.W. Greenberg and H.J. Lipkin, Nucl. Phys. A370, 349 (1981).
7. D. Robson, AIP Conf. Proc. 110, 100 (1984).
8. C.H. Horowitz et al., AIP Conf. Proc. 110, 312 (1984).
9. K. Holinde, Phys. Rep. 68, 122 (1981).
10. M. Harvey and J. Le Tourneus, Nucl. Phys. A424, 419 (1984).
11. M. Encinosa, private communication.
12. M. Lacombe et al., Phys. Rev. C21, 861 (1980).
13. R.A. Arndt et al., Phys. Rev. C15, 1002 (1977); Phys. Rev. D28, 97 (1983).
14. R.D. McKeown, Phys. Rev. Letts. 56, 1452 (1986).

K^+-N, K^{+*}-N CHANNEL COUPLING IN A QUARK POTENTIAL MODEL*

Roy K. Campbell**
Florida State University, Tallahassee, FL 32306

ABSTRACT

The K^+-N interaction was calculated with a quark potential model using the resonating group method (RGM). For the central interaction the $2\hbar\omega$ components of the nucleon and kaon wave functions were included in the RGM kernels. The total isospin I=0 s-wave phase shifts are in good agreement with experimental results. The I=1 interaction does not exhibit enough repulsion. The K^+-N channel was coupled to the K^{+*}-N channel via quark exchange. The channel coupling gives effectively a more repulsive interaction for both I=1 and I=0.

The K^+-N system has several features which make it a good prototype for studying hadron-hadron interactions at the quark level. One-pion exchange is forbidden, there is no $q\bar{q}$ annihilation and there is no tensor term in the K^+-N interaction.

The quark-quark potential and hadron wave functions used are those of Stanley and Robson.[1] Their phenomenological potential consists of parametrized short range one-gluon exchange type terms and a linear confinement term. The parameters of this model were fixed by the meson spectrum and the baryon spectrum was then calculated using a complete set of three body oscillator basis states.

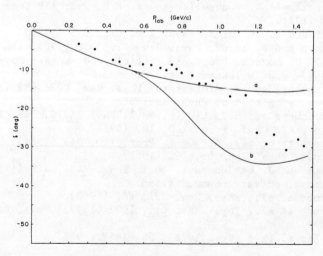

Fig. 1. S-wave phase shifts, I=0.
  a) single channel b) with channel coupling

* Supported in part by the U.S. Department of Energy.
** Present Address: Southwestern Adventist College, Keene, TX 76059.

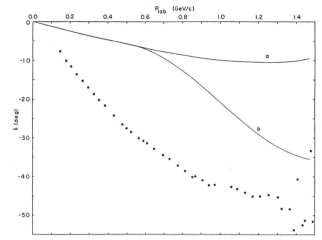

Fig. 2. S-wave phase shifts, I=1.
a) single channel b) with channel coupling

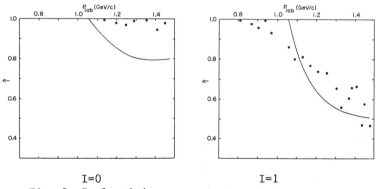

Fig. 3. Inelasticity parameters.

The resonating group method (RGM)[2] was applied to the $4q\bar{q}$ K^+-N system. The RGM wave function was first approximated by a single channel two cluster wave function. The s-wave phase shifts are given in figures 1 and 2 and compared to the phase shift analysis of Martin and Oades.[3]

The coupling of the K^+-N channel to the K^{+*}-N channel was calculated using the multichannel RGM. The resulting phase shifts are given in figures 1 and 2. The inelasticity parameters are given in figure 3.

REFERENCES

1. D.P. Stanley and D. Robson, Phys. Rev. D 21, 3180 (1980).
2. Y.C. Tang, M. Lemere and D.R. Thompson, Phys. Rep. 47, 167 (1978).
3. B.R. Martin and G.C. Oades, contribution submitted to the IV Conf. on Baryon Resonances (1980).

ANOMALOUS DIMENSIONS OF MULTIQUARK BOUND STATES*

Chueng-Ryong Ji[a]

Stanford Linear Accelerator Center
Stanford University, Stanford, California, 94305

and

Institute of Theoretical Physics, Department of Physics,
Stanford University, Stanford, California 94305

ABSTRACT

The evolution of six-quark color-singlet state distribution amplitudes is formulated as an application of perturbative quantum chromodynamics to nuclear wave functions. We present a general method of solving the evolution equation for multiquark bound states and predict the asymptotic Q^2 slope for the deuteron charge form factor as a result.

Because of the asymptotic freedom property of quantum chromodynamics (QCD), a perturbative analysis of strong interaction processes should be rigorous when the momentum transfer q is much larger than the QCD scale parameter Λ_{QCD} and the value of the strong coupling constant α_s becomes small. Applying light-cone perturbation theory,[1] amplitudes for exclusive processes are given by factorized forms. For example, hadronic form factors at asymptotic high $Q^2 \equiv -q^2$ are generically given by $\int \Phi(x) T_H(x,y) \Phi(y)[dx][dy]$, with the longitudinal momentum fractions x_i of the quarks in the hadron. Therefore, detailed analyses for exclusive processes require knowledge of the valence-quark distribution amplitude $\Phi_H(x_i, Q)$ of hadrons.[1,2] Since the QCD theory requires new degree of freedoms to form multiquark systems which do not exist in the conventional nuclear theory (hidden-color states), it is important to construct a basis to calculate the multiquark bound state wavefunctions. The multiquark bound state wavefunction should be completely antisymmetric in the total color(C), spin(S), iso-spin(T), and orbit(O) quantum space. As an explicit example, a specific representation for the asymptotic six quark systems with the Young symmetry $f_T = (33), f_{CS} = (222)_C \times (6)_S$, and $f_O = (6)$ ($T = 0$, $S = S_Z = 3$, and S-wave) is given by[3]

---

* Work supported in part by the Department of Energy, contract DE-AC03-76SF00515 and in part by National Science Foundation Grant PHY-85-08735.

[a] Present address: Department of Physics, Brooklyn College, Brooklyn, New York 11210.

$$\Phi_d(x_i,q) = \frac{a_0}{48\sqrt{5}}\Big[-\epsilon_{ijk}\epsilon_{lmn}\left(\epsilon_{ad}\epsilon_{be}\epsilon_{cf} + \epsilon_{ae}\epsilon_{bd}\epsilon_{cf} + \epsilon_{ad}\epsilon_{bf}\epsilon_{ce} + \epsilon_{af}\epsilon_{bd}\epsilon_{ce}\right)$$

$$+ \epsilon_{ijl}\epsilon_{kmn}\left(\epsilon_{ac}\epsilon_{be}\epsilon_{df} + \epsilon_{ac}\epsilon_{bf}\epsilon_{de} + \epsilon_{ae}\epsilon_{bc}\epsilon_{df} + \epsilon_{af}\epsilon_{bc}\epsilon_{de}\right)$$

$$- \left(\epsilon_{ijm}\epsilon_{kln} + \epsilon_{ijn}\epsilon_{klm}\right)\left(\epsilon_{ac}\epsilon_{bd}\epsilon_{ef} + \epsilon_{ad}\epsilon_{bc}\epsilon_{ef}\right)$$

$$+ \left(\epsilon_{ikm}\epsilon_{jln} + \epsilon_{ikn}\epsilon_{jlm} + \epsilon_{jkm}\epsilon_{iln} + \epsilon_{jkn}\epsilon_{ilm}\right)\epsilon_{ab}\epsilon_{cd}\epsilon_{ef}$$

$$- \left(\epsilon_{ikl}\epsilon_{jmn} + \epsilon_{jkl}\epsilon_{imn}\right)\left(\epsilon_{ab}\epsilon_{ce}\epsilon_{df} + \epsilon_{ab}\epsilon_{cf}\epsilon_{de}\right)\Big]$$

$$\times a_i^\dagger(1) b_j^\dagger(2) c_k^\dagger(3) d_l^\dagger(4) e_m^\dagger(5) f_n^\dagger(6)$$

$$\times x_1 x_2 x_3 x_4 x_5 x_6 \left(ln\frac{Q^2}{\Lambda^2}\right)^{-\gamma_0}, \qquad (1)$$

where the indices $i, j, \cdots, n$ and $a, b, \cdots, f$ are the color (r, y, b) and isospin (u, d) indices, respectively. The ϵ_{ijk}'s and ϵ_{ab}'s are the completely antisymmetric Cartesian tensors of $SU(3)_C$ and $SU(2)_T$. The coefficient a_0 represents the six quark amplitude at the origin and can be calculated after the complete six-quark wavefunction is given. However, the variation of the amplitude at the asymptotic high Q^2 region is mainly determined by the leading anomalous dimension γ_0. Anomalous dimensions of exclusive hadron amplitudes are given by solving the QCD evolution equation.

The evolution of the amplitude for simpler hadrons such as quark–antiquark meson[1,4] and three quark baryon[1] systems have already been formulated and solved. While these conventional hadrons have only one color singlet representation, six–quark systems considered here have five independent color singlet representations. The formulation of the evolution equation for totally antisymmetric multiquark states is not trivial even though it is a natural extension of the three-quark case. We have presented a general method for solving the QCD evolution equations which govern relativistic multi–quark wave functions.[5] We have also applied it to a four–quark toy system in $SU(2)_C$ and derived some constraints on the effective force between two baryons.[6] However, since the antisymmetric representation of a multiquark wave function must be constructed explicitly, it is hard in practice to solve the multi-quark evolution equation.

We avoid this problem by exploiting the permutation symmetry of the evolution kernel.[3] Each eigensolution of a multiquark state satisfies a kernel equation of the form

$$K|\phi_A\rangle = \gamma|\phi_A\rangle, \qquad (2)$$

where K, γ and $|\phi_A\rangle$ represent the kernel, the eigenvalue (e.g., anomalous dimensions) and the eigenfunction which is given by a linear combination of the antisymmetric representations respectively. The most important observation in this formulation is that the kernel K is a linear combination of the operators Θ_{fY} in color space each of which has a definite Young symmetry f with Yamanouchi labels Y;

$$K = \sum_{fY} K_{fY} \Theta_{fY} . \tag{3}$$

For simplicity (but without loss of generality), let us consider a six-quark case as an example of multiquark systems. The six-quark system has five orthogonal color singlet states $|(222)\alpha\rangle$ with $\alpha = 1, 2, \ldots, 5$ and the evolution kernel becomes 5×5 matrix:

$$K_{\alpha\beta} = \langle (222)_C \alpha | K | (222)_C \beta \rangle . \tag{4}$$

An essential simplification can be obtained by replacing $K_{\alpha\beta}$ with K_{fY} such that

$$\begin{aligned} K_{\alpha\beta} &= \left\langle (222)_C \alpha \left| \sum_{fY} K_{fY} \Theta_{fY} \right| (222)_C \beta \right\rangle \\ &= \sum_f \sum_Y \langle (222)_C \alpha, fY | (222)_C \beta \rangle K_{fY} \end{aligned} \tag{5}$$

where the possible f which gives the non-zero Clebsch-Gordon coefficient[7] is only (6) or (42). For example, the leading anomalous dimension of the deuteron state can be obtained by following procedures. Projecting out a certain state which has common C, T and O representations, we get a set of equations for spin states. Since the kernel of each equation has a definite symmetry and its explicit representation is known, we can determine relative weighting factors among the independent equations by counting the number of spin annihilation terms in the kernel. The only equation which we have to solve explicitly is the equation which has the symmetric kernel $K_{(6)}$. After taking into account the relative weighting factor we get the leading anomalous dimension for a deuteron state,

$$\gamma_0 = \frac{3}{4} \frac{C_F}{\beta} \quad \text{for} \quad S_Z = 0 , \tag{6a}$$

$$= \frac{7}{8} \frac{C_F}{\beta} \quad\quad\quad = \pm 1 . \tag{6b}$$

Using the result of Eq. (6), one can calculate the asymptotic deuteron form factor $F_d(Q^2)$. The QCD prediction for the asymptotic Q^2-behavior of the deuteron reduced form factor[8] $f_d(Q^2)$ defined by

$$f_d(Q^2) = \frac{F_d(Q^2)}{F_N^2\left(\frac{Q^2}{4}\right)}$$

is given by

$$f_d(Q^2) \sim \frac{\alpha_s(Q^2)}{Q^2}\left(\ell n \frac{Q^2}{\Lambda^2}\right)^{C_F/2\beta} . \tag{11}$$

The deuteron state which has the leading anomalous dimension is related to the NN, $\Delta\Delta$, and hidden color (CC) physical bases, for both the $(TS) = (01)$ and (10) cases with Young symmetry of $\{33\}$, by the formula[9]

$$\psi_{[6]\{33\}} = \sqrt{\frac{1}{9}}\,\psi_{NN} + \sqrt{\frac{4}{45}}\,\psi_{\Delta\Delta} + \sqrt{\frac{4}{5}}\,\psi_{CC} .$$

The fact that the six-quark state is 80 percent hidden color at small transverse separation implies that the deuteron form factors cannot be described at large Q^2 by meson–nucleon degrees of freedom alone.

ACKNOWLEDGMENT

This work has been done in collaboration with Stanley J. Brodsky. I would like to thank J. Dirk Walecka for many helpful discussions.

REFERENCES

1. G. P. Lepage and S. J. Brodsky, Phys. Rev. **D22**, 2157 (1980).
2. C.-R.Ji, in Proceedings of the Workshop on Nuclear Chromodynamics, Santa Barbara, August 1985, edited by S. Brodsky and E. Moniz, pp. 60-75; S. J. Brodsky and C.-R. Ji, Prog. Part. Nucl. Phys. **13**, 299 (1984), edited by A. Faessler, and references therein.
3. C.-R. Ji and S. J. Brodsky, SLAC–PUB–3148 (1985)(to be published in Phys. Rev. **D**).
4. V. N. Baier and A. G. Grozin, Nucl. Phys. **B192**, 476 (1981).
5. S. J. Brodsky and C.-R. Ji, Phys. Rev. **D33**, 1951 (1986).
6. S. J. Brodsky and C.-R. Ji, Phys. Rev. **D33**, 1406 (1986).
7. M. Hamermesh, *Group Theory* (Addison-Wesley, Reading, Mass., 1962).
8. S. J. Brodsky, C.-R. Ji and G. P. Lepage, Phys. Rev. Lett. **51**, 83 (1983); S. J. Brodsky and C.-R. Ji, Phys. Rev. **D33**, 2653 (1986).
9. M. Harvey, Nucl. Phys. **A352**, 301 (1981); **A352**, 326 (1981).

DIRAC VS. SCHROEDINGER APPROACH IN PROTON-NUCLEUS SCATTERING: A SIMPLE MODEL

M. Thies

Free University, Amsterdam, The Netherlands

The striking success of the Dirac approach[1] to proton-nucleus scattering does <u>not</u> necessarily imply that relativity is important in nuclear physics. To illustrate this point, we consider a simple, exactly soluble model. We start from the Dirac equation with local scalar (S) and vector (V) potentials, after elimination of the "lower components":

$$(\vec{\sigma}\cdot\vec{p}(E+m+S-V)^{-1}\vec{\sigma}\cdot\vec{p} + S+V) u_k = (E-m) u_k \qquad (1)$$

Typical values are S=-400 MeV, V=+350 MeV at r=0. The model consists in assuming that the cancellation in S+V is exact (S+V=0) and going to the 0 energy limit. Eq.(1) can then be written as

$$(p^2 + \vec{\sigma}\cdot\vec{p}\frac{2mU}{(1-2mU)}\vec{\sigma}\cdot\vec{p}) u_k = k^2 u_k \qquad (U=\frac{V-S}{4m^2}) \qquad (2)$$

The effective potential can be decomposed into spin-orbit, effective mass and $N\bar{N}$ pair term (Z-graph, Fig.1), using the identity

$$\vec{\sigma}\cdot\vec{p}\frac{U}{(1-2mU)}\vec{\sigma}\cdot\vec{p} = \frac{1}{r}\frac{dU}{dr}\vec{\sigma}\cdot\vec{L} + \vec{p}\cdot U\vec{p} + \vec{\sigma}\cdot\vec{p}\frac{2mU^2}{(1-2mU)}\vec{\sigma}\cdot\vec{p} \qquad (3)$$

The Lippmann-Schwinger equation for the elastic T-matrix,

$$T = \vec{\sigma}\cdot\vec{p}\frac{U}{(1-2mU)}\vec{\sigma}\cdot\vec{p}\,(1+G_o T) \qquad (4)$$

can be solved algebraically in the 0 energy limit, with the result

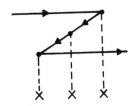

Fig.1 : Example of a "Z-graph" appearing in the Dirac approach to proton-nucleus elastic scattering.

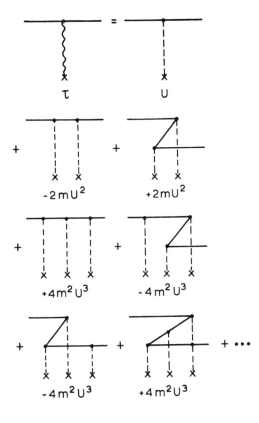

Fig.2 : Perturbative solution of the model problem up to order U^3, illustrating the exact cancellation of all higher order terms at 0 energy. In the corresponding non-relativistic case, all the non-Z-graphs survive.

$$T = \vec{\sigma}\cdot\vec{p}\ U\ \vec{\sigma}\cdot\vec{p} \qquad \text{(Dirac)} \qquad (5)$$

Here, we have used

$$\vec{\sigma}\cdot\vec{p}\ G_o\ \vec{\sigma}\cdot\vec{p} = \frac{2mp^2}{k^2-p^2+i\varepsilon} \xrightarrow[k^2=0]{} -2m \qquad (6)$$

Remarkably, T is simpler than the effective potential it has been derived from, see eq. (4). In particular, being linear in U, T does not involve any $N\bar{N}$ pair terms. All "Z-graphs" have been cancelled against higher order iterations of the spin-orbit and effective mass potentials, as can easily be verified order by order (Fig. 2).

In the non-relativistic approach, one usually starts from the Schroedinger equation with "almost local" potentials

$$(-\vec{\nabla}\cdot\frac{1}{2m^*}\vec{\nabla} + \frac{1}{r}\frac{dU_{LS}}{dr}\vec{\sigma}\cdot\vec{L} + U_c)\ u_k = \frac{k^2}{2m} u_k \qquad (7)$$

To come as close as possible to our model problem, let us choose

$$U_c=0,\ U_{LS}= \frac{1}{2m^*} - \frac{1}{2m} = U,\qquad k^2 \to 0 \qquad (8)$$

Then, the non-relativistic Lippmann Schwinger equation becomes

$$T = \vec{\sigma}\cdot\vec{p}\ U\ \vec{\sigma}\cdot\vec{p}\ (1+G_o T) \qquad (9)$$

and differs from the "relativistic" one, eq. (4), only by the absence of pair terms. The solution of (9) for $k^2 \to 0$ is

$$T = \vec{\sigma}\cdot\vec{p}\ \frac{U}{(1+2mU)}\ \vec{\sigma}\cdot\vec{p} \qquad \text{(Schroedinger)} \qquad (10)$$

There are non-vanishing multiple scattering terms, namely all the non-Z-graphs of Fig. 2. Note however that T has the unusual feature of being (almost) local. This is a consequence of eq. (6) which shows that the propagator reduces to a δ-function in r-space, at 0 energy. Hence, the term of order U^n in eq. (10) requires n target nucleons to be simultaneously within the range of the NN spin-orbit force (~ 0.4 fm) of the projectile. Such processes will be strongly affected by correlations and nucleon substructure effects, and are presumably spurious in this simple form. Therefore, it may be safest to delete them by applying the "Lorentz-Lorenz correction", well known from pion physics[2]. Here, it simply yields the replacement

$$U \longrightarrow \frac{U}{(1-2mU)} \qquad (11)$$

which, in turn, leads back to the Born approximation T-matrix (5), i.e. the exact Dirac result.

This trivial example shows that the virtue of the Dirac approach is not that it does *more*, but rather that it does *less* than the Schroedinger approach, cf. eqs. (5) and (10). Various physical mechanisms for suppressing the "δ-function" type iterations of the LS-potential are conceivable, and one is not forced to resort to very large potentials. The sensitivity to short-range processes may in fact be grossly exaggerated by the use of almost local potentials. A more detailed discussion of these problems and applications of similar ideas to higher energies, inelastic scattering and the Dirac-Hartree approach can be found in ref. 3.

REFERENCES

1. Proc. LAMPF Workshop on Dirac Approaches to Nuclear Physics, J.R.Shepard et al. eds. (Los Alamos 1985).
2. M. and T.E.O.Ericson, Ann.of Phys. 36, 323 (1966).
3. M.Thies, Phys. Lett. 162B, 255 (1985) and 166B, 23 (1986), Nucl. Phys. A (1986), to be published.

Dudley's Dilemma: Magnetic Moments in Relativistic Theories

J. A. McNeil
Department of Physics and Atmospheric Science
Drexel University
Philadelphia, Pennsylvania 19104
and
Department of Physics
University of Pennsylvania
Philadelphia, Pennsylvania 19104

Abstract

In 1975 L. Dudley Miller showed how the basic phenomenology of the major shell and spin-orbit splittings constrained the relativistic scalar/vector structure model to values of the potentials incompatible with the observed magnetic moments of nuclei one nucleon away from closed shell [1]. In this talk the resolution of this problem is presented from three different perspectives. First a self-consistent Landau-Migdal approach is used to define the single particle isoscalar current in infinite nuclear matter. The constraint of self-consistency provides a vector suppression factor to the single particle current which returns the current to its nonrelativistic form and resolves the problem. The same suppression factor is shown to follow as well from either a consideration of gauge invariance or (equivalently) the relativistic random phase approximation. Local density approximation calculations of isoscalar magnetic moments of nuclei one nucleon away from closed shell recover the Schmidt values, thus resolving this longstanding problem.

In this talk I want to present some recent work which finally rids the magnetic moment monkey from the back of relativistic structure theories. Long ago L. Dudley Miller [1] showed how the strong scalar and vector potentials characteristic of relativistic approaches to nuclear physics induced an enormous enhancement in the Dirac current with disasterous consequences for the predicted magnetic moments. On the other hand the (nonrelativistic) Schmidt values for nuclei one particle away from closed shell agreed well with the isoscalar magnetic moments (the isovector moments are strongly influenced by the isovector meson exchange currents and therefore constitute a special problem for either the relativistic or nonrelativistic approaches). The disagreement between the isoscalar Schmidt values and the relativistic shell model or Hartree predictions has long been an embarassment for proponents of the relativistic approach. This problem stands in singular contrast to an otherwise quite successful theory. The recent successful developements using relativistic approaches to scattering at intermediate energies are but the most recent examples [2]. The resolution of the magnetic moment problem, as we shall see, turns out to be better condensed matter physics. In this talk I will present three ways of looking at the problem which all give the same result: the vector interaction suppresses the current back to its nonrelativistic value. The first approach, following Matsui [3], uses a Landau-Migdal method to define the single particle current in a self-consistent way. This leads to Matsui's vector suppression factor. An alternative approach is to invoke gauge invariance (Ward identity [4]) to arrive at the same suppression factor. The third approach is to use the relativistic random phase approximation (RRPA) to calculate the transverse polarization insertion. The RRPA was first studied by S. A. Chin,[5], and recently used by Kurasawa and Suzuki, [6], to examine relativistic effects in quasi-elastic electron scattering. Carefully taking the zero momentum transfer limit of the transverse polarization insertion gives the same result as the other treatments. The methods are not independent, each but another way of treating the linear response of the system to the external electromagnetic field in a self-consistent manner.

Reference [7] provides a comprehensive review of the relativistic many body problem. Much of what appears here was worked out in a collaboration with R. D. Amado, C. J. Horowitz, M. Oka, J. R. Shepard, and D. A. Sparrow [8].

Before starting let me point out that one does not have to go to the real (i.e. finite) nucleus to see the symptoms of the magnetic moment problem. The large enhancements to the Dirac current responsible for the problem are already present in the simpler case of infinite nuclear matter. The approach taken therefore is to focus on infinite nuclear matter, solve the problem there and then apply the result to finite nuclei using a local density approximation. The details of the final answer will of course depend on this last approximation, but the physics of the resolution to the dilemma will not. We begin therefore with the statement of the problem for which we need a lightning review of Walecka's mean field model of infinite nuclear matter [7] and the Landau Fermi liquid treatment of T. Matsui [3]. We use the conventions of Bjorken and Drell [9]. Walecka's simplest relativistic model for nuclear matter consists of three fields, the nucleon field, Ψ, and two isoscalar meson fields, σ, a Lorentz scalar and ω, a Lorentz vector. The Lagrangian density is given by:

$$\mathcal{L} = -\tfrac{1}{4}(\partial_\mu V_\nu - \partial_\nu V_\mu)^2 + \tfrac{1}{2}m_\omega^2 V_\mu V^\mu + \tfrac{1}{2}\partial_\mu \Phi \partial^\mu \Phi - \tfrac{1}{2}m_\sigma^2 \Phi^2 \\ + \overline{\Psi}(i\gamma_\mu \partial^\mu - m)\Psi + g_\sigma \overline{\Psi}\Psi\Phi - g_\omega \overline{\Psi}\gamma_\mu \Psi V^\mu \qquad (1)$$

where V_μ is the vector (ω) field with mass m_ω, Φ is the scalar (σ) field with mass m_σ, and Ψ is the nucleon field with mass m. Isospin indices are suppressed. For bulk (long wavelength) properties and low-lying excitations it is sufficient to average the meson source terms yielding the following static mean fields:

$$\overline{\Phi} = -\frac{g_\sigma}{m_\sigma^2}\langle\overline{\Psi}\Psi\rangle \qquad (2)$$

$$\overline{V}_\mu = -\frac{g_\omega}{m_\omega^2}\langle\overline{\Psi}\gamma_\mu\Psi\rangle \qquad (3)$$

where the brackets indicate the ground state expectation value. The nucleon field equation then becomes the self-consistent Dirac equation:

$$(i\gamma_\mu \partial^\mu - m - \Sigma)\Psi = 0 \qquad (4)$$

where, letting $\lambda_{\sigma,\omega} = g_{\sigma,\omega}^2/m_{\sigma,\omega}^2$,

$$\Sigma = \lambda_\omega \gamma^\mu j_{B\mu} - \lambda_\sigma \rho_s \tag{5}$$

with

$$j_{B\mu} = \langle \overline{\Psi} \gamma^\mu \Psi \rangle = (\rho_0, \vec{j}_B) \tag{6}$$

$$\rho_s = \langle \overline{\Psi} \Psi \rangle \tag{7}$$

One defines the effective mass, m^*,

$$m^* = m - \lambda_\sigma \rho_s. \tag{8}$$

The Dirac equation has the following quasiparticle solutions in an arbitrary frame:

$$\Psi_i - \left(\frac{E_i^* + m^*}{2E_i^*} \right)^{\frac{1}{2}} \left(\begin{array}{c} 1 \\ \frac{\vec{\sigma} \cdot (\vec{k}_i - \lambda_\omega \vec{j}_B)}{E_i^* + m^*} \end{array} \right) \chi e^{ik_i z_i} \tag{9}$$

where χ is the Pauli 2-spinor and

$$E_i^* = \left[(\vec{k}_i - \lambda_\omega \vec{j}_B)^2 + m^{*2} \right]^{\frac{1}{2}} \tag{10}$$

Note that this is in fact a self-consistency condition in that the solution for Ψ_i involves Ψ_i itself. For infinite nuclear matter filled to Fermi momentum k_F, the baryon density and current, and scalar density are given by:

$$\rho_0 = \sum_i n_i \tag{11}$$

$$\vec{j}_B = \sum_i n_i \frac{(\vec{k}_i - \lambda_\omega \vec{j}_B)}{E_i^*} \tag{12}$$

$$\rho_s = \sum_i n_i \frac{m^*}{E_i^*} \tag{13}$$

where $n_i = \theta(k_F - |\vec{k}_i|)$. Actually the Fermi distribution in the moving frame is elliptical not spherical, but since in the end we will go to the nuclear rest frame, $\vec{j}_B \to 0$, it turns out not to matter for the purposes of this work. The energy density for a zero temperature Fermi liquid filled to Fermi momentum k_F is:

$$\mathcal{E} = \frac{1}{2}\frac{g_\omega^2}{m_\omega^2}(\rho_0^2 + \vec{j}_B \cdot \vec{j}_B) + \frac{1}{2}\frac{m_\sigma^2}{g_\sigma^2}(m - m^*)^2 + \sum_i n_i E_i^* \qquad (14)$$

Note that the energy density is stationary under variations in m^* and $\vec{j}_B$ and ρ_0 is fixed by normalization. Fixing the ratios, g_σ/m_σ and g_ω/m_ω, to the binding energy per particle and saturation density, yields scalar and vector potentials of roughly -450 MeV and +400 MeV respectively. These values are consistent with the major shell and spin-orbit splitting in finite nuclei. Roughly speaking, the nonrelativistic central potential obtained from these scalar/vector strengths is proportional to the sum, about -50 MeV. The spin-orbit potential on the other hand is proportional to the difference, or +850 MeV (when the nuclear geometry factors are included this results in an equivalent Schrödinger spin-orbit potential of -20 MeV). It is seldom appreciated just how very strong the spin-orbit force in nuclei really is, and it is this large force which drives the difference in the scalar and vector potentials to such large values.

One consequence of the large scalar/vector potentials is the enhancement of the lower component of the Dirac 4-spinor relative to its value in free space. Writing out the upper and lower components,

$$\Psi_i = \begin{pmatrix} u_i \\ l_i \end{pmatrix}, \qquad (15)$$

one can use the Dirac equation to find the lower component in terms of the upper. In the nuclear rest frame one finds:

$$l_i = \frac{-i\vec{\sigma} \cdot \vec{\nabla}}{E + m + S - V} u_i \qquad (16)$$

where S and V are the scalar and vector potentials respectively. At nuclear matter densities the lower component is enhanced by about 70% over its value in free space. This alteration of the upper/lower ratio is a new degree of freedom in relativistic approaches in that this ratio is implicitly assumed to be fixed at the free value in nonrelativistic treatments. Allowing for an independent lower component adds the additional degree of freedom which has been shown to be essential to predicting spin observables in intermediate energy proton-nucleus elastic scattering at intermediate energies. There are alternative, but equivalent, ways of expressing this. If one restricts oneself to a free basis, the inclusion of this new degree of freedom requires, by completeness, the addition of negative energy states relative to that basis. Therefore, from this view one is led to the so-called Z-graph explanation of the success of the Dirac approach. It should be emphasised however that the nucleon which happens to find itself in a strong scalar/vector environment always has positive definite energy.

We come now to Dudley's dilemma. The enhancement of the lower components in relativistic models leads directly to a corresponding enhancement in the Dirac current, $\vec{\gamma}$, contribution to the magnetic moment. The anomalous (tensor) contribution to the magnetic moment is not significantly altered since its matrix elements are dominated by the upper components of the wavefunction which are essentially the same as their nonrelativistic counterparts. Specifically for the relativistic models one finds for the isoscalar magentic moment of a valence nucleon of total angular momentum j:

$$\frac{\mu}{\mu_0} = \begin{cases} \frac{1}{2}(j+\frac{1}{2})R_{lj} + \kappa_0 & ; j = l + \frac{1}{2} \\ \frac{1}{2}\frac{j}{j+1}(j+\frac{1}{2})R_{lj} - \kappa_0 & ; j = l - \frac{1}{2} \end{cases} \quad (17)$$

where κ_0 is the isoscalar anomalous moment,

$$R_{lj} = \int_0^\infty dr r^2 \frac{|u_{lj}|^2}{1 + S/m} + O(|l|^2/|u|^2), \quad (18)$$

and terms involving derivatives of the scalar potential have been neglected. In the weak field limit, $S/m \to 0$, $R_{lj} \to 1$ and the magnetic moments reduce to their (purely geometric) Schmidt values. It is the large and negative value of S which is responsible for Dudley's dilemma. Table 1 shows the Dirac current contribution to the isoscalar magnetic moments of nuclei one nucleon away from closed shell. The relatively small enhancement in the A=17 system is due to the relatively weaker binding of the valence nucleon in that system. We may summarize Dudley's dilemma as follows: relativistic scalar/vector models cannot simultaneously fit both the spin-orbit splitting and the magnetic moments.

The resolution of the problem begins with the recognition that the single independent particle model implicit in the previous calculations is too radical for this dense strongly interacting context. We need better condensed matter physics. There are three essentially equivalent ways of constructing the proper many body effective current: a Landau-Migdal self-consistent Fermi liquid approach, an appeal to gauge invariance (Ward identity), and the relativistic random phase approximation (RRPA). We will treat each briefly in turn.

We have already worked out the mean field (Hartree) quasiparticle solutions in the Walecka model. It remains to determine the single quasiparticle properties in a self-consistent manner. According to Landau and Migdal, the single quasiparticle quantity can be obtained from the corresponding bulk property by determining how the bulk property changes when that particle is removed. For example from the total energy density, Eq.(14), the single particle energy is defined by:

$$\varepsilon_i \equiv \frac{\delta \mathcal{E}}{\delta n_i} = E_i^\star + \lambda_\omega \rho_0 \tag{19}$$

Despite the complicated density dependences in $\mathcal{E}(m^\star, \vec{j}_B)$ the simplicity of this result follows from the stationary nature of $\mathcal{E}$ under variations in $m^\star$ and $\vec{j}_B$. This definition implicitly assumes that the system has time to rearrange itself to adjust to the quasiparticle's absense. This definition is only appropriate therefore for energies near the Fermi surface. In similar fashion one defines the single quasiparticle current as the change in the total baryon current when the particle is removed:

$$\vec{\tilde{j}}_i \equiv \frac{\delta \vec{j}_B}{\delta n_i} \tag{20}$$

This not only accounts for the missing particle but also for the current due to the rest of the medium adjusting to that particle, i.e. the backflow.

Using the self-consistent $\vec{j}_B$, Eq.(6), one finds in the nuclear rest frame:

$$\tilde{\vec{j}}_i = \left(\frac{\delta \vec{j}_B}{\delta n_i}\right)_{\vec{j}_B \to 0}$$

$$\tilde{\vec{j}}_i = \frac{\vec{k}_i}{(k_i^2 + m^{*2})^{\frac{1}{2}}} \frac{1}{1 + \varsigma_\omega} \quad (21)$$

where

$$\varsigma_\omega = \lambda_\omega \sum_j n_j \frac{(\frac{2}{3}k_j^2 + m^{*2})}{(k_j^2 + m^{*2})^{\frac{3}{2}}} = \frac{2\lambda_\omega k_F^2}{3\pi^2 E_F^*} \quad (22)$$

where E_F^* is the Fermi energy. The first term is the usual relativistically enhanced (m/m^*) current responsible for Dudley's dilemma. However, we see that the self-consistency condition yields a suppression factor first noted by Matsui for the interparticle strong magnetic interaction[3]. It is easy to show that this is the current which couples to the electromagnetic field [8]. Table 1 shows the Dirac current contribution to the isoscalar magnetic moments for nuclei one nucleon away from closed shell using the self-consistently determined current, Eq.(21). In applying this result to finite nuclei we have used a local density approximation. The suppression factor reduces the Dirac current contribution back to the Schmidt value. The same cancellation which gives the central potential of -50 MeV is at work here as well. In Table 2 we include the anomalous moment and compare with data. It is not surprising that the suppressed current yields essentially identical results as the successful Schmidt values. The dilemma L. D. Miller first noted is resolved.

We wish to understand this result from the perspective of gauge invariance. The importance of gauge invariance in constraining currents in the general interacting context has been emphasised by W. Bentz, et. al. [4]. We will apply the constraint of gauge invariance to the Walecka model specifically. Including a weak external electromagnetic field, minimally coupled to the nucleons to insure gauge invariance, modifies the self-energy of the ith particle:

$$\Sigma_i^{(A)} = \Sigma(..., k_j^\mu \to k_j^\mu + eA^\mu, ...) + e\gamma_i^\mu A_\mu \quad (23)$$

Dirac Current Contribution to Isoscalar Magnetic Moments			
Mass No.	Schmidt	$\bar{\gamma}$	$\tilde{\bar{\gamma}}$
15	.167	.277	.180
17	1.50	1.63	1.48
39	.60	.88	.64
41	2.0	2.3	2.0

Table 1:
The Dirac current contributions to the isoscalar magnetic moments with ($\tilde{\bar{\gamma}}$) and without ($\bar{\gamma}$) the vector suppression factor, Eq.(21), (calculated in local density approximation) are compared with the corresponding Schmidt values.

Isoscalar Magnetic Moments				
Mass No.	Exp	Schmidt	$\bar{\gamma}$	$\tilde{\bar{\gamma}}$
15	.218	.187	.297	.203
17	1.414	1.440	1.569	1.415
39	.706	.636	.918	.673
41	1.918	1.940	2.237	1.930

Table 2:
Two relativistic isoscalar magnetic moment calculations of closed shell plus or minus one nucleon are shown along with the experimental values and the Schmidt values. All theoretical values include the anomalous moment without modification from its free space value.

where the minimal coupling rule is indicated for the *internal* momenta of the Hartree loop. The electromagnetic vertex function is defined by the linear response of the self-energy to the external field:

$$e\Gamma_i^\mu = \left(\frac{\partial \Sigma_i^{(A)}}{\partial A_\mu}\right)_{A_\mu \to 0} \qquad (24)$$

For the purposes of the magnetic moments we focus on the space like component, l. Writing out the sums in Σ and using the minimal coupling substitution to rewrite the A-derivatives in terms of k-derivatives, we have:

$$\Gamma_i^l = \gamma_i^l + \left\{ \lambda_\omega \vec{\gamma}_i \cdot \sum_j \frac{\partial}{\partial k_{jl}} n_j \frac{(\vec{k}_j - \lambda_\omega \vec{j}_B)}{E_j} - \lambda_\sigma \sum_j \frac{\partial}{\partial k_{jl}} n_j \frac{m^\star}{E_j} \right\}_{\vec{j}_B \to 0} \qquad (25)$$

where $\vec{j}_B$ is set to zero at the end to yield the rest frame result. Diagramatically this amounts to connecting a photon line to every possible nucleon line in the self-energy. We shall return to this point later. The k-derivatives acting on scalar functions of k_j^2 or $\vec{j}_B \cdot \vec{k}_j$ give terms which sum to zero in the nuclear rest frame. The only surviving terms are those arising from the baryon current, $\vec{j}_B$, yielding:

$$\Gamma_i^l = \gamma_i^m \left\{ \delta_m^l - \frac{\lambda_\omega}{1 + \varsigma_\omega} \sum_j \frac{n_j}{E_j} (\delta_m^l - \frac{k_j^l k_{jm}}{E_j^2}) \right\} \qquad (26)$$

Performing the sum recovers the Landau-Migdal result:

$$\Gamma_i^l = \frac{\gamma_i^l}{1 + \varsigma_\omega} \qquad (27)$$

One can therefore consider the suppression factor as arising from a consideration of gauge invariance.

Yet another useful perspective on this problem is provided by the relativistic random phase approximation (RRPA) [5] [6]. Besides recovering the previous results, the RRPA provides a natural means for calculating renormalized vertices at finite momentum transfers. Go back to the gauge invariance arguments. The mean field result is equivalent to the Hartree

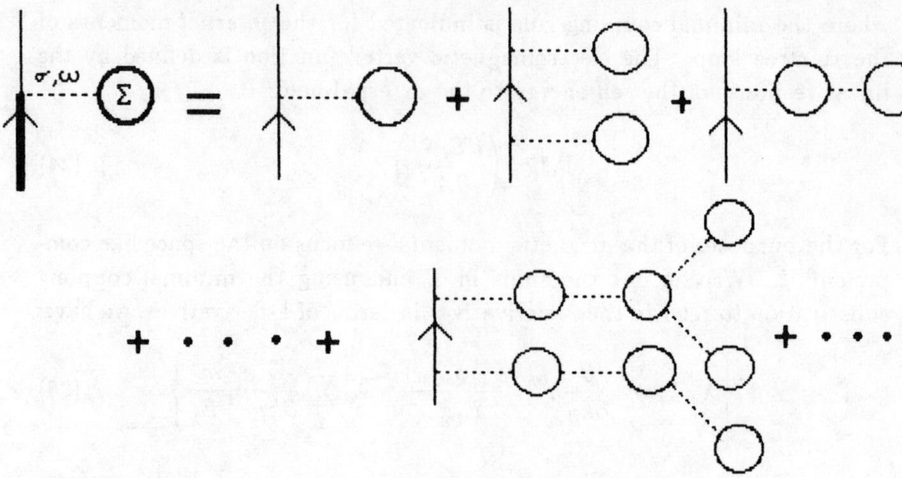

Figure 1: Hartree Green's function with self-energy insertion Σ.

self-energy insertion shown in Figure 1. As we saw in the preceeding section, gauge invariance can be expressed in terms of derivatives with respect to the momenta of the nucleons within the self-energy loop (Ward identity). This amounts to attaching a photon line to all possible nucleon lines in the self-energy loop which for transverse photons generates the diagrams shown in Figure 2 which in turn we recognize as the RRPA ring sum. The details of RRPA have been worked out by S.A. Chin [5]. The relevant quantity we need here is the transverse polarization insertion, Eq.(5.32) in Ref.[5].

$$\Pi^{ll} = -\frac{32g_\omega^2}{(2\pi)^3} \int_0^{k_F} \frac{dk k^2}{2E_k} \int_{-1}^{1} d\chi \frac{\frac{1}{2}q^2 \vec{k} \cdot \vec{k}(1-\chi^2) + (q \cdot k)^2}{q^4 - 4(q \cdot k)^2} \quad (28)$$

There is an additional factor of 2 here because we are dealing with nuclear matter, instead of neutron matter. The full transverse vertex is determined by the polarization insertion:

$$\Gamma_i^l = \frac{\gamma_i^l}{1 - \Pi^{ll} D_\omega} \quad (29)$$

where D_ω is the ω-meson propagator. For the magnetic moment we need the $|\vec{q}| \to 0$ then $q_0 \to 0$ limit which once again reproduces the current

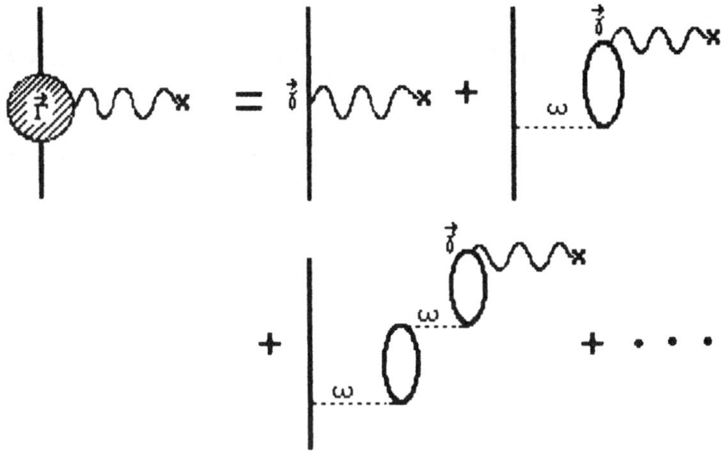

Figure 2: Transverse relativistic RPA vertex

operator with the Matsui suppression factor, Eq.(27):

$$\lim_{q_0 \to 0} \left\{ \lim_{|\vec{q}| \to 0} \Gamma_i^l \right\} = \frac{\gamma_i^l}{1 + \zeta_\omega} \qquad (30)$$

The added advantage to the RRPA, as already mentioned, is that it allows a natural extention to finite momentum transfers. A study of the impact of this class of medium modifications to electromagnetic form factors in the relativistic context is currently underway.

To summarize, we have seen how the strong scalar/vector potentials of relativistic models are well constrained by the spin-orbit and major shell splittings. This basic phenomenology leads to large and oppositely signed scalar and vector potentials. These large potentials in turn enhance the lower components and lead to anomalously large current contributions to the magnetic moments, Dudley's dilemma. The resolution to the dilemma lies in a slightly more sophisticated treatment of the many body problem. In this talk I have reviewed three approaches to the problem (Landau-Migdal, gauge invariance, and RRPA) which all lead to the same result: the medium suppresses the electromagnetic transverse vertex through the vector interaction. The renormalized current is within binding energy corrections of the free current; so it is no surprise that local density calculations using this current recover the nonrelativistic Schmidt values thus resolving the longstanding problem with magnetic moments.

We thank G. E. Brown for useful conversations and for bringing Ref.[3] to our attention. We thank W. Bentz for bringing Ref.[4] to our attention. This work was supported in part by the National Science Foundation and the Department of Energy.

Bibliography

[1] L. D. Miller, Ann. Phys. **91**, 40 (1975).

[2] J. A. McNeil, J. R. Shepard, and S. J. Wallace, Phys. Rev. Lett. **51**, 1066 (1982).

[3] T. Matsui, Nuc. Phys. **A370**, 365 (1981).

[4] W. Bentz, A. Arima, H. Hyuga, K. Shimizu, and K. Yazaki, Nuc. Phys. **A436**, 593 (1984), and private communication.

[5] S. A. Chin, Ann. Phys. **108**, 301 (1977).

[6] H. Kurasawa and T. Suzuki, preprint RIFP-613, July 1985. See also G. E. Brown, **Proceedings of the Indiana Workshop on Nuclear Physics**, 1985, to be published.

[7] J. D. Walecka, Ann. Phys. **83**, 491 (1974), and B. D. Serot and J. D. Walecka, "The Relativistic Nuclear Many-Body Problem", **Advances in Nuclear Physics**, J. W. Negele and E. Vogt, Plenum Press, New York, New York. See also M. R. Anastasio, L. S. Celenza, W. S. Pong and C. M. Shakin, Phys. Rep. **100**, 327 (1983) and references therein.

[8] J. A. McNeil, R. D. Amado, C. J. Horowitz, M. Oka, J. R. Shepard, and D. A. Sparrow, to be published in Phys. Rev. C.

[9] Bjorken and Drell, "Relativistic Quantum Field Theory", McGraw-Hill, New York, New York, 1964.

(n,p) REACTION STUDIES

W.P. Alford*
TRIUMF, 4004 Wesbrook Mall, Vancouver, B.C., Canada V6T 2A3

ABSTRACT

The study of (n,p) reactions on nuclei at intermediate energies is a promising new tool in the investigation of isovector excitations, particularly those involving spin-flip transitions. Initial results from a new charge exchange facility at TRIUMF are discussed in the context of earlier (n,p) measurements and complementary studies with other probes.

INTRODUCTION

Current interest in the (n,p) reaction has been stimulated greatly by results of recent studies of the inverse (p,n) reaction. Since about 1980, it has been shown that the (p,n) results for even-A targets are accounted for satisfactorily by a simple impulse approximation reaction model and that at energies above about 100 MeV the cross section at small momentum transfer is dominated by the central isovector spin-flip effective interaction, exciting Gamow-Teller transitions.[1] It has also been shown that the cross section at small momentum transfer provides a quantitative measure of Gamow-Teller strength,[2] and this knowledge has been used to study the Gamow-Teller giant resonance for targets ranging over the whole periodic table. Finally, the comparison[3] of measured Gamow-Teller strengths with the sum rule prediction

$$S^- - S^+ = \Sigma B_f^- - \Sigma B_f^+ = 3(N-Z)$$

with $$B^{\pm} = \frac{1}{2J_i+1} |<f\|\sigma\tau^{\pm}\|i>|$$

has revealed the problem of missing strength which has generated much interest in the possibility of Δ-isobar components in low-lying nuclear states. In addition to the GT giant resonance, other spin resonances[4] are observed in the (p,n) reaction. Thus measurements of (n,p) cross sections are needed, in part to determine the GT strength B^+ as a component of the sum rule. It should also be noted that on targets with a neutron excess the (n,p) reaction excites only states of isospin $T_f=T_i+1$. Since many spin-flip resonances observed in the (p,n) reaction are expected to be excited at lower excitation energy and with less spreading in the (n,p) reaction, it is anticipated that (n,p) studies will yield interesting new information about these resonances.

The investigation of (n,p) reactions has been severely limited due to experimental difficulties, and only a few scattered measurements[5] have been reported until quite recently. Over the past few

---

*permanent address: Physics Department, University of Western Ontario, London, Ontario, Canada

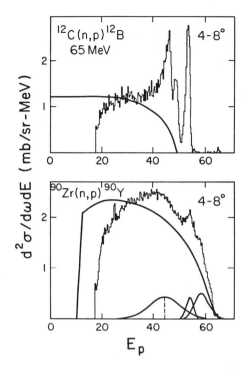

Fig. 1. (n,p) cross sections at 65 MeV for targets of ^{12}C and ^{90}Zr. An estimate of the continuum background is indicated on each spectrum.

years the group at U.C. Davis have developed a facility which is now yielding interesting results[6,7] on isovector giant resonances. Typical spectra at 65 MeV are shown in Fig. 1. In ^{12}C, both the GT and spin dipole states are strongly excited. In ^{90}Zr, the data indicate the presence of both the monopole and quadrupole isovector giant resonances.

The interpretation of these results is complicated by two factors. At 65 MeV both the spinflip and non-spinflip interaction are of comparable strength, and both types of states are excited with comparable cross sections. The reaction mechanism is also more complicated with substantial non-direct contributions to the cross section in interesting regions of the spectrum. These are indicated by the curves on Fig. 1.

THE TRIUMF CHARGE EXCHANGE FACILITY

During the past year a new facility for both (p,n) and (n,p) measurements has been commissioned at TRIUMF and some initial results from this system will be described as an indication of future possibilities.

The system is based on the existing medium resolution spectrometer (MRS) which is used to detect protons arising from the (n,p) reaction in a suitable target. A schematic layout of the system is shown in Fig. 2. Neutrons are produced in the primary target T_{pn}, and the proton beam is then deflected through an angle of 20° by a magnet (not shown) and directed into a beam dump. Neutrons incident on the secondary target T_{np} give rise to the protons which are then detected by the MRS particle detection system. For (p,n) measurements, T_{pn} is moved downstream by 92 cm to be over the MRS pivot and T_{np} is moved the same distance closer to the MRS entrance. In this case T_{np} is some hydrogenous material such as CH_2 and the protons arise as recoil particles from elastic scattering of the neutrons. For (n,p) measurements, T_{pn} the neutron production target is normally ^{7}Li while T_{np} at the MRS pivot is the material to be studied. The target T_{np} actually consists of up to six layers separated by wire chambers which

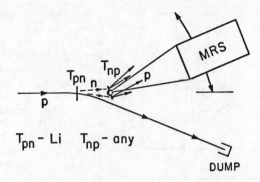

Fig. 2. Schematic diagram of the TRIUMF charge exchange spectrometer.

serve to define the origin of each proton entering the MRS. With this system it is possible to use total target thicknesses up to about 1 g/cm^2 while maintaining resolution better than 1 MeV. A typical spectrum obtained with the system is shown in Fig. 3. This is for ^{12}C(n,p)^{12}B at 200 MeV, and the results are seen to be quite comparable to those available for (p,n) time-of-flight measurements at this energy. It is also worth noting that with this system one of the target layers is usually CH$_2$. This permits a measurement of cross sections relative to that for hydrogen, without requiring a direct measurement of neutron flux or MRS detector efficiency.

Initial studies with this facility have been directed towards measurements of G-T strength in a variety of nuclei. For targets of ^{6}Li, ^{12}C and ^{13}C, the strength B$^+$ is known from β-decay measurements and the results serve to check the ratio $\sigma(0°)/B^+$ for the (n,p) reaction at 200 MeV. As shown in Table I, the preliminary results are quite consistent for the (n,p) reaction and generally agree with those derived from the (p,n) reaction.

Table I. The normalized cross section per unit GT strength extrapolated to q=0 for (p,n) and (n,p) reactions on light nuclei.

Target	$\dfrac{\sigma_{pn}(q=0)^a}{B^-}$	$\dfrac{\sigma_{np}(q=0)^b}{B^+}$
^{6}Li	9.0 ± 0.7	9.9 ± 0.7
^{12}C	9.1 ± 1.0	9.5 ± 0.7
^{13}C	13.2 ± 2.9	10.0 ± 0.7

^{a}mb/sr (cm) [Ref. 8].
^{b}mb/sr (cm), results of preliminary analysis with an assumed $\sigma_{np}(0°)$ = 12.4 mb/sr (cm) for the H(n,p)n reaction.

In another measurement, we have studied the ^{26}Mg(n,p)^{26}Na reaction to obtain the strength S$^+$ for ^{26}Mg. These results complement

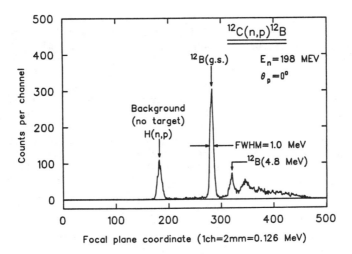

Fig. 3. Spectrum from $^{12}C(n,p)^{12}B$ at 0°, E_p = 200 MeV. The primary target was ^7Li and the ^{12}C target consisted of four layers of graphite, each 93 mg/cm² in thickness. The peak labelled H results from hydrogen in the windows of the target chamber.

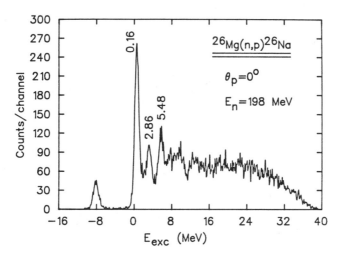

Fig. 4. Spectrum from $^{26}Mg(n,p)^{26}Na$ at 0°, E_p = 200 MeV. The groups at excitation energies of 0.16, 2.86 and 5.48 MeV show the forward peaking characteristic of GT transitions.

existing measurements[9] of ^{26}Mg(p,n) and will provide a test of the full G-T sum rule, without requiring assumptions about the magnitude of S^+. The 0° spectrum is shown in Fig. 4, and displays three distinct groups at excitation energies of 0.16, 2.86, and 5.48 MeV. A shell model calculation which provides a good fit to the G-T strength distribution B^- for ^{26}Mg has been used[10] to calculate the B^+ distribution, and the predicted strengths are indicated by the vertical bars on Fig. 5. In this calculation the G-T operator has been renormalized by a factor of 0.6 relative to the free-nucleon value. A theoretical spectrum generated by spreading the calculated G-T strength with the observed line shape in these measurements is also shown in Fig. 5, along with the measured 0° spectrum. The agreement is fair, though not as good as for the (p,n) results, probably reflecting the fact that the predicted strength for $T_>$ states is very sensitive to details of the model Hamiltonian. At 10°, the spectra show other groups at low excitation, indicating the importance of higher multipole transitions. Figure 6 shows the measured spectrum at 10°, superimposed on a suitably normalized 0° spectrum. It is clear that the extraction of G-T strength at excitation energies above 3 MeV will require careful analysis.

A test of the G-T sum rule was also carried out in the ^{54}Fe(n,p) reaction at 300 MeV. At $\theta=0°$ the cross section below $E_x=8$ MeV (see Fig. 7) was observed to be 16 mb. Data at larger angles show that only a small fraction of this cross section in this region of excitation results from L>0 multipoles. The resulting estimate of GT strength observed in this measurement is $5.0 < S^+ < 6.7$. Although the energy distribution of the GT strength agrees with shell model calculations by Bloom and Fuller[11] and by Muto,[12] the strength is reduced compared to the predicted $S^+_{theory} \approx 9.2$. Combining the (n,p) results

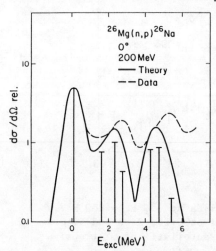
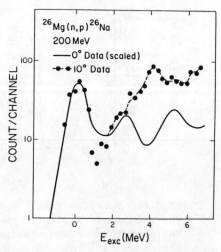

Fig. 5. Comparison of the 0° cross section for ^{26}Mg(n,p)^{26}Na with the cross section predicted for G-T transitions only.

Fig. 6. Spectra at 0° and 10° normalized to the same intensity for the 0.16 MeV group to show the appearance of groups with L>0 at low excitation.

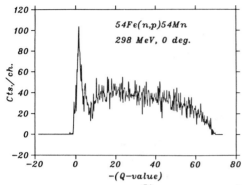

Fig. 7. Spectrum for ^{54}Fe(n,p) at 0° and 300 MeV. Most of the cross section in the peak at low excitation represents G-T transition strength.

with previous[13] (p,n) data, S^- = 7.8±1.9, it is seen that the experimental sum rule value, $S^- - S^+$ = 2.0±2.7, is smaller than 3(N-Z) = 6. This is an indication that a substantial fraction of G-T is spread out to higher excitation energies. The (n,p) results at E_x < 8 MeV are also of considerable astrophysical interest[11] since electron capture GT transitions in the N=28, Z=28 region play an important role in the pre-supernova stage of stellar evolution.

Perhaps the most careful effort to account for the missing strength in (p,n) measurements has been carried out for ^{90}Zr. The ^{90}Zr(p,n) reaction has been studied at 200 MeV,[14] with measurements extending to 18.7° in angle and 70 MeV in excitation energy in the final nucleus. These results have been analyzed by Osterfeld[15] et al. in an attempt to identify the missing strength. Wave functions for excited nuclear states were calculated with the RPA assuming p-h excitations up to 3 ℏω as well as Δ-isobar excitations in states from 1s up to 1j. Reaction cross sections were then calculated with the impulse approximation, assuming only spin-flip transitions with multipolarities L = 0 to 4. The conclusion of this analysis is that the data are consistent with the calculations assuming no Δ-isobar excitations, so that the G-T sum rule would be satisfied by S^- = 3(N-Z), provided that S^+ = 0. A determination of S^+ would thus be of great importance in establishing the possible importance of isobar excitations.

We have measured the ^{90}Zr(n,p) cross section at 200 MeV and angles out to 17°. Preliminary data at 0° and 6° are shown in Fig. 8. There is no obvious concentration of G-T strength in the 0° spectrum, while the spectrum at 6° indicates the excitation of higher multipoles. The detailed analysis of these results has just begun and quantitative conclusions will have to await results of that analysis.

In a further search for G-T strength in heavy nuclei we have measured ^{208}Pb(n,p) at 200 MeV also. The spectrum at 0° is shown in Fig. 9. The prominent peak at an excitation of 5 MeV persists at angles out to 9°, indicating substantial contributions from transitions with L>0. Once again, careful analysis will be required to establish whether any GT strength can be identified. It is interesting to note that the spin dipole resonance is predicted to occur at low excitation[16] in ^{208}Tℓ, with strengths for the 0^-, 1^- and 2^- components as indicated by the vertical bars on the figure. Thus it is likely that much if not all of the strength in the 0° peak arises from the spin dipole resonance.

As these examples have shown, it is now feasible to carry out measurements of the (n,p) reaction at intermediate energies with negligible background and resolution of 1 MeV or better. As to the

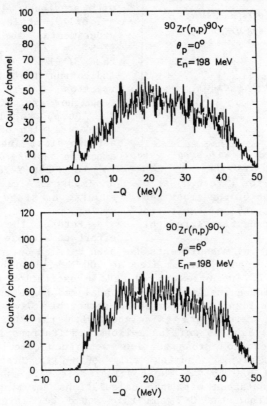

Fig. 8. Spectra for ^{90}Zr(n,p) at 200 MeV and angles of 0° and 6°.

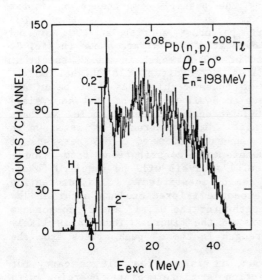

physics of such measurements, it appears that studies of the higher spin multipoles may be as important and interesting as the study of G-T excitations has been in the (p,n) reaction.

Fig. 9. Zero-degree spectrum of ^{208}Pb(n,p) at 200 MeV. The vertical bars indicate relative magnitudes of the transition strengths calculated for components of the spin-dipole resonance (Ref. 12).

COMPLEMENTARY PROBES

Two other reactions which should complement (n,p) measurements should also be noted, the (d,^{2}He) and ($\bar{p},\bar{n}$) reactions.

Studies of the (d,^{2}He) reaction have been reported both at low energy[17] (55 MeV) and at intermediate energy[18] (650 MeV). The deuteron ground state is $^3S_1$, so that if the two protons in ^{2}He are detected in a $^1S_0$ state, then the reaction involves both spin- and isospin-flip, and is similar to (n,p) at intermediate energies. In fact it is found that the excitations resulting from (d,^{2}He) are similar to those observed for the (n,p) reaction on ^{6}Li and ^{12}C. The detailed analysis of (d,^{2}He) results is complicated by distortion effects, however, and quantitative comparisons with (n,p) are not yet available. Two aspects of the (d,^{2}He) reaction promise to be of great importance at intermediate energies, however. It has recently been shown that polarization transfer in the ($\vec{p},\vec{n}$) reaction[19] can be used to identify spin flip strength, both for well-defined resonances and for continuum states. The ($\vec{d},^2$He) reaction with tensor polarized deuterons is predicted[20] to show significant analyzing power, which should also serve to identify spin flip transitions corresponding to (n,p) excitations. This is a very interesting possibility, though no actual measurements have been reported to date. Finally, it is found that the (d,^{2}He) reaction produces strong excitation of the Δ-isobar region,[18] analogous to what is observed in (p,n)[21] and (^{3}He,t)[22] reactions. It is probable that measurements in this region will be much easier with the (d,^{2}He) than with the (n,p) reaction.

The ($\bar{p},\bar{n}$) reaction involves the same isospin quantum numbers as (n,p) and may be expected to produce similar nuclear excitations. Initial observations of ($\bar{p},\bar{n}$) are planned for the near future.[23] It is expected that initial results will be important mainly in extending our knowledge of the antinucleon-nucleus interaction, but future work may provide an interesting complement to (n,p) studies.

ACKNOWLEDGEMENTS

The development of the TRIUMF charge exchange facility, and the use of the facility to obtain the results quoted here has involved the close collaboration of a number of individuals. These include R. Abegg, A. Celler, D. Frekers, P.W. Green, O. Häusser, R. Helmer, R. Henderson, K. Hicks, K.P. Jackson, C.A. Miller, M. Vetterli and S. Yen. We also wish to thank the technical staff of TRIUMF for their efforts in the construction of the facility, and the operations staff for their help during experimental runs. The Director of TRIUMF and his associates provided strong encouragement to the project and substantial financial support. The work was also supported by the Natural Sciences and Engineering Research Council of Canada.

REFERENCES

1. J. Rapaport et al., Phys. Rev. C 24, 335 (1981).
2. C.D. Goodman et al., Phys. Rev. Lett. 44, 1755 (1980).
3. C.D. Goodman in Spin Excitations in Nuclei, ed. F. Petrovich et al. (Plenum, New York, 1984), p. 143.

4. C. Gaarde et al., Nucl. Phys. A422, 189 (1984).
5. D.F. Measday and J.N. Palmieri, Phys. Rev. 161, 1071 (1967).
6. F.P. Brady et al., J. Phys. G 10, 363 (1984).
7. T.D. Ford et al., preprint 1986.
8. T. Taddeucci, private communication.
9. R. Madey et al., preprint 1986.
10. B.A. Brown, private communication.
11. S.D. Bloom and G.M. Fuller, Nucl. Phys. A440, 511 (1985).
12. K. Muto, Nucl. Phys. A451, 418 (1986).
13. J.R. Rapaport et al., Nucl. Phys. A410, 371 (1985).
14. C. Gaarde et al., Nucl. Phys. A369, 258 (1981).
15. F. Osterfeld et al. Phys. Rev. C 31, 372 (1985).
16. F. Krmpotic et al., Nucl. Phys. A342, 497 (1980).
17. D.P. Stahel et al., Phys. Rev. C 20, 1680 (1980).
18. C. Gaarde, private communication.
19. T.N. Taddeucci et al., Phys. Rev. C 33, 746 (1986).
20. C. Wilkin and D.V. Bugg, Phys. Lett. 154B, 243 (1985).
21. B.E. Bonner et al., Phys. Rev. C 18, 1418 (1978).
22. D. Contardo et al., Phys. Lett. 168B, 331 (1986).
23. A.I. Yavin, private communication.

THREE-BODY FORCES

B.M.K. Nefkens
University of California. Los Angeles, CA 90024

The existence of a three-body force is required by any theory of the strong interaction based on the exchange of virtual quanta. The crucial unknown is the relative strength of the two- and three-body forces.

An international symposium was held at the end of April in Washington D.C. on "The Three-Body Force in the Three-Nucleon System." Its purpose was to review the current status of the nuclear three-body force and to identify top priorities for theoretical and experimental work for fostering maximum progress in the next five years. The subject matter of the symposium was divided into four regimes:

1) the bound-state properties of the A=3 nuclei: binding energies, electromagnetic moments and radii, form factors;
2) the long-wavelength region, $\lambda >$ tritium-diameter, where nuclear collective effects are important;
3) the intermediate energy region where πN and $\pi\pi$ resonance formation such as the delta and rho play a significant role;
4) the high energy region, $\lambda < 0.4$ fm, where the new quark degrees of freedom are expected to be important.

There is an impressive body of high quality, quantitative, theoretical and experimental work on the A=3 system that has been performed in the past five years. The theory for regions 1) and 2) is based mainly on a nonrelativistic Hamiltonian of the form

$$H = \sum_i (T_i + V_i) + \sum_{i<j} V_{ij} + \sum_{i<j<k} V_{ijk},$$

where V_{ij} is the two-body interaction potential and V_{ijk} is the optional three-body one. The three nucleons in ^{3}H are weakly bound and move nonrelativistically since the binding energy is less than 1% of the rest mass. Various Fadeev and variational techniques for computing bound state properties of ^{3}H and ^{3}He using different models of the two-nucleon force all give mutually consistent results good to 50 KeV in the binding energy. The bottom line of all calculations is the same: the tritium is underbound by about 1 MeV, the rms charge radius is too large, the ^{3}H asymptotic normalization constants are too small, and finally the n-d spin doublet scattering length is too large by a factor of two, see Table I.

To fix up the shortcomings of the two-body force calculations the three-body force is invoked which for this purpose is defined[1] as a force that depends in an irreducible way on the simultaneous coordinates of three nucleons when only nucleon degrees of freedom are taken into account. A number of different theoretical approaches have been used to describe the three-body force. All include two-meson exchange with an intermediate delta, Fig. 1c. Some models use the boson exchange and others a more

Table I Comparison of Two-Body Force Calculations with Experimental Data in the A=3 System

	Experiment	2 BF calculations	
		RSC (34ch.)	AV14
1. Triton binding energy (MeV)	8.48	7.35	7.67
2. ^{3}He rms charge radius (fm)	1.93±0.03	2.04	2.02
3. Asympt. norm const D_2(fm^2) ^{3}H → nd	-0.279±0.012	-0.243	
4. n-d doublet scatt. length $^2a_{nd}$(fm)	0.65±0.04	1.31	

phenomenological approach to characterize the two-nucleon potential. The three-body force calculations show improvements but by no means are all the discrepancies with the data gone. For instance, the two-pion-three-body force calculation of the Tuscon-Melbourne group does increases the binding energy of tritium by 1.5 MeV which is a little too much, the rms charge radius is now a bit too small, the asymptotic normalization constants are improved as are the e.m. form factors. The Hanover group specializing in the NN-NΔ coupled-channels calculation finds an increase in the ^{3}H binding energy of only 0.3 MeV. The conclusions about the first regime are: the static properties of ^{3}H and ^{3}He cannot be described adequately by standard two-body forces. The existing three-body force models are not yet sufficiently sophisticated to account entirely for the discrepancies but they reduce them.

There are some noteworthy points in this regime.
a) The rms charge and magnetic radii of the A=3 nuclei have been determined to almost 2% accuracy.
  $r_c(^3H) = 1.83 ± 0.05$ fm, $r_c(^3He) = 1.93 ± 0.03$ fm,
  $r_m(^3H) = 1.80 ± 0.03$ fm, $r_m(^3He) = 1.93 ± 0.07$ fm .
b) The charge and magnetic form factors of tritium are now measured[2] out to $q^2 = -25$ fm^{-2}, see Figs. 2 and 3.
c) For the first time a determination has been made of the proton matter form factor of tritium using π^+ - ^{3}H elastic scattering. This is possible because of certain unique properties of pion scattering in the Δ-region. The agreement with the magnetic form factor is remarkable, Fig. 4. We now have the opportunity for a direct measurement of meson currents in ^{3}H, it requires a new more precise experiment.

In the long wavelength region there are several beautiful, new experiments[4,5] on polarized neutron-deuteron scattering which turn out to be impressive triumphs for the Fadeev calculations see Figs. 5 and 6. An attractive place to look for evidence for

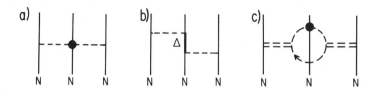

Fig. 1. Examples of a three-nucleon force due to meson exchange.

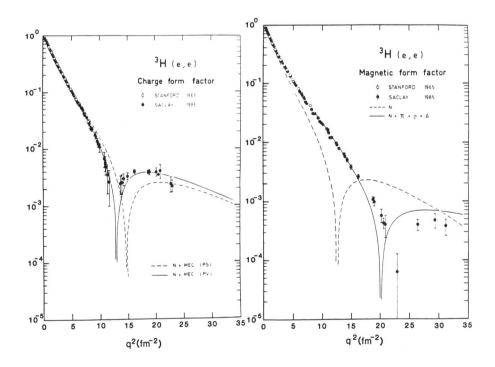

Fig. 2. Charge form factor of tritium, measured at Saclay, Ref. 2. The dashed and solid curves are calculations by Hajduk, Sauer, and Strueve.

Fig. 3. Same as Fig. 2 but for ^{3}H magnetic form factor.

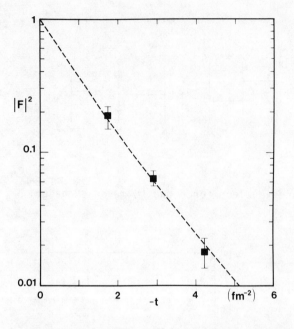

Fig. 4.
^{3}H proton matter form factor measurement reported in Ref. 3. The dashed line is the ^{3}H magnetic form factor.

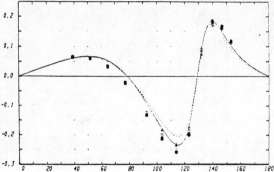

Fig. 5.
nd analyzing power at 27.5 MeV, Ref. 4. Solid line: Fadeev calculation, dashed line: PEST 4 potential.

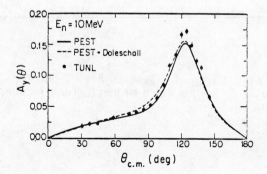

Fig. 6.
Vector analyzing power measurements and calculations for pd → pd at 10 MeV, Ref. 5.

the three-body force appears to be the "star" and the "collinear" geometry of three-nucleon continuum final states reactions. Electromagnetic reactions such as ^{3}He (γ,pp) n and D($\vec{n}$,γ) ^{3}He also deserve detailed attention when coupled with good calculations.

In the resonance region the pion and delta degrees of freedom can manifest themselves directly. Among the new experiments we single out the herculean job done in the pd elastic scattering reaction by the UCLA II group and their collaborators.[6] They are measuring 34 different spin variables which is more than sufficient to determine the 12 independent amplitudes that characterize pd → pd. Detailed calculations by M. Bleszynski et al.[7] show that the deuteron tensor polarization is especially sensitive to the contact term which in turn is equivalent to a three-body interaction. Two examples are shown in Fig. 7 and 8, details are given in Ref. 7.

Very interesting possibilities are presented by p+d → ^{3}H + π^+ and $\gamma+^3$He ⇄ p+d measurements in the Δ resonance region. New measurements do not agree with calculations based on two nucleon interactions. The agreement is particularly bad in the backward direction, see Fig. 9 where kinematic conditions require all three nucleons to participate in the reactions. These two-body reactions involving a pion or a photon are potentially very fertile ground when polarization variables are measured to probe the three-body force; careful calculations are needed to interpret the experiments.

The fourth and final regime is that of quarks and gluons. There has been obtained as yet little experimental data at sufficiently high energy except for exploratory electron scattering data from SLAC; theoretical discussions outnumber experiments. The possibilities for seeing three-body force effects are large. It is good to remind oneself that the antisymmetrization of three identical, colored quarks in a nine quark bag necessarily has the effect of a three-body force, see Fig. 10. The first exploratory calculations of the effect of quark clusters on the static properties of tritium and helium-three are beginning to appear, e.g., it has been reported[8] that a cluster with radius of 0.5 fm could change the binding energy of ^{3}H by 0.1 MeV. Some experiments in which one expects to see quark degrees of freedom are listed in Table II while experiments to test three-body force involving quarks are given in Table III. The conclusion regarding this regime is easy: start the experimental program.

The overall conclusion regarding the three-body force is as follows: The two-body force cannot explain the observations in any of the four regimes. The discrepancies are most blatant or most interesting depending on one's point of view when polarizations are measured. The three-body force calculations in their present form help to reduce the discrepancies but they are not adequate. The next five years are full of exciting possibilities for experimentalists and theorists alike.

We gratefully acknowledge the contributions of Ben Gibson and Frank Gross who provided me with copies of their summary talks of the "International Symposium on the Three-Body Force in the Three-Nucleon System."

This work was supported in part by the U.S. D.O.E.

724

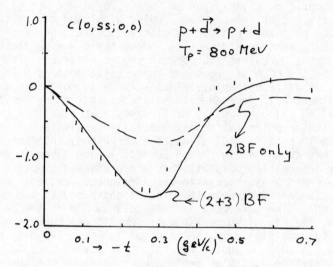

Fig. 7. Tensor analyzing power in p$\vec{d}$ → pd at 800 MeV (UCLA II collaboration).

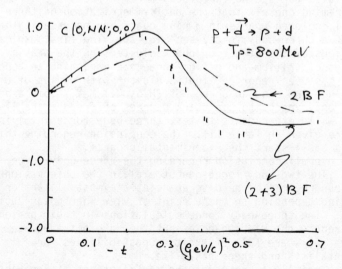

Fig. 8. Same as Fig. 7.

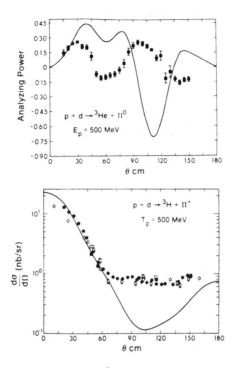

Fig. 9. Measurements of pd → ^{3}H + π^0 by Cameron at al., compared with calculations by Laget presented at this meeting.

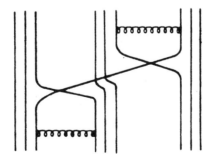

Fig. 10. The antisymmetrization of a 9-quark wavefunction introduces matrix elements which depend on three coordinates and hence play the role of a three-body force.

Table II   Experiments to Test Quark Degrees of Freedom

a. Neutron electric form factor using longitudinally polarized electrons
$$\vec{e} + d\uparrow \rightarrow e + d$$
$$\vec{e} + {}^3He\uparrow \rightarrow e + {}^3He$$

b. Electric quadrupole transition strengths of the radiative decay of the delta resonance, $\Delta \rightarrow N\gamma$
$$\gamma + N \rightarrow \gamma + N \text{ (including all spin variables)}$$
$$\vec{e} + N \rightarrow e' + N'$$
$$\vec{e} + {}^3He\uparrow \rightarrow e + {}^3He$$

c. Deep inelastic scattering, $x > 1$
$$e + {}^3He \rightarrow e' + x$$
$$e + {}^3He \rightarrow e' + y$$

Table III   Experiments to Test for the Three-Body Force with Quark Degrees of Freedom

1. Electric and magnetic form factors of ^3H and ^3He at large momentum transfer.

2. ^3He spectral functions in reactions such as
$$p + {}^3He \rightarrow p' + x$$
$$p + {}^3He \rightarrow p + p + d$$
$$e + {}^3He \rightarrow e' + x$$
$$e + {}^3He \rightarrow e' + d + y$$
$$\pi^{\pm} + {}^3He \rightarrow \pi^{\pm} + p + --$$
$$\gamma + {}^3He \rightarrow p + n + p$$
The same for ^3H instead of ^3He.

References

1. J. Friar, B. Gibson, G. Payne, Ann. Rev. Nucl. Part. Sci. **34**, 403 (1984).
2. F. Juster et al., Phys. Rev. Lett. **55**, 2261 (1985).
3. B.M.K. Nefkens, Contribution to the Symposium on the Three-Body Force in the Three-Nucleon System, The George Washington University, 1986, (Three-Body Force Symposium) and to be published.
4. H.O. Kluges et al., Contr. to the Three-Body Force Symposium.
5. W. Tornow et al., Contr. to the Three-Body Force Symposium.
6. D. Adams et al., Contr. to the Three-Body Force Symposium.
7. M. Bleszynski, et al., Contr. to the Three-Body Force Symposium.
8. K. Maltman, Contr. to the Three-Body Force Symposium.

MULTIPOLE MATRIX ELEMENTS FOR ^{208}Pb

D. K. McDaniels, J. Lisantti
University of Oregon, Eugene, OR 97403

L. W. Swenson, X. Y. Chen
Oregon State University, Corvallis, OR 97330

F. E. Bertrand, E. E. Gross, D. J. Horen, C. Glover,
R. Sayer, B. L. Burks
Oak Ridge National Laboratory, Oak Ridge, TN 37831

O. Häusser, K. Hicks
Simon Fraser University, Burnaby, B. C. V5A 1S6

ABSTRACT

New measurements of inelastic proton scattering to low-lying states of ^{208}Pb at 200 and 400 MeV are reported. Deformation lengths extracted from angular distributions for the 3^- (2.614 MeV), 5^-_1 (3.198 MeV), 5^-_2 (3.209 MeV), 2^+ (4.086 MeV) and 4^+ (4.324 MeV) states are in good accord with values extracted at other incident proton energies. The fact that the deformation lengths are independent of incident proton energy within experimental uncertainty provides support for the validity of the collective DWBA for medium energy proton scattering to strongly excited states. Advantage is taken of this to extract statistically more precise values of the ratio of neutron to proton multipole matrix elements (M_n/M_p). Different methods of determining the appropriate average value of M_n/M_p are discussed.

RESULTS AND DISCUSSION

New high resolution measurements to several low lying states of ^{208}Pb have been made at 200 and 400 MeV. Angular distributions around the first diffraction peak have been made for the states at 2.614 MeV (3^-), 3.198 MeV (5^-_1), 3.709 MeV (5^-_2), 4.086 MeV (2^+) and 4.324 MeV (4^+). Deformation lengths have been extracted with the collective DWBA code ECIS79 and compared with results from other (p,p') experiments. Within uncertainties the results for these strongly excited states in ^{208}Pb are consistent with the picture of a deformation length that is independent of the energy of the incident proton, in agreement with our earlier work.[1]

We now utilize the deformation length data to evaluate M_n/M_p, the ratio of neutron and proton transition multipole moments. These latter are interesting in that they are quite sensitive to details of nuclear structure models. The multipole moment ratio is expressed in terms of the deformation lengths by the relation

$$\frac{M_n}{M_p} = [\frac{N}{Z}] [\frac{\delta_H}{\delta_p} + \frac{Z}{(b^p_n/b^p_p)N} \{\frac{\delta_H}{\delta_p} - 1\}] . \qquad (1)$$

In this equation δ_n and δ_p are nuclear structure deformation lengths, δ_H is the value measured in our (p,p') experiment and $b^p{}_n$ and $p^p{}_p$ represent the interaction between the incident proton and the target nucleons. We have evaluated these using the Love-Franey[3] representation of the nucleon-nucleon force.

Our values for $\tilde{M}_n/\tilde{M}_p = [M_n/M_p][N/Z]$ are summarized in the second column of Table 1. This contains the average value of our results calculated using the measured values of δ_H at 200, 334 and 400 MeV. The appropriate force ratio $b^p{}_n/b^p{}_p$ was calculated at each energy. The $\tilde{M}_n/\tilde{M}_p$ values were corrected for different neutron and proton matter densities.[3] For comparison, we tabulate in the third column the average value of $\tilde{M}_n/\tilde{M}_p$ calculated from literature values of δ_H at lower energies between 60 and 135 MeV. In the last column the numerical results at 800 MeV[3] are listed.

Table 1. Multipole Matrix Elements ($\tilde{M}_n/\tilde{M}_p$) for ^{208}Pb

STATE	Present (200-400)	Low Energy	800 MeV
3^-	1.04±0.04	1.06±0.04	1.12±0.03
2^+	1.00±0.04	1.01±0.03	1.28±0.05
4^+	0.95±0.07	0.93±0.12	1.10±0.06
5^-_1	0.93±0.11	1.03±0.08	1.18±0.07
5^-_2	0.72±0.10	0.89±0.06	1.10±0.07

Within uncertainties the values obtained by us in the 200- to 400-MeV energy range agree with those calculated using the measured lower energy deformation lengths. However, the values found from the 800-MeV measurements are 20% higher than the other results, on the average. Since $\tilde{M}_n/\tilde{M}_p$ is a property of the target nucleus, it cannot vary with energy. Also, the difference cannot be due to the wrong choice of the force ratio. For the 3^- state for example, even if the force ratio were infinite (rather than 0.82 as used) $\tilde{M}_n/\tilde{M}_p$ would still be 1.09. Thus, we conclude that a small normalization problem exists either for the 800-MeV data, or for the rest of the measurements at other energies.

REFERENCES

1. D. K. McDaniels, J. Lisantti, J. Tinsley, I. Bergqvist, L. W. Swenson, F. E. Bertrand, E. E. Gross, and D. J. Horen, Phys. Lett. 162B, 277 (1985).

2. M. A. Franey and W. G. Love, Phys. Rev. C31, 488 (1985); W. G. Love and M. A. Franey, Phys. Rev. C24, 1073 (1981).

3. M. M. Gazzaly, N. M. Hintz, G. S. Kyle, R. K. Owen, G. H. Hoffman, M. Barlett, and G. Blanpied, Phys. Rev. C25, 408 (1982).

THE CHOU-YANG MODEL AND p-^{4}He ELASTIC SCATTERING FROM 45-393 GeV

Mujahid Kamran
Centre for High Energy Physics, University of the Punjab, Lahore-20
Pakistan

I.E.Qureshi
Nuclear Physics Division, PINSTECH, P.O.Nilore, Islamabad, Pakistan

ABSTRACT

The Chou-Yang model is applied to p-^{4}He elastic data in the 45-393 GeV region. The proton electric form factor due to Borkowski et al is used. The model gives a good overall description of the data. Comparison with Glauber theory is made and aspects of amplitude structure described.

Elastic p-α data[1] at high energies exhibits features which are similar to those exhibited by hadron-hadron scattering such as rising total cross-sections, shrinkage of diffraction peak, minima in dσ/dt, etc. Recently[2] it has been pointed out that the p$\alpha \to$ pα data exhibits scaling. It is therefore natural to apply models of hadron-hadron scattering to hadron-nucleus and nucleus-nucleus scattering. In this paper we apply the Chou-Yang model[3] to pα scattering. An application of this model to α-nucleus (A$\geq$4) scattering was made in ref.4 where it was pointed out that the success of the model is comparable to that of the Glauber theory, the latter being comparatively tedious and difficult to apply to larger nuclei.

The formulae of the model[3] are ($q^2=-t$)

$$T = i\int\{1-e^{-\Omega}\}J_0(b\sqrt{-t}) \; bdb \tag{1a}$$

$$\Omega = K\int G_\alpha(t) \; G_E(t) \; J_0(bq) \; qdq \tag{1b}$$

$$d\sigma/dt = \pi|T|^2 \; \text{GeV}^{-4} \tag{2}$$

$$\sigma_{tot} = 4\pi\int\{1-e^{-\Omega}\}bdb \; \text{GeV}^{-2} \tag{3}$$

G_α is the α-form factor and G_E is the proton electric form factor. For G_α we use the expression quoted in ref.5 whereas for G_E we use the four pole fit quoted in ref.6 as this pays close attention to the small -t pp data. Glauber and Velasco[7] have recently pointed out that this fit to G_E is favoured by the Collider $\bar{p}p$ data in the context of a generalised Chou-Yang model. The constant K is determined by the total cross-section.

In fig.1 we show a sample comparison of the model with experiment. We find that all energies in the 45-393 GeV range the model gives an excellent description of $d\sigma/dt$ upto the dip. However the dip results from a zero of the amplitude and therefore $d\sigma/dt$ vanishes in contradiction with experiment. This calls for the introduction of a real part of the amplitude but the model ignores the real part which is a major short coming. Furthermore we find that the model predicts a height of the secondary maximum which is somewhat below that of the data. This problem seems to be opposite to the one the model encounters in pp and $\bar{p}p$ scattering where the secondary maximum is higher than the data[8]. At 393 GeV the model predicts another dip around $-t \simeq 0.93$ where the amplitude changes sign once again. A comparison with the results of Glauber theory for the same process is shown in the figure. Glauber theory predicts another dip around $-t \simeq 0.7$ at high energies where as the data[9] do not reveal such a dip. Currently data extend upto $-t \simeq 0.72$.

REFERENCES

1. A.Bujak et al, Phys.Rev.D23, 1895 (1981).
2. S.Patel, Pramana 25, 685 (1985).
3. T.T.Chou and C.N.Yang, Phys.Rev.168, 1594 (1968).
4. Y.Li and S.Lo, Aust.J.Phys.37, 255 (1984).
5. C.R.Otterman et al, Nucl.Phys.A436, 688 (1985).
6. G.Borkowski et al, Nucl.Phys.B93, 461 (1975).
7. R.Glauber and J.Velasco, Phys.Lett.147B, 189 (1984).
8. T.Fearnley, CERN-EP/85-137 (Sep.9, 1985).
9. W.Bell et al, Phys.Lett.117B, 131 (1982).

FIGURE

Comparison of the model predictions (solid curve) with 393 ($\sqrt{s}$=54)GeV data (open circles). The solid curve corresponds to σ_{tot}=125.93mb (K=33.21GeV^{-2}). Also the $\sqrt{s}$=88 GeV data points from ref.9 alongwith the predictions of Glauber theory (broken curve) with inelastic shadow corrections included, are shown as solid squares. For $\sqrt{s}$=88 GeV σ_{tot}=130±20mb.

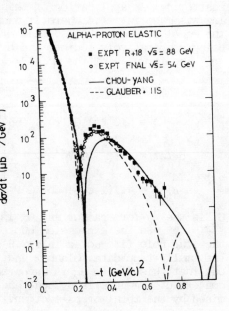

HADRON SPECTROSCOPY

Coordinators: J. F. Dohoghue
             A. McDonald

INTERACTING BOSON MODEL: SELECTED RECENT DEVELOPMENTS*

A. B. Balantekin[†]
Physics Division, Oak Ridge National Laboratory
Oak Ridge, TN 37831 U.S.A.

ABSTRACT

The Interacting Boson Model is briefly reviewed. Recent applications of this model to the low-lying collective magnetic-dipole excitations and to the spectra of ^{195}Ir are described.

INTRODUCTION

Interacting Boson Model is an algebraic approach to even-even nuclei formulated in terms of bosonic variables with emphasis being placed on the symmetries associated with the quadrupole collectivity.[1] In this model, low-lying energy levels are generated as states of a system of N bosons carrying angular momentum L = 0 or L = 2. N is taken to be the number of correlated pairs of valence neutrons or pairs of valence protons. The pairs with L = 0 are similar to Cooper pairs in a gas of electrons. Introducing creation and annihilation operators for six types of bosons, b_i, $b_i^\dagger$, one obtains a bosonic realization of the U(6) algebra, $G_{ij} = b_i^\dagger b_j$. The one-boson terms of the Hamiltonian are written as a linear combination of the U(6) generators and the boson-boson interaction terms as a combination of quadratic products of these generators:

$$H = \Sigma \, \epsilon_{ij} \, G_{ij} + \Sigma \, u_{ijk\ell} \, G_{ij} \, G_{k\ell} \qquad (1)$$

The coefficients in the above equation are free parameters to be fixed separately for each nucleus, subject to the restriction that the total angular momenta of the bosons is conserved. Eigenstates of the Hamiltonian are constructed by the application of boson creation operators on a vacuum N-times

$$b_{i_1}^\dagger \, \cdots \cdots \, b_{i_N}^\dagger | \, 0 \rangle \qquad (2)$$

i.e., one forms the completely symmetric representations of U(6). In general, such states are labeled by the boson number N and by the labels of all the representations of all possible subalgebras, included in the completely symmetric representation of U(6). Hence,

---

*Research sponsored by the Division of Nuclear Physics, U.S. Department of Energy under contract DE-AC05-84OR21400 with Martin Marietta Energy Systems, Inc.

[†]Eugene P. Wigner Fellow.

for most nuclei, the diagonalization of the Hamiltonian must be performed numerically. However, the popularity that the Interacting Boson Model has enjoyed in recent years is mostly facilitated by the existence of analytically solvable limits.

Analytical solutions are possible when the free parameters take special values such that the first term in Eq. (1) can be written as a combination of linear Casimir operators and the second term as a combination of quadratic Casimir operators of the algebra U(6) and its subalgebras. In such cases the quantum numbers are given by the labels associated with the algebras in a particular chain starting with U(6) and ending with SO(3), and the Hamiltonian is said to possess a dynamical symmetry. This situation is reminiscent of the Gell-Mann-Okubo SU(3) mass formula

$$E(I, I_3, Y) = E + a Y + b \left[(Y^2/4) - I(I+1) \right]. \quad (3)$$

which is associated with the group chain

$$SU(3) \supset SU(2) \times U(1) \supset SO(2) \times U(1). \quad (4)$$

There are three possible dynamical symmetries for the Interacting Boson Model Hamiltonian with the second group in the chain being SU(5), SU(3), or SO(6).

The existence of dynamical symmetries of the Interacting Boson Model Hamiltonian implies the existence of analytical expressions not only for energy eigenvalues, but also for the eigenstates. Consequently, to establish whether or not a particular dynamical symmetry is physically realized, it is not sufficient to compare energy spectra. It is essential to test predictions for the electromagnetic transition rates, which are calculated using the wavefunctions. The dynamical symmetries of the Interacting Boson Model provide a simple framework to analyze and classify the experimental data and have been successfully used in this context during the last ten years.[1] In this talk, I highlight two of the recent developments in the applications of the Interacting Boson Model.

II. LOW-LYING COLLECTIVE MAGNETIC-DIPOLE EXCITATION MODE

There is a second version of the Interacting Boson Model, which is much richer in structure than the version discussed in the Introduction. If no distinction is made between bosons which are correlated pairs of neutrons and bosons which are correlated pairs of protons, then the wavefunction given in Eq. (2) contains only the completely symmetric representations of U(6). However, if such a distinction is made, the states which are partially antisymmetrized between neutron and proton bosons should also be experimentally observable. This new version of the model is called the neutron-proton Interacting Boson Model.

Semiclassically, a deformed nucleus can be visualized as an ellipsoid inside of which protons and neutrons are uniformly distributed. When such a nucleus is placed in an external field, the proton ellipsoid translationally oscillates around the neutron ellipsoid. The direction of the vibration can be either parallel or perpendicular to the major axis of the deformed nucleus. Consequently, a double-humped strength distribution can be experimentally observed. This is the well-known Goldhaber-Teller mode. For two separate ellipsoidal rigid bodies, which are slightly tilted away from each other, a different kind of oscillation is also possible where the two shapes wobble around the same axis[2] as depicted in Fig. 1. Such a motion would excite a $J^{\pi}=1^+$ state through the orbital current part of the M1 operator. One would expect that the energy of such a state is comparable to the energy of the E1 giant resonance (the Goldhaber-Teller mode).

A collective magnetic-dipole excited 1^+ state has been discovered[3] in electron scattering first over ^{156}Gd targets and later other nuclei in this region. The excitation energy in this mass region was found to be approximately 3 MeV. Recently (γ,γ') experiments on a series of Gd isotopes has confirmed[4] the energies and more importantly the spin of these states. This excitation energy is much lower than one would expect for the wobble mode discussed in the previous paragraph, where the entire nuclear system participates. However, the mixed symmetry states of the neutron-proton Interacting Boson Model in the SU(3) limit give rise[5] to a low-lying $K^{\pi}=1^+$ band, where the M1 matrix element connecting the

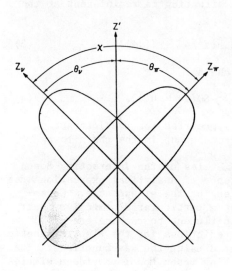

Fig. 1. Classical picture of the collective 1^+ configuration

ground state and the bandhead is very large. Since in this model only the valence nucleons contribute to the wobble motion, as opposed to the entire nuclear system, the excitation energy would be relatively low. Furthermore, one can show[6] that in the classical limit of the neutron-proton Interacting Boson Model Hamiltonian these 1^+ states can be regarded as small amplitude oscillations of the angle between the two symmetry axes of the deformed valence neutrons and protons. Consequently, one can interpret the experimentally observed 1^+ state in ^{156}Gd as a neutron-proton antisymmetric state in the Interacting Boson Model framework.

After the initial discovery in ^{156}Gd, the new collective isovector magnetic dipole mode has been observed in a number of nuclei.[7] One should stress that this new mode is not an isolated example, but the prototype of an entire new class of states which occur at that energy and have collective properties.[8] Since these states are neutron-proton antisymmetric, their experimental study should yield a better understanding of the neutron proton forces in the collective models.

III. DESCRIPTION OF THE NUCLEUS ^{195}Ir IN THE FRAMEWORK OF U(6/4) SUPERSYMMETRY

In order to study odd-even nuclei, where one of the nucleons is unpaired, in a framework similar to the Interacting Boson Model, fermionic degrees of freedom should be explicitly introduced. Such an extended model is the Interacting Boson-Fermion Model.[9] If the angular momentum of the fermion has some specific value, and furthermore if the boson-boson and boson-fermion interactions have particular forms, one can also obtain analytical expressions for the energies and transition rates. An experimentally relevant example is provided when the bosonic core is described by the O(6) dynamical symmetry of the Interacting Boson Model and when j = 3/2 for the unpaired fermion, since j = 3/2 is the angular momentum in the SU(2) decomposition of the lowest-dimensional spinor representation of O(6). In this scheme, bosonic states (i.e. those in the even-even nucleus) are transformed by the tensor representations of O(6) and fermionic states (i.e. those in the neighboring odd-even nucleus) are transformed by the spinor representations, which are associated with the universal covering group Spin(6). Such symmetries were termed as Bose-Fermi symmetries.[10]

In principle, the spectra of the even-even and the odd-even nuclei can be fitted separately. However, in certain cases the parameters appearing in the Hamiltonian and in the electromagnetic transition operators take the same value for both kinds of nuclei. In such cases, a supersymmetry is realized. Using the isomorphism Spin(6) ~ SU(4), the scheme described in the previous paragraph can be embedded in the supergroup SU(6/4). This supersymmetry has been observed[11] in the Os-Ir region where a number of nuclei are placed in the same supermultiplet (Fig. 2). There are other examples of approximate supersymmetries in nature, one of them being the equivalence of the slopes of the Regge trajectories for mesons and baryons. Indeed, the U(N/M) type supergroups were first introduced in an effort[12] to explain this equivalence. The prominence of the nuclear supersymmetries, however, is that their well-defined predictions can be thoroughly tested in the laboratory. In the rest of this talk, I describe such a recent test.

There is a simpler description of the odd-even nuclei, called the weak coupling model. In this model the angular momentum of the fermion is directly coupled to the states L = 0, 2, .. in the even-even core and the resulting states are split by a J(J+1) interaction. In the Spin(6) scheme the 3/2 member of the 3/2 x 2 multiplet is at an energy higher than that is expected from the simple-minded weak coupling picture. Since this state is generated by the quadrupole interaction between the bosons and the fermion, it does not belong to the same Spin(6) representation as the ground state. Consequently, a precise measurement of the position of this state is requisite to

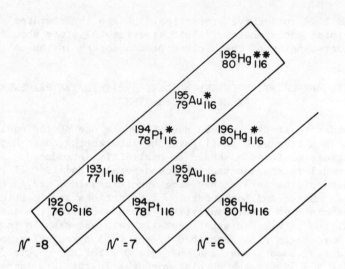

Fig. 2. Supersymmetric multiplets in the Os,Pt region.

establish the applicability of the Spin(6) Bose-Fermi symmetry.

Based on previous work in this region, ^{195}Ir is expected to be a good example of Spin(6) symmetry. The low-lying positive parity states and their electromagnetic de-excitations have recently been studied[13] in the ^{193}Ir$(2n,\gamma)^{195}$Ir reaction. Using the observed E2 transition, the 3/2 state mentioned in the preceding paragraph has been located at 286 keV. In ^{195}Ir, the ground state is located in the σ=13/2 representation of Spin(6), and the 286-keV 3/2 state is the head of the σ=11/2 representation. Furthermore, using the parameters obtained for the nucleus ^{194}Os, one can predict the spectra of ^{195}Ir by invoking the U(6/4) Spin(6) supersymmetry. The resulting predictions are compared with the experimental level scheme in Fig. 3, taken from Cizewski, et al.[13] The salient feature of this figure is the remarkable regularity in the repetition of the σ sequences, which is a unique signature of the supersymmetry scheme.

I would like to thank J. Cizewski for communicating her recent results prior to publication.

REFERENCES

1. For recent reviews see A. Arima and F. Iachello, Adv. Nucl. Phys. 13, 139 (1984); A.E.L. Dieperink and G. Wenes, Ann. Rev. Nucl. Part. Sci. 35, (1985).
2. N. Lo Iudice and F. Palumbo, Phys. Rev. Lett. 41, 1532 (1978).
3. D. Bohle et al., Phys. Lett. 137B, 27 (1984).
4. U.E.P Berg et al., Phys. Lett. 149B 59 (1984).
5. F. Iachello, Nucl. Phys. A358, 89c (1981); A.E.L. Dieperink, Prog. in Part. and Nucl. Phys. 9, 121 (1983).
6. A. B. Balantekin and B. R. Barrett, Phys. Rev. C32, 288 (1985); S. Pittel and J. Dukelsky, ibid. 335 (1985); R. Bijker, ibid.

1442 (1985); N. R. Walet, P. J. Brussaard, and A.E.L. Dieperink, Phys. Lett. 163B, 1 (1985).
7. For a recent review, see A. Richter, in Nuclear Structure 1985, ed. by R. Broglia, G. B. Hagemann, and B. Herskind (Elsevier-North-Holland, Amsterdam, 1985).
8. F. Iachello, Phys. Rev. Lett. 53, 1427 (1984).
9. For a recent review, see O. Scholten, Prog. in Part. and Nucl. Phys. 14, 189 (1985).
10. F. Iachello, Phys. Rev. Lett. 44, 772 (1980).
11. A. B. Balantekin, I. Bars, and F. Iachello, Phys. Rev. Lett. 47, 19 (1981).
12. H. Miyazawa, Phys. Rev. 170, 1586 (1968).
13. J. A. Cizewski, et al., Yale preprint YALE-3074-854a (1986).

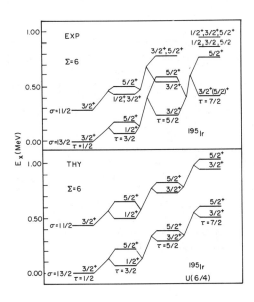

Fig. 3. Comparison between supersymmetry predictions and positive-parity states in ^{195}Ir (from Ref. 13).

FUNDAMENTAL INTERACTION STUDIES IN NUCLEI

W. C. Haxton
Institute for Nuclear Theory, Department of Physics, FM-15
University of Washington, Seattle, WA 98195

ABSTRACT

There is a growing class of nuclear and atomic physics experiments in which the special properties of many-body systems are exploited in tests of fundamental symmetries. Three representative topics are discussed: parity nonconservation in the NN interaction, time-reversal-odd nuclear moments, and matter-enhanced neutrino oscillations.

INTRODUCTION

Many particle physicists are uneasy over the prospect that the mass scales governing physics beyond the standard model may not be directly accessible in accelerator experiments. If this is the case, this new physics may prove very elusive. However, two alternatives to standard accelerator experiments hold great promise. Theoreticians have become increasingly adept at analyzing the ultimate high-energy experiment, the big bang. Similarly, many experimentalists believe that evidence for grand unification may appear at low energies in the guise of rare events forbidden in the standard model. Experiments have been mounted to search for monopoles, proton decay, neutrinoloss double beta decay, new forces of intermediate range, electric dipole moments of the neutron, electron, and various atoms, axions, neutrino masses, neutrino oscillations, muon-number- and lepton-number-violating muon and kaon decays, etc. It is clear from many of the talks at this conference that the alliance of particle, nuclear, and atomic physicists exploring this frontier is rapidly growing. In this talk I would like to discuss three classes of experiments from nuclear and atomic physics that have yielded some interesting results this past year.

PARITY NONCONSERVATION

The first problem I will discuss is, in a sense, a "loose end" of the standard electroweak theory. While elegant experiments have been performed to test the predictions of the standard model for leptonic and semileptonic interactions, the hadronic weak interaction has been more elusive. At low energies it can be studied only when the strong and electromagnetic interactions are forbidden by a symmetry principle, such as flavor conservation. However, in the standard model the neutral-current contributions to $\Delta S = 1$ and $\Delta C = 1$ weak processes are greatly suppressed. Therefore neutral weak interactions can only be probed in flavor-conserving processes, where parity nonconservation must serve as the filter to isolate the relevant observables. The NN interaction is the single practical example.

The weak NN interaction is a sum of isoscalar, isovector, and isotensor terms. The charged current contribution to the $\Delta I = 1$ amplitudes is suppressed by $\sin^2 \theta_c$, where θ_c is the Cabibbo angle, while the neutral current contribution is unsuppressed. Thus this component of the weak NN interaction should provide a direct measure of the neutral current interactions of quarks. In the conventional single-meson-exchange description of the weak NN interaction, where one meson-nucleon vertex is governed by the strong interaction and the second by the weak, the isovector amplitude is dominated by pion exchange. The neutral current is expected to increase the weak pion-nucleon coupling f_π from the Cabibbo angle (~ 0.5) to approximately 12.

The experimental task of isolating the pion exchange amplitude is a difficult one. At low energies there are six independent S-P amplitudes (if one considers the long-range pion-exchange and short-range vector-meson exchange contributions to the $^1S_0$ - $^3P_0$ amplitude to be distinct). Ideally one would make six independent measurements in the NN system to determine these amplitudes. To date only one definitive result, the helicity dependence of the $\vec{p}+p$ cross section, has been obtained. (See Adelberger's talk for a discussion of other possible parity-violating observables in the pp and pn systems.)

Another approach to this problem is the extraction of the weak meson-nucleon couplings from measurements of the parity mixing of nuclear states. The success of this approach depends in part on our ability to determine the many-body matrix elements of the weak NN potential. I will discuss only a single example, ^{18}F. A more complete discussion of parity violation in nuclei can be found in Ref. 1.

A diagram of the relevant levels in ^{18}F (and of the parity doublets that have been studied in the neighboring nuclei ^{19}F and ^{21}Ne) is shown in Fig. 1. Since the separation ΔE of the $J^\pi T = 0^-0$ and 0^+1 levels is only 39 keV, one can reasonably assume that the parity admixture in the 0^-0 state is due to mixing with the 0^+1 state. This admixture can be measured by observing the circular polarization P_γ of the gamma ray emitted in the $0^-0 \to 1^+0$ (g.s.) decay:

$$P_\gamma = \frac{2}{\Delta E} Re \left[\frac{<1^+0\|M1\|0^+1>}{<1^+0\|E1\|0^-0>} <0^+1|V_{PNC}^{\Delta I=1}|0^-0> \right] \quad (1)$$

This formula illustrates a number of attractive features of the doublet. The definite isospin of the nuclear states isolates a single component, $V_{PNC}^{\Delta I=1}$, of the weak NN potential. While the parity-allowed E1 transition is isospin-forbidden, the parity-forbidden M1 transition is unusually strong, 10.3 ± 1.5 W.u. Thus the matrix element ratio $<M1> / <E1> \approx 110$ amplifies the effect of the parity admixing. A further amplification occurs because of the small energy denominator separating the doublet states. Thus one expects $P_\gamma \sim 10^{-3}$, four orders of magnitude greater than the scale for parity nonconserving observables in the NN system.

A final remarkable aspect of the ^{18}F doublet involves a relation between the PNC matrix element in Eq. (1) and the β decay of ^{18}Ne.[2,3] The ^{18}Ne ground state is the isospin analog of the 1.042 MeV 0^+1 state in ^{18}F. Thus the first-forbidden transition leading to the 1.081 MeV 0^-0 state in ^{18}F links the same states that appear in Eq. (1). In the long wavelength limit this transition proceeds through the axial charge operator. In addition to the familiar one-body axial charge operator

$$F_A \sum_{i=1}^{A} \frac{\vec{\sigma}(i) \cdot \vec{p}(i)}{M} \tau_- \quad (2)$$

there is an important two-body contribution that is also of order (v/c). This operator, which is determined by current algebra and PCAC, is identical, apart from an isospin rotation and overall coupling constants, to the pion-exchange piece of $V_{PNC}^{\Delta I=1}$.

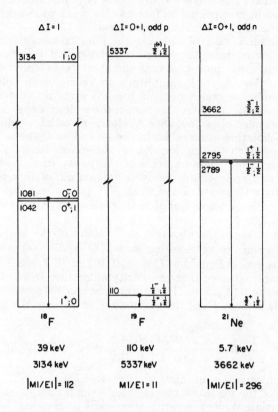

Figure 1. Parity-mixed doublets in light nuclei. The transitions displaying the amplified PNC effect are indicated. The quantities ΔE and $\Delta E'$ are the smallest and next smallest energy denominators governing the parity mixing. The quantities shown in the bottom row are "amplification factors."

More importantly, the exchange current, when averaged over core nucleons, becomes an effective one-body operator whose form is also given by Eq. (2). That is, the exchange current, to an excellent approximation, renormalizes the one-body coupling constant F_A. Thus, even though the matrix elements of the one- and two-body operators are difficult to calculate, the ratio of these matrix elements should be very insensitive to nuclear physics uncertainties. Thus an estimate of $< V_{PNC}^{\Delta I} >$ can be obtained from the measured[4] first-forbidden β decay rate and the calculated ratio of one- and two-body matrix elements. The more detailed arguments given in Ref. 2 indicate that this procedure is remarkably reliable.

This result permits one to extract from an experimental measurement of

P_γ (and thus $<V_{PNC}^{\Delta I=1}>$) a reliable value for the weak pion-nucleon coupling f_π. The beautiful experiments of the Queen's University/Princeton/Caltech[5] and Florence[6] groups, when combined with earlier measurements[7], yield $P_\gamma = (0.8 \pm 3.9) \times 10^{-4}$ and $|<V_{PNC}^{\Delta I=1}>| \leq 0.09$ eV. The corresponding limit on f_π is a factor of five smaller than the standard model best value calculated by Desplanques, Donoghue, and Holstein.[8]

This result is consistent with other measurements of PNC ($A_L(\vec{p}+p)$, $A_L(\vec{p}+{}^4\text{He})$, $A_L(\vec{p}+d)$, $A_\gamma({}^{19}\text{F})$), with the exception of the circular polarization measurement in ^{21}Ne. However, the interpretation of the ^{21}Ne result depends entirely on shell model estimates of the nuclear matrix element. Even if one accepts the ^{21}Ne constraint, a global fit[1] of weak meson-nucleon couplings to PNC observables indicates that the neutral current enhancement of f_π is less than that predicted by theory. Perhaps the isospin dependence of V_{PNC} will prove to be as surprising as the $\Delta I = 1/2$ rule in $\Delta S = 1$ interactions.

T-ODD NUCLEAR MOMENTS

The observation of CP-violation in neutral kaon decays[9], coupled with the CPT theorem, suggests elementary particles may have non-zero "odd" static moments (C1/E1, M2, C3/E3,...) forbidden by time reversal symmetry. Although no non-zero odd moments have been detected, the sensitive limit on the electric dipole moment (edm) of the neutron[10] ($d_n < 4 \cdot 10^{-25} e - cm$) has provided important constraints on theoretical models incorporating CP violation.

Recently very powerful techniques for measuring the edm's of neutral atoms have been developed, yielding, in the case of the ^{199}Hg experiment of Fortson et al., a limit of approximately $10^{-26} e - cm$.[11] The importance of atomic experiments as tests of the electron edm has long been appreciated. However, with the exquisite precision of recent measurements, these experiments are also emerging as a competitor to the neutron edm measurements in probing possible CP-violating interactions between quarks.

As a probe of nuclear edm's, atomic experiments must contend with Schiff's theorem: for a point nucleus, there is no term in the interaction energy of a neutral atom in an external field that is linear in the edm.[12] The perfect shielding limit is not realized in Nature due to the finite nuclear size: the interaction energy is proportional to the r^2-weighted moment of the difference between the normalized nuclear charge and edm distributions. Thus heavy nuclei are favored. There are also terms in the interaction energy that depend on the M2 nuclear moment[13,16] and on relativistic corrections to the electron-nucleus Coulomb interaction.[12]

Henley and I discussed the possibility that the sensitivity of atomic edm experiments to the parameters governing CP violation might, in favorable cases, be much greater than one would expect from single-particle estimates.[14] This suggestion was originally made by Feinberg[12], who pointed out that nearly degenerate ground-state opposite-parity doublets are found in certain rare earth and actinide nuclei. In such cases enhanced T-odd nuclear moments can result from the mixing of the doublet states by the CP-nonconserving NN interaction. We calculated the moments that would result if the QCD θ parameter were the dominant source of CP-violation. In this case the CP-violating NN interaction is due to single pion exchange, where one πNN vertex is CP-violating (and proportional to θ) and the other is governed by the strong interaction. The edm's due to this many-body effect are, in favorable cases, 100 times that due to the unpaired valence nucleon. In one extraordinary case, the 200 eV parity doublet in ^{229}Pa, the enhancement is 10^4. I will not discuss this work further in the written version of my talk because adequate references already exist.[15]

Similar work was done by Khriplovich[13,16] and his collaborators for the Kobayashi Maskawa CP-violating phase in the quark mass matrix. Khriplovich found that the CP-violating KM NN interaction was dominated by a kaon-exchange diagram. The KNN CP-violating vertex was estimated by evaluating a penguin diagram. (The πNN vertex for the θ parameter is fixed by current algebra.[17]) These calculations yield the surprising result that the KM NN interaction, for a fixed value of the neutron edm, is roughly two orders of magnitude stronger than the θ-parameter NN interaction. The calculations of Ref. 13 suggest that the Seattle result for $d_A(^{199}\text{Hg})$ may be about an order of magnitude more restrictive than the neutron edm measurement in limiting the size of the KM phase. Clearly this result will motivate careful re-examinations of the particle theory, nuclear theory, and atomic theory needed to interpret the atomic edm measurements.

MATTER-ENHANCED OSCILLATIONS OF SOLAR NEUTRINOS

Mikheyev and Smirnov[18] have discovered a mechanism by which electron neutrinos produced in the solar core could be efficiently converted into neutrinos of a different flavor. This mechanism depends on an effective density-dependent electron neutrino mass arising from the weak charged-current interactions with solar electrons, a phenomenon first discussed by Wolfenstein.[19] The Mikheyev-Smirnov-Wolfenstein mechanism provides a plausible particle physics solution to the solar neutrino puzzle: neutrino oscillations governed by small vacuum mixing angles can produce the needed factor-of-three suppression of the ^{37}Cl counting rate.[20,21]

Numerical studies of the effect of matter-enhanced oscillations on the counting rates of the ^{37}Cl and ^{71}Ga experiments have been carried out by Mikheyev and Smirnov[18], Rosen and Gelb[22], Hampel and Kirsten[23], and others. However, most of the physics underlying these detailed calculations can be understood in a simple model that exploits the analogy between the MSW mechanism and the phenomenon of adiabatic level crossing in atomic collisions. I would like to briefly describe this approach. More detailed discussions can be found in the papers of Bethe[24] and Haxton.[25]

In the adiabatic approximation one describes the evolution of the neutrino wave function in terms of the stationary eigenstates of a time-independent Hamiltonian evaluated for the appropriate instantaneous electron density. Consider the case of two-state mixing where the ν_e, in vacuum, is composed primarily of the lighter mass eigenstate and the "ν_μ" of the heavier. Then adiabatic propagation of the neutrino through the sun will lead to a large $\nu_e \to \nu_\mu$ amplitude provided the ν_e, when produced in the solar core, was initially the heavy eigenstate of the instantaneous core Hamiltonian. Such level crossing[24] occurs if

$$\beta \rho(t=0) > \cos 2\theta \tag{3}$$

where β and ρ are the dimensionless quantities $\beta = 2E/\delta m^2 R_s$ and $\rho(t) =$

$\sqrt{2}\, G_F\, \eta_e(t) R_s$, with E the neutrino energy, $\delta m^2 = m_2^2 - m_1^2 > 0$, R_s the sun's radius, and $\eta_e(t)$ the instantaneous electron density at time t; t is measured in units of R_s and thus runs from 0 to 1 for a ν_e produced at the sun's center. An additional constraint comes from the requirement of adiabatic propagation. For small vacuum mixing angles θ, the oscillation frequency moves through a pronounced minimum at a critical density that determines the level-crossing point

$$\beta\rho(t_c) \equiv \beta\rho_c = \cos 2\theta \tag{4}$$

The adiabatic condition relates the allowed mixing angles to the rate of change of the density

$$\tan^2 2\theta \gtrsim \frac{\sin^3 2\theta}{\left[(\beta\rho - \cos 2\theta)^2 + \sin^2 2\theta\right]^{3/2}} \left|\frac{1}{\rho_c^2}\frac{d\rho}{dt}\right| \tag{5}$$

a constraint that becomes particularly stringent at t_c

$$\gamma \equiv \frac{\tan^2 2\theta}{\left|\frac{1}{\rho^2}\frac{d\rho}{dt}\right|_{t_c}} \gtrsim 1 \tag{6}$$

The term adiabatic conversion will be used to describe neutrino oscillations when both the level-crossing condition (Eq. (3)) and the adiabatic condition (Eq. (6)) are satisfied.

The constraints imposed by Eqs. (3) and (6) are plotted in Fig. 2 as a function of the two relevant parameters, $\delta m^2/E$ and $\sin^2 2\theta$. Above the horizontal line no level crossing occurs: even in the solar core the electron neutrino is dominantly the lighter of the two instantaneous mass eigenstates. Below the diagonal line (defined by $\gamma = 1$) the propagation of the neutrino in the vicinity of the crossing point is nonadiabatic. The region of adiabatic conversion lies entirely inside these boundaries. (The adiabatic condition will be discussed more fully below.)

As Davis has measured a *nonzero* signal of 2.0 ± 0.3 SNU, some of the solar neutrinos that contribute to the ^{37}Cl counting rate are not converted to ν_μ. Bethe noted that Eq. (3) permits one to exclude low-energy neutrinos from adiabatic conversion for an appropriate choice of δm^2. Bethe found $\delta m^2 \sim (0.008\text{eV})^2$, corresponding to a critical neutrino energy $E_c^B \sim 6$ MeV, provides the correct suppression of the experimental signal. Eq. (6) then requires $\sin^2 2\theta \gtrsim 8 \cdot 10^{-4}$. This solution is one of two found in the numerical studies of the MSW mechanism. Bethe's analysis identifies this solution with the boundary in the $\delta m^2 - \sin^2 2\theta$ plane where the level-crossing requirement for adiabatic conversion fails for low-energy neutrinos.

The second solution found in the numerical studies was identified in Ref. 25 with the adiabatic boundary ($\gamma = 1$) in Fig. 2. For a fixed angle θ, the condition $\gamma = 1$ defines a density that I will denote by $\rho^{\gamma=1}(\theta)$. Now note that the denominator on the right-hand side of Eq. (6) is a monotonically increasing function of t, running from 0 at $t = 0$ to 1 at $t \sim 0.9$ for a neutrino produced at the sun's center. Thus, for fixed θ, Eq. (6) is satisfied by any δm^2 and E that lead to $\rho_c > \rho^{\gamma=1}(\theta)$; i.e.,

$$\frac{\delta m^2 R_s}{2E} > \frac{\rho^{\gamma=1}(\theta)}{\cos 2\theta}. \qquad (7)$$

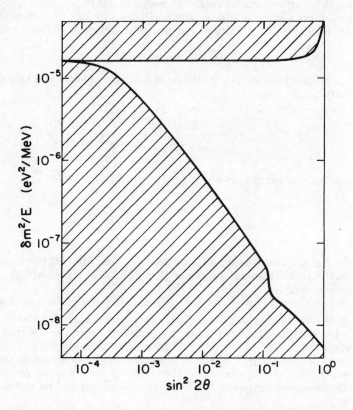

Figure 2. The shaded areas define those values of $\delta m^2/E$ and $\sin^2 2\theta$ that fail to satisfy the constraints imposed by Eq. (3) (upper region) and by the condition of adiabatic propagation (Eq. (6), lower region). The conversion of solar $\nu_e \to \nu_\mu$ is highly efficient in the unshaded region. Solutions consistent with Davis's experiment lie along the boundaries for an appropriate choice of an average E.

For our present "back-of-the-envelope" purposes we will assume that the condition $\gamma = 1$ defines a sharp boundary between the adiabatic and nonadiabatic regions. (This approximation will be improved below.) Thus, for fixed δm^2, neutrinos of energy E *above* a critical value will not satisfy Eq. (7) and thus not undergo adiabatic conversion. Davis's result is consistent with an absence of neutrinos with energies below 10.4 MeV. Thus a second solution to the solar neutrino problem is obtained from Eq. (7)

$$\delta m^2 = 5.9 \cdot 10^{-9} \text{eV}^2 \, \frac{\rho^{\gamma=1}(\theta)}{\cos 2\theta} \qquad (8)$$

so that δm^2 is determined as a function of θ. For the spectrum of neutrinos sampled by the ^{37}Cl detector, this solution connects smoothly to Bethe's solution at $\delta m^2 \sim (0.008 \text{ eV})^2$ and $\sin^2 2\theta \sim 8 \cdot 10^{-4}$, but runs to much smaller δm^2 and large mixing angles. Values satisfying Eq. (8) for a range of mixing angles are given in the third column of Table 1 and are very similar to the results of detailed numerical calculations. Thus solutions to the ^{37}Cl puzzle are uniquely defined as those δm^2 and $\sin^2 2\theta$ that lie along the two boundaries of the region of adiabatic conversion, an intuitively appealing result.

$\sin^2 2\theta$	$\rho_c^{\gamma=1}$	$\delta m^2 (\text{eV}^2)$	$\delta m_{LZ}^2 (\text{eV}^2)$	$\sin^2 2\theta$	$\rho_c^{\gamma=1}$	$\delta m^2 (\text{eV}^2)$	$\delta m_{LZ}^2 (\text{eV}^2)$
8E-4	1.1E4	6.4E-5	7.3E-5	3E-2	3.3E2	2.0E-6	9.8E-7
1E-3	9.8E3	5.5E-5	6.1E-5	6E-2	1.6E2	9.9E-7	3.2E-7
3E-3	3.6E3	2.1E-5	1.1E-5	1E-1	8.4E1	5.2E-7	1.8E-7
6E-3	1.9E3	1.1E-5	5.4E-6	3E-1	2.0E1	1.4E-7	8.6E-8
1E-2	1.1E3	6.6E-6	3.2E-6				

Table I. Values of δm^2, $\rho_c^{\gamma=1}(\theta)$, and $\sin^2 2\theta$ of the adiabatic-boundary solution for Eq. (8) (δm^2) and for the calculation that includes the width of the adiabatic boundary and presence of hybrid solutions for small $\sin^2 2\theta$ (δm_{LZ}^2). The right-hand side of Eq. (3) was taken from a smooth fit to the tabulated density profile of Ref. 21. $\rho_c^{\gamma=1}$ can be compared to the sun's central density, $\rho(0) = 2.8\text{E}4$. (EX = 10^x).

The level-crossing solution discussed by Bethe leads to a reduced flux of higher energy neutrinos and thus, in various radiochemical experiments, mimics the signal expected of a nonstandard solar model. The critical densities for this solution are achieved in the solar core. For the adiabatic-boundary solution the strongest suppression occurs (except for the small-angle hybrid solutions discussed below) for the low-energy neutrinos. Thus a distinctively low counting rate will result in the gallium experiment. The critical densities become progressively lower (i.e., the crossing point occurring nearer the surface) as one moves away from Bethe's solution to smaller δm^2.

It is rather easy to improve the simple arguments above to account for the fact that the boundaries represent transition regions. The behavior near the level-crossing boundary is governed by the overlap of ν_e with the heavy-mass eigenstate $\nu_H(t)$

$$<\nu_e|\nu_H(t)> = \frac{1}{\sqrt{2}} \left[1 \pm \frac{1}{\sqrt{1+\alpha^2}}\right]^{1/2} \quad (9)$$

where $\alpha(t) = \sin 2\theta/(\beta\rho(t) - \cos 2\theta)$ and the + (-) sign is taken for α positive (negative). At the level-crossing boundary $\alpha(t=0)$ is infinite. However, for small $\sin 2\theta$, modest variations in $\delta m^2/E \propto \beta^{-1}$ drive $\alpha(t=0)$ to small values and the overlap to ~ 0 (above the boundary) or to ~ 1 (below boundary). The representation of the transition region by a sharp boundary in $\delta m^2/E$ is thus a good approximation for small $\sin 2\theta$, and Bethe's E_c^B is well defined.

The behavior near the diagonal (adiabatic) boundary is determined by the probability of remaining on the heavy mass trajectory while crossing the critical point. This is given approximately by the Landau-Zener[26] factor $1 - e^{-\pi\gamma/2}$. This result, which is familiar from level-crossing problems in atomic physics, is derived by approximating the off-diagonal mixing matrix element (and therefore $n_e(t)$) by a linear function of t near the crossing point. The Landau-Zener factor is an excellent smooth approximation to the numerical boundary profiles given, effectively, as a function of $\delta m^2/E$ in Figs. 1 of the paper by Rosen and Gelb. A single parameter γ governs the physics of the adiabatic region, the boundary region, and the highly nonadiabatic region below the diagonal line in Fig. 2.

There is an explicit quadratic dependence of γ on $\delta m^2/E$ and an implicit dependence through $d\rho/dt_c$. This dependence is much more gentle than that of Eq. (9), so that the notion of sharp cutoff energy at 10.4 MeV is an oversimplification. However, by folding the neutrino spectrum with the appropriate Landau-Zener factors, the arguments leading to Eq. (3) can be generalized to account for the width of the adiabatic boundary. The results are given in Table 1 as δm_{LZ}^2. Near the intersection with Bethe's solution (i.e., for small $\sin^2 2\theta$) low-energy neutrinos (^{7}Be, pep, CNO) fail to satisfy the level-crossing condition and thus contribute substantially to the ^{37}Cl capture rate. Excluding these small $-\sin^2 2\theta$ hybrid solutions, one must reduce the naive coefficient in Eq. (8) from 5.9 to approximately 2.5 to reproduce on average the last column of Table 1.

Efficient adiabatic conversion of solar ν_e's to ν_μ's will occur throughout much of the unshaded region in Fig. (1). One quickly realizes the implications of the MSW mechanism in view of the *nonzero capture rate* for the ^{37}Cl experiment. Davis's result excludes, in the case of two-state mixing, a substantial portion of the $\sin^2 2\theta - \delta m^2$ plane corresponding, roughly, to the region bounded by the two solutions to the ^{37}Cl puzzle and by the line $\sin^2 2\theta \sim 0.4$ (a requirement imposed by Eq. (9) as $\rho \to 0$). The excluded masses and mixing angles are totally unexplored in terrestrial experiments.

Clearly the proposed Ga experiment will provide a crucial test of neutrino physics. The Ga detector is sensitive to a flux that is distinctively solar, the low energy pp neutrinos (0.42 MeV endpoint) produced copiously in the ppI cycle. The predicted standard-solar-model capture rate is 122 SNU. The minimum astronomical rate consistent with the assumption of steady-state hydrogen burning is 78 SNU.[27] For $\sin^2 2\theta \gtrsim 10^{-2}$ the adiabatic-boundary solution yields a very suppressed pp neutrino flux. Counting rates far below the minimum astronomical value would result, providing a compelling argument for neutrino masses and mixing. As one continues along this solution to somewhat smaller angles so that the pp neutrinos begin to cross the level-crossing boundary in Fig. 2, the Ga counting rate increases. These hybrid solutions are special in that both $\sin^2 2\theta$ and δm^2 could be determined from the results of the Ga and Cl experiments. For still smaller values of $\sin^2 2\theta$ the entire pp flux lies above the region of adiabatic conversion, guaranteeing that the Ga counting rate exceeds the minimum astronomical value. As in the case of Bethe's solution, an experiment capable of measuring the spectrum of ^{8}B neutrinos would be needed to distinguish oscillations from nonstandard solar physics.

Any substantial counting rate in the Ga experiment will unambiguously rule out a large region in δm^2 and $\sin^2 2\theta$ where adiabatic conversion of the pp neutrinos must take place. This region is similar in size and shape to that tested by the ^{37}Cl experiment but is shifted, as the effective E in $\delta m^2/E$ is about an order of magnitude smaller, to lower δm^2. Adiabatic conversion for δm^2 as small as 10^{-8} eV^2 can take place. For those larger δm^2 that are also tested in the ^{37}Cl experiment, the adiabatic region for the Ga experiment will extend to somewhat smaller mixing angles.

I thank E. Adelberger, H. Bethe, E. Fischbach, N. Fortson, B. Heckel, S. Koonin, F. Raab, S.P. Rosen, and L. Wilets for helpful discussions. This work was supported in part by the U.S. Department of Energy.

REFERENCES

1. E.G. Adelberger and W.C. Haxton, Ann. Rev. Nucl. Part. Sci. 35, 501 (1985).
2. W.C. Haxton, Phys. Rev. Lett. 46, 698 (1981).
3. C. Bennett, M.M. Lowry, and K. Krien, Bull. Am. Phys. Soc. 25, 486 (1980).
4. E.G. Adelberger, M.M. Hindi, C.D. Hoyle, H.E. Swanson, R.D. Von Lintig, and W.C. Haxton, Phys. Rev. C 27, 2833 (1983).
5. H.C. Evans et al., Phys. Rev. Lett. 55, 791 (1985).
6. M. Bini, T.F. Fazzini, G. Poggi, and N. Taccetti, Phys. Rev. Lett. 55, 795 (1985).
7. C.A. Barnes et al., Phys. Rev. Lett. 40, 840 (1978); P.G. Bizetti, T.F. Fazzini, P.R. Maurenzig, A. Perego, and G. Poggi, Lett. Nuovo Cimento 29, 167 (1980); G. Ahrens, W. Harfst, J.R. Kass, E.V. Mason, and M. Schober, Nucl. Phy. A 390, 486 (1982).
8. B. Desplanques, J.F. Donoghue, and B.R. Holstein, Ann. Phys. (NY) 124, 449 (1980).
9. J.H. Christenson, J.W. Cronin, V.L. Fitch, and R. Turley, Phys. Rev. Lett. 13, 138 (1964).
10. I.S. Alterev et al., Phys. Lett. 102B, 13 (1981); N.F. Ramsey, Rep. Prog. Phys. 45, 95 (1982). The present Grenoble experiment may improve this limit by an order of magnitude (B. Heckel, private communication).
11. E.N. Fortson, B. Heckel, J. Jacobs, S. Lamoreaux, F.J. Raab, and T.G. Vold, to be published; T.G. Vold, F.J. Raab, B. Heckel, and E.N. Fortson, Phys. Rev. Lett. 52, 2229 (1984).
12. L.I. Schiff, Phys. Rev. 132, 2194 (1963); G. Feinberg, Trans. N.Y. Acad. sci. 38, 6 (1977).
13. V.V. Flambaum, I.B. Khriplovich, and O.P. Sushkov, Zh. Eksp. Teor. Fiz. (JETP) (1984).
14. W.C. Haxton and E.M. Henley, Phys. Rev. Lett. 51, 1937 (1983).
15. In addition to Ref. 14, see W.C.Haxton, in *Nuclear Shell Models*, ed. M. Vallieres and B.H. Wildenthal, p. 471 (World Scientific, 1985).
16. I.B. Khriplovich, Zh. Eksp. Teor. Fiz. 71, 51 (1976) (JETP 44, 25 (1976)).
17. R.J. Crewther, P. DiVecchia, G. Veneziano, and E. Witten, Phys. Lett. 88B, 123 (1979) and 91B, 487 (1980).

18. S.P. Mikheyev and A. Yu. Smirnov, Yad. Fiz. 42, 1441 (1985).
19. L. Wolfenstein, Phys. Rev. D 16, 2369 (1978).
20. J.N. Bahcall, B.T. Cleveland, R. Davis, Jr., and J.K. Rowley, Astrophys. J. 292, 279 (1985).
21. J.N. Bahcall, W.F. Huebner, S.H. Lubow, P.D. Parker, and R.K. Ulrich, Rev. Mod. Phys. 54, 767 (1982).
22. S.P. Rosen and J. Gelb, submitted to Phys. Rev. D (1986).
23. T. Kirsten, talk presented at the Int. Symp. on Beta Decay and Neutrino Masses, Osaka (1986).
24. H. Bethe, Phys. Rev. Lett. 56, 1305 (1986).
25. W.C. Haxton, submitted to Phys. Rev. Lett. (1986).
26. L.D. Landau, Physik. Z. Sowjetunion 2, 46 (1932); C. Zener, Proc. Roy. Soc. A137, 696 (1932).
27. J.N. Bahcall, in *Solar Neutrinos and Neutrino Astronomy*, ed. M.L. Cherry, W.A. Fowler, and K. Lande, AIP Conf. Proc. No. 126, p. 60 (New York, 1985).

ARE NUCLEONS AND DELTAS DEFORMED?*

M. Bourdeau, R. Davidson and Nimai C. Mukhopadhyay+
Department of Physics, Rensselaer Polytechnic Institute
Troy, New York 12181

ABSTRACT

We investigate the extraction and modelling of the electromagnetic multipoles in the transition $N \to \Delta(1232)$. We show that the extracted E2 multipole amplitude indicates that nucleons and deltas are not spherical. We study these deformations in various models and discuss possible experiments to test these theoretical inferences.

The question as to whether nucleons and deltas are deformed has been with us for more than twenty years[1]. In the simple SU(6) and SU(6)w models, as well as in the original MIT bag model, they are spherical and the electric quadrupole (E2) transition amplitude between these states is identically zero. Thus, any significant departures from this vanishing value would indicate departures from sphericity. In the modern versions[2] of the quark shell models, and in the soliton (Skyrmion) models[3] of baryons, deformed hadrons are distinct possibilities. We shall study here implications of various quark shell models on the size of the transition E2, longitudinal/scalar quadrupole amplitudes as functions of photon mass squared. These could be tested through future experiments at the new electron machines, like CEBAF.

We first note that our analysis[4] of the available data on the magnetic dipole (M1) and E2 amplitudes for the photoproduction of pions around the delta region, in the framework of a phenomenological $\gamma N \Delta$ interaction and background, together with constraint of unitarity via Watson's theorem has yielded the E2 and M1 resonance amplitude ratio (EMR) to be $(-1.5 \pm 0.2)\%$. While this number seems small, this is considerably larger than the prediction of the Isgur-Karl shell model (IK). It is the same order of magnitude as indicated by the soliton models[3]. This answers our question posed in the title of this contribution in the affirmative.

We can now use the quark shell models to investigate structures of these hadrons further. In the model of Glashow-Vento-Baym-Jackson[2] (GVBJ), the nucleon and delta wave functions are

---

*Contribution presented at the Second Conference on the Interaction between Particle and Nuclear Physics; Chateau Lake Louise, Canada (Session: Hadron Spectroscopy; May 27, 1986.)
+Supported by the U.S. Department of Energy (Contract #DE-AC02-83ER40114-A003).

$$|N\rangle = \sqrt{1-\alpha} \, |S\rangle_s + \sqrt{\alpha} \, |D\rangle_M, \qquad (1)$$

$$|\Delta\rangle = \sqrt{1-3\beta} \, |S\rangle_s + \sqrt{2\beta} \, |D\rangle_s - \sqrt{\beta} \, |D\rangle_M, \qquad (2)$$

where α is fixed from the nucleon axial-vector coupling constant to be ~22%. We can use the extracted EMR (or just E2 amplitude) to fix β. This comes to be 35%. Note that there is an implicit sign ambiguity in (2), which could yield a different value of β.

In the IK model, the $|N\rangle$ and $|\Delta\rangle$ are given in terms of a larger basis, with the coefficients of the wave functions determined from the diagonalization of the quark effective hamiltonian. The EMR calculated at the photon point is obtained to −0.31%, about a fifth of the "experimental" value.

At this point, we recognize that the E2 amplitude for <u>real</u> photon transtion $N \leftrightarrow \Delta$ can be calculated by two different ways. The usual "current" approach, used above, or by a different "charge" approach[5], by which one calculates S2, the scalar quadrupole multipole amplitude; this is equal to the E2 one at real photon point. Using this approach, the IK wave function yield EMR = −0.6%, while the GVBJ model gives 3.2%, with a wrong sign! This difference persists if we calculate the longitudinal quadrupole multipole L2 by the "current" way and by "charge" way: $L2 = \frac{k_o}{k} S2$, k_o energy, k the momentum for photons. Future experiments using polarized electron and nucleons should be able to measure this multipole.

Thus, the discrepancy of the IK model with the inferred EMR at photon point may not be as serious as it looks in the "current" approach.

1. C. Becchi and G. Morpurgo, Phys. Lett. <u>17</u>, 352 (1965).

2. N. Isgur and G. Karl, Phys. Rev. D<u>18</u>; 4187 (1975); S.L. Glashow, Physica <u>96A</u>, 27 (1979); V. Vento, G. Baym and A.D. Jackson, Phys. Lett. <u>102B</u>, 97 (1981).

3. C. Nappi, Lewes Conf. (1985).

4. R. Davidson, N.C. Mukhopadhyay and R. Wittman, Phys. Rev. Lett. <u>56</u>, 804 (1986).

5. D. Drechsel and M.M. Giannini, Phys. Lett. <u>143B</u>, 329 (1984).

SPIN OBSERVABLES AT INTERMEDIATE ENERGIES:
A TOOL IN VIEWING THE NUCLEUS

J. B. McClelland
Los Alamos National Laboratory, Los Alamos, NM   85745

In this paper I attempt to summarize some of the advances made in intermediate nuclear physics through measurements of spin observables, notably in the range of bombarding energies from 100 to 1000 MeV. I leave the discussion of the important nucleon-nucleon (NN) measurements to other speakers. Relative to measurements of cross section, spin observables offer a highly selective filter in viewing the nucleus. Their general utility is found in their sensitivity to particular nuclear transitions and is further augmented by their simple connections to the NN force. The advantage of higher energies is apparent from the dominance of single-step mechanisms even at large energy losses where general nuclear spin responses may be made. Experimentally, this is an energy range where efficient, high-analyzing-power polarimeters can be coupled with high resolution detection techniques.[1]

The first experiment to measure a complete set of spin observables for the elastic scattering of protons from a nucleus[2] provided the impetus for a Dirac description of the scattering process.[3] An apparent failure of the nonrelativistic KMT treatment of intermediate-energy proton elastic scattering data for cross sections and, most noticeably, analyzing powers had already been extensively investigated looking at numerous corrections in order to resolve the discrepancies. Furthermore, it was believed that the data were driven by the geometries of the nucleus such that the third independent observable for elastic scattering, the spin-rotation parameter, Q, would be predicted from the other two, cross section and analyzing power. The data for Q turned out to be in total disagreement with this prediction and not explained by the standard KMT analysis. Predictions of Q using the Dirac phenomenology, however, provided excellent agreement with the data.

As can be seen in Fig. 1, only small differences in the cross section are seen between a more recent relativistic impulse approximation[9] (solid curve) and nonrelativistic impulse approximation (dashed curve) predictions, whereas the analyzing power (or polarization P) and the spin rotation parameter (Q) are both qualitatively and quantitatively different. The underlying physics is quite different. The Dirac approach includes processes such as virtual pair production and annihilation in the field of the nucleus not present in nonrelativistic dynamics. The 500-MeV data markedly favor the Dirac treatment. It should be pointed out, however, that spin rotation data at other energies and on other targets are not in as good agreement, but it is precisely these type of data that are likely to shed light on this issue.

A great deal of effort has gone into planning a spin-transfer experiment from a polarized nuclear target (^{13}C) where the relativity of the target may be tested in hope of finding identifiable differences in the nonrelativistic and relativistic approaches to nuclear physics.

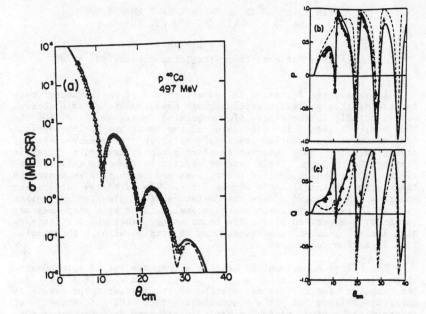

Fig. 1. ^{40}Ca(p,p) scattering at 500 MeV with relativistic (solid curve) and nonrelativistic calculations for cross section, analyzing power, and spin rotation parameter from Ref. 4.

A direct connection can be made between spin observables and the squared moduli of the coefficients of the effective NN scattering amplitude given by

$$M(q) = A + B\sigma_{1n}\sigma_{2n} + C(\sigma_{1n} + \sigma_{2n}) + E\sigma_{1q}\sigma_{2q} + F\sigma_{1p}\sigma_{2p} , \quad (1)$$

where 1(2) denotes the target (projectile) nucleon and the unit vectors $(\hat{n},\hat{q},\hat{p})$ are in the $\vec{K}\times\vec{K}'$, $\vec{K}-\vec{K}'$, and $\hat{q}\times\hat{n}$ directions, with K(K') the relative momentum in the NN system before(after) collision. For unnatural parity transitions, it has been shown[5,6] that in the static limit

$$I_o = (C^2 + B^2 + F^2)X_T^2 + E^2 X_L^2 , \quad (2.1)$$

$$I_o D_{nn} = (C^2 + B^2 - F^2)X_T^2 - E^2 X_L^2 , \quad (2.2)$$

$$I_o D_{pp} = (C^2 - B^2 + F^2)X_T^2 - E^2 X_L^2 , \quad (2.3)$$

$$I_o D_{qq} = (C^2 - B^2 - F^2)X_T^2 + E^2 X_L^2 , \quad (2.4)$$

$$I_o D_{no} = I_o D_{on} = 2X_T^2 \text{Re}(BC^*) , \quad (2.5)$$

$$I_o D_{qp} = -I_o D_{pq} = 2X_T^2 \text{Im}(BC^*) . \quad (2.6)$$

where $X_L^2(X_T^2)$ is the static longitudinal(transverse) form factor. One can see from Eq. 2 that if the nuclear structure is known (i.e. X_L^2 and X_T^2), the q dependence of the effective NN interaction may be mapped out by measuring a complete set of spin observables to discrete final states at several momentum transfers. Although Eqs. 2 are strictly valid in the plane wave impulse approximation (PWIA), full distorted wave (DWIA) calculations have shown that distortion or details of the transition density have little effect on the spin observables for a transition dominated by a single multipolarity near the peak of the associated form factor. Thus, Eqs. 2 are expected to still be valid under these conditions.

The first complete set of spin observables at intermediate energy for inelastic scattering used the two lowest 1^+ states in ^{12}C at 500 MeV to map the q dependence of the individual coefficients of the NN spin-dependent interaction for both isospin channels.[7] The results were consistent with the free NN amplitudes. Further measurements are needed to improve the accuracy of these results as well as extending them to larger q by choosing states of higher multipolarity. In principle one can be divorced from uncertainties in nuclear structure by doing similar measurements in quasi-free scattering.[8] It is no longer possible to make the isospin decomposition in (p,p') directly, but similar measurements will soon be possible in the (p,n) reaction, which is purely isovector in nature.[9] The combination of (p,p') and (p,n) would be complimentary and both would require only modest energy resolution.

Spin observables have also been shown to be more sensitive to convection ($\vec{J}$) and composite ($\vec{J} \times \vec{\sigma}$) currents than unpolarized cross sections alone.[10] Observables such as $\sigma(P-A)$ and $\sigma(D_{ls}+D_{sl})$ have been found to be most useful in detecting and confronting composite currents. Nonrelativistic and relativistic theories all contain these currents at some level of approximation, although the relativistic treatment gives rise to these currents in a more natural way through the lower component.

As an example of the selectivity and sensitivity of spin observables to particular nuclear transitions, consider Fig. 2, which is the spectrum of inelastic states in ^{12}C at 397 MeV from 7 to 23 MeV in excitation. This is seen in the top portion of the figure. The spectrum is dominated by the natural parity $\Delta S=0$ transitions at 7.6 and 9.6 MeV. General symmetry properties of the scattering amplitude imply that for transitions involving spin-parity transfer of $J^\pi=0^-$, $D_{NN}=-1$, and for transitions involving $J^\pi=0^+$, $D_{NN}=+1$. In general a positive value of D_{NN} is a signature of $\Delta S=0$ strength, while $\Delta S=1$ transitions yield a negative or zero value of D_{NN}. This is seen directly in the bottom portion of Fig. 2 for the spin-flip cross section, $d\sigma/d\Omega \cdot S_{NN}$, where $S_{NN}=(1-D_{NN})/2$ is the transverse spin-flip probability. The natural parity $\Delta S=0$ transitions in the top spectrum are completely suppressed in the spin-flip cross section. Only the unnatural parity $\Delta S=1$; 1^+ and 2^- and natural parity $\Delta S=1$; 2^+ states persist.

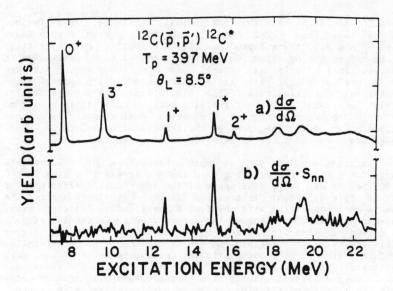

Fig. 2. ^{12}C(p,p) scattering at 397 MeV showing yield spectrum (top) and spin-flip cross section (bottom) from Ref. 28.

This selective technique of picking out only the spin-flip strength has been applied to continuum studies using the (p,p') and (p,n) reactions, in particular in the investigation of the Gamow-Teller (GT) and closely related M1 missing strength problem. The proportionality between intermediate energy (p,n) 0 cross section and beta decay transition strengths has provided a direct means of measuring GT strength in a wide range of nuclei.[11,12] The surprising feature, of course, has been the apparent lack of GT strength. Less than two thirds of the predicted strength based on an essentially model-independent sum rule has systematically been observed.[13] Explanations of this effect range from conventional nuclear mixing to delta-isobar admixtures to the nuclear wave functions. Resolution of this problem seems to rest with experiments that are sensitive to thinly distributed GT strength in the continuum. Cross-section angular distributions are primarily sensitive to the orbital angular momentum transfer ΔL rather than the total angular momentum transfer ΔJ. D_{NN} is sensitive to the spin transfer ΔS, hence it provides information on $\Delta J = \Delta L + \Delta S$. Its simple prediction and interpretation make it a powerful tool for these types of investigation. Figure 3 shows a ^{90}Zr(p,n) cross section and polarization transfer cross section at 160 MeV. The 0^+ ($\Delta L=0, \Delta S=0$) isobaric analog state (IAS) is seen with its corresponding $D_{NN}=1$. Virtually all of the remaining cross section corresponds to $\Delta S=1$ transitions as evidenced by its $D_{NN} \leq 0$. The observed value for D_{NN} in the region of the giant GT resonances further demonstrates that it is essentially all GT in nature, without other contributions. RPA calculations have been done for ^{90}Zr(p,n) cross section at

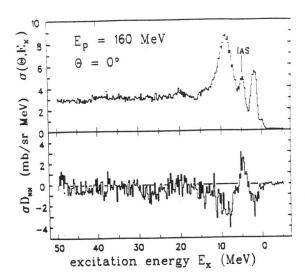

Fig. 3. ^{90}Zr(p,n) reaction at 160 MeV showing cross section (top) and polarization transfer cross section (bottom) from Ref. 29.

200 MeV[14,15] and have found no evidence for a need to include quenching due to delta-isobars with the proviso that the S_{β^+} strength obtained from the (n,p) reaction (as yet unmeasured) is small. However, this same calculation predicts only a small amount of 1^- and 2^+ natural parity strength in the 0° cross section in contradiction of the (p,n) results at 160 MeV. Further analysis of these types of spin transfer measurements as well as their angular distributions is certainly needed, as well as (n,p) and higher energy data where the delta region can be investigated directly.[9]

A similar program exists in (p,p') addressing the question of M1 giant resonances.[16] These resonances have been reportedly seen in (p,p') spectra on a variety of medium-to-heavy nuclei.[17] With only cross-section data available, the assignment of M1 is based primarily on the characteristic ΔL=0 angular distribution and on the centroid and width of the distributions. These giant resonances are not systematically observed in back-angle electron scattering, presumably sensitive to M1 strength. If these giant resonances are indeed M1 in nature, they should exhibit large ΔS=1 strength over their region of excitation. Spin-flip cross sections appear to be an ideal tool for resolving this controversy.

Equation 2 may also be viewed as a way of getting nuclear structure information if the effective interaction is known. An investigation of the nuclear continuum using polarized protons [18,19] uses this approach to relate high energy (200 GeV) deep inelastic lepton scattering (DILS) to nucleon scattering at intermediate energies (500 MeV). The connection between the two is through explanations of the European Muon Collaboration (EMC) effect and their implications for inclusive proton scattering. The EMC effect is the apparent modification of the nucleon structure function, or quark distribution, when the nucleon is embedded in the nuclear medium. This is reflected as an enhancement of the ratio of F_2 structure functions for a heavy nucleus compared to deuterium as measured in DILS at small values of the scaling variable x--scattering from the sea quarks--and a corresponding depletion at large x--scattering from the valence quarks. Most explanations of this effect either invoke a dynamic rescaling or increased confinement size of a nucleon within the nucleus or a more conventional approach involving enhancement of the pion field within the nucleus. The second has a certain intuitive appeal since each nucleon is surrounded by a cloud of pions contributing to DILS (as a $q\bar{q}$ pair of sea quarks). When embedded in the nuclear medium, the net number of pions per nucleon may increase, either due simply to exchange of pions with neighboring nucleons providing binding to the system, or by a nuclear many-body enhancement of the πNN vertex through an attractive NN interaction. It has been further suggested that dynamic rescaling simply mimics the pionic effects.

In both scattering processes the enhancement of the πNN vertex arises in the same way, and any explanation of the EMC effect invoking enhanced pion fields within the nucleus must confront the lower-energy hadron scattering data.

One model of the pionic enhancement uses the spin-isospin responses of Alberico, Ericson, and Molinari (AEM)[20,21] which is calculable for quasi-free scattering in infinite nuclear matter. The separation between spin-longitudinal and spin-transverse provides additional selectivity to the $\vec{\sigma} \cdot \vec{q}$ and $\vec{\sigma} \times \vec{q}$ parts of the residual interaction given by

$$V_L^{res}(q,\omega) \sim \frac{f_\pi^2}{m_\pi^2} \left[g' - \frac{q^2}{q^2 + m_\pi^2 - \omega^2} \right] \quad (3.1)$$

$$V_T^{res}(q,\omega) \sim \frac{f_\pi^2}{m_\pi^2} \left[g' - \left(\frac{f_\rho^2/m_\rho^2}{f_\pi^2/m_\pi^2}\right) \frac{q^2}{q^2 + m_\rho^2 - \omega^2} \right] \quad (3.2)$$

(to within vertex form factors). Since $m_\rho \approx 5.5 m_\pi$, these two pieces of the interaction have very different q-dependences as seen in Fig. 4a along with the corresponding response functions (4b) for $g'=.7$ at 1.75 fm^{-1}, the momentum transfer corresponding to the largest enhancement in this model. Figure 4(c) is the ratio of spin-longitudinal to spin-transverse response functions in this model. It is this proposed attractive behavior of V_L^{res} which enhance the pion field in the nucleus giving rise to the EMC effect.[22] It should be noted that other models of V^{res} do not exhibit this attractive behavior[23] and hence would not predict any excess pions.

(a)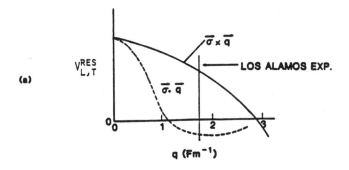

(b)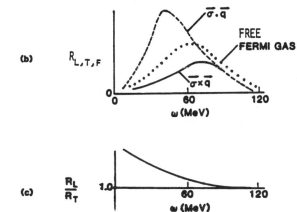

(c)

Fig. 4. (a) Longitudinal and transverse particle-hole interaction in the AEM model described in the text. (b) Response functions at $q = 1.75 \text{fm}^{-1}$ for interactions shown above. (c) Ratio of longitudinal to transverse response functions.

The proton experiment consists of precise determination of the complete set of polarization-transfer coefficients for 500 MeV inclusive scattering from Pb, Ca, and ^{2}H at a momentum transfer of $q = 1.75 \text{fm}^{-1}$. The spin-longitudinal and spin-transverse spin-flip probabilities are constructed using

$$IS_L = 1/4 \, [1 - D_{NN} + (D_{SS} - D_{LL})\sec\theta_{lab}] \, , \qquad (4.1)$$

$$IS_T = 1/4 \, [1 - D_{NN} + (D_{SS} - D_{LL})\sec\theta_{lab}] \, . \qquad (4.2)$$

For free NN scattering these combinations will isolate pure spin-longitudinal and spin-transverse couplings of the two nucleon system. For intermediate energy nucleon-nucleus interactions the following prescription is used:

$$IS_L = I^{NN} S_L^{NN} R_L(q,\omega) N_e \quad , \qquad (5.1)$$

$$IS_T = I^{NN} S_T^{NN} R_T(q,\omega) N_e \quad , \qquad (5.2)$$

$$I = I^{NN} R(q,\omega) N_e \quad , \qquad (5.3)$$

where NN refers to the nucleon-nucleon values. N_e is the effective number of participating nucleons. The spin-longitudinal, transverse, and total response functions per nucleon in the A-nucleon system are defined as

$$R_L(q,\omega) = |<q,\omega| f(\vec{r}) \vec{\sigma} \cdot \vec{q} e^{i\vec{q} \cdot \vec{r}} |o>|^2 \quad , \qquad (6.1)$$

$$R_T(q,\omega) = |<q,\omega| f(\vec{r}) \vec{\sigma} \times \vec{q} e^{i\vec{q} \cdot \vec{r}} |o>|^2 \quad , \qquad (6.2)$$

$$R(q,\omega) = \frac{C^2 + B^2 + F^2}{I^{NN}} R_T + \frac{E^2}{I^{NN}} R_L + \frac{A^2 + C^2}{I^{NN}} R_0 \quad , \qquad (6.3)$$

It should be noted that R_L is new nuclear structure information not available in (e,e') or (π,π') scattering. Taking the same approach as in the EMC experiment, the isospin-averaged values for S_L^{NN} and S_T^{NN} were experimentally determined by quasi-free scattering from ^{2}H in order to eliminate any uncertainties in the phase shift values of these quantities. From these data the ratio of $R_L(q,\omega)/R_T(q,\omega)$ was extracted by

$$S_L^{Pb}/<S_L^D> = R_L(q,\omega)/R(q,\omega) \quad , \qquad (7.1)$$

$$S_T^{Pb}/<S_T^D> = R_T(q,\omega)/R(q,\omega) \quad , \qquad (7.2)$$

where A refers to the heavy target, D to the deuterium data, and < > implies an average over ω.

Even at the level of the individual spin-flip probabilities, S_L and S_T for the heavy target show no difference from ^{2}H, and hence no effect is seen in in the ratio of responses derived from them. (Recall that the response functions are normalized to unity for free NN scattering).

Several corrections must be made to the proton data, however, before the level of sensitivity to the predicted enhancement of R_L can be determined. It is expected that a density correction must be made to account for the surface peaking of protons scattering from the nucleus at these energies. Two methods were employed, a local Fermi gas approximation where the interaction profile is calculated using a detailed Intranuclear Cascade code[24] and the Semi-Infinite Slab model,[25] which accounts well for medium-energy p-nucleus continuum data.[26] Both yield essentially identical results for the ratio of R_L/R_T. It also demonstrates that the Ca data provide as good a density profile as Pb at these energies.

Secondly, the calculated ratio is purely isovector. Corrections must be made for the mixed isospin contributions for (p,p') scattering. This is accomplished using the isospin decomposition of the NN interaction from the 500-MeV phase-shift solution of Arndt. The results for $q = 1.75\text{fm}^{-1}$ are (in terms of the coefficients in Eq. 1)

$$E^2_{T=1}/E^2_{T=0} = 3.62 \quad (8.1)$$

$$E^2_{T=1}/E^2_{T=0} = 1.15 \quad (8.2)$$

The longitudinal interaction is dominantly isovector but the transverse consists of nearly equal mixtures of both isospins. We define

$$\tilde{R}_L = \frac{1}{4.62}(3.62 R_L^{T=1} + R_L^{T=0}) \quad (9.1)$$

and

$$\tilde{R}_T = \frac{1}{2.15}(1.15 R_T^{T=1} + R_T^{T=0}) \quad (9.2)$$

and all isoscalar responses are assumed to be the free, non-interacting functions. The calculated ratios of $\tilde{R}_L/\tilde{R}_T$ are shown in Fig. 5 along with the data for the quasi-free experiment. The different curves represent different values of g'. Recent analysis of the EMC data requires $g' = 0.55$ in order to fit the low-x region,[27] in disagreement with the proton data. In fact the data

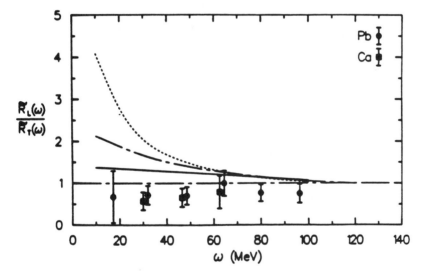

Fig. 5. Comparison of theory and proton scattering data for the ratio $\tilde{R}_L/\tilde{R}_T$. Calculations are for values of $g' = 0.55$ (dotted), $g' = 0.7$ (dot-dashed), and $g' = 0.9$ (solid).

favor a large value of $g' = 0.9$ at this momentum transfer. It should be noted that most of our knowledge of g' comes from $q = 0$ and the q-dependence is essentially undetermined.

Many other possible sources of "theoretical error" have been investigated. These include verifying the validity of the approximations at small ω, coupling of longitudinal and transverse modes in a finite nucleus, distortion effects, and differential range effects for one-π and one-ρ exchange potentials. These are detailed in Ref. 19. None of these effects seem to account for the lack of enhancement in the proton data predicted by those pion models used to explain the EMC effect. Such comparisons could not be made, however, without spin-transfer measurements.

It seems clear that spin observables at intermediate energies will provide the required detailed information needed to address important fundamental questions in nuclear physics today. These programs are relatively new, but have already made significant impact. Many laboratories are now pursuing them with great vigor and enthusiasm.

The work and ideas presented here include those of many of my colleagues and collaborators. I would especially like to acknowledge helpful discussions with Joel Moss, Tom Carey, Terry Taddeucci, and Charles Glashausser.

REFERENCES

1. J. B. McClelland et al., "A Polarimeter for Analyzing Nuclear States in Proton-Nucleus Reactions Between 200 and 800 MeV," Los Alamos National Laboratory Report No; LA-UR-84-1671 (May 1984); T. N. Taddeucci, et al., Nucl. Instrum. Methods A241, 448 (1985).
2. A. Rahbar et. al., Phys. Rev. Lett. 47, 1811 (1981).
3. Proceedings of the LAMPF Workshop on Dirac Approaches to Nuclear Physics, Los Alamos National Laboratory document LA-10438-C (May 1985).
4. S. J. Wallace, "Development of the Relativistic Impulse Approximation," Proceedings of the LAMPF Workshop on Dirac Approaches to Nuclear Physics, Los Alamos National Laboratory document LA-10438-C (May 1985).
5. J. M. Moss, Phys. Rev. C26, 727 (1982).
6. E. Bleszynski, M. Bleszynski, and C. A. Whitten, Jr., Phys. Rev. C26, 2063 (1982).
7. J. B. McClelland et. al., Phys. Rev. Lett. 52, 98 (1984).
8. C. Horowitz and M. Iqbal, "Relativistic Effects on Spin-Observables in Quasielastic Proton Scattering," to be published.
9. J. B. McClelland, et al., "Development Plan for the Nucleon Physics Laboratory Facility at LAMPF," Los Alamos National Laboratory report LA-10278-MS (February 1986).
10. W. G. Love and Amir Klein, "Nuclear Currents in Inelastic Scattering: Relativistic and Nonrelativistic Approaches," Proceedings of the LAMPF Workshop on Dirac Approaches to Nuclear Physics, Los Alamos National Laboratory document LA-10438-C (May 1985); D. A. Sparrow, et al., Phys. Rev. Lett. 54, 2207 (1985).

11. C. D. Goodman et. al., Phys. Rev. Lett. 44, 1755 (1980).
12. T. N. Taddeucci et. al., Phys. Rev. C25, 1094 (1982).
13. C. D. Goodman and S. D. Bloom, in "Spin Excitations in Nuclei," editied by F. Petrovich (Plenum, New York); C. Gaarde, J. S. Larsen, and J. Rapaport, ibid.; J. Rapaport, in The Interaction Between Medium Energy Nucleons in Nuclei-1982, edited by H. O. Meyer, AIP Conference Proceedings No. 97 (American Institute of Physics, New York, 1983).
14. F. Osterfeld, D. Cha, and J. Speth, Phys. Rev. C31, 372 (1985).
15. Amir Klein, W. G. Love, and N. Auerbach, Phys. Rev. C31, 710 (1985).
16. C. Glashausser, Comments Nucl. Part. Phys. 14, 39 (1985).
17. See, for example, C. Djalali, Proceedings Int. Conf. Highly Excited States and Nucl. Struc., Orsay, 1983 (Editions du Physique, Paris, 1984),; and A. Richter, Nucl. Phys. A374, 177c (1982).
18. Los Alamos Experiment #741, T. A. Carey, K. W. Jones, J. B. McClelland, J. M. Moss, L. B. Rees, N. Tanaka, and A. D. Bacher (1984).
19. T. A. Carey et. al., Phys. Rev. Lett. 53, 144 (1984); L. B. Rees et. al., "Continuum Polarization Transfer in 500 MeV Proton Scattering and Pionic Collectivity in Nuclei," Los Alamos National Laboratory report LA-UR-86-35 (December 1985), accepted for publication Phys. Rev. C.
20. W. M. Alberico, M. Ericson, and A. Molinari, Nucl. Phys. A379, 429 (1982).
21. W. M. Alberico, M. Ericson, and A. Molinari, Phys. Rev. C30, 1776 (1984).
22. M. Ericson and A. W. Thomas, Phys. Lett. 128B, 112 (1983).
23. J. Speth et. al., Nucl. Phys. A343, 382 (1980); M. Rho, Annu. Rev. Nucl. Part. Phys. 34 (1984); G. E. Brown.
24. Calculations performed with the Los Alamos version of ISABEL by Y. Yariv and Z. Fraenkel.
25. G. F. Bertsch and O. Scholten, Ann. Phys. 157, 255 (1984).
26. J. M. Moss et. al., Phys. Rev. Lett. 48, 789 (1982).
27. D. Stump, G. F. Bertsch, and J. Pumplin in "Hadron Substructure in Nuclear Physics," American Institute of Physics, New York (1984), p. 339.
28. S. J. Seestrom-Morris et. al., Phys. Rev. C26, 2131 (1982).
29. T. N. Taddeucci et. al., Phys. Rev. C33, 746 (1986).

SPECTROSCOPY WITHOUT QUARKS:
A SKYRME-MODEL SAMPLER

Marek Karliner and Michael P. Mattis
SLAC, bin 81, Stanford, CA 94305

ABSTRACT

A potpourri of Skyrme-model results for the meson-baryon system is surveyed.

This talk is devoted to a study of meson-nucleon scattering in skyrmion models of the nucleon.[1,2] We shall focus on the characteristic energy-range of the baryon resonances, typically 1.5-2.5 GeV. This is well beyond the point where the chiral Lagrangians are conventionally applied; nor is it known at present how to apply QCD directly in this domain. Thus it is especially interesting to see what insights emerge in this regime from skyrmion physics. The results presented here will only be valid to leading order in $1/N_c$, where N_c is the number of colors of the underlying gauge theory.[3]

The object of our investigations will be effective Lagrangians (Skyrme's included) of the form

$$\mathcal{L}_\pi = \frac{f_\pi^2}{16} \text{Tr}\,(\partial_\mu U \partial^\mu U^\dagger) + \cdots. \tag{1}$$

The leading term is the usual 2-flavor or 3-flavor nonlinear sigma model, depending on whether $U \in SU(2)$ or $U \in SU(3)$. The dots stand for higher-derivative terms, which are not usually exploited in traditional soft-pion physics. Nevertheless, they are needed to stabilize a soliton, or "skyrmion," whose topological charge (following Skyrme) is interpreted as baryon number. The standard identification of the pion field in (1) in the baryon-number-0 sector of the 2-flavor theory is via:

$$U(x) = \exp(\frac{2i}{f_\pi} \vec{\pi}(x) \cdot \vec{\sigma}). \tag{2}$$

Thus the pions can be thought of as "small fluctuations" about the trivial vacuum $U(x) \equiv 1$.

It is a straightforward procedure to introduce additional fields into (1) in such a way as to preserve chiral invariance.[4] In particular, the traditional approach to studying the coupling of pions to the nucleon isodoublet N is to set

$$\mathcal{L}_{\pi N} = \frac{f_\pi^2}{16} \text{Tr}\,(\partial_\mu U \partial^\mu U^\dagger) + \bar{N}(i\gamma^\mu \mathcal{D}_\mu - m)N + g_A \mathcal{D}_\mu \vec{\pi} \cdot \bar{N}\vec{\tau}\gamma^\mu \gamma^5 N. \tag{3}$$

Here $\mathcal{D}$ is the covariant derivative appropriate to the manifold G/H, where $G = SU(2)_L \times SU(2)_R$ and $H = SU(2)_{\text{isospin}}$. From this Lagrangian, all soft-pion theorems pertaining to the πN interaction, such as Weinberg's calculation of the S-wave scattering lengths,[5] can be derived.

It is the moral of this talk that the purely mesonic Lagrangian (1) contains *at least* as much information as does (3)! Not only does (1) properly encompass soft-pion physics, as Schnitzer has shown,[6] but in addition—well beyond the soft-pion regime—it yields surprisingly accurate predictions concerning the spectrum of nucleon and Δ resonances and the qualitative behavior of the large majority of πN and $\overline{K}N$ partial-wave amplitudes.

The study of meson-nucleon scattering in skyrmion models involves splitting the Goldstone field $\vec{\pi}$ into two pieces: a spatially-varying c-number piece, i.e., the skyrmion, and a fluctuating piece, which we identify with physical mesons.[7] The skyrmion will be assumed to be of the hedgehog form:

$$U_\circ(\vec{x}) = \exp(iF(r)\hat{r}\cdot\vec{\sigma}). \qquad (4)$$

Calculating the T-matrix then reduces to a problem of potential scattering, from which partial-wave phase-shifts can be extracted in the usual manner. In addition, it is necessary to fold in a little group theory, as we now describe.

For simplicity, let us focus on the non-strange processes $\pi N \to \pi N$, $\pi N \to \pi\Delta$ and $\pi\Delta \to \pi\Delta$. The quantum numbers needed to describe such processes are: the initial and final pion angular momenta L and L'; the initial and final spin (or isospin) representation of the baryon s and s', which equal $\frac{1}{2}$ for nucleons and $\frac{3}{2}$ for Δ's; and the total pion-baryon isospin and angular momentum $\mathbf{I}$ and $\mathbf{J}$. The T-matrix describing such processes can then be shown to be:[9,10]

$$T(\{LsIJ\} \to \{L's'I'J'\}) = \delta_{II'}\delta_{I_zI'_z}\delta_{JJ'}\delta_{J_zJ'_z}(-1)^{s'-s}$$
$$\times \sqrt{(2s+1)(2s'+1)} \sum_{K=L-1}^{L+1} (2K+1) \begin{Bmatrix} K I J \\ s'L'1 \end{Bmatrix} \begin{Bmatrix} K I J \\ sL1 \end{Bmatrix} T_{KL'L}. \qquad (5)$$

The expressions in curly brackets are $6j$-symbols. The quantities $T_{KL'L}$, which are functions of pion energy, are the "reduced amplitudes" of the model, obtainable numerically from a phase-shift analysis about the skyrmion. The Kronecker δ's express the reassuring fact that $\mathbf{I}$ and $\mathbf{J}$ are conserved in these models, as they ought to be.

Although, in Eq. (5), K plays the part of a dummy index, it actually has an interesting physical interpretation. Specifically, K can be viewed as the vector sum of the pion's angular momentum and isospin in the unphysical frame in which the pion scatters, not from a nucleon, but rather from an unrotated hedgehog soliton. This frame is "unphysical" in that a nucleon properly corresponds to a *rotating* hedgehog soliton in the skyrmion approach.[2]

Pleasingly, all the model-dependence in (5) arising from the details of the terms indicated by dots in the Lagrangian (1) is subsumed in the reduced amplitudes $T_{KL'L}$; the $6j$-symbols, in contrast, follow purely from the assumed hedgehog symmetry of the skyrmion. Equation (5) is thus analogous to the Wigner-Eckart theorem in that a large number of physical matrix elements (the T's) are expressed in terms of a substantially smaller set of reduced matrix elements (the T_K's) weighted by appropriate group-theoretical coefficients.

One can carry the analogy further by finding those special linear combinations (analogous to the Gell-Mann-Okubo formula) for which the model-dependent right-hand side of (5) cancels out; the net result will be a set of energy-independent linear relations between physical scattering amplitudes that serve as a test of the applicability of skyrmion physics to the real world. For example, for $\pi N \to \pi N$, one can for each value of $L \geq 1$ solve for the two independent isospin-$\frac{3}{2}$ amplitudes (*i.e.*, with $J = L \pm \frac{1}{2}$) as linear combinations of the two isospin-$\frac{1}{2}$ amplitudes![9,10] How well do these relations work as applied to the experimental scattering data? Consider the case of the F waves (*i.e.*, $L = 3$). Figures 1a and 1b depict the real-world F_{35} and F_{37} amplitudes,[11] respectively (indicated by solid lines), compared with the linear combinations of the F_{15} and F_{17} amplitudes (dotted lines) to which they should correspond, if Eq. (5) holds true.[9] The degree of agreement is impressive.

Similar relations hold[9] for $\pi N \to \pi \Delta$. Figure 1c displays the FP_{15} amplitude compared with the appropriate multiple of the FP_{35} amplitude predicted by Eq. (5). One can even find relations between $\pi N \to \pi N$ and $\pi N \to \pi \Delta$ amplitudes; a typical such prediction is illustrated in Fig. 1d. In both cases, although the sizes of the two comparison curves are not in especially close accord, the disagreement is on the order of the typical error-bars of the data. We should point out that the agreement in the *signs* of the amplitudes is in itself a completely nontrivial result.

It is straightforward to generalize Eq. (5) to the case when the initial and/or final meson has arbitrary spin and isospin; one need only replace the $6j$ symbols by $9j$ symbols.[12] This allows us to treat in the skyrmion framework such experimentally accessible processes as $\pi N \to \rho N$ and $\pi N \to \omega N$. A typical relation for $\pi N \to \rho N$ is illustrated in Fig. 1e. Alternatively, one can generalize Eq. (5) to incorporate *strangeness*, which entails interpreting U in Eq. (1) as an $SU(3)$ matrix.[13,14] This enables us to study KN and $\overline{K}N$ scattering as well. In this context, one can obtain predictions relating $\pi N \to \pi N$ to $\overline{K}N \to \overline{K}N$,[15] as exemplified in Fig. 1f.

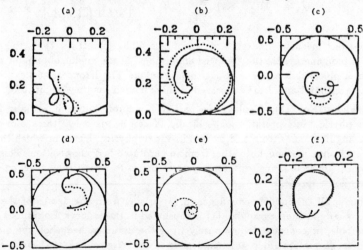

Fig. 1. Model-independent linear relations between experimental amplitudes. (a)[9] $\pi N \to \pi N$: F_{35} vs. $\frac{1}{7}F_{15} + \frac{6}{7}F_{17}$. (b)[9] $\pi N \to \pi N$: F_{37} vs. $\frac{9}{14}F_{15} + \frac{5}{14}F_{17}$. (c)[9] $\pi N \to \pi \Delta$: FP_{15} vs. $\sqrt{10}FP_{35}$. (d)[9] $\pi N \to \pi N$ vs. $\pi N \to \pi \Delta$: $\frac{1}{\sqrt{2}}(P_{11} - P_{13})$ vs. $\frac{5}{4}PP_{33} - \frac{1}{8}PP_{11}$. (e)[12] $\pi N \to \rho N$: $-\frac{1}{2}SD_{31}$ vs. SD_{31}. (f)[15] $\pi N \to \pi N$ vs. $\overline{K}N \to \overline{K}N$: $\frac{1861}{2395}F_{15} - \frac{10661}{14370}F_{37}$ vs. $F_{05} - \frac{1952}{1437}F_{17}$. In all cases, the first-named amplitude is depicted by a solid line, the second by a dotted line. As usual, the graphs depict the T-matrices in the complex plane.

We should reiterate that all the curves in Fig. 1 are drawn from experiment. In other words, they are examples of *model-independent* predictions of skyrmion physics. (As far as we know, no such relations follow from $SU(6)$, or from the quark model.) In our eyes, Fig. 1 is compelling evidence for the validity of the hedgehog-soliton picture of the nucleon.

Of course, one can also adopt a *model-dependent* approach to Eq. (5). This involves specifying an effective meson Lagrangian, calculating the $T_{KL'L}$'s numerically, reconstructing the complete partial-wave T-matrix using (5), and then comparing with experiment.[8,10] Let us focus on the venerable Skyrme model, in which the first term of Eq. (1) is supplemented by the soliton-stabilizing term $\frac{1}{32e^2}\text{Tr}[(\partial_\mu U)U^\dagger, (\partial_\nu U)U^\dagger]^2$.

Figure 2 presents the partial-wave T-matrix of the Skyrme model, juxtaposed with experiment. The overall degree of accord is quite striking. The exceptions are in the lower partial waves, the S_{31}, P_{11}, P_{33} and D_{35} channels especially. We have argued at length elsewhere[8,14] that the poor agreement in these channels can be attributed to our failure to factor out the rotational and translational zero-modes of the skyrmion in our formalism. As such, they represent failures, not of the Skyrme model *per se*, but rather of our treatment of it. We are confident that a higher-order $1/N_c$ analysis would dramatically improve the agreement in these channels. Fortunately, in higher partial waves, which do not mix with the zero-modes, there is no such problem.

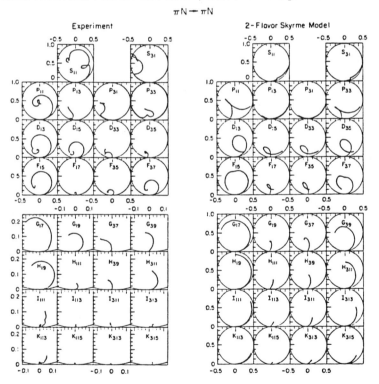

Fig. 2. $\pi N \to \pi N$: comparison between 2-flavor Skyrme model and experiment (from Ref. 14).

The most intriguing feature of Fig. 2 is how the Skyrme model reproduces the pattern of size alternation evident in the experimental curves for each value of $L \geq 1$. For example, the F_{15} and F_{37} curves are much bigger than the F_{17} and F_{35} amplitudes. This "big-small-small-big" pattern finds a natural group-theoretic explanation in the context of skyrmion physics.[9] According to Eq. (5), the physical scattering amplitudes can be expressed as linear combinations of reduced amplitudes. In the F-wave sector, one finds, for example:

$$\mathbf{T}(F_{15}) = \frac{5}{9} \cdot \mathcal{T}_{233} + \frac{4}{9} \cdot \mathcal{T}_{333}, \tag{6a}$$

$$\mathbf{T}(F_{17}) = \frac{1}{4} \cdot \mathcal{T}_{333} + \frac{3}{4} \cdot \mathcal{T}_{433}, \tag{6b}$$

$$\mathbf{T}(F_{35}) = \frac{5}{63} \cdot \mathcal{T}_{233} + \frac{5}{18} \cdot \mathcal{T}_{333} + \frac{9}{14} \cdot \mathcal{T}_{433}, \tag{6c}$$

$$\mathbf{T}(F_{37}) = \frac{5}{14} \cdot \mathcal{T}_{233} + \frac{3}{8} \cdot \mathcal{T}_{333} + \frac{15}{56} \cdot \mathcal{T}_{433}. \tag{6d}$$

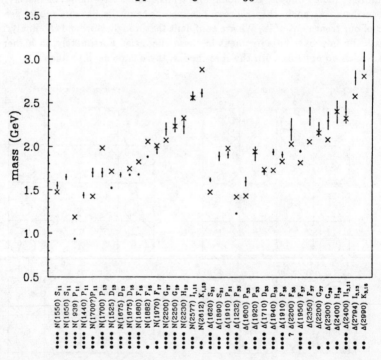

Fig. 3. Spectrum of N and Δ resonances: Skyrme model *vs.* experiment (from Ref. 14). The experimental masses (with uncertainties) are indicated by dots, the Skyrme-model masses by crosses. The Skyrme-model values for m_N and m_Δ are obtained from Eq. (9) of Ref. 2, using our "best fit" parameters $\{e = 4.79, f_\pi = 150 \, \text{MeV}\}$.

Note that, whereas the F_{17} and F_{35} amplitudes receive the bulk of their contributions from T_{433}, the F_{15} and F_{37} amplitudes are composed primarily of T_{233} and T_{333}. The big-small-small-big pattern can thus be simply explained by the assumption that T_{433} is small compared with T_{233} and T_{333}, etc. Indeed, this turns out to be true numerically in the Skyrme model; presumably it is true as regards the optimal effective Lagrangian of Nature, as well.

Figure 3 presents the spectrum of nucleon and Δ resonances of the Skyrme model compared with experiment. The Skyrme-model masses are the results of a least-squares fit to the data, with all resonances weighted equally. (Recall that the model has two "free parameters," f_π and e.) The agreement with Nature is surprisingly good—better than 7% on average up to 3 GeV. Interestingly, two resonances, the $N(1882)$ F_{15} and the $\Delta(2350)$ F_{37}, are not present in the 2-flavor Skyrme model but only in the 3-flavor model; other than this, the two versions yield identical spectra.[16]

Figure 4 displays both the 2-flavor and 3-flavor Skyrme-model amplitudes for the process $\pi N \to \pi \Delta$, indicated by dotted and solid curves, respectively. In general, the agreement with experiment is excellent. It is especially pleasing that the model correctly predicts a negative DD_{13} amplitude, in stark contrast to all the other PP, DD and FF channels. Likewise, in the model as in Nature, the FF_{15} amplitude curves around much more than its three F-wave counterparts. We should also note that—unlike $SU(6)$—the Skyrme model correctly predicts the signs of the $\pi N \to \pi \Delta$ amplitudes in channels in which the pion's angular momentum jumps by 2 units (not depicted).

Finally, Figs. 5-7 present a sampler of results corresponding to the processes $\overline{K}N \to \overline{K}N$, $\pi N \to K\Lambda$, and $\overline{K}N \to \pi\Lambda$, in which the 3-flavor Skyrme model is compared with experiment. In general, the agreement is quite satisfactory. (As one would expect, the accord is much less good in the S- and P-wave sectors, which we have not depicted, due to mixing with the zero-modes of the skyrmion.) Although the Skyrme-model graphs are often much too large, as in Fig. 6, the *relative* sizes and signs between amplitudes are well rendered by the model. In particular, for $\overline{K}N \to \overline{K}N$, note that the D_{03}, F_{05} and G_{07} amplitudes dominate their counterparts in both the model and experiment; we have dubbed this the "big-small-small-small" pattern, in analogy with the big-small-small-big pattern discussed earlier. Likewise the model successfully predicts a "down-up" pattern of sign alternation for $\pi N \to K\Lambda$ and $\overline{K}N \to \pi\Lambda$. Not surprisingly, these patterns, too, find a natural group-theoretic explanation in the skyrmion framework, just as in the case of the big-small-small-big pattern.

In closing, we would like to express our wonderment that so much detailed structure of the meson-nucleon T-matrix—much of it in reasonable accord with Nature—can emerge from a simple meson Lagrangian with no explicit quark or nucleon fields.

This work was supported by the Dept. of Energy, contract DE-AC03-76SF00515.

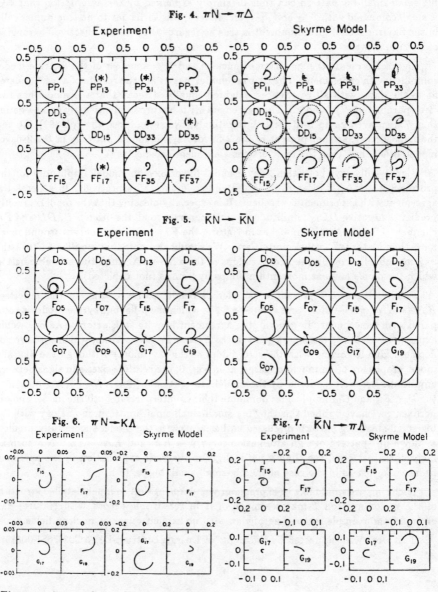

Fig. 4. $\pi N \to \pi \Delta$

Fig. 5. $\bar{K}N \to \bar{K}N$

Fig. 6. $\pi N \to K\Lambda$

Fig. 7. $\bar{K}N \to \pi\Lambda$

Figures 4-7 are taken taken from Ref. 14

REFERENCES

1. T. H. R. Skyrme Proc. Roy. Soc. **A260** (1961) 127; Nucl. Phys. **31** (1962) 556.
2. G. Adkins, C. Nappi, and E. Witten, Nucl. Phys. **B228**, 552 (1983).
3. The parameters of the effective Lagrangians have well defined scaling properties as functions of N_c. (For a detailed exposition of this issue see Ref. 2 and references therein). Consequently in the large-N_c limit certain terms in the meson equations of motion are formally supressed and these equations can therefore be linearized in a consistent fashion.[8,9]
4. S. Coleman, J. Wess, B. Zumino, Phys. Rev. **177** (1969) 2239; C. Callan, S. Coleman, J. Wess, B. Zumino, ibid. 2247.
5. S. Weinberg, Phys. Rev. Lett. **17**, 168 (1966).
6. H. Schnitzer, Phys. Lett. **139B**, 217 (1984); Nucl. Phys. **B261** (1985) 546.
7. In truth, not all fluctuations correspond to *bona fide* mesonic excitations: some serve only to rotate or translate the skyrmion, and should therefore be viewed as *baryonic* degrees of freedom. These *zero-modes* mix only with S-, P- and D-wave pions, and tend to spoil the results in these channels.
8. M. P. Mattis and M. Karliner, Phys. Rev. **D31** (1985) 2833.
9. M. P. Mattis and M. Peskin, Phys. Rev. **D32** (1985) 58.
10. A. Hayashi, G. Eckart, G. Holzwarth, and H. Walliser, Phys. Lett. **147B**, 5 (1984).
11. We shall follow the customary notation for partial-wave amplitudes: $\pi N \to \pi N$ and $\pi N \to K\Lambda$ amplitudes are indicated by $L_{2I,2J}$, $\overline{K}N \to \overline{K}N$ and $\overline{K}N \to \pi\Lambda$ amplitudes by $L_{I,2J}$, and $\pi N \to \pi\Delta$ and $\pi N \to \rho N$ amplitudes by $LL'_{2I,2J}$, where L and L' are the incoming and exiting pion orbital angular momenta, and I and J are the total meson-baryon isospin and angular momentum.
12. M. P. Mattis, Phys. Rev. Lett. **56** (1986) 1103.
13. M. P. Mattis, *Aspects of Meson-Skyrmion Scattering*, SLAC-PUB. 3795, in S. Brodsky and E. Moniz, eds., Proceedings of the 1985 ITP Workshop on Nuclear Chromodynamics, World Scientific Press.
14. M. Karliner and M. P. Mattis, *πN, KN and $\overline{K}N$ Scattering: Skyrme Model vs. Experiment*, SLAC-PUB-3901, 1986.
15. M. Karliner, *How chiral solitons relate $\overline{K}N$ and πN scattering*, SLAC-PUB 3958, 1986.
16. M. Karliner and M. P. Mattis, Phys. Rev. Lett. **56** (1986) 428.

LIGHT QUARK PAIRS IN HEAVY QUARKONIA*

Nina Byers and Vasilis Zambetakis
Department of Physics, University of California, Los Angeles, CA 90024

ABSTRACT

We report results of a study of potential models which explicitly include light quark pair production vacuum polarization effects. Fits to the $c\bar{c}$ and $b\bar{b}$ data including leptonic and hadronic widths of $\psi''(3772)$ give 0.31 GeV^2 for the slope of the linear potential (string tension), a value significantly higher than the value of 0.22 GeV^2 obtained without explicit account taken of light quark pair production. This fit to the data is obtained in a model in which the confining potential is Lorentz vector. We are not able to find parameter values to fit the data in models in which the confining potential is scalar.

There is little doubt that there are virtual quark pairs in the chromodynamic fields of a heavy quark and antiquark. The Compton wave lengths of light quarks are comparable to the radii of low mass charmonium and upsilon states. Therefore virtual light quark pairs can have large effects in these states. Furthermore light quark pair production accounts for the OZI allowed decays. These decay channels as virtual intermediate states give significant and sometimes large mass shifts and configuration mixings. Estimates of such effects in charmonium were first made by Eichten, Gottfried, Kinoshita, Lane, and Yan.[1] They showed that the level shifts are nearly spin independent. In this case level spacings can be described by single channel potential models with potential parameters that implicitly contain quark pair production effects. Single channel models, however, do not give large enough configuration mixing ($^3D_1$ - $^3S_1$ mixing) to account for the observed leptonic width of the $\psi''(3772)$. The work of Eichten et al. showed, on the other hand, that if light quark pairs are explicitly included one might be able to account for the observed properties of $\psi''(3772)$.

Perturbative QCD cannot describe light quark pair production at large quark-antiquark separations where confining forces become strong. It is therefore necessary to describe such pair production phenomenologically using, for example, a Wigner-Weisskopf formalism as was done by Eichten et al.. In this report we present results of a study of models similar to that of Eichten et al. in which the quark-antiquark potentials are Coulomb-like (one gluon exchange) at small distances and linearly confining at large distances, and in which pair production occurs according to usual rules of quantum field theory assuming that the potentials are due to scalar and/or time-like vector exchange. We find that the parameters of the potential are significantly changed when quark pair production is explicitly included; for example, the string tension (slope of the linear potential) increases by 40%.[2]

We took pair production from both one gluon exchange and the confining potential into account. We found that the inclusion of one gluon pair production changes the results of Eichten et al.[1] significantly. One gluon pair production by itself is small but its interference with pair production from the confining piece is important. Further we studied the two cases: (i) the confining potential time-like vector and (ii) scalar, and we refer to case (i) as the confining gluon model and to case (ii) as scalar confinement. In the confining gluon case, we took the gluon propagator to vanish like $1/q^2$ as $q^2 \to \infty$ (Coulomb-like) and to have the $1/q^4$ behavior as $q^2 \to 0$ of linear confinement. For scalar confinement we took the confining interaction to be a scalar exchange mechanism.

There are many models which have potentials which are Coulombic at small distances and are linearly confining at large distances.[3,4] For low mass states they all give essentially

---

*Work supported in part by Department of Energy, contract DE-AT03-81ER40024.

the same results because the wave functions of these states average out the details of the potentials. All these models can be characterized by the three parameters which give (i) the strength of the short distance Coulombic interaction; (ii) the slope of the linearly rising potential; and (iii) the zero of energy, thus interpolating between small and large distances. We define three such parameters using the potential of Eichten et al.; viz.,

$$V(r) = -\kappa/r + C + r/a^2 . \qquad (1)$$

The interpolation constant C requires some discussion. In the confining gluon case, pair production amplitudes do not depend on C because they depend on the derivative of the potential. However in the scalar confinement case, pair production amplitudes are sensitive to the value of the constant C. In this case they depend on the potential, not its derivative. In this case both C and the linear piece give large effects which tend to cancel. The result is a relatively small difference of two large quantities. Similar cancellations also occur in mass shifts from $(v/c)^2$ corrections to potential models with scalar confinement.[5] Fits to the observed spectrum of charmonium with the linear potential scalar seem to require that the interpolating piece C is also scalar.[4]

Assuming (1) and the confining gluon model, we found a set of values for the parameters κ, a, and C from a fit to the charmonium masses as shown in Table 1. This fit was obtained using the constituent quark masses chosen by Eichten et al. and the same decay channels as they used; namely, the S-wave charmed mesons D, D^*, F, and F^*. Assuming κ decreases as is usual owing to asymptotic freedom, we took the parameter values determined by the fit to the charmonium masses and calculated the upsilon mass spectrum. The results are shown in Table 2.

State	ΔM	$c\bar{c}$ M	exp.
ψ	100.9	3096.9	3096.9±0.1
ψ'	309.0	3678.0	3686.0±0.1
$\chi^c_{c.o.g.}$	251.3	3495.4	3524.9±1.3

Table 1. Mass shifts and masses for $c\bar{c}$ states; $\Delta M = M_0 - M$ where M_0 is the so-called bare mass which is an eigenvalue of the two-body Hamiltonian with potential energy (1). For the above fit the parameters of the potential are $\kappa = 0.49$, $C = -0.968$ GeV, and $a = 1.80$ GeV^{-1}.

We also used these parameter values to compute the ratio of e^+-e^- cross sections R in the $\psi''(3772)$ resonance region.[7] With these parameters and inclusion of pair production from the one gluon exchange term, we obtained a hadronic width of about 28 MeV and leptonic width of about 0.2 keV, in agreement with measured values.

The above results indicate that one can account for the gross $c\bar{c}$ and $b\bar{b}$ spectroscopy quite well with a confining gluon model. However a Breit-Fermi extension of this model does not yield a fit to the fine structure splitting of the ^3P_J masses; a fit to the fine structure splitting requires scalar confinement.[4]

State	ΔM	$b\bar{b}$ M	exp.
Υ	33.1	9460.2	9460.1±0.2
Υ'	182.5	10001.8	10023.4±0.4
Υ''	273.9	10364.1	10355.5±0.5
$\chi^b_{c.o.g.}$	126.5	9909.0	9900.6±0.4
$\chi^{b'}_{c.o.g.}$	241.1	10264.3	10260.9±0.8±6

Table 2. Calculated mass shifts and masses for $b\bar{b}$ states.[2]

We studied the scalar confinement case with light quark pair productions explicitly included.[7] We were not able to find parameter values for the potential (1) which gave a fit to the data (Tables 1, 2, and the ψ'' widths).

We conclude, therefore, that simple relativistic extensions of potential models in which the potential is a scalar and/or vector exchange potential cannot account both for the spin dependence of the quark-antiquark force and also light quark pair productions. One can take the view that the spin dependence of the interquark force is not simply derivable from a Breit-Fermi extension of the nonrelativistic model. In this case, we view our results as indicating validity for the confining gluon model.

Recently Adler[8] has reported that QCD pairing model calculations give concrete realization of chiral symmetry breaking and the accompanying Nambu-Goldstone pion provided the instantaneous potential has a confining piece and this piece is Lorentz vector. So, if one is prepared to forego obtaining in a simple way the spin dependence of the quark-antiquark force, the confining gluon model may be a correct description of confinement.

REFERENCES

1. E. Eichten, K. Gottfried, T. Kinoshita, K. D. Lane, and T.-M. Yan, Phys. Rev. D17, 879 (1978).
2. V. Zambetakis, Ph.D. thesis, UCLA/86/TEP/2.
3. J. L. Richardson, Phys. Lett. 82B, 272 (1978); W. Buchmüller, S.-H. H. Tye, Phys. Rev. D 24, 132 (1981).
4. A. B. Henriques, B. H. Kellet, and R. G. Moorhouse, Phys. Lett. 64B, 85 (1976); R. L. McClary and Nina Byers, Phys. Rev. D 28, 1692 (1983); S. N. Gupta, S. F. Radford, and W. W. Repko, Phys. Rev. D 26, 3305 (1982).
5. R. McClary, Ph.D. thesis, UCLA, 1982.
6. N. Byers and V. Zambetakis (to be published).
7. N. Byers, H. Grotch, V. Zambetakis, UCLA preprint (in preparation).
8. Stephen L. Adler, "Gap Equation Models for Chiral Symmetry Breaking," Progress in Theoretical Physics (Supplement), dedicated to Professor Yoichiro Nambu.

NUCLEAR-LIKE STATES OF QUARK MATTER

T. Goldman, K. E. Schmidt[†] and G. J. Stephenson, Jr.

Theoretical Division, Los Alamos National Laboratory,
Los Alamos, NM 87545

ABSTRACT

We show that, in a world with only one flavor of light quark, QCD suggests that the low energy states of quark matter are similar to nuclei, but are <u>not</u> well represented as collections of baryons. Except for the existence of open nucleon channels, the same would be true for the actual, two-light-flavor world.

INTRODUCTION

Why does nuclear physics have to be that way? More precisely, is the surprising utility of the point nucleon approximation in nuclear physics due to general properties of QCD, or due to some accident? We cannot really know the range of validity of this approximation until we know its origins in the theory of the strong interactions. We describe here, in the context of a model of QCD, cases in which QCD is virtually unaltered, but in which the point baryon is a poor approximation for describing nuclear-like states.

A ONE FLAVOR WORLD

Suppose only one flavor of quark were light. Then 3 quarks in the same space state would also be symmetric in flavor, since they must be overall antisymmetric, and antisymmetric in color to form a color singlet. Thus, they form a delta as the lowest energy state. (A spin 1/2 state may be formed by radial or orbital excitation of one of the quarks, but this is a higher mass state -- an N^*.) Note that QCD has suffered only a tiny change in its beta-function, and is virtually unaltered -- it still confines, with almost precisely the same strength and range, according to conventional ideas wherein the gluon self-interaction dominates the theory.

What do nuclei look like in this world? For simplicity, we first consider the analog of the deuteron.

We model QCD with a linear (scalar) confining potential

$$V = 0.9 \text{ GeV/fm} \times (r-0.57 \text{ fm}) \tag{1}$$

which is consistent with the potential analyses of Buchmuller and Tye[1]. A three quark system in this potential has an rms radius of ~0.8 fm. Including the color-magnetic spin-splitting effect of the current-current interaction of these (massless) quarks allows us to fix the effective QCD coupling constant by fitting to the delta

---

[†]Present address: Courant Institute, New York, NY 10012

mass. For massless gluon exchange, we obtain α_c ~ 1.0. Analysis of the gluon field equations as fluctuations about a mean field responsible for the confining potential, however, suggests that these additional exchanged gluons suffer a confinement effect also, which is representable as a mass term. For a Yukawa value of ~400 MeV, we find α_c ~ 1.5. Although the reasonable values in (1) and (2) have been chosen to fit the real world spectrum, we expect only very small changes in the one flavor world. In all of these calculations, we have used a Monte Carlo integration technique so that we could apply it to the geometrically more complicated problems below.

Our model for the "deuteron" is then to build a two well potential with a separation R between the minima. The value of the potential at any point is given by (1) where r is the distance to the nearer minimum. (A section through both minima shows a W-shaped potential.) This is similar to what would be found for the two meson case by averaging the model of Lenz et al.[2] Naturally a symmetric wavefunction gives the lowest energy, due to tunnelling, as we have noted before.[3] We load this potential with 6 quarks, saturating the color and spin degrees-of-freedom.

We then find a total binding energy from the effective of quark delocalization of 528 MeV at an energy minimum occurring at $R \simeq 1.1$ f. However, the color-magnetic spin-spin interaction must also be included. Although some of the pairs of quarks are now in color-6 and spin-zero representations, they still produce the same sign interaction as the color-$\bar{3}$, spin-one pairs. The current-current interaction has three times the weight as in one delta (due to the larger number of interacting pairs) but the current-current spatial matrix element is smaller. The net effect of all of this, since the color-magnetic spin interaction is repulsive in the delta, is to reduce the binding energy, to 114 MeV, and to shift the preferred size to $R \simeq 1.4$ f. This is still a very strongly bound state. The total energy of the system rises sharply with decreasing R; all of the binding energy is lost by R ~ 1 f. Thus the system shows a strong core repulsion: this is primarily due to the color magnetic interaction.

TWELVE QUARKS

How much does this result depend upon the number of quarks in the system? To study this, we consider a 4-well system, with minima at the corners of a tetrahedron. Again, R is the separation between minima, and the value of the potential is given by (1) referred to the nearest minimum. We restore the second light flavor, and saturate the spatial wavefunction symmetric over all 4 wells with twelve quarks. The color magnetic interaction involves more pair combinations, but produces the same factor of three as above. We find a binding energy of 1060 MeV at R just under 1.6 f.

Once again the system is strongly bound, with respect to 4 deltas. Unlike the previous case, however, we must close the nucleon channel "by hand", as the system is ~120 MeV above the energy of 4 nucleons. (It can just barely decay into a pion plus

^{4}He; otherwise it must decay by relatively slow gamma emission.) As can be seen from the one flavor case, this need not mean that any drastic change has been implicitly made in the strong interactions. Once again, we also find that the binding energy vanishes if the system is compressed to R ~ 0.7 f, showing a strong core replusion.

We have also tried to examine the effect of spatial correlations in this system. We have done this with a single parameter ϵ. For $\epsilon = 0$, the quark wavefunctions are those for individual deltas centered in each well. For $\epsilon = 1$, we return to the symmetric wavefunction used above. We also kept track of which pairs of quarks came from the same "delta", and which pairs were from different ones. This allowed us to follow the system to four separate deltas at $\epsilon = 0$ and large R. Not surprisingly, the energy of the system was higher at all R and ϵ than at the minimum found above for $\epsilon = 1$; this was due to increases in both localization and color-magnetic energy as ϵ was reduced. Furthermore, it occcured even though no effort was made to include repulsive effects, due to the Pauli principle, from overlapping components of different deltas.

CONCLUSIONS

By closing the nucleon channel in two ways which do not affect the strong interaction, we have shown that nuclear-like states are formed in our model of QCD. There are local density fluctuations such as one would find from localized baryons, and short distance repulsive effects which suggest density and binding energy saturation in the "nuclear matter" limit. However, the minimum energy states are more properly described as quark matter: The quark wavefunctions extend over the entire "nucleus", and there are no local deltas, although the overlap with same is not small. In the A=4 case, the large binding energy makes it clear that this system is not an approximate sum of point baryons. It also suggests that there may be a significant 4-delta component of ^{4}He. Finally, note that this system has an enormous EMC effect, as the quark wavefunctions are very different from those for 4 incoherently summed baryons.

We conclude that nuclear physics didn't have to be that way, and that the nature of QCD is not the determining factor for the point nucleon approximation. Two light flavors are essential.

We wish to thank Kim Maltman on for stimulating discussions the question of the quark structure of nuclei.

REFERENCES

1. W. Buchmuller and S.-H. H. Tye, Phys. Rev. D24, 132 (1981).
2. F. Lenz, et al., "Quark Confinement and Hadronic Interactions", SIN preprint 85-09.
3. T. Goldman and G. J. Stephenson, Jr., Phys. Lett. 146B, 143 (1984).

RECENT MARK III RESULTS ON
J/ψ HADRONIC AND RADIATIVE DECAYS

WILLIAM S. LOCKMAN
Representing the MARK III Collaboration*
*Institute for Particle Physics University of California
Santa Cruz, California 95064*

ABSTRACT

Recent results from the MARK III experiment on J/ψ radiative and hadronic decays are presented, based on an analysis of 5.8×10^6 J/ψ decays collected at the SPEAR e^+e^- storage ring. The status of the $\xi(2230)$ is reviewed and the result of a search for a new rare decay mode of the $\xi(2230)$ is presented. The complete set of $\eta_c \to 1^{--}1^{--}$ decays is discussed. In this channel, the η_c decays preferentially to final states containing strange quarks. In the decays J/$\psi \to \{\gamma, \omega, \phi\}X, X = \{K\overline{K}, \pi\pi, K\overline{K}\pi, \eta\pi\pi\}$, J/$\psi \to K^*K\pi$ and J/$\psi \to \rho\eta\pi$, the vector mesons are used to probe the quark content of the recoil system. Possible evidence for $\theta(1720)$, but no indication for the $\iota(1460)$ is seen in these final states.

1. INTRODUCTION

Since its first observation in 1974, the J/ψ resonance has continued to be a unique laboratory for the study of hadron dynamics. The J/ψ is produced with high rate and little background, and has well defined spin, parity and charge conjugation ($J^{PC} = 1^{--}$). Since the initial state contains no u, d or s quarks and the mass of the J/ψ lies below threshold for open charm production, the most prominent decays of the J/ψ shown in Figure 1 are OZI suppressed.[1] In the case of J/ψ hadronic decays (Figures 1a and b), these properties provide a unique way to study light quark spectroscopy. Applying the OZI rule to the final state, the singly OZI (SOZI) suppressed diagram shown in Figure 2a should be greatly enhanced relative to the doubly OZI (DOZI) suppressed diagram of Figure 2b.[2] In decays of the type J/$\psi \to \phi X$ and J/$\psi \to \omega X$, the dominance of the SOZI diagram implies that the $\phi(s\overline{s})$ and $\omega(\frac{1}{\sqrt{2}}(u\overline{u}+d\overline{d}))$ can be used to probe, respectively, the strange and nonstrange quark content of the state X.

The radiative J/ψ decay mode (Figure 1c) is considered to a likely channel to find bound states of two gluons, or "glueballs".[3] In fact, many new states, e.g., $\iota(1460)$, $\theta(1720)$, $\xi(2230)$ and pseudoscalar structures decaying into $\omega\omega$ and $\rho\rho$ have been found in this mode.[4] By comparing the ω and ϕ recoil systems in

---

* Caltech: D.Coffman, G.Dubois, G.Eigen, J.Hauser, D.Hitlin, C.Matthews,Y.Zhu. Santa Cruz: D.Dorfan, F.Grancagnolo, R.Hamilton, C.Heusch, L.Köpke, W.Lockman, R.Partridge, H.Sadrozinski, A.Seiden, M.Scarlatella, T.Schalk, S.Watson, A.Weinstein, R.Xu. Illinois: J.Becker, G.Blaylock, J.Brown, B.Eisenstein, T.Freese, G.Gladding, C.Simopoulos, I.Stockdale, B.Tripsas, J.Thaler, A.Wattenberg, W.Wisniewski. SLAC: T.Bolton, K.Bunnell,R.Cassell, D.Coward, K.Einsweiler, D.Favart, U.Mallik, R.Mozley, A.Odian, R.Schindler, W.Stockhausen, W.Toki, F.Villa, S.Wasserbaech, N.Wermes, D.Wisinski, G.Wolf. Washington: T.Burnett, V.Cook, A.Duncan, A.Guy, P.Mockett, B.Nemati.

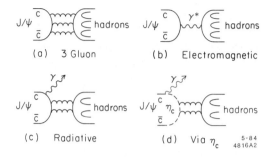

Fig. 1. Lowest order diagrams for J/ψ decay: (a) Three gluon annihilation; (b) electromagnetic decay proceeding through $c\bar{c}$ annihilation into a virtual photon; (c) radiative decay into a final state of one photon and two gluons; (d) magnetic dipole transition to η_c.

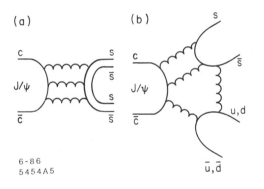

Fig. 2. (a) Singly OZI (SOZI) suppressed diagram; (b) doubly OZI (DOZI) suppressed diagram.

hadronic J/ψ decays with the spectra seen in radiative decays, more information about the quark/gluon content of these new states may be obtained.

The J/ψ decay (Figure 1d) to the η_c has been seen in numerous hadronic channels. Since the hadronization of the η_c proceeds to lowest order through the two gluon intermediate state, the study of the η_c decay pattern may also further our understanding of gluonia decays.

In section 2, the status of the $\xi(2230)$ and a search for $\xi(2230) \to \omega\phi$ are presented. In section 3, the $\eta_c \to 1^{--}1^{--}$ decays are discussed. In the remaining sections, the quark content of various resonances are probed in two-body J/ψ hadronic decays. In section 4, the properties of the $K\bar{K}$ and $\pi\pi$ systems recoiling against γ, ω and ϕ are studied to probe the quark content of the isoscalars, f, f', S^* and θ. In section 5, the $K\pi$ system recoiling against the K^* and the $\eta\pi$ system recoiling against the ρ are discussed to understand the nature of the $K^*(1430)$, A_2 and $\delta(980)$. In the last section, the $K\bar{K}\pi$ and $\eta\pi\pi$ systems recoiling against the γ, ω and ϕ are examined to study the $\eta(1275)$, $D(1285)$, $E(1420)$ and $\iota(1460)$.

2. STATUS OF THE $\xi(2230)$

The $\xi(2230)$, a narrow, massive state seen in radiative J/ψ decays, was first observed by the MARK III collaboration in 1983, in a data sample of 2.7×10^6 J/ψ decays.[5] To verify the existence of the $\xi(2230)$, the MARK III collected another 3.1×10^6 J/ψ events in 1985. Applying somewhat looser cuts than in its previous analysis, the MARK III observed the $\xi(2230)$ in its K^+K^- decay mode in both the 1983 and 1985 data samples with a combined significance of 4.5 s.d. and with the following resonance parameters: $m = (2230 \pm 6 \pm 14)$ MeV, $\Gamma = (26^{+20}_{-16} \pm 17)$ MeV, and $B(J/\psi \to \gamma\xi(2230)) \times B(\xi(2230) \to K^+K^-) = (4.2^{+1.7}_{-1.4} \pm 0.8) \times 10^{-5}$.[6] The $K^0_s K^0_s$ decay mode of the $\xi(2230)$ was also observed with a statistical significance of 3.6 s.d. and the following resonance parameters:

$m = (2232 \pm 7 \pm 7)$ MeV, $\Gamma = (18^{+23}_{-15} \pm 10)$ MeV, and $B(J/\psi \to \gamma\xi(2230)) \times B(\xi(2230) \to K^0_s K^0_s) = (3.1^{+1.6}_{-1.3} \pm 0.7) \times 10^{-5}$.[6]

In 1984, the DM2 collaboration at the DCI machine logged 8.6×10^6 J/ψ events. The DM2 detector was similar to the MARK III apparatus in its drift and shower chamber coverage and resolution, but had substantially poorer time of flight resolution (500 psec) compared to the MARK III detector (200 psec). No signal was seen near 2230 MeV in either the K^+K^- or $K^0_s K^0_s$ channel. Assuming a width of 25 MeV and a mass of 2230 MeV, an upper limit on the product branching ratio: $B(J/\psi \to \gamma\xi(2230)) \times B(\xi(2230) \to K^+K^-) < 2.4 \times 10^{-5}$(95% C.L.) was obtained.[7] A small excess of events was observed in the $K^0_s K^0_s$ spectrum near 2185 MeV. Using these events, an upper limit on the product branching ratio $B(J/\psi \to \gamma\xi(2230)) \times B(\xi(2230) \to K^0_s K^0_s) < 3.1 \times 10^{-5}$(95% C.L.) was obtained, again assuming a 25 MeV width.[7]

One likely interpretation of the $\xi(2230)$ is that it is an $L = 3$ $s\bar{s}$ meson with $J^{PC} = 2^{++}$.[8] In this case, the branching fractions to $K^*\bar{K} + c.c.$ and $K\bar{K}$ are roughly equal, the $\phi\phi$ mode is roughly a factor of seven smaller, and the $\pi\pi$ mode is absent. Currently, the $J/\psi \to \gamma\phi\phi$ decay mode is being investigated for possible structure and the upper limits on the $K^*\bar{K}$ and $\pi\pi$ branching ratios presented in reference 6 are consistent with these predictions.

Another possibility is that the ξ could be a hybrid, or meikton state.[9] In this case, the $\xi(2230)$ would have prominent decay modes to $K^*\bar{K}^*$ and $\omega\phi$, the latter mode being OZI-suppressed for a $q\bar{q}$ meson. A search for the $\xi(2230)$ in the $J/\psi \to \gamma\omega\phi$ channel has been carried out, and the $\omega\phi$ invariant mass distribution is shown in the upper part of Figure 3. Fitting the events near 2.22 GeV with a flat background and a gaussian resolution function, the upper limit $B(J/\psi \to \gamma\xi(2230)) \times B(\xi(2230) \to \omega\phi) < 5.9 \times 10^{-5}$ at 90% C.L. is obtained for the production of the $\xi(2230)$.

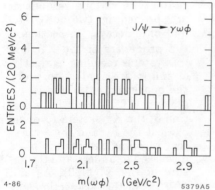

Fig. 3. $\omega\phi$ mass distribution from the reaction $J/\psi \to \gamma\omega\phi$ (upper plot). Background estimate from ω-sidebands (lower plot).

3. $\eta_c \to 1^{--}1^{--}$ DECAYS

The measurement of the $J/\psi \to \gamma\omega\phi$ decay mode now completes the study of the decay sequence $J/\psi \to \gamma\eta_c, \eta_c \to 1^{--}1^{--}$. Figure 3 shows no hint of an η_c signal decaying into $\omega\phi$, leading to an upper limit $B(J/\psi \to \gamma\eta_c) \times B(\eta_c \to \omega\phi) < 1.3 \times 10^{-5}$ (90% C.L.). From this upper limit and previous MARK III measurements,[10] one obtains a complete set of $\eta_c \to 1^{--}1^{--}$ branching ratios, which are presented in table 1. Also shown are the reduced branching ratios $\tilde{B}$[11] normalized to the $\phi\phi$ reduced branching ratio, and the corresponding SU(3) prediction. If SU(3) symmetry were exact, the reduced branching ratios $\tilde{B}(\eta_c \to \phi\phi)$, $\tilde{B}(\eta_c \to \rho^0\rho^0)$, $\tilde{B}(\eta_c \to \omega\omega)$ and $\frac{1}{2}\tilde{B}(\eta_c \to K^{*0}\bar{K}^{*0})$ would all be equal. The

Table 1. $\eta_c \to 1^{--}1^{--}$ decays.

Decay	$B(\eta_c \to X)$	$\frac{\tilde{B}(\eta_c \to X)}{\tilde{B}(\eta_c \to \phi\phi)}$	SU(3)
$\phi\phi$	0.8 ± 0.2	1	1
$K^*\overline{K}^*$	0.9 ± 0.5	0.85 ± 0.47	4
$\rho\rho$	< 1.4	< 1.08	3
$\omega\omega$	< 0.31	< 0.24	1
$\omega\phi$	< 0.10	< 0.10	0

$\eta_c \to 1^{--}1^{--}$ branching ratios in percent.[10] The reduced branching ratios ($\tilde{B}$) and the SU(3) predictions are normalized to the $\phi\phi$ decay mode.

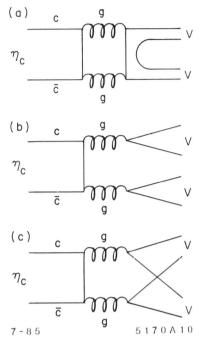

Fig. 4. Lowest order graphs for $\eta_c \to 1^{--}1^{--}$ decay. In (b), final state interactions (e. g., soft gluon exchange, not shown) convert the outgoing particles into color-singlets.

observed $\eta_c \to 1^{--}1^{--}$ decay rates appear to increase with the number of strange quarks in the final state, in contrast to the SU(3)-breaking pattern seen in $J/\psi \to 1^{--}0^{-+}$ decays,[12] where the reduced branching ratios decrease with the number of strange quarks in the final state. To achieve a better understanding of the SU(3) breaking, the relative contributions of the lowest order $\eta_c \to 1^{--}1^{--}$ decay diagrams, shown in Figure 4, must be estimated. The $\rho\rho$, $\omega\omega$ and $\phi\phi$ decays can be produced by all three mechanisms, $K^*\overline{K}^*$ by (a) and (c), while $\omega\phi$ can arise only from (b). If graph (b) dominates, then one predicts that $\tilde{B}(\eta_c \to \omega\phi)/\tilde{B}(\eta_c \to \phi\phi) = 0.60 \pm 0.05$.[13] The measured value of this ratio is much smaller. In addition, a substantial $K^*\overline{K}^*$ decay rate is seen, implying that no single graph is dominant. Clearly, both more theoretical work as well as more precise measurements of the decay rates are needed in order to understand the SU(3) breaking pattern exhibited by the $\eta_c \to 1^{--}1^{--}$ decays.

4. $K\overline{K}$ AND $\pi\pi$ SYSTEMS RECOILING AGAINST γ, ω AND ϕ

In this section we compare the properties of the $K\overline{K}$ and $\pi\pi$ systems recoiling against the γ, ω and ϕ in J/ψ decays. The masses, widths and product branching ratios for the various structures described in sections 4-6 are listed in table 2.

Figure 5a shows the mass distribution of the K^+K^- system recoiling against the radiative photon. In addition to the $\xi(2230)$ discussed earlier, there is clear evidence for the production of the f' and $\theta(1720)$. From the rates in table 2, the θ decays preferentially to final states containing strange quarks.

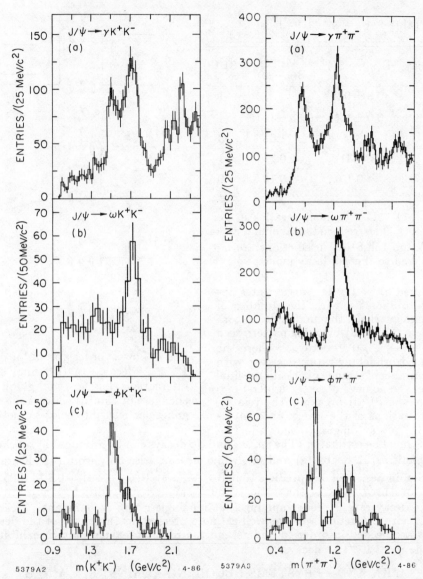

Fig. 5. K^+K^- mass distributions.   Fig. 6. $\pi\pi$ mass distributions.

In Figure 5b, the mass spectrum of the $K\overline{K}$ system recoiling against the ω shows clear evidence for a structure having mass and width values similar to the $\theta(1720)$. No evidence for f' production is seen in this channel.

In Figure 5c, the mass distribution of the K^+K^- system recoiling against the ϕ shows a strong f' signal and also a shoulder on the high side of the f'. Fitting the f' and shoulder to two noninterfering Breit-Wigner amplitudes yields standard f' resonance parameters for the first peak. The second Breit-Wigner mass is about 50 MeV lower than the θ mass, while the width of the $\theta(1720)$

and this structure are consistent with each other. When the two Breit-Wigner amplitudes are allowed to interfere, the branching ratio for the higher mass structure decreases by a factor of ≈ 2.5.

Figure 6a shows the invariant mass spectrum of the $\pi^+\pi^-$ system recoiling against the radiative photon. The large ρ-peak, as well as most of the background at large $\pi\pi$ mass come from the $J/\psi \to \rho^\pm\pi^\mp$ and $J/\psi \to \rho^0\pi^0$ channels, where $\rho^\pm \to \pi^\pm\pi^0$, $\rho^0 \to \pi^+\pi^-$ and the soft photon from an asymmetric decay of the π^0 is lost. In addition to the prominent f peak, a shoulder on the high side of the f is visible. This shoulder, which is absent in the $\omega\pi\pi$ channel, could be due to f-f' interference or possibly related to the box-like structure seen in the DOZI channel $J/\psi \to \phi\pi\pi$. Also evident is a structure at (1713 ± 15) MeV whose mass and width are compatible with the θ. Finally, the interpretation of a third structure at a mass of (2086 ± 15) MeV is not yet clear.

The mass spectrum of the $\omega\pi\pi$ channel shown in Figure 6b is dominated by the f. Also present is a broad enhancement around 500 MeV which has also been seen by the DM2 collaboration.[14] Its nature is not yet understood. Finally, a small signal near 1000 MeV, also seen by DM2, may be the $S^*(975)$.

In Figure 6c, the mass spectrum of the $\pi\pi$ system recoiling against the ϕ shows many interesting features: a clear $S^*(975)$ peak, a "box-like" structure in the 1.1-1.5 GeV region which may contain the f, $\epsilon(1300)$ or another new structure, and an enhancement near 1750 MeV. The presence of the S^* in this channel suggests that it has a large strange quark content. Lying below $K\overline{K}$ threshold, its prominent decay mode is to $\pi\pi$. A coupled channel Breit-Wigner parameterization provides a good fit to the S^* shape.[15]

In summary, the f and f' dominate their respective singly OZI suppressed channels, $\omega\pi\pi$ and $\phi K\overline{K}$, and are largely absent in their doubly OZI suppressed channels, $\phi\pi\pi$ and $\omega K\overline{K}$. Assuming that the enhancements seen near 1730 MeV are all due to the θ, one concludes that the θ is more SU(3) symmetric in its production and decay than either the f or f'. In addition, the ratio of radiative to hadronic production is larger for the θ than for either the f or f'.

5. SYSTEMS RECOILING AGAINST K* AND ρ

In the previous section, the isoscalar states ω and ϕ were used to probe the nonstrange and strange quark content of systems decaying to $\pi\pi$ and $K\overline{K}$. In this section, the study of quark correlations is extended to include systems recoiling against the $I = \frac{1}{2}$ and $I = 1$ states, the K* and ρ. In the case of the K*, the $J/\psi \to K^+\pi^-K^-\pi^+$ decay was studied. Figure 7a displays the correlation between the $K^-\pi^+$ and $K^+\pi^-$ invariant masses. The bands due to K^{*0} and $\overline{K^{*0}}$ production are evident. Within each band, a clustering of events near 1.4 GeV is also visible, corresponding to the decay $J/\psi \to K^{*0}\overline{K}^*(1430)^0$. For this process, the branching ratio is $B(J/\psi \to K^{*0}\overline{K}^*(1430)^0 + c.c.) = (56.2 \pm 4.0 \pm 8.4) \times 10^{-4}$.

Figure 7b shows the mass spectrum of the $K^-\pi^+$ system recoiling against the K^{*0}. A clear peak near 890 MeV indicates a substantial decay rate for the SU(3) violating electromagnetic decay $J/\psi \to K^{*0}\overline{K}^{*0}$.

Figure 8a shows the combined $\eta\pi^\mp$ and $\eta\pi^0$ mass distributions for events recoiling against the $\rho^\pm$ and ρ^0, respectively, where $\rho^\pm \to \pi^\pm\pi^0$ and $\rho^0 \to \pi^+\pi^-$. A strong A_2 signal, near 1320 MeV, is seen, together with a much weaker signal near the nominal $\delta(980)$ mass. In Figure 8b, the $\eta\pi$ mass spectrum for events where the recoil $\pi\pi$ mass lies outside the ρ region indicates that the A_2 signal is correlated with a ρ. The measured decay rate is $B(J/\psi \to \rho A_2) = (118 \pm 8 \pm 29) \times 10^{-4}$.

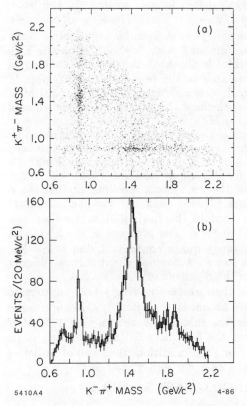

Fig. 7. (a) Mass of $K^+\pi^-$ versus $K^-\pi^+$. (b) Mass spectrum of $K^-\pi^+$ system recoiling against K^{*0}.

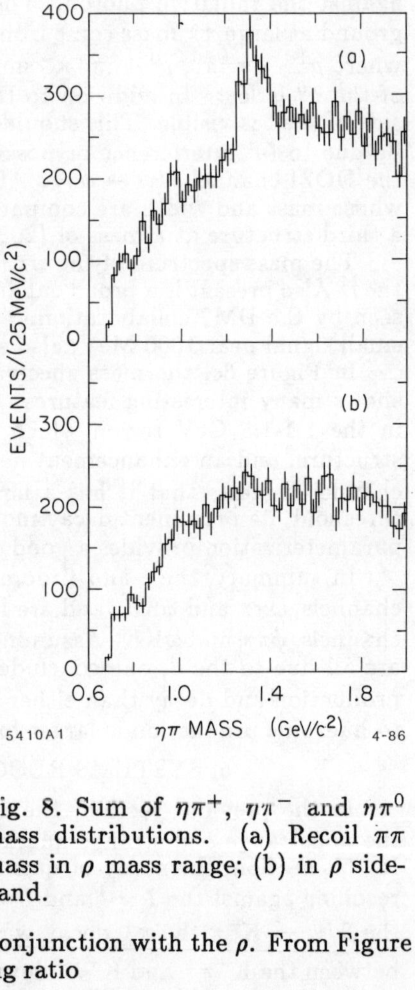

Fig. 8. Sum of $\eta\pi^+$, $\eta\pi^-$ and $\eta\pi^0$ mass distributions. (a) Recoil $\pi\pi$ mass in ρ mass range; (b) in ρ sideband.

There is little evidence for a δ signal in conjunction with the ρ. From Figure 8a, the upper limit on the product branching ratio

$$B(J/\psi \to \rho\delta) \times B(\delta \to \eta\pi) < 4.4 \times 10^{-4} (90\% \text{ C.L.}) \qquad (1)$$

is approximately 25 times smaller than the branching ratio for $J/\psi \to \rho A_2$ and $J/\psi \to \rho\pi$ decays, indicating that the $\delta(980)$ may not be the isotriplet member of the 0^{++} $q\bar{q}$ multiplet. Other than $q\bar{q}$ states, the S^* and δ could be four-quark states[16] or $K\bar{K}$ molecules.[17]

Assuming that the S^* is the $s\bar{s}$ member of an ideally mixed $q\bar{q}$ multiplet implies that $B(J/\psi \to \omega S^*) \approx 0$, and neglecting phase space and SU(3) breaking effects, $B(J/\psi \to \rho\delta) = 3 \times B(J/\psi \to \phi S^*)$.[13]

The expected $J/\psi \to \rho\delta$ rate is smaller if the S* and δ are $K\overline{K}$ molecules. In this case, one expects that $B(J/\psi \to \rho\delta) = \frac{3}{2} \times B(J/\psi \to \phi S^*) = 3 \times B(J/\psi \to \omega S^*)$ in the SU(3) symmetric limit.[18] Given the measurement $B(J/\psi \to \phi S^*) = (4.4 \pm 0.6 \pm 1.0) \times 10^{-4}$, the upper limit presented in equation (1), and the hint of a small S* signal in the $J/\psi \to \omega\pi\pi$ channel, the $K\overline{K}$ molecule model predictions seem more consistent with the data than the $q\bar{q}$ model.

In the case of the tensor nonet, the $q\bar{q}$ model predicts that $B(J/\psi \to \rho A_2) = 3 \times B(J/\psi \to \omega f') = 3 \times B(J/\psi \to \phi f') = \frac{3}{2} \times B(J/\psi \to K^{*0}\overline{K}^*(1430)^0 + c.c.)$.[13] SU(3) breaking strongly suppresses the $B(J/\psi \to \phi f')$ rate, where $B(J/\psi \to \rho A_2) \approx 16 \times B(J/\psi \to \phi f') \times B(f' \to K\overline{K})$, while the other predictions agree reasonably well with the data. If the S* and δ were $q\bar{q}$ states, a similar suppression might also be expected for the $J/\psi \to \phi S^*$ rate. However, the $K\overline{K}$ molecule prediction would be relatively immune to these SU(3) breaking effects, since both final states contain the same number of s-quarks from the "sea".

6. $K\overline{K}\pi$ AND $\eta\pi\pi$ SYSTEMS RECOILING AGAINST γ, ω AND ϕ.

The 0^{-+} $\iota(1460)$, shown in Figure 9a, is produced with the largest branching ratio in radiative J/ψ decays, and is considered to be a prime "glueball" candidate. Seen by previous e^+e^- experiments,[19] this state has a mass and width of $(1456 \pm 5 \pm 6)$ MeV and $(95 \pm 10 \pm 15)$ MeV, respectively,[20] and decays primarily to (low mass $K\overline{K})\pi$. Another state, the E(1420), also has a $K\overline{K}\pi$ decay mode and mass similar to the ι. First seen in $\bar{p}p$ annihilations at rest, its spin-parity was determined to be 0^-.[22] Subsequent observations of the E(1420) in other hadroproduction experiments showed the state to have $J^{PG} = 1^{++}$ and a dominant K*K decay mode.[21] In the latest measurements, a $J^{PG} = 0^{-+}$ E was seen to couple to $K\overline{K}\pi$[23] and to $\eta\pi\pi$[25] through a $\delta\pi$ intermediate state. The observation of a 0^{-+} E has led to speculation that the E and ι are the same state. By examining the $K\overline{K}\pi$ and $\eta\pi\pi$ spectra recoiling against γ, ω and ϕ in J/ψ decays, the production and decay properties of the E and ι can be determined within one experiment.

In Figure 9b, a prominent peak at 1444 MeV is visible in the summed mass spectra from the $K^\pm K_s^0 \pi^\mp$ and $K^+K^-\pi^0$ systems recoiling against the ω. This object appears to be formed out of nonstrange quarks. Its mass lies midway between the E and ι masses and its width, $24 < \Gamma < 84$ MeV (90% C.L.), is barely compatible with that of the ι. This state decays more strongly to K*K and less strongly to (low mass $K\overline{K})\pi$ than the ι does. Given the large backgrounds, a precise spin determination is difficult; however, the data favor a non-zero spin assignment for this state. Based on these observations, this state is most likely not the ι. However, it does have many features in common with the hadroproduced E resonance, including its width, decay properties and spin. If this state is the E, then the absence of a signal in the $K^\pm K_s^0 \pi^\mp$ system recoiling against the ϕ (Figure 9c) implies that the E is not the $s\bar{s}$ member of an ideally mixed 1^{++} nonet.

In Figures 10a-b, the mass spectra of the $\eta\pi\pi$ systems recoiling against the γ and ω show structure near 1400 MeV. At least one $\eta\pi$ combination was required to have a mass within 50 MeV of the $\delta(980)$ to enhance the peaks which were seen in the raw distributions. The peak at 1395 MeV in the $\gamma\eta\pi\pi$ channel has a mass which is about 25 MeV lower than the E mass. There is no evidence for the ι in this channel.

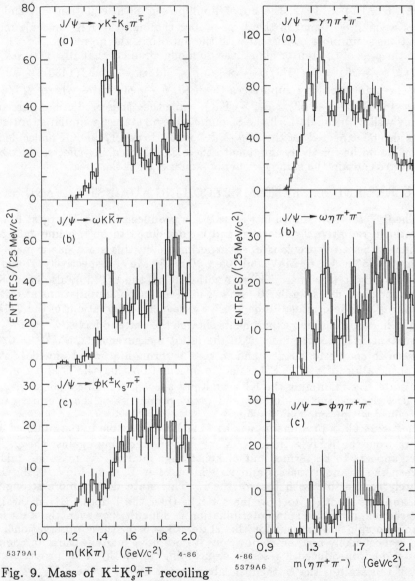

Fig. 9. Mass of $K^\pm K^0_s \pi^\mp$ recoiling against (a) γ and (c) ϕ; (b): sum of $K^\pm K^0_s \pi^\mp$ and $K^+ K^- \pi^0$ recoiling against ω. Backgrounds estimated from ω and ϕ sidebands have been subtracted in (b) and (c), respectively.

Fig. 10. Mass of $\eta\pi\pi$ recoiling against (a) γ, (b) ω and (c) ϕ. At least one $\eta\pi$ mass combination is required to be in the δ mass region (a,b).

The structure seen near 1420 MeV in the $\eta\pi\pi$ system recoiling against the ω has a mass and width consistent with the E(1420). Whether this enhancement is the 0^{-+} state seen by Ando et al.,[25] or the 1^{++} E, has not yet been determined. However, the ratio $B(J/\psi \to \omega K\overline{K}\pi)/B(J/\psi \to \omega\eta\pi\pi) = \frac{7}{9}$ is inconsistent with recent measurements of the 1^{++} E decay rates.[26]

In Figures 10a-c, the mass spectra of the $\eta\pi\pi$ systems recoiling against the γ, ω and ϕ all show structures near 1285 MeV. Based on their masses and widths, it is tempting to assume that these states are due to the 1^{++} D(1285). Traditionally, the D is classified as the $\frac{1}{\sqrt{2}}(u\overline{u} + d\overline{d})$ member of a nearly ideally mixed 1^{++} nonet and its presence in the ϕ recoil spectrum with $B(J/\psi \to \phi D) \approx 0.4 \times B(J/\psi \to \omega D)$ is therefore somewhat surprising. While the 1^{++} nonet may not be ideally mixed, another possibility is that the 0^{-+} $\eta(1275)$,[24,25] a radial excitation of the η, contributes to these spectra. A spin-parity determination of the observed E- and D-like structures is clearly needed to clarify the structure of the 1^{++} and radially excited 0^{-+} nonets.

Finally, the broad peaks in the 1500-2000 MeV range may be related to the pseudoscalar structures seen in the $\gamma\rho\rho$ and $\gamma\omega\omega$ final states, but such an interpretation cannot be verified until a spin-parity analysis in the $\eta\pi\pi$ channel has been completed.

The author would like to thank Drs. M. Chanowitz, N. Isgur and W. Dunwoodie for many useful discussions. This work was supported in part by the Department of Energy, under contract numbers DE – AC03-76SF00515, DE – AC02-76ER01195, DE – AC03-81ER40050, DE – AM03-76SF00034, and by the National Science Foundation.

Table 2. Isoscalar Resonance Parameters.

Object	Seen in $J/\psi \to$	Mass (MeV)	Width (MeV)	Product BR (units of 10^{-4})
f'	γK^+K^-	$1525 \pm 10 \pm 10$	85 ± 35	$6.0 \pm 1.4 \pm 1.2$
f'	ϕK^+K^-	1520 (fixed)	75 (fixed)	$6.4 \pm 0.6 \pm 1.6$
f'	ωK^+K^-	1520 (fixed)	75 (fixed)	< 1.2(90% C.L.)
θ	γK^+K^-	$1720 \pm 10 \pm 10$	130 ± 20	$9.6 \pm 1.2 \pm 1.8$
θ	$\gamma \eta \eta$	$1670 \pm 50^\dagger$	$160 \pm 80^\dagger$	$3.8 \pm 1.6^\dagger$
θ	$\gamma \pi^+\pi^-$	1713 ± 15	130 (fixed)	$2.4 \pm 0.6 \pm 0.5$
θ? (incoh. fit)	ϕK^+K^-	$1671^{+15}_{-17} \pm 10$	$126^{+60}_{-40} \pm 15$	$3.4^{+1.0}_{-0.8} \pm 0.9$
θ?	ωK^+K^-	$1731 \pm 10 \pm 10$	$110^{+45}_{-35} \pm 15$	$4.5^{+1.2}_{-1.1} \pm 1.0$
$S^*(975)$	$\gamma \pi^+\pi^-$	975 (fixed)	35 (fixed)	< 0.7(90% C.L.)
$S^*(975)$	$\phi \pi^+\pi^-$	coupled channel parametrization		$3.4 \pm 0.5 \pm 0.8$
f	$\gamma \pi^+\pi^-$	1269 ± 13	180 (fixed)	$11.5 \pm 0.7 \pm 1.9$
f	$\omega \pi^+\pi^-$	1277 (fixed)	182 ± 10	$49.3 \pm 2.5 \pm 12.5$
D?	$\omega \eta \pi^+\pi^-$	$1283 \pm 6 \pm 10$	$14^{+19}_{-14} \pm 10$	$4.3 \pm 1.2 \pm 1.3$
$D?, \eta(1275)$?	$\phi \eta \pi^+\pi^-$	$1283 \pm 6 \pm 10$	$24^{+20}_{-14} \pm 10$	$1.6^{+0.6}_{-0.5} \pm 0.4$
$\iota(1460)$	$\gamma K^\pm K^0_s \pi^\mp$	$1456 \pm 5 \pm 6$	$95 \pm 10 \pm 15$	$50 \pm 3 \pm 8$
$\iota(1460)$	$\gamma K^+ K^- \pi^0$	$1461 \pm 5 \pm 5$	$101 \pm 10 \pm 10$	$49 \pm 2 \pm 8$
$\iota(1460)$	$\phi K^\pm K^0_s \pi^\mp$	1460 (fixed)	95 (fixed)	< 1.8(90% C.L.)
E?	$\omega K \overline{K} \pi$	$1444 \pm 5^{+10}_{-20}$	$40^{+17}_{-13} \pm 10$	$6.8^{+1.9}_{-1.6} \pm 1.7$
E?	$\phi K^\pm K^0_s \pi^\mp$	1420 (fixed)	52 (fixed)	< 1.1(90% C.L.)
E?	$\gamma \eta \pi^+\pi^-$	≈ 1390	≈ 50	−
E?	$\omega \eta \pi^+\pi^-$	$1421 \pm 8 \pm 10$	$45^{+32}_{-23} \pm 15$	$9.2 \pm 2.4 \pm 2.8$

Parameters of structures seen in the $K\overline{K}$, $\pi\pi$, $K\overline{K}\pi$ and $\eta\pi\pi$ systems recoiling against the γ, ω and ϕ. The product branching ratios are corrected for unobserved decay modes. Unless otherwise noted, the data are MARK III measurements.

† Reference 27.

REFERENCES

1. S. Okubo, Phys. Lett. **5**, 165(1963); G. Zweig, 1964 (unpublished); J. Iizuka, Prog. Theor. Phys. Suppl. **37-38**, 21 (1966).
2. The suppression of the doubly OZI suppressed diagram is experimentally motivated by the observation that $B(J/\psi \to \omega f) > 41 \times B(J/\psi \to \omega f') \times B(f' \to K\bar{K})$ in the MARK III data.
3. S. J. Brodsky et al., Phys. Lett. **73B**, 203(1978);
   K. Koller and T. Walsh, Nucl. Phys. **B140**, 449(1978).
4. R. M. Baltrusaitis et al., Phys. Rev. Lett. **55**, 1723 (1985);
   R. M. Baltrusaitis et al., Phys. Rev. **D33**, 1222(1986).
5. K. Einsweiler, Proceedings of the International Europhysics Conference on High Energy Physics, Brighton, 1983;
   D. Hitlin, Proceedings of the International Symposium on Lepton and Photon Interactions, Cornell, 1983; W. Toki, SLAC-PUB-3262, Nov. 1983.
6. R. M. Baltrusaitis et al., Phys. Rev. Lett. **56**, 107(1986).
7. G. Szklarz, Proceedings of the XXI Rencontre de Moriond, March, 1986.
8. S. Godfrey et al., Phys. Lett. **141B**, 439 (1984).
9. M. S. Chanwitz and S. R. Sharpe, Phys. Lett. **132B**, 413 (1983).
10. R. M. Baltrusaitis et al., Phys. Rev. Lett. **52**, 2126 (1984);
    R. M. Baltrusaitis et al., Phys. Rev. **D33**, 629 (1986).
11. $\tilde{B} = B/P_V^3$, where P_V is the 1^{--} momentum in the η_c rest frame.
12. R. M. Baltrusaitis et al., Phys. Rev. **D32**, 2883 (1985).
13. H. Haber and J. Perrier, Phys. Rev. **D32**, 2961 (1985).
14. J. E. Augustin et al., Orsay Preprint LAL 85-27, July, 1985.
15. A relativistic extension of the Flatté parameterization (S. M. Flatté, Phys. Lett. **63B**, 224(1976)) has been used.
16. R. L. Jaffe, Phys. Rev. **D15**, 267 (1977).
17. J. Weinstein and N. Isgur, Phys. Rev. Lett. **48**, 659 (1982);
    J. Weinstein and N. Isgur, Phys. Rev. **D27**, 588 (1983).
18. A. Seiden, Processes Related to Two Photon Physics, presented at the 7th Int. Workshop on Photon-Photon Collisions, Paris, France, April, 1986.
19. D. Scharre et al., Phys. Lett. **97B**, 329 (1980);
    C. Edwards et al., Phys. Rev. Lett. **49**, 259 (1982).
20. J. Richman, *A Study of the ι in Radiative J/ψ Decays*, CALT-68-1231, December, 1984.
21. C. Dionisi et al., Nucl. Phys. **B169**, 1 (1980);
    T. Armstrong et al., Phys. Lett. **146B**, 273 (1984).
22. P. Baillon et al., Nuovo Cimento **50A**, 393 (1967).
23. S. Chung et al., Phys. Rev. Lett. 55, 779 (1985).
24. M. Stanton et al., Phys. Rev. Lett. **42**, 346 (1979).
25. A. Ando et al., KEK Preprint 86-8, May, 1986.
26. O. Villalobos Baillie, Proc. Int. Europhysics Conf. on High Energy Physics, Bari, Italy, July, 1985. In this experiment, a 1^{++} E is seen with $B(E \to \eta\pi\pi) = (0.12 \pm 0.08) \times B(E \to K_s^0 K^\pm \pi^\mp)$.
27. E. D. Bloom and C. W. Peck, Ann. Rev. Nucl. Part. Sci **33**, 143 (1983).

HADRON STRUCTURE FROM LATTICE QCD

R.M. Woloshyn
TRIUMF, Vancouver, Canada, V6T 2A3

INTRODUCTION

Hadron electromagnetic form factors and radiative transition amplitudes, extracted from the vector current three-point functions, are a useful probe of hadron structure.[1-4] They can be calculated in lattice QCD in a colour gauge invariant way and, ultimately, are directly comparable to experimental results. In this paper some recent lattice results for the pseudoscalar meson electric form factor and the vector meson decay amplitude are presented.

THREE-POINT FUNCTION

A conserved vector current can be constructed from the Lagrangian for fermions on the lattice using the Noether procedure. For Wilson's form of the fermion lattice action the current is

$$j_\mu(x) = -i\kappa[\bar{\psi}(x)(1+\gamma_\mu)U_\mu(x)\psi(x+a_\mu) - \bar{\psi}(x+a_\mu)(1-\gamma_\mu)U_\mu^\dagger(x)\psi(x)]. \quad (1)$$

With appropriate charge assignments the electromagnetic current can then be constructed. The interpolating fields for pseudoscalar and vector mesons are taken to be $O_0(x)=\bar{\psi}(x)\gamma_5\psi(x)$, $O_1^{(i)}=\bar{\psi}(x)\gamma_i\psi(x)$.

To get the pseudoscalar meson electric form factor the lattice three-point function

$$A = \sum_{\vec{x}} e^{-i\vec{p}'\cdot\vec{x}} < O_0(x) \sum_{\vec{z}} e^{i\vec{q}\cdot\vec{z}} j_4(z) O_0^\dagger(0)> \quad (2)$$

is calculated. At large time separations $0 \ll t_x \ll t_z$ this is proportional to the matrix element

$$<0^-\vec{p}'|j_4(0)|0^-\vec{p}> = \frac{(E_{\vec{p}'}+E_{\vec{p}})}{2\sqrt{E_{\vec{p}'}E_{\vec{p}}}} F(q) . \quad (3)$$

For radiative decay the three-point function

$$\tilde{A} = \sum_{\vec{x}} e^{-i\vec{p}'\cdot\vec{x}} <O_0(x) \sum_{\vec{z}} e^{i\vec{p}'\cdot\vec{z}} j_1(z) O_1^{(3)\dagger}(0)> \quad (4)$$

is calculated. This yields the matrix element $<0^-,\vec{p}'|j_1(0)|1^-,\vec{0}>$.

ELECTRIC FORM FACTOR

Calculations were done in a model for quenched lattice QCD using SU(2) colour at one value of the coupling constant (β=2.3). The lattice size was 10×20×10×10. Form factors were calculated at one value of momentum transfer ($q=\pi/10$, the smallest non-zero momentum value available). The form factor was parametrized by $F(Q)=(1+Q^2/\lambda^2)^{-1}$ and the charge radius was calculated from the derivative of $F(Q^2)$ at $Q^2=0$.

Form factors and charge radii obtained in the Wilson scheme for different values of the hopping parameter are shown in Fig. 1. As expected larger quark masses result in smaller radii. The parameter λ^2 scales with the vector-meson mass squared (Fig. 2), reminiscent of vector dominance ideas.

A comparison of results obtained using the Wilson and staggered fermion schemes is given in Fig. 3. Pseudoscalar (π) and vector-meson (ρ) masses were used to fix the lattice scales, which, however, came out to be quite different in the two fermion schemes. At smaller mass the Wilson and staggered results seem to be compatible.

RADIATIVE VECTOR-MESON DECAY

The matrix element of the M1-radiative decay $V \to P\gamma$ can be written in the form

$$\langle 0^-, p' | j_\mu | 1^-, p \rangle = \frac{(-ie)g\,\epsilon_{\mu\rho\sigma\tau}P_\rho P'_\sigma S_\tau}{2\sqrt{E_0(p')E_1(p)}\,M_1} . \qquad (5)$$

The transition strength g is related to the decay width (in the rest frame, $\vec{p}=0$)

$$\Gamma = \frac{1}{3}\alpha g^2 \frac{p'^3}{M_1^2} . \qquad (6)$$

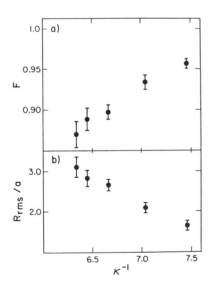

Fig. 1. Electric form factor and charge radius versus inverse of the hopping parameter.

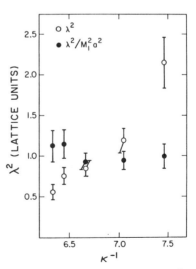

Fig. 2. Behavior of the λ^2 parameter versus the inverse hopping parameter.

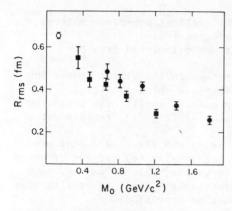

Fig. 3. Comparison of charge radii using Wilson (●) and staggered (■) fermions.

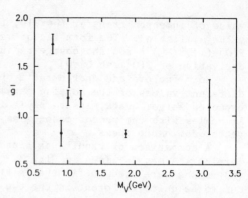

Fig. 4. Lattice (●) and experimental (▲) results for the reduced radiative transition strength.

It should be borne in mind that the radiative decay amplitude is calculated off-shell, i.e., at an unphysical value of the momentum. Even for twenty lattice sites the smallest nonzero momentum is still too large to allow on-shell energy and momentum conservation.

So far, calculations have been done using the Wilson fermion scheme. Staggered fermions present a complication due to the fact that species doubling has not been completely eliminated. The lowest mass vector and pseudoscalar mesons (reducible to local lattice operators) do not carry the same Susskind flavour and thus are not connected by the M1-multipole operator. The calculation therefore involves operators which can not be reduced to local form. Reliable results would require very high statistics.

Preliminary results for the reduced transition strength (eliminating isospin and quark fractional charge factors) are shown in Fig. 4. Also shown are transition strengths extracted from the measured $\rho \to \pi\gamma$, $\phi \to \eta\gamma$ and $\psi \to \eta_c\gamma$ decay widths. In general the lattice results are lower than the experimental values, a trend also observed for purely hadronic coupling constants.[5]

ACKNOWLEDGEMENT

Some of this work was done in collaboration with A.M. Kobos and W. Wilcox. This work is partially supported by the Natural Sciences and Engineering Research Council of Canada.

REFERENCES

1. W. Wilcox and R.M. Woloshyn, Phys. Rev. D32, 3282 (1985).
2. R.M. Woloshyn and A.M. Kobos, Phys. Rev. D33, 222 (1985).
3. R.M. Woloshyn, Phys. Rev. D, in press.
4. W. Wilcox and K.F. Liu, Phys. Lett., in press.
5. S. Gottlieb et al., Nucl. Phys. B263, 704 (1986).

LATTICE GAUGE THEORY AS A NUCLEAR MANY-BODY PROBLEM

G. J. Mathews, S. D. Bloom, and N. J. Snyderman
University of California, Lawrence Livermore National Laboratory
Livermore CA, 94550

ABSTRACT

We discuss the conceptual connection between lattice quantum chromodynamics and a nuclear many-body problem. We begin with an illustrative example of how the O(3) nonlinear sigma model in (1+1) dimensions can be computed with a nuclear shell-model code with a speed which is competitive with other approaches. We then describe progress toward the implementation of this technology in lattice SU(2) Yang-Mills gauge theory.

INTRODUCTION

Since this is a conference on the intersections between particle and nuclear physics, we should begin with a discussion of how it is that lattice gauge theory occurs at such an intersection. From a particle theorist's point of view, lattice gauge theory, as proposed by Wilson,[1] is the only formulation of quantum chromodynamics which is amenable to numerical solution and thus, is a vehicle in which to analyze the theory of strong interactions. From a nuclear physicst's view point, such an ability to obtain numerical results in QCD will allow for a more fundamental understanding of nucleonic structure and interactions.

There is, however, another reason why this particle-physics problem is of interest to a nuclear physicist. That reason will be the main focus of this paper. It is that, at a computational level, the Hamiltonian formulation[2] of QCD lattice gauge theory is very similar to a problem which has been at the center of nuclear physics for many years; namely, the determination of the eigenstates of a nonrelativistic interacting many-body system of coupled angular momenta. In this paper, we develop this analogy and show, as an example, how our existing nuclear shell-model code can be applied, without modification, to a simple lattice theory, the O(3) nonlinear sigma model. This model has many properties in common with QCD. We also report on our progress toward adapting this approach to the more realistic SU(2) Yang-Mills pure gauge theory in (3+1) dimensions.

O(3) NONLINEAR SIGMA MODEL AND THE NUCLEAR SHELL MODEL

To illustrate the analogy between lattice QCD and a nuclear many-body problem, it is instructive to begin with a simpler lattice theory in which this equivalence is more transparent. The O(3) nonlinear sigma model in (1+1) dimensions has fundamental similarities to the SU(2) gauge theory in (3+1) dimensions. For example, it has a set of massless perturbative degrees of freedom (Goldstone bosons), the analogue of gluons in Yang-Mills gauge theories, which do not appear in the spectrum of the theory. In both cases, only massive hadronic particles appear in the spectrum. The theory is also similar to SU(2) in that it is asymptotically free, has instantons, and is classically scale invariant with the consequent dimensional transmutation in the quantum field theory. Furthermore, the lattice O(3) model is computationally quite close to the SU(2) lattice gauge theory. As we shall see, both problems are excercises in the coupling of many angular momenta and the Hamiltonian matrix elements are, in both cases, linear combinations of 3n-j symbols.

The Lagrangian density for the nonlinear sigma model is,

$$L = (1/g^2)\partial_\mu \vec{\phi} \cdot \partial^\mu \vec{\phi} \quad , (\mu = 0,1) \tag{1}$$

where, the field variable, $\vec{\phi}$, is constrained by $\vec{\phi} \cdot \vec{\phi} = 1$. Thus, $\vec{\phi}$ is a three-component unit spin which can be parameterized by the spherical polar angles, θ and φ; $\vec{\phi} = (\sin\theta\cos\varphi, \sin\theta\sin\varphi, \cos\theta)$. The discretized Hamiltonian[3] corresponding to equation (1) can be written,

$$Ha = (g^2/4)\sum_n \vec{L}_n^2 + (2/g^2)\sum_n (1 - \vec{\phi}_n \cdot \vec{\phi}_{n+1}) \tag{2}$$

where, a, is the lattice spacing and n labels the sites on a one-dimensional lattice. The quantum coordinates are θ and φ, and their conjugates give the kinetic-energy differential operator, $\vec{L}^2$ in Eq. (2), whose eigenfunctions are the spherical harmonics.

$$\vec{L}_n^2 Y_{\ell m}(\theta_n, \varphi_n) = \ell(\ell+1) Y_{\ell m}(\theta_n, \varphi_n) \tag{3}$$

The interaction term in equation (2) simply becomes,

$$\vec{\phi}_n \cdot \vec{\phi}_{n+1} = \frac{4\pi}{3}\sum (-1)^m Y_{\ell,m}(\theta_n, \varphi_n) Y_{\ell,-m}(\theta_{n+1}, \varphi_{n+1}) \tag{4}$$

Therefore, by selecting a basis of angular momentum eigenstates one can immediately see an analogy with a nuclear shell-model problem, i.e. the Hamiltonian can be written,

$$H = \sum_i T_i + \sum_{i,j=i+1} V_{i,j} \tag{5}$$

where the "single-particle" orbitals correspond to sets of angular momentum eigenstates at each lattice site. The "single-particle" energies, T_i, now are $\ell(\ell+1)$, and the "two-body" potential corresponds to an interaction between adjacent lattice sites. This two-body potential is a simple dipole-dipole interaction, very similar to the quadrupole force often applied in shell-model calculations.

We have diagonalized the nonlinear sigma model Hamiltonian (Eq. 2) using the Livermore system of vectorized Lanczos-method shell-model codes[4]. Figure 1 compares our results for a 5 site lattice in a basis constructed by 3 operations of H on the strong-coupling ground state (ℓ = 0 on all sites). This corresponds to a basis of up to 3707 m-scheme basis vectors. This figure shows the lattice mass gap (essentially the excitation energy of the L = 1 first excited state relative to the ground state in units of the lattice spacing) as a function of the coupling constant, ($2/g^2$). Our results compare well with previous calculations based upon the Monte-Carlo functional integral[5] or a Hamiltonian variational calculation[6]. Even though we utilize a much smaller lattice we are able to obtain reasonable values for some distance into the weak-coupling regime. Using the Lanczos method in this way, seems to offer the advantage of fairly rapid convergence to the exact mass gap for a given basis. This calculation was run in about 100 sec of CDC 7600 cpu time per data point. Our good agreement may be somewhat fortuitous, however, since the results depend upon chosing the right combination of lattice and basis sizes. Nevertheless, these results have encouraged us to apply these same techniques to the more challanging Yang-Mills theories.

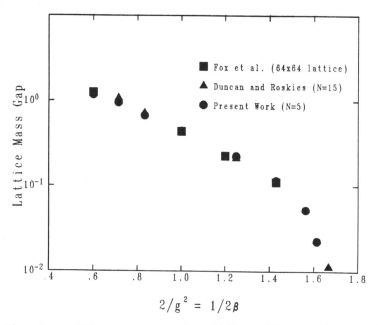

Fig. 1 Lattice mass gap for O(3) nonlinear sigma model in (1+1) dimensions. Results from our Lanczos-method shell-model code are compared with other techniques.

SU(2) LATTICE GAUGE THEORY IN (3+1) DIMENSIONS

Unlike QED without the electron field, which is a theory of free photons, the SU(2) gauge theory without quark fields is an interacting theory of closed-loop electric flux vorticies (glueballs). In the presence of quark fields, this electric flux string can end on quark charges, making quark-antiquark mesons. (In the SU(3) theory, "Y" configurations of flux are also possible, which confine three quarks.)

In the lattice theory, on each link there is the field variable, exp(igAa), where the vector potential, $A = \vec{A}\cdot\vec{\tau}/2$, is a 2x2 matrix. The product of these operators around a closed loop on the lattice is the exponential of the magnetic flux through the loop. The vector potential becomes a set of angular variables, $\vec{\theta} = g\vec{A}a$, which are the quantum coordinates. In the gauge where the scalar potential is zero, the electric fields become canonical momenta and the Hamiltonian can be written,[2]

$$Ha = (g^2/2)\sum_\ell \vec{J}^2 + (2/g^2)(2 - \sum_p tr U_1 U_2 U_3^\dagger U_4^\dagger) \qquad (6)$$

where, $\vec{J}^2$ is a kinetic-energy differential operator whose eigenfunctions are the Wigner D- matrices, and eigenvalues are $j(j+1)$, where $j = 0, 1/2, 1, 3/2, \cdots$. The interaction term, $tr U_1 U_2 U_3^\dagger U_4^\dagger$, is a contraction of four D-matricies around a closed loop of four adjacent links on the lattice, a placquette. The basis states correspond to the excitation of angular momenta of links on the lattice for different numbers of placquettes under the addtional constraint (Gauss' law) that the the angular momenta of the links intersecting a given site must couple to zero.

Because of the large number of states associated with the implementation of a four-body operator, it is not feasible to treat this problem with an m-scheme code as we did with the nonlinear sigma model discussed in the previous section. With the constraint of Gauss' law, imposed by Clepsch-Gordan coefficients, the m's can be summed over, so the basis states are completely specified by the angular momenta on the links of the lattice. Therefore, the Hamiltonian SU(2) lattice gauge theory can also be thought of as a many-body system of coupled angular-momentum states. The main distinction from a nuclear many-body problem will be that the interaction term is a four-body operator.

Since the Lanczos algorithm[4] is based upon an iterative scheme involving successive operations of the Hamiltonian on a given start vector (the noninteracting strong-coupling ground state), we have found it convenient to rewrite the algorithm in terms of a nonlinear iterative combination of the diagonal matrix elements of the strong coupling ground state for powers of the Hamiltonian. These matrix elements can then be decomposed as a sum of products of the matrix elements of the Hamiltonian between arbitrary intermediate states. These intermediate matrix elements can be conveniently reduced (via angular-momentum coupling algebra) to the product of a Racah coefficient and two 9-j symbols around the four corners of a placquette. The problem then reduces to the calculation of these coupling coefficients.

A major simplification can be introduced by exploiting translational invariance on the lattice. The most efficient exploitation of this degeneracy has been proposed by Horn and Weinstein[7,8]. The problem of computation of $<H^n>$ can be reduced to the calculation of only terms which have a linear dependence on the volume, i.e. an overall translational degeneracy of N, where N is the number of placquettes on the lattice. These linear contributions correspond to connected graphs, (intersecting placquettes). The usual calculation of connected graphs involves excluded volume factors which are extremely difficult to compute. Horn and Weinstein[7,8] have found a way to circumvent this problem such that only manifestly connected graphs occur in the computation of the matrix elements. This approach can lead to an enormous simplification of the problem. For example, the calculation of $<H^4>$ on a 3x3x3 lattice requires 7695 intermediate states. This calculation can be reduced, however, to the calculation of 3 manifestly connected graphs. We are currently in the process of implementing this algorithm for the calculation of high powers of $<H^n>$. We estimate that with these simplifications we should be able to calculate to order $<H^{20}>$.

At present we have available matrix elements of up to $<H^{10}>$ (basis of H^5) computed by Duncan and Roskies[6] using a different technique. This limited basis, however, only allows for 5 Lanczos iterations on the strong-coupling ground state start vector. Although with each iteration we are approaching closer to the weak coupling regime, many more Lanczos iterations, within this fixed basis, will be required before we approach the weak coupling regime. With a basis of up to H^{10}, and applying enough Lanczos iterations in this basis, we expect to reach the scaling regime necessary to obtain reliable calculations of the hadronic mass spectrum.

ACKNOWLEDGMENT

Work performed under the auspices of the U. S. Department of Energy by the Lawrence Livermore National Laboratory under contract number W-7405-ENG-48.

REFERENCES

1. K. G. Wilson, Phys. Rev. D10, 2435 (1974).
2. J. Kogut and L. Susskind, Phys. Rev., D11, 395 (1975).
3. C. J. Hammer, J. B. Kogut, and L. Susskind, Phys. Rev., D19, 3091 (1979).
4. G. J. Mathews, S. D. Bloom, K. Takahashi, G. M. Fuller, and R. F. Hausman, in "Nuclear Shell Models", M. Vallieres and B. H. Wildenthal, eds., (World Scientific; Singapore) (1985) pp. 447-457.
5. G. Fox, R. Gupta, O. Martin, and S. Otto, Nucl. Phys., B205, 188 (1982).
6. A. Duncan and R. Roskies, Phys. Rev., D31, 364, (1985).
7. D. Horn and M. Weinstein, Phys. Rev., D30, 1256 (1984).
8. D. Horn, M. Karliner, and M. Weinstein, Phys. Rev., D31, 2589 (1985).

WIGNER-WEYL AND NAMBU-GOLDSTONE REALIZATIONS

OF

CHIRAL SYMMETRY IN QUANTUM CHROMODYNAMICS

R. Acharya
Physics Department, Arizona State University, Tempe, Arizona 85281

and

P. Narayana Swamy
Physics Department, Southern Illinois University, Edwardsville,
Illinois 62026.

Abstract

We employ the gauge technique developed by Salam, in conjunction with the Schwinger-Dyson equation for the quark propagator to study the Wigner-Weyl and Nambu-Goldstone realizations of chiral symmetry in Quantum Chromodynamics at zero temperature. We demonstrate that due to the presence of the transverse piece of the quark gluon vertex, the chirally symmetric Wigner-Weyl solution is <u>not</u> realized.

Quantum Chromodynamics [1] (QCD) is a theory of strong interactions based on three colors and N_F flavors. The Goldstone [2] nature of the pion, proposed some time ago by Weinberg [3], forms the basis of current understanding of the special role of pion in QCD. In the absence of explicit symmetry breaking via current quark mass terms, the chiral $SU_L(N_F) \times SU_R(N_F)$ symmetry is expected to be realized in the Nambu-Goldstone (NG) fashion [2], resulting in $N_F^2 - 1$ massless pseudoscalar mesons.

Standard continuum field theory provides a convenient framework in which to investigate the problem of chiral symmetry breakdown [4-9]. Recent studies [10,11] have relied on a nonperturbative treatment of the dynamical content of QCD described by the Schwinger-Dyson equations and the Ward-Takahashi (WT) identities satisfied by the vertex functions. Such nonperturbative investigations employ a "solution" of the WT identities, a solution that specifies the longitudinal part of the vertex functions developed by the work of Ball and Zachariasen [7,12] in the context of the gauge technique [13]. The Schwinger-Dyson equation then leads to an equation for the dynamical mass of the quark, analogous to the gap equation obtained by Lane [14] in his work on the Goldstone realization of chiral symmetry and asymptotic freedom. Such equations are generally believed to admit two solutions: the broken symmetry solution where the dynamical mass $M(p^2) \neq 0$, with $M(0)$ representing the strength of chiral symmetry breaking and the symmetric solution $M(p^2) = 0$, corresponding to unbroken symmetry. It is usually argued that the symmetric solution corresponds to lower energy or that it does not satisfy stability criteria. The problem of eliminating the symmetric solution, however, has never really been resolved to everyone's satisfaction.

It is the purpose of this note to reexamine the problem of NG realization of chiral symmetry in QCD in its most general form. Our investigation has a goal which is twofold. One is to analyze the

Schwinger-Dyson equation for the quark propagator and WT identity for the quark gluon vertex function without any questionable assumptions and without any approximations that may cloud the issues. Any conclusions we arrive at would thus be truly model independent. As a result of this analysis we shall obtain the equation which determines the dynamical quark mass. Although we are unable to establish that this equation possesses nontrivial solutions, we derive it nonetheless in a form that is suitable for further investigation.

Following Mandelstam [15], we may conveniently carry out the analysis in the covariant Landau gauge, ignoring the ghost fields. This is justified since one does not expect chiral symmetry phenomena to arise from ghosts. The quark gluon proper vertex function Γ^α_μ will consequently obey the naive WT identity

$$q^\mu \Gamma^\alpha_\mu(p,p+q) = \tfrac{1}{2}\lambda^\alpha S^{-1}(p+q) - S^{-1}(p)\tfrac{1}{2}\lambda^\alpha \qquad (1)$$

where λ^α are the color SU(3) matrices, $\alpha = 1,2,\ldots,8$ and $S^{-1}(p)$ is the inverse of the renormalized quark propagator with the standard decomposition

$$S^{-1}(p) = \not{p} - \Sigma(p) = A(p^2)\not{p} - M(p^2),$$
$$S(p) = F(p^2)\not{p} + G(p^2). \qquad (2)$$

The Schwinger-Dyson equation for $S^{-1}(p)$ reads [1]

$$S^{-1}(p) = \not{p} - (2\pi)^{-4}\int d^4k\, \gamma_\mu \tfrac{1}{2}\lambda^\alpha D^{\mu\nu}_{\alpha\beta}(p-k) S(k)\Gamma^\beta_\nu(p,k). \qquad (3)$$

We shall often suppress the color indices by employing the convention

$$\Gamma^\alpha_\mu(p,p') = \tfrac{1}{2}\lambda^\alpha \Gamma_\mu(p,p') \qquad (4)$$

as a means to eliminate the λ^α matrices.

It has been shown by Pagels [4] that the inverse of the quark propagator admits an unambiguous separation of the quark mass into a current part and a constituent part, thus

$$M(p^2) = M_E(p^2) + M_D(p^2) , \qquad (5)$$

where the subscripts signify explicit and dynamical symmetry breaking. We recall that chiral symmetry is implied by [4] $m_o(\Lambda) \equiv 0$ for all values of the cut-off parameter Λ, where m_o is the bare quark mass. Then the renormalized current quark mass $m = Z_m^{-1} m_o = 0$, where Z_m renormalizes the mass operator $\bar{q}q$. Since we are interested in the case when current quark mass is absent, we may in fact set $M_E = 0$ and thus not distinguish between M and M_D. In this case, the assertion of spontaneous chiral breaking implies that one is dealing with the infrared dominated "regular" solution spelled out by Lane[14], Pagels[4] and Politzer[16] with the soft asymptotic behavior $M_D \sim p^{-2} (\ln p^2)^c$, characteristic of a dynamically generated mass. Since ultraviolet softness is a general feature of a field theory where the mass scale is introduced through dynamical breaking of scale invariance, we choose here to work with the regular solution. This is in accord with experiment, as lucidly discussed by Pagels[17] and by Pagels and Stokar[18].

The WT identity in Eq.(1) may now be solved in the standard fashion and yields the familiar expression[7,11]

$$\Gamma_\mu(p,p') = \tfrac{1}{2}(A+\tilde{A})\gamma_\mu + (M-\tilde{M})(p'^2-p^2)^{-1}(\gamma_\mu \not{p} + \not{p}' \gamma_\mu)$$
$$- \tfrac{1}{2}(A-\tilde{A})(p'^2-p^2)^{-1}\left[2\not{p}\gamma_\mu\not{p}' + (p'^2+p^2)\gamma_\mu\right]$$
$$+ \Gamma_\mu^T(p,p') , \qquad (6)$$

where $\tilde{A} = A(p'^2)$, $\tilde{M} = M(p'^2)$ and $p' = p+q$. The transverse piece explicitly displayed here satisfies the condition

$$(p'-p)^\mu \Gamma_\mu^T(p,p') = 0, \qquad (7)$$

and exists by virtue of gauge invariance. The general form of the full gluon propagator in the Landau gauge, consistent with Slavnov-Taylor identity is[1]

$$D^{\mu\nu}{}_{\alpha\beta}(q) = g^2 \delta_{\alpha\beta} \left[q^\mu q^\nu/q^2 - g^{\mu\nu}\right] q^{-2} D(q^2) \qquad (8)$$

which we insert in Eq.(3), valid for all values of p. We may take advantage of the resulting simplification by setting p=0 in Eq.(3). Accordingly we have

$$S^{-1}(0) = g^2 \int d^4k \, D(k^2)\gamma_\mu \, k^{-2} \left[g^{\mu\nu} - k^\mu k^\nu/k^2\right] S(k)\Gamma_\nu(0,k). \quad (9)$$

Here we have absorbed all constant factors including the invariant $\lambda^\alpha \lambda^\alpha$ in g^2. Following Cornwall [8], we may employ the effective quark mass M(0) defined by

$$M(0) = M(p^2=0) = -S^{-1}(0) \quad (10)$$

which measures the 'strength' of chiral symmetry breakdown, not to be identified with either the current quark mass m or the constituent mass. By hypothesis of breakdown of chiral symmetry, $S^{-1}(0) \neq 0$. This of course implies $S(0) < \infty$.

We first need to compute $\Gamma_\nu(0,k)$ in order to deal with Eq.(9). From Eq.(6), we obtain after some algebra,

$$\Gamma_\nu(0,k) = A(k^2)\gamma_\nu + k^{-2}\left[M(0)-M(k^2)\right] \not{k} \, \gamma_\nu + \Gamma_\nu^T(0,k). \quad (11)$$

Substituting in Eq.(9) and performing the necessary algebra with the γ matrices, we may simplify by employing the familiar result [19]

$$\int d^4k \not{k} \, H(k^2) = 0 \quad (12)$$

which follows for an arbitrary function $H(k^2)$ from angular integration, and obtain

$$M(0) = g^2 \int d^4k \, D(k^2)\gamma_\mu \, k^{-2}\left[(k^\mu k^\nu/k^2 - g^{\mu\nu})\right]$$
$$\left[\gamma_\nu F(k^2)M(0) + S(k)\Gamma_\nu^T(0,k)\right]. \quad (13)$$

After further simplification we thus derive

$$M(0) = -3g^2 M(0) \int d^4k \, k^{-2} D(k^2) F(k^2)$$
$$+ g^2 \int d^4k \, k^{-2} D(k^2)\gamma_\mu \left[k^\mu k^\nu/k^2 - g^{\mu\nu}\right] S(k) \, \Gamma_\nu^T(0,k). \quad (14)$$

It is difficult to effect further reduction of this equation without some knowledge of the transverse part of the vertex function, which is an unknown arbitrary quantity. The best we can do is to utilize an invariant decomposition

$$\Gamma_\nu^T(0,k) = F_1(k^2) + k_\nu F_2(k^2) + \gamma_\nu \not{k} F_3(k^2) + \not{k} k_\nu k^{-2} F_4(k^2) \quad (15)$$

where $F_i(k^2)$ are arbitrary invariant functions of k^2. The transversality condition, Eq.(7) relates F_4 and F_3 to the other two and hence we obtain

$$\Gamma_\nu^T(0,k) = (\gamma_\nu - \not{k} k_\nu/k^2)F_1(k^2) + (k_\nu - \gamma_\nu \not{k})F_2(k^2). \quad (16)$$

The requirement that the decomposition be devoid of kinematical singularities leads to specific restrictions [12]. For instance, it is seen that $F_1(0) = 0$ and that F_2 must be regular at $k^2=0$. If we utilize Eqs.(12) and (16), we can again simplify the second term on the right hand side of Eq.(14). As a consequence we obtain

$$M(0) = -3g^2 M(0) \int d^4k \, k^{-2} \, D(k^2) F(k^2)$$
$$-3g^2 \int d^4k \, k^{-2} \, D(k^2) (GF_1 + k^2 FF_2). \quad (17)$$

By carrying out the angular integrations and absorbing constants in g_1^2, we may now rewrite this equation in the form

$$M(0) = -g_1^2 M(0) \int_0^\infty dx \, D(x) F(x)$$
$$-g_1^2 \int_0^\infty dx \, D(x) \left[G(x) F_1(x) + x F(x) F_2(x) \right]. \quad (18)$$

We cannot of course reduce this any further without model dependent assumptions about F_1 and F_2. We can, however, derive an important conclusion.

The most important conclusion stems from the observation that Eq.(18) does not in general admit the chirally symmetric solution $M(p^2) = 0$. Since $G(x) = M(x)/(A^2 x - M^2)$, it follows that the chirally symmetric solution is allowed by Eq.(18) if and only if

$$\int_0^\infty dx \, x \, D(x) F(x) F_2(x) = 0. \quad (19)$$

This condition is not likely to be satisfied since F remains non-zero in the chiral Wigner-Weyl limit; F_2 being one of the invariants in Eq.(16), is in general completely arbitrary and hence nonvanishing. In other words, theoretically, this alternative is not "natural" because a quantity that has no reason to vanish in general is indeed different from zero [20]. Hence we conclude that the chiral symmetric solution is ruled out in QCD.

It is interesting to observe that Eq.(18) is the QCD analog of the Nambu-Jona-Lasinio consistency condition[2]. There is, however, an all important difference. Because of the presence of the transverse part Γ_ν^T in the second integral, Eq.(18) is not an eigenvalue condition for g^2. This should be so, since, as emphasized by Weinberg[21], "a small change in the gauge coupling corresponds to a general change of mass scale and cannot shift a massless bound state away from zero mass". If chiral breaking solutions exist, they should after all occur for a range of coupling.

A knowledge of the infrared behavior of the propagator and the quark gluon vertex function was not needed and hence not employed in our demonstration that the chirally symmetric solution is ruled out. One may therefore make the observation that chiral symmetry breakdown may not necessarily be linked with confinement. Thus although confinement may imply chiral breaking[4,5] it should be recognized that chiral symmetry breakdown may occur in nonconfining theories as well[9]. In other words, quark confinement is a sufficient condition for chiral symmetry breakdown but not a necessary one. While we have ruled out the solution corresponding to unbroken chiral symmetry, we have not really established the existence of a solution, $M(p^2) \neq 0$ of Eq.(18). This problem indeed requires a specific model for the transverse piece form factors F_1 and F_2.

We should point the gauge invariant nature of our analysis. We have verified that our calculations remain valid if carried out in an arbitrary covariant gauge in which Eq.(8) is replaced by

$$D^{\mu\nu}{}_{\alpha\beta}(q) = g^2 \delta_{\alpha\beta} \left[(1-\alpha)q^\mu q^\nu/q^2 - g^{\mu\nu}\right] q^{-2} D(q^2) \qquad (20)$$

where α is a constant gauge parameter. Explicit calculation shows that the only change that occurs is the replacement of factor 3 in Eq.(17) et seq by the factor $(3+\alpha)$. Hence our conclusions remain valid in all covariant gauges.

REFERENCES

1. W.Marciano and H.Pagels, Phys.Rep.36C, 137 (1978).
2. Y.Nambu, Phys.Rev.Lett.4,380(1960); Y.Nambu and G.Jona-Lasinio, Phys.Rev.122,345(1961); J.Goldstone,A.Salam and W.Weinberg,Phys.Rev.127,965(1962).
3. S.Weinberg,Lectures on Elementary Particles and Quantum Field Theory,Brandeis University Summer Institute 1970, MIT Press (Cambridge, Mass)1970, p.283.
4. H.Pagels,Phys.Rev.D19,3080(1979).
5. C.Callan,R.Dashen and D.Gross, Phys.Rev.D17,2717(1978); D.G. Caldi, Phys.Rev.Lett.39, 121(1977).
6. C.Callan,R.Dashen and D.Gross,Phys.Rev.D19,1826(1979).
7. J.S.Ball and F.Zachariasen, Phys.Lett.106B,133(1981).
8. J.M.Cornwall,Phys.Rev.D22,1452(1980).
9. J.Finger and J.Mandula, Nuc.Phys.B199,168 (1982); A.Patrasciouiu and M.Scadron, Phys.Rev.D22,2054(1980); M.Peskin in Recent Advances in field theory and statistical mechanics,editors J.Zuber and R.Stora , N.Holland Pub.Co.(1984)p.219.
10. R.Acharya and P.Narayana Swamy, Phys.Rev.D26,2797(1982).
11. R.Acharya and P.Narayana Swamy, Nuovo Cimento 86A,157 (1985); Z.Phys.C--Particles and fields 28,463(1985).
12. J.S.Ball and T.Chiu, Phys.Rev.D22,2542(1980); M.Baker,J.Ball and F.Zachariasen, Phys.Rev.31, 2575 (1985).
13. A.Salam, Phys.Rev.130, 1287 (1963); R.Delburgo and P.West, J.Phys.A10, 1049 (1977).
14. K.Lane, Phys.Rev.D10, 2605(1974).
15. S.Mandelstam, Phys.Rev.D20, 3223 (1979).
16. H.Politzer, Nuc.Phys.B117, 397 (1976); also V.Elias and M.Scadron, Phys.Rev.D30, 647 (1984).

17. H.Pagels, Phys.Rev.D21, 2336(1980).
18. H.Pagels and S.Stokar, Phys.Rev.D20, 2947 (1979).
19. See e.g., S.Schweber, An Introduction to relativistic quantum field theory, Harper and Row Pub.Co.(New York)1964, p.521.
20. J.Gasser and H.Leutwyler, Phys.Rep.C87, 78 (1982).
21. S.Weinberg, Phys.Rev.D13, 974 (1976).

HEAVY ION PHYSICS

Coordinators: O. Hansen
J. W. Harris

DIRECT LEPTON PRODUCTION IN HADRONIC AND NUCLEAR COLLISIONS AT LOW PT AND LOW PAIR MASS

G. Roche(*)
Lawrence Berkeley Laboratory, Berkeley, CA 94720

ABSTRACT

A brief review of the direct production of single leptons and dileptons in hadronic collisions at low pt and low pair mass is given. Several open questions are pointed out and the universality of lepton production is briefly discussed. The Bevalac Di-Lepton Spectrometer project is presented.

INTRODUCTION

Lepton production has been extensively studied in hadronic collisions from the GeV region up to the highest available energies at Serpukhov, Fermilab and CERN. In ultra-relativistic Heavy Ion collisions, it is considered as a possible signature of the Quark-Gluon Plasma formation. A typical mass spectrum of lepton pairs produced at high energies is shown in Fig. 1. It exhibits peaks corresponding to the various meson resonances and a continuum. The high mass continuum is well interpreted by the Drell-Yan hard quark-antiquark annihilation process, while the low mass continuum, sometimes referred to as "the anomalous dilepton continuum", is not fully understood. Lepton production has also been investigated in inclusive experiments. Here again, there remain questions about the lepton yield at low pt compared to the inclusive pion one, after subtracting known contributions. The copious single lepton yield at low pt and the low pair mass anomaly are frequently considered in connection, even though the issue is not yet completely clear.

I will mostly restrict the talk to low pt and low pair mass direct lepton production (direct leptons are those not produced in the decay of known particles or resonances). There exist previous reviews on the subject(2). I will first summarize existing experimental data and models on dileptons, trying to point out the basic envisionned processes, and then go to single leptons which better leads to the possibility

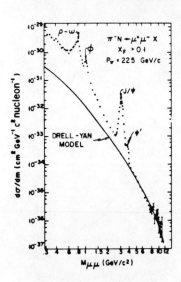

Fig. 1. Mass spectrum of dimuons measured by Chicago-Princeton group and contribution of the Drell-Yan process(1).

of the existence of some universal process from the few GeV domain to the highest energies. Finally, I will present the Bevalac Di-Lepton Spectrometer(DLS) project and conclude with some remarks on the nucleus-nucleus case for which no experimental data exist yet.

THE LOW MASS DILEPTON CONTINUUM

A sample of experimental works on low mass dileptons is contained in ref. 3 to 10. All of these data exhibit an excess of low mass pairs compared to the decay of vector mesons and is not compatible with the Drell-Yan mechanism in a wide range of incident energies and for various hadronic projectiles. A first example was shown in Fig. 1. A second one from Mikamo et al.(3) is given in Fig. 2 with comparisons to some models that we are going to discuss below. The slopes of the mass distributions seem to depend upon the nature of the projectile, proton or pion, and somewhat upon the incident energy, but the characteristics of the cross sections are not well established because of the various kinematical regimes and, most of the time, lack of statistics. In contrast to these data, two CERN-ISR experiments(11,12) report the low mass pair continuum to be in agreement with the Dalitz decay of known mesons or the semi-leptonic decay of charmed particle pairs.

The models that are used to interpret the low mass continuum can be classified into two broad groups, inner bremsstrahlung and soft parton annihilation models, both with or without radiative corrections. Subclasses could be identified depending on the constituants involved in the process and the treatment of the interaction.

Bremsstrahlung models(13) consider the internal conversion of vir-

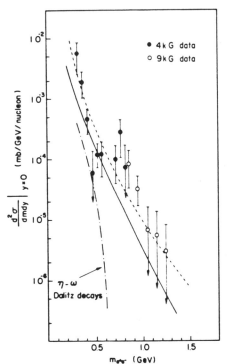

Fig. 2. Dielectron mass spectrum from the KEK experiment at 13 GeV/c. The solid curve is the prediction of Shuryak(17) and the dotted curve is a fit to the model of Kinoshita et al.(18). The dot-dashed curve is the estimated background due to Dalitz decay.

tual photons produced by the bremsstrahlung of charged constituants created during the collision. They suffer difficulties in suppressing real photons and getting the right amount of internal conversion. Besides, the mass distribution of the order of 1/m for small masses does not agree with experimental data and yields a wrong e/μ ratio.

The soft parton models are somewhat more successful. They were initiated by Bjorken and Weisberg[14] who take into account the presence of the many wee partons and antipartons created during the collision process and which may reannihilate into leptons before the products of the collision emerge. Latter works consider meson-meson subprocesses [15] or go to the quark level [16]. Shuriak[17] considers that dileptons are produced through a created quark-gluon plasma with an initial temperature of the order of 0.5 GeV at high incident energies. The phenomelogical model constructed by Kinoshita et al.[18] to unify low mass low pt dilepton and low pt hadron productions is very useful to compare data over the entire incident energy range. It has been found to agree quite well with various experimental data with slightly different parameters.

For the purpose of the DLS design (see a latter section), I have studied the validity of the Kinoshita et al. model that I will refer to as KSS. I checked the model as a scaling law for pion production over a wide range of incident energies. Using the scaling radial variable $x0=E/Emax$ introduced by Taylor et al.[19], where E is the energy of the detected particle in the center-of-momentum frame and Emax the maximum energy kinematically available to the detected particle in the c.m. frame, the following simple relationship was fit to high energy data:

$$E \frac{d\sigma}{dp^3} = C e^{-\lambda E_t} (1-x_0)^\alpha$$

where E_t is the tranverse kinetic energy, $\lambda = 6$ GeV^{-1}, $\alpha = 3.5$, and C=85 and 55 mb/(GeV)^2 for positive and negative pions, respectively. This relationship turned out to work quite well all the way down to the GeV region. In the same way, using the radial scaling variable and Mikamo et al. parameters[3], the dielectron cross section can be written as:

$$E \frac{d\sigma}{dm\,dp^3} = \frac{C}{m^4} (1-x_0)^\alpha \frac{\lambda^2 e^{-\lambda E_t}}{2(\lambda m+1)}$$

with C=1.3 10^{-5} mb.GeV^3 and the other constants being the same as above. I will discuss the agreement of the model with single electron data in the next section.

At this stage, the actual source of low mass dilepton might be a superposition of several processes, depending on the incident energies, the projectile particles or the kinematics involved. More experimental data on mass, y and pt distributions, with good statistics, are needed to settle the question.

SINGLE LEPTONS
AND THE UNIVERSALITY OF LEPTON PRODUCTION

The yield of single direct leptons is expressed in terms of lepton to pion production ratio as a function of xf or pt. Fig. 3 is a typical representation of the production ratio as a function of pt, here for the CERN-ISR data (20). The e/π ratio is independent of the incident energy and kinematical region (within errors), with a constant magnitude of about 10^{-4} and a rise at low pt. Other works support the same conclusion for other projectiles and for both positive and negative electrons. The figure also shows various known contributions which cannot account for the measured electron yield. The data on single muons are less abundant but they support the same production magnitude without any rise at small pt. This point is important because it may favor the bremsstrahlung origin of the low pt single leptons. Finally, two low energy measurements at 256 and 800 MeV (21) have found no evidence for direct electrons at the 10^{-6} level. This raises the question of a possible threshold effect somewhere below 10 GeV/c.

Most of the works on the low mass dilepton continuum have compared their results to single lepton data (see for instance ref. 3, 5 and 6) by integrating their pair cross sections over the proper kinematical range. These calculations support the hypothesis of the low mass dilepton continuum and the low pt single leptons being of the same origin. However, there is a major difficulty in connecting dileptons and single leptons in the fact that dilepton cross sections have not been measured below 100 or 200 MeV. With a steep dependance at the lowest measured masses, like $1/m^3$, an average assumption, the problem is to find a realistic low mass cut in the integration that would simulate the unknown cross section. Blockus et al.(5) lean on the argument that most of the single-arm spectrometer experiments require one and only one particle in the

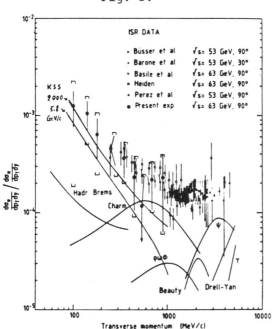

Fig. 3.

spectrometer. This requirement amounts to a cut on mass which is taken at 20 MeV. They note that the computed e/π ratio from the pair cross section would be reduced by a factor of 2 if the mass cut were increased to 60 MeV. Adams et al.(6) use a mass cut of 100 MeV in their calculation. I have studied the question with the KSS model for an incident momentum of 5.8 GeV/c. There is almost an order of magnitude in change when the cut is varied from 50 to 100 MeV, the curves being mostly translated parallel to the vertical axis.

As a consequence, the connection of single electron and dielectron is not straightforward. I mentionned the need for more data. I would then suggest that dielectron measurements might be more useful for the understanding of the processes because they carry more detailed informations. In my opinion, the meaning of the mass cut discussed above is a modification in the dielectron cross section behaviour which should get to a peak or at least to a sharp change in slope at low mass. The peak or change in slope suggested by the mass cuts should occur in between 20 and 100 MeV. It is then important to investigate the pair cross section below about 200 MeV which would have a definitive impact on the understanding of the low mass dilepton continuum. To this regard, the problem is different for dimuons which have an inherent threshold of 210 MeV.

I have compared the e/π data with the KSS model and found a good agreement with a mass cut of 100 MeV. As an example, I have plotted a curve on the CERN-ISR data of Fig. 3 (the paper states that contributions from low mass electron pairs less than 100 MeV are eliminated from the data). On the same figure is plotted a curve computed at 5.8 GeV/c incident momentum which show some decrease in magnitude and a slightly steeper slope. This observation might explain the Los Alamos(21) null result by a strong reduction in phase space when going down to 800 MeV. Here again, more measurements in the GeV region would be needed to settle the question of a possible threshold.

Finally, I am getting to the universality of the single lepton data. It is absolutely striking that experimental results (here the ratio l/π) lie on the same curve over the whole range in energy from 10 Gev up to 2000 GeV (equivalent CERN-ISR incident energy), for different projectile particles and kinematical regimes. In my opinion, it must be a universal basic process which is the source of the phenomenon. It would be very unlikely that contributions from different processes would add up to give an unique behaviour. I would like to quote Cerny et al.(16) who say "if the hadronic origin of the LM (low mass) dileptons is confirmed, then the LM dilepton production could become the simplest hadronic process and as such could be useful for the analysis of more complicated processes of hadron production".

THE BEVALAC DI-LEPTON SPECTROMETER PROJECT

The DLS Collaboration(22) has undertaken an electron pair production study in the few GeV incident energy domain in p-nucleus and nucleus-nucleus collisions at the Bevalac, in order to answer some of the questions mentionned in the previous sections. It also seeks for specific interests in Heavy Ion collisions as will be indicated below. A single electron experiment was first performed(23), the data analysis being still in progress.

The DLS experimental setup is shown in Fig. 4. It consists of a segmented target (about 5 segments) and two symmetric arms, each including a large aperture dipole magnet, two scintillator hodoscopes which provide accurate time of flight information and offer the possibility of using the system in a single arm trigger mode, two segmented gas Cerenkov counters working at one atmosphere for electron identification, and three or four drift chamber stacks for tracking. The conical scattering chamber provides a minimum amount of material along the particle trajectories into the spectrometer together with the use of a multiplicity array detector positionned around the target. Several lead glass blocks behind the detectors will be used for the Cerenkov counter calibration (pion rejection power measurement).

The DLS program will first be exploratory, with data tacking in p+Be collisions at 4.9 and 2.1 GeV, then going to Ne+Ne and Ca+Ca at 2.1 GeV/A, and Fe+Fe at 1.7 GeV/A. The developpement of the program will depend on the first results and might consist in going up in energy or up in mass (therefore, down in energy). The system has been designed to provide suitable acceptance and resolution in mass, y and pt. First detector test runs were just completed and turned out to be quite successful. The data acquisition runs will begin in the fall.

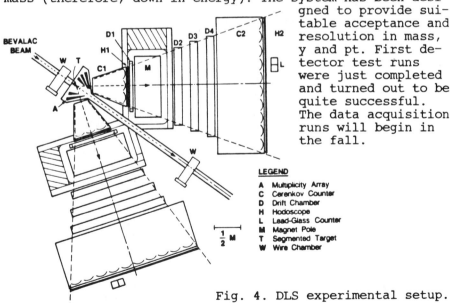

LEGEND
A Multiplicity Array
C Cerenkov Counter
D Drift Chamber
H Hodoscope
L Lead-Glass Counter
M Magnet Pole
T Segmented Target
W Wire Chamber

Fig. 4. DLS experimental setup.

Fig. 5. Production yield of dielectrons at Bevalac energies(24) from baryon-baryon cascading (solid) and pion-pion annihilation (dashed).

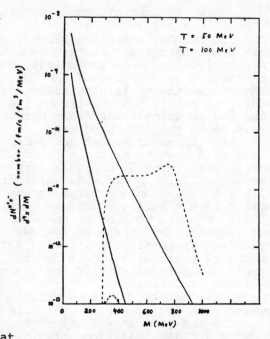

J. Kapusta(24) has recently presented a first estimate of the dielectron characteristics at Bevalac energies in nucleus-nucleus collisions. Fig. 5 gives the expected production yield per unit volume and per unit mass from pion-pion annihilation and baryon-baryon cascading bremsstrahlung for two temperatures of the fire ball and for a baryon density of twice the normal nuclear density. The curves show that the pion-pion annihilation would be undetectable at low temperature while it would start to show up at higher temperature. Besides, the annihilation contribution shape depends upon the effective pion mass in the nuclear medium.

The study should bring useful informations on the first stage of the nucleus-nucleus collision, owing to the general property of virtual photons to interact only weakly with the environment, in contrast to hadrons.

CONCLUSION

The low pt single lepton and low mass dilepton productions probably involve some basic and universal process(es) of hadronic interactions but there are still several open questions that must be answered before a good understanding is achieved. I will summarize these questions in three groups:
- the existence of a threshold in single lepton production; data are needed in the few GeV incident energy domain;
- a quantitative connection between single leptons and dileptons still remains to be verified; more lower pt single lepton data and measurements of the dilepton cross section at lower masses would be useful;
- higher statistics experiments on dilepton production, in particular at lower incident energies, would improve our knowledge of the cross section characteristics and our understanding of the low mass continuum origin.

In ultra-relativistic Heavy Ion collision, it is obvious that a good knowledge of the hadronic case must be achieved before trying to extract any possible Quark-Gluon Plasma signature from the low mass dilepton background. On the other hand, M. Danos(25) has described a possible scenario of Quark-Gluon Plasma formation in which plasma "seeds" are initially produced in the hot compressed medium and would then grow or die, depending on the amount of cooling in the system. In this event and according to Shuryak 's quark-gluon plasma picture(17), the low mass dileptons might be at the onset of a larger plasma formation.

ACKNOLEDGMENTS

This work was supported by the Director, Office of Energy Research, Division of Nuclear Physics of the Office of High Energy and Nuclear Physics of the U.S. Department of Energy under Contract DE-AC03-76SF00098.

FOOTNOTES AND REFERENCES

(*) On leave from the University of Clermont-Ferrand, France.
(1) A.J.S. Smith, Moriond Workshop on Lepton Pair Production, Les Arcs-Savoie-France, January 25-31, 1981.
(2) S. Mikamo, 1979 INS Symposium on Particle Physics in the GeV Region, Tokyo, November 21-23, 1979; H.J. Specht, Quark Matter '84, Helsinki, June 17-21, 1984.
(3) A. Maki et al., Phys. Lett. 106B, 423 (1981); S. Mikamo et al., Phys. Lett. 106B, 428 (1981).
(4) K. Bunnel et al., Phys. Rev. Lett. 40, 136 (1978); B. Haber et al., Phys. Rev. D22, 2107 (1980).
(5) D. Blockus et al., Nucl. Phys. B201, 205 (1982).
(6) M.R. Adams et al., Phys. Rev. D27, 1977 (1983).
(7) J. Ballam et al., Phys. Rev. Lett. 41, 1207 (1978).
(8) D.M. Grannan et al., Phys. Rev. D18, 3150 (1978).
(9) K.J. Anderson et al., Phys. Rev. Lett. 37, 799 (1976).
(10) M. Kasha et al., Phys. Rev. Lett. 36, 1007 (1976).
(11) A. Chilingarov et al., Nucl. Phys. B151, 29 (1979).
(12) J.H. Cobb et al., Phys. Lett. 78B, 519 (1978).
(13) N.S. Craigie et al., Nucl. Phys. B141, 121 (1978).
(14) J.D. Bjorken et al., Phys. Rev. D13, 1405 (1976).
(15) T. Goldman et al., Phys. Rev. D20, 619 (1979).
(16) V. Cerny et al., Phys. Rev. D24, 652 (1981).
(17) E.V. Shuryak, Phys. Lett. 78B, 150 (1978).
(18) K. Kinoshita et al., Phys. Rev. D17, 1834 (1978).
(19) F.E. Taylor et al., Phys. Rev. D14, 1217 (1976).
(20) T. Akesson et al., Phys. Lett. 153B, 419 (1985).
(21) A. Browman et al., Phys. Rev. Lett. 37, 246 (1976).
(22) LBL: G. Claesson, D. Hendrie, G. Krebs, E. Lallier, H. Matis, J. Miller, T. Mulera, C. Naudet, H. Pugh, G. Roche, L. Schroeder, A. Yegneswaran; UCLA: J. Carrol, J. Gordon, G. Igo, S. Trentalange, Z.F. Wang, J. Bystricky (on leave from IN2P3, Fr.); JHU: T. Hallman, L. Madansky; LSU: J.F. Gilot, P. Kirk, S. Christo; NWU: D. Miller; Clermont-Fd: G. Landaud.
(23) J. Carroll, Bull. Am. Phys. Soc. 30, 1271 (1985).
(24) J. Kapusta, Bevalac Users' Meeting, LBL, April 18-19, 1986.
(25) M. Danos, Workshop on SPS Fixed-target Physics, CERN, December 6-10, 1982.

SUBTHRESHOLD PRODUCTION OF STRANGE HADRONS IN RELATIVISTIC HEAVY ION COLLISIONS.

S. Trentalange, S. Carlson, J.B. Carroll,
J. Gordon, G. Igo, and Zhi-Fu Wang
University of California at Los Angeles, Los Angeles CA 90024

T. Hallman
Johns Hopkins University, Baltimore MD 21218

P.N. Kirk
Louisiana State University, Baton Rouge LA 70803

A. Shor
Brookhaven National Laboratory, Upton NY 11973

B.J. Keay, G. Krebs, P. Lindstrom, T. Mulera, and V. Perez-Mendez
Lawrence Berkeley Laboratory, Berkeley CA 94720

ABSTRACT

We report on preliminary measurements of subthreshold antikaons produced in relativistic heavy ion collisions. Comparison with hadrochemical and transport models show that while the total level of antikaon production may be accounted for, important difficulties remain in describing their momentum and angular distributions.

INTRODUCTION

This talk is intended to be an update on the present collaborative effort at the Bevalac to study relativistic heavy ion collisions by sifting through the debris of these reactions for extremely rare events. One such class of events is subthreshold particle production. A nuclear process is considered to be subthreshold if it is kinematically forbidden under the assumption that the individual nucleons all translate rigidly with some average velocity. These processes occur, of course, because the momenta of the nucleons are partitioned according to the Fermi distribution. The energy of several nucleons may also be shared by the nature of the reaction process, $i.\,e.$, either by clustering or secondary reactions. Thus, subthreshold particle production may be a powerful tool in distinguishing among reaction mechanisms. To give a simple example, consider the production of kaons through the three reactions:

$$NN \rightarrow NNK^+K^- \quad E_{thresh} = 2.6 \text{ GeV/A}.$$

$$NN \to NN\Lambda K \quad E_{thresh} = 1.6 \text{ GeV/A}.$$

$$(NN)N \to (NN\pi)N \to NN(\pi N) \to NN\Lambda K \quad E_{thresh} < 1 \text{ GeV/A}.$$

The vast differences in threshold energies for these reactions make them easily identifiable since the production cross section as a function of bombarding energy will fall rapidly below the threshold for the relevant reaction.

Since 1981 our collaboration has attempted more and more complex measurements of antikaon production in nuclear systems. The first experiment consisted of a single measurement, $d^2\sigma/dpd\Omega$ for $\theta_{cm} = 0°$, $p_{kaon} = 1.0$ GeV/c in the system Si+Si at 2.1 GeV/nucleon. In the next experiment (1983) we measured the cross section at 0° as a function of kaon kinetic energy. These results[1] are depicted in fig. 1, and may be summarized as follows: the invariant differential cross section is approximately an exponential function of the kaon center-of-mass kinetic energy. The decay constant of this exponential is about 90 MeV (slightly higher than that for pions). If one assumes isotropy in the CM system, the total cross section is about 1 μbarn. The absence of structure in the cross section would appear to rule out certain exotic production mechanisms such as the ϕ-bremsstrahlung model of K.H. Muller[2].

In the third phase of this experimental program, completed in November, 1985, we have measured the kaon momentum spectra at 0° in Si+Si as a function of bombarding energy down to 1.4 GeV/nucleon. We have also investigated the A-dependence at 2.1 GeV/A by measurements in the systems C+C, Si+Si, and Ca+KCl. Although we have not as yet reduced these data to absolute cross sections, we have obtained preliminary K^-/π^- ratios. These results have yet to be corrected for target absorption or pion-kaon acceptance differences, however, they are accurate to about 30%. First I will describe the experimental apparatus and design, then I will return to the results and model calculations.

EXPERIMENTAL APPARATUS

A schematic diagram of the apparatus is given in fig. 2. Projectile nuclei strike the target ahead of dipole magnet S1M7, and the products of the reaction enter a momentum-recombining secondary beamline. The momentum resolution of the system is $\Delta p/p = 2\%$, while the acceptance is 6 msr.

A trajectory thru the entire spectrometer is ensured by requiring a coincidence of all scintillation counters. Particles are then identified by precise time-of-flight measurements with ultra-fast scintillators/phototubes at S1 and S2, followed by rejection of the pion background with an array of efficient aerogel and gas Cerenkov detectors. Pion "leak-thru" for these requirements

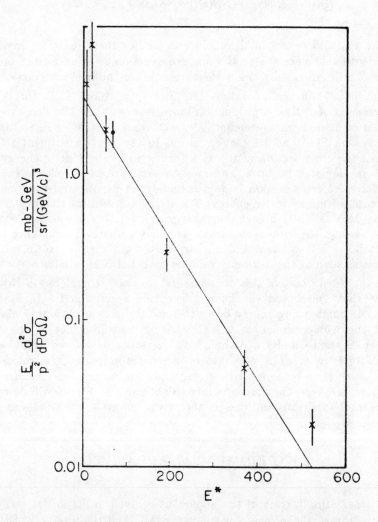

Fig. 1. Invariant cross section for the production of K^- at $0°$ in Si + Si at 2.1 GeV/nucleon as a function of the K^- kinetic energy in the nucleus–nucleus center-of-mass frame. The solid line represents a fit to an exponential with a slope parameter of 91 MeV.

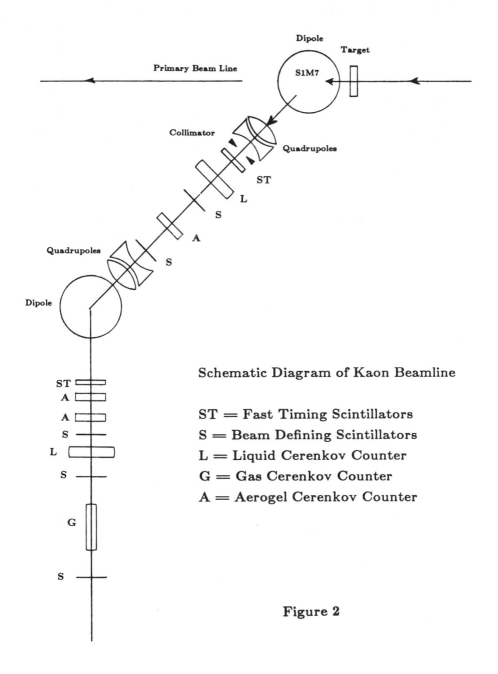

Figure 2

is estimated to be less than 10^{-6}. To search for rare particles, kaons can, in turn, be rejected with a pair of liquid Cerenkov counters.

By taking great care in the construction of this system, we have virtually eliminated the background and sources of systematic error which were so troublesome in the previous two experiments. For example, the pion TOF was found to be Gaussian over more than 3 orders of magnitude, with a slewing-corrected width of $\sim$75 picoseconds. This allows us to establish the position of its centroid with an error of only $\sim$0.2 picosecond, and that of the kaon to within $\sim$5 picosconds. Background, after applying Cerenkov rejection, was confined to two events, each of which had a time-of-flight consistent with that of an antiproton. A more precise analysis of these events, including a study of their pulse heights in each counter, is now underway.

The quantity measured in our system is the K/π ratio at 0° for spectrometer momenta between 1.0 and 1.9 GeV/c. Absolute kaon differential cross sections are obtained by correcting these ratios for particle decay in flight, then multiplying by previously measured pion cross sections.

RESULTS

Early model calculations were often in serious disagreement with the data for both kaon and antikaon production[1,3]. For example, a single collision fermi momentum calculation, where the kaons are produced in elementary baryon-baryon collisions, predicts a kaon cross section which is too low by a factor of 20. At the other extreme, models which assume complete chemical and thermal equilibrium give both kaon and antikaon cross sections which are too large by a factor of 20. Of course, neither assumption is realistic. Nobody expects chemical equilibrium in a system as small as Si+Si, or that such strongly interacting particles will suffer only single collisions. As we may anticipate, better agreement is found among models which try to ascertain the degree of chemical and thermal equilibrium attained in the collision.

There are two models on which I will concentrate. The first is the Hadrochemical theory of Barz et al [4]. They divide the collision into three phases. During the first phase, two spherical nuclei collide and a number of nucleons undergo a primary collision which scatter these nucleons and produce new particles. The particles which have participated in this interaction are assumed to thermalize in the first collision and form a spherical firecloud. The second phase follows the chemical and hydrodynamical evolution of this firecloud. A large number of elementary processes are allowed, the most relevant for our discussion being reactions of the type $\pi Y \longleftrightarrow NK$, where $Y=(\Lambda, \Sigma)$, since this was found to be the source of most of the antikaons. Finally, the third phase occurs when the firecloud expands to the point where the time between collisions is as large as the characteristic expansion time, and all reactions are assumed to cease.

The second model is that of Schurmann and Zwermann[5]: Semi-analytic Transport Theory. In this model the authors derive analytic expressions for the convolution of the number and momentum distributions with the elementary cross sections for each type of particle. The only allowed reactions are of the type BB → BYK, with B=(N,Δ) and Y=(Λ, Σ). This type of calculation is essentially a multiple collision fermi momentum model. As an additional check, this model was shown to produce good fits to other particle production data, such as high energy proton momentum/angle distributions, as well as the energy dependence of Λ production at 3.6 GeV/nucleon.

In fig. 3 we plot our preliminary results for the ratio of the K/π total cross sections as a function of bombarding energy for the system Si+Si. On this same graph we display the results of the both the hadrochemical model and semi-analytic transport theory. The agreement is much closer than the factors of twenty quoted previously, and we notice that the data is bracketed by the theoretical predictions. Thus there is real hope, at first glance, that slight refinement of these models will enable them to predict the total production cross sections.

Unfortunately, these models have great difficulty in reproducing the momentum/ angular distributions for both kaons and antikaons. Starting with the transport theory, it is found that the calculated momentum spectra fall off much faster than the data. A natural question to ask, then, is whether this excess of high energy kaons could be produced by scattering as they emerge from the reaction. Investigation has shown that this is a very large effect; in fact, far too many high energy kaons are produced in this way, and the comparision with data becomes worse. In addition, the resulting kaon and antikaon angular distributions would be highly anisotropic, in contradiction with K$^+$ data at 2.1 GeV/nucleon[6]. At the present time it is unclear how this model can be modified to reproduce the momentum spectra.

In the hadrochemical model, isotropy of the firecloud in the center-of-mass is assumed, a fact in agreement with data, but artificially imposed. In order to study the details of kaon production in a more realistic framework, Barz and Iwe have carried over some of the ideas of the hadrochemical model into a transport calculation[7]. As in transport theory they find that kaon rescattering produces a large anisotropy, however, it is not as large as that calculated by Schurmann and Zwermann. Barz and Iwe point out that scattering may be less important than previously thought because the majority of kaons arise from pion interactions late in the collision, during the expansion phase. Both models make predictions about the A-dependence of the total cross sections, however, our present data do not cover a large enough variation in mass to make a meaningful decision about which model is correct. We find the K/π ratio scales very weakly, approximately as $A^{0.2}$.

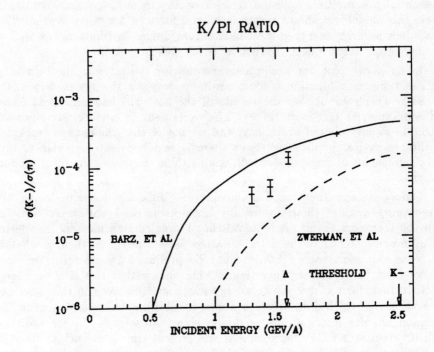

Figure 3

K/π total cross section ratios vs bombarding energy for the system Si+Si. Model calculations are explained in the text.

Since future progress in this field is likely to hinge on the crucial question of kaon scattering, we have planned more experiments this year which will bear directly on this point. The first is a measurement of the K⁻ momentum differential cross section at $\theta_{cm} = 90°$. This would be the most direct evidence possible for or against kaon scattering. Second in importance is our improvement of our A-dependence measurements. At an energy of 2.1 GeV/A, we are limited (budget-wise) to beams lighter than argon; at lower energies, projectiles as heavy as lanthanum are available. Finally, we will continue to explore lower bombarding energies, down to the 0.8-1.0 GeV/A region where the rapid variation of the excitation function will place severe constraints on model calculations.

REFERENCES

1) E. Barasch, et al, Physics Letters **161B** (1985) 265.
2) K.H. Muller, Nucl. Phys. **A395** (1983) 509.
3) K.A. Olive, Phys. Lett. **95B** (1980) 355.
4) H.W. Barz, et al, Z. Phys. **A311** (1983) 311.
5) W. Zwermann and B. Schurmann, Phys. Lett. **145B** (1984) 315.
6) S. Schnetzer, et al, Phys. Rev. Lett. **49** (1982) 989.
7) H.W. Barz and H. Iwe, Phys. Lett. **143B** (1984) 55.

PRODUCTION OF METASTABLE STRANGE-QUARK DROPLETS IN RELATIVISTIC HEAVY-ION COLLISIONS

Gordon L. Shaw
Physics Department, University of California, Irvine, CA 92717

ABSTRACT

Of enormous consequence is the theoretical possibility that not only may multiquark S droplets with large strangeness be metastable, but large extended S matter might be absolutely stable. If indeed, the energy/baryon < 938 MeV for S matter, then this would provide the ultimate energy source. Thus the detection of the metastable S droplets in relativistic heavy-ion collisions would be of great interest. We have calculated the production probability of S droplets in heavy-ion collisions by fragmentation of quarks following formation of hot quark-gluon droplets. We proposed a very sensitive detection scheme (for the present fixed target heavy-ion BNL and CERN facilities) in which an S droplet interacts in a secondary target to produce many Λ's, a striking, readily identified signature. Here, we also discuss detectable consequences of neutron stars being S matter with essentially no crust, in particular, the possibility of a pulsed $\bar{\nu}_e$ flux from fast pulsars.

INTRODUCTION

Considerable theoretical interest[1-6] has focused on the intriguing possibility that not only may nuclear-density multiquark droplets (S droplets) with large strangeness be very long-lived[3] but large bodies of strange matter (S matter) might be absolutely stable[5,6]. Roughly, neglecting the strange-quark mass, m_s, the number of strange quarks n_s should be $\approx A$, the baryon number, since $n_s = n_u = n_d$ in order to lower the Fermi energies of the u and d quarks. Chin and Kerman[3] calculated (using the MIT bag model) that S droplets with $A \gtrsim 10$ and $n_s/A \gtrsim 0.8$ would be metastable with lifetimes $\tau_s \gtrsim 10^{-4}$ sec. They proposed producing S droplets in relativistic heavy-ion collisions. Recently Witten[5] suggested that extended S matter with $n_s/A \sim 1$ might be absolutely stable, i.e., the energy/A < 938 MeV. Farhi and Jaffe[6] then, using a Fermi-gas model with α_c QCD corrections, showed that for a range of parameters (α_c and m_s) S matter could be stable. It seems <u>certain</u> that in the foreseeable future QCD calculations will not be accurate enough to give a definitive prediction on the stability of S matter. Thus experimental tests are crucial and must be done since the consequences are of such enormity. A (positively charged) large A stable chunk of S matter would be the ultimate energy source; it would readily gobble up neutrons, increasing its size and giving off energy. Liu and I[7] calculated the production probability of S droplets in relativistic heavy-ion collisions. We proposed a very sensitive detection scheme (for the present fixed target relativistic heavy-ion facilities at BNL and CERN) in which an S droplet produced in the primary collision interacts in a secondary target to produce many Λ's The study of the properties of metastable S droplets should shed considerable light on the stability of S matter.

As noted by Witten[5], a consequence of the stability of S matter would be that "neutron" stars are really S matter with essentially no crust. Benford, Silverman and I[8] have shown that a fast strange matter pulsar would emit an observable ν flux. Metastable S drops stripped from the pulsar (and accelerated) by electrodynamic fields yield a $\bar{\nu}_e$ flux by sequential decays of the S drops. These decays may be rapid enough to see the pulsar frequency in the neutrino signal in future, large detectors. Possible relevance to Cygnus X-3 is discussed. If indeed the contraversial secondary muons observations[9-12] from Cygnus X-3 are correct, then some new physics would appear to be involved. A good candidate might be a long-lived neutral S droplet.[13,14] In fact we suggest that light, long-lived, neutral S droplets might be coming from all fast pulsars.

PRODUCTION OF S DROPLETS IN HEAVY-ION COLLISIONS

The process for S droplets production was considered in two steps: We assumed that in a relativistic heavy-ion collision with a large A fixed target nuclei, a high density, hot quark-gluon (qg) droplet[15] (or perhaps two) is formed with probability P_{qg}. Note that P_{qg} could be fairly small, and also that a real phase transition to a qg plasma is not necessary for our proposal. In the second step, we calculated[7] the emission of mesons and baryons in several processes from the hot, expanding qg droplet leading to an S droplet. By far the dominant process was the emission of a high momentum u or d quark from the surface of the qg droplet which fragments and picks up an $\bar{s}$ quark leaving behind an s from an $\bar{s}s$ pair. We found that the probability P of producing the metastable S droplet with $n_s/A \geq .8$ from the initial qg droplet is quite model dependent, but highly favoring small A droplets where it may be very large. Although very many particles will be produced in each primary interaction, and even if the probability of producing an S droplet, $P_{qg} \cdot P$, is extremely small, it should be readily detected in a secondary collision. Our scheme is summarized in Fig. 1.

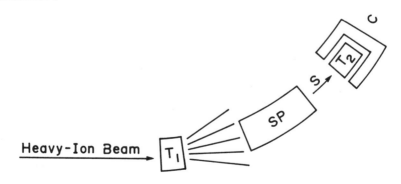

Fig. 1   Schematic diagram of a proposed[7] experiment to detect the quasistable S droplets. The relativistic heavy-ion beam hits a <u>thin</u> Pb target T_1 designed so that produced S droplets suffer no secondary collisions. From the huge number of particles produced in each primary collision, the spectrometer SP would tend to separate out S droplets by passing fragments with $A \gtrsim 10$ and Z/A small or negative. The observation of multiple Λ's, in counters C, emitted in the interaction of an S droplet in target T_2 would be a striking, readily observed signature.

Particles produced in the initial collision in (a thin Pb) target T_1 pass through a spectrometer SP that tends to separate S droplets by passing fragments with baryon number $A \gtrsim 10$ and small or negative electric-charge ratio Z/A. The secondary collision of the relativistic S droplet in T_2 will have center-of-mass energies $\gg$ the binding energy of the S droplet and will then disintegrate into a huge number of hypersons. The observation of multiple Λ decays would be a readily detected and striking signature. Such an experiment should be extremely sensitive. For example, the BNL heavy-ion facility will have a 16 GeV/A beam with $\sim 10^{10}$ oxygen ions per pulse, giving $\sim 10^{14}$ per day. Thus sensitivities of detecting an S droplet produced per 10^{10} collisions should be possible.

PULSED $\bar{\nu}_e$ BEAMS FROM STRANGE MATTER PULSARS

If indeed S matter is stable, Witten[5] noted that "neutron" stars would really be S matter (one huge bag of u, d and s quarks) with essentially no crust. Benford, Silverman and I[8] then proposed that a fast S matter pulsar would emit an observable, possibly pulsed $\bar{\nu}_e$ beam. Metastable S droplets stripped from the pulsar (and accelerated) by electrodynamic fields yield mainly $\bar{\nu}_e$'s by sequential decays.

A pulsar rapidly rotates (period P) and carries its huge magnetic field B with it. As a result, a surface gap (height h) in charge density <u>must exist</u>, and the pulsar must emit a current to supply the corotation charge. This outward flowing ion beam (charge Z, baryon number A) excavates matter at a rate[16]

$$F \approx 2 \times 10^{-3} A(ZP)^{-1}(B/10^{12}G) \text{ g cm}^{-2}\text{s}^{-1} \qquad (1)$$

from the polar cap. It is understood[17] that a TeV electron-positron cascade is produced in the charge gap and these electrons hitting the polar cap excavate the ions. We[8] propose that this self-consistent picture[16,17] could liberate S droplets (along with more conventional ions) without necessarily breaking them up. (A TeV electron hitting the polar cap could produce a high energy pion which rapidly distributes its energy in a small volume). These S droplets would perhaps have $30 \lesssim A \lesssim 10^4$, some with lifetimes $\gg 10^{-4}$s. The charge density gap will accelerate the ions to a maximum energy[18]

$$E_{max} \sim 3 \times 10^{15} \, Z(B/10^{12}G)(P/10 \text{ ms})^{-\frac{5}{2}} (h/10 \text{ km}) \text{ eV}. \qquad (2)$$

These accelerated high-energy S droplets would undergo a chain of decays. Silverman[19] has shown that for a range of parameters consistent with Farhi and Jaffe[6] that an S droplet by the end of its decay sequences emits roughly a number of $\bar{\nu}_e \sim A$. Thus a S matter fast pulsar should be an incredible source of high-energy ($\gtrsim 10$ GeV) $\bar{\nu}_e$'s. It also appears that the decay times of S droplets might be such that the $\bar{\nu}_e$ beam displays the pulsar frequency. This would be a great help in rejecting a signal from background. As a bonus, note that separate measurement in future, large detectors of μ ane e events from the (mainly) $\bar{\nu}_\mu$ beam would be an extremely sensitive test of neutrino oscillations. Not only would the appearance of $\bar{\nu}_\mu$'s test Δm^2 to $\sim 10^{-16}$ eV^2, but would be simultaneously sensitive to small mixing angles. A larger, fully instrumented version of the presently planned DUMAND[20] would be necessary for this accuracy.

CONCLUSION

One of the most exciting conjectures[5] in physics is that S matter might be absolutely stable, i.e., have energy/A < 938 MeV. A chunk of stable S matter would then provide the <u>ultimate</u> energy source. Although, compatible[6] with QCD, no forseeable calculations can settle the question of whether S matter is stable. Thus experimental tests are crucial. We have suggested two relevelant experiments: We have proposed[7] that metastable S droplets be produced in reltivistic heavy-ion collisions at the present BNL and CERN fixed target facilities, both beginning operation by the end of this year. The observation of multiple Λ's emitted in the interaction of an S droplet in the <u>secondary</u> target (see Fig. 1) would be a striking, readily identified signature. A second proposal[8] is that an S matter pulsar would have S droplets <u>stripped</u> from its polar cap (and <u>accelerated</u>) by electrodynamic fields. These high-energy S droplets would undergo a sequence of decays yielding a measurable $\bar{\nu}_e$ flux in future, large detectors. We further note the possibility of there existing a very-long lived ($\gtrsim 1$ year) electrically neutral, light S droplet and it being produced in these sequential decays of the S droplets. This might be relevant to Cygnus X-3. If the contraversial[9-12] secondary muon observations from Cygnus X-3 are correct, then new physics seems to be needed. A good candidate might be a long-lived neutral S droplet.[13,14] We strongly suggest that it would be much larger A than the proposed A=2 H particle.[14] Finally, we suggest that such light, long-lived, ($\gtrsim 1$ year) neutral S droplets may be coming from all fast pulsars (no companion, as for Cygnus X-3, is necessary).

This work was supported in part by the National Science Foundation.

REFERENCES

1. A. R. Bodmer, Phys. Rev. $\underline{D4}$, 1601 (1971).
2. B. Freedman and L. McLerran, Phys. Rev. $\underline{D17}$, 1109 (1978).
3. S. A. Chin and A. K. Kerman, Phys. Rev. Lett. $\underline{43}$, 1292 (1979).
4. A. K. Mann and H. Primakoff, Phys. Rev. $\underline{D22}$, 1115 (1980).
5. E. Witten, Phys. Rev. $\underline{D30}$, 272 (1984).
6. E. Farhi and R. L. Jaffe, Phys. Rev. $\underline{D30}$, 2379 (1984).
7. H.-C. Liu and G. L. Shaw, Phys. Rev. $\underline{D30}$, 1140 (1984).
8. G. L. Shaw, G. Benford and D. J. Silverman, Phys. Lett. $\underline{169B}$, 275 (1986).
9. M. L. Marshak et al., Phys. Rev. Lett. $\underline{54}$, 2079 (1985); $\underline{55}$, 1965 (1985).
10. G. Battistoni et al., Phys. Lett. $\underline{155B}$, 465 (1985).
11. Frejus Collab., 19th Intern. Cosmic ray Conf. (U. C. San Diego, August 1985).
12. J. Oyama et al. Phys. Rev. Lett. $\underline{56}$, 991 (1986).
13. M. V. Barnhill, T. K. Gaisser, T. Stanev and F. Halzen, Nature $\underline{317}$, 409 (1985).
14. G. Baym, E. W. Kolb, L. McLerran, T. P. Walker and R. L. Jaffe, Phys. Lett. $\underline{160B}$, 181 (1985).
15. For a general review see "Quark matter formation and heavy ion collisions," Edt. by M. Jacob and J. Tran Thanh Van, Physics Reports $\underline{88}$, 321 (1982).
16. M. Ruderman, in: Pulsars, Intern. Astronomical Union Symp. No. 95, eds. W. Sieber and R. Wielebinski (Reidel, Dordrecht, 1980) p. 87; M. A. Ruderman and P. G. Sutherland, Astrophys. J. $\underline{196}$ 51 (1975).
17. J. Arons, in Positron-Electron Pairs in Astrophysics, ed. M. L. Burns, A. K. Harding and R. Ramaty, Am. Institute of Physics, 1983, p. 163.
18. P. A. Sturrock, Astrophys. J. $\underline{164}$, 529 (1971); E. T. Scharlemann, J. Arons and W. M. Fawley, Astrophys. J. $\underline{222}$, 297 (1978).
19. D. J. Silverman, to be published. Also see M. S. Berger, Radioactivity in strange quark matter, MIT Report June, 1986.
20. DUMAND proposal (Hawaii, November 1982).

POSITRONS FROM HEAVY IONS: A PUZZLE FOR PHYSICISTS

Berndt Müller
Institut für Theoretische Physik, Joh.Wolfg.Goethe-Universität
D-6000 Frankfurt am Main, West Germany

INSTRUCTIONS

Cut out the pieces and try to combine them into a square. If you do not succeed, perform a new experiment or invent a new model.

THE DATA

Narrow lines have been found in positron spectra from collisions of a variety of very heavy ions, with combined nuclear charges ranging from $Z_1+Z_2=163$ (Th+Ta) up to 188 (U+Cm). The line positions are confined to a rather narrow energy interval, about 320-380 keV in spectra taken by the EPOS (GSI-Yale-Heidelberg-Frankfurt-Main) collaboration[1,4] and about 230-330 keV in those measured by the Munich-GSI collaboration[2,3]. The observed line-widths are always of the order of 75 keV, agreeing with expectations for Doppler broadening of a mono-energetic line emitted from the center-of-mass system[4] The line intensities are also similar in all cases, of order 10 μb/sr (see Fig. 1, left part). Nuclear pair-conversion processes seem to be ruled out as origin, both from the line-width argument[4], and from simultaneous measurement of photon and electron spectra[1,2] in the energy region between 1 and 1.5 MeV.

More recently, coincident peaks in electron spectra at precisely (within 20 keV) the same energy have been discovered[5]. The coincidence peaks seems to be able to account for the intensity of the positron lines observed in singles spectra, and they fall into the same energy interval. The narrowness of the peak in the sum-energy spectrum indicates back-to-back emission of e^+ and e^- with cancelling Doppler shift. In the same system (Th+Th) the line position seems to *shift as a* function of bombarding energy from 310 keV to 370 KeV.

For more details and discussion of the data the reader is referred to the talk given by J. Greenberg at this conference.

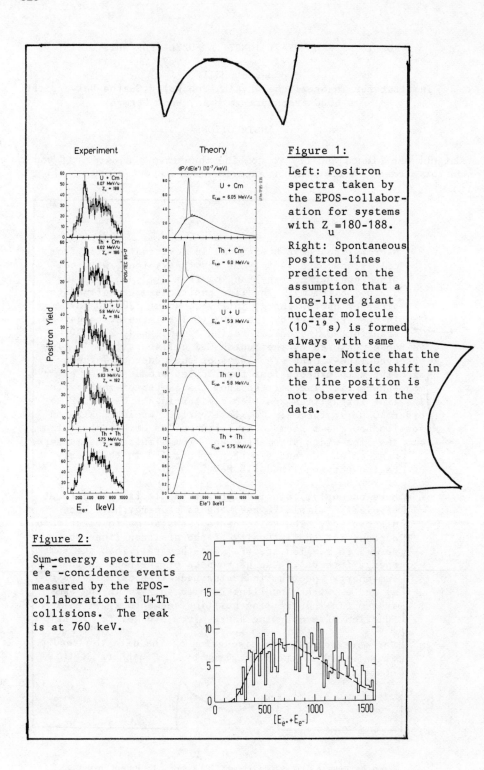

Figure 1:

Left: Positron spectra taken by the EPOS-collaboration for systems with Z_u=180-188.

Right: Spontaneous positron lines predicted on the assumption that a long-lived giant nuclear molecule (10^{-19}s) is formed always with same shape. Notice that the characteristic shift in the line position is not observed in the data.

Figure 2:

Sum-energy spectrum of e^+e^--concidence events measured by the EPOS-collaboration in U+Th collisions. The peak is at 760 keV.

THEORY(I): SPONTANEOUS VACUUM DECAY IN SUPERCRITICAL FIELDS

When two very heavy nuclei ($Z_1+Z_2>175$) are brought close together the K-shell electrons, which form a quasi-atomic orbit around both nuclei, become bound by more than 1 MeV. Their state then becomes part of the Dirac sea of negative energy electron states, and the Dirac vacuum acquires a non-zero charge. When the K-shell state is unoccupied prior to this, a vacancy is introduced into the Dirac sea which spontaneously escapes as a positron within about $t_{sp} \approx 10^{-19}$s (see ref. 6 for details). In a heavy-ion collisions at 6 MeV/u the K-shell electrons are ionized with a probability of about 5 percent during the approach of the nuclei. A narrow positron line will then develop at energy $E_p = E_K - 2m_e$, where E_K is the K-shell binding energy, if the collision time is comparable to t_{sp}. This can occur, if the nuclear potential contains an attractive pocket when the nuclei come into touch (insert in left part of Fig. 3), allowing them to form a giant nuclear molecule.[7,8] When the time delay is calculated in the framework of resonance scattering theory, a narrow positron line at about 200 keV is predicted[9] (Fig. 3), which compares favorably with the U+U data of the Munich-GSI group (Fig. 4). The peak energy is somewhat too low, but note that this could be corrected by assuming the nuclei to overlap to some extent, forming a deformed giant nucleus. Unfortunately, the model fails miserably in predicting the observed dependence of the peak energy with combined nuclear charge Z_u (see Fig. 1). The peak should vanish somewhere between $Z_u=175...180$, and must be absent for all lower Z_u, in contrast to observation: the line only shifts by less than 100 keV between 163 and 188. In principle, a peak can still be predicted for Z<175, if energy transfer between the nuclei and the positron is taken into account as in the Raman effect[9]. However, the near Z-independence of the peak energy is difficult to understand. Pair-conversion processes in the giant atom have been suggested as possible source, but then a dominant spontaneous decay line should also be there[10]. At present, supercritical QED theory cannot provide a consistent picture of all combined data, but it cannot be ruled out as explanation for some of the data, in particular for the heaviest systems. Also, it does not explain the existence of the e^+e^- - coincidence events.

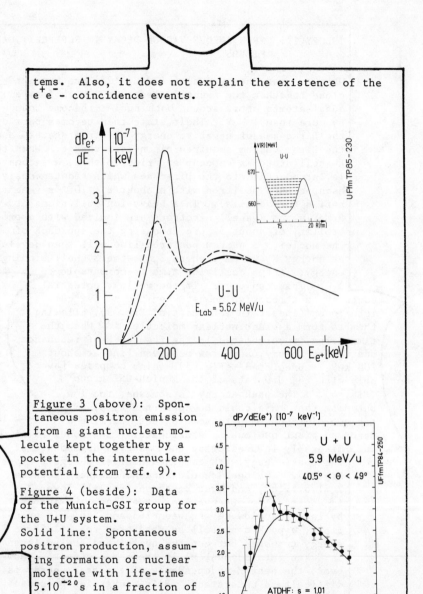

Figure 3 (above): Spontaneous positron emission from a giant nuclear molecule kept together by a pocket in the internuclear potential (from ref. 9).

Figure 4 (beside): Data of the Munich-GSI group for the U+U system.
Solid line: Spontaneous positron production, assuming formation of nuclear molecule with life-time $5 \cdot 10^{-20}$s in a fraction of 1.2×10^{-3} of all events.

THEORY (II): A NEW NEUTRAL PARTICLE

A new neutral (most likely pseudoscalar) boson of mass 1.7-1.8 MeV, which decays into an e^+e^--pair has been suggested as possible source of the positron lines and the coincidence events[11,12,4,5]. However, in addition to providing no obvious explanation for the observed slight variation in line position (!), this mechanism faces serious difficulties:

(a) Precision experiments, e.g. leptonic (g-2)-values, Lamb shift, Delbrück scattering, neutron scattering, rare pion and kaon decays, set extremely stringent bounds on the coupling constants of such a light boson to electrons, quarks and the electromagnetic field. On the basis of these limits, estimates of the production cross-sections in perturbation theory fail to reproduce the data by orders of magnitude[11-17].

(b) Conventional models predict production of the particle with a broad energy spectrum. However, to explain the narrowness of the separate e^+e^- lines the particle would have to decay almost at rest (v/c < 0.05), and no mechanism is known which could stop the particles inside the target. Suggestions that the particles could be emitted mono-energetically from a long-lived giant nucleus do not seem to work quantitatively [13,18].

(c) More exotic models, either beyond the standard gauge model[19] or on the basis of hypothetical highly localized e^+e^- bound state[20,21] have been suggested but, so far, their consistency with other data and their ability to explain the GSI observations have not been proven.

(d) Other experiments, e.g. looking at nuclear and particle decays and beam dump experiments, have failed to detect the existence of a new light neutral boson.

Krauss and Wilczek[22] observed that the failure of perturbative production mechanisms to explain the shape of the observed spectra could possibly be remedied, if the new particles could be produced in a bound state around the nuclei, because then they would move slowly. It turns out that the failure to reproduce the required intensity could be cured by the same mechanism, if the binding would be sufficient strong to significantly lower the effective mass of the particle, thus enhancing the production cross-section[23]. This would occur quite naturally in (viable) axion models because, in addition to the linear pseudoscalar coupling, there exists a quadratic coupling to the scalar quark density, of the form $(m_q/f_a^2)\bar{\Psi}\Psi\phi^2$, where m_q is the (effective) quark mass and f_a is the axion decay constant. The nuclei would then act as deep potential wells for the axion field, which reinforce their effect when they approach each other. If a threshold axion bound state exists for a single Uranium nucleus, the effective axion mass becomes zero at 2a = 35 fm distance[23] (see Fig. 5). Spontaneous

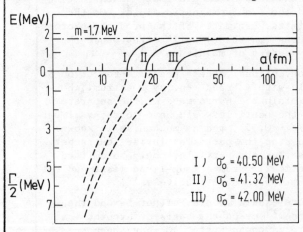

Figure 5: Axion bound states in the vicinity of two uranium nuclei. 2a is distance between the nuclei, σ_o is the depth of the potential wells. Γ is the decay width of the supercritical axion bound state.

axion pair production would occur when the nuclei come closer on a time-scale of less than 10^{-21}s. In this way axions could be abundantly produced and emitted almost at rest at the end of the collision. A nuclear time-delay might strongly enhance the creation rate. However, it is doubtful whether the interaction strength σ_o can be sufficiently large without violation of other experimental limits.

CONCLUSIONS

The positron lines and electron-positron coincidences observed at GSI probably represent the most intriguing set of data in nuclear and particle physics today. Unfortunately, the spontaneous decay of the QED vacuum in strong fields apparently has not been found, so far. But note that this does not necessarily imply a contradiction to theory, because the prediction of a positron line signalling spontaneous pair production relies on the existence of long nuclear contact times ($\gtrsim 10^{-19}$s). It may well be that nature is not kind enough to provide us with such long-lived giant nuclear molecules.

At present, the experimental data could be best explained by the pair-decay of a neutral boson of mass around 1.7 MeV that is produced preferentially with low momentum in heavy ion collisions. To cope with the existence of several line positions, several particles or a particle with internal structure and low-energy excited states are needed. To account for the measured cross-sections a complex, non-linear production mechanism must be invoked, since all models with linear couplings seem to fail. Formation of a bound state around the nuclei during the collision, which may even become unstable against boson condensation at small separations, would be one such non-linear mechanism that might also solve the problems of low-momentum production and very strong beam-energy dependence. We should not be too surprised if the data signal the change in a vacuum other than the electrodynamic one.

One may also be tempted to associated collective solid state effects with the observed e^+e^- signal. In doing so, however, one must keep in mind that excitations in condensed matter of the required properties (energy, e^+e^--coupling) are unknown. Discovery of such states would have far-reaching implications.

It is obvious that more refined experiments are required to establish the source of the GSI positron events. In particular, the 'new particle' explanation - although rather attractive - must be viewed with healthy skepticism until it is confirmed by an invariant mass measurement.

REFERENCES

1. J. Schweppe, et al., Phys. Rev. Lett. $\underline{51}$ (1983) 2261
2. M. Clemente, et al., Phys. Lett. $\underline{137B}$ (1984) 41
3. H. Tsertos, et al., Phys. Lett. $\underline{162B}$ (1985) 372
4. T. Cowan, et al., Phys. Rev. Lett. $\underline{54}$ (1985) 1761
5. T. Cowan, et al., Phys. Rev. Lett. $\underline{56}$ (1986) 444
6. W. Greiner, J. Rafelski, B. Müller, Quantum Electrodynamics of Strong Fields (Springer, Heidelberg 1985)
7. J. Rafelski, B. Müller, W. Greiner, Z. Phys. $\underline{A285}$ (1978) 49
8. J. Reinhardt, et al., Z. Phys. $\underline{A303}$ (1981) 173
9. S. Schramm, et al., Z. Phys. $\underline{A323}$ (1986) 275
10. P. Schlüter, et al., Z. Phys. $\underline{A323}$ (1986) 139
11. A. Schäfer, et al., J. Phys. $\underline{G11}$ (1985) L69
12. A. B. Balantekin, et al., Phys. Rev. Lett. $\underline{55}$ (1985) 461
13. A. Chodos, L.C.R. Wijewardhana, Phys. Rev. Lett. $\underline{56}$ (1986) 302
14. J. Reinhardt, et al., Phys. Rev. $\underline{C33}$ (1986) 194
15. A. Schäfer, et al., Mod. Phys. Lett. $\underline{A1}$ (1986) 1
16. K. Lane, Phys. Lett. $\underline{169B}$ (1986) 97
17. M. Suzuki, Berkeley preprint UCB-PTH 85/54, LBL-21565
18. B. Müller, J. Reinhardt, Phys. Rev. Lett. $\underline{56}$ (1986) 2108
19. A. Schäfer, et al., GSI preprint 86-12
20. B. Müller, et al., J. Phys. $\underline{G12}$ (1986) L109
21. C. Y. Wong, R.L. Becker, Oak Ridge preprint
22. L. Krauss, F. Wilczek, preprint NSF-1TP-86-18
23. B. Müller, J. Rafelski, to be published

ACKNOWLEDGEMENT

I thank all my collaborators for their contributions to the results presented here. I am particularly indebted to J. Reinhardt and A. Schäfer for stimulating remarks.

THE NUCLEAR MATTER EQUATION OF STATE
FROM RELATIVISTIC HEAVY IONS TO SUPERNOVAE

John W. Harris
Nuclear Science Division, Lawrence Berkeley Laboratory
University of California, Berkeley CA 94720 USA

INTRODUCTION

From a few microseconds after the beginning of the universe, when quarks became confined in nuclear particles, until the explosive end of the nuclear lifecycle of large stars, the nuclear matter equation of state plays an important role in governing processes that are fundamental to understanding the world around us. Until recently the study of the nuclear equation of state was confined to temperatures and densities near those of ground state nuclei. In relativistic collisions of nuclei high temperatures and densities are attained providing a unique opportunity to study, in a somewhat controlled manner, nuclear matter under extreme conditions.

In this presentation the relationship between relativistic nucleus-nucleus collisions and the nuclear equation of state will be discussed. The connection between observables measured in the experiments and thermodynamic variables used to describe the system will be made. Through this connection a semi-empirical nuclear equation of state is extracted from the data. The resulting equation of state will be discussed in terms of nuclear matter calculations, neutron star stability and supernova collapse.

RELATIVISTIC NUCLEUS-NUCLEUS COLLISIONS,
EXPERIMENTAL OBSERVABLES AND THERMODYNAMIC VARIABLES

In collisions of two heavy nuclei at the Bevalac where velocities approach the speed of light, three distinct stages of the reaction are predicted - compression, high density and expansion. On the microscopic level binary nucleon-nucleon collisions initially dominate as the two nuclei collide. As the nuclei interpenetrate, successive nucleon-nucleon collisions occur in the interaction region. In macroscopic terms this results in conversion of part of the energy of relative motion into internal degrees of freedom while also increasing the temperature and density of the interaction region. This first stage of the collision is the compression stage. The dynamics of the collision is easily seen in predictions of the intranuclear cascade model[1] shown in Fig.1 for central (small impact parameter) collisions of La + La at 1.0 GeV/n. Plotted logarithmically are the a) baryon density, b) number of baryon-baryon collisions per unit time and c) number of produced particles (pions and delta resonances) in the system as a function of time in the collision. The number of created particles represents part of the energy that is transformed from relative motion into other degrees of freedom. All three variables in Fig.1 are observed to increase rapidly during the compression stage. At a somewhat later time, in the middle of the collision process, the baryon density reaches a maximum before rapidly decreasing exponentially. This is the high density stage of the collision.

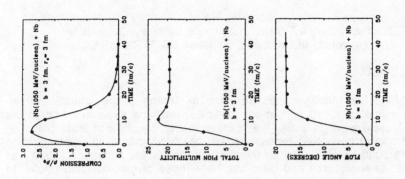

Fig.2. Time dependence of the baryon density, pion multiplicity and flow angle from VUU calculations.[5]

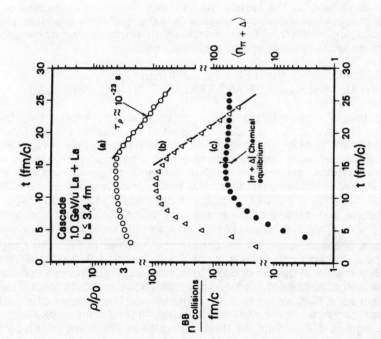

Fig.1. Time dependence of the baryon density, number of baryon-baryon collisions per fm/c and number of pions + deltas from the intranuclear cascade.[1,3]

Here the baryon-baryon collision rate peaks at approximately 90 collisions per fm/c and the number of created particles also reaches a maximum. The exponential decrease in density, the rapid decrease in the collision rate, and a saturation of the abundance of produced particles characterizes the third stage, namely expansion.

To understand the behavior of nuclear matter at high temperatures and densities, it is important to study central collisions. From rate estimates[2] it has been shown[3] that both thermal and chemical equilibrium of pions, nucleons and deltas should be attained in the high density stage of the collision. Thermal equilibrium continues late into the expansion stage while chemical equilibrium, which determines the number of deltas and pions, cannot be maintained in the expansion. The final number of pions in the system is determined at the time of high density and should not change with expansion. This is observed in Fig.1 as the constancy of the pion multiplicity from the time of highest density onwards. A prediction for the pion multiplicity from a chemical equilibrium model[4] is also shown and agrees with the cascade result. Intuitively, the large number of baryon-baryon collisions in the high density stage strongly suggests a large amount of "mixing" which leads to the predicted equilibrium conditions. Results of another microscopic model Vlasov-Uehling-Uhlenbeck (VUU)[5] which also includes potential effects in the form of a mean field are displayed in Fig.2 for the system Nb + Nb at incident energy E_{lab}=1.05 GeV/n and 3 fm impact parameter. Again there is a high density region after which time the total pion multiplicity decreases slightly then remains constant. In addition, the flow angle is established at this time. The flow angle refers to the direction of maximum energy flow in an event and like the pion multiplicity provides information on the high density stage of the reaction. In addition to the pion multiplicity and flow angle, the entropy has been shown[6] to remain constant from the beginning of the expansion onward. The entropy can best be extracted from a determination of the nuclear cluster concentrations,[7] however neither comprehensive measurements of the physical observables (ratios of p, d, t, ^{3}He, ^{4}He, ...) nor a consistent method of interpretation in terms of the thermodynamic variable (entropy) have yet been made. In the following, only results from the pion production and flow of matter will be presented.

PION PRODUCTION

The sensitivity of pion production to the nuclear equation of state was initially pointed out by several authors.[8,9] The first quantitative results[10] were obtained by using hydrodynamics and are shown in Fig.3. The pion multiplicity is significantly lower for the stiff (K_0 = 300 MeV) equation of state, which has higher compressional energy at a given baryon density, than for the softer one (K_0 = 100 MeV). Displayed in Fig.4 are the experimentally observed ratios of pions to participant nucleons as a function of incident energy taken in the Streamer Chamber[11] at the Bevalac. These ratios rise monotonically with energy and are identical for both the Ar + KCl and La + La systems. Also shown are the prediction of the chemical model andpredictions from two intranuclear

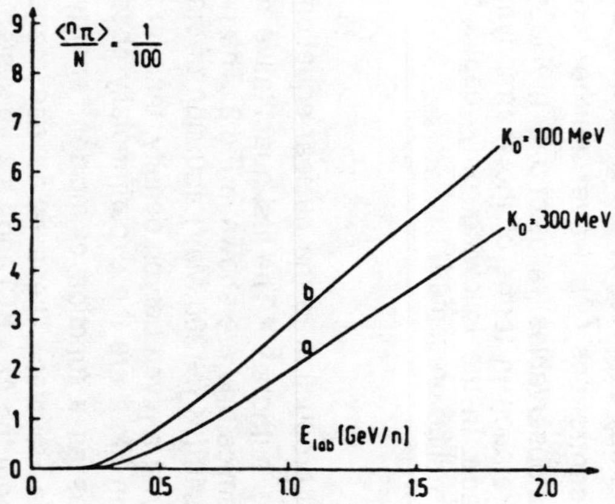

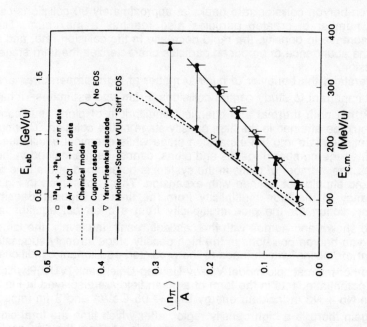

Fig.3. Relative pion multiplicity as a function of incident laboratory energy assuming relatively a) stiff and b) soft equations of state.[10]

Fig.4. Ratio of pions to participant nucleons as a function of incident energy.[3,13]

cascade models,[1,12] all of which are similar and overpredict the observed pion/participant ratios. These models do not incorporate potential degrees of freedom as manifested in the equation of state. In a simple approximation[13] the total energy available in the system can be partitioned into the kinetic energy and the potential energy degrees of freedom. The kinetic energy is available for particle production while the potential energy is not. Thus, the noninclusion of potential degrees of freedom in the purely thermal models (chemical and cascade) lead to all the energy being available for pion production, thus the overprediction of the pion multiplicity. With only kinetic degrees of freedom, the results of VUU calculations without an equation of state (not shown) are very similar to those of the thermal models. However, when potential degrees of freedom are included, in the form of a stiff equation of state, the VUU predicts the observed pion multiplicies as seen in Fig.4.

COLLECTIVE FLOW OF MATTER

Recent observations of sidewards flow of matter have provided further evidence for the necessity to include potential degrees of freedom, in descriptions of relativistic nucleus-nucleus collisions. Both the Plastic Ball[15] and Streamer Chamber[16] groups have observed sidewards flow. A systematic study[17] of the energy and mass dependence of sidewards flow has just been completed using the Plastic Ball. The mass and multiplicity dependence of the data are displayed in Fig.5. Plotted as a function of fractional multiplicity, which is inversely related to the impact parameter, are the distributions of flow angles for three systems Ca + Ca, Nb + Nb and Au + Au at fixed incident energy. The flow angle is defined as the angle of the major axis of an event ellipsoid with respect to the beam axis. This angle is determined from a sphericity analysis where the ellipsoid represents the shape of an event in a (weighted) momentum or velocity space. The flow angle is observed to increase with the mass of the incident system and the fractional charged-particle multiplicity, i.e. the centrality of the collision. The observation of finite flow angles is a strong indication for collective sidewards flow in these events. Models which do not include potential degrees of freedom, such as the intranuclear cascade, do not predict the large flow angles observed.

A more sensitive global event analysis technique is the transverse momentum (p_t) analysis.[18] Displayed in Fig.6a are the mean p_t per particle projected onto the event reaction plane as a function of the particle's rapidity for Streamer Chamber data. As observed in the sphericity analysis for the heavier systems, a definite sidewards flow is observed for the lighter Ar + KCl system from this analysis. Using this technique, the sidewards flow appears in the form of a mean transverse momentum boost which is opposite in the forward and backward hemispheres of the c.m. system. As displayed in Fig.6b, the intranuclear cascade predicts much lower values for the mean p_t projected onto the reaction plane. VUU with the stiff equation of state, used to describe the pion multiplicities, predicts fairly well the mean p_t in the reaction plane as shown in Fig.6c. When a soft equation of state is incorporated into the VUU code the mean p_t is underpredicted as seen in Fig.6d. The actual forms of these

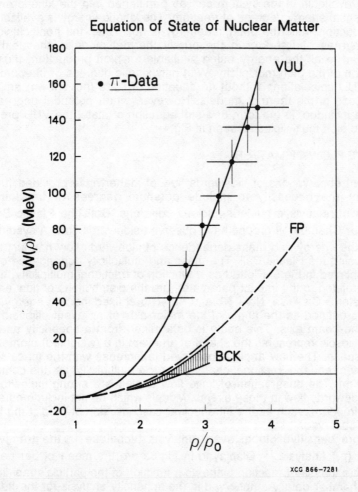

Fig.7. Internal energy per baryon as a function of nuclear density extracted from pion multiplicity data (points).[3] Also displayed are the "stiff" VUU,[5] FP[20] and BCK[21] equations of state. See text for details.

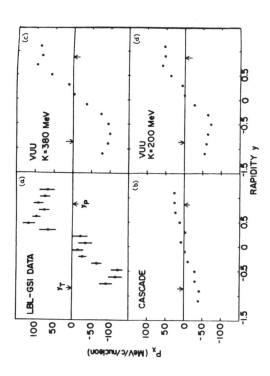

Fig.6. Transverse momentum per nucleon projected into the reaction plane[18] as a function of rapidity for 1.8 GeV/n Ar + KCl a) data, b) cascade and VUU model with c) stiff and d) soft equations of state.[5]

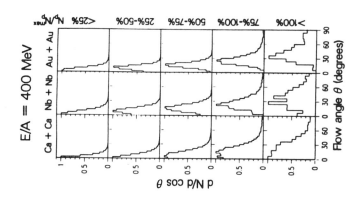

Fig.5. Flow angular distributions for three systems at fixed incident energy as a function of fractional multiplicity.[17]

equations of state will be presented below. Together the pion multiplicity data and the flow data support a stiff nuclear equation of state in relativistic nucleus-nucleus collisions.

THE NUCLEAR MATTER EQUATION OF STATE

A semi-empirical equation of state has been extracted[3,13] from the pion multiplicity data. If the total available c.m. energy is partitioned into thermal and potential energies, the thermal energy can easily be found using the "thermal" models (chemical and cascade) in Fig.4. The potential energy is the difference between the total c.m. energy of the experiment and the thermal energy which is the energy necessary in the "thermal" models to predict the observed pion multiplicity. These potential energies as a function of c.m. energy are represented by arrows in Fig.4. The relationship between the potential energy per nucleon and the nuclear density will be referred to as the nuclear equation of state. To derive an equation of state from the potential energies of Fig.4 for the chemical model, the densities were found by assuming one dimensional shock compression. The resulting equation of state[3] incorporating the density dependent Fermi energy[19] is shown in Fig. 7. A previous analysis[13] using the intranuclear cascade densities determines an equation of state which lies within the error bars of the chemical approach. Also shown in Fig.7 are the FP,[20] BCK[21] and VUU[14] equations of state. The VUU curve is the stiff equation of state which successfully predicts both the pion multiplicities and the flow data. The FP equation of state has been used successfully in nuclear matter calculations and BCK in supernova simulations. Both are considerably softer than that determined from the data and the stiff equation of state used in VUU calculations. In fact, the soft equation of state used in the VUU calculations of Fig.6d above is similar to the FP curve. The discrepancy between the various equations of state that are used for nuclear matter, supernova explosions and relativistic nucleus-nucleus interaction calculations may be understood by investigating the regions of temperature and density that the calculations address. Nuclear matter calculations are sensitive to the nuclear equation of state at zero temperature near the saturation density of nuclear matter ρ_0. Neutron star structure and supernova explosions are governed by the equation of state at low temperatures, $T \sim 15$ MeV, and densities $\rho \sim 2\rho_0$. On the other hand, relativistic nucleus-nucleus collisions occur at high temperatures, $T \sim 60 - 100$ MeV, and densities $\rho \sim 3\text{-}4\rho_0$. The behavior of the nuclear equation of state under these drastically different conditions must still be understood. The equation of state is expected to be stiffer at higher temperatures[22] due to a decrease in the effective mass of the nucleons at high T and a decrease in the strength of the attractive sigma-interaction coupled with an increase in strength of the repulsive omega-interaction.

ACKNOWLEDGEMENTS

I would like to acknowledge the other members of the GSI/LBL Streamer Chamber Group without whom this presentation would not have been possible: D. Bangert, J. Brannigan, R. Bock, R. Brockmann, A. Dacal, C. Guerra, G. Odyniec, M.E. Ortiz, H.G. Pugh, W. Rauch, R.E. Renfordt, A. Sandoval, D. Schall, L.S. Schroeder, R. Stock, H. Ströbele, J. Sullivan, L.Teitelbaum, M.L. Tincknell and K.L. Wolf.

REFERENCES

1. J. Cugnon, T. Mitzutani and J. Vandermuelen, Nucl.Phys. A352 (1981) 505; J. Cugnon, D. Kinet and J. Vandermuelen, Nucl. Phys A379 (1982) 553.
2. A. Z. Mekjian, Nucl. Phys. A312 (1978) 491; S. Das Gupta and A.Z. Mekjian, Phys. Rep. 72 (1981) 131.
3. J.W. Harris et al., Phys. Lett. 153B (1985) 377; J.W. Harris and R. Stock, Proc. of Seventh Oaxtepec Symposium on Nuclear Physics, Notas de Fisica, UNAM, Vol.7 (1984) p.61 and Lawrence Berkeley Laboratory Report LBL-17054
4. R. Hagedorn and J. Ranft, Suppl. Nuovo Cimento 6 (1968) 169;
J. I. Kapusta, Phys. Rev. C16 (1977) 1493.
5. H. Kruse, B.V. Jacak and H. Stöcker, Phys. Rev. Lett. 54 (1985) 289;
J.J. Molitoris, D. Hahn and H. Stöcker, MSU Preprint MSUCL-530,June 1985.
6. G. Bertsch and J. Cugnon, Phys. Rev. C24 (1981) 2514.
7. P. Siemens and J.I. Kapusta, Phys. Rev. Lett. 43 (1979) 1486.
8. G.F. Chapline et al., Phys. Rev. D8 (1973) 4302.
9. M.I. Sobel et al., Nucl. Phys. A251 (1975) 502.
10. H. Stöcker, W. Greiner and W. Scheid, Z. Phys. A286 (1978) 121.
11. GSI/LBL Streamer Chamber Collaboration: D. Bangert, R. Bock, R. Brockmann, A. Dacal, C. Guerra, J.W. Harris, G. Odyniec, M.E. Ortiz, H.G. Pugh, W. Rauch, R.E. Renfordt, A. Sandoval, D. Schall, L.S. Schroeder, R. Stock, H. Ströbele, J. Sullivan, M.L. Tincknell and K.L. Wolf.
12. Y. Yariv and Z. Fraenkel, Phys. Rev C20 (1979) 2227; C24 (1981) 488; and Y. Yariv, private communication.
13. R. Stock et al., Phys. Rev. Lett. 49 (1982) 1236.
14. J.J. Molitoris and H. Stöcker, private communication.
15. H.A. Gustafsson et al., Phys. Rev. Lett 52 (1984) 1590.
16. R.E. Renfordt et al., Phys. Rev. Lett. 53 (1984) 763; D. Keane, " Collective Flow and the Stiffness of Compressed Nuclear Matter", presented at this conference.
17. H.G. Ritter et al., Nucl. Phys. A447 (1985) 3c; K.G.R. Doss et al., Lawrence Berkeley Laboratory Preprint LBL-21464 (1986).
18. P. Danielewicz and G. Odyniec, Phys. Lett. 157B (1985) 146.
19. M. Sano et al., Phys. Lett. 156B (1985) 27.
20. B. Friedman and V.R. Pandharipande, Nucl. Phys. A361 (1981) 502.
21. E. Baron, J. Cooperstein and S. Kahana, Nucl. Phys. A440 (1985) 744.
22. E. Baron et al., SUNY - Stony Brook Preprint 1985.

COLLECTIVE FLOW AND THE STIFFNESS OF COMPRESSED NUCLEAR MATTER[‡]

D. Keane, D. Beavis[*], S.Y. Chu, S.Y. Fung, W. Gorn, Y.M. Liu,
G. VanDalen, and M. Vient
Department of Physics, University of California,
Riverside, California 92521

J.J. Molitoris and H. Stöcker
Institut für Theoretische Physik, Goethe Universität,
D-6000 Frankfurt am Main, West Germany

ABSTRACT

Sideward flow parameters for Ar + KCl at 1.2 and 1.8 A GeV are compared with VUU model predictions for different nuclear equations of state. This model indicates that compressional effects are needed to account for the observed flow. Various uncertainties hinder a precise determination of the compressibility; our current best estimate lies somewhere between the so-called "medium" and "stiff" equations of state.

Theoretical estimates of the peak density attained during the compressional phase of relativistic nucleus-nucleus collisions are typically in the range 2 to 4 times normal nuclear matter density. Model simulations indicate that certain final state observables reach a maximum at about the same time as maximum density, and remain essentially unchanged during the subsequent stages of the collision process. Collective sideward flow is one such observable, and shows promise of providing valuable information about the equation of state of the compressed nuclear matter. In this study, published sideward flow parameters for charged particle exclusive Ar + KCl data from the Bevalac streamer chamber are compared with several sets of model predictions, each corresponding to a different incompressibility coefficient ("stiffness") of the high density nuclear matter.

The transverse momentum method[1] is now widely accepted[2-5] as the most useful parametrization of collective sideward flow; for experimental data, this approach involves estimating the reaction plane for each event using

$$\underline{Q} = \Sigma_\nu w_\nu \underline{p}^t_\nu, \quad w_\nu = \pm 1 \text{ for baryons with rapidity } y_{cm} \gtrless \pm \delta$$
$$= 0 \text{ otherwise.}$$

where $\underline{p}^t_\nu$ is the transverse momentum per nucleon for the νth track. The quantity $\langle p^x \rangle (y)$ is the mean component of transverse momentum per nucleon in the estimated reaction plane:

$$p^{x'}_\nu = \underline{p}_\nu \cdot \underline{Q}_\nu / Q_\nu, \qquad \underline{Q}_\nu = \Sigma_{\mu \neq \nu} w_\mu \underline{p}^t_\mu$$

The condition $\mu \neq \nu$ removes the "self correlation" term from the above scalar product, and thereby avoids one type of finite multiplicity distortion. The component in the true reaction plane, p^x, is systematically larger than the component in the estimated plane, $p^{x'}$, hence

$$\langle p^x \rangle = \langle p^{x'} \rangle / \langle \cos\Phi \rangle; \quad \langle \cos\Phi \rangle \simeq \langle wp^{x'} \rangle \{\langle W^2 - W \rangle / \langle Q^2 - \Sigma(wp^t)^2 \rangle\}^{1/2}$$

where $W = \Sigma|w|$. Particles close to mid-rapidity are least likely to be correlated with the event reaction plane, but contribute unwanted statistical fluctuations; a non-zero setting of the parameter δ minimizes these fluctuations. In the context of an event-generating model, the true reaction plane can immediately be obtained from the initial orientation of the nuclei, and hence a more direct calculation of $\langle p^x \rangle$ is possible.

The model[6] used in this study is a microscopic simulation based on the Vlasov-Uehling-Uhlenbeck[7] (VUU) equation. It proceeds in terms of a cascade of binary collisions between nucleons, deltas and pions according to the experimental scattering cross sections for free particles, corrected by a factor for Pauli blocking. The isospin of each particle is explicitly incorporated. The dependence on the equation of state enters via the acceleration of nucleons in the nuclear mean field. It is assumed that the local potential, U, is determined by the nucleon density within a radius of 2 fm, with a functional form $U(\rho) = a\rho + b\rho^c$. The parameter c fixes the incompressibility, K, and the remaining two parameters are constrained by nuclear equilibrium conditions. c = 7/6 corresponds to K = 200 MeV, termed "medium" stiffness; this is the value inferred from nuclear monopole vibrations, which probe densities very close to equilibrium. c = 2 corresponds to K = 380 MeV, and implies a "stiff" equation of state. The VUU code also contains provision for a special case where only the attractive part of the mean field interaction is present; this case is labeled K = 0.

Streamer chamber data for Ar + KCl have previously been compared with VUU model predictions[2,4], but since these predictions were not filtered to simulate the experimental sample selection criteria or the detector inefficiencies, only preliminary estimates of the stiffness parameter, K, were possible. Figure 1 shows VUU predictions for 1.2 A GeV Ar + KCl, before and after the streamer chamber filter. The main sources of distortion are energy loss and absorption in the target, reduced efficiency for certain kinematic regions near target rapidity, and smearing due to statistical particle identification in the target and mid-rapidity regions.[8] In addition, we normally misidentify A > 4 projectile fragments as hydrogen isotopes or ^{3}He, depending on their rigidity. For relatively high multiplicity events, A > 4 projectile fragments make a significant contribution only at the most forward rapidities, and so we avoid comparing the model with experiment in this region.

Figure 2 shows experimental results for 1.2 A GeV Ar + KCl in

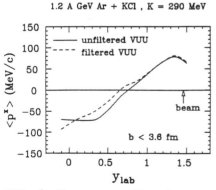

FIG. 1: Transverse flow for the VUU model, averaged over all impact parameters b < 3.6 fm.

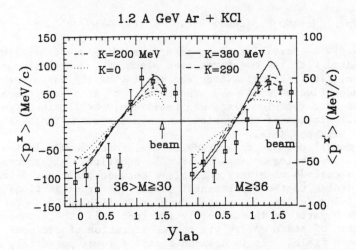

FIG. 2: Mean transverse momentum/nucleon in the reaction plane, as a function of rapidity, for 571 streamer chamber events with charged multiplicity $M \geq 30$. The curves are the filtered VUU model predictions for 4 different equations of state.

two multiplicity bins.[4] The plotted sample contains 571 events with charged multiplicity > 30; this cut selects just over 20% of the inelastic cross section. In the simplest geometrical model, this corresponds to impact parameters $b < 3.6$ fm, as plotted in Figure 1. However, for comparison with the experimental data, we have simulated collisions with $b \leq 5.1$ fm, and have selected the appropriate fraction of these events on the basis of filtered multiplicity. In Figure 2, we show the resulting VUU predictions for four different values of K. A total of 5400 events were simulated for $K = 290$ MeV, and 2000 events for each of the other three cases; the statistical uncertainties on the corresponding VUU predictions are about 0.45 and 0.75 times the experimental error bars, respectively. It appears that sensitivity to K improves at the higher multiplicities, and is also better in the forward rapidity region. Confining our attention to the range $1.0 > y_{lab} > 1.5$, where the overall detector efficiency is high and there is useful sensitivity to K, our preliminary conclusion is that $K = 0$ (attractive mean field only) can be excluded with reasonable confidence, and that the data on balance favor values in the range $K = 200$ to 300 MeV. Some controversy[9-11] still surrounds the question of whether the attractive part of the mean field interaction is sufficient to account for the experimental flow results; the present findings confirm that the VUU approach requires a contribution from the repulsive interaction.

Figure 3 shows experimental results for 1.8 A GeV Ar + KCl;[1] this sample contains 495 central trigger events, corresponding to about 10% of the inelastic cross section. For K = 200 and 380 MeV, we have generated 2000 VUU model collisions with $b \leq 4.2$ fm, and have selected about one third of these events using a simulated central trigger. The resulting predictions for $y_{lab} > 1.2$ are shown

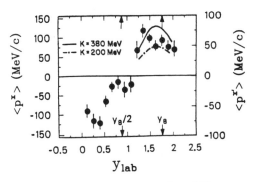

FIG. 3: As Fig. 2, but for 495 central trigger events, from ref. 1.

in Figure 3. The experimental data favor an incompressibility K somewhere between the 2 plotted predictions - say, K ~ 290 ± 50 MeV. In particular, the simulated central trigger leads to $\langle p^x \rangle$ values about 20% higher than for a sample selected by impact parameter alone (b < 2.4 fm);[2] the latter prescription overestimates the stiffness.

In conclusion, we have compared sideward flow data for Ar + KCl at 1.2 and 1.8 A GeV with appropriately filtered VUU model predictions. The results indicate that in the context of this model, a significant contribution from compressional potential energy is needed to account for the observed data. The VUU code correctly describes the increase in transverse flow parameters between 1.2 and 1.8 A GeV. Attempts to quantify the stiffness of the nuclear equation of state at high density are subject to several uncertainties, some of which are not yet well understood; however, the streamer chamber data for Ar + KCl appear to favor incompressibilities in the "medium" to "stiff" range.

‡ Work supported by the U.S. Department of Energy.
* Present address: Brookhaven National Laboratory, Upton, NY 11973.
[1] P. Danielewicz and G. Odyniec, Phys. Lett. 157B, 146 (1985).
[2] J.J. Molitoris and H. Stöcker, Phys. Rev. C 32, 346 (1985).
[3] J.J. Molitoris and H. Stöcker, Phys. Lett. 162B, 47 (1985).
[4] D. Beavis, S.Y. Chu, S.Y. Fung, W. Gorn, D. Keane, Y.M. Liu, G. VanDalen, and M. Vient, Phys. Rev. C 33, 1113 (1986).
[5] K.G.R. Doss, H.Å. Gustafsson, H.H. Gutbrod, K.H. Kampert, B. Kolb, H. Löhner, B. Ludewigt, A.M. Poskanzer, H.G. Ritter, H.R. Schmidt, and H. Wieman, LBL preprint.
[6] G. Bertsch, H. Kruse, and S. Das Gupta, Phys. Rev. C 29, 673 (1984); H. Kruse, B.V. Jacak, and H. Stöcker, Phys. Rev. Lett. 54, 289 (1985).
[7] E.A. Uehling and G.E. Uhlenbeck, Phys. Rev. 43, 552 (1933).
[8] D. Beavis, S.Y. Chu, S.Y. Fung, W. Gorn, A. Huie, D. Keane, J.J. Lu, R.T. Poe, B.C. Shen, and G. VanDalen, Phys. Rev. C 27, 2443 (1983); M. Vient, U.C. Riverside report FPF5 85-8.
[9] Y. Kitazoe, H. Furutani, H. Toki, Y. Yamamura, S. Nagamiya, and M. Sano, Phys. Rev. Lett. 53, 2000 (1984); Phys. Rev. C 29, 828 (1984).
[10] B. Schürmann and W. Zwermann, Phys. Lett. 158B, 366 (1985); Phys. Rev. C 33, 1668 (1986).
[11] J.J. Molitoris, H. Stöcker, H.Å. Gustafsson, J. Cugnon, and D. L'Hôte, Phys. Rev. C 33, 867 (1986).

PHASE TRANSITIONS IN QCD FROM THE LATTICE PERSPECTIVE *

R. V. Gavai[1,2]
Physics Department
Brookhaven National Laboratory
Upton, NY 11973
U S A

ABSTRACT

A brief survey of the latest results in the field of finite temperature lattice quantum chromodynamics (QCD) is presented with special emphasis on simulations that incorporate the dynamical fermions. For the theory with four light flavours, three different apparently inequivalent methods yield results in quantitative agreement with each other. Recent attempts to obtain nonperturbative estimates of the velocity of sound in both the hadronic and quark-gluon phase are summarised along with the results.

INTRODUCTION

Numerical simulations of the lattice formulation of Quantum Chromo Dynamics (QCD) have offered us a unique tool to probe the nonperturbative domain of the theory. Indeed, a host of the static properties of the theory such as the hadron spectrum can now be quantitatively obtained from essentially the first principles. It therefore appears appropriate to ask whether one can obtain the predictions of QCD at finite temperatures and/or densities using these techniques. In particular, one would like to know how the system changes phases from our present spontaneously broken chiral symmetric and colour confined (hadronic) state to the chirally symmetric, colour deconfined (quark-gluon plasma) state, as temperature/density is increased. Apart from satisfyng a theoretical curiousity, these considerations could be relevant to the studies of early stages of our Universe and perhaps the proposed relativistic heavy ion experiments at the Brookhaven National Laboratory and CERN.

It may be emphasized here that unlike many other approaches which too attempt to answer the same question, the lattice QCD approach is free of arbitrary assumptions or parameters. The only parameters that enter these calculations are the quark masses and the scale of the theory Λ_{QCD}, all of which can be fixed by calculating the hadron masses using the same methods. Furthermore, the very existence of a phase transition can be investigated since one can study a given physical observable over the entire temperature domain of interest.

[1]Address after November 1, 1986, and permanent address: Theory Group, Tata Institute of Fundamental Research, Homi Bhabha Road, Bombay 400005 INDIA

[2]Work supported by the U. S. Department of Energy under contract No. DE-AC-02-76CH00016

The new results in this field in the past year or so can be broadly put in three catagories: i) thermodynamics without dynamical quarks but on large lattices, ii) thermodynamics with dynamical quarks, and iii) attempts to obtain the equation of state non-perturbatively. After a bit of controversy in the beginning, a consistent picture seems to be now emerging about the predictions of QCD with light, dynamical quarks, as I hope to show you later. One may therefore look forward in future to more precise calculations on bigger lattices and with more physical observables of interest. For the non-zero chemical potential or the finite baryon density case the situation is, however, far from clear. Already in the early stages it was recognised that new conceptual problems arise as chemical potential is introduced. Due to further technical difficulties, almost no realistic calculation exists so far in this area. I intend to present the results of two groups towards the end, leaving the conclusions about them to you. Before I do so, I would like to give a brief outline of how one obtains the thermodynamics of QCD from first principles. The interested reader may find it more rewarding to fill in the details from literature[1] elsewhere.

The thermodynamic observable of a theory can be obtained from its partition function Z by using the usual formulae:

$$Z = \mathrm{Tr}\, \exp\left(-(H-\mu N)/T\right) \qquad (1)$$

Here H is the Hamiltonian of the theory, T and μ are the temperature and the chemical potential of the system respectively, and N is the conserved baryon density. Tr denotes the sum over all physical states of the theory. What we wish to do is to substitute the Hamiltonian of QCD in the equation above and evaluate various physical quantities and order parameters as a function of temperature (or chemical potential): a phase transition will show up as a non-analytic behaviour. The lattice approach to do this consists of three major steps. First, one rewrites the partition function as a sum over all possible classical paths with given boundary conditions. Each path in the sum is weighted by the exponential of its action. In the case of QCD, Z is given by

$$Z = \int_{bc} DA_\mu\, D\Psi\, D\bar\Psi \exp\left(-\int_0^{1/T} dt \int d^3x\, L_{QCD}(\bar\Psi,\Psi,A_\mu; g^2,\mu,m_i)\right). \qquad (2)$$

A_μ, $\bar\Psi$, Ψ in eq. (2) are the gluon and antiquark-quark fields, and L_{QCD} is the usual QCD lagrangian (with an additional $\mu\bar\Psi\gamma_0\Psi$ term). bc denote the (anti) periodic boundary conditions in the temperature direction for the (antiquark-quark) gluon fields.

The second step consists of introducing a space-time lattice so as to simplify (and even define) the complicated functional integrals in eq. (2). If the lattice has N_β sites in the temperature direction and a_β is the lattice spacing in that direction then the temperature $T = (N_\beta a_\beta)^{-1}$. Analogous quantities in the three spacial directions determine the volume $V = N_\sigma^3 a_\sigma^3$. Associating the fermion fields $\Psi, \bar\Psi$ with the lattice sites $n = (n_1, n_2, n_3, n_4)$ and the gluon fields U_n^μ with the oriented bonds of the lattice, Z takes the form

$$Z = \int \prod_{\substack{bc \\ \mu}} dU_n^\mu \prod_n d\psi(n) \, d\bar\psi(n) \, \exp\left(-S(\psi,\bar\psi,U_n^\mu;g^2,\mu,m_i)\right). \tag{3}$$

The lattice action S is chosen by demanding a) gauge invariance on the lattice and b) appropriate classical continuum limit, i.e. $\lim_{a\to 0} S(\psi,\bar\psi,U_n^\mu) = \int dt \, d^3x \, L_{QCD}$. $U_n^\mu \sim \exp(ia\cdot g \, A_\mu(n))$ in this limit. The third step is comprised of numerical evaluation of $\langle\Theta\rangle$ for any physical observable Θ with respect to the Z in eq. (3) so that results relevant to the continuum theory can be obtained by numerically taking the limit of vanishing lattice spacing. Asymptotic freedom of QCD tells us how the bare coupling must change as the lattice spacing $a \to 0$:

$$a\Lambda_L = (b_0 g^2)^{-b_1/2b_0^2} \exp\left(-1/2 \, b_0 g^2\right) \left[1 + O(g^2)\right], \tag{4}$$

where $b_0 = 33-2N_f/48\pi^2$, $b_1 = 153-19N_f/384\pi^4$ and N_f is the number of massless flavours in the theory. Using Monte Carlo techniques, Creutz[2] showed that eq. (4) holds true for $N_f = 0$ on rather small lattices and rather large g^2. In the asymptotic scaling region, where the above equation is satisfied, one can obtain continuum results for any physical quantity of interest using eq. (4).

THERMODYNAMICS WITHOUT AND WITH DYNAMICAL FERMIONS

The anticommuting nature of the fermion variables $\psi,\bar\psi$ in eq. (3) makes it difficult to apply the above procedure in a straightforward manner to obtain the thermodynamics of QCD. One finds it usually convenient to carry out the fermionic integrals explicitly since S in (3) is typically $S = S_G(U) + \sum_{n,n'} \bar\psi(n) Q_{nn'} \psi(n')$. This leads to the following expression for Z:

$$Z = \int \prod_{\substack{bc \\ \mu}} dU_n^\mu \, \exp\left(-S_G(U)\right) \cdot \det Q(U). \tag{5}$$

In a typical calculation, Q is a square matrix of dimension ~ 6000, and one needs to evaluate det Q about 1-10 million times. ($N_\sigma = 8$, $N_\beta = 4$ was assumed). Even with clever tricks which make use of the properties of Q such a calculation would need about three years on even a CRAY-XMP.

The early calculations[3] were therefore done by dropping the determinant altogether, which can be thought of as the heavy quark limit of our world. There it was found that QCD had a strong first order deconfinement phase transition with a latent heat of about 1 GeV/fm^3. With the advent of the giant supercomputers these calculations have now been extended to rather large lattices. Table I shows the latest results[4] obtained on $19^3 \times$ and 8, 10, 12, and 14 lattices. If the continuum physics is indeed being simulated on these lattices, then T_c/Λ_L should be independent of the lattice size.

The last three rows show this to be the case. The last column of $T_c/\sqrt{\sigma}$, where σ is the string tension, gives us a hope that continuum physics may be setting on even earlier.

Table I. Compilation of Tc and $\sqrt{\sigma}$ values.

Lattice size	$6/g_c^2$	T_c/Λ_L	$\sqrt{\sigma}/\Lambda_L$	$T_c/\sqrt{\sigma}$
$10^3 \times 4$	5.70	76.0	133	0.57
$16^3 \times 6$	5.93	65.5	106	0.61
$19^3 \times 8$	6.02	54.5	104	0.52
$17^3 \times 10$	6.15	50.5	102	0.50
$19^3 \times 12$	6.32	50.9	104.5	0.49
$19^3 \times 14$	6.47	51.7	106	0.49

Research efforts in the past couple of years or so have been concentrated on the question of making the calculations above more realistic by considering lighter quarks. Based on simple models that exploit only the symmetry aspects, one can argue what one would expect as the quark mass, m_q, is gradually lowered. These expectations are summarised in Fig. 1. The first order phase transition in the quenched or quarkless QCD is denoted there by T_c on the $m_q = \infty$

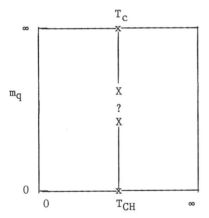

Fig. 1 Expected phase diagram of QCD

line. As m_q is lowered, one expects a line of first order phase transitions along which the latent heat decreases. The end of this line will be marked by a point where latent heat vanishes. It is not clear whether a second order line continues beyond this point. Starting from the other end, $m_q = 0$, one expects a chiral symmetry restoring phase transition there: we believe that chiral symmetry is broken in our world ($\langle\bar{\psi}\psi\rangle \neq 0$, $m_\pi^2 \to 0$ as $m_q \to 0$ etc.) and it can be shown[5] that at sufficiently high temperatures it must be restored.

If this phase transition is also of first order, then as m_q is increased from zero, one expects an analogous scenario as that for the deconfinement phase transition.

The interesting question, of course, is about the positions of the two end points. Are they close to each other? or perhaps overlapping? Could one have two types of phase transitions for some value of m_q? These and other similar questions of details can only be answered after a good scheme to approximate the fermion determinant det Q is found. There are lots of proposals for such schemes in the literature, quite a few of which have been already used to study the full QCD thermodynamics, often leading to unfortunately confusing, sometimes even contradictory, results. In many cases the source of such a confusion is the method used to incorporate the fermion effects. Thus, one needs to test the methods rather thoroughly before drawing any firm conclusions, which is being done only recently.

Using the so-called pseudo-fermion method[6], Karsch and I[7] have studied the full QCD with three light dynamical flavours. Figures 2 and 3 show the energy density ε, and the order parameters $\langle\bar{\psi}\psi\rangle$ and $\langle L \rangle$ as a function of $\beta = 6/g^2$ or equivalently temperature on an $8^3 \times 4$ lattice. $\langle\bar{\psi}\psi\rangle$, the chiral condensate, can be thought of as a direct measure of the constituent mass of the quarks while $\langle L \rangle$ can be loosely described at the deconfinement order parameter: $\langle L \rangle \simeq 0$ corresponds to a confining phase, and $\langle L \rangle \neq 0$ to a deconfined phase.

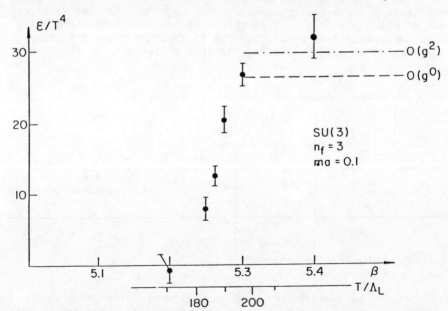

Fig. 2   Energy density as a function of β ($=6/g^2$) for QCD with three flavours of mass 0.1 on an $8^3 \times 4$ lattice. The temperature scale has been obtained by assuming the validity of eq. (4).

One sees clearly that all these quantities undergo a rapid variation in a small range of temperature. The energy density jumps from a small value (~ 0.) to a value corresponding to that of an ideal gas of quarks and gluons. The constituent mass of the quarks becomes very small at the phase transition, and deconfinement seems to take place coincident with chiral symmetry restoration. We made attempts to look for the characteristic two state signal of a first order phase transition with negative results. While our results agree with the previous results[8] obtained on smaller lattices and with lesser statistics quantitatively, both sets of authors obtained $\beta_c \sim 5.25$, ref. 8 did find a first order phase transition. Our study suggests that the nature of the phase transition is sensitively dependent on the choice of physics independent parameters pertaining to the method. In particular, the signal observed in ref. 8 was washed out in higher statistics studies. Nonetheless, a safe conclusion would perhaps be that even if the phase transition in the full theory is indeed of first order, the latent heat (and similar discontinuities) is much smaller than was estimated in the earlier studies of quenched

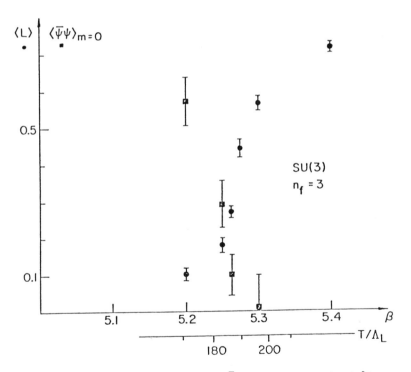

Fig. 3 The order parameters $\langle L \rangle$ and $\langle \bar{\psi}\psi \rangle_{m=0}$ versus $\beta(=6/g^2)$ and T/Λ_L. All the input parameters are the same as in Fig. 2 except that an additional quark mass of 0.075 was used to obtain a linear extrapolation to $\langle \bar{\psi}\psi \rangle_{m=0}$.

(or heavy quark) QCD. Phenomenological implications of this conclusion could be significant, especially in the studies of space-time evolution of the plasma and the experimental signatures to detect the plasma. Our lattice was perhaps not big enough to allow a good estimate of T_c in MeV. A rough estimate can, however, be obtained by using the recent spectroscopic calculations[9] to set the scale Λ_L: $T_c \sim 200$-250 MeV.

Figure 4 exhibits $\langle\bar{\Psi}\Psi\rangle$ and $\langle L\rangle$ as a function of β, calculated by using three different apparently inequivalent methods: the Langevin method[10], the microcanonical method[11], and the pseudofermion method[12]. All calculations were done on an $8^3 \times 4$ lattice for four light flavours of mass 0.1 (in lattice units). A good quantitative agreement is evident over the entire temperature range studied. This is indeed very encouraging, and leads one to believe that these results are perhaps stable and reliable.

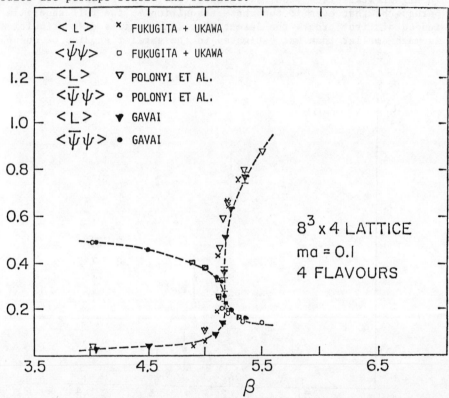

Fig. 4 $\langle L\rangle$ and $\langle\bar{\Psi}\Psi\rangle_{ma=0.1}$ as a function of β. All the data have been obtained for QCD with four light flavours of mass 0.1 on an $8^3 \times 4$ lattice. The dashed curves are drawn to guide the eye. Data from refs. 10, 11, and 12.

VELOCITY OF SOUND IN LATTICE GAUGE THEORIES

A popular approach to obtain a detailed space-time picture of the quark-gluon plasma produced in the relativistic heavy ion collisions is to employ the equations of relativistic hydrodynamics. One needs an equation of state to solve these equations. It is therefore of some interest to know the velocity of sound in the quark-gluon "fluid" in the non-perturbative region around the phase transition. Since $V_s^2 = (\partial P/\partial \varepsilon)_S$, one can use the lattice approach to obtain it. Unfortunately, both ε and P are hard to obtain since both are dominated by two terms of about the same magnitude but opposite sign. It turns out[13], however, that one can relate it to another quantity which involves less of these uncertainties, and some preliminary results[13,14] have thus been obtained.

First, let us consider what one expects from simple considerations. Approximating the confined phase at low temperatures by a nonrelativistic ideal gas of hadrons, it is simple to obtain $V_s^2 = \gamma T/m_H$ where m_H is the (effective) hadron mass. At large temperatures, one ought to have an ideal relativistic gas of quarks and gluons, and hence $V_s^2 \to 1/3$. In the absence of a phase transition, one expects thus the dashed line to represent V_s^2 as a function of T in fig. 5. At the phase transition point V_s^2 goes to zero, and the solid line in fig. 5 then would depict the behaviour of V_s^2.

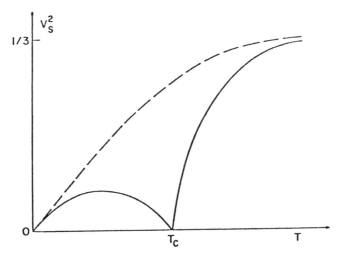

Fig. 5 A schematic representation of theoretical expectations for V_s^2 as a function of T.

Figure 6 shows the lattice evaluation of V_s^2 in the case of quenched QCD which is consistent with these naive expectations. Estimating the glueball mass from the low temperature relation above, one obtains m_G = 900 MeV which is certainly in the right ballpark.

This calculation has now been performed with the dynamical fermions, and one again obtains a similar picture. Since the effective m_H is

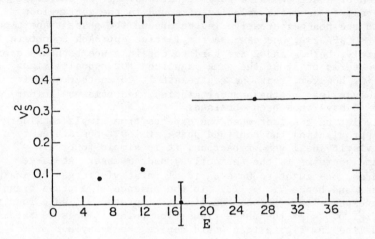

Fig. 6 The velocity of sound squared versus ε for the quenched QCD on an $8^3 \times 4$ lattice.

then expected to decrease, one should see the height of the maximum in the confined phase increase appreciably, which is what one finds in the Monte Carlo simulations[14].

CONCLUSIONS

We now have good approximation schemes to include the fermion determinant, which seem to lead to results that are independent of these methods. All of them predict a rapid variation in $\langle \bar{\psi}\psi \rangle$ and $\langle L \rangle$ that may be characteristic of a coincident chiral and deconfinement phase transition. One has now begun to obtain quantities of phenomenological interest, such as the velocity of sound, using the lattice approach, and the first set of results in this area appear quite encouraging.

This work was supported by the U. S. Department of Energy under contract No. DE-AC-02-76CH00016.

REFERENCES

1) For a review of lattice QCD: M. Creutz, L. Jacobs, and C. Rebbi, Phys. Rep. 93, 201 (1983);
   For finite temperature lattice QCD: J. Cleymans, R. V. Gavai, and E. Suhonen, Phys. Rep. 130, 217 (1986);
   B. Svetitsky, Phys. Rep. 132, 1 (1986).

2) Creutz, M., Phys. Rev. Lett. 43, 553 (1979);
   _____ Phys. Rev. D21, 2308 (1980).

3) Kajantie, K., Montonen, C., and Pietarinen, E., Z. Phys. C9, 253 (1981);
   Kogut, J. et al., Phys. Rev. Lett. 50, 393 (1983);
   Çelik, T., Engels, J., and Satz, H., Phys. Lett. 125B, 411 (1983)

4) Karsch, F. and Petronzio, R., Phys. Lett. 139B, 403 (1983);
   Gottlieb, S.A., et al., Phys. Rev. Lett. 55, 1958 (1985);
   Christ, N.H. and Terrano, A.E., Phys. Rev. Lett. 56, 111 (1986);
   for $\sqrt{\sigma}$ values, see Barkai, D., Moriarty, K.J.M., and Rebbi, C., Phys. Rev. D30, 1293 (1984).

5) Tomboulis, E.T. and Yaffe, L.G., Phys. Rev. Lett. 52, 2115 (1984)

6) Fucito, F., Marinari, E., Parisi, G., and Rebbi, C., Nucl. Phys. B180, [FS2] 369 (1981).

7) Gavai, R.V., and Karsch, F., Nucl. Phys. B261, 273 (1985).

8) Fucito, F., Solomon, S., and Rebbi, C., Nucl. Phys. B248, 397 (1984).

9) Fucito, F., Moriarty, K.J.M., Rebbi, C., and Solomon, S., "The hadronic spectrum with dynamical fermions", Brookhaven Preprint, BNL 37546.

10) Fukugita, M. and Ukawa, A., "Deconfining and Chiral Transitions of finite temperature Quantum Chromodynamics in the presence of dynamical quark loops", Kyoto preprint, RIFP-642.

11) Polonyi, J., Wyld, H.W., Kogut, J.B., Shigemitsu, J., and Sinclair, D.K., Phys. Rev. Lett. 53, 644 (1984).

12) Gavai, R.V., Nucl. Phys. B269, 530 (1986).

13) Gavai, R.V. and Gocksch, A., Phys. Rev. D33, 614 (1986).

14) Redlich, K. and Satz, H., Phys. Rev. D (1986), in press.

OBSERVATION OF KNO SCALING IN THE NEUTRAL ENERGY
SPECTRA FROM αα AND pp COLLISIONS AT ISR ENERGIES

Michael J. Tannenbaum
Brookhaven National Laboratory
Upton, New York 11973

ABSTRACT

Neutral transverse energy spectra in pp and αα interactions are analyzed in terms of the Wounded Nucleon Model. Analysis of the αα spectrum by application of a multiple nucleon-nucleon collision mechanism is conveniently performed when a Gamma Distribution is used to represent the pp spectral shape. The Wounded Nucleon Model provides a reasonable description of the αα spectrum for the first 3 orders of magnitude, but completely fails to account for the slope of the high energy tail of the distribution. However, the pp and αα spectra can both be fit to the same Gamma Distribution when scaled by their respective mean values and thus exhibit KNO scaling.

This talk describes the work of the Brookhaven-CERN-Michigan State-Oxford-Rockefeller collaboration at the CERN ISR. These results have been published earlier this year[1]. The measurements were made during the second ISR light ion run in August 1983 with colliding beams of proton-proton, d-d and αα, all at the same nucleon-nucleon center of mass energy, $\sqrt{s_{NN}}$ = 31 GeV. The measurements were published[2] in 1984, with further details presented at Steamboat Springs[3]. The present work is a detailed analysis of the data in terms of the "Wounded Nucleon Model"[4], together with the empirical observation that the pp and αα spectra exhibit KNO[5] scaling over the 10 orders of magnitude range of measured cross section. This empirical observation is completely unexpected theoretically.

Some may find it disquieting that this early step in the quest for the quark-gluon-plasma in relativistic heavy ion collisions produced a completely unexpected result. Much can be learned in this regard by reviewing the progress of high energy strong interaction physics over the past twenty years[6].

In the 1960's high energy physics studied mainly exclusive reactions in which all particles on individual events are fully kinematically reconstructed. Most of our well known particles were discovered in this way. A typical example is cascade (Ξ^-) production in a bubble chamber.

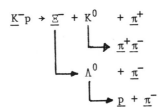

All the underlined charged tracks are measured and the event fully kinematically reconstructed, so that the masses of the $K^0 \Lambda^0$ and Ξ^- could be determined. Such measurements became nearly impossible when the center-of-mass energies rose above 20 GeV in the early 1970's. Then, single particle inclusive measurements became the rage.

A single particle inclusive reaction involves the measurement of just one particle coming out of a reaction. Inclusive reaction measurements at the CERN ISR in the early 1970's produced very surprising results: an unexpectedly large yield of particles with large transverse momentum. The CCR[7] and CCRS[8] groups led the way in making systematic measurements of the reaction

$$p + p \rightarrow \pi^0 + \text{anything}$$

as a function of the transverse momentum, p_T, of the π^0 and the C-M energy, $\sqrt{s}$, of the pp collision over the full ISR energy range (Figure 1). The solid line illustrated the $\exp(-6 \, p_T)$ dependence of the invariant cross section observed at lower p_T. The availability of a consistent set of data covering a wide range of the variables p_T and $\sqrt{s}$ enabled an empirical relationship between the C.M. energy and transverse momentum dependences of the large p_T phenomena to be determined. This is illustrated in Figure 2 in which all the data are plotted, multiplied by p_T^n, as a function of the scaled variable $p_T/\sqrt{s}$. A universal behavior is seen. In this case, the theoreticians were enormously helpful, having suggested[3] that this form of scaling would occur if the π^0 were produced as the result of large momentum transfer quasielastic scattering of pointlike constituents inside the proton. Unfortunately, the value of the parameter $n = 8.24$ favored by the measurements did not agree with the value $n = 4$ favored by the theorists. In fact the theorists were correct,[10,11] as was later determined by measurements at even higher transverse momenta[12,13]. However the scaling shown in Figure 2, with the parameter $n = 8$, spawned a whole new class of theories[14], which were eventually ruled out by subsequent measurements[15].

In order to track down the curious behavior exhibited by the single particle inclusive measurements, experimenters in the mid '70's started making two particle inclusive measurements, and

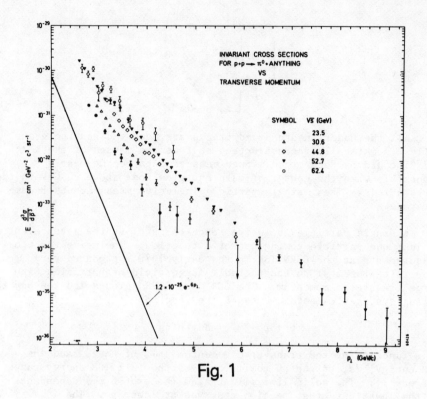

Fig. 1

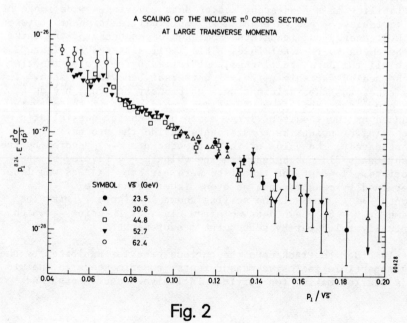

Fig. 2

eventually in the late '70's and early '80's, multiparticle inclusive measurements, in which many but not all particles from an interaction are measured. One of the most beautiful results from two particle inclusive measurements is that of the CCOR collaboration[16] at the ISR who measured

$$p + p \rightarrow \pi^0 + \pi^0 + \text{anything},$$

with the two π^0's roughly opposite in azimuth. If the π^0's were the fragments of jets produced by hard-scattering of the constituents of the proton, then the condition of requiring both of the π^0's to have large transverse momentum selects events in which both jets consist mainly of a strongly leading π^0. In this picture, the mass of the dipion system is a measure of the center-of-mass energy of the constituents, the net rapidity of the π^0 pair gives the transformation to the constituent C.M. system and the angle of the dipion axis in this system corresponds very closely to the C.M. scattering angle, θ^*, of the constituents (Figure 3). These data provided the first direct measurement of the constituent scattering angular distribution and were in excellent agreement with the predictions of QCD, including the variation of the coupling constant and the structure functions with momentum-transfer. These results were all subsequently (brilliantly) confirmed at the CERN SppS collider.

It is interesting to note as a digression that the era of discovery of hard-scattering of constituents of the proton spanned a period of about 15 years, from 1968 when proton substructure was first discovered at SLAC to 1983 when the $W^{\pm}$ and Z^0 particles and unbiased jet structure was discovered at the CERN collider in excellent accord with QCD predictions. The first papers applying QCD quantitatively to hadron collisions appeared only in 1957-78[10,11]. Meanwhile, in the period 1972-1982, the properties of hard-scattering and jet structure in hadron collisions were laboriously mapped out at the CERN ISR, starting from the original discovery of the large yield of particles with large transverse momentum in 1972[17].

The decade of the 1980's ushered in the era of multiparticle inclusive measurements. These were stimulated by the desire to detect and study the jets from hard scattering, with an unbiased trigger. One of the most popular measurements is the transverse energy flow, or "E_T" emitted from a hadron collision. In this type measurement, the transverse energy, $E_{Ti} = E_i \sin\theta_i$, is summed over all the particles emitted on an event into a fixed but large solid angle, typically $\Delta\phi = 2\pi$, $\Delta y = \pm 1$, where y is the rapidity in the C.M. system. A typical E_T spectrum is shown in Figure 4. Contrary to early expectations[18], E_T spectra are dominated by "soft collisions". The transverse energy is made up of a structureless cloud of low transverse momentum (~ 0.40 GeV/c) particles[19,20]. Jets are swamped.

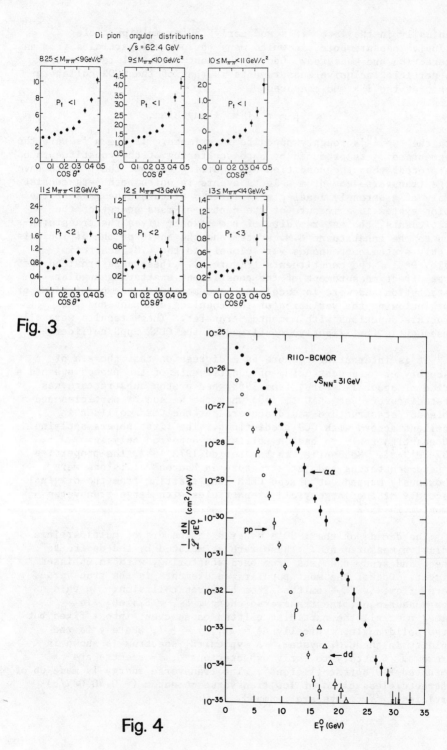

Fig. 3

Fig. 4

This completes the review of the progress of high energy strong interaction particle physics and brings us back to the original subject, the unexpected observation of KNO scaling in the neutral energy spectra from pp and $\alpha\alpha$ interactions at nucleon-nucleon C.M. energy $\sqrt{s_{NN}} = 31$ GeV. The spectrum of total neutral energy emitted in the central region was measured using an electromagnetic shower counter which detected, but did not separately resolve, the photons from the decays of π^0 and η^0 particles ($\pi^0 \to \gamma\gamma$, $\eta^0 \to \gamma\gamma$). A sum was made of all the neutral energy observed in the detector for each event having a vertex with at least 2 charged tracks, and a frequency distribution of this quantity was tabulated, normalized by the integrated luminosity. The center-of-mass acceptance in which the neutral energy was detected covered 90% of 2π in azimuth with an average rapidity acceptance inside this region of $\Delta y = \pm 0.9$ about $y = 0$. Only the statistical errors are plotted. (Figures 4,5). The systematic errors are small for the spectra of E^0_{TOT} shown in Figure 5. In this detector, E^0_{TOT} and the transverse neutral energy, E^0_T, are proportional, $E^0_T \sim 0.87 \, E^0_{TOT}$, but the systematic errors for the measurement of E^0_T are larger[2]. Hence only the E^0_{TOT} spectra are used for further analysis. A full description of the experiment[1] and details of the apparatus[2,3] have been given elsewhere.

The quantity E^0_{TOT} is a multiparticle measure, like multiplicity, being the sum over many particles in the detector on an event-by-event basis. Thus E^0_{TOT} is additive in the case of multiple nucleon-nucleon (N-N) collisions. The E^0_{TOT} spectrum in $\alpha\alpha$ interactions is conventionally analyzed as the result of multiple independent N-N collisions, with each N-N collision producing the observed E^0_{TOT} spectrum of pp interactions. The observed pp spectrum is treated as the probability function for the collision of two nucleons:

$$f_1(E) = \frac{1}{\sigma^{pp}_{in}} \left(\frac{d\sigma}{dE}\right)_{pp}$$

where $\sigma^{pp}_{in} = \int_0^{\sqrt{s}} (d\sigma/dE)_{pp} \, dE$ and the $f_1(E)$ is the differential probability for the emission of energy E in dE in our detector for a single nucleon-nucleon collision. Then $f_n(E_0)dE_0$, the probability of observing E_0 in dE_0 for n such collisions overlapped, is given by the n-fold convolution of $f_1(E)$. This is easy to understand by writing $f_n(E_0)$ in the recursive form:

$$f_n(E_0) = \int_0^{E_0} dE \, f_1(E) \, f_{n-1}(E_0 - E)$$

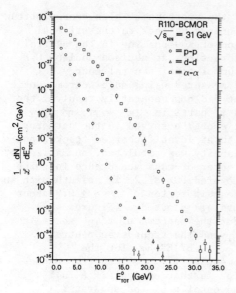

Fig. 5

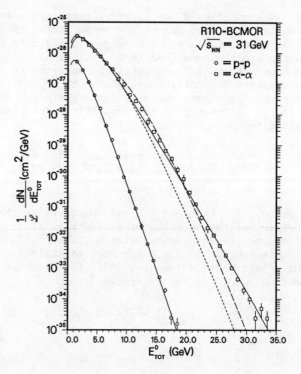

Fig. 6

where $0 \leq E_0 \leq n \sqrt{s_{NN}}$. Here, E_0 represents the energy emitted on n collisions. The first term inside the integral is the probability of emitting energy E on one collision and the second term is the probability of emitting a total of $E_0 - E$ on $n - 1$ collisions.

It should be noted that all the n-collision spectra are normalized to unity; only the shape is determined. They can be renormalized to the probability or cross section for n simultaneous nucleon-nucleon collisions, if this is known from a model. Alternatively, the data can be used to fit for the n-collision cross sections σ_n, $n = 1,2,...m$ as parameters:

$$\frac{1}{L} \frac{dN^{\alpha\alpha}}{dE^0_{TOT}} = \sum_{n=1}^{m} \sigma_n f_n(E^0_{TOT}) . \qquad (1)$$

The analysis is greatly facilitated by using a Gamma Distribution to represent the N-N probability distribution $f_1(E)$. The pp spectrum in Figure 5 is fit to the function

$$f_1(E) = \frac{\alpha}{\Gamma(p)} (\alpha E)^{p-1} e^{-\alpha E} \qquad (2)$$

for the parameters p and α. It should be noted that $p > 0$, $\alpha > 0$, $f_1(E)$ is normalized to unity over the range $0 \leq E \leq \infty$ and that $\Gamma(p)$ is the gamma function of p, $= (p - 1)!$ if p is an integer. The function fits the pp data very well with the result:

$\alpha = 1.41 \pm 0.01$ GeV^{-1} $p = 2.50 \pm 0.06$ $<E> = p/\alpha = 1.77 \pm 0.03$ GeV

$\sigma^{pp}_{in} = 13.1 \pm 0.3$ mb $\chi^2 = 24.6/15$ d.o.f.

The energy spectrum produced by n simultaneous independent N-N collisions, for $n = 1,2,3,...$, is then simply given by the function:

$$f_n(E) = \frac{\alpha}{\Gamma(np)} (\alpha E)^{np-1} e^{-\alpha E} \qquad (3)$$

with the values of α and p determined above. All the $f_n(E)$ remain normalized to unity.

The "Wounded Nucleon Model"[4] has been used successfully by the AFS group[20,21] to relate the central multiplicity distributions ($|y| < 0.8$) of pp, αp, and αα collisions at $\sqrt{s_{NN}} = 31$ GeV. The main distinctive feature of the "Wounded Nucleon Model" is that it counts the number of struck nucleons geometrically, using measured nucleon density distributions; and that a nucleon contributes only once to the production of particles no matter how many times (≥ 1) it is successively struck. The AFS group[21] determined the relative cross sections for n nucleon-nucleon collisions per αα interaction:

$$r_n \equiv \frac{\sigma_n}{\sigma_{in}^{\alpha\alpha}}.$$

The results agree very well with the α-particle geometry. These results can be applied directly in Equation (1), with Equations (2) and (3) used for $f_n(E)$, if n is allowed to take on half-integer as well as integer values. For instance, n = 2-1/2 corresponds to 5 wounded nucleons (w): w = 2n.

The results of the fits are shown in Figure 6. The line in the pp data is the fit to Equation (2) given above. The dotted line on the αα data is the best fit using the n-collision probabilities derived by the AFS collaboration[21] from their charged multiplicity data. This fits the E^0_{TOT} spectrum out to 10 GeV, or about 1-1/2 orders of magnitude down in cross section. The dashed line on the αα data allows the σ_n, n = 1,1-1/2,2,...4, to take on the values which give the best fit to the data (see Table 1). It is clear from Table 1 that this fit is obtained by saturating the data with the largest number of wounded nucleons allowed, leaving the probability for 1 and 2 pp collisions unchanged from the AFS values. This curve fits the E^0_{TOT} spectrum out to 20 GeV, or over 4-1/2 orders of magnitude. Note that both fits have essentially the same shape in the region beyond 15 GeV which is dominated by 8 wounded nucleons, the maximum allowed in this model; whereas the slope of the data is much flatter. In summary, the "Wounded Nucleon Model" works well for the first 90% of the αα cross section, and can fit the data with unreasonable parameters over half the measured range of E^0_{TOT}, but leads to the wrong functional form for the high energy tail of the data over a 6 order-of-magnitude range of cross section.

A surprising result was obtained by fitting the αα spectrum to a single Gamma Distribution (Equation 2) with all parameters free. This is the solid line on the αα data in Figure 6. The fit is quite reasonable over the whole E^0_{TOT} range, with parameters:

$\alpha = 0.822 \pm 0.005$ GeV^{-1} p = 2.48 ± 0.05   <E> = p/α = 3.01 ± 0.05 GeV

$\sigma_{in}^{\alpha\alpha} = 124 \pm 1.5$ mb $\chi^2 = 139.7/30$ d.o.f.

The fact that the parameter p is the same within errors for both the pp and αα data means that the E^0_{TOT} spectra for pp and αα interactions measured in this experiment obey KNO scaling[5]. This can be seen mathematically by rewriting Equation 2 in terms of the scaled variable $z \equiv E/\langle E \rangle = p/\alpha$. The distribution depends only on the parameter p:

$$\langle E \rangle f_1(E) = \psi(z) = \frac{p}{\Gamma(p)} (pz)^{p-1} e^{-pz} \qquad (4)$$

A much more graphic illustration of the KNO scaling is obtained by making a KNO-plot of both the pp and αα data (Figure 7). The E^0_{TOT} spectra plotted in this way for pp and αα interactions are nearly indistinguishable over 10 orders of magnitude! In particular, the tails of the pp and αα distributions are the same, when measured in units of the average value of the energy $\langle E^0_{TOT} \rangle$, in each case.

The KNO formulation seems to account beautifully for the shape of the pp and αα spectra but does not provide any guidance for the relationship between $\sigma^{\alpha\alpha}_{in}$ and σ^{pp}_{in} or $\langle E^0_{TOT} \rangle^{\alpha\alpha}$ and $\langle E^0_{TOT} \rangle^{pp}$. However, these gross features of data are accounted for very well in the geometrical framework[4]:

$$\langle w \rangle \equiv 2 \langle E^0_{TOT} \rangle^{\alpha\alpha}/\langle E^0_{TOT} \rangle^{pp} = 2 \times (1.70 \pm 0.04)$$

$$\langle n \rangle \equiv 16 \, \sigma^{pp}_{in}/\sigma^{\alpha\alpha}_{in} = 1.69 \pm 0.04.$$

Also the AFS r_n parameters[21] give directly

$$\langle w \rangle/2 = 1.61.$$

These values are also in excellent agreement with the values 1.74 ± 0.06 and 2.0 ± 0.2 for $\langle w \rangle/2$ and $\langle n \rangle$ derived from full solid angle multiplicity measurements in pp and αα interactions at the ISR[22].

Previous authors[21,23,24] have examined the question of KNO scaling in nuclear collisions, but have always favored the wounded-nucleon, or other similar multiple independent-collision model, to explain the data. In general, these conclusions were based on data spanning only 2 to 3 orders-of-magnitude in cross section. An expanded plot of the first 3 orders-of-magnitude of Figure 7 is shown in Figure 8, together with the KNO fit, Equation 4 with p = 2.48. On this scale, the KNO curve fits the pp data very well, but the αα data systematically miss the curve. This effect is also clear in Figures 6 and 7. The αα data are systematically higher than the KNO fit over the range $12 \leq E^0_{TOT} \leq 20$ GeV, and for the lowest data

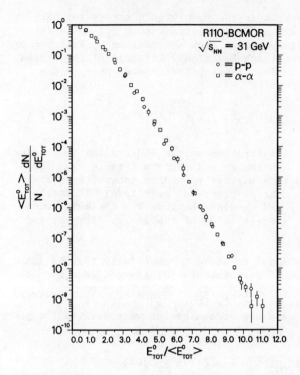

Fig. 7

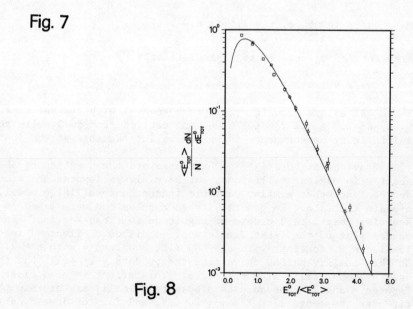

Fig. 8

point. Nevertheless, it is clear from Figure 7 that over the 10 orders-of-magnitude for which the $E^0{}_{TOT}$ spectra in the rapidity range $|y| < 0.9$ have been measured for pp and $\alpha\alpha$ interactions, the most reasonable representation of the data is a single Gamma Distribution, Equation 4, exhibiting KNO scaling for the pp and $\alpha\alpha$ distributions at the same value of nucleon-nucleon c.m. energy, $\sqrt{s_{NN}}$ = 31 GeV, with integrated observed $\alpha\alpha$ cross section and $\langle E^0{}_{TOT}\rangle^{\alpha\alpha}$ related to the pp values by the wounded-nucleon or other multiple-collision model.

It is interesting to note that multiplicity distributions in pp and p̄p interactions do not, in general, exhibit KNO scaling[25,26]. However, all the available pp and p̄p charged multiplicity distributions for $\sqrt{s}$ above 10 GeV can be fit by the scaled Negative Binomial Distribution[25,26], which is the same as Equation 4 in the continuum limit. Thus, a simple representation exists which describes all this "soft" multiparticle physics in terms of systematic variations of the mean value and the parameter p for all measured distributions. However, the underlying explanation for this apparent simplicity remains a mystery[27,28] at the present time.

In conclusion, the high energy tail of the $E^0{}_{TOT}$ spectrum in $\alpha\alpha$ interactions cannot be obtained from the pp spectrum using the extreme-independent-collision hypothesis fundamental to the wounded nucleon mode. In itself this is not surprising since the very simplistic asumptions of the model omit any multiple scattering or coherence between the single particle interactions. The truly surprising result of the data presented is that the KNO scaling of the pp spectrum fits the $\alpha\alpha$ data over ten decades including the high energy tail. In a sense KNO scaling implies the opposite of independent particle interactions. Any simple independent particle model for the $\alpha\alpha$ reaction allows the energy release in each N-N interaction to be combined in a random way with that of another N-N interaction. Thus the use of convolution integrals. However, convolutions of a function do not have the same shape as the original function and cannot obey KNO scaling. In terms of the Gamma distribution (Eq. 3), convolutions leave the parameter α the same and let p increase to np. Whereas the observed scaling leaves p unchanged but varies α by the mean value. The implication, the, of the excellent fit to the data in Figure 7 is that the N-N collisions involved in the $\alpha\alpha$ scattering must be, unexpectedly, highly correlated.

TABLE 1

Parameters of Wounded Nucleon Fit

n	w	AFS r_n Fixed r_n(percent)	Best Fit r_n Free r_n(percent)
1	2	44.7	47.7 ± 8.4
1-1/2	3	19.9	14.1 ± 12
2	4	16.0	28.4 ± 6.4
2-1/2	5	10.2	0 ± 0.8
3	6	5.97	0 ± 0.1
3-1/2	7	2.65	0 ± 0.1
4	8	0.54	9.7 ± 0.3

$\sigma_{in}^{\alpha\alpha}$ (mb) 134 ± 2   129 ± 4

χ^2/dof 1984/32   1044/26

REFERENCES

1. A.L.S. Angelis, G. Basini, H.J. Besch, R.E. Breedon, L. Camilleri, T.J. Chapin, C. Chasman, R.L. Cool, P.T. Cox, Ch. Von Gagern, C. Grosso-Pilcher, D.S. Hanna, P.E. Haustein, B.M. Humphries, J.T. Linnemann, C.B. Newman-Holmes, R.B. Nickerson, J.W. Olness, N. Phinney, B.G. Pope, S.H. Pordes, K.J. Powell, R.W. Rusack, C.W. Salgado, A.M. Segar, S.R. Stampke, M. Tanaka, M.J. Tannenbaum, P. Thieberger and J.M. Yelton, Phys. Lett. 168B, 158 (1986).
2. A.L.S. Angelis, et al., Phys. Lett. 141B, 140 (1984).
3. M. Tanaka, A.L.S. Angelis, et al., AIP Conf. Proc. Vol. 123, Ed. R. Mischke (AIP, New York, 1984) pp 762-769.
4. A. Bialas, A. Bleszynski and W. Czyz, Nucl. Phys. B111, 461 (1976).
5. Z. Koba, H.B. Nielsen and P. Olesen, Nucl. Phys. B40, 317 (1972).
6. This comparison was pointed out by Leon Van Hove at the Quark Matter '86 Conference.
7. CERN-Columbia-Rockefeller Collaboration: F.W. Busser, et al., Phys. Lett. 46B, 471 (1973).
8. CERN-Columbia-Rockefeller-Saclay Collaboration: F.W. Busser, et al., Nucl. Phys. B106, 1 (1976).
9. S.M. Berman, J.D. Bjorken and J.B. Kogut, Phys. Rev. D4, 3388 (1971).
10. R.F. Cahalan, K.A. Geer, J. Kogut and Leonard Susskind, Phys. Rev. D11, 1199 (1975).

11. R.D. Field, Phys. Rev. Lett. 40, 997 (1978); R. Cutler and D. Sivers, Phys. Rev. D16, 679 (1977); ibid D17, 196 (1978); B.L. Combridge, J. Kripfganz and J. Ranft, Phys. Lett. 70B, 234 (1977); A.P Contogouris, R. Gaskell and S. Papadopoulos, Phys. Rev. D17, 2314 (1978); J.F. Owens, E. Reya and M. Gluck, Phys. Rev. D18, 1501 (1978); J.F. Owens and J.D. Kimel, ibid 3313 (1978).
12. A.G. Clark, et al., Phys. Lett. 74B, 267 (1978).
13. A.L.S. Angelis, et al., Phys. Lett. 79B, 505 (1978).
14. J.F. Gunion, S.J. Brodsky and R. Blankenbecler, Phys. Rev. D6, 2652 (1972); Phys. Lett. 42B, 461 (1972); Phys. Rev. D12, 3469 (1975); ibid D18, 900 (1978).
15. H.J. Frisch, et al., Phys. Rev. Lett. 44, 511 (1980).
16. CERN-Columbia-Oxford-Rockefeller Collaboration: A.L.S. Angelis, et al., Nucl. Phys. B209, 284 (1982); J. De Physique 43, C3-134 (1982).
17. For the flavor, see Maurice Jacob's Eulogy for the ISR, TH. 3807-CERN, January 1984 (unpublished).
18. W.J. Willis, BNL 17522, ISABELLE-Physics Prospects, pp 207-234, Brookhaven National Laboratory, Upton, NY (1972).
19. C. De Marzo, et al., Nucl. Phys. B211, 375 (1983).
20. H. Gordon, et al., Phys. Rev. D28, 2736 (1983).
21. T. Åkesson, et al., Phys. Lett. 119B, 464 (1982).
22. W. Bell, et al., Phys. Lett. 128B, 349 (1983); and references therein.
23. J. Hüfner and B. Liu, Z. Phys. C27, 283 (1985).
24. W.Q. Chao and H.J. Pirner, Z. Phys. C14, 165 (1982).
25. G.J. Alner, et al., Phys. Lett. 160B, 193 (1985).
26. G.J. Alner, et al., Phys. Lett. 160B, 199 (1985).
27. P. Carruthers, Proc. Quark Matter'84, K. Kajantie, Ed., Springer-Verlag, Berlin (1985), pp 93-100.
28. A. Giovannini and L. Van Hove, Z. Phys. C30, 391 (1986).

HIGH p_T $\pi°$ PRODUCTION IN $\alpha\alpha$, dd, and pp COLLISIONS AT THE CERN ISR*

BNL-CERN-Michigan State-Oxford-Rockefeller (BCMOR) Collaboration[1]

Mitsuyoshi Tanaka
AGS Department, Brookhaven National Laboratory
Upton, New York 11973

ABSTRACT

The invariant cross sections for inclusive high p_T $\pi°$ production in $\alpha\alpha$, dd, and pp collisions have been measured in the central rapidity (y) region at the same center-of-mass energy per nucleon pair, $\sqrt{s_{NN}}$ = 31 GeV. The transverse momentum range covers 3.5 < p_T < p_T < 9.0 GeV/c. The cross section ratio R ($\alpha\alpha$/pp) shows clearly the anomalous nuclear enhancement with <R> = 23.8 + 0.5 whereas the ratio R(dd/pp) shows no such enhancement and is roughly constant at <R> = 3.7 + 0.1. The data will be compared with theoretical predictions.

INTRODUCTION

The interest in studying high energy heavy ion collisions has been recently increased in conjunction with the anomalous nuclear enhancement as well as the theoretical expectations of observing a new state of matter, the quark-gluon plasma.[2] In August 1983 the second and last light ion run took place at $\sqrt{s_{NN}}$ = 31 GeV at the CERN ISR. The new $\alpha\alpha$ data are 15 times statistically better than the previous data (1980)[3] and the dd data are available for the first time, allowing a detailed comparison between light ion and pp interactions a the highest energy available at accelerators.

In this paper we report new results for inclusive high $p_T\pi°$ production in $\alpha\alpha$ and dd collisions for p_T > 3.5 GeV/c, and compared with pp collisions at the same $\sqrt{s_{NN}}$. The main purpose of this experiment is to investigate the P_T dependence of the anomalous nuclear enhancement which was first observed by the Chicago-Princeton Group in proton-nucleus collisons.[4] Results for the transverse energy spectra E_T of neutral particles ($\pi°$) have been already published elsewhere.[6]

EXPERIMENTAL PROCEDURE

The experiment was performed at the CERN ISR with the R110 detector which was basically an electromagnetic calorimeter with a system of drift chambers in a magnetic field of 1.4 T. Measurements of the invariant cross sections of $\pi°$ were made using the two lead-glass modules and the drift chambers. Each module consisted of an array of 168 blocks (17 radiation lengths (X_0) thick) proceeded by an array of 34 blocks (4 X_0 thick). The rms energy resolution was

---

*Work performed under the auspices of the U.S. Department of Energy.

$(4.3/\sqrt{E} + 2)$ % where E in GeV. The trigger required an energy deposit above a given threshold, localized in either one of the two lead-glass modules. The data were taken with various trigger thresholds. Integrated luminosities used in this analysis are 91, 400, and 502 nb^{-1} for $\alpha\alpha$, dd and pp interactions, respectively.

After an off-line checking of the trigger condition, events were required to have at least three charged tracks with a reconstructed vertex inside a defined intersection region. Neutral clusters were defined as a distribution of energy which occupied at most 3*3 adjacent blocks in the back array plus the associated front array blocks. In addition, cuts were made on the ratio of energy deposited in the front and back arrays in order to reduce background from charged hadrons. For each cluster an estimated energy loss in the iron and magnetic coils was added and the center-of-mass (c.m.) transverse momentum p was calculated assuming a π° coming from the vertex since the individual γ-rays from π° decays were not resolved. In order to insure a uniform acceptance and to avoid possible trigger inefficiencies near the threshold, the final corrected transverse momentum was required to be well above the nominal threshold, and the c.m. rapidity y and azimuth angle ϕ were required to be in fiducial region; $\|y\| < 0.40$ and $\|\phi\| < 0.33$ for the inside module, $\|y\| < 0.55$ and $\|\phi\| < 0.43$ for the outside module.

Since there were no significant systematic differences in the data from the inside and outside modules, the invariant cross sections from both the modules were combined, treated as a single detector.

RESULTS AND CONCLUSIONS

Figure 1 shows the invariant cross sections $Ed^3\sigma/dp^3$ for π° production in $\alpha\alpha$, dd, and pp collisions as a function of p_T. The data were fitted for each type of beam by the empirical expression

$$E \frac{d^3\sigma}{dp^3} = A \ p_T^{-n} \ e^{-bx_T}$$

where $x_T = 2p_T/\sqrt{s}$ and A, n and b are parameters to fit. The results of these fits are shown as the solid lines in the figure. They are in excellent agreement with the data. The quoted errors are purely statistical. It should be noted that as for our previous publications, the cross sections here are not for reconstructed π°'s, but for single clusters which are consistent with the decays $\pi^\circ \to \gamma\gamma$, $\eta \to \gamma\gamma$. No correction was applied to the data for the η contribution. Both the pp and $\alpha\alpha$ data are in reasonable agreement with the previously published data. Since the pp comparison run was very close in time to the light ion run, it is expected that any remaining systematic errors will be cancelled by measuring the cross section ratios of these data. There is an overall systematic uncertainty of 5% in the absolute p_T scale which will be cancelled by measuring the ratios.

In Figure 2, the cross section ratios of the $\alpha\alpha$ and dd data to the fitted pp data, $R(\alpha\alpha/pp)$ and $R(dd/pp)$, are shown as a function

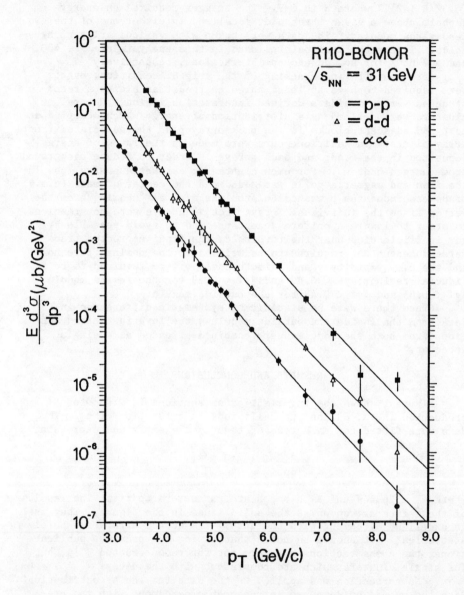

Figure 1 - The invariant inclusive cross section for π° production in $\alpha\alpha$, dd and pp collisions as a function of p_T at $\sqrt{s_{NN}}$ = 31 GeV. The solid lines are the best fits to the empirical expression described in the text.

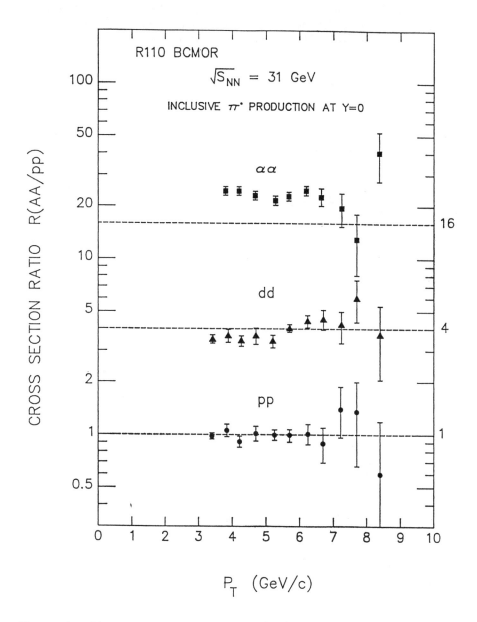

Figure 2 - The cross section ratios R(αα/pp) and R(dd/pp) of the αα and dd data to the fitted pp data as a function of p_T at $\sqrt{S_{NN}}$ = 31 GeV, as well as the ratio of the pp data to the fit.

of p_T, as well as the ratio of the pp data to the fit. The ratio $R(\alpha\alpha/pp)$ shows no p_T dependence in the measured range $3.7 < p_T < 9.0$ GeV/c but exhibits clearly the nuclear enhancement ($> A^2 - 16$) with $\langle R \rangle = 23.8 \pm 0.5$.

On the other hand, the ratio $R(dd/pp)$ shows no clear evidence for the anomalous enhancement and is consistent with remaining constant with $\langle R \rangle = 3.7 \pm 0.1$ over the range $3.2 < p_T < 9.0$ GeV/c, though the possible increase of R with p_T cannot be ruled out.

A number of theoretical models with three different mechanisms[6] have been proposed to account for the nuclear enhancement observed in $\alpha\alpha$ collisions at high p_T, depending on which of the following incredients are responsible for the enhancement: i) the nuclear structure functions, ii) the multiple parton scattering in the nucleus, and iii) the recombination of the high p_T scattered partons into a meson. Our results of the ratio $R(\alpha\alpha/pp)$ are consistent with predictions of a multiple parton scattering model[7] and also of a QCD parton model[8] with assumption that quarks and gluons undergo multiple scattering, as well as of a recently proposed recombination model.[9] There are presently no theoretical predictions available for the ratio $R(dd/pp)$.

ACKNOWLEDGMENTS

We wish to thank the ISR Division for the superb operation of the ISR, Robert Gros for his technical help and Marie-Ann Huber for her painstaking data reduction duties. This research was supported in part by the U.S. Department of Energy under Contract No. DE-AC02-76CH00016.

REFERENCES

1. The BCMOR Collaboration members are: A.L.S. Angelis, G. Basini, H.-J. Besch, R.E. Breedon, L. Camilleri, T.J. Chapin, C. Chasman, R.L. Cool, P.T. Cox, Ch. von Gagern, C. Grosso-Pilcher, D.S. Hanna, P.E. Haustein, B.M. Humphries, J.T. Linnemann, C.B. Newman-Holmes, R.B. Nickerson, J.W. Olness, N. Phinney, B.G. Pope, S.H. Pordes, K.J. Powell, R.W. Rusack, C.W. Salgado, A.M. Segar, S.R. Stampke, M. Tanaka, M.J. Tannenbaum, P. Thieberger and J.M. Yelton.
2. See e.g., Quark Matter '83, Proceedings of the Third International Conference on Ultra-relativistic Nucleus-Nucleus Collisions, BNL 1983, Eds. T.W. Ludlam and H.E. Wegner, Nucl. Phys. A418 (1984).
3. A.L.S. Angelis, et al., Phys. Lett. 116B (1982) 379.
4. J.W. Cronin, et al., Phys. Rev. D11 (1975) 3105.
5. BCMOR Collaboration, A.L.S. Angelis, et al., Phys. Lett. 141B (1984) 140, Phys. Lett. 168B (1986) 158, M. Tanaka, et al., AIP Conf. Proc. 123, R. Mischke, Ed., AIP, NY (1984) 762-769.
6. See e.g., M.A. Faessler, Phys. Report 115 (1984) 68-76.
7. U.P. Sukhatme and G. Wilk, Phys. Rev. D25 (1982) 1978.
8. M. Lev and B. Peterson, Z. Phys. C21 (1983) 155.
9. T. Ochiai, preprint RUP-85-9 (October 1985), Rikkyo University.

FINITE TEMPERATURE MEAN FIELD CALCULATIONS WITH ISOBARS

H. G. Miller and R. M. Quick, Theoretical Physics Division, NRIMS, CSIR, Pretoria, South Africa

J. P. Vary, Physics Department, Iowa State University, Ames, IA 50011, U.S.A.

W. Fabian, Frankfurt, W. Germany

ABSTRACT

Finite Temperature Hartree Fock (FTHF) calculations have been performed in ^{4}He in a quasiparticle basis which contains nucleons as well as $\Delta(1232)$ isobars. For temperatures less than 10 MeV, the isobar admixture probabilities are negligibly small.

It has long been recognized that the $\Delta(1232)$ resonance plays an important role in pion condensation and is a crucial ingredient in the formation of giant Gamow Teller states. Calculations of such phenomena inevitably assume that the uncorrelated nuclear ground state does not contain any admixtures of Δ's. One would probably not expect Δ's to be present in the nuclear ground state obtained via a mean field calculation at T=0. In spite of the fact that a number of the matrix elements involving Δ degrees of freedom are large, due to larger values of some of the basic coupling constants, and spin-isospin factors, they do not appear to be large enough to compensate for the mass difference between the nucleon and the Δ. However, at finite temperatures, isobar admixture probabilities in mean field calculations will ultimately become nonzero when enough thermal energy is available to excite additional degrees of freedom in the nucleon. The crucial question is at what temperatures does this occur in realistic calculations.

In the present work a FTHF calculation has been performed in ^{4}He in which the HF quasiparticles contain admixtures of nucleons and Δ's. The present preliminary calculation has been performed in a model space with 200 single particle (s.p.) basis states. The nucleon sector contains 40 s.p. basis states that are in the 0s, 0p and 1s-0d shells. The Δ sector contains the corresponding 160 s.p. basis states. The Hamiltonian describing the system is

$$H = (T_{rel} + V_{eff})_{NN} + \Delta M + T_{lab} + V_{\Delta N} + V_{\Delta\Delta}$$

where ΔM gives the mass difference between the nucleon and Δ. The HF quasiparticle states are assumed to have good $T_z = \pm 1/2$. In the nucleon sector, the Hamiltonian operator contains an effective potential which is described in ref. 1. In the Δ sector, the bare interactions arising from one pion and one rho exchange used in earlier deuteron calculations[2] have been employed. In order to avoid the singularities in both of the transition potentials a dipole regularization factor, which depends on the regularization parameter, Λ, has been used. In the nucleon sector the relative kinetic energy operator has been used while, for simplicity, the laboratory kinetic energy operator has been employed in the Δ sector.

In principle, one should renormalize the transition potentials to take into account the truncation of the model space. Rather than renormalize, we have performed preliminary calculations for different values of the regularization constant, Λ. For $\Lambda=5$ fm^{-1}, the percentage admixture of isobars in the deuteron is roughly 1%[2]. Increasing Λ makes the transition potentials less repulsive and increases the isobar admixture probability in the deuteron[2]. Perhaps a more suitable way to determine Λ and also eventually to determine the values of the effective coupling constants in the transition potentials would be to calculate the bound state of the deuteron in the model space and require that isobar admixtures were similar to those obtained in the standard calculations.

In the present calculation the ^{4}He HF ground state is only axially symmetric and is not an eigenstate of the parity operator. A similar type of HF ground state has been found previously in ^{16}O.[3] No isobar admixtures were present in the HF ground state in ^{4}He for $\Lambda=5$, 10, and 20 fm^{-1}. At finite temperatures up to 10 Mev, the temperature at which a deformed to spherical phase transitions in ^{4}He occurs, the isobar admixture probability remains zero for the aforementioned choices of the regularization constant. This indicates that the Δ degrees of freedom probably will not influence nuclear shape transitions in the static mean field calculations at normal densities. Reliable calculations at higher temperatures will require a much larger model space. Future calculations will attempt to determine the role of Δ degrees of freedom as a function of pressure as well as temperature.

This work was supported in part by the U.S. Department of Energy under Contract No. DE-AC02-82ER40068, Division of High Energy and Nuclear Physics.

REFERENCES

1. G. Bozzolo and J. P. Vary, Phys. Rev. C31, 1909 (1985).
2. H. Arenhövel, Z. Phys A275, 189 (1975).
3. H. G. Miller and J. P. Vary, Phys. Lett. B150, 11 (1985).

PRODUCTION MECHANISM FOR A LIGHT PSEUDOSCALAR BOSON AND THE ANOMALOUS POSITRON SPECTRUM

M.S. Zahir and D.Y. Kim
Department of Physics and Astronomy, University of Regina
Regina, Canada S4S 0A2

ABSTRACT

To explain the anomalous positron peak observed at the GSI, a detailed QED calculation is done for the bremsstrahlung mechanism to produce a pseudoscalar boson ϕ of mass $\simeq 1.6$ MeV which subsequently decays into e^+e^- pair. The coupling g_p of ϕ with nucleons is obtained as $g_p < 10^{-1}$ which is much smaller than the value ($g_p^2 / 4\pi \simeq 10^{-4}$) obtained from the semiclassical calculation of the same mechanism. It is also shown that the strong interaction induced bremsstrahlung mechanism allows a more acceptable value for g_p ($\simeq 10^{-3}$). Finally to explain the narrow width of the peak, a model to produce ϕ nearly at rest is also outlined.

INTRODUCTION

In heavy ion collisions with very high-Z nuclei at the GSI (Gesellschaft fur Schwerionenforschung), positrons with puzzling kinetic energy spectrum has been observed for the first time. Later on, the EPOS group performed the coincident experiment in which they again found similar spectra both for e^+ and e^-. The striking feature of these spectra can be listed as: (1) The peaks are extremely narrow ($\simeq 70$ keV), (2) In the data for single spectrum,[1] the position of the e^+ peak is at 220 - 310 keV (TUM - GSI) and 310 - 310 keV (EPOS) whereas in the coincident experiment[2], the peaks are at 380 keV (e^+) and 375 keV (e^-), (3) The narrow anomalous peaks are observed only when the beam energy is about 6.0 MeV/nucleon which is just the Coulomb barrier, (4) The peak position does not change with total charge $Z_3 = Z_1 + Z_2$ (Z = 180 -188) but it seems to change with beam energy[3].

Because of the observation (4), the origin of the peaks cannot be attributed to the decay of the QED vacuum (theory predicts a strong dependence on Z). As a result, theoretical attempts are based on a hypothesis about the production of a light neutral boson ϕ (mass $m_\phi \simeq 1.6 - 1.8$ MeV) decaying into e^+e^- pairs. The various model calculations can be classified mainly into two groups. (1) Nuclear pseudoscalar bremsstrahlung: (a) semi-classical approach[4], (b) linear QED calculation[5] with an effectve Lagrangian. (2) Production of ϕ from the electromagnetic fields of resonating nuclear complex in a semiclassical approach. Although none of the above mechanisms can explain the narrow spectrum, they all set limits on the coupling parameter. This talk refers to calculations belonging to group 1(b). At first, we analyze the mechanism in terms of bremsstrahlung induced by electromagnetic interaction. We assume a model in which the two ions [(A_1,Z_1) and (A_2, Z_2)] barely come into contact with each other and form a super-nucleus $(A_1+ A_2, Z_1+ Z_2)$ with life time $\simeq 10^{-20}$ sec. Inside the super-nucleus, each incident proton interacts with the strong Coulomb field of the target emitting a ϕ whose life time is $\simeq 10^{-10}$ sec. Then, as the ions recede from each other, the ϕ's subsequently decay into e^+e^- pairs (Fig.1).

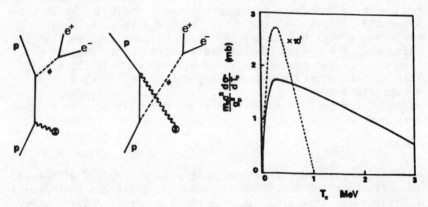

Fig.1 Production of ϕ and subseqeunt decay into e^+e^- in strong Coulomb field.

Fig.2 The positron spectrum versus T_e. The solid curve is for $T_p \simeq 6.0$ MeV and dashed curve is for $T_p \simeq 2.0$ MeV.

THE POSITRON ENERGY SPECTRUM

The calculated energy spectrum of the positrons is[5]

$$\frac{d\sigma}{dT_e} = [4(1 - 4m^2_e/m^2_\phi)]^{-1/2} \int_{w^-}^{w^+} \frac{dw}{\sqrt{w^2 - m^2_\phi}} \left(\frac{d\sigma^B}{dw}\right) \qquad (1)$$

where $d\sigma^B/dw$ is the production cross section for ϕ alone and is given by

$$\frac{d\sigma^B}{dw} = \frac{(Z_1Z_2e^2g_p)^2\sqrt{(w^2 - m^2_\phi)[(E_1- w)^2 - m^2]}}{32\ \pi^3 \sqrt{E_1^2-m^2}\, E_1(E_1- w)[E_1(E_1- w) - m^2]} \log \frac{4[E_1(E_1-w)-m^2]}{m^2_\phi} \qquad (2)$$

and

$$w^\pm = \frac{m^2_\phi}{2m^2_e}\left[E \pm \sqrt{E^2 - m^2_e}\left(1 - \frac{4m^2_e}{m^2_\phi}\right)^{1/2}\right] \qquad (3)$$

Here, m_e, m_ϕ, m and E, w, E_1 are the mass and total energies of the positron, ϕ and initial proton respectively. Also, $E = T_e + m_e$, $E_1 = T_p + m$ with T_e, T_p being the kinetic energies of the positron and proton respectively. The predicted positron spectrum is shown in Fig.2 for two values of T_p. The prediction is broader than the experiment but interestingly as T_p is reduced (the ϕ's are slower, too) the width of the spectrum becomes narrower.

Again the integrated cross-section is

$$\sigma_I = \int \frac{d\sigma}{dT_e} dT_e = (Z_1^2 Z_2^2 e^4) \ g_p^2 \ (5.13 \times 10^{12}) \ \text{MeV}^{-2} \qquad (4)$$

Let Z_1, $Z_2 = 92$. We compare σ_I with the experimental value[3] of 200 μb and obtain $g_p \lesssim 10^{-7}$.

RESULTS AND DISCUSSION

The limits on g_p obtained from a calculation of the nuclear bremsstrahlung mechanism in a semiclassical approach[4] is $g_p^2/4\pi \simeq 10^4$. The reason behind this difference of a few orders of magnitude is that the semiclassical calculations give a smaller value for the cross section; hence require a larger value for g_p to agree with the experiment. However, a more restrictive limit on g_p is obtained from other theoretical and experimental considerations[4] such as anomalous magnetic moment of electrons and hyperfine splitting of atomic lines etc. These considerations require $g_p \lesssim 10^{-2} - 10^{-4}$. If we take $g_p \simeq 10^{-3}$ as 'acceptable', then the total integrated cross section of the bremsstrahlung mechanism induced by electromagnetic interaction is smaller than the observed one by a factor of 10^{-4}. However, in that case, is it not possible for the same mechanism to be induced by strong interaction? To estimate this, one needs to determine the number of nucleons effectively playing a role in the process. Since electromagnetic interaction is long ranged, all the protons of both ions were involved whereas in the case of strong interaction, only those protons and neutrons lying within a distance of 2-3 fm of the surfaces of the interacting ions are effective. Therefore, for the total cross section, in eq.(4), one needs to replace the factor $(Z_1^2 Z_2^2 e^4)$ by $(A_1^2 A_2^2)_{eff} \cdot g_{st}^2$ where $g_{st}^2/4\pi = 14$. In a simple calculation for $U^{238} - U^{238}$, one can obtain[5] $(A_1 A_2)_{eff} = Z_1 Z_2/5$ and thus, the value of σ_I is enhanced by a factor $\simeq 10^5$ over the case when the bremsstrahlung is induced by electromagnetic interaction. And it is just about the right factor needed for the theory to agree with the experiment with g_p as small as 10^{-3}. So, we conclude that the interaction between ions has to be strong to agree with the experiment and at the same time have an acceptable value for the parameter. At this point, it can be mentioned that the position of the positron peak is controlled by m_ϕ alone. Therefore, if more than one peak is observed, it would then definitely encourage further speculations about compound ϕ or more than one ϕ.

MOMENTUM DISTRIBUTION OF φ

Of course, even after all this, one is still left with the task of explaining the narrowness of the peak. In Fig. 2, we only showed that, if the average energy of the protons can be made smaller inside a supernucleus (i.e. then the φs will have smaller energy too), a narrower positron peak results. We think it would require a detailed calculation to investigate how this is possible and therefore it will be considered beyond the scope of this talk. We only would like to illustrate how small kinetic energy of φ directly affects the width of the peak from an alternative standpoint. Assuming that the actual mechanism to produce φ is not known, then in eq.(1), one can take for $d\sigma^B/dw$ a Boltzmann like distribution. So, we rewrite eq.(1) as[5]

$$\frac{dN}{dT_e} = [4(1 - 4m_e^2/m_\phi^2)]^{-1/2} \int_{k_-}^{k_+} \frac{dk}{k} \frac{dN^\phi}{dk} \quad (5)$$

where $k_\pm = (w_\pm^2 - m_\phi^2)^{1/2}$, k is the magnitude of the three momenta of φ and $d\sigma/dT_e = \sigma_{cl} dN/dT_e$; $\sigma_{cl} = 12.6$ b. For dN^ϕ/dk is

$$\frac{dN^\phi}{dk} = N_0 (\beta/\pi)^{3/2} k^2 \exp(-k^2 \beta)$$

It can be shown numerically[5] that eq.(5) gives the observed data-both the position and narrow width ($\simeq$ 70 keV) of the peak and the cross section with No = 4.0 x 10^{-5} and $\beta^{-1/2} = 0.2 m_e$ and $m_\phi \simeq 1.62$ MeV. This gives the average K.E. of φ, $\langle w \rangle = 3/4 [1/m_\phi \beta] = 5$ keV or the velocity $\simeq 0.08$ c. Therefore, the major problem in front of us now is to explain how φs can be produced nearly at rest inside the supernucleus. We have no definite answer as yet for that puzzle although we would like to present some numbers to ponder before we draw our conclusions.

The binding energy of U^{238} is about 7.6 MeV/u. However, if one extrapolates the binding energy curve or uses a semiempirical mass formula, one can derive that the binding energy of the supernucleus ($U^{238} + U^{238}$) is about 5.5 MeV/u. Therefore a part of the kinetic energy of the incident nucleons will be required to supply this extra energy for all the nucleons of the supernucleus. The energy required is $(A_1 + A_2) \times (7.6 - 5.5)$ MeV the energy available is 6.0 x A_1 MeV. Therefore the energy left per incident nucleon is about 1.8 MeV. This means, in bremsstrahlung mechanism (induced by strong interaction), the nucleons have $T_p \simeq$ 1.8 MeV effectively. And that is enough to produce a φ with $m_\phi \simeq 1.6 - 1.7$ MeV nearly at rest.

CONCLUSIONS

We found that a detailed QED calculation lowers the limit on g_p by a few orders of magnitude compared to the semiclassical results. Nevertheless, we conclude that the bremsstrahlung mechanism induced by electromagnetic interaction requires an unacceptably large value for $g_p (\approx 10^{-1})$. Thus it can be accepted that none of the conventional mechanisms (as long as induced by e-m interaction) for ϕ production is compatible with the experimental observation. We found, however, that the alternative production mechanism, the bremsstrahlung type process induced by <u>strong interaction</u> (e.g. pp → ppϕ, pn → pnϕ) seems to provide a cross section, that agrees well with the observed one. Although, we seem to have solved one of the two puzzles, we need to investigate further to explain how ϕ s can be produced nearly at rest in a supernucleus.

Of course, the very question of how ϕ fits into the rest of the particle world at a fundamental level is still a big mystery. A few attempts have been made to answer that question. For example, whether the axion could be related to the particle in question. Any possibility of identifying the ϕ particle with axion[7] have been ruled out, at least with the standard axion. However, both theoretical[8] and experimental[9] works on a similar light boson have been carried out independent of and prior to the GSI discovery. Whether they can be related to light pseudoscalar boson in question requires futher investigations. Finally, it is necessary to confirm the existence of the particle in some other experiments dealing with different processes. In order to explore this possibility, photo and electroproduction processes of the ϕ particle have been discussed[10].

REFERENCES

1. J. Schweppe, et al.,Phys. Rev. Lett., 51, 2261 (1983).
   M. Clemente, et al.,Phys. Lett., 137B, 41 (1984).
   T. Cowan, et al.,Phys. Rev. Lett.,54, 1761 (1985).
2. T. Cowan, et al.,Phys. Rev. Lett., 56, 444 (1986).
3. J. S. Greenberg, Anomalous Positron Production. In this proceeding and references therein
4. B. Muller, Positron Lines - New Particle or Decaying Vacuum, In this proceeding and references therein
5. D.Y. Kim and M.S. Zahir, Regina Preprint URTP 86-03, submitted to Phys. Rev. D. (1986)
6. A. Chodos and L.C.R. Wijewardhana, Phys. Rev. Lett. 56, 302 (1986); K. Lane, Phys. Lett. 169B, 97 (1986)
7. A.B. Balantekin,et al. Phys. Rev. Lett., 55, 461 (1985).
   F. Calaprice, Restrictions on Axions from Nuclear Internal Pair Decays. In this Proceeding.
8. D.Y. Kim and S.I.H. Naqvi, Lett. Nuovo Cim., 35, 79 (1982).
9. L. Greenberg, et al., Lett Nuovo Cimento, 36, 221 (1983).
10. D.Y. Kim, S.R. Valluri and M.S. Zahir, Ann. Phys(Leipzig) in press.

HYPERNUCLEAR PHYSICS

Coordinators: R. E. Chrien
                R. H. Dalitz

THEORETICAL ASPECTS OF DIBARYONS

M. E. Sainio
Swiss Institute for Nuclear Research (SIN)
CH-5234 Villigen, Switzerland

ABSTRACT

Some aspects of the dibaryon spectroscopy are reviewed.

INTRODUCTION

The question of dibaryons, i.e. resonances or bound states with the baryon number B=2, has been widely discussed in recent years, for a review see e.g. Ref. 1. However, the existence of long-lived dibaryons has not yet been established experimentally. Thus there remains a number of questions to be answered to help the experiments to identify these objects.

In the standard approach to strong interaction, QCD, with SU(3) color gauge symmetry mesons are $q\bar{q}$ configurations and baryons three quark systems q^3. For these valence quark configurations there is a unique way of coupling to get an overall color singlet state and so the color content of ordinary ground state hadrons is trivial. The situation is qualitatively different for systems which contain more than three quarks, e.g. $q^2\bar{q}^2$, $q^4\bar{q}$ or q^6. These configurations correspond to mesons, baryons and dibaryons respectively. They are possibly exotic, i.e. carrying unusual quantum numbers. Of course, other exotic hadrons can be built adding gluons.

For the six quark system (q^6) there is a number of possibilities for the color coupling. By $(q^N)_c$ we denote a N-quark cluster coupling to the color representation $\underline{c}$. Then the following hidden color configurations are possible:

$$(q^5)_{3^*}-(q)_3; \quad (q^3)_8-(q^3)_8$$
$$(q^4)_3-(q^2)_{3^*}; \quad (q^4)_{6^*}-(q^2)_6.$$

Of course, the configuration

$$(q^3)_1-(q^3)_1$$

is also possible. It does not represent hidden color and is governed by the conventional dynamics through color singlet exchanges. The essential question is now if these group theoretical possibilities are realized in nature and what could be the observable consequences. If realized the hidden color configurations are likely to obey selection rules different from ordinary singlet-singlet states. In particular, they could lead to narrow resonances or even to bound states with respect to the strong interaction. On the level of theoretical predictions all these questions are, however, still very open.

In the next sections the different approaches to calculate the spectra of B=2 states are discussed. The main emphasis will be in the sector of genuine six-quark states, but some features of the color singlet dynamics will be discussed as well.

DIBARYON MASSES AND WIDTHS

Most predictions for the spectrum of six-quark states come from the MIT bag model, for a review see Ref. 2. In such a model the masses of hadronic states can be written as a sum of four terms representing kinetic energy, volume energy, zero-point energy and center-of-mass corrections, and the color magnetic splitting [3]. The one-gluon-exchange color magnetic term is after averaging over the interaction strengths proportional to

$$\Delta = -\sum_{i>j} \frac{1}{4}(\sigma\lambda)_i \cdot (\sigma\lambda)_j \qquad (1)$$

$$= -\frac{1}{4} N(10-N) + \frac{1}{3} s(s+1) + F_f^2 + \frac{1}{2} F_c^2$$

where N is the number of quarks, s is the spin, and F_f^2 and F_c^2 are the quadratic Casimir operators for SU(3) flavor and color respectively.

Once the four parameters, bag constant B, zero point energy Z_0, strong coupling constant α_c and the strange quark mass m_s, have been fixed using ground state hadron spectra one can make predictions for the q^6 sector assuming that all parameters retain their values. In this sort of framework Jaffe proposed [4] the existence of a stable dihyperon (strangeness S=-2), the H particle, with a mass 80 MeV below the $\Lambda\Lambda$ threshold ($2m_\Lambda$=2231 MeV). The H dihyperon is a flavor singlet $J^P=0^+$, I=0 s-state (uuddss). The color magnetic force is maximally binding for the lowest dimensional flavor representation (Δ=-6). The most extensive calculations within the MIT bag model have been done by the Nijmegen group [5]. They have extended the model to string-like bags with the mass formula

$$M = [M_0^2 + (1/\alpha)\ell]^{1/2} + M_{mag} \qquad (2)$$

where M_0 is calculated in the spherical bag approximation (without color magnetic term) and $1/\alpha$ is the slope of the trajectory depending on the color charges of the two clusters at the ends of the string. The relative orbital angular momentum between the clusters is denoted by ℓ. The color magnetic splitting is taken to be

$$M_{mag} = m_1\Delta_1 + m_2\Delta_2 \qquad (3)$$

where m_1 and m_2 are strengths which depend upon the number of nonstrange and strange quarks in the cluster, and Δ_1 and Δ_2 are given by Eq. (1). A large number of states with different cluster structure is predicted [5], see Table I. However, various modifications, including center-of-mass [6] and pionic corrections [7,8], have been discussed to this simple picture. For the lowest mass dibaryons the c.m. corrections raise the mass by about 70-90 MeV from the original MIT value [6]. In this case the H particle would be just unbound. The inclusion of pionic degrees of freedom raises the masses of dibaryons as well. For the H dihyperon the effect is about 70-80 MeV [7,8] and so H could be weakly bound or unbound. Aerts and Rafelski have taken [9] a more phenomenological approach in treating the color magnetic splitting and allowed for two more free parameters in describing the interaction strengths between different quark species. This approach raises the masses for all multi-quark states, for the H particle the effect is 60-90 MeV.

By imposing confinement to the q^6 states the bag model, however, neglects the fact that quarks may regroup to color singlet clusters inside the bag and should as such not be confined. This problem was discussed by Jaffe and Low who proposed [10] a method, the P-matrix formalism, to take into account the coupling to open channels. In this scheme the multi-quark bag model states are not supposed to be interpreted as physical states, poles of the S-matrix, but should be taken as primitives, poles of the P-matrix. The P-matrix is essentially the logarithmic derivative of the channel wave functions at a surface which divides the space into internal and external regions governed by different dynamics. The poles of the P-matrix correspond to a vanishing wave function at the boundary r=b. A simple Ansatz for this boundary assumes the following connection to the bag radius R [11]

$$b \simeq 1.37 \left(\frac{N}{N_1 N_2}\right)^{1/2} R \qquad (4)$$

where N is the number of quarks separating into two hadrons with N_1 and N_2 quarks respectively. In a comprehensive review of the formalism and applications see Ref. 11. The coupling to open channels causes a hadronic shift, i.e. a difference between the S-matrix and P-matrix pole positions. Typically it is expected to be of the order of the resonance width [11]. In the nucleon-nucleon scattering the P-matrix analysis is complicated by the opening of pion production channels. The hadronic shift for the H dihyperon primitive has been estimated in Refs. 12 and 13. The shift is of the order of 150-200 MeV more binding and so, if the primitive mass lies below the $\Sigma\Sigma$ threshold (2386 MeV), the H particle is expected to be bound [13].

There is no bag model calculation which would incorporate all corrections discussed above. Thus we are left with a quite large, at least 100 MeV, uncertainty of the mass predictions.

More limited predictions for dibaryon masses are obtained from other models. In the non-relativistic quark model the model hamiltonian is taken to be

Table I The lowest mass dibaryon states from the bag model calculation of the Nijmegen group [5]. The pionic decay thresholds are 2.02, 2.19 and 2.37 GeV for the Y=2,1,0 states respectively

Quark configuration	Orbital state	Spin	Isospin	Mass (GeV)
Y=2				
$(q^6)_1$	$L^P=0^+$	1	0	2.16
$(q^4)_3-(q^2)_{3*}$	$L^P=1^-$	1	0	2.11
	$L^P=1^-$	0	1	2.20
Y=1				
$(q^6)_1$	$L^P=0^+$	1	1/2	2.17
$(q^4)_3-(q^2)_{3*}$	$L^P=1^-$	0	1/2	2.11
	$L^P=1^-$	1	1/2	2.15
Y=0				
$(q^6)_1$	$L^P=0^+$	0	0	2.16
	$L^P=0^+$	1	0,1	2.35
$(q^4)_3-(q^2)_{3*}$	$L^P=1^-$	0	0,1	2.30

$$H = \sum_{i=1}^{6}(m_i + p_i^2/2m_i) + \sum_{i<j}(V_{conf}^{ij} + V_{hyp}^{ij}) \quad (5)$$

where, in addition to the mass and kinetic energy terms, there are terms representing two-body confining potential and the hyperfine splitting due to one-gluon-exchange (Eq. (1)). Here, like in the bag model, the six-quark spectra are qualitatively determined by the hyperfine interaction. Large attraction is found e.g. in the $\Delta\Delta$ channel where the (isospin, spin) = (0,3) state could lie below the $N\Delta\pi$ threshold [14]. In the strangeness S=-2 channel Oka et al. [15] find attraction in the SU(3)-flavor singlet state. However, the $\Lambda\Lambda$ system is not bound, but, instead, there is a sharp coupled channel resonance at E=27 MeV, i.e. just below the $N\Xi$ threshold. The SU(3) flavor symmetry breaking effects in the hyperfine interaction for the H dibaryon have recently been studied in detail by Rosner [16]. These effects reduce the attraction by about 100 MeV, but the H particle still remains bound by about 50 MeV in the model considered.

Recently some estimates of the masses of the B=2 states have been published in the framework of topological chiral soliton models, for reviews see Refs. 17 and 18. In the Skyrme model the Lagrangian is taken to be [18]

$$L = \frac{f_\pi^2}{4} \text{Tr}[\partial_\mu U \partial^\mu U^+] + \frac{1}{32e^2}\text{Tr}[(\partial_\mu U)U^+, (\partial_\nu U)U^+]^2 \quad (6)$$

where $U(x)$ is the chiral field (a $N_f \times N_f$ matrix of unit determinant), f_π is the pion decay constant and e is a dimensionless parameter. In this model a conserved topological current

$$B_\mu = \frac{1}{24\pi^2} \varepsilon_{\mu\alpha\beta\gamma} \text{Tr}[U^+\partial_\alpha U)(U^+\partial_\beta U)(U^+\partial_\gamma U)] \quad (7)$$

can be constructed. The associated generalized charge

$$B = \int d^3x \, B_o \quad (8)$$

is the winding number which characterizes the soliton solutions. It can be interpreted as the baryon number. The solutions are constructed using a spherically symmetric Ansatz which for the B=2 states is based on the SO(3) subgroup of flavor SU(3) [19]. After quantization the masses for dibaryons can be calculated. When explicit SU(3) flavor symmetry breaking effects are included Balachandran et al. [20] obtain the value 1030 MeV for the SU(3) singlet (the H particle). To fit the baryon octet and decuplet a f_π value is needed which disagrees qualitatively with the experimental value. For the parameter e in Eq. (6) the value 6.49 is obtained. The problem in the model seems to be that the zero point corrections are not small. If an overall constant is added to the baryon masses and its size is fixed by using the experimental value for f_π in addition to the baryon multiplets, another estimate for the H mass is obtained, m_H=2100 MeV, assuming that the same overall constant is applicable for both dibaryons and baryons. In addition to the SU(3) singlet dibaryon octet and decuplet masses are estimated [20] with the values 300-350 MeV and 600-700 MeV above the singlet respectively. Another chiral model with the Skyrme term replaced by the ω meson has been used to estimate the H mass [21]. The problems appear to be very similar to the ones discussed above in the context of the Skyrme model. There [21] the best estimate is the result in the SU(3) limit which would make H bound by about 200 MeV. A Skyrme model calculation has been performed also by Jaffe and Korpa [22]. The resulting dibaryon masses are roughly 600 MeV below the bag model predictions [4]. In addition, they [22] analyze the role of the Wess-Zumino term which can be added to the Skyrme Lagrangian, and demonstrate that it selects the same relation between color and flavor quantum numbers for the Skyrme dibaryon as found in the quark model. The relation of the Skyrme model and the quark model with a variable number of colors has been discussed for dibaryons in Ref. 23. Baryon number B=2 states have also been discussed with specific SU(2) Ansätze, see e.g. Refs. 24 and 25, or within the SU(4) flavor extension [26]. To summarize, the H dihyperon seems to be bound in the chiral soliton models. However, uncertainties are large and, if the typical 30 % precision in the single baryon sector can serve as any guide, better than that accuracy can not be expected.

The techniques of lattice QCD have also been used to estimate the splitting between the H mass and the mass of two Λ particles [27]. These calculations indicate that the H particle is not bound, but the uncertainties are still large.

Very little work has been done to estimate the decay widths of dibaryons. Qualitative estimates can be obtained by comparing the color magnetic attraction (Eq. (1)) in different quark cluster configurations. For the non-strange sector the bag model primitives lie above the NNπ threshold and so they would mostly acquire a large width in a P-matrix calculation [11]. Branching ratio of dibaryon decay into two different pionic channels has been estimated by Grein et al. [28] with the result that the πd channel is very small in comparison with the πNN continuum. For the strangeness S≠0 systems the lowest lying states may become stable with respect to pionic decay due to the stronger color magnetic attraction. Aerts and Dover have estimated [29] the widths of the $^1P_1$(2112) and ^3P_J(2152) (I=1/2) states [5] in a model which consists of an oscillator wave function overlap and a space-spin-color-flavor recoupling coefficient. The results from this model indicate that these states with $(q^4)_3$-$(q^2)_{3*}$ cluster structure could be very narrow, $\Gamma \lesssim 10$ MeV. The weak decay of ΛΛ bound states has been discussed in Ref. 30.

DIBARYONS AND COLOR SINGLET DYNAMICS

In much of the discussion above it was assumed that dibaryons are related to the hidden color configurations of the baryon-baryon system. Such configurations would be "genuine" six-quark dibaryons. In practice, however, it may not be possible to separate them from B=2 resonances which involve conventional dynamics based on color singlet meson exchanges. There are threshold effects, like the excitation of the Δ-resonance in nucleon-nucleon scattering or the ΣN cusp in ΛN interaction, which can confuse the analysis of multi-quark states. These phenomenological aspects have been extensively discussed e.g. in Ref. 1 where the emphasis is in the nonstrange sector, and in Ref. 31 which deals with the $S \neq 0$ baryon-baryon system. If one would observe a very narrow B=2 resonance, it would be a good candidate for a genuine six-quark state, but, even that would not constitute a proof for hidden color. Maybe it will turn out, that dibaryons can be established only as a part of a coherent picture of hadron-hadron interaction.

CONCLUSIONS

The central issue in the search for multi-quark hadrons is color confinement. The question is, if confinement is realized in some particular way for multi-quark states or, if any states beyond $q\bar{q}$ and q^3 appear as physical states. Unfortunately, the theoretical models used to estimate the masses of the q^6 states cannot be considered reliable in a quantitative sense. In addition, numerous other effects, less directly related to the color degree of freedom, confuse the experimental analysis considerably. In particular, many candidates for six-quark states seem to be related to inelastic threshold effects [1]. Recently Lenz et al. [32] have discussed some models for the s-wave q^2-$\bar{q}^2$ system where narrow resonances may appear close to thresholds due to quark exchange, if the color hidden ground state is nearby. In addition to these threshold resonances the model also produces wider resonances related to the color hidden spectrum.

From a practical point of view adding strange quarks to the system to look for B=2 states could be helpful, because the number of available strong decay channels is reduced due to the additional color-magnetic attraction which shifts the spectrum downwards. Specific possibilities have been discussed in Ref. 33.

ACKNOWLEDGMENTS

I have greatly benefitted from discussions with my colleagues at the SIN Theory Group. In addition, I would like to thank Milan Locher for reading the manuscript.

REFERENCES

1. M.P. Locher, M.E. Sainio and A. Švarc, Adv. Nucl. Phys. (to appear).
2. A.W. Thomas, Adv. Nucl. Phys. $\underline{13}$, 1 (1984).
3. T. DeGrand et al., Phys. Rev. $\underline{D12}$, 2060 (1975).
4. R.L. Jaffe, Phys. Rev. Lett. $\underline{38}$, 195 (1977); (E) $\underline{38}$, 617 (1977).
5. P.J. Mulders, A.T. Aerts and J.J. de Swart, Phys. Rev. $\underline{D21}$, 2653 (1980).
6. K.F. Liu and C.W. Wong, Phys. Lett. $\underline{113B}$, 1 (1982).
7. P.J. Mulders and A.W. Thomas, J. Phys. $\underline{G9}$, 1159 (1983).
8. K. Saito, Prog. Theor. Phys. $\underline{72}$, 674 (1984).
9. A.T.M. Aerts and J. Rafelski, Phys. Lett. $\underline{148B}$, 337 (1984).
10. R.L. Jaffe and F.E. Low, Phys. Rev. $\underline{D19}$, 2105 (1979).
11. B.L.G. Bakker and P.J. Mulders, Adv. Nucl. Phys. (to appear).
12. A.M. Badalyan and Yu. A. Simonov, Sov. J. Nucl. Phys. $\underline{36}$, 860 (1982).
13. B.O. Kerbikov, Sov. J. Nucl. Phys. $\underline{39}$, 516 (1984).
14. K. Maltman, Nucl. Phys. $\underline{A438}$, 669 (1985).
15. M. Oka, K. Shimizu and K. Yazaki, Phys. Lett. $\underline{130B}$, 365 (1983).
16. J.L. Rosner, Phys. Rev. $\underline{D33}$, 2043 (1986).
17. A.P. Balachandran, Syracuse preprint SU-4222-314 (1985).
18. R. Vinh Mau, Acta Physica Austriaca, Suppl. XXVII, 91 (1985).
19. A.P. Balachandran et al., Phys. Rev. Lett. $\underline{52}$, 887 (1984).
20. A.P. Balachandran et al., Nucl. Phys. $\underline{B256}$, 525 (1985).
21. S.A. Yost and C.R. Nappi, Phys. Rev. $\underline{D32}$, 816 (1985).
22. R.L. Jaffe and C.L. Korpa, Nucl. Phys. $\underline{B258}$, 468 (1985).
23. H. Gomm, F. Lizzi and G. Sparano, Phys. Rev. $\underline{D31}$, 226 (1985).
24. H. Weigel, B. Schwesinger and G. Holzwarth, Phys. Lett. $\underline{168B}$, 321 (1986).
25. E. Braaten and L. Carson, Phys. Rev. Lett. $\underline{56}$, 1897 (1986).
26. F. Hussain and M.S. Sri Ram, Phys. Rev. Lett. $\underline{55}$, 1169 (1985).
27. P.B. Mackenzie and H.B. Thacker, Phys. Rev. Lett. $\underline{55}$, 2539 (1985).
28. W. Grein, K. Kubodera and M.P. Locher, Nucl. Phys. $\underline{A356}$, 269 (1981).
29. A.T.M. Aerts and C.B. Dover, Phys. Lett. $\underline{146B}$, 95 (1984).
30. M.I. Krivoruchenko and M.G. Shchepkin, Sov. J. Nucl. Phys. $\underline{36}$, 769 (1982).
31. R.H. Dalitz, Nucl. Phys. $\underline{A354}$, 101c (1981).
32. F. Lenz et al., Ann. Phys. (to appear).
33. C.B. Dover, Nucl. Phys. $\underline{A450}$, 95c (1986).

FUTURE DIBARYON RESEARCH AT BNL

Philip H. Pile
Brookhaven National Laboratory, Upton, New York, 11974

ABSTRACT

An experiment designed to search for a strangeness -1 dibaryon is discussed. The experiment will use the ^{3}He(K^-, π^+)nD_S reaction and is expected to begin near the end of the year at the BNL AGS using the existing LESB-I beamline and hypernuclear spectrometer system. Planned searches for a strangeness -2 dibaryon are also discussed. The strangeness -2 dibaryon searches will however require the construction of a new 1.8 GeV/c kaon beam line at the AGS.

INTRODUCTION

The success of the quark model in describing the baryon and meson mass spectra has led to calculations which predict that certain systems, with the quarks of two baryons, may be stable with respect to strong decay or at least have narrow decay widths. In particular, most bag models predict that although non-strange dibaryon states will have large decay widths, strangeness -1 (S = -1) dibaryons may have narrow decay widths and one S = -2 dibaryon may be stable with respect to strong decay. There has not been an experiment to date with the required sensitivity or the necessary freedom from background processes which can prove or disprove the existence of dibaryon states. Planned experiments at the Brookhaven AGS which will attempt to identify S = -1 and S = -2 dibaryon states will be discussed.

S = -1 DIBARYONS

Theoretical predictions of the absolute masses of dibaryons have rather large uncertainties. However, the uncertainty in the difference in mass between dibaryons with a given strangeness is not so large. Figure 1 shows the result of a calculation by Mulders, Aerts and deSwart.[1]

Two S = -1 dibaryons are predicted to have rather narrow decay widths. Both are isospin 1/2, $(Q)^4 \times (Q)^2$ clusters, in a relative P state. The state designated as D_t at 2150 MeV/c^2 is formed with the initial baryons in a spin triplet state and the D_s state at 2110 MeV/c^2 with the initial baryons in a spin singlet state. The $(Q)^6$ configuration at 2170 MeV/c^2 is expected to be unstable with respect to quark tunneling and therefore will not have a narrow decay width. All other dibaryon states are predicted to lie above the $\Lambda N\pi$ threshold and will have very broad decay widths.

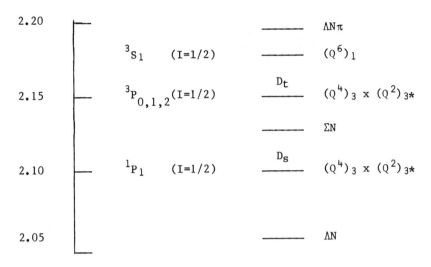

Fig. 1. Bag model prediction of S = -1 dibaryon states. Only the states labeled D_t and D_s are predicted to have narrow decay widths. The vertical scale is mass in GeV/c^2.

A search for the D_t dibaryon state shown in Fig. 1 is discussed by Piekarz elsewhere in these proceedings and in ref. 2. Piekarz presented data which suggest that the state exists with a 2140 MeV/c^2 mass. If the D_t state does indeed exist with this mass then the spin 0 partner D_s state should be seen about 40 MeV/c^2 lower in mass. Although the ^{2}H(K$^-$,π^-)nD_s reaction could produce the dibaryon, a spin flip would be required and D_s signal would be buried in the Λp quasifree missing mass tail. The (K$^-$,π^-) reaction on ^{3}He is a better choice since the spin 0 proton pair is available to allow the pp(K$^-$,π^+)D_s reaction to occur without spin flip. The production cross section then should be comparable to the D_t formation cross sections in the ^{2}H(K$^-$,π^-)D_t reaction. The use of a ^{3}He target does, however, lead to a three-body final state. If the spectator neutron in the final state is left undetected the missing mass signature for the D_s formation will have a rather broad width of about 25 MeV/c^2.

Unlike the D_t dibaryon, the D_s state has only one channel in which to decay--the ΣN channel is energetically forbidden and only the Λn channel is open. Also unlike the D_t, the D_s formation is only through intermediate Σ^- production whereas the D_t intermediate states involve both Λ and Σ states. Furthermore, since the (K$^-$,π^+) reaction on ^{3}He cannot produce Λ's in a single step the D_s signal should appear in a background-free region.

The S = 1 dibaryon collaboration[3] is planning a search for the D_s dibaryon at the end of 1986. A 25 cm long liquid ^{3}He target is now being constructed by the AGS cryogenic group. The search will be made with a 870 MeV/c kaon momentum and about a 20 degree reaction angle. The experimental sensitivity is expected to be about 150 nD_s events per 100 hours per μb/sr. The BNL LESB-I kaon beam line and hypernuclear spectrometer system will be used for the search. The target design will allow a scintillator hodoscope barrel to be placed around it, as was done for the D_t search,[2] but the hodoscope should not be essential to the success of the experiment as was the case for the D_t search. The hodoscope can be used to suppress the Σ^-pn quasifree background at the expense of D_s signal.

Detection of the spectator neutron will be necessary if a width measurement of the D_s state is to be made. Current K^- beam intensities available at the AGS make neutron detection in coincidence with the (K^-,π^-) tag difficult due to the low (few/100 hours) coincidence rates. This part of the experiment may be possible with future improvements in the AGS beam intensity.

S = -2 DIBARYONS

The only dibaryon that is predicted to be stable with respect to strong decay is the S = -2 "H" dibaryon predicted by Jaffe in 1977.[4] There have been no searches to date which have had the required sensitivity to rule out the existence of the H dibaryon. Some of the properties of the H are spin and isospin 0, strangeness 0, Q^6 quark structure in relative S state, and the quarks of two lambdas.

The H dibaryon was characterized by Jaffe using the MIT bag model. Although considerable fluctuations in the predicted mass of the H have occurred using improved bag model calculations, the various corrections have tended to cancel each other. For example coupling to open channels relaxes the requirement for a spherical bag and decreases the H mass by 100-200 MeV while adding the proper treatment of center of mass motion tends to increase the mass by about 100 MeV. Other models such as the Skyrme or lattice gauge also predict the existence of the H particle.

Jaffe predicted the H mass to lie about 80 MeV/c^2 below the $\Lambda\Lambda$ mass. He, however, pointed out that the uncertainty in the parameters of his bag model would allow the H mass to lie anywhere between the $\Lambda\Lambda$ mass and the ΛN mass. Furthermore, although less probable, his model does not rule out an H mass that lies between the $\Sigma\Sigma$ and NN masses. Since Jaffe's original article, careful theoretical estimates[5] have expanded the most favored region to extend from about 40 MeV/c^2 below the $N\Lambda$ mass to about 20 MeV/c^2 above the $\Lambda\Lambda$ mass.

There is currently a conditionally approved experiment[6] (contingent upon the construction of a new beamline) designed to search for the H dibaryon using the two step process:

$K^- + p \rightarrow \Xi^- + K^+$ then $\Xi^- + d \rightarrow n + H$

The essential features of the experiment are:

1. The (K^-,K^+) reaction on a liquid hydrogen target is used to create a Ξ^- beam.

2. The Ξ^- energy is degraded with tungsten slabs and is brought to rest in a liquid deuterium target.

3. The Ξ^- d atoms formed will collapse onto an H particle plus a monoenergetic neutron some fraction of the time.

4. The monoenergetic neutron is detected and represents the "signal" for H formation.

The principle disadvantage of this process is that the branching ratio for the Ξ^- d atom to form an H particle decreases rapidly with decreasing H mass. The sensitivity of the experiment does however extend from about 20 MeV above the $\Lambda\Lambda$ threshold to about 100 MeV/c^2 below the $\Lambda\Lambda$ threshold. About one half of Jaffe's most favored region will be covered by this experiment.

The H particle can be produced in a single step using the reaction:

$$^3He(K^-,K^+)nH$$

This reaction was discussed in detail by Aerts and Dover.[7] The essential features of this reacion are:

1. The (K^-,K^+) reaction is used on a ^{3}He target to produce the H and a spectator neutron in a single step.

2. The (K^-,K^+) momentum difference spectrum provides the information necessary to identify the H Particle, even without detection of the spectator neutron.

The production mechanism for this reaction is illustrated in Fig. 2 below. The reaction cross section is predicted to increase

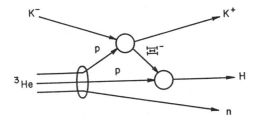

Fig. 2. Schematic of the reaction leading to H formation with the ^{3}He(K^-,K^+)nH reaction.

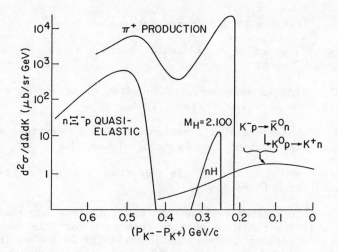

Fig. 3. Predicted dibaryon signal (nH) and 3 background sources.

with decreasing H mass (opposite to the Ξ^- atom case). However, due to Ξ^- quasielastic background the sensitivity of this approach does not extend above the $\Lambda\Lambda$ threshold.

The predicted[7] ^{3}He (K^-,K^+)nH momentum difference signature for nH production is shown in Fig. 3. The nH signal will appear with about a 30 MeV/c width (FWHM). The sharp peaking of the nH momentum difference spectrum is due to the dynamics of the (K^-,K^+) reaction. The relative momentum between the Ξ^- and the proton (see Fig. 2) must be small in order for the H to be formed. The maximum K^+ momentum and hence smallest momentum transfer to the Ξ^- will be preferred. Also shown in the figure are three of the background sources which must be contended with. The quasielastic Ξ^- production background presents a barrier to the observation of dibaryon masses above the $\Lambda\Lambda$ threshold.

Background reactions that are not sufficiently suppressed could appear to simulate a dibaryon signal. The background reactions will, however, be measured with separate triggers. The shape and positions of the sources will then be well characterized and any "leak through" will be easily identified. It should be noted that the 30 MeV/c expected width of the nH peak will be about a factor of 2 more narrow than the width of any of the background processes. Good momentum resolution will be necessary to properly characterize the shape of the various spectra.

Two H particle searches using a ^{3}He target have been proposed at the AGS. One by the proponents of the Ξ^-d atom experiment[8] (Proposal 836), and one by the S = -1 dibaryon collaboration.[9] Proposal 836 would use the low resolution large solid angle dipole proposed for the Ξ^- atom H search as the K^+ spectrometer while Proposal 830 would use the present medium resolution hypernuclear spectrometer reconfigured to handle the higher momentum required.

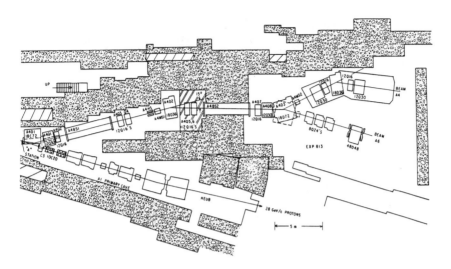

Fig. 4. Proposed 1-2 GeV/c kaon beam line for the AGS as it would be configured if located at the "A" target station. At the end of the beam line two different experimental setups are shown.

Both of these proposed experiments will be sensitive to H masses from about the ΛΛ mass to below the NN mass.

At the present time there does not exist a beamline with a high enough momentum and sufficient intensity to complete either the Ξ^- atom H search or the ^{3}He H searches. The experiments will require the construction of a new beamline at the AGS optimized for about a 1.8 GeV/c momentum. A 1-2 GeV/c kaon beam line has been proposed[10] and is shown in Fig. 4 located at the AGS "A" target station. The 31 meter beamline incorporates third order corrections in the quadrupoles and two ExB velocity selection stages. The beamline promises to deliver above 10^6 K$^-$ per AGS beam spill with a beam purity approaching 1:1. Shown at the end of the beam line are two proposed H particle search experimental setups. The beam right momentum dispersed side being for the Ξ^- atom experiment and on the beam left side is a reconfigured version of our present Moby Dick hypernuclear spectrometer system suitable for the ^{3}He H particle search (Proposal 830) and Ξ and Λ hypernuclear studies.

CONCLUSIONS

The present S = -1 dibaryon search using the ^{2}H(K$^-$,π$^-$) reaction in an attempt to identify the D_t dibaryon will be complemented by use of the ^{3}He(K$^-$,π$^+$) reaction to search for the D_s dibaryon state. The meaningful search for a S = -2 H dibaryon at the AGS will depend upon the construction of a new 1.8 GeV/c kaon beam line. The two proposed methods of producing the H dibaryon (using either Ξ^- atoms or a ^{3}He target) are complementary in their regions of sensitivity and together promise to provide a clear answer to the question of existence of the H in the mass range from 20 MeV/c^2 above the $\Lambda\Lambda$ mass to below the NN mass.

ACKNOWLEDGEMENT

Research supported by USDOE contract DE-AC02-76CH00016.

REFERENCES

1. P. J. C. Mulders, A. T. M. Aerts, and J. J. deSwart, Phys. Rev. D21 (1980) 265.

2. H. Piekarz, Nucl. Phys. A450 (1986) 85c.

3. Brandeis, BNL, MIT, Osaka, Houston, Texas, Vassar collaboration, AGS Experiment 820, "Search for S = -1 Dibaryon Resonances (Ds) in the Mass Region (2050-2130) MeV Using the Reaction ^{3}He(K$^-$,π^+) nDs."

4. R. L. Jaffe, Phys. Rev. Lett. 38 (1977) 195.

5. G. Baum, E. W. Kolb, L. McLerran, T. P. Walker, R. L. Jaffe, Phys. Lett. 160B (1985) 180; M. M. Waldrop, Science 231 (1986) 336.

6. CMU, BNL, Erlanger-Nurnberg, Freiburg, Houston, Kyoto Sangyo, New Mexico, Pittsburgh, CEN Saclay, Vassar, AGS Experiment 813, "Search for a Strangeness -2 Dibaryon."

7. A. T. M. Aerts and C. B. Dover, Phys. Rev. D 28 (1983) 450.

8. CMU, BNL, Erlanger-Nurnberg, Freiburg, Kyoto Sangyo, New Mexico, Pittsburgh, CEN Saclay, Vassar collaboration, AGS Proposal 836, "Search for a Strangeness -2 Dibaryon Using a ^{3}He Target."

9. Brandeis, BNL, LANL, MIT, Osaka, Houston, Rutgers, Texas, Vassar collaboration, AGS Proposal 830, "A Search for the H-Particle Using the ^{3}He(K$^-$,K$^+$)nH Reaction."

10. P. H. Pile, Nucl. Phys. A450 (1986) 517c; R. E. Chrien, "A 1-2 GeV/c Beam Line for Hypernuclear and Kaon Research", BNL 36082.

"THE REACTIONS $K^-d \to N\Lambda\pi$ AND $N\Sigma\pi$"

M. Torres[*] and R.H. Dalitz
Department of Theoretical Physics
Oxford University, Oxford

A. Deloff
Institute for Nuclear Studies
Hoza 69, Warsaw 00-681, Poland

ABSTRACT

The reactions $K^-d \to N\Lambda\pi$ and $N\Sigma\pi$ are calculated using a three-body multichannel formalism and two-body interactions between all pairs of particles. Quantitative agreement with the data on relative rates and spectra is found using acceptable two-body potentials. The results imply that there is no unstable bound state (2nd sheet pole above ΛN threshold) in the $(\Sigma N, \Lambda N)$ system but do require a virtual state (4th sheet pole) near the ΣN threshold.

INTRODUCTION

There exists a substantial body of data on the reactions $K^-d \to N\Lambda\pi$ and $N\Sigma\pi$[1,2] at low energy, which call for interpretation in terms of two-body hadronic interactions. The main purpose of the work on K^-d reactions is to throw additional light on the study of YN interactions. Of particular interest is the striking peak that has keen observed in the Λp mass distribution for the $K^-d \to p\Lambda\pi^-$ reaction near the ΣN threshold. This peak, has been interpretated as either the result of a cusp effect or as a ΣN bound state embedded in the ΛN continuum. The K^-d reactions are also sensitive to the other pair-wise interactions, namely; πN, NN and $\bar{K}N \leftrightarrow Y\pi$. At low energy, the multiple scattering effects are of prime importance for the K^-d system[3,4]. We report here a Faddeev calculation for the $K^-d \to NY\pi$ reactions that differs in some important details and results from previous work[3].

FORMALISM

We calculate the reaction amplitudes required by using the Faddeev equations for a multichannel set of three-particle systems. We label the three particles in the order $(1,2,3) = (N, N$ or $Y, \bar{K}$ or $\pi)$. Using the Faddeev decomposition the three-body t-matrix is given by

---

[*] Now at Instituto de Física
Apdo. Postal 20-364, 01000 México, D.F. México.

$$T = T_1 + T_2 + T_3$$

One term for each spectator particle. For scattering of particle 3 (Kaon) from the bound state of particle 1 and 2 the Faddeev equations are

$$T_i = (1-\delta_{i3})t_i + t_i G_o \sum_{j \neq i} T_j$$

where t_i is the two-body t-matrix for the interaction between particles j and k ($j \neq k \neq i$) and G_o is the free three-particle propagator.

Our calculation have assumed K^--capture from an S-orbit about deuterium. We use non-relativistic kinematics and we also neglect all mass differences whitin isoboric multiplets. One further approximation is to represent all the two-body interactions by one-term separable potentials. We chose the potential parameters such that the calcualted scattering and reaction cross sections give a good fit to the experimental data available[4]. We consider s-wave interactions between all pairs of particles in each channel, except in the πN case for which the important p-wave πN ↔ Δ interaction is included and for the addition of a p-wave YN resonance. For the $^3S_1$ YN interactions we found two solutions that fit the two body data fairly well: i) the model YN1 that has no ΛN resonance but has a 4th sheets S-matrix pole not far from the ΣN threshold and ii) the model YN2 that has an unstable bound state (UBS) such that the phase shift $\delta_{\Lambda N}$ passes upward through 90° at 0.45 Mev below the ΣN threshold. We finally notice that the two-body data do not determine in general the signs of the off-diagonal potentials[4]. For a given YN potential, we have two situations denoted by YN<$\pm$>(for positive or negative value of the ΛN-ΣN potential) that produce the same two-body results but that give different solutions for the three-body problem.

RESULTS AND CONCLUSIONS

The total reaction rates and branching ratios to the NYπ final states are listed on table I for the models YN1 <±> and YN2 <±>. Rather good agreement with the data is found for YN1<+>, well within the experimental errors.

The calculated m(Λp) spectrum is compared with Tan's on Fig. 1(a) for the cases YN1<±>, with the low-momentum cut-off for the final proton in this data. The agreement is good for YN1<+> below the dominating threshold cusp and also above except for the shoulder at about 2140 Mev. The calculated amplitude in the NΛπ channel is the sum of

Table I   Reaction Rates for $K^-d \to NY\pi$.   The data are mostly from ref. [1], the starred values being from ref. [2]

	$A(K^-d)$ fm.	$p\Lambda\pi^-$ $=2(n\Lambda\pi^0)$	$n\Sigma^+\pi^-$	$p\Sigma^0\pi^-$ $=p\Sigma^-\pi^0$	$n\Sigma^0\pi^0$	$n\Sigma^-\pi^+$	Total Rate	Σ^-/Σ^+	R_{pn}
YN1<+>	$-1.34 + i1.04$	1.34 (20%)	1.51 (23%)	0.24 (3.7%)	1.29 (19.5%)	1.31 (19.8%)	6.61	1.03	11.5
YN1<->	$-1.41 + i1.00$	0.91 (14.3%)	1.82 (28.6%)	0.28 (4.4%)	1.38 (21.8%)	1.28 (19.4%)	6.36	0.83	10.8
YN2<+>	$-1.33 + i1.09$	2.13 (31%)	0.86 (12%)	0.27 (3.8%)	0.99 (14.2%)	1.38 (19.8%)	6.95	1.92	8.4
YN2<->	$-1.43 + i1.00$	1.71 (24.6%)	1.07 (15.4%)	0.25 (3.6%)	1.02 (14.6%)	1.21 (17.4%)	6.37	1.36	9.05
Data	–	–	21.5% ±3	21.3% ±3	3.5% ±1.5	17.8% ±2	20.2% ±4	–	1.15±0.4 10.3±3.0 1.16±0.05* 11.0±0.8*

two Faddeev amplitudes (T_1 and T_3 which interfere in the final spectrum. For the YN1<+> solution, they interfere constructively below the ΣN threshold, but destructively above this threshold. For the YN1<-> solution the situa-

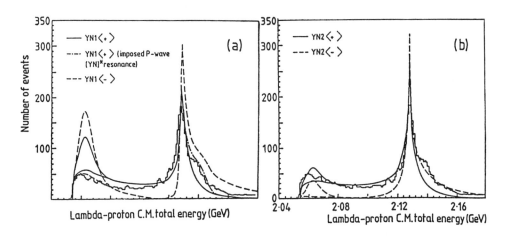

Fig. 1. The $m(\Lambda p)$ spectra for the models YN1<$\pm$> and YN2<$\pm$> are compared with the data of Tan². Below 2080 Mev, the lower branch of each curve gives the rate when only events with proton recoil momenta $\geq$ 75 Mev/c are counted. The effect of an imposed YN resonance is shown for YN1<+>.

tion is reversed and the constructive interference above the threshold produces a gentle shoulder centred about 2144 Mev, but it is not prominent and our spectrum does not agree other wise with the data above threshold and disagrees strongly below threshold.

We also present the results when a p-wave ΛN-ΣN resonance centred at 2140 Mev is included, the m(Λ p) spectrum is much improved around the shoulder, cf. Fig. 1(a). Fig. 1(b) shows the m(Λp) spectra for the YN2<$\pm$> models that contain a ΣN UBS. Its presence increases the intensity on the lower energy wing of the cusp and diminishes it on the upper, a trend not favoured by the data.

In fig. 2 we present the neutron momentum distribution for n$\Sigma^+\pi^-$. The strong $\pi N \rightarrow \Delta$ excitation gives an important contribution in this channel. We observe that the models YN1<$\pm$> led to an acceptable fit to the data. However, this agreement does not persist when there is a YN UBS.

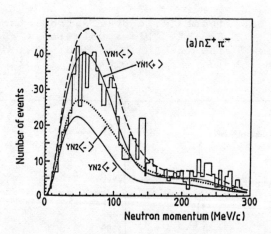

Fig. 2. Neutron momentum spectra for YN1<$\pm$> and YN2 <$\pm$> models for the $K^-d \rightarrow n\Sigma^+\pi^-$ reaction is compared with the data of Tan^2.

Finally, we emphasize that for the K^-d reactions at low energy only calculations that solve the Faddeev equations can be given credence. We find that such calculations do account for the data in considerable detail, in terms of forces that fit the known scattering and reaction data in the ($\bar{K}N$, πY) and YN systems. From our results we conclude that there is no UBS for the YN system, but the YN amplitudes near the ΣN threshold have a strong pole in the 4th Rieman sheet.

REFERENCES

1. L.W. Alvarez, UCRL rept. No. 9354 (1960).
2. T.H. Tan, Phys. Rev. Letters 23, 395 (1969); Phys. Rev. D7, 600 (1973).
3. G. Toker, A. Gal and J.M. Eisenberg, Phys.Lett. 88B (1979) 235; Nucl. Phys. A362, 405 (1982).
4. M. Torres, R.H. Dalitz and A. Deloff, to appear in Phys. Letters B.

K-NUCLEON SCATTERING AND THE CLOUDY BAG MODEL

B.K. Jennings
TRIUMF, 4004 Wesbrook Mall, Vancouver, B.C., Canada V6T 2A3

ABSTRACT

The cloudy bag model (CBM) has been applied with considerable success to low energy meson-nucleon scattering. In this talk I will describe in particular calculations for kaon-nucleon and antikaon-nucleon scattering. The main emphasis will be on s-waves with special attention paid to the antikaon-nucleon system in the isospin zero channel where the $\Lambda(1405)$ is important. In the CBM the $\Lambda(1405)$ is an antikaon-nucleon bound state and I show that this interpretation is consistent with the antikaon-nucleon scattering in the region of the $\Lambda(1670)$ and $\Lambda(1800)$ although ambiguities in the phase shift analysis prevent a definite conclusion.

INTRODUCTION

In recent years the cloudy bag model (CBM)[1] has been used extensively to study meson-nucleon scattering with considerable success being achieved in the pion-nucleon,[1-8] kaon-nucleon[8,10] and antikaon-nucleon[8,9] systems. In the present talk, for reasons of time and space, I will deal mainly with the kaon-nucleon and antikaon-nucleon system although there is still interesting work being done on the pion-nucleon system.[11] Despite the large mass of the kaon it has been shown[8] that it is useful to start with a model with SU(3) symmetry. I will first set up the formalism, and use kaon-nucleon scattering to check the model. Then I will turn to the antikaon-nucleon system and look at the s-wave isospin zero channel where the $\Lambda(1405)$ plays a large role. The nature of the $\Lambda(1405)$ has been in dispute for many years with early claims that it is a antikaon-nucleon bound state[12] and more recent claims that it is a more or less normal three quark state,[13,14] although it stands out in quark models as one state whose energy is particularly hard to fit. It has been notoriously difficult to get a clear signature that either description is correct. The CBM, however, comes down firmly on the side of the $\Lambda(1405)$ being predominately an antikaon-nucleon bound state with some small admixture of three quark states. In the present work I show that this interpretation is also consistent with the phase shift analysis in the region of the $\Lambda(1670)$ and $\Lambda(1800)$ if these states are assumed to be almost pure SU(3) states. Unfortunately, the ambiguities in the phase shift analysis prevent a definite conclusion but raise the hope that more data on the scattering in this region may help resolve the problem.

THE FORMALISM

The CBM is a quark model that restores chiral symmetry through the introduction of the pion (or pseudoscalar octet mesons) as an elementary field. The CBM is distinguished from other chiral bag models in that the mesons are allowed inside the bag and treated perturbatively. Both these characteristics make the CBM more calculationally tractable. I start from the Lagrangian:[1]

$$\mathcal{L} = (i\bar{q}\slashed{\partial}q - B)\theta_V - \frac{1}{2}\bar{q}\,e^{i\gamma_5\vec{\lambda}\cdot\vec{\phi}/f}\,q\,\delta_s + 1/2(D_\mu\phi)^2 - \frac{1}{2}m^2\phi^2 \quad . \tag{1}$$

Here $q(x)$ and $\vec{\phi}(x)$ are the quark and meson-octet fields. B is the phenomenological bag energy, f is the meson-octet decay constant, and λ_i are the SU(3) Gell-Mann matrices. The function θ_V is one inside the bag and zero outside, while δ_s is a surface delta function. For a static spherical bag (which I assume) these functions reduce to $\theta(R-r)$ and $\delta(R-r)$. As has been stressed[2-4] this Lagrangian is not a good starting point for doing a perturbative calculation since the surface coupling of the meson to the quarks guarantees[3] that the mode sum inherent in a perturbative calculation is poorly convergent. To cure this problem I follow Ref. 6 and make the transformation:

$$q_w = \exp(i\gamma_5\vec{\lambda}\cdot\vec{\phi}/2f) \tag{2}$$

which gives the following Lagrangian:[6]

$$\mathcal{L} = (i\bar{q}\slashed{\partial}q - B)\theta_V - 1/2\,\bar{q}q\,\delta_s + \frac{1}{2}(D_\mu\phi)^2 - \frac{1}{2}m^2\phi^2$$
$$+ 1/2f\,\bar{q}\,\gamma^\mu\gamma_5\,\lambda\cdot q(D_\mu\vec{\phi})\theta_V - \frac{1}{2}m^2\phi^2 \tag{3}$$

where we have dropped the subscript 'w' and the D's denote the various covariant derivatives. Although Eq. (1) and Eq. (3) are formally identical and all physical results must be the same regardless of which Lagrangian is used Eq. (3) is much better for real calculations since the soft pion results appear in a much more transparent way (see Ref. 3 and 15). In short, I believe that Eq. (1) is obsolete. The interaction terms in this Lagrangian can be written:

$$\mathcal{L}_s = \frac{1}{2f}\,\bar{q}\,\gamma^\mu\gamma_5\,\lambda\cdot q(D_\mu\phi)\theta_V \tag{4}$$

and

$$\mathcal{L}_c = -\frac{\theta_V}{4f^2}\,\bar{q}\,\lambda\gamma^\mu q\cdot\vec{\phi}\times D_\mu\vec{\phi} \quad . \tag{5}$$

The first of these is the covariant derivative of the meson dotted into the axial current of the quarks while the second is $\vec{\phi}\times D_\mu\vec{\phi}$ dotted with the vector current. It must be stressed that these

terms are not unique to the CBM but are a consequence of the present
realization of chiral symmetry. They would also arise in chiral
potential models (either realtivistic or nonrelativistic). Note
that the time part of the vector current involves just the normalization integral of the quarks so all that can change from one
chiral model to another is the form factor. It is the time part of
the vector current that plays the most important role in my calculation so the qualitative conclusions are more general than the derivation through the CBM might suggest. I now expand the Lagrangian
in powers of the meson fields and to lowest order the covariant
derivatives become just the ordinary derivatives.

The details of how a calculation is done are given in considerable detail in Ref. 8. (They are also discussed to some extent in
any of Ref. 1-11.) Only a brief sketch will be given here. The
interaction terms are evaluated between bag model wave functions.
The time part of the contact term (Eq. 5) gives a central potential
which serves as a driving term in a Lippmann-Schwinger equation.
The spatial part of the contact term contributes only a spin-orbit
potential which I do not include since in this talk I will consider
only s-wave scattering. Two terms involving one meson field must be
combined with an energy denominator to give a bare resonance. This
bare resonance is then also used as a driving term in the Lippman-
Schwinger equation. Both the location and couplings of the resonances will be renormalized by the iteration in the Lippman-
Schwinger equation. In the Lippman-Schwinger equation relativistic
kinematics are used for the mesons, and the baryon is allowed to
recoil.

THE RESULTS

As already mentioned I restrict my attention to s-waves. For
pion-nucleon scattering the contact term evaluated in first Born
approximation gives[8] the correct scattering lengths, although considerably more work[4,7,11] is required to get the energy dependence
of the phase shifts correct. It is in fact necessary to include
terms of higher order in the meson fields from the Lagrangian.
These terms are expected to be somewhat less important for the kaon-
nucleon system but there is evidence that they do contribute and
work is now in progress to calculate them.[16] It is not expected
that they will change our results qualitatively.

In Fig. 1 (see Ref. 8) I show the results for s-wave kaon-
nucleon scattering. A radius of 1.0 fm has been chosen for the bag.
In the S11 channel the CBM does a good job of describing the data
including the scattering length (see Ref. 10). In the S01 channel
it gives a very small scattering length. Experimentally the sign of
the scattering length is not well determined but its magnitude is
small. It can be seen from the figure that we are missing repulsion
which may arise from the higher order terms in the Lagrangian which
I have so far neglected. Perhaps the main conclusion to be drawn
from the S01 channel is the need for better data. From the kaon
scattering we see that the CBM gives repulsion where repulsion is

needed and small results where the experimental results are small although there is a clear signal of missing ingredients in the S01 channel where the present model gives very small results.

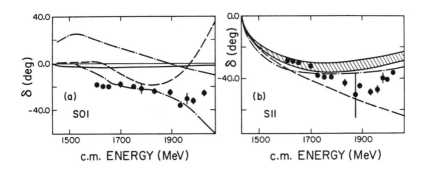

Fig. 1. Phase shifts for (a) S01 and (b) S11 partial waves. My results[8] are given by the solid curve for the S01 partial wave and by the dashed region (95 MeV < f < 105 MeV) for S11. The dots are Hashimoto's phase-shift analysis,[17] while the dashed curves are the phase shift analysis of Arndt and Roper[18] for I=1 and Corden et al.[19] for I=0. The dash-dot curves are the meson exchange model results of Davis et al.[20]

I now turn to the problem of real interest, the antikaon-nucleon in the S01 partial wave. The first point to be made is that in this channel the contact term is large and attractive. It is attractive for both the antikaon-nucleon and the pion-sigma. There is also a coupling between these two channels. I do not believe that it is an accident that the CBM gives attraction precisely where the quark models[13,14] have the greatest difficulty fitting the data and give the state too high an energy. However, it is at this point that difficulties arise. There are two things one can visualize happening. One is that the attraction simply moves the lowest three quark state down in energy while leaving its structure more or less untouched. The second possibility is that the attraction creates a new state, an antikaon-nucleon bound state while leaving the three quark state or states relatively untouched. From numerical studies it appears quite difficult to realize the first scenario. The problem arises when one tries to lower a resonance past a threshold. As the attraction is increased the state does move down in energy but at some point a new state will appear. This is what happens in the CBM. The situation is a bit more complicated in real life because there is not just a single three quark state but three of them and, moreover, if there is a bound state one of the three quark states is not seen experimentally.

To check our model I compare against two types of data, first there is the pion-sigma mass spectrum measured by Hemingway[21] for $k^-p \to \pi^+\pi^-(\Sigma^-\pi^+)$. This gives us information on the pion-sigma cross section below the kaon-nucleon threshold. Second there are the antikaon-nucleon phase shift analysis.[22-24] Unfortunately there are two different analysis and they disagree quite significantly in precisely that partial wave and energy range I am interested in. In Fig. 2, I show the pion-sigma mass distribution of Hemingway[21] along with the center of mass momentum times the square of the pion-sigma scattering amplitude[8] (the normalization is arbitrary) for two sets of parameters. In set A (solid line) R=1.0 fm and f=120 MeV while for set B (dashed line) R=1.1 fm and f=110 MeV. In these calculations one bare quark state was included. For the first set of parameters its mass was 1630 MeV while for the second it was 1650 MeV. It will be noted that for each set of parameters it was necessary to increase f and/or the bag radius from that used in the kaon-nucleon calculation. Both these changes have the effect of reducing the attraction so it is almost certain that even if I include the extra repulsion that the I=0 kaon scattering suggested I would still get an antikaon-nucleon bound state, i.e. I have attraction to spare. Both sets of parameters give the asymmetric shape suggested by the data although our resonance is too wide. The Λ(1405) is not a good Breit-Wigner resonance and one should treat calculations that assume it is with suspicion. In the present calculation the exact value of the bare energy of the three quark state is not of much importance although it is necessary to have a bare state somewhere around 1600 to 1700 MeV to get the bound state with our choice of the other parameters.

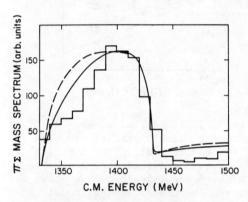

Fig. 2. The pion-sigma mass distribution.[8] The histogram is data from Ref. 21. The theoretical curves are the center of mass momentum times the square of the pion-sigma scattering amplitude. The solid curve corresponds to parameter set A and the dashed curve to parameter set B.

Much effort has been put into trying to decide if the Λ(1405) is an antikaon-nucleon bound state or a three quark state and we have two models which seem to contradict each other on this point, namely the CBM and the nonrelativistic quark model.[13,14] To check the CBM further I have calculated the antikaon-nucleon elastic scattering and the antikaon-nucleon to pion-sigma reaction amplitude from threshold to 1840 MeV. These are shown in Figs. 3 and 4, respectively, along with the data from Ref. 23. The bag radius and meson decay constant are the same as in set A above but we include three bare quark states not just one. The form factors are taken to be the same for each of the three states. The masses and structure of the three states are chosen to fit the scattering data.

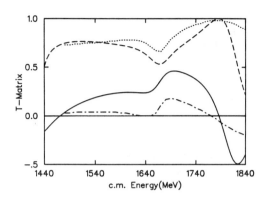

Fig. 3. The antikaon-nucleon scattering amplitude for elastic scattering. The real and imaginary parts are shown by the solid and dashed curves respectively. The phase shift analysis of Ref. 23 is given by the dash-dotted curve (real part) and the dotted curve (imaginary part).

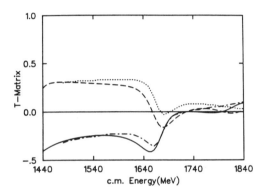

Fig. 4. The antikaon-nucleon to pion-sigma reaction amplitude. The real and imaginary parts are shown by the solid and dashed curves respectively. The phase shift analysis of Ref. 23 is given by the dash-dotted curve (real part) and the dotted curve (imaginary part).

The lowest of our quark states is taken to be predomenately an SU(3) singlet state with a 10% admixture of the spin singlet SU(3) octet state and to have a mass of 1643 MeV. The second state is the orthogonal mixture to this and has a mass of 1773 MeV, while the highest state was taken to be a pure spin triplet SU(3) octet state with a mass of 1913. With this choice the highest state decouples from the antikaon-nucleon channel which would explain why it is not seen experimentally. It plays little role if it is not too close to the other states and its mass is quite arbitrary. As expected from its couplings it affects mainly the inelastic cross section and even there it does very little. For the lowest state the two SU(3) model states are added with the opposite phase to that of lowest state of Ref. 13 (see also Ref. 25,26). The mixing angles are not well determined by my fitting procedure. The error seen in the elastic amplitudes at higher energy is due to the increasing number of open inelastic channels I have not included. (I have included only the antikaon-nucleon and pion-sigma channels.) Including these three states instead of just the one has very little effect on the amplitudes below the kaon-nucleon threshold (i.e. on Fig. 2).

The first conclusion from Figs. 3 and 4 is that it is possible to fit the scattering in this energy range even if the Λ(1405) is a bound state. The second point is that I have chosen to fit the phase shift analysis of Gopal et al.[23] Within the CBM it is much more difficult to fit the analysis of Martin et al.[24] and by trial and error I was unable to get a good fit, although this does not prove it is impossible. On the other hand the nonrelativistic quark model is much more consistent with the analysis of Martin. For example, Martin's analysis gives a large coupling of the $\Lambda(1800)$ to the sigma-pi channel, while in the analysis of Ref. 23 it almost decouples from that channel. The nonrelativistic quark model predicts a large coupling[25,26] and is thus inconsistent with Ref. 23. To oversimplify things, Martin's analysis is consistent with the following scenario: the three nonrelativistic quark model states are all that are present but one of them is shifted down in energy, possibly by the attraction we have seen. The analysis of Gopal et al.[23] on the other hand has couplings that are inconsistent with that scenario but is consistent with the $\Lambda(1405)$ being a bound state and the higher resonances being almost pure $SU(3)$ states. Thus we see that data in this higher energy region would put severe constraints on models of the $\Lambda(1405)$.

CONCLUSION

In conclusion we have seen that the CBM is quite capable of describing the kaon and antikaon scattering. The CBM also suggests very strongly that the $\Lambda(1405)$ is an antikaon-nucleon bound state and I have shown that this is consistent with the scattering data up to the region of the $\Lambda(1800)$, provided the analysis of Ref. 23 is correct. It is very important that the ambiguities in the phase shift analysis in the S01 partial wave be resolved.

ACKNOWLEDGMENT

The following people: E.A. Veit, A.W. Thomas, R.C. Barrett, O.V. Maxwell and E.D. Cooper, have been involved in this work at various times and their contribution is hereby acknowledged. N. Isgur is thanked for useful discussions and E.D. Cooper is thanked for useful discussions and for reading the manuscript. The hospitality of the physics department at the University of Toronto where part of this work was done is also acknowleged. Financial support from the Natural Sciences and Engineering Research Council of Canada is gratefully acknowledged.

REFERENCES

1. A.W. Thomas, S. Théberge, and G.E. Miller, Phys. Rev. D24, 216 (1981); D22, 2838 (1980); D23, 2106(E) (1981).
2. E.A. Veit, B.K. Jennings, and A.W. Thomas, Phys. Rev. D33, 1859 (1986).
3. B.K. Jennings and O.V. Maxwell, Nucl. Phys. A422, 589 (1984).
4. E.D. Cooper and B.K. Jennings, Phys. Rev. D33, 1509 (1986).
5. A.W. Thomas, Adv. Nucl. Phys. 13, 1 (1984).

6. A.W. Thomas, J. Phys. G7, L283 (1981).
7. B.K. Jennings, E.A. Veit, and A.W. Thomas, Phys. Lett. 148B, 28 (1984).
8. E.A. Veit, B.K. Jennings, A.W. Thomas, and R.C. Barrett, Phys. Rev. D31, 1033 (1985).
9. E.A. Veit, B.K. Jennings, R.C. Barrett, and A.W. Thomas, Phys. Lett. 137B, 415 (1984).
10. E.A. Veit, A.W. Thomas, and B.K. Jennings, Phys. Rev. D31, 2242 (1985).
11. E.D. Cooper, B.K. Jennings, P. Guichon, and A.W. Thomas, work in progress.
12. R.H. Dalitz and S.F. Tuan, Ann. Phys. (N.Y.) 10, 307 (1960).
13. N. Isgur and G. Karl, Phys. Rev. D18, 4187 (1982).
14. S. Capstick and N. Isgur, University of Toronto preprint, UTPT-85-34 (1985).
15. G. Kalbermann and J. Eisenberg, Phys. Rev. D28, 66 (1983); D28, 71 (1983).
16. R.C. Barrett and B.K. Jennings, work in progress.
17. K. Hashimoto, Phys. Rev. C29, 1377 (1984).
18. R.A. Arndt and L.D. Roper, Scattering analysis interactive dial-in (SAID) from VPI&SU Report No. KN83-1 (unpublished).
19. M.J. Corden, G.F. Cox, D.P. Kelsey, C.A. Lawrence, P.M. Watkins, O. Hamon, J.M. Levy, and G.W. London, Phys. Rev. D25, 720 (1982).
20. A.C. Davis, W.N. Cottingham, and J.W. Alcock, Nucl. Phys. B111, 233 (1976); B102, 173 (1976).
21. R.J. Hemingway, Nucl. Phys. B253, 742 (1985).
22. G.P. Gopal, in Proceedings of the 4th International Conference on Baryon Resonances (Toronto 1980); edited by N. Isgur, p. 159.
23. G.P. Gopal et al., Nucl. Phys. B119, 362 (1977).
24. B.R. Martin et al., Nucl. Phys. B126, 266 (1977); B126, 285 (1977); B127, 34 (1977).
25. J.W. Darewych, R. Koniuk, and N. Isgur, Phys. Rev. D32, 1765 (1985).
26. R. Koniuk and N. Isgur, Phys. Rev. D21, 1868 (1980).

RELATIVISTIC WAVE FUNCTIONS IN KAON PHOTOPRODUCTION FROM NUCLEI *

C. Bennhold and L.E. Wright
Ohio University, Athens, Ohio 45701

ABSTRACT

Photoproduction of positively charged kaons from light nuclei is used to investigate the effect of relativistic bound state wave functions. These nuclear and hypernuclear shell model states are obtained by solving the Dirac equation with scalar and time-like vector Woods-Saxon potentials, which have been adjusted to fit binding energies. The calculations have been performed in momentum space to treat non-localities from the production operator straightforwardly and to include fermi motion. Significant relativistic effects are found in the full non-local calculation, but appear much reduced in a local frozen nucleon approximation.

INTRODUCTION

Kaon photoproduction from nuclei has great prospects of becoming an important research tool in hypernuclear physics,[1] since it offers a "clean" way to study the nuclear interior through the weakly interacting photon and K^+ and preferentially excites unnatural parity and high spin states, complementing the reactions (K^-,π^-) and (π^+,K^+).[2] Since the process (γ,K^+) involves large momentum transfers to the nuclear system, it also appears to be a good candidate for examining bound state relativistic effects, which have provided a superior description of proton-nucleus scattering as well as a natural explanation of the nuclear spin-orbit splitting.

DISCUSSION

Our analysis is performed in a relativistic impulse approximation framework (R.I.A.). Hence, the production process in the nuclear medium is described by the free photoproduction operator which is based on first-order Feynman diagrams and include low-lying baryon and meson resonances.[3] Since we work in momentum space, non-localities arising from the propagation of the various intermediate states can be included and fermi motion is treated properly as well.[4] Thus, the operator is used without any local or non-relativistic approximations and is assumed to be a one-body operator in the nuclear many-body context. This allows the separation of the matrix element into single particle matrix elements, that contain the dynamics of the process, and nuclear structure matrix elements. For the present investigation we employ pure single-particle single-hole states. The single particle matrix element in momentum space assuming plane wave kaons is given by:

* Supported in part by DOE Grant DE-AC02-79ER10397-07.

$$\langle \alpha, K^+ | t | \beta, \gamma \rangle = \int d^3p \, \bar{\psi}_\alpha(\vec{p}') \, t(\vec{p},\vec{p}',\vec{k},\vec{q}) \, \psi_\beta(\vec{p}) \, , \qquad (1)$$

where α and β denote the lambda and proton orbitals, respectively; t is the free production operator, and $\psi_{\alpha,\beta}$ are Dirac bound state wave functions that have been Fourier transformed into momentum space.[5] These relativistic shell model wave functions are solutions of the Dirac equation with large scalar and vector potentials:

$$[\not{p} - m - S(r) - \gamma^0 V(r)] \, \psi_K^\mu(\vec{r}) = 0 \qquad (2)$$

The potentials are of Woods-Saxon shape and for ^{12}C we used a diffuseness $a = 0.65$ fm and a radius $R = 1.27 \, A^{1/3}$ fm. The depths of the potentials are adjusted to fit the s- and p-shell binding energies. For the proton we used $S_p = -430$ MeV and $V_p = 360$ MeV,[6] while the lambda potentials are about 1/3 as strong with $S_\Lambda = -144$ MeV and $V_\Lambda = 116$ MeV.[7]

To compare our analysis with standard non-relativistic calculations, we use the plane wave relation between upper and lower components of the Dirac spinor and identify the upper component as a solution of a conventional Schrödinger equation with central and spin-orbit potentials.

Fig. 1 displays some results of the non-local calculation for different states in the reaction $^{12}C(\gamma,K^+)^{12}_\Lambda B$. The large differences between the relativistic and non-relativistic model come mostly from the lower component of the proton wave function.[5] Through the large scalar and vector potentials the lower component is enhanced by about a factor of 1.7 in the nuclear interior.[6] Since the reaction involves a large change in rest masses coupled with large momenta and energies, the process projects out high momentum components of the wave functions and the lower component plays an important role. In fig. 2 we show results for the local frozen nucleon approximation. In this case relativistic effects are much reduced compared to the full calculation. All of these results were obtained with only the Born terms of the full relativistic operator of ref. 3.

CONCLUSION

Although relativistic effects are large and should be observable, ambiguities involving energies or off mass shell effects arising from the R.I.A. may not be completely negligible. Also kaon distortion effects may lower the cross sections but are not expected to alter the conclusion that relativistic effects are important. It is also quite clear that, as in the case of pion photoproduction,[4] non-local effects are important.

REFERENCES

1) S.S. Hsiao and S.R. Cotanch, Phys. Rev. C28, 1668 (1983).
2) C.B. Dover and G.E. Walker, Phys. Rep. 89, 1 (1982).
3) R.A. Adelseck, C. Bennhold and L.E. Wright, Phys. Rev. C32, 1681 (1985).
4) L. Tiator and L.E. Wright, Phys. Rev. C30, 989 (1984).
5) C. Bennhold and L.E. Wright, to be published.
6) J.R. Shepard et al., Phys. Rev. C29, 2243 (1984).
7) R. Brockmann and W. Weise, Nucl. Phys. A355, 365 (1981).

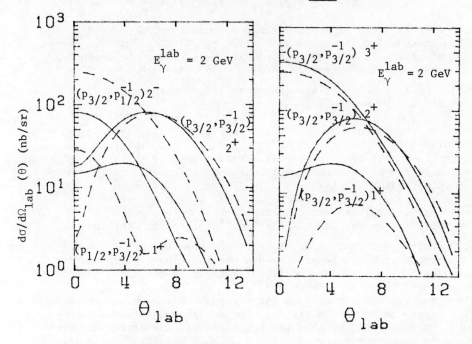

FIG. 1 Non-local calculation with —— relativistic, ---- non-relativistic

FIG. 2 Local calculation with —— relativistic, ---- non-relativisitic.

Hypernuclear Experiments at KEK

Junsei Chiba

National Laboratory for High Energy Physics
Oho-machi, Tsukuba-gun, Ibaraki 305, Japan

Abstract

Hypernuclear experiments at the National Laboratory for High Energy Physics (KEK, Japan) are reported with emphasis on the recent results of the (Stopped K^-,π) spectroscopy for Σ hypernuclei. The stopped-K method was proved to be powerful to study hypernuclear physics especially for Σ hypernuclei. An extended experiment with an upgraded detector system is now under running. Another hypernuclear experiment using the (π^+,K^+) reactions is also programmed and will run in this year.

Until a few years ago, experimental hypernuclear physics was studied mainly by means of the recoilless (K^-,π^-) reactions. In this reaction, populations of produced hypernuclear states are highly selective, namely, substitutional states are dominant due to small momentum transfers in the reaction. This is, in some sense, an advantage of the method because the selectivity largely simplifys momentum spectra of outgoing pions so that peaks in the spectra can be easily identified. However, the advantage of the selectivity turns to be a disadvantage for more detailed studies of hypernulei, since populations of non-substitutional states are very small. While it is very important to study non-substitutional states, especially hypernuclear ground states, in order to investigate ΣN interactions as well as behaviours of nucleons in the nuclear deep interior. Coincidence measurements $(K^-,\pi^-\gamma)$ may reveal such states, but limited kaon intensity and small γ-ray detection efficiency make it difficult to carry out such experiments. New experimental methods were eagerly awaited.

Recently the $(Stopped K^-,\pi)$ spectroscopy for hypernuclear study was revived at KEK. This method was almost forgotten for a long period after the first measurement at CERN[1] where they showed tiny peaks, corresponding to Λ hypernuclear states, in a huge background. The KEK experiment was a byproduct of K^+ decay experiment searching for heavy neutrinos and ran only for a short time (∼ 80 hrs). The successful results indicated that this method might be useful if we could apply a proper tagging method to suppress the huge continuum background.

Since the result of the KEK experiment has been already published[2,3], I will describe only a guideline of the experiment. Pions emitted after K^- stopped in a stack of thin plastic scintillators were measured in a magnetic spectrometer. The momentum acceptance of the spectrometer (120 ∼ 270 MeV/c) was suitable to detect pions from the formation of Σ hypernuclei. Together with $\pi^\pm$ in the spectrometer, π^0's were measured in an array of NaI(Tl) detectors. Detection of π^0 was found to play an essencial role in suppressing the huge background for Σ^- hypernuclear studies. When a Σ^- hypernucleus is formed, it decays mainly through $\Sigma^- p \to \Lambda n$ process. Since thus produced Λ is energetic (∼40MeV) it is emitted in the free space and decays into $n\pi^0$ (36%). This is the only process in which π^0 is emitted in coincidence with π^+. Thus π^0-tagging works effectively to select Σ^- hypernuclear events.

Fig. 1 shows π^+ momentum spectrum in hypernuclear mass scale after applying the π^0-tagging. Three peaks seem to correspond to Σ^- hypernuclear states. Although configuration assignments of these states can not be made only from this data, some interesting information can be drawn out by comparing with in-flight (K^-,π^+) experiments at CERN[4]. The peak A was assigned to a substitutional state of the $p_{3/2}$ orbital. The peak location agreed well with the CERN data. For the peak B, configuration of $(p_{3/2}^{-1})_p(p_{1/2})_\Sigma$ is most probable. If above assignment is correct, spin-orbit splitting of Σ^- is deduced to be about 5 MeV. There are a lot of theoretical predictions on the splitting. The observation provided a nice test to investigate the origin of the spin-orbit forces. See Ref.2 for more detailed discussions. A Λ hypernuclear state was also found in

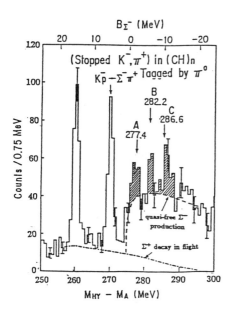

Fig. 1 Momentum distribution of π^+ from stopped K^- in $(CH)_n$ in hypernuclear mass scale after the π^0-tagging was applied. Three peaks A, B and C correspond to Σ^- hypernuclear states. The position of the peak A well agreed to the result of in-flight (K^-,π^+) spectroscopy at CERN.

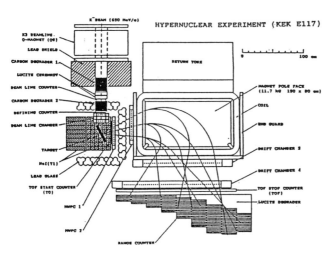

Fig. 2 Plan view of experimental setup of hypernuclear experiment by the stopped-K method. It is currently running at KEK.

π^- spectra but the momentum acceptance was not suitable for Λ hypernuclei in the previous experiment.

An extension of the (stopped K^-,π) spectroscopy (E117) started a few weeks ago at KEK with an upgraded detector system shown in Fig. 2. The momentum acceptance is wide enough (120 ~ 300 MeV/c) to cover both Λ and Σ hypernuclei. The targets being prepared are ^3He, ^4He, ^6Li, ^7Li, ^9Be, ^{12}C and heavier ones such as ^{40}Ca. The experiment will last at the end of this year. Fruitful data will come out soon. For further stopped-K^- experiments (E130), a large scale spectrometer system with a superconducting troidal magnet is being built. Also another hypernuclear experiment with use of the (π^+,K^+) spectroscopy has been approved (E150), encouraged by the BNL experiment[5], and will run in this year.

This year seems to be a "Hypernuclear year" at KEK, and what will come in next year in addtion to a "TRISTAN year"?

References

1. M.A. Faessler *et al.*, *Phys.Lett.* **46B** (1973) 468.

2. T. Yamazaki *et al.*, *Nucl.Phys.* **450A** (1986) 1c.

3. T. Yamazaki *et al.*, *Phys.Rev.Lett.* **54**(1985) 102.

4. R. Bertini *et al.*, *Phys.Lett.* **136B** (1984) 29.

5. C. Milner *et al.*, *Phys.Rev.Lett.* **54** (1985) 1237.

QUASIFREE PROCESS IN HYPERNUCLEAR FORMATION

T. Kishimoto*,**
Department of Physics
University of Houston
Houston, Texas

The apparent observation of narrow structures in Σ-hypernuclei has caused much interest in the Σ-nucleon interaction in nuclei[1,2,3,4,5]. The observation contradicts the naive belief that the width of a Σ-level should be large because of the strong conversion process $\Sigma + N \to \Lambda + N$ in nuclei. Since the structures lie in the unbound region, a large escape width is required in addition to the conversion width. The implications of narrow states give rise to interesting new physics, and it is important to have in hand a proper account of the quasifree process which underlies the presumed states.

Previous authors[6] have used the Fermi gas model for quasifree spectral calculations. In this paper, the process has been treated with the pole graph method.[7] For light nuclei the strength of the pickup reaction is concentrated in a few discrete states, and it is not appropriate to treat these states statistically. In the present analysis the initial and final states are treated explicitly.

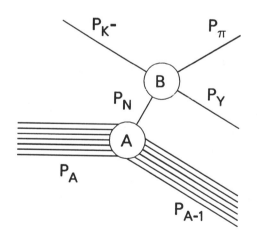

Fig. 1.

The reaction $^A_Z(K^-,\pi Y)^{A-1}$ (Z or Z-1) is shown graphically with momenta for particles.

*Present address Brookhaven National Laboratory, Upton, New York.
**On leave from Tokyo Institute of Technology, Tokyo, Japan.

The process is schematically shown in Fig. 1. An expression for the reaction T matrix is as follows:

$$T_{fi} \propto \int d^3 p_N dE_N \frac{M_A M_B}{(p_N^2 - 2M_N E_N - i\eta)} \times \delta^3(p_A - p_{A-1} - p_N) \times$$

$$\delta^3(p_K + p_N - p_\pi - p_Y) \times \delta(\omega_A - \omega_{A-1} - \omega_N) \times \delta(\omega_K + \omega_N - \omega_\pi - \omega_Y)$$

Plane waves are assumed for all particles. The vertex function at M_A is obtained by fitting the momentum distribution of p shell nucleons observed in the (e,e'p) knockout reaction[8] within the framework of this analysis. One very important feature of the analysis is the explicit inclusion of the momentum-dependent elementary amplitudes. The presence of the strong Y*(1520) resonance affects the quasifree spectrum substantially for incident kaons near 400 MeV/c. The amplitude is taken from the Gopal phase shift analysis.[9] Figure 2 shows the effect of including this resonance in the quasifree calculation. Figure 3 shows the $^{16}O(K^-, \pi^+)$ reaction taken at BNL (713 MeV/c) and CERN (450 MeV/c). The gross structure at 713 MeV/c is well reproduced

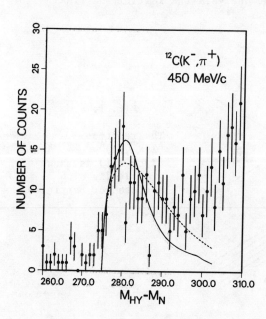

Fig. 2

Calculated quasifree spectra are shown with experimental data for the $^{12}C(K^-, \pi^+)$ reaction at p_{K^-}=450 MeV/c (CERN)[2]. A dashed line shows quasifree spectrum assuming constant amplitude. A solid line is a calculation including elementary amplitude.

by assuming knockout from the $P_{1/2}$ and $P_{3/2}$ shells. In the case of ^{16}O at 450 MeV/c, the narrowness of the quasifree peaks is due both to the smaller momentum transfer and to the effect of the resonance. This spectrum had been previously interpreted as an indication of L-S splitting in the Σ-nucleus interaction. The present work indicates that the quasifree background must be more accurately determined, and better statistical accuracy obtained, before strong conclusions about the reality of Σ-states can be drawn.

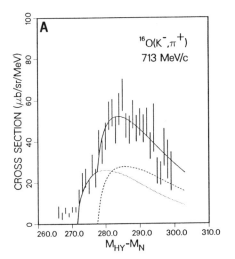 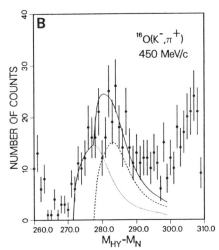

Fig. 3. Calculated quasifree spectra are shown with experimental data for the ^{16}O target: A) $^{16}O(K^-,\pi^+)$ at $p_K=713$ MeV/C (BNL))[4]; B) $^{16}O(K^-,\pi^+)$ at $p_K=450$ MeV/c (CERN))[3]. Dotted and dashed lines represent a quasifree process of $P_{1/2}$ and $P_{3/2}$ protons. The sum of the $P_{1/2}$ and $P_{1/2}$ contributions are represented by solid lines.

References

1. R. Bertini et al., Phys. Lett. B90 (1980) 375.
2. R. Bertini et al., Phys. Lett. B136 (1984) 29.
3. R. Bertini et al., Phys. Lett. B158 (1985) 19.
4. H. Piekarz et al., Phys. Lett. B110 (1982) 428.
5. T. Yamazaki et al., Phys. Rev. Lett. 54 (1985) 102.
6. R. H. Dalitz and A. Gal, Phys. Lett. B64 (1976) 154.
7. I. S. Shapiro, Nucl. Phys. 28 (1961) 244.
8. M. Bernheim et al., Nucl. Phys. A375 (1982) 381.
9. G. P. Gopal et al., Nucl. Phys. B119 (1977) 362.

CORRELATIONS IN HYPERNUCLEI

R. Büttgen, K. Holinde[+] and B. Holzenkamp
Institut für Kernphysik der KFA Jülich,
D-5170 Jülich W. Germany

J. Speth[*]
Los Alamos National Laboratory
Los Alamos, NM 87545

ABSTRACT

We investigate the relationship between the free ΛN-interaction and the effective interaction inside of a nucleus. These polarization effects are taken into account within a generalized Brueckner G-matrix. Within this approximation we calculate the Landau-parameters in the ΛN-system and compare the correlated ΛN- and NN-wavefunctions in nuclear matter.

G-MATRIX FOR THE ΛN-SYSTEM

The most essential input to modern calculations of nuclear structure and nucleon-nucleus scattering is the nucleon-nucleon (NN) interaction. Because strong short-range correlations are introduced by a repulsive core in certain NN-states the interactions used for these problems are necessarily effective ones which are qualitatively different from the free NN-interaction. An understanding of the relationship between these effective interactions and their free counterparts has been one of the major objectives of nuclear physics over the past twenty years. To a good first approximation the effective NN-interaction (G) may be related to the free NN potential via the Bethe-Goldstone equation. In the following we report on results of such calculations for the ΛN-system. The G-matrix allows us to calculate, e.g. the binding

---

[+]also: Institut für Theoretische Kernphysik, Universität Bonn
[*]Permanent address KFA Jülich - Supported in part by NATO Grant RG-85/0093

energy of the Λ-particle in nuclear matter, the effective mass of the Λ, and the effective ΛN-interaction. In the present calculations we used as the free ΛN-potential the generalized Bonn-potential described in Ref. 1 and emphasize especially the Landau parameters for the ΛN-system that are directly connected with the excited states of Λ-hypernuclei.

In analogy with the treatment of the free ΛN-interaction we introduce a direct G-matrix, $G_{NΛ,NΛ}$, and an exchange G-matrix $G_{ΛN,NΛ}$. The single particle potential U of a Λ in nuclear matter is calculated in the usual self-consistent way:

$$U^Λ(k_Λ) = \sum_{n\epsilon F} \langle Λn|G_{ΛN,ΛN}(E_Λ+E_n)|Λn\rangle + \langle nΛ|G_{NΛ,ΛN}(E_Λ+E_n)|Λn\rangle \quad (1)$$

where the sum runs over all occupied nucleon states n. For the Fermi momentum $k_F=1.36$ [fm^{-1}] we obtain $m_Λ^*/m_Λ = 0.87$ $U_0^Λ = -28.5$ MeV. The parameters that have been used for the ΛN-interaction correspond to Nijmegen model D (Ref. 2). We can compare our results with those in Ref. 3 and 4 where they obtained $m_Λ^*/m_Λ = 0.78$ and $U_0^Λ = -39.3$ MeV, and $U_0^Λ = -37.6$ MeV, respectively. In both cases the Nijmegen potential[2] (model D) has been used. We believe that the differences are connected with the approximations made by the Nijmegen group which have been discussed (in Ref. 1).

As in the NN-case we can obtain the operator structure of the ΛN effective interaction $G_{ΛN}$ directly from the G-matrix elements.[5] In the Landau limit, where one assumes that all particles in the initial and final states are on the Fermi sphere, the effective interaction $\tilde{G}_{ΛN}$ depends only on the (direct) momentum transfer q and the exchange momentum transfer Q. As the dependence on the latter is in general very weak, one obtains for the central part of the effective interaction the familiar form:

$$\tilde{G}_{ΛN}(q,k_F) = F(q,k_F) + G(q,k_F)\vec{\sigma}_1 \cdot \vec{\sigma}_2.$$

As one example we show in Fig. 1 the spin independent effective interaction as a function of the momentum transfer q. The dashed curve indicates the bare potential. The dashed-dotted curve is the

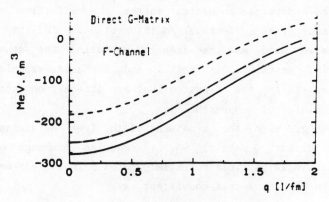

Fig. 1 Spin independent interaction.

result of a G-matrix calculation in which we neglect the coupling between the direct- and exchange-G matrix. The full line corresponds to the results obtained from the solution of the coupled Bethe-Goldstone equations. The interaction in the direct spin-independent channel is very attractive. However, from our experience in the NN-case[6] we expect large corrections in this channel due to higher order and relativistic-effects that will both reduce the attraction appreciably. The strength of the force in all the other cases is weak and indicates that the energy of excited Λ-hypernuclei will essentially be given by the unperturbed single-particle energies.

COMPARISON OF THE CORRELATIONS IN NN- and ΛN-SYSTEMS

From the conventional nuclear structure calculations one knows that the effective interaction in nuclei differs qualitatively from the free nucleon-nucleon interaction due to the large polarization effects of the surrounding nucleons. In the following, we compare correlated and uncorrelated NN- and ΛN-system's in free space as well as in nuclear matter. For this reason, we calculate the corresponding two-particle wavefunctions of the NN- and ΛN-system.

For two interacting particles in nuclear matter, we obtained the partial waves $\Psi_{LL'}^{JS}(r)$ from the G-matrix using the equation[7]

$$\Psi_{LL'}^{JS}(r) = j_L(q_0 r) \delta_{LL'} + \int_0^\infty \frac{Q(k) \, G_{LL'}^{JS}(Z|k\, q_0) j_{L'}(kr)}{Z - E(k)} k^2 \, dk \qquad (2)$$

where $Q(k)$ is the Pauli operator. The relative two-particle wavefunctions shown in the following are obtained by summing over the L'. Figure 2 shows the two-particle wavefunction as a function of the relative distance between the two particles for the free scattering case of the ΛN-system (upper part) and the NN-system (lower part) in the relative $^3S_1$ state. For comparison we also show in all figures the uncorrelated plane wave (dashed curve). The huge difference between the two systems shown in Fig. 2 is due to the difference in the tensor force. The tensor interaction in the free NN-system is strong and attractive whereas in the ΛN-system it is very weak[1].

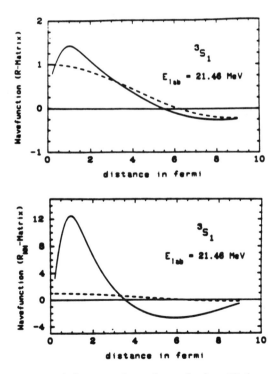

Fig. 2 Two particle wavefunction of the ΛN (upper) and NN (lower) system in free space. The dashed lines denote the unperturbed wavefunction.

The correlated two-particle wavefunctions of the NN- and ΛN-system in nuclear matter are shown in Fig. 3. The starting moment of q_0=100 MeV corresponds to the starting energy of E_{Lab} = 21.46 MeV in the free case. Here we observe that in nuclear matter the differences between the two-particle wavefunctions of the NN- and ΛN-system are much smaller compared with free scattering. One notices the effect of an attractive interaction which is somewhat larger for the NN system. The $^1S_0$-wavefunctions show a similar behavior. The much stronger free NN-interaction obviously gives rise to much larger polarization effects. Therefore, the ΛN- and NN-interaction which are very different in the free space are quite similar in nuclear matter.

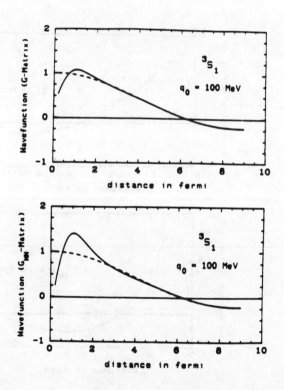

Fig. 3 Two-particle relative wavefunctions of the ΛN-(upper) and NN-(lower) system in nuclear matter.

REFERENCES

1. R. Büttgen, K. Holinde, B. Holzenkamp and J. Speth, Nucl. Phys. A450 (1986) 403c.
2. M. M. Nagels, "Baryon-Baryon Scattering in a One-Boson Exchange Potential Model," Proefschrift (1975).
3. H. Bando et al., Progress of Theoretical Physics (Supplement) No. 81 (1985).
4. J. Rosynek and J. Dabrowski, Phys. Rev. $\underline{20}$ C 1612 (1979).
5. K. Nakayama, S. Krewald, J. Speth and W. G. Love, Nucl. Phys. A431, 419 (1984).
6. K. Nakayama, S. Krewald and J. Speth, Phys. Lett. $\underline{145B}$, 310 (1984).
7. M. I. Haftel and F. Tabakin, Nucle. Phys. $\underline{A158}$, 1 (1970).

THE ETA-MESIC NUCLEUS: A NEW NUCLEAR SPECIES?

L. C. Liu and Q. Haider
Isotope and Nuclear Chemistry Division
Los Alamos National Laboratory, Los Alamos, NM 87545, USA

ABSTRACT

The strong-interaction dynamics of the ηN system can cause the η meson to be captured into nuclear orbitals in nuclei with mass numbers greater than 10. We present our prediction for what should be the experimental signature of the formation of an η-mesic nucleus and describe briefly a forthcoming experiment at Brookhaven. The scientific significance of η-mesic nuclei is also discussed.

INTRODUCTION

Searching for new forms of matter constitutes one of the most rewarding endeavors in science because every successful discovery inevitably generates quantum leaps in our understanding of nature. Familiar examples in nuclear physics are the uses of Λ- and Σ-hypernuclei to advance our knowledge of the ΛN and ΣN interactions. While meson clouds and meson-exchange currents have for long been an integral part of the theoretical description of known nuclear systems, formation of a mesonic nucleus by capturing a physical meson into a <u>nuclear</u> orbital has not yet been discovered. However, the existence of a mesic nucleus cannot be readily investigated with pions or kaons. The low-energy s-wave pion-nucleus interaction is weak and repulsive. Although the p-wave pion-nucleus interaction is attractive, its strength depends critically on the local pion momentum that theoretical estimates indicate is very small. The K^+ meson is not suitable because the low-energy K^+N interaction is repulsive. Although the low-energy K^-N interaction is attractive, the use of a stopped K^- beam could be hampered by the presence of the Coulomb interaction that would cause K^- to form, preferentially, the mesic atom. Furthermore, because of strangeness conservation, K^- will be produced in nuclear reactions only through K^+K^- pair-production that has a small cross section. We show that the η-mesic nucleus can exist[1] and can be studied experimentally.

There are currently extensive theoretical and experimental efforts devoted towards achieving a better understanding of the η and η' mesons. Studying η-nucleon interactions can yield additional important information on the nature of the η meson. Because it is nearly impossible to produce an η beam, the nucleus provides a natural laboratory for such investigations.

EVIDENCE AND SEARCH FOR ETA-MESIC NUCLEUS

Recent coupled-channel analysis[2] of πN→πN, πN→ππN, and πN→ηN reactions indicates that pionic η production on a nucleon in the threshold region proceeds mainly through the formation of the $N^*(1535)(S11)$ resonance and that the low-energy ηN interaction is attractive with a scattering length (0.28 + 0.20i) fm. It is easy to prove that this attractive interaction is a necessary consequence of the fact that the threshold for ηN scattering is below $N^*(1535)$, the lowest resonance in the s-wave channel.

Our recent theoretical study[1] indeed indicates that nuclear bound states of the η meson can exist in nuclei with mass numbers A > 10. Such a bound system, termed the η-mesic nucleus, is caused by the strong interaction between the electrically neutral η meson and all the nucleons in the nucleus. In Fig. 1, we present the predicted A-dependence of the binding energies of η. The underlying dynamics that gives rise to this systematics has been discussed in Ref. 1.

As can be seen from Fig. 1, in light nuclei the η is bound in the s-state for which the bound-state wave function is peaked at low momenta. The production of a low-momentum η meson will, therefore, favor the formation of the η-mesic nucleus. Because of the kinematics of the πN→ηN reaction, a low-momentum η is associated with a high-momentum nucleon in the laboratory frame. We conclude that the reaction $A(\pi^+,p)_\eta B$ leading to the emission of a high-energy outgoing proton is favorable to the search for the η-mesic nucleus. An experiment along this line of consideration will soon be carried out at Brookhaven.[3] Pions of momentum ~740 MeV/c will be used to bombard three nuclear targets: ^{12}C, ^{16}O, and ^{28}Si. The reaction kinematics of the experiment is illustrated in Fig. 2, where we also show, as an example, the predicted proton energy spectrum for the reaction $^{16}O(\pi^+,p)X$. The two peaks situated at outgoing proton kinetic energies 185 and 158 MeV correspond, respectively, to the formation of the mesic nucleus $^{15}O_\eta$ after the removal of the 1p- and 1s-shell neutron in the ^{16}O target.[4] The background at T_p > 185 MeV is due mainly to (π,π xp) and (π,2π xp) reactions. The background at lower proton energies contains also a component arising from the (π,η xp) reactions. To facilitate the determination of systematic errors, a nuclear target of ^7Li that cannot bind an η meson will also be used.

The calculated widths of η bound-states are ~10 to 13 MeV in nuclei with A < 30 and are ~15 to 20 MeV for nuclei with higher A. About 95% of the width is due to the decay of the η-mesic nucleus with the emission of one pion and one nucleon. The remaining 5% is related to the emission of two pions. The η absorption, being strongly suppressed, does not contribute significantly to the width. Details of medium effects on the binding energies and widths of the η-mesic nucleus can be found in Ref. 4. We stress that the order of magnitude of the above-mentioned widths represents an upper limit. This is because our calculations were based on the η interacting with one and/or two nucleons at a time. It is conceivable that the large wave length of the bound η meson could favor genuine η-3N interaction. In such a case, the η will share its energy and momentum with three nucleons and the pion emission will be completely blocked by the Pauli principle. As a result, widths of the order of keV may be possible. Consequently, if the widths observed at Brookhaven are comparable to the energy resolution of the experiment, then a better measurement of the widths should be seriously considered in the future.

IMPLICATIONS OF ETA-MESIC NUCLEUS

The magnitude of the η binding energies depends on the magnitude of the ηNN* coupling constant. Since N*(1535) is not a ground state of the quarks, the ηNN* coupling constant depends not only on the quark content of the η meson but also on the quark excitation mechanism in N*. Thus, studies of η-mesic nuclei will improve our understanding of the η and N*. It is further noteworthy that the η-mesic nuclear levels correspond to an average excitation energy of ~540 MeV (Fig. 3), to be compared with an average excitation energy of ~200 MeV associated with the Λ- and Σ-hypernuclei. We anticipate that the existence of nuclear bound states with such high excitation energies will trigger many new fundamental studies in nuclear structure theory.

We thank the INC Division at Los Alamos for constant encouragement. One of us (LCL) wishes to express his deep gratitude to Drs. H. O. Funsten, R. E. Chrien, P. Pile, and C. B. Dover for stimulating discussions on various aspects of the Brookhaven experiment. This work was supported by the US DOE.

REFERENCES

1. Q. Haider and L. C. Liu, Phys. Lett. 172B, 257 (1986).
2. R. S. Bhalerao and L. C. Liu, Phys. Rev. Lett. 54, 865 (1985).
3. Brookhaven National Laboratory AGS experiment # 828, spokesmen L. C. Liu, H. O. Funsten, and R. E. Chrien.
4. L. C. Liu and Q. Haider, Los Alamos Preprint # LA-UR-86-1185 (submitted to Phys. Rev. C).

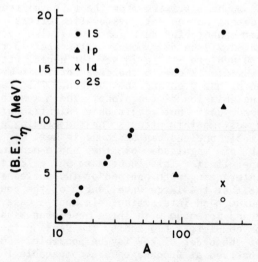

Fig. 1: Binding energies of η.

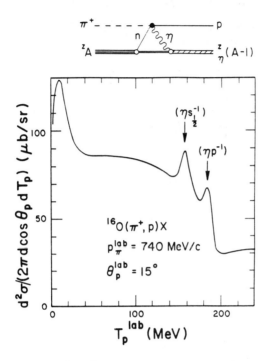

Fig. 2: The proton energy spectrum.

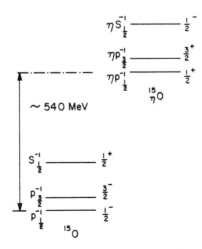

Fig. 3: The level schemes of ^{15}O and $^{15}_{\eta}$O.

WEAK DECAY OF Λ HYPERNUCLEI

John J. Szymanski
Carnegie Mellon University, Pittsburgh, PA 15213

ABSTRACT

The isospin structure of the $\Lambda N \to NN$ non-mesonic weak interaction is discussed as one aspect of hypernuclear weak decay. Measurements of the hypernuclear lifetime and neutron stimulated fraction for $^{12}_{\Lambda}C$ and $^{11}_{\Lambda}B$ are reported. Preliminary results for $^{4,5}_{\Lambda}He$ are presented.

INTRODUCTION

A particle stable hypernucleus will decay electromagnetically to its ground state. Such a hypernucleus must subsequently decay via a strangeness-changing weak interaction. The aim of our experimental studies is a better understanding of these nonleptonic weak interactions.

The predominant decay modes are of two types: mesonic modes, where $\Lambda \to p\pi^-(\Gamma_{\pi^0})$ or $\Lambda \to n\pi^0(\Gamma_{\pi^0})$ and non-mesonic modes, where $\Lambda p \to np$ (Γ_p, proton stimulated) or $\Lambda n \to nn$ (Γ_n, neutron stimulated). The mesonic modes are analogous to free Λ decay, but are suppressed in hypernuclei because of phase space restrictions and Pauli blocking of the final state nucleons. Conversely, the non-mesonic modes open up in hypernuclei, and are the dominant decay modes for all but the lightest hypernuclear systems. An experimental attraction of the non-mesonic modes is their ease of identification because of the energetic nucleons in the final state. The sum of all four partial rates gives the total decay rate for a particular hypernucleus. Semi-leptonic decays are ignored here.

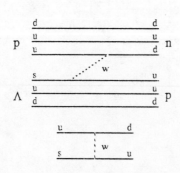

Fig. 1. Quark level description of proton stimulated decay.

There are several aspects to the problem of calculating hypernuclear weak decay rates. I would like to concentrate on the weak Hamiltonian required for non-mesonic rate calculations. At the quark level, the lowest order diagram which describes the Γ_p transition is shown in figure 1. The basic exchange diagram is also shown in figure 1. At zero momentum transfer, the Hamiltonian for this process is[1]

$$\mathcal{H}_{V-A} = \frac{G_F}{\sqrt{2}} \sin\theta_c \cos\theta_c \, \bar{u} \gamma_\mu (1-\gamma_5) s \, \bar{d} \gamma^\mu (1-\gamma_5) u + cc \quad (1)$$

It is well known that hyperon decays indicate that $\Delta I=1/2$ transitions are greatly enhanced with respect to $\Delta I=3/2$ transitions. The Hamiltonian in eqn (1), however, has both $\Delta I=1/2$ and $\Delta I=3/2$ components. The derivation of a Hamiltonian giving the "$\Delta I=1/2$ rule" is still an open question, although several avenues look promising.[2]

To what level are $\Delta I=1/2$ transitions dominant in non-leptonic interactions within a nuclear medium, i.e., the non-mesonic decays? Leonard Kisslinger will discuss one approach to this question in the next talk. I will now discuss an alternative approach put forth by Dalitz and his collaborators.[3]

Assuming the Λ-N interaction is via s-wave, the possible non-mesonic transitions are shown in table I. The rate, R_{NS}, for the Λ interacting with a specific nucleon, N, in a specific initial spin state, S, is a sum of those transitions indicated in table I.

Table I. Allowed non-mesonic transitions[3]

Initial state	Final state	T_{final}	Λ+p	Λ+n	Parity Change
$^1S_0$	$^1S_0$	1	R_{p0}	R_{n0}	+1
$^1S_0$	$^3P_0$	1	R_{p0}	R_{n0}	-1
$^3S_1$	$^3S_1$	0	R_{p1}	-	+1
$^3S_1$	$^1P_1$	0	R_{p1}	-	-1
$^3S_1$	$^3D_1$	0	R_{p1}	-	+1
$^3S_1$	$^3P_1$	1	R_{p1}	R_{n1}	-1

R_{NS}: N = n,p initial nucleon; S = 0,1 spin

The $\Delta I=1/2$ rule implies $R_{n0} = 2R_{p0}$ and $R_{n1} \leq 2R_{p1}$. Experimentally, we hope to measure Γ_p and Γ_n for $^4_\Lambda$He and $^5_\Lambda$He. These experimentally measurable rates can be expressed in terms of the spin-isospin rates as shown in table II.

Table II. Experimental rates in terms of elementary rates[3]

	Γ_p	Γ_n	
$^4_\Lambda$He	$1/6\, \rho_4(3R_{p1} + R_{p0})$	$1/6\, \rho_4(2 R_{n0})$	ρ = average nucleon density at the
$^5_\Lambda$He	$1/8\, \rho_5(3R_{p1} + R_{p0})$	$1/8\, \rho_5(3R_{n1} + R_{n0})$	Λ position.

RECENT CALCULATIONS

McKellar and Gibson[4] describe the non-mesonic rate in nuclear matter as the interference between pion and rho mesons exchanged between weak and strong vertices. Oset and Salcedo[5] consider pion exchange only, but include effects due to the interaction of the pions with the nuclear medium. They calculate the rates for several finite nuclei and nuclear matter. Dubach et al.[6] use $SU(6)_{weak}$ to relate weak vertices in a meson exchange model, which is described elsewhere in this volume. Kisslinger et al.[7] use the Hybrid Quark-Hadron model to calculate the non-mesonic rates, which is also described elsewhere in these proceedings.

EXPERIMENTAL TECHNIQUE

The BNL experiment 759 and 788 collaboration has been studying the weak decay of $^{12}_\Lambda C$, $^{4}_\Lambda He$ and $^{5}_\Lambda He$.[8-10] Final results for $^{12}_\Lambda C$ will be discussed in this report. Data were taken on $^{5}_\Lambda He$ in June, 1985 and preliminary results are reported below.

The experimental goals are 1) to measure excitation energy spectra with and without decay product tags, 2) to measure hypernuclear lifetimes, 3) to detect protons and neutrons from non-mesonic decay and charged pions from mesonic decay.

The experiments use the $^{A}Z(K^-,\pi^-)^{A}_\Lambda Z$ reaction to produce and tag hypernuclear states. The K^- and π^- are momentum analyzed in the BNL Hypernuclear Spectrometer system. A 4 gm/cm^2 scintillator target was exposed to a 800 MeV/c K^- beam (2×10^5 K^-/sec) to produce $^{12}_\Lambda C$. A 5.87 gm/cm^2 ^{6}Li target was used to produce $^6_\Lambda Li$, the ground state of which decays via proton emission to $^5_\Lambda He$. The $^4_\Lambda He$ studies will use a liquid ^{4}He target to produce $^4_\Lambda He$ directly.

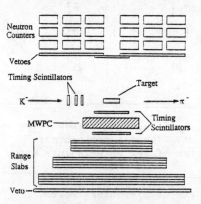

Fig. 2. Experimental apparatus.

Two detector systems surround the production target to detect decay products (see Figure 2). Above the target is a neutron counter array which measures neutron energies by time-of-flight. Below the target sits a range-time spectrometer capable of stopping protons up to ~120 MeV and pions up to ~50 MeV. The first two elements of the spectrometer are high-quality timing scintillators which give the decay time of the hypernucleus. A multi-wire proportional chamber gives particle trajectory information.

$^4_\Lambda He$ AND $^5_\Lambda He$ RESULTS

Figure 3 is an excitation energy spectrum with a decay proton coincidence requirement. The state at ~2 MeV excitation is identified as the $^6_\Lambda Li$ (ground state) which decays to p + $^5_\Lambda He$, thus giving energetic protons from $\Lambda p \to np$ in $^5_\Lambda He$. The state at ~20 MeV excitation is then identified as the $n(S_{1/2}^{-1})\Lambda(S_{1/2})$ substitutional state, which must also decay to a stable hyper-nucleus to give energetic protons. The separation of these two states is comparable to the 18.3 MeV separation measured by Bertini et al.[11] Also note that the $n(p_{3/2}^{-1})\Lambda(p_{3/2})$ state seen in the Bertini data is strongly suppressed by the proton coincidence requirement, implying this state decays by Λ emission.

Figure 4 is a preliminary $^4_\Lambda$He excitation spectrum showing the $n(S_{1/2}^{-1})\Lambda(S_{1/2})$ $^4_\Lambda$He (ground state). The $^4_\Lambda$He excitation spectrum has no decay coincidence requirement.

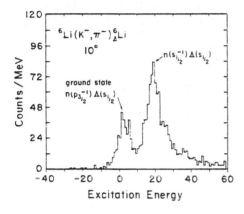

Fig. 3. Excitation energy spectrum with decay proton coincidence.

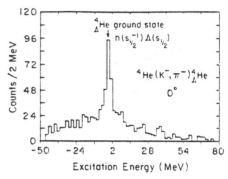

Fig. 4. $^4_\Lambda$He excitation energy spectrum.

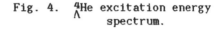

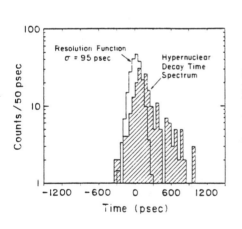

Fig. 5. Time spectra for $^5_\Lambda$He.

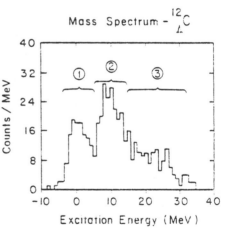

Fig. 6. $^{12}_\Lambda$C excitation energy spectrum with proton coincidence.

The production and decay times of hypernuclei are measured directly by two scintillators in the K⁻ beam and the two previously mentioned timing scintillators in the range spectrometer. An important characteristic of these counters is their time resolution function, which we measure throughout the experiment using the ^{6}Li(π,p)X reaction. In Figure 5 the resolution function for the entire two week run is plotted along with the proton time-of-flight spectrum from $^5_\Lambda$He decays. The resolution function has a width σ = 95 psec and has been scaled to match the area of the decay spectrum. The decay spectrum clearly shows the time shift due to the lifetime of $^5_\Lambda$He.

$^{12}_\Lambda$C RESULTS

The excitation energy spectrum for $^{12}_\Lambda$C in coincidence with decay protons is shown in figure 6. Region 1 is the particle stable $^{12}_\Lambda$C (ground state), region 2 is the $^{12}_\Lambda$C n$(p_{3/2}^{-1})\Lambda(P_{3/2})$ state, which is known[12] to decay to p + $^{11}_\Lambda$B, and region 3 is in the vicinity of the $^{12}_\Lambda$C n$(S_{1/2}^{-1})\Lambda(S_{1/2})$ region with unknown breakup. The weak decay rates for these three regions are given in table III where Γ_Λ is the free Λ decay rate. The errors in the lifetime and n measurements are dominated by statistics, whereas the errors in Q^- are mainly due to uncertain background subtraction from the ground state region of the coincident pion excitation energy spectrum. The results for n have also been corrected for nuclear multiple scattering and charge exchange with an intranuclear cascade calculation[13] (correction ~20%). Using $\Gamma_{\pi^0}/\Gamma_{\pi^-}$ = 7/6 as calculated by Gal[14] and by Bando et al.[15] and the $^{12}_\Lambda$C results for Γ/Γ_Λ and Q^- we find (Γ_{nm} = total nonmesonic rate = $\Gamma_n + \Gamma_p$)

$$\Gamma_{nm} = (1.14 \pm .20)\Gamma_\Lambda . \qquad (2)$$

Table III

State	hypernucleus	Γ/Γ_Λ	$n = \dfrac{\Gamma_n}{\Gamma_{nm}}$	$Q^- = (\Gamma_{nm})/\Gamma_{\pi^-}$
g.s.	$^{12}_\Lambda$C	1.25±0.18	0.57 $^{+0.14}_{-0.23}$	22 $^{+43}_{-12}$
$p^{-1}p$	$^{11}_\Lambda$B	1.37±0.16	0.51 $^{+0.11}_{-0.15}$	
$s^{-1}s$	unknown	1.31±0.2		Γ_Λ = free Λ decay rate

DISCUSSION

Preliminary results for $^5_\Lambda$He indicate $\Gamma_{\pi^-}/\Gamma_p \sim 1$ or greater. Compare this to $^{12}_\Lambda$C where non-mesonic decay modes totally dominate the mesonic modes. This is consistent with the expected suppression of the mesonic modes for heavier hypernuclei.

The total measured non-mesonic rate can be compared with the calculated non-mesonic decay rates for infinite nuclear matter obtained by McKellar and Gibson[4] ($\Gamma_{nm}/\Gamma_\Lambda$ = 2.3) and by Dubach et al.[6] ($\Gamma_{nm}/\Gamma_\Lambda$ = 1.23), and for the hypernucleus $^{12}_\Lambda$C, obtained by Kisslinger et al[7] ($\Gamma_{nm}/\Gamma_\Lambda$ = 1.28) and by Oset and Salcedo[5] ($\Gamma_{nm}/\Gamma_\Lambda$ = 1.5). The calculations do quite well at reproducing the experimental non-mesonic decay rate. Two points should be made, however: a. Rates calculated in nuclear matter may be reduced significantly, as seen by Kisslinger et al. (Γ_{nm}(nuclear matter)/$\Gamma_{nm}(^{12}_\Lambda$C) = 2.3) and by Oset and Salcedo (Γ_{nm}(nuclear matter)/$\Gamma_{nm}(^{12}_\Lambda$C) = 1.3). b. Dubach is the only author to report a value for Γ_p/Γ_n (= 2.9 $\Rightarrow$ n ~ .25). This ratio is another sensitive test for calculations. As a final conclusion the $^{12}_{\Lambda}$C results are consistent with the $\Delta I=1/2$ rule, which implies $\Gamma_n \leq 2\Gamma_p$.

CONCLUSIONS

There are several interesting aspects of hypernuclear weak decay which have been explored in various calculations. This report emphasizes the sensitivity of the non-mesonic decay rates to the isospin structure of the effective weak Hamiltonian.

The experimental situation is promising. New results exist for $^{11}_{\Lambda}$B and $^{12}_{\Lambda}$C. Final results for $^{5}_{\Lambda}$He are forthcoming, and the $^{4}_{\Lambda}$He experiment is approved and waiting to run. Because of these new measurements I want to encourage calculations of $^{4}_{\Lambda}$He and $^{5}_{\Lambda}$He rates and refinements in the $^{12}_{\Lambda}$C calculations.

REFERENCES

1. N. Cabibbo, Phys. Rev. Lett. **10**, 531 (1963).
2. F.J. Gilman and M.B. Wise, Phys. Rev. **D20**, 2392 (1979), and references therein.
3. R.H. Dalitz and L. Liu, Phys. Rev. **116**, 1312 (1959); R.H. Dalitz and G. Rajasekharan, Phys. Lett. **1**, 58 (1962); M.M. Block and R.H. Dalitz, Phys. Rev. Lett. **11**, 96 (1963).
4. B.H.J. McKellar and B.F. Gibson, Phys. Rev. **C30**, 322 (1984).
5. E. Oset and L.L. Salcedo, Nucl. Phy. **A443**, 704 (1985); E. Oset and L.L. Salcedo, Nucl. Phys. **A450**, 371 (1986).
6. J. Dubach, Nucl. Phys. **A450**, 71 (1986); B. Desplanques, J. Donoghue and B.R. Holstein, Ann. Phys. **124**, 449 (1980).
7. D.P. Heddle and L.S. Kisslinger, preprint Carnegie Mellon University, 1985; C.Y. Cheung, D.P. Heddle and L.S. Kisslinger, Phys. Rev. **C27**, 335 (1985).
8. AGS Proposal E-759, The Weak Decay Modes of Hypernuclei; R. R. Grace, P.D. Barnes, R.A. Eisenstein, G.B. Franklin, C. Maher, R. Rieder, J. Seydoux, J. Szymanski, W. Wharton, S. Bart, R.E. Chrien, P. Pile, Y. Xu, R. Hackenburg, E.V. Hungerford, B. Bassalleck, M. Barlett, E.C. Milner and R.L. Stearns, Phys. Lett. **55**, 1055 (1985).
9. AGS Proposal E-788, The Four Fermion Weak Interaction and the Decay of $^{4}_{\Lambda}$He and $^{5}_{\Lambda}$He, P.D. Barnes, G.B. Franklin, G. Diebold, R. Grace, D. Hertzog, C. Maher, B. Quinn, J. Seydoux, J. Szymanski, S. Bart, R.E. Chrien, P. Pile, R. Sutter, E.V. Hungerford, T. Kishimoto, L.G. Tang, B. Bassalleck and R.L. Stearns, 1983.
10. P.D. Barnes, Nucl. Phys. **A450**, 43 (1986).
11. R. Bertini et al., Nucl. Phys. **A368**, 365 (1981).
12. A. Montwill et al, Nucl. Phys. **A234**, 413 (1974); T. Cantwell et al., Nucl. Phys. **A236**, 445 (1974).
13. R. Grace, The Weak Decay of P-Shell Lambda Hypernuclei (Ph.D. dissertation, Carnegie Mellon Univ., 1985).
14. A. Gal, private communication.
15. H. Bando and H. Takaki, Prog. of Theor. Phys. **72**, 106 (1984); Phys. Lett. **150B**, 409 (1985).

LIFETIMES FOR NON MESONIC DECAY IN HYPERNUCLEI

Leonard S. Kisslinger
Carnegie-Mellon University, Pittsburgh, PA 15213

ABSTRACT

The nonmesonic lifetime in the HQH Model is closely related to the $\Delta I = 1/2$ rule, and we obtain a good fit to the $^{12}_{\Lambda}C$ lifetime with a quark Hamiltonian which satisfies this rule. Hypernuclear structure with an accurate shell model is particularly crucial for proton- vs. neutron-stimulated Λ decays.

I. INTRODUCTION

In the free (mesonic) decay of the Λ, $\Lambda \to N+\pi$, which has a width $\Gamma_{free} \simeq 2.5\times10^{-12}$ MeV, the nucleon receives a kinetic energy of only 5 MeV. Since the depth of the Λ potential in typical hypernuclei is greater than 30 MeV, the mesonic decay is inhibited. The dominant decay processes are nonmesonic two-baryon weak processes

$$\Lambda+p \to n+p \quad \text{(proton stimulated)}$$
$$\Lambda+n \to n+n \quad \text{(neutron stimulated)} , \quad (1)$$

in which each final nucleon has a center of mass momentum of about 415 MeV/c, which is greater than the nuclear Fermi momentum. In this paper I discuss the Hybrid Quark-Hadron (HQH) Model description of the nonmesonic Λ-hypernuclear decays[1] and give our present results.

Recently, the lifetimes for Λ-decays in $^{12}_{\Lambda}C$ have been measured in a BNL experiment, which John Szymanski describes in these proceedings.[2] The most important results for our purposes are

$$\Gamma^{nonmesonic}/\Gamma_{free} = 1.14 \pm 0.2$$
$$\Gamma_n / \Gamma_p \approx 1.3 \text{ (neutron/proton stimulated widths)} \quad (2)$$

There are both short-distance and long distance aspects of this process. The relative center of mass momentum of the two final nucleons in Eq. (1) is

$$q = |\vec{k}_1 - \vec{k}_2|/2 \approx 2.1 \text{ fm}^{-1} \quad (3)$$

where $\vec{k}_1$ and $\vec{k}_2$ are the final nucleon momenta. In the Born approximation, neglecting final state distortions, the width for $(\Lambda N)_i \to NN$ is given by

$$\Gamma^{nonmesonic} \propto \left| \int d^3r \, e^{-i\vec{q}\cdot\vec{r}} \, V \, \Psi_i(\vec{r}) \right|^2$$
$$\propto \left| \sum_{\ell} \int d\Omega \, i^{\ell} \sqrt{2\ell+1} \, Y_{\ell 0}(\Omega) \int dr \, r^2 \, j_{\ell}(qr) V \, \Psi_i(\vec{r}) \right|^2 \quad (4)$$

in a hadronic model. From this we observe

i) For low ℓ's <u>short-distance phenomena</u> are very important.
ii) Large ℓ's are enhanced.
iii) Since the average NN separation is large, $\bar{r}_{NN} \approx 1.8$ fm, large distances and hypernuclear structure are expected to be very important.

II. THEORETICAL MODELS OF WEAK HADRONIC/NUCLEAR INTERACTIONS

A. Quark vs. Meson Exchange Models: HQH Model

The Standard Model of weak leptonic interactions has been remarkably successful. The theory of weak interactions in hadrons and nuclei is far more complicated. Although at very high momentum transfers the quark weak interactions are expected to be very similar to weak leptonic interactions, at low energies the strong interaction corrections are very large and not yet understood. At low energy the form of the effective weak quark-quark interaction is

$$\mathcal{H}_{qq}^{W(eff)} = \frac{G}{\sqrt{2}} J^{W\lambda} J_{\lambda}^{W}, \qquad (5)$$

where J_{λ}^{W} is the weak quark current. However, for ΛN separation $r_{\Lambda N} \gtrsim 1.0$ fm, this model cannot be applied directly, since the quarks in the two baryons are in separate clusters and one does not have an adequate quark model at the present time.

For long distance nuclear interactions the pion exchange model is known to be excellent, and for $r_{\Lambda N} \gtrsim 1.0$ fm the one pion exchange weak interaction, $\mathcal{H}_\pi^W$, is reasonably well known. However, meson exchange models introduce heavy meson exchange to describe the short-range properties, with ranges $\lesssim 1/4$ fm. At these short distances the very concept of a meson exchange potential is in question. Moreover, the parameters for these potentials are not well-known. Therefore we feel that the short-range weak nuclear interactions should be described by quark interactions rather than meson exchange Λ-N interactions.

The Hybrid Quark-Hadron (HQH) Model of weak nuclear interactions introduces a projection operator in coordinate space, with explicit quark degrees of freedom for interbaryon separation $r < r_0$ and hadronic degress of freedom for $r > r_0$. In this model the weak $\Lambda N \to NN$ process is represented as $\mathcal{H}_\pi^W$, the weak one pion exchange potential, for $r > r_0$ and the effective weak quark Hamiltonian, $\mathcal{H}_{qq}^W$, for $q < r_0$. A typical weak matrix element is given by

$$\langle f | \mathcal{H}^W | i \rangle = \int_{r > r_0} \Psi_f^*(r_1) \mathcal{H}_\pi^W \Psi_i(r) + \langle NN(6q) | \mathcal{H}_{qq}^W | \Lambda N(6Q) \rangle \qquad (6)$$

B. Effective Weak Quark Interaction

The effective $\Delta S = 1$ weak quark interaction arising from W and Z gauge boson exchange with strong interaction corrections calculated from QCD perturbation theory is of the form, with Q_i local operators,

$$\mathcal{H}_{qq}^{W(eff)}(\Delta S=1) = \frac{G}{\sqrt{2}} \sin\theta_c \cos\theta_c \sum_{i=1}^{6} C_i Q_i, \qquad (7)$$

with five independent operators. The coefficients C_i for $i = 1-6$ have been calculated from one-loop gluonic diagrams[3] and "penguin" diagrams[4] using renormalization group methods. In the notation of Gilman and Wise,[4] only two of the operators have large coefficients, so

$$\mathcal{H}_{qq}^{W(eff)}(\Delta S=1) \cong C_1 Q_1 + C_2 Q_2 , \qquad (8)$$

with $Q_1 = \bar{s}\gamma^\mu(1-\gamma_5)d\bar{u}\,\gamma_\mu(1-\gamma_5)u$ and $Q_2 = \bar{s}\,\gamma^\mu(1-\gamma_5)u\,\bar{d}\gamma_\mu(1-\gamma_5)u$. The values of the coefficients are $C_1 = 1.51(1.45)$ and $C_2 = -0.856$ (-1.05) from Gilman and Wise (Schifman et al.).[4]

There are important experimental constraints for the coefficients C_i of Eq. (7). The most important is the $\Delta I = 1/2$ rule. Recognizing that Q_1-Q_2 is a $\Delta I = 1/2$ operator and that Q_1+Q_2 is a $\Delta I = 3/2$ operator, the theoretical ratio of $\Delta I = 1/2$ to $\Delta I = 3/2$ amplitudes is[4]

$$R^{theory} = (A_{1/2}/A_{3/2})^{theory} = (C_1-C_2)/(C_1+C_2) = 3.6 \qquad (9a)$$

while R^{exp} depends on the particular process, but

$$R^{exp} \approx 20 . \qquad (9b)$$

There are a number of other constraints which have been discussed in specific models,[5,6] but the constraint (9b) is most important for our present work. One should note, however, that if one simply picks the renormalization point to change R^{theory} to 20, there can be a problem[6] with the CP violation parameter $\epsilon'/\epsilon \propto \langle\pi\pi(I=0)|Q_6|K^0\rangle$.

III. HQH MODEL FOR NONMESONIC Λ-HYPERNUCLEAR DECAY

The HQH Model[7] has been applied for the Λ-hypernuclear lifetime in a preliminary calculation of Λ-hypernuclear matter[8] and a detailed calculation[1] for $^{12}_\Lambda C$. A brief description follows:

A. Interior (six-quark) Region

The initial bound state is taken as shell model hypernuclear state. The interior region of the particular ΛN cluster undergoing decay is taken as

$$|\phi_{\Lambda N}^{6q}\rangle = \sqrt{P^{6q}}\,|(S_{1/2})^4\,[\alpha(S_{1/2})^2 + \beta(P_{3/2})^2 + \gamma(P_{1/2})^2$$
$$+ \frac{\delta}{\sqrt{2}}(P_{3/2}P_{1/2} + P_{1/2}P_{3/2})$$
$$\alpha^2 + \beta^2 + \gamma^2 + \delta^2 = 1 \qquad (10)$$
$$P^{6q} = 1 - \int_{r>r_0}|\Psi_{\Lambda N}|^2\,dV .$$

The most important configuration is $(S_{1/2})^6$, and we explore sensitivity to quark configuration mixing through the parameters α, β, γ and δ. The interior part of the final NN scattering states are obtained[7] by using probability current conservation to determine the amplitudes of the interior six-quark. I.e., the interior six-quark wave functions are given by

$$\phi_{SLJI}^{6q}(E) = A(E)_{SLJI} \phi_{SLJI}^{SLJI}(6q) , \qquad (11)$$

with the amplitudes $A(E)_{SLJI}$ calculated from the experimental phase shifts and a potential which can approximately give these phase shifts.

The resulting six-quark contribution to the nonmesonic width is

$$\Gamma_{6quark}^{nonmesonic} / \Gamma_{free} = 2.73(C_1+C_2)^2 P_{\Lambda N}^{6q} F(\alpha,\beta,\gamma,\delta)$$

$$= 2.73 \, P_{\Lambda N}^{6q} \, F(\alpha,\beta,\gamma,\delta) \begin{cases} 1.0 & \text{Cabibbo} \\ 0.42 & \text{strong interaction correction} \\ 0.14 & \text{fit } \Delta I = 1/2 \text{ rule} \end{cases}, \qquad (12)$$

where $F(\alpha,\beta,\gamma,\delta)$ is a known function calculated from the wave function[s] Eqs. (10)-(11).

B. Exterior region--weak pion exchange

The weak one pion exchange potential is taken as the static limit derived from the strong πNN and the weak $\pi N\Lambda$ interaction Hamiltonians

$$\mathcal{H}_{int}^S = iG_S \, \bar{\Psi}_N \gamma_5 \, \vec{\tau} \cdot \vec{\phi}_\pi \, \Psi_N$$
$$\mathcal{H}_{int}^W = iG_W \, \bar{\Psi}_N (1+\lambda\gamma_5) \vec{\tau} \cdot \vec{\phi}_\pi \, \Psi_\Lambda \qquad (13)$$

resulting in

$$V(r) = \left[V_0(r)\vec{\sigma}_1 \cdot \vec{\sigma}_2 + V_1(r)\vec{\sigma}_2 \cdot \vec{r} + V_2(r)S_{12}\right] \vec{\tau}_1 \cdot \vec{\tau}_2 , \qquad (14)$$

with the $V_i(r)$ Yukawa forms (see ref. 1 for details).

C. Nuclear Matter.

In nuclear matter the ΛN cluster is in a ^1S or ^3S state. For the initial state we use the Bethe-Goldstone radial wave function with correlations.[9] Thus

$$|\bar{\Psi}_i^{\Lambda N}\rangle = |R_i(kr)X(L=0,S=0,1)\rangle$$

$$R_i(kr) = \left[j_0(kr) + 1.07 \sin(kr_c)S_i(1.63r) / kr\right]\theta(r-r_c) , \qquad (15)$$

and the final wave function is given by an eikonal model. Our results for the interior (six-quark) and exterior (pionic) regions are as follows:

$$\Gamma_{6quark}^{nonmesonic} / \Gamma_{free} = \begin{cases} 5.2 & \text{Cabibbo} \\ 2.2 & \text{strong interaction correction} \\ 0.73 & \text{fit } \Delta I = 1/2 \text{ rule} \end{cases},$$

$$\Gamma_\pi / \Gamma_{free} = 0.77$$

resulting in

$$\Gamma^{nonmesonic}_{(nuclear\ matter)} / \Gamma_{free} = \begin{cases} 10.0 & \text{Cabibbo} \\ 5.5 & \text{strong interaction corrections.} \\ 3.0 & \text{fit } \Delta I = 1/2 \text{ rule} \end{cases}$$

If we consider the neutron- vs. proton-stimulated widths, it turns out that for the one pion exchange model the $j_2(qr)$ Bessel function with $^3S \to {}^3D$ dominates, and from isospin considerations[10]

$$\Gamma_n / \Gamma_p \ll 1$$

In comparison with the results of Grace et al.,[2] clearly nuclear matter calculations are inadequate for the study of lifetimes in finite nuclei.

D. Finite nuclei: $^{12}_\Lambda C$

The initial Λ-hypernucleus state is known to be well-described by weak coupling

$$\left| \Psi(^{12}_\Lambda C) \right> = \left| (S_{1/2})^4 (P_{3/2})^7 S^\Lambda_{1/2} \right>. \qquad (16)$$

To get the probabilities for the various ΛN clusters, we expand via spectroscopic factors

$$\left| \Psi(^{12}_\Lambda C) \right> = \sum_{J_f I_f} \Big\{ \sqrt{S_p\{[J_f I_f] P_{3/2}\}} \left| [J_f I_f] P^{(p)}_{3/2} S^{(\Lambda)}_{1/2} \right> $$
$$+ \sqrt{S_p\{[J_f I_f] S_{1/2}\}} \left| [J_f I_f] S^{(p)}_{1/2} S^{(\Lambda)}_{1/2} \right> \qquad (17)$$
$$+ \sqrt{S_n\{[J_f I_f] P_{3/2}\}} \left| [J_f I_f] P^{(n)}_{3/2} S^{(\Lambda)}_{1/2} \right> $$
$$+ \sqrt{S_n\{[J_f I_f] S_{1/2}\}} \left| [J_f I_f] S^{(n)}_{1/2} S^{(\Lambda)}_{1/2} \right> \Big\}$$

We find that the largest spectroscopic factors are

$$S_n\{[2,1](P^{(n)}_{3/2} S^{(\Lambda)}_{1/2})\} = 2.5$$
$$S_p\{[3,0](P^{(n)}_{3/2} S^{(\Lambda)}_{1/2})\} = 1.75 \qquad (18)$$

One sees that there are large relative P-wave components in the initial state, in contrast to nuclear matter (Eq. (15)), and that there are large P→P transitions via $j_2(qr)$ as well as p.v. P→D transitions. Thus from Eq. (18) one can expect $\Gamma_n \approx \Gamma_p$.

Our results for the width are[8]

$$\Gamma^{nonmesonic}_{6q}(^{12}_\Lambda C) / \Gamma_{free} = 0.24 \quad \text{(fit } \Delta I = 1/2 \text{ rule)}$$
$$\Gamma^{nonmesonic}_\pi (^{12}_\Lambda C) / \Gamma_{free} = 0.41$$
$$\Gamma^{nonmesonic}(^{12}_\Lambda C) / \Gamma_{free} = 1.31$$

in good agreement with experiment.

IV. CONCLUSIONS

In conclusion we would like to emphasize

- The study of lifetimes for nonmesonic decay of Λ-hypernuclei is rich in particle/nuclear physics.

- The HQH model has now been developed so that the six-quark amplitudes are known. In that model the six-quark cluster process is directly related to aspects of the weak Hamiltonian related to the $\Delta I = 1/2$ rule.

- A weak effective Hamiltonian for which the renormalization group technique[3,4] is altered to fit the $\Delta I = 1/2$ rule gives the best fit to the $^{12}_{\Lambda}C$ experiment.

- A detailed finite-nucleus treatment is necessary, particularly for the O.P.E. part.

- The neutron- vs. proton-stimulated widths are particularly useful quantities for study.

This work was supported in part by NSF Grant PHY84-0655.

REFERENCES

1. D.P. Heddle and L.S. Kisslinger, Phys. Rev. **C33**, 608 (1986); D.P. Heddle, Carnegie Mellon University Ph.D. Dissertation (1985).
2. R. Grace et al., Phys. Rev. Lett. **55**, 1055 (1985); R. Grace, Carnegie Mellon University Ph.D. Dissertation (1986).
3. M.K. Gaillard and B.W. Lee, Phys. Rev. Lett. **33**, 108 (1974); G. Altarelli and L. Maiani, Phys. Lett. **52B**, 351 (1974).
4. M.A. Shifman, A.I. Vainstein, and V.I. Zacharov, Nucl. Phys. **B120**, 316 (1977); F.J. Gilman and M.B. Wise, Phys. Rev. **D20**, 2392 (1979).
5. J. Donoghue, E. Golowich, W. Ponce, and B. Holstein, Phys. Rev. **D21**, 186 (1980).
6. F.J. Gilman and J.S. Hagelin, Phys. Lett. **126B**, 111 (1983); P. Ginsparg and M.B. Wise, Phys. Lett. **127B**, 265 (1983).
7. L.S. Kisslinger, Phys. Lett. **112B**, 307 (1982); E.M. Henley, L.S. Kisslinger and G.A. Miller, Phys. Rev. **C28**, 1277 (1983).
8. G.Y. Cheung, D.P. Heddle and L.S. Kisslinger, Phys. Rev. **C27**, 335 (1983).
9. J.B. Adams, Phys. Rev. **150**, 1611 (1967).
10. M.M. Block and R.H. Dalitz, Phys. Rev. Lett. **11**, 96 (1963).
11. D.J. Millener et al., Phys. Rev. **C31**, 4991 (1985); private communication.

THE NON-MESONIC DECAY OF HYPERNUCLEI

J. Dubach
University of Massachusetts, Amherst, Mass. 01003

ABSTRACT

The $\Lambda N \to NN$ non-mesonic decay mode of hypernuclei is examined in a model that employs a weak, strangeness-changing meson exchange potential in a Fermi gas model of the hypernucleus. Results are in good agreement with recent experiments for A=11 and A=12 hypernuclei. The calculations suggest that the ratio of proton-stimulated to neutron-stimulated decays and the ratio of parity-violating to parity-conserving decays are particularly sensitive to the details of the model. Finally, preliminary results using a more sophisticated shell model of the hypernucleus are discussed.

INTRODUCTION

Cheston and Primakoff[1] recognized in 1953 the potential interest in studying the decay modes of hypernuclei. The free-space "mesonic" decay, $\Lambda \to N\pi$, imparts very little energy (< 6 MeV) and very little momentum (∿ 100 MeV/c) to the nucleon. Thus, a lambda bound deeply inside a hypernucleus which decays by this mode will produce a nucleon which must remain well bound in the daughter nucleus. However, the relevant nucleon levels are already occupied for the most part and the available final state phase space is considerably reduced. The mesonic decay is therefore expected to be strongly suppressed due to this "Pauli-blocking" effect. Explicit calculations[2,3,4] verify these expectations and show how rapidly the supression increases with increasing size of the hypernucleus. By A=12, the suppression is a factor of 6 to 10 and by A=208, it is more than 1000. Thus, we expect the lifetime of large (A > 4) hypernuclei to be significantly longer than the lifetime of the free Λ (260 psec) unless there are other competing processes.

Cheston and Primakoff[1] suggested that the more interesting two-hadron non-mesonic decay mode $\Lambda N \to NN$ might be important and even dominate the process. Here one allows the pion produced in the free-space decay mode to become virtual and to be absorbed via a strong interaction with a nucleon inside the hypernucleus. The resulting additional energy (i.e., the mass of the pion) leaves the two final state nucleons in the continuum with energies of about 70 MeV each and eliminates the Pauli-blocking suppression. The dominance of this non-mesonic mode, as shown in Fig. 1, was observed in a series of experiments[5] in the 1960's and 1970's. The measured ratios of Fig. 1 along with the theoretical expectations for the suppression of the mesonic mode suggest that the non-mesonic decay rate is roughly constant and about equal to the free-space value, at least for 5 < A < 20.

Direct mesaurement of the absolute non-mesonic rate proved

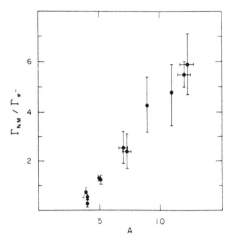
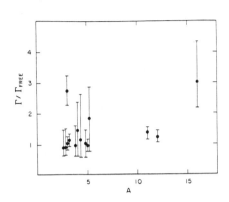

Fig. 1 Selected experimental data for Γ_{NM}/Γ_π as a function of hypernuclear mass number.

Fig. 2 Experimental values for the hypernuclear decay rate.

difficult and the early results[6] were often inconsistent or had very large error bars as shown in Fig. 2. Only recently, improved accelerator beams and sophisticated electronic timing techniques have allowed for the accurate measurement[7] of the non-mesonic decay rate for A=11 and A=12. These data are also shown in Fig. 2. As expected, these measured rates are roughly equal to the free-space mesonic decay rates. The experiments of Figs. 1 and 2 thus confirm the dominance of the non-mesonic decay modes.

THEORETICAL MODEL

In collaboration with L. delaTorre, J.F. Donoghue, and B.R. Holstein[4,8], I have evaluated the non-mesonic decay rate in a model which allows for not only virtual pion exchange but also the heavier ρ, ω, η, K, and K^* mesons as diagrammed in Fig. 3a and 3b. Here one constructs a $\Lambda N \to NN$ $\Delta S=1$ weak "transition potential"

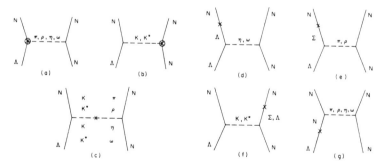

Fig. 3 Meson exchange diagrams used to evaluate the transition potential in the present model.

in analogy with weak parity-mixing nucleon-nucleon
interactions[9]. The full parity-conserving weak vertices are
calculated assuming a pole model as depicted in diagrams (c)-(g) of
Fig. 3 where the cross indicates a weak meson→meson or baryon→
baryon transition amplitude. Using an $SU(6)_W$ symmetry, enforcing
the $\Delta I=1/2$ rule, and using PCAC, the meson→meson amplitudes can be
related to the physical $K \to \pi\pi$ decay rate. In the same fashion,
baryon baryon amplitudes can be related to the free-space Λ and Σ
decays. With similar use of the appropriate symmetries all
parity-conserving baryon-baryon-meson couplings may be related to
the Λ-p-π and Σ-p-π couplings. Further details may be found in
Refs. 4 and 8.

For simplicity, we evaluate this transition potential in a
Fermi gas model of the hypernucleus. The Λ is taken to be at
rest. Only initial ΛN relative S-states are considered while the
outgoing nucleon pair are allowed to be in relative L=0, 1, or 2
states. Hadron-hadron correlations will play an important role in
these calculations. We have examined a number of possible forms
for such correlations but here only show results using a
phenomenological form[10] for the initial state correlation and
allowing the final state nucleon pair to interact via a Reid soft
core potential.

RESULTS

The results of this model are summarized in Table I. The
rates are given in units of the free non-mesonic rate for the six
partial-wave channels considered. A number of variations of the
model are presented; the columns are labelled according to the
meson exchanges included. The first two columns compare the
results for pion-exchange only with and without the hadron-hadron
correlations. These clearly play an important role, reducing the
total rate by more than a factor of two. (In fact, this process
was once proposed[11] as an important one for getting at the
hadron-hadron correlations.) (The tensor part of the Reid
soft-core interaction mixes the $^3S_1$ and $^3D_1$ final states
and splits the uncorrelated $^3D_1$ into two "mixed" states.)

Table I. Decay rates for variations of the present model.

	π no corr	π corr	$\pi+\rho$	$\pi,\rho,\eta,\omega,K,K^*$
$^1S_0 \leftarrow {}^1S_0$	.01	--	.001	.001
$^3P_0 \leftarrow {}^1S_0$	.156	.037	.052	.018
$^3P_1 \leftarrow {}^3S_1$	.312	.117	.113	.456
$^1P_1 \leftarrow {}^3S_1$	.468	.128	.100	.110
$^3S_1 \leftarrow {}^3S_1$	.01	.789	.589	.202
$^3D_1 \leftarrow {}^3S_1$	2.93	.751	.693	.444
Total	3.89	1.82	1.55	1.23

These results are in basically good agreement with other pion-only calculations[13]. Including all of the meson exchanges reduces the rate somewhat further to a final value of 1.23 in good agreement with the measured rates[12] of 1.25±0.18 (A=12) and 1.37±0.16 (A=11). We have also considered the role of "correlated" two-pion exchanges as mocked up in a scalar σ exchange, but these effects are small[4].

While the addition of the heavier mesons provides only a modest decrease in the total rate, it does considerably modify the distribution of this rate among the individual partial-wave channels. In particular the $^3P_1 \leftarrow {^3S_1}$ rate is increased by about a factor of four while the mixed $^3S_1/^3D_1$ rates are reduced by more than a factor of two. Processes which emphasize the various partial-wave rates in different ways may be quite sensitive to the addition of the heavier mesons. In particular, a separation of the total rate into parity-violating and parity-conserving components or into "proton-stimulated" and "neutron-stimulated" components should be especially revealing. The calculated ratios:

$$\frac{\Gamma_{pv}}{\Gamma_{pc}} = \frac{\Gamma(^3P_0 \leftarrow {^1S_0}) + \Gamma(^3P_1 \leftarrow {^3S_1}) + \Gamma(^1P_1 \leftarrow {^3S_1})}{\Gamma(^1S_0 \leftarrow {^1S_0}) + \Gamma(^3S_1 \leftarrow {^3S_1}) + \Gamma(^3D_1 \leftarrow {^3S_1})}$$

$$\frac{\Gamma_{\Lambda p \rightarrow np}}{\Gamma_{\Lambda n \rightarrow nn}} = \frac{3\Gamma_{T=0} + \Gamma_{T=1}}{2\Gamma_{T=1}}$$

(1)

where

$$\Gamma_{T=0} \equiv \Gamma(^1P_1 \leftarrow {^3S_1}) + \Gamma(^3S_1 \leftarrow {^3S_1}) + \Gamma(^3D_1 \leftarrow {^3S_1})$$

$$\Gamma_{T=1} \equiv \Gamma(^1S_0 \leftarrow {^1S_0}) + \Gamma(^3P_0 \leftarrow {^1S_0}) + \Gamma(^3P_1 \leftarrow {^3S_1})$$

are shown in Table II for the variations of the model considered above. Here we see a dramatic change by factors of 4 when the heavier mesons are included. The final result for the ratio of proton- to neutron-stimulated rates, 2.9, is not in particularly good agreement with the experimental[12] value of about .8 (with an error bar encompassing .4 to 1.9), but it offers significant improvement in that regard over the pion-only results.

Table II. The ratios of Eq. 1 in the present model.

	Γ_{pv}/Γ_{pc}	$\Gamma_{\Lambda p \rightarrow np}/\Gamma_{\Lambda n \rightarrow nn}$
π (no corr)	0.31	11.2
π (corr)	0.18	16.6
$\pi+\rho$	0.21	13.1
$\pi,\rho,\eta,\omega,K,K^*$	0.90	2.9

DISCUSSION

Our theoretical predictions for the total rates are in very good agreement with the most recent experiments. Given the relatively large uncertainties due to the role of hadron-hadron correlations, we cannot expect to judge the accuracy of the model in much detail. However, the measurement of the proton/neutron ratio offers a better opportunity to make such a judgment. While these experimental values are not completely explained theoretically, they do seem to support the role of the heavier meson exchanges included in our model. A measurement of the parity-violating/parity-conserving ratio should provide an "orthogonal" test of this role.

Of course, the most obvious flaw in the theoretical analysis is the use of a Fermi gas model to describe the structure of hypernuclei with $A \simeq 12$. This should be particularly true for predictions of the (partial-wave dependent) ratios. Recently, M. Kimura and I have begun to re-examine these calculations within the context of a shell model description of the hypernucleus. One must make a number of simplifying assumptions in order to make the problem tractable. To date we have included only pion exchange and are still in the process of testing our various assumptions. Nonetheless, I can report the following qualitative conclusions. The shell model predictions for the rates will be (somewhat) smaller than for the Fermi gas model (if you like, it is simply appropriate to use a smaller Fermi momentum for A=12). This is consistent with other calculations that have estimated this effect in various ways[13]. Our preliminary (pion-only) results show the proton/neutron ratio is a bit smaller than that predicted for nuclear matter, but it is still greater than 5. However, the most important reason for examining this model is to get a realistic assessment of the role of initial ΛN pair relative P-states, particularly in the various ratio predictions. Our preliminary results suggest that the initial P-states contribute less than 20% of the total rate. The proton/neutron ratio is somewhat smaller for these initial P-states, but their relative unimportance yields little change in the total proton/neutron ratios. It will, of course, be most interesting to see the effect of the heavier meson exchanges in this model.

REFERENCES

1. W. Cheston and H. Primakoff, Phys. Rev. 92, 1537 (1953).
2. H. Bando and H. Takaki, Prog. Th. Phys. 72, 106 (1984).
3. H. Bando and H. Takaki, Phys. Lett. 150B, 409 (1985).
4. J. Dubach, Nucl. Phys. A450, 71c (1986).
5. A. Montwill, P. Moriarty, D.H. Davis, T. Pniewski, T. Sobczak, O. Adamovic, U. Krecker, G. Coremans-Bertrand, and J. Sacton, Nucl. Phys. A234, 413 (1974); and references therein.
6. K. J. Nield, T. Bowen, G. D. Cable, D. A. DeLise, E. W. Jenkins, R. M. Kalbach, R. C. Noggle, and A. E. Pifer, Phys.

Rev. C13, 1263 (1976); G. Keyes, J. Sacton, J. H. Wickens, and M. M. Block, Nucl. Phys. B67, 269 (1973); G. Bohm, J. Klabuhn, U. Krecker, F. Wysotzki, G. Bertran-Coremans, J. Sacton, J. Wickens, D. H. Davis, J. E. Allen, and K. Garbowska-Pniewska, Nucl. Phys. B23, 93 (1970); R. E. Phillips and J. Schneps, Phys. Rev. 180, 1307 (1969); G. Keyes, M. Derrick, T. Fields, L. G. Hyman, J. G. Fetkovich, J. McKenzie, B. Riley, and I. T. Wang, Phys. Rev. D1, 66 (1970); R. J. Prem and P. H. Steinberg, Phys. Rev. 136, B1803 (1964).

7. R. Grace, P. D. Barnes, R. A. Eisenstein, G. B. Franklin, C. Maher, R. Rieder, J. Seydoux, J. Szymanski, W. Wharton, S. Bart, R. E. Chrien, P. Pile, Y. Xu, R. Hackenburg, E. Hungerford, B. Bassalleck, M. Barlett, E. C. Milner, and R. L. Stearns, Phys. Rev. Lett. 55, 1055 (1985).
8. L. de la Torre, J. F. Donoghue, J. Dubach, and B. R. Holstein, to be published.
9. B. Desplanques, Contr. to 8th Int. Workshop on Weak Interactions and Neutrinos, Javea, Spain (1982).
10. B. H. J. McKellar and B. F. Gibson, Phys. Rev. C30, 322 (1984).
11. J. B. Adams, Phys. Rev. 156, 1611 (1967).
12. P. D. Barnes, Nucl. Phys. A450, 43c (1986).
13. See references quoted in Ref. 4.

WEAK DECAYS OF THE H DIBARYON

Eugene Golowich
University of Massachusetts, Amherst, MA 01003

ABSTRACT

We posit the existence of a flavor, color, and spin singlet six-quark configuration called the H dibaryon. A wave function for this particle is written down, and its weak decays for a range of possible H-masses are computed. Both $\Delta S=2$ and $\Delta S=1$ transitions are considered.

INTRODUCTION

John Donoghue, Barry Holstein, and I have undertaken a study of weak decays of a particle called the H-dibaryon. We were motivated to do so in part by astrophysical considerations. The astronomical system Cygnus X-3 has been identified as the source of radiation which on occasion is remarkable for its intensity. This radiation might have a hadronic component which produces muons in the Earth's atmosphere and mantle. The muons are then detected in deep mine proton decay experiments.[1] It has been conjectured that the hadronic component consists of H-dibaryons.[2] According to this premise, the Cygnus X-3 system contains a compact object consisting of degenerate quark matter with macroscopic numbers of strange quarks. Pieces of this matter, presumably H-dibaryons, are energetically hurled into space by the violent physical processes of matter accretion from the other member of the binary system. The proper lifetime of the H dibaryon must be about 10 years to allow propagation over the 3.7×10^4 lt-yrs distance to the Earth.

A second motivation for considering weak decays of the H comes from an attempt to expand our understanding of hadron spectroscopy. The phenomenological description of hadrons is generally expressed in terms of "valence configurations" of quarks and gluons. Aside from baryons (three quarks) and mesons (quark-antiquark), not much is known about other possible arrangements. In this language, the H would be a six-quark configuration behaving as a singlet under color, spin, and flavor transformations. A measure of the importance of this issue is demonstrated by the forthcoming experimental program at Brookhaven National Laboratory to detect the H-dibaryon.

THE CALCULATION

As a working hypothesis, we assume that the H-dibaryon exists as a bound system of six quarks (uuddss) and that to a first approximation, color and spin correlations can be neglected in the wave function. Further, we consider the full range of possible mass values, $2M_\Lambda > M_H > 2M_N$. For masses exceeding $2M_\Lambda$ the H is unbound, whereas for masses below $2M_N$ the H is strictly stable. The mass plays a crucial role in determining how the H decays weakly. For $M_\Lambda + M_N > M_H > 2M_N$, only the $\Delta S=2$ transition $H \rightarrow NN$ is allowed.

For $M_\Lambda + M_N > M_H > 2M_N$, the $\Delta S=1$, $\Delta I=1/2$ decay $H \to \Lambda N$ dominates, whereas for $2M_\Lambda > M_H > M_\Sigma + M_N$ the additional $\Delta S=1$ mode $H \to \Sigma N$ can occur with both $\Delta I=1/2$, $3/2$ amplitudes possible.

The calculation proceeds in three stages: (i) write down the H-wave function in terms of quark creation operators, (ii) compute the relevant weak decay amplitude (we find it convenient to do this in the bag model framework), and (iii) use the P-matrix formalism to relate the ensuing quark configuration to dibaryon scattering states, and thus obtain the associated decay rate.

There is not room to detail these steps here. They have been presented elsewhere.[3] However, a few comments are in order. The potentially onerous problem of constructing the H wavefunction is circumvented by noting that, according to Jaffe, it is the only color, spin, and flavor singlet state consisting of six ground-state quarks.[4] Thus any such state must be acceptable. Indeed, we have shown that several apparently different forms represent in fact identical quantum states. Computation of the H weak-decay amplitude is entirely standard, and we refer the reader to the literature.[5] For the final part of the calculation, we use Mulders' P-matrix analysis of nucleon-nucleon scattering to fix parameters relating six-quark bag states to dibaryon scattering states.[6] This step is perhaps less familiar than the others. It is instructive to divide it into two parts. First note that the weak transition induces an H-pole in each P-matrix channel of the scattering dibaryons. To see this simply diagonalize the mass matrix of the coupled-channel system, where the off-diagonal elements arise from the weak amplitudes. Next relate the H resonance in the S-matrix to the corresponding poles in the P-matrix. The NN, ΛN, ΣN dibaryon scattering states of interest to us are components of a 27-plet of SU(3). Thus we use SU(3) symmetry to relate the experimentally fixed NN pole residue of the P-matrix to its ΛN, ΣN counterparts. However, we incorporate SU(3) breaking when relating the P-matrix mass parameter in the NN channel to that in the $\Lambda N, \Sigma N$ channels.

RESULTS AND SUMMARY

Although the computed lifetimes depend on the mass value assumed for the H, it is nonetheless possible to arrive at some relatively general statements about our results.

For all but the very lightest H masses, the $\Delta S=2$ mode has lifetimes of order 10^6 sec., i.e., several days. This is hundreds of times smaller than the 10 yr. value needed to support the Cygnus X-3 hypothesis described earlier. Thus even if the H were bound so tightly as to obey $M_H < M_\Lambda + M_N$, the estimated lifetime would appear to rule out (or at the very least place rather severe constraints on) any attempt to interpret the hadronic component of cosmic radiation emanating from Cygnus X-3 as the H dibaryon.

The $\Delta S=1$ decays turn out to be of comparable interest. For the mass range $M_\Sigma + M_N > M_H > M_\Lambda + M_N$, only the $\Delta I=1/2$ transition $H \to \Lambda N$ occurs and we find a lifetime of order 10^{-7} sec. It is natural to wonder why this is so much longer than hyperon lifetimes $O(10^{-10}$ sec.). The point is that this weak transition occurs via a $\Delta I=1/2$ nonleptonic Hamiltonian which belongs to a 27-plet, and not to an octet, of operators. Such an interaction is induced by QCD radiative effects and thus has a coefficient substantially smaller than the octet operator.

A heavier H can decay into ΣN as well as ΛN. Perhaps surprisingly, we find that the $\Delta I=3/2$ mode dominates the $\Delta I=1/2$ mode. In this mass range $2M_\Lambda > M_H > M_\Sigma + M_N$ the total lifetime is of order 10^{-8} sec. The reason for $\Delta I=3/2$ dominance lies with the relatively large ratio of $\Delta I=3/2$ to $\Delta I=1/2$ coefficients in the weak Hamiltonian. We thus predict that $H \to \Sigma N$ decays, if discovered, would be the first baryonic nonleptonic transitions to grossly violate the famous $\Delta I=1/2$ rule. The root cause for this is the 27-plet symmetry structure of the dibaryon final states, which forces the weak Hamiltonian to transform in like manner when coupling to the SU(3) singlet H.

Our findings for the $\Delta S=1$ decays of the H have real significance for attempts to experimentally detect the H. One should not assume that the H will decay like other baryons. Rather, account must be made of the comparatively long H propagation length before decay. Otherwise the H dibaryon will likely pass through the detection chamber before decaying, and thus avoid detection!

REFERENCES

1. M. L. Marshak et al., Phys. Rev. Lett. 54, 2079 (1985).
2. G. Baym, R. L. Jaffe, E. Kolb, L. McLerran, and T. P. Walker, Phys. Lett. 160B, 181 (1985).
3. J. F. Donoghue, E. Golowich, and B. R. Holstein, Phys. Lett. B (in press); "Weak Decays of the H-Dibaryon", Univ. of Massachusetts preprint UMHEP-253 (1986).
4. R. L. Jaffe, Phys. Rev. Lett. 38, 195 (1977).
5. E.g., see J. F. Donoghue, E. Golowich, and B. R. Holstein, Phys. Rep. 131, 319 (1986).
6. P. J. Mulders, Phys. Rev. D28, 443 (1983); D26, 3039 (1982).

KAON DECAY PHYSICS

Coordinators: D. A. Bryman
M. P. Schmidt

KMC Unitarity and $K^+ \to \pi^+ \nu \bar{\nu}$

William J. Marciano
Department of Physics
Brookhaven National Laboratory
Upton, NY 11973

ABSTRACT

Experimental constraints on the Kobayashi-Maskawa-Cabibbo (KMC) matrix are surveyed and shown to provide a test of the standard model at the level of its $O(\alpha)$ radiative corrections. The three generation prediction $BR(K^+ \to \pi^+ \nu \bar{\nu}) \simeq (0.35 \sim 3) \times 10^{-10}$ is reviewed and the potential for enhancement up to $\approx 3 \times 10^{-9}$ due to the fourth generation mixing is described.

The three generation charged current quark mixing matrix (hereafter referred to as the KMC matrix)

$$V = \begin{pmatrix} V_{ud} & V_{us} & V_{ub} \\ V_{cd} & V_{cs} & V_{cb} \\ V_{td} & V_{ts} & V_{tb} \end{pmatrix} \quad (1)$$

is constrained by unitarity. Indeed, the orthonormality conditions

$$\sum_i |V_{ij}^* V_{ik}| = \sum_i |V_{ji}^* V_{ki}| = \delta_{jk} \quad (2)$$

provide a powerful consistency check on the standard model. Any apparent deviation from unitarity which could not be accounted for by hadronic effects or electroweak radiative corrections would be evidence for new physics.

The present "best" values for directly measured KMC elements are given below along with some brief comments.

$|V_{ud}| = 0.9740 \pm 0.0012$: This value is obtained from the superallowed beta decay $^{14}O \to {}^{14}Ne^+\nu$ after accounting for rather large $\simeq 3.7\%$ electroweak radiative corrections [1]. Recent developments are the inclusion of higher order leading short-distance $O(\alpha^n \ln^n(m_W/m_p))$, $n = 2, 3 \ldots$ corrections and axial-vector induced $O(\alpha)$ effects [2]. Also employed is a new calculation of the $Z\alpha^2$ Coulomb corrections ($Z = 7$ for this decay) by Sirlin and Zucchini [3].

$|V_{us}| = 0.221 \pm 0.002$: This value was determined by Leutwyler and Roos [4] from their analysis of K_{e3} and hyperon beta decays. It should be noted that more recent hyperon decay data further improves agreement between the $|V_{us}|$ obtained by these two methods [5].

$|V_{cd}| = 0.207 \pm 0.024$ and $|V_{cs}| = 0.95 \pm 0.14$: These values are extracted from CDHS [6] dimuon data on charm production in deep-inelastic $\overset{(-)}{\nu}$ scattering. They have been updated [5] by the use of a larger $c \to \mu$ branching ratio (based on MARK III results) and a new strange quark sea distribution (based on CCFRR data).

$|V_{ub}| \leq 0.009$ and $|V_{cb}| = 0.043^{+0.006}_{-0.008}$: These quantities are based [5] on a world average $\tau_b = 1.1 \times 10^{-12}s$ and $\Gamma(b \to u)/\Gamma(b \to c) \leq 0.08$ using the theoretical analysis of B decay by Grinstein, Wise and Isgur [7]. If one wishes to be more conservative, $\Gamma(b \to u)/\Gamma(b \to c) < 0.25 \to |V_{ub}| < 0.016$ should be employed [8].

From the above values, one finds for the first row in Eq. (1)

$$|V_{ud}|^2 + |V_{us}|^2 + |V_{ub}| = 0.9975 \pm 0.0025 \qquad (3)$$

which is in very good accord with the three generation prediction of 1. That must be viewed as a significant quantum loop triumph for the standard model, since without radiative corrections [1] one would have obtained 1.034 (an apparent violation of unitarity). The result in Eq. (3) can be used to limit heavy neutrino mixing, bound the scale of compositeness [9], constrain supersymmetry mass scales [10], etc. In the case of a fourth generation of quarks (referred to as $t'\&b'$), there is still room for relatively large $V_{t'd}$ mixing at the level of about 0.05. The second row gives

$$|V_{cd}|^2 + |V_{cs}|^2 + |V_{cb}|^2 = 0.95 \pm 0.27 \qquad (4)$$

which is consistent with unitarity but much less constraining.

Three generation unitarity can also be used to bound the degree of t quark mixing. Allowing for a 2σ variation in $|V_{cb}|$, one finds

$$|V_{tb}| > 0.998 \qquad (5a)$$
$$|V_{ts}| < 0.056 \qquad (5b)$$
$$|V_{td}| < 0.056 \qquad (5c)$$
$$|V_{ts}^*V_{td}| < 0.0016 \qquad (5d)$$

The constraint on $|V_{ts}^*V_{td}|$ in Eq. (5d) implies a suppression of virtual t quark loop effects in kaon decay. For example, in the case of $K^+ \to \pi^+\nu\bar{\nu}$, one finds [11] (for $m_t \simeq 40 - 70 GeV$) employing Eq. (5d)

$$BR(K^+ \to \pi^+\nu\bar{\nu}) \simeq (0.35 \sim 3) \times 10^{-10} \qquad (6)$$

That range of predictions is well below the present bound[12]

$$BR(K^+ \to \pi^+ \nu \bar{\nu}) < 1.4 \times 10^{-7} \quad \text{(Exp.Bound)} \qquad (7)$$

but close to the capability of a BNL experiment[13] now underway which is sensitive to $\simeq 2 \times 10^{-10}$. I might remark that for a class of models in which induced $Zd\bar{s}$ couplings are the main contribution to $K^+ \to \pi^+ \nu \bar{\nu}$ (as in some supersymmetric models[14]) one can use the constraint on such a coupling from the $K_L \to \mu^+ \mu^-$ rate[11,15,16] to infer

$$BR(K^+ \to \pi^+ \nu \bar{\nu}) \lesssim 6 N_\nu \times 10^{-11} \qquad (8)$$

where N_ν is the number of neutrino species. This shows that at about the level of 2×10^{-10}, $BR(K^+ \to \pi^+ \nu \bar{\nu})$ becomes a better constraint on flavor changing neutral current effects than $K_L \to \mu^+ \mu^-$. Of course, it is also much cleaner theoretically.

The predicted range in Eq. (6) depends strongly on three generation unitarity and the suppression of $|V_{ts}^* V_{td}|$ implied by the long b lifetime. It can be loosened by the introduction of a fourth generation of quarks which automatically invalidates the bound on $|V_{ts}^* V_{td}|$ in Eq. (5d) and replaces it with the less restrictive $K_L \to \mu^+ \mu^-$ constraint[11,15,16]

$$|\text{Re} \sum_{i=u,c,t,t'} V_{is}^* V_{id} C(m_i^2/m_W^2)| \lesssim 2 \times 10^{-3} \qquad (9a)$$

$$C(x) = \frac{x}{4} - \frac{3}{4}\frac{x}{x-1} + \frac{3}{4}(\frac{x}{x-1})^2 \ln x \qquad (9b)$$

(The bound in Eq. (9a) can even be relaxed by about a factor of 2 if one allows for so-called "maximal" destructive interference with long distance dispersive contributions[11].) Barring a subtle cancellation, this implies for $m_t \simeq 40 GeV$, $|V_{ts}^* V_{td}| < 0.01$. In such a scenario, one finds from the general formula of Lim and Inami[16] (neglecting charged lepton masses)

$$BR(K^+ \to \pi^+ \nu \bar{\nu}) \simeq 1.5 N_\nu \times 10^{-5} | \sum_{i=u,c,t,t'} V_{is}^* V_{id} D(m_i^2/m_W^2)|^2 \qquad (10a)$$

$$D(x) = \frac{x}{4} - \frac{3}{4}\frac{x}{1-x} + \frac{1}{8}[1 + \frac{3}{(1-x)^2} - \frac{(4-x)^2}{(1-x)^2}]x \ln x \qquad (10b)$$

that the t quark contribution can be quite large. Indeed, for $m_t \simeq 40 GeV$, $BR(K^+ \to \pi^+ \nu \bar{\nu})$ can be about as large as 3×10^{-9} before one starts to conflict with the $K_L \to \mu^+ \mu^-$ constraint in Eq. (9). (It could even be somewhat larger if destructive interference occurs in the $K_L \to \mu^+\mu^-$ amplitude[11].) As a concrete example, consider the following four generation mixing matrix[5]

$$\begin{pmatrix} .9742 & .2230 & .0060 & .0342 \\ -.2150 & .9623 & .0490 & -.1592 \\ -.069c_4 & .15c_4 - .04s_4 e^{i\delta} & -.068c_4 + .996s_4 e^{i\delta} & .985c_4 + .06s_4 e^{i\delta} \\ -.069s_4 & .15s_4 + .04c_4 e^{i\delta} & -.068s_4 - .996c_4 e^{i\delta} & .985s_4 - .06c_4 e^{i\delta} \end{pmatrix} \quad (11)$$

$$c_4 \equiv \cos\theta_4, \quad s_4 \equiv \sin\theta_4.$$

in which one angle and one phase have been left arbitrary while the other parameters have been fixed by unitarity and phenomenology assuming relatively large fourth generation mixing. That sample mixing gives

$$BR(K^+ \to \pi^+ \nu\bar{\nu}) \simeq 6 \times 10^{-9}(0.08 + c_4^2 D(m_t^2/m_W^2) + s_4^2 D(m_{t'}^2/m_W^2))^2 \quad (12)$$

which for $c_4^2 \gg s_4^2$ and $m_t \simeq 40 GeV$ implies

$$BR(K^+ \to \pi^+ \nu\bar{\nu}) \simeq 3 \times 10^{-9} \quad \text{(Example in Eq.(11))} \quad (13)$$

That is an order of magnitude larger than the maximum three generation prediction in Eq. (6). It has also been pointed out by U. Türke[17] that t' loop effects with $m_{t'} \simeq 200 GeV$ may lead to an enhanced $BR(K^+ \to \pi^+ \nu\bar{\nu})$ in four generation models. I might note that in scenarios such as Eq. (11) where CP violation in the kaon system comes from heavy quark loops, ϵ'/ϵ turns out generally to be very small[17] $\simeq 10^{-4}$.

In conclusion, unitarity has proven to be a powerful constraint on quark mixing. It already tests the standard model at the quantum loop level and leads to the fairly definite prediction $BR(K^+ \to \pi^+ \nu\bar{\nu}) < 3 \times 10^{-10}$ for three generations. An experimental finding of $BR(K^+ \to \pi^+ \nu\bar{\nu}) > 3 \times 10^{-10}$ would, therefore, most likely suggest relatively large fourth generation mixing, an exciting possibility. That large a branching ratio should be observable at a new BNL experiment[13] which begins taking data in late 1986 and will start to have results some time in 1987.

Acknowledgment

This work was supported by the U.S. Department of Energy under Contract DE-AC02-76CH00016.

References

1. A. Sirlin, Rev. Mod. Phys $\underline{50}$, 573 (1978);
   V.T. Koslowsky et al. in "Proceedings of the Seventh Int. Conf. on Atomic Masses and Fundamental Constants", edited by O. Klepper (Darmstadt, 1984), p. 572..
2. W.J. Marciano and A. Sirlin, Phys. Rev. Lett. $\underline{56}$, 22 (1986).
3. A. Sirlin and R. Zucchini, NYU preprint (1986).
4. H. Leutwyler and M. Roos, Z. Phys. $\underline{C25}$, 91 (1984).
5. W.J. Marciano and Z. Parsa, 1986 Annual Review of Nucl. and Part. Science.
6. H. Abromowicz et al., Z. Phys. $\underline{C15}$, 19 (1982).
7. B. Grinstein, M. Wise, and N. Isgur, Phys. Rev. Lett. $\underline{56}$, 298 (1986).
8. R. Wilson, 1986 Vanderbilt Conference talk.
9. W.J. Marciano and A. Sirlin, work in progress.
10. R. Barbieri, C. Bouchiat, A. Georges and P. le Doussal, Nucl. Phys. $\underline{B269}$, 253 (1986).
11. J. Ellis and J. Hagelin, Nucl. Phys. $\underline{B217}$, 189 (1983);
    F.J. Gilman and J. Hagelin, Phys. Rev. $\underline{133B}$, 443 (1983).
12. Part. Data Table, Rev. Mod. Phys. $\underline{56}$ (1984).
13. L. Littenberg, private communication.
14. S. Bertolini and A. Masiero, NYU preprint, March (1986).
15. R. Shrock and M. Voloshin, Phys. Lett. $\underline{87B}$, 375 (1979).
16. T. Inami and C.S. Lim, Prog. Theor. Phys. $\underline{65}$, 297 (1981).
17. U. Türke, E. Paschos, H. Usler, Nucl. Phys. $\underline{B285}$, 313 (1985);
    A.A. Anselm et al., Phys. Lett. $\underline{156B}$, 102 (1985);
    X.-G. He and S. Pakvasa, Phys. Lett. $\underline{156B}$, 236 (1985).

RARE DECAYS

T. Numao
TRIUMF, 4004 Wesbrook Mall, Vancouver, B.C., Canada V6T 2A3

ABSTRACT

Two experimental efforts to understand the generation problem are discussed. A recent search for the lepton-flavor-nonconserving reaction $\mu^- + \text{Ti} \to e^- + \text{Ti}$ has yielded a new upper limit of 4×10^{-12} (90% C.L.). The status of the experiment BNL787 to search for the rare kaon decay mode $K^+ \to \pi^+\nu\bar{\nu}$ is also discussed.

INTRODUCTION

One of the major problems of elementary particle physics is how to understand the existence of multiple generations. Each generation is just a repetition of the first and has universal coupling constants for electro-magnetic and weak interactions. They are strongly supported by the measurements of the deviations of electron and muon g-factors from 2,[1] the branching ratio of the $\pi-e\nu/\pi-\mu\nu$ decay[2] and the decay characteristics of the τ-lepton.[3] There seems to be no difference between the lepton generations except for masses. Being a single generation model, the standard W-S-G theory doesn't answer the question of this redundancy. Two approaches aimed at learning more about this problem are reported here: the search for μ-e conversion,[4] in which lepton number violation is looked for, and the search for $K^+ \to \pi^+\nu\bar{\nu}$,[5] in which the number of generations is counted.

μe CONVERSION

Present situation

Many models beyond the standard model predict lepton-number violating transitions through mixing of massive neutrinos or through new particles such as flavour non-conserving Higgs scalars and leptoquarks.[6,7] The range of the prediction of the branching ratio for lepton flavor violating processes such as $\mu \to e\gamma$, and $\mu A \to eA$ reactions is from 10^{-9} to zero depending on the masses and couplings of neutrinos and other new particles. Experimentally, no evidence of such transitions has been observed. The current upper limit (90% C.L.) of the branching ratio for each process is,

$\mu^+ \to e^-\gamma$	4.9×10^{-11}	(Ref. 8)
$\mu^+ \to e^+e^+e^-$	2.4×10^{-12}	(Ref. 9)
$\mu^-A \to e^-A$	1.6×10^{-11} (in Ti)	(Ref.10)

According to many theoretical calculations, coherent neutrinoless μ-e conversion, in which the nucleus remains in the ground state, may be enhanced over other lepton number violating processes because of the Pauli blocking effect in the nucleus. The signature of coherent μ-e conversion is a single peak at the energy approximately $E_e = m_\mu c^2 - B$, where m_μ and B are the muon mass and the binding energy of the muonic atom, respectively.

$\mu^- + Ti \rightarrow e^- + Ti$ experiment

A 73 MeV/c μ^- beam from the TRIUMF M9 channel was stopped at a typical rate of 10^6 s^{-1} in a 20 cm long shredded natural Ti target with a density of 0.1 g cm^{-3}. The raw μ^- beam contained a similar flux of pions and an order of magnitude more electrons. An RF particle separator in the beam channel reduced the pion and electron contaminations to $\pi/\mu \approx 10^{-4}$ and $e/\mu \approx 10^{-2}$.

The detection system was based on a hexagonal time projection chamber (TPC). The titanium target was mounted on the TPC axis surrounded by inner trigger counters consisting of six plastic scintillators and a cylindrical wire chamber. The outer trigger counters consisted of six planar wire chambers each sandwiched between two plastic scintillators. On the top and sides of the magnet were two layers of drift chambers and a layer of plastic scintillators for detecting cosmic ray initiated events. The magnetic field of 9 kG prevented most $\mu \rightarrow e\nu\bar{\nu}$ decay electrons from reaching the outer trigger counters. An electron trigger, defined as a coincidence of both inner counter layers and at least two of the three outer layers, was accepted for ~ 1 µs after a muon stops in the target.

Analysis and Results

The initial ~5×10^7 events on tape were fitted to a helix and reduced with loose cuts. Background events due to cosmic rays and beam pions were identified by the surrounding cosmic ray counters and beam scintillators, respectively. The remaining $\mu \rightarrow e$ candidates and $\mu \rightarrow e\nu\bar{\nu}$ events above momentum P=70 MeV/c, approximately 3×10^4 events, were further analyzed with more stringent cuts. No candidate events were observed in the 96.5 to 106 MeV/c region where coherent μ-e conversion events are expected. Most of the events at this stage were from the bound decay of muons in an atomic orbit. A typical example of good events is shown in Fig. 1. The estimated acceptance for overall

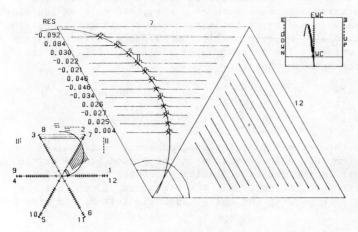

Fig. 1. A typical good event from the $\mu^- \rightarrow e\nu\bar{\nu}$ decay.

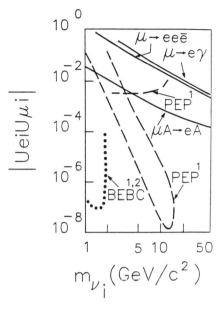

Fig. 2. Limits on the mixing matrix $|U_{e_i}U_{\mu_i}|$ as a function of neutrino mass m_{ν_i}, as obtained from BEBC[12] (dotted line), PEP[12] (dashed line) and the present analysis on the upper limits for $\mu \to e$ conversion, $\mu \to e\gamma$[8] and $\mu \to e e\bar{e}$[9] (solid lines). The superscripts indicate that the matrix is $|U_{e_i}|^2$ (1) or $|U_{\mu_i}|^2$ (2).

cuts was ~6%. Based on the absence of candidate events in 10^{13} stopped muons and assuming the Poisson distribution, a preliminary upper limit R for coherent μ-e conversion is

$$R < 4 \times 10^{-12} \quad (90\% \text{ C.L.}) \ .$$

Assuming only left handed interactions and a massive neutrino for the third generation, Altarelli et al.[6] derived analytical expressions for branching ratios for $\mu \to e\gamma$ and $\mu \to e e\bar{e}$ decays and of neutrinoless $\mu \to e$ conversion. The branching ratios were related to the neutrino mixing matrix $|U_{e_i}U_{\mu_i}|$. In Fig. 2, the upper bounds of the matrix from the $\mu \to e$ conversion, $\mu \to e\gamma$ and $\mu \to e e\bar{e}$ measurements[8,9] are shown by a solid line. A recent beam dump result (a dotted line) and analysis on ee^+ data at PEP (dashed lines) are also shown in Fig. 2 for comparison.[11,12]

$K^+ \to \pi^+ \nu\bar{\nu}$ DECAY

Motivation

The phase space of the decay $K^+ \to \pi^+ \nu\bar{\nu}$ is roughly proportional to the number of neutrino generations. This, in principle, allows counting of the number of neutrino generations from the decay rate. The diagrams, which are responsible for the decay, are shown in Fig. 3. At the present time there are several unknown parameters needed to accurately estimate the decay rate. The mass of the top quark has a large effect on the decay amplitude through the term proportional to $U_{ts}^* U_{td} (m_t/m_w)^2 \ln(m_w/m_t)^2$, where U_{ts} and U_{td} are quark mixing matrices. The bounds for the decay rate were estimated as a function of a top mass using the mixing matrix limit obtained from the decay $K^0 \to \mu\bar{\mu}$ and the CP violation parameter ε.[13,14] For a top quark mass of 40 GeV/c^2

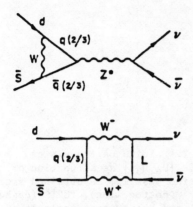

Fig. 3. Diagrams responsible for the decay $K^+ \to \pi^+ \nu \bar{\nu}$.

the lower and upper bounds on the decay $K^+ \to \pi^+ \nu \bar{\nu}$ are 10^{-11} and 4×10^{-11} per neutrino generation, respectively.

As shown in the diagrams of Fig. 3, the decay is induced only by second order interactions due to GIM suppression. Since other second order decays such as $K^+ \to \pi^+ ee$ and $K^+ \to \mu \bar{\mu}$ involve a long range force which is comparable or greater than the weak interaction effect, the study of the decay $K^+ \to \pi^+ \nu \bar{\nu}$ becomes the cleanest test of higher order weak interactions in the standard model. In addition, the same experiment offers other interesting features. Unobserved particles are not limited to neutrinos but also other neutral particles such as axions, photinos, etc. The presence of these new particles indicates new physics beyond the standard model.

Upper limits on the decay $K^+ \to \pi^+ \nu \bar{\nu}$ were previously set by Cable et al.[15] at 5×10^{-7} and by Asano et al.[16] at 1.4×10^{-7}. The BNL787 experiment[5] is designed to improve the existing limit by three orders of magnitude.

Detector

The detector will have a large solid angle and a capability of good background rejection from $K^+ \to \mu^+ \nu$, $K^+ \to \mu^+ \nu \gamma$, $K \to \pi^+ \pi^0$, etc. Figure 4 shows the assembly of the detector in a magnetic field of 1 Tesla. A kaon beam from the BNL LESBI channel is stopped in a 256-element fiber scintillation target surrounded by a cylindrical drift chamber. Pions from the decay $K^+ \to \pi^+ \nu \bar{\nu}$ are detected by the drift chamber and a scintillation-counter range stack divided into 21 layers each of 24 sectors. Momenta, energies and stopping ranges are determined with these elements of the detector. Each sector of the range stack has 2 wire chambers in the middle to provide the position and the angle of the pion track and to help obtain accurate range measurements. The decay sequence $\pi \to \mu \to e$ will be observed by a transient digitizer to identify $K^+ \to \pi^+ x$ decays from $K^+ \to \mu^+ x$ decays. Each range stack counter will be viewed by a photomultiplier on both ends to provide the positions along the counter for the pion and the muon in the sequential $\pi \to \mu \to e$ decay. Outside the range stack is a photon detector of a 1 mm Pb and 5 mm scintillator sandwich. The

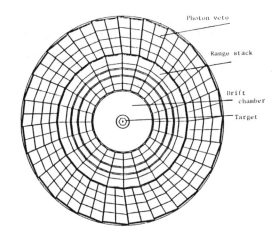

Fig. 4. The end view of the BNL787 detector.

forward and backward endcap regions also have a photon detector. Inefficiency of π^0 detection is estimated to be better than 10^{-5}.

Many background sources are expected through misidentification of muons as pions and through inefficiencies of the photon detectors and the pion beam detectors. A large amount of efforts were spend on Monte Carlo calculations. All backgrounds are estimated to be less than 10^{-10} in terms of kaon branching ratio.

REFERENCES

1. J. Bailey et al., Nucl. Phys. B150, 1 (1979).
2. D.A. Bryman et al., Phys. Rev. D33, 1211 (1986).
3. E. Fernandez et al., Phys. Rev. Lett. 54, 1624 (1985).
4. TRIUMF TPC Collaboration, S. Ahmad, G. Azuelos, D.A. Bryman, A. Burham, E.T.H. Clifford, M. Hasinoff, J.A. Macdonald, T. Numao, J-M. Poutissou, TRIUMF; M.S. Dixit, C.K. Hargrove, H. Mes, NRC; P. Depommier, R. Poutissou, Univ. de Montréal; and M. Blecher, VPI
5. BNL787, BNL-Princeton-TRIUMF Collaboration.
6. G. Altarelli et al., Nucl. Phys. B125, 285 (1977).
7. O. Shanker, Nucl. Phys. B206, 253 (1982).
8. R.D. Bolton et al., LAMPF LAUR-86-446 (1986).
9. W. Bertel et al., Nucl. Phys. B260, 1 (1985).
10. D.A. Bryman et al., Phys. Rev. Lett. 55, 465 (1985).
11. A.M. Cooper-Sarker et al., Phys. Lett. 160B, 207 (1985).
12. F.J. Gilman and S.H. Rhie, Phys. Rev. D32, 324 (1985).
13. J. Ellis and J.S. Hagelin, Nucl. Phys. B217, 189 (1983).
14. F.J. Gilman and J.S. Hagelin, SLAC-PUB-3226 (1983).
15. G.D. Cable et al., Phys. Rev. D8, 3807 (1973).
16. Y. Asano et al., Phys. Lett. 107B, 159 (1981).

SEARCH FOR RARE MUON AND PION DECAY MODES
WITH THE CRYSTAL BOX DETECTOR

L.E. Piilonen, R.D. Bolton, J.D. Bowman, M.D. Cooper, J.S. Frank,
A.L. Hallin,[a] P. Heusi,[b] C.M. Hoffman, G.E. Hogan, F.G. Mariam,
H.S. Matis,[c] R.E. Mischke, D.E. Nagle, V.D. Sandberg, G.H. Sanders,
U. Sennhauser,[d] R. Werbeck, and R.A. Williams
Los Alamos National Laboratory, Los Alamos, New Mexico 87545

S.L. Wilson,[e] R. Hofstadter, E.B. Hughes, and M. Ritter[f]
Stanford University, Stanford, California 94305

D. Grosnick and S.C. Wright
University of Chicago, Chicago, Illinois 60637

V.L. Highland and J. McDonough
Temple University, Philadelphia, Pennsylvania 19122

ABSTRACT

New experimental upper limits for the branching ratios of the lepton-family-number nonconserving decays $\mu^+ \to e^+\gamma$ and $\mu^+ \to e^+\gamma\gamma$ are presented. A new determination of γ, the ratio of pion axial-vector to vector form factors, from radiative pion decay is also reported. These results are from data taken with the Crystal Box detector at LAMPF.

RARE MUON DECAYS

No process violating conservation of lepton family number, like $\mu^+ \to e^+\gamma$, $\mu^+ \to e^+e^+e^-$, $\mu \to e^+\gamma\gamma$, and $\mu^-A \to e^-A$, has ever been seen. Such processes are forbidden in the standard model[1] of electroweak interactions, but are allowed in many extensions to this model.[2] The existing experimental upper limits for the transition rates impose model-dependent constraints on the theoretical parameters, like mixing angles or gauge-boson masses, that describe such processes.

Prior to the Crystal Box, the best experimental upper limits of the branching ratios (90% C.L.) for $\mu^+ \to e^+\gamma$ and $\mu^+ \to e^+\gamma\gamma$ were[3,4]

$$B_{e\gamma} \equiv \frac{\Gamma(\mu \to e\gamma)}{\Gamma(\mu \to all)} \leq 1.7 \times 10^{-10} \quad \text{and} \quad B_{e\gamma\gamma} \equiv \frac{\Gamma(\mu \to e\gamma\gamma)}{\Gamma(\mu \to all)} \leq 8.4 \times 10^{-9}.$$

We report here improved limits for $B_{e\gamma}$ and $B_{e\gamma\gamma}$ from data taken with the Crystal Box detector in the stopped muon channel at the Clinton P. Anderson Meson Physics Facility (LAMPF).

The Crystal Box detector,[5] shown in Fig. 1, consists of 396 NaI(Tℓ) crystals, 36 plastic scintillation hodoscope counters, and a cylindrical 8-plane stereo drift chamber[6] surrounding a thin planar polystyrene target where muons decay at rest. The position resolution of each wire is 330 μm. The NaI(Tℓ) position resolution is 4.7 cm. The NaI(Tℓ) energy resolution for positrons and photons is ~ 7%. The timing resolution of the NaI(Tℓ) is 1.2 ns and of the scintillators is 0.29 ns. (All resolutions are FWHM.)

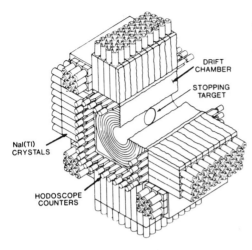

Fig. 1. The Crystal Box

The data for $\mu^+ \to e^+\gamma$, $\mu^+ \to e^+e^+e^-$, and $\mu^+ \to e^+\gamma\gamma$ were collected concurrently. Data written on magnetic tape for each candidate rare-decay event included timing and energy information from all hodoscope counters and from those NaI(Tℓ) crystals with at least 0.1 MeV deposited energy, and timing information from the drift-chamber cells that were hit.

The apparatus acceptances for the rare decay modes were determined with a Monte Carlo simulation, based on the shower code EGS3,[7] that accurately reproduced the response of the detector to photons, positrons, and electrons.

The $\mu^+ \to e^+\gamma$ data analysis is presented first. The signature for a $\mu^+ \to e^+\gamma$ decay at rest is a positron and a photon back-to-back, in time coincidence, with $E_e = E_\gamma = 52.8$ MeV. The hardware trigger for $\mu^+ \to e^+\gamma$ required a coincidence within ±5 ns of a "positron quadrant" and an opposite "photon quadrant", with at least 30 MeV deposited in the NaI(Tℓ) of each quadrant. A positron quadrant had a hodoscope counter signal and one or more NaI(Tℓ) discriminator signals. A photon quadrant had no hodoscope counter signal and at least one NaI(Tℓ) discriminator signal. The trigger selected $\sim 10^7$ candidate from $\sim 10^{12}$ muon decays. The detector resolutions for energies, times, and directions were sufficient to identify backgrounds from $\mu^+ \to e^+\nu\bar{\nu}\gamma$ and random coincidences.

The offline data reduction retained for subsequent analysis all $\mu^+ \to e\gamma$ events and an appreciable number of $\mu^+ \to e^+\nu\bar{\nu}\gamma$ events and random coincidences. The remaining 17 073 events satisfied $|\Delta t_{e\gamma}| < 5$ ns, $\theta_{e\gamma} \geq 160°$, $E_e \geq 44$ MeV, and $E_\gamma \geq 40$ MeV. Fig. 2a shows $\Delta t_{e\gamma}$, the photon-positron relative timing, for a subset of these events. The broad distribution is due to random coincidences, while the prompt peak is due to $\mu^+ \to e^+\nu\bar{\nu}\gamma$ and possibly $\mu^+ \to e^+\gamma$.

The $\mu^+ \to e^+\gamma$ content was found by maximizing the likelihood

$$L(n_{e\gamma}, n_{IB}) = \prod_{i=1}^{N} \left[\frac{n_{e\gamma}}{N} P(\vec{x}_i) + \frac{n_{IB}}{N} Q(\vec{x}_i) + \frac{n_R}{N} R(\vec{x}_i) \right] \quad (1)$$

with respect to the parameters $n_{e\gamma}$, n_{IB}, and $n_R = N - n_{e\gamma} - n_{IB}$ that estimated the number of $\mu^+ \to e^+\gamma$, $\mu^+ \to e^+\nu\bar{\nu}\gamma$, and random events in the total sample of N events. P, Q, and R were the probability distributions for $\mu^+ \to e^+\gamma$, $\mu^+ \to e^+\nu\bar{\nu}\gamma$, and random events, respectively. The vector $\vec{x}$ had components $\theta_{e\gamma}$, $\Delta t_{e\gamma}$, E_e, and E_γ.

Fig. 3 shows the normalized likelihood function. It peaks at $n_{e\gamma} = 0$ and $n_{IB} = 3470 \pm 80 \pm 300$ events. The latter agrees well

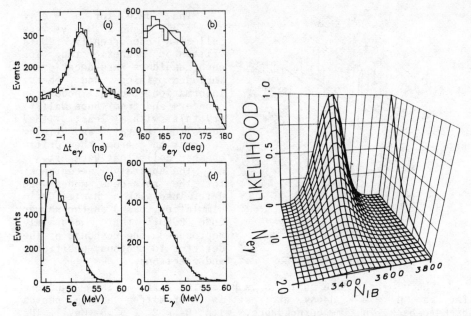

Fig. 2. Histograms of eγ candidates   Fig. 3. μ→eγ likelihood

with the 3960 ± 90 ± 200 $\mu^+ \rightarrow e^+\nu\bar{\nu}\gamma$ events expected in the data. The likelihood function distribution implies $n_{e\gamma} < 11$ events (90% C.L.). Using the number of muons stopped, 1.35×10^{12}, during the live time of the experiment, the apparatus acceptance for $\mu^+ \rightarrow e^+\gamma$, 0.305, and the detection efficiency, 0.545, we obtain $B_{e\gamma} < 4.9 \times 10^{-11}$. Fig. 2a-d shows the agreement between the data (histogrammed) and the best mix of $\mu^+ \rightarrow e^+\nu\bar{\nu}\gamma$ and randoms (smooth) as determined by the likelihood analysis.

The preliminary analysis of ~ 60% of the $\mu^+ \rightarrow e^+\gamma\gamma$ data is presented next. The signature for a $\mu^+ \rightarrow e^+\gamma\gamma$ decay at rest is a positron and two photons in time coincidence emerging with zero net momentum and $E_{tot} = E_e + E_{\gamma 1} + E_{\gamma 2} = 105.6$ MeV. The three particles could strike two quadrants of the Crystal Box and fire the eγ trigger discussed above, or could strike three quadrants and fire the eγγ trigger. The $\mu^+ \rightarrow e^+\gamma\gamma$ trigger required a time coincidence within ±5 ns of a positron quadrant and two photon quadrants, with at least 70 MeV deposited in the NaI(Tℓ) calorimeter.

The eγγ trigger recorded ~ 10^6 candidates from ~ 10^{12} muon decays. In addition, ~ 10^5 candidates were found in the eγ-triggered events, where the positron and one photon occupied the same quadrant. Fig. 4 shows the relative timing distribution for some of these events, the majority being backgrounds from triple random coincidences or two-particle prompt events in random coincidence with a third particle (e.g., $\mu^+ \rightarrow e^+\nu\bar{\nu}\gamma + \gamma$).

The offline analysis removed most of the double- and triple-random coincidences while retaining all of the $\mu^+ \rightarrow e^+\gamma\gamma$ events, assuming the most general local interaction for the

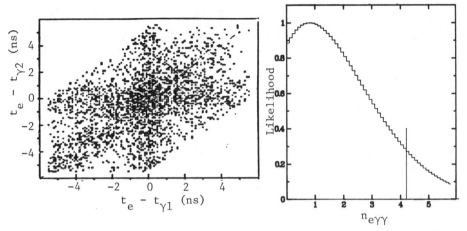

Fig. 4. Scatter plot of eγγ candidates   Fig. 5. μ→eγγ likelihood

$\mu^+ \to e^+\gamma\gamma$ matrix element.[8] Events with one particle showering and appearing as two hits in the trigger in coincidence with another particle were removed by energy and opening-angle cuts.

The number of $\mu^+ \to e^+\gamma\gamma$ events in the remaining sample of 15 events was estimated by maximizing the likelihood

$$L(n_{e\gamma\gamma}) = \prod_{i=1}^{N} \left[\frac{n_{e\gamma\gamma}}{N} P(\vec{x}_i) + \frac{n_R}{N} R(\vec{x}_i) \right] \quad (2)$$

with respect to the parameters $n_{e\gamma\gamma}$ and $n_R = N - n_{e\gamma\gamma}$ that estimated the number of $\mu^+ \to e^+\gamma\gamma$ and background events in the sample of N events. P and R were the probability distributions for $\mu^+ \to e^+\gamma\gamma$ and background, respectively. The components of $\vec{x}$ were E_{tot}, $\Delta t = 2t_e - t_{\gamma 1} - t_{\gamma 2}$, $p_{\parallel} = |\vec{p}_a + \vec{p}_b + \vec{p}_c \times \hat{p}_{ab}|$ and $\cos\alpha = \hat{p}_c \cdot \hat{p}_{ab}$, where $\vec{p}_a$ and $\vec{p}_b$ were the momenta most nearly perpendicular to each other, $\hat{p}_{ab}$ was the unit vector normal to the p_a-p_b plane, and $\vec{p}_c$ was the third particle's momentum.

The likelihood function distribution in Fig. 5 implies $n_{e\gamma\gamma} < 4.2$ (90% C.L.). Using the number of muons stopped, 8.2×10^{11}, during the live time of the experiment, the apparatus acceptance for $\mu^+ \to e^+\gamma\gamma$, 0.071, and the detector efficiency, 0.501, we obtain $B_{e\gamma\gamma} < 1.5 \times 10^{-10}$ (90% C.L.).

RADIATIVE PION DECAY

The low-energy behavior of QCD, the strong-interaction component of the standard model, is extremely difficult to determine. Nevertheless, there are serious attempts to calculate low-energy parameters such as the ratio $\gamma \equiv F_A/F_V$ of the pion weak axial-vector to vector form factors.[9] This ratio can be measured in the radiative decay of the pion, $\pi^+ \to e^+\nu_e\gamma$, where the decay rate is determined by the coherent admixture of an amplitude sensitive to the strong interaction (γ dependent[10]) and an amplitude that accounts for QED corrections to the decay $\pi^+ \to e^+ \nu_e$.

The results of two previous measurements[1f,12] are ambiguous because the experiments detected photons and positrons in a region of phase space where the term in the decay rate proportional to $(1+\gamma)^2$ dominates. The weighted averages of γ are $\gamma = 0.41 \pm 0.06$ or $\gamma = -2.36 \pm 0.06$. We report here data that resolves the ambiguity in the measurement of γ.

Pions passed through a CH_2 degrader and a segmented scintillation beam counter, then stopped and decayed in a planar CH_2 target in the Crystal Box. The trigger required a coincidence within ±5 ns of a positron quadrant and an opposite photon quadrant; these signals had to appear within 50 ns of a signal from the beam counters. The trigger recorded ~ 10^7 candidates from 4×10^{10} pion decays. Events from $\mu^+ \to e^+ \nu \bar{\nu} \gamma$ were eliminated by requiring $E_e + E_\gamma + |\vec{p}_e + \vec{p}_\gamma| > 115$ MeV. Events from $\pi^+ \to e^+ \nu_e \gamma$ and random coincidences in the shaded region of Fig. 6 with $105° \leq \theta_{e\gamma} < 180°$ were retained for subsequent analysis.

The ratio of the number of prompt events in the cross-hatched region of Fig. 6 to the number of pion stops (corrected for $\pi^+ \to e^+ \nu_e \gamma$ detection efficiency) gave the measured branching ratio for $\pi^+ \to e^+ \nu_e \gamma$ in this region. Comparing this measurement to the branching ratio calculated as a function of γ, we find $\gamma = 0.22 \pm 0.15$ or $\gamma = -2.13 \pm 0.15$, in agreement with the previous measurements.

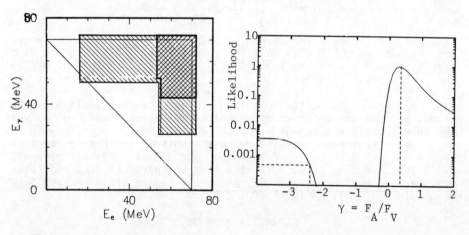

Fig. 6. $\pi \to e\nu\gamma$ Dalitz plot   Fig. 7. $\pi \to e\nu\gamma$ likelihood

The distribution of events in the shaded region of Fig. 6 was used to resolve the ambiguity in γ. We maximized the likelihood

$$L(n_{e\nu\gamma}, \gamma) = \prod_{i=1}^{N} \left[\frac{n_{e\nu\gamma}}{N} P_\gamma(\vec{x}_i) + \frac{n_R}{N} R(\vec{x}_i) \right] \quad (3)$$

by varying the ratio γ and the parameters $n_{e\nu\gamma}$ and $n_R = N - n_{e\nu\gamma}$ that estimated of the number of $\pi^+ \to e^+ \nu_e \gamma$ and random events in the sample of N events. P_γ and R were the probability distributions for $\pi^+ \to e^+ \nu_e \gamma$ (for a particular value of γ) and randoms, respectively. The coordinates of $\vec{x}$ were $\theta_{e\gamma}$, $\Delta t_{e\gamma}$, E_e, and E_γ.

Fig. 7 shows the normalized likelihood as a function of γ. The positive value for γ is favored over the negative value by a likelihood ratio of 2175 to 1. Thus we obtain the unique solution $\gamma = 0.25 \pm 0.12$. The new world average is $\gamma = 0.39 \pm 0.06$.

SUMMARY

Using data from μ^+ decays in the Crystal Box detector, we have obtained improved upper limits on the branching ratios of the lepton-family-number nonconserving decays $\mu^+ \to e^+\gamma$ and $\mu^+ \to e^+\gamma\gamma$ of $B_{e\gamma} < 4.9 \times 10^{-11}$ and $B_{e\gamma\gamma} < 1.5 \times 10^{-10}$, respectively. Using data from π^+ decays, we have resolved the ambiguity in the measurement of γ, the ratio of pion axial-vector to vector form factors, in favor of the positive value, with a new world average of $\gamma = 0.38 \pm 0.06$.

We acknowledge the extraordinary assistance from the many people at each of our institutions and from the operations staff at LAMPF. This work was supported in part by the U.S. Department of Energy and the National Science Foundation.

REFERENCES

a. Now at Physics Dept., Princeton University, Princeton, NJ 08544
b. Now at ELEKTROWATT Ing. AG., Zurich, Switzerland
c. Now at Lawrence Berkeley Laboratory, Berkeley, CA 94720
d. Now at SIN, CH-5234 Villigen, Switzerland
e. Now at Los Alamos National Laboratory, Los Alamos, NM 87545
f. Now at Lockheed Missiles and Space Company, Palo Alto, CA 94304

1. S.L. Glashow, Nucl. Phys. 22, 579 (1961); S. Weinberg, Phys. Rev. Lett. 19, 1264 (1967); A. Salam, in Elementary Particle Theory: Relativistic Groups and Analyticity (Nobel Symposium No. 8), ed. N. Svartholm (Almqvist and Wiksells, Stockholm, 1968), p. 367.
2. C.M. Hoffman, in Fundamental Interactions in Low-Energy Systems, ed. P. Dalpiaz et al., (Plenum Press, New York, 1985), p. 138; P. Herczeg and T. Oka, Phys. Rev. D29, 475 (1984); D.E. Nagle, Comments on Nuclear and Particle Physics XI, 277 (1983).
3. W.W. Kinnison et al., Phys. Rev. D25, 2846 (1982).
4. G. Azuelos et al., Phys. Rev. Lett. 51, 164 (1983).
5. R.D. Bolton et al., Phys. Rev. Lett. 56, 2461 (1986) and references therein.
6. R.D. Bolton et al., Nucl. Instr. and Methods 241, 52 (1985).
7. R.L. Ford and W.R. Nelson, Stanford Linear Accelerator Center No. SLAC-210, 1978 (unpublished).
8. J.D. Bowman et al., Phys. Rev. Lett. 41, 442 (1978).
9. B.R. Holstein, Phys. Rev. D33, 3316 (1986).
10. D.A. Bryman et al., Phys. Rep. 88, 151 (1982).
11. A. Stetz et al., Nucl. Phys. B4, 189 (1978).
12. A. Bay et al., Proceedings of Tenth International Conference on Particles and Nuclei, Heidelberg (1984).

LEPTON NUMBER AND LEPTON FLAVOR VIOLATION IN SUSY MODELS[+]

G. K. Leontaris, K. Tamvakis and J. D. Vergados
The University of Ioannina, Theoretical Physics Division, Ioannina-Greece

ABSTRACT

We examine the consequences on lepton number and lepton flavor non-conservation of softly broken supersymmetric extensions of the standard model which arise in spontaneously broken supergravity theories.

Lepton number (L) and lepton flavor (L_f) appear to be conserved in all experiments performed up to this date. Such a conservation follows automatically in the standard model[1] built in as a global symmetry of the Lagrangian. This, however, is not completely satisfactory since according to present day dogma only local symmetries are admitted as exact symmetries of nature.

It is easy to achieve both L and L_f non-conservation if one goes beyond the standard model by enlarging the standard symmetry $G_S = SU(3) \times SU(2) \times U(1)$ and/or enriching the particle content of the theory[2]. Thus the introduction of the right-handed neutrino naturally[2] leads to a non-zero Dirac mass, mixing between the neutrinos and L_f non-conservation. Furthermore the introduction of additional Higgs particles (e.g. the isosinglet and/or isotriplet) leads to a Majorana mass and both L and L_f non-conservation [2]. All such processes are, however, suppressed [2-4], if the neutrinos are much lighter or much heavier than m_W =80GeV. L and L_f non-conservation can also be achieved via intermediate Higgs scalars [4] but then one is faced with the uncertainties regarding the Higgs sector.

In this talk we will examine the consequences on L and L_f non-conservation which may arise in the supersymmetric extensions[5] (SUSY) of the standard model with the particle content given in table I. Our results will be applicable in a variety of models with spontaneously broken supergravity. Typical diagrams which lead to L and L_f violation are shown in fig. 1. Due to lack of time we will skip all details about SUSY[6] models and present the results relevant to us.

In the absence of supersymmetry breaking the mass matrices[6] for the s-leptons

$$m_\ell^2: \begin{array}{c|c} & \tilde{\ell} \\ \hline \tilde{\ell}^* & m^+ m \end{array} \quad (\text{charged}) \qquad (m_{\tilde{\nu}})^2: \begin{array}{c|cc} & \tilde{\nu} & \tilde{N} \\ \hline \tilde{\nu}^* & m_D^+ m_D & m_D^+ M \\ \tilde{N}^* & M^+ m_D & M^+ M \end{array} \quad (neutral)$$

are simply related to the well known matrices of the usual leptons

[+] Presented at the Conference by J. D Vergados

Table I: The various particles which appear in the supersymmetric extension of the standard model

NAME	SPIN	SYMBOL	NAME	SPIN	
Gluons	1	$g^i, i=1,\ldots,8$	Gluinos	½	$\tilde{g}^i \; i=1,\ldots,8$
Gauge Bosons	1	$W^\pm, Z$	Gauginos	½	$\tilde{W}^\pm, \tilde{Z}^0$
Photon	1	γ	Photino	½	$\tilde{\gamma}$
Quarks	½	$Q = \begin{pmatrix} u^i \\ d^i \end{pmatrix}_L, u_L^{ic}, d_L^{ic}$ $i = 1,2,3$	Squarks	0	$\tilde{Q} = \begin{pmatrix} \tilde{u}^i \\ \tilde{d}^i \end{pmatrix}, \tilde{u}_i^c, \tilde{d}_i^c$ $i = 1,2,3$
Leptons	½	$L = \begin{pmatrix} \nu_i \\ e_i^- \end{pmatrix}_L, e_{iL}^c$ $i = 1,2,3$	Sleptons	0	$\tilde{L} = \begin{pmatrix} \tilde{\nu}_i \\ \tilde{e}_i^- \end{pmatrix}, \tilde{e}_i^c$ $i = 1,2,3$
Higgses	0	$H = \begin{pmatrix} h_i^+ \\ h_i^0 \end{pmatrix} \; i=1,2$ $\bar{H} = \ldots$	Higgsinos	½	$\tilde{H} = \begin{pmatrix} \tilde{h}_i^+ \\ \tilde{h}_i^0 \end{pmatrix} i=1,2$ $\tilde{\bar{H}} = \ldots$

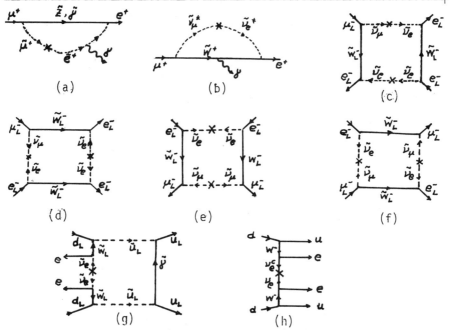

Figure 1. Typical new diagrams in SUSY models.

	ℓ_R^0
$\bar{\ell}_L^0$	m

	ν_R^{0c}	N_R
$\bar{\nu}_L^0$	0	m_D
$\bar{N}_L^{0c}$	m_D^T	M

as required by supersymmetry.

After the kind of SUSY breaking alluded to above the fermion mass matrix is unaffected at the tree level but the s-lepton matrices become [6]

	$\tilde{\ell}$	$\tilde{\ell}^*$
$\tilde{\ell}^*$	$m^+m + m_{3/2}^2 \mathbf{1}$	$A m_{3/2} m$
$\tilde{\ell}$	$A^* m_{3/2} m^+$	$m^+m + m_{3/2}^2 \mathbf{1}$

	$\tilde{\nu}$	$\tilde{N}$	$\tilde{\nu}^*$	$\tilde{N}^*$
	$m_{\tilde{\nu}}^2 + m_{3/2}^2 \mathbf{1}$	0	0	$A^+ m_{3/2} m_D^+$
	0	$A^+ m_{3/2} \tilde{m}_D^*$	$A m_{3/2} m_D^T$	$B m_{3/2} M^+$
	$A m_{3/2} m_D$	$B m_{3/2} M$	$(m_{\tilde{\nu}}^2)^+ + m_{3/2}^2 \mathbf{1}$	

where $m_{3/2}$ = supersymmetry breaking scale, identified with the mass of gravitino, and A and B are constants of order unity.

From the structure of the charged s-lepton matrix we immediately see that the eigenvalues are $m_i^2 = m_{3/2}^2 (1+\mu_i^2+A\mu_i)$ and $(m'_i)^2 = m_{3/2}^2 (1+\mu_i^2-A\mu_i)$ with $\mu_i = m_i/m_{3/2}$. The mixing matrix is proportional to the usual charged lepton mixing matrix. Thus the current of fig. 1a is diagonal in flavor space and does not contribute. This is analogous to the familiar case of the neutral currents. The neutral s-lepton mass matrix is obviously more complicated. In general it can lead to flavor mixing currents since the mixing matrix is no longer expected to be proportional to S_e. In the special case,

$$|m_{ij}| \ll m_{3/2} \ll |M_{ij}|$$

favored by gauge theories one can show [6] that

$$(\tilde{m}_i)^2 = m_{3/2}^2 (1+\lambda_i) \quad , \quad (\tilde{m}'_i)^2 = m_{3/2}^2 (1+\lambda'_i) \quad , \quad \lambda_i, \lambda'_i \ll 1$$

Thus if we consider the flavor violating or F-vertices of fig. (1b) and (1c) we obtain [6]

$$\tilde{\nu}_e \text{---} \times \text{---} \tilde{\nu}_\mu = \sum_i U_{ei} U_{\mu i}^* \frac{1}{p^2 - \tilde{m}_i^2} = \xi_{\tilde{\nu}_e \tilde{\nu}_\mu} m_{3/2}^2 / (p^2 - m_{3/2}^2)^2$$

where $\xi_{\tilde{\nu}_e \tilde{\nu}_\mu^*} = \Sigma U_{ej} U\mu_j^* \lambda_j$ (we made an expansion in powers of λ_j. The leading term vanishes due to the GIM mechanism). We immediately see that $\xi_{\tilde{\nu}_e \tilde{\nu}_\mu^*}$ is small so long as $\lambda_j \ll 1$. In fact in the special case of two generations one can show [2] that

$$|\xi_{\tilde{\nu}_e \tilde{\nu}_\mu}| \lesssim \frac{m_{\nu_\mu}^2 - m_{\nu_e}^2}{m_{3/2}^2} \approx 10^{-18} \quad \text{for} \quad m_{\nu_\mu} = 10 m_{\nu_e} \approx 100 eV, \quad m_W \approx m_{3/2} \approx 100 GeV$$

Clearly such processes are unobservable.

Let us consider the " Majorana like " or M-vertices of fig. 1d-1g. Then one can show [6] that

$$\tilde{\nu}_e \to \times \to \tilde{\nu}_e^* = \frac{1}{2} \sum_i U_{ei}^2 \left\{ \frac{1}{p^2 - \tilde{m}_i^2} - \frac{1}{p^2 - \tilde{m}_i'^2} \right\} \approx \xi_{\tilde{\nu}_e \tilde{\nu}_e^*} \frac{m^2_{3/2}}{(p^2 - m^2_{3/2})^2}$$

$$\xi_{\tilde{\nu}_e \tilde{\nu}_e^*} \approx \frac{1}{2} \sum_j U_{ei}^2 (\lambda_j - \lambda_j')$$

We see that this amplitude is small if $\lambda_j \approx \lambda'_j$. In the special case of one generation one finds that $\lambda_j = -\lambda_j' \approx (2A-B)(m^2/(Mm_{3/2}))$ i.e.

$$\xi_{\tilde{\nu}_e \tilde{\nu}_e^*} = (2A-B) \, m^2/m_{3/2} M = (2A-B) m_{\nu_e}/m_{3/2} \approx 10^{-9}$$

Since $\xi_{\tilde{\nu}_\mu \tilde{\nu}_\mu^*}$ and $\xi_{\tilde{\nu}_e \tilde{\nu}_\mu^*}$ are also expected to be small the $\mu \to 3e$ and M-M̄ oscillations are going to be equally suppressed in this mechanism as well. The neutrinoless ββ-decay (fig. 1h) can also occur in SUSY models (fig. 1g). The corresponding lepton violating parameter is

$$\eta_{\tilde{u}} = 8(4\pi\alpha) \left(\frac{m_{\tilde{w}}}{m_w}\right)^4 \frac{m_{3/2} \, \tilde{m}_{\gamma} \, m_p}{(m_{\tilde{u}})^4} \quad \xi_{\tilde{\nu}_e \tilde{\nu}_e^*} = 5 \times 10^{-14}$$

(we used $m_w = m_{\tilde{w}} \approx m_{\tilde{u}} \approx 100$ GeV and $m_{\tilde{\gamma}} = 5$ GeV). This process is also unobservable since the present experimental limit[2] on η (for heavy indermediate particles) is $|\eta| < 1 \times 10^{-7}$.

In conclusion we can say that in S̄USY models with soft symmetry breaking emerging from spontaneously broken supergravity with the particle content of table I lepton number and lepton flavor violating processes are greatly suppressed. The reason is that even if the s-neutrinos acquire favorable masses ($\sim M_W$) they remain essentially degenerate. As a result the GIM mechanism continues to be effective.

REFERENCES

1. S. L. Glashow, Nucl. Phys. 22, 597 (1961); S. Weinberg, Phys. Rev. Lett. 19, 267 (1967); A. Salam, In Elementary Particle Theory: Rel. Groups and Analyticity (Nobel Symposium No 8) Ed. by S. Svartholm and Wikells, Stockholm, 1967 p. 367.
2. For a recent review and complete set of references see e.g. J. D. Vergados, Phys. Rep. 133, 1 (1986)
3. See also G. K. Leontaris and J. D. Vergados, Double ββ-decay and Lepton Flavor non-conservation, these proccedings.
4. G. K. Leontaris, K. Tamvakis and J. D. Vergados, Phys. Lett. 162B, 373 (1985).
5. For a Review of supersymmetric models see e.g. H. p. Niles, Phys. Rep. 110, 1 (1984).
6. G. K. Leontaris, K. Tamvakis and J. D. Vergados, Phys. Lett. 171B, 412 (1986).

STATUS OF THE STANDARD ELECTROWEAK MODEL IN MUON DECAY

D. P. Stoker
The Johns Hopkins University, Baltimore, Md. 21218

ABSTRACT

The results of muon decay experiments are reviewed and are shown to be consistent with both the (V-A) interaction of the standard electroweak model and a scalar-pseudoscalar interaction coupling to left-handed charged fermions and right-handed neutrinos. The latter case is, however, strongly disfavored by the measured rate for inverse muon decay. It is concluded that the (V-A) interaction accounts for >87% (>79%) of the normal muon decay rate with 68% (90%) confidence. Upper limits on other possible couplings are given.

INTRODUCTION

The standard electroweak model, based on the gauge group $SU(2)_L \times U(1)$, predicts μ decay parameters differing from those of a "V-A" contact interaction by negligible terms of order[1] $(M_\mu/M_W)^2$. Our knowledge of the decay interaction is based primarily on measurements of the longitudinal and transverse e^+ polarizations and the e^+ energy spectrum in both unpolarized and polarized μ^+ decay. Non-observation of the neutrinos, and in particular their helicities, implies an incomplete determination of the interaction structure.

REVIEW OF MUON DECAY MEASUREMENTS

The muon decay parameters describe the momentum spectrum (ρ, η), the asymmetry (ξ, δ), and the longitudinal (ξ', ξ'') and transverse $(\alpha, \alpha', \beta, \beta')$ polarizations of the e^- in the decay $\mu^- \to e^- \nu_\mu \bar{\nu}_e$. Experiments performed in the 1960's found the parameters ρ, η, ξ, δ, and ξ' to be in good agreement with their V-A values.

Recent experiments at SIN have measured the transverse[2] and longitudinal[3] e^+ polarizations by means of the spin dependence of Bhabha scattering and annihilation in flight. The results $P_L = 0.998 \pm 0.045$, $P_{T1} = 0.016 \pm 0.023$ and $P_{T2} = 0.007 \pm 0.023$ are in agreement with the V-A values of 1, 0 and 0. The energy dependence of P_{T1} and P_{T2} imply[2] the decay parameters $\alpha/A = 0.015 \pm 0.052$, $\beta/A = 0.002 \pm 0.018$, $\alpha'/A = -0.047 \pm 0.052$ and $\beta'/A = 0.017 \pm 0.018$ in agreement with the V-A values of 0. T-invariance also requires $P_{T2} = \alpha'/A = \beta'/A = 0$. The value $\xi'' = 0.65 \pm 0.36$ was deduced[3] from the P_L measurements ($\xi'' = 1$ for V-A).

The development of "surface" beams has made available μ^+ beams with polarization $P_\mu = 1$. Their low momentum (29 MeV/c) has allowed the use of thinner stopping targets resulting in smaller energy-loss straggling and scattering of the emitted e^+. Surface μ^+ beams have been used in new experiments at TRIUMF to measure η, $\xi P_\mu \delta/\rho$, and δ.

TRIUMF E134, using an axial-focussing spectrometer to analyze e^+ with p>6MeV/c from stopped surface μ^+, has yielded the preliminary value[4] $\eta = 0.11 \pm 0.09 \pm 0.05$. A value $\eta = 0.011 \pm 0.085$ has been obtained[2,5] indirectly from the e^+ transverse polarization measurements at SIN.

The ξ parameter is measured as the product ξP_μ. Neglecting radiative corrections the energy integrated forward-backward asymmetry is $\xi P_\mu/3$. A new measurement of ξ in progress at SIN (see paper of R. von Dincklage) proposes an order of magnitude improvement over the previous best determination[6] using $\pi^+ \to \mu^+ \to e^+$ decay in photoemulsion in a pulsed 14T field which gave $\xi P_\mu = 0.975 \pm 0.015$. The SIN experiment has the novel feature of using both decay in flight and at rest of μ^+ from π^+ decay in flight.

TRIUMF E185 has measured the endpoint decay rate for e^+ emitted from surface μ^+ opposite their polarization direction relative to the unpolarized μ^+ decay rate. This quantity is $1 - \xi P_\mu \delta/\rho$ and vanishes in the V-A limit if $P_\mu = 1$. The final result[7] of the endpoint rate analysis is $\xi P_\mu \delta/\rho = 0.99863 \pm 0.00046 \pm 0.00075$. The same apparatus also measured the μSR signal amplitude of e^+ emitted with $p > 46$MeV/c from spin-precessed surface μ^+ and found[8] $\xi P_\mu \delta/\rho = 0.9984 \pm 0.0016 \pm 0.0016$. Since there may be unknown sources of μ^+ depolarization the combined result $\xi P_\mu \delta/\rho = 0.99860 \pm 0.00082$ may be expressed as the 90% confidence limit $\xi P_\mu(\pi) \delta/\rho > 0.99753$, where $P_\mu(\pi)$ is the intrinsic μ^+ polarization in π^+ decay.

TRIUMF E247 has measured the decay asymmetry of e^+ emitted with $p > 19$MeV/c using the μSR mode of the E185 apparatus. The preliminary result[9] $\delta = 0.748 \pm 0.005$ is more precise than the current world average $\delta = 0.7751 \pm 0.0085$.

The primary input to the world average value $\rho = 0.7517 \pm 0.0026$ remains the spark chamber spectrometer measurement of Peoples[10].

Recent precise measurements[11,12] of the $\mu^\pm$ mean-lives have lead to the improved world averages[5] $\tau_\mu = (2.197033 \pm 0.000038)\mu$s and μ^+/μ^- mean-life ratio 1.000029 ± 0.000078 (CPT theorem predicts 1).

The CHARM collaboration has measured[13] the rate for inverse muon decay $\nu_\mu + e^- \to \mu^- + \nu_e$ to be $S = 0.98 \pm 0.12$ relative to the V-A prediction.

Recent measurements[14] of the μ^+ momentum from π^+ decay at rest gave $M(\nu_\mu)^2 = -0.163 \pm 0.080$MeV^2/c^4 and a 90% confidence limit $M(\nu_\mu) < 0.25$MeV/c^2. A new high precision measurement[15] of the π^-/e^- mass ratio from pionic X-rays leads to the re-evaluation, assuming $M(\pi^-) = M(\pi^+)$, of $M(\nu_\mu)^2 = -0.097 \pm 0.072$MeV^2/c^4 and $M(\nu_\mu) < 0.27$MeV/c^2.

The measurement $\xi P_\mu(\pi)\delta/\rho > 0.99753$ together with the model independent constraint $|\xi\delta|/\rho \leq 1$ and angular momentum conservation gives the 90% confidence limit on the ν_μ helicity in π^+ decay $h(\nu_\mu) < -0.99753$.

LIMITS ON NON-(V-A) COUPLINGS

General analyses of muon decay have been made[2,16] using the parameters in Table I as input. The parameter values in Table I are the world average values[5] except that the new value of $\xi P_\mu \delta/\rho$ (not used in refs. 2,16) and the new preliminary value of δ (not used in ref. 2) are included here.

The general four-fermion contact interaction contains 10 complex coupling constants. The "helicity projection form"[17,18] of the interaction uses fermion fields of definite chirality. Following ref. 16 the coupling constants are denoted by $g^\gamma_{\epsilon\mu}$, where $\gamma = S$

(scalar, pseudoscalar), V (vector, axial-vector), T (tensor) and ϵ,μ=L,R are the e^- and μ^- chiralities. The neutrino helicities, determined by γ, ϵ and μ, are given in Table II. A purely (V-A) interaction corresponds to only g^V_{LL} non-zero. The coupling constants are related to the definitions of the decay parameters by[16]

$$Q_{RR} = (1/4)|g^S_{RR}|^2 + |g^V_{RR}|^2 \qquad = 2(b+b')/A$$

$$Q_{LR} = (1/4)|g^S_{LR}|^2 + |g^V_{LR}|^2 + 3|g^T_{LR}|^2 \qquad = [(a-a')+6(c-c')]/2A$$

$$Q_{RL} = (1/4)|g^S_{RL}|^2 + |g^V_{RL}|^2 + 3|g^T_{RL}|^2 \qquad = [(a+a')+6(c+c')]/2A$$

$$Q_{LL} = (1/4)|g^S_{LL}|^2 + |g^V_{LL}|^2 \qquad = 2(b-b')/A$$

$$B_{LR} = (1/16)|g^S_{LR} + 6g^T_{LR}|^2 + |g^V_{LR}|^2 \qquad = (a-a')/2A$$

$$B_{RL} = (1/16)|g^S_{RL} + 6g^T_{RL}|^2 + |g^V_{RL}|^2 \qquad = (a+a')/2A$$

$$I_\alpha = (1/4)g^V_{LR}(g^S_{RL}+6g^T_{RL})^* + (1/4)g^{V*}_{RL}(g^S_{LR}+6g^T_{LR}) = (\alpha+i\alpha')/2A$$

$$I_\beta = (1/2)g^V_{LL}g^{S*}_{RR} + (1/2)g^{V*}_{RR}g^S_{LL} \qquad = -2(\beta+i\beta')/A$$

The constraints $Q_{RR},\ldots,B_{RL} > 0$, $Q_{RR}+Q_{LR}+Q_{RL}+Q_{LL}=1$, $B_{LR} < Q_{LR}$, $B_{RL} < Q_{RL}$, $B_{LR}B_{RL} > |I_\alpha|^2$, $Q_{LL}Q_{RR} > |I_\beta|^2$ are model independent.

The $Q_{RR},\ldots,Q_{LL}$ are the fractional decay rates to final states with the indicated electron and muon chiralities. B_{LR} and B_{RL} contain interference terms between the scalar and tensor couplings which give final states with the same chiralities (cf. Table II).

The decay parameters have the values:

$$\rho = (3/4) - (B_{LR}+B_{RL}) + (1/4)(Q_{LR}+Q_{RL})$$

$$\xi = 1 - 2Q_{RR} - (10/3)Q_{LR} + (4/3)Q_{RL} + (16/3)(B_{LR}-B_{RL})$$

$$\delta = (3/4)[1 - 2Q_{RR} - (7/3)Q_{LR} + (1/3)Q_{RL} + (4/3)(B_{LR}-B_{RL})]/\xi$$

$$\xi' = 1 - 2Q_{RR} - 2Q_{RL}$$

$$\xi'' = 1 - (10/3)(Q_{LR}+Q_{RL}) + (16/3)(B_{LR}+B_{RL})$$

The limits shown in Table II were obtained by Monte Carlo integration of the joint probability density implied by the parameter values in Table I. Although the interaction couples overwhelmingly to left-handed μ^- and e^- ($Q_{LL} > 0.964$ at 68% C.L.), the non-observation of the neutrino helicities means that the normal muon decay data cannot distinguish between the (V-A) coupling g^V_{LL} and the scalar-pseudoscalar coupling g^S_{LL}. In the extreme case all normal muon decay data is consistent with $|g^S_{LL}|=2$ and $|g^V_{LL}|=0$. However, the two couplings may be distinguished by the inverse muon decay data. Neglecting the other couplings, the inverse muon decay rate normalized to the (V-A) prediction (S=0.98±0.12) is[16]

$$S = |g^V_{LL}|^2(1-\epsilon) + (3/64)|g^S_{LL}|^2 \epsilon$$

Table I. Values of the muon decay parameters used in analysis.

Decay Parameter	V-A Value	Experimental Value	Comments
ρ	3/4	0.7517 ±0.0026	
δ	3/4	0.7502 ±0.0043	Ref. 9 value included.
$\xi P_\mu \delta/\rho$	1	0.99860±0.00082	Ref. 7 value. Table II limits assume P_μ=1.
ξ'	1	0.998 ±0.042	
ξ''	1	0.65 ±0.35	
α/A	0	0.015 ±0.052	
β/A	0	0.002 ±0.018	
α'/A	0	-0.047 ±0.052	
β'/A	0	0.017 ±0.018	
$\rho(\alpha/A,\beta/A)=\rho(\alpha'/A,\beta'/A)$		-0.894	Correlation (ref. 2)

Table II. 68% (90%) confidence limits on non-(V-A) couplings.

$10^3 \times (Q_{\epsilon\mu}$ and $B_{\epsilon\mu})$ 68% (90%) C.L.	Coupling	Handedness of ν_μ ν_e		Coupling: 68% (90%) C.L. $10^3 \|g^\gamma_{\epsilon\mu}\|$
Q_{RR} < 0.75 (1.1)	g^S_{RR}	L	R	< 55 (66)
	g^V_{RR}	R	L	< 28 (33)
Q_{LR} < 2.8 (4.0)	g^S_{LR}	L	L	<107 (127)
B_{LR} < 2.6 (3.7)	g^V_{LR}	R	R	< 51 (61)
	g^T_{LR}	L	L	< 31 (37)
Q_{RL} < 32 (47)	g^S_{RL}	R	R	<358 (434)
B_{RL} < 9.3 (13)	g^V_{RL}	L	L	< 97 (114)
	g^T_{RL}	R	R	<103 (125)
Q_{LL} > 964 (948)	g^S_{LL}	R	L	<719 (918)
	g^V_{LL}	L	R	>933 (888)

where $\varepsilon=h(\nu_\mu)+1$ is the deviation of the ν_μ helicity in π^+ decay from -1. Since from the previous discussion $\varepsilon<0.0025$ the second term may be neglected so that $|g^V_{LL}|^2=S$ and, from the definition of Q_{LL}, $|g^S_{LL}|^2\leq 4(1-S)$. The limits on $|g^V_{LL}|$ and $|g^S_{LL}|$ in Table II were obtained after excluding the unphysical S>1 region.

The limits obtained for the couplings imply that the (V-A) interaction accounts for >87% (>79%) of the total muon decay rate with 68% (90%) confidence. The largest possible admixture is the coupling g^S_{LL} which may account for as much as 13% (68% C.L.) of the total rate. A stronger limit on $|g^S_{LL}|$ requires a more precise measurement of the inverse muon decay rate.

Finally it is noted that in the context of manifest left-right symmetric models[19], based on the gauge group $SU(2)_L\times SU(2)_R\times U(1)$, the new measurement of $\xi P_\mu\delta/\rho$ yields[7] limits on both the mass of a predominantly right-handed gauge boson W_2 and the left-right mixing angle ζ. The 90% confidence limits, valid if any ν_R coupling to W_R has a mass $M(\nu_R)<6MeV/c^2$, are $M(W_2)>432GeV/c^2$ for any ζ; $M(W_2)>514GeV/c^2$ for $\zeta=0$; $-0.050<\zeta<0.035$ for any $M(W_2)$; and $|\zeta|<0.035$ for infinite $M(W_2)$.

REFERENCES

1. T. D. Lee and C. N. Yang, Phys. Rev. 108, 1611 (1957).
2. H. Burkard et al., Phys. Lett. 160B, 343 (1985).
3. H. Burkard et al., Phys. Lett. 150B, 242 (1985).
4. R. Bossingham, private communication.
5. M. Aguilar-Benitez et al., Review of Particle Properties, Phys. Lett. 170B, (1986).
6. V. V. Akhmanov et al., Sov. J. Nucl. Phys. 6, 230 (1968).
7. A. E. Jodidio, Ph.D. Thesis, LBL report LBL-21084 (1986).
8. D. P. Stoker et al., Phys. Rev. Lett. 54, 1887 (1985).
9. B. Balke et al., LBL report LBL-18320.
10. J. Peoples, Nevis Cyclotron Report No. 147 (1966).
11. G. Bardin et al., Phys. Lett. 137B, 135 (1984).
12. K. L. Giovanetti et al., Phys. Rev. D29, 343 (1984).
13. F. Bergsma et al., Phys. Lett. 122B, 465 (1983).
14. R. Abela et al., Phys. Lett. 146B, 431 (1984).
15. B. Jeckelmann et al., Phys. Rev. Lett. 56, 1444 (1986).
16. W. Fetscher, H.-J. Gerber, and K. F. Johnson, Phys. Lett. 173B, 102, (1986).
17. F. Scheck, Leptons, Hadrons and Nuclei, North-Holland, Amsterdam (1983).
18. K. Mursula and F. Scheck, Nucl. Phys. B253, 189 (1985).
19. J.C. Pati and A. Salam, Phys. Rev. D10, 275 (1974);
    R.N. Mohapatra and J.C. Pati, Phys. Rev. D11, 566, 2558 (1974);
    G. Senjanovic and R.N. Mohapatra, Phys. Rev. D12, 1502 (1975);
    G. Senjanovic, Nucl. Phys. B135, 334 (1979).

REVIEW OF THE DECAY OF THE TAU LEPTON

Patricia R. Burchat

Santa Cruz Institute of Particle Physics
University of California, Santa Cruz CA 95064

ABSTRACT

The predictions of the standard electroweak model for the τ lepton decay modes are discussed and compared to experimental measurements. The measured τ lifetime, final state charged lepton momentum spectra, relative branching fractions and quantum numbers of hadronic final states all agree well with the standard model. However, the sum of the measured branching fractions is only $(89.7 \pm 2.0)\%$. Most of the unidentified decay modes appear to result in only one charged particle in the final state.

INTRODUCTION

Since its discovery in 1975 at the e^+e^- storage ring SPEAR,[1] the τ lepton has been extensively studied. The threshold behaviour of the cross section for τ-pair production was used to establish that the τ is a spin-$\frac{1}{2}$ particle with a mass[2] of $(1784 \pm 3)\,MeV/c^2$. Thus, the τ is massive enough to decay to final states which include mesons. According to the standard model, the τ decays via the weak charged current as shown in Figure 1. Since there are five diagrams which contribute approximately equally (two in which the W^- couples to a lepton and its corresponding neutrino, and three in which the W^- couples to a quark-antiquark pair in each of the three different colors), the branching fractions are expected to be approximately 20% each for the leptonic decays $\tau^- \to \nu_\tau e^- \bar{\nu}_e$ and $\tau^- \to \nu_\tau \mu^- \bar{\nu}_\mu$ and about 60% for the semi-leptonic decays. More precise theoretical calculations generally predict the partial width for a particular τ decay mode but not the absolute magnitude of the branching fraction. In the following sections, I review the theoretical predictions for decay widths by Y. S. Tsai,[3] and by F. J. Gilman and Sun Hong Rhie.[4] I also review the status of published measurements of branching fractions for each decay mode and compare these values with the theoretical predictions.

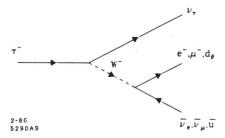

Figure 1. Feynman diagram for the decay of the τ lepton.

LEPTONIC DECAY MODES

As in muon decay, the energy spectrum of the charged lepton in the final state of purely leptonic decays can be used to study the nature of the $\tau - \nu_\tau - W$ coupling. Assuming only V and A couplings, the energy spectrum of the electron (or muon) in the τ center of mass frame is given by $N(x)\,dx = 4x^2(3(1-$

$x) + 2\rho(\frac{4x}{3} - 1) + r(x))dx$ where $x = E/E_{max}$ is the lepton energy relative to its maximum possible value E_{max}, $\rho = \frac{3}{4}\frac{g_L^2}{g_L^2 + g_R^2}$ is the Michel parameter, and $r(x)$ takes into account radiative corrections. g_L and g_R are the couplings to a left- and right-handed ν_τ, respectively. For $V - A$ coupling, ρ is equal to 3/4; for $V + A$, it is zero; and for pure V or pure A, it is 3/8. The charged lepton energy spectrum is harder for large values of ρ. Measurements of the ρ parameter have been published by only two experiments. The DELCO experiment[5] used 594 candidate $\tau^- \to \nu_\tau e^- \bar{\nu}_e$ decays to measure $\rho = 0.72 \pm 0.10 \pm 0.11$. The CLEO experiment[6] has recently used 699 candidate $\tau^- \to \nu_\tau e^- \bar{\nu}_e$ decays and 727 candidate $\tau^- \to \nu_\tau \mu^- \bar{\nu}_\mu$ decays to measure $\rho = 0.71 \pm 0.09 \pm 0.03$. The combined result is $\rho = 0.71 \pm 0.09$. This measured ρ parameter value excludes a $V + A$ coupling and disfavours pure V or A couplings. A comparison with the muon decay measurement[7] $\rho = 0.7517 \pm 0.0026$ demonstrates the lack of precision in the τ decay measurement due largely to a lack of statistics.

The decay width for $\tau^- \to \nu_\tau e^- \bar{\nu}_e$ with the electron mass neglected is $\Gamma(\tau^- \to \nu_\tau e^- \bar{\nu}_e) = \frac{G_F^2 m_\tau^5}{192\pi^3}$ according to the standard model with $V - A$ coupling of universal strength at the $\tau - \nu_\tau - W$ vertex. The decay width for $\tau^- \to \nu_\tau e^- \bar{\nu}_e$ can also be related to the tau lifetime through the branching ratio: $B(\tau^- \to \nu_\tau e^- \bar{\nu}_e) = \tau_\tau \times \Gamma(\tau^- \to \nu_\tau e^- \bar{\nu}_e)$. The average measured τ lifetime[8] of $(2.85 \pm 0.18) \times 10^{-13}$ s, a tau mass of $1784\, MeV/c^2$, and a combination of the above two equations results in $B(\tau^- \to \nu_\tau e^- \bar{\nu}_e) = (17.9 \pm 1.1)\%$. The average measured branching fraction for $\tau^- \to \nu_\tau e^- \bar{\nu}_e$ is $(17.8 \pm 0.5)\%$. The agreement between theory and experiment is excellent but the errors on the τ lifetime are still relatively large.

Taking into account the mass of the muon, we predict $\frac{\Gamma(\tau^- \to \nu_\tau \mu^- \bar{\nu}_\mu)}{\Gamma(\tau^- \to \nu_\tau e^- \bar{\nu}_e)} = 1 - 8y + 8y^3 - y^4 - 12y^2 \ln y = 0.972$, where $y = (\frac{m_\mu}{m_\tau})^2$. The average measured branching fraction for $\tau^- \to \nu_\tau \mu^- \bar{\nu}_\mu$ is $(17.1 \pm 0.6)\%$. Therefore, the ratio of the measured branching fraction for $\tau^- \to \nu_\tau \mu^- \bar{\nu}_\mu$ to $\tau^- \to \nu_\tau e^- \bar{\nu}_e$ is 0.96 ± 0.04. Again, there is good agreement with theory.

When joint measurements of the branching fractions for $\tau^- \to \nu_\tau e^- \bar{\nu}_e$ and $\tau^- \to \nu_\tau \mu^- \bar{\nu}_\mu$ which assume $e - \mu$ universality are included in the world average, the measured branching fraction for $\tau^- \to \nu_\tau e^- \bar{\nu}_e$ is $(17.9 \pm 0.4)\%$. I will normalize measured branching fractions for other decay modes to this measurement for comparison with theoretically predicted decay rates.

SEMI-LEPTONIC DECAY MODES

The standard model leads to the following predictions for the semi-leptonic decay modes.
- The hadronic portion of the decay amplitude has the following G-parity: $G = +$ for the vector current and $G = -$ for the axial current. Currents which have the wrong G parity are called second class currents (SCC) and are not easily incorporated in the standard model.[9] The scalar term ($J^P = 0^+$) is zero both because it is a second class current and by the conserved vector current hypothesis (CVC).
- The rates for decays involving the strange quark are related to those

involving the down quark by the Cabibbo angle.
- From $\mu - \tau$ universality, the rate for $\tau^- \to \nu_\tau \pi^-$ is related to that for $\pi^- \to \bar\nu_\mu \mu^-$ and the rate for $\tau^- \to \nu_\tau K^-$ is related to that for $K^- \to \bar\nu_\mu \mu^-$.
- According to CVC, the rates for τ decays involving the charged weak vector current are related to rates for similar interactions involving the neutral electromagnetic vector current.

The hadronic states to which the charged weak current is expected to couple in τ decay are shown in the upper two sections of Table 1. The hadronic resonances to which the charged weak current is *not* expected to couple are shown in the bottom section of the table. These resonances correspond to second class currents.

Table 1. Hadronic states to which the charged weak current is expected to couple in τ decay (upper two sections) and to which the charged weak current is *not* expected to couple (lower section).

Description	J^{PG}	Resonance	Major Final State
Allowed,	0^{--}	π^-	π^-
non-strange	1^{+-}	$A^-(1270)$	3π
	1^{-+}	$\rho^-(770)$	2π
		$\rho^-(1250), \rho^-(1600)$	4π
Allowed,	0^-	K^-	K^-
strange	1^+	Q^-	$K^-\pi^+\pi^-(\pi^0)$
	1^-	K^{*-}	$K\pi$
Not Allowed	0^{+-}	$\delta^-(980)$	$\eta\pi$
(SCC)	1^{++}	$B^-(1235)$	$\omega\pi \to 4\pi$

I begin with the pseudoscalar hadronic final states. The decay width for $\tau^- \to \nu_\tau \pi^-$ is given by $\frac{\Gamma(\tau^- \to \nu_\tau \pi^-)}{\Gamma(\tau^- \to \nu_\tau e^- \bar\nu_e)} = \frac{(f_\pi \cos\theta_c)^2}{m_\tau^2} 12\pi^2 \left(1 - \frac{m_\pi^2}{m_\tau^2}\right)^2$. Both $\tau^- \to \nu_\tau \pi^-$ and $\pi^- \to \bar\nu_\mu \mu^-$ involve the coupling of the axial current to the pion. Therefore the coupling factor $f_\pi \cos\theta_c$ can be determined from the measured pion lifetime since the branching fraction for $\pi^- \to \bar\nu_\mu \mu^-$ is 100%. This results in $\frac{\Gamma(\tau^- \to \nu_\tau \pi^-)}{\Gamma(\tau^- \to \nu_\tau e^- \bar\nu_e)} = 0.607$. The world average measured branching fraction for $\tau^- \to \nu_\tau \pi^-$ is $(10.9 \pm 1.4)\%$. Therefore, the relative branching fraction of $\tau^- \to \nu_\tau \pi^-$ to $\tau^- \to \nu_\tau e^- \bar\nu_e$ is measured to be 0.61 ± 0.08 in good agreement with theory.

Both $\tau^- \to \nu_\tau K^-$ and $K^- \to \bar\nu_\mu \mu^-$ involve the coupling of the axial current to the kaon. Therefore the coupling factor $f_K \sin\theta_c$ can be determined from the dominant decay of the charged kaon. This results in $\frac{\Gamma(\tau^- \to \nu_\tau K^-)}{\Gamma(\tau^- \to \nu_\tau e^- \bar{\nu}_e)} = 0.0395$. The world average measured branching fraction for $\tau^- \to \nu_\tau K^-$ is $(0.67 \pm 0.17)\%$ and the branching fraction relative to $\tau^- \to \nu_\tau e^- \bar{\nu}_e$ is 0.037 ± 0.009, again in good agreement with theory.

Now I will discuss the axial-vector hadronic final states. The lowest lying non-strange, isospin-1 resonance with $J^{PG} = 1^{+-}$ is the $A^-(1270)$ whose dominant decay mode is to $\rho\pi$. Several experiments have studied $3\pi^\pm$ final states in τ decay to determine the intermediate state; the spin-parity, mass and width of the resonance; and the branching fraction for $\tau^- \to \nu_\tau \pi^- \pi^+ \pi^-$. The Argus collaboration[10] has presented a preliminary analysis of the $3\pi^\pm$ final state. This study contains more than twice as many events as the other analyses combined. They determine that the probability of the $3\pi^\pm$ system containing a component which is not $\rho\pi$ is less than 10% at the 90% confidence level. A Dalitz plot analysis is used to determine the spin-parity of the three pion state. All experiments find that the data are well represented by a $J^P = 1^+$ distribution with the $\rho\pi$ in a relative s-wave. In the Mark II analysis,[12] the data are compared to incoherent mixtures of the dominant 1^+ s-wave and small amounts of the other allowed spin-parity states. Upper limits at the 95% confidence level for these contributions are found to be 18% for the 0^- hypothesis and 29% for the 1^+ d-wave. Alternatively, the best fit population in combination with the 1^+ s-wave is $(10 \pm 5)\%$ for 0^- and $(16 \pm 8)\%$ for the 1^+ d-wave. The measured values of the mass and width of the resonance are listed in Table 2. The apparent discrepancy between the various measurements is due to the use of parameterizations which differ by up to three powers of mass for the three pion mass distribution. The parameterization is not unique because of arbitrary form factors at both the $W - A$ and $A - \rho - \pi$ vertices. The world average measured branching fraction for $\tau^- \to \nu_\tau \pi^- \pi^+ \pi^-$ is $(6.6 \pm 0.4)\%$. The decay mode $\tau^- \to \nu_\tau \pi^- 2\pi^0$ is difficult to study because of the large number of photons from the neutral pions. However, isospin conservation imposes the following important constraint: $B(\tau^- \to \nu_\tau \pi^- 2\pi^0) \leq B(\tau^- \to \nu_\tau \pi^- \pi^+ \pi^-)$.[4] If the three pion system is completely dominated by the $A(1270)$ the branching fractions to one and three charged pions are identical.

The lowest lying strange $J^P = 1^+$ resonances are $Q(1280)$ and $Q(1400)$ which decay through various intermediate states to $K^- \pi^+ \pi^- (\pi^0)$. The theoretical estimate for the branching fraction for $\tau^- \to \nu_\tau K^- \pi^+ \pi^- (\pi^0)$ is 0.11% and the measured value, based on eight events from the DELCO experiment,[11] is $(0.22^{+0.16}_{-0.13})\%$.

Finally, we arrive at the vector hadronic final states. The lowest lying non-strange isospin-1 resonance with $J^{PG} = 1^{-+}$ is $\rho(770)$ whose dominant decay mode is to two pions. Studies of the two pion final state of the tau show that it is consistent with pure $\rho(770)$. CVC can be used to relate the strength of the charged weak vector current coupling to $\pi^- \pi^0$ to that of the neutral electromagnetic vector current. The strength of the latter coupling is measured by the cross section $\sigma(e^+ e^- \to \gamma \to \pi^+ \pi^-)$. Using the measured cross sections from $e^+ e^-$ storage rings, Gilman and Rhie calculate $\frac{\Gamma(\tau^- \to \nu_\tau \pi^- \pi^0)}{\Gamma(\tau^- \to \nu_\tau e^- \bar{\nu}_e)} = 1.23$. The

Table 2. The measured values of the $A(1270)$ mass and width. The first three rows correspond to analyses of $3\pi^{\pm}$ final states in τ decay. The last row corresponds to the Particle Data Group[7] values from hadronic scattering.

Experiment	Mass (MeV/c^2)	Width (MeV/c^2)
ARGUS (preliminary)[10]	1046 ± 11	521 ± 27
DELCO[13]	1056 ± 30	476 ± 140
Mark II[12]	1194 ± 20	462 ± 70
PDG[7]	1275 ± 28	316 ± 45

world average measured branching fraction for $\tau^- \to \nu_\tau \pi^- \pi^0$ is $(22.1 \pm 1.2)\%$. Relative to the branching fraction for $\tau^- \to \nu_\tau e^- \bar{\nu}_e$, this is 1.23 ± 0.06, in good agreement with theoretical predictions.

The lowest lying strange $J^P = 1^-$ state is the K^{*-}. The decay rate for $\tau^- \to \nu_\tau K^{*-}$ can be related to that for $\tau^- \to \nu_\tau \rho^-$ by a phase space factor, the Cabbibo factor $\tan^2 \theta_C$, and $SU(3)$ breaking sum rules. Gilman and Rhie predict $\frac{\Gamma(\tau^- \to \nu_\tau K^{*-})}{\Gamma(\tau^- \to \nu_\tau e^- \bar{\nu}_e)} = 0.079$. The world average measured branching fraction for $\tau^- \to \nu_\tau K^{*-}$ is $(1.4 \pm 0.3)\%$. Relative to the branching fraction for $\tau^- \to \nu_\tau e^- \bar{\nu}_e$, this is 0.08 ± 0.02, again in agreement with theoretical predictions.

The higher mass non-strange vector resonances which are allowed in tau decay are $\rho(1250)$ and $\rho(1600)$. The $\rho(1250)$ was first seen in photo-production at the SLAC Hydrogen Bubble Camber[14] in the final state $\omega \pi$. Some evidence for the $\rho(1250)$ was also seen in $\pi \pi$ phase shift experiments and in photoproduced $e^+ e^-$ final states.[7] However, the Omega photon collaboration[15] has done a spin-parity analysis of the $\omega \pi$ state and finds that is dominantly $J^P = 1^+$ indicating that it corresponds to the $B(1235)$, so the status of the $\rho(1250)$ is not firm.

Coupling of the tau to these higher mass resonances is expected to result in four pions in the final state. The rate for $\tau^- \to \nu_\tau (4\pi)^-$ can be related to the cross section for $e^+ e^- \to \gamma \to 4\pi$ assuming CVC. Gilman and Rhie have used recent data from $e^+ e^-$ storage rings to predict $\frac{\Gamma(\tau^- \to \nu_\tau \pi^- 3\pi^0)}{\Gamma(\tau^- \to \nu_\tau e^- \bar{\nu}_e)} = 0.055$ and $\frac{\Gamma(\tau^- \to \nu_\tau \pi^- \pi^+ \pi^- \pi^0)}{\Gamma(\tau^- \to \nu_\tau e^- \bar{\nu}_e)} = 0.275$. The world average measured branching fraction for $\tau^- \to \nu_\tau \pi^- \pi^+ \pi^- \pi^0$ is $(4.8 \pm 0.3)\%$ or, relative to the branching fraction for $\tau^- \to \nu_\tau e^- \bar{\nu}_e$, 0.27 ± 0.02, in good agreement with the prediction. The decay $\tau^- \to \nu_\tau \pi^- 3\pi^0$ is difficult to study because of the large number of photons in the final state but the branching fraction is predicted to be 1.0%.[4]

The Argus collaboration[16] has recently studied the $3\pi^{\pm} \pi^0$ final state in τ decay and they observe a clear ω signal in the neutral three pion mass combination $\pi^-\pi^+\pi^0$. A spin-parity analysis shows that this $\pi^-\omega$ resonance is consistent with pure $J^P = 1^-$. The $\rho(1600)$ decays through the intermediate state $A(1270)\pi$ and not through $\omega\pi$. Hence, the signal may be evidence for the $\rho(1250)$. However, the mass spectrum of the $\omega\pi$ signal is very broad but the statistics are still quite low. The preliminary branching fraction for this mode is about $(1.4 \pm 0.3 \pm 0.5)\%$.

INCLUSIVE BRANCHING FRACTIONS

The inclusive branching fractions of the τ to final states containing one, three and five charged particles have been measured in many experiments. The results are $B_1 = (86.8 \pm 0.3)\%$, $B_3 = (13.1 \pm 0.3)\%$, and $B_5 = (0.14 \pm 0.04)\%$.[8] The sums of measured exclusive branching fractions discussed in the previous sections are $(77.5 \pm 2.0)\%$ * and $(12.1 \pm 0.7)\%$ for modes resulting in one and three charged particles in the final state, respectively. Therefore, $(10.3 \pm 2.0)\%$ of the decay rate is as yet unidentified and most of this results in one charged particle in the final state.

CONCLUSIONS

The measured τ lifetime and relative branching fractions agree well with the predictions of the standard electroweak model. However, $(10.3 \pm 2.0)\%$ of the decay rate is still unidentified. The τ lepton provides a unique environment for the study of $J^{PG} = 1^{-+}$ and $J^{PG} = 1^{+-}$ hadronic resonances.

REFERENCES

1. M. L. Perl et al., Phys. Rev. Lett. **35**, 1489 (1975); M. L. Perl et al., Phys. Lett. B **63**, 466 (1976); M. L. Perl et al., Phys. Lett. B **70**, 487 (1977).
2. R. Bradelik et al., Phys. Lett. B **73**, 109 (1978); W. Bacino et al., Phys. Rev. Lett. **41**, 13 (1978); W. Bartel et al., Phys. Lett. B **77**, 331 (1978).
3. Y. S. Tsai, Phys. Rev. D **4**, 2821 (1971).
4. F. J. Gilman and Sun Hong Rhie, Phys. Rev. D **31**, 1066 (1985).
5. W. Bacino et al., Phys. Rev. Lett. **42**, 749 (1979).
6. S. Behrends et al., Phys. Rev. D **32**, 2468 (1985).
7. Review of Particle Properties, Phys. Lett. B **170** (1986).
8. For recent reviews of published measurements of the τ lifetime and branching fractions see P. R. Burchat, Stanford University Ph. D. thesis and SLAC Report No. 292 (1986); K. K. Gan, 1985 Annual Meeting of the Division of Particles and Fields of the APS, Eugene, Oregon, August 12-15, 1985 and Purdue University report PU-85-539 (1985).
9. E. D. Commins and P. H. Bucksbaum, *Weak Interactions of Leptons and Quarks*, p. 169. Cambridge University Press, 1983.
10. Argus collaboration, private communication (1986).
11. G. B. Mills et al., Phys. Rev. Lett. **54**, 624 (1985).
12. W. B. Schmidke et al., submitted to Phys. Rev. Lett. (1986).
13. W. Ruckstuhl et al., Phys. Rev. Lett. **56**, 2132 (1986).
14. J. Ballam et al., Nucl. Phys. B **76**, 375 (1974).
15. M. Atkinson et al., Nucl. Phys. B **243**, 1 (1984).
16. R. Ammar, presented at the General Meeting of the American Physical Society, Washington DC, April 28 - May 1, 1986.

---

* Here I assume that $B(\tau^- \to \nu_\tau \pi^- 2\pi^0)$ is equal to $B(\tau^- \to \nu_\tau \pi^- \pi^+ \pi^-)$ as implied by isospin rules if the 3π state is pure $A(1270)$ resonance, and that $B(\tau^- \to \nu_\tau \pi^- 3\pi^0) = 1.0\%$ which is the result of using CVC and e^+e^- cross section measurements to 4π final states.

STUDY OF CP VIOLATION IN A TAGGED NEUTRAL KAON BEAM

Abstract
Details are presented of an experiment to study symmetry violation in all decay modes of neutral kaons. The method makes use of initially pure K^0 and $\bar{K}^0$ beams produced in $\bar{p}p$ annihilations at rest at LEAR. The detector and trigger are described, and some indication of the scope of the physics given.

Basle CERN CEN-Saclay DAP-Athens Democritos ETH-Zurich Fribourg Liverpool SIN Stockholm Thessaloniki Collaboration.

Presented by J. R. Fry   Liverpool University

0.1 Introduction

CP Violation is detected whenever a particle and its antiparticle decay at a different rate into the same final state. Alternatively, CP violation may be detected because of the interference between the amplitudes for decay into a particular final state of the particle and its antiparticle. These are of course related: the time-dependent asymmetry, $A\pm$, shown in fig 3a is obtained from the time-dependent decay rate for the K^0 and $\bar{K}^0$ as:

$A\pm(t)=(R(K^0 \to \pi^+\pi^-)-R(\bar{K}^0 \to \pi^+\pi^-))/(R(K^0 \to \pi^+\pi^-)+R(\bar{K}^0 \to \pi^+\pi^-))$

CP violation has so far only been observed in the two pion decay mode of the K^0, where it is explained on the basis of mass-mixing, or CP impurity, in the observed weak eigenstates at the level of $\varepsilon = 2.27 \; 10^{-3}$ Theoretical calculations, however, suggest that CP violation should occur in the decay amplitudes, and the Kobayashi-Maskawa 'standard' model predicts the ratio of ε'/ε to be a few times 10^{-3} It is therefore extremely important that high precision measurements are undertaken of the two pion final states to challenge the theoretical models and test whether the standard model is adequate to explain CP violation.

This experiment proposes to study symmetry violation in all K^0 decay modes in the same experiment with tightly controlled systematics. In particular we plan to:

- measure ε'/ε with a precision of $2 \; 10^{-3}$
- measure the phase difference ($\phi_{+-} - \phi_{00}$) to ± 1.5 ° as a test of CPT.
- measure η_{+-0} and η_{000} to $6 \; 10^{-4}$ and thereby observe CP violation in K^0 short-lived decay modes.

- investigate the semi-leptonic decay modes with a view to testing the $\Delta Q = \Delta S$ rule and whether CP violation is 'compensated' by T violation, as would be expected if CPT were a good symmetry.

Other experiments will achieve some of these goals before us; thus NA31 and E731 plan to measure ε'/ε to 10^{-3} within a year, and E621 will probably detect CP violation in the K^o(short) decay mode within a similar time. All such experiments are very difficult, and it is clear that any challenge to theoretical models will require the most detailed understanding of systematic errors and backgrounds. The fact that our experiment is so different from all others can only help in this respect.

0.2 Detector Configuration and Triggering.

It is proposed to study neutral kaon decays by selecting the particular final state $K^{\pm}\pi^{\mp}K^o$ formed in $\bar{p}p$ interactions at rest. A beam intensity of $2 \cdot 10^6$ antiprotons per second is required, to give a total of 10^{13} interactions, corresponding to the production of a total of $4 \cdot 10^9$ neutral kaons.

The detector is built inside a solenoid of length 3m and inner diameter 2m, having an axial magnetic field of strength 0.5 T uniform to 2%. A gaseous hydrogen target of radius 10 cm is used to minimise regeneration effects, and this is surrounded by a proportional chamber of wire spacing 1 mm and six drift chambers of wire spacing 1 cm, the whole assembly having a mass equivalent to less than 1% of a radiation length (fig 2). On-line, flash TDC's give drift chamber precision of 1 mm in r-ϕ, while the off-line precision is 200 μm. Outside the wire chambers, two cylindrical layers of tubes, run in limited streamer mode, enable the hit position of a charged particle to be determined to 2-3 cm by using the difference in propagation time of the signal to the two ends of the tube. Using this information together with the approximate position of the interaction, a road may be defined for the particle track, and information from two of the cathode planes of the wire chambers used to obtain z-coordinates with high precision.

In order to tag the charged kaons, and hence label the neutral kaon, a Cerenkov-Scintillator sandwich is used. To give an acceptable trigger a charged kaon must be entirely contained within one of the 32 sectors, must give light in the scintillator at front and back, and must give no light in the Cerenkov cell. Two Cerenkov designs are currently under study. One involves two independent perspex cells filled with water or FC72 and viewed at each end by photomultiplier tubes (fig 2), while the other consists of a thin FC72 radiator with a quartz window and uses a wire chamber with TMAE to detect the Cerenkov radiation. Outside this is an electromagnetic calorimeter of 6 radiation lengths comprising

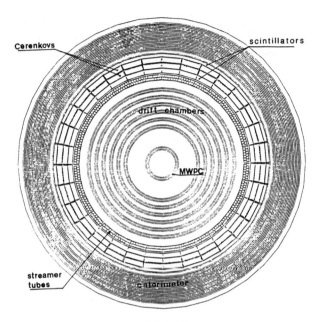

FIGURE 2

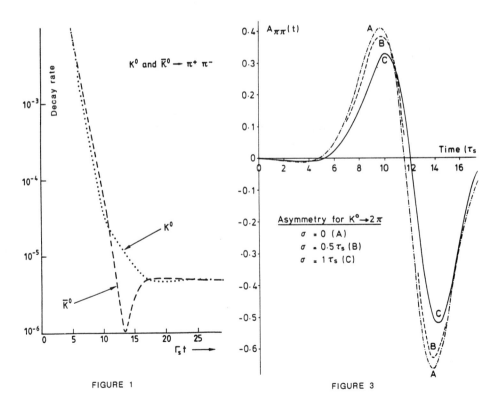

FIGURE 1

FIGURE 3

19 layers. Each layer consists of 4x4.5 mm tubes run in limiting streamer mode with cathode planes above and below at angles of ±30 °, separated by 1.5 mm lead sheets encased in thin stainless steel. A total of 20 000 tubes and 60 000 cathode strips are involved. Monte Carlo simulations based on test data indicate that the position of a photon shower can be obtained to ±5 (±15) mm for photons which do not (do) interact before reaching the calorimeter.

Finally, the barrel-detector is sealed by two end caps whose function is to veto events where charged or neutral particles are detected, thus ensuring that ALL particles in a triggered event exit into the barrel.

The importance of having good spatial resolution for the calorimeter is illustrated in fig 3, where the effect on the measured time-dependent asymmetry of uncertainty in the proper lifetime of the neutral kaon is shown. The dominant contribution to this uncertainty in the case of neutral decay modes of the kaon comes from the directional (and hence positional) uncertainty of the photon showers. In consequence the value of η_{oo} which is extracted from the 2 π^o decay mode is insufficiently accurate to be used for the calculation of ε'/ε. The phase information is affected to a much smaller degree. For charged decay modes the uncertainty on the vertex position is of course very small and does not significantly affect the accuracy with which one can analyse the asymmetry data.

Purpose built hard-wired trigger processors are being designed and tested to select the final state $K^{\pm}\pi^{\mp}K^o$ and the K^o decay products on-line. The first stage requires a charged particle multiplicity of 2 or 4 in the inner scintillation barrel and a multiplicity of 2 or 4 in the MWPC, to ensure that all charged particles are within the geometrical acceptance of the detector, and at least one of these particles must be a kaon candidate. A track follower combines information from the scintillators, streamer tubes and wire chambers, and by using look-up tables calculates the transverse momentum, azimuthal angle and z component of momentum for each track. At this stage a momentum cut is applied to the kaon candidates and only events with one such candidate are retained. A second processor calculates the missing mass to the ($K^{\pm}\pi^{\mp}$) combination(s) and a cut is applied. The master trigger is defined by using pulse height information from the Cerenkov and time of flight information from the scintillators to tighten up the kaon definition. The momentum resolution is calculated to be 2-3% (dominated by Coulomb scattering) and the resolution on the neutral kaon mass to be ±50 Mev/c². The geometrical acceptance is 25%

A third processor treats the photon data and obtains the number of showers together with the coordinate of the shower foot and an estimate of its energy based on a count of the number of tubes which have fired. For those candidate $K^o \rightarrow \pi^o\pi^o$ decays, the K^o decay vertex is calculated on-line using an algorithm

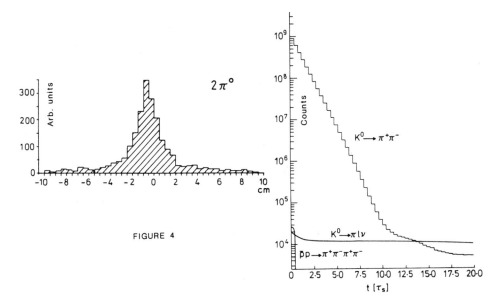

FIGURE 4

FIGURE 5

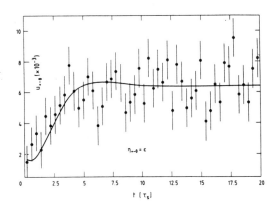

FIGURE 6

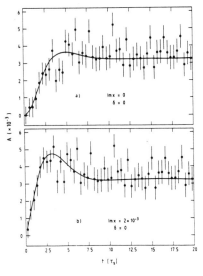

FIGURE 7

which makes use only of the shower positional information. Fig 4 shows the distribution obtained from Monte Carlo studies based on test data obtained from the calorimeter test-modules. A significant reduction in the data to be written to tape is effected by cutting 2π decay modes with K^0 lifetimes smaller than about $5\tau_s$ by demanding the decay vertex to lie more than 10 to 15 cm from the production vertex. Identification of the decay mode relies on both charged and photon multiplicity information, and in addition energy conservation is useful to positively select the $\pi^+\pi^-$ decay mode and as a consistency check on the other pion channels. Additional cuts on the opening angle between the two charged decay tracks help to select the semi-leptonic mode.

An additional trigger is defined which is identical to the real trigger in all respects EXCEPT that the Cerenkov is required to give light for the 'pseudo' kaon candidate. This pseudo trigger is very useful in studying backgrounds from $\bar{p}p$ annihilation final states, such as $\bar{p}p \to 4\pi$ which would mimic our master trigger with 2π decay mode of the K^0 in the event of a misidentified charged kaon, but its major use is in the validation of our measurement for ε'/ε, as discussed below.

0.3 Physics Analysis.

The value of ε'/ε is calculated from the $\pi^+\pi^-$ and $\pi^0\pi^0$ data using the integral asymmetry, $I = \int_0^T A(t)dt$, where $T \gg \tau_s$. Assuming CPT, then $I_{+-} = 2\text{Re}(\varepsilon+2\varepsilon')$ and $I_{00} = 2\text{Re}(\varepsilon-4\varepsilon')$, so that $\varepsilon'/\varepsilon = (I_{+-}/I_{00} -1)/6$. Background is small (fig. 5) and systematic effects cancel to a very high order. Moreover, systematics can be checked directly using the pseudo-trigger, where the measured value of ε'/ε must be zero (strong interaction).

CP violation in the K^0 decay is measurable through the interference at short times in the 3π decay modes. Fig. 6 shows the expected asymmetry, U_{+-0}, on the assumption that it equals η_{+-}. The study of both 3π decay modes is unique to this experiment.

Measurement of the $\pi^\pm \mp \nu$ decay mode allows us to measure T violation in the mass matrix by detailed balance, to measure the K^0 and $\bar{K}^0$ mass difference with improved accuracy, to test the $\Delta S = \Delta Q$ rule with increased precision and to investigate CP violation in the decay amplitude. Fig. 7 shows the time-dependent asymmetry parameter relevant to the last two points, indicating the clear improvement possible in this experiment.

SEARCH for T and CPT VIOLATION in the NEUTRAL KAON SYSTEM

B. Lee Roberts, Department of Physics, Boston University, Boston, MA 02215, On behalf of the The University of Birmingham; Boston University; University of Coimbra; Technische Hogeschool, Delft; University and Jozef Stefan Institute, Ljubljana; University of New Mexico; University of Oxford; College of William and Mary; Institut für Mittelenergiephysik, E.T.H., Zürich; and Phys. Institut der Universität, Zürich Collaboration, N.W. Tanner - Spokesman - CERN PSSC Proposal P88.

Abstract

The LEAR cooled $\bar{p}$ beam will be used to produce tagged $K^0, \bar{K}^0$ in equal numbers through the reactions: $\bar{p} + p \to K^0 + K^- + \pi^+$ or $\bar{p} + p \to \bar{K}^0 + K^+ + \pi^-$ (total branching ratio 0.42%). The sign of the charged kaon determines whether a K^0 or a $\bar{K}^0$ was produced. Since the physical states of the neutral kaon system are the K_L, K_S there will be transitions between K^0 and $\bar{K}^0$ and vice versa. Eventually the neutral kaon will decay. By measuring the charge of electron in the K_{e3} decays: $\bar{K}^0 \to e^- + \pi^+ + \bar{\nu}$; $K^0 \to e^+ + \pi^- + \nu$, one can determine whether the particle which decayed was K^0 or $\bar{K}^0$. By studying the time evolution of the $K_{\ell 3}$ decays, where the initial state at $t = 0$ is pure K^0 or $\bar{K}^0$, it is possible to measure the CP violation parameter ε, the CPT violation parameters $\Delta \lambda_0$, y_ℓ and the $\Delta S = -\Delta Q$ parameters x_ℓ, $\bar{x}_\ell$ (which can also reflect CPT violation). This experiment would be the first experiment to search directly for T violation in the neutral kaon system where CP has been shown to be violated.

The neutral kaon system has been a rich source of theoretical and experimental discoveries in the past 30 years. The unusual properties of the $(K^0, \bar{K}^0)$ system were first pointed out by Gell-Mann and Pais[1]. At that time, two physical states, $|K_1^0>, |K_2^0>$ were believed to exist which had the properties under CP, $CP|K_1^0>= +1|K_1^0>$ and $CP|K_2^0>= -1|K_2^0>$, which allowed K_1^0 to decay into two pions, but K_2^0 could only decay into three pions. The discovery[2] that the long-lived K_2^0 decayed into two pions with a branching ratio of $\sim 2 \times 10^{-3}$, implied a CP impurity in the physical eigenstates so that the pure CP eigenstates, $|K_1^0>$ and $|K_2^0>$, do not represent the physical eigenstates.[3,4]

The physical eigenstates, $|K_S>, |K_L>$ can be defined most generally as[4,5]

$$|K_S> = \frac{1}{\sqrt{2}} \left\{ (1 + \varepsilon + \Delta) |K^0> + (1 - \varepsilon - \Delta) |\bar{K}^0> \right\} \qquad (1a)$$

and

$$|K_L> = \frac{1}{\sqrt{2}} \left\{ (1 + \varepsilon - \Delta) |K^0> - (1 - \varepsilon + \Delta) |\bar{K}^0> \right\} \qquad (1b)$$

where the states $|K^0>$ and $|\bar{K}^0>$ are orthogonal eigenstates of H_{strong} and H_{em}, which are assumed to be invariant under CP, T, and CPT. The (complex) quantities ε and Δ are small parameters which represent the breakdown of CP invariance.[4,5] Their significance is:

CP invariance requires $\varepsilon = \Delta = 0$
CPT invariance requires $\Delta = 0$
T invariance requires $\varepsilon = 0$

The neutral kaon system is the only place where we have discovered CP violation. Unlike P and C which are violated at the maximum level in the weak interaction this CP violation is minimal. Everything we know currently about CP violation can be summarized by the small number[6] $|\varepsilon| = (2.27 \pm 0.02) \times 10^{-3}$. It is expected that the observed CP violation is accompanied by a compensating T violation such that CPT invariance still holds. Since T violation has never been observed directly is is of interest to try to measure it in the K^0 system. While it is expected from field theories that CPT is a good symmetry, with the introduction of superstring theories additional motivation has been given to this experiment. It is not clear whether such theories require CPT invariance.[7,8]

In terms of the mass matrix, the parameters ε and Δ are given by[4,5]

$$\varepsilon = \frac{(-2\Im M_{12} + i\Im\Gamma_{12})}{2(\gamma_S - \gamma_L)} \quad \text{and} \quad \Delta = \frac{i(M_{11} - M_{22}) + \frac{1}{2}(\Gamma_{11} - \Gamma_{22})}{2(\gamma_S - \gamma_L)} \quad (2)$$

The indices $(1,2)$ indicate $(K^0, \bar{K}^0)$ and $\gamma_{S,L} = iM_{S,L} + \frac{1}{2}\Gamma_{S,L}$ are the eigenvalues of the matrix $iM + \frac{1}{2}\Gamma$. One notes that the CPT violating parameter Δ is proportional to both the $K^0, \bar{K}^0$ mass difference and lifetime difference. Higher order terms in ε and Δ will be neglected in this discussion.

The isospin amplitudes for $\pi\pi$ decay are defined by

$$A(K^0 \to (\pi\pi)_I) = A_I e^{i\delta_I}; \quad A(\bar{K}^0 \to (\pi\pi)_I) = \bar{A}_I e^{i\delta_I} \quad (3)$$

where δ_I is the s-wave $\pi\pi$ scattering phase shift for isospin I at the energy of the K mass. In the limit of CPT invariance one can choose $\bar{A}_0 = A_0 =$ real and $\bar{A}_2 = A_2^*$. If CPT invariance does not hold, the phase can be chosen such that the ratio of the isospin zero amplitudes, $\bar{A}_0/A_0 =$ real, even in the presence of CPT violation in the $K \to \pi\pi$ amplitudes.[5]

One can define a (real) number λ_0 which represents a measure of the CPT-violation in the isospin $= 0$ decay amplitude to $\pi\pi$, and a generalized epsilon parameter ε_0 by

$$\lambda_0 = \left(\frac{\bar{A}_0 - A_0}{\bar{A}_0 + A_0}\right) \quad \text{and} \quad \varepsilon_0 = \varepsilon - \Delta - \lambda_0 \quad (4)$$

The parameter λ_0 is real because of the phase convention, but the corresponding quantity for $I = 2$, λ_2, is complex. In the limit of CPT invariance ε_0 is equal to ε. It is important to note that in much of the literature ε is used instead of ε_0 under the assumption that CPT is a good symmetry.

Two additional parameters, η_{+-} and η_{00} which describe the amplitudes for CP violation are given by

$$\eta_{+-} = \frac{<\pi^+\pi^-|T|K_L>}{<\pi^+\pi^-|T|K_S>} \quad \text{and} \quad \eta_{00} = \frac{<\pi^0\pi^0|T|K_L>}{<\pi^0\pi^0|T|K_S>} \quad (5)$$

The parameters η_{+-}, η_{00} are related to the ε parameters through

$$\eta_{+-} = \varepsilon_0 + \varepsilon' \quad \text{and} \quad \eta_{00} = \varepsilon_0 - 2\varepsilon' \quad (6)$$

where ε' is a measure of the amplitude for a CP violating transition to the $I = 2$ $(\pi\pi)$ final state. Eliminating ε' from Eq. 6 and using the definition of ε_0 yields

$$\frac{1}{3}(2\eta_{+-} + \eta_{00}) = \varepsilon - \Delta - \lambda_0 \quad (7)$$

Since ε' is rather small compared with ε_0, we can neglect its contribution in Eq. (6) and within this approximation η_{+-} and η_{00} should be equal. Using the definition $\eta = |\eta|e^{i\phi}$ the experimental values[9] are $|\eta_{+-}| = (2.274 \pm 0.022) \times 10^{-3}$, $\phi_{+-} = (44.6 \pm 1.2)°$; and $|\eta_{00}| = (2.33 \pm 0.08) \times 10^{-3}$, $\phi_{00} = (54 \pm 5)°$. The phase angles differ by almost two standard deviations which is the largest indication of a deviation from CPT invariance in the $K^0\bar{K}^0$ system.[10]

The central focus of the proposed experiment is measurement of the semileptonic decays. The four possible decays and corresponding amplitudes are

$$K^0 \to \ell^+ + \pi^- + \nu_\ell \qquad f_\ell \qquad \bar{K}^0 \to \ell^- + \pi^+ + \bar{\nu}_\ell \qquad \bar{f}_\ell \qquad (8a)$$

$$K^0 \not\to \ell^- + \pi^+ + \bar{\nu}_\ell \qquad x_\ell f_\ell \qquad \bar{K}^0 \not\to \ell^+ + \pi^- + \nu_\ell \qquad \bar{x}_\ell \bar{f}_\ell \qquad (8b)$$

where the decays indicated by $\not\to$ are forbidden by the $\Delta S = \Delta Q$ rule. The quantities x_ℓ and $\bar{x}_\ell$ are measures of the $\Delta S = -\Delta Q$ decay amplitudes, and the amplitudes f_ℓ ($\bar{f}_\ell$) can be written as $f_\ell = F_\ell(1 - y_\ell)$, and $\bar{f}_\ell = F_\ell(1 + y_\ell)$. The parameter y_ℓ is a CPT violating fractional difference in the $\Delta S = \Delta Q$ amplitudes.

In these expressions CPT invariance has not been assumed. The $\Delta S = \Delta Q$ rule is expected to hold to order 10^{-14} in the standard model, but it need not be true for the purpose of this experiment, since the experiment is quite sensitive to the level of violation of this rule. With CPT invariance and no Coulomb interaction in the final state one has $\bar{f}_\ell = f_\ell$ and $\bar{x}_\ell = x_\ell$

It is useful to define two ratios which can be measured experimentally. (Two additional ratios, γ_ℓ and α_ℓ, are defined by Tanner and Dalitz.[5]) The notation is as before, viz. that a bar implies an initial $\bar{K}^0$, the sign $\pm$ indicates $\ell^\pm$, and $R_{\ell\pm}(t)$ and $\bar{R}_{\ell\pm}(t)$ are the transition rates for the semi-leptonic decays along the particle trajectory as a function of proper time.

The lepton decay asymmetry δ_ℓ is defined as

$$\delta_\ell(\infty) = \frac{\Gamma_L(\ell^+) - \Gamma_L(\ell^-)}{\Gamma_L(\ell^+) + \Gamma_L(\ell^-)} = \frac{R_{\ell+}(\infty) - R_{\ell-}(\infty)}{R_{\ell+}(\infty) + R_{\ell-}(\infty)} = \frac{\bar{R}_{\ell+}(\infty) - \bar{R}_{\ell-}(\infty)}{\bar{R}_{\ell+}(\infty) + \bar{R}_{\ell-}(\infty)} \qquad (9)$$

where $\Gamma_L(\ell^\pm)$ is the total K_L semi-leptonic decay rate for $K_L \to \pi^- e^\pm \nu(\bar{\nu})$. The lepton asymmetry δ_ℓ is measured to be[9] $(3.30 \pm 0.12) \times 10^{-3}$. This asymmetry does not depend on whether the initial state was K^0 or $\bar{K}^0$ at $t = 0$ since it is characteristic of the K_L and not how it was formed. The infinite value of t implies $t \gg \frac{1}{\Gamma_S}$ To lowest order $\delta_\ell(\infty)$ is given by

$$\delta_\ell(\infty) = 2\Re(\varepsilon - \Delta) - \Re(2y_\ell + x_\ell - \bar{x}_\ell). \qquad (10)$$

If both CPT invariance and $\Delta S = \Delta Q$ hold in the $K_{\ell 3}$ decay, then the second term on the rhs of Eq. (10) is zero. The imaginary part of $\varepsilon - \Delta$ is given by the imaginary part of Eq. (7).

Of particular interest is β_ℓ which is defined as

$$\beta_\ell(t) = \frac{\bar{R}_{\ell+}(t) - R_{\ell-}(t)}{\bar{R}_{\ell+}(t) + R_{\ell-}(t)} \qquad \beta_\ell(\infty) = 4\Re\varepsilon - \Re(2y_\ell + x_\ell - \bar{x}_\ell). \qquad (11)$$

The time dependence of $\beta_\ell(t)$ is given by[5]

$$\beta_\ell(t) = 4\Re\varepsilon - 2\Re y_\ell - \Re(x_\ell - \bar{x}_\ell)\left(\frac{sinh(t\bar{\Gamma})}{cosh(t\bar{\Gamma}) - cos(t\delta M)}\right) + \Im(x_\ell + \bar{x}_\ell)\left(\frac{sin(t\delta M)}{cosh(t\bar{\Gamma}) - cos(t\delta M)}\right)$$

where δM is the $K_S - K_L$ mass difference and $\bar{\Gamma}$ is the average of the K_S, K_L widths. From this expression one can see that in the absence of $\Delta S = -\Delta Q$ transitions and if CPT holds ($x = \bar{x}$, $y_\ell = 0$) then β_ℓ will be independent of time and given by

$\beta_\ell(\infty) = 4\Re\varepsilon \approx 0.65\%$. In Fig. 1 the time dependence is shown for a value of $\Im x_\ell$ which is an order of magnitude smaller than the current limit.

The tagged production determines whether one has K^0 or $\bar{K}^0$ in the initial state at $t = 0$. If the $\Delta S = \Delta Q$ rule holds, x_ℓ and $\bar{x}_\ell$ are zero and and a decaying K^0 is identified by an ℓ^+ and a $\bar{K}^0$ by an ℓ^-. In this case the ratio $\beta_\ell(t)$ is related to the difference between the two rates $K^0 \to \bar{K}^0$ and $\bar{K}^0 \to K^0$. This T invariance test stems from a suggestion by Kabir,[11] that a non-zero value for β_ℓ implies a deviation from reciprocity, and thus represents a violation of T invariance. This test was impossible to realize experimentally until a tagged source of K^0, $\bar{K}^0$ became available, since the $K_{\ell 3}$ branching ratios quoted in the literature[12] are <u>not</u> normalized to the production rate, but rather the $K_L \to \pi \ell \nu$ rates are normalized to the $K_S \to \pi^+\pi^-$ rates in the same experiment. Absolute branching ratios for the $K_{\ell 3}$ decays are not known. If CPT or $\Delta S = \Delta Q$ were to be violated in the $K_{\ell 3}$ decays, this violation could contribute to the lepton asymmetry $\beta_\ell(t)$ and Kabir's argument for T violation breaks down.

The proposed experiment employs a liquid hydrogen target, at the center of a 2 m diameter solenoid, where the LEAR antiproton beam is stopped in $\sim 1 mm^3$. Decay regions are defined by cylindrical chambers, the K_S decay region covering out to a radius of $2 < r < 10$ cm and the K_L decay region covering $10 < r < 50$ cm. Only 12% of the K_L decay in this region. A typical semileptonic decay event is shown in Fig. 2.

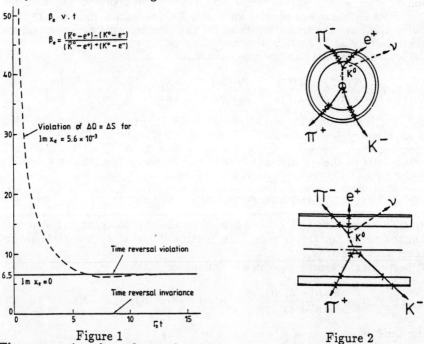

Figure 1
The proper time dependence of $\beta_\ell(t)$

Figure 2
A possible event topology

In addition to the semi-leptonic decays, the π^+, π^- decays will be measured to determine the ratio

$$D_{\pi 2} = \frac{\int_0^{t_{max}} R_{\pi 2} dt - \int_0^{t_{max}} \bar{R}_{\pi 2} dt}{\int_0^{t_{max}} R_{\pi 2} dt + \int_0^{t_{max}} \bar{R}_{\pi 2} dt} = 2\Re(\varepsilon - \delta) + f(\eta_{+-})$$

where $f(\eta_{+-})$ is an interference term. The experimental sensitivity to the various parameters is given in Table I.

Table I

Parameters which can be measured in the proposed experiment assuming the geometry mentioned above and 10^{12} stopped antiprotons. The symmetry which is violated for a non-zero value of the parameter is listed with the parameter. The theory column gives the value predicted if CPT and $\Delta S = \Delta Q$ are not violated. Note that in listing the current value, no assumption of CPT invariance is made. An exponent of 10^{-4} has been suppressed for all numbers in the table.

Parameter	Experimental Source	Statistical Error	Theory	Present Value
$\Re\varepsilon$ - CP,T	$\gamma_\ell, D_{\pi 2}, \eta_{+-}$	0.5	16	< 400
$\Im\varepsilon$ - CP,T	$K_{e3}, \delta_\ell, \eta_{+-}$	6	16	–
$\Re\Delta$ - CP,CPT	$\gamma_\ell, D_{\pi 2}, \eta_{+-}$	0.5	0	< 300
$\Im\Delta$ - CP,CPT	K_{e3}, δ_ℓ	6	0	–
$\frac{M-\bar{M}}{M_L-M_S}$ - CP,CPT	$\Re$ and $\Im\Delta$	6	0	–
$\frac{\Gamma-\bar{\Gamma}}{\Gamma_L-\Gamma_S}$ - CP,CPT	$\Re$ and $\Im\Delta$	1†	0	< 300†
$\Re y_\ell$ - CP,CPT	$D_{\pi 2}, \eta_{+-}, \delta_\ell$	2	0	< 300
$\beta_\ell(\infty)$ - T	$K_{e3}, D_{\pi 2}, \eta_{+-}, \delta_\ell$	4	65	–
$\frac{1}{2}\Im(x_\ell + \bar{x}_\ell)$ - $\Delta S, \Delta Q$	$K_{e3}, \gamma_\ell, \delta_\ell$	8	0	-40 ± 260
$\frac{\Gamma^+-\Gamma^-}{\Gamma^++\Gamma^-}$ - CP, CPT	K^+, K^-	1	0	< 15

†Assuming unitarity.

The time evolution of the ratios presented above for the semi-leptonic and $\pi\pi$ decays determines all the parameters except ε'. In principle, since ε' is a second order term, if higher orders in Δ and ε were retained in the calculation the sensitivity could be seen. However, the statistics needed to realize this sensivity are prohibitive.

References

1. M. Gell-Mann and A. Pais, Phys. Rev. **97**, 1387 (1955).

2. J.H. Christenson, J.W. Cronin, V.L. Fitch and R. Turlay, Phys. Rev. Lett. **13**, 138 (1964).

3. V.L. Fitch, Rev. Mod. Phys. **53**, 367 (1981).

4. J.W. Cronin, Rev. Mod. Phys. **53**, 373 (1981). As pointed out in Ref. 5, the signs attached to $\Im x$ and $\Im \bar{x}$ are wrong, but there are no significant consequences.

5. N.W. Tanner and R.H. Dalitz to be published.

6. L. Wolfenstein, to be published in Ann. Rev. Nucl. and Part. Sci.

7. E. Witten, private communication.

8. S.L. Glashow, private communication.

9. Particle Data Group, Rev. Mod. Phys. **56** S1 (1984).

10. V.V. Barmin et al., Nucl. Phys. **B247**, 293 (1984).

11. P.K. Kabir, Phys. Rev. **D2**, 540 (1970).

12. Y. Cho et al., Phys. Rev. **D1**, 3031 (1970), G. Burgun et al., Nucl. Phys. **B50**, 194 (1972), B.R. Webber et al., Phys. Rev. **D3**, 64 (1971) and W.A. Mann et al., Phys. Rev. **D6**, 137 (1972).

CP VIOLATION: PAST AND PRESENT*

Barry R. Holstein
University of Massachusetts at Amherst 01003

ABSTRACT

After over twenty years of work, there still exist at least four classes of viable models of CP violation. Fortunately present and near future experiments appear capable of narrowing the field.

DISCUSSION

Over two decades have past since the discovery of CP violation[1] and during this period we have accumulated increasingly precise data on CP nonconserving effects in the K_L-K_S system

$$\frac{\langle \pi^+\pi^-|H_w|K_L\rangle}{\langle \pi^+\pi^-|H_w|K_S\rangle} = \eta_{+-} = \varepsilon + \varepsilon'$$

$$\frac{\Gamma(K_L \to \ell^+\nu\pi^-) - \Gamma(K_L \to \ell^-\bar{\nu}\pi^+)}{\Gamma(K_L \to \ell^+\nu\pi^-) + \Gamma(K_L \to \ell^-\bar{\nu}\pi^+)} = \delta = 2\mathrm{Re}\,\varepsilon \quad (1)$$

$$\frac{\langle \pi^0\pi^0|H_w|K_L\rangle}{\langle \pi^0\pi^0|H_w|K_S\rangle} = \eta_{00} = \varepsilon - 2\varepsilon'$$

Using the experimental results[2]

$$|\eta_{+-}| = (2.27 \pm 0.03) \times 10^{-3}$$

$$\phi_{+-} = 44.6 \pm 1.2° \quad (2)$$

$$\delta = (3.30 \pm 0.12) \times 10^{-3}$$

and[3]

$$\left|\frac{\eta_{00}}{\eta_{+-}}\right|^2 = 1.028 \pm 0.032 \pm 0.014$$

and (3)

$$0.990 \pm 0.050$$

we find

---

*Research supported in part by the National Science Foundation

$$|\varepsilon| = 2.3 \times 10^{-3}$$

$$\frac{\varepsilon'}{\varepsilon} = (-4.6 \pm 5.3 \pm 2.4) \times 10^{-3} \tag{4}$$

and

$$(+1.6 \pm 8.0) \times 10^{-3}$$

In analyzing these results we shall need the expressions derived on the basis of CPT invariance[4]

$$\varepsilon = \sqrt{\frac{1}{2}} e^{i\theta} (\frac{m'}{\Delta m} + \frac{ImA_0}{ReA_0}) \qquad \theta = \tan^{-1}\frac{2 \Delta m}{\Gamma_s} = 44°$$

$$\varepsilon' = \sqrt{\frac{1}{2}} e^{i\theta'} \omega(\frac{ImA_2}{ReA_2} - \frac{ImA_0}{ReA_0}) \qquad \theta' = \delta_2 - \delta_0 + \frac{\pi}{2} = 48 \pm 8° \tag{5}$$

Here

$$\langle(\pi\pi)_I | H_w | K^0\rangle = A_I e^{i\delta_I} \qquad I = 0,2 \tag{6}$$

and ω is the small parameter which gives the strength of the $\Delta I = \frac{3}{2}$ transition

$$\omega = \frac{ReA_2}{ReA_0} = 0.045 \tag{7}$$

while m', Δm are the Imaginary, Real components of the K^0-$\bar{K}^0$ mixing matrix element.

Of course, theorists have not been idle during this time and at this point, although some models have been ruled out, there still exist four classes of model which are viable:
1) Superweak[5] - in which is postulated the existence of a tiny ($\sim 10^{-9}$) $\Delta S=2$ interaction which mixes the K^0 and $\bar{K}^0$ to lowest order in perturbation theory as opposed to the usual weak interaction which contributes to K^0-$\bar{K}^0$ only in second order.
2) KM Model[6] - in which the phase parameter δ in the usual representation of the KM matrix is nonvanishing. However, even if δ is sizable any weak effect must arise proportional to $s_1 s_2 s_3 s_\delta \ll 1$.
3) Higgs Model[7] - in which one has at least three Higgs doublets arranged so that neutral Higgs couplings are flavor conserving. In such a model CP violation can arise from charged Higgs exchange between quark fields.
4) Left-Right Models[8] - based on the gauge group $SU(2)_L \times SU(2)_R \times U(1)$ and in which the weak interaction is produced both by exchange of left-handed gauge bosons $W_L^\pm$ and their right-handed counterparts $W_R^\pm$. These latter couple to the quark fields with possibly complex couplings which give rise to CP violation.

Note that only the simple KM model does not involve some sort of new physics. Thus confirmation of one of these of other possibilities or showing that the KM picture is invalid is a very exciting prospect!

In addition there have been generated a large number of null re-

sults in other systems,
  i) strong interactions, e.g.[9]

$$\left|\frac{Amp(CP=-)}{Amp(CP=+)}\right| < 5 \times 10^{-4}$$

ii) electromagnetic decays, e.g.[10]

$$^{131}Xe^* \rightarrow {}^{131}Xe + \gamma \qquad \sin\eta = (1.1 \pm 1.1) \times 10^{-3} \qquad (9)$$

iii) semileptonic weak transitions, e.g.[11]

$$^{19}Ne \rightarrow {}^{19}F + e^+ + \nu_e \qquad D = (0.7 \pm 6) \times 10^{-4} \qquad (10)$$

However, these vanishing effects are all anticipated on the basis of models i,...,iv (although some T violation in semileptonic weak transitions could be permitted in the left-right models). A measurement which <u>will</u> prove to be significant in probing such models in the upper bound on the neutron electric dipole moment[12]

$$|d_n| \lesssim 10^{-25} \text{ e cm} \qquad (11)$$

Although, as stated above, <u>any</u> of these models are consistent with present experimental data, two of the models - KM and Higgs - are currently being somewhat challenged. In the case of the Higgs model, early calculations indicated that[13]

$$\frac{ImA_0}{ReA_0} \gg \frac{m'}{\Delta m} \qquad (12)$$

so that

$$\left|\frac{\varepsilon'}{\varepsilon}\right|_{Higgs} \simeq \omega \qquad (13)$$

and

$$d_n \sim 10^{-24} \text{ e cm} \qquad (14)$$

However, recently it was pointed out that chiral properties of the $K \rightarrow 2\pi$ Lagrangian were not properly imposed in the earlier calculation and that a more reliable estimate is[15]

$$\left|\frac{\varepsilon'}{\varepsilon}\right|_{Higgs} \sim 0.005 \qquad (15)$$

Also a valence quark calculation of the neutron EDM in the Higgs model has given[16]

$$d_n^{Higgs} \gtrsim 3 \times 10^{-26} \text{ e cm} \qquad (16)$$

although the influence of heavy quark intermediate states is not included here. Thus the Higgs model predictions are well within the capability of ongoing experiments which seek to measure ε'/ε to be a part in 10^3 and d_n to the 10^{-27} e cm level.

In the case of the KM model, any CP violating amplitude must be proportional to

$$\sigma = s_1 s_2 s_3 s_\delta \qquad (17)$$

Thus experiments in the b-quark system which provide upper bounds on s_2, s_3 give[17]

$$|\sigma| < 3 \times 10^{-4} \tag{18}$$

Neglecting long-distance contributions and keeping only the box diagram we predict

$$\varepsilon = \exp(i\tfrac{\pi}{4}) \frac{\sigma}{s_1} BF(m_t, s_2(s_2 + s_3 c_\delta)) \tag{19}$$

where F is a known function of the indicated quantities and

$$B = \frac{\langle \bar{K}^0 | \bar{s}\gamma_\mu (1+\gamma_5) d \; \bar{s}\gamma^\mu (1+\gamma_5) d | K^0 \rangle}{\tfrac{16}{3} F_K^2 m_K^2} \tag{20}$$

is the so-called B parameter, which is calculated from $K \to 2\pi$ decay data using PCAC and SU(3) to be[18]

$$B \simeq 0.33 \tag{21}$$

If m_t is in the vicinity of 40 GeV then the KM model is threatened by values $\lesssim 0.5$ and could be nearly ruled out for values $B \sim 1/3$. Although chiral loop studies by Bijnens et al. suggest that B may well be significantly larger than this[19] - say $B \sim 0.66$ - other calculations of chiral loop effects in semileptonic hyperon decay[20] may signify that this is a considerable overestimate. Thus while the KM model is possibly challenged by the size of ε, current theoretical technology does not permit a definitive conclusion to be drawn. Ongoing experiments on the neutron EDM should not see an effect here since theoretical estimates yield $d_n \lesssim 10^{-30}$ e cm.[21] On the other hand the KM model yields a lower bound[22]

$$\frac{\varepsilon'}{\varepsilon} \gtrsim 3 \times 10^{-3} \tag{22}$$

so that experiments to a part in 10^3 should find a nonzero result.

In the case of the superweak model the prediction for the two ongoing experiments is

$$\frac{\varepsilon'}{\varepsilon} = d_n = 0 \tag{23}$$

so that obtaining an effect in either experiment would signal the death-knell of this model.

In the case of the left-right symmetric model there exist a large number of additional parameters - three mixing angles and six phases - so that there is no real predictive power. Thus nearly any result for ε'/ε and d_n can be comfortably accommodated. Detailed predictions may be found in ref. 23.

Fortunately the ε'/ε and d_n measurements are not the only ones which should soon bear fruit. Experiments are also underway to look at the muon polarization in $K_L \to \mu^+\mu^-$ and to measure the three-pion analogue of η_{+-}.

$$\frac{\langle \pi^+\pi^-\pi^0|H_w|K_S\rangle}{\langle \pi^+\pi^-\pi^0|H_w|K_L\rangle} = \eta_{+-0} \tag{24}$$

I do not have time to explore these modes in detail, but I do wish to present one piece of encouraging new to those endeavoring to look at η_{+-0}. If the KM model is the cause of CP violation then current algebra-PCAC studies have been able to relate CP violation in the $K \to 3\pi$ channel to that seen in $K \to 2\pi$ 24

$$\frac{1}{\varepsilon}|\eta_{+-0} - \varepsilon| = 2\varepsilon'/\varepsilon \gtrsim 6 \times 10^{-4} \tag{25}$$

However, we feel that this estimate is unduly pessimistic. In fact, if one includes next order chiral terms, one can have

$$\frac{1}{\varepsilon}|\eta_{+-0} - \varepsilon| \sim 2 \frac{m_K^2}{\Lambda^2} \frac{\varepsilon'}{\omega\varepsilon} \gtrsim 3 \times 10^{-2} \tag{26}$$

where $\Lambda \sim 1$ GeV is the chiral scale parameter. This is still not a large effect. However, it is a bit more encouraging than the simple estimate in eqn. 25.

Thus we are at a critical point of time in the subject of CP violation in that these ongoing experiments are now at the level where they should be able to confirm or rule out at least some of these models.

REFERENCES

1. C. R. Christenson et al., Phys. Rev. Letters 13, 138 (1964).
2. Particle Data Group, Rev. Mod. Phys. 56, 51 (1984).
3. R. H. Bernstein et al., Phys. Rev. Letters 54, 1631 (1985); J. K. Black et al., Phys. Rev. Letters 54, 1628 (1985).
4. L. Wolfenstein in Theory and Phenomenology in Particle Physics, A. Zichichi, ed. (Academic Press, New York, 1969), p. 218.
5. L. Wolfenstein, Phys. Rev. Letters 13, 180 (1964).
6. M. Kobayashi and M. Taskawa, Prog. Theo. Phys. 49, 652 (1973).
7. T. D. Lee, Phys. Rev. D8, 1226 (1973) and Phys. Rept. 96, 143 (1979); S. Weinberg, Phys. Rev. Letters 37, 657 (1976).
8. R. N. Mohapatra and J. C. Pati, Phys. Rev. D11, 566 (1975); P. Herczeg, Phys. Rev. D28, 200 (1983); D. Chang, Nucl. Phys. B214, 435 (1983).
9. E. Blanke et al., Phys. Rev. Letters 51, 355 (1983).
10. J. L. Gimlett et al., Phys. Rev. C25, 1567 (1982).
11. A. Hallin et al., Phys. Rev. Letters 52, 337 (1984).
12. I. S. Altarev et al., Phys. Letters 102B, 13 (1981); J. M. Pendlebury et al., Phys. Letters 136B, 327 (1984).
13. A. Sanda, Phys. Rev. D23, 2647 (1981); N. G. Deshpande, Phys. Rev. D23, 2654 (1981); J. F. Donoghue, J. S. Hagelin and B. R. Holstein, Phys. Rev. D25, 195 (1982).
14. S. Weinberg, ref. 7.
15. J. F. Donoghue and B. R. Holstein, Phys. Rev. D32, 1152 (1985).
16. G. Beall and N. G. Deshpande, Phys. Letters 132B, 427 (1983).

17. This value will rise somewhat if the b→u/b→c ratio rises above 5%.
18. J. F. Donoghue, E. Golowich and B. R. Holstein, Phys. Letters 119B, 412 (1982).
19. J. Bijnens et al., Phys. Rev. Letters 53, 2367 (1984).
20. J. F. Donoghue and B. R. Holstein, Phys. Letters 160B, 173 (1985).
21. N. G. Deshpande, G. Eilam and W. L. Spence, Phys. Letters 108B, 42 (1982); D. Nanopoulos, A. Yildiz and P. Cox, Phys. Letters 87B, 53 (1979) and Ann. Phys. (N.Y.) 127, 126 (1980); E. Golowich and B. R. Holstein, Phys. Rev. D26, 182 (1982).
22. The strict value depends upon the top quark mass. The value given here is for $m_t \sim 40$ GeV. F. J. Gilman and J. S. Hagelin, Phys. Letters 126B, 111 (1983).
23. P. Herczeg, ref. 8.
24. L.-F. Li and L. Wolfenstein, Phys. Rev. D21, 178 (1980).

A PROGRESS REPORT ON THE MEASUREMENT OF ε'/ε AT FERMILAB

G. J. Bock
Fermilab, Batavia, IL 60510

P. Cabeza, R. Coleman, R. Daudin, P. Debu, M.Dung, G.Gollin, G.Grazer,
Y.Hsiung, P. Jarry, K.Nishikawa, J.Okamitsu, R. Patterson, B. Peyaud,
K. Stanfield, R. Stefanski, E. Swallow, R. Turlay, Y. Wah, M. Woods,
B. Winstein, R. Winston, T. Yamanaka

ABSTRACT

An ongoing experiment at Fermilab intends to measure the ratio of the CP violation parameters ε'/ε to a precision of .001. The experimental technique and apparatus are described. Sources of systematic error are discussed and some preliminary data from the test run in 1985 is shown. The status of the analysis is described and plans for the upcoming final data run in March 1987 are given.

INTRODUCTION

Fermilab E731 a collaboration of physicists from five laboratories and universities (Chicago, Elmhurst, Fermilab, Princeton, and Saclay) is an experiment intended to study CP violation through observations of the decays of neutral K mesons to two pions. More specifically the experiment addresses the issue of whether or not CP violation is confined to $\Delta S = 2$ interactions.

Neutral kaon phenomenology has been discussed by several speakers at this conference.[1] Only the highlights will be reviewed here. Recall that the physical neutral K states ($K_{L,S}$) are almost, but not quite, the CP eigenstates (K_1[CP+1], K_2[CP-1]). Specifically the K_L is: $K_L \sim K_2 + \varepsilon K_1$ where $\varepsilon \approx 0.002$. The CP mixture of the K^0 states results from a small T violation in the $\Delta S = 2$ process $K^0 <\!-\!-\!> K^0$ allowed by second order weak interactions. To date all observations of CP violation can be accounted for by the single parameter ε describing this imbalance. This is the prediction of the superweak model. There is also a parameter describing $\Delta S = 1$, or "milliweak" CP violation, ε', which is predicted to be small but non-zero in many models of the electrweak interactions. The usual definitions :

$$\eta_+ = A(K_L \to \pi^+\pi^-)/A(K_S \to \pi^+\pi^-),$$

$$\eta_0 = A(K_L \to \pi^0\pi^0)/A(K_S \to \pi^0\pi^0),$$

expressed in terms of ε and ε' are:

$$\eta_{+-} = \varepsilon + \varepsilon', \text{ and } \eta_{oo} = \varepsilon - 2\varepsilon'$$

Since the phases of ε and ε' are nearly equal[1], we see that a test of the equality of $\eta_{+-} = \eta_{oo} = \varepsilon$ is a test for $\Delta S = 1$ CP violation and we can write:

$$\varepsilon'/\varepsilon \approx 1/3\, (1 - \eta_{+-}/\eta_{oo}) \qquad (1)$$

A number of calculations predict $\varepsilon'/\varepsilon > .002$ [1]. Motivated partially by the successes of the standard model and its natural incorporation of CP violation, two recently completed experiments[2,3] and two experiments underway [4,5] address the question of whether all the CP vioaltion is confined to this imbalance or whether there may be some "milliweak" CP violation in the decay itself.

EXPERIMENTAL TECHNIQUE

This experiment, like its predecesssor Fermilab E617 (Chicago/Saclay), makes use of the double beam technique: two distinct K_L beams are simultaneously incident on a spectrometer designed to detect two pion decays from the K_L. A thick boron carbide regenerator is located in one beam providing a source of K_S. Decays from this "regenerated beam are used to normalize the CP violating decays detected in the neighboring vacuum beam. There are two modes of data collection. In the charged mode the detector records $K^o \sim > \pi^+\pi^-$ decays, in the neutral mode, $K^o \sim > \pi^o\pi^o$. In each mode the decay region and regenerator are the same and data is taken simultaneously from both the regenerated and vacuum beams. The charged mode is run separately from the neutral mode, there are slight changes to the apparatus depending on the mode. Four integrated decay rates are measured in this experiment and combined to produce the following quantity:

$$\left|\eta_{+-}/\eta_{oo}\right|^2 = \{ (N^V_{+-}/N^r_{+-})/(N^V_{oo}/N^r_{oo})\} \equiv R_{+-}/R_{oo} \qquad (2)$$

where $N^{V(r)}_{+-}$ is simply the number of $\pi^+\pi^-$ events collected from the vacuum (regenerated) beam. The neutral mode data is defined similarly. Combining (1) and (2) we see that

$$\varepsilon'/\varepsilon = 1/6\, (1 - R_{+-}/R_{oo})$$

To determine ε'/ε to .001 (statistically) would require 75K events in each mode. In practice, of course, regenerated events are more easily obtained than vacuum events and charged data much easier than neutral, and these types of experiments are limited by the number of neutral, vacuum events.

SYSTEMATIC EFFECTS

Several systematic effects make the experiment more difficult than the above idealized description might imply. However some potentially troublesome systematic effects are eliminated in the double beam technique. Since both K_L and K_S decays are collected simultaneously in each mode, the measurement of R_{+-}/R_{00} does not directly depend on maintaining precise calibration over long periods of time. The most obvious potential problem might be the lead glass gains. However while a gain change of 1-2% would change the acceptance for neutral events to high order the <u>ratio</u> of acceptances will not change. The technique is similarly insensitive to beam rate (accidentals) inefficiencies.

There are several sources of systematic effects which do not cancel in this type of experiment. There is background under the mass peaks in both the charged and neutral mode data. This background results from semileptonic K decays in the charged mode data and $K\to 3\pi^0$ in the neutral mode sample. This correction is insignificant in the regenerated beam sample because of the large ratio of K_S/K_L behind the regenerator, but must be made to each of the vacuum beam samples. The regenerated beam samples require a different correction. The experiment is normalized to coherently (0 momentum transfer) regenerated events. There are two other processes which produce K_S in the boron carbide and their contributions to the sample must be subtractted. The largest of these two is the contribution from inelastically produced or regenerated events. Indeed this correction was the largest source of systematic error from E617. Although insignificant in E617, a small correction may be necessary to account for elastic or diffractive regeneration, in which a K_L is transformed into a K_S during an elastic scatter in the regenerator.

There are geometrical acceptance differences between the vacuum and regenerated beam events which must be understood. This will amount to about a 4% correction (averaged over all momentum) and will require a good understanding of the spectrometer by monte carlo calculation. This acceptance difference arises beacause of the different longitudinal decay distributions of the K_L and K_S decays. We rely on the known distribution of the $K_L \to \pi^+\pi^-$ and $\pi^0\pi^0$ decays to check our understanding. Additionaly we collected during actual data taking approximately 50 times more $K_L\to 3\pi^0$ than $K_L\to\pi^0\pi^0$. These events also have a known decay distribution and provide an invaluable check on the acceptance and performance of the spectrometer.

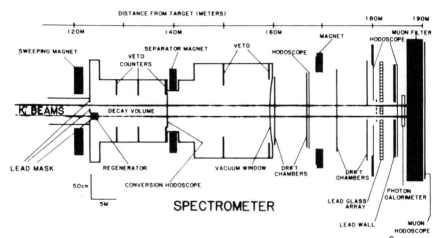

Fig. 1. The E731 spectrometer. The twin K^0 beams are shown entering the detector from the left. Distances shown are from the MC primary beam production target.

BEAMLINE AND DETECTOR

The result of E617 was ε'/ε = -0.0046 ± 0.0053 ±.0024 where the first error is statistical and the second, systematic. In order to achieve a precision of 0.001 it is clear that improvements had to be made in the systematics as well as statistics. A statistical limitation of the earlier experiment was the anomalously high flux of soft neutral (not kaons) particles in the beams which produced high singles rates throughout the spectrometer. For E731 a new long lived neutral beam has been constructed with special attention paid to clean collimation and sweeping which significantly reduces the soft beam halos found in E617. A new primary proton beam for the experiment allows for variable production targetting angles about a nominal 5 mr for optimizing the neutron/kaon content of the beams. The new beam also takes full advantage of the 800 GeV protons and high duty cycle of the Tevatron to provide a beam with five times more useable (100 GeV region) neutral kaon

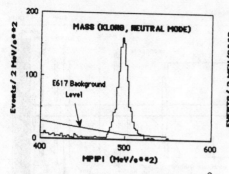
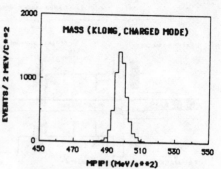

Fig. 2. Reconstructed K_L^0 invariant mass for the neutral (4 photon) mode. The background level from the 1982 experiment is superimposed.

Fig. 3 Reconstructed K_L^0 invariant mass for the charged mode.

fluxes per hour than ever before at Fermilab. A typical spill during the last run had 1E12 800 GeV protons per 20 second pulse each minute and about 3E6 K_L/beam.

The spectrometer shown in Figure 1 has greater rate capability, a factor of 6 more geometrical acceptance, and much better background suppression capabilities than the E617 spectrometer. The 800 element lead glass calorimeter used in E617,was restacked in a circular array which, together with an increase in the gap of the analysis magnet, and shortening the distance from the decay region to the lead glass, gave a factor of six increase in geometrical acceptance over E617. A new high rate 2000 wire drift chamber system (resolution better than 150 microns) was constructed to handle the higher fluxes and improve resolution needed for background supression. The conversion hodoscope consisted of a 1mm thick scintillator plane, and a 0.1 radiation length sheet of lead (for neutral mode running only), followed by another scintilator plane.The entire assembly was also in vacuum. The neutral trigger required the conversion of a single photon from the K decay. The electron pair is tracked back to the conversion plane to calculatet he kaon's transverse momentum. This is then used to determine whether ithe eventcame from the vacuum or regenerated beam. Occasionally, a neutron interaction in the lead sheet triggers the spectrometer. A π^0 from such an interaction can be deteced in the lead glass calorimeter and serves as an added source of energy calibration during the data taking. Background suppression was improved by including veto counters wtihin and without the vacuum chamber and a segmented, regenerator including hodoscopes for identifying the inelastic events in the K_S beam. The regenerated moves from one beam to the other on a pulse by pulse basis to cancel any small asynnetries in the apparatus .

1985 TEST RUN

The analysis of the 1985 data is now in progress. During the 1985 run approximately 1500 $K_L \to 2\pi^0$ were collected per (good) week. That is over four times better than the previous experiment even though there were significant deadtime losses. Approximately 8000 neutral mode and 35,000 charged mode K_L decays have been recorded along with about 24,000 neutral and 100,000 K_S decays. Additionaly we have almost 400,000 $K_L \to 3\pi^0$ decays taken along with the data for calibration purposes and acceptance checks.

Special momentum analyzed electron pair calibration data has also been taken during the run. Currently the lead glass resolution calculated from these electron pair data is 1.5% +4.5%/$\sqrt{E}$ to be compared with the 2% + 6% /$\sqrt{E}$ from the 1982 data.

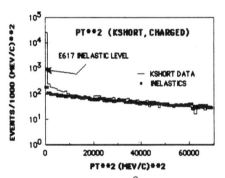

Fig. 4. The p_t^2 distribution for K_s data. The coherent peak is seen in the first bin. See text.

Figures 2 and 3 show the K_L neutral and charged mode mass distributions from a sample of the 1985 data. Significant improvement in the background of the neutral mode data has been made using the new photon veto counters. The successful suppression of the inelastic correction in the K_S data is shown in Figure 4. Also shown overlayed on Fig. 4 is the p_t^2 distribution for inelastic events identified by the regenerator hodoscopes. An order of magnitude improvement in the suppression of inelastic events uncovers some elastic (diffractive) regeneration.

FUTURE WORK

Currently the large task of careful data and monte carlo comparisons for the acceptance calculation is underway. There are enough events from the 1985 sample to provide a (statistical) measurement of ε'/ε to about ± 0.003. While that analysis is in progress, work is underway to ready the experiment for the final data run next March. The experiment will take data at the design intensity of 3E12 protons per pulse. Changes to the spectrometer will include a completely new Fastbus data aquisition sytem, and a hardware trigger cluster finder , and a new conversion hodoscope to handle the higher beam (and data) rates. We hope to collect our goal of 100,000 neutral K_L events during the next run, while maintaining our improved systematics.

REFERENCES

1. B. Holstein, D. Wilkinson, B.Winstein, Invited Talks at the "2nd Conference on the Intersections of Particle and Nuclear Physics", Lake Louise, Canada, \ May 1986. See also J. Cronin, Rev. Mod. Phys. 53,371(1981) and references therein for a complete discussion.
2. R. Bernstein, et al., Phys. Rev. Letters 54, 1631 (1985).
3. J.K. Black, et al., Phys. Rev. Letters 54, 1628 (1985).
4. G. Gollin, et al. Fermilab Proposal E731, (unpublished).
5. M.Calvetti, Invited Talk at the "2nd Conference on the Intersections of Particle and Nuclear Physics", Lake Louise, Canada, May 1986.

NEUTRINO PHYSICS

Coordinators: W.-Y. Lee
S. P. Rosen

DOUBLE BETA DECAY IN ^{82}Se, ^{128}Te AND ^{130}Te

M. K. Moe
University of California, Irvine, CA 92717

ABSTRACT

The results of a ^{82}Se double beta decay experiment in a time projection chamber are presented, together with a brief review of the geochemical measurements on ^{128}Te and ^{130}Te. The TPC experiment has shown the ^{82}Se half life to be $> 1.0 \times 10^{20}$ years at 68% confidence, consistent with geochemical data, but in conflict with nuclear shell model calculations. The tellurium controversy continues, with new geochemical measurements being proposed.

THE ^{82}Se EXPERIMENT

A double beta decay search with a time projection chamber (TPC) containing 14 grams of selenium enriched to 97% isotope 82 has been operating for 5426 hours in its present configuration.[1,2]

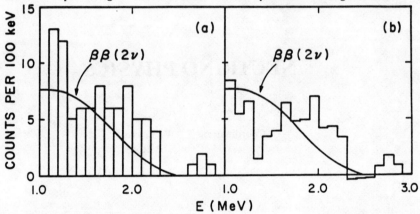

Fig. 1 (a) The electron sum energy spectrum of the 88 double beta decay candidates found in 5426 live hours. The theoretical $\beta\beta(2\nu)$ spectrum is normalized to the geochemical rate of Kirsten. (b) The same, with identified background subtracted.

Helices are fit to the electron tracks, and from these fits, the energies and opening angles of the electrons are derived. Conversion electrons from ^{207}Bi, and ^{208}Tl sources serve as calibration lines. The sum energy spectrum for 88 double beta decay candidate events is shown in Fig. 1a. A threshold at 1.1 MeV excludes background from beta decay with internal conversion in ^{214}Pb. This contaminant is deposited on the surface of the double beta decay source by decaying ^{222}Rn in the TPC gas. Tests have shown that radon, in turn, comes from the Be-Cu wire used for the grid, cathode, and field wires of the chamber.

Also eliminated from the spectrum are ^{214}Bi events, similar in origin to ^{214}Pb, but of higher energy. Here the alpha particle from the polonium daughter flags the false double beta events.

Superimposed on the data is the theoretical spectrum[3] for the two neutrino mode ($\beta\beta(2\nu)$) normalized to $(1.26\pm.04)\times10^{20}$ yr, the geochemical rate reported by Kirsten.[4] The TPC data do not give a good fit to the expected $\beta\beta(2\nu)$ spectrum, the most probable explanation being the presence of residual background.

Certain background processes are known to be present from manifestations of associated activity. For example, the thorium decay chain includes a rapid beta-alpha sequence from the decay of ^{212}Bi and its 0.3 μs daughter. These events have a distinctive appearance in the TPC, and their rate can be measured. With the thorium series activity thus determined, the background contribution from the series' most troublesome member, ^{208}Tl, can be calculated to be between 5% and 9% of the 88 candidates. The energy spectrum of the ^{208}Tl background is known from an injection of the parent ^{220}Rn into the TPC. (No permanent contamination results, because the radon daughters all decay away in a few days.) Following the radon injection, strong conversion lines from ^{208}Tl decay can be seen in the spectrum of single electrons belonging to false double beta decay pairs. The absence of conversion lines in the corresponding singles spectrum of the 88 double beta candidates confirms the lack of strong ^{208}Tl background.

Likewise, background spectra from Möller scattering and double Compton scattering can be found. The input for these calculations is the spectrum of single electrons (not members of pairs) originating at the double beta decay source. The intensities of the identified backgrounds have been tabulated elsewhere.[5]

A composite of these calculated background spectra can be subtracted from the double beta candidates of Fig. 1a leaving the spectrum of Fig. 1b. This, too, is a poor fit to the expected $\beta\beta(2\nu)$ spectrum. Possible explanations are 1) Double beta decay does not have the expected $\beta\beta(2\nu)$ shape, 2) The geochemical half life is too short, 3) The background is not understood. The last possibility must be eliminated before the other two can be considered seriously. Therefore a new TPC is being constructed without the radioactive wire and other materials suspected of causing background. Depending on the outcome, we may also consider relocating the experiment in an underground laboratory.

Meanwhile, a half life limit for ^{82}Se can be found by subtracting the estimated background (26 counts) from the 88 candidates, and assigning the remainder to double beta decay. The result is $T_{\frac{1}{2}}(2\nu) > 1.0 \times 10^{20}$ yr at 68% c.l. The sum energy for a $\beta\beta(0\nu)$ ground-state transition in ^{82}Se is 3.0 MeV. The probability that a neutrinoless event would be detected in a 300 keV window centered on 3.0 MeV is estimated to be 0.21 ± 0.02. The absence of counts in this window during 7106 hours yields $T_{\frac{1}{2}}(0\nu) > 9 \times 10^{21}$ yr at 68% c.l., or about 3 times the previous published limit[6] for ^{82}Se.

^{128}Te AND ^{130}Te EXPERIMENTS

The tellurium isotopes ^{128}Te and ^{130}Te are both unstable against double beta decay to the corresponding isotopes of xenon. Although still out of reach of direct counting experiments, they are accessible to the geochemical method, with some controversy about ^{128}Te whose long half life places it close to the threshold of detectability. The on-again, off-again history of ^{128}Te double beta decay detection is summarized in Table IV of reference 7.

Much of the interest in these isotopes stems from the 1968 argument of Pontecorvo,[8] that uncertainties in nuclear matrix

Table I. Comparison of Heidelberg and Missouri data*

^{i}Xe	Atmosphere	Extracted Xe Heidelberg	Missouri	Neutron capture	Fission
124	0.35	0.36			
126	0.33	0.32			
128	7.136	7.152 ± 0.036	13.4 ± 0.2	^{127}I(n,γ)	
	fission corrected	(7.175 ± 0.043)			
129	98	147	234	^{128}Te(n,γ)	
130	15	424.86 ± 1.69	10,822 ± 32		
	fission corrected	(426.25 ± 1.7)			
131	79	113	133	^{130}Te(n,γ)	Yes
132	=100	=100	=100		Yes
134	39	39	39		Yes
136	33	33	33		Yes

* All but isotope 128 and 130 values have been rounded. For complete data see refs. 9 and 13.

elements could be overcome by considering the ratio of half lives of pairs of similar nuclei. In the case of $T_{\frac{1}{2}}(^{128}Te)/T_{\frac{1}{2}}(^{130}Te)$, if the matrix elements are assumed equal, then this ratio is, within limits, just the ratio of phase spaces, which for these isotopes of quite different transition energies is strongly dependent on the decay mode. Phase space having greater dependence on energy for ββ(2ν) gives rise to a half life ratio much larger for ββ(2ν) than for ββ(0ν). Thus, under the equal matrix element assumption, a measurement of the ratio of half lives is a test for ββ(0ν). A ratio measurement is also attractive experimentally because any errors from gas diffusion or incorrect dating cancel out.

The positive ^{128}Te result from a measurement at the University of Missouri at Rolla by Hennecke, Manuel, and Sabu[9] triggered a sequence of theoretical papers[10,11,12] interpreting the half life ratio in terms of lepton number violation and finite neutrino mass. Later, Kirsten, Richter, and Jessberger of the Max-Planck-Institute in Heidelberg reported a conflicting result in their paper, "Rejection of Evidence for Nonzero Neutrino Rest Mass from Double Beta Decay."[13]

The Missouri group appeared to capitulate in a 1986 paper by Richardson, Sinha, and Manuel,[14] where they say, "our results indicate that the value reported by Hennecke, et. al. is an upper limit and thus lend credence to the lower value reported by Kirsten, et. al."

However, Prof. Manuel[15] still believes that radiogenic ^{128}Xe has been detected and he is proposing a new measurement. Let us look at the data from the two groups as shown in Table I. Xenon is released in stepwise heating of the ores, with the bulk of the radiogenic xenon appearing at the melting point. It is the xenon from this melt fraction that is displayed in the table. The abundances of atmospheric and extracted xenon isotopes are normalized to 100 for ^{132}Xe. Neutron capture reactions and spontaneous fission are indicated where they contribute to the isotopic excess. A striking difference in the Heidelberg and Missouri data is the strength of the ^{130}Xe excess, being far greater in the Missouri sample. This large release of radiogenic ^{130}Xe implies that ^{128}Xe should also be stronger in the Missouri data, and indeed the Missouri ^{128}Xe is measured as nearly twice the atmospheric abundance, compared to the Heidelberg value which shows no significant excess over the atmospheric level.

Nevertheless, the superior statistics of the Heidelberg data for ^{128}Xe exclude the Missouri result. To make the comparison properly, one should use the fission-corrected Heidelberg numbers. Although the shielded isotopes ^{128}Xe and ^{130}Xe are not produced by fission, there is a fission contribution to ^{132}Xe, to which the other isotopes are normalized. Subtracting a small fission component from ^{132}Xe therefore increases ^{128}Xe and ^{130}Xe slightly, as shown in parentheses in the table. The fission corrected Heidelberg excess ^{128}Xe above atmospheric is 0.039 ± 0.043, consistent with zero, and in clear conflict with the equivalent Missouri value of 0.240 ± 0.008 (that is, the Missouri excess scaled down to Heidelberg on the basis of the ^{130}Xe excesses in the two laboratories). ^{128}Xe can arise from sources other than double beta decay, namely from neutrons on ^{127}I, and from spurious behavior of the apparatus known as the memory effect. Such effects can only increase the measured ^{128}Xe excess, so the lower value, in this case Heidelberg's, has to be considered more realistic.

Prof. Manuel argues,[15] however, that the Heidelberg result is very sensitive to the fission correction, and that a proper accounting of the uncertainties would increase the ^{128}Xe limit that Heidelberg allows. He also argues that the Xe excesses in the melt fraction should be measured relative to the release in the preceding heating step, rather than relative to air. Such a procedure would yield a positive excess of ^{128}Xe from the

Heidelberg data. Profs. Kirsten[16] and Manuel do not agree on these points, but we can anticipate some useful discussion at the upcoming International Symposium on Nuclear Beta Decays and Neutrinos in Osaka, where both Kirsten and Manuel will speak.

To summarize the experimental tellurium situation at present, there is general agreement within a factor of 2 or 3 on the half life for ^{130}Te, but qualitative disagreement on whether ^{128}Te double beta decay has been detected. We do not have compelling experimental evidence for lepton number nonconservation, and the Heidelberg group does not believe that the geochemical technique can be pushed further. The burden is on Missouri or some other laboratory to demonstrate otherwise.

REFERENCES

1. M. K. Moe, A. A. Hahn, and H. E. Brown, in The Time Projection Chamber, edited by J. A. Macdonald, AIP Conference Proceedings No. 108 (American Institute of Physics, New York, 1984) p. 37.
2. M. K. Moe, A. A. Hahn, and S. R. Elliott, in Proceedings of the Conference on Neutrino Mass and Low Energy Weak Interactions, Telemark, Wisconsin, 1984, edited by Vernon Barger and David Cline (World Scientific, Singapore, 1985).
3. H. Primakoff and S. P. Rosen, Rep. Prog. Phys. 22, 121 (1959).
4. T. Kirsten, Talk at International Symposium on Nuclear Beta Decays and Neutrino, Osaka (1986).
5. S. R. Elliott, A. A. Hahn, and M. K. Moe, Phys. Rev. Lett. 56, 2582 (1986).
6. B. T. Cleveland, et. al., Phys. Rev. Lett. 35, 757 (1975).
7. W. C. Haxton and G. J. Stephenson, Jr., Prog. Part. Nucl. Phys. 12, 409 (1984).
8. B. Pontecorvo, Phys. Lett. 26B, 630 (1968).
9. E. W. Hennecke, O. K. Manuel, and D. D. Sabu, Phys. Rev. C 11, 1378 (1975).
10. D. Bryman and C. Picciotto, Rev. Mod. Phys. 50, 11 (1978).
11. M. Doi, et. al., Phys. Lett. 103B, 219 (1981).
12. W. C. Haxton, G. J. Stephenson, Jr., and D. Strottman, Phys. Rev. D25, 2360 (1982).
13. T. Kirsten, H. Richter, and E. Jessberger, Phys. Rev. Lett. 50, 474 (1983).
14. J. F. Richardson, B. Sinha, and O. K. Manuel, Nucl. Phys. A453, 26 (1986).
15. O. K. Manuel, private communication.
16. T. Kirsten, private communication.

^{76}Ge ββ-DECAY EXPERIMENTS AND THEIR ANALYSES, AN UPDATE

F.T. Avignone and Harry Miley
University of South Carolina, Columbia, SC 29208

R.L. Brodzinski and J.H. Reeves
Pacific Northwest Laboratory, Richland, Washington 99352

ABSTRACT

The status, progress and projections of ^{76}Ge ββ-decay experiments is reviewed. Sources of radioactive background, progress in their elimination and projections of future background levels are discussed. Existing data from the lowest background experiments are combined to extract a "world limit" on the Majorana mass of the electron neutrino, $<M_\nu> \lesssim 2.2$ eV.

INTRODUCTION

The renaissance of interest in ββ-decay in the late 1970's was mainly driven by the importance of the properties of neutrinos to the structure of theories of Grand Unification. There are a number of recent comprehensive reviews in the literature[1], all with somewhat different emphases. The interesting connection between 0ν ββ-decay and fundamental particle theory is that it can be engendered by Majorana neutrino mass or explicit right handed neutrino couplings to hadrons, or both. There are many realistic scenarios in which Majorana neutrino masses on the order of tenths to tens of eV arise. This entire domain should be accessible to 0ν ββ-decay experiments in the future.

There are interesting models in which neutrinos have a wide range of masses. In the SU(3)XSU(3)XSU(3) trinification model of de Rújula, Georgi and Glashow[2], for example, there exist scenerios in which $m(\nu_\tau) \sim$ keV, $m(\nu_\mu) \sim$ eV and $m(\nu_e) \sim 10^{-4}$ eV.[3] In another recent example, Mohapatra[4] has proposed a mechanism to explain small ν-masses in E_6 supersymmetric theories which predicts $m(\nu) \sim 1$ eV. Since such results depend on assumptions and estimates, to fix the parameters, we might conclude that there are indeed a number of reasonable scenarios in which $m(\nu_e)$ might lie between 0.1 eV and 10 eV.

GENERAL THEORETICAL CONSIDERATIONS IN 0ν ββ-DECAY

The semi-leptonic interaction for the β-decay of a d-quark is written as follows:

$$H = \frac{-G_F \cos\theta_c}{\sqrt{2}} \{ j_L^\mu [J_{L\mu}^+ + \eta_{LR} J_{R\mu}^+] + j_R^\mu [\eta_{RL} J_{L\mu}^+ + \eta_{RR} J_{R\mu}^+] \} \quad (1)$$

where j_L^μ (j_R^μ) are the components of left (right)-handed leptonic currents, $J_{L\mu}^R$ ($J_{R\mu}$) are the components of the left (right)-handed hadronic currents, and η_{LR}, η_{RL} and η_{RR} are the relative weightings of the various currents.

The ββ-decay rate, for the 0ν mode, can be expressed in the following general form:

$$\omega(yr^{-1}) = \alpha_1(yr^{-1})\{x^2+\alpha_2 y^2+\alpha_3 z^2+\alpha_4 xy+\alpha_5 xz+\alpha_6 yz\}. \quad (2)$$

The parameters x, y and z are proportional to $\langle m_\nu \rangle$, η_{RL} and, η_{RR} respectively. The quantities $\{\alpha_i\}$ depend on the nuclear and atomic wave functions and relevant matrix elements. Values of these parameters, from recent calculations, are given Table I.

TABLE I. Numerical values for the parameters α_i in equation (2) for the ββ-decay of ^{76}Ge.

	Ref. (5)	Ref. (6)	Ref. (7)
α_1	1.08×10^{-13} yr^{-1}	1.12×10^{-13} yr^{-1}	2.18×10^{-13} yr^{-1}
α_2	1.44	0.92	1.04
α_3	0.65	114	7016
α_4	-0.45	-0.33	-0.36
α_5	-0.38	-6.68	82.0
α_6	-1.34	-0.68	-0.82

NUCLEAR STRUCTURE CALCULATIONS

There are a number of recent nuclear structure calculations of ββ-decay matrix elements. The most extensive are the weak-coupling shell model calculations of Haxton, Stephenson and Strottman[5], who treated ^{76}Ge, ^{82}Se, ^{128}Te and ^{130}Te.

The Osaka group included the p-wave[6] effect as well as a weak magnetism correction to the nuclear current.

The recent paper by Tomoda, Faessler, Schmid and Grümmer,[7] describes the initial and final nuclear states by angular-momentum-and particle-number-projected Hartree-Fock-Bogoliubov wave functions. In this approach the energy was minimized after projection. The p-wave contribution was included as well as relativistic corrections to the nuclear current, including weak magnetism.

Another recent set of calculations, of the ββ-decays of ^{128,130}Te, ^{82}Se and ^{76}Ge, was published by Grotz and Klapdor[8]. Right handed couplings were neglected and the nuclear matrix elements were calculated in the framework of particle-number-projected BCS wave functions with a two body interaction including pairing, double Gamow-Teller, and quadrupole-quadrupole terms.

Earlier, Klapdor and Grotz[9] concluded that strong cancellations, from Gamow-Teller and quadrupole-quadrupole correlations, reduced the 2ν matrix elements of the Te isotopes by more than a factor of 10. These cancellations were not found to effect 0ν decay significantly; hence, may explain the discrepancy between the shell model predictions[5] and the geochronological half lives[10], while supporting the existence of large matrix elements in 0+ → 0+, 0ν ββ-decay. Recent calculations by Vogel and Fisher[11] include pairing, static quadrupole deformation, spin-isospin polarization and the Δ_{33} isobar admixtures. Their 2ν ββ-decay rate for ^{76}Ge is larger than the shell model value[5], and they predict faster than observed rates for ^{82}Se, ^{130}Te and ^{150}Nd. Their calculation has not removed the discrepancy with experimental 2ν ββ-decay lifetimes, and they conclude that it will be difficult to interpret searches for 0ν ββ-decay until that discrepancy is removed.

^{76}Ge DOUBLE BETA DECAY EXPERIMENTS AND RESULTS

The clever introduction of Ge detectors to search for 0ν ββ-decay was due to Fiorini and his co-workers[12]. A number of improved experiments have been initiated but only the most recent ones are discussed[13]. The experiment of Ejiri et al. is a coincidence experiment, and its main goal is the detection of 0+ → 2+, 0ν ββ-decay. The others are single counting experiments with ultimate goals of reducing the background and increasing the volume of Ge to increase the sensitivity. Our 135cm^3 protoype, for example, has been operating in the Homestake goldmine for more than two years in a variety of shielding configurations. The details of our background reduction are given elsewhere[14]. This effort will be pursued in search of the crucially important 2ν ββ-decay.

The experiment was first operated underground with no shielding between the low background, recently mined, lead and the copper cryostat. A large bremsstrahlung background resulted from the β-decay of ^{210}Bi which follows the β-decay of ^{210}Pb present in the lead shielding bricks themselves. Later, a copper inner liner was installed which was 7.3 cm thick on the sides and 7.6 cm thick on the ends. This reduced the background in the 250 keV region by about a factor of 13. It is interesting to note that the experiment is sensitive enough to detect the 1125 keV line from the decay of ^{65}Zn present in the Ge crystal due to the cosmic ray-neutron-generated reaction, ^{70}Ge(n,α2n)^{65}Zn. This line is the sum of the 1115.5 keV transition in ^{65}Cu, following the electron capture of ^{65}Zn, and the Cu x-rays. In addition cosmogenically produced isotopes of Mn, Fe and Co are also observed in the Cu liner (see Table II).

Available data through mid 1986 have been combined to obtain a "world limit" on $T_{\frac{1}{2}}$ of 0ν ββ-decay. A maximum likelihood analysis was used, and the results are given in Table III.

TABLE II. Primary reactions of cosmic ray neutrons with the copper liner and the equilibrium concentrations of the product isotope.

Reaction	Specific Activity dpm/kg	Reaction	Specific Activity dpm/kg
$^{63}Cu(n,\alpha 2n)^{58}Co$	0.05	$^{63}Cu(n,\alpha 4n)^{56}Co$	0.006
$^{63}Cu(n,2\alpha 2n)^{54}Mn$	0.02	$^{63}Cu(n,\alpha p)^{59}Fe$	0.004
$^{63}Cu(n,\alpha)^{60}Co$	0.01	$^{63}Cu(n,\alpha 3n)^{57}Co$	<0.002

The number given in the second column of Table III, in counts/keV/yr/10^{23} ^{76}Ge atoms, provides a figure of merit of the radiopurity of each experiment.

TABLE III. Summary of the results of maximum likelihood analyses of spectra in the region of 2041 keV from various ^{76}Ge $\beta\beta$-decay experiments.

Experiment	(BG/Nt)X10^{23}	Nt(10^{23} yr)	$T_{\frac{1}{2}}^{0\nu}$ limit (10^{23} yr)
PNL/USC	0.40	4.07	1.40
MILANO	0.68	2.64	0.74
GUELPH	0.50	6.56	2.10
CALTECH	0.50	2.08	0.55
UCSB/LBL	0.35	9.19	2.50
TOTAL	0.45	24.45	4.90

A complete analysis of the data must include all possible combinations of the terms in equation (2). The values of $<m_\nu>$, x, y and z for $T_{\frac{1}{2}} = 10^{23}$y are given in Table IV. Allowing only one parameter (x, y or z) to be non zero yields slightly more optimistic limits. For example, the three values corresponding to $<m_\nu>$ = 4.9 eV, 4.2 eV and 3.4 eV in Table IV are 4.1 eV, 4.0 eV and 2.9 eV when the interferences are neglected. The calculations of Grotz and Klapdor yield $<m_\nu>$ = 1.6 eV for $T_{\frac{1}{2}} = 10^{23}$y.

The most conservative nuclear stucture calculations in Table IV combined with the half life limit 4.9 x 10^{23}yr leads to $<m_\nu> \lesssim 2.2$ eV with a 68% level of confidence.

POSSIBLE FURTHER BACKGROUND REDUCTION

The dramatic reduction in primordial radioactivity in the present detectors, compared to those of earlier detectors, is impressive (see Figure 1). Steps are being pursued to achieve further reduction. In the PNL/USC detector the copper liner was recently replaced by 448 year old lead. The background has been substantially reduced in the energy region of maximum probability for 2ν $\beta\beta$-decay. In addition, all of the γ ray lines below 1461 keV appear to have been eliminated or greatly reduced.

TABLE IV. Experimental limits on $<m_\nu>$, x, y and z, with the interference terms of eq. 2 included, for $T_{\frac{1}{2}}^{0\nu} = 10^{23}$yr.

	Ref. (5)	Ref. (6)	Ref. (7)
$<m_\nu>$max	4.9 eV	4.2 eV	3.4 eV
x(max)	9.5×10^{-6}	8.3×10^{-6}	6.6×10^{-6}
y(max)	1.1×10^{-5}	8.4×10^{-6}	5.6×10^{-6}
z(max)	1.6×10^{-5}	8.1×10^{-7}	7.7×10^{-8}

Our spectrum has a broad peak at about 5.2 MeV followed by a significant continuum. A similar peak was observed in the LBL/UCSB detector[13] at 5.1 MeV and has been attributed to a Doppler-broadened line produced by the reaction $^{28}Si(n,n\gamma)^{28}Si$. Our line at 5.2 MeV is attibuted to α-particles from ^{210}Po on the surface of a solder connection near the Ge crystal. Evidence supporting this interpretation can be obtained from the study of a high-gain spectrum shown in Figure 2. The existence of Pb, K and L x rays demonstrates that the source is inside of the cryostat. In addition, the Sn x rays are most certainly from the solder while the presence of the 46.5 keV γ ray following the β-decay of ^{210}Pb is extremely convincing. Finally the 5.3 MeV alpha peak activity decays with the expected 138 day half life of ^{210}Po. This implies that the ^{210}Po is brought to the surface of the solder when melted. The solder has now been eliminated; and, the alpha peak does not appear to be present. More counting time will be needed to accurately assess the impact of this change.

Important insight into various sources of background on 0ν $\beta\beta$-decay experiments can be gained by comparing the backgrounds in the PNL/USC experiment with those of the UCSB/LBL effort. We particularly refer to their data acquired underground and published in 1986.[13] This experiment had roughly the same background in the 2041 keV region as our data as seen by reference to Table III. The UCSB/LBL spectrum, as well as other published spectra, show definite γ ray lines at 185.72, 911.07, 1001.03, 1460.75, and 2614.47 keV associated with the decays of ^{235}U, ^{228}Ac, ^{234m}Pa, ^{40}K and ^{208}Tl. The only one of these lines in our spectrum is that of the 1460.75 keV γ ray from the decay of ^{40}K (see Table V). The count rates under the peaks of the UCSB/LBL spectrum have been graphically integrated and compared to the rates in our experiment, renormalized to an equivalent Ge volume. These results are presented in Table VI. The rate of the 2614 keV γ ray peak is more than 39 times higher in the UCSB/LBL spectrum than the renormalized limit on the corresponding rate in the PNL/USC experiment. The continuum background rates at 2041 keV are the same for both experiments; hence, one concludes that the dominant source of background in this energy region is not from ^{208}Tl in our detector. There are no peaks between 1460 keV and the degraded, broad α-peak at ~5.2 MeV in our spectrum; therefore, this peak and its associated continuum must be the dominant source of counts in the 2041 keV region.

TABLE V Comparison of primordial radioactivity levels in the background of the Ge spectrometer before and after rebuilding with radiopurity selected materials.

Primordial Radionuclide	Gamma ray energy (keV)	Count rate before rebuilding (c/hr)	Count rate after rebuilding (c/hr)	Improvement factor
^{235}U	185.72	73.0	<0.0029	>25000
^{228}Ac(^{232}Th)	911.07	9.0	<0.0012	>7500
^{234m}Pa(^{238}U)	1001.03	3.4	<0.00080	>4300
^{40}K	1460.75	22.0	0.014	1600
^{208}Tl(^{228}Th)	2614.47	1.0	<0.00068	>1500

TABLE VI Comparison of primordial radioactivity levels in the backgrounds of the PNL/USC and UCSB/LBL detectors. The PNL/USC rates are scaled up by the fiducial volume ratio 5.26.

Primordial Radionuclide	Gamma ray energy (keV)	Count rate UCSB/LBL (c/hr)	Count rate x 5.26 PNL/USC	Ratio
^{235}U	185.72	0.36	<0.015	>23
^{228}Ac(^{232}Th)	911.07	0.14	<0.0063	>23
^{234m}Pa(^{238}U)	1001.03	0.07	<0.0042	>16
^{40}K	1460.75	3.36	0.074	45
^{208}Tl(^{228}Th)	2614.47	0.14	<0.0036	>39

The primordial background in the other detectors might well be due to the presence of aluminum or of aluminized Mylar commonly used as a multi-layer infrared reflector. Although the elmination of this material introduces complications in maintaining low liquid nitrogen consumptions, it is a source of background.

The present world effort, in its second generation of experiments, has just completed the equivalent of a 0.4 year count with a 1770 cm³ detector with (BG/Nt) x 10^{23} = 0.45. The most optimistic interpretation of the likelihood function yields $T_{1/2} \gtrsim 4.9 \times 10^{23}$ yr. The average background was 10 counts/keV. Ignoring the slight dip in the spectrum at 2041 keV the half life becomes 3.3 x 10^{23} years. Using this figure and continuing to count for 3.7 years, with the present world volume of Ge, limits of $T_{1/2} \gtrsim 10^{24}$ years and $<m_\nu> \lesssim 1.6$ eV can be reached. With 10 years of counting these limits can be extended to $T_{1/2} \lesssim 1.7 \times 10^{24}$ years and $<m_\nu> \lesssim 1.2$ eV, while 20 years of counting would only increase the sensitivity to $T_{1/2} \gtrsim 2.3 \times 10^{24}$ years and $<m_\nu> \lesssim 1.0$ eV. We are clearly near the point of diminishing returns with respect to learning new physics utilizing detectors of the present levels of background. Further background reductions of two orders of magnitude in the 2041 keV region are required and are achievable.

A distinctive feature of Ge detectors, is that they are ideal for identifying the background and, to a degree, its location.

An important product of the effort to observe 0ν $\beta\beta$-decay of ^{76}Ge is the development of new low background technology which will undoubtedly play an important role in basic as well as applied research programs.

This work was supported by the U.S. Department of Energy under Contract No. DE-AC06-76RLO 1830 and the National Science Foundation under Grant No. PHY-8405654.

REFERENCES

1. H. Primakoff and S.P. Rosen, Ann. Rev. Nucl. Part. Sci 31, 145 (1981); W.C. Haxton and G.J. Stephenson, Jr., Progress in Particle and Nuclear Physics 12, 409 (1984); M.G. Shchepkin Sov. Phys. Usp. 27, 555 (1984); Masuru Doi, Tsuneyuki Kotani and Eijchi Takasugi, Progress in Theoretical Phys. Suppl. 83, 1 (1985); J. Vergados, Phys. Repts. 133, 1 (1986); Paul Langacker, B. Sathiapalan and Gary Steigman, Nucl. Phys. B266, 699 (1986).
2. A. de Rújula, H. Georgi, and S. L. Glashow, in Fifth Workshop on Grand Unification, edited by K. Kang, H. Fried and P. Frampton (World Scientific, Singapore, 1984), p. 88.
3. See Babu, He and Pakvasa, Phys. Rev. D 33, 763 (1986).
4. Rabindra N. Mohapatra, Phys. Rev. Lett. 56, 561 (1986).
5. W.C. Haxton, G.J. Stephenson, Jr., and D. Strottman, Phys. Rev. D25, 2360 (1981).
6. M. Doi, T. Kotani and E. Takasugi, Osaka preprint OS-Ge 85-07, March 1985.
7. T. Tomoda, A. Faessler, K.W. Schmid and F. Grümmer, Nucl. Phys. (1986).
8. K. Grotz and H.V. Klapdor, Phys. Lett. 153B, 1 (1985).
9. H.V. Klapdor and K. Grotz, Phys. Lett. 142B, 323 (1984);
10. T. Kirsten, H. Richter and E. Jessberger, Phys. Rev. Lett. 50, 475 (1983).
11. P. Vogel and P. Fisher, Phys. Rev. C32, 1362 (1985).
12. E. Fiorini, A. Pullia, G. Bertolini, F. Cappellani, and G. Restelli, Phys. Lett. 25B, 602 (1967); Lett. Nuovo Cimento 3, 149 (1970); Nuovo Cimento 13A, 747 (1973).

13. F.T. Avignone, III, R.L. Brodzinski, D.P. Brown, J.C. Evans, Jr., W.K. Hensley, J.H. Reeves and N.A. Wogman, Phys. Rev. Lett. 54, 2309 (1985); E. Bellotti, O. Cremonesi, E. Fiorini, C. Liguori, A. Pullia, P. Sverzellati and L. Zanotti, Phys. Lett. 146B, 450 (1984); A. Forster, H. Kwon, J.K. Markey, F. Boehm and H.E. Henrikson, Phys. Lett. 138B, 301 (1984); J.J. Simpson, P. Jagam, J.L. Campbell, H.L. Malm and B.C. Robertson, Phys. Rev. Lett. 53, 141 (1984); H. Ejiri, N. Takahashi, T. Shibata, Y. Nagai, K. Okada, N. Kamikubota T. Watanabe, T. Irie, Y. Itoh and T. Nakamura, Nucl. Phys. A448, 271 (1986); D.O. Caldwell, R.M. Eisberg, D.M. Grumm, D.L. Hale, M.S. Witherell, F.S. Goulding, D.A. Landis, N.W. Madden, D.F. Malone, R.H. Pehl and A.R. Smith, Phys. Rev. Lett. 54, 281 (1985); Phys. Rev. D33, 2737 (1986); G. Aardsma, P. Jagam, B.C Robertson and J.J. Simpson (April 1986, to be published).

14. R.L. Brodzinski, D.P. Brown, J.C. Evans, Jr., W.K. Hensley, J.H. Reeves, N.A. Wogman, F.T. Avignone, III and H.S. Miley, Nucl. Instr. and Meth. A239, 207 (1985).

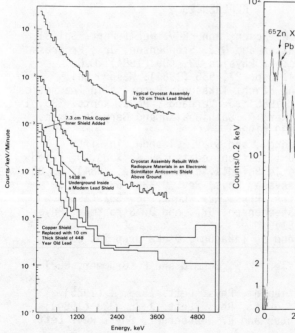

Figure 1. Background spectra of the PNL/USC, 135cm³ prototype Ge spectrometer.

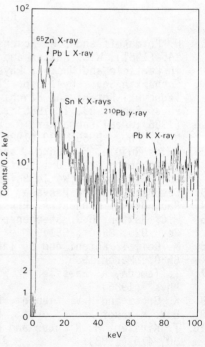

Figure 2. Ten weeks of data from the low energy portion of the Ge detector spectrum.

NEUTRINOLESS DOUBLE β-DECAY AND LEPTON FLAVOR VIOLATION[+]

G.K. Leontaris and J.D. Vergados
The University of Ioannina, Department of Physics, Ioannina, Greece

ABSTRACT

We discuss the phenomenological inplications of gauge theories on lepton number and lepton flavor non-conservation. In particular we compare neutrinoless double β-decay to muon violating processes.

1. INTRODUCTION

Lepton number (L) and lepton flavor (L_f) appear to be conserved in all experiments performed thus far [1]. Such a conservation is understood in the context of the phenomenologically successful standard model as a consequence of a global symmetry of the Lagrangian. This does not completely settle the issue, however, since only (local) gauge symmetries are accepted as exact. Beyond the standard model one can achieve both L and L_f non-conservation. The most familiar mechanism is that which involves massive intermediate neutrinos. Thus the presence of Dirac mass terms which enter when the right handed neutrino exists, leads to L_f violation. Furthermore the presence of Majorana mass terms, which connect a neutrino with an antineutrino, may lead to ΔL=2 transitions. In any case the leptonic currents are no longer diagonal but they can take the form [2]

$$j^L_\mu = 2(\bar{e}_L \gamma_\mu U^{(11)} \nu_L + \bar{e}_L \gamma_\mu U^{(12)} N_L + h.c.), \quad j^R_\mu = 2(\bar{e}_L \gamma_\mu U^{(21)} \nu_R + \bar{e}_R \gamma_\mu U^{(22)} N_R + h.c.) \quad (1.1)$$

The conjugate currents in a CP conserving theory are easily expressed [2] as:

$$(j^L_\mu)^c = 2(\bar{\nu}_R e^{i\alpha}(U^{(11)})^T e^c_R + \bar{N}_R e^{i\varphi}(U^{(12)})^T \gamma_\mu e^c_R) + h.c.$$
$$(j^R_\mu)^c = 2(\bar{\nu}_L e^{i\alpha}(U^{(21)})^T e^c_L + \bar{N}_L e^{i\varphi}(U^{(22)})^T \gamma_\mu e^c_L) + h.c. \quad (1.2)$$

Less familiar mechanisms, involve exotic Higgs scalars [3] or even supersymmetric partners of known particles [4]

2. GROSS FEATURES OF L AND L_f VIOLATING PROCESSES IN NUCLEI

The oldest lepton violating process related to the nature of the neutrino (Majorana or Dirac) is neutrinoless (0ν) ββ-decay

$$(A,Z) \to (A,Z\pm 2) + e^{\mp} + e^{\mp}, \quad e^-_b + (A,Z) \to (A,Z-2) + e^+ \quad (2.1)$$

which together with the allowed (2ν) ββ-decay

$$(A,Z) \to (A,Z\pm 2) + e^{\mp} + e^{\mp} + \{{}^{2\bar{\nu}_e}_{2\nu_e}, \quad e^-_b + (A,Z) \to (A,Z-2) + e^- + \nu_e + \tilde{\nu}_e \quad (2.2)$$

are the only decay modes of some otherwise absolutely stable nuclei [2]. One finds that

[+] Presented at the Conference by J.D. Vergados

$$T_{1/2}(0\nu) = K_{0\nu}/|\eta|^2|ME|^2_{0\nu} \quad , \quad T_{1/2}(2\nu) = K_{2\nu}/|ME|^2_{2\nu} \qquad (2.3)$$

where η is a suitably defined lepton violating parameter and $|ME|^2$ are the corresponding nuclear matrix elements. The quantities $K_{0\nu}$ and $K_{2\nu}$ are functions of (A,Z) which can take the following values [4]

$$1.5 \times 10^{13} y < K_{0\nu} < 2.5 \times 10^{17} y \quad , \quad 2.5 \times 10^9 y < K_{2\nu} < 3.3 \times 10^{26} y \qquad (2.4)$$

Thus if η, which contains all information about the gauge models, is not much smaller than 10^{-5} the present experimental limit [1], detection of 0ν $\beta\beta$-decay is within the capabilities of present experiments.

Another interesting process is the (μ^-, e^+) conversion [2,5]

$$\mu_b^- + (A,Z) \to e^+ + (A,Z-2)^*, \quad T_{1/2}(\mu^-, e^+) = K'_{0\nu}/|\eta'|^2|ME'|^2_{0\nu} \qquad (2.5)$$

where

$$K'_{0\nu}(\mu^-, e^+) = 2.2 \times 10^{10} y \; A^{2/3}/(Z^4_{eff}/Z) \quad \text{or} \quad 5 \times 10^7 y < K'_{0\nu} < 6 \times 10^8 y \qquad (2.6)$$

for nuclei ranging ^{58}Ni to ^{12}C. Even though this process is 10^{10} faster than its sister (e^-, e^+) conversion, it must compete against ordinary muon capture $\mu^- + A \to A^* + \nu_\mu$. Thus one gets a branching ratio

$$R = \Gamma(\mu^- \to e^+)/\Gamma(\mu^- \to \nu_\mu) \simeq 1.5 \times 10^{-21} |\eta'|^2 |ME'|^2/[Z(1.62(Z/A)-0.62)] \qquad (2.7)$$

where η' is the corresponding lepton violating parameter. Thus for $|\eta'| \simeq |\eta| \leq 10^{-5}$ this process is unobservable even if transitions to all final nuclear states are considered $|ME'|^2 \approx 0.1 Z^2 = 100$. (The present experimental limit is $R < 9 \times 10^{-9}$, see A. Baderscher ref. 1).

Among the various L_f violating processes the most interesting for this audience is the process

$$\mu^- + (A,Z) \to e^- + (A,Z)^* \qquad (2.8)$$

For ground state transitions we get [2]

$$R = \frac{\Gamma(\mu^- \to e^-)}{\Gamma(\mu^- \to \nu_\mu)} = |\tilde{\eta}_\alpha|^2 \mathcal{P}_\alpha \frac{E_e P_e}{m_\mu^2} \frac{h_\alpha(A,Z)}{Z(1.62(Z/A)-0.62)} |F_{ch}(q^2)|^2 \qquad (2.9)$$

In the special case of ^{58}Ni with $F_{ch}(-m_\mu^2) = 0.5$ we obtain

$$R = \begin{cases} 2 \times 10^{-21} |\tilde{\eta}_\nu^L|^2 \\ 2 \times 10^{-2} |\tilde{\eta}_N^L|^2 \end{cases} (L-L) \qquad R = \begin{cases} 1 \times 10^{-8}(|\tilde{\eta}_\nu^+|^2+|\tilde{\eta}_\nu^-|^2) \\ 3 \times 10^{-2}(|\tilde{\eta}_N^+|^2+|\tilde{\eta}_N^-|^2) \end{cases} (L-R) \qquad (2.10)$$

where the upper row refers to light neutrino while the lower to heavy ones.

Comparison with the corresponding branching ratios [2] for $\mu \to e\gamma$

$$R(\mu \to e)/R(\mu \to e\gamma) = \begin{cases} 12 \\ 2 \times 10^{-3} \end{cases} (L-L), \quad R(\mu \to e)/R(\mu \to e\gamma) = \begin{cases} 14 \\ 1.4 \times 10^{-2} \end{cases} (R-L) \quad (2.11)$$

shows that (μ^-, e^-) is favorable for light neutrinos.

3. L AND L_f VIOLATING PARAMETERS

It is now straightforward to compute the amplitude for L and L_f violating processes. Typical diagrams are shown in figs. (1) and (2).

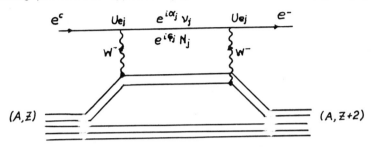

Fig. 1. Neutrinoless ββ-decay mediated by Majorana neutrinos. The relevant submatric for the coupling depends in the type of leptonic current (L or R) and the neutrino type (v_j or N_j). The diagram for (μ^-, e^+) is analogous.

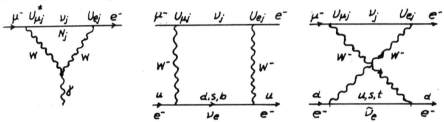

Fig. 2. Typical diagrams which lead to flavor change. Diagram (a) can also lead to (μ, e) and $\mu \to e^+e^-$ if the photon is virtual. Diagrams with W replaced by Higgs scalars must also be considered when appropriate.

We are now in a position to obtain expressions for the lepton violating parameters. We distinguish the following cases
i) Both leptonic currents of figs. (1) and (2) are left-handed (j_L-j_L). Then for light neutrinos ($m_j \ll m_W$) we get

$$\eta_\nu = \eta_\nu^L = \langle m_\nu \rangle / m_e \quad , \quad \langle m_\nu \rangle = \sum_j (U_{ej}^{(11)})^2 e^{i\alpha_j} m_j \quad (\text{ov } \beta\beta\text{-decay}) \quad (3.1a)$$

$$\eta' = \eta_\nu'^L = \sum_j U_{ej}^{*(11)} U_{\mu j}^{*(11)} e^{-i\alpha_j} m_j \quad (\mu^- \to e^+) \quad (3.1b)$$

$$\tilde\eta = \tilde\eta_\nu^L = \sum_j U_{ej}^{(11)} U_{\mu j}^{*(11)} (m_j/m_e)^2 \quad (\mu \to e\gamma,\ \mu \to e^-,\ \mu \to 3e\ \text{etc.}) \quad (3.1c)$$

Note the presence of CP eigenvalues in (3.1a) and (3.1b) which distinguish between Majorana and Dirac particles, and the unfavorable mass dependence in the case of (3.1c). In the case of heavy neutrinos ($m_j \gtrsim m_W$) one finds

$$\eta = \eta_N^{(L)} = \sum_j (U_{ej}^{(12)})^2 e^{i\varphi_j} (m_p/m_j) \quad (m_p: \text{proton mass}) \quad (3.2a)$$

$$\eta' = \eta_N^{'(L)} = \sum_j U_{ej}^{*(12)} U_{\mu j}^{(12)} e^{-i\varphi_j} m_p / M_j \qquad (3.2b)$$

$$\tilde{\eta} = \tilde{\eta}_N^{(L)} = \sum_j U_{ej}^{(12)} U_{\mu j}^{(12)} e^{-i\varphi_j} (m_W/M_j)^2 [a \ln(M_j/m_W)^2 + b] \qquad (3.2c)$$

where a and b are constants of order unity [2] which depend on the details of the graph of fig. (2). Note that U^{12} is expected to be small.

ii) Both leptonic currents are right-handed (j_R-j_R). The light neutrino contribution is now negligible. For heavy neutrinos one gets

$$\eta = \eta_N^{(R)} = (\epsilon^2 + \kappa^2) \sum_j (U_{ej}^{(22)})^2 e^{i\varphi_j} m_p / M_j \qquad (3.3a)$$

$$\eta' = \eta_N^{'(R)} = (\epsilon^2 + \kappa^2) \sum_j U_{ej}^{*(22)} U_{\mu j}^{*(22)} e^{-i\varphi_j} m_p / M_j \qquad (3.3b)$$

$$\tilde{\eta} = \eta_N^{(R)} = (\epsilon^2 + \kappa^2) \sum_j U_{ej}^{(22)} U_{\mu j}^{(22)} (m_W/M_j)^2 [a \ln(M_j^2/m_W^2) + b - a \ln(\epsilon^2/(\kappa^2 \epsilon^2))] \qquad (3.3c)$$

where $\kappa = (m_W/m_{WR})^2$ and $\varepsilon = \tan\zeta$, $\zeta = w_L$-w_R mixing angle. The suppression now is due to mass of the vector boson which mediates the right-handed interaction and/or the small w_L-w_R mixing.

iii) The leptonic current is j_L-j_R type. In this case the helicities are such that one picks the momentum instead of the mass of the intermediate neutrinos. One finds

$$\eta = \eta_{RL} = \sqrt{\kappa^2 + \epsilon^2} \sum_j U_{ej}^{(11)} U_{\mu j}^{*(21)} e^{i\alpha_j}, \quad \eta' = \eta'_{RL} = \sqrt{\kappa^2 + \epsilon^2} \sum_j U_{ej}^{*(11)} U_{\mu j}^{*(11)} e^{-i\alpha_j} \qquad (3.4a)$$

$$\tilde{\eta} = \tilde{\eta}_\nu^\pm = (\kappa \text{ or } \epsilon) \sum_j (U_{ej}^{(11)} U_{\mu j}^{*(21)} \pm U_{ej}^{(21)} U_{\mu j}^{*(11)}) m_j / m_e \qquad (3.4b)$$

$$\tilde{\eta} = \tilde{\eta}_N^\pm = (\kappa \text{ or } \epsilon) \sum_j (U_{ej}^{(12)} U_{\mu j}^{*(22)} \pm U_{ej}^{(22)} U_{\mu j}^{*(12)}) m_W / M_j \qquad (3.4c)$$

We note the favoralbe explicit dependence on the neutrino mass but we caution that this may be offset by the smallness of U^{21} and U^{12}. (The heavy neutrino contribution in η and η' is negligible).

4. INTERMEDIATE HIGGS SCALARS AND SUPERSYMMETRIC PARTICLES

One can have L and L_f violation, even if the neutrino mass mechanism is suppressed, via intermediate Higgs scalars [2]. Typical L_f violating diagrams are shown in fig. 3.(ov) $\beta\beta$-decay and (μ^-,e^+) cannot occur directly since the exotic Higgs scalars (S^-, S^{--} and T^{--}) do not directly couple to the quarks. They can proceed however, via the couplings of such Higgs with the ordinary Higgs isodoublets or with the right-handed vector bosons [2,5]. Thus it is natural to expect that L_f will be favored over L. The contribution of supersymmetric particles is discussed elsewhere in these proceedings [4].

5. CONCLUSIONS

The most conventional L and L_f violating mechanisms involve the neutrino mass. Since the neutrinos produced in weak interactions are linear combinations of mass eigenstates the mass combination measured

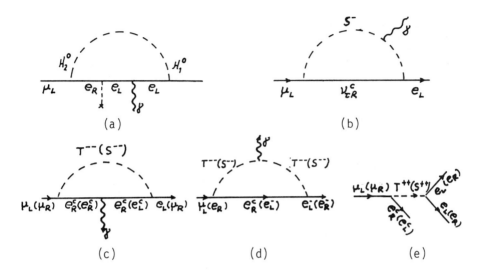

Fig. 3. μ → eγ decay if there exist two different isodoublets (a), a singly charged isosinglet (b) (which always changes flavor) and doubly charged isosinglet (S⁻⁻) and isotriplet (T⁻⁻) (c) and (d). In the last case μ → ee⁺e⁻ can proceed faster at the tree level (e).

in various experiments are [2].
i) Decay experimenys producing flavor α measure

$$\sum_j |U_{\alpha j}|^2 m_j \quad \text{or} \quad |U_{\alpha j}|^2 \text{ and } m_j$$

depending in whether the masses can be resolved.
ii) (ov) ββ-decay measures the quantity

$$|\langle m_\nu \rangle| = |\sum_j U_{ej}^2 e^{i\alpha_j} m_j|$$

Thus it is possible that [1] $\langle m_\nu \rangle = 0$ even if the neutrinos are massive Majorana particles.
iii) Neutrino oscillations in principle can disentangle [2] the neutrino mixing from the mass and measure the quantity

$$\delta m^2 \frac{L}{E_\nu} \quad , \quad \delta m^2 = (m_K^2 - m_\ell^2)$$

where E_ν = neutrino energy and L = source - detector distance.
The present limits in lepton violating parameters as well as the predictions of some models are given in table I. From the table we see that the most likely process for observing lepton non-conservation is the (ov) ββ-decay. The best candidate for observing lepton flavor non-conservation appears to be the neutrino oscillation since nature allows a variaty of L/E_ν so that the smallness of δm² can be overcome. Admittedly these experiments are very hard and they must be done against pessimistic theoretical predictions. We hope however that,

since such experiments are truly fundamental and the theoretical predictions not completely variable, they will continue unhindered.

Table I The Lepton violating parameters in some gauge models which are completely analyzed in ref. 2. The experimental limits are obtained (a) from (on) ββ-decay, T. Kirsten et al (b) from (μ⁻, e⁻) Bryman et al, (c) from $\mu \to e\gamma$ Kinnison et al (ref. 1).

model / Parameter	Witten	SO(10)	M - S$_{(A)}$	M - S$_{(B)}$	Experimental limit	
η_ν^L	2×10^{-6}	4×10^{-9}	1.5×10^{-6}	4.0×10^{-6}	1.0×10^{-6}	(a)
η_N^L	5×10^{-14}	6.0×10^{-34}	2.0×10^{-18}	-	6.0×10^{-7}	(a)
η_N^R	1.0×10^{-7}	2×10^{-11}	3.0×10^{-12}	4.0×10^{-7}	6.0×10^{-7}	(a)
η_{RL}	5.0×10^{-7}	6×10^{-12}	-	-	1.0×10^{-6}	(a)
$\tilde{\eta}_\nu^L$	1.0×10^{-7}	1.4×10^{-13}	1.3×10^{-7}	2.1×10^{-5}	1.0×10^{-5}	(b)
$\tilde{\eta}_N^L$	-	2.5×10^{-38}	9.4×10^{-22}	-	3.0×10^{-5}	(c)
$\tilde{\eta}_N^{(R)}$	-	0	1.8×10^{-11}	4.2×10^{-7}	3.0×10^{-5}	(c)
$\tilde{\eta}_\nu^{(+)}$	5.0×10^{-12}	1.6×10^{-17}	2.2×10^{-22}	8.9×10^{-6}	4.0×10^{-2}	(b)
$\tilde{\eta}_\nu^{(-)}$	5.0×10^{-12}	2.0×10^{-17}	0	0	4.0×10^{-2}	(b)
$\tilde{\eta}_N^{(+)}$	-	2.0×10^{-24}	5.8×10^{-12}	-	3.0×10^{-7}	(c)
$\tilde{\eta}_N^{(-)}$	-	1.8×10^{-24}	0	-	3.0×10^{-7}	(c)

REFERENCES
1. F.T. Avignone, Phys. Lett. 54, 2309 (1985);T.Kirsten, et al, Phys. Rev.Lett. 50,474 (1973);W. Bertl et al, Nucl.Phys. B260,1 (1985); D.A. Bryman et al, Phys.Rev.Lett. 55,465 (1985) and Private Communication;W. Kinnison et al, Phys.Lett. D25, 2846 (1986); A. Badertscher et al, Nucl. Phys. A377, 406 (1982).
2. J.D. Vergados, Phys. Rep. 133, 1 (1986)
3. S.T. Petkov, Phys. Lett. 115B, 401 (1982)
   G.K.Leontaris,K.Tamvakis and J.D. Vergados,Phys.Lett.162B,153 (1986)
4. G.K. Leontaris, K. Tamvakis and J.D. Vergados, Phys. Lett. 171B, 412 (1986); See also these proceedings.
5. G.K. Leontaris and J.D. Vergados, Nucl. Phys. B224, 137 (1983).

UPPER LIMIT ON THE NEUTRINO MASS FROM THE LOS ALAMOS FREE MOLECULAR GASEOUS TRITIUM EXPERIMENT

D.A. Knapp, T.J. Bowles, M.P. Maley, R.G.H. Robertson, and J.F. Wilkerson,
(presented by D.A. Knapp)
Physics Division
Los Alamos National Laboratory, Los Alamos, NM 87544

Abstract

The mass of the electron antineutrino can be determined by measurement of the shape of the beta spectrum of tritium near the 18.6 keV endpoint. A measurement of this spectrum has been undertaken at the Los Alamos National Laboratory, and initial results have been obtained, allowing an upper limit of 29.7 eV to be set on the neutrino mass.

Introduction

The possibility of the existence of a rest mass for the electron antineutrino is a topic of considerable current interest. The mass can be determined from measurements of the beta spectrum of tritium near the endpoint. The effect of the neutrino mass extends over a region from the endpoint to an energy about $2m_\nu c^2$ below the endpoint. There are several other processes that have a similar effect on the spectrum, and therefore must be accurately quantified in order to determine the neutrino mass. These include the atomic or molecular final states, solid-state effects, and energy loss in the source. The system for which these effects are best known is gaseous tritium, which has a calculable final state distribution and no solid-state effects. Therefore, an experiment to measure the beta spectrum of gaseous tritium has been constructed at the Los Alamos National Laboratory[1].

Apparatus

The use of a gaseous source engenders significant complications in the apparatus used to measure the beta spectrum of tritium. The Los Alamos apparatus is shown in Fig. 1. It consists of a source region, a differential pumping and electron extraction region, and an iron-free Tret'yakov-type toroidal beta spectrometer. The tritium decays in a long aluminum tube filled with tritium gas, which is floated at a high negative potential. The electrons are confined by a longitudinal magnetic field, which extracts them from the source without letting them touch the walls. The electrons are accelerated as they leave the source tube, which eliminates the background from tritium in any other part of the system and allows the electron energy seen by the rest of the apparatus to remain constant. The initial beta energy is selected by changing the accelerating voltage on the source tube. Differential pumping reduces the partial pressure of tritium at the entrance to the spectrometer to less than 10^{-10} Torr. The electrons are focused to

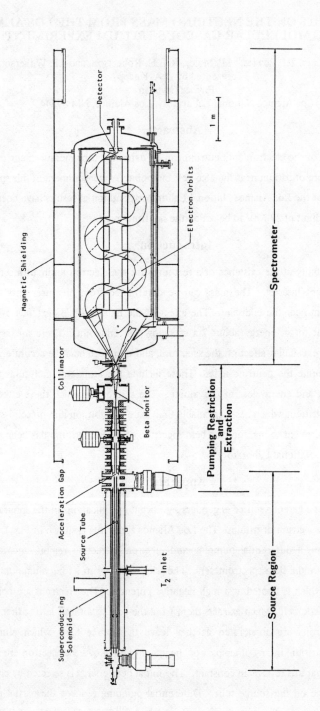

Fig. 1. Cross-sectional view of the Los Alamos Gaseous Molecular Tritium Experiment.

a collimator at the entrance to the spectrometer by means of nonadiabatic transport through a rapidly changing magnetic field. A solid-state detector located in this region detects betas originating between the source volume seen by the spectrometer and the walls of the source tube, thus allowing an accurate determination of the source intensity. The electrons are detected by a position and energy sensitive proportional counter located at the image of the spectrometer. The spectrometer resolution with a gaseous source can be measured using ^{83m}Kr, a short-lived (1.8 h) conversion-line source with an electron energy of 17.84 keV.

With an integrated source strength of 8×10^{15} tritium molecules per cm^2, the count rate observed in the detector is about .15 per second in the last 100 eV of the beta spectrum. The background rate is about 2×10^{-4} counts per second per eV. The spectrometer has an energy resolution near 35 eV for 26 keV electrons coming from the source.

Data

The apparatus has been used to take three data runs of about 72 hours each. The data in each case cover an energy range of about 16500 to 19000 eV. The data points were taken in a random order, for 600 sec. each. In order to track any long-term variation in the efficiency of the beta monitor, shorter calibration runs were taken at a fixed beta energy after each two data points. The events recorded in the spectrometer detector were gated by position and energy. After the data run was completed, each data point was normalized by using the beta monitor count rate observed during the taking of that point. A typical beta spectrum is shown in Fig. 2.

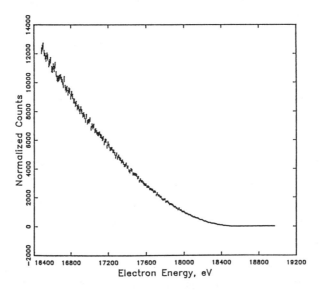

Fig. 2. A typical beta spectrum obtained from the apparatus.

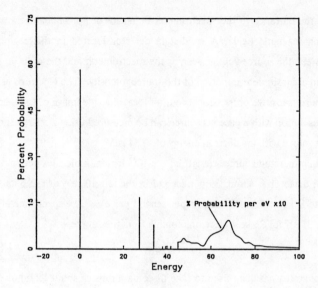

Fig. 3. The final-state spectrum following the beta decay of molecular tritium.

Data Analysis

Because of the low background rate and the relatively small number of counts near the endpoint of the beta spectrum, the neutrino mass obtained from the data is extremely sensitive to the estimation parameter used. The conventional χ^2 is not appropriate since the data is not distributed normally. Instead, a modified version of the Poisson maximum likelihood estimator was used. This estimator produces an unbiased estimate of the neutrino mass, since it preserves the area of the data in the fitted spectrum.

The final state distribution for molecular tritium has been calculated[2], and it is shown in Fig. 3. For this analysis, the continuum was approximated by 6 states which accurately reproduce the first 12 spectral moments of the actual distribution. The energy loss spectrum of the electrons in the gaseous source can be divided into two parts: a part resulting from single scatters of electrons traversing the source, and a part resulting from multiple scatters of electrons emitted in a direction such that they are trapped by inhomogeneities in the magnetic field of the source. A planned gradient in the field of the source was not operational during the taking of this data, so the fraction of electrons trapped was substantial. The energy loss spectrum for the singly scattered electrons was calculated from data obtained by Wellenstein et al.[3] and Geiger[4]. The spectrum for the trapped electrons was calculated by a Monte Carlo program. The spectrometer resolution was measured using the ^{83m}Kr source, for which the shakeoff spectrum was calculated

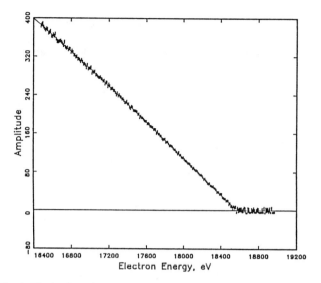

Fig. 4. Kurie plot of a typical beta spectrum, with a fit to $m_\nu = 0$ shown.

using the method of Levinger[5]. This source was also used for the energy calibration of the spectrometer; however, since the energy of the conversion line is only known to an accuracy of about 20 eV, the actual T-^{3}He$^+$ mass difference could not be determined very well.

Using all these corrections, and including the effects of systematic errors, each spectrum was fit to a neutrino mass. A Kurie plot of one of the data runs, together with the fit for zero neutrino mass, is shown in Fig. 4. The three spectra are consistent with each other and with a neutrino mass of zero, and an upper limit can be set on the mass:

$m_\nu < 29.7$ eV   (95% C.L.)

and

$m_\nu < 25.4$ eV   (90% C.L.).

This limit is still dominated by statistical uncertainties. It is expected that the ultimate limit obtainable by this experiment should be about 10 eV.

References

1. J.F. Wilkerson, T.J. Bowles, D.A. Knapp, M.P. Maley, and R.G.H. Robertson, paper presented at the Moriond conference, Jan. 1986.
2. R. Martin and J. Cohen, Phys. Lett. 110A, 95 (1986).
3. R. Ulsh, H. Wellemstein, and R. Bonham, J. Chem. Phys. 60, 103 (1974).
4. J. Geiger, Z. Phys. 181, 413 (1964).
5. J. Levinger, Phys. Rev. 90, 11 (1953).

AN UPPER LIMIT FOR THE ELECTRON ANTINEUTRINO MASS

M. Fritschi, E. Holzschuh, W. Kündig, J. W. Peterson*,

R. E. Pixley, and H. Stüssi

Physics Institute, University of Zürich, 8001 Zürich Switzerland

ABSTRACT

The endpoint region of the tritium β-spectrum has been measured with 27 eV resolution, using a toroidal field magnetic spectrometer, modified with a radial, electrostatic retarding field around the source. The tritium activity was implanted into an thin layer of carbon. The $\bar{\nu}_e$-mass determinded is consistent with zero with an upper limit of 18 eV, which includes instrumental and statistical uncertainties as well as uncertainties due to the energy loss in the source and the final electronic states. As an illustration Fig. 1 shows a portion of the endpoint region for one of the four spectra measured. The data are fitted with an assumed mass of 0 and 35 eV. In Fig. 2 the normalized deviations for one of the spectra are plotted for the two cases.

Our result is in strong contradiction with the ITEP result claiming evidence for a nonzero mass of 20 eV $\leq m_\nu \leq$ 45 eV with a central value of 35 eV[1]. We see no possible source of error in our experiment large enough to account for this discrepancy. A paper describing the experiment has been submitted to Physics Letters. Some details on the spectrometer and the source preparation may be found in the Proceedings of the Moriond Workshop[2].

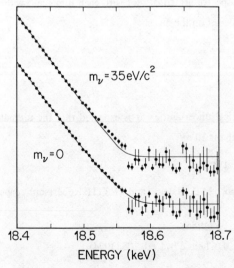

Fig. 1

Section of the data in the form of a Kurie-plot near the endpoint and the best fits for m = 0 and 35 eV (the free parameters are: the background B, the normalization A, the backscattering parameter α, and the endpoint E_o

* Present address: CERN, 1211 Geneva, Switzerland

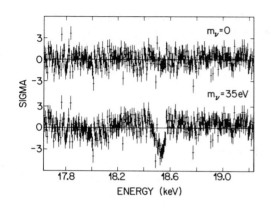

Fig. 2

Plot of the difference between the fitted function and the data, divided by the standard deviation with two fixed values of m_ν. The free parameters B, A, α, and E_0 being fitted. The χ^2 for the shown fits with 318 degrees of freedom are 344.6 and 518.0 respectively. For the total dataset with 1370 degrees of freedom one gets 1370.2 and 1773.2 respectively.

REFERENCES

1. V. A. Lubimov et al. Phys. Lett. B94, 266 (1980); Sov. Phys. JETP 54, 616 (1981); and Phys. Lett. B159, 217 (1985).
2. Proceedings of the Moriond Workshop 1984 and 1986, ed. by J. Tran Thanh Van (Edition Frontières, Paris).

Coherent Scattering of Neutrinos and Antineutrinos by Quarks in a Crystal

J. Weber
University of California, Irvine, California 92717

and

University of Maryland, College Park, Maryland 20742

ABSTRACT

An earlier paper[1] presented a new method for observation of neutrinos and antineutrinos. Coherent scattering is employed for tightly bound nuclei in a nearly perfect crystal. Total cross sections are large.

In this paper the theory is extended, for a crystal with nuclei described by the modern theory of quarks.

Experimental data are presented.

INTRODUCTION

The theory of coherent elastic scattering of tightly bound nuclei was presented[1], with the total cross sections proportional to the square of the number of scatterers. The theory is extended, for a crystal with nuclei composed of up and down quarks.

QUARK MODEL CROSS SECTIONS

For unpolarized scatterers, the S matrix gives a total cross section σ_{TOTAL}, with

$$\sigma_{TOTAL} = \frac{G_W^2 E_\nu^2}{4\pi \hbar^4 c^4} \left(.387 N_U - .693 N_D\right)^2 \quad (1)$$

N_U is the total number of up quarks, N_D is the total number of down quarks, in a macroscopic crystal which is assumed to be perfect and "infinitely stiff." E_ν is the neutrino or antineutrino energy, G_W is the weak interaction coupling constant.

A torsion balance has been employed to measure the total cross section for tritium antineutrinos incident on a 13 gram crystal. The measured cross section is

σ_{TOTAL}(measured) = 2.05 ± 0.25 cm^2

[1] J. Weber Phys. Rev. C 31, 4, 1468(1985). This research was supported in part by the Advanced Research Projects Agency.

For higher energy antineutrinos the finite stiffness and crystal imperfections are expected to give smaller cross sections. For antineutrinos from a 20 megawatt nuclear reactor, a 100 gram crystal is observed to have a cross section

$$\sigma_{TOTAL}(\text{measured}) \sim 0.2 \text{cm}^2$$

The reactor antineutrino value is approximate and tentative. Search is continuing for systematic errors.

Search is being carried out for solar neutrinos. For two crystals with equal mass but different Debye temperatures, a positive result is expected for the Eötvös experiment, because of momentum transfer by solar neutrinos.

COHERENT SCATTERING OF LOW ENERGY NEUTRINOS FROM MACROSCOPIC OBJECTS

R. C. Casella
Reactor Division, Institute for Materials Science and Engineering
National Bureau of Standards, Gaithersburg, Md. 20899

ABSTRACT

It is known since the work of Freedman that neutral-current scattering of neutrinos from the nucleons in a nucleus can be coherent, leading, for sufficiently long wavelengths, to cross sections which are proportional to the square of the nuclear baryon number. When extended to macroscopic objects containing N nuclei, it has recently been reported[2], on the one hand, that coherent cross sections proportional to N^2 have been observed, and on the other, that coherence can occur only for wavelengths comparable to the sample size. I find that coherent scattering can indeed occur on the scale of an entire crystal for incident neutrino wavelengths comparable to the internuclear separation or less, but that the cross section remains linear in N. Hence coherence cannot explain the reported observation of a macroscopic force exerted on a crystal by reactor antineutrinos.

THE COHERENT CROSS SECTION

For sufficiently small momentum transfers, $|q^2| \ll M_{Z^o}^2$, neutral current neutrino processes follow from the standard-model Hamiltonian

$$H = (G/\sqrt{2})[\bar{\nu}\gamma^\mu(1-\gamma^5)\nu][\bar{q}(C_V\gamma^\mu + C_A\gamma^\mu\gamma^5)q] , \qquad (1)$$

where $C_V = I_3 - 2Q\sin^2\theta_W$ and $C_A = -I_3$, G is the Fermi coupling and q is an elementary weak isodoublet: $q = (\nu,\ell)$, (u,d'), (p,n), or (A_+,A_-). The first entries are the leptons and quarks of each family. For smaller momentum transfers the nucleon isodoublet is treated as elementary, while for sufficiently low energies we may treat the entire nucleus as elementary with $A_\pm$ the $I_3 = \pm 1/2$ components of q. When, in the latter case, vector coupling dominates, the scattering amplitude $a(E,\theta)$ in the neutrino-nucleus c.m. frame satisfies

$$a(E,\theta) = (1/\sqrt{2})(GE/2\pi)[Z(1 - 4\sin^2\theta_W) - N_n]\cos(\theta/2). \quad (2)$$

Here E and θ are the incident energy and scattering angle in the c.m. frame and θ_W is the weak angle. This leads[1] to cross sections which are quadratic in baryon number $A = Z + N_n$. Weber[2] has considered coherent neutrino scattering from an entire crystal, obtaining cross sections proportional to N^2 for crystals containing N nuclei and has reported observing a macroscopic force produced by the scattering of reactor antineutrinos via this process. His analysis has been criticized by Butler[3] who states that coherent scattering and N^2 enhancements can occur only for neutrino wavelengths comparable to the linear dimension of the sample.

I consider coherent neutral current scattering from crystals by neutrinos with wavelengths of the order of and considerably smaller than the internuclear separation. The process is analogous to the coherent scattering of X-rays or thermal neutrons from crystals. I calculate directly the net force exerted on a crystal by the coherent scattering of every neutrino from the N nuclei, assuming phase coherence is maintained even for neutrino wavelengths much smaller than the lattice constant, but larger than nuclear dimensions. Letting b denote the constant multiplying the quantity $E \cos(\theta/2)$ in Eq.(2) for $a(E,\theta)$ and F_z, the net force exerted by a current-density source distribution $J_z(E)$ of (anti)neutrinos, I find[4]

$$F_z = (4\pi/3) \, N \, e^{-2W} \, b^2 \int^{E_c} dE \, E^3 \, J_z(E). \qquad (3)$$

Here E_c is a cut-off energy above which the scattering becomes incoherent and $\exp(-2W)$ is the temperature dependent Debye-Waller factor due to thermal vibrations of the nuclei. F_z is linear in N. Thus, while coherence can occur in the scattering of neutrinos of wavelengths much smaller than the sample size, an N^2 enhancement is possible only for ultra-low-energy neutrinos with wavelengths of order the linear dimensions of a macroscopic object. The latter part of this statement is in accord with the observation of Butler[3] and precludes the possibility of crystal-wide coherence in the scattering of antineutrinos from fission reactors leading to the production of a force of sufficient magnitude to be observed at a level of sensitivity requiring an N^2 enhancement. Accounting for recoil of either the entire crystal or of each nucleus, i.e., transforming Eq.(2) to the appropriate rest-frame coordinates, cannot introduce an additional factor N into Eq.(3) for F_z. Only when wavelengths are of order the crystal size, allowing the crystal to be treated as an elementary particle (ignoring the relative phases associated with internal coordinates), can F_z be proportional to N^2.

CONCLUSION

The reported observation[2] of a macroscopic force exerted on a crystal by antineutrinos from a fission reactor cannot be explained by an N^2 enhancement at the crystalline level of the nuclear cross-section due to coherence, at least by extrapolating known physics.

REFERENCES

1. D. Z. Freedman, Phys. Rev. D, 9, 1389 (1974).
2. J. Weber, Phys. Rev. C, 31, 1468 (1985) and this conference.
3. M. N. Butler, preprint, Cal. Inst. Tech. (1985).
4. It is planned to publish a more detailed report of the present analysis.

Neutrino-Oscillation Experiments
at Reactors

Z.D. Greenwood
Physics Department
University of California, Irvine
Irvine, CA 92717
USA

ABSTRACT

A general review of neutrino-oscillation experiments at reactors is given. Preliminary results of two measurements at 18.2m and 23.7m from a production reactor at the Savannah River Plant are discussed.

Motivations

Double beta decay, beta decay, and neutrino oscillation experiments yield information on the following fundamental questions: Are neutrinos Dirac or Majorana particles? Are neutrinos massive, and if so, is there mixing between neutrinos of different types? A positive neutrino oscillations test would indicate that the ν_e, ν_μ and ν_τ are massive but not mass eigenstates. Rather these weak states (ν_e, ν_μ, ν_τ) are a linear superposition of mass eigenstates (ν_1, ν_2, ν_3).

This idea, which first appeared as theoretical speculation in the late fifties (see the review article by Bilenky and Pontecorvo[1]), is of current interest to particle physicists because of certain Grand Unified Theories, and to astrophysicists and cosmologists because it could (1) offer a solution to the disparity of the Homestake Solar Neutrino Experiment results with the Standard Solar Model and (2) address the missing mass (in the universe) problem.

Reactors as Neutrino Sources

Nuclear reactors yield about six electron-antineutrinos per fission. Therefore, when a reactor is operating at 2000 MW, one obtains a flux of about $10^{13}/cm^2/sec$ at a distance of 20m. Because reactor neutrino spectra essentially go to zero above 10 MeV. Should the $\bar{\nu}_e$ change to another type, the new neutrino will be below the inverse beta threshold, resulting in a deficiency in the $\bar{\nu}_e$ signal. Thus, neutrino oscillation experiments at reactors are usually disappearance experiments. The current generation of experiments use the inverse beta decay reaction:
$\bar{\nu}_e + p \rightarrow e^+ + n$.

Also, because of the low energies of the neutrinos, the inverse beta reaction is nonrelativistic resulting in the simple relationship

$$E_{e+} = E_{\bar{\nu}} - 1.8 \text{ MeV}.$$

Therefore, the measured positron spectrum together with the known interaction cross section, yields the neutrino source spectrum.

Neutrino Oscillations Phenomonology

For simplicity we consider the case of oscillations between two neutrino types:

$$\bar{\nu}_e = \nu_1 \cos\theta + \nu_2 \sin\theta$$

$$\bar{\nu}_x = \nu_2 \cos\theta - \nu_1 \sin\theta$$

where $\bar{\nu}_x$ represents any neutrino other than $\bar{\nu}_e$.

One finds the probability of finding $\bar{\nu}_e$ at distance d from the source

$$P(d) = \frac{1}{2}(\sin^2 2\theta) [1-\cos(2.5\Delta m^2 d/E_\nu)]$$

where $\Delta m^2 = |m_1^2 - m_2^2|c^4$. The wavelength for oscillations is

$$\lambda(m) = 2.5 \, E_\nu \text{ (MeV)}/\Delta m^2 \text{ (eV}^2\text{)}.$$

The expected inverse beta reaction rate is given by

$$Y(L,E,\Delta m^2, \sin^2 2\theta) = \frac{N_p \varepsilon \int \sigma(E_\nu) r(E_{obs}, E_\nu) S(E_\nu) (P(d) dE_\nu dV_r dV_d}{4\pi d^2}$$

where N_P = proton number, ε is the detector efficiency, σ is the inverse beta decay cross section, $S(E_\nu)$ is the neutrino spectrum, and r is the detector response function. Because of the finite sizes of the reactor core and detector, the rate involves an integration over the reactor volume V_r and detector volume V_d.

A straight forward method of searching for neutrino oscillations at a reactor is to measure Y at multiple reactor-detector distances and analyze the ratios of the rates. This approach has the advantage of cancelling many uncertainties associated with efficiencies and cross sections. However, one also must have a good knowledge of the reactor power source and of the reactor and detector geometries.

Detector Sensitivities

At the ILL Grenoble, Bugey, and Savannah River Plant (SRP) facilities, one can place detectors ≤ 18m. from the reactor for high statistics measurements yielding information on quite small $\sin^2 2\theta$. At SRP and Gosgen, experiments can be located ≥ 50m. from the reactor which means a nominal lower sensitivity of

$$\Delta m^2(eV^2) = 2.5\ E_\nu(MeV)/d(m) \simeq .02\ eV^2.$$

Because of the finite sizes of reactor and detector, oscillations for λ less than some minimum dimensions wash out. Experiments which compare the results two or more distances are insensitive above

$$\Delta m^2(eV^2) = 2.5\ E_\nu\ (MeV)/R(m.) \simeq 2.0\ eV^2.$$

The Caltech - SIN - TU Munich Collaboration

This group has measured at 8.76m from the ILL Grenoble reactor and at 37.9m, 45.9m, and 64.7m. from a Gosgen power reactor.[2] The detector (see Figure 1) employed consists of planes of scintillation counters between which ^{3}He MWPC's are sandwiched. A signature for a good event requires a correlation in time and position between a positron and neutron. Pulse shape discrimination is employed when viewing the target scintillator. The results of these experiments are consistent with the expected flux and no oscillations.

The Grenoble - LAPP Annecy Collaboration

The detector used by this group at Bugey,[3] France is essentially identical to that used at the Gosgen experiment. They were able to place the detector at 13.6m. and 18.3m. from a power reactor yielding a measurement with much higher statistics than the Gosgen experiment. They observe a 2σ deviation from the expected value of the ratio of integrated yields at the two positions. In terms of two neutrino oscillations, the results are compatible with $\Delta m^2 \sim 0.2\ eV^2$ and $\sin^2 2\theta \sim 0.25$. These results are barely compatible with the Gosgen three position limits.

The UCI Neutrino Group

This group has just completed a two position measurement (18m and 24m from a production reactor) at the Savannah River Plant (SRP). A description of this experiment follows.

1. The SRP Reactor as a Neutrino Source

Most fissions at the SRP production reactor are of ^{235}U. ^{239}Pu and ^{238}U, which have different neutrino spectra, produce 8% and 4% of the fissions respectively.

The reactor shuts down at regular intervals, allowing frequent updates of the measured background.

These advantages are absent at power reactors where significant and varying fractions of ^{239}Pu and ^{238}U exist and the reactors are run for lengthy periods preventing background measurement.

2. Description of the UCI Detector

A schematic of the detector is given in Figure 1. In the center

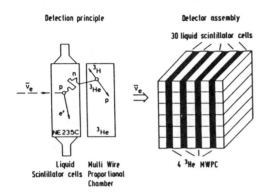

Figure 1. A Schematic of the ILL, Gosgen, and Bugey Detection Schemes

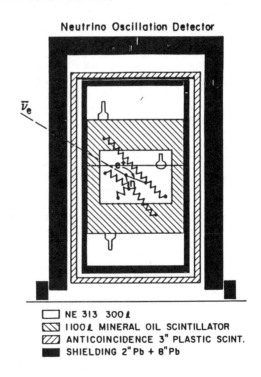

Fig. 2. Schematic of the UCI detector

is the target volume: a liquid scintillator loaded with gadolinium. This scintillator allows particle discrimination by pulse shape. The Gd readily absorbs neutrons and yields about 8 MeV in gamma deexitation energy. Surrounding the target are two volumes of mineral oil based scintillator.

Surrounding these detectors are two inches of low background lead, a three-inch thick plastic scintillator anticoincidence and eight inches of lead.

When an electron antineutrino interacts with a target proton, a positron and neutron are created. The target detector registers a prompt pulse proportional to the positron kinetic energy plus absorbed radiation from positron annihilation. Most cosmic ray induced events are eliminated by the surrounding detector used in anticoincidence. The neutron, after thermalizing, is captured by a Gd nucleus with a mean time of 10 μs. The "delayed" portion of a signature is the gamma energy from Gd deexitations deposited in the target and surrounding liquid detectors.

3. UCI Experimental Data

In Figure 3, we present a comparison of the data taken at 18.2m and preliminary data taken at 23.7m. These data are unnormalized and represent 38,000 events at the close position and 19,000 at the far position.

After correcting for reactor power differences and distance, the rates in the two locations can be compared as a function of energy (Figure 4). A ratio of unity is expected in the absence of oscillations. These data are used to set limits on Δm^2 and $\sin^2 2\theta$. The ratio of the integrated rates above an observed energy of 1.8 MeV is 1.038 ± 0.0137 (statistics) ± 0.02 (systematic).

The Kurchatov Institute of Atomic Energy

This group has completed a measurement at 18m from the Rovno reactor in the USSR. The detector they designed is similar to the UCI detector. They have collected 15,000 events.

By comparing their measured spectral shape to a predicted shape and their measured integral rate to a predicted rate, they derive exclusion regions.

Summary of Multi-Position Experiments

The allowed region of the Bugey experiment and the exclusion regions of the Gosgen and SRP experiments are shown in Figure 5. The SRP curve is a 99.7% confidence contour; the Gosgen 90%; and the Bugey 95%.

As for the future, the Bugey group is repeating their original experiment and preparing a new experiment. The UCI group is preparing for a measurement at 52m.

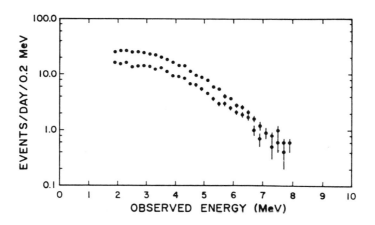

Fig. 3. Comparison of the Unnormalized Spectra Measured at 18.2m and 23.7m at SRP.

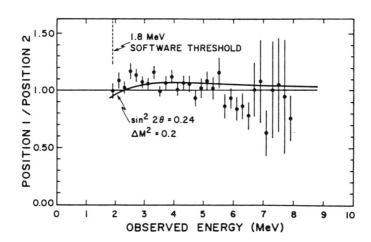

Fig. 4. The ratio of the Integrated Rates Above E_{obs} = 1.8 MeV.

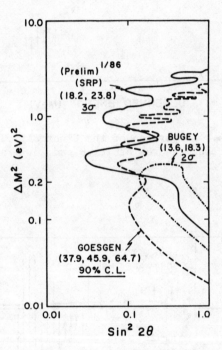

Fig. 5. Summary of Multi-position Experiment Results.

This work is supported in part by the U.S. Department of Energy.

References
1. Bilenky et al., Phys. Rep. C41, 225 (1978).
2. Kwon, H., et al., Phys. Rev. D., 24, 1097, Sept. (1981). Vuilleumier, J.L., et al., Physics Letters, 114B, 298, Nov. (1984). Gabathuler et al., Physics Letters, 138B, 449, April (1984).
3. Cavaignac, J.F., et al., Physics Letters, 148B, 387, Nov. (1984).
4. Afonin, A. et al, JETP Lett. Vol. 42, No. 5, (1985) 285.

RECENT RESULTS ON $\nu_e e^-$ SCATTERING*

R.C. Allen, V. Bharadwaj,[a] G.A. Brooks,[b] H.H. Chen, P.J. Doe,
R. Hausammann,[c] W.P. Lee, H.J. Mahler,[d] M.E. Potter,
A.M. Rushton,[e] and K.C. Wang[f]
University of California, Irvine, California 92717

T.J. Bowles, R.L. Burman, R.D. Carlini, D.R.F. Cochran, J.S. Frank
E. Piasetzky,[g] and V.D. Sandberg
Los Alamos National Laboratory, Los Alamos, New Mexico 87545
and

D.A. Krakauer and R.L. Talaga
University of Maryland, College Park, Maryland 20742

Presented by W. P. Lee

Department of Physics
University of California
Irvine, California 92717

ABSTRACT

The latest results from the neutrino-electron elastic scattering experiment at the LAMPF beam stop are presented. Based on the data sample collected from September 1983 to December 1985, we observed 121 ± 25 events consistent with $\nu_x e^-$ scattering, of which 99 ± 25 events are assigned to $\nu_e e^-$ scattering. The resulting cross section agrees with standard electroweak theory and rules out the constructive interference between weak charged-current and neutral-current interactions.

INTRODUCTION

With the absence of observed structure for leptons, the study of neutrino electron elastic scattering provides a stringent testing ground for the validity of the electroweak theory by Weinberg,[1] Salam[2] and Glashow[3] (WSG). In contrast to $\nu_\mu e^-$, $\bar{\nu}_\mu e^-$ and $\bar{\nu}_e e^-$ scattering,[4] which had been studied extensively at high energy accelerators and the reactor, the first observation of $\nu_e e^-$ scattering was reported only very recently.[5] Unlike $\overset{(-)}{\nu}_\mu e^-$ scattering, the $\nu_e e^-$ reaction can proceed via both the weak charged-current (CC) and neutral-current (NC) interactions. Measurement of the cross section can therefore provide fundamental information on the interference between these two interactions.[6]

---

*Work supported in part by the U.S. National Science Foundation under grant No. PHY-8501559 and by the U. S. Department of Energy.

The beam stop at LAMPF (Los Alamos Meson Physics Facility) is a unique ν_e source. Stopped π^+ decay followed by stopped μ^+ decay leads to the production of equal numbers of ν_e, ν_μ and $\bar{\nu}_\mu$ with well defined energy spectra below 53 MeV. Production of $\bar{\nu}_e$ is highly suppressed ($\lesssim 10^{-3}$) because of the absorption of π^- and μ^- by nuclei in the beam stop.

At a LAMPF proton current of 1 mA, the neutrino flux is about $4 \times 10^7/cm^2$-sec at a distance of 9 meters from the beam stop. The kinematical signature of $\nu_x e^-$ scattering is such that, with a 20 MeV detection threshold, the recoil electron will be confined to a forward angular cone of $10°$.

THE EXPERIMENT

The central neutrino detector, with a sensitive mass of 15 metric tons and dimensions 3m x 3m x 3.5m, is a fine-grained sandwich system arranged in 40 closely-packed identical layers, and is located in a "Neutrino House" well shielded from cosmic-rays as well as most beam-associated backgrounds. Each sandwich layer consists of a plane of plastic scintillator and a polypropylene flash-chamber module (FCM) for recording energy and track information respectively. Each FCM contains 10 panels alternating vertically and horizontally. Each panel consists of 520 flash tubes, which typically operate at an average efficiency of 55%.

From the outside inwards, the cosmic ray shields include some 900 gm/cm^2 of inert material to attenuate hadrons, followed by a multiwire proportional counter (MWPC) system to veto and tag incoming charged particles, (mostly muons), and then, another inert shield of 100 gm/cm^2 to attenuate and convert gamma rays. In addition, a steel shield 6.3m thick is located between the beam stop and the detector system to suppress beam-associated neutrons by a factor of about 10^{15}. With all these shields, a neutrino signal to background ratio of better than 1 to 1 is achieved at our final data analysis level. Details of this detector system and its performance will be provided elsewhere.[7]

During data taking, the on line trigger is defined by a coincidence between at least 3 adjacent scintillator planes with an energy deposition between 1 and 16 MeV per plane, and no veto signal from the MWPC system. The trigger initiates the discharge of the HV pulsers for the FCM's and subsequent acquisition of data via CAMAC from the various detector components scanning a time period of 32 µS before and 64 mS after the trigger. The pre-trigger and post-trigger information is used in the off-line analysis for identifying stopped-muon decay and ν_e capture by ^{12}C respectively. Data was taken during LAMPF beam spill (beam On) as well as between spills (beam Off) to allow cosmic-ray background subtraction. The typical system trigger rate is less than 0.1/s with a dead-time of about 15%. In addition, special triggers were arranged to take cosmic-ray through-going muons and stopped-muon decay events periodically for calibration and monitoring purposes.

DATA REDUCTION

The data sample presented here is based on a total beam exposure of 3.49 A-hr of protons in the beam stop, and a beam Off/On live-time ratio of 3.88. Offline data reduction was conducted in 3 stages. First, events with substantial activity before and during the trigger or without reconstructable tracks were removed. Second, a fiducial volume cut was applied and some energy and dE/dx constraints were imposed to enhance the electron samples. Finally, events with relatively short track lengths were also rejected.

In Fig. 1a, the $\cos\theta_e$ distribution of all events after the third stage of data reduction is plotted for both beam-On and beam-Off. Here θ_e is the reconstructed angle relative to the incident neutrino direction. The forward angular cone, defined as $\cos\theta_e > 0.960$, contains 164.9 ± 20.0 beam-associated events as shown in Fig. 1b. Beam-associated background was estimated by using events in the $0.96 > \cos\theta_e > 0.84$ region and correcting for, based on a Monte Carlo simulation, the small fraction of $\nu_x e$

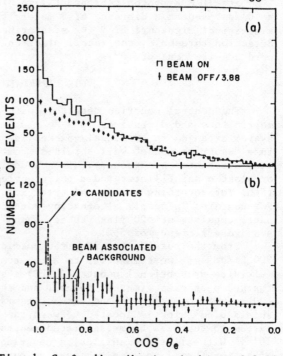

Fig. 1 $\cos\theta_e$ distribution in bins of 0.02.
(a) For beam-on and normalized beam-off events.
(b) For the beam-associated events.

scattering events in this angular range. This was found to be 44.2 ± 14.2,[8] and subtracting this from the 164.9 ± 20.0 events above resulted in 120.7 ± 24.5 events remaining in the forward angular cone which we assign to neutrino electron scattering.

MONTE CARLO SIMULATION

To understand better the detector response to the various physical processes of interest and to evaluate the detection efficiency, a detailed Monte Carlo simulation was carried out. This simulation is based on the EGS-4 (Electron-Gamma-Shower) code released recently.[9] The simulation took into account light attenuation in the scintillators, PMT photo-electron statistics, online trigger requirements, and detailed tracking in the FCM's. The simulated events were translated into the on line data format and then passed through the off line analysis programs. Aspects of the detector not simulated, such as the pretrigger activity in the scintillators or the performance of the MWPC veto system, were examined separately using the measured cosmic-ray muon data. In all, a total systematic error of 6% was assigned to the absolute detection efficiency for $\nu_x e^-$ scattering.

In Fig. 2, comparisons are made of the energy and the angular distributions between Monte Carlo and measured stopped-muon decay data. The agreement as shown provides confidence in our understanding of the detector performance.

RESULTS

Knowledge of the total neutrino flux is based on an earlier neutrino source intensity calibration experiment by H.H. Chen et. al. at LBL (Lawrence Berkeley Laboratory) using 720 MeV incident protons.[10]

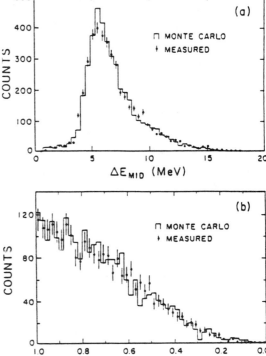

Fig. 2 Comparison of measured and Monte Carlo events for the stopped μ-decay process.
(a) Distribution of ΔE_{mid}, the energy deposited in the middle scintillator slabs.
(b) $\cos\theta_e$ distribution.

Scaling that measured result to the LAMPF beam stop proton energy of 765 MeV and accounting for differences in the beam stops, we arrived at a total neutrino exposure of $(6.88 \pm 0.83) \times 10^{14}/cm^2$ for the present experiment.

To extract the number of $\nu_e e^-$ events, the contribution from $\overset{(-)}{\nu}_\mu e^-$ scattering must be estimated. Using the world-averaged measured cross sections and the corresponding detection efficiencies as determined from our Monte Carlo simulation, we assign 7.1 ± 1.8 and 14.2 ± 3.1 events to $\nu_\mu e^-$ and $\bar{\nu}_\mu e^-$ scattering respectively. Subtracting these from the total $\nu_x e^-$ events observed, 99.4 ± 24.8 events remain, which are assigned to $\nu_e e^-$ scattering. This result agrees with the WSG electroweak theory with

$$\sin^2\theta_w = 0.24 \pm {}^{0.09}_{0.10} \text{ (stat)} \pm {}^{0.05}_{0.06} \text{ (syst)}$$

and a $\nu_e e^-$ cross section of

$$\sigma(\nu_e e^-)/E_\nu = [9.8 \pm {2.7 \atop 2.6}\text{(stat)} \pm {1.5 \atop 1.6}\text{(syst)}] \times 10^{-45} \text{ cm}^2/\text{MeV}.$$

In Fig. 3, the present $\nu_e e^-$ result is compared with the expected number of $\nu_e e^-$ events under various CC and NC interference assumptions. As shown, our result, in good agreement with WSG expectation, rules out constructive interference by more than 4 standard deviations and suggests strongly the existence of CC and NC interference.

We have also directly compared the 120.7 ± 24.5 $\nu_x e^-$ events with the WSG prediction to determine

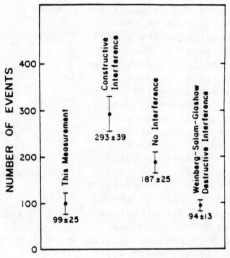

Fig. 3 Comparison of observed ν_e events with expectations involving several Nc and CC interference possibilities.

$$\sin^2\theta_w = 0.24 \pm {0.06 \atop 0.09}\text{(stat)} \pm {0.04 \atop 0.06}\text{(syst)}$$

PLANS AND EXPECTATIONS

We plan to continue taking data in 1986 to increase our beam exposure by about 50%. Combined with analysis improvements currently under investigation we anticipate reducing the statistical error to the 16% level. The present overall systematic error is dominated by the neutrino flux uncertainty of 12%. A new experiment is in progress at LAMPF which will reduce this uncertainty to 6%.[11] We anticipate that with these additional efforts, the $\nu_e e^-$ cross section will be determined to about 18% by the end of 1986.

REFERENCES

(a) Now at Fermilab, Batavia, Il 60510
(b) Now at Baylor College of Medicine, Houston, TX 77004
(c) Now at University of Geneva, Geneva, Switzerland
(d) Now at Cerberus A.G., Mannedorf, Switzerland
(e) Now at Argonne National Laboratory, Argonne, Il 60439
(f) Now at Rockwell International, Thousand Oaks, CA 91360
(g) Also at Tel-Aviv University, Ramat Aviv, Israel 69978
1. S. Weinberg, Phys. Rev. Lett. 19, 1264 (1967).

2. A. Salam, in Elementary Particle Theory: Relativistic Groups and Analyticity (Nobel Symposium No. 8), edited by N. Svartholm (Almqvist and Wiksell, Stockholm, 1968), p. 367.
3. S.L. Glashow, Nucl. Phys. 22, 579 (1961).
4. For $\nu_e e^-$ scattering, F. Reines, H.S. Gurr, and H.W. Sobel, Phys. Rev. Lett. 37, 315 (1976). For $\bar{\nu}_\mu e^-$ scattering, see L.A. Ahrens et. al., Phys. Rev. Lett. 54, 18 (1985) and references within that paper.
5. R.C. Allen et. al., Phys. Rev. Lett. 55, 2401 (1985).
6. B. Kayser, E. Fischbach, S.P. Rosen, and H. Spivack, Phys. Rev. D20, 87 (1979).
7. P.J. Doe at. al., Nucl. Inst. and Methods (to be published).
8. The error quoted here includes some additional systematic uncertainties associated with the shape and the angular region over which the beam associated background is estimated.
9. W.R. Nelson, H. Hirayama, and D.W.O. Rogers, SLAC-Report-265 (1985).
10. H.H. Chen, J.F. Lathrop, R. Newman, and J.C. Evans, Nucl. Instrum. Methods, 160, 393 (1979).
11. Experiment 866 at LAMPF.

Neutrino-Nucleus Interactions:

charged current interactions at Line E of LAMPF

The E764 Collaboration:

University of California, Riverside,
Riverside, CA 95251

University of California, Los Angeles
Los Angeles, CA 90024

Los Alamos National Laboratory,
Los Alamos, NM 87545

University of New Mexico,
Albuquerque, NM 87131

Temple University,
Philadelphia, PA 19122

University of Iowa,
Iowa City, IA 52242

Valparaiso University,
Valparaiso, IN 46383

Presented by R. Fisk
Valparaiso University

ABSTRACT

This paper describes the progress and capabilities of E764 at LAMPF's Line E decay-in-flight neutrino source to measure the $\nu_\mu + {}^{12}C \rightarrow \mu^- + X$ cross section. A small amount of data has been taken and analyzed and, although no physics conclusions can be made with this amount of data, techniques to separate signal from background have been successfully developed. From 7×10^{18} protons on target, we have observed 16 charged current candidates, 12% of which we estimate are background. By August 1986 we expect to have 25 to 50 times more events.

I. THE PHYSICS

The study of charged current neutrino interactions on carbon with energies near the muon production threshold (see figures 1 and 2) over a wide range of q^2 (energy transfers from 0 to 100 MeV) is of interest for several reasons. First, it offers to test the Fermi Gas Model with neutrinos on neutrons, whereas only electrons on protons have tested such models in the past. Second, it will help us learn to what extent the charged current cross section arises from specific nuclear states as opposed to the continuum. Third, it will probe the axial vector structure of the nucleus (which electron scattering alone is unable to do), which may be quite different from that obtained by summing the contributions of individual nucleons.[1,2]

II. THE REACTION

The cross section for $\nu_\mu + {}^{12}C \rightarrow \mu^- + X$ will be measured. There are three physical processes of interest that are used to identify the above final state. These processes occur over three easily separated time intervals. The signal from the muon produced by this interaction in the detector occurs tens of nanoseconds after the protons hit the target. This time interval is called the "prompt interval." The muon then stops and decays with a lifetime of 2.2 μsec. To observe this decay, the detector is activated during a so-called "muon decay interval" which lasts up to 20 μsec after the prompt interval. Finally, in 10% of the cases, the "X" in the above reaction is ^{12}N which beta decays with a lifetime of 11 msec. To observe this decay, the detector is activated during an "N12 interval" which lasts up to 50 ms after the prompt interval. The presence of "prompt" and "muon decay" (and perhaps "N12") signals in spatial proximity within the detector, then, identifies the event as a charged current candidate.

III. THE EXPERIMENTAL ARRANGEMENT

The decay-in-flight neutrino source consists of a water target, a 12 m pion decay channel and a tungsten beam stop followed by 8 m of iron and concrete shielding. The detector is positioned 21 m from the target. The central portion of the detector is a cylinder, 214 cm high and 160 cm in diameter, housing 26 "honeycomb" modules filled with liquid scintillator providing a fiducial mass of 4.5 tons. A 9 inch phototube looks into each end of each module. Outside the central detector is a veto consisting of 2.5 tons of liquid scintillator and eight 2 inch photo-tubes. Surrounding this is 5 radiation lengths of lead shot. Outside the lead absorber is an additional veto which surrounds the entire assembly except at the bottom. The overall veto inefficiency has been measured to be 2.6×10^{-5}.

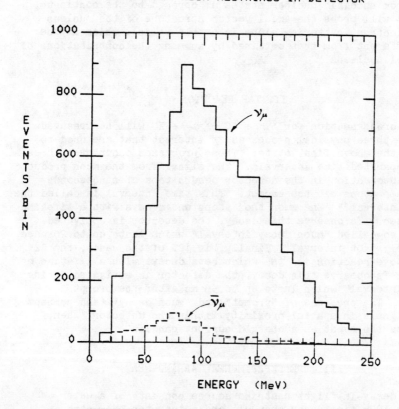

Fig. 1 -- Monte Carlo generated energy spectrum for muon neutrinos and antineutrinos which strike the detector.

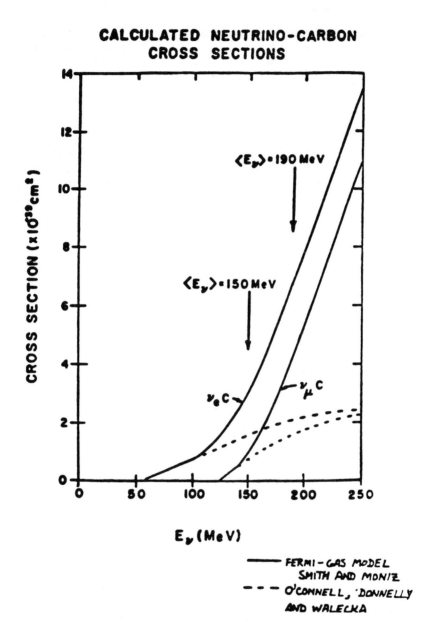

Fig. 2 -- Calculated neutrino-carbon cross sections. For descriptions of the two models used in this graph, see references 1 and 2.

IV. THE PERFORMANCE OF THE DETECTOR

The performance of the detector is exemplified in figure 3 where we see the track produced by a stopped cosmic muon. "P" represents a signal in the prompt interval while "D" represents a signal in the decay interval. The letter size indicates the amount of energy detected in each module. From TDC information, we can also ascertain the height of the track and the relative timing of each module. The latter is displayed by a string of asterisks next to each "P" or "D"; one asterisk per nanosecond. A study of cosmic muons determined that our resolution in energy is $\Delta E/E$ = 12%, our resolution in height is Δz = 30 cm, and our resolution in time is Δt = 0.5 ns. Our estimated energy resolution was also verified by looking at the energy released when muons which stopped in our detector decayed. Figure 4 shows an excellent agreement between the observed decay energy and that predicted using our energy resolution in an EGS-4 Monte Carlo.

V. DATA ANALYSIS

Table I shows the fate of our initial 2762 "beam neutral" events as they underwent various data cuts. A "beam neutral" event consists of signals in both the prompt and decay time intervals with no coincident signals in the vetoes. Most of these events were actually background neutrons. Since most of the neutrons arrived at the detector later than the neutrinos, many were eliminated by cutting on a time of flight measurement (requiring our proton beam to have been given to us in bursts.) Further cuts were made on the spatial proximity of the prompt and decay signals (in x, y and z) and on the total energies within the two time intervals. After the data sample had been reduced to a reasonable number of candidates, further cuts were imposed on an event-by-event basis. Figure 5 shows a typical event rejected by these "scanning" cuts. This event was rejected because the length of the prompt track was too long for a 50 MeV muon and, further, the "muon" was travelling in the wrong direction to have caused the signal seen in the decay region. Such scanning cuts reduced our number of charged current candidates to 21. These same cuts were also applied to a sample of "cosmic neutrals" and "cosmic stopped muons." The "cosmic neutrals", identical to "beam neutrals" but taken between beam bursts, fared roughly the same fate as did our beam-induced neutrals when the same cuts were applied. However, when the live-time for the acceptance of these cosmic-induced events was taken into account, we found the background of such events in our data sample to be insignificant. A sample of "cosmic stopped muon" events, identical to "cosmic neutrals" but <u>with</u> a coincident veto signal, were collected and analyzed. These events should have passed our cuts, since they involve the travel and decay of muons, just as do charged current events. All but 25% of the stopped muons passed our cuts, giving us an estimate of the fraction of true charged current events lost in our analysis.

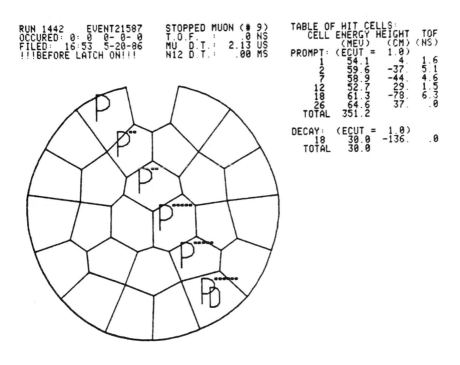

Fig. 3 -- The track of a muon which stopped and decayed in the detector.

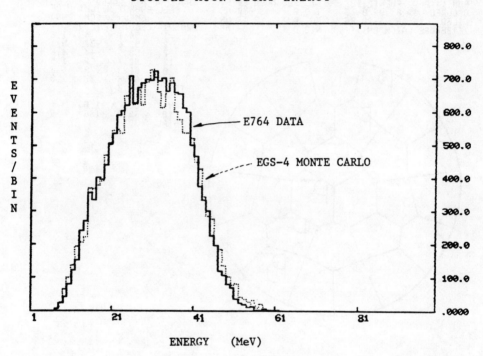

Fig. 4 -- The decay energy distribution of cosmic muons which stopped and decayed in the detector.

TABLE I: REDUCTION OF THE DATA

(7×10^{18} protons on target)

CUT	BEAM NEUTRAL	COSMIC NEUTRAL	COSMIC STOPPED MU
Events w/ trigger in muon decay interval	2762	532*	9671
"P"'s and "D"'s in same or adjacent cells	472	145	9564
Fiducial volume and T.O.F. cuts	200	-	-
$\lvert Z_P - Z_D \rvert < 40$ cm	66	90	7323
$E_D < 60$ MeV	47	86	7255
$E_P < 100$ MeV	42	32†	-
"Scanning" cuts (muon in correct direction, w/ proper range, clean track, etc.)	21	-	-

$$* \quad 532 \times \frac{347 \text{ sec.}}{161000 \text{ sec.}} = 1.1 \text{ event}$$

$$\dagger \quad 32 \times \frac{347 \text{ sec.}}{161000 \text{ sec.}} = 0.07 \text{ event}$$

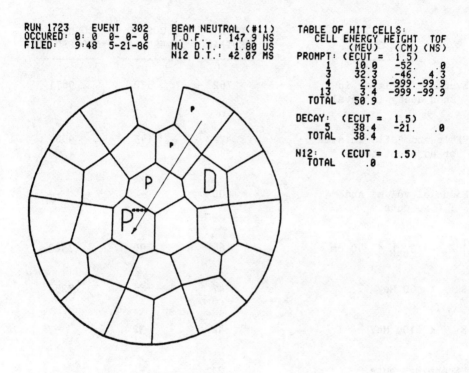

Fig. 5 -- An event rejected by our scanning cuts. The length of the prompt track is too long for a 50 MeV muon and the "muon" is travelling in the wrong direction to have caused the signal in the decay interval.

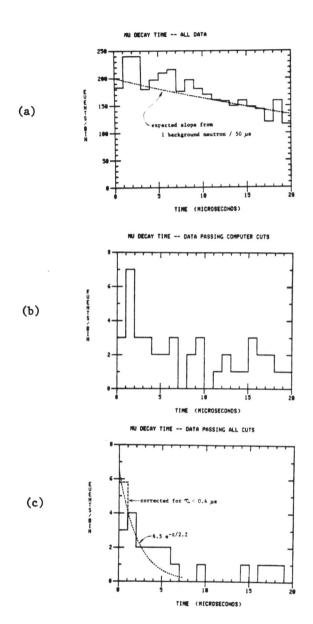

Fig. 6 -- The decay time distributions for (a) all events, (b) those events passing our computer cuts, (c) those events passing all cuts. Notice the predominance of events at low decay times for those events passing our cuts.

TABLE II: BACKGROUND ESTIMATION

	$\tau_\mu < 7$ μsec	$\tau_\mu > 7$ μsec
# events passing cuts	16	5
random background consistent with 5 events > 7 μsec	3.1 ± 1.4	5
"non-adjacent" events	31	102
multiplied by 5/102	1.5 ± 0.7	5

BACKGROUND ESTIMATE: Of the 16 events, 2 ± 1 are background.

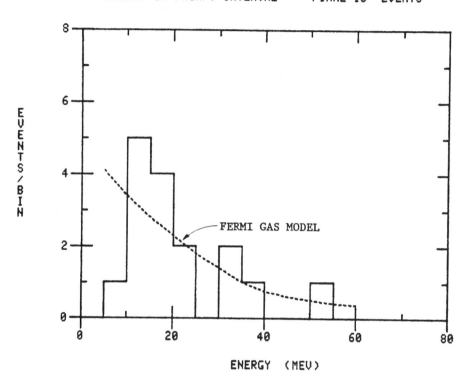

Fig. 7 — The prompt energy distribution of our final 16 charged current candidates. If specific nuclear states make a significant contribution to the cross section, we should see an enhancement in the number of events at the upper end of this spectrum. By August 1986 we expect to have 25 to 50 times more events.

VI. DECAY TIME DISTRIBUTIONS

Figure 6a shows the time distribution of all events in the decay interval. The slope is consistent with background neutrons arriving in our detector at a rate of one every 50 μsec. Figure 6b shows the same distribution for those 42 events which passed our computer cuts. Notice the peak at low decay times, indicating the richness of true charged current events. After our scanning cuts (figure 6c), the distribution is sharply peaked at low decay times, consistent with most of these events being charged current induced.

VII. BACKGROUND ESTIMATION

Table II shows two methods for estimating the background in our final 21 charged current candidates. From figure 6c, we see 16 events have times less than 7 μsec and 5 events have times greater than 7 μsec. Assuming the latter 5 events to be all background, we can estimate how many of the first 16 events are background by assuming a background shape like that of figure 6a. Another way to make this same estimate is to take a sample of events whose prompt and decay signals are in non-adjacent cells, assuring us these are background neutrons. We can then look at the ratio of these events with times less than and greater than 7 μsec. Multiplying this ratio by our 5 events gives us another estimate of our background. In each case, we estimate that roughly 2, or 12%, of our 16 candidates are background.

VIII. CONCLUSION

With the statistics we expect to gather this summer (25 to 50 times more events), we should be able to measure a cross section which will help choose between the cross section predictions from the various models. Plotting our charged current candidates vs. energy in the prompt interval (see figure 7) should give us additional information for this comparison. If specific nuclear states make a significant contribution, the number of events having energetic muons (and therefore less energy transferred to the nucleus) should be enhanced.

REFERENCES

1. R. A. Smith and E. J. Moniz, Nucl. Phys. <u>B43</u>, 605 (1972).

2. J. S. O'Connell, T. W. Donnelly and J. D. Walecka, Phys. Rev. <u>C6</u>, 719 (1972).

NON-ACCELERATOR PHYSICS

Coordinators: D. S. Ayres
 D. A. Sinclair

ULTRA-HIGH ENERGY POINT SOURCES OF COSMIC RAYS

J. W. Elbert
Department of Physics, University of Utah
Salt Lake City, Utah 84112

ABSTRACT

Over 40 air shower observations of 12 point sources are consistent with the hypothesis that the primary cosmic ray particles are 10^{12}-10^{16} eV γ-rays. However, a few observations involving a muon signal from Cygnus X-3 cannot be interpreted as due to γ-rays or other known particles. This problem is described as well as the techniques, data, and prospects in this rapidly growing field.

INTRODUCTION

This article will give a brief summary of the results of work on "γ-rays" and muons coming from unresolved (<1°) angular regions of the sky. The word γ-rays is in quotes because there is not direct evidence that the interstellar particles which produce the small subclass of air showers from point sources are γ-rays. The interstellar or primary particles described here are in the TeV (10^{12} eV) and PeV (10^{15} eV energy range. The distances of the point sources are usually a number of kiloparsecs (1 pc= 3.26 light years). There is an interstellar magnetic field of about 2.5 μgauss and magnetic deflection ensures a nearly isotropic distribution of charged cosmic rays. Therefore, signals from point sources superimposed on the nearly isotropic background are due to neutral particles.

It has been assumed that the neutral primary particles from the point sources are γ-rays. Unlike hadronic showers, few muons are expected from γ-ray showers. But nucleon decay searches at Soudan[1] and Mt. Blanc[2] have independently reported muon signals pointing toward the X-ray source Cygnus X-3. In both cases, the muon flux depends on the phase of the 4.8 hour X-ray period of Cygnus X-3.

If the primary particles producing the muon signal are not γ-rays, but are neutral, what are they? The primaries can't be neutrons because Cygnus X-3 is over 10 kpc away, requiring about 10^5 lifetimes for ~10 TeV neutrons. The primaries are not neutrinos because the flux would have to be too high[3], and the muon signal does not exhibit the weak zenith angle dependence expected from neutrinos. The fact that a periodic signal is present implies that the particles are extremely relativistic. Given the approximate energy of the primary particles, this implies that their mass is less than a few GeV/c^2.

A comparison of the Cygnus X-3 air shower flux to the muon flux implied by the Soudan and Mt. Blanc experiments yields a surprising result. The muon flux is much too high if the air showers are produced by γ-rays. Even hadronic air showers would fall far short of producing the observed muon flux. Halzen[4] points out that the fluxes might be reconciled if the primary particles have small cross sections so they usually would not interact in the atmosphere. The cross section should not be too small, either, or the muon flux would have a weaker zenith angle dependence than observed in the Mt. Blanc experiment. The allowed cross sections on nucleon targets are $10\mu b < \sigma < 1 mb$. If the particles don't have large masses or extremely low cross sections, why aren't they observed in accelerator experiments?

Although the muon signals from point sources raise exciting nuclear physics questions, the field of TeV and PeV astronomy is also exciting and rapidly evolving. As shown in Table 1, about a dozen

Table 1 - Reported TeV and PeV sources

Number of Observations	Object	Type	Reference Number
16	Cygnus X-3	binary pulsar	5-20
11	Crab Nebula	fast pulsar	21
3	Hercules X-1	binary pulsar	22
3	Galactic Plane	our galaxy	23
?	Vela Pulsar	fast pulsar	24
2	4U115+63	binary pulsar	25
1	LMC X-4	binary pulsar	26
1	Vela X-1	binary pulsar	27
1	Centaurus A	radio galaxy	28
1	Andromeda galaxy(M31)	nearby galaxy	29
1	CG195+4(Geminga)	binary pulsar?	30
1	PSR 1953+29	fast pulsar	31
1	PSR1937+21	fast pulsar	32

point sources (plus the galactic plane) have been reported. A casual search turned up 45 detections or indications of these objects. Most sources fall into two classes: rapidly rotating pulsars and binary X-ray sources. Hillas[33] maintains that Cygnus X-3 may accelerate enough nuclei to supply all the <100 PeV cosmic rays in the galaxy. The point sources may be the long sought cosmic ray accelerators.

γ-RAY OBSERVATIONS

The basic sensitivity limitation of the air shower work is the fluctuations in the number of observed showers from the nearly uniform cosmic ray background. The flux sensitivity level for detecting showers above an energy E can be described by $\Sigma = EF$, where F is the flux needed to give a 4σ detection. For Cygnus X-3, Σ (in units of eV cm^{-2}s^{-1}) is only weakly dependent on E for a detector of area A, resolution solid angle Ω, and operating time t. It is easy to show that for a background integral cosmic ray spectrum I,

$$\Sigma \sim 4E \sqrt{\frac{I\Omega}{At}} \qquad (1)$$

In order to detect a low flux, a low value of $\sum$ is required. Eqn.1 illustrates the need for good angular resolution (so that Ω is small) together with large effective areas and long operating times.

Reported detections of Cygnus X-3 are listed in Table 2. The $\sum$ values are approximately given by the products of the reported fluxes and the threshold energies. In the TeV region the $\sum$'s range from 18-89 eV cm^{-2}s^{-1}. The PeV results vary from 11-1300 eV cm^{-2}s^{-1}.

Table 2
Cygnus X-3 TeV and PeV γ-Ray Observations

E_γ (TeV)	#σ	$\sum$ (eV cm^{-2}s^{-1})	Phase	Years	Method*	Ref.
0.5	4.2	40	0.6	'81	Ch	5
0.8	4.4	37	0.63	'83	Ch	6
1	3.6	89	(sporadic)	'80	Ch	7
1	4.4	18	0.625	'81,'82	Ch	8
2	~3.8	60	0.16	'72,-'75	Ch	9
2	3.5	30	0.75	'80	Ch	10
3	~4	24	0.18	'77	Ch	11
30	2.8	126†	0.63	'80-'83	P	12
32	3.7†	173†	0.25,068	'85	Ch	13
300	3.6	18†	0.6	'84-'85	P	14
320	3.4†	128†	0.68	'85	Ch	13
500	5.0	1300	0.6	'76-'77	Ch	15
1,000	3.5	320	0.25	'83	Ch	16
1,000	--	45	0.27,0.63	'84	P	17
1,000	2.9	11	0.6	'81-'84	P	18
2,000	>4.4	148	0.2	'76-'80	P	19
3,000	4.0	45	0.25	'78-'83	P	20
10,000	--	110	0.2	'76-'80	P	19

*Ch:Cherenkov detector,   P:Particle counter array,   †:preliminary

Two basic methods are used to detect air showers in this work. One is to detect Cherenkov light emitted by particles in the air shower. The other is to use an array of particle detectors to sample the shower as it passes through ground level. The effective area of a single optical detector is roughly 5×10^4 m^2. The solid angle, Ω, corresponds to about a $2°$ diameter cone. The yearly viewing time during clear, moonless nights while a source is near to overhead is about 100 hours. For E=1 TeV, Eqn. 1 yields $\sum$=36 eV cm^{-2}s^{-1}. If all the signal falls in one of 10 phase bins of the 4.8 hour period, the result is about 11 eV cm^{-2}s^{-1}. This shows how $\sum$ values like those in Table 2 are attained.

The alternative detector, the air shower array, typically consists of a large number of ~1 m^2 scintillation detectors which sample the particle density at many points on a grid. The total

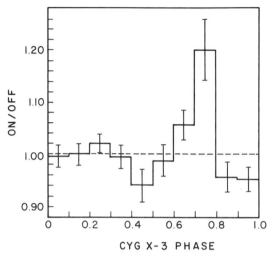

Fig. 1. The ratio of the event rates in the Cygnus X-3 direction (ON) to a background region (OFF) as a function of the 4.8 hour phase. (From Ref. 10)

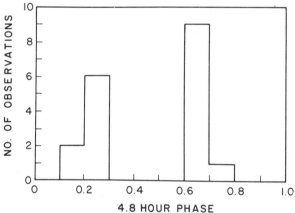

Fig. 2. Phases of the 4.8 hour period in which signals have been reported for Cygnus X-3.

number of particles is calculated and the shower energy is estimated. The shower direction is obtained by accurate timing. With 1 ns time resolution and 10 m grid spacing, 0.5° angular resolution is obtainable. An array can observe a source for ~1500 hours in a year. An example of a next generation array[34] would consist of 1000 detectors covering an area of 300 m x 300 m. At 0.3 PeV, and for a signal in 1 of 10 phase bins, $\sum$ would be about 3 eV $cm^{-2} s^{-1}$. In addition, if a very large muon detector is built below the air shower array, it should improve $\sum$ by an additional order of magnitude. D. A. Sinclair's talk describes such a system.

Figure 1 is an example of a Cygnus X-3 result. For each event the number of elapsed 4.8 hour cycles since a time at which the phase was 0 was calculated. This number (modulo 1) was the event's phase. The phases at which peaks have been reported are shown in Figure 2. Peaks are observed near 0.25 and 0.65. The agreement of a large fraction of all northern hemisphere experimental groups on these phases is one of the convincing features of these observations.

Figure 3 shows the Cygnus X-3 spectrum. The dashed lines show two fits from Ref. 19 to the data above 3.5 x 10^7 eV. Both fits were proportional to $E^{-1.1}$. The measurements are scattered widely. Different techniques, with relative normalization uncertainties, were used in the PeV studies. The Cygnus X-3 flux is time dependent as well.[6] Thus, part of the variation is real.

AIR SHOWER STUDIES WITH MUON DETECTORS

The first reported PeV observation of Cygnus X-3 was by Samorski and Stamm,[19] using the Kiel air shower detector. There was a 4.4σ excess of showers from the direction of Cygnus X-3. In addition, a peak was observed in the 4.8 hour phase distribution. A muon detector was used with the air shower array. The expected result was that showers displaying the "γ-ray" signal would have about a factor of 30 fewer low energy muons than the ordinary cosmic ray showers. The result is shown in Figure 4. The "on source" showers were from the direction of Cygnus X-3 and were chosen from the 4.8 hour phase region in which the signal occurred. The "off source" showers were not from Cygnus X-3. The very surprising result was that there was little difference between the muon densities of the two sets of showers. None of the "on source" showers have extremely low muon densities.

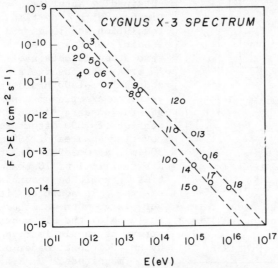

Fig. 3. Cygnus X-3 integral spectrum. Points 1-10 and 12-17 are from Ref. 5-14 and 15-20, respectively. Points 11 and 18 are from Ref. 13 and 19, respectively.

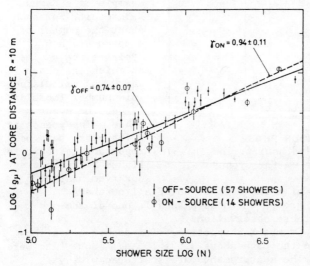

Fig. 4. Muon densities in the Kiel observations of Cygnus X-3.

Some evidence contradicts the Kiel result. At the Akeno air shower array a small signal from Cygnus X-3 was detected by selecting showers with less than 0.03 of the normal muon density.[18] At Tien Shan, Kirov et al.[35] obtained a 4.2σ signal from the Crab Nebula by selecting showers with less than 11% of the average muon density. The signal was not seen without the muon cut. These results indicate that a muon-poor component exists.

DEEP UNDERGROUND MUON OBSERVATIONS

The unexpected muon signals in deep underground experiments were mentioned above. Figure 5 shows the result from the Soudan detector. The dashed line is the expected background. Excess events seen in the phase region from 0.65-0.9 amount to 60 events and a 3.5σ peak. The significance is reduced, however, since there was no a priori reason to look for an excess in this broad phase region. The talk by D. S. Ayres at this meeting discusses some of the Soudan results.

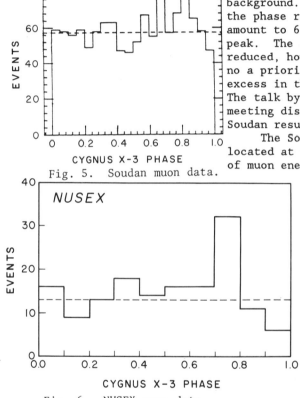

Fig. 5. Soudan muon data.

Fig. 6. NUSEX muon data.

The Soudan detector is located at a depth which, in terms of muon energy losses, is equivalent to 1800 meters of water. In Mt. Blanc, there is another nucleon decay detector, called NUSEX, which is almost 3 times deeper. Figure 6 shows the results from over two and a half years of operation. The data are from a 10° by 10° region centered on Cygnus X-3. In the phase interval 0.7-0.8 there were 32 events observed over a background of 13.0 ± 0.2. The chance probability for this is less than 10^{-4}. The angular resolution of the detector is about 1°, but the bin size was adjusted to give the largest signal.

Figure 7 is the distribution of the events relative to Cygnus X-3. There is no narrow enhancement near Cygnus X-3. The broad region in which the signal appears exceeds the width arising from detector resolution and muon multiple scattering. Conceivably, the muons could be deflected by transverse momentum from the decay of a very massive particle. Then the lower energy muons at Soudan would be spread more than those at NUSEX. This is apparently not the case.

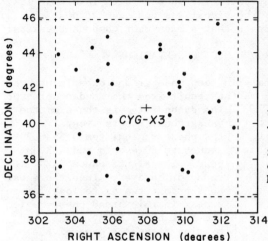

Fig. 7. NUSEX event directions relative to Cygnus X-3.

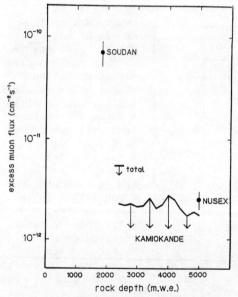

Fig. 8. The Kamiokande muon flux limits compared with the Soudan and NUSEX fluxes. The crooked line shows upper limits for phase 0.7-0.8. The limit for all phases is marked "total".

The Frejus detector has been built at about the same depth as NUSEX. F. Raupach[36] recently reported on a search for a Cygnus X-3 signal in that detector. No signal was found. Figure 8 shows the limits from Kamiokande[37], a nucleon decay detector in Japan. Although Kamiokande is deeper (2400 m.w.e.) than Soudan, the results tend to contradict both Soudan and NUSEX.

CONCLUDING REMARKS

The experimental situation is likely to improve in the next few years. I described a proposed, very sensitive air shower array above to be located at Dugway, Utah. Another array that should gather crucial data is project "Cygnus" at Los Alamos, described in the talk by J. Goodman. These projects will study low energy muons and measure the total number of shower particles. The deep muon situation should be clarified by the large new detectors such as Homestake, Soudan II, and Frejus. The especially difficult problems caused by the apparent variability of Cygnus X-3 may be reduced by simultaneous observatons by suitable pairs of detectors. Examples are the Dugway array with the "Cygnus" project and NUSEX with Frejus.

ACKNOWLEDGEMENTS

This work was supported by the U.S. NSF grants PHY-8201089 and PHY-8515265.

REFERENCES

1. M. L. Marshak et al., Phys. Rev. Lett. 54, 2079 (1985).
2. G. Battistoni et al., Phys. Lett. 155B, 465 (1985).
3. T. K. Gaisser and Todor Stanev, Phys. Rev. Lett. 54, 2265 (1985).
4. Francis Halzen, University of Wisconsin Report MAD/PH/260, 1985.
5. R. C. Lamb, et al. Nature 296, 543 (1982).
6. M. F. Cawley et al., Harvard-Smithsonian Center for Astrophysics preprint 2133 (1985).
7. V. P. Fomin et al., in Proc. 17th ICRC, Paris, 1 28 (1981).
8. J. C. Dowthwaite et al., in Proc. 19th ICRC, La Jolla 1,79('85).
9. A. A. Stepanian, et al.in Proc. 15th ICRC, Plovdiv, 1, 135 (1977).
10. S. Danaher et al., Nature 289, 568 (1981).
11. J. B. Mukanov et al., in Proc. 16th ICRC,Kyoto, 1, 143 (1979).
12. C. Morello, G. Navarra, and S. Vernetto, in Proc. 18th ICRC, Bangalore, 1, 127 (1983).
13. J. W. Elbert et al., University of Utah preprint UUHEP-86/4.
14. V. V. Alexeenko et al., in Proc. 19th ICRC, La Jolla, 1,91('85).
15. C. L. Bhat, M. L. Sapru, and H. Razdan, Bhabha Atomic Research Center (Sprinagar, India) preprint (1985).
16. R. M. Baltrusaitis et al., Ap.J. 297, 145 (1985).
17. A. Lambert et al., in Proc. 19th ICRC, La Jolla, 1, 71 (1985).
18. T. Kifune et al., Ap. J. 301, 230 (1986).
19. M. Samorski and W.Stamm, Ap.J. 268, L17 (1983).
20. J. Lloyd-Evans et al., Nature 305,784 (1983).
21. J. Boone et al., Ap. J. 285, 264 (1984).
22. M. F. Cawley et al., in Proc. 19th ICRC, La Jolla, 1, 119 (1985).
23. J. C. Dowthwaite et al., Astron. Ap. 142, 55 (1985).
24. S. K. Gupta et al., in Very High Energy Gamma Ray Astronomy, edited by P. V. Ramana Murthy and T. C. Weekes (Tata Institute for Fundamental Research, Bombay, 1982), p. 282.
25. R. C. Lamb and T. C. Weekes, Harvard Smithsonian Center for Astrophysic preprint 2239 (to be published in Ap. Letters).
26. R. J. Protheroe and R. W.Clay, Nature 315, 205 (1985).
27. R. J. Protheroe, R.W. Clay, and P. R. Gerhardy, Ap. J 280,L47 (1984).
28. J. Grindlay et al., Ap. J. 197, L9 (1975).
29. J.C. Dowthwaite et al., Astron. Ap. 136. L14(1984).
30. Yu. L. Zyskin and D. B. Mukanov, Sov. Astr. Lett. 9, 117 (1983).
31. P. M. Chadwick et al., University of Durham preprint, 1985.
32. M. P. Chantler et al., University of Durham preprint, 1984.
33. A. M. Hillas, Nature 312, 50 (1984).
34. J. W. Cronin, private communication.
35. I. N. Kirov et al., in Proc. 19th ICRC, La Jolla, 1, 135 (1985).
36. F. Raupach in a talk at the Moriond Astrophysics Meeting at Les Arcs, France, March, 1986.
37. Y. Oyama et al., Phys. Rev. Lett. 56, 991 (1986).

Cygnus Experiment at Los Alamos

B.L. Dingus, J.A. Goodman, S.K. Gupta, R.L. Talaga,
C.Y. Chang and G.B. Yodh
University of Maryland, College Park, MD. 20742

R.W. Ellsworth
George Mason University, Fairfax. VA. 22030

R.D. Bolton, R.L. Burman, K.B. Butterfield, R. Cady, R.D. Carlini,
J.S. Frank, W. Johnson, D.E. Nagle, V.D. Sandberg, and R. Williams
Los Alamos National Lab., Los Alamos. NM. 87545

J. Linsley
University of New Mexico, Albuquerque. NM. 87131

H.H. Chen, R.C. Allen, P.J. Doe, and W.P. Lee
University of California, Irvine, CA. 92717

Abstract

The Cygnus experiment at Los Alamos National Laboratory has been designed to study, with high angular accuracy, point sources of gamma rays above 10^{14} eV. The experimental detector consists of an air shower array to observe gamma ray showers and a shielded, large area track detector to study the muon content of the showers. In this paper we present preliminary data from the array and describe its performance.

Introduction

The study of ultra high energy cosmic ray gamma rays provides an exciting new window to explore the origin of high energy cosmic rays (A.M.Hillas 1984). The interest in this field has grown steadily since the first reports of observations of point sources (Stepanian et al 1972, Ramanmurthy and Weekes 1982, A.A.Watson 1985). The reports of the Kiel group(Samorsky and Stamm 1983) showed strong evidence for a signal from Cygnus X-3, which showed phase correlation with the orbital period as seen in X-rays. This result also suggested that the observed showers had a muon content much higher than would be expected from gamma ray induced showers. Since then, other experiments using both air shower techniques and underground muon detectors have reported results of varying statistical significance. If showers from Cygnus X-3 are muon rich relative to expectation they may indicate the existance of new particle physics phenomena either at the source or in the interactions of high energy gamma rays with atmospheric nuclei(Barnhill et al 1985) None of these experiments (Keil excepted) has shown a significant signal without the use of phase analysis.

The Cygnus experiment at Los Alamos National Laboratory was designed to search for the presence of point sources of ultra high energy gamma rays and to study the muon content of their air showers. This detector was designed to have an angular accuracy of better than 1°, in order to improve the signal to background ratio of this experiment. The location of this experiment was chosen for several reasons. First, the presence of a working fine grained track detector (E225) which could be used to detect muons in a clear and unambiguous manner. Second, the detector is located at an altitude of 7000' which allows this experiment to observe showers produced from lower energy gamma rays than previous air shower experiments done at greater depths. Third, the facilities of a major laboratory were available to facilitate the construction and operation of this experiment.

EAS Detector Design

The EAS detector consists of 64 counters placed in an array of radius 60 meters with a typical separation of 14 meters. Each counter contains a scintillator, approximately 1 square meter by 8 cm thick, with a 2 inch photomultiplier tube positioned 70 cm above the scintillator. Single minimum ionizing particles selected by small scintillator paddles result in a timing resolution of standard deviation 1.6 ns and give about 20 photoelectrons in the photomultiplier tube.

EAS Trigger

Every counter is used in making the trigger decision as well as giving pulse height and timing information. The basic trigger requires that a given number of counters must fire their individual discriminators within a 300 ns interval, the time for an EAS with a zenith angle of 45° to the array. The discriminators responsible for the trigger are also used to determine the timing, so the threshold is set very low, about 1/10 of a minimum ionizing particle, in order to fire the discriminator on the earliest photoelectron.

A software cut is implemented to elliminate non-analyzable showers before they are recorded on magnetic tape and reduces our data taking rate by about a factor of five. Figure 1 shows the online display of a typical event which passes the software criteria.

Muon Detector

Muon information is also recorded for every trigger. The muon detector was designed for studying the elastic scattering of accelerator produced neutrinos with electrons (E225) and is currently being used for that purpose (Allen, 1985). A multiplexing circuit allows both the neutrino and the air shower experiment to use the detector simultaneously. The detector is shielded above by $1700 g/cm^2$ of steel and concrete. Two components of the detector–the multiwire proportional chambers (MWPCs) and the flash chamber calorimeter– are used to determine the muon content and direction in showers. The MWPCs surround the detector with 4 layers on all 6 walls except the floor which is 1 layer. Each MWPC is typically 520 cm long×20 cm wide×5 cm thick and the horizontal area is 36 m^2. The muon number can be determined exactly for small numbers and systematically for higher densities.

The 208,000 flash chambers cover a volume of $305 \times 305 \times 348 cm^3$ and have sufficient resolution to determine the muon direction and number as can be seen from the example of Figure 2. Simulations of a fitting algorithm show that 95% of the tracks can be reconstructed to within 0.5° . However, the flash chambers can not be triggered for all EAS which come at a rate of 1.2 Hz.

Smart Trigger

The rate at which it is reasonable to fire the flash chambers in the E225 calorimeter is 0.02 hz. It was necessary therefore to reduce the rate of triggers to the E225 detector by a factor of greater than 50. A computer controlled hardware trigger was devised to allow only events that come from a specified direction to trigger E225. This smart trigger was needed because the flash chambers must be fired within a microsecond of particles traversing it. This does not allow for timing information to be digitized or processed. The direction of a suspected source moves 1° in four minutes, thus it requires that the trigger be continually updated. This is accomplished by the use of ECLine Camac programable logic delays. The gates for this trigger are set to a width 8 ns and the relative delays of each counter are set to zero for showers coming from the source direction. A multiplicity coincidence level is set and only showers coming near the desired direction are accepted. In figure 3 we show a plot of the sky (right ascension and declination) the crosses are showers which pass the smart trigger criteria over a 24 hour period in which three sources were being watched, Cygnus X-3, the Crab, and Herc X-1. This trigger selects events in a cone of half angle 9° and gives a trigger every 2 min when a source is overhead.

Monte Carlo

Monte Carlo calculations were performed to study the design and triggering conditions of the array. These calculations included proton and gamma ray induced showers. The primary energies were selected from a spectrum and the cores of these showers were throw over the area of the detector and the surrounding areas. The hadron component of these showers was simulated fully. Electromagnetic showers were followed down to 500 GeV where Approximation B was used for longitudinal development. The radial disrtibution of each subshower was computed. Muon densities were computed for annular rings for each shower. Differing trigger conditions where imposed and rates for each were computed.

The results of the simulations show that with a trigger requirement of 10 counters each with greater than 2 equivalent minimum ionizing particles the effective threshold for proton showers is 10^{14}ev while gamma ray induced showers have a threshold of 2×10^{14}ev. The muon simulation showed that the E225 detector contained at least one muon for 80% of proton induced triggers. This number is consistent with the observed number of 75% of showers having one or more muons in data.

The Monte Carlo was also used to study reconstruction algorithms. These simulations showed that with a timing resolution of 2ns (for large signals) it was possible to obtain a resolution of better than $.65°^2$ for showers which passed our threshold.

Reconstruction and Resolution of Events

Event direction is computed by fitting the shower arrival direction. This fitting is done by minimizing χ^2 of the arrival time distribution. The showers are seen *in data* to have a curvature. The observed value of this curvature ($\sim 10ns/60m$) is included in the fits. Counters are weighted in this fit so that counters with larger signals are given more significance. An estimate of the resolution of these counters can be obtained by studying the distribution of the quantity $\frac{\chi^2}{\nu} \times \sigma^2$ where ν is the number of degrees of freedom. This quantity should have the average value of σ^2. For our showers this yields a value of $\sigma = 2ns$ for greater than 3 particles signal.

A test of the random reconstruction error in the array can be obtained from data by the following procedure: Counters are divided in to two groups - odd and even numbered counters. Each group of counters is used independently to reconstruct the arrival direction. The space angle between these directions is found to have a median value of $< 1°$. This predicts a resolution for the combined array of $< 0.75°^2$. Work is still being done to improve this resolution.

Three independent tests of pointing accuracy are being under taken. First, the arrival direction of muons, detected in the E225 flash chambers, is being compared to the air shower data. Very preliminary results indicate reasonable agreement between the two directions, consistent with the expected multiple scattering angle of the muons in the shielding.

Second, a small Cherenkov array has been deployed at the experiment site. This will be used to determine shower direction and energy in way which is systematically different than the air shower method. Tests have been made with these counters and it is planned to take data shortly.

Third, we plan to use our existing data to study the shadow of the moon using ordinary hadron data. If our accuracy is better than $(1°)^2$ then we should see a substantial reduction of data in bins which contain the moon. This may also work with the sun. We hope to have results shortly on this technique.

Status

The experiment has been operational since early March 1986 with more than 40 scintillation counters and the MWPC information. More than 10^6 events have been recorded. As of May 15, 60 counters are deployed, and by the end of summer nearly 100 detectors will compose an expanded array of radius 90 m. The flash chamber multiplexing scheme is operational, and data will be recorded on a regular basis following the completion of routine detector maintenance by June 26.

An additional detector (E645) which can give muon information, both number and direction, is coming on line this summer to study the oscillations of accelerator produced neutrinos (Smith, 1985). This detector, consisting of liquid scintillator and drift tubes, has an horizontal area of 56 m^2 and an overburden of 3000 g/cm^2. Liquid scintillator information has been successfully recorded for EAS triggers.

References

A. M. Hillas, Nature, **312**, 50, 1984.

A. A. Stepanyan et al., Nature Phys. Sci. **239**, 40,1972.

B. M. Vladirmirski,A. M. Gal'per, B. I. Luchkov and A. A. Stepanyan, Usp. Fiz. Nauk,**145**, 255. 1985, [Soviet Phys. Usp.**28**,153, (1985)].

P. V. Ramanamurthy and T. C. Weekes, editors, Proc. of the Int. Workshop on Very High Energy Gamma Ray Astronomy, Ootacamand, India, 1982.

A.A.Watson, Rapporteur paper presented at the 19th International Cosmic Ray Conference, La Jolla 1985,(to be published). This paper gives a complete review of UHE gamma ray experiments.

M. Samorski and W. Stamm, Ap. J. Lett.,**268**, 117, 1983.

M. V. Barnhill, T. K. Gaisser, T. Stanev and F. Halzen, Nature, **317**, 409, 1985.

R. C. Allen et al, Phys. Rev. Lett. **55**, 2401, 1985.

E. S. Smith et al, "Search for Neutrino Oscillations at LAMPF" presented at Moriond Conference on Massive Neutrinos and Astrophysics, Jan. 1985.

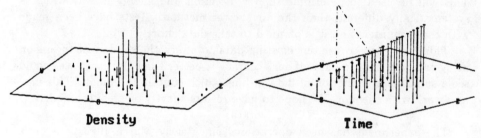

Fig 1. Density profile and arrival time of an EAS event.

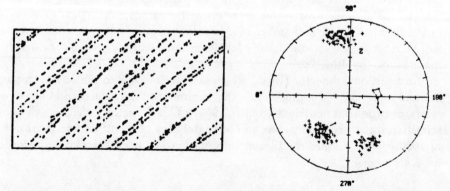

Fig 2. Side view of flash chambers.   Fig 3. RA and Dec for smart trigger.

NEW EVIDENCE FROM SOUDAN 1 FOR
UNDERGROUND MUONS ASSOCIATED WITH CYGNUS X-3

D.S. Ayres
High Energy Physics Division
Argonne National Laboratory
for the
Soudan 1 Collaboration[*]

ABSTRACT

The Soudan 1 experiment has obtained additional evidence for underground muons associated with the x-ray pulsar Cygnus X-3. We report the preliminary analysis of data recorded during the October 1985 radio outburst of Cygnus X-3, which show a significant excess of muons for a narrow range of Cygnus X-3 phases.

Several observers have reported cosmic-ray air showers originated by 1 to 1000 TeV primaries associated with the x-ray pulsar Cygnus X-3.[1] Both the direction (declination 40.8°, right ascension 307.6°) and flux modulation characteristics (4.8 hour period) of Cygnus X-3 are used to associate the showers with this source. The primaries must be stable (Cygnus X-3 is at least 37,000 light years away), electrically neutral (to avoid deflection by the galactic magnetic field), light (to preserve the phase signature), and must initiate showers in the atmosphere (to be detected). The photon is the only known particle which satisfies the criteria.

During 1981-3, the 30-ton Soudan 1 proton-decay detector acquired a large data sample of cosmic-ray muons which had penetrated its 2000 ft rock overburden (surface energy > 650 GeV). Analysis of these data by methods similar to those used by air-shower experiments have shown evidence for a substantial flux of underground muons associated with Cygnus X-3.[2] This flux is several hundred times that expected from the photon primaries associated with the source, leading to a serious dilemma which seems to require the existence of a new phenomenon. Either multi-TeV photons have an unexpected new interaction which produces high-energy muons, or the primary is a new, neutral, light, stable particle. While the NUSEX experiment has also reported underground muons associated with Cygnus X-3,[3] other experiments have reported flux limits lower than Soudan 1 and NUSEX.[4]

The Soudan 1 detector recorded 340,000 cosmic-ray muon events between February 1985 and February 1986. This 0.4 year live-time data sample is about 40% of the 1981-3 sample. We report here the preliminary results from the analysis of the new data. The previous evidence for underground muons associated with Cygnus X-3 relied on coherence with the precisely known 4.8-hour periodicity of the source.[2] Events pointing to within 3° of the source direction were binned according to the phase determined from the arrival time and the known ephemeris of Cygnus X-3.[5] The resulting phase plot showed an excess of events with phases between 0.65 and 0.90, consistent with the phases of air-shower observations. The new data show no evidence for an excess of events in any phase interval. Similarly, no signal is seen in the phase plot for muons which arrived during high-rate periods (within 30 minutes of another muon).[2]

From 3-13 October 1985, Cygnus X-3 produced the most intense radio outburst ever recorded for this source.[6] The phase plot for the 102 muon events recorded between 24 September and 7 October is shown in Fig. 1 . The signal has been enhanced by choosing this two-week period about a week earlier than the peak in radio emissions, by increasing the angular acceptance from a 3° to a 5° half-angle cone, and by choosing phase bins of 0.025 instead of 0.05. The background shown in the figure was obtained from 23 5° cones centered at the same declination as Cygnus X-3, but at other right ascensions spaced at 15° intervals. Figure 2 shows the rate of muons per week in the 0.725-0.750 phase interval selected from Fig. 1. The 16 events recorded from 24 September to 21 October is much larger than the 6.3 events expected from the analysis of the 23 background directions during the same four weeks; the Poisson probability for such an occurrence is 0.05%. The probability of such a fluctuation occurring in one of the 10 phase bins of our 1981-3 peak is ten times this, or 0.5%.

In order to better assess the statistical significance of this result, we have performed a search for other bursts of muons in

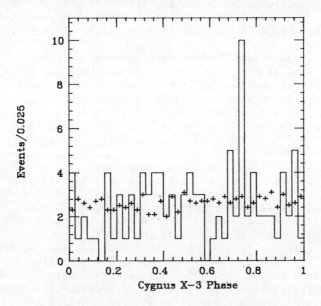

Fig. 1. Cygnus X-3 phase plots for events between 24 September and 7 October 1985, within 5° of the source direction. The solid histogram is for the Cygnus X-3 direction, and the dashed histogram is for 23 the off-source directions.

narrow bins of Cygnus X-3 phase and time. The number of events in each of 4600 bins of phase (0.005 bin width) and time (2 week bin width) was calculated for the 1985-6 data sample. Figure 3 is a histogram of the occupation frequency of these two-dimensional phase-time bins, for both the Cygnus X-3 direction and the 23 background directions. The average number of events per bin is 0.3; the numbers of events expected from the Poisson distribution with this average are shown in the figure. The phase-time bin with 6 muons has the phase range of 0.740 to 0.745 and the time range of 24 September to 7 October. The probability of such an occurrence is consistent with the 0.5% derived from Fig. 2. Data from the 23 background directions had no bursts with more than five muons in any phase-time bin.

Observation of underground muons with Soudan 1 will continue until the much larger Soudan 2 experiment begins operation. With several other large underground experiments also operating, the question of whether Cygnus X-3 emissions include primaries which produce underground muons should be resolved within the next few years.

This work was supported by the U.S. Department of Energy and the Graduate School of the University of Minnesota. The experiment has been conducted with the cooperation of the State of Minnesota, Department of Natural Resources, particularly the staff at Tower-Soudan State Park.

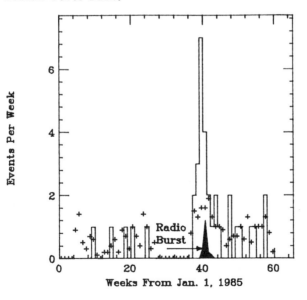

Fig. 2. Events recorded per week within the phase interval 0.725 to 0.750. The solid histogram is for events within 5° of the Cygnus X-3 direction, and the dashed histogram is obtained from the 23 off-source directions. Weeks with zero background events indicate times when the detector was off.

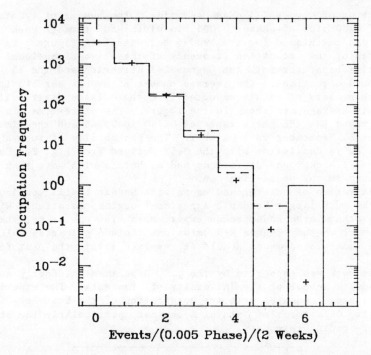

Fig. 3. Frequency of occurrence of different bin populations in two-dimensional phase-time plots, for on-source (solid) and off-source (dashed) data during 1985-6. The average bin population is 0.3, and the points show the expected Poisson distribution for this average. The one bin with six events occurred during the Cygnus X-3 radio outburst.

REFERENCES

*The Soudan 1 collaboration consists of J. Bartelt, H. Courant, K. Heller, S. Heppelmann, T. Joyce, M. Marshak, E. Peterson, K. Ruddick, and M. Shupe at the University of Minnesota, and D. Ayres, J. Dawson, T. Fields, E. May, L. Price, and K. Sivaprasad at Argonne National Laboratory.

1. J. Lloyd-Evans et al., Nature 305, 784 (1983) and other references listed in this paper.
2. M. Marshak et al., Phys. Rev. Lett. 54, 2079 (1985), and 55, 1965 (1985).
3. G. Battistoni et al., Phys. Lett. 155B, 465 (1985).
4. Y. Oyama et al., Phys. Rev. Lett. 56, 991 (1986), and unpublished reports at the 7th Workshop on Grand Unification, Toyama, Japan, April 1986.
5. M. van der Klis and J.M. Bonnet-Bidaud, Astron. Astrophys. 95, L5 (1981).
6. M. Waldrop, Science 231, 336 (1986), news article.

The Search for Solar Neutrinos at Kamioka

Presented by B. Cortez[†]

(for Kamiokande II Collaboration)
California Institute of Technology
Pasadena, California 91125

ABSTRACT

The Kamiokande II[1] collaboration upgraded the Kamioka nucleon decay experiment to allow a search for solar neutrinos. The ^{8}B solar neutrinos are detected via neutrino-electron scattering yielding electrons above the Cherenkov threshold in water. The current threshold of 10 MeV gives an expected signal of 0.1 events/day, about an order of magnitude below the background. This background is due to a combination of uranium dissolved in the water, photons generated in the rock, and beta decays from long-lived isotopes caused by the breakup of ^{16}O by cosmic ray muons traversing the detector. Each background can be reduced in the coming months and allow the threshold to be reduced to 7 MeV.

Introduction

The detection of neutrinos is an important way to study processes taking place in the interior of stars, since their spectrum tells you which nuclear reactions are occurring and at what rate. The pioneering experiment of R. Davis et al.[2] in the Homestake Goldmine was the first attempt to measure these neutrinos from the sun. Their ^{37}Cl detector could measure individual ^{37}Ar atoms produced through the charged current reaction $\nu_e + {}^{37}\text{Cl} \rightarrow e^- + {}^{37}\text{Ar}$, above a threshold of 0.8 MeV. This threshold is too large to detect the vast majority of solar neutrinos, produced by the $pp \rightarrow de^+\nu_e$ reaction in the sun. However, the highest energy neutrinos, produced by ^{8}B in the sun, are detected with large efficiency, accounting for ~75% of the expected rate, according to calculations

---
[†]Supported by Robert A. Millikan Research Fellowship and DOE Contract No. DE-AC03-81-ER40050.

of Bahcall et al.[2] The Davis experiment saw only 1/3 as many neutrinos as expected, which is consistent with complete absence of the ^{8}B neutrinos. This constitutes the solar neutrino puzzle.

The Detector

The Kamioka nucleon decay experiment is a large tank with 3kT of purified water, surrounded by 948 20" photomultiplier tubes (PMT). Charged particles emitting Cherenkov light are easily detected down to a threshold of ~15 MeV. The background due to cosmic ray muons or terrestrial radioactivity is negligible at this level, compared to a signal of contained events from cosmic ray induced neutrinos of 0.3 events per day, allowing searches for nucleon decay at the level of a few events per year.

Recently, the collaboration, based at the University of Tokyo, was expanded to include the University of Pennsylvania and Caltech. An upgrade reduced the threshold to 5 MeV, to begin the search for the ^{8}B solar neutrinos. First, new discriminator and multihit timing electronics allow the trigger to be based simply on the number of PMT in coincidence. The existing trigger has evolved from a simple summed pulse height trigger that could not be lowered below 30 MeV to a more complicated scheme requiring adjacent energy deposition that worked to 13 MeV, still too large to look for solar neutrinos. The discriminator operates at a threshold of 0.3 photoelectron (pe). The timing measurement enables reconstruction of the vertex and direction of low energy events, to help reject background. Finally, an active, instrumented water shield is placed around the detector to attenuate the entering photon background and easily veto entering or exiting tracks. The sum of the output of all the PMT's in the detector and in the anticounter is recorded using separate flash ADC's to look before and after the event. Figure 1 shows the upgraded detector.

The Signal

The solar neutrinos scatter off the electrons in the water. The cross section is very small, ~10^{-44} cm^2. On average, half of the neutrino energy goes to the scattered electron, which lights up 4 PMT per MeV. At a threshold of 7 MeV, we expect 0.5 events per day at the Davis flux. The energy spectrum drops rapidly to a maximum energy of 14 MeV. The energy calibration is critical, because 10% systematic errors at the threshold can translate into 30% errors in the rate. There is no calibration source readily available at the relevant energy of 5-15 MeV, so we now rely on other methods. The EGS program simulates electrons stopping in water and calculates how much Cherenkov light is given off. The simulation is tied to the data by comparing the light generated for decay electrons from stopping muons and reconstructing the Michel spectrum up to 53 MeV. An additional check comes from studying high energy cosmic ray muons traversing the detector.

The multiple scattering of these low energy electrons, which averages about 10° per mm of track, is important for a proper simulation. The Cherenkov ring is modified by the multiple scattering, affecting the vertex and direction resolution. Figure 2 shows the tracks of 20 simulated 10 MeV/c electrons. By combining the timing and topology, we are able to reconstruct the vertex to a mean error of 1.5m and the direction to 35° for events with the solar neutrino energy spectrum above a threshold of 20 PMT.

The signal, due to neutrino electron scattering, has the kinematic property that the electron follows the neutrino to within 20°. By measuring the electron direction and comparing it to the direction of the sun, we should be able to reduce the background by an order of magnitude and verify that the detected particles do indeed come from the sun.

Data Reduction

The upgraded experiment has been taking data since December 1985, with an on-line threshold of 20 PMT in 100ns coincidence, giving an event rate of ≤1 Hz. Some of this, 0.3 Hz, is due to cosmic ray muons traversing the detector and the remainder are triggers below 20 MeV. We reduce the data to a manageable level by the following steps:

1) Require at least 25 PMT in coincidence. The trigger is 50% efficient at 20 PMT, but is ~100% efficient at 25 PMT so we don't have to worry about trigger efficiency calculation.

2) The anticounter must be quiet during the event. This throws out entering or exiting particles.

3) The anticounter must be quiet for 20 μsec before each event, using the flash ADC. This will eliminate cosmic ray muons which lit up the anticounter, but slowed down below Cherenkov threshold as they entered the detector and did not cause a trigger. Then 2 μsec later a muon decay electron could be generated at the edge of the detector,

with no apparent cause.

4) No trigger can occur in the last 100 μsec. This removes all muon decay electrons.

5) The maximum number of pe is 100. This corresponds to ~25 MeV, well above the solar neutrino endpoint of 14 MeV.

6) The maximum pe in one PMT must be less than 10. This cuts out events near the edge of the detector.

7) Find the vertex of the event and require that it is more than 2m in from the plane of the PMT's. This eliminates entering photon background from radioactivity in the rock walls or the steel tank.

These requirements are highly efficient at saving solar neutrino interactions and removing background. Ninety events/day remain above 25 PMT, and 3.8 events/day remain above 35 PMT, at least an order of magnitude larger than the expected signal of 0.1 events/day. Figure 3 shows one event passing these criteria. These events come from the following background sources.

1) There is 0.5×10^{-9} uranium in the water by weight. The ^{238}U has a decay chain that leads to ^{214}Bis with a beta decay endpoint of 3.26 MeV. The rate of decay is 2×10^4 Hz or 1.5×10^9/day. It is likely that the dominant background from 3-7 MeV is due to ^{214}Bis decays.

2) The ^{238}U can undergo spontaneous fission with a lifetime of 10^{16} years. This leads to a rate of 400/day. On average, 7 MeV/fission is released into photons. Occasionally ($<10^{-2}$) more than 7 MeV goes into one photon which could look similar to solar neutrino events. To reduce backgrounds 1 and 2, we must filter the uranium and other heavy elements out of the water. An ion exchange resin can remove a factor of 40 of the uranium in each pass, taking a month to cycle through the water. To remove the ^{214}Bis decays, we must also eliminate radium with a 1600 year lifetime which decays to radon (3.8 days) to bismuth. More sophisticated filtering techniques are under construction.

3) The rock surrounding the detector produces high energy gammas. These photons are attenuated going through the 1.5m water shield (40 cm interaction length), but several hundred per day will reach the inner detector, mostly interacting at the outside. Occasionally, a photon will reach the fiducial volume before interacting or the vertex reconstruction will be off and move an event into the fiducial volume.

4) The high energy cosmic ray muons traversing the detector can cause the ^{16}O nucleus to break up into various isotopes, some of which are relatively long lived and have energetic beta decays. The worst backgrounds are due to ^8B and ^8Li, with endpoints of 14 MeV and lifetimes of 0.8 sec. To remove this background, we must reconstruct the path of every cosmic ray muon and throw out every possible event that occurs in the next few seconds within a 2m cylinder centered around the track. This background can be used as an additional calibration since it has a known spectrum in the region of interest (5-15 MeV).

5) Stopping μ^- can be captured on ^{16}O. This creates ^{16}N with a 7 sec lifetime and beta decay endpoint of 10 MeV. We must find each stopping muon, and for those events without a decay, we need to throw out all events in the next 30 sec which occur within 2m of the endpoint. We expect ~1 event per day above a 7 MeV threshold.

Even after some of these background events filter through, we have an additional handle to separate out the signal. We reconstruct the direction of the event and compare to the sun. The solar neutrino induced electrons should point back to the sun with a mean error of 35°. We can get an additional factor of 5 to 10 rejection of isotropic background by this method.

Conclusion

The Kamiokande II detector can operate down to a threshold of 7 MeV. The background is still large, but methods are in progress to remove it. Even with a background ten times larger than the signal, the angular requirement pointing to the sun will produce an effect of five standard deviations in one year of running. A measurement of the solar neutrino flux from ^8B at the level of the Davis experiment is possible within the year.

References

1. The Kamiokande II Collaboration includes:
    B. G. Cortez, California Institute of Technology; K. Takahashi, KEK; K. Miyano, U. of Niigata; E. W. Beier, L. Feldscher, S. B. Kim, A. K. Mann, F. M. Newcomer, R. Van Berg, W. P. Zhang, U. of Pennsylvania; and K. Hirata, T. Kajita, T. Kifune, M. Koshiba, M. Nakahata, Y. Oyama, N. Sato, T. Suda, A. Suzuki, M. Takita, Y. Totsuka, Univ. of Tokyo.

2. J. N. Bahcall, B. T. Cleveland, R. Davis and J. K. Rowley, *Astrophysical Journal* **292**, L79 (1985).

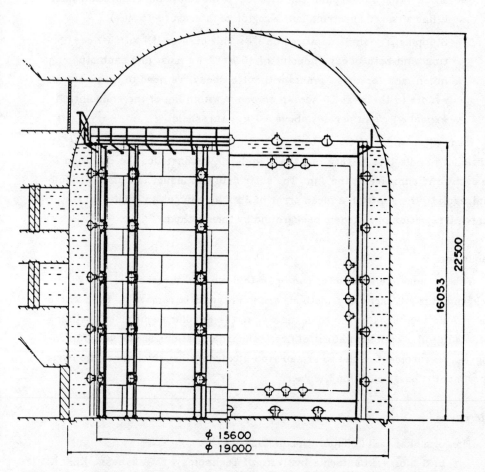

Figure 1. The Kamiokande II detector. Dimensions are in millimeters.

Figure 2. Representation of the trajectories of twenty Monte Carlo generated 10 MeV/c electrons in water. The initial direction of all electrons is the same.

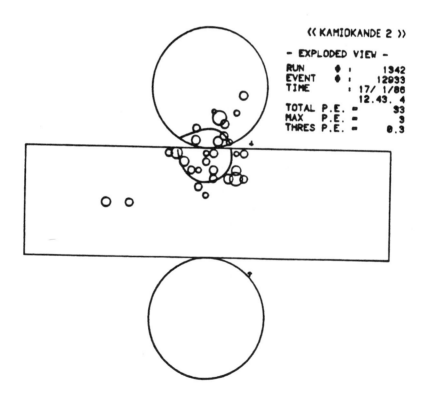

Figure 3. A candidate electron event with energy approximately 10 MeV. For a detailed description of the figure see the text.

A D_2O CERENKOV DETECTOR FOR SOLAR NEUTRINOS

E.D. Earle[1], G.T. Ewan[2], H.W. Lee[2], H.-B. Mak[2],
B.C. Robertson[2], R.C. Allen[3], H.H. Chen[3], P.J. Doe[3],
D. Sinclair[4], W.F. Davidson[5], C. Hargrove[5],
R.S. Storey[5], C. Aardsma[6], P. Jagam[6], J.J. Simpson[6],
E.D. Hallman[7], A.B. McDonald[8], A.L. Carter[9], and
D. Kessler[9].

[1] Chalk River Nuclear Laboratories, Chalk River, Ontario, K0J 1J0, Canada
[2] Queen's University, Kingston, Ontario, K7L 3N6, Canada
[3] University of California, Irvine, CA, 92717, USA
[4] University of Oxford, Oxford, OX1 3NP, UK
[5] National Research Council of Canada, Ottawa, Ontario, K1A 0R6, Canada
[6] University of Guelph, Guelph, Ontario, N1G 2W1, Canada
[7] Laurentian University, Sudbury, Ontario, P3E 2C6, Canada
[8] Princeton University, Princeton, NJ 08544, USA
[9] Carleton University, Ottawa, Ontario, K1S 5B6, Canada

ABSTRACT

The construction of a solar neutrino observatory utilizing 1000 Mg of D_2O in a nickel mine near Sudbury is proposed. The rationale for such an observatory, the characteristics of the detector and its sensitivity are discussed. Technical details that have been solved and several that remain are outlined. Technical measurements and designs are in progress with a view to preparing detailed costing estimates.

INTRODUCTION

We have proposed[1] that a D_2O Cerenkov detector be installed in a nickel mine near Sudbury, Ontario. This detector would measure the incident direction, energy distribution and flux of electron neutrinos above 5 MeV partially through the neutrino electron scattering reaction, (ν_e, e)

$$\nu_e + e \rightarrow \nu_e' + e'$$

but primarily via the inverse β-reaction, (ν_e, d)

$$\nu_e + d \rightarrow p + p + e - 1.44 \text{ MeV}.$$

and, for the first time, permit direct spectroscopic measurements to be made of solar core reactions.

In addition, if certain backgrounds can be made small, the total neutrino flux could be measured[2] via neutrino disintegration of deuterium, (ν_x, d)

$$\nu_x + d \to \nu_x' + n + p - 2.22 \text{ MeV}$$

with the free neutron detected by measuring the 6.25 MeV γ-ray following neutron capture in deuterium. The primary objectives of the experiment are to measure the solar $^8B(\nu_e)$ spectrum and the total solar $^8B(\nu_x)$ flux above 2.22 MeV. The detector will also be sensitive to neutrinos produced by cosmic rays, solar flares, collapsing stars and mode independent proton decay.

The ν_e flux above 2 MeV is dominated by ^8B decay in the sun and is predicted[3] to be 4×10^6 cm^{-2}s^{-1} by the standard solar model. The Cl-Ar radiochemical experiment[4] in the Homestake mine has, over a 15 year period, measured a flux which is at least a factor of two lower than this prediction. In fact, the detection threshold for the Cl-Ar experiment is 0.82 MeV making the Cl detector sensitive to other ν_e's, in particular $^7Be(\nu_e)$, so that the experimental result is consistent with a $^8B(\nu_e)$ flux equal to zero but, more realistically, 1/3 of the standard solar model prediction. Various solutions to this solar neutrino problem (SνP) have been proposed including a large number of non-standard solar models. These models introduce difficult new astrophysical problems and have not been generally accepted. Of particular interest to the nuclear physicist is the possibility that the standard solar model is correct and that the ν_e's have decayed or oscillated during their flight to earth. The interest in neutrino oscillations has recently received added impetus from the realization[5] that neutrino oscillations can be greatly enhanced in the sun through the mechanism of matter oscillations[6] as distinct from the more familiar vacuum oscillations. How these matter oscillations might affect the experimental ν_e flux in proposed experiments has been discussed by Rosen and Gelb[7].

Two Ga-Ge radiochemical experiments[8,9] have been funded and will complement the Cl-Ar experiment. These experiments can detect ν_e's above 0.23 MeV and, in the standard solar model, will be sensitive to the pp(ν_e), the dominant energy producing solar reaction, and $^7Be(\nu_e)$'s. A ν_e flux measurement with a Ga detector will give valuable additional information contributing toward resolving the SνP but it may not enable us to distinguish conclusively between oscillations and solar models as the explanation for the Cl results. On the other hand a measurement of the shape of the $^8B(\nu_e)$

energy spectrum could determine if neutrino oscillations are the cause of the SνP and, if so, could identify probable values for the mass difference, δm^2, and mixing angle, $\sin^2 2\theta$, between the different neutrino species. A 3000 Mg H_2O Cerenkov detector built (Kamioka, Japan) for proton decay studies has been upgraded[10] to detect $^8B(\nu_e)$ via the (ν_e,e) reaction and should provide a direct check on the Cl results. However the sensitivity and signal to noise ratio of this detector may not be sufficient to measure the $^8B(\nu_e)$ spectral shape. If matter oscillations in the sun's core are the reason for the SνP this H_2O detector may not detect any solar neutrinos.

The proposed D_2O Cerenkov detector, operating via the (ν_e,d) reaction, is nearly 10x more sensitive than a detector with the same volume of H_2O and, because of the rock overburden (2000 m) and construction details, it will have a very good signal to background ratio. In addition, its sensitivity to the (ν_x,d) reaction may permit a measurement of the ν_x flux thereby checking solar models independently of neutrino oscillations.

THE SNO DETECTOR

A sketch of the proposed D_2O observatory is shown in Figure 1. It will contain 1000 Mg of 99.8% D_2O enclosed in an acrylic tank. The 10.5 m diameter by 10.5 m high acrylic tank will be surrounded by 3.5 m of H_2O and 1 m of low background concrete. Twenty-four hundred 50 cm diameter Hamamatsu photomultipliers, mounted uniformly in the H_2O 2.5 m from the acrylic tank, will provide 40% coverage of the D_2O. The detector cavity will be constructed 2000 m underground in the INCO nickel mine at Creighton, Ontario and will be located in the non ore-bearing, norite rock. Phototube noise will be reduced by cooling the water to 10°C. The ν_e energy resolution at 7 MeV is about 15%. Photomultiplier timing resolution is 8 nsec, resulting in a spatial resolution of <1 m after vertex reconstruction.

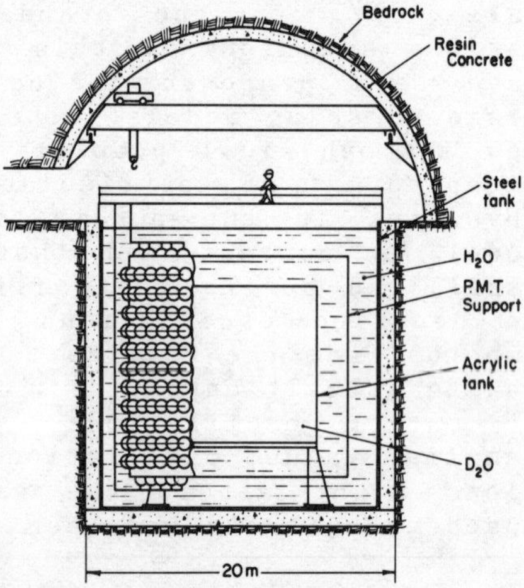

Fig. 1 Sketch of SNO Detector

The proposal is supported by the results of numerous calculations, measurements and discussions. These include:

a) Extensive Monte Carlo calculations on the performance of a D_2O Cerenkov detector[11] and a comparison between the performance predictions and the experimental experience with H_2O Cerenkov detectors e.g. at Kamioka[10] and IMB[12]. The conclusions about energy, spatial and angular resolution, about the number of phototubes in coincidence per event and about predicted pattern recognition are based on these calculations.

b) Light attenuation in D_2O. In the wavelength region of interest for the D_2O Cerenkov detector the attenuation length was measured[13] to be ≥ 30 m in a D_2O sample supplied by Atomic Energy of Canada Limited (AECL).

c) Phototube timing measurements. The timing characteristics of three Hamamatsu 50 cm diameter phototubes have been measured[14] and found to be consistent with their specifications[15].

d) Availability of D_2O. AECL has on hand, for future CANDU reactor sales, sufficient D_2O for SNO. The ^3H content of this D_2O is typically 300 μCi kg^{-1}, too high for the SNO detector[16]. However, Ontario Hydro has D_2O for its reactors with a ^3H content <0.05 μCi kg^{-1}. The loan of 1000 Mg of AECL's D_2O has been discussed and, in principle, it and the exchange of D_2O between Ontario Hydro and AECL can be arranged. The time between a signed CANDU reactor sale and its D_2O loading is seven years.

e) Underground location. The Creighton mine management has been very cooperative during numerous discussions and underground measurements. Their detailed plans for the construction of a 20 m diameter cavity at the 2000 m level in the norite and at a location 250 m from their present mining operations will depend on rock testing on location. These tests will be performed after a 250 m access tunnel has been drilled to the proposed site.

f) Acrylic tank. A manufacturer of large acrylic aquariums[17] foresees no major technical difficulties in constructing and assembling a suitable D_2O container in the underground cavity.

g) Cosmic ray background. At 2000 m the muon flux is negligible, about 24 identifiable high energy muons per day. It is more than 100x lower than at Kamioka and at least 20x lower than at Gran Sasso.

h) Background measurements[18] in the mine. The γ-ray and neutron fluxes in the norite have been measured and the U and Th content of the rock assayed. The amount of concrete and H_2O required to shield the D_2O from these backgrounds is as in Figure 1.

i) Background measurements of detector materials. The U and Th content of the phototube glass and components has been measured[19] by neutron activation techniques and samples of acrylic have been found[20] to contain Th at the 4×10^{-11} g/g concentration. The Th content of a 10 ℓ D_2O sample was determined[21] to be $(4\pm1) \times 10^{-14}$ g/g by removing the Th from the D_2O by adsorption on silica-immobilized 8-hydroxyquinoline and then measuring the concentrate by inductively coupled plasma mass spectrometry. The concentrations of Th in the acrylic and D_2O contribute a negligible background for the $^8B(\nu_e)$ spectrum measurement but are both about a factor of four too high for a $^8B(\nu_x)$ measurement.

The (ν_x,d) reaction measures the total neutrino flux above 2.2 MeV and tests the standard solar model independently of neutrino oscillations. The (ν_e,d) reaction measures the $^8B(\nu_e)$ spectrum, the shape of which may be affected by neutrino oscillations[7]. The ratio of the (ν_e,d) rate to the (ν_x,d) rate tests for neutrino oscillations independently of solar models. The theoretical[22] (ν,d) and (ν,e) cross sections have been verified by experiments[23]. Our calculated rates for each reaction are listed in Table 1. It is assumed that the $^8B(\nu_e)$ flux is 2×10^6 cm^{-2}s^{-1}, that the $^8B(\nu_x)$ flux is 4.6×10^6 cm^{-2}s^{-1} and that the detection threshold is 7 MeV. The detection efficiency for the (ν_e,d) reaction is relatively better than for the (ν_e,e) reaction because all of the available energy is transmitted to the electron in the (ν_e,d) reaction whereas in the (ν_e,e) reaction it is shared with the scattered ν_e. The detection efficiency for the (ν_x,d) reaction is poor because 50% of the neutrons created in 99.8% D_2O leave the acrylic tank before capture and 50% of the remainder are captured in the 0.2% H_2O rather than in the D_2O. The detection efficiency for the (ν_x,d)

Table 1 Reaction rates and sensitivity

Reaction	Events (day^{-1})	Events x detector efficiency (day^{-1})	Flux sensitivity (cm^{-2} s^{-1})
ν_e,e	3.5	0.7	3×10^5
ν_e,d	12	8	7×10^4
ν_x,d	14	3	4×10^5

reaction could be improved by a factor of three by doping the D_2O with Gd thereby capturing the neutrons before they escape from the tank or are captured in H_2O.

The flux sensitivities (Table 1) are estimates of the minimum ^8B neutrino flux that could be observed with confidence (3σ) in a period of six months running with D_2O after running a similar time with H_2O. All backgrounds associated with the (ν_e,e) and (ν_e,d) reactions are included and total less than two events per day from γ-rays external to the D_2O not rejected after reconstruction.

In addition to the two background events per day from external γ-rays, the (ν_x,d) reaction has an internal background due to the presence of ^{232}Th and its daughters which decay via the 2.614 MeV first excited state of ^{208}Pb. These 2.614 MeV γ-rays cause deuterium photodisintegration which produces neutrons indistinguishable from those from deuterium neutrino disintegration. Th in the acrylic at 10^{-11} g/g and in the D_2O at 10^{-14} g/g will each produce one background event per day, events not included in the 4×10^5 value listed in Table 1. It is expected that acrylic without dye and ultraviolet absorber will be lower in Th than the 4×10^{-11} g/g measured[20]. It is also expected that on-line filtration systems will reduce the Th content of the water to 10^{-14} g/g.

These projected and measurable concentrations of Th will not solve the background problems if Th is in disequilibrium with its daughters, in particular ^{228}Ra. Procedures to determine the concentration of Th daughters are being investigated for both the acrylic and D_2O. The ability of the detector to detect (ν_x,d) events is, at the present time, uncertain.

THE $^8B(\nu_e)$ SPECTRUM

The good event rate and detection efficiency of the D_2O Cerenkov detector as compared to existing detectors indicates that a measurement of the $^8B(\nu_e)$ energy spectrum will be practical. Doping the D_2O with ^{10}B will remove any residual background from neutrons due to deuterium disintegration. Rosen and Gelb[7] have indicated the importance of this spectrum as a means of choosing between oscillations and the solar model as the cause of the SνP and also as a means of distinguishing between different sets of oscillation parameters.

The predicted spectrum in our D_2O detector from a $^8B(\nu_e)$ flux of 2×10^6 cm^{-2} s^{-1} is shown in Figure 2 as a function of detected photo electrons, assuming 40% surface coverage. Eighty photoelectrons will be detected from a 10 MeV electron. The Monte Carlo calculated resolution function has been folded with the expected $^8B(\nu_e)$ energy spectrum for no neutrino oscillations, for matter oscillations in the body of the sun ($\delta m^2 = 10^{-6}$ eV^2, $\sin^2 2\theta = 0.04$) and for matter oscillations in the sun's core ($\delta m^2 = 10^{-4}$ eV^2, $\sin^2 2\theta = 0.04$). A measured $^8B(\nu_e)$ spectrum similar to the $\delta m^2 = 10^{-4}$ curve

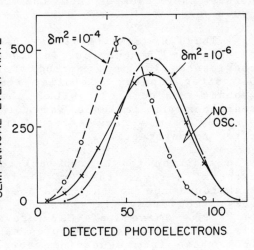

Fig. 2 Predicted $^8B(\nu_e)$ Spectra

will suggest that matter oscillations in the sun's core is the reason for the SνP. If the spectrum is similar to the other curves then it will be more difficult to distinguish between matter oscillations in the body of the sun and non-standard solar models. In such a case the Ga radiochemical experiment will prove crucial since it is in this region of δm^2 that a near zero value for the pp(ν_e) flux is predicted from matter oscillations.

SUMMARY

The SNO detector can measure the $^8B(\nu_e)$ energy spectrum and, combined with the results of a Ga radiochemical experiment, can resolve the SνP. If backgrounds internal to the detector can be reduced by removing Th and its daughters then the detector can independently resolve the SνP and check for neutrino oscillations with a sensitivity unavailable to experiments utilizing earth bound sources. An engineering design study has been commissioned to verify that a large cavity at 2000 m can be constructed, to determine that 1000 Mg of D_2O can be handled realistically and to obtain detailed construction costs. A program to determine methods of further reducing the internal background is also in progress.

1) G.T. Ewan et al., SNO-85-3, NRCC, Ottawa, Canada, (July 1985)
2) H.H. Chen, Phys. Rev. Lett. $\underline{55}$ 1534(1985)
3) J.N. Bahcall, AIP Conf. Proc. $\underline{126}$ 60 (1985)
4) J.K. Rowley, B.T. Cleveland and R. Davis, Jr., AIP Conf. Proc. $\underline{126}$ 1 (1985)
5) S.P. Mikheyev and A. Yu. Smirnov, in Proceedings of the Tenth International Workshop on Weak Interactions, Savonlinna, Finland, 16-25 June 1985 (unpublished).
6) W.L. Wolfenstein, Phys. Rev. D. $\underline{17}$ 2369 (1978)
7) S.P. Rosen and J.M. Gelb, Los Alamos preprint (March 1986) and these proceedings.
8) W. Hampel, AIP Conf. Proc. $\underline{126}$ 162(1985)
9) I.R. Barabanov et al., AIP Conf. Proc. $\underline{126}$ 175 (1985)
10) B.G. Cortex, preceding paper at this conference.
11) R.C. Allen, H.H. Chen, P.J. Doe and K. Roemheld, SNO-85-5, UCI-Neutrino #162 (Nov. 1985).
12) H.S. Park et al., Phys. Rev. Lett. $\underline{54}$ 22(1985)
13) L.P. Boivin, W.F. Davidson, R.S. Storey, D. Sinclair and E.D. Earle, Appl. Opts. $\underline{25}$ 877(1986)
14) H.-B. Mak, private communication to SNO collaboration.
15) H. Kume et al., Nucl. Instr. $\underline{205}$ 443(1983)
16) E.D. Earle, AECL Progress Report, PR-P-140 (1985).
17) Reynolds and Taylor Inc., Santa Ana, California.
18) H.-B. Mak et al., Queen's Annual Report (1985).
19) G. Aardsma, P. Jagam and J.J. Simpson, private communication (see also Ref. 1).
20) G. Aardsma, P. Jagam and J.J. Simpson, private communication to SNO collaboration.
21) J.W. McLaren, R.E. Sturgeon, S.N. Willie and W.F. Davidson, NRCC private communication and CAP Congress, (June 1986).
22) F.J. Kelly and H. Uberall, Phys. Rev. Lett. $\underline{16}$ 145(1966)
S.D. Ellis and J.N. Bahcall, Nucl. Phys. $\underline{A114}$ 636(1968)
A. Ali and C.A. Domenguely, Phys. Rev. D. $\underline{12}$ 3673(1975)
23) R.L. Allen et al., Phys. Rev. Lett. $\underline{55}$ 2401 (1986)
S.E. Willis et al., Phys. Rev. Lett. $\underline{44}$ 522(1980)
E. Pasierb et al., Phys. Rev. Lett. $\underline{43}$ 96(1979)

A NEW FORCE IN NATURE?

Ephraim Fischbach[1,2], Daniel Sudarsky[2], Aaron Szafer[2], Carrick Talmadge[2] and S. H. Aronson[3,4]

1 — Physics Department, University of Washington, Seattle, WA 98195
2 — Physics Department, Purdue University, West Lafayette, IN 47907
3 — Physics Department, Brookhaven National Laboratory, Upton, NY 11973
4 — CERN - EP Division, 1211 Geneva 23, Switzerland

ABSTRACT

We review recent experimental and theoretical work dealing with the proposed fifth force. Further analysis of the original Eötvös experiments has uncovered no challenges to our original assertion that these data evidence a correlation characteristic of the presence of a new coupling to baryon number or hypercharge. Various models suggest that the proposed fifth force could be accomodated naturally into the existing theoretical framework.

INTRODUCTION

In a recent reanalysis[1] of the classic paper of Eötvös, Pekár, and Fekete[2] (EPF), we have raised the possibility that there exists a new intermediate-range force in Nature whose source is baryon number or hypercharge. The publication of Ref. 1 has stimulated a considerable amount of theoretical and experimental interest and, as we anticipate the arrival of the first experimental results in the not-too-distant future, it is appropriate to use the occasion of this conference to summarize the present status of the putative "fifth force". In what follows we discuss i) the motivation that led to Ref. 1, ii) unresolved uncertainties and questions regarding the various anomalous results, iii) various criticisms of our work and our replies to them, iv) attempts to understand the EPF data in terms of conventional physics, v) new experiments, and vi) theoretical models.

MOTIVATION

The motivation for reexamining the EPF paper came from the suggestion that anomalous results in both the $K^\circ - \overline{K}^\circ$ system[3] and in geophysical determinations of the Newtonian constant of gravity[4] could be accounted for by assuming that both arose from a new coupling of the form

$$U_Y(r) = +f^2 B_1 B_2 \frac{e^{-r/\lambda}}{r}. \qquad (1)$$

Here $U_Y(r)$ is the potential energy of the two point objects having hypercharge or baryon numbers $B_{1,2}$ separated by a distance $r = |\vec{r}_1 - \vec{r}_2|$. The strength f and range λ can be inferred (at least approximately) from the geophysical data and are given by[4,5]

$$f^2 = (8 \pm 3) \times 10^{-39} e^2 \quad \text{(Gaussian Units)},$$
$$\lambda \approx 200 \, \text{m}.$$
(2)

For later purposes it is important to emphasize that although $\lambda \approx 200$ m is perhaps the "best value" implied by the geophysical data, the actual constraints on λ are relatively weak and come at short distances from laboratory experiments,[6] and at large distances from satellite determinations[7] of the local acceleration of gravity $\vec{g}$. Taken together we can guess that any value in the range $10 \, \text{m} \lesssim \lambda \lesssim 1500 \, \text{m}$ would be compatible with what is presently known. If the geophysical data represent a real effect (see below), and arise from $U_Y(r)$ in (1), then the locus of values allowed in the $f^2 - \lambda$ plane is shown by the indicated curve in Fig. 1.

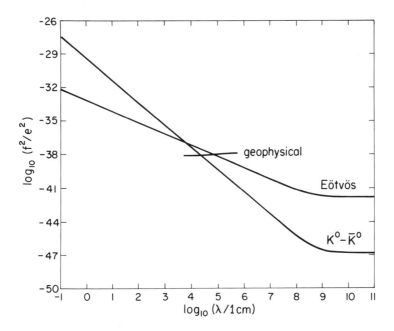

Fig. 1: Constraint curves arising from geophysical, EPF, and kaon experiments.

One can likewise plot the constraint curve implied by the $K^\circ - \overline{K}^\circ$ data of Ref. 3, if we assume that the anomalous energy-dependences found there are also a real effect (see below), which can be attributed to the same $U_Y(r)$ in (1). We see from Fig. 1 that the curves implied by the geophysical and $K° - \overline{K}°$ data intersect, which means that the anomalous effects observed in these systems could be compatible with the same values of f^2 and λ. Although the modelling that goes into these curves is somewhat crude at present, it is sufficient to indicate the possibility of a common origin of these effects, and motivated a search for other systems where the presence of $U_Y(r)$ could show up. Among various possibilities that were examined, the elegant experiments at the University of Washington by Vold, et al.,[8] on the electric dipole moment of $_{54}Xe^{129}$ were of particular interest, since they are capable of detecting external perturbations which produced energy shifts as small as $\sim 10^{-22}$ eV. It turns out, however, that this system and others are not as sensitive as might be thought initially, for reasons that will be discussed elsewhere. However, our analysis suggested that the original EPF experiment could be sufficiently sensitive to detect the presence of $U_Y(r)$ in (1): If we assume that EPF could measure fractional acceleration differences between two samples as small as 1×10^{-9}, then an effect observed at this level would be compatible with the values of f^2 and λ indicated by the corresponding curve in Fig. 1. Remarkably, the Eötvös curve is seen to pass near the intersection of the geophysical and $K° - \overline{K}°$ constraint curves, thus suggesting that the EPF experiments could also have detected the coupling in Eqs. (1) and (2). The curves in Fig. 1 provided the specific impetus for the reanalysis of the EPF experiment in Ref. 1.

UNCERTAINTIES IN THE RESULTS

Having now found[1] in the EPF data the very effect we were searching for, it is tempting to conclude that the geophysical and $K° - \overline{K}°$ data which motivated this search are correct, in the sense of representing real effects. However, such a conclusion would be premature at present, for the following reasons. The geophysical evidence for the non-Newtonian contribution from $U_Y(r)$ in (1)-(2) comes from a comparison of the laboratory value G_0 of the Newtonian constant of gravity, and the value G_1 determined from various measurements below the surface of the Earth. These are given by

$$G_0 = (6.6720 \pm 0.0041) \times 10^{-8}\,\text{cm}^3\,\text{g}^{-1}\,\text{s}^{-2},$$
$$G_1 = (6.720 \pm 0.024) \times 10^{-8}\,\text{cm}^3\,\text{g}^{-1}\,\text{s}^{-2}, \tag{3}$$

where for G_1 we use the Hilton mine determination of Stacey, et al.,[4] which is probably the best of the geophysical values. We see that G_1 differs from G_0 by approximately twice the quoted error, where the error on G_1 is largely an estimate of the inaccuracy of density measurements. Although the magnitude of the discrepancy is not large, its significance lies in the fact that all such determinations of G_1 give values which are greater than G_0. Nonetheless the problem remains that the primitively measured experimental quantity is proportional to the product $G_1\bar{\rho}$, where $\bar{\rho}$ is the local density of the Earth. Hence a small underestimate of the *in situ* density $\bar{\rho}$ would make G_1 appear systematically larger than its actual value, and hence great care must be taken to properly understand the sample densities. This problem has been discussed in some detail by Stacey and collaborators,[4] and remains a possible explanation of the discrepancy between G_0 and G_1.

The situation with respect to the $K^\circ - \overline{K}^\circ$ data is also problematical for a number of reasons. The curve shown in Fig. 1 comes from the 3σ limit for the slope parameter $b_\eta^{(2)}$ in Ref. 3,

$$b_\eta^{(2)} \leq 0.57 \times 10^{-6}, \qquad (4)$$

under the assumption that an energy-dependence of $|\eta_\pm|$ was actually seen at the upper limit. One difficulty is that a recent high energy experiment by Coupal, et al.,[9] has found a result consistent with no energy-dependence of $|\eta_\pm|$. On the other hand, an earlier experiment by Birulev, et al.,[10] and an independent analysis[11] of some of the same data that led to the results quoted in Ref. 3, are both consistent with the suggestion of an anomalous energy-dependence. At present it is not clear whether one (or more) of these experiments/analyses is (are) incorrect, or whether the discrepancies arise from some subtle differences in the way these experiments were configured. For example, the experiment of Coupal, et al.,[9] took place above ground, whereas the earlier Fermilab data were taken in a beam line $\sim (3-5)$ m below the surface of the Earth. Conceivably this difference in the nearby matter distribution could account for all or part of the differences in the corresponding values of $|\eta_\pm|$. Hopefully ongoing experiments to measure $|\eta_\pm|$ and τ_S at high energies will resolve the question of an energy-dependence in the near future. Finally, even if an effect were definitely confirmed, understanding it quantitatively would not be trivial since the matter distribution relevant for the $K^\circ - \overline{K}^\circ$ system is genuinely 3-dimensional, whereas it is effectively 2-dimensional for the EPF experiment.

Despite the uncertainties in the geophysical and $K^\circ - \overline{K}^\circ$ data, and other uncertainties in the EPF data to which we will turn shortly, the observation that effects appear in all three systems, which are at least roughly compatible with the same potential $U_Y(r)$ in (1) – (2), must be taken seriously. Nonetheless it could very well develop in the end that all of these effects are spurious, and that $U_Y(r)$ does not exist. We will then be left to ponder the question of how three experiments which are so different, which were performed by different groups at different places and times, and which depend on such totally unrelated technologies, could all have yielded effects which could be simulated by the same potential $U_Y(r)$.

The preceding discussion leads naturally to the EPF experiment itself, which is an attempt to measure the quantity $\Delta a_\perp/g$ for two dissimilar objects. Here $\Delta a_\perp = (a_1 - a_2)_\perp$ is the acceleration difference of the two objects in a direction perpendicular to the net acceleration field $(\vec{a}_c + \vec{g})$, where $\vec{a}_c$ is the centrifugal acceleration due to the rotation of the Earth, and $g = |\vec{g}| = 980\,\text{cm}\,\text{s}^{-2}$.

$\Delta a_\perp$ depends not only on the composition of the samples being compared, but also on the presumed source responsible for the acceleration difference. If the source is taken to be the Earth itself (viewed as a rigid rotating sphere), acting through the potential $U_Y(r)$ in (1)-(2), then

$$\Delta \vec{a} = \Delta(B/\mu)\vec{y}. \qquad (5)$$

Here $\vec{y}$ is the hypercharge field of the Earth, and $\Delta(B/\mu) = (B_1/\mu_1 - B_2/\mu_2)$, where $\mu_{1,2}$ are the masses of the samples expressed in terms of $m(_1H^1) = 1.00782519(8)$ u. Following the publication of Ref. 1, it was shown by us[12] and others[4,13,14] that the result in (5), which formed the basis for the analysis in Ref. 1, had to be augmented to take into account the presence of the local matter distribution, such as buildings and mountains. For the values of f^2 and λ in (2), these can give the largest contributions to $\Delta a_\perp$, even though the Earth as a whole is the dominant source of the local gravitational field. If we add to (5) the hypercharge contributions $y'(y'')$ from a building(basement) located at an azimuthal angle $\phi'(\phi'')$ relative to the apparatus, then the quantity $\Delta\kappa$ quoted by EPF becomes[12]

$$\Delta\kappa \equiv \frac{\Delta a_\perp}{-g\sin\beta} = -\Delta\left(\frac{B}{\mu}\right)\left\{\frac{y}{g} + \frac{y'}{g}\frac{\cos(\phi'+\beta)}{\sin\beta} - \frac{y''}{g}\frac{\cos(\phi''+\beta)}{\sin\beta}\right\}, \qquad (6)$$

where $\beta \cong \beta_1 \cong \beta_2$ is the angle a plumb line makes (for mass 1,2) with the vertical. Numerically β is given by

$$\beta \cong \tan\beta = \frac{a_c\sin\theta}{g} \cong \frac{1}{581}, \qquad (7)$$

with $\theta \cong 45°$ being the latitude at which the EPF experiment was performed. The presence of the factor $\sin\beta$ in (6) enhances the contribution from y' and y'' relative to that from y, with the result that y' and y'' largely determine both the sign and magnitude of the expression in { } in (6). It is important to point out that the signs of the second and third terms in { } depend on the location of the building and its basement relative to the apparatus, but since this information is provided in the EPF paper, we have evaluated (6) for the appropriate configuration. Note that the relative sign of y'' and y' is a consequence of the fact that the "missing mass" from the basement acts as a "hole" in the otherwise uniform matter distribution of the Earth. It turns out that for typical buildings $y'' > y' > y$, and that as a consequence the sign of the expression in { } is determined by the third term. Our failure to take into account the contributions from y' and y'' in Ref. 1, led to a legitimate criticism raised by Thodberg[15] and by Hayashi and Shirafuji.[16] They correctly pointed out that the naive model of Eq. (5), when taken in conjunction with the EPF data, suggests a force which is attractive rather than repulsive. As we observed,[12] however, when y' and y'' are included neither the sign nor magnitude of the expression in { } in (6) can be ascertained without a detailed knowledge of the local matter distribution in the vicinity of the experiment. For this reason, one cannot at present establish with certainty whether the EPF data imply an attractive or repulsive force. However, if the simple picture that underlies Eq. (6) is correct, along with our estimates of y' and y'', then the EPF data would in fact correspond to a repulsive force, in agreement with the suggestion of the geophysical data. In the original analysis of Ref. 1 the sign of the slope $\Delta\kappa/\Delta(B/\mu)$ was simply fixed to correspond to a repulsive force.

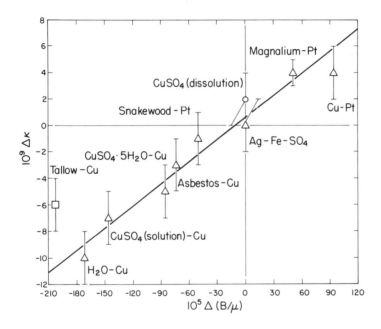

Fig. 2: Updated plot the measured values of $\Delta\kappa$ quoted by EPF versus the calculated values of $\Delta(B/\mu)$. The solid line represents a fit to all data points denoted by a triangle, which are considered to be "reasonably good" data points. The tallow-Cu datum is denoted by a square to indicate an uncertainty in the value of (B/μ) for tallow. The dissolution of copper sulfate is represented by a circle because this datum is not discussed by EPF in any detail, and therefore must be regarded as being somewhat tentative.

It follows from Eq. (6) that when the values of $\Delta\kappa = \kappa_1 - \kappa_2$ quoted by EPF are plotted as a function of the calculated values of $\Delta(B/\mu)$, the result should be a straight line passing through the origin. Fig. 2 shows an updated version of the figure published in Ref. 1, which contains three additional data points: snakewood-Pt, tallow-Cu, and the dissolution of $CuSO_4 \cdot 5H_2O$ in water. As we discuss elsewhere, B/μ for snakewood was determined from a chemical analysis of several samples of snakewood that we obtained, and tallow-Cu is the result of our "best guess" as to the composition of what EPF identified as *talg*. EPF present a result for the dissolution of copper sulfate, without any further discussion, and we simply quote their result as given. Our best fit to the data of Fig. 2 (*i.e.*, including all points denoted by triangles) gives

$$\Delta\kappa = \gamma\Delta(B/\mu) + \delta,$$
$$\gamma = (+5.60 \pm 0.71) \times 10^{-6}, \quad \delta = (+0.64 \pm 0.62) \times 10^{-9}, \tag{8}$$
$$\chi^2 = 2.2 \quad (6 \text{ degrees of freedom}).$$

We see from Fig. 2 and Eq. (8) that the fitted lines passes through the origin as it should which, as we discuss elsewhere, is not generally the case when $\Delta\kappa$ is plotted against other variables. From (6) the slope γ is given by

$$\gamma = -\left\{\frac{y}{g} + \frac{y'}{g}\frac{\cos(\phi' + \beta)}{\sin\beta} - \frac{y''}{g}\frac{\cos(\phi'' + \beta)}{\sin\beta}\right\}. \tag{9}$$

Using Ref. 2 and information provided to us[17] concerning the location of various buildings (and their basements) at the time the EPF experiment was performed, along with the parameters in Eq. (2), we have evaluated a generalized version of (9) numerically and found

$$\gamma \approx +1.7 \times 10^{-6}. \tag{10}$$

The agreement between (8) and (10) is reasonably good, particularly in view of the large uncertainty in λ, and in view of the fact that we have had to make educated guesses as to the densities of various buildings. It is clear in any case that including the effects of local mass anomalies in (9) substantially improves the agreement between the calculated and measured values of the slope γ, compared to what was found in Ref. 1 (where only y was taken into account).

CRITICISMS AND REPLIES

Following the publication of Ref. 1, a number of criticisms were raised by various authors. In addition to the issue of the sign and magnitude of γ which we have already discussed, questions were also raised about various details of our reanalysis and about the compatibility of the EPF results and our interpretation of them with other experiments. These questions, and our replies to them, have now been published,[5,12,15,16,18-22] and from what we know at present, there are no outstanding criticisms which challenge the basic assertions contained in Ref. 1: a) That the EPF data exhibit a correlation between $\Delta\kappa$ and $\Delta(B/\mu)$, and b) that the slope $\gamma = \Delta\kappa/\Delta(B/\mu)$ agrees reasonably well with the value expected from the geophysical data, given the uncertainties we have already discussed. By way of summary we outline some of the salient points made in the various papers.

  i) A number of authors[5,18-20] pointed out that if the putative fifth force arises from a coupling to hypercharge, then one can derive stringent limits on the product of $f^2\lambda^2$ from the experimental limits that can be inferred from the decay rate $\Gamma(K^\pm \to \pi^\pm \gamma_Y)$, where γ_Y denotes the "hyperphoton". The suggestion that γ_Y couples to a quantum number other than the baryon number B, which is all that is demanded by the EPF data, comes from the $K^\circ - \overline{K}^\circ$ data: If ordinary matter can influence the energy-dependence of the $K^\circ - \overline{K}^\circ$ parameters through the exchange of a virtual γ_Y, then the decays $K^\pm \to \pi^\pm \gamma_Y$ into real hyperphotons must also be allowed, and in fact are enhanced for reasons that were originally discussed by Weinberg.[23] $\Gamma(K^\pm \to \pi^\pm \gamma_Y)$ can be obtained experimentally from existing limits on $K^\pm \to \pi^\pm X$, where X is any missing neutral such as an axion or γ_Y. It can be predicted theoretically by standard techniques, given a knowledge of f and λ, and hence a comparison of theory and experiment constrains $f^2\lambda^2$. In deriving the experimental constraint, it is important to bear in mind that the energy scale set by the hyperphoton mass $m_Y = \lambda^{-1} \cong 10^{-9}$ eV is as far removed from the usual hadron scale ($e.g.$, $m_\pi \cong 10^8$ eV), as the latter is from the presumed Grand Unification scale ($m_{\rm GUT} \cong 10^{24}$ eV). Hence there is no particular reason for supposing that the dynamics at the scale of the m_Y have any obvious relation to ordinary hadron dynamics. For this reason we generalize the notion of hypercharge by assuming that γ_Y couples not to the usual quantity $Y = B + S$ (where S is strangeness), but to an appropriate linear combination of all possible flavors (including charm, truth, beauty, lepton number, ...). In the simplest picture, we retain only B and S in the form of the generalized hypercharge Y_θ given by[5]

$$Y_\theta = \sqrt{2}\cos\theta_Y B + \sqrt{2}\sin\theta_Y S. \tag{11}$$

Combining the experimental limits from Asano, $et\ al.$,[24] with a theoretical model for $K^\pm \to \pi^\pm \gamma_Y$ then leads to the constraint

$$\tan^2\theta_Y \left(\frac{f^2}{G_0 m_p^2}\right)\left(\frac{\lambda}{1\,\rm m}\right)^2 \lesssim 4.7, \tag{12}$$

where m_p is the proton mass. For the usual hypercharge $\tan\theta_Y = 1$, and Eq. (12) implies that $\lambda \lesssim 25$ m, using f^2 from (2). This is much smaller than the "best" value for λ quoted in (2) but, as we have already noted, even a value as small as 25 m would not be inconsistent with the geophysical data. Additionally, there is no compelling reason why $\tan\theta_Y$ must be unity. More stringent limits for the $K \to \pi\gamma_Y$ modes should be available in the forseeable future from an experiment at Brookhaven and, when combined with the other relevant data, should be able to set tighter limits on f, λ, and θ_Y (or its generalization).

ii) In Ref. 1 we discussed one more system (in addition to the EPF, geophysical, and $K^\circ - \overline{K}^\circ$ experiments) which, with a modest increase in sensitivity, could detect the presence of U_Y. This is the comparison of satellite and terrestrial determinations of the local acceleration of gravity $\vec{g}$, which has been discussed by Rapp.[7] Subsequently Neufeld[14] and Nussinov[25] called attention to an experiment by Kreuzer,[26] which is also at the edge of setting a nontrivial constraint on U_Y. In fact an experiment somewhat similar to Kreuzer's is underway at the University of Washington by Davisson.[27]

iii) Keyser, Niebauer, and Faller[21] (KNF) raised the question of why we excluded the data of Renner,[28] who repeated the EPF experiment with a modified version of the original apparatus in 1935. The reason for this, as we noted in Ref. 1, was that Roll, Krotkov, and Dicke[29] (RKD) found various inconsistencies in Renner's analysis and results. Renner's data are discussed in much greater detail in our longer paper, but since KNF in no way refute the arguments of RKD, one must be very wary in using these data. Moreover, even if Renner's data were free of problems, it would still be incorrect to directly compare them to the EPF data since the two experiments were carried out in different places.[17] In the presence of nearby sources whose effects are comparable to that of the Earth, two experiments (such as EPF and Renner) at different locations could obtain correlations of $\Delta\kappa$ and $\Delta(B/\mu)$ which differ in magnitude and even in sign. In the case of these two experiments, the differences in local mass distributions are not well-known. This would prevent our combining or comparing their results even if the serious problems with Renner's analysis did not exist.

iv) KNF and others also raised a question concerning the brass vials in which some of the samples were contained. For example, in the asbestos-Cu comparison, both the asbestos and the Cu standard were contained in similar brass vials. EPF assumed that the effects of the vials simply cancelled out, which for present purposes is completely legitimate, since $(B/\mu)_{\text{brass}} \cong (B/\mu)_{\text{Cu}}$. One can show in a straightforward way that one obtains virtually identical results for γ and δ in (8), irrespective of whether one views the comparisons as being asbestos-Cu or (asbestos + vial$_1$) − (Cu + vial$_2$).

v) Among the other questions raised by KNF, the RaBr$_2$-Pt comparison deserves a brief discussion since the KNF paper is confusing on this point. This datum has not been included in our analysis because of the uncontrolled systematics introduced by the heating of the sample due to the radium.[22] KNF correctly point out that since a very small sample of RaBr$_2$ (in a glass vial) was placed in a brass container to carry out the RaBr$_2$-Pt comparison, one can for present purposes ignore both the glass and the RaBr$_2$ compared to the brass. The comparison is then effectively brass-Pt $\approx$ Cu-Pt, since B/μ for brass is very nearly the same as that for Cu. KNF then claim that the result one infers in this way for $\Delta\kappa$(Cu-Pt) disagrees with that quoted by EPF for their direct Cu-Pt comparison, which would then imply that the EPF methodology is suspect. However, in carefully checking the final EPF results by starting with their quoted scale deflections, we found in their paper a misprint in the *sign* for this datum, whereas all of the other EPF results are correct. When this sign is corrected the two data agree within their quoted errors, so this comparison becomes a nice consistency check on the EPF results. The corrected RaBr$_2$-Pt datum is shown in Fig. 1 of KNF, however, it appears that the original version of their text has not been modified to reflect the corrected value of $\Delta\kappa$.

The preceding discussion leads naturally to the question of the overall consistency of the EPF data, as illustrated above by the comparison of brass-Pt and Cu-Pt. This question is of interest because if some of the data are excluded, it might be possible to explain the EPF results in terms of conventional physics, as we discuss below. The data that are the leading candidates to be excluded are those involving Pt, which were taken with their earliest methodologies. In addition to the brass-Pt and Cu-Pt comparison given above, one can compare $\Delta\kappa(\text{CuSO}_4 \cdot 5\text{H}_2\text{O-Cu})$ and $\Delta\kappa(\text{Cu-Pt})$. These measurements were carried out using EPF's Method III and Method II respectively, and are interesting because CuSO$_4 \cdot$ 5H$_2$O and Pt have very similar values of B/μ:

$$(B/\mu)_{\text{Pt}} = 1.00801, \qquad (B/\mu)_{\text{CuSO}_4 \cdot 5\text{H}_2\text{O}} = 1.00809. \qquad (13)$$

This is a reflection of the distinctive shape of B/μ as a function of atomic number, namely the fact that B/μ peaks at Fe and is smaller at either end of the Periodic Table. Since both of the samples in (13) were directly compared to Cu, it follows that the fractional acceleration differences $\Delta\kappa$ should be very nearly the same in both cases. Using the data in Table I of Ref. 1 we find

$$\frac{\Delta(B/\mu)_{\text{Cu-Pt}}}{\Delta(B/\mu)_{\text{CuSO}_4 \cdot 5\text{H}_2\text{O-Cu}}} = \frac{+94 \times 10^{-5}}{-86 \times 10^{-5}} = -1.1, \qquad (14a)$$

$$\frac{\Delta\kappa_{\text{Cu-Pt}}}{\Delta\kappa_{\text{CuSO}_4 \cdot 5\text{H}_2\text{O-Cu}}} = \frac{(+4 \pm 2) \times 10^{-9}}{(-5 \pm 2) \times 10^{-9}} = -0.8 \pm 0.5. \qquad (14b)$$

The results in (14a,b) agree quite well, and in fact are in even better agreement when one uses the values of $\Delta\kappa$ recalculated (without rounding off) from their scale deflections. We can thus infer that EPF were able to achieve a high degree of internal consistency between the results of Method II and III. More generally, our analysis of the EPF paper indicates that, in every way that we could check, the EPF data are internally consistent as well as being consistent with the hypothesis that the observed effects arise from a coupling to baryon number or hypercharge. The question is whether there are alternative explanations of the EPF results, to which we turn next.

ALTERNATIVE EXPLANATIONS OF THE EPF DATA

We consider the question of whether the observed correlation between $\Delta\kappa$ and $\Delta(B/\mu)$ might be attributable to something other than an external baryon number or hypercharge field, as we have been assuming. The possibility that this correlation is simply the result of a statistical fluctuation, is extremely remote. This leaves two other classes of possibilities to be explored: i) Can the EPF data be explained in terms of more conventional physics as arising from a systematic effect? ii) If not, is the new interaction whose existence they imply necessarily a coupling to baryon number?

We will show elsewhere that the answer to ii) above is that if the EPF results do in fact represent new physics, then the only known quantum number that accounts for the data is B. [A coupling to the combination $(B + \epsilon L)$, where L is lepton number and ϵ is of order 10^{-3}, would also be allowed.] The critical question is then i). At the present, there is only one model that we know of which as a serious chance of explaining the EPF data in conventional terms. This is the "thermal gradient" model of Chu and Dicke[30] (CD), which can be described briefly as follows: If horizontal thermal gradients were present in the EPF apparatus, they could produce a gentle "breeze" which could exert a force on the samples being compared. Since the samples, or the containers they were in, had different physical dimensions, the forces exerted on opposite sides of the apparatus would not be equal, and a net torque could result. In practice, this would lead to a correlation between $\Delta\kappa$ and $\Delta(1/\rho)$ or ΔS, where ρ is the density of the sample and S is the cross-sectional surface area of the sample or its container. This is a very nice model from a pedagogical view, since it clearly illustrates how a systematic effect can appear to depend on a property of the samples, such as $1/\rho$. Moreover, this model is sufficiently realistic that we can use it to explore the necessary properties that any model must have to account for the EPF data.

It is useful to picture any acceptable alternative to the baryon number or hypercharge hypothesis as satisfying two separate criteria: i) To start with, the mechanism in question must have the property that the torque it produces would switch sign when the EPF apparatus is rotated through 180°. ii) The torque produced by the alternative mechanism must not only be composition dependent, but must vary from one material to another in a manner that (at least approximately) simulates a variation with B/μ.

The significance of the first criterion, when applied to the CD model, is to suggest that if a thermal gradient (or breeze) is to simulate the effects of a matter distribution (*e.g.*, a mountain or building), it must be similarly constant on average both spatially and temporally relative to the apparatus. However, it is unclear how any likely heat source (*e.g.*, a window or radiator) would always produce a gradient with the *same sign*, independent of time of day or year over the period of months or years that the experiment took place. The challenge to the CD model arising from the second criterion is to adequately explain the Pt data, which tend not to fit that well irrespective of how the model is formulated. As CD correctly point out, there is a clear suggestion of a correlation between $\Delta\kappa$ and ΔS or $\Delta(1/\rho)$ evidenced by the double torsion balance data alone, particularly for those comparisons carried out using EPF's Method III, which happen not to involve Pt. The reason for this is illuminating, and provides an insight into other possible correlations as well. If we examine a plot B/μ as a function of atomic number Z, we see that for the EPF samples B/μ is an approximately monotonically increasing function of Z, providing we exclude Pt. It follows that since B/μ does in fact correlate with κ, so will any other nearly monotonically changing variable, provided the Pt data are excluded. Hence it is precisely these data which test the characteristic shape of B/μ as a function of Z, and which thus discriminate between a correlation with B/μ and some other variable. It follows that the suggestion of an approximate correlation between $\Delta\kappa$ and $\Delta(1/\rho)$ is not suprising, since $(-1/\rho)$ is also an approximately increasing function of Z for the substances studied by EPF. In summary, the CD model is very clever and is sufficiently promising to warrant more detailed study, should ongoing experiments fail to confirm the original EPF results. The questions that it must address more fully (if possible) are mechanisms for producing a spatially and temporally constant thermal gradient over a long term, and the behavior of the Pt data. In this context, we recall the previous discussion of the overall consistency of the EPF results involving Pt, which suggests that these data should be included on an equal footing with the rest of the EPF data.

NEW EXPERIMENTS

Numerous experiments have been proposed to test various implications of the fifth force, and several of these are presently under way. Space limitations prevent a detailed discussion of the current experimental situation, which will be presented elsewhere, and hence we will simply list the ongoing experiments that we know of, along with others that have recently been suggested. Since the experimental situation is constantly changing, we apologize in advance to any individuals whose contributions we have neglected.

Laboratory Eötvös experiments
- D. F. Bartlett [Colorado]
- C. W. F. Everitt and P. Worden [Stanford]
- J. Faller and P. Keyser [JILA, Colorado]
- G. Luther [NBS, Maryland]
- R. Newman and P. Nelson [U. California, Irvine]

Eötvös experiments near a mountain
- E. Adelberger, B. Heckel, F. Raab, and C. Stubbs [U. Washington, Seattle]
- P. Boynton, D. Crosby, P. Ekstrom, and A. Szumilo [U. Washington, Seattle]
- G. Edwards [U. Washington, Seattle]
- R. Newman and P. Nelson [U. California, Irvine]
- P. Thieberger [BNL]

Repetitions of the Galileo experiment
- V. Cavasinni, E. Iacopini, E. Polacco, and G. Stefanini [CERN/Pisa]
- J. Faller and T. Niebauer [JILA, Colorado]
- S. Richter [Harris Corp., Florida]
- A. Sakuma [BPIM, Paris]

Other experiments
- E. Amaldi, R. Bizzarri, A. Degasperis, G. Muratori, G. V. Pallotino, G. Pizzella, F. Ricci, and C. Rubbia [Rome and CERN] — An Eötvös experiment using a gravity-wave detector and external rotating masses.
- N. Beverini, *et al.*[Pisa, LANL, Rice, Texas A&M, Genoa, Kent State, Case Western Reserve, CERN, NASA/Ames] — Comparison of p and p̄ masses at LEAR in a vertical drift tube.
- R. Davisson [U. Washington, Seattle] — Variant of the Kreuzer experiment.
- G. Gabrielse, X. Fei, K. Helmerson, H. Kalinowsky, W. Kells, S. Rolston, R. Tjoelker, and T. A. Trainor [Washington, Mainz, Fermilab] — Comparison of p and p̄ in a Penning Trap at LEAR.
- H. J. Paik [Maryland] — Measurement of the Laplacian of the gravitational field over various distance scales.
- C. C. Speake and T. J. Quinn [BPIM, Paris] — Beam balance comparison of two masses.

In addition to the experiments cited above, the gravity measurements of Stacey and collaborators are, of course, continuing. Other experiments have also been proposed[31] utilizing satellite technology, and these may be worth exploring if indications from terrestrial experiments confirm the presence of the fifth force.

THEORETICAL MODELS

Despite the fact that the very existence of a fifth force is yet to be convincingly demonstrated, one can speculate on how such a force might fit in with currently accepted ideas. This is, of course, a bit difficult to do given how little we know at present, but various observations suggest that a coupling with the properties suggested by the available data, might find a natural place in the company of existing theories.

In characterizing the fifth force in field theoretical terms, we assume at the outset that it has at least one component which is described by a massive spin-1 (vector) field. A vector field is the natural choice to account for the fact that this force is repulsive, as suggested by the geophysical data. There may, of course, be more than one vector field, as well as spin-0 (scalar) and spin-2 (tensor) fields, and hence the number of theoretical possibilities is quite large. One of the more interesting conjectures is that the fifth force is evidence for supersymmetric theories,[25,32,33] some of which predict the existence of particles similar to γ_Y which couple weakly to ordinary matter. If the new field is a vector, then its coupling strength f and mass $m_Y = \lambda^{-1}$ characteristically appear in the combination f^2/m_Y^2 and, for the values suggested in (2), this ratio is characteristic of a hadron scale. For example, in Ref. 5 it was noted that the branching ratio for decays into γ_Y, compared to electromagnetic decays, is of order

$$\frac{f^2}{m_Y^2} \frac{m_K^2 - m_\pi^2}{2e^2} = 7 \times 10^{-4}. \tag{15}$$

It follows that for interactions involving $W^\pm$ and Z this ratio can be of order unity, and this observation has been made by Nussinov[25] and by Bars and Visser.[32] Following Nussinov we note, for example, that from (2)

$$\frac{f^2}{m_Y^2} \simeq \frac{1}{(131\,\text{GeV})^2}. \tag{16}$$

This means that if we assume that γ_Y obtains its mass through some Higgs mechanism [so that $m_Y = f\langle\phi\rangle$] then $\langle\phi\rangle \simeq 240\,\text{GeV}$, which is appropriate to the standard model,[25] comes close to satisfying (16). These and other related observations[34-37] suggest that there may be an underlying connection between the electroweak interaction and the fifth force.

Another class of theories incorporating the fifth force focuses on the similarity between its strength and that of the gravitational interaction:

$$\frac{f^2}{G_0 m_p^2} \simeq 10^{-2}. \tag{17}$$

In these theories the fifth force arises naturally as a modification of the usual field equations of General Relativity.[38] An example is the nonsymmetric gravitation theory of Moffat,[39] which is interesting in that it makes a number of rather distinctive predictions. One of these is that the dominant influence in the EPF experiment is the Earth as a whole, rather than the nearby buildings or mountains as would be the case for the potential of Eqs. (1) and (2). Should the EPF results turn out to be a real effect, then this prediction can easily be tested in several of the ongoing experiments being conducted near mountains. Another gravity-like theory which incorporates the fifth force, has been described recently by Mészáros.[40]

Should the fifth force turn out to be real, then much more theoretical effort will be needed to sort out its specific properties. However, the preceding examples point to an even more exciting possibility: If the values of f and λ are such that the fifth force can be naturally united with both the electroweak and gravitational interactions, then perhaps it could provide the basis for unifying these interactions. In fact if we combine (16) and (17), and use m_π^2 in place of

m_p^2 to estimate the strength of the gravitational interaction, then we find

$$G_0 \langle \phi \rangle^2 \approx \frac{2m_Y^2}{m_\pi^2}. \tag{18}$$

This is an interesting relation in that it connects the parameters G_0, $\langle \phi \rangle$, and m_Y, which arise from the gravitational, electroweak, and fifth forces respectively. Thus the fifth force, far from being an unwanted intruder into the present theoretical picture, could instead be the missing link in our quest for a unified picture of the known interactions.

REFERENCES

1. E. Fischbach, D. Sudarsky, A. Szafer, C. Talmadge, and S. H. Aronson, Phys. Rev. Lett. **56**, 3(1986).

2. R. v. Eötvös, D. Pekár, and E. Fekete, Ann. Phys. (Leipzig) **68**, 11 (1922); R. v. Eötvös, D. Pekár, and E. Fekete, in *Roland Eötvös Gesammelte Arbeiten*, Edited by P. Selényi (Akadémiai Kiado, Budapest, 1953) pp. 307-372; the latter paper has been translated into English by J. Achzehnter, M. Bickeböller, K. Bräuer, P. Buck, E. Fischbach, G. Lübeck, and C. Talmadge, University of Washington preprint 40048-13-N6.

3. S.H. Aronson, G.J. Bock, H.Y. Cheng, and E. Fischbach, Phys Rev. Lett. **48**, 1306 (1982); Phys. Rev. **D28**, 476 (1983) and **D28**, 495 (1983); E. Fischbach, H.Y. Cheng, S.H. Aronson, and G.J. Bock, Phys. Lett. **116B**, 73 (1982).

4. F. D. Stacey and G. J. Tuck, Nature **292**, 230 (1981); S. C. Holding and G. J. Tuck, Nature **307**, 714 (1984); F. D. Stacey, Sci. Prog. Oxf. **69**, 1 (1984); S. C. Holding, F. D. Stacey, and G. J. Tuck, Phys. Rev. **D33**, 3487 (1986); F. D. Stacey, G. J. Tuck, G. I. Moore, S. C. Holding, B. D. Goodwin, and R. Zhou, University of Queensland preprint, 1986 (submitted to Rev. Mod. Phys.).

5. S. H. Aronson, H.-Y. Cheng, E. Fischbach, and W. Haxton, Phys. Rev. Lett. **56**, 1342 (1986).

6. D. Long, Nature **260**, 417 (1976); D. R. Mikkelsen and M. J. Newman, Phys. Rev. **D16**, 919 (1977); V. I. Panov and V. N. Frontov, Zh. Eksp. Teor. Fiz. **77**, 1701 (1979) [Sov. Phys. JETP **50**, 852 (1979)]; H.-T. Yu, *et al.*, Phys. Rev. **D20**, 1813 (1979); H. Hirakawa, K. Tsubono, and K. Oide, Nature **283**, 184 (1980); R. Spero, J. K. Hoskins, R. Newman, J. Pellam, and J. Schultz, Phys. Rev. Lett. **44**, 1645 (1980); G. W. Gibbons and B. F. Whiting, Nature **291**, 636 (1981); H. A. Chan, M. V. Moody, and H. J. Paik, Phys. Rev. Lett. **49**, 1745 (1982); Y. Ogawa, K. Tsubono, and H. Hirakawa, Phys. Rev. **D26**, 729 (1982); Y. T. Chen, A. H. Cook, and A. J.

F. Metherell, Proc. R. Soc. Lond. **A394**, 47 (1984); J. K. Hoskins, R. D. Newman, R. Spero, and J. Schutlz, Phys. Rev. **D32**, 3084 (1985).

7. R. H. Rapp, Geophys. Res. Lett. **7**, 35 (1974); Bull. Geod. **51**, 301 (1977); Report of Special Study Group No. 5.39 of the International Association of Geodesy, "Fundamental Geodetic Constants (August 1983)", in *Travaux de l'Association Internationale de Geodesie* (International Association of Geodesy, Paris, 1984) Vol. 27.

8. T. G. Vold, F. J. Raab, B. Heckel, and E. N. Fortson, Phys. Rev. Lett. **52**, 2229 (1984).

9. D. P. Coupal, et al., Phys. Rev. Lett. **55**, 566 (1985).

10. V. K. Birulev, et al., Nucl. Phys. **B115**, 249 (1976).

11. A. Gsponer, D. Sc. thesis, Swiss Federal Institute of Technology, Zurich, 1978, (unpublished) p. 49.

12. C. Talmadge, S. H. Aronson, and E. Fischbach, in *XXI Rencontre de Moriond*, "Perspectives in Electroweak Interactions and Unified Theories", Les Arcs, March 9-16, 1986; E. Fischbach, D. Sudarsky, A. Szafer, C. Talmadge, and S. H. Aronson, Phys. Rev. Lett. **56**, 2424 (1986).

13. P. Thieberger, Phys. Rev. Lett. **56**, 2347 (1986); A. Bizzeti, University of Florence Report No. 86, 1986 (to be published); M. Milgrom, Institute for Advanced Study Report, 1986 (to be published).

14. D. A. Neufeld, Phys. Rev. Lett. **56**, 2344 (1986).

15. H. H. Thodberg, Phys. Rev. Lett. **56**, 2423 (1986).

16. K. Hayashi and T. Shirafuji, Kitasato University Report No. 86-2, 1986 (to be published).

17. J. Barnóthy, J. Németh, and P. Király, private communication.

18. M. Suzuki, Phys. Rev. Lett. **56**, 1339 (1986).

19. C. Bouchiat and J. Iliopoulos, Phys. Lett. **169B**, 447 (1986).

20. M. Lusignoli and A. Pugliese, Phys. Lett. **171B**, 468 (1986).

21. P. T. Keyser, T. Niebauer, and J. E. Faller, Phys. Rev. Lett. **56**, 2425 (1986).

22. E. Fischbach, D. Sudarsky, A. Szafer, C. Talmadge, and S. H. Aronson, Phys. Rev. Lett. **56**, 2426 (1986).

23. S. Weinberg, Phys. Rev. Lett. **13**, 495 (1964).

24. Y. Asano, et al., Phys. Lett. **107B**, 159 (1981) and **113B**, 195 (1982); T. Shinkawa, D. Sc. thesis, University of Tokyo, 1982 (unpublished).

25. S. Nussinov, Phys. Rev. Lett. **56**, 2350 (1986).

26. L. B. Kreuzer, Phys. Rev. **169**, 1007 (1968); J. J. Gilvarry and P. M. Muller, Phys. Rev. Lett. **28**, 1665 (1972); W.-T. Ni, Ap. J. **181**, 939 (1973).

27. R. Davisson, private communication.

28. J. Renner, Matematikai és Természettudományi Értesitő **53**, 542 (1935).
29. R. H. Dicke, Sci. Am. **205**, 84 (1961); P. G. Roll, R. Krotkov, and R. H. Dicke, Ann. Phys. (N.Y.) **26**, 442 (1964).
30. S. Y. Chu and R. H. Dicke, 1986 preprint (submitted to Phys. Rev. Lett.).
31. Y. Avron and M. Livio, Ap. J. **304**, L61 (1986); M. P. Silverman, (submitted to Europhys. Lett.); J. G. Hills, Ap. J. (to be published).
32. I. Bars and M. Visser, USC preprint 86/05 (February, 1986).
33. P. Fayet, Phys. Lett. **171B**, 261 (1986) and Phys. Lett. **172B**, 363 (1986).
34. R. V. Wagoner, Phys. Rev. **D1**, 3209 (1970); J. O'Hanlon, Phys. Rev. Lett. **29**, 137 (1972).
35. Y. Fujii, Nature Phys. Sci. **234**, 5 (1971); Ann. Phys. (N.Y.) **69**, 494 (1972); Phys. Rev. **D9**, 874 (1974); Gen. Rel. Grav. **13**, 1147 (1981); Prog. Theoret. Phys. **75**, XXX (1986), to be published.
36. A. D. Sakharov, Sov. Physics Doklady **12**, 1040 (1968).
37. S.L. Glashow, Harvard University preprint HUTP-86/A012, (Contribution to the 1986 Moriond Workshop on Neutrino Masses, Tignes, France) January, 1986.
38. M. Gasperini, Univ. of Torino preprint (February, 1986); H. Peng, University of Alabama preprint UAHDP-261 (February, 1986); H. Dehnen and F. Ghaboussi, Konstanz University preprint, March 1986 (accepted for publication in Z. f. Physik); T. Goldman, R. J. Hughes, and M. M. Nieto, Phys. Lett. **171B**, 217 (1986).
39. J. W. Moffat, P. Savaria, and E. Woolgar, University of Toronto Preprint (March, 1986); J. W. Moffat, University of Toronto Preprint (May, 1986).
40. A. Mészéros, Charles University Preprint (accepted for publication by Ap. J.) 1986.

ACOUSTIC DETECTION OF LOW-ENERGY RADIATION

C.J. Martoff, B. Cabrera, B. Neuhauser
Department of Physics, Stanford University
Stanford, CA 94305

ABSTRACT

The feasibility of detecting nuclear radiation by the phonon flux produced is discussed. This method should permit construction of very large detectors because of the long phonon mean free paths in commercial crystals. Such detectors may be expected to have energy resolutions for gamma radiation similar to conventional semiconductor diode detectors, and time resolutions of the order of 0.1 microsecond. With appropriate readout schemes, phonon focussing should permit event topology reconstruction within monolithic crystals, with spatial resolution of a few mm. Aside from their obvious potential in radiochemical analysis and nuclear physics, such detectors could be well suited to measurements of neutrino interactions and other rare processes involving small energy deposits.

Existing types of radiation detectors rely directly or indirectly upon the production of free electrical charge in matter exposed to a radiation flux. The energy required to produce one electron-hole pair depends on the material, but is in the range of a few electron volts. This energy per pair, plus the statistics of the cascade governing the number of pairs produced, determine the energy sensitivity and energy resolution of ionization detectors. At present, an excellent Si/Li detector with state-of-the-art electronics may achieve an energy sensitivity of the order of KeV, with resolution of a few percent.

A detector based on quanta with much lower energy than electron-hole pairs would be expected to achieve improved energy sensitivity and resolution. The capability of detecting events with only a few KeV energy deposited permits some very interesting experiments like measurement of the coherent nuclear scattering of neutrinos, to be discussed below.

The characteristic energy of phonons produced in radiation interaction with matter are of the order of 10^{-3} electron volts. Theories of the energy deposition process in insulators(9) predict that as much as 75% of the energy is in the form of prompt phonons. Thus if an appropriate transducer can be found, phonon detection would appear to be a very attractive method of radiation detection. As will be discussed below, appropriate transducers do exist. Furthermore, crystalline anisotropies should lead to phonon focusing effects, which would permit spatial reconstruction of events occurring within monolithic large volume detectors.

Consider the mechanism of energy deposition by neutrinos in matter. The fundamental interactions of neutrinos with electrons and quarks (nucleons) include charge-changing and neutral current processes (1). Interactions with free electrons lead to a roughly flat spectrum of recoil electron energies, extending from zero to the incident neutrino energy.(1) In dense matter, interesting solid state and atomic coherence effects may arise (2,3) but these do not alter this qualitative discussion. Neutral-current scattering of low energy neutrinos from nuclei turns out to be enhanced (4) over the incoherent sum of nucleon cross sections, by a coherence factor of the neutron number N. This occurs because the relevant neutral current matrix element is spin independent and the low energy neutrino wave has constant phase through the nuclear volume. The nuclear recoil spectrum is a linearly decreasing function of energy with a maximum energy given approximately by $(2E_\nu)^2/2M$ for 180° scattering. A 1 KeV maximum Si recoil energy requires 3.6 MeV neutrinos, available from a nuclear reactor(5) or the solar ^8B decay.(6)

Interestingly, the coherent nuclear scattering cross section is independent of neutrino flavor (electron muon, etc.). However, the electron-neutrino scattering cross sections are flavor dependent (due to the absence of charged current interactions for low energy muon neutrinos). This difference could in principle be exploited in a neutrino oscillation experiment to provide an oscillation independent normalization, as was attempted by Reines et al.(7)

We have estimated interaction rates per unit detector mass from neutrino sources of interest, specializing to the case of Si detectors and imposing an energy deposition threshold of 1 KeV (Table I). The calculations neglected atomic effects (1,2). Because of the easy availability of high purity, high quality crystals, we are working with silicon while developing transducers and other necessary technology. The acoustic detection method should work with any insulating or semiconducting crystalline material.

The detectability of the interactions in Table I depends upon: 1) the efficiency of prompt phonon production by radiation interactions and the ability of these phonons to propagate across a reasonably sized detector without scattering or degradation; 2) the availability of sensitive transducers to convert phonons into electrical signals; and 3) the backgrounds due to other radiations striking the detector. Each of these points will be briefly discussed below.

PROMPT PHONON PRODUCTION

Interaction of nonionizing radiation in the MeV energy range, with an insulating material produces fast ionization electrons ($v \sim c$) or slow atomic recoils ($v \ll \alpha c$). The electron primaries then Møller scatter, producing further ionization, initiating a cascade of secondaries which may themselves have enough energy to produce ionizations, etc. (8,9)

Simulations of the ionization cascade have been undertaken(9) to explain the observed Fano factors, which determine the energy resolution of conventional semiconductor detectors. Production of phonons is believed to occur in the ionization events themselves, as well as during the very fast relaxation to the conduction band edge by secondaries below the threshold for further ionization. The fraction of the primary energy going into prompt phonons is quite strongly coupled to the derived Fano factors. The results indicate that approximately half of the primary energy goes into phonons promptly, with the remainder going into potential energies of electrons and holes. This is also consistent with the experimental fact that the average radiation energy ε needed to produce one electron-hole pair in a semiconductor is much larger than the band gap E_g (for Si, ε=3.6 ev while E_g = 1.1 ev). At least 70% of a totally absorbed x-ray's energy has been experimentally observed to appear as heat within less than a millisecond (10).

Under the assumption of energy closure between phonons and electron hole pairs, and independence of the average phonon energy from the details of the cascade, the Fano fluctuations in electron hole pair number will lead to corresponding fluctuations in the phonon number. The relative fluctuation of phonon number $\delta N/N$ would then be reduced from the relative fluctuations in electron-hole number, by a factor E_g/ϵ. (Recombination of electrons and holes in pure Si is very slow at low temperatures, and will not contribute appreciably to the phonon number).

For slow atomic recoils, energy loss by ionization becomes negligible (below $\sim$30 KeV for Si in Si)(9). Nearly all the energy is dissipated as phonons and lattice damage. The temperature of the phonon-emitting region and hence the average phonon wavelength, can be estimated from range data to be a few tens of degrees Kelvin. Once these long wavelength phonons have been produced in pure crystals, they propagate with mean free paths greater than the order of several cm, if the bulk temperature is kept low enough. Temperatures below 4 K prevent scattering of the radiation-produced phonons by thermal background phonons.(11)

PHONON FOCUSING

A bonus comes from the crystalline anisotropy: we illustrate this again for the case of Si. Phonons initially generated isotropically in propagation vector $\vec{k}$ space, will be highly anisotropic in group velocity direction. This is illustrated in Figure 1, taken from the experimental work of Northrop and Wolfe on Ge.(12) Figure 2 shows our Monte Carlo calculations of phonon intensities at the [100] faces of a Si cube due to a point source of phonons within the crystal. The focusing results in $\sim$90% of the phonon energy being concentrated into $\sim$10% of the total area. Event positions within the crystal may be reconstructed from the focusing pattern and independently from relative arrival times of the phonons at opposing sides of the crystal (for a 1 kg cube, 7.5 cm on a side, the spread of arrival times is only a few tenths of a microsecond compared to transit times of several microseconds). This should permit fairly powerful event topology discrimination, with spatial resolution of the order of mm, and the ability to distinguish events with "multiple hits".

TRANSDUCERS

There is an extensive literature of phonon spectroscopy.(13) Several transducers have been used, including thin film graphite thermistors;(14) thin film Al superconducting transition edge detectors;(12) and superconducting tunnel junctions(13), shown schematically in Figure 3. The SIS tunnel junctions appear to be superior for the present purpose, having very high sensitivity (phonons are converted to electron hole pairs at >1:1 ratio), high impedance, and stable operation. The energy sensitivity of small (.1 x .1 mm) Sn/SnO/Sn junctions has been shown to be more than sufficient for direct detection of 5 KeV x-rays interacting in the junctions themselves.(15) The threshold phonon energy for detection with SIS tunnel junctions is typically less than 10 milli-eV or 1 K, (twice the superconducting energy gap of the junction material). Although our initial experimental work at Stanford was with graphite films at ultralow temperature (Figure 4), we are now intensively working on novel designs for large area, low capacitance SIS junctions of various materials, to be

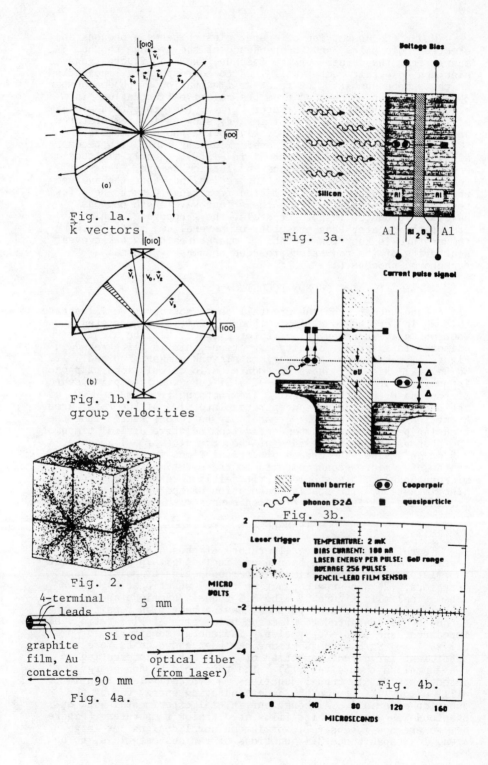

Fig. 1a. $\vec{k}$ vectors

Fig. 1b. group velocities

Fig. 2.

Fig. 3a.

Fig. 3b.

Fig. 4a.

Fig. 4b.

used at 0.1 - 1.3 K . Fortunately, excellent facilities and considerable expertise is available to us through the Stanford Center for Integrated Systems.

BACKGROUNDS

Serious attention must be given to the many sources of background which will interfere with low rate mesurements as indicated in Table I. The interfering backgrounds are rather different for the energetic electron recoils than for the few KeV atomic recoils from coherent neutral current nuclear scattering.

The subject of ultralow backgrounds is of course quite complex. Only a few points of particular interest with respect to our work will be mentioned here.

Great strides have been made recently by several groups (after prodigious efforts!) in reduction of backgrounds in semiconductor diode detectors for Ge double beta decay experiments (16-18). Although the radionuclide content of Si is likely to be different from Ge, the Ge results are at least indicative of what may be achievable.

The lowest continuum background count rates reported in these deep underground experiments are a few counts/Kg day KeV at 10-20 KeV, and about .01 count/Kg day Kev around 2 McV. Similar rates in a silicon acoustic detector of a few Kg mass, would permit a first measurement of coherent neutral current neutrino-nucleus scattering at a high flux reactor.

The expected rate from solar neutrinos is 10^5 times less. Ninety per cent of the solar neutrino interaction rate comes from neutrino-electron scattering, with energy deposits of hundreds of KeV or more.(19) The background in this energy region would need to be reduced by a factor of roughly 10^4 from tne Ge results, to become comparable to the signal. Event topology discrimination should permit considerable background reduction in an acoustic detector, as should the capability of defining a fiducial volume entirely within the monolithic Si crystal.

No appreciable content of ^{32}Si ($t_{1/2}$=280 yr) should be present in silicon prepared from appropriately chosen deep-mined ores. This isotope is produced by cosmic ray spallation of atmospheric argon and/or heavy impurities in the Si ore. Deep-mine sources of inorganic silicon with very low heavy impurity concentrations are known and are commercially exploited for electronics manufacture (20). These ores should have an "effective age" for ^{32}Si decay comparable to that of inorganic carbon samples for ^{14}C decay, for which an experimental lower limit is 50,000 years (21). This leads to the expectation of insignificant remaining ^{32}Si activity. Our plans include measurements of the isotopic composition of various silicon samples to check this argument.

Other radionuclides produced by cosmic ray spallation of the silicon itself can result in appreciable counting rates (on the solar-neutrino scale!). Our calculations show that the largest rate will be due to tritium, at about 3.5 disintegrations/Kg day for each 90 days exposure of silicon to cosmic rays at sea level. Sodium-22 and ^{7}Be are also produced at lower activities.

The continuum beta spectra of tritium and ^{22}Na would interfere with solar neutrino measurements. One possible solution would be to prepare detector material for this purpose underground, from ore to finished detector. The monochromatic gamma from ^{7}Be daughter decay, like any monochromatic radiation totally absorbed within the crystal, is not an interfering background.

References
1. E.D. Commins, P. Bucksbaum, Weak Interactions of Leptons and Quarks, Cambridge, 1983, p.124.
2. L.M. Sehgal, M. Wanninger, Phys Lett B 171, 107 (1986).
3. A.A. Varfolomeev, Sov J Nuc Phys 31, 655 (1980).
4. D.Z. Freedman et al, Ann Rev Nuc Sci 27, 167 (1977).
5. N. Avignone, Phys Rev D 2,2609 (1970).
6. J.N. Bacall, Rev Mod Phys 50, 88 (1978).
7. F. Reines et al, Phys Rev Lett 45, 1307 (1980). In this work the neutral current dissociation of a deuteron into a neutron plus a proton was used as a normalization for the flavor dependent charged current dissociation into two neutrons plus a positron. See however the conflicting final results of F. Boehm et al, Phys Lett 92B, 310 (1980).
8. W. Van Roosbeck, Phys Rev 139A, 1702 (1965). and C. Klein, IEEE Trans Nuc Sci 15, 220 (1968).
9. R.S. Nelson, The Observation of Atomic Collisions in Crystalline Solids, American Elsiever, 1968 p 35 ff and J. Lindhard, M. Scharff, Phys Rev 124, 128 (1961).
10. D. Mc Cammon et al, J Appl Phys 56, 1263 (1984).
11. J.M. Ziman, Principles of the Theory of Solids, Cambridge, 1972, pp 239 ff.
12. G.A. Northrop, J.P. Wolfe, Phys Rev B 22, 6196 (1980).
13. For a review see W. Eisenmenger, Physical Acoustics XII, Academic Press, 1976 ,p 79.
14. W. Knaak, M. Meissner, Proc. LT17, U. Eckhern et al, eds., Elsevier, 1984, p 667.
15. D. Twerenbold, Europhysics Lett 1, 209 (1986).
16. F.T. Avignone et al, Phys Rev Lett 54, 2309 (1985).
17. A. Forster et al, Phys Lett 138B, 301 (1984).
18. F.S. Goulding et al, LBL Report 18043 (1984) and D.O. Caldwell et al, Phys Rev Lett 54, 281 (1985).
19. B. Cabrera et al, Phys Rev Lett 55,25 (1985).
20. W.E. Ver Plank, Calif. Div Mines & Geol Bull 187, 58 (1966).
21. See for example H.E. Gove ed., Proc. First Conf. Radiocarbon Dating with Accelerators, Rochester, 1978.

Figure Captions:
1. (Reproduced from Ref. 12). a. Intersection of constant frequency surface with (001) plane in $\vec{k}$ space for slow TA phonon mode in Ge. b. Group velocities corresponding to $\vec{k}$ vectors shown in a. The surface contains folds, e.g. along (010).
2. Monte Carlo results for point source of phonons at x/a=.7, y/a=.5, z/a=.4 in Si cube of side "a" (70,000 points, isotropic $\vec{k}$'s)
3. a. Schematic of phonons incident on SIS tunnel junction. b. Energy level diagram for quasiparticle tunneling in a.
4. a. First Si phonon detector built at Stanford. b. Oscilloscope trace from ballistic phonons following laser excitation of a.

Table I. Neutrino interaction rates in Si detector (1 KeV threshold imposed).

High flux reactor	10^{13} ν/cm^{2}sec	100 counts/Kg day
Solar neutrinos	few 10^{11} ν/cm^{2}sec	$1.5 \cdot 10^{-3}$ /Kg day
100 KT fission weapon at 1 Km range	few 10^{14} ν/cm^{2}blast	10^{-2} /Kg blast

MODERN IMPLICATIONS OF NEUTRON β-DECAY

S. J. Freedman
Argonne National Laboratory, Argonne, IL 60439-4843

ABSTRACT

The ratio of the axial-vector to vector coupling constant for neutron decay is best determined by recent neutron β-asymmetry experiments. The new value of g_A/g_V resolves serious discrepancies in neutron lifetime measurements. Predictions of theories using g_A/g_V as input are seriously modified.

It is surprising that improving our knowledge of the rather mundane process of neutron β-decay remains a critical element in several important investigations. It is also embarrassing that despite fifty years since the neutron's discovery we still have an inadequate empirical understanding of one of its most fundamental characteristics - its lifetime. The problem is aggravated by serious discrepancies among the four direct lifetime measurements with the smallest reported errors. Thus we have four numbers for the neutron lifetime:[1] τ=(1013±26), (919±14), (877±8) and (937±18) sec. The spread makes the neutron lifetime from direct measurement more questionable than the lifetime of the Ω^-.

A useful parameterization of neutron beta decay is the six coupling constants associated with the matrix elements of the hadronic current, with the usual notation we have,

$$\langle n|J_\mu|p\rangle = \bar{u}_n(g_V\gamma_\mu + ig_M\sigma_{\mu\nu}q_\nu + g_S q_\mu)u_p$$
$$+ \bar{u}_n(g_A\gamma_5\gamma_\mu + ig_T\gamma_5\sigma_{\mu\nu}q_\nu + g_P\gamma_5 q_\mu)u_p , \qquad (1)$$

and $q=(p_n-p_p)$ is the momentum transfer. By convention the overall Fermi strength for neutron β-decay is written as $G_\beta = G\cos\theta_c g_V$, where G is the universal strength of the weak interaction which is accurately measured in muon decay, and $\cos\theta_c$ is the cosine of the Cabibbo angle. Since the weak vector current is conserved the vector coupling constant g_V is also the quark vector coupling (aside from conventions about radiative corrections), and equal to unity in the standard model. Experiments with neutrons could in principal help us determine the Cabibbo angle and measure all the neutron coupling constants. However, at present almost nothing we understand about the size of the small "induced" terms in the nucleon weak current has been aided by studies of neutron β-decay. Nevertheless, numerous experimental studies of nuclear β-decay and μ-capture indicate there is no contradictions with expectations based on the conserved-vector-current hypothesis (CVC), the partially-conserved axial-vector-current theory (PCAC), and the absence of second-class currents. Neutron decay experimental sensitivities are just at the level where the effect of g_M (the weak magnetism effect) is a correction comparable to the experimental error; CVC implies that $g_M=(\mu_p-\mu_n-1)/2M_n$. The best determination of $\cos\theta_c$ now comes from nuclear β-decay experiments. Null experiments have failed to find any indication that time reversal symmetry is violated in neutron β-decay.

The principal utility of neutron β-decay experiments has been the determination of g_A/g_V. Unlike the vector current, the axial-vector current is not conserved, thus g_A is expected to change inside nuclei, and g_A/g_V must be measured with free neutron decay. We shall first discuss the present state of our knowledge of g_A/g_V the implications, and then briefly discuss some expectations of future neutron β-decay experiments.

In terms of the experimental observables so far utilized and assuming time reversal invariance the decay angular correlation in neutron decay is given by the expression:

$$dW = [1 + \frac{a\vec{p}_e\cdot\vec{p}_\nu}{E_e E_\nu} + \frac{A\vec{J}\cdot\vec{p}_e}{\langle J\rangle E_e} + \frac{B\vec{J}\cdot\vec{p}_\nu}{\langle J\rangle E_\nu}] \, dW(p_e)d\Omega_e d\Omega_\nu \qquad (2)$$

Neglecting small radiative corrections, induced couplings, and with $\lambda=|g_A/g_V|$, we have in the allowed approximations:

$$a \approx a_o = \frac{1+\lambda^2}{1+3\lambda^2} \qquad \text{for the β-ν correlation,} \qquad (3)$$

$$A \approx A_o = \frac{-2\lambda(\lambda-1)}{1+3\lambda^2} \qquad \text{for the β-asymmetry,} \qquad (4)$$

and

$$B \approx B_o \frac{2\lambda(\lambda+1)}{1+3\lambda^2} \qquad \text{for the ν-asymmetry.} \qquad (5)$$

The neutron fτ-value is given by:

$$f\tau = \frac{2\pi^3\hbar^7}{G^2 g_V^2 \cos^2\theta_c m_e^5 c^4 (1+3\lambda^2)} \qquad (6)$$

As we have noted the value of $G\cos\theta_c$ can be measured in nuclear β-decay, notably in $0^+\to 0^+$ super-allowed Fermi decays. An analysis due to Wilkinson[2] determines the neutron Fermi factor including the effects of the induced couplings and radiative corrections; using the corrected fτ-value of super-allowed decays he obtains, $\tau=(5188.8\pm2.4)/(1+3\lambda^2)$.

Until recently the lifetime measurements determined g_A/g_V with the smallest reported errors, although this feature is negated by the systematic disagreement of the experiments. In any case, the best determination of g_A/g_V now comes from a recent measurement of the β-asymmetry.

The new experiment (a Heidelberg/ANL/ILL collaboration) employs a cold beam of polarized neutrons from the 57 MW high flux

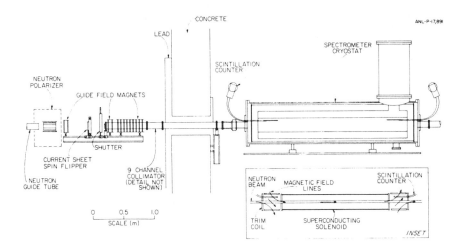

Fig. 1 Experimental arrangement of PERKEO

reactor operated by the Institute Laue-Langevin in Grenoble, France. Figure 1 shows the experimental arrangement. The polarized neutron beam (P≈97%) enters a 1.7 m long superconducting solenoidal spectrometer (called PERKEO) with an internal field of ≈ 15 kgauss. The neutron polarization is along the beam direction and it is reversed periodically with a current sheet spin reverser. Electrons from neutron decay are constrained by the magnetic field to helical paths with diameters less than 1 cm. Trim coils at the ends of the apparatus distort the field and the electrons are guided to plastic scintillator counters outside of the region of the neutron beam. The scintillators are calibrated with conversion electron sources. Backscattered electrons are trapped in the spectrometer and eventually deposit all their energy in the countners; timing information is used to determine the counter struck first.

The β-asymmetry as a function of kinetic energy is measured by observing the count rate asymmetry for the two polarization states. Figure 2 shows the result. The principal energy dependence is caused by the factor of v/c in Eq. 2. To obtain A_o (Eq. (4)) the data is fitted to the energy dependence from v/c, weak magnetism and the measured detector response. The amplitude from the fit is corrected for the measured polarization, spin flip efficiency, and an adjustment of about 10% to account for the magnetic mirror effect (if the initial direction of the decay electron is toward a region of increasing magnetic field there is a finite reflection probability). Table 1 tabulates the results of the new experiment along with previous β-asymmetry measurements. This is the first experiment to measure the asymmetry without counting electron recoil-proton coincidences and the first to measure the energy dependence of the asymmetry. Previous experiments were statistics

Table I Characteristics and Results of β-asymmetry Measurements

Year	Reference	Polarization (%)	Typical Count Rate (sec^{-1})	Asymmetry	g_A/g_V
1960	Burgy et al.[1]	84±7	0.015	−0.114(19)	−1.257(50)[†]
1961	Clark & Robson[4]	89±5	−	−0.090(50)	−1.20(12)[†]
1969	Christensen et al.[5]	87±3	0.013	−0.115(9)	−1.260(23)[†]
1971	Erozolimskii et al.[6]	77±2	0.055	−0.120(10)	−1.273(27)[†]
1975	Krohn & Ringo[7]	79±1.5	0.117	−0.111(8)	−1.249(22)[†]
1976	Erozolimskii et al.[6]	73.3±2	0.106	−0.112(5)	−1.257(14)[†]
1984	Bopp et al.[8]	96.7±0.7	200	−0.118(3)[*]	−1.270(9)
1985	Bopp et al.[9]	97.4±0.5	200	−0.1146(19)[*]	−1.262(5)

[*] The asymmetries in these experiments are corrected to correspond to the β-asymmetry in the allowed approximation (A_o).

[†] The value of g_A/g_V is obtained by correcting A for induced effects and recoil over the reported energy range by the method described in Ref. 7.

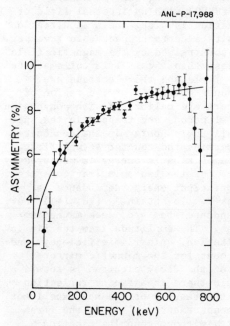

Fig. 2 β-asymmetry vs energy from 150 hours of running hours with PERKEO.

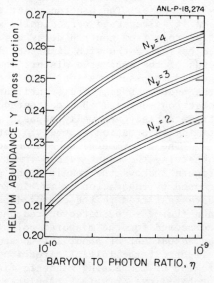

Fig. 3 The abundance of primordial ^{4}He vs proton to baryon ratio for different numbers of species of light neutrinos.

limited but the new experiments obtains a count rate more than three orders of magnitude higher than before; the final error is almost entirely systematic, from subtracting background, measuring the polarization, calibrating and accounting for the magnetic mirror effect.

The asymmetry measurements are consistent and we are justified in taking the weighted average, obtaining $g_A/g_V = -1.262\pm0.004$. The predicted neutron lifetime is (898 ± 5) sec. Previous $\beta-\nu$ correlation experiments obtain $g_A/g_V = -1.256\pm0.015$, in good agreement. Measurements of the ν-asymmetry are rather insensitive to g_A/g_V.

Theoretical calculations of g_A/g_V have been attempted for some time. The early calculations were based on PCAC often improved with various corrections. More recently, calculations have been made in various models of quark confinement and in the Skyrme model. Some of the results are consistent with experiment but the theories are not well enough developed to make any definite conclusions from the level of agreement. Eventually g_A/g_V should provide a useful tool for finding the correct theory for low energy limit of the quark interaction.

In the generalized Cabibbo model, the magnitude of g_A/g_V should coincide with F+D from semileptonic hyperon decay. However, a recent measurement[10] gave F+D=1.18(2) in rather poor agreement.

The value of g_A/g_V is a critical input for the calculation of the solar neutrino flux. In the standard solar model the rate for the fundamental hydrogen burning process is determined by g_A/g_V and the weak coupling strength. The most recent calculation using the new value of g_A/g_V gives 5.8 ± 2.2 SNU (3σ) for the predicted ^{37}Cl capture rate.[11] Although the reduction from previous estimates because of the revised g_A/g_V is about 1 SNU the resulting decrease in error makes the discrepancy with the measured value of 2.2 ± 0.3 SNU even more serious.

The g_A/g_V is in necessary input in calculations of the primordial ^{4}He abundance based on big bang nucleosynthesis.[12] Until recently the neutron lifetime uncertainties were regarded as the largest error in the calculation. Figure 3 shows the predicted ^{4}He mass fraction, Y, as a function of the baryon to photon ratio, η, with the new value of g_A/g_V. The three curves in the figures correspond to calculations with 2, 3 and 4 species of light neutrinos (or other light particles). Some suggested bounds[13] on $\eta(\eta>2\times10^{-10})$ and Y(Y<0.25) seem to indicate $N_\nu \leq 4$. Prevous analysis with an effectively longer neutron lifetime as input seemed to favor no more than 3 light particles.

The experimental problem with the directly measured lifetime is still the most serious issue in neutron decay. Data for a new direct lifetime determinations by two separate "in beam" techniques has already been collected by a new Heidelberg/ANL/ILL collaboration using the PERKEO spectrometer.[14] The measurement is expected to give the lifetime to about 1%. Hopefully the new measurement will help to resolve the present conflict. A new measurement using a "bottle" to confine the neutrons promises to avoid the basic limitations of in beam methods, perhaps leading to an order of magnitude more precision.[15] Combined with a comparable improvement in the β-asymmetry the neutron data could provide a more precise determination of $\cos\theta_c$. This possibility is particularly exciting because of recent indications that the more general Kobayashi-Moskawa mixing matrix may fail to satisfy unitarity.[16] Using $\theta_c(=\theta_1)$ determined from neutron decay would relieve uncertainties from radiative corrections and nuclear physics effects.

Finally, more sensitive neutron decay correlation experiments will be able to test important principles that constrain the induced form factors without the uncertainties of nuclear corrections, and to improve limits on time symmetry violation. For example, a sensitive measurement of the energy dependence of a and A would provide an unambiguous test of weak magnetism and the absence of second class currents. Time reversal symmetry is now tested to a part in 10^3 in neutron decay, but techniques available now could push the limit another order of magnitude.

This work supported by the U.S. Department of Energy, Nuclear Physics Division, under contract W-31-109-ENG-38.

REFERENCES

1. For a complete review of neutron β-decay experiments see J. Byrne, Rep. Prog. Phys. 45, 115 (1982).
2. D. H. Wilkinson, Nucl. Phys. A377, 424 (1982).
3. M. T. Burgy, et al., Phys. Rev. 20, 1829 (1960).
4. M. A. Clark and J. M. Robson, Can. J. Phys. 39, 13 (1961).
5. C. J. Christensen, Phys. Lett. 28B, 411 (1969); this reference combines the result of Ref. 3. Table I includes only the 1969 result.
6. B. G. Erozolimskii et al., Sov. J. Nucl. Phys. 30, 356 (1979); this reference supercedes earlier papers; B. G. Erozolimskii et al., JETP Lett. 13, 252 (1971); JETP Lett. 23, 663 (1976).
7. V. E. Krohn and G. R. Ringo, Phys. Lett. 55B, 175 (1975); this reference combines the results of Refs. 3 and 4. Table I includes only the 1975 result.
8. P. Bopp et al., Journal de Phys. C3, 21 (1984).
9. P. Bopp et al., Phys. Rev. Lett. 56, 919 (1986).
10. M. Bourquin et al., Z. Phys. C21, 27 (1983)/
11. J. N. Bahcall et al., Astro. J. 292, 179 (1985).
12. K. A. Olive et al., Astro. J. 246, 557 (1981).
13. D. A. Dicus et al., Phys. Rev. D26, 2694 (1982).
14. D. Dubbers, NSB Spec. Pub. 711, 54 (1986).
15. W. Mampe, NBS Spec. Pub. 711, 59 (1986).
16. W. J. Marciano and A. Sirlin, Phys. Rev. Lett. 56, 22 (1986).

COMPARISON OF BETA-RAY POLARIZATIONS IN FERMI
AND GAMOW-TELLER TRANSITIONS AND $SU(2)_L \times SU(2)_R \times U(1)$ MODELS

P. Herczeg
Theoretical Division, Los Alamos National Laboratory
Los Alamos, New Mexico 87545

ABSTRACT

We analyze the information on $SU(2)_L \times SU(2)_R \times U(1)$ models provided by measurements of the ratio of beta-ray polarizations in Fermi and Gamow-Teller transitions.

$SU(2)_L \times SU(2)_R \times U(1)$ models[1] are attractive extensions of the standard model of the electroweak interactions. A characteristic feature of these models is the presence of right-handed charged currents. Among the sensitive probes of right-handed currents are the longitudinal polarizations (P_e^F, P_e^{GT}) of electrons (or positrons) in pure Fermi or Gamow-Teller β-decays[2]. An approach followed in recent and in ongoing experiments[3] involves a comparison of P_e^F and P_e^{GT} for positrons of the same energy. The present experimental result $P_e^F/P_e^{GT} = 0.986 \pm 0.038$ (J. van Klinken et al., Ref. 3) implies the limit

$$|r|_{expt} < 10^{-2} \quad (90\% \text{ confidence}) \qquad (1)$$

for the quantity $r \equiv \{(P_e^F/P_e^{GT}) - 1\}/8$. The experiments under way[3] plan to improve the accuracy for P_e^F/P_e^{GT} by 1-2 orders of magnitude. The result (1) was interpreted so far only in the framework of manifestly left-right symmetric models. In the investigation reported here we study the implications of (1) for more general $SU(2)_L \times SU(2)_R \times U(1)$ models, including the most general one which allows for CP-violation, unequal left- and right-handed quark mixing angles, and mixing in the leptonic sector. In each case we compare the corresponding constraints on the parameters with constraints provided on them by other data. Below we give a brief account of our results[4].

For allowed decay, ignoring recoil-order terms, higher-forbidden contributions and electromagnetic effects, r is given in $SU(2)_L \times SU(2)_R \times U(1)$ models by[5]

$$r \simeq -\tilde{v}_e \text{Re} \eta_{RR} \eta_{RL}^*, \qquad (2)$$

where $\eta_{ik} \equiv a_{ik}/a_{LL}$ (i = L,R; k = L,R); a_{ik} are the coupling constants of the $\Delta S = 0$ semileptonic Hamiltonian (a_{RL} is associated with the term involving the right-handed leptonic and the left-handed quark current, etc.). In terms of the parameters of the most general model

$$r \simeq \tilde{v}_e \, t_\theta \zeta_g \cos(\alpha+\omega), \qquad (3)$$

where $t_\theta = t\cos\theta_1^R/\cos\theta_1^L$, $t = g_R^2 m_1^2/g_L^2 m_2^2$, $\zeta_g = g_R\zeta/g_L$, α and ω are CP-violating phases in the right-handed quark mixing matrix and in W_L-W_R mixing, respectively; m_1, m_2 are the masses of W_1, W_2; ζ is the W_L-W_R mixing angle; $\tilde{v}_e = v_e/u_e$, $u_e = \Sigma_i' |U_{ei}|^2$, $v_e = \Sigma_i' |V_{ei}|^2$, U and V are the left-handed and right-handed leptonic mixing matrices, respectively; the summation in u_e and v_e is over the neutrino states produced in the decay (see Ref. 6).

(A) <u>Models with negligible leptonic mixing.</u> Except when the right-handed muon-neutrino is too heavy to be produced in π-decay, in which case the conclusions of case (B) below apply, the best limit on t_θ comes from measurements of the quantity R (see Ref. 6), related to the positron momentum spectrum end point in polarized μ-decay[6]. Combined with the best limit on ζ_g, provided by the ρ-parameter, one obtains $|r| < 1.3 \times 10^{-3}$. We note that as $\cos(\alpha+\omega) = 0$ is not ruled out, an upper limit on r does not constrain $t_\theta \zeta_g$. In models where $\theta_i^R = \theta_i^L$, the $K^\circ \to \bar{K}^\circ$ amplitude, nonleptonic K-decays and the D-coefficient in β-decay yield the limit $|r| \lesssim 2 \times 10^{-5}$ (see Ref. 6). For the general case the bound $|r| \lesssim 2 \times 10^{-4}$ follows from R, nonleptonic K-decays and the D-coefficient.

(B) <u>Models with leptonic mixing.</u> Except for the case when all the neutrinos can be produced in the decay, muon decay does not provide constraints on $t\sqrt{\tilde{v}_e}$ or $t_\theta\sqrt{\tilde{v}_e}$. The best limit on r from leptonic and semileptonic processes is obtained combining ρ_{expt} and Gamow-Teller β-decay data[7], yielding $|r| < 6 \times 10^{-3}$. As $\tilde{v}_e \lesssim 1$, in models where $\theta_i^R = \theta_i^L$ the limit on $|r|$ from data involving nonleptonic transitions is the same as in case (A). In the general case only the constraint from nonleptonic K-decays and the D-coefficient is applicable, implying only the limit $|r| \lesssim 4 \times 10^{-3}$.

To conclude, in some special versions of $SU(2)_L \times SU(2)_R \times U(1)$ models either muon-decay data, or constraints which include information from non-leptonic transitions, or both of these, provide an upper limit on r which is more stringent than the limit from the present experimental result on P_e^F/P_e^{GT}. However in $SU(2)_L \times SU(2)_R \times U(1)$ models where θ_i^R and θ_i^L are unrelated and v_μ ($\equiv \Sigma_i' |V_{\mu i}|^2$) is arbitrary the limit on r from the direct measurement is comparable to the limit on r implied by other data.

ACKNOWLEDGEMENTS

I would like to thank Professors J. Deutsch, A. Rich and M. Skalsey for informative conversations. This work was performed under the auspices of the U. S. Department of Energy.

REFERENCES

1. J. C. Pati and A. Salam, Phys. Rev. Lett. <u>31</u>, 661 (1973); Phys. Rev. <u>D10</u>, 275 (1974); R. N. Mohapatra and J. C. Pati, Phys. Rev. <u>D11</u>, 566, 2558 (1975);, G. Senjanović and R. N. Mohapatra, Phys. Rev. <u>D12</u>, 1502 (1975).
2. M. A. Bég et al., Phys. Rev. Lett. <u>38</u>, 1252 (1957).
3. M. Skalsey et al., Phys. Rev. Lett. <u>49</u>, 708 (1982); A. Rich and M. Skalsey, private communication; J. van Klinken et al., Phys. Rev. Lett <u>50</u>, 94 (1983); J. Deutsch, private communication.
4. A detailed account will be given in a forthcoming article. Some preliminary remarks on P_e^F/P_e^{GT} are contained in P. Herczeg, in <u>Neutrino Mass and Low-Energy Weak Interactions</u>, Telemark 1984, edited by V. Barger and D. Cline, World Scientific Publishing Co., 1985, p. 288.
5. We assume that the effect of the masses of the neutrinos produced in the decay can be neglected.
6. P. Herczeg, Los Alamos National Laboratory preprint LA-UR-85-2761, submitted to Nucl. Phys. B.
7. J. van Klinken, F. W. J. Koks and H. Behrens, Phys. Lett. <u>79B</u>, 199 (1978).

STATUS OF THE SOUDAN 2 NUCLEON DECAY EXPERIMENT[*]
Edward N. May
For the Soudan 2 Collaboration
Argonne National Laboratory, Argonne, Illinois 60439

ABSTRACT

Soudan 2 is an 1100-ton fine-grained tracking calorimeter now being constructed to search for nucleon decay. It is distinguished by superior background rejection for a wide variety of decay modes and by a trigger threshold low enough for 100 MeV neutrino-induced muons. A new laboratory large enough for an eventual 3300 ton experiment has been constructed in Minnesota's Soudan mine at a depth equivalent to 2200 meters of water. The detector will be assembled from 256 5-ton calorimeter modules, utilizing corrugated steel sheets and plastic drift tubes to obtain a rather isotropic honeycomb geometry. Data collection will begin in the fall of 1986 and assembly of the 1100 tons will be completed in 1988. Installation progress and the performance of the first 5-ton modules will be described.

INTRODUCTION

The Soudan 2 experiment is being built by a collaboration of five institutions from the USA (Argonne, Minnesota, Tufts U) and the United Kingdom (Oxford, Rutherford). The philosophy of the detector is to use a dense fine-grained tracking calorimeter to detect and measure many properties of nucleon decay events predicted by grand unified theories.[1] Candidate events will be identified by their unique topological properties, compared to the principal backgrounds induced by the interaction of atmospherically produced neutrinos. Soudan 2 is unique in being able to identify particle types and measure energies of each track resulting from a decay or neutrino interaction. This results from the design features which give a measurement of the deposited ionization as a function of track length for each track stopping in the detector. Unique kinematic reconstruction of most recorded events will be possible. The low threshold of the design permits highly efficient searches for decay modes containing many particles in the final state.

THE DETECTOR

The basic detector element of the experiment is shown in Fig. 1. Formed steel sheets 1.6 mm thick are stacked vertically to give a hexagonal close-packed array of holes filled by resistive plastic (hytrel) tubes (ID=15 mm, OD=16 mm, length=1 m)

---

[*]Work supported by the U.S. Department of Energy, Division of High Energy Physics, Contract W-31-109-ENG-38 and the U. K. Science and Engineering Research Council.

and enclosed in mylar sheets for insulation. This is shown in the upper inset of Fig. 1. A -10KV electric potential is applied to the central of 17 copper strips to grade the electric field to zero at each tube end. This provides a uniform axial drift field to transport ionization electrons from charged tracks at up to the full 50 cm depth within the module. The transported ionization is detected at each tube end by vertical proportional wires (64) backed by horizontal cathode strips (256) as shown in the lower inset of Fig. 1. The drift velocity using 80/20 Ar/CO_2 is 0.8 cm/usec. Full 3D readout is obtained from the wire and strip locations and multi-hit drift time digitalization. In addition the charge content of each hits is measured with 6-bit flash ADC. A detailed description of the electronics is given in Ref. 2. The tracking resolution is 4 × 3 × 2 mm RMS. The threshold will be 100 MeV kinetic energy for charged particles. The module shown in Fig. 1 is 1m × 1m × 2.5m and mass 5 ton giving an average density of 2 gm/cm^3. Two hundred and fifty six will be arranged as 2 high × 8 wide × 16 long to provide detector of 1100 metric tons active mass. This main detector will be entirely surrounded by an active shield constructed of a 2-layer array of extruded aluminum proportional tubes.

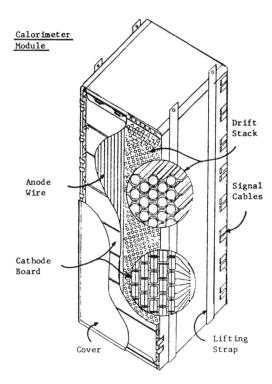

Fig. 1

MODULE TESTING

As part of the module construction and before installation at the mine site, each module is tested to verify that it meets minimum performance standards. These tests are performed in a cosmic ray teststand on the surface using an external trigger to provide stiff cosmic ray muon tracks. All 640 anode/cathode channels are digitized and readout. Fitted tracks are then used to verify the drift velocity, efficiency, gain uniformity, position resolution, etc. In addition, imbedded Fe^{55} radioactive sources are monitored to obtain an absolute gain calibration and a high precision measurement of the charge resolution. A detailed report of module performance is given in Ref. 3.

THE SITE

The detector will be located on the 27th level of the Soudan Iron Mine, Tower-Soudan State Park, Soudan, Minnesota, USA. Latitude 48°N, Longitude 92°W. The depth of the rock overburden is 700 m, equivalent to 2200 m of water. The rate of cosmic ray muons traversing the main detector will be 0.5 Hz. A new cavity of dimensions 14 m × 72 m × 11 m (high) has been excavated. This size is sufficient to accommodate 3300 metric tons of detector, should later expansion be desirable. A steel framework and crane system for mounting the detector modules has been installed. The cavity is currently being outfitted as a physics laboratory with heat, light, power, and enclosures for data acquisition electronics and computers.

SCHEDULE

The first elements of the detector and shield will be shipped to the mine during the summer of 1986. A half wall (8 modules) and overhead shield will be completely installed by the fall of 1986. Sufficient electronics will be installed to test the trigger and observe cosmic rays. We will operate two walls (32 modules, 150 tons) by the firsts of the year 1987. This will provide the first useful fiducial volume for physics. At the time the production rate will be two per week, which will allow the complete 1100 tons to be in operation in the mine by the end of 1988. During the year 1988, approximately 9 modules will be taken to an accelerator laboratory to be calibrated in neutrino and charged particle beams.

1. H. Georgi and S. Glashow, Phys. Rev. Lett. 32, 438 (1974); H. Georgi, H. Quinn and S. Weinberg, Phys. Rev. Lett. 33, 451 (1974).
2. J. D. Dawson et al., Proceedings of the 1985 IEEE Nuclear Science Symposium, NS-33, 106 (1986).
3. J. Hoftiezer et al. Proceedings of the 1985 IEEE Nuclear Science Symposium, NS-33, 181 (1986).

SEARCHES FOR MONOPOLES: PAST, PRESENT AND FUTURE

S.P. Ahlen
Physics Department
Boston University, Boston, MA 02215

ABSTRACT

This paper will summarize the status of experiments searching for Grand Unified Theory magnetic monopoles. Emphasis will be placed on those types of experiments which rely on ionization/excitation techniques (another paper in these proceedings deals with superconductive detection techniques). Special attention will be given to a description of the MACRO experiment and its monopole detection capabilities.

INTRODUCTION

In recent years, there has been a great deal of interest in the possibility of detecting supermassive magnetic monopoles which may have been produced in the early moments of the universe. Such particles are a natural consequence of Grand-Unified Theories (GUT's) and Kaluza-Klein theories.[1] If they exist, they are likely to have a charge given originally by Dirac, $g = 137e/2$, a mass in the range 10^{16} to 10^{19} GeV/c^2, and a velocity relative to the Earth greater than or equal to $10^{-3}c$. Their abundance cannot be predicted accurately (standard cosmologies produce an abundance of monopoles which is known to be too great; inflationary models of the universe allow production of monopoles with a very wide range of possible abundances). However, a useful reference flux for monopoles is the Parker Limit, which, for a monopole mass of 10^{16}GeV/c^2 is:

Parker Limit (10^{16} GeV/c^2) = 5×10^{-16}/(cm^2 sr s).

Straightforward arguments imply that if monopoles existed at fluxes greater than the Parker Limit, they would "short circuit" the magnetic field of our Galaxy, which is known not to be the case. Thus, to provide information on monopole abundance which goes beyond that which can be inferred from such basic astrophysical arguments, detectors which search for magnetic monopoles need to be quite large (of the order of 10,000 m^2 sr, for experiments running for several years).

Since there is little guidance from theory as to the abundance of monopoles produced in the early universe, no one knows how far beyond the Parker Limit to probe to have a realistic expectation of detecting monopoles. However, there is a rather remarkable coincidence which suggests that the Parker flux may be a natural scale to achieve with experiments in the near future. The coincidence concerns the cosmic mass density of monopoles. For the GUT mass scale (10^{16} GeV/c^2), monopoles would account for all of the "missing mass" in the universe (i.e. their density would correspond to the mass density required to close the universe) if they were distributed uniformly throughout the universe at the Parker flux. Thus, by probing a factor of 10 beyond the Parker Limit, one rules out monopoles as a significant form of dark matter in the universe if none are observed.

INDUCTION TECHNIQUE

To probe at the level of the Parker Limit is not easy. Induction experiments are attractive in the simplicity of their interpretation. A magnetically charged particle passes through a superconducting loop of wire. The induced electric field (analogous to the magnetic field set up by a moving electric charge) produces a current in the loop. Monopoles having the Dirac charge produce signals which can be calculated accurately. An attractive feature of induction devices is that they respond to monopoles having any velocity. However, it is quite difficult to construct these detectors with great enough area to approach the Parker Limit. As of early 1986, the 90% confidence level (C.L.) flux limit on monopoles from all induction experiments was $3 \times 10^{-12}/(\text{cm}^2 \text{ sr s})$.

TRACK ETCH TECHNIQUE

Other means of detecting monopoles do exist however. One possibility, which uses naturally occurring mica crystals, could detect monopoles at quite low flux levels, assuming the following sequence of events: 1) monopoles enter the earth with zero or negative electric charge; 2) the monopoles capture nuclei into bound states in the Earth's crust through magnetic-dipole magnetic-monopole interactions; 3) the bound pair passes intact through several kilometers of Earth crust, eventually going through an underground sample of muscovite mica; 4) the monopole-nucleus composite produces a trail of crystal defects in the crystal at the same frequency as would an infinitely heavy isotope of the captured nucleus having the same velocity as the monopole-nucleus composite; 5) the trail of defects remains intact (i.e. thermal annealing does not proceed at an appreciable rate) over nearly a billion years; 6) the trail of defects, when immersed in a solution of hydrofluoric acid, will be etched into a pit on the surface of the mica; and 7) the etch pit will be visible through an optical microscope.

Experiments have been done using this technique.[2,3] The monopole flux limit from the most recent of these is about $10^{-17}/(cm^2 \text{ sr s})$, about 100 times below the Parker Limit. The monopole velocity range covered by this experiment was from $3 \times 10^{-4}c$ to $3 \times 10^{-3}c$. The mica technique is quite powerful for searching for monopoles due to the enormous collecting power of billion year old detectors. However, in view of the large number of assumptions required, one must be careful to avoid over-interpretting negative results from this type of experiment. For, example, if monopoles catalyze baryon decay, it is unlikely that a monopole-nucleus composite could survive long enough to produce a track in mica. If monopoles are produced as positively charged dyons, or if they capture a proton prior to entering the Earth, they will be unable to capture a heavy nucleus due to Coulomb repulsion. Another potential limitation is the difficulty of predicting the shape of an etch pit produced by a slowly moving monopole-nucleus composite, so that they may not be easy to observe with optical microscopes. Finally, the lack of sensitivity to high velocity monopoles (which is due to the fact that the nuclear component of stopping power decreases with increasing velocity) would make it difficult to observe monopoles at the low end of the mass range, as these would be accelerated to a considerable extent by galactic magnetic fields.

HELIUM TECHNIQUE

A detection technique which in a sense would provide a check on the mica technique is that suggested by Drell et al.[4] This technique is based on the large cross section for excitation of helium by low velocity monopoles due to level mixing. Transfer of helium excitation energy to the ionization of additive molecules by the Penning effect could be detected through standard gas amplification methods. Unfortunately, if monopoles carry positive electric charge, they have a much higher velocity threshold than if they are electrically uncharged. So if monopoles enter the Earth with a bound proton, they may be difficult to detect by this means.

Experiments using this method have been done recently. One of these[5] has reported a flux limit (90% C.L.) of $1.44 \times 10^{-13}/(cm^2 \text{ sr s})$ for monopoles moving faster than $7 \times 10^{-4}c$.

SCINTILLATION TECHNIQUES

In my opinion, the best technique for searching for monopoles is that which uses thick slabs of organic scintillator to detect monopole induced molecular excitation of fluor molecules through the emission of photons, which are subsequently detected by photomultiplier tubes. An experiment of this type using 1 slab of acrylic scintillator, and having a collecting power of approximately 20 m^2 sr has been reported.[6] There are several advantages of this technique over the ones mentioned above: 1) It is easy, and relatively inexpensive to costruct very large area arrays of scintillator; 2) It has been established that scintillators are excited by singly charged particles moving at low velocities ($4 \times 10^{-4}c$). This has been done by detecting the light emission from scintillators due to neutron induced proton recoils.[7] Preliminary data from an experiment recently performed at the High Flux Beam Reactor at Brookhaven have shown that a finite amount of light is emitted by protons as they slow from an initial energy of 60 eV to an energy of 30 eV. Thus, monopoles, which have electron scattering cross sections comparable to those for singly charged electric particles, will also probably excite scintillators. There is no need to form heavy nucleus composites; 3) The scintillator signal gets larger for increased velocity, so that the thick-slab technique has a much broader region of velocity sensitivity than does the mica technique; 4) Organic scintillator will respond at low velocities (below $10^{-3}c$) to proton-monopole pairs, whereas mica and helium will not; 5) The thick slab produces a phototube signal which is unique for slowly moving massive particles (a very long pulse, of the order of a microsecond, as opposed to signals from more conventional particles which are spread out over tens of nanoseconds). This allows for excellent rejection of background events; 6) The scintillator technique is sensitive to slow, supermassive, electrically charged particles (such as advocated by Wen and Witten[8]). The induction, mica, and helium techniques are not.

The thick slab experiment mentioned above provided a flux limit (90% C.L.) of $9 \times 10^{-13}/(cm^2 \text{ sr s})$. The Baksan scintillation experiment[9] has attained a flux limit of $2 \times 10^{-15}/(cm^2 \text{ sr s})$ (90% C.L.). The sensitivity of this experiment to monopoles is probably limited to velocities greater than $10^{-3}c$ due to the manner in which the phototube signals were processed.

The logical step to take next in the search for monopoles is to carry the thick-slab-Baksan technique to a detector an order of magnitude larger than Baksan, which has a collecting power of 1850 m^2 sr, and to develop the detector electronics in such a way to be able to detect monopoles with velocity below the Baksan velocity threshold. Such an experiment, involving a collaboration of physicists in the United States and Italy [10], has now been approved and will be constructed in the Gran Sasso laboratory over the next several years. The experiment is called MACRO, for Monopole, Astrophysics and Cosmic Ray Observatory.

MACRO

The MACRO detector will be installed in Hall B of the Gran Sasso Laboratory, which is under the Gran Sasso Mountain approximately 80 miles east of Rome, Italy. The main part of the detector is a horizontal structure consisting of two layers of liquid scintillation counters, ten layers of plastic streamer (Iarocci) tubes and a sandwhich of plastic track-etch detectors. The scintillators will be 25 cm thick, and will provide energy loss and timing information, while the streamer tubes will provide tracking and ionization information. The scintillators will search for monopoles through the thick-slab technique, while the streamer tubes will utilize the Drell et al.-Penning effect to search for monopoles. For additional redundancy, the CR-39 track detectors will also be available to search for monopoles, possibly at velocities below that which either the scintillators or streamer tubes are sensitive. The collecting power of MACRO will be approximately 12,000 m^2 sr. It will be configured so that it can also be used to detect upward going neutrinos (thus making it the world's first high energy neutrino telescope which is capable of observing point sources in our Galaxy), and to measure properties of downward going bundles of muons (which provide important information on the properties of high energy cosmic rays, and on high energy particle interactions). The first supermodule (12% of the full detector) will be installed by the summer of 1987. The full detector will probably be complete by 1989.

REFERENCES

1. See papers in Monopole '83, edited by J.L. Stone (Plenum, New York, 1984).

2. P.B. Price, Shi-Lun Guo, S.P. Ahlen, and R.L. Fleischer, Phys. Rev. Lett. $\underline{52}$, 1265 (1984).

3. P.B. Price and M.H. Salamon, Phys. Rev. Lett. $\underline{56}$, 1227 (1986).

4. S.D. Drell et al., Phys. Rev. Lett. $\underline{50}$, 644 (1983).

5. T. Hara et al., Proc. 19th Inter. Cosmic Ray Conf. (La Jolla) $\underline{8}$, 218 (1985).

6. T.M. Liss, S.P. Ahlen, and G. Tarle, Phys. Rev. D $\underline{30}$, 884 (1984).

7. S.P. Ahlen, T.M. Liss, C. Lane, and G. Liu, Phys. Rev. Lett. $\underline{55}$, 181 (1985).

8. Xiao-Gang Wen and Edward Witten, Nucl. Phys. $\underline{B261}$, 651 (1985).

9. E.N. Alexeyev et al., Proc. 19th Inter. Cosmic Ray Conf. (La Jolla) $\underline{8}$, 250 (1985).

10. MACRO Collaboration; B.Barish (Caltech), E. Iarocci (Frascati) co-spokesmen; other members being C.DeMarzo, O.Enriquez, N.Giglietto, F.Posa (Bari); M.Attolini, F.Baldetti, G. Giacomelli, F.Grianti, A.Margiotta, P.Serra (Bologna); S.Ahlen, G.Auriemma, A.Ciocio, M.Felcini, D. Ficenec, A.Marin, J.Stone, L.Sulak, W.Worstell (Boston); D.Imel, C.Lane, G.Liu (Caltech); R.Steinberg (Drexel); G.Battistoni, H.Bilokin, C.Bloise, P.Campana, V.Chiarella, A.Grillo, A.Marini, A.Rindi, F.Ronga, L.Satta, M.Spinetti, S.Torres L.Trasatti, V.Valente (Frascati); R.Heinz, S.Mufson, J.Petrakis, J.Reynoldson (Indiana); D.Crary, M.Longo, J.Musser, G.Tarle (Michigan); C.Angelini, A.Baldini, C.Bemporad, A.Cnops, V.Flaminio, G.Giannini, R.Pazzi, B.Saitta (Pisa); M.DeVincenzi, E.Lamanna, G.Martellotti, S.Petrera, L.Petrillo, P.Pistilli, G.Rosa, A.Sciubba, M.Severi (Roma); P.Green, R.Webb (Texas A&M); M.Arneodo, G.Borreani, P.Giubellino, F.Marchetto, A.Marzari, S.Palestini, L.Ramello (Torino); B.Laubis, D.Solie, P.Trower (Virginia Tech); Proc. 19th Inter Cosmic Ray Conf. (La Jolla) $\underline{8}$, 228 (1985).

A Search for Anomalously Heavy Isotopes of Low Z Nuclei

D. Nitz and D. Ciampa
Department of Physics, University of Michigan, Ann Arbor, MI 48109

T. Hemmick, D. Elmore, P.W. Kubik, S.L. Olsen, and T. Gentile
Dept. of Physics and Astronomy, University of Rochester, Rochester, NY 14627

H. Kagan and P. Haas
Department of Physics, Ohio State University, Columbus, OH 43210

P.F. Smith
Rutherford Laboratory, Chilton, Oxon, UK

ABSTRACT

We present preliminary results of a search for anomalously heavy isotopes of certain light elements using an electrostatic charged particle spectrometer in conjunction with the MP tandem accelerator facility at the Nuclear Structure Research Laboratory of the University of Rochester. New limits for the existence of anomalous, heavy isotopes (100–10,000 amu) in ordinary, terrestrial Li, Be, B and F samples and enriched H^2, C^{13}, and O^{18} samples are reported.

INTRODUCTION

It is a curious fact that, in spite of the large variety of elementary particles that have been observed and hypothesized during the past 40 years, it appears that all stable matter can be explained as various combinations of neutrons, protons, and electrons – particles well known to physicists since the 1930's. Big-Bang cosmology implies that all types of particles were present in large numbers during the earliest moments of creation. Thus, particles of virtually any mass that have lifetimes comparable to the age of the universe ($\approx 10^{10}$ years) should exist today as remnants of the Big-Bang. Various calculations have been performed[1,2] which yield estimates of 10^{-12}–10^{-10} for the concentration of anomalously heavy isotopes in nature. There are a several potential candidates among the particles that are commonly considered in high energy physics possessing the required level of stability. In technicolor theories[3], for example, the lightest "techni-baryon", a technicolor singlet state of 3 technicolor quarks, is expected to have a lifetime of $\approx 10^{16}$ years. Supersymmetric theories[4] predict that all fermions have bosonic partners and vice versa; the lightest of these is expected to be stable. Particles of these types, if charged, should be observable in matter. Positively charged particles would have similar chemical properties to hydrogen and would appear in nature as an isotope of hydrogen with an anomalous mass. Negatively charged particles would bind to ordinary nuclei, changing a nucleus of atomic number Z into one with atomic number Z-1 and anomalous mass.

We have constructed an all-electrostatic beam line for the University of Rochester Nuclear Structure Research Laboratory (NSRL) MP tandem electrostatic accelerator to search systematically for such components of matter. An electrostatic beam line transports ions independently of their masses, an essential feature since the masses of the ions for which we are searching are not known. We can enhance the relative selection of these ions by tuning the beam and appropriately configuring the detector.

Searches for massive isotopes of hydrogen have been reported[5,6,7], the most sensitive being that by Smith and co-workers[8]. Using electrolysis followed by analysis in a time-of-flight mass spectrometer, they were able to establish concentration limits of $<10^{-28}$ per nucleon in ordinary water for isotopes in the mass range 8–1200 amu. Searches for anomalous mass isotopes of heavier nuclei have had less sensitivity. One of the most sensitive to date is that of Turkevich et al.[9], which places limits on the natural abundance of >100 amu carbon-like nuclei at less than 1 per 10^{15} nucleons. Other less sensitive searches covering more limited mass regions have been reported for helium[10], lithium[10], beryllium[10], oxygen[11], and sodium[12].

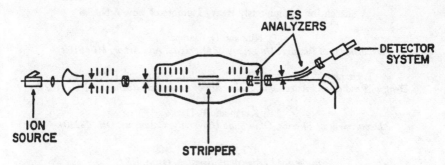

Fig. 1. A plan view of the apparatus including the injection system into the accelerator.

THE ELECTROSTATIC SPECTROMETER

A diagram of the accelerator, spectrometer, and detection system is shown in Fig. 1. It has been described in detail in ref. 13. A cesium ion source, which can be used to sample virtually any type of solid material and many gasses, feeds into the accelerator directly through electrostatic lenses. The sample beam is accelerated to 5 MeV in charge state 1^- to the terminal where it is stripped to charge state 1^+ in a 5 μg/cm² carbon foil. At the high energy end of the accelerator, the 10 MeV beam is focused with an electrostatic quadrupole doublet and deflected 1.3° off axis by a pair of electrostatic plates within the accelerator pressure vessel. Outside the machine, the beam passes through a removable 5 μg/cm² foil stripper and a set of adjustable slits at the focus of the quadrupole. This is followed by a high resolution 20° electrostatic analyzer, an electrostatic quadrupole doublet, and a small magnet that can be energized to sweep out low mass (<100 amu) ions. A slit located at the focus of the second quadrupole system is followed by a microchannel plate time-of-flight transmission detector, a gas pressure cell to range out unwanted ions, and a gas ionization counter to measure the $\delta E/\delta x$, range, and total energy of the ions.

We periodically verified the mass independence of the beam transport by measuring the yields from a specially prepared BeCuAu alloy. Small variations in the optimum beam tunes for the three elements were identified as being due to small differences in the energy losses in the two stripper foils and to small residual magnetic fields at the low energy end of the accelerator. These effects, while significant for masses of a few amu, have little effect for the high mass, low-Z ions that are the subject of the search reported here.

For the hydrogen search we introduced a small magnetic field just downstream of the ion source. This eliminated light particles from the beam but had negligible effects on particles with masses ≥100 amu.

For the other searches, we exploit the fact that heavy ions, in passing through a thin foil, don't have as many orbital electrons stripped as do light ions of the same energy. As a general rule, only those electrons that have orbital speeds less than that of the ion are stripped. This means, for example, that a 10 MeV 1000 amu "isotope" of beryllium will strip to charge state 1^+ ≈½ of the time, while normal Be^9 will usually strip completely to charge state 4^+, emerging in charge state 1^+ with a probability of 10^{-4}. By tuning for charge state 1^+ after both the terminal and high energy stripper foils, we achieve a rejection factor of ≈10^7 for normal ions. Normal ions that do pass through the system are rejected in the detector, since a heavy version of any particular element will have a distinctly different $\delta E/\delta x$ and range.

RESULTS

We measured a variety of samples, including normal beryllium, lithium, boron, and fluorine, and specially enhanced samples of oxygen, carbon, and hydrogen. We used a commercially available sample of O^{18}, the preparation of which effectively enhances the concentration of heavy isotopes relative to O^{16} by a factor of 416, and a carbon sample prepared in the C^{13} separator at Los Alamos in which the concentration of possible heavier isotopes was enhanced by $\approx 10^5$.

In our searches for heavy hydrogen, we used several samples: sea water from a depth of 3 km (no enhancement), a commercially available deuterium sample (enhancement factor = 6600), and samples of lake water enriched at the Rutherford Laboratory (enhancement factor 10^6–10^9).

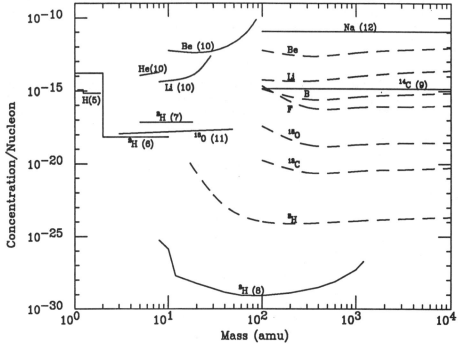

Fig. 2. Concentration limits (90% confidence level) for the existence of heavy isotopes in matter. Our preliminary results are shown as dashed lines, and previously published results as solid lines.

The sensitivity for detection of heavy isotopes is given by the expression:

$$S = \frac{n_0 \times \gamma_m}{Y_1 \times Y_2 \times \varepsilon_t \times \varepsilon_m \times I \times f \times \Delta t \times A \times \gamma_s} \tag{1}$$

Here S is the measured sensitivity, Y_1 and Y_2 are the stripping yields for the heavy isotope into charge state 1^+ at the first and second stripping foils, ε_t is the transmission efficiency of the heavy isotope, ε_m is the mass dependent correction for the magnetic sweeping, I is the average current from the ion source, f is the ratio of specific ion current to the total current, Δt is the total live time accumulated, A is the mass number of sample nuclei, γ_s is the enrichment of the sample, γ_m is a mass dependent enhancement factor due to molecules from the source, and n_0 is the minimum number of events required to define a signal. In the present preliminary analysis we have assumed $\varepsilon_m = 1$, we have taken the heavy isotope transmission efficiency to be 10%, consistent with our measurements of the beam transmission for normal ions, and have considered only the masses ≥ 100 amu where the effect of the magnet sweeping is insignificant. We require ≥ 1 event ($n_0 = 2.3$ for 90% CL upper limit) to define a positive signal in the heavy hydrogen search. For the other samples, we require ≥ 3 events ($n_0 = 5.3$ for 90% CL upper limit) within a region taken to be ≥ 3 times wider than our FWHM resolution in the total ionization energy measurement and 4 $\delta E/\delta x$

measurements. In Fig. 2, we show our preliminary results as dashed lines, and previously published results as solid lines. We expect that further refinement of our analysis of these same data will enhance the concentration limits and extend our mass range.

CONCLUSIONS

It is not expected that geological fractionation processes will dramatically change the concentration of heavy isotopes in normal matter[14]. In light of the large discrepancy between these measurements and the predictions of refs. 1 and 2, it would appear safe to rule out the existence of stable charged particles in the mass range 100–10,000 amu.

ACKNOWLEDGEMENTS

We gratefully acknowledge Dr. K. Nishiizuni for providing the deep sea water sample, Prof. J. Bigeleisen for providing the enriched carbon samples and for helpful discussions regarding the chemical history of heavy isotopes, and Dr. J.D. Bowman for the loan of the time-of-flight detectors. We thank Mr. T. Haelen of the University of Rochester's 130" cyclotron laboratory for his help in designing and installing the apparatus. We also wish to express our appreciation to the staff of the NSRL for their aid in the operation of the experiment. Mr. N. Conard helped with chemical preparation of the samples and Mr. R. Teng helped maintain the ion source and detector. This work has been supported by the United States Department of Energy and National Science Foundation.

REFERENCES

1. S. Wolfram, Phys. Lett. **82B** (1979) 65; N. Isgur and S. Wolfram, Phys. Rev. **D19**, (1979) 234.
2. C.B. Dover et al., Phys. Rev. Lett. **42**, (1979) 1117.
3. R. Cahn and S. Glashow, Science **213**, (1981) 607.
4. J. Wess and B. Zumino, Nucl. Phys. **B70**, (1974) 39.
5. R.N. Boyd et al., Phys. Lett. **72B**, (1978) 484.
6. R. Muller et al., Science **196**, (1977) 521.
7. T. Alvager and R.A. Naumann, Phys. Lett. **24B**, (1967) 647.
8. P.F. Smith et al., Nucl. Phys. **B206**, (1982) 333.
9. A. Turkevich et al., Phys. Rev. **D30**, (1984) 1876.
10. J. Klein et al., Symp. on Accelerator Mass Spectrometry, Argonne National Lab. (1981) 136.
11. R. Middelton et al., Phys. Rev. Lett. **43**, (1979) 429.
12. W.J. Dick et al., Phys. Rev. Lett. **53**, (1984) 431.
13. D. Elmore et al., Nucl. Instr. and Meth., **B10**, (1985) 738.
14. J. Bigeleisen, private communication.

SPIN PHYSICS

Coordinators: J. B. Roberts
W. T. H. Van Oers

POLARIZED PROTONS AT THE AGS AND HIGH $P_\perp^2$ SPIN EFFECTS

Thomas Roser
Randall Laboratory of Physics, The University of Michigan
Ann Arbor, MI 48109

ABSTRACT

At the AGS an intense polarized proton beam was for the first time successfully accelerated to 22 GeV/c, preserving a polarization of $\sim 45\%$. Using this polarized beam a surprisingly strong energy dependence of the spin-spin correlation parameter A_{nn} in high $p_\perp^2$ elastic proton-proton scattering was discovered.

A study of interactions based on measurements of cross sections and decay rates always involves averaging over the possible spin configurations of the system. To avoid that, and thus explore the interactions between pure quantum states, we have to study spin effects. In fact the correct description of spin effects is one of the most critical tests of any dynamical theory of hadronic interactions. Since the experimental hardware requirements are extraordinary, however, spin experiments mostly are second generation experiments and, more often than not, reveal unexpected properties of the interaction which are difficult to explain with the established theories.

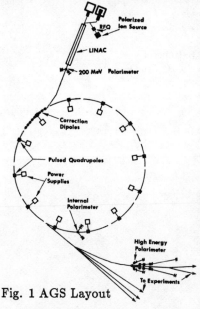

Fig. 1 AGS Layout

I will report here the investigations of double spin effects in elastic proton-proton scattering conducted from December 1985 to February 1986 at the AGS, Brookhaven Natl. Lab., as well as the results of commissioning of polarized proton beam at the AGS.

Fig. 1 shows the major hardware items installed for the polarized proton beam project in a collaboration of BNL, Michigan, Rice, Argonne and Yale. The pulsed polarized H^- ion source now operates at $25\mu A$. After acceleration through a RFQ and the Linac the polarization is for the first time measured at 200 MeV using elastic $p-^{12}C$ scattering. At the following injection into the AGS main ring the 2 electrons are stripped off the H^- ions, leaving the bare vertically polarized protons to be accelerated.

As the energy $E = \gamma m_p$ of the protons in the main ring increases to the maximum energy of about 22 GeV achieved during this run the protons encounter about 40 depolarizing resonances. The protons are depolar-

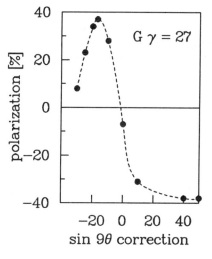
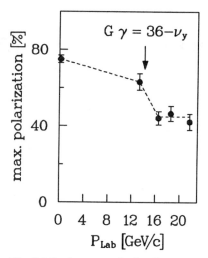

Fig.2 Polarization vs. amplitude of $sin9\theta$ correction (in arbitrary units)

Fig.3 Maximum polarization vs. beam momentum

ized whenever they see horizontal magnetic field components with a frequency equal to their precession frequency. The horizontal fields of the focussing elements in the ring cause 'intrinsic resonances' with the resonance condition $G\gamma = nP \pm \nu_y$ where G is the anomalous magnetic moment of the proton, ν_y the vertical betatron tune ($\sim$ 8.75 at the AGS), P the periodicity of the machine (12 at the AGS) and n an integer. However, since the strength of the resonance depends on how fast the protons cross the resonance, depolarization can be prevented by rapidly shifting the tune as the protons approach the resonance condition. In this way we successfully jumped 5 intrinsic resonances ($G\gamma = 0 + \nu_y, 12 + \nu_y, 36 - \nu_y, 24 + \nu_y$ and $48 - \nu_y$) with 10 fast pulsed quadrupole magnets which are capable of shifting the tune within $1.6\mu sec$.

Misalignements of the main bending magnets also lead to horizontal field components. The resonance condition for these 'imperfection resonances' is $G\gamma = n$. We passed about 35 imperfection resonances using 96 small pulsed dipole magnets to generate the appropriate harmonic corrections at the energies $E = nm_p/G$. It turns out that imperfection resonances are considerably stronger in the neighbourhood of the strongest intrinsic resonances ($G\gamma = 0 + \nu_y, 36 - \nu_y$) and that these resonances are more easily corrected using a harmonic which beats against an integer multiple of the periodicity of the AGS (e.g. the 9th harmonic for $G\gamma = 27$ since $9 = 3 \times 12 - 27$) than with their primary harmonic. Fig. 2 shows the dependence of the polarization to the amplitude of the sine component of the 9th harmonic correction at $E = 27m_p/G$. It also appeared that a 20% polarization loss near 14 GeV/c was caused by interference between the $G\gamma = 36 - \nu_y$ intrinsic resonance and the

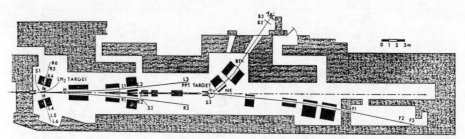

Fig.4 Layout of external polarimeter and spectrometer

$G\gamma = 27$ imperfection resonance. Decreasing the tune using slow quadrupole magnets increased the separation of these two resonances and indeed reduced the loss of polarization.

During these commissioning studies the polarization was monitored with two polarimeters. The internal polarimeter uses as a target a 0.1 mm nylon string inside the AGS ring and measures the left-right asymmetry in the p-nucleon scattering at $p_\perp^2 \sim .15 \ (GeV/c)^2$. The external polarimeter is located in one of the extracted proton beam lines and measures the left-right asymmetry in proton-proton elastic scattering at $p_\perp^2 \sim .3 \ (GeV/c)^2$. Fig. 3 shows the maximum polarization reached during the last run.

I will now turn to our experiment on spin effects on elastic proton-proton scattering, a collaboration of Michigan, BNL, Maryland, MIT, Notre Dame, Texas A&M and ETH (Zurich). The target and the spectrometer are located immediately down stream of the external polarimeter as shown in Fig. 4 The polarized proton beam with an intensity of up to 8×10^9 protons per 2.2 sec pulse is scattered from the University of Michigan polarized proton target. This target contains NH_3 beads which were irradiated at the MIT Bates Linac with a total dose of $5 \times 10^{16} \ e^-/cm^2$ to produce radicals with spin-unpaired electrons. The beads are cooled to $0.5°K$ by a $^3He - \ ^4He$ evaporation refrigerator in a 2.5 T magnetic field. A 70 GHz microwave system drives the polarizing transitions and a 107 MHz NMR system continuously monitors the polarization of the hydrogen protons. The maximum target polarization was about 70% and the average polarization for our data run was $51 \pm 3\%$.

Elastic scattering events are detected in a double arm forward-backward spectrometer using 8 channel scintillation counter hodoscopes. Measuring the number of elastic events for the 4 pure initial spin states $N(beam, target)$ allows us to calculate the spin-spin correlation parameter A_{nn}:

$$A_{nn} = \frac{1}{P_B P_T} \frac{N(\uparrow\uparrow) - N(\uparrow\downarrow) - N(\downarrow\uparrow) + N(\downarrow\downarrow)}{N(\uparrow\uparrow) + N(\uparrow\downarrow) + N(\downarrow\uparrow) + N(\downarrow\downarrow)}$$

where P_B and P_T are the beam and target polarizations respectively.

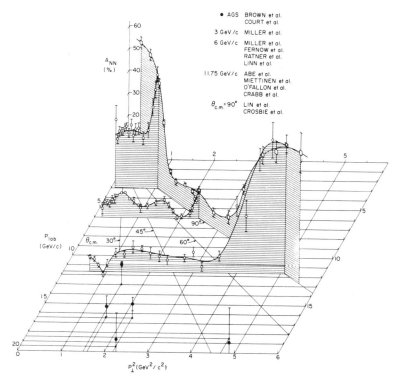

Fig. 5 A_{nn} measurements at high beam momentum

Measurements of A_{nn} at high momentum transfer were first performed with the polarized beam of the ZGS. Fig. 5 shows a compilation of A_{nn} measurements with a beam momentum greater than 3 GeV/c[1]. A striking result of the ZGS data is that at $p_\perp^2 \approx 3.5\ (GeV/c)^2$ A_{nn} starts to increase to a value of about 55% at $p_\perp^2 = 5.1\ (GeV/c)^2$ [at $p_{lab} = 11.75\ GeV/c$ this corresponds to $\theta_{cm} = 90°$]. In fact this value of $p_\perp^2$ is believed to be the onset of the region where the scattering is dominated by direct interaction of the constituents of the protons. As can be seen in fig. 5 the sharp increase of A_{nn} occurs at the same $p_\perp^2$ if, instead of the incident beam momentum, the center-of-mass scattering angle is kept constant at 90°. This confirms that the large values for A_{nn} are indeed related to hard scattering and not just a reflection of the high degree of symmetry of the 90° scattering of identical particles.

With the polarized beam of the AGS we are now in the position to study A_{nn} at high $p_\perp^2$ at center-of-mass angles considerably different from 90°. Our main data run was taken at an incident beam momentum of 18.5 GeV/c at $p_\perp^2 = 4.7\ (GeV/c)^2$ ($\theta_{cm} = 49°$). The result $A_{nn} = -2 \pm 16\%$ is shown in fig. 6 together with lower energy ZGS measurements at the same $p_\perp^2$.

The observed surprisingly sharp decline of A_{nn} as a function of the incident beam energy for fixed $p_\perp^2$ indicates that being in a hard scattering region is not sufficient for large values of A_{nn}. It clearly is of great importance to extend the A_{nn} measurements to higher values of $p_\perp^2$ (and thus closer to $\theta_{cm} = 90°$) at fixed beam momentum as well as to higher beam momentum at fixed $p_\perp^2$.

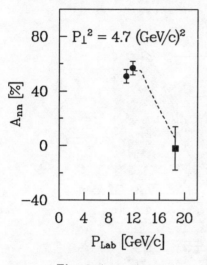

Fig. 6 A_{nn} vs. p_{Lab}

Today the most favoured candidate for a dynamic theory of hadronic phenomena is without doubt quantum chromodynamics (QCD). In this high $p_\perp^2$ region it is believed that a perturbative calculation should be applicable. This assumption is supported by the success of the prediction that e.g. the cross section of pp elastic scattering goes like s^{-10}. For $\theta_{cm} = 90°$ a value of 1/3 is predicted for A_{nn} with possible oscillations around this value as suggested by the observed oscillations of the cross section around the s^{-10} behaviour[2]. However it is not yet clear how this fixed-angle prediction is related to our observed fixed-$p_\perp^2$ behaviour.

It certainly became clear that with the availability of a high energy high intensity polarized proton beam at the AGS a new and unique region for precision tests of our understanding of the dynamics of hadrons has opened up.

This work was supported by a research contract from the U.S. Department of Energy.

REFERENCES

1. D.Miller et al., Phys. Rev. D16, 7 2016 (1977); S.L.Linn et al., Phys. Rev. D26, 3 550 (1982); K.A.Brown et al., Phys. Rev. D31, 11 3017 (1985); R.C.Fernow et al., Phys. Lett. 52B, 2 243 (1974); L.G.Ratner et al., Phys. Rev. D15, 3 604 (1977); A.Lin et al., Phys. Lett. 74B, 3 273 (1978); E.A.Crosbie et al., Phys. Rev. D23, 3 600 (1981); K.Abe et al., Phys. Lett. 63B, 2 239 (1976); H.E.Miettinen, Phys. Rev. D16, 3 549 (1977); J.R.O'Fallon et al., Phys. Rev. Lett. 39, 12 733 (1977); D.G.Crabb et al., Phys. Rev. Lett. 41, 19 1257 (1978); G.R.Court et al., to be published
2. S.J.Brodsky, C.E.Carlson and H.J.Lipkin, Phys. Rev. D20, 2278 (1979)
G.F.Wolters, Phys. Rev. Lett. 45, 776 (1980)
G.R.Farrar, Phys. Rev. Lett. 56, 1643 (1986)

EFFECTS ON SPIN ASYMMETRIES OF SPECIAL EFFECTS AT 90°*

Harry J. Lipkin
High Energy Physics Division, Argonne National Laboratory
Argonne, Illinois 60439

Hadron and quark exchange contributions introduce additional spin asymmetries in hadron elastic scattering which are a maximum at 90° but go away at small angles, where such exchanges always involve one backward scattering with a much higher momentum transfer than the direct forward scattering with only gluon exchanges. Double-flip and nonflip contributions to $\sigma_{\uparrow\uparrow}$ are incoherent, while direct and exchange contributions to each are coherent. For $\sigma_{\uparrow\downarrow}$ the direct nonflip and exchange double-flip amplitudes are coherent and vice versa. At 90° even if the direct amplitude is spin independent and dominantly nonflip, the nonflip coherence in $\sigma_{\uparrow\uparrow}$ can introduce a factor of 2 in $\sigma_{\uparrow\uparrow}/\sigma_{\uparrow\downarrow}$, while small double-flip amplitudes which are coherent with nonflip in $\sigma_{\uparrow\downarrow}$ but incoherent in $\sigma_{\uparrow\uparrow}$ can introduce further large factors.

Recent measurements of spin effects in elastic scattering of polarized protons[1] show a very small value for the asymmetry parameter $A_{nn} = -2 \pm 16\%$ at $P_t^2 = 4.7$ (GeV/c)^2. This is in sharp contrast with the previous value near 60% obtained at the same value of P_t but at the lower energy of 12 GeV/c. However, the 12 GeV/c measurement was taken at a scattering angle of 90°, while the 18.5 GeV/c measurement was taken at a scattering angle of 49° in the center of mass system.

The difference between 90° and 49° is crucial because of both symmetry and dynamical effects. For the scattering of identical particles, any measurement sums the scattering at an angle θ and an angle π-θ, and the sum may be coherent or incoherent, depending upon the particular spin states involved. At high momentum transfer, where the scattering amplitude is expected to fall off rapidly with increasing scattering angle, this effect should be negligible at 49°, since the scattering amplitude at 131° is expected to be very much smaller. At 90°, however, the effects of particle interchange are a maximum and can easily produce large spin asymmetries even if the

---

*Work supported by the U.S. Department of Energy, Division of High Energy Physics, Contract W-31-109-ENG-38.

**1985-86 Argonne Fellow on Leave from Weizmann Institute of Science, Rehovot, Israel.

asymmetries at angles like 49° are very small. Such effects have been shown to produce a value of $A_{nn} = 1/3$ at 90° for completely spin-independent scattering which gives $A_{nn}=0$ at small angles.[2] We shall see here that small additional spin-dependent amplitudes that have a negligible effect at small angles can give $A_{nn} = 0.6$ at 90°.

There are also important differences in QCD dynamics between 90° and small angle scattering. The dominant mechanisms for hadron scattering in QCD are gluon exchange and quark exchange. The question of which is more important in different kinematic regions remains open. Single-gluon exchange between color singlet hadrons is forbidden in QCD; any scattering without quark exchange must involve at least two exchanged gluons. The question at high momentum transfers where all interactions are suppressed is whether the price paid for an additional gluon is greater than the price paid for a quark exchange.

At small angles, one might expect gluon exchange without quark exchange to dominate, because any quark exchange involves a "backward scattering" at the quark level with a much larger momentum transfer than forward scattering. At 90°, however, there is no difference between forward and backward; quark exchange involves the same momentum transfer at the quark level as pure gluon exchange without quark exchange. Thus there may be a transition between quark-exchange dominance and gluon-exchange dominance at some intermediate scattering angle.

In order to unscramble these effects and obtain a true picture of the dynamics of spin dependence, it is necessary to separate out the 90° coherence effects from energy dependence, and look for possible transitions between quark exchange dominance and gluon exchange dominance. This can be done by measurements taken at different energies either all at 90° or all at small angles where coherence effects are negligible. Measurements of angular distributions at a given energy would also reveal the existence of peculiar behavior at 90° and possible transitions. Measurements of final state polarization may also be relevant as discussed below.

The peculiar conditions at 90° can be seen explicitly as follows. Consider the scattering of protons transversely polarized normal to the scattering plane; i.e. transversity eigenstates. The physics is particularly simple in the transversity basis, where the number of contributing amplitudes is small. Since angular momentum and parity conservation require transversity conservation modulo 2, there are only nonflip and double-flip transitions; single transversity flip is forbidden.

If the incident protons have parallel transverse spins, the scattered protons also have parallel transverse spins, either nonflipped or double-flipped. In either case interchanging the two protons gives the same final state and it is impossible to distinguish between scattering by θ and by θ-π. The two amplitudes are coherent and must be added before they are squared. If the incident protons have antiparallel transverse spins, the scattered protons also have antiparallel transverse spins, but interchanging the two scattered protons gives a different final state with both transversities flipped. Thus for antiparallel transversities, the nonflip amplitude at angle θ leads to the same final state and is coherent with the double-flip amplitude at angle θ-π and vice versa. Thus:

$$\sigma_{\uparrow\uparrow}(\theta) = |N_{\uparrow\uparrow}(\theta) + N_{\uparrow\uparrow}(\theta-\pi)|^2 + |D_{\uparrow\uparrow}(\theta) + D_{\uparrow\uparrow}(\theta-\pi)|^2 \quad (1a)$$

$$\sigma_{\uparrow\downarrow}(\theta) = |N_{\uparrow\downarrow}(\theta) + D_{\uparrow\downarrow}(\theta-\pi)|^2 + |D_{\uparrow\downarrow}(\theta) + N_{\uparrow\downarrow}(\theta-\pi)|^2 , \quad (1b)$$

where $\sigma_{\uparrow\uparrow}$ and $\sigma_{\uparrow\downarrow}$ denote the cross sections and $N_{\uparrow\uparrow}$, $D_{\uparrow\uparrow}$, $N_{\uparrow\downarrow}$ and $D_{\uparrow\downarrow}$ denote the nonflip and double-flip amplitudes respectively for the states of parallel and antiparallel transverse spins.

The angular distribution of the scattering is expected to be forward peaked. Thus at small angles, where all backward scattering by an angle θ-π can be neglected, the relations (1) simplify to give

$$\sigma_{\uparrow\uparrow}(\theta) = |N_{\uparrow\uparrow}(\theta)|^2 + |D_{\uparrow\uparrow}(\theta)|^2 \quad (2a)$$

$$\sigma_{\uparrow\downarrow}(\theta) = |N_{\uparrow\downarrow}(\theta)|^2 + |D_{\uparrow\downarrow}(\theta)|^2 \quad (2b)$$

$$\frac{\sigma_{\uparrow\uparrow}(\theta)}{\sigma_{\uparrow\downarrow}(\theta)} = \frac{|N_{\uparrow\uparrow}(\theta)|^2 + |D_{\uparrow\uparrow}(\theta)|^2}{|N_{\uparrow\downarrow}(\theta)|^2 + |D_{\uparrow\downarrow}(\theta)|^2} \equiv S(\theta) . \quad (2c)$$

The quantity $S(\theta)$ defined by Eq. (2c) gives a direct measure of the spin dependence of the average of the nonflip and double-flip cross sections.

At 90 degrees, where $\theta = \pi-\theta$, and we assume equal amplitudes with a positive relative phase at θ and -θ, the relations (1) simplify to give

$$\sigma_{\uparrow\uparrow}(\theta) = 4|N_{\uparrow\uparrow}(\theta)|^2 + 4|D_{\uparrow\uparrow}(\theta)|^2 \quad (3a)$$

$$\sigma_{\uparrow\downarrow}(\theta) = 2|N_{\uparrow\downarrow}(\theta) + D_{\uparrow\downarrow}(\theta)|^2 \quad (3b)$$

$$\frac{\sigma_{\uparrow\uparrow}(90^0)}{\sigma_{\uparrow\downarrow}(90^0)} = 2 \frac{|N_{\uparrow\uparrow}(90^0)|^2 + |D_{\uparrow\uparrow}(90^0)|^2}{|N_{\uparrow\downarrow}(90^0) + D_{\uparrow\downarrow}|^2} = 2 \frac{(1+\epsilon^2)}{(1+\epsilon)^2} S(90^0) \quad (4a)$$

where

$$\varepsilon = \frac{D_{\uparrow\downarrow}(90^0)}{N_{\uparrow\downarrow}(90^0)} \; . \tag{4b}$$

This result (4) shows that the ratio $\sigma_{\uparrow\uparrow}/\sigma_{\uparrow\downarrow}$ at 90° contains an additional factor $2(1 + \varepsilon^2)/(1 + \varepsilon)^2$ which is completely unrelated to the spin dependence of the scattering amplitude and depends only on the coherence effects between the nonflip and double-flip antiparallel amplitudes.

The fact that effects completely unrelated to spin dependence arise at 90° is clearly seen in the simplest case of spin independent scattering, where

$$N_{\uparrow\uparrow}(\theta) = N_{\uparrow\downarrow}(\theta) \tag{5a}$$

$$D_{\uparrow\uparrow}(\theta) = D_{\uparrow\downarrow}(\theta) \; . \tag{5b}$$

In this case, for small angles

$$\frac{\sigma_{\uparrow\uparrow}(\theta)}{\sigma_{\uparrow\downarrow}(\theta)} = S(\theta) = 1 \tag{6a}$$

and the asymmetry parameter A_{nn} is then

$$A_{nn} = 0 \tag{6b}$$

as expected for spin-independent scattering.

However, for angles near 90°,

$$\frac{\sigma_{\uparrow\uparrow}(90^0)}{\sigma_{\uparrow\downarrow}(90^0)} = \frac{2(1 + \varepsilon^2)}{|1 + \varepsilon|^2} \; . \tag{7a}$$

This can vary rather wildly, depending upon ε, even though $S(\theta) = 1$. For $\varepsilon \sim 0$, the asymmetry parameter A_{nn} is

$$A_{nn} = +1/3 \; . \tag{7b}$$

From these specific cases, we see that it is risky to compare the values of A_{nn} at small angles with those at 90°. A value near zero at small angles can increase to a value of 1/3 at 90° for reasons of permutation symmetry which have no implications for the dynamics.

Note that the value of A_{nn} at 90° is very sensitive to small double-flip amplitudes which can interfere destructively in the antiparallel cross section (2b). Consider, for example, the case where $\varepsilon = -0.1$ so that the double-flip cross section is only 1% of the nonflip cross section. Then

$$\frac{\sigma_{\uparrow\uparrow}(90^U)}{\sigma_{\uparrow\downarrow}(90^U)} = 2.5 \; S(90^U) \; , \tag{8a}$$

If $S(\theta) = 1$, as would be the case if the nonflip and double-flip amplitudes are both spin-independent, i.e. they satisfy Eqs. (5), the asymmetry parameter A_{nn} is

$$A_{nn} = +0.43 \; . \tag{8b}$$

This is very different from the value (7b) obtained without the small double-flip amplitude.

The observed value at 90° of $A_{nn} = 0.6$ is obtained, with $A_{nn} = 0$ at small angles, if $S(\theta) = 1$ and $\varepsilon = -0.27$; i.e. the double-flip cross section is 7% of the nonflip cross section for antiparallel spins and the amplitudes have a negative relative phase. Thus it is possible to have $A_{nn}=0$ at small angles and 60% at 90° with only a small double-transversity-flip amplitude in addition to an amplitude which gives $A_{nn} = +1/3$ at 90°.

Since most models based upon QCD use helicity amplitudes, and helicity is conserved in high-momentum scattering processes with emission and absorption of vector gluons, it is interesting to see the role of helicity conservation in these considerations. We first note that helicity conservation is confused by exchange effects of two kinds, hadron exchange and constituent exchange. In proton-proton scattering where the initial protons have opposite helicities, a given detector will observe protons of both helicities, even if helicity is conserved, because of the exchange of the two protons. This effect can be very large at 90°, but is expected to be small at small angles like 49°.

If the dominant mechanism for the scattering is constituent interchange, then even if helicity is conserved at the quark level, the exchange of quarks with opposite helicity between the two protons can give an effective helicity flip at the proton level. This effect is not expected to be strongly dependent on angle or particularly strong at 90°, since it does not depend upon the particles being identical. Quark exchange can give an effective helicity flip even in meson-baryon scattering, where there is no question of identical particles.[3] Note, however that all these exchange effects can produce only a double helicity flip and never a single helicity flip, as the overall helicity of the two-particle system must be conserved. Note also that the effective double-helicity-flip due to exchange can only occur for initial states of opposite helicity. If both particles have the same helicity, there is no way to obtain a double flip by any exchange.

It is convenient to use the amplitudes which have simple symmetry properties for the scattering of identical particles developed using the H-spin formalism[2,4] because the scattering matrix is nearly diagonal in this basis, and independently discovered in another context[5] because they behave simply under reflection about 90°.

$$N = \phi_s = (\phi_1 - \phi_2)/2 , \qquad (9a)$$

$$S = \phi_t = (\phi_1 + \phi_2)/2 , \qquad (9b)$$

$$L = \phi_T = (\phi_3 - \phi_4)/2 , \qquad (9c)$$

$$V = \phi_\tau = (\phi_3 + \phi_4)/2 , \qquad (9d)$$

$$F = \phi_5 , \qquad (9e)$$

where N,S,L,V and F are the notation of ref. 2, ϕ_s, ϕ_t, ϕ_T and ϕ_τ are the notation of ref. 5, and

$$\phi_1 = \langle ++|++\rangle \qquad (10a)$$

$$\phi_2 = \langle ++|--\rangle \qquad (10b)$$

$$\phi_3 = \langle +-|+-\rangle \qquad (10c)$$

$$\phi_4 = \langle +-|-+\rangle \qquad (10d)$$

$$\phi_5 = \langle ++|+-\rangle \qquad (10e)$$

are conventional helicity amplitudes[5].

In terms of these amplitudes, the asymmetry parameter A_{nn} is given by

$$A_{nn} = (|S|^2 + |L|^2 + 2|F|^2 - |N|^2 - |V|^2)/ \\ (|S|^2 + |L|^2 + 2|F|^2 + |N|^2 + |V|^2). \qquad (11)$$

For the case where helicity is conserved except for exchange effects,

$$\phi_2 = 0 \qquad (12a)$$

$$\phi_5 = 0 , \qquad (12b)$$

but ϕ_4, which is an apparent double-flip amplitude, can occur as a result of exchange effects. In the notation (9) these conditions become

$$N = S \tag{13a}$$

$$F = 0 \tag{13b}$$

and

$$A_{nn} = (|L|^2 - |V|^2)/(2|N|^2 + |L|^2 + |V|^2) \ . \tag{14}$$

The five amplitudes (9) were chosen to have simple properties under reflection about 90°. The amplitudes N, S, and L are even and need not vanish at 90°; the amplitudes V and F are odd and vanish at 90°. Thus

$$A_{nn}(90°) = (|L|^2)/(2|N|^2 + |L|^2) \ . \tag{15}$$

When H-spin[2,4] is conserved, as in the constituent interchange model,

$$N = S = L \ , \tag{16a}$$

and

$$A_{nn} = (1/3) - (4/3)(|V|^2)/(3|L|^2 + |V|^2) \leqslant 1/3 \tag{16b}$$

$$A_{nn}(90°) = 1/3 \ . \tag{16c}$$

The value (16c) is in disagreement with the large value of A_{nn} observed at 90°. indicating that H-spin is not conserved and that the constituent interchange model does not give the complete amplitude for the process. If we still assume that helicity is conserved in the process, we can obtain the value $A_{nn}(90°) = 0.6$ by setting

$$|L|^2 = 3|N|^2 \ . \tag{17a}$$

At angles far from 90°, we can obtain $A_{nn} = 0$ by setting

$$V = L \ . \tag{17b}$$

Note that the condition (17b) implies the vanishing of the double-helicity-flip amplitude ϕ_4, which is nonvanishing in the constituent interchange model, where it results from interchange of quarks with opposite helicity between the two protons. If the standard SU(6) wave function is used for the proton, the constituent interchange model gives for small angles

$$V = (3/31)L \ . \tag{17c}$$

This is very different from the value (17b) and predicts that A_{nn} is within 2% of 1/3.

We thus see that there are two different ways to get agreement with the present experimental results. A small transversity-flip amplitude added to a spin-independent transversity conserving amplitude can give agreement with experiment, as shown by eqs. (8). However these amplitudes violate helicity conservation even at the constituent level. Alternatively, helicity conservation at the hadron level (which means no interchange of quarks with opposite helicity between hadrons) can give agreement with the recent low asymmetry at 49° while the coherent effect at 90° can give the high value of A_{nn} observed if the helicity amplitude for oppsite opposite helicity scattering at 90° is greater than the amplitude for the same helicity scattering by a factor of $\sqrt{3}$.

In order to determine what is really happening, experimental data are needed both at 90° and at smaller angles for the same values of momentum transfer, or there should be measurements of the transversity flip. Only in this way will it be possible to disentangle large effects depending upon transverse momentum from comparatively small effects enhanced at 90° by these peculiar coherence effects.

REFERENCES

1. G. R. Court et al., Michigan preprint UM HE 86-03, submitted to Phys. Rev. Letters
2. Harry J. Lipkin, in School on Intermediate and High Energy Physics, Edited by W. R. Falk, TRIUMF report, p. 1
3. H.J. Lipkin, Phys. Rev. Lett. 53, (1984) 2075.
4. S. Brodsky, C. Carlson and H. J. Lipkin, Phys. Rev. D20, (1979) 2278.
5. H. Spinka, Phys. Rev. D30, (1984) 1461.

HIGH ENERGY POLARIZED BEAMS FROM HYPERON DECAYS§

David G. Underwood
Argonne National Laboratory, Argonne, Illinois 60439

ABSTRACT

The use of various ways to utilize lambda decays to obtain polarized beams of protons and antiprotons is emphasized. Examples described are the Fermilab polarized beam, now under construction, and the use of similar techniques at other energies. Beam transport, spin precession and reversal systems, and polarimeters are also discussed.

INTRODUCTION

The parity violating decays of hyperon and their anti-particles are a potentially quite useful source of polarized nucleons, other hyperons, and their anti-particles. The decay length and c.m. momentum from lambda decay allow two simple and clean ways of selecting polarized protons (or antiprotons) in a beamline. Exploitation of these possibilities in the polarized beam at Fermilab will allow two-spin experiments in a new energy regime as well as untried or improved single spin experiments. There is a plan for a similar beam to be built at the future 3 TeV accelerator, UNK. Specific techniques developed for the 200-400 GeV region can be approximately scaled to other energies such as 8.9 GeV or 16 TeV with simple formulas to determine their appropriateness at these energies.

BASICS

There are two simple ways to obtain polarized proton beams from lambda decays by using properties of beam transport systems. We call these the longitudinal and transverse modes. The polarization comes from the parity violating decays of lambdas. Spin direction is almost unchanged in transforming from the lambda center-of-mass frame to the proton frame.

The simplest way to get a tertiary beam of polarized protons is to use the high energy end point of the lambda production spectrum as proposed at CERN in 1977.[1-4] In this longitudinal mode, high momentum protons are chosen which necessarily come from forward decays of lambdas near the end point of the lambda production spectrum. This is similar to the way a polarized muon beam is produced by matching beam transmittance to accept only decays which are forward in the laboratory. This method does not work for $\bar{p}$ from $\bar{\Lambda}^0$ because the spectrum does not fall off fast enough in X_F. The proton polarization direction can be reversed or transformed to transverse by systems of spin precessors (snakes),[5] but only one polarization direction at a time is available.

In the transverse mode,[2-4] we are interested in a Λ°'s ($\overline{\Lambda}^0$'s) which decay near 90° to their path in the laboratory. If we look at p's ($\overline{p}$'s) from decays some distance from the production target, there is a correlation between polarization and effective proton source position (see Fig. 1). There is a virtual source around the target with large radially directed polarization farthest from the center (approximately 1 cm) and no useable polarization in the center. We use magnetic sweeping to eliminate both direct particles from the target and particles from decays close to the target which have a virtual source which is too small to deal with.

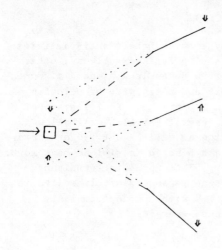

In a practical beam there is a balance among total flux, polarized flux, sweeping length, target size, and chromatic abberations. The size and polarization distribution within the spot are optimum and most clearly defined only for decays at a fixed distance from the target. Although the flux is highest from decays closest to the target, the image of the virtual spot will be the smallest from these, and furthermore will be smeared by the production target size and chromatic abberations in the beamline.

Fig. 1 Virtual source from lambda decay.

A beam transport system (Fig. 2) can make a real image of this virtual source at a focus. We can then use tagging or collimation to select a particular polarization. The tagging has to deal only with those products of Λ° ($\overline{\Lambda}^0$) or K° decay which will be transmitted to the experiment by a tuned beamline. In our case at Fermilab, tagging has the advantage over collimation in that the useful beam is doubled and opposite polarizations are available simultaneously which should reduce systematic errors. With a wide momentum band beam (we may use ± 7%), the focus of the virtual image will be at different z for different momenta. This can be handled in the tagging for every particle by using two tagging hodoscopes separated in z, using some momentum tagging and encoding the correlations in ECL memory look-up units.

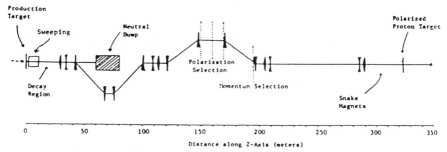

Fig. 2 Arrangement of beamline.

TRANSPORT

A high intensity proton beam from lambda decay will have large divergence, large effective source size, and large momentum spread. A beam designed for reasonable intensity will have spin rotations of up to 90° in the quadrupoles and up to 360° in the dipoles. It is possible to transport such a beam with very little depolarization.

The beam transport involves global cancellation of spin precession by quadrupoles (Fig. 3), and local cancellation of spin precession by dipoles (Fig. 2).

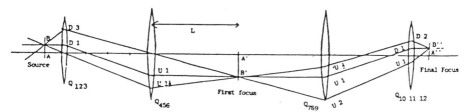

Fig. 3 Spin precession in quadrupoles. The field integral on any ray is zero.

The basic point in cancelling spin precession is that the field integral must be zero. This means that a ray at the final focus much have the same direction it had before entering the transport system. The overall transverse matrix is then $\begin{pmatrix} 1 & 0 \\ 0 & 1 \end{pmatrix}$ in each dimension. The beam we are building is mirror symmetric about an intermediate focus with a transfer matrix of $\begin{pmatrix} -A & 0 \\ 0 & -\frac{1}{A} \end{pmatrix}$ at that focus.

In order for the global cancellation of precession in the quadrupoles to occur, the spin directions entering one set of quadrupoles must be the same as the directions exiting the preceding set. Precessions by bends must cancel between quadrupole sets. Net bends between quadrupoles can lead to depolarization because rotations about orthogonal axes do not commute.

SCALING

If we are building a beamline to use transverse mode at a particular momentum, we can scale the beam length to match the lambda decay length. The spot size for tagging will be a constant, roughly 3. cm across, depending on the specifics. Once we have this scaling in length and approximately constant $p_\perp$ distribution for lambdas, we can use the same quadrupoles and transport dipoles at any momentum (e.g. 8.9 GeV and 16 TeV). This ignores the facts that magnet lengths are a fraction of the beam length at low momentum and that we might not want a single-momentum beam.

The sweeping magnets do not scale in the same way at all. If we assume the Fermilab scheme and say that the primary beam must be bent to an angle greater than lambda production angles in a distance less than a lambda decay length, the field required in the sweeper goes roughly as

$$B_{sweep} \text{ (kg)} = \frac{1000}{P_{pri} \text{ (GeV)} \cdot X^2_F \text{ sec}}$$

in this example we need 1 kg at the SSC and 1500 kg at 8.9 GeV. For a polarized $\bar{p}$ source for a storage ring, we might not sweep but use a careful acceptance at the image of the virtual spot to avoid the primary beam and limit the source emittance. (The source emittance is still too large for existing accumulators, approximately 100π MM-MR vs. 20π MM-MR.)

POLARIMETERS

We are developing two new kinds of polarimeters for the Fermilab beam. One based on the Primakoff effect[6-8] (Fig. 4) and one based on the Coulomb-nuclear interference.[8] The analyzing power for purely hadronic elastic scattering at small t is too small to be useful. Both of the new polarimeters use interactions at $t \approx 10^{-3}$, and require precise reconstruction of beam and outgoing tracks. The analyzing power for the Primakoff should be over 70% for a selected sample of events, and about 5% for the Coulomb nuclear. Further details are given in the References.

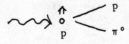

Data exists in the s channel for 620 MeV unpolarized γ on polarized proton ($M_{p\pi} \approx 1.4$ GeV).

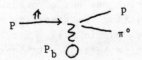

We will use the t channel with the photon from the Coulomb field of lead.

Fig. 4 Measuring polarization using the Primakoff effect.

REFERENCES

1. P. Dalpiaz et al., CERN/ECFA/72/4. Vol. I, p. 284; CERN proposal SPSC/p. 87, July 1977.
2. D. Underwood et al., A Polarized Beam for the M-3 Line (Fermilab), December 1977.
3. D. Underwood, Hyperon Beams as a Source of Polarized Protons, AIP Conf. Proc. 51, 318 (1978).
4. G. Shapiro and D. Underwood, Some Possibilities for Doing Polarized p,p̄ Experiments at the SSC, Proc. of the SSC Fixed Target Workshop, The Woodlands, Texas, Jan. 1984, APS-DPF and Houston Area Research Center.
5. D. Underwood, A Survey of Eight Magnet Spin Precession Snakes, Nucl. Instr. and Meth. 173, 351 (1980).
6. D. G. Underwood, A Method of Measuring the Polarization of High Momentum Proton Beams, ANL-HEP-PR-77-56 (1977).
7. B. Margolis and G. H. Thomas, One Photon Exchange Processes and the Calibration of Polarization of High Energy Protons, AIP Conf. Proc. 42, 173 (1978).
8. K. Kuroda, Proton Polarimeters at High Energy, LAPP (Annecy) Preprint and AIP Conf. Proc. 95, 618 (1982).

---

§ Work supported by the U.S. Department of Energy, Division of High Energy Physics, Contract W-31-109-ENG-38.

SPIN EFFECTS IN EXCLUSIVE REACTIONS AT HIGH $P_\perp$*

Yousef Makdisi †
AGS Department, Brookhaven National Laboratory
Upton, New York 11973

ABSTRACT

Production and decay angular distributions from the process $\pi^-p \to \rho^-p$ at 90° in the center-of-mass are presented. A large spin flip amplitude is observed, the ramifications of which are noted in the context of the known theories.

It is becoming quite clear that spin is an important facet of the High Energy scattering domain. The interest in spin effects is growing as experiments probe further into hadron interactions at shorter distances with the expectation that this will lead to a better understanding of quark interaction dynamics.

Large polarizations have been observed in inclusive and exclusive reactions and the effects persisted at relatively large momentum transfer.[1] It should be noted that common wisdom had predicted these effects to be small and to vanish as energies got higher and reactions more violent.

In departure from discussing experiments using polarized beams or targets, we describe the spin effects observed in an exclusive two body reaction $\pi^-p \to \rho^-p$ which was part of a general program to measure cross sections and decay angular distributions of a large class of exclusive reactions at the kinematic limit, namely 90° in the center-of-mass system.

These experiments were carried out at the Brookhaven AGS where the beam energy is a good match between reaching a reasonably high $P_\perp$ and attaining good statistical accuracy in a manageable running period since exclusive cross sections fall sharply with beam energy. This compromise sets the experiment in a $P_\perp$ domain ranging from 2-2.5 GeV/c or an equivalent t of 8-14 $(GeV/c)^2$ where arguments are aplenty on whether perturbative QCD is applicable.

The experiment has been described elsewhere.[2] Briefly, we utilize a single arm spectrometer to measure the scattered baryon and side chambers to track the outgoing meson and its decay into $\pi^-\pi^0$ of which only the π^- is observed. Two related reactions are discussed here: the elastic π^-p and the exclusive ρ meson production. While the elastics serve as a guide, the decay angular distributions of the ρ form the bulk of this paper.

Data were collected at two beam energies 10 and 13.4 GeV/c respectively. The missing mass spectra are shown in Fig. 1(a,b) where the elastic peak is shaded. The elastics sample contained 1150 events at 10 GeV/c and 230 events at 13.5 GeV/c resulting in average differential cross sections $d\sigma/dt$ of 1.69 ± 0.20 $nb(GeV/c)^2$

---

*Work performed under the auspices of the U.S. Dept. of Energy.
†Reporting on behalf of the "Exclusive Collaboration" of AGS Exp. 755 (BNL/Minn/SE Mass).

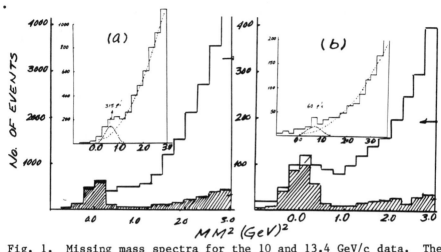

Fig. 1. Missing mass spectra for the 10 and 13.4 GeV/c data. The shaded regions represent the elastics cuts and the insets show the fit to the ρ mass and background.

and 0.23 ± 0.03 nb/(GeV/c)^2 respectively. These points fall in line with the power low dependence of S^{-8} in $d\sigma/dt$ at fixed angles. The exponent being the number of constituent valence quarks participating in the reaction reduced by 2 from dimensional counting arguments. This dependence was first predicted by Brodsky and Farrar[3] and has been borne out by πp, pp and γp experiments. If this is an indication of hard scattering, we note that this sets in for beam momenta as low as 5 GeV/c. The corresponding ρ cross sections are $1.18 \pm .27$ and $.15 \pm .05$ nb/(GeV/c)^2 respectively. These two points follow a scaling power of 7.1 ± 1.8.

In the language of valence quark diagrams, both elastic scattering and exclusive ρ production can proceed via one or all of the diagrams shown in Fig. 2. Other reactions can proceed via a subset of these. Thus, the importance of measuring exclusive reactions.

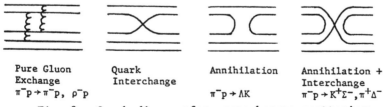

Pure Gluon Exchange
π⁻p → π⁻p, ρ⁻p

Quark Interchange

Annihilation
π⁻p → ΛK

Annihilation + Interchange
π⁻p → K⁺Σ⁻, π⁺Δ⁻

Fig. 2. Quark diagrams for meson-baryon scattering.

Spin dependence, under certain assumptions, could readily separate the first diagram from the rest. In the world of quarks, with nearly massless fermions, one can equate chirality and helicity. Quark gluon coupling being purely vector will result in strict helicity conservation. Any helicity flip amplitudes are expected to be small and of the order of constituent quark masses divided by their respective energies. In hadron scattering, under perturbative assumptions, individual quarks interact perturbatively conserving helicity. Thus, helicity is preserved for the whole reaction.

The helicity amplitudes are proportional to the angular distribution matrix elements of the decaying rho meson. Assuming parity conservation, these distributions for a spin 1 particle are given by

$$(4\pi/3) W(\theta,\phi) = r_{0,0} \cos^2\theta + r_{1,1} \sin^2\theta - r_{1,-1}\sin^2\theta \cos 2\phi - 2 (\text{Re } r_{1,0})\sin 2\theta \cos \phi \qquad (1)$$

when θ and ϕ are the polar and azimuthal angles in the center-of-mass system of the rho meson.

A few results can be pointed out readily:

a) Helicity Conservation reduces the off diagonal elements to zero. The expression becomes

$$4\pi/3 \, W(\theta,\phi) = r_{1,1} + (r_{0,0} - r_{1,1}) \cos^2\theta \qquad (2)$$

with no ϕ dependence. Figure 3 (a,b) show the ϕ angular distributions of the selected ρ events. There is a striking similarity between the two sets of data. The ϕ dependence is not flat but consistent with $\sin^2\phi$ lending evidence to some helicity nonconservation in the reaction.

b) If pure gluon exchange is assumed then only $r_{0,0}$ will be nonzero.

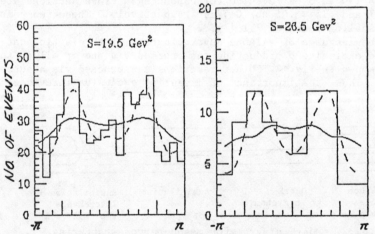

Fig. 3. ϕ (CM Helicity Frame). Fits to the ϕ decay angular distribution for ρ mass $.25 < p < 1.0$ GeV^2. The solid line represents the acceptance.

The decay angular distributions were fit with a sum of spherical harmonics $Y_M^L(\theta,\phi)$ with the series cut off at $L = 2$. Table I

has the corresponding density martrix elements for the two beam
energies.

TABLE I

		Helicity Conserving		Non Conserving	
		$r_{0,0}$	$r_{1,1}$	$r_{1,-1}$	$Re(r_{1,0})$
10 GeV	(ρ mass)	.07 ± .21	.46 ± .11	.29 ± .07	.05 ± .04
	(>ρ mass)	.31 ± .21	.35 ± .11	-.05 ± .07	-.02 ± .04
13.4 GeV	(ρ mass)	1.00 ± .34	.00 ± .17	.26 ± .18	.03 ± .15
	(>ρ mass)	.25 ___	.37 ___	.02 ___	.02 ___

The 13.4 GeV/c results are still preliminary. Additional analysis is underway to determine the sensitivity of the results to the shape of the background.

The experiment lacked sensitivity to the $r_{0,0}$ and $r_{1,1}$, namely cos θ terms, but was very sensitive to the ϕ terms. The fact that $r_{1,-1}$ is large is a good indication that helicity is not conserved and the spin flip term is substantial. This term is consistent with zero for masses above that of the rho. While the 10 GeV data is statistically significant, the 13.4 GeV/c data consolidate these findings.

Where does this leave us with respect to the various theoretical interpretations? The lack of helicity conservation negates the pure gluon exchange picture in favor of a mixture of the above diagrams since quark exchange or annihilation would allow helicity nonconservation. Cross sections will serve to assess the relative contributions of these diagrams to the scattering process.

G. Farrar[4] combined these data along with, A_n measurement at 28 GeV and A_{nn} data at 11.75 GeV/c in pp elastic scattering[1] to estimate that the higher twist amplitudes in this exclusive process to be ~ 30% compared to leading terms. Helicity nonconservation arises from the interference between the leading and nonleading twist amplitudes.

Nardulli, Preparata and Soffer[5] use a meson exchange model to obtain the observed angular distributions namely the θ and ϕ dependence. But our data differ from their predictions of the ratios of certain exclusive cross sections.

One last comparison of our results with ρ meson data taken at 6 GeV/c and up to t ~ 1 $(GeV)^2$ from Gordon et al.[6] are presented in Fig. 4. It is remarkable that the density matrix elements are similar over such a wide gap in t. This probably indicates that soft scattering is still present there even at t ~ 9 $(GeV)^2$. These results can only point to the fact that more experimental data are needed as input and that theoretical understanding is still lacking.

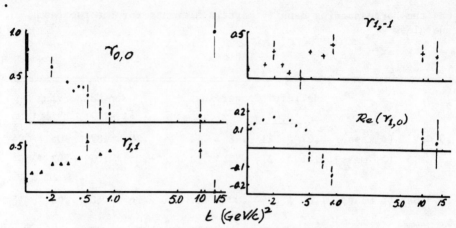

Fig. 4. The density matrix elements vs. t.

REFERENCES

1. See for example: K. Heller, Inclusive Hyperon Polarization: A Review. Proc. of the 6th Intern. Symp. on High Energy Spin Physics, J. Soffer, Editor, Marseille, France, 1984; P.R. Cameron, et al. Measurement of the Analyzing power for $p + p\uparrow \to p + p$ at $p_\perp^2 = 6.5$ $(GeV/c)^2$, UM-HE 85-17; E.A. Crosbie, et al., Phys. Rev. D23, 600 (1981).

2. G. Blazey et al. Hard Scattering with Exclusive Reactions. Phys. Rev. Lett. 55, No. 18, 1820 (1985).

3. S. Brodsky and G. Farrar. Phys. Rev. Lett. 31, 1153 (1973).

4. G. Farrar, Phys. Rev. Lett. 56, 1643 (1986).

5. G. Nardulli, G. Preparata, J. Soffer, Phys. Rev. D31, 626 (1985).

6. H. Gordon, et al., Phys. Rev. D8, 779 (1973).

SPIN OBSERVABLES IN PROTON-NEUTRON SCATTERING AT INTERMEDIATE ENERGY§

H. Spinka
Argonne National Laboratory, Argonne, IL 60439

ABSTRACT

A summary of np elastic scattering spin measurements at intermediate energy is given. Preliminary results from a LAMPF experiment to measure free neutron-proton elastic scattering spin-spin correlation parameters are presented. A longitudinally polarized proton target was used. These measurements are part of a program to determine the neutron-proton amplitudes in a model independent fashion at 500, 650, and 800 MeV. Some new proton-proton total cross sections in pure helicity states ($\Delta\sigma_L(pp)$) near 3 GeV/c are also given.

INTRODUCTION

Although there have been a large number of new measurements of spin parameters for pp elastic and inelastic reactions below T_{lab} = 1100 MeV in the last few years, the np data base remains sparse above about 500 MeV.[1] There are a significant number of np differential cross section and polarization data, especially at backward angles. These include recent results from the polarized neutron beam at Saclay.[2,3] However, there are only a few measurements of other spin parameters. This article describes recent np elastic scattering measurements from LAMPF, as well as the outlook for a complete determination of the five complex isospin-0 (I=0) amplitudes in the near future. Finally some intriguing new information has become available from a final analysis of past ZGS data on $\Delta\sigma_L(pp)$ (the total cross section difference between antiparallel and parallel longitudinally polarized beam and target), which are also included.

For this paper, the notation (beam, target; forward scattered, recoil) will be used to express the elastic scattering spin observables. The spin directions are $\vec{N}$ (up and normal to the horizontal scattering plane), $\vec{L}$ (longitudinal, along the incident beam direction or the outgoing particle directions), and $\vec{S} = \vec{N} \times \vec{L}$. An "O" denotes that the spin direction is not measured.

np ELASTIC SCATTERING MEASUREMENTS

The free np elastic scattering data base presently contains many different spin parameters and energies below about 500 MeV, with a large body of data from TRIUMF a few years ago.[4-6] Above 500 MeV, there are numerous differential cross section and polarization results. In addition, there is a set of older measurements from

DUBNA near 600 MeV (see Ref. 1), a few spin transfer parameter data in the charge exchange peak at 800 MeV,[7] and some preliminary polarized beam - polarized target results up to 665 MeV (C_{NN} = (N,N;0,0), see Ref. 8).

Several groups have attempted to extract np spin parameters from pd quasielastic scattering measurements. For example, a number of D_{ij} = (i,0;j,0) parameters, as well as differential cross sections and polarizations, were obtained at $\theta_{c.m.} \lesssim 70°$ and 800 MeV using the LAMPF High Resolution Spectrometer (HRS - Refs. 9, 10). The pp spin parameter data extracted from these experiments agree reasonably well with free pp measurements from other experiments. This strongly suggests that the np spin parameters extracted similarly agree with free np results. Recently, these measurements were extended to larger $\theta_{c.m.}$ values in another beamline by the same group[11], but preliminary results are not yet available. Data are also being analyzed for D_{NN} from free np scattering at 800 MeV and large $\theta_{c.m.}$ values.[12]

In addition to the pd experiments above, the $\vec{p} + d \rightarrow \vec{n} + p + p$ reaction with the outgoing neutron along the incident proton direction is used to generate the LAMPF polarized neutron beam. The spin transfer parameters K_{LL} = (L,0;0,L) and K_{NN} = (N,0;0,N) have been measured at three beam energies.[13,14]

Finally, total cross section differences for pd interactions

$$\Delta\sigma_L(pd) = \sigma^{Tot}(\rightleftarrows) - \sigma^{Tot}(\rightrightarrows)$$

were measured some time ago at the Argonne ZGS.[15] Using Glauber theory relations,[16,17] Fermi motion and deuteron D-state corrections, and results from dispersion relations,[18,19] the values of $\Delta\sigma_L$(np) were extracted; the results were reasonably featureless. Then the (Fermi-smeared) $\Delta\sigma_L$(pp) data were subtracted to yield $\Delta\sigma_L$ (I=0). These derived results showed considerable structure because of the $\Delta\sigma_L$(pp) subtraction. Free np measurements are being planned for SIN, Saclay and LAMPF to check the true value of $\Delta\sigma_L$ (I = 0). The observed structure, if confirmed, might suggest the presence of dibaryon resonances.[19,20]

A three year program at LAMPF to measure free np spin-spin correlation parameters for $\theta_{c.m.} \sim 50°$ to 160° and T_{lab} = 500, 650 and 800 MeV recently completed data taking. The spins of both the polarized neutron beam and the polarized proton target were in the horizontal scattering plane. The spin parameters measured were C_{SS} = (S,S;0,0), $C_{LS} = C_{SL}$ = (L,S;0,0) and C_{LL} = (L,L;0,0). Interference with the polarized target magnet coils prevented detection of both outgoing particles in coincidence. Consequently, only the recoil proton was measured in a magnetic spectrometer.

Preliminary results over part of the angular range are shown in Fig. 1. Fairly good agreement with phase shift predictions can be seen at 500 MeV, where a variety of different spin parameters were previously measured at TRIUMF and elsewhere, and also at 650 MeV. Refinements to the values shown in Fig. 1 are expected soon for improved magnetic field integrals, background subtraction and cuts on the target projections and the particle mass.

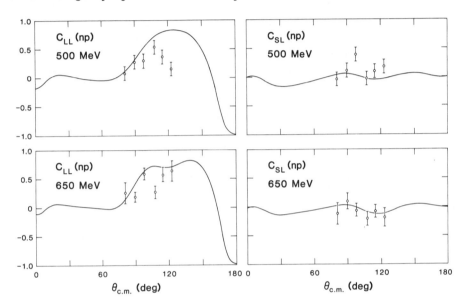

Fig. 1 Preliminary results for C_{LL} and C_{SL} for free np elastic scattering at 500 and 650 MeV. The curves are phase shift predictions. The quoted errors are presently dominated by uncertainties in the background subtraction.

Basically the same apparatus used for measurements of C_{SS}, C_{SL} and C_{LL} was also employed in a short experiment to determine the polarization of free np scattering at 800 MeV and small $\theta_{c.m.}$. These data were collected late in 1985 and they are presently in the process of being analyzed.[12]

In the past few years, model independent amplitude analyses[21] for pp elastic scattering have been achieved for data from SIN,[22] LAMPF,[23] and the ZGS[24]. Results from a single experimental group dominated the data base in each of these cases. Based on experience with the pp system at LAMPF, phase shift predictions generally did not fit data for new spin observables very well until a model independent amplitude analysis could be performed. In the np case, this is likely to be true as well.

Using this experience as a guide, the number of additional np measurements needed to obtain unique phase shifts can be obtained. It has been shown[25] that measurements with polarized beam and/or polarized target ($d\sigma/d\Omega$, P, C_{NN}, C_{SS}, C_{SL}, C_{LL}) at $\theta_{c.m.}$ and $\pi - \theta_{c.m.}$, and knowledge of the pp amplitudes at $\theta_{c.m.}$, allows the model independent determination of the I=0 amplitudes up to discrete ambiguities. The absolute phase at each angle is anchored to the phase of the pp amplitudes. Therefore, the free np elastic data on C_{LL}, C_{SL} and C_{SS} from this experiment and existing data on the other three parameters, will allow an I=0 amplitude determination at 500 and 650 MeV at about 10 angles between $\theta_{c.m.}$ = 50° and 90°. (The 800 MeV C_{NN} results will be measured this summer, permitting the reconstruction of the amplitudes at 800 MeV as well.) At each energy, there will be four discrete solutions as a function of angle. The existence of other spin parameter results, described above, may allow a unique solution at each energy.

Nevertheless, it is expected that additional measurements at 650 and especially 800 MeV may be required to clearly establish the I=0 amplitudes. Experimental difficulties with producing monoenergetic neutron beams and with the detection of neutrons probably leads to considerably larger systematic errors for np than for pp scattering. Differences in data sets from various experiments on polarization, total and backward differential cross sections outside quoted errors are probably caused by these experimental difficulties. Thus, it will be important to overdetermine the amplitudes, providing consistency checks on the data. For example, it is considered important to measure $K_{ij} = (i,0;0,j)$ for free np scattering at LAMPF for this purpose. The new high intensity polarized ion source planned for LAMPF should make such experiments feasible over the full angular range.

Additional experiments on free np scattering at SIN[26], Saclay[27], and TRIUMF[28] are expected to begin later this year. These include $\Delta\sigma_L(np)$, as well as $C_{ij}(np)$, etc. The neutron beam energies will be T_{lab} = 400 to 1150 MeV at Saclay and below 600 MeV at SIN and TRIUMF.

NEW $\Delta\sigma_L(pp)$ RESULTS

Final results for $\Delta\sigma_L(pp)$ from the ZGS at several new energies have revealed intriguing energy behavior. The experiment was similar to that reported previously.[29-31] The results are shown in Fig. 2 along with some of the earlier data. The errors plotted are purely statistical; systematic uncertainties are estimated to be 6% of $\Delta\sigma_L$. Coulomb-nuclear interference corrections are not included. It should be noted that the Saclay measurement[32] near 3.25 GeV/c is consistent with the new data.

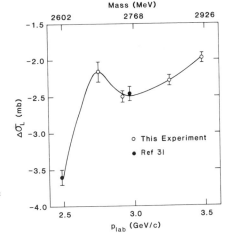

Fig. 2

The $\Delta\sigma_L(pp)$ dependence on P_{lab} from 2.5 to 3.5 GeV/c from the new measurements reported and from Ref. 31. The errors shown are statistical only. The curve is to guide the eye.

Although the statistics are not compelling, a new peak structure is indicated near 2.75 GeV/c. If it is due to a resonance, the mass is around 2700 MeV with a width less than 80 MeV. Earlier indications of structure near this mass include pp spin-spin correlation parameter data at $\theta_{c.m.} = 90°$ ($k^2 C_{LL} d\sigma/d\Omega$[33] and $k^2 C_{NN} d\sigma/d\Omega$[34], where k is the c.m. momentum) and polarization for the pp → dπ reaction from Saclay.[35] Cloudy Bag model predictions[36,37] include a narrow s-wave state near 2700 MeV, which may be consistent with the new structure. It is highly desirable to measure additional $\Delta\sigma_L(pp)$ data in finer steps near 2.75 GeV/c to clarify the energy dependence in this energy region.

ACKNOWLEDGEMENTS

I wish to thank my many colleagues from Argonne, LAMPF, University of Montana, New Mexico State University, Texas A&M University, and Washington State University for useful discussions on np scattering, and especially to T. Shima and M. Rawool. I would also like to express my gratitude to M. Barlett, L. Greeniaus, R. Hess and F. Lehar for useful communications.

§ Work supported by the U.S. Department of Energy, Division of High Energy Physics, Contract W-31-109-ENG-38.

REFERENCES

1. R. A. Arndt et al., Phys. Rev. D28, 97 (1983).
2. J. Bystricky et al., Nucl. Phys. A444, 597 (1985).
3. G. A. Korolev et al., Phys. Lett. 165B, 262 (1985).
4. A. S. Clough et al., Phys. Rev. C21, 988 (1980).
5. D. Axen et al., Phys. Rev. C21, 998 (1980).
6. R. K. Keeler et al., Nucl. Phys. A377, 529 (1982).
7. R. D. Ransome et al., Phys. Rev. Lett. 48, 781 (1982).
8. T. S. Bhatia et al., Fifth International Symposium on Polarization Phenomena in Nuclear Physics, Santa Fe, New Mexico, 1980 - A.I.P. Conference Proceedings #69, p. 123.
9. M. L. Barlett et al., Phys. Rev. C27, 682 (1983).
10. M. L. Barlett et al., Phys. Rev. C32, 239 (1985).
11. M. L. Barlett, private communication.
12. G. Glass, private communication.
13. P. J. Riley et al., Phys. Lett. 103B, 313 (1981).
14. J. S. Chalmers et al., Phys. Lett. 153B, 235 (1985).
15. I. P. Auer et al., Phys. Rev. Lett. 46, 1177 (1981).
16. C. Sorensen, Phys. Rev. D19, 1444 (1979).
17. G. Alberi et al., Phys. Rev. D20, 2437 (1979).
18. W. Grein and P. Kroll, Nucl. Phys. B157, 529 (1979).
19. W. Grein and P. Kroll, Nucl. Phys. A377, 505 (1982).
20. K. Hashimoto, Y. Higuchi, and N. Hoshizaki, Prog. Theor. Phys. 64, 1678 (1980); K. Hashimoto and N. Hoshizaki, ibid. 1693 (1980).
21. G. R. Goldstein and M. J. Moravcsik, Ann. Phys. (N.Y.) 142, 219 (1982) and references therein.
22. E. Aprile et al., Phys. Rev. Lett. 46, 1047 (1981).
23. M. J. Moravcsik et al., Phys. Rev. D31, 1577 (1985).
24. I. P. Auer et al., Phys. Rev. D32, 1609 (1985).
25. H. Spinka, Phys. Rev. D30, 1461 (1984).
26. R. Hess, private communication.
27. F. Lehar, private communication.
28. L. Greeniaus, private communication.
29. I. P. Auer et al., Phys. Lett. 67B, 113 (1977).
30. I. P. Auer et al., Phys. Lett. 70B, 475 (1977).
31. I. P. Auer et al., Phys. Rev. Lett. 41, 354 (1978).
32. J. Bystricky et al., Phys. Lett. 142B, 130 (1984).
33. I. P. Auer et al., Phys. Rev. Lett. 48, 1150 (1982).
34. A. Yokosawa, Phys. Rep. 64, 64 (1980) and A. Lin et al., Phys. Lett. 74B, 273 (1978).
35. R. Bertini et al., Proceedings of the Sixth International Symposium on Polarization Phenomena in Nuclear Physics, Osaka, Japan, 1985 (J. Phys. Soc. Japan, 55, (1986), Suppl. p. 976-977).
36. E. L. Lomon, Nucl. Phys. A434, 139C (1985); P. LaFrance and E. L. Lomon, to be published.
37. P. J. Mulders and A. W. Thomas, J. Phys. G9, 1159 (1983).

PARITY AND TIME-REVERSAL VIOLATION IN NUCLEI AND ATOMS

E.G. Adelberger
Nuclear Physics Laboratory, GL-10
University of Washington
Seattle, WA 98195 U.S.A.

ABSTRACT

Two topics are briefly reviewed: the parity (P)-violating NN interaction and the time-reversal (T) and P-violating electric moments (EDM's) of atoms. The $\Delta I=1$ P-violating NN amplitude dominated by weak $\pi^{\pm}$ exchange is found to be appreciably smaller than bag-model predictions. This may be a dynamical symmetry of flavor-conserving hadronic weak processes reminiscent of the $\Delta I=1/2$ rule in flavor-changing decays. General principles of experimental searches for atomic EDM's are discussed. Atomic EDM's are sensitive to electronic or nuclear EDM's and to a P-and-T-violating electron-quark interaction. Even though the experimental precision is still ~ 10^4 times worse than counting statistics, the recent results have reached a sensitivity to nuclear EDM's which rivals that of the neutron EDM data. Further significant improvements can be expected.

INTRODUCTION

In this talk I will give an experimentalist's quick overview of P and T violation in nuclear and atomic systems. I will discuss the P-violating NN interaction, concentrating on the motivation for these studies and a review of results which have appeared since the last Intersections between Particle and Nuclear Physics Meeting at Steamboat Springs, Colorado. I conclude with a short discussion of permanent electric dipole moments (EDM's) of atoms. Here I discuss a few basic ideas, stress the opportunities for significant results and mention some recent work. Because of time limitations I have chosen to neglect several very interesting topics which should have been included under my title. There is no real discussion of theory because W.C. Haxton has covered some of this. I will not discuss the problems of systematic errors in high precision experiments because V. Yuan and M. Simonius will deal with this issue in their talks. I omit a discussion of CP violation in neutron scattering since J.D. Bowman will talk on this subject. Finally I will not deal with the topic of P-violation in atoms.

THE P-VIOLATING NN INTERACTION

Haxton and I recently completed an extensive review[1] of this topic. Consequently, I shall focus on new results and refer the reader to our review for more details.

Since we have a very successful standard theory of electroweak interactions and quarks, why should one undertake the very difficult measurements of P-violating hadronic interactions? I see two motives. First, in principle, one could learn something new about the fundamental weak interaction. We suspect that the standard theory is not complete and requires extensions. One possible extension is that there are extra Z^o weak bosons. If these "extra" Z's have very small couplings to leptons they would have probably escaped detection in conventional particle physics experiments. On the other hand, if the extra Z's coupled with normal or greater strength to the quarks then they would have detectable consequences in the P-violating NN interaction. It turns out that the flavor-conserving P-violating hadronic interaction provides a unique window on Z^o exchange between the quarks. We can only study weak interactions among hadrons when the strong interaction is suppressed by a symmetry. Normally this suppression is due to flavor violation as in the decay $\Lambda \rightarrow N\pi$. However, the neutral current contribution to flavor-changing processes is highly suppressed by the GIM mechanism. To probe the neutral current hadronic interaction we must study flavor-conserving processes which violate parity — of which the P-violating NN interaction is the only existing practical example.

The second motive is more applied. Even if we had a complete and exact theory of electroweak interactions of quarks we still would not know how to compute the weak interactions of hadrons because we do not have exact knowledge of the structure of the hadrons. This problem of computing weak interaction matrix element between hadronic states is both important and interesting. Important, for example, because it limits our ability to interpret the lovely ϵ'/ϵ data in kaon decay, and interesting because of the appearance of dynamical symmetries (e.g., the $\Delta I=1/2$ rule) which are still not well understood. By studying a new class (flavor-conserving) of hadronic weak interactions we may get new insight into the problem of weak matrix elements.

Under strong isospin (I) rotations, the P-violating NN interaction is expected to transform as a mixture of $\Delta I=0$, 1, and 2 with the $\Delta I=1$ contribution being especially sensitive to Z^0 exchange. The standard theoretical analysis of the P-violating NN interaction employs a one-meson exchange model in which one of the meson (M)-nucleon-nucleon vertices is a weak process. Bag-model estimates of the weak MNN vertices by Desplanques, Donoghue and Holstein[2] (DDH) predict that the $\Delta I=0$, 1, and 2 components of the P-violating NN interaction should have roughly equal strengths.

How do these predictions compare to the experimental results? Unfortunately, the data is sparse because of the difficulty of performing sufficiently precise experiments. In principle, one could completely determine the low-energy P-violating NN interaction from six different measurements in the NN system, involving p+p, n+p and n+n initial states. However, the predicted P-violating effects range between $\sim 10^{-7}$ to $\sim 10^{-8}$ so that today definite effects have been seen only in the p+p system. An impressively precise new SIN result for $A_L(pp)$ (the longitudinal analyzing power) in p+p at $E_p^{lab} \approx 45$ MeV is being reported at this conference by M. Simonius.[3] The SIN result is shown in Fig. 1,

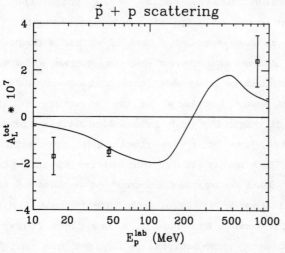

Fig. 1. Measured values of A_L in $\vec{p}+p$ scattering. The curve is a P-violating meson exchange theory calculation described in Ref. 1.

along with previous results from the Los Alamos tandem and LAMPF. Also shown is a prediction for A_L (see Ref. 1 for details) based on the DDH estimates of the MNN vertices. The agreement is encouragingly good. The other new datum in the NN system is an ILL measurement[4] of A_γ(np) (the asymmetry when polarized cold neutrons are captured by protons). This quantity is particularly interesting because it is sensitive primarily to F_π, the $\Delta I=1$ amplitude for weak $\pi^\pm$ exchange. (Note that pp scattering is completely insensitive to F_π in lowest order. The experimental value[4] $A_\gamma = (-4.7 \pm 4.7) \times 10^{-8}$ is consistent with the prediction (see Ref. 1) $A_\gamma = -5 \times 10^{-8}$. Blayne Heckel has proposed[5] measuring A_L(np) for cold neutrons on a parahydrogen target. It seems that a measurement of A_L(np), which is mainly sensitive to the ρ and ω exchange amplitudes, can be made with enough precision to test the theory.

On the other hand, it is currently impossible to do an adequately precise A_L(nn) measurement and quite difficult to make a large improvement in the errors on A_γ(np). However, one can study nn and np scattering indirectly, by investigating parity impurities in nuclear states. The most interesting cases of nuclear parity mixing are the "parity doublets" - nuclei with closely spaced pairs of levels of the same spin and opposite parity. Some important examples of parity doublets are shown in Fig. 2. These examples all have the property that one member of the doublet has a much longer decay lifetime than the other. As a result pseudoscalar observables associated with the decay of the longer lived states will be considerably enhanced - because the small energy dominator and the fact that a little admixture of a rapidly decaying level has a relatively big effect.

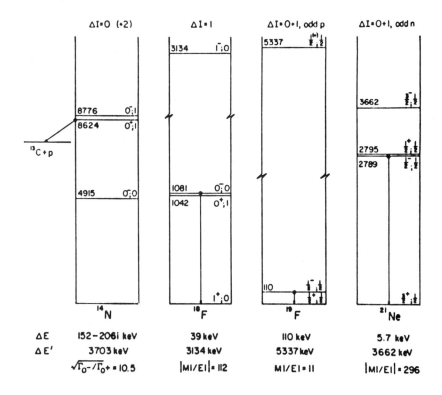

Fig. 2. Parity doublets in light nuclei. The transitions displaying the amplified P-violating effect are indicated. The quantities ΔE and $\Delta E'$ are the smallest and next smallest energy denominators governing the parity mixing. The quantities in the bottom row are "amplification factors."

The doublets also act as "isospin filters", the various doublets isolate specific isospin components of the P-violating NN interaction. For example the ^{18}F doublet involves the mixing of I=1 and I=0 levels; it probes weak $\pi^{\pm}$ exchange which has $\Delta I=1$. Therefore $P_\gamma(^{18}F)$ and $A_\gamma(np)$ both are sensitive to the same underlying physics. They provide a dramatic illustration of the gain of the nuclear P-violating amplifier. The DDH "best value" predictions for the observables are $A_\gamma(np)=-5\times10^{-8}$ and $|P_\gamma(^{18}F)|=1.5\times10^{-3}$ - which corresponds to a gain of 3×10^4. In this case the gain is also nearly "noise-free". The ^{18}F nuclear matrix element needed to compute $P_\gamma(^{18}F)$ from F_π is determined by the measured $^{18}Ne \rightarrow ^{18}F(0^-)$ β decay rate (see Ref. 1).

Recently groups at Queens University[6] and Firenze[7] have published results for $P_\gamma(^{18}F)$ with considerably smaller errors than the previous world average. To obtain this precision the experimenters had to develop high-rate data acquisition systems and produce targets that would survive under prolonged intense bombardments. When the new ^{18}F results are added to previous data the new world average is $P_\gamma=(1.2\pm3.8)\times10^{-4}$, which corresponds to $|F_\pi|<3\times10^{-7}$.

The $A_\gamma(^{19}F)$ and $A_L(\vec{p}+^4He)$ data (both ^{19}F and p+^{4}He are odd proton systems) yield similar and consistent constraints on a linear combination of $\Delta I=1$ and $\Delta I=0$ amplitudes. The $P_\gamma(^{21}Ne)$ result (an odd neutron system) corresponds to a different linear constraint on the $\Delta I=1$ and $\Delta I=0$ amplitudes. These constraints are shown in Fig. 3. The $\pm1\sigma$ constraints do not overlap. However, the ^{21}Ne constraint is shown as dashed because it was not possible to "calibrate" the shell model matrix elements against corresponding first forbidden β decay rates. I therefore discount the ^{21}Ne constraint and provisionally conclude that F_π is considerably smaller and F_0 somewhat larger than the DDH best values.

How can we test these conclusions? It seems difficult to make significant improvements in either the $A_\gamma(np)$ or $P_\gamma(^{18}F)$

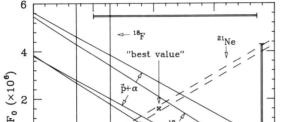

Fig. 3. Analysis of the P-violation in the light systems $p+^4He$, ^{18}F, ^{19}F and ^{21}Ne in terms of F_0 and F_π as defined in Ref. 1. The linear ±1σ con-straints imposed by the experimental results are shown. The DDH "best value" is shown as a cross and the DDH "reasonable ranges" are shown as double bars on the top and right.

experiments. A possible test would be to measure a pure ΔI=0 transition that imposes a horizontal constraint in Fig. 3. The mixing[8] of I=1 and 0^+ and 0^- levels in ^{14}N (see Fig. 2) provides just such a constraint. One could then check for internal consistency among the ^{18}F, ^{19}F, p+α and ^{14}N results. We are currently trying to study the P-mixing in ^{14}N by a novel technique — measuring $A_L(\vec{p}+^{13}C)$ at the elastic-scattering resonance corresponding to the narrow 0^+ state.[9] The P-violating effect predicted using the DDH best values and Haxton's matrix element[9] is shown in Fig. 4. We hope to have results within a year.

What conclusions can be drawn at this point?

1) There are no strikingly large anomalies — cases where the measured effects are much larger than can be accounted for by the standard model. Previous anomalies, such as the earlier $P_\gamma(np)$ and $A_\gamma(nd)$ results have been superceded by more precise data which is in general agreement with expectations. Hence there is nothing that lends strong support to the existence of "extra" Z^0's.

2) There is a suppression of F_π (the ΔI=1 amplitude for weak $\pi^\pm$ exchange) compared to bag model calculations. Donoghue and

Holstein[10] have considered the implications of small values for F_π and argued that it teaches us about quark masses and SU(3) breaking effects. It would be good to update their analysis.

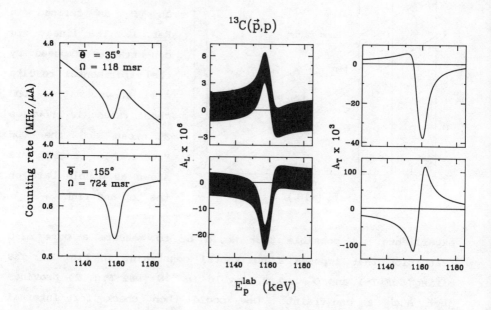

Fig. 4. Predicted counting rates, P-violating A_L, and P-conserving A_T for the 0^+ resonance in ^{13}C+p. Predictions are based on the DDH "best values." The solid region in the A_L plot represents a $\pm 1\sigma$ band where σ is the statistical error after running for 1 μA-day of integrated beam.

3) The one-boson exchange model of the P-violating NN interaction at low and intermediate energies appears to work quite well (see Fig. 1). This is consistent with the successes of boson-exchange models in many other areas of nuclear physics.[11]

4) It would be very interesting to extend the A_L(pp) results to considerably higher energies where the boson exchange model must break down and explicit quark calculations are required.

ATOMIC EDM'S

Although T-violation was observed in the kaon system 22 years ago[12] we still do not understand its origin – largely because T-violation has not yet been seen in any other system. It is well known that the existing limits on a permanent EDM of the neutron [which would violate both P and T (time reversal)] provides a sensitive constraint on theories of T-violation. Recent advances in experimental technique have opened the possibility that searches for T-violating EDM's of atoms will ultimately be a more sensitive probe of the origin of T-violation than the neutron EDM data. In this section I will briefly discuss recent work on atomic EDM's by E.N. Fortson, B.R. Heckel, F. Raab, and their students at Seattle.

Atomic EDM's are measured by polarizing atoms confined in a cell by optical pumping (sometimes in conjunction with spin exchange). A weak magnetic field B_z and a strong parallel or anti-parallel electric field E_z are applied to the cell. The atoms, which were initially polarized perpendicular to the z-axis, precess around z. If the atoms have a T-violating EDM there will be a small change in the precession frequency when the sign of E_z is changed. Since the precession frequencies can be measured very precisely the resulting errors on the EDM are very small.

The Seattle experimenters are working on two types of experiments:

1) Odd-Z atoms with $^2S_{1/2}$ electronic structure. Here one polarizes the electron and measures a $\vec{\sigma}\cdot\vec{E}$ correlation. Such experiments are primarily sensitive to an EDM of the electron or to a scalar-pseudoscalar P-and-T violating electron-quark interaction. Examples of this are Rb and Cs.

2) Even-Z atoms with $^1S_0$ electronic structure and spin 1/2 nuclei. Here one polarizes the nucleus and measures a $\sigma_{nucleus}\cdot\vec{E}$ correlation. Such experiments are primarily sensitive to a nuclear EDM or to a tensor-pseudotensor P-and-T violating electron-quark interaction. Examples of this are ^{129}Xe and ^{199}Hg.

In order to understand the sensitivity of atomic EDM's to the underlying T-violating interactions one must appreciate the phenomenon of atomic shielding of EDM's. Suppose the atomic electrons were non-relativistic (i.e., subject only to Coulomb forces). Then, since electron and nucleus have no net acceleration, the net E fields on the electron and on the nucleus must vanish. Therefore, the applied external $\vec{E}$ field will be shielded exactly and the EDM observable must vanish. When the electrons are relativistic (i.e., magnetic forces become appreciable) the perfect shielding breaks down. It turns out that the observable effects of e^- or nuclear EDM's grow roughly like Z^3 or Z^2 depending on the origin of the atomic EDM. In other words, an intrinsic electronic or nuclear EDM produces a much larger atomic EDM in a heavy atom than in a light atom.

The Seattle group's first EDM result was for the $^1S_0$ atom ^{129}Xe. In 1984 they published a value[13] $\mu e=(-0.3\pm1.1)\times10^{-26}$ ecm. I.B. Khriplovich[14] has developed a model for estimating the EDM's expected in complex nuclei if the Kobayashi-Maskawa 6-quark mixing is responsible for the T-violation seen in kaon decay. According to Khriplovich[14] "this [existing ^{129}Xe] limit approaches by its implications the limit on the neutron EDM." However Haxton and McKellar[15] suspect that Khriplovich's calculation overestimates the nuclear EDM's by a factor of ten or more.

Recently, the Seattle group has been working on the $^1S_0$ atom ^{199}Hg. Their preliminary experimental upper limits[16] on the atomic EDM are comparable to their published results for ^{129}Xe. However because of the higher Z of the atom these results correspond to a ~ 10 fold improvement on the nuclear EDM and an ~ 20 fold improvement on the electronic EDM compared to ^{129}Xe. The group has also investigated $^2S_{1/2}$ atoms. Their preliminary[16] result for Rb corresponds to an electronic EDM $\mu_e \leq 6\times10^{-25}$ ecm which is a 5-fold improvement over the previous e^- EDM limit set by Player and Sandars.[17] The Seattle group is now preparing to run with Cs which, because of its higher Z, should have ~ 5 times more sensitivity than Rb. According to Fortson[16] "if we put Cs in the cell and run overnight, we should get the electron EDM to 1×10^{-25} ecm."

I would like to conclude this topic with some general observations. It turns out that one can measure $\sigma_{nucleus}\cdot\vec{E}$ correlations much more precisely than $\sigma_e\cdot\vec{E}$ correlations because nuclear spins retain their coherence for a much longer time ($\tau_{nucleus} \sim 10^3$s) than electronic spins ($\tau_e \sim 10^{-1}$s). This large difference occurs because $\mu_e >> \mu_N$ and because the electrons shield the nuclear spin from many of the perturbing effects of collisions. As a result Fortson et al.[16] believe that they can ultimately set a better limit on the electron EDM from $^1S_0$ atoms rather than from $^2S_{1/2}$ atoms even though an e^- EDM contributes to the atomic EDM in $^1S_0$ atoms only via a higher order effect of the hyperfine interaction!

We can anticipate considerable improvements in the precision of these very elegant atomic experiments. The fundamental signal-to-noise limitations (counting statistics) permit improvements over existing results by a factor of roughly 10^4, once systematic effects are understood and eliminated. But even the existing results have reached a marvelous level of precision. To borrow an analog from Norman Ramsey, if the ^{129}Xe atom were blown up to the size of the earth, the bulge of positive charge in one hemisphere would be known to be less than 0.06 Å.

References

1. E.G. Adelberger and W.C. Haxton, Ann. Rev. Nucl. Part. Sci. 35, 501 (1985).
2. B. Desplanques, J.F. Donoghue, and B.R. Holstein, Ann. Phys. 124, 449 (1980).
3. M. Simonius, private communication.
4. R. Wilson et al., The Investigation of Fundamental Interactions with Cold Neutrons, ed., G.L. Greene, NBS

Special Publ. 711 p. 85.

5. B. Heckel, The Investigation of Fundamental Interactions with Cold Neutrons, ed., G.L. Greene, NBS Special Publ. 711 p. 90.
6. H.G. Evans et al., Phys. Rev. Lett. 55, 791 (1985).
7. M. Bini et al., Phys. Rev. Lett. 55, 795 (1985).
8. E.G. Adelberger, P. Hoodbhoy, and B.A. Brown, Phys. Rev. C 30, 456 (1984); Phys. Rev. C 33, 1840 (1986).
9. E.G. Adelberger et al., Annual Report, Univ. Washington, Nuclear Physics Lab, 1985, 1986 (unpublished).
10. J.F. Donoghue and B.R. Holstein, Phys. Rev. Lett. 46, 1603 (1981).
11. G.E. Brown and M. Rho, Comm. Nucl. Part. Phys. 15, 245 (1986).
12. J.H. Christenson et al., Phys. Rev. Lett. 13, 138 (1964).
13. T.G. Vold et al., Phys. Rev. Lett 52, 2229 (1984).
14. V.V. Flambaum, I.B. Khriplovich and D.P. Sushkov, Phys. Lett. B 162, 213 (1985).
15. B.H.J. McKellar and W.C. Haxton (private communication).
16. E.N. Fortson (private communication).
17. M.A. Player and P.G.H. Sandars, J. Phys. B 3, 1620 (1970).

PARITY VIOLATION IN PROTON-PROTON SCATTERING AT INTERMEDIATE ENERGIES

V. Yuan[*], H. Frauenfelder, and R. W. Harper[**]
University of Illinois, Urbana, IL 61801

J. D. Bowman, R. Carlini, D. W. MacArthur, R. E. Mischke, and D. E. Nagle
Los Alamos National Laboratory, Los Alamos, NM 87545

R. L. Talaga
University of Maryland, College Park, MD 20742

A. B. McDonald
Princeton University, Princeton, NJ 08540

ABSTRACT

Results of a measurement of parity nonconservation in the $\vec{p}$-p total cross section at 800-MeV are presented. The dependence of transmission on beam properties and correction for systematic errors are discussed. The measured longitudinal asymmetry is $A_L = (+2.4 \pm 1.1(\text{statistical}) \pm 0.1(\text{systematic})) \times 10^{-7}$. A proposed experiment at 230 MeV is discussed.

INTRODUCTION

An experiment has been carried out to search for parity nonconservation (PNC) in the scattering of 800 MeV longitudinally-polarized protons from an unpolarized hydrogen target. In the experiment, PNC appears as a small helicity dependence in the total cross section when the helicity of incoming protons is reversed. A longitudinal asymmetry A_L is defined: $A_L = (\sigma_+ - \sigma_-)/(\sigma_+ + \sigma_-)$ where $\sigma_+(\sigma_-)$ is the total cross section for positive (negative) helicity protons on the target.

PNC arises from an interference between the strangeness-conserving weak interaction and the strong interaction. Meson-exchange calculations[1-3] of the parity-violating amplitudes give predictions that agree with experimental measurements[4,5] at energies of 15 and 45 MeV. However, at higher energies (6 GeV), the meson-exchange predictions[6] do not explain the large experimental result.[7] Other theoretical models,[8,9] designed to be valid in the high-energy region, have predictions that agree with the 6-GeV results. In the intermediate energy region of 800-MeV, it is not clear which existing models correctly describe PNC effects. I will discuss results from a parity measurement at 800 MeV that is the highest energy $\vec{p}$-p measurement to date, achieving a sensitivity in its result comparable to that of the low-energy experiments.

EXPERIMENTAL SETUP

The experiment was performed with the Clinton P. Anderson Meson Physics Facility (LAMPF) polarized H^- beam. Polarized H^- ions were produced in a Lamb-shift-type ion source.[10] Neutral hydrogen atoms, initially polarized in the spin-filter region of the source, had their polarization reversed at 30 Hz by a weak magnetic field. Beam pulses were of 500-μs duration with a 120-Hz repetition rate. The proton-beam intensity ranged from 1 to 5 nA, and average polarization was 70%.

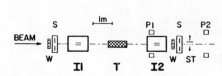

FIG. 1. Experimental setup

The layout of the apparatus is shown in Fig. 1. The transmission of protons through a one-meter long liquid-hydrogen target was measured by integrating ion chambers, I1 and I2, located upstream and downstream of the target. Detector noise due to nuclear spallation reactions in ion-chamber surfaces was a significant factor in determining the final statistical sensitivity of the measurement. To reduce spallation effects, special ion chambers[11] were used.

A major concern of the experiment were changes in beam properties that occur synchronous to the helicity reversal (beam systematics). Such changes could give rise to a spurious PNC signal. Detectors used to monitor the beam properties on a pulse-by-pulse basis are shown in Fig. 1. Integrating multi-wire ion chambers,[12] W, monitored beam position and size. A four-arm polarimeter, P1, used the LH_2 target as an analyzer to measure net transverse polarization (T_{pol}). The upstream ion chamber of the transmission measurement recorded intensity variations of the incident beam. Lastly, a second polarimeter utilized a narrow moving target, ST, to sample the transverse-polarization distribution across the beam profile every two minutes. The first moment of transverse polarization across the beam profile, C_{pol}, can result in an unwanted contribution to Z even if both $T_{pol} = 0$ and the beam is on the symmetry axis.[4,13] A dual-loop feedback system (not shown) stabilized the beam in position as well as angle.

The helicity-dependent fractional change in transmission, Z, was determined from the analog difference of the I1 and I2 signals. For each group of four beam pulses (quad), the quantity $Z_q = (\bar{T}_+ - \bar{T}_-)/(\bar{T}_+ + \bar{T}_-)$ was calculated, where $\bar{T}_+(\bar{T}_-)$ is the average transmission for a pair of +(-) helicity pulses. The helicity reversal pattern for each quad was + - - + in order to reduce the effects of drifts and to remove 60-Hz effects. After the accumulation of ~10^5 quads, an average, Z, and a statistical uncertainty in Z were computed from from the individual Z_q and their fluctuations. The longitudinal asymmetry A_L can be calculated from the transmission asymmetry:

$$A_L = Z/(P \ln T)$$

where P is the beam polarization and T is the average transmission of the target. For this experiment a $P = 0.7$ and $T = 0.85$ resulted in a value of $1/(P \ln T)$ of -8.8. Hence to reach a sensitivity in A_L of 10^{-7}, Z had to be measured to nearly one part in 10^8.

BEAM SYSTEMATICS

To correct for beam systematics, the sensitivity of Z to each beam systematic was determined in a separate measurement. Subsequently, beam properties were recorded during the transmission measurement, and corrections to Z were later applied in the off-line analysis. Pulse-by-pulse corrections were made for beam intensity, position, and size. Corrections for T_{pol} were made for each quad. Corrections for C_{pol} and for unwanted electrical couplings were applied on a run-by-run basis.

Contributions from transverse polarization were minimized by locating the beam along the symmetry-axis of the transmission detectors. The servo-loop system maintained the beam on the symmetry axis during data taking. The sensitivity of Z to intensity modulations was determined using an apparatus[14] consisting of a set of stripper grids that were moved in and out of the H⁻ beam path to produce a 10% intensity modulation at 30 Hz. At each transmission detector, position scans were performed to measure the sensitivity of Z to position. Small corrections for size variations were calculated from the quadratic components in the position dependence of Z. 30-Hz electrical pickup in the difference signal was minimized by using a 15-Hz digital signal to transmit the helicity-reversal information from the polarized source to the experiment. Care was also taken to insert either optical or analog isolators in all important signal paths.

To cancel contributions to A_L from changes not correlated to the beam helicity, the experiment was run for equal time periods in two different operating configurations (N and R)[15] of the spin filter[10] in the polarized source. Data taken in the two configurations can be combined in two ways: $PNC = (Z_N - Z_R)/2$ measures the longitudinal asymmetry while canceling some systematic effects; $HI = (Z_N + Z_R)/2$ is expected to be zero and serves as a test for unidentified systematic errors. Unidentified systematic errors were also addressed by analyzing the data using a shift in the quad grouping that eliminates helicity dependence from the calculated A_L. The resultant value, A_L (shift), was consistent with zero.

The final PNC and HI values of A_L, along with corrections made to A_L for each systematic, are given in Table I. Within each data run, pulse-to-pulse fluctuations in A_L are smaller for the corrected data than for the uncorrected data. This decrease occurs because correlations between Z and various beam systematics are removed by the corrections. In addition, testing the data from all runs for the hypothesis that $HI = 0$ and that PNC has a definite value produces a χ^2 value for corrected data which is nearly a factor of 2 smaller than that for uncorrected data.

TABLE I. Results for A_L and beam-systematic corrections. Both statistical and systematic contributions to uncertainty are given.

Quantity	PNC($\times 10^7$) Value	Stat	Sys	HI($\times 10^7$) Value	Stat	Sys
Corrections to A_L						
Position	0.3	0.3	0.1	-2.7	0.3	0.4
Intensity	-0.8	0.5	0.1	7.7	0.5	0.8
Size	-0.1	0.0	0.1	0.2	0.0	0.1
Polarization	<0.1	0.0	0.0	<0.1	0.0	0.0
C_{pol}	-0.1	0.4	0.0	-0.2	0.4	0.0
Electrical pickup	0.0	0.0	0.0	0.6	0.0	0.0
A_L (uncorrected)	3.0	1.2		-5.0	1.2	
A_L (corrected)	2.4	1.1	0.1	0.2	1.1	0.9
A_L (shift)	-0.7	1.1		-0.3	1.1	

The measured longitudinal asymmetry at 800 MeV is $A_L = (+2.4 \pm 1.1(\text{statistical}) \pm 0.1(\text{systematic})) \times 10^{-7}$. The corrected HI result is consistent with zero.

FUTURE EXPERIMENTS

An improved experiment by the Simonius group is being conducted at SIN, and new results will be presented in another talk at this conference. No independent verification of the large ZGS result exists. Therefore, an experiment utilizing the 30-GeV polarized beam at Brookhaven would be desirable. However, at this time a proposal has yet to be submitted.

In the intermediate energy range, a $\vec{p}$-p experiment at 230 Mev has been proposed at TRIUMPF by a Manitoba/ Alberta/ TRIUMPF/ LANL/ Washington/ Irvine collaboration. At this energy the $^1S_0 - {}^3P_0$ PV transition amplitude does not contribute to A_L because of kinematics and strong-interaction phases.[16] As a result other PV amplitudes dominated by ρ-meson exchange become significant, and there is promise for a determination of h_ρ. Finally, a letter of intent has been submitted by the TRIUMPF collaboration to perform a 2.8 GeV measurement at Saturne.

We acknowledge the contributions of the entire LAMPF staff. This work was supported by contract NSF PHY81-03317 and the U. S. D. O. E. One of us (RLT) acknowledges a grant from the General Research Board of the University of Maryland.

REFERENCES

(*)Present address: Los Alamos National Laboratory, Los Alamos, NM 87545
(**)Present address: Ohio State University, Columbus, OH 43210

1. V. R. Brown, E. M. Henley, and F. R. Krejs, Phys. Rev. C $\underline{9}$, 935 (1974). Note that the asymmetry defined there is $2A_L$ as here defined.
2. M. Simonius, Nucl. Phys. $\underline{A220}$, 269 (1974)
3. B. Desplanques, J. F. Donoghue, and B. R. Holstein, Ann. Phys. $\underline{124}$, 449 (1980).
4. D. E. Nagle et al., in High Energy Physics with Polarized Beams and Targets, ed. G. H. Thomas, AIP Conference Proceedings No. 51, (American Institute of Physics, New York, 1978), p. 224.
5. R. Balzer et al., Phys. Rev. C $\underline{30}$, 1409 (1984).
6. E. M. Henley and F. R. Krejs, Phys. Rev. D $\underline{11}$, 605 (1975)
7. N. Lockyer et al., Phys. Rev. Lett. $\underline{45}$, 1821 (1980).
8. T. Goldman and D. Preston, Nucl. Phys. $\underline{B217}$, 61 (1983).
9. G. Nardulli and G. Preparata, Phys. Lett. $\underline{117B}$, 445 (1982).
10. J. McKibben, in Polarization Phenomena in Nuclear Physics, ed. G. G. Ohlsen et al., AIP Conference Proceedings No. 69, (American Institute of Physics, New York, 1980), p. 830.
11. J. D. Bowman et al., Nucl. Inst. Meth. $\underline{216}$, 399 (1983).
12. D. W. MacArthur et al., Nucl. Inst. Meth. (to be published).
13. M. Simonious et al., Nucl. Inst. Meth. $\underline{177}$, 471 (1980).
14. D. W. MacArthur, Nucl. Inst. Meth. $\underline{A243}$, 281 (1986).
15. R. W. Harper et al., Phys. Rev. D $\underline{35}$, 1151 (1985)
16. M. Simonius, in Interaction Studies in Nuclei, ed. H. Jochim and B. Ziegler, (North Holland, Amsterdam, 1975), p. 3.

TEST OF TIME REVERSAL SYMMETRY WITH RESONANCE NEUTRON SCATTERING

J. David Bowman
Los Alamos National Laboratory
Los Alamos, New Mexico 87545

It may be possible to search for a time-reversal-odd, parity-odd interaction between nucleons in a nucleus with a sensitivity of 10^{-5} of the weak interaction between nucleons. It has been shown experimentally that large parity violating effects are present in the scattering of cold and epithermal neutrons from nuclei. The possibility of using this phenomenon to carry out sensitive tests for time reversal symmetry violation has been discussed theoretically by several authors.[1,2,3,4,5] In this contribution I will discuss the possibility of searching for time-reversal symmetry violation in the scattering of epithermal neutrons from the Los Alamos spallation neutron source from nuclei. I will discuss both statistical and systematic errors.

How large might one expect time reversal violation to be? There are two indications, CP violation in K decays and searches for an electric dipole moment of the neutron. In K mixings[6] CP-odd mixing amplitudes are measured to be 2×10^{-3} of CP-conserving decay amplitudes. Since the product TCP is believed to be conserved this amplitude ratio may be taken as the size of time reversal symmetry violation. Two types of theories are used to describe CP violation in the kaon system; milliweak theories in which CP-violating mixing between the K^0 and $\bar{K}^0$ takes place via a $\Delta S = 1$ interaction acting in second order, or super weak theories in which the mixing takes place via a $\Delta S = 2$ interaction acting in first order. For milliweak theories, CP-odd effects may be as large in other systems as they are in kaon decays. For super weak theories CP violation will be unobservably small outside of the kaon system. The present limit on the electric dipole moment of the neutron, d, is $d < 6 \times 10^{-25}$ e cm.[7] The neutron electric dipole moment is zero if either time reversal symmetry or parity symmetry holds. Following Wolfenstein[8] we take the nuclear magneton divided by the speed of light as an estimate of the size of d to be expected in the absence of any inhibiting symmetries. As an estimate of the reduction due to parity symmetry we take

$$\frac{G_F m_q^2}{4\pi} = 10^{-7} \quad ;$$

then

$$d = \frac{\mu_n}{c} \frac{G_F m_q^2}{4\pi} R < 6 \times 10^{-25}$$

where R is the fraction of the weak force which is time-reversal odd, G_F is the Fermi constant, m_q is a quark mass, and μ_n is the nuclear magneton. Then $R < 6 \times 10^{-4}$. These qualitative estimates make a search for R with a sensitivity of 10^{-5} attractive.

The energy dependence of the cross section for the scattering of epithermal neutrons from nuclei exhibits a rich structure of resonances. The density of states having a given set of quantum numbers is 0.1 eV and typical widths are 0.1 eV. Resonances are characterized by the orbital angular momentum of the neutron relative to the target. S-wave resonances have have cross sections as large as 10^{-20} cm^2, possible angular momenta of $J + 1/2$ and $J - 1/2$ (where J is the angular momentum of the target ground state) and the same parity as the ground state. For S-wave resonances, gamma decay and neutron decay compete. P-wave resonances have smaller cross sections, 10^{-24} cm^2. Angular momenta range from $J - 3/2$ to $J + 3/2$, their parity is opposite to the ground state, and gamma decay dominates.

Large parity-violating admixtures of S-wave resonances into P-wave resonances have been observed in the scattering of polarized neutrons from unpolarized targets.[9] In pulsed reactor experiments the neutron energy is measured by time of flight. The 0.9 eV resonance in ^{139}La shows a helicity dependence of 0.07. Because the density of states is high and the S-wave scattering amplitudes are much larger than P-wave amplitudes, mixing matrix elements of 10^{-3} eV can result in such large effects. Mixing matrix elements extracted from observed helicity dependences range from 0.38 to 3.0 MeV.[7]

The characteristics of the spallation neutron source at the LAMPF proton storage ring represent a thousand-fold improvement over the Dubna pulsed reactor where the above measurements were done. The neutron spectrum has an approximately 1/E shape above thermal energies with a flux of 5×10^7 neutrons per second into 0.1 msr. The 1 microsecond pulse width allows the resolution of resonances up to 100 eV with a 10-meter flight path.

The studies of parity violation used the observable $\sigma \cdot k$ where σ is the neutron polarization and k is the neutron momentum. The quantity $\sigma \cdot k$ is parity-odd, time-reversal-even. The study of time reversal requires a polarized target with polarization vector I. The observable is $\sigma \cdot I \times k$ which is both parity-odd and time-reversal-odd. There are three effects which may be studied: a dependence of the total cross section of a P-wave resonance on $\sigma \cdot I \times k$, the development of a neutron polarization perpendicular to both I and k when an unpolarized neutron beam passes through a polarized target, and the precession of the neutron polarization about $I \times k$. Adelberger[5] has discussed the third of these in the context of cold neutrons. For epithermal any of these effects may be used as probes of time-reversal symmetry violation. I will concentrate on the the first, although the choice of the effect depends on factors such as resonance parameters, as well as considerations of systematic and statistical errors.

The statistical sensitivity is excellent. For a one interaction length ^{139}La target, with the same type of polarimeter as used in the Dubna experiments, a statistical error of 3×10^{-7} eV in the time-reversal odd-mixing matrix element can be obtained in a 10^5 second run at the proton storage ring. For ^{139}La the measured mixing matrix element is 1.28×10 MeV.[7] The ratio R of time-reversal-odd, parity-odd matrix element to time-reversal-even parity-odd matrix element is measured with a statistical sensitivity of 2.3×10^{-4}. A 10^7 run, which appears feasible, would obtain a 10 times smaller statistical sensitivity.

An attractive feature of the type of time-reversal searches under consideration is the absence of final-state interaction effects. Kabir[1] has shown that final-state interactions do not falsify a time-reversal-odd mixing.

There are a number of interactions that change the direction of the neutron. In designing an experiment to search for a change in the neutron spin resulting from time-reversal symmetry violation it is important that these other interactions are not confused with time-reversal symmetry violation. The interactions that change the neutron spin are:

1. Magnetic fields. Magnetic fields result from at least three sources: external fields used to polarize the target, magnetization fields in ferromagnetic target materials, and the magnetic fields resulting from the polarization of the target nuclei.
2. Abragam rotation.[10] This source of spin rotation is associated with the polarization of the target nuclei, but is not a magnetic field. Due to the spin-orbit force the real part of the neutron scattering amplitude is different for the neutron spin parallel or antiparallel to the target polarization. This coherent effect causes the neutron spin to precess around the target polarization direction.
3. Parity-violating $\sigma \cdot k$ interaction. As discussed above the existence of a parity-violating $\sigma \cdot k$ interaction causes the total neutron scattering cross section to depend on the neutron helicity. In addition the neutron polarization will precess about k.

It is important to use thick targets in order to achieve good statistical accuracy. For a strong P-wave resonance a target thickness of 10^{22} nuclei/cm² would be appropriate. The use of thick targets means that the three sources cited above act over long distances and that spin rotation angles may be large. One must therefore deal with the problem of keeping the neutron spin direction perpendicular to both I and k on the average. An attractive approach to the above problems is the use of internal fields in ferromagnetic alloys to polarize the target nuclei. When a rare-earth atom is alloyed with iron the magnetic field at the point of the rare-earth nucleus may be much larger than the bulk magnetization of the iron. Fields for dilute substitutional alloys are typically hundreds of kilogauss. The magnetic field that acts on the neutron spin would be the bulk field, while the magnetic field that polarizes the target nucleus would be the much larger internal field.

The figure shows an approach to the use of these ideas. Magnetized plates of sample alloy are alternated with plates of iron. The iron plates are magnetized in the opposite direction to the target plates. The flux return path is within each plate, so that the fields outside of the plates are small. The function of the dummy iron plates is to undo the neutron spin rotation that takes place in the target plates.

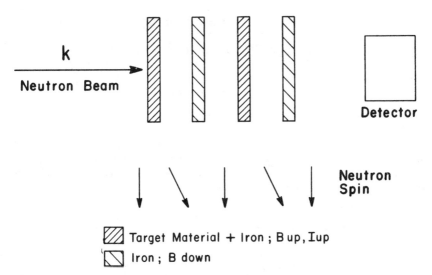

The first task in a search for time-reversal-odd effects is to make a survey of parity violation in nuclear levels. A good candidate level for time reversal would show large parity violation, a cross section larger than the potential scattering background, have an energy of a few electron volts, and be polarizable. The survey of parity violation would itself produce interesting results. A knowledge of the size, distribution, and signs of parity-violating weak matrix elements between states would lead to a better understanding of the questions of nuclear structure relevant to the interpretation of both parity violation and time-reversal violation.

REFERENCES

1. P. K. Kabir, Phys. Rev. D25, 2013 (1982).
2. V. E. Bunakov and V. P. Gudkov, Journal de Physique 45, 3 (1984).
3. V. E. Bunakov and Gudakov, Nucl. Phys. A401, 93 (1982).
4. L. Stodolsky, preprint (1985).
5. E. G. Adelberger, Intersection Between Particle and Nuclear Physics (1984), R. E. Mischke, ed., p. 300.
6. E. D. Commins and P. H. Bucksbaum, Weak Interactions of Leptons and Quarks, Cambridge University Press, Cambridge (1983), p. 262.
7. I. S. Altarev et al., Phys. Lett. B102, 13 (1981).
8. L. Wolfenstein, Nucl. Phys. B77, 375 (1974).
9. V. P. Alfimenkov et al., Pis'ma Zh. Eksp. Teor. Fiz. 35, 42 (1982), [JETP Lett. 35, 51 (1982)].
10. V. G. Baryshevskii and M. I. Podgoretskii, J. Exptl. Theoret. Phys. 47, 1050 (1964).

SPIN DEPENDENCE OF ρ° PRODUCTION IN $\pi^+ n_\uparrow \to \pi^+ \pi^- p$

M. Svec
Dawson College and McGill University, Montreal, P.Q., Canada
A. de Lesquen, L. van Rossum
Centre d'Etudes Nucléaires, Saclay, France

ABSTRACT

First measurements of $\pi^+ n_\uparrow \to \pi^+ \pi^- p$ on polarized target at 6 and 12 GeV/c provide information on momentum-transfer evolution of mass dependence of $\pi^+ \pi^-$ P-wave production amplitudes. Large and unexpected spin effects are observed.

First measurements of $\pi^+ n_\uparrow \to \pi^+ \pi^- p$ and $K^+ n_\uparrow \to K^+ \pi^- p$ on polarized target were performed by the Saclay group at CERN-PS at 6 and 12 GeV/c [1-3]. The experiment yields 15 spin density matrix elements for dimeson masses $m \lesssim 1$ GeV where only S- and P-waves contribute. These observables and $d^2\sigma/dmdt \equiv \Sigma$ can be expressed in terms of 2 S-wave and 6 P-wave recoil nucleon transversity amplitudes. In our normalization

$$d^2\sigma/dmdt = (|S|^2 + |\overline{S}|^2 + |L|^2 + |\overline{L}|^2 + |U|^2 + |\overline{U}|^2 + |N|^2 + |\overline{N}|^2)\Sigma$$

where $A = S,L,U,N$ and $\overline{A} = \overline{S},\overline{L},\overline{U},\overline{N}$ are normalized amplitudes with recoil transversity "down" and "up", respectively. The S-wave amplitudes are S and $\overline{S}$. The P-wave amplitudes $L,\overline{L}$ have dimeson helicity $\lambda = 0$ while $U,\overline{U},N,\overline{N}$ are combinations with $\lambda = \pm 1$. In terms of t-channel nucleon helicity amplitudes A_n, $n = 0,1$

$$A = (A_0 + iA_1)/\sqrt{2} \quad \text{and} \quad \overline{A} = (A_0 - iA_1)/\sqrt{2}$$

for $A = S,L,U$. For $N,\overline{N}$ the signs are opposite. In $\pi N \to \pi^+ \pi^- N$, the nonflip amplitudes S_0,L_0,U_0 exchange "A_1", the flip amplitudes S_1,L_1,U_1 exchange "π", and N_0,N_1 exchange "A_2". Amplitudes S and N are invariant under rotation from s- to t-channel frame.

Linear combinations of the measured s.d.m.e. give upper and lower bounds on the normalized moduli $|A|^2$ and $|\overline{A}|^2$ in each (m,t) bin[4]. The bounds are most restrictive on the moduli of P-wave amplitudes. The momentum-transfer evolution of their mass dependence is shown in the Figure. Production of ρ° with $\lambda = 0$ proceeds mostly with recoil nucleon transversity "up" and is decreasing with t. ρ° with $\lambda = \pm 1$ produced mostly via "A_2" exchange with recoil transversity "down" which increases with t. Unnatural exchange production of ρ° with $\lambda = \pm 1$ is small for $|t| < 0.5$. The naive expectation of resonant peak at ρ° of similar shape for all t in all P-wave moduli is contradicted by the observation of oscillation-like t-evolution of amplitudes $|L|^2, |\overline{L}|^2$ and $|N|^2, |\overline{N}|^2$ in the ρ° mass

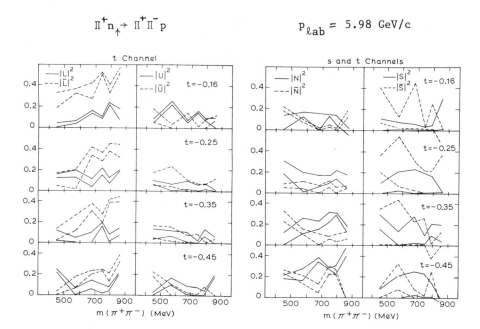

region. This behaviour reflects the zero structure and relative phases of the nucleon helicity amplitudes. The large difference $|L|^2 \neq |\bar{L}|^2$ indicates the presence of a strong A_1 exchange. The equality $|N|^2 \approx |\bar{N}|^2$ at $t \approx -0.45$ is observed also in KN CEX where it is due to zero in nonflip A_2 exchange amplitude[5]. This zero is confirmed also in $K^+ n_\uparrow \to K^+ \pi^- p$[4]. Complete account of this work, including estimate of error, bounds on cosines of certain relative phases, and the results for $K^+ n_\uparrow \to K^+ \pi^- p$ will appear elsewhere.

This work was supported by Fonds F.C.A.R., Ministère de l'Education du Québec, Canada and, in part, by Commissariate à l'Energie Atomique, CEN-Saclay, France.

REFERENCES

1. A. de Lesquen et al., CEN-Saclay Report, DPhPE 82-1 (1982).
2. A. de Lesquen et al., Phys. Rev. D32, 21 (1985).
3. A. de Lesquen et al., to appear.
4. M. Svec, A. de Lesquen, L. van Rossum, to appear.
5. M. Fujisaki et al., Nucl. Phys. B152, 232 (1979).

SPIN TRANSFER IN HYPERON PRODUCTION

Jeffrey Kruk[*]
T. W. Bonner Nuclear Laboratories
Rice University
P. O. Box 1892, Houston, TX 77251

ABSTRACT

The polarization transfer coefficient D_{NN} provides a sensitive test of present quark recombination models of hyperon production. In this paper we present preliminary results from a recent measurement of D_{NN} in the reaction $\vec{p}Be \to \vec{\Lambda}X$ at the Brookhaven AGS using the newly available polarized proton beam at incident momenta of 13 and 18 GeV/c.

It has long been known that inclusively produced hyperons are polarized, but the fundamental mechanism responsible for this polarization is still unknown.[1,2] DeGrand et al. have developed a very general framework for parametrizing the spin dependence of quark recombinations in baryon to baryon transitions, and have shown that the polarization transfer coefficients D are independent of the individual spin-dependent amplitudes in the limit of exact SU(6) symmetry. For the case of Λ production by protons, they predict both the analyzing power A and D to be zero. Λ's can also be produced indirectly through the decays of Σ^0's, for which D is predicted to be +2/3. Σ^0's are produced at about one third the rate of directly produced Λ's, and Λ's from Σ^0 decay have a polarization of -1/3 that of the Σ^0, so the observed Λ's are expected to have $A \approx 0.1$ and $D \approx -0.1$.

The experimental layout is shown in fig.1. Λ's are produced in the beryllium target, some of which decay between scintillators S4 and S5 to a pion and a proton. The pions and protons are momentum analyzed in the Brookhaven MultiParticle Spectrometer. Protons are identified by the segmented threshold Cerenkov counter C7 and hodoscope H7. The acceptance of the system covered a range $-.4 \lesssim X_F \lesssim .6$ and $.5 \leq P_T \leq 2$ GeV/c. The distribution of reconstructed Λ masses had a mean value of $M_\Lambda = 1115.7$ MeV/c^2 and a width of 2 MeV/c^2. The beam polarization was measured with a polarimeter consisting of a CH_2 target and horizontal and vertical scintillator telescopes, which was calibrated against the Univeristy of Michigan polarimeter. The average beam polarization was 60% at 13 GeV/c and 40% at 18 GeV/c. About 20% of the 18 GeV/c data have been analyzed. Fig.2 shows some previous data on the polarization of inclusively produced Λ's,[3] along with our present results. Our results are in good agreement with the previous data. The analyzing power is shown in fig.3. The mean value of $-.012 \pm .005$ is not consistent with the small positive A expected from Λ's arising from Σ^0 decay. D_{NN} is shown in fig.4. The mean value of $-.055 \pm .035$ is consistent with the predictions of the model of DeGrand et al. for Λ's produced in part by Σ^0 decay. Analysis of the remaining data will permit a study of the X_F and P_T dependence of D_{NN} and A.

*Work supported by the U.S. Dept. of Energy. E817 Collaboration: S.D. Baker, B.E. Bonner, J.A. Buchanan, J.M. Clement, M.D. Corcoran, I.M. Duck, N.M. Krishna, J.W. Kruk, H.E. Miettinen, R.M. Moss, G.S. Mutchler, F. Nessi-Tedaldi, M. Nessi, G.C. Phillips, J.B. Roberts, P.M. Stevenson, S.R. Tonse, Rice; A. Birman, S.U. Chung, R.C. Fernow, H. Kirk, S. Protopopescu, Brookhaven; Z. Bar-Yam, J. Dowd, W. Kern, E. King, Southeastern Massachusetts; B.W. Mayes, L.S. Pinsky, U. Houston; T. Hallman, L. Madansky, Johns Hopkins.

1. B. Andersson et al. Phys.Lett. 85B, 417 (1979).
2. T.A. DeGrand et al. Phys. Rev. D32, 2445.
3. K. Heller, High Energy Spin Physics-1982, AIP Conf.Proc. No.95, G. Bunce, ed. (AIP, NY, 1983), p. 320.

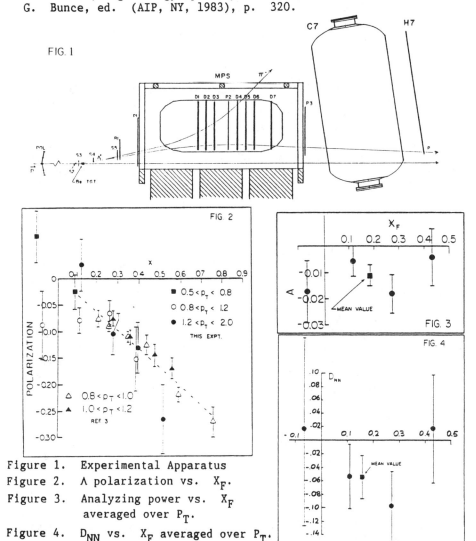

Figure 1. Experimental Apparatus
Figure 2. Λ polarization vs. X_F.
Figure 3. Analyzing power vs. X_F averaged over P_T.
Figure 4. D_{NN} vs. X_F averaged over P_T.

CHARGE SYMMETRY BREAKING IN THE n-p SYSTEM

L.G. Greeniaus*
TRIUMF, 4004 Wesbrook Mall, Vancouver, B.C. V6T 2A3

ABSTRACT

The status of charge symmetry breaking in the n-p system and related reactions is reviewed. Theoretical predictions and recent experimental results for the reaction d+d → ^{4}He+$\pi°$ are summarized, as are planned measurements in the same reaction and n+p → d+π. Experiments to measure the difference between the neutron and proton analyzing powers in n-p elastic scattering are described and results of the TRIUMF experiment are compared to theory.

INTRODUCTION

Recognizing that symmetries apply to the forces between interacting particles has helped in our understanding of those forces. Probing how these symmetries are broken reveals important details about the particle interactions. Isospin conservation is a good example. It was the first "internal" symmetry postulated for the nucleon-nucleon system[1]. At present, it is not clear if isospin conservation is a fundamental symmetry or an accident due to the smallness of the up and down quark masses. Extensive reviews of tests of isospin invariance of the NN force can be found in references 2-4.

Conservation of isospin in the strong interaction implies that the Hamiltonian is invariant under ARBITRARY rotations in isospin space.

$$[H_{nn}, T^2] = 0 \qquad [H_{nn}, T_3] = 0 \qquad (1)$$

This CHARGE INDEPENDENCE of nuclear forces forbids transitions between states of different isospin (T^2 is a good quantum number), and requires that in the same space and spin states the nn, pp and np forces be identical. Unlike parity or time reversal invariance, violation of isospin invariance is relatively easy to observe in nuclear systems. This is because isospin is not conserved by the electromagnetic (EM) or weak forces. Since EM effects are large, typically of the order of the fine structure constant, interpretation of isospin violation in the strong force is often ambiguous and difficult to establish. However, results such as the difference in the low energy nn and np scattering lengths indicate that charge independence in the NN force is violated to a small degree.

CHARGE SYMMETRY is a less restrictive variation of isospin invariance. It requires that the strong force be invariant for 180° rotations about any axis perpendicular to the charge axis (3-axis). The charge symmetry operator and its action on T_3 can be written:

$$P_{cs} = e^{i\pi T_2} \qquad P_{cs} |T_3\rangle = (-1)^{T+T_3} |-T_3\rangle \qquad (2)$$

---

* Permanent address: Nuclear Research Centre, Alberta, Edmonton, Alberta, CANADA, T6G 2N5

That is, it effectively transforms the quantum number T_3 into $-T_3$. This symmetry implies that the nn and pp forces are identical, but does not require that the np force be equal to the nn or pp forces.

Up to the present there has been no clear evidence for charge symmetry breaking (CSB) in the nucleon-nucleon force. CSB effects that are not explained by direct EM corrections have been observed, but either the interpretation is ambiguous or has some origin other than the NN force. In $\pi^{\pm}$ - d total cross section measurements, the energy dependent effect has large EM corrections and the remainder appears to originate in the mass differences of the charge states of the Δ-isobar[5]. The deviation from unity of the super-ratio obtained from $\pi^{\pm}$ scattering from ^{3}H and ^{3}He is not satisfactorily explained.[6,7]

Tests of CSB in the free NN system are potentially fruitful for several reasons. Firstly, the strong interaction amplitudes are (relatively) well understood. Secondly, the many body complications that arise in nuclei are avoided. Finally, for the free np system, the magnitude of EM effects is minimized, thereby making observable strong CSB effects easier to interpret. Tests of CSB in the NN system will allow us to test new aspects of meson exchange forces ($\Delta T=1$ terms), or allow us to probe "indirect" EM effects -- for example photon exchanges between quarks, or simultaneous photon - meson exchanges between nucleons. They may even provide us with a window to probe the SU(2) symmetry breaking intrinsic to the strong interaction which is manifested by the u-d quark mass difference.

CLASSIFICATION OF CSB TESTS

Tests of CSB can be divided into three convenient classes according to the properties of the system with respect to the charge symmetry operator.

1. Tests involving systems that are not eigenstates of P_{cs}

Class 1 tests require a comparison of two or more separate measurements made with mirror systems. Examples are: (i) comparison of the low energy scattering lengths a_{nn} and a_{pp} (equal within experimental and theoretical uncertainties), (ii) the 81±29 keV difference in the ^{3}H and ^{3}He binding energies (may be within the limits of theoretical uncertainties in the EM corrections)[8], or (iii) the $\pi^{\pm}$ - d total cross section differences mentioned earlier. In general, these comparisons have at least one system with two interacting charged particles. This results in large EM corrections compared to the strong interaction CSB effects.

2. Tests involving systems that are eigenstates of P_{cs}

For systems with $T_3 = 0$, charge symmetry prohibits even-odd isospin mixing even if charge independence is violated. Some of the immediate consequences of this are: (i) reactions like d+d → ^{4}He+π° are forbidden, (ii) reactions like n+p → d+π have $\sigma(\theta)$ symmetric about 90° in the center of mass, and the vector polarization observables are anti-symmetric about 90°, and (iii) polarization variables in n-p elastic scattering are restricted as discussed later.

3. Tests involving CSB along with other symmetries

G-parity breaking occurs in the mixing of the T=0 and T=1 mesons. The decay $\eta° \to 3\pi$ is interpreted via $\pi°-\eta°$ mixing. Similarly $\omega° \to 2\pi$ is interpreted via $\rho°-\omega°$ mixing. These measurements lead to constraints on the u-d mass difference. They also enter into the CSB tests that will be described below. Charge symmetry combined with time reversal invariance is used in comparing the neutron polarization with the proton analyzing power in the reaction $p+^3H \to n+^3He$.

THE n+p → d+π° REACTION

Charge independence requires that
$$2\, d\sigma/d\Omega\, [np \to d\pi°] = d\sigma/d\Omega\, [pp \to d\pi+]. \qquad (3)$$
This relation has been tested at the 5% level. Charge symmetry only requires that the differential cross section be symmetric about 90° in the center of mass. Cheung has calculated[9] the difference in the cross sections for deuterons emitted forward and backward in the c.m. If one defines the forward-backward asymmetry as
$$A_{fb}(\theta) = 2\, \frac{\sigma(\theta) - \sigma(\pi-\theta)}{\sigma(\theta) + \sigma(\pi-\theta)}, \qquad (4)$$
Cheung's calculations indicate that A_{fb} is a maximum (in magnitude) near 0° with a value $A_{fb} = -0.7\%$ that is nearly independent of energy between 400 and 800 MeV. The best measurement to date is by Hollas et al.[10] at 795 MeV who integrated A_{fb} from 0° to 90° and obtained $A_{fb} = -(0.30 \pm 1.00)\%$. Cheung's angle averaged prediction is −0.22%

This reaction is interesting because the CSB effect is dominated by $\pi-\eta$ mixing. The contribution depends on both the coupling constant $g_{\eta NN}$ and the $\pi-\eta$ mixing element. The former is uncertain to at least a factor of three, the latter to 10-20%. There are very few ways to access these quantities directly in experiments. Therefore the use of CSB effects as a probe appears very promising. At TRIUMF there is a proposal to determine A_{fb} at 0° with a precision of 0.1%[11]. A comparison measurement with the $p+p \to d+\pi^+$ reaction will be made to reduce systematic errors.

THE d+d → ⁴He+π° REACTION

The $d+d \to {}^4He+\pi°$ reaction is closely related to the $NN \to d\pi$ reaction. Production of a $\pi°$ involves large momentum transfer between the centers of mass of the deuterons. Quasi-free NN interactions are likely to play a large role because it is difficult to get that momentum transfer from the tail of the deuteron wave function alone. Cheung[12] has calculated the 0° cross section in the $\Delta(33)$ region, in the range 500 MeV $< T_d <$ 700 MeV where $NN \to N\Delta \to NN\pi$ is dominant. The $dd \to {}^4He\pi°$ cross section peaks at 600 MeV and 60% of the cross section comes from $\pi-\eta$ mixing, the largest remaining part coming from one photon exchange. The cross section is in the range 0.01 to 0.1 pb/sr and decreases as the energy increases beyond 600 MeV.

Measurements of d+d → ^{4}He+π° have been made recently at Saclay at incident deuteron energies of 800 MeV and 1350 MeV[13]. The ^{4}He was detected in the SPES-4 spectrometer in coincidence with at least one photon from the π° decay. The major experimental problems came from the 2π° production background. The preliminary results are:

800 MeV - $d\sigma/d\Omega$ (100° cm) < 0.8 pb/sr at a 90% confidence level based on a possible 7 events which are consistent with background

1350 MeV - $d\sigma/d\Omega$(77° cm) < 5 pb/sr at a 68% confidence level based on 1 event in the π° region.

These results are to be compared to the previous best limit of 19 pb/sr[14]. A new generation of experiments is planned at Saclay to lower this limit to the 0.1 pb/sr level or better[15]. Model independent predictions[16] at 1950 MeV indicate a d+d→^{4}He+π° cross section of 0.12 ± 0.05 pb/sr at 146° cm. This calculation is based on the observed cross section for dd→^{4}Heη and includes only contributions from π-η and π-η' mixing.

CSB TESTS IN n-p ELASTIC SCATTERING

Charge symmetry imposes restrictions on polarization observables in the n-p system. If charge symmetry is valid, then the analyzing power A_n for polarized neutrons scattered from unpolarized protons, equals the analysing power A_p when the protons are the polarized particles; Then $A_n(\theta) = A_p(\theta) = A(\theta)$. If there are CSB forces in the n-p system we can have $\Delta A = A_n - A_p \neq 0$. In an experiment where the left-right scattering asymmetries are measured with beam and target polarized SEPARATELY normal to the scattering plane one has

$$\varepsilon_B - \varepsilon_T = A(\theta) [P_B - P_T] + \frac{1}{2} \Delta A [P_B + P_T]. \quad (5)$$

A similar expression is obtained if the scattering asymmetry $\varepsilon_{\pm}+$ is measured with the beam and target polarized simultaneously in opposite directions. The only difference is that the spin correlation C_{nn} enters in a term $(1-C_{nn} P_B P_T)$ as a divisor to (5) above. Since the CSB term ΔA is small, errors in measuring beam and target polariztions can be important. In the region where A = 0, it is not necessary to know P_B or P_T accurately. This makes the region of the analyzing power crossover a favourable one for experimental determination of ΔA. When beam and target are polarized simultaneously, knowledge of the ratio P_B/P_T is needed where $A \neq 0$. The only other experimentally accessible CSB effect in the n-p system is the spin correlation parameter difference $(C_{zx}-C_{xz})$.

Two experimental groups are attempting to determine ΔA, one at IUCF[4] and the other at TRIUMF[17,18]. These experiments have been described in some detail in earlier conferences. The IUCF experiment is performed at a neutron beam energy of 188 MeV incident on a "spin refrigerator" polarized proton target. The detection system consists of two symmetric wire chamber, scintillation counter telescopes which detect the proton and neutron in coincidence. The counters are placed so that they cover the angular range 60° < θ_{cm} < 120°. The crossover angle occurs at 96° at 188 MeV. The experiment uses simultaneously polarized beam and target. The goal is to measure $\Delta A(\theta)$ at six points with a precision of ± 0.001. The IUCF group have obtained some preliminary data. They are now working on controlling systematic errors which arise from polarization components that are transverse to the scattering plane, and are improving the magnitude of the polarization attainable with their target.

The TRIUMF experiment was performed with a 477 MeV incident neutron beam energy incident on a polarizable hydrogen target of the frozen spin type. In this experiment measurements were made with beam and target polarized in separate stages of the experiment. Protons and neutrons were detected in coincidence in telescopes that subtended the angle range $60° < \theta_{cm} < 80°$. Symmetric detector systems were placed to observe both left and right scattered events. The experiment was designed as a null experiment to accurately determine the difference in the crossover angle of the neutron and proton analyzing powers. Only the beam and target polarizations were varied during the experiment, a technique which allows most systematic errors to be eliminated in comparing the neutron and proton analyzing power results. The value of ΔA is obtained only at the crossover angle. While less ambitious than the IUCF experiment, it probably allows for much closer control over systematic errors. The result obtained is[19]

$$\Delta A = A_n(\theta_0) - A_p(\theta_0) = 0.0037 \pm 0.0017 \ (\pm 0.0008 \text{ systematic}).$$

The systematic error is a worst case estimate, and analysis is continuing to reduce it.

Significant theoretical progress has been made recently and new calculations have been made available by three groups[20,21,22]. Ge and Svenne (GS), and Miller, Thomas and Williams (MTW) perform non-relativistic potential calculations, while Iqbal, Thaler and Woloshyn have performed a relativistic calculation where the np mass difference enters via the propagator and in the external wave functions. ITW have not evaluated the EM or $\rho-\omega$ contributions. They find important effects when π, σ and ω exchanges are included in their calculation. Controversy over the sign of the term due to the np mass difference in OPEP has been resolved and calculations have been done for two pion exchange, ρ exchange, $\rho-\omega$ mixing, short range quark effects as well as one photon exchange and OPE. A very condensed synopsis of these calculations is given in the table. For more information the forthcoming publications should be consulted.

Table 1
Contributions to $\Delta A \times 10^4$ from various calculations

Term	477 MeV			188 MeV		
	MTW	GS	ITW	MTW	GS	ITW
EM	6	22	–	10	9	–
OPEP	34	60		6	13	
TPEP	−0.4	–	14	−0.5	–	8
ρ	9±3	–		−1.2±0.4	–	
$\rho-\omega$	5±1.2	−3	–	3±0.8	4	–
TOTAL	54±4	84	25	17±1	27	21
Experiment		37±17(±8)				

This first evidence for CSB effects in the np system has already lead to tests of new aspects of the OPEP and possible speculation about sensitivity to the ρNN coupling constants. Clearly the dust must settle in the theoretical arena. It will be interesting to see the development of this topic when results of the IUCF experiment are available as well as a new proposed experiment at TRIUMF at 350 MeV.[23]

References

1. W. Heisenberg, Zeits. F. Physik 77, 1 (1932)
2. E.M. Henley in Isospin in Nuclear Physics, ed. by D.H. Wilkinson, p16 (1969 North-Holland).
3. E.M. Henley and G.A. Miller, Mesons in Nuclei, ed. by M.R. Rho and D.H. Wilkinson, Vol I, p406, (1979 North-Holland).
4. S.E. Vigdor, Int. Conf. on Current Problems in Nuclear Physics, Crete, Greece, 1985.
5. E. Pedroni et al, Nucl. Phys. A300 321 (1978).
6. B.M.K. Nefkens et al, Phys. Rev. Lett. 52 735 (1984).
7. S. Barshay and L.M. Sehgal , Phys. Rev. C31, 2133 (1985)
8. P. Langacker and A.D. Sparrow, Phys. Rev. C25, 1194 (1982).
9. C.Y. Cheung, et al., Phys. Rev. Lett. 43 1215 (1979) and Nucl. Phys. A348 365 (1980).
10. C.L. Hollas et al., Phys. Rev. C24, 1561 (1982).
11. L.G. Greeniaus et al, TRIUMF experiment #368.
12. C.Y. Cheung, Phys. Lett 119B, 47 (1982).
13. F. Plouin and L. Goldzahl, private communication. Saclay experiment #105, F.Plouin, spokesman.
14. J. Banaigs et al, Phys. Lett. 53B, 390 (1974).
15. F. Plouin and B. Preedom are spokesmen for the future Saclay experiments.
16. S.A. Coon and B.M. Preedom, Phys. Rev. C33, 605 (1986).
17. R. Abegg et al., Nucl. Instrum. Methods A234, 11 (1985).
18. R. Abegg et al., Proc. 6th Int. Symp. on Polarization Phenomena in Nuclear Physics, Osaka, 1985, 369.
19. R. Abegg et al., accepted for publication in Phys. Rev. Lett., June 1986
20. L. Ge and J.P. Svenne, Phys. Rev. C33, 417 (1986), and erratum (in press, June 1986).
21. G.A. Miller, A.W. Thomas and A.G. Williams, accepted for publication in Phys Rev. Lett. June 1986.
22. J. Iqbal, J. Thaler and R. Woloshyn, private communication
23. TRIUMF Experiment #369, R. Tkachuk, L.G. Greeniaus and W.T.H. van Oers, cospokesmen.

SPIN OBSERVABLES IN PROTON DEUTERON ELASTIC SCATTERING[#]
E. Bleszynski[+], M. Bleszynski[+] and T. Jaroszewicz[*]
[+]Physics Department, University of California, Los Angeles, CA90024
[*]Lyman Laboratory of Physics, Harvard University, Cambridge, MA02138

Abstract

Some interesting aspects of spin observables in medium energy proton-deuteron scattering are disussed. It is argued that measurements of spin observables in this reaction can provide new information on the three-body (contact) interactions. Such an information is not available from measurements of NN observables and can be obtained from the analysis of the 12 amplitudes which describe p-d elastic scattering. We discuss the sensitivity of the spin observables to such three-body interactions in the framework of a realistic model in which each of these amplitudes is separated into 1) " trivial part", involving single and double scattering terms, unambiguosly expressible in terms of physical nucleon-nucleon (NN) scattering amplitudes and 2) the "nontrivial part" which can not be expressed in terms of two-body interactions, and represents contributions generated by three-body forces. We find that the agreement between the data and the model predictions is considerably improved when such three-nucleon interaction are included. The presence of these three-body forces, of the type of contact interactions, is also required by the high energy asymptotic behavior of the amplitudes involved, and can be, to some extent, determined from soft-pion theorems. Much more information on them should become available from the systematic analysis of the spin structure of the p-d scattering amplitude and constrains given by the data on spin observables.

***

The purpose of this talk it to discuss what new information can be extracted from measurements of spin observables in proton-deuteron elastic scattering at intermediate energies.

With the recent advances in experimental technique it has become possible to measure spin observables involving various combinations of the target and beam polarizations in the intitial and final states. There is considerable amount of current activity on both experimental[1,2] and theoretical[3,4,5] sides of this reaction and new experiments in LAMPF, Saturne, TRIUMF, IUCF and KEK accelerators are being planned.

The spin structure of the proton-deuteron collision matrix is very rich. We deal with the collision of two objects of spin 1/2 and spin 1 respectively. From the general symmetry arguments it follows that the collision matrix for the elastic proton-deuteron scattering is composed of 12 independent complex amplitudes and can be decomposed suitably in terms of products of the projectile spin $\vec{\sigma}$, the deuteron spin $\vec{J}$ and quadrupole moment operators $Q_{ik} = 1/2(J_i J_k + J_k J_i) - 2/3 \delta_{ik}$ as follows[3]:

$$F = F_o^o + F_y^o J_y + F_{xx}^o Q_{xx} + F_{yy}^o Q_{yy} + F_x^x \sigma_x J_x + F_{xy}^x \sigma_x Q_{xy} +$$
$$+ F_o^y \sigma_y + F_y^y \sigma_y J_y + F_{xx}^y Q_{xx} \sigma_y + F_{yy}^y Q_{yy} \sigma_y + F_z^z J_z \sigma_z + F_{yz}^z Q_{yz} \sigma_z .$$

Here the x-axis is parallel to the momentum transfer and the y-axis is normal to the scattering plane. These 12 independent, complex amplitudes $F_0^0,\ldots,F_{yz}^z$ determine the p-d collision matrix. In order to deterime the p-d collision matrix up to an arbitrary phase, it is therefore necessery to perform 23 independent measurements. The relations between the spin observables which we denote as C and the 12 subamplitudes are given by:

$$C(\alpha,i;\beta,k) = \text{Tr}\{F\,\sigma_\alpha\,O_i\,F^+\sigma_\beta\,O_k\}/\text{Tr}\{F\,F^+\},$$

where $O_i = 1, J_i, Q_{mn}$, $m,n=x,y,z$.

A number of spin observables has been already measured in the rangle of energies 500- 1000 MeV[1] . The most extensive set of spin observables exists at 800 MeV. These include C(0,SS;0,0), C(0,NN;0,0), C(0,N ;0,0), C(N,0 ;N,0), C(S,0 ;S,0), C(L,0 ;L,0), C(S,0 ;L,0), C(L,0 ;S,0), C(S,SS;0,0), C(S,L ;0,0), C(S,S ;0,0), C(L,L ;0,0), C(N,SS;0,0), C(N,NN;0,0), C(N,N ;0,0), C(S,S ;0,0), C(L,L ;0,0), C(S,N ;S,0), C(N,N ;N,0), C(L,N ;L,0), C(S,N ;L,0), C(L,N ;S,0), C(S,NN;S,0), C(N,NN;N,0), C(L,NN;L,0), C(S,NN;L,0), C(L,NN;S,0), in the typical range of momentum trasfers squared $0 < -t < .8$ $(\text{GeV/c})^2$. The letters L and N appearing as two first arguments denote, in accordance with the Madison convention, the directions parallel to the initial beam momentum and normal to the scattering plane. The letter S appearing as the first argument corresponds to the direction forming the right-handed coordinate frame (N,L,S). The letters N,L and S appearing as the third or fourth arguments of the C observables correspond to the right handed coordinate frame with L taken along the scattered beam momentum. With the recent data from Saturne, which are currently analyzed[1] the total number of observables at 800 MeV will be 35). We note that some of above mentioned observables are not linearly independent.

Spin observables are attractive because the offer a possibility of selective study of various effects which contribute with different strenghts to p-d 12 transition amplitudes. Each of the 12 amplitudes F_k^i describing the p-d scattering depends in general on different ingredients of the reaction theory, like different spin components of the NN amplitude or different deuteron formfactors. Thus, for example the observables C(0,SS;0,0) and C(0,NN,0,0) depend mainly on the D-wave component of the deuteron wave function. Also by taking suitable linear combinations of the observables $C(\alpha,0;\beta,0)$ it is possible to construct quantities which depend selectively on the double spin flip components of the NN amplitude[6].

However, the most interesting aspect of investigating spin effects in proton-deuteron scattering is the potential possibility of finding the effects effects which cannot be explained in terms of a superposition of the underlying two-nucleon interactions but, might require inclusion of other degrees of freedom, like three-body forces. Such effects contribute to the double collision term and turn out to be strongly spin dependent.

The double collision term plays a very important role, in particular for momentum transfers $-t \geq .3$ $(\text{GeV/c})^2$. In order to illustrate its importance, in Fig. 1 we display the data for the observable C(0,SS;0,0), for 800 MeV incident proton energy ,together with the theory predictions obtained by using the complete p-d collision amplitude (solid lines) and the single collision approximation (dashed lines). The contributions from the single and double collision terms enter with different strenghts to each of the amplitudes F_k^j.

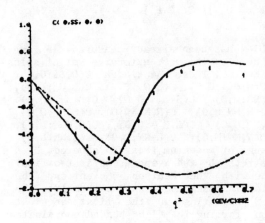

Fig 1. The observable C(0,SS;0,0) (measured by the UCLA collaboration) compared with the model predictions without (dashed lines) and with the double collision term (solid lines).

From the inspection of the structure of the 12 amplitudes describing p-d elastic scattering[3] it follows that 4 of them: the amplitudes F_{yy}^0, F_{yy}^y, F_{xy}^x, and F_{yz}^z have a very attractive property –they do not contain the single scattering contributions.

In order to extract the information on these 4 amplitudes a complete set of of 23 measurements should be performed. Current accuracy of the data and the fact that some of them cover very limited ranges of momentum transfers does not allow us yet to obtain such an information. With the increasing supply of new data[1] such a task should be possible in the near future.

The present theoretical effort[3] is focused on using the spin observables to extract information on nontrivial three-nucleon information, not available from two-nucleon scattering data. Our approach consists of separating each of the 12 amplitudes describing p-d elastic scattering into sums "trivial" and a "nontrivial" components:

$$F_k^j = F_k^{j,trivial} + F_k^{j,nontrivial}$$

The "trivial" parts are constructed from the single and double scattering terms, unambiguosly expressible in terms of physical nucleon-nucleon (NN) scattering amplitudes. The uncertainties in the existing phase shift analyses of the NN amplitudes are believed to be very small in the range of energies up to 800 MeV[7]. Consequently the "trivial" parts of the 12 amplitudes can be evaluated quite accurately.

The "nontrivial" parts can not be expressed in terms of two-body interactions, and require some additional dynamical input. Each amplitude F_k^j can be, in general, written, as a sum of invariant amplitudes A_i (free of kinematical singularities), multiplied by appropriate spin kinematical factors. The amplitudes A_i are functions of several invariants, the most relevant being $s=(k_i^1+p_1-p_1')^2$ and the crossed- channel variable $u=(k_i^1+p_2-p_2')^2$, where k_i^1 is the projectile initial momentum and p_1, p_1', and p_2, p_2' are the initial and scattered momenta of the two target nucleons respectively.

From general analycity and unitarity principles we know that the amplitudes A_i must have poles at $s=m^2$ an $u=m^2$ (m is the nucleon mass), corresponding to the nucleon intermediate states, e.g.,

$$A_i(s,u,\ldots) \simeq \frac{B_{i_2} B_{i_1}}{m^2-s} \text{ for } s \simeq m^2.$$

Clearly the residues of these poles factorize into products of factors B_{i_1} and B_{i_2} expressible in terms of physical elastic scattering amplitudes[2]. The model involving the minimum amount of dynamical input is just this: invariant amplitudes are assumed to be pure pole terms, with residues determined from (2→2) nucleon scattering. We refer to it as to the "two-body model" or as to the "trivial part"

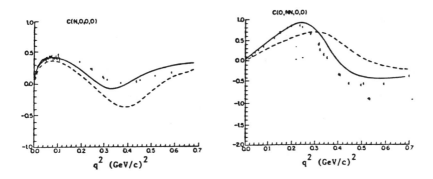

Fig.2 The data for deuteron vector and tensor analyzing powers compared with the model predictions with 2-body interactions (dashed lines) and those in which 3-body interactions were included (solid lines).

Although the invariant amplitudes are pure poles, the poles may be cancelled by spin kinematical factors, and non-pole terms may, and in general will, appear in the full amplitude F. It is these non-pole terms in F which can be interpreted as resulting from the propagation of the virtual negative-energy part of the (antinucleon states). The reason for such an interpretation is obvious: the negative energy part of the propagator has no pole at $s=m^2$ (this argument can be made more precise to show that there is a one-to-one correspondence between pole (non-pole) parts of F and positive-(negative)- energy intermediate states. Consequently, significant virtual antinucleon effects are un anavoidable consequence of the "two-body model".

We find that the model based on two-body interactions only is inadequate for two reasons:

(i) From the unitarity and analycity it follows that the invariant amplitudes must have, beside the poles, inelastic cuts in s and u, resulting from resonances and multiparticle intermediate states. The corresponding s- and u- discontinuities have a high energy behavior which, generally requires subtractions in dispersion relations for $A_i(s,u,...)$. The subtraction constants indroduce extra non-pole terms which cannot be determined from the unitarity alone, and certainly not from (2→2) -nucleon scattering data. In this sense these non-pole terms represent independent dynamical information characteristic of the (3→3) nucleon amplitude, and can be therefore interpreted as genuine three-body effects.

(ii) The model based on solely two-body NN interactions turns out to be in a quite poor agreement with experiments on spin observables. In Fig 2. we present the data for 2 spin observables: C(N,0,0,0) and C(0,NN,0,0) together with the curves resulting from the calulation in which only two-body interactions were taken into account (dashed curves) and those in which three-body interactions were added (solid curves). (The data have been measured by the UCLA collaboration.)

In the above mentioned calculation[5] the p-d scattering amplitude was assumed to be given as a sum of multiple scattering terms representing the single and the double collision processes. Both term in the series can be expressed in terms of : (a) vertex functions describing projectile-target nucleon interaction, (b) projectile propagator and (c) target wave function (together with the relativistic impulse aproximation prescription).

The double collision term we can be written schematically as:

$$F_d = \frac{-1}{2m\pi^2} \int d^3p \; \langle e^{i[(\vec{k}_i+\vec{k}_f)/2 - \vec{p}]\cdot\vec{r}} \; F_{NN}(\vec{k}_f,\vec{p}) \; \frac{1}{\slashed{p}-m} \; F_{NN}(\vec{p},\vec{k}_i) \rangle$$

where $\langle...\rangle$ denotes the average taken with the deuteron wave function, $(\slashed{p}-m)^{-1}$ is the projectile Dirac propagator and the NN vertices describing the first and the second collision of the projectile with the target nucleons are assumed to have the following forms:

$$F_{NN}(\vec{p}',\vec{p}) = \sum_{n=1}^{5} (1+\xi_n \frac{\not{p}'-m}{2m}) \hat{O}_n F_n(s,t) (1+\xi_n \frac{\not{p}-m}{2m}).$$

In the above expression $\hat{O}_n$, $n=1,5$ are the operators acting in the four dimensional Dirac spaces of the projectile constructed from the scalar, vector, tensor, pseudoscalar and axial operators. The target nucleons are described in terms of a nonrelativistic wave function. The subamplitudes F_n, $n=1,5$ are functions of two independent Lorentz scalars $t = -(\vec{p}-\vec{p}')^2$, where $\vec{p}$ and $\vec{p}'$ are the initial and final momenta of the projectile, and $s = (p+p')^2$. The on-shell (i.e. on mass shell) part of the above amplitude is uniquely determined by being consistent with a phase shift analysis of NN scattering data. Terms involving ξ_i do not contribute to the on-mass-shell NN scattering, but contribute to the double collision term in which the integration over the intermediate projectile momentum covers all values. Each term $\xi_n(\not{p}-m)$ entering into expression for F_{NN} generates a nonpole term in the amplitude F_d ("nontrivial part") which can be interpreted as three-body contact interactions.

We conclude that the three- body forces understood in the above sense are very important in p-d scattering. We stress that their existence has been established in a model independent way, since the two-body interactions were taken into account without approximations, i.e. from the phase shift compilation of the NN data.

As a future program we plan to perform a systematic phenomenological analysis of the spin structure of the elastic and inelastic p-d scattering amplitudes and of the constrains given by the data on spin observables, in order to extract more information on the non-trivial genuine three-body part of the (3→3)-nucleon amplitude.

REFERENCES

1. G. Igo, UCLA collaboration, private communication.
2. M. Bleszynski, in AIP Conference Proceedings 128,(1984) 220.
3. G. Alberi, M. Bleszynski, and T. Jaroszewicz, Ann. Phys. 142 (1982) 299.
4. D. L. Adams and M. Bleszynski Phys. Lett. 136B (1984) 10
5. E. Bleszynski and M. Bleszynski, and T. Jaroszewicz, to be published.
6. M. Bleszynski Phys. Lett., 106B (1981) 42.
7. R. A. Arndt and L. D. Roper, Phys. Rev D 25 (1982) 2011.

RECENT TOPICS ON THE SPIN PHYSICS AT KEK

Akira Masaike
Department of Physics, Kyoto University
Kyoto, Japan
and
National Laboratory for High Energy Physics
Tsukuba, Ibaraki, Japan

ABSTRACT

Some of the recent topics on the polarization phenomena at KEK are discribed. They include the experiments which are being carried out using the proton synchrotron, the booster synchrotron and other facilities of KEK. The phase shift analysis of nucleon-nucleon scattering is also mentioned briefly.

INTRODUCTION

There are many topics concerning the spin at KEK, because the spin observables are related to the fundamental properties of elementary particles and nuclei. In this report I will be able to mention only a few projects, which are now going on or recently got interesting results. Serveral important topics have to be omitted because of the limit of the time.

First, I will try to explain the slow neutron polarization using the polarized proton filter and its applications.

Then I will show a new Hoshizaki's phase shift analysis of nucleon-nucleon scattering, which includes recent experimental data.

The project of the polarized proton beam at KEK will be mentioned.

SLOW NEUTRON POLARIZATION AND ITS APPLICATION

i) SLOW NEUTRON POLARIZATION USING POLARIZED PROTON FILTER

The neutron beam with the energy between 10^{-2} eV and 10 MeV could be polarized by the polarized proton filter, because the cross section for the neutron-proton scattering in the singlet state is substantially larger than that in the triplet state. Such an idea to produce polarized slow neutron beam has several advantages over the traditional methods, especially in obtaining epithermal polarized neutrons. High intensity polarized beam can be obtained using this method. An experiment of the polarization of slow neutron using an transversely polarized proton filter was made and it showed the usefulness of the filter as an effective polarizer or analyzer.[1]

Recently N. Hoshizaki and the author suggested that the neutron can be polarized in the direction of the beam using a longitudinally polarized proton filter.[2] The degree of the polarization is nearly the same as that in the case using the transversely polarized proton filter in the neutron energy lower

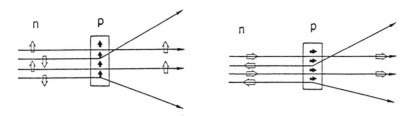

Fig. 1. Schematic view of the slow neutron polarization with the polarized proton filter.

than 1 MeV and higher in the energy higher than 1 MeV.

An experiment of longitudinally polarized filter was carried out using a pulsed neutron source (KENS) at KEK booster synchrotron.[3] In this experiment, the same technique as for the polarized proton target in high energy physics experiments was used. The protons in a polycrystalline sample of ethylene glycol containing a few percent of Cr^V complexes were dynamically polarized by means of microwave of 70 GHz in the magnetic field of 2.5 T at the temperature of 0.4 K. A superconducting solenoid and a horizontal He^3 cryostat were used, which were installed in a neutron beam line. Typical polarization of neutrons at 50 meV was 70%. The results show that the longitudinally polarized neutron will be quite useful for several applications.

Recently neutron beam intensity was much improved, using a He^3-He^4 heat exchanger which avoided the absorption of the neutron in the He^3.[4]

ii) MEASUREMENT OF T AND P VIOLATIONS USING POLARIZED SLOW NEUTRONS

A few years ago, E. G. Adelberger pointed out the possibility of the test of T-invariance using the resonance scattering of the polarized slow neutron on the polarized nuclear target.[5] If the component of T-violation exists, longitudinally polarized neutron spin rotates in the plane including the direction of the target spin, because the $\vec{\sigma} \cdot (\vec{k} \times \vec{I})$ term relates to the T-violation, where $\vec{\sigma}$ is the neutron spin, k is the neutron wave number and $\vec{I}$ is the spin of target nucleus. The effect of the final state interaction can be avoided, if the 0°-scattering is measured.

Recently, a large parity violation at the 0.75 eV neutron resonance of lanthanum-139 was observed in an experiment on the transmission of polarized neutrons.[6] It was pointed out that the mixing of the P-wave and S-wave causes the parity-violation, since it is a P-wave resonance. It is natural to assume that in the case of large P-violation T-violation is also large.

At KEK an experiment of the test of P-violation is being carried out in the reaction of slow neutron resonance absorption on the Xe-129 using the polarized neutrons produced by the method mentioned in §2, (i).[7] The Xe-129 has a P-wave resonance at 9.4 eV. Furthermore, it can be polarized by the optical pumping method. Therefore, the Xe-129 is one of the candidates to use for a test of T-violation.[8] The resonance peak of 9.4 eV was clearly

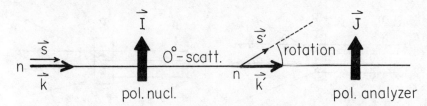

Fig. 2. Test of the T-violation by rotation of the spin direction of longitudinally polarized neutron after 0°-scattering off the transversely polarized nucleus.

observed in the first test run.

We hope that the test of T-violation can be made in the measurement of the spin rotation of the neutron by the resonance absorption of the neutron on the polarized nuclear target of Xe using a good polarization analyzer.

iii) THE NEUTRON SCATTERING FROM CLUSTER OF POLARIZED PROTONS

One of the applications of the neutron scattering from the polarized proton is the measurement of microscopic inhomogeneity of the dynamic polarization.

The cross section of the elastic scattering of the slow neutron from the polarized nucleus is

$$\frac{d\sigma}{d\Omega} = \left|\sum_\ell b_\ell \exp(ik, R_\ell)\right|^2 + \left|\sum_\ell B_\ell P_\ell \exp(ik, R_\ell)\right|^2 + \sum_\ell \left\{\frac{I_\ell + 1}{I_\ell} - P_\ell^2\right\} B_\ell^2$$

where b_ℓ is the coherent scattering length of the nucleus, B_ℓ is the "polarization" scattering length, I_ℓ is the nuclear spin and P_ℓ is the nuclear polarization. If the nuclei are polarized, the second term of the equation increases, while the third term decreases. If the size of the polarized hydrogen cluster is larger than the neutron wave length, the small angle scattering is enhanced.

The molecule of the dicyclohexyl-18-crown-6 ($C_{20}H_{32}O_6$) doped with EHBA-Cr^V was put in the fully deuterated propanediol. The small angle scattering amplitude was found to be enhanced, if the proton in the molecule of $C_{20}H_{32}O_6$ was polarized. The scattering amplitudes on deuterons and other nuclei around $C_{20}H_{32}O_6$ were very small. We calculated the size of the molecule from the measurement. The size measured by this method is about 400Å. It is consistent with the result from X-ray diffraction.

This method will be applicable for the wide field of molecular biology, and for the study of polymers and other large molecules.

N-N SCATTERING AND NEW HOSHIZAKI'S PHASE SHIFT ANALYSIS

In 1985 several experimental results of the proton-proton

scattering were reported. Some of them are crucial for the phase shift analysis.

The measurement of A_{yy} and A_{zz} in the Coulomb interference region is important to determine the real part of the forward double-spin-flip amplitudes. The experiment was performed with LAMPF-HRS and KEK-spin-frozen-Target at 1.46 GeV/c and 1.28 GeV/c.[10] Angular range was 4.5°∿30° c.m.

In addition to the results of A_{yy} and A_{zz} in the Coulomb interference region, new results[11] of $\Delta\sigma_L^{yy}$ from LAMPF, of D, R, A, R' and A' parameters from LAMPF[12], of A_{yy} and A_{zz} parameters from Saclay[13], and of $\Delta\sigma_L$ (inelastic) from ANL[14] are included in the new version (May, 1986) of Hoshizaki's energy independent PSA.[15]

Here we show the Argand plots of $^1D_2$, $^3F_3$, $^3P_0$, $^3P_1$ and $^3P_2$. The plots of $^1D_2$ and $^3F_3$ are almost the same as the versions of 1984 and of 1985, and confirm the resonance-like loopings in the both waves. On the other hand, no resonance-like-loop was found in the $^3P_0$, $^3P_1$ and $^3P_2$ waves. Possible existence of a resonance in $^3P_0$, $^3P_1$ or $^3P_2$ was suggested previously, because the

$k^2\left[(1+A_{zz})\frac{d\sigma}{d\Omega}\right]_{90°}$ and $k^2\left[(A_{yy}-A_{zz})\frac{d\sigma}{d\Omega}\right]_{90°}$ curves show structures and these quantities contain spin-triplet partial waves in $J = L \pm 1$.[16] Further analysis is required for these waves.

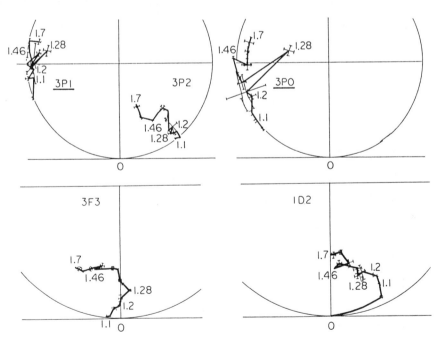

Fig. 3. Argand plots of $^1D_2$, $^3F_3$, $^3P_0$, $^3P_1$ and $^3P_2$ waves from Hoshizaki's PSA.

PRESENT STATUS OF POLARIZED PROTON BEAM AT KEK

The project of the acceleration of the polarized beam has been carried out for the last few years.[17]

A polarized proton ion source using a new technique of the optical pumping was successfully operated in 1983.

The acceleration of the beam in the 500 MeV booster synchrotron was tested in 1984 and it was found that the polarization of the beam can be maintained during the acceleration using the method of the spin flip at energies of strong intrinsic depolarization resonance (260 MeV) and imperfection resonance (180 MeV).

After the long shut down period of the proton synchrotron, the test of the acceleration in the main ring of the 12 GeV synchrotron started. In March of 1986 the depolarization in the first strong intrinsic resonance ($G = 12 - \nu_z$) at 2.08 GeV was found to be avoided by the tune jump method.[18] In this experiment the beam polarization was monitored by a polarimeter in the tunnel of the main ring which could be operated during the acceleration. The method of the polarization measurement during the acceleration will be able to be applied to several purposes.

The number of strong depolarization resonances in the main ring is about 10. The spin flip method can be applied to some of these resonaces. Although the present value of the polarization at 2.5 GeV is rather low ($\sim 25\%$), the value will hopefully increase by tuning parameters of the accelerator.

REFERENCES

1. S. Hiramatsu et al. : J. Phys. Soc. Jpn. <u>45</u> (1978) 949.
2. N. Hoshizaki and A. Masaike : Jpn. J. Appl. Phys. <u>25</u> (1986) L244.
3. S. Ishimoto et al. : Jpn. J. Appl. Phys. <u>25</u> (1986) L246.
4. Y. Masuda et al. : to be published.
5. E. G. Adelberger : J. Phys. Soc. Jpn. <u>54</u> (1985) Suppl. I, 6.
6. V. P. Alfimenkov et al. : Nucl. Phys. A <u>398</u> (1983) 93.
7. Y. Masuda : private communication.
8. K. Morimoto et al. : proposed experiment at KEK.
9. M. Ishida et al. : KENS-Report 1984.
10. M. Gazzaly et al. : Proc. 6th Int. Symp. Pol. Phenom. in Nucl. Phys., Osaka, (1985) 814.
11. W. P. Madigan et al. : Phys. Rev. D. <u>31</u> (1985) 966.
12. M. L. Barlett et al. : Phys. Rev. C. <u>32</u> (1985) 239.
13. J. Bystricky et al. : Nucl. Phys. B <u>258</u> (1985) 483.
14. I. P. Auer et al. : Phys. Rev. Lett. <u>51</u> (1983) 1411.
15. N. Hoshizaki : private communication.
16. G. R. Burleson et al.; Nucl. Phys. B <u>213</u> (1983) 365.
17. S. Hiramatsu : Proc. 6th Int. Sym. High Energy Spin Physics (Marseille)(1984), C2-511.
18. S. Hiramatsu and K. Imai : private communication.

TENSOR ANALYZING POWER IN πd ELASTIC SCATTERING

G.R. Smith
TRIUMF, 4004 Wesbrook Mall, Vancouver, B.C., Canada V6T 2A3

Two years ago at Steamboat Springs, I reviewed the status of spin observables in πd elastic scattering. In the meantime, much new information has been gathered, and the overall picture has changed dramatically.

One of the original motivations for studying spin observables in πd elastic scattering was the observation of resonances in nucleon-nucleon scattering in $\Delta\sigma_L$ and $\Delta\sigma_T$. The existence of these highly inelastic resonances has been more firmly established in the meantime, both through further measurements and more detailed phase-shift analyses (the situation is reviewed in Ref. [1]). Their width, however, is similar enough to that of the delta that their internal structure cannot be ascribed definitely to six quark bags. Such an assignment would rely heavily on the observation of a narrow (10-20 MeV) resonance, of which only one potential candidate has been seen. It is with respect to this candidate, observed in the πd elastic channel, that most of the experimental attention has been focused recently.

As long ago as 1982, a group from ETH Zurich claimed to have seen[2] evidence for a dibaryon resonance in measurements of the tensor polarization, t_{20}(lab), in πd elastic scattering at SIN. The evidence consisted of violent angular distributions of t_{20}(lab) at T_π=134 MeV, which flattened out at neighboring bombarding energies as close as 15-20 MeV. Their results for t_{20}(lab) at all energies were mostly positive, with almost maximal t_{20}(lab) values at 120° and 150° at 134 MeV. Theoretical predictions, on the other hand, were mostly negative and had a smooth dependence on energy and angle.

From the beginning, these results were highly controversial, not only because of the exciting physics they implied, but because they were in complete disagreement with t_{20}(lab) measurements[3] made with a similar apparatus by a group consisting mainly of people from ANL at LAMPF. The LAMPF results were entirely negative, and had no violent energy or angular dependence. The results of a third experiment,[4] recently carried out at TRIUMF, were in agreement with the LAMPF data. However, the ETH group at SIN had in the meantime reproduced[5] their earlier results with a completely redesigned polarimeter.

All of the above experiments employed the same experimental technique for measuring t_{20}(lab), namely, recoil deuterons from πd elastic scattering were analyzed in a second scattering with a ^{3}He cell polarimeter via the ^{3}He(d,p)^{4}He reaction. Obviously, a completely different, independent experimental approach was required to resolve the controversy.

The most ideal solution would be to measure the tensor analyzing power, T_{20}(cm), in a single scattering experiment with a tensor polarized deuteron target. Such an experiment has just been completed at TRIUMF. As it resolves unambiguously the controversy laid out above, I shall dwell on it for the majority of this presentation.

The (spherical) tensor analyzing power is obtained from the following simple expression, valid when the target magnetic field is aligned parallel to the incident beam momentum:

$$T_{20}(\text{cm}) = \frac{\sqrt{2}}{P_{zz}} \left(\frac{\sigma_p}{\sigma_0} - 1 \right) \quad (1)$$

where P_{zz} is the target tensor polarization (Cartesian), and $\sigma_p(\sigma_0)$ the relative πd elastic differential cross section measured with the target polarized (unpolarized). This observable can then be related to the tensor polarization of the recoil deuteron, t_{20} (lab), measured in the double scattering experiments via the relation

$$t_{20}^{\text{lab}}(\theta_d) = T_{20}^{\text{cm}}(\theta_d) d_{00}^2(\theta_d) + 2T_{21}^{\text{cm}}(\theta_d) d_{10}^2(\theta_d) + 2T_{22}^{\text{cm}}(\theta_d) d_{20}^2(\sigma_d) \quad (2)$$

where the $d_{ik}^i(\theta_d)$ are the usual Wigner d functions.

The TRIUMF tensor analyzing power measurement employed a tensor polarized deuteron target consisting of frozen 1 mm diameter deuterated butanol beads contained in a teflon basket measuring 16×16×5 mm^3. The basket was immersed in ^{3}He/^{4}He in the mixing chamber of a dilution refrigerator. The polarizing field of 2.5 T was provided by a superconducting split pair solenoid with a magnetic field axis along the incident beam direction. The average target tensor polarization (P_{zz}) achieved was 0.085±0.008. This value was checked with three independent techniques. The first technique involved a comparison of the area of the dynamically polarized NMR signal to the area of the thermal equilibrium NMR signal. The second technique used the asymmetry of the dynamically polarized NMR signal to deduce P_z. Both of these techniques utilized the relationship between the vector polarization, P_z, and P_{zz} given by $P_{zz} = 2 - \sqrt{4 - 3P_z^2}$. The third technique involved a direct measurement of P_{zz} by utilizing the known tensor analyzing power at 90° cm in the πd to 2p reaction. All three techniques gave consistent results for the target tensor polarization.

The detection system used for the πd elastic scattering measurements of the tensor analyzing power consisted of an array of plastic scintillator telescopes which measured the TOF, energy loss, and total energy of deuterons in coincidence with scattered pions. Six angles could be measured simultaneously in this arrangement. The experimental arrangement is shown in Fig. 1. A typical spectrum of deuteron TOF vs deuteron E+DE is shown in Fig. 2. The polygon encloses the πd elastic events. The other band identifies protons arising from πp quasielastic reactions and absorption. The deuterons are cleanly separated from these other particles.

A direct comparison of T_{20}(cm) with t_{20}(lab) requires a knowledge of T_{21} and T_{22}. As these observables have not yet been measured, calculated values of T_{21} and T_{22} from Garcilazo[6] are used in Eq. (2) with the measured T_{20}(cm) in order to compare to the earlier double scattering measurements of t_{20}(lab). The conversion is required because the t_{20}(lab) measurements are done in a coordinate system in which the z-axis is along the direction of the outgoing deuteron momentum in the lab system. The T_{20}(cm) measurements, on the other hand, are performed in a coordinate system in which the z-axis is along the incident beam direction, which is the same for cm and lab frames. The results are

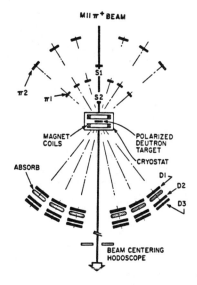

Fig. 1. The experimental layout is shown, with the pion beam incident from the top. Each of six identical arms consists of a pion telescope (π1 and π2) in coincidence with an associated deuteron telescope (D1, D2 and D3). Deuteron energy loss is measured in D1. The absorber thickness is adjusted so that deuterons stop in D2 at a given angle. The D3 counter supplies a veto signal. The incident beam is counted directly with the coincidence of S1 and S2.

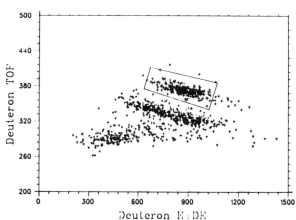

Fig. 2. A typical two-dimensional spectrum of deuteron TOF (vertical axis) vs deuteron E+DE is shown for a pion bombarding energy of 180 MeV. The polygon encloses the πd elastic events.

shown in Fig. 3, together with the earlier double scattering data. The TRIUMF tensor analyzing powers are in agreement with the tensor polarizations from LAMPF and TRIUMF, but they are not consistent with the tensor polarizations of the SIN experiment.

The effect of the uncertainties in the magnitude of T_{21} and T_{22} on the conversion from T_{20}(cm) to t_{20}(lab) can be assessed as follows. A model independent comparison can be made by using the maximum physically possible bounds on T_{21} and T_{22} in Eq. (2). The ensuing band of allowable t_{20}(lab) determined from the T_{20}(cm) data and these bounds is completely negative for cm angles greater than 145 degrees, where the SIN t_{20}(lab) data reach positive values as high as +0.6.

In summary, the dibaryon candidate suggested by the SIN t_{20}(lab) data has turned out to be a phantom. At this point in time, therefore,

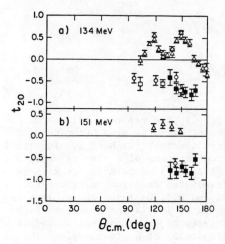

Fig. 3. The tensor analyzing power ($T_{20}(cm)$) angular distributions (solid squares) are compared to the existing tensor polarization ($t_{20}(lab)$) data from SIN (Ref. [2]) (open triangles), from LAMPF (Ref. [3]) (open circles), and from TRIUMF (Ref. [4]) (open square), at 134 MeV (a), and 151 MeV (b). The conversion of $T_{20}(cm)$ to $t_{20}(lab)$ is done by admixing calculated $T_{21}(cm)$ and $T_{22}(cm)$ values to the measured $T_{20}(cm)$ (see text).

no narrow resonances exist which might be construed as dibaryon candidates in the πd elastic channel. Although discrepancies remain between three body calculations and measured values of the vector[7] and tensor analyzing powers, they are not serious enough to mandate the addition of dibaryon resonances. The discrepancies in both iT_{11} and T_{20} seem to indicate rather that the presently accepted splitting of the P_{11} πN partial wave (which accounts for pion absorption) into nucleon pole and non-pole parts is not being done correctly. The parameterization of pole and non-pole parts is difficult since their energy dependence and off-shell behaviour are unknown. Together they cancel almost completely to give the small on-shell P_{11} phase. But in the three body calculations, the Pauli principle forbids the pole contribution for certain partial waves, thus the non-pole part has a major impact in these cases. The sensitivity to the P_{11} term in the calculations may be gauged by a comparison to calculations in which this term is left out. As Fig. 4 demonstrates, calculations[8] in which pion absorption is left out entirely agree with the T_{20} data much better than the full calculation[8] which includes the P_{11} and pion absorption. Finally, calculations[6] with reduced contributions from the P_{11} partial wave lie in between the other two, as expected.

If one assumes that the P_{11} splitting is at the source of the discrepancies between theoretical calculations and experimental data, how does one reconcile this with the fact that several resonances seem firmly established in the nucleon-nucleon channel? This question has been addressed in Ref. [9]. It was found that if a resonance of spin J couples with lower pion orbital angular momentum ℓ_π=J-1 to the πd channel, then only minimal effects arise in the vector and tensor analyzing powers. If, on the other hand, upper coupling dominates ℓ_π=J+1, then the effects on iT_{11} and T_{20} are much more pronounced. The observed behaviour of iT_{11} and T_{20} are, therefore, consistent with the nucleon-nucleon results if indeed lower coupling dominates. The absence of oscillations in iT_{11} and T_{20} may have an even more trivial explanation. Recent calculations[10] suggest that dibaryon coupling to the πd channel is smaller than first expected.

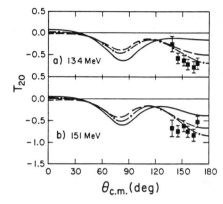

Fig. 4. The tensor analyzing power data at 134 MeV (a) and 151 MeV (b) are compared to three body Faddeev calculations. The solid curves (full calculation) and dash-dot curves (no P_{11} rescattering and no absorption) are from Blankleider and Afnan (Ref. [8]). The dashed curves are from Garcilazo (Ref. [6]) (reduced contribution from pion absorption.

For the future, measurements of T_{21} and T_{22} in πd elastic scattering, together with more complete measurements of T_{20}, will form the basis for a more reliable phase shift analysis than has been possible until now. Hopefully, these new observables may also shed some light on the P_{11} problem. Together with information on spin observables being gathered in the absorption and breakup channels, the new πd elastic scattering data will provide a solid foundation on which to test the three body calculations of these fundamental reactions. However, as no narrow resonances have yet been established in the πd channel, it seems unlikely that these data will provide evidence for exotic six quark states. More promising on theoretical grounds would be the investigation of the strangeness -1 and -2 channels, for which several potential dibaryon candidates have already been observed. Such a search would be greatly facilitated by the existence of a kaon factory.

REFERENCES

1. D.V. Bugg, Prog. Part. Nucl. Phys. 7, 47 (1981);
   D.V. Bugg, Nucl. Phys. A416, 227c (1984);
   C. Lechanoine-Leluc, J. Phys. (Paris) Coll. C2, 46, 399 (1985).
2. J. Ulbricht et al., Phys. Rev. Lett. 48, 311 (1982);
   W. Gruebler et al., Phys. Rev. Lett. 49, 444 (1982);
   V. Koenig et al., J. Phys. G9, L211 (1983).
3. R.J. Holt et al., Phys. Rev. Lett. 43, 1229 (1979);
   R.J. Holt et al., Phys. Rev. Lett. 47, 472 (1981);
   E. Ungricht et al., Phys. Rev. Lett. 52, 333 (1984);
   E. Ungricht et al., Phys. Rev. C31, 934, (1985).
4. Y. M. Shin, et al., Phys. Rev. Lett. 55, 2672 (1985).
5. Swiss Institute for Nuclear Research Annual Reports (NL18), (1984).
6. H. Garcilazo, Phys. Rev. Lett. 53, 652 (1984).
7. G.R. Smith et al., Phys. Rev. C29, 2206 (1984).
8. B. Blankleider and I.R. Afnan, Phys. Rev. C24, 1572 (1981).
9. K. Kubodera and M.P. Locher, Phys. Lett. 87B, 169 (1979);
   K. Kubodera et al., J. Phys. G6, 171, (1980).
10. W. Grein et al., Nucl. Phys. A356, 269, (1981);
    H.G. Dosch and E. Ferreira, Phys. Rev. C32, 496 (1985).

Radiative Capture of Polarized Protons and Deuterons
by the Hydrogen Isotopes*

W.K. Pitts
Indiana University Cyclotron Facility
Bloomington, Indiana 47405, USA

ABSTRACT

Several measurements of polarized light ion radiative capture at the IUCF have either been completed or are being analyzed. Polarized deuteron capture was used to study the D-state components of the ^{3}He and ^{4}He wavefunctions with the reactions ^{1}H$(d,\gamma)^3$He and ^{2}H$(d,\gamma)^4$He; the tensor analyzing powers observed in these reactions have been found to be very sensitive to the D-state components of the residual nucleus. Polarized proton capture was used to study the "quasi-deuteron" effect and its signature in the analyzing power for the reaction ^{2}H$(p,\gamma)^3$He. Both measurements illustrate the importance of spin observables in elucidating details of nuclear structure or reaction mechanisms in few nucleon systems.

INTRODUCTION

My charge today is to review recent radiative capture experiments in light nuclei at the IUCF. My involvement in this program was limited to E 234, a study of the reaction ^{2}H$(d,\gamma)^4$He. The results and interpretation presented here for E 207, ^{2}H$(p,\gamma)^3$He, are the work of M.A. Pickar and his collaborators.

E 234: TENSOR ANALYZING POWERS AND D-STATE COMPONENTS

Experiment E 234 was a study of the reactions ^{1}H$(d,\gamma)^3$He and ^{2}H$(d,\gamma)^4$He with tensor polarized deuterons. The tensor analyzing powers for these reactions are given (to a very good appproximation) by interference terms involving the D-state component of the residual helium nucleus. The major "contaminant" in the tensor analyzing power is the deuteron D-state which contributes no more than a few percent of the tensor analyzing power. This process should be one of the most reliable methods for determination of the ^{4}He D-state component[1]. The ^{2}H$(d,\gamma)^4$He reaction has some other interesting features. With identical spin-one bosons in the entrance channel and a 0^+ ground state all the odd electric multipoles are forbidden (assuming only central nuclear forces). Since ^{4}He is self conjugate, isospin selection rules result in magnetic transitions weaker than expected, with the M1 in particular reduced by a factor of about 100. The E2 multipole is dominant resulting in a $\sin^2 2\theta$ shape for this cross section[2]. Additional simplifications arise from the isoscalar nature of the reaction, resulting in a suppression of some meson exchange currents (especially those involving the Δ-isobar).

At lower energies (E_d=10 MeV) this simple picture is not totally correct. A recent measurement[3] found that M2 radiation must be added to complete the picture. At these low energies there are strong initial state interactions due to excited states of ^{4}He and it is likely that this large M2 component is resulting from an initial state resonance[4]. At our energy (E_d=95 MeV) these initial state interactions should be much less important.

To isolate the radiative capture events the photon and the helions were detected in coicidence (Fig. 1). The photons were identified in lead glass Cherenkov detectors while plastic scintillator telescopes cleanly separated radiative capture events from other processes. The final spectra were very clean, with less than 1% background under the ^{1}H(d,γ)^{3}He peak (Fig. 2). A more complete description is contained in a recent IUCF Annual Report[5].

About half of the ^{2}H(d,γ)^{4}He data have been analyzed. The relative cross section (normalized to $\sin^2 2\theta$ at 52°) is shown in Fig. 3. At 90° the cross section is not zero, and even if the detector acceptance is taken into account the cross section is still an order of magnitude larger than that expected for a pure $\sin^2 2\theta$ distribution. At very low energies (E_d=2 MeV) this enhancement can be attributed to D-state effects[6] but at these higher energies it is not clear which physical process is filling the minimum.

The data analysis of the ^{1}H(d,γ)^{3}He reaction is nearly complete. Our data are shown in Fig. 4. Systematic errors (to within an overall luminosity uncertainty) are included for the cross section while only statistical errors are shown for the analyzing powers. The overall uncertainty in the analyzing power data is 5% from the uncertainty in the the polarimeter calibration. Also plotted are the results of a PWBA calculation[7] for this process. This model includes the E1 and E2 multipoles with a ^{3}He wavefunction generated from a Faddeev calculation[8] of ^{3}H with the Reid soft core potential. A value of η (the asymptotic D/S state normalization ratio) of -0.029 was derived by comparison to the asymptotic Hankel functions at 6 fm. This value of η is low; a recent calculation[9] including three body forces gave an η of -0.043. The overall agreement with our data is quite good, especially for the tensor analzying power A_{yy}. This value for η also results in a good agreement with another measurement[7] of A_{yy} at a photon center of mass energy of 15 MeV. A calculation[10] at this energy using a full Faddeev treatment of both the entrance and exit channels also finds that the expected value of η overpredicts A_{yy}, with a disagreement of 20%. A puzzling feature of these calculations is that the results are qualitatively similar for distinctly different ^{3}He configurations. In the PWBA calculation[7] all of the tensor analyzing power is given by the projection of ^{3}He into a deuteron and a proton in a relative D-state. In the full Faddeev treatment only 9% of the tensor analyzing power arises from this partition while 89% is from the partition into a two nucleon cluster (relative L=1) orbited by the third nucleon in a relative L=1 orbit about the other two. These are distinctly different partitions and it is not clear why these different configurations give such similar results.

The measured vector analyzing power A_y is small and is within two standard deviations of zero at most angles. It is expected to be non-zero because of distortions in the incoming channel[7].

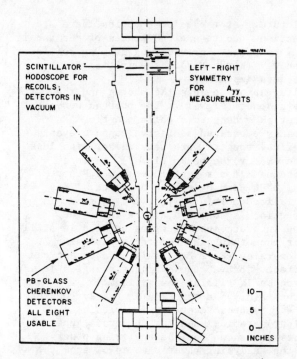

Fig. 1. Experimental apparatus for the (d,γ) measurements.

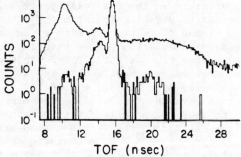

Fig. 2. Typical time of flight spectrum for the $^1H(d,\gamma)^3He$ reaction, with and without radiative capture cuts. FWHM for ^3He peak is 500 psec.

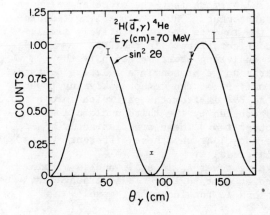

Fig. 3. Preliminary relative cross section for the $^2H(d,\gamma)^4He$ reaction, normalized at 52° to a $\sin^2 2\theta$ distribution.

1227

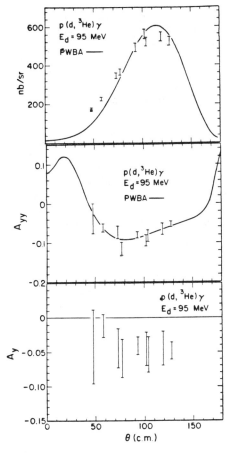

Fig. 4. Preliminary results for the $^1H(d,\gamma)^3He$ reaction

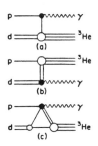

Fig. 5. Terms contributing to $^2H(p,\gamma)^3He$. A and B are the Born terms, C is the triangle term.

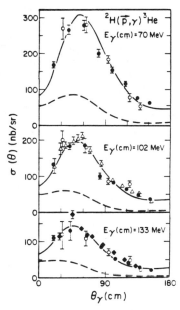

Fig. 6. Cross section for $^2H(p,\gamma)^3He$. Data from:
● IUCF E 207   ○ Ref. 14
◆ Ref. 15   △ Ref. 16

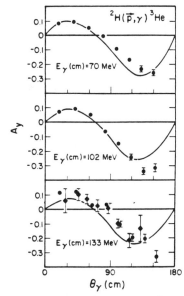

Fig. 7 Analyzing power for $^2H(p,\gamma)^3He$, data as Fig. 6

E 207: THE "QUASI-DEUTERON" MECHANISM IN ^{2}H$(p,\gamma)^3$HE

The goal of experiment E 207 was to investigate the influence of the quasi-deuteron mechanism or the "triangle" diagram (fig. 5c) in proton capture processes. It was known that the Born terms (fig. 5a and 5b) alone resulted in satisfactory fits to the data at low energies[11] while inclusion of the triangle term[12] was needed to explain existing cross section data at photon energies of about 100 MeV. Analyzing powers were also to be measured in the hope that as the quasi-deuteron mechanism became more dominant the analyzing power would resemble the analyzing power in the basic n$(p,\gamma)^2$H reaction.

The cross section data are shown in fig. 6 with calculations using the amplitude of Prats[12]. A calculation with only the Born terms (dashed curve) results in a poor fit to the data, especially for the backward angles and higher momentum transfers. When the triangle term is added (solid curve) the calculation agrees with the data very well. Fig. 7 shows the analyzing power data. The predicted A_y from Partovi's calculation[13] of n$(p,\gamma)^2$H at the same center of mass photon energy is also plotted. Good agreement is seen at almost all angles and energies. The conclusion must be that the triangle term is the dominant term at these momentum transfers.

ACKNOWLEDGEMENTS

Many people were involved in these experiments. M.A. Pickar, H.J. Karwowski, J.D. Brown, J.R. Hall, M. Hugi, R.E. Pollock, V.R. Cupps, M. Fatyga and A.D. Bacher participated in E 207. H.O. Meyer, L.C. Bland, J.D. Brown, R.C. Byrd, M. Hugi, H.J. Karwowski, P. Schwandt, A. Sinha, J. Sowinski and I.J. van Heerden collaborated on E 234. Valuable discussions with F.D. Santos are greatly appreciated.

REFERENCES

* Supported in part by the National Science Foundation
1. T.E.O. Ericson and M. Rosa-Clot, Annu. Rev. Nucl. Part. Sci. 1985, 35, 271 (1985)
2. H.R. Weller, J. Phys. Soc. Jpn. 55, suppl. 113 (1986)
3. S. Mellema et al, Phys. Lett. 166B, 282 (1981)
4. F.D. Santos, personal communication
5. 1985 IUCF Annual Report 1985 (p. 46)
6. J.L. Langenbruneer et al, Bull. Am. Phys. Soc. 31, 786 (1986)
7. A. Arriaga and F.D. Santos, Phys. Rev. C29, 1945 (1984)
8. T. Sasakawa and T. Sawada, Phys. Rev. C19, 2035 (1979)
9. S. Ishikawa and T. Sasakawa, Phys. Rev. Lett. 56, 317 (1986)
10. J. Joudan et al, Phys. Lett. 162B, 269 (1985)
11. B.A. Craver et al, Nucl. Phys. A276, 237 (1977)
12. F. Prats, Phys. Lett. 88B, 23 (1979)
13. F. Partovi, Ann. Phys. (NY) 27, 79 (1964)
14. N.M. o'Fallon et al, Phys. Rev. C5, 1926 (1972)
15. J.M. Cameron et al, Nucl. Phys. A424, 549 (1984)
16. J.P. Didelez et al, Nucl. Phys. A143, 602 (1970)

A SPIN-SPLITTER FOR ANTIPROTONS IN LOW ENERGY STORAGE RINGS

Y. Onel[1], A. Penzo[2] and R. Rossmanith[3]

[1] DPNC, University of Geneva, Geneva, Switzerland
[2] INFN and University of Trieste, Italy
[3] CERN, Geneva, Switzerland. On leave from DESY, W. Germany

Abstract: A method to polarize antiprotons in LEAR or other storage rings via Stern-Gerlach effect is suggested. The method can be tested with protons circulating in the same machine.

INTRODUCTION

Even with the advent of LEAR, and its $\bar{p}$ beams of superior quality, the perspectives for obtaining a complete description of the N-$\bar{N}$ amplitudes are limited, mainly because of the lack of polarized antiprotons.

To overcome this deficiency, several methods have been suggested for polarizing antiprotons. Most of these methods, like scattering on hydrogen or nucleon targets (1) differential absorption on a polarized target (2) or polarization of antihydrogen (3), are of limited practical value as they involve severe losses of intensity and modest final beam polarization. A much more appealing scheme, which does not involve, as in the previous ones, a large reduction of flux in the course of polarization, has been recently proposed (4) for high energy machines and is based on the spatial separation of circulating particles with opposite spin directions, by means of the Stern-Gerlach effect in the field gradients of quadrupoles in the machine. Although the effect of the field gradient on the magnetic moment is extremely small compared to the effects on the charge of the circulating particle, the choice of specific coherence conditions between the orbit and spin motions, allow a constructive build-up of spacial separation, leading to a mechanism of self-polarization in storage rings.

In order to achieve a build up over many revolutions, the spin tune and betatron frequency must be, modulo 2π, identical. This can be expressed by

$$a \cdot \gamma = Q \pm n \qquad \ldots (1)$$

where $a=(g-2)/2(=1.793$ for protons), $\gamma = (1-\beta^2)^{-1/2}$ is the Lorentz factor, n is an integer.

The field of a quadrupole magnet is described to first order by

$$B_x = b \cdot y$$
$$B_y = b \cdot x \qquad \ldots(2)$$

in a coordinate system where x and y are the transverse horizontal and vertical axis respectively and z the axis along the design orbit. The force acting on a magnetic dipole is

$$F = \nabla(\vec{\mu} \cdot \vec{B}) = b(\mu_x y + \mu_y x) \qquad \ldots(3)$$

and the deflections of the particles in the x-z and y-z planes are, respectively

$$x = b\mu_y L/p\beta$$
$$y = b\mu_x L/p\beta \qquad \ldots(4)$$

where μ_x and μ_y are the projections of the magnetic moment along the x and y directions, L is the length of the quadrupole and p is the momentum of the beam. We note that there is no deflection when the spin points in the direction of the beam. When the spin points into the horizontal plane the particle is vertically deflected. When the spin points into the vertical direction the spin is horizontally deflected.

In the following we present a scheme suitable for low energy storage rings, which can be, under certain circumstances, quite simple. For concreteness, we discuss a possible layout for LEAR.

THE LAYOUT FOR LEAR

For LEAR, a machine designed for low energies, the polarizing element is concentrated: the so-called spin-splitter (5). This spin splitter consists of two quadrupoles and a solenoid in between

placed in one straight section of the ring. The quadrupoles have opposite polarity. In between the quadrupoles a solenoid rotates the spin by 180 degrees around the direction of particle motion.

The characteristics of this configuration are the following:

a.) The existing machine lattice is not changed and the spin-splitter configuration can be made transparent th the rest of the ring for the orbit motion (see end of this chapter). Furthermore large separations between particles with opposite spin can be obtained by using strong superconducting quadrupoles.

b.) The spin tune is affected in a characteristic way by the spin splitter because the solenoid behaves as a "siberian snake" for the low energy storage ring. Each second revolution the spin aims for each particle into the same direction. The noninteger part of the spin tune is therefore 1/2 and independent on the energy of the particle.

c.) The solenoid would introduce a coupling between the horizontal and vertical plane which would affect the beam behaviour in the machine.

This effect can be most simply cured by rotating the quadrupoles of the spin-splitter by 45° (skew quadrupoles) and adjusting their fields with respect to the solenoid field in order to minimize such coupling.

THE BUILD UP TIME FOR POLARIZATION

The separation time can be deduced from the formula (4); μ is $1,41 \cdot 10^{-23}$ erg/Gauss[7]. Going to low energies the separation speed increases. If a momentum of 200 MeV/c is assumed and a gradient of 20 Tm^{-1}, 1m long, the deflection per turn is $42 \cdot 10^{-15}$ rad and for 1 second ($7,8 \cdot 10^5$ turns) $3,35 \cdot 10^{-8}$ rad. Particles with opposite spin direction are separated with a speed of 2.5mm/hour (a quarter betatron wavelength of 10m was assumed). As the beam in LEAR has vertical dimensions of this order, this separation speed seems to be quite adequate. The energy for applying the build up of polarization has to be chosen as a compromise between separation rate and emmittande growth; at 600 MeV/c, for example, the separation speed is three quarters of the speed at 200 MeV/c but the emittance is a factor of two smaller.

Before any further consideration of the possible implementation of the described scheme, it seems of great interest to perform a test with protons stored in LEAR, in order to establish the feasibility of the proposed method (5).

REFERENCES

1. R. Birsa et al., Phys. lett. 155B (1985) 437

2. H. Dobbeling et al., Proposal P92, CERN/PSCC/85-80

3. K. Imai, Preprint KUNS 801, July 1985

4. T.O. Niinikoski and R. Rossmanith, CERN-EP/85-46(1985)
   T.O. Niinikoski and R. Rossmanith, Proc. of the Workshop on Polarized Protons at SSC and on polarized Antiprotons, Eds. A.D. Krisch, A.M.T. Linn and O. Chamberlain, to be published in AIP Conf. Proc. Series (1986)

5. Y. Onel, A. Penzo, R. Rossmanith, CERN/PSCC/86-22, PSCC/M258.

MEASUREMENT OF THE VECTOR ANALYSING POWER IN THE πd-BREAKUP.

W. Gyles, E.T. Boschitz, H. Garcilazo, E.L. Mathie,
C.R. Ottermann and G.R. Smith
Kernforschungszentrum Karlsruhe, Institut für Kernphysik and
Institut für Experimentelle Kernphysik der Universität Karlsruhe,
7500 Karlsruhe, Federal Republic of Germany

S. Mango and J.A. Konter,
Schweizerisches Institut für Nuklearforschung, 5234 Villigen,
Switzerland.

R.R. Johnson
TRIUMF, University of British Columbia, Vancouver, B.C.,
Canada V6T 2A3 .

ABSTRACT

The vector analysing power, iT_{11}, has been measured for the $\pi^+ d \to \pi^+ pn$ reaction in a kinematically complete experiment. The dependence of iT_{11} on the momentum of the proton has been obtained for 35 pion-proton angle pairs at pion bombarding energies of 180 MeV, 228 MeV and 294 MeV. The data are compared with a relativistic Faddeev calculation.

INTRODUCTION AND EXPERIMENT

In the πNN system, those reactions with 3-body final states have been largely neglected compared to the 2-body reactions. While the additional experimental and theoretical difficulties may be intimidating, these reactions are an integral part of the πNN system and must be included in tests of a consistent 3-body theory. Also, it is possible that non-conventional dynamics (not included in the 3-body theory) may show themselves preferably in these reactions[1]. An advantage of the 3-body final state is that different aspects of the reaction dynamics are emphasised in different parts of the phase space. This allows one to concentrate on particular graphs, while supressing the effects of others. This experiment was performed on the πM3 beamline at SIN. The polarized target and detector arrangement have been described in detail elsewhere[2,3].

RESULTS AND DISCUSSION

The iT_{11} were calculated from σ^+ and σ^-, the cross sections with positively and negatively, vertically polarized target, and σ^0, the background cross section, using
$$\frac{\sigma^+ - \sigma^-}{\sqrt{3}\,(P^-\sigma^+ + P^+\sigma^- - (P^++P^-)\sigma^0)}$$
The cross sections are the threefold differential cross sections $d\sigma/d\Omega_\pi\, d\Omega_p\, dp_p$. A sample of the data at 228 MeV is shown in Figure 1. The points where the momenta and angles are equal to those

of free πp scattering are indicated by asterices and those where the invariant masses of the final state proton and pion are equal to the Δ(1232) mass are indicated by arrows. The solid lines are the results of Faddeev calculations and the dotted lines the results of impulse approximation calculations.

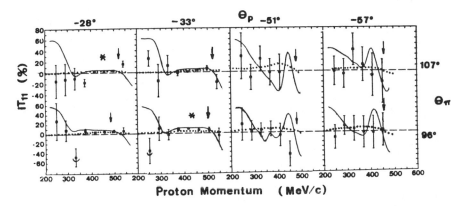

Fig. 1. Sample of iT_{11} at T_π = 228 MeV. Solid lines are Faddeev calculations and dotted lines are impulse approximation.

The agreement between the calculations and the data is generally good. The positive sign of iT_{11} near πp scattering kinematics reflects the positive sign of the πp polarization parameter. Indeed, exactly on πp kinematics the iT_{11} values at the 3 energies studied correspond very closely to the πp polarization values, P, if we use the relationship $iT_{11}=P/\sqrt{3}$. For large values of the final state proton momentum, the D state of the deuteron is sampled more. The spin of the proton is opposite to the spin of the deuteron in the D state and the sign of iT_{11} is therefore negative here. At low proton momenta at angles where the neutron and proton are close together, the calculated iT_{11} rise due to the final state interaction between the proton and neutron. The data suggest this rise, but are not sufficiently precise to allow a detailed comparison with the calculation in these regions. An interesting feature of the calculations is the sharp peak in the region of the Δ(1232) at large pion and proton angles. At these angles, the neutron recoils with large momentum and at a large angle to both the pion and the proton. Hence, any rescattering with the neutron is minimised here, and this is just the region therefore, where any effects of the Δ would be expected to be clearest. More precise data in this region should be possible with the higher target polarizations now available.

REFERENCES

1) W. Grein, K. Kubodera and M.P. Locher, Nucl. Phys. A356, 269 (1981)
2) G.R. Smith et al., Phys. Rev. C29, 2206 (1984).
3) G.R. Smith et al., Phys. Rev. C30, 980 (1984).

2nd CONFERENCE ON THE INTERSECTIONS BETWEEN
PARTICLE AND NUCLEAR PHYSICS
Lake Louise, Canada - May 26-31, 1986

PLENARY LECTURES

Monday - May 26, 1986

	Chairman-L. Rosen, Los Alamos
8:30 a.m.	E. W. Vogt, TRIUMF
	Welcome
8:45 a.m.	A. D. Krisch, Michigan
	Introduction
9:00 a.m.	D. H. Wilkinson, Sussex
	Symmetry Tests in Particle and Nuclear Physics
9:50 a.m.	Coffee
	Chairman-R. E. Taylor, SLAC
10:20 a.m.	S. E. Koonin, Caltech
	QCD and Lattice Gauge Theory
11:00 a.m.	B. D. Winstein, Chicago/Stanford
	CP Phenomenology
11:40 a.m.	Lunch
	Chairman-L.C. Teng, Fermilab
8:30 p.m.	H. A. Grunder, CEBAF
	Superconducting Accelerator Technology
9:10 p.m.	A. M. Sessler, Berkeley
	Future Accelerator Technology
9:50 p.m.	M. K. Craddock, Univ. of British Columbia/TRIUMF
	High Intensity Proton Accelerators
10:30 p.m.	End

Tuesday May 27, 1986

	Chairman-V. W. Hughes, Yale
8:30 a.m.	R. G. Arnold, SLAC/American Univ.
	EM Structure of Nuclei
9:10 a.m.	P. D. Barnes, Carnegie-Mellon
	Searches for Strange Matter
9:50 a.m.	Coffee
	Chairman-A. Masaike, Kyoto
10:20 a.m.	G. P. Lepage, Cornell
	Bound Between Perturbative and Non-perturbative QCD
11:00 a.m.	J. S. Greenberg, Yale
	Anomalous Positron Production
11:40 a.m.	Lunch

Wednesday May 28, 1986

	Chairman-M. Bardon, U.S. N.S.F.
8:30 a.m.	R. G. H. Robertson, Los Alamos
	Neutrino Physics Review
9:10 a.m.	A. Gal, Hebrew University/TRIUMF
	Overview of Hypernuclear Interactions
9:50 a.m.	P. G. Langacker, Pennsylvania
	The Standard Model and Beyond
10:30 a.m.	End
10:45 a.m.	Excursion to Columbia Icefield

Thursday May 29, 1986

	Chairman-T. A. O'Halloran, Illinois
8:30 a.m.	E. L. Berger, Argonne
	The EMC Effect
9:10 a.m.	M. Simonius, ETH, Zurich
	High Precision Parity Tests
9:50 a.m.	Coffee
	Chairman-E. J. Moniz, MIT
10:20 a.m.	G. Karl, Guelph
	Quarks in Spectroscopy
11:00 a.m.	T. W. Ludlam, Brookhaven
	Relativistic Heavy Ion Collisions
11:40 a.m.	Lunch
	Chairman-W. A. Wallenmeyer, U.S. DOE
8:30 p.m.	J. D. Jackson, SSC
	The SSC Project
9:10 p.m.	L. S. Schroeder, Berkeley
	Relativisitic Heavy Ion Facilities
9:50 p.m.	F. E. Mills, Fermilab
	Cooling of Stored Beams
10:30 p.m.	End

Friday May 30, 1986

	Chairman-J. M. Cameron, Alberta
8:30 a.m.	M. Oka, Pennsylvania
	Quark Model of the N-N Interaction
9:10 a.m.	E. Leader, Birkbeck, London
	High Energy Spin Physics
9:50 a.m.	Coffee
	Chairman-J. P. Schiffer, Argonne
10:20 a.m.	C. B. Dover, Brookhaven
	Antiproton-Proton Interactions
11:00 a.m.	D. H. Perkins, Oxford
	Non-Accelerator Experiments
11:40 a.m.	Lunch

Saturday, May 31, 1986

	Chairman-H. D. Holmgren, SURA
8:30 a.m.	Accelerator Physics Summary
8:50 a.m.	Antiproton Physics Summary
9:10 a.m.	Electron and Muon Physics Summary
9:30 a.m.	Hadron Scattering Summary
9:50 a.m.	Coffee
	Chairman-G. Fidecaro, CERN
10:20 a.m.	Hadron Spectroscopy Summary
10:40 a.m.	Heavy Ion Physics Summary
11:00 a.m.	Hypernuclear Physics Summary
11:20 a.m.	Kaon Decay Physics Summary
11:40 a.m.	Lunch
	Chairman-M. H. Macfarlane, Indiana
2:00 p.m.	Neutrino Physics Summary
2:20 p.m.	Non-Accelerator Physics Summary
2:40 p.m.	Spin Physics Summary
3:00 p.m.	R. Hofstadter, Stanford
	Closing Remarks
4:00 p.m.	End

Second Conference on the Intersections Between
Particle and Nuclear Physics
Lake Louise, Canada
May 26-31, 1986

PARALLEL SESSIONS

Monday - May 26, 1986

2:30-4:10 p.m.	Room	4:40-6:20 p.m.	Room
Accelerator	Sun	Antiproton	430
Electron and Muon	Tom Wilson	Hadron Scattering	Sun
Heavy Ion	Agnes	Hypernuclear	Agnes
Kaon Decay	430	Neutrino	330
Non-Accelerator	330	Spin	Tom Wilson

Tuesday - May 27, 1986

2:30-4:10 p.m.	Room	4:40-6:20 p.m.	Room
Accelerator	Tom Wilson	Antiproton	Sun
Hadron Spectroscopy	Agnes	Hadron Scattering	Tom Wilson
Heavy Ion	Sun	Hypernuclear	Agnes
Kaon Decay	430	Neutrino	330
Non-Accelerator	330	Spin	430

Thursday - May 29, 1986

2:30-4:10 p.m.	Room	4:40-6:20 p.m.	Room
Electron and Muon	Tom Wilson	Antiproton	330
Hadron Spectroscopy	Agnes	Hadron Scattering	Tom Wilson
Heavy Ion	Sun	Hypernuclear	Agnes
Neutrino	430	Spin	430
Non-Accelerator	330	Electron and Muon	Sun

Friday - May 30, 1986

2:30-4:10 p.m.	Room	4:40-6:20 p.m.	Room
Accelerator	430	Antiproton	Sun
Hadron Scattering	Tom Wilson	Electron and Muon	Tom Wilson
Hadron Spectroscopy	Agnes	Heavy Ion	430
Hypernuclear	330	Kaon Decay	Agnes
Spin	Sun	Non-Accelerator	330

ACCELERATOR PHYSICS

Monday, May 26, 2:30 - 4:10 p.m.   R. E. Pollock - Chairman

2:30   J. P. Schiffer, Argonne
       On the Possibility of Achieving a Condensed Crystalline
       State in Cooled Particle Beams

2:55   G. R. Young, Oak Ridge
       Storage Rings for Heavy Ion Atomic and Nuclear Physics

3:20   D. Larson, Univ. of Wisconsin
       Toward Multi-GeV Electron Cooling

3:45   P. A. Thompson, Brookhaven National Lab
       Status of Magnet System for RHIC

4:00   Discussion

Tuesday, May 27, 2:30 - 4:10 p.m.   P. Koehler - Chairman

2:30   J. Jowett, CERN/SLAC
       Does the Transition to Chaos Determine the Dynamic
       Aperture?

2:55   B. E. Norum, Univ. Virginia
       Polarized Electrons for Intermediate Energy Nuclear
       Physics

3:20   F. S. Dietrich, Lawrence Livermore Lab
       An Exploratory Gas Target Experiment at PEP Using the
       TPC/2γ Facility

3:45   V. W. Hughes, Yale
       Plans for a New Precision Measurement of the Muon g-2
       Value

Friday, May 30, 2:30 - 4:10 p.m.   R. E. Pollock - Chairman

2:30   J. Griffin, Fermilab
       Use of the FNAL Antiproton Source for Experiments

2:55   G. A. Smith, Pennsylvania State Univ.
       Is There a Need for a LEAR-Type Facility in North America?

3:20   H. A. Thiessen, Los Alamos
          LAMPF-II 1986: Matching the Machine to the Physics
          Requirements

3:45   K. Y. Ng, Fermilab
          Microwave Instability Criterion for Overlapped Bunches

4:00   Discussion

<u>Saturday, May 31, 8:30 a.m.</u>

       P. Koehler, Fermilab
          Accelerator Physics Summary

ANTIPROTON PHYSICS

Monday, May 26, 4:40 - 6:20 p.m.   G. A. Smith - Chairman

4:40   J. A. Niskanen, Helsinki
        $\bar{N}N$ Strong Interaction Physics (Theory)

5:05   S. M. Playfer, Pennsylvania State Univ.
        $\bar{N}N$ Annihilation Interactions (Experiment)

5:30   P. Barnes, Carnegie-Mellon
        $\bar{N}N$ Non-Annihilation Interactions (Experiment)

5:55   R. Ransome, Rutgers
        $\bar{N}N$ Spin Physics

Tuesday, May 27, 4:40 - 6:20 p.m.   G. A. Smith - Chairman

4:40   T. Goldman, Los Alamos
        Experimental Evidence for Quantum Gravity?

4:50   R. E. Brown, Los Alamos
        Measurement of the Gravitational Acceleration of the
        Antiproton

5:15   B. Bassalleck, Univ. New Mexico
        Measurement of the Antineutron-Proton Annihilation Cross
        Section Extremely Close to $\bar{N}N$ Threshold

5:25   Y. Onel, Geneva
        Spin Physics at LEAR

5:35   G. Marshall, Univ. Victoria/TRIUMF
        Antiproton-Proton Annihilation into Collinear Charged
        Pions and Kaons

5:45   S. M. Playfer, Pennsylvania State Univ.
        Search for Narrow Lines in Gamma Spectra from $\bar{p}p$

5:55   L. Pinsky, Univ. of Houston
        Search for Narrow States in $\bar{n}p$ Annihilation
        Near $\bar{N}N$ Threshold

Thursday, May 29, 4:40 - 6:20 p.m.   D. Axen - Chairman

4:40   R. Cester, Torino
        Heavy Quark Physics in $\bar{N}N$ Interactions

5:05   H. Poth, Karlsruhe
          Physics with Antihydrogen

5:30   H. Koch, Karlsruhe
          Antiprotonic Atoms

5:55   J. Donoghue, Univ. Massachusetts
          Tests of CP Invariance in $\bar{p}p$ Interactions

<u>Friday, May 30, 4:40 - 6:20 p.m.</u>   D. Axen - Chairman

4:40   J. Peng, Los Alamos
          $\bar{N}$-Nucleus Interactions

5:05   P. McGaughey, Los Alamos
          Annihilation in $\bar{N}$-Nucleus Interactions

5:30   W. Gibbs, Los Alamos
          Production of High Energy Density in
          $\bar{N}$-Nucleus Interactions

5:55   L. Pinsky, Univ. Houston
          Summary of Fermilab Low Energy Antiproton Workshop

6:10   N. Auerbach, Univ. Tel Aviv
          Charge Exchange Reactions with Antinucleons

<u>Saturday, May 31, 8:50 a.m.</u>

          D. Axen, Univ. British Columbia
          Antiproton Physics Summary

ELECTRON AND MUON PHYSICS

Monday, May 26, 2:30 - 4:10 p.m.   S. Kowalski - Chairman

2:30   G. Taylor, CERN
       A Review of the EMC Effect

2:55   D. Day, Univ. Virginia
       Scaling in Nuclei

3:20   L. Lapikas, NIKHEF-K
       Nuclear Medium Effects on the Photon-Proton Vertex
       Investigated with the (e,e'p) Reaction

3:45   R. Dymarz, Univ. Alberta
       Scaling in Inclusive (e,d) Scattering and the $\Delta\Delta$-Component
       of the Deuteron Wave Function

4:00   S. Cotanch, North Carolina State Univ.
       $^{16}O(\gamma,p)^{15}N$ Reaction in the Δ-Region

Thursday, May 29, 2:30 - 4:10 p.m.   S. Kowalski - Chairman

2:30   D. Beck, MIT
       Electromagnetic Structure of ^3H and ^3He

2:55   P. Bosted, American Univ./SLAC
       New Measurements of the Deuteron Magnetic Form Factor at
       High Q^2

3:20   F. C. Khanna, Univ. Alberta
       Tensor Polarization of the Deuteron in the Elastic (e,d)
       Scattering

3:35   G. Dodson, MIT
       Coherent Photoreactions in Nuclei: (γ,π°) and (γ,γ) on
       ^{12}C and ^4He

4:40   R. von Dincklage, SIN
       Integral Asymmetry Parameter of Muon Decay

Thursday, May 29, 4:40 - 6:20 p.m.   T. Kinoshita - Chairman

4:40   J. Sapirstein, Notre Dame
       Present Status of QED

5:04   V. Hughes, Yale
       Experimental Possibilities in Muon Electrodynamics: The
       LAMPF Workshop on Fundamental Muon Physics

5:30   R. Arnowitt, Northeastern University
       Supersymmetry Corrections to Muon (g-2)

5:55   A. Olin, TRIUMF/Univ. Victoria
       Thermal Muonium in Vacuum

Friday, May 30, 4:40 - 6:20 p.m.   T. Kinoshita - Chairman

4:40   N. Byers, UCLA
       Leptonic Decay Rates of Heavy Leptonium States

4:55   S. Mintz, Florida International Univ.
       Muon and Neutrino Production in Proton - ^{12}C Scattering

5:10   F. Gross, CEBAF
       Research Program at CEBAF

5:35   B. A. Mecking, CEBAF
       Large Acceptance Magnetic Detector for Photonuclear
       Physics at CEBAF

5:50   V. Burkert, CEBAF
       Polarization in Electron Scattering Experiments at CEBAF

Saturday, May 31, 9:10 a.m.

       S. Kowalski, MIT
       Electron and Muon Physics Summary

HADRON SCATTERING

Monday, May 26, 4:40 - 6:20 p.m.   N. Isgur - Chairman

4:40   B. Wicklund, Argonne
       Phase Shift Analysis of NN → NNπ and Dibaryons

5:10   M. Banerjee, Univ. Maryland
       Chiral Bag Model of the NN Interaction

5:40   D. Riska, Helsinki
       Model Independent NN Exchange Currents

6:10   K. Seth, Northwestern Univ.
       A Precision Test of Charge Independence

Tuesday, May 27, 4:40 - 6:20 p.m.   G. Igo - Chairman

4:40   J. Peng, Los Alamos
       The (π,η) Reaction

5:00   H. Baer, Los Alamos
       Pion Nucleus Reactions Above the Δ Resonance

5:20   M. Moinester, Tel Aviv
       Pion Absorption on the Diproton

5:40   W. Kluge, Karlsruhe
       Coulomb Nuclear Interference in Pion Scattering

6:00   B. Nefkens, UCLA
       The Splitting of the Roper Resonance

6:10   M. Bleszynski, UCLA
       The Sequential Nature of Pion Double Charge Exchange

Thursday, May 29, 4:40 - 6:20 p.m.   N. Isgur - Chairman

4:40   H. Lipkin, Weizmann/Argonne/Fermilab
       Models of Multiquark States

5:10   K. Maltman, Los Alamos
       Exotic 6q and 5q$\bar{q}$ States

5:35   D. Robson, Florida State Univ.
       Flux Link Picture of Multiquark States

6:00   R. K. Campbell, Southwestern Adventist College
       KN and K*N Channel Coupling in a Quark Potential Model

<u>Friday, May 30, 2:30 - 4:10 p.m.</u>   G. Igo - Chairman

2:30   M. Thies, Vrije Univ.
       Q and P in Proton Elastic Scattering

2:50   J. A. McNeil, Drexel Univ.
       Relativistic Treatment of Nuclear Magnetic Moments

3:10   P. Alford, TRIUMF/Western Ontario
       (n,p) Studies

3:30   B. Nefkens, UCLA
       Three Body Forces

3:50   C. Horowitz, MIT
       Relativity in Elastic and Quasi-Elastic Proton-Nucleus
       Scattering

4:00   D. K. McDaniels, Oregon Univ.
       Multipole Matrix Elements for ^{208}Pb

<u>Saturday, May 31, 9:30 a.m.</u>

       N. Isgur, Univ. Toronto
       Hadron Scattering Summary

HADRON SPECTROSCOPY

Tuesday, May 27, 2:30 - 4:10 p.m.   J. Donoghue - Chairman

2:30   A. B. Balantekin, Oak Ridge
        The Interacting Boson Model

3:00   F. Calaprice, Princeton Univ.
        Searches for Axions and Light Scalar Bosons in Nuclear
        Decay Processes -- A Summary

3:30   W. Haxton, Univ. Washington
        Fundamental Interaction Studies in Nuclei

4:00   N. Mukhopadyay, RPI
        Are Nucleons and Deltas Deformed?

Thursday, May 29, 2:30 - 4:10 p.m.   J. Donoghue - Chairman

2:30   J. McClelland, Los Alamos
        Spin Observables at Intermediate Energy - A Tool in
        Viewing the Nucleus

3:00   M. Mattis, SLAC
        Hadron Spectroscopy without Quarks: A Skyrme Model
        Sampler

3:30   N. Byers, UCLA
        The Effects of Light Quarks on Heavy Quarkonia

3:50   T. Goldman, Los Alamos
        Nuclear-like States of Quark Matter

Friday, May 30, 2:30 - 4:10 p.m.   J. Donoghue - Chairman

2:30   W. Lockman, U.C. Santa Cruz
        Recent MARK III Results in J/Ψ Hadronic and Radiative
        Decays

3:00   M. Chanowitz, LBL
        Issues in Spectroscopy of Gluonic States

3:30   R. Woloshyn, TRIUMF
        Hadron Structure from Lattice QCD

Saturday, May 31, 10:20 a.m.

        J. Donoghue, Univ. Massachusetts
        Hadron Spectroscopy Summary

HEAVY ION PHYSICS

Monday, May 26, 2:30 - 4:10 p.m.   J. Harris - Chairman

2:30   G. Roche, LBL
       Universality of Lepton Production in pp, pA and AA
       Collisions

3:10   S. Trentalange, UCLA
       Subthreshold Production of Strange Hadrons in Relativistic
       Heavy Ion Collisions

3:50   G. Shaw, U. C. Irvine
       Production of Metastable Strange Quark Droplets in
       Relativistic Heavy Ion Collisions

Tuesday, May 27, 2:30 - 4:10 pm.   J. Harris - Chairman

2:30   B. Müller, Univ. Frankfurt
       Positron Lines - New Particle or Decaying Vacuum?

3:10   J. Harris, LBL
       The Nuclear Matter Equation of State from Relativistic
       Heavy Ions to Supernovae

3:50   D. Keane, U. C. Riverside
       Collective Flow and the Stiffness of Compressed Nuclear
       Matter

Thursday, May 29, 2:30 - 4:10 p.m.   J. Harris - Chairman

2:30   R. Gavai, Brookhaven
       Phase Transitions from Lattice QCD Perspectives

3:05   M. Tannenbaum, Brookhaven
       Observation of KNO Scaling in the Neutral Energy Spectra
       from $\alpha\alpha$ and pp Collisions at ISR Energies

3:40   H. Miettinen, Rice Univ.
       Proton-Nucleus Collisions at Large Transverse Energies

3:55   M. Tanaka, Brookhaven
       High p_T π° Production in $\alpha\alpha$, dd and pp Collisions at the
       CERN ISR

Friday, May 30, 4:40 - 6:20 p.m.   J. Harris - Chairman

4:40   J. Boguta, Berkeley
           Topological Excitations in Nuclei

5:20   H. G. Miller, NRIMS, South Africa
           Finite Temperature Calculations with Isobars

5:40   M. S. Zahir, Univ. Regina
           Production Mechanism for a Light Pseudoscalar Boson
           (m = 1.6 MeV) and Anomalous Positron Spectra

Saturday, May 31, 10:40 a.m.

         J. Harris, LBL
             Heavy Ion Physics Summary

HYPERNUCLEAR PHYSICS

Monday, May 26, 4:40 - 6:20 p.m.   R. Chrien - Chairman

4:40   M. Sainio, SIN
        Dibaryons: A Review

5:15   H. Piekarz, Brandeis Univ.
        Search for an S = -1 Dibaryon

5:45   P. Pile, Brookhaven
        Future Plans for Dibaryon Searches at the AGS

Tuesday, May 27, 4:40 - 6:20 p.m.   R. Dalitz - Chairman

4:40   M. Torres-Labansat, Univ. Mexico
        New Calculations for $D(K^-,\pi^-)$ at Rest

5:15   B. K. Jennings, TRIUMF
        Kaon-Nucleon Interactions

5:50   C. Bennhold, Ohio Univ.
        Relativistic Wave Functions for Kaon Photoproduction in
        Nuclei

Thursday, May 29, 4:40 - 6:20 P.M.   R. Chrien - Chairman

4:40   J. Chiba, KEK
        Recent Σ-Hypernuclear Studies with Stopped Kaons at KEK

4:55   T. Kishimoto, Univ. Houston
        Quasifree Processes in Hypernuclei

5:15   E. V. Hungerford, Univ. Houston
        Final State Interaction Effects in (K,π) Reactions

5:40   J. Speth, Los Alamos
        Strangeness in Nuclei

6:00   L.-C. Liu, Los Alamos
        The Eta-Mesic Nucleus: A New Nuclear Species?

Friday, May 30, 2:30 - 4:10 p.m.   R. Dalitz - Chairman

2:30   J. Szymanski, Carnegie-Mellon Univ.
           Hypernuclear Lifetimes and Decay Modes

3:00   L. Kisslinger, Carnegie-Mellon Univ.
           Weak Decays in a Quark Model

3:20   J. Dubach, Univ. Massachusetts
           Non Mesonic Weak Decay in Hypernuclei

3:45   E. Golowich, Univ. Massachusetts
           Decay of the H Dibaryon

Saturday, May 31, 11:00 a.m.

       R. Chrien
           Hypernuclear Physics Summary

KAON DECAY PHYSICS

Monday, May 26, 2:30 - 4:10 p.m.   M. P. Schmidt - Chairman

2:30   W. Marciano, Brookhaven
          Constraints on the K-M Matrix

2:55   T. Numao, TRIUMF
          Rare Decays

3:20   L. Piilonen, Los Alamos
          LAMPF Crystal Box

4:00   J. Vergados, Univ. Ioannina
          Lepton Number and Lepton Flavor Violation in SUSY Models

Tuesday, May 27, 2:30 - 4:10 p.m.   T. Numao - Chairman

2:30   D. Stoker, John Hopkins Univ.
          Status of the Electroweak Model in Muon Decay

2:55   P. Burchat, U.C. Santa Cruz
          Review of the Decays of the Tau Lepton

3:20   J. Fry, Univ. Liverpool
          CP Violation in Experiments at LEAR

3:45   B. Roberts, Boston Univ.
          Search for T and CPT Violation in the Neutral Kaon System

Friday, May 30, 4:30 - 6:20 p.m.   M. P. Schmidt - Chairman

4:30   B. Holstein, Univ. Massachusetts
          CP Violation

4:55   M. Calvetti, Univ. Pisa
          ε' Measurement at CERN

5:20   G. Bock, Fermilab
          Progress Report on the Precision Measurement
          of $|\eta_{00}/\eta_{+-}|$ at Fermilab

5:45   G. Thomson, Rutgers
          New Results on η_{+-0}

Saturday, May 31, 11:20 a.m.

          M. P. Schmidt, Yale
          Kaon Decay Physics Summary

NEUTRINO PHYSICS

Monday, May 26, 4:40 - 6:20 p.m.   S. Rosen - Chairman

4:40   W. Haxton, Univ. Washington
            Double Beta Decay and Nuclear Theory

5:05   M. Moe, U.C. Irvine
            Double Beta Decay of ^{82}Se and ^{130,128}Te

5:30   F. Avignone, Univ. Southern California
            Double Beta Decay of ^{76}Ge

5:55   J. Vergados, Univ. Ioannina
            Double Beta Decay and Lepton Flavor Violation

Tuesday, May 27, 4:40 - 6:20 p.m.   S. Rosen - Chairman

4:40   D. Knapp, Los Alamos
            Los Alamos Tritium Beta Decay Experiment

5:10   W. Kundig, Univ. Zürich
            Zürich - SIN Tritium Beta Decay Experiment

5:40   J. Weber, Univ. California, Irvine
            New Method for Weak Interactions

5:50   R. Casella, NBS
            Coherence in the Scattering of Low Energy Neutrinos from
            Macroscopic Objects

6:10   G. Karl, Guelph
            Some Remarks on the 17 keV Neutrino

Thursday, May 29, 2:30 - 4:10 p.m.   W.-Y. Lee - Chairman

2:30   T. O'Halloran, Univ. Illinois
            Neutrino Oscillations at Accelerators

2:50   Z. Greenwood, U.C. Irvine
            Neutrino Oscillations at Reactors

3:10   W. P. Lee, U.C. Irvine
            Neutrino-electron Scattering

3:30   R. Fisk, Valparaiso Univ.
            Neutrino Nucleus Interactions

Saturday, May 31, 2:00 p.m.

    B.H.J. McKellar, Univ. Melbourne
    Neutrino Physics Summary

NON-ACCELERATOR PHYSICS

<u>Monday, May 26, 2:30 - 4:10 p.m.</u>   D. Sinclair - Chairman

2:30   J. W. Elbert, Univ. Utah
       Ultra-high Energy Point Sources of Cosmic Rays

3:10   J. Goodman, Univ. Maryland
       The Cygnus Experiment at Los Alamos

3:35   D. Ayres, Argonne
       New Evidence from Soudan 1 for Underground Muons
       Associated with Cygnus X-3

3:50   D. Sinclair, Univ. Michigan
       A New Gamma-ray Telescope

<u>Tuesday, May 27, 2:30 - 4:10 p.m.</u>   D. Ayres - Chairman

2:30   S. P. Rosen, Los Alamos
       Amplification of Neutrino Oscillations in the Sun

3:00   B. G. Cortez, Caltech
       Search for Solar Neutrinos at Kamioka

3:30   E. D. Earle, Chalk River Nuclear Lab
       A D_2O Cerenkov Detector for Solar Neutrinos

<u>Thursday, May 29, 2:30 - 4:10 p.m.</u>   D. Ayres - Chairman

2:30   E. Fischbach, Univ. Washington
       A New Force in Nature?

2:55   J. C. Martoff, Stanford Univ.
       Acoustic Detection of Neutrinos

3:25   S. J. Freedman, Argonne
       Modern Implications of Neutron Beta Decay

3:50   P. Herczeg, Los Alamos
       Constraints on General Left-Right Symmetric Electroweak
       Models from Muon Decay and Beta Decay

Friday, May 30, 4:40 - 6:20 p.m.   D. Sinclair - Chairman

4:40   B. G. Cortez, Caltech
            The Status of Proton Decay

5:00   E. N. May, Argonne
            Status of the Soudan 2 Nucleon Decay Experiment

5:15   S. P. Ahlen, Boston Univ.
            Searches for Monopoles: Past, Present, and Future

5:45   D. F. Nitz, Univ. Michigan
            A Search for Anomalously Heavy Isotopes of Low Z Nuclei

6:00   Discussion

Saturday, May 31, 2:20 p.m.

        D. Sinclair, Univ. Michigan
            Non-Accelerator Physics Summary

SPIN PHYSICS

Monday, May 26, 4:40 - 6:20 p.m.   W. van Oers - Chairman

4:40   T. Roser, Univ. Michigan
        Spin Effects in Proton-Proton Elastic Scattering at High $P_\perp^2$

5:05   D. G. Underwood, Argonne
        High Energy Polarised Beams from Hyperon Decays

5:30   Y. I. Makdisi, Brookhaven
        Spin in Exclusive Processes at High Momentum Transfer

5:55   H. Spinka, Argonne
        Spin Observables in Neutron-Proton Scattering at Intermediate Energies

Tuesday, May 27, 4:40 - 6:20 p.m.   J. Roberts - Chairman

4:40   E. W. Adelberger, Univ. Washington
        Parity Violation and C. P. Violation in Nuclear Systems

5:10   V. Yuan, Los Alamos
        Parity Violation in Proton-Proton Scattering

5:35   J. D. Bowman, Los Alamos
        A Test of Time Reversal Invariance in Resonance Neutron Scattering

6:00   M. Svec, McGill Univ.
        Spin Dependence of ρ^0 and K^0 Resonance Production at 6 GeV/c

6:10   J. Kruk, Rice Univ.
        Spin Transfer in Hyperon Production

Thursday, May 29, 4:40 - 6:20 p.m.   W. van Oers - Chairman

4:40   L. G. Greeniaus, TRIUMF/Univ. Alberta
        Tests of Charge Symmetry in the Neutron-Proton System

5:10   R. A. Arndt, Virginia Polytechnic Inst.
        Pion-nucleon and Kaon-nucleon Phase Shifts Analyses

5:35   M. Bleszynaski, UCLA
       Spin-Dependent Observables in p-d Elastic Scattering at
       Intermediate Energies

6:00   A. Masaike, Kyoto
       Spin Physics at KEK

<u>Friday, May 30, 2:30 - 4:10 p.m.</u>   J. Roberts - Chairman

2:30   G. Smith, TRIUMF
       Vector and Tensor Analyzing Powers in π-d Elastic
       Scatterng

2:55   W. K. Pitts, IUCF
       Radiative Capture of Polarized Protons and Deuterons by
       the Hydrogen Isotopes

3:20   Y. Onel, Univ. Geneva
       Antiproton Spin Physics at LEAR

4:00   W. Gyles, Karlsruhe
       Measurement of the Vector Analyzing Power in πd Breakup

<u>Saturday, May 31, 2:40 p.m.</u>

       J. B. Roberts, Rice Univ.
       Spin Physics Summary

CONFERENCE PARTICIPANTS

Raghunath Acharya	Arizona State University
Eric G. Adelberger	University of Washington
I. R. Afnan	Flinders University
Mohammad M. Agbareia	Toronto, Canada
Lewis Agnew	Los Alamos National Laboratory
Steven Ahlen	Boston University
W. P. Alford	University of Western Ontario
J. C. Allred	Los Alamos National Laboratory
Jonas Alster	Tel Aviv University
Richard A. Arndt	Virginia Polytechnic Institute
Raymond G. Arnold	Stanford Linear Accelerator Center
Richard L. Arnowitt	Northeastern University
Naftali Auerback	Tel Aviv University
Frank T. Avignone	University of South Carolina
D. A. Axen	University of British Columbia
D. S. Ayres	Argonne National Laboratory
Mark G. Bachman	University of Texas
Helmut Baer	Los Alamos National Laboratory
A. Baha Balantekin	Oak Ridge National Laboratory
Manoj K. Banerjee	University of Maryland
Marcel Bardon	National Science Foundation
Peter D. Barnes	Carnegie-Mellon University
Bernd Bassalleck	University of New Mexico
Douglas H. Beck	Massachusetts Institute of Technology
Cornelius Bennhold	Ohio University
Edmond L. Berger	Argonne National Laboratory
Aron M. Bernstein	Massachusetts Institute of Technology
James Birchall	University of Manitoba
Elizabeth Bleszynski	UCLA
Mark Bleszynski	UCLA
Gregory J. Bock	Fermi National Accelerator Laboratory
J. Boguta	Lawrence Berkeley Laboratory
Bernardino Bosco	Universita di Firenze
Peter Bosted	American University, SLAC
Theodore Bowen	University of Arizona
Charles R. Bower	Indiana University
J. David Bowman	Los Alamos National Laboratory
James Bradbury	Los Alamos National Laboratory
Ronald E. Brown	Los Alamos National Laboratory
Stanley G. Brown	Physical Review Letters
Patricia R. Burchat	Stanford Linear Accelerator Center
Volker Burkert	CEBAF
M. P. Bussa	I.N.F.N. Torino
Nina Byers	UCLA

Frank P. Calaprice	Princeton University
Mario Calvetti	University of Pisa
John Cameron	University of Alberta
Roy K. Campbell	South Western Adventist College
Russell C. Casella	National Bureau of Standards
Rosanna Cester	Istituto di Fisica Torino
Kai Chang	California Institute of Technology
Michael S. Chanowitz	Lawrence Berkeley Laboratory
Xiao-Yan Chen	TRIUMF
Chi-Yee Cheung	University of Colorado
Junsei Chiba	KEK
C. R. Ching	Academia Sinica
R. E. Chrien	Brookhaven National Laboratory
Timothy Chupp	Harvard University
Edward D. Cooper	TRIUMF
Martin D. Cooper	Los Alamos National Laboratory
B. G. Cortez	California Institute of Technology
Stephen R. Cotanch	North Carolina State University
Geoffrey R. Court	University of Michigan
M. K. Craddock	TRIUMF
Hall Crannell	Catholic University
John F. Dawson	University of New Hampshire
Donal B. Day	University of Virginia
Frank S. Dietrich	Lawrence Livermore National Laboratory
George Dodson	MIT/Bates
Thomas W. Dombeck	Los Alamos National Laboratory
John F. Donoghue	University of Massachusetts
Carl B. Dover	Brookhaven National Laboratory
John F. Dubach	University of Massachusetts
Rafal Dymarz	University of Alberta
E. Davis Earle	Chalk River Nuclear Laboratory
J. W. Elbert	University of Utah
Willie R. Falk	University of Manitoba
M. Farkhondeh	MIT/Bates
Mirek Fatyga	Indiana University
Harold W. Fearing	TRIUMF
G. Fidecaro	CERN
Rudolf H. Fiebig	Florida International University
Ephraim Fischbach	University of Washington
Randy Fiske	Valparaiso University
Daniel H. Fitzgerald	Los Alamos National Laboratory
Sherman Fivozinsky	U. S. Department of Energy
Stuart J. Freedman	Argonne National Laboratory
John Fry	University of Liverpool

Avraham Gal	TRIUMF
Siegfried Galster	KFK Karlsruhe
Rajiv Gavai	Brookhaven National Laboratory
Donald F. Geesaman	Argonne National Laboratory
M. Gering	University of Witwatersrand
William R. Gibbs	Los Alamos National Laboratory
D. R. Gill	TRIUMF
Charles M. Glashausser	Rutgers University
George Glass	Los Alamos National Laboratory
R. J. Glauber	Harvard University
T. Goldman	Los Alamos National Laboratory
Eugene Golowich	University of Massachusetts
Jordan Goodman	University of Maryland
Jack S. Greenberg	Yale University
L. Gordon Greeniaus	TRIUMF/University of Alberta
Zeno D. Greenwood	University of California, Irvine
James E. Griffin	Fermi National Accelerator Laboratory
David P. Grosnick	Los Alamos National Laboratory
Franz L. Gross	CEBAF
Geraldine J. Grube	Champaign, IL, USA
Hermann A. Grunder	CEBAF/SURA
William Gyles	Universität Karlsruhe
H. Haghbin	Shiraz University
Aksel L. Hallin	Princeton University
Emil L. Hallin	University of Saskatchewan
John W. Harris	Lawrence Berkeley Laboratory
Michael D. Hasinoff	TRIUMF
Wick C. Haxton	University of Washington
Richard M. Heinz	Indiana University
Peter Herczeg	Los Alamos National Laboratory
N. Hessey	Birmingham University
Kenneth H. Hicks	TRIUMF
David L. Hill	Valutron N.V.
John R. Hiller	University of Minnesota
Tso Hsiu Ho	Academia Sinica
Robert Hofstadter	Stanford University
Harry D. Holmgren	SURA
Barry Holstein	University of Massachusetts
Charles J. Horowitz	Massachusetts Institute of Technology
Vernon W. Hughes	Yale University
Ed V. Hungerford	University of Houston
G. J. Igo	UCLA
Mohammed Javed Iqbal	TRIUMF
N. Isgur	University of Toronto

J. D. Jackson	SSC/Berkeley
B. K. Jennings	TRIUMF
Chueng R. Ji	Stanford University
Mikkel B. Johnson	Los Alamos National Laboratory
John Jowett	Stanford Linear Accelerator Center
Mujahid Kamran	University of the Punjab
Y. Karant	Lawrence Berkeley Laboratory
Gabriel Karl	University of Guelph
Declan Keane	University of California, Riverside
F. C. Khanna	University of Alberta
D. V. Kim	University of Regina
T. Kinoshita	Cornell University
Tadafumi T. K. Kishimoto	University of Houston
Leonard S. Kisslinger	Carnegie-Mellon University
Wolfgang Kluge	KFK, Karlsruhe
David Knapp	Los Alamos National Laboratory
Helmut K. Koch	KFK, Karlsruhe
Peter F. M. Koehler	Fermi National Accelerator Laboratory
S. E. Koonin	California Institute of Technology
Stanley Kowalski	Massachusetts Institute of Technology
Alan D. Krisch	University of Michigan
J. W. Kruk	Rice University
Walter Kundig	University of Zürich
Ray Kunselman	University of Wyoming
Paul G. Langacker	University of Pennsylvania
Louis L. Lapikas	NIKHEF-K
Delbert J. Larson	University of Wisconsin
Donald M. Lazarus	Brookhaven National Laboratory
Elliot Leader	Birkbeck College, London
Wen-Piao Lee	University of California, Irvine
Peter Lepage	Cornell University
B. Joseph Lieb	George Mason University
V. Gordon Lind	Utah State University
Harry J. Lipkin	Weizmann Inst. of Sci./Argonne Nat. Lab
Jerry Lisantti	University of Oregon
Lon-chang Liu	Los Alamos National Laboratory
William S. Lockman	Stanford Linear Accelerator Center
Thomas W. Ludlam	Brookhaven National Laboratory
William M. MacDonald	University of Maryland
Malcolm H. Macfarlane	Indiana University
Hay Boon Mak	Queen's University
Y. I. Makdisi	Brookhaven National Laboratory
A. Malecki	INFN Laboratori Nazionali
Kim Maltman	Los Alamos National Laboratory

William Marciano	Brookhaven National Laboratory
Glen M. Marshall	University of Victoria/TRIUMF
J. C. Martoff	Stanford University
Akira Masaike	Kyoto University
Grant J. Mathews	Lawrence Livermore National Laboratory
Michael Mattis	Stanford Linear Accelerator Center
Edward N. May	Argonne National Laboratory
Morgan May	Brookhaven National Laboratory
Bill W. Mayes	University of Houston
James S. McCarthy	University of Virginia
John B. McClelland	Los Alamos National Laboratory
David K. McDaniels	University of Oregon
Joseph P. McDermott	University of Colorado, Boulder
Arthur B. McDonald	Princeton University
Patrick L. McGaughey	Los Alamos National Laboratory
Bruce H. J. McKellar	University of Melbourne
Robert D. McKeown	Caltech
James A. McNeil	Drexel University
Bernhard Mecking	CEBAF
Hannu E. Miettinen	Rice University
H. G. Miller	NRIMS, CSIR
J. P. Miller	Boston University
Frederick E. Mills	Fermi National Accelerator Laboratory
Stephan L. Mintz	Florida International University
Michael Moe	University of California, Irvine
Murray Moinester	Tel Aviv University
E. J. Moinz	Massachusetts Institute of Technology
George C. Morrison	University of Birmingham
Joel M. Moss	Los Alamos National Laboratory
Steven A. Moszkowski	UCLA
Jean Mougey	CEBAF
Nimai C. Mukhopadhyay	Rensselaer Polytechnic Institute
Berndt Müller	F. W. Goethe Universität
Gordon S. Mutchler	Rice University
Swamy P. Narayana	Southern Illinois University
Bernard M. K. Nefkens	UCLA
King-Yuen Ng	Fermi National Accelerator Laboratory
Jouni A. Niskanen	University of Helsinki
David F. Nitz	University of Michigan
Lee Northcliffe	Texas A & M University
Blaine E. Norum	University of Virginia
Toshio Numao	TRIUMF
Thomas A. O'Halloran	University of Illinois
Makoto Oka	University of Pennsylvania
Arthur Olin	TRIUMF
Rigobert Olszewski	University of British Columbia
Yasar Onel	Université de Genève

David S. Onley   Ohio University
Clemens Ottermann   Universität Karlsruhe

Shelley A. Page   University of Manitoba
Jen-Chieh Peng   Los Alamos National Laboratory
Victor Perez-Mendez   Lawrence Berkeley Laboratory
D. H. Perkins   University of Oxford
William R. Phillips   Argonne National Laboratory
C. Picciotto   University of Victoria
Henryk Piekarz   Brandeis University
Leo Piilonen   Los Alamos National Laboratory
Fulvia Pilat   Los Alamos National Laboratory
Philip H. Pile   Brookhaven National Laboratory
Lawrence S. Pinsky   University of Houston
Karl Pitts   Indiana University Cyclotron Facility
Stephen M. Playfer   Pennsylvania State University
R. E. Pollock   Indiana University
Helmut H. Poth   CERN
Barry M. Preedom   University of South Carolina
Melvin A. Preston   University of Saskatchewan

J. P. Ralston   University of Kansas
W. D. Ramsay   TRIUMF
Ronald D. Ransome   Rutgers University
Lazarus G. Ratner   Brookhaven National Laboratory
Pradosh K. Ray   Tuskegee University
Richard S. Raymond   Brookhaven National Laboratory
Clarence R. Richardson   U. S. Department of Energy
Dan O. Riska   Helsinki University
B. Lee Roberts   Boston University
Jabus B. Roberts   Rice University
R. G. Hamish Robertson   Los Alamos National Laboratory
Donald Robson   Florida State University
Guy R. Roche   Lawrence Berkeley Laboratory
Louis Rosen   Los Alamos National Laboratory
Simon P. Rosen   Los Alamos National Laboratory
Thomas Roser   University of Michigan
Csada M. Rozsa   Harshaw/Filtrol

M. Sainio   SIN
Vern Sandberg   Los Alamos National Laboratory
Jonathan R. Sapirsten   University of Notre Dame
John P. Schiffer   Argonne National Laboratory
Michael P. Schmidt   Yale University
L. S. Schroeder   Lawrence Berkeley Laboratory
Peter Schwandt   Indiana University
Urs J. Sennhauser   SIN
Andrew M. Sessler   Lawrence Berkeley Laboratory

Kamal K. Seth	Northwestern University
Gordon Shaw	University of California, Irvine
James R. Shepard	University of Colorado
Helmy S. Sherif	University of Alberta
B. Shin	University Saskatchewan
P. J. Siemens	Texas A & M University
Markus Simonius	ETH Zürich
Daniel Sinclair	University of Michigan
D. Sivers	Argonne National Laboratory
Gerald A. Smith	Pennsylvania State University
Gregory R. Smith	TRIUMF
Jack Smith	SUNY
Paul A. Souder	Syracuse University
Josef Speth	Los Alamos National Laboratory
Brian M. Spicer	University of Melbourne
Harold M. Spinka	Argonne National Laboratory
S. Standil	University of Manitoba
Gerard J. Stephenson, Jr.	Los Alamos National Laboratory
David P. Stoker	Stanford Linear Accelerator Center
M. Svec	McGill University
L. Wayne Swenson	Oregon State University
John Szymanski	Carnegie-Mellon University
Mitsuyoshi Tanaka	Brookhaven National Laboratory
Michael J. Tannenbaum	Brookhaven National Laboratory
Geoffrey N. Taylor	Oxford/CERN
Richard E. Taylor	Stanford Linear Accelerator Center
Lee C. Teng	Fermi National Accelerator Laboratory
Kent M. Terwilliger	University of Michigan
Michael Thies	Vrije Universiteit
Henry A. Thiessen	Los Alamos National Laboratory
Pat A. Thompson	Brookhaven National Laboratory
Gordon Thomson	Rutgers University
D. R. Toi	Newport News, VA, USA
M. Torres-Labansat	Univ. Nac. Autonama De México
Stephen Trentalange	UCLA/Lawrence Berkeley Laboratory
Christoph Tschalaer	SIN
David G. Underwood	Argonne National Laboratory
Jack L. Uretsky	Elmhurst College
Michel Vallieres	Drexel University
Olin van Dyck	Los Alamos National Laboratory
W. T. H. van Oers	University of Manitoba
J. P. Vary	Iowa State University
John D. Vergados	University of Ioannina
E. W. Vogt	TRIUMF
R. von Dincklage	ETH Zürich

William A. Wallenmeyer	U.S. Department of Energy
Joseph Weber	University of California, Irvine
Morton S. Weiss	Lawrence Livermore National Laboratory
Roy R. Whitney	CEBAF
A. B. Wicklund	Argonne National Laboratory
Denys H. Wilkinson	University of Sussex
Harvey B. Willard	National Science Foundation
Bruce Winstein	University of Chicago/Stanford
Richard Woloshyn	TRIUMF
S. S. M. Wong	University of Toronto
Ming-Jen Yang	University of Chicago
Alva F. Yano	California State University, Long Beach
Fleur B. Yano	California State Univ., Los Angeles
Paul F. Yergin	Rensselaer Polytechnic Institute
Glenn R. Young	Oak Ridge National Laboratory
Vincent Yuan	Los Alamos National Laboratory
Muhammad S. Zahir	University of Regina
Ben Zeidman	Argonne National Laboratory

CONFERENCE STAFF

Sheila Dodson	Massachusetts Institute of Technology
Kathy Einfeldt	Brookhaven National Laboratory
Lorraine King	TRIUMF
Margaret Lear	TRIUMF
Frances Mann	Lawrence Berkeley Laboratory
Roberta Marinuzzi	Los Alamos National Laboratory
Sue Marsh	University of Michigan
Karen Poelakker	Los Alamos National Laboratory

AUTHOR INDEX

A

Aardsma, G., 1094
Acharya, R., 796
Adelberger, E. G., 1177
Adney, J. R., 366
Adrian, S., 640
Ahlen, S. P., 1137
Ahmad, S., 456
Alford, W. P., 710
Allen, R. C., 1050, 1078, 1094
Amsler, C., 456
Anderson, D. R., 366
Arends, J., 640
Armenteros, R., 456
Armstrong, T., 445, 460
Arndt, R. A., 650
Arnold, R. G., 83
Arnowitt, R., 582
Aronson, S. H., 1102
Artuso, M., 627
Auerbach, N., 520
Auld, E., 456
Avignone, F. T., 1017
Axen, D. A., 307, 456
Ayres, D. S., 1083

B

Bailey, D., 456
Balantekin, A. B., 732
Balestra, F., 526
Barlag, S., 456
Barlow, D., 627
Barnes, P. D., 99, 418
Bassalleck, B., 445
Batusov, Y. A., 526
Beavis, D., 844
Beck, D. H., 547
Beer, G., 456
Bendiscioli, G., 526
Bennhold, C., 914
Berger, E. L., 165
Bertrand, F. E., 727
Bharadwaj, V., 1050
Bizot, J.-C., 456

Bleszynski, E., 1208
Bleszynski, M., 644, 1208
Bloom, S. D., 791
Bock, G. J., 1004
Bolton, R. D., 966, 1078
Boschitz, E. T., 1232
Bossolasco, S., 526
Bosted, P. E., 554
Botlo, M., 456
Bourdeau, M., 749
Bowles, T. J., 1031
Bowman, J. D., 966, 1189, 1194
Breivik, F. O., 526
Briscoe, W., 640
Brodzinski, R. L., 1017
Brooks, G. A., 1050
Brown, R. E., 436
Burchat, P. R., 981
Burkert, V., 604
Burks, B. L., 727
Burman, R. L., 1050, 1078
Bussa, M. P., 526
Busso, L., 526
Butterfield, K. B., 1078
Büttgen, R., 924
Byers, N., 770

C

Cabeza, P., 1004
Cabrera, B., 1119
Cady, R., 1078
Campbell, R. K., 686
Caria, M., 456
Carlini, R. D., 1050, 1078, 1189
Carlson, S., 814
Carroll, J. B., 814
Carter, A. L., 1094
Casella, R. C., 1040
Cester, R., 468
Chang, C. Y., 1078
Chen, H. H., 1050, 1078, 1094
Chen, X. Y., 727
Chiba, J., 917
Chrien, R. E., 325

Chu, C, 460
Chus, S. Y., 844
Ciampa, D., 1143
Clement, J., 445, 460
Cline, D. B., 366
Cochran, D. R. F., 1050
Coleman, R., 1004
Comyn, M., 456
Cooper, M. D., 966
Cortez, B., 1081
Cotanch, R., 542
Cottingham, J., 371
Craddock, M. K., 63

D

Dahl, P., 371
Dahme, W., 456
Dalitz, R. H., 901
Danby, G., 382
Daudin, R., 1004
Davidson, R., 749
Davidson, W. F., 1094
Debu, P., 1004
Delcourt, B., 456
de Lesquen, A., 1198
Deloff, A., 901
Dhawan, S. K., 382
Dietrich, F. S., 378
Dingus, B. L., 1078
Disco, A., 382
Doe, P. J., 1050, 1078, 1094
Donoghue, J. F., 313, 495
Doser, M., 456
Dover, C. B., 272
Dubach, J., 946
Duch, K.-D., 456
Dung, M., 1004
Dymarz, R., 540, 559

E

Earle, E. D., 1094
Eichon, A., 640
Elbert, J. W., 1070
Elinon, C., 460
Ellsworth, R. W., 1078
Elmore, D., 1143
Engelage, J., 640
Erdman, K., 456
Ewan, G. T., 1094

F

Fabian, W., 877
Falomkin, I. V., 526
Farley, F. J. M., 382
Feld-Dahme, F., 456
Fernow, R., 371
Ferrero, L., 526
Fetscher, W., 646
Fischbach, E., 1102
Fisk, R., 1056
Frank, J. S., 966, 1050, 1078
Frauenfelder, H., 1189
Freedman, S. J., 1125
Fritschi, M., 1036
Fry, J. R., 987
Fung, S. Y., 844
Furic, M., 445, 460

G

Gal, A., 127
Garber, M., 371
Garcilazo, H., 1232
Gastaldi, U., 456
Gavai, R. V., 848
Gentile, T., 1143
Gerber, H. J., 646
Ghosh, A., 371
Gibbs, W. R., 505
Glauber, R., 644
Glover, C., 727
Goldman, T., 434, 773
Gollin, G., 1004
Golowich, E., 952
Goodman, J. A., 1078
Goodzeit, C., 371
Gordon, J., 814
Göring, K., 646
Gorn, W., 844
Graessle, S., 640
Grazer, G., 1004
Greenberg, J. S., 112
Greene, A., 371
Greeniaus, L. G., 1202
Greenwood, Z. D., 1042
Griffin, J. E., 391
Grosnick, D., 966
Gross, E. E., 727
Grunder, H. A., 37
Guaraldo, C., 526

Gupta, S. K., 1078
Gyles, W., 1232

H

Haas, P., 1143
Haatuft, A., 526
Hahn, H., 371
Haider, Q., 930
Hallin, A. L., 966
Hallman, E. D., 1094
Hallman, T., 814
Halsteinslid, A., 526
Hargrove, C., 1094
Harper, R. W., 1189
Harris, J. W., 318, 835
Hartline, B. K., 37
Hartman, K., 460
Hausammann, R., 1050
Häusser, O., 727
Haxton, W. C., 738
Heel, M., 456
Hemmick, T., 1143
Herczeg, P., 1131
Herrera, J., 371
Heusi, P., 966
Hicks, A., 445, 460
Hicks, K. 727
Highland, V. L., 966
Hoffman, C. M., 966
Hofstadter, R., 346, 966
Hogan, G. E., 966
Holinde, K., 924
Holstein, B. R., 998
Holzenkamp, B., 924
Holzschuh, E., 1036
Horen, D. J., 727
Howard, B., 456
Hsiung, Y., 1004
Hughes, E. B., 966
Hughes, R. J., 434
Hughes, V. W., 382, 575
Hungerford, E., 460

I

Igo, J., 814
Isgur, N., 310

J

Jackson, J., 382
Jacobsen, T., 526
Jagam, P., 1094
Jaroszewicz, T., 1208
Jarry, P., 1004
Jeanjean, J., 456
Jennings, B. K., 906
Ji, C.-R., 688
Johnston, R. R., 1232
Johnson, W., 1078
Jowett, J. M., 374

K

Kagan, H., 1143
Kahn, S., 371
Kalinowsky, H., 456
Kamran, M., 729
Karl, G., 200
Karliner, M., 762
Kayser, F., 456
Keane, D., 844
Keay, B. J., 814
Kelly, E., 371, 382
Kessler, D., 1094
Khanna, F. C., 540, 559
Kim, D. Y., 879
Kim, G., 640
Kirk, P. N., 814
Kishimoto, T., 460, 921
Kisslinger, L. S., 940
Klein, U., 646
Klempt, E., 456
Kluge, W., 646
Knapp, D. A., 1031
Koch, H., 490
Koehler, P. F. M., 303 25
Konter, J. A., 1232
Krakauer, D. A., 1050
Krebs, G., 814
Krienen, F., 382
Kruk, J., 445, 460, 1200
Kuang, Y., 382
Kubik, P. W., 1143
Kündig, W., 1036

L

Laa, C., 456
Landua, R., 456

Langacker, P., 142
Lapikás, L., 535
Larson, D. J., 366
Leader, E., 257
Lee, H. W., 1094
Lee, W. P., 1050, 1078
Leontaris, G. K., 972, 1025
Lewis, R. A., 445, 460
Lindstrom, P., 814
Linsley, J., 1078
Lipkin, H. J., 647, 1153
Lisantti, J., 727
Liu, L. C., 930
Liu, Y. M., 844
Lochstet, W., 445, 460
Lockman, W. S., 776
Lodi-Rizzini, E., 526
Lowenstein, D. I., 445, 460
Lubell, M., 382
Ludeking, L. D., 542

M

MacArthur, D. W., 1189
Maggiora, A., 526
Maher, C. J., 418
Mahler, H. J., 1050
Mak, H.-B., 1094
Makdisi, Y., 1166
Malenfant, J., 595
Maley, M. P., 1031
Maltman, K., 672
Mango, S., 1232
Marciano, W. J., 956
Mariam, F. G., 966
Marshall, G. M., 456
Marston, P., 382
Martoff, C. J., 1119
Masaike, A., 1214
Mathews, G. J., 791
Mathie, E. L., 1232
Matis, H. S., 966
Matthäy, H., 646
Mattis, M. P., 762
May, E. N., 1134
Mayes, B., 445, 460
McClelland, J. B., 751
McDaniels, D. K., 727
McDonald, A. B., 1094, 1189
McDonough, J., 966
McKellar, B. H. J., 339

McNeil, J. A., 696
Mecking, B. A., 601
Melnikoff, S. O., 378
Metzler, M., 646
Miklebost, K., 526
Miley, H., 1017
Miller, H. G., 877
Mills, F. E., 226
Mintz, S. L., 597
Mischke, R. E., 966, 1189
Moe, M. K., 1012
Moinester, M. A., 636
Molitoris, J. J., 844
Morgan, G., 371
Moss, B., 445, 460
Mukhopadhyay, N. C., 749
Mulera, T., 814
Müller, B., 827
Mutchler, G., 445, 460

N

Nagle, D. E., 966, 1078, 1189
Narayana Swamy, P., 796
Nath, P., 582
Nefkens, B. M. K., 640, 719
Neuhauser, B., 1119
Ng, K.-Y., 401
Nguyen, H., 456
Nieto, M. M., 434
Nishikawa, K., 1004
Niskanen, J. A., 406
Nitz, D., 1143
Numao, T., 961

O

Ohashi, Y., 640
Oka, M., 238
Okamitsu, J.
Olin, A., 587
Olsen, J., 526
Olsen, S. L., 1143
Onel, Y., 449, 1229
Orth, H., 382
Ottermann, C. R., 1232

P

Panzieri, D., 526
Parker, B., 627
Patterson, R., 1004

Pedroni, E., 646
Peng, J. C., 499, 630
Penzo, A., 1229
Perez-Mendez, V., 814
Perkins, D. H., 293
Peterson, J. W., 1036
Peyaud, B., 1004
Piasetzky, E., 1050
Piilonen, L. E., 966
Pile, P. H., 894
Pinsky, L., 445, 460, 515
Piragino, G., 526
Pitts, W. K., 1224
Pixley, R. E., 1036
Plate, S., 371
Playfer, S. M., 412, 458
Pontecorvo, G. B., 526
Poth, H., 445, 480
Potter, M. E., 1050
Prévot, N., 456
Prodell, A., 371, 382

Q

Quick, R. M., 877
Qureshi, I. E., 729

R

Rahman, A., 354
Ransome, R. D., 429
Reeves, J. H., 1017
Riedlberger, J., 456
Riska, D. O., 618
Ritter, M., 966
Roberts, B. L., 993
Robertson, B. C., 1094
Robertson, L., 456
Robertson, R. G. H., 115, 1031
Robson, D., 676
Roche, G., 806
Roser, T., 1148
Rossmanith, R., 1229
Rotondi, A., 526
Rushton, A. M., 1050

S

Sabev, C., 456
Sadler, M., 640
Sainio, M. E., 886

Salvini, P., 526
Sampson, W., 371
Sandberg, V. D., 966, 1050, 1078
Sanders, G. H., 966
Sapirstein, J., 567
Sapozhnikov, M. G., 526
Sayer, R., 727
Schaefer, U., 456
Schiffer, J. P., 354
Schmidt, K. E., 773
Schmidt, M. P., 334
Schneider, R., 456
Schneider, W., 371
Schreiber, O., 456
Schroeder, L. S., 205
Seftor, C., 640
Sennhauser, U., 966
Sessler, A. M., 53
Seth, K. K., 627
Shaw, G. L., 814
Shor, A., 814
Shutt, R., 371, 382
Simonius, M., 185
Simpson, J. J., 1094
Sinclair, D., 1094
Smith, G. A., 395, 445, 460
Smith, G. R., 1219, 1232
Smith, P. F., 1143
Snyderman, N. J., 791
Sober, D., 640
Sorensen, S. O., 526
Souadranayagam, R., 627
Speth, J., 924
Spinka, H., 1171
Stanfield, K. 1004
Stefanski, R., 1004
Stephenson, Jr., G. J., 773
Stöcker, H., 844
Stoker, D. P., 976
Stokes, W., 382
Storey, R. S., 1094
Straumann, U., 456
Stüssi, H., 1036
Sudarsky, D., 1102
Sundquist, M. L., 366
Svec, M., 1198
Swallow, E., 1004
Swenson, L. W., 727
Szafer, A., 1102
Szymanski, J. J., 934

T

Talaga, R. L., 1050, 1078, 1189
Talmadge, C., 1102
Tamvakis, K., 972
Tanaka, M., 872
Tang, L., 460
Tannenbaum, M. J., 858
Tanner, N. W., 993
Taragin, M., 640
Tarrh, J., 382
Taylor, G. N., 530
Thies, M., 692
Thiessen, H. A., 400
Thompson, P. A., 371
Torres, M., 901
Tosello, F., 526
Trentalange, S., 814
Truöl, P., 456
Turlay, R., 1004

U

Underwood, D. G., 1161

V

Van Bibber, K. A., 378
van Rossum, L., 1198
VanDalen, G., 844
Vary, J. P., 877
Vergados, J. D., 972, 1025
Vient, M., 844
Vinson, L. S., 445
Vogel, G., 382
Vonach, H., 456
von Dincklage, R. D., 561
von Witsch, W., 445, 460

W

Wah, Y., 1004
Wanderer, P., 371
Wang, K. C., 1050
Wang, Z.-F., 814
Weber, J., 1038
Werbeck, R., 966
White, B., 456
Wicklund, A. B., 609
Wiedner, U., 646
Wilkerson, J. F., 1031
Wilkinson, D., 1
Willen, E., 371
Williams, R. A., 966, 1078
Williams, W., 382
Wilson, S. L., 966
Winstein, B., 24
Winston, R., 1004
Wodrich, W. R., 456
Woloshyn, R. M., 788
Woods, M., 1004
Wright, L. E., 914
Wright, S. C., 966

X

Xue, Y., 445, 460

Y

Yamanaka, T., 1004
Yodh, G. B., 1078
Young, G. R., 357
Yuan, V., 1189

Z

Zahir, M. S., 879
Zambetakis, V., 770
Zenoni, A., 526
Ziegler, M., 456
Ziock, H., 640

AIP Conference Proceedings

		L.C. Number	ISBN
No. 1	Feedback and Dynamic Control of Plasmas – 1970	70-141596	0-88318-100-2
No. 2	Particles and Fields – 1971 (Rochester)	71-184662	0-88318-101-0
No. 3	Thermal Expansion – 1971 (Corning)	72-76970	0-88318-102-9
No. 4	Superconductivity in d- and f-Band Metals (Rochester, 1971)	74-18879	0-88318-103-7
No. 5	Magnetism and Magnetic Materials – 1971 (2 parts) (Chicago)	59-2468	0-88318-104-5
No. 6	Particle Physics (Irvine, 1971)	72-81239	0-88318-105-3
No. 7	Exploring the History of Nuclear Physics – 1972	72-81883	0-88318-106-1
No. 8	Experimental Meson Spectroscopy –1972	72-88226	0-88318-107-X
No. 9	Cyclotrons – 1972 (Vancouver)	72-92798	0-88318-108-8
No. 10	Magnetism and Magnetic Materials – 1972	72-623469	0-88318-109-6
No. 11	Transport Phenomena – 1973 (Brown University Conference)	73-80682	0-88318-110-X
No. 12	Experiments on High Energy Particle Collisions – 1973 (Vanderbilt Conference)	73-81705	0-88318-111-8
No. 13	π-π Scattering – 1973 (Tallahassee Conference)	73-81704	0-88318-112-6
No. 14	Particles and Fields – 1973 (APS/DPF Berkeley)	73-91923	0-88318-113-4
No. 15	High Energy Collisions – 1973 (Stony Brook)	73-92324	0-88318-114-2
No. 16	Causality and Physical Theories (Wayne State University, 1973)	73-93420	0-88318-115-0
No. 17	Thermal Expansion – 1973 (Lake of the Ozarks)	73-94415	0-88318-116-9
No. 18	Magnetism and Magnetic Materials – 1973 (2 parts) (Boston)	59-2468	0-88318-117-7
No. 19	Physics and the Energy Problem – 1974 (APS Chicago)	73-94416	0-88318-118-5
No. 20	Tetrahedrally Bonded Amorphous Semiconductors (Yorktown Heights, 1974)	74-80145	0-88318-119-3
No. 21	Experimental Meson Spectroscopy – 1974 (Boston)	74-82628	0-88318-120-7
No. 22	Neutrinos – 1974 (Philadelphia)	74-82413	0-88318-121-5
No. 23	Particles and Fields – 1974 (APS/DPF Williamsburg)	74-27575	0-88318-122-3
No. 24	Magnetism and Magnetic Materials – 1974 (20th Annual Conference, San Francisco)	75-2647	0-88318-123-1

No. 25	Efficient Use of Energy (The APS Studies on the Technical Aspects of the More Efficient Use of Energy)	75-18227	0-88318-124-X
No. 26	High-Energy Physics and Nuclear Structure – 1975 (Santa Fe and Los Alamos)	75-26411	0-88318-125-8
No. 27	Topics in Statistical Mechanics and Biophysics: A Memorial to Julius L. Jackson (Wayne State University, 1975)	75-36309	0-88318-126-6
No. 28	Physics and Our World: A Symposium in Honor of Victor F. Weisskopf (M.I.T., 1974)	76-7207	0-88318-127-4
No. 29	Magnetism and Magnetic Materials – 1975 (21st Annual Conference, Philadelphia)	76-10931	0-88318-128-2
No. 30	Particle Searches and Discoveries – 1976 (Vanderbilt Conference)	76-19949	0-88318-129-0
No. 31	Structure and Excitations of Amorphous Solids (Williamsburg, VA, 1976)	76-22279	0-88318-130-4
No. 32	Materials Technology – 1976 (APS New York Meeting)	76-27967	0-88318-131-2
No. 33	Meson-Nuclear Physics – 1976 (Carnegie-Mellon Conference)	76-26811	0-88318-132-0
No. 34	Magnetism and Magnetic Materials – 1976 (Joint MMM-Intermag Conference, Pittsburgh)	76-47106	0-88318-133-9
No. 35	High Energy Physics with Polarized Beams and Targets (Argonne, 1976)	76-50181	0-88318-134-7
No. 36	Momentum Wave Functions – 1976 (Indiana University)	77-82145	0-88318-135-5
No. 37	Weak Interaction Physics – 1977 (Indiana University)	77-83344	0-88318-136-3
No. 38	Workshop on New Directions in Mossbauer Spectroscopy (Argonne, 1977)	77-90635	0-88318-137-1
No. 39	Physics Careers, Employment and Education (Penn State, 1977)	77-94053	0-88318-138-X
No. 40	Electrical Transport and Optical Properties of Inhomogeneous Media (Ohio State University, 1977)	78-54319	0-88318-139-8
No. 41	Nucleon-Nucleon Interactions – 1977 (Vancouver)	78-54249	0-88318-140-1
No. 42	Higher Energy Polarized Proton Beams (Ann Arbor, 1977)	78-55682	0-88318-141-X
No. 43	Particles and Fields – 1977 (APS/DPF, Argonne)	78-55683	0-88318-142-8
No. 44	Future Trends in Superconductive Electronics (Charlottesville, 1978)	77-9240	0-88318-143-6
No. 45	New Results in High Energy Physics – 1978 (Vanderbilt Conference)	78-67196	0-88318-144-4
No. 46	Topics in Nonlinear Dynamics (La Jolla Institute)	78-57870	0-88318-145-2

No. 47	Clustering Aspects of Nuclear Structure and Nuclear Reactions (Winnepeg, 1978)	78-64942	0-88318-146-0
No. 48	Current Trends in the Theory of Fields (Tallahassee, 1978)	78-72948	0-88318-147-9
No. 49	Cosmic Rays and Particle Physics – 1978 (Bartol Conference)	79-50489	0-88318-148-7
No. 50	Laser-Solid Interactions and Laser Processing – 1978 (Boston)	79-51564	0-88318-149-5
No. 51	High Energy Physics with Polarized Beams and Polarized Targets (Argonne, 1978)	79-64565	0-88318-150-9
No. 52	Long-Distance Neutrino Detection – 1978 (C.L. Cowan Memorial Symposium)	79-52078	0-88318-151-7
No. 53	Modulated Structures – 1979 (Kailua Kona, Hawaii)	79-53846	0-88318-152-5
No. 54	Meson-Nuclear Physics – 1979 (Houston)	79-53978	0-88318-153-3
No. 55	Quantum Chromodynamics (La Jolla, 1978)	79-54969	0-88318-154-1
No. 56	Particle Acceleration Mechanisms in Astrophysics (La Jolla, 1979)	79-55844	0-88318-155-X
No. 57	Nonlinear Dynamics and the Beam-Beam Interaction (Brookhaven, 1979)	79-57341	0-88318-156-8
No. 58	Inhomogeneous Superconductors – 1979 (Berkeley Springs, W.V.)	79-57620	0-88318-157-6
No. 59	Particles and Fields – 1979 (APS/DPF Montreal)	80-66631	0-88318-158-4
No. 60	History of the ZGS (Argonne, 1979)	80-67694	0-88318-159-2
No. 61	Aspects of the Kinetics and Dynamics of Surface Reactions (La Jolla Institute, 1979)	80-68004	0-88318-160-6
No. 62	High Energy e^+e^- Interactions (Vanderbilt, 1980)	80-53377	0-88318-161-4
No. 63	Supernovae Spectra (La Jolla, 1980)	80-70019	0-88318-162-2
No. 64	Laboratory EXAFS Facilities – 1980 (Univ. of Washington)	80-70579	0-88318-163-0
No. 65	Optics in Four Dimensions – 1980 (ICO, Ensenada)	80-70771	0-88318-164-9
No. 66	Physics in the Automotive Industry – 1980 (APS/AAPT Topical Conference)	80-70987	0-88318-165-7
No. 67	Experimental Meson Spectroscopy – 1980 (Sixth International Conference, Brookhaven)	80-71123	0-88318-166-5
No. 68	High Energy Physics – 1980 (XX International Conference, Madison)	81-65032	0-88318-167-3
No. 69	Polarization Phenomena in Nuclear Physics – 1980 (Fifth International Symposium, Santa Fe)	81-65107	0-88318-168-1
No. 70	Chemistry and Physics of Coal Utilization – 1980 (APS, Morgantown)	81-65106	0-88318-169-X

No. 71	Group Theory and its Applications in Physics – 1980 (Latin American School of Physics, Mexico City)	81-66132	0-88318-170-3
No. 72	Weak Interactions as a Probe of Unification (Virginia Polytechnic Institute – 1980)	81-67184	0-88318-171-1
No. 73	Tetrahedrally Bonded Amorphous Semiconductors (Carefree, Arizona, 1981)	81-67419	0-88318-172-X
No. 74	Perturbative Quantum Chromodynamics (Tallahassee, 1981)	81-70372	0-88318-173-8
No. 75	Low Energy X-Ray Diagnostics – 1981 (Monterey)	81-69841	0-88318-174-6
No. 76	Nonlinear Properties of Internal Waves (La Jolla Institute, 1981)	81-71062	0-88318-175-4
No. 77	Gamma Ray Transients and Related Astrophysical Phenomena (La Jolla Institute, 1981)	81-71543	0-88318-176-2
No. 78	Shock Waves in Condensed Mater – 1981 (Menlo Park)	82-70014	0-88318-177-0
No. 79	Pion Production and Absorption in Nuclei – 1981 (Indiana University Cyclotron Facility)	82-70678	0-88318-178-9
No. 80	Polarized Proton Ion Sources (Ann Arbor, 1981)	82-71025	0-88318-179-7
No. 81	Particles and Fields –1981: Testing the Standard Model (APS/DPF, Santa Cruz)	82-71156	0-88318-180-0
No. 82	Interpretation of Climate and Photochemical Models, Ozone and Temperature Measurements (La Jolla Institute, 1981)	82-71345	0-88318-181-9
No. 83	The Galactic Center (Cal. Inst. of Tech., 1982)	82-71635	0-88318-182-7
No. 84	Physics in the Steel Industry (APS/AISI, Lehigh University, 1981)	82-72033	0-88318-183-5
No. 85	Proton-Antiproton Collider Physics –1981 (Madison, Wisconsin)	82-72141	0-88318-184-3
No. 86	Momentum Wave Functions – 1982 (Adelaide, Australia)	82-72375	0-88318-185-1
No. 87	Physics of High Energy Particle Accelerators (Fermilab Summer School, 1981)	82-72421	0-88318-186-X
No. 88	Mathematical Methods in Hydrodynamics and Integrability in Dynamical Systems (La Jolla Institute, 1981)	82-72462	0-88318-187-8
No. 89	Neutron Scattering – 1981 (Argonne National Laboratory)	82-73094	0-88318-188-6
No. 90	Laser Techniques for Extreme Ultraviolt Spectroscopy (Boulder, 1982)	82-73205	0-88318-189-4
No. 91	Laser Acceleration of Particles (Los Alamos, 1982)	82-73361	0-88318-190-8
No. 92	The State of Particle Accelerators and High Energy Physics (Fermilab, 1981)	82-73861	0-88318-191-6

No. 93	Novel Results in Particle Physics (Vanderbilt, 1982)	82-73954	0-88318-192-4
No. 94	X-Ray and Atomic Inner-Shell Physics – 1982 (International Conference, U. of Oregon)	82-74075	0-88318-193-2
No. 95	High Energy Spin Physics – 1982 (Brookhaven National Laboratory)	83-70154	0-88318-194-0
No. 96	Science Underground (Los Alamos, 1982)	83-70377	0-88318-195-9
No. 97	The Interaction Between Medium Energy Nucleons in Nuclei – 1982 (Indiana University)	83-70649	0-88318-196-7
No. 98	Particles and Fields – 1982 (APS/DPF University of Maryland)	83-70807	0-88318-197-5
No. 99	Neutrino Mass and Gauge Structure of Weak Interactions (Telemark, 1982)	83-71072	0-88318-198-3
No. 100	Excimer Lasers – 1983 (OSA, Lake Tahoe, Nevada)	83-71437	0-88318-199-1
No. 101	Positron-Electron Pairs in Astrophysics (Goddard Space Flight Center, 1983)	83-71926	0-88318-200-9
No. 102	Intense Medium Energy Sources of Strangeness (UC-Sant Cruz, 1983)	83-72261	0-88318-201-7
No. 103	Quantum Fluids and Solids – 1983 (Sanibel Island, Florida)	83-72440	0-88318-202-5
No. 104	Physics, Technology and the Nuclear Arms Race (APS Baltimore –1983)	83-72533	0-88318-203-3
No. 105	Physics of High Energy Particle Accelerators (SLAC Summer School, 1982)	83-72986	0-88318-304-8
No. 106	Predictability of Fluid Motions (La Jolla Institute, 1983)	83-73641	0-88318-305-6
No. 107	Physics and Chemistry of Porous Media (Schlumberger-Doll Research, 1983)	83-73640	0-88318-306-4
No. 108	The Time Projection Chamber (TRIUMF, Vancouver, 1983)	83-83445	0-88318-307-2
No. 109	Random Walks and Their Applications in the Physical and Biological Sciences (NBS/La Jolla Institute, 1982)	84-70208	0-88318-308-0
No. 110	Hadron Substructure in Nuclear Physics (Indiana University, 1983)	84-70165	0-88318-309-9
No. 111	Production and Neutralization of Negative Ions and Beams (3rd Int'l Symposium, Brookhaven, 1983)	84-70379	0-88318-310-2
No. 112	Particles and Fields – 1983 (APS/DPF, Blacksburg, VA)	84-70378	0-88318-311-0
No. 113	Experimental Meson Spectroscopy – 1983 (Seventh International Conference, Brookhaven)	84-70910	0-88318-312-9

No. 112	Particles and Fields – 1983 (APS/DPF, Blacksburg, VA)	84-70378	0-88318-311-0
No. 113	Experimental Meson Spectroscopy – 1983 (Seventh International Conference, Brookhaven)	84-70910	0-88318-312-9
No. 114	Low Energy Tests of Conservation Laws in Particle Physics (Blacksburg, VA, 1983)	84-71157	0-88318-313-7
No. 115	High Energy Transients in Astrophysics (Santa Cruz, CA, 1983)	84-71205	0-88318-314-5
No. 116	Problems in Unification and Supergravity (La Jolla Institute, 1983)	84-71246	0-88318-315-3
No. 117	Polarized Proton Ion Sources (TRIUMF, Vancouver, 1983)	84-71235	0-88318-316-1
No. 118	Free Electron Generation of Extreme Ultraviolet Coherent Radiation (Brookhaven/OSA, 1983)	84-71539	0-88318-317-X
No. 119	Laser Techniques in the Extreme Ultraviolet (OSA, Boulder, Colorado, 1984)	84-72128	0-88318-318-8
No. 120	Optical Effects in Amorphous Semiconductors (Snowbird, Utah, 1984)	84-72419	0-88318-319-6
No. 121	High Energy e^+e^- Interactions (Vanderbilt, 1984)	84-72632	0-88318-320-X
No. 122	The Physics of VLSI (Xerox, Palo Alto, 1984)	84-72729	0-88318-321-8
No. 123	Intersections Between Particle and Nuclear Physics (Steamboat Springs, 1984)	84-72790	0-88318-322-6
No. 124	Neutron-Nucleus Collisions – A Probe of Nuclear Structure (Burr Oak State Park - 1984)	84-73216	0-88318-323-4
No. 125	Capture Gamma-Ray Spectroscopy and Related Topics – 1984 (Internat. Symposium, Knoxville)	84-73303	0-88318-324-2
No. 126	Solar Neutrinos and Neutrino Astronomy (Homestake, 1984)	84-63143	0-88318-325-0
No. 127	Physics of High Energy Particle Accelerators (BNL/SUNY Summer School, 1983)	85-70057	0-88318-326-9
No. 128	Nuclear Physics with Stored, Cooled Beams (McCormick's Creek State Park, Indiana, 1984)	85-71167	0-88318-327-7
No. 129	Radiofrequency Plasma Heating (Sixth Topical Conference, Callaway Gardens, GA, 1985)	85-48027	0-88318-328-5
No. 130	Laser Acceleration of Particles (Malibu, California, 1985)	85-48028	0-88318-329-3
No. 131	Workshop on Polarized ^{3}He Beams and Targets (Princeton, New Jersey, 1984)	85-48026	0-88318-330-7
No. 132	Hadron Spectroscopy–1985 (International Conference, Univ. of Maryland)	85-72537	0-88318-331-5

No. 133	Hadronic Probes and Nuclear Interactions (Arizona State University, 1985)	85-72638	0-88318-332-3
No. 134	The State of High Energy Physics (BNL/SUNY Summer School, 1983)	85-73170	0-88318-333-1
No. 135	Energy Sources: Conservation and Renewables (APS, Washington, DC, 1985)	85-73019	0-88318-334-X
No. 136	Atomic Theory Workshop on Relativistic and QED Effects in Heavy Atoms	85-73790	0-88318-335-8
No. 137	Polymer-Flow Interaction (La Jolla Institute, 1985)	85-73915	0-88318-336-6
No. 138	Frontiers in Electronic Materials and Processing (Houston, TX, 1985)	86-70108	0-88318-337-4
No. 139	High-Current, High-Brightness, and High-Duty Factor Ion Injectors (La Jolla Institute, 1985)	86-70245	0-88318-338-2
No. 140	Boron-Rich Solids (Albuquerque, NM, 1985)	86-70246	0-88318-339-0
No. 141	Gamma-Ray Bursts (Stanford, CA, 1984)	86-70761	0-88318-340-4
No. 142	Nuclear Structure at High Spin, Excitation, and Momentum Transfer (Indiana University, 1985)	86-70837	0-88318-341-2
No. 143	Mexican School of Particles and Fields (Oaxtepec, México, 1984)	86-81187	0-88318-342-0
No. 144	Magnetospheric Phenomena in Astrophysics (Los Alamos, 1984)	86-71149	0-88318-343-9
No. 145	Polarized Beams at SSC & Polarized Antiprotons (Ann Arbor, MI & Bodega Bay, CA, 1985)	86-71343	0-88318-344-7
No. 146	Advances in Laser Science–I (Dallas, TX, 1985)	86-71536	0-88318-345-5
No. 147	Short Wavelength Coherent Radiation: Generation and Applications (Monterey, CA, 1986)	86-71674	0-88318-346-3
No. 148	Space Colonization: Technology and The Liberal Arts (Geneva, NY, 1985)	86-71675	0-88318-347-1
No. 149	Physics and Chemistry of Protective Coatings (Universal City, CA, 1985)	86-72019	0-88318-348-X